WILLIAM F. MAAG LIBRARY
YOUNGSTOWN STATE UNIVERSITY

Organometallic Chemistry

Volume 15

A Specialist Periodical Report

Organometallic Chemistry
Volume 15

A Review of the Literature Published during 1985

Senior Reporters
E. W. Abel, *Department of Chemistry, University of Exeter*
F. G. A. Stone, *Department of Inorganic Chemistry, University of Bristol*

Reporters
D. A. Armitage, *Kings College, University of London*
M. Bochmann, *University of East Anglia*
B. J. Brisdon, *University of Bath*
D. A. Edwards, *University of Bath*
P. G. Harrison, *University of Nottingham*
J. A. S. Howell, *University of Keele*
W. E. Lindsell, *Heriot-Watt University*
D. R. Russell, *University of Leicester*
A. K. Smith, *University of Liverpool*
T. R. Spalding, *University College, Cork*
J. L. Wardell, *University of Aberdeen*
J. W. Wilson, *University of Ulster at Coleraine*
M. J. Winter, *University of Sheffield*
A. H. Wright, *University of Nottingham*

The Royal Society of Chemistry
Burlington House, London, W1V 0BN

ISBN 0-85186-631-X
ISSN 0301-0074

Copyright © 1987
The Royal Society of Chemistry

All Rights Reserved
No part of this book may be reproduced or transmitted
in any form or by any means – graphic, electronic,
including photocopying, recording, taping or
information storage and retrieval systems – without
written permission from The Royal Society of Chemistry

Printed in Great Britain at the Alden Press, Oxford,
London and Northampton

Foreword

Publication of Volume 15 of this series of Specialist Periodical Reports is something of a landmark. We continue to make every effort to meet effectively our two main objectives, of being both monitoring and archival in nature. Whilst the main stucture of the Volumes has been preserved, the loss and gain of some chapter titles and the changes in relative chapter lengths reflect some of the qualitative and quantitative changes of emphasis in the subject over the years. Our reporters have responded positively to the severe restrictions of space, and we would ask readers to accept a consequential concentrated presentation.

E.W. Abel
F.G.A. Stone

Contents

CHAPTER 1 Group I: The Alkali and Coinage Metals
By J.L.Wardell

1	Alkali Metals	1
	1.1 General	1
	1.2 Alkyl Derivatives	1
	1.3 Functionally Substituted Alkyl Derivatives	1
	1.4 Aryl Derivatives	2
	1.5 Benzyl Derivatives	3
	1.6 Other Derivatives	3
2	Copper, Silver and Gold	5
	2.1 π-Complexes	5
	2.2 σ-Complexes	6
3	Bibliography	8
	References	10

CHAPTER 2 Group II: The Alkaline Earths and Zinc and its Congeners
By J.L. Wardell

1	Beryllium	14
2	Magnesium	14
3	Zinc and Cadmium	16
4	Mercury	17
	4.1 General	17
	4.2 Methylmercury Species	17
	4.3 Other Derivatives	19
5	Bibliography	20
	References	21

CHAPTER 3 Boron with the Exception of the Carbaboranes
By J.W. Wilson

1	Introduction	24
2	Books and Reviews	24
3	Uses of Organoboranes and Organoborates in Organic Synthesis	24
4	Preparations and Reactions of Organoboranes	29
5	Physical Data	34
	References	35

CHAPTER 4 Carbaboranes, including their Metal Derivatives
By T.R. Spalding

1	Introduction, Review Articles and Theoretical Aspects	39
2	Carbaborane Synthesis, Characterisation and Reactions	39
3	Metallacarbaboranes excluding Cage-Bonded Compounds	43
4	Cage-Bonded Metallacarbaboranes	43
	References	54

CHAPTER 5 Group III: Aluminium, Gallium, Indium and Thallium
By P.G. Harrison

1	Structure, Spectroscopy and Bonding	58
2	Reactions and Applications in Organic Synthesis	73
	References	79
	Bibliography	81

CHAPTER 6 Group IV: The Silicon Group
By D.A. Armitage

1	Introduction	84
2	The Carbon – Metalloid Bond	85
3	Catenation	98
4	Hydrogen Derivatives	101

	5	Radicals and Metal Derivatives	102
	6	Nitrogen Derivatives	104
	7	Phosphorus Group Derivatives	108
	8	Oxygen Derivatives	109
	9	Sulphur, Selenium and Tellurium Derivatives	114
	10	Halogen Derivatives	117
	11	Complexes	118
		References	119

CHAPTER 7 Group V: Arsenic, Antimony and Bismuth
By J.L. Wardell

	1	Tervalent Compounds	138
		1.1 Metal – Metal Bonded Species	138
		1.2 Triorganometallic Species	138
		1.3 Other Compounds	139
	2	Quinquevalent Compounds	139
	3	Uses in Organic Synthesis	141
	4	Bibliography	141
		References	142

CHAPTER 8 Metal Carbonyls
By B.J. Brisdon

	1	Introduction	144
	2	General and Theoretical Studies	144
	3	Chemistry of Metal Carbonyls	146
		3.1 Mononuclear Carbonyl Derivatives	146
		3.2 Binuclear Carbonyl Derivatives	147
		3.3 Polynuclear Carbonyl Derivatives	147
	4	Cluster Carbonyls containing C, N, or S	149
	5	Metal Carbonyl Hydrides	150
	6	Metal Carbonyl Halides	151
		References	151

CHAPTER 9 Organometallic Compounds Containing Metal – Metal Bonds
By W.E. Lindsell

1 Introduction 156
  1.1 Reviews 156
  1.2 Theoretical Studies 156
  1.3 Physical Studies 157
  1.4 Surface Bound Species 157

2 Compounds with Homonuclear Transition Metal Bonds 158
  2.1 Titanium, Zirconium and Hafnium 158
  2.2 Niobium 158
  2.3 Chromium, Molybdenum and Tungsten 158
  2.4 Manganese and Rhenium 160
  2.5 Iron 163
  2.6 Ruthenium and Osmium 165
  2.7 Cobalt 169
  2.8 Rhodium and Iridium 170
  2.9 Nickel 172
  2.10 Palladium and Platinum 172
  2.11 Copper and Gold 173

3 Compounds with Heteronuclear Transition Metal Bonds 173
  3.1 Binuclear Complexes 175
  3.2 Tri- and Higher Nuclearity Complexes 177

4 Compounds containing Bonds between Transition and Main Group Metals 182
  4.1 Lithium 182
  4.2 Group II 182
  4.3 Gallium 182
  4.4 Group IV 182
  4.5 Bismuth 183

References 183

CHAPTER 10 Ligand Substitution Reactions of Metal and Organometal Carbonyls with Group V and VI Donor Ligands
By D.A. Edwards

1 Introduction and Reviews 197

2 Papers of General Interest 197
  2.1 Nitrogen Donor Ligands 199
  2.2 Phosphorus and the Heavier Group V Donor Ligands 200
  2.3 Group VI Donor Ligands 202

	3	Groups IV and V	202
	4	Group VI	203
		4.1 Carbonyl Complexes of Cr^0, Mo^0 and W^0	203
		4.2 Carbonyl Complexes of Mo^{II} and W^{II}	207
		4.3 Cyclopentadienyl and Arene Complexes	208
	5	Group V	209
		5.1 Carbonyl, Carbonyl Halide and Related Complexes	209
		5.2 Cyclopentadienyl and Other Complexes	213
	6	Group VIII: Iron, Ruthenium and Osmium	215
		6.1 Iron, Ruthenium, and Osmium Carbonyl Complexes	215
		6.2 Cyclopentadienyl Complexes	218
	7	Group VIII: Cobalt, Rhodium and Iridium	219
		7.1 Carbonyl Complexes	219
		7.2 Cyclopentadienyl Complexes	221
	8	Group VIII: Nickel, Palladium and Platinum	221
		References	222

CHAPTER 11 Complexes Containing Metal – Carbon σ–Bonds of the Groups Scandium to Manganese, Including Carbenes and Carbynes
By Mark J. Winter

1	Introduction	229
2	Groups 3 (Sc, Y, and La), Lanthanides and Actinides	229
3	Group 4 (Ti, Zr, and Hf)	231
4	Group 5 (V, Nb, and Ta)	237
5	Group 6 (Cr, Mo, and W)	238
6	Group 7 (Mn, Tc, and Re)	251
	References	257

CHAPTER 12 Complexes Containing Metal – Carbon σ–Bonds of the Groups Iron, Cobalt and Nickel
By A.K.Smith

1	Introduction	265
2	Reviews and Articles of General Interest	265
3	Metal – Carbon σ–Bonds involving Group III Metals	265

		3.1 The Iron Triad	265
		3.2 The Cobalt Triad	273
		3.3 The Nickel Triad	279
	4	Carbene and Carbyne Complexes of the Group VIII Metals	286
		4.1 The Iron Triad	286
		4.2 The Cobalt Triad	290
		4.3 The Nickel Triad	292
	5	Bibliography	292
		References	293

CHAPTER 13 **Metal – Hydrocarbon π–Complexes, Other than π–Cyclopentadienyl and π–Arene Complexes**
By James A. Howell

A	Reviews		301
B	Allyl Complexes and Complexes Derived from Monoolefins		301
	1	Ni, Pd and Pt	301
	2	Co, Rh and Ir	306
	3	Fe, Ru and Os	308
	4	Cr, Mo and W	310
	5	Other Metals	311
C	Complexes Derived from Unconjugated Dienes		315
	1	Ni, Pd and Pt	315
	2	Co, Rh and Ir	315
	3	Other Metals	315
D	Complexes Derived from Conjugated Dienes		318
	1	Fe, Ru and Os	318
		i. Acyclic Dienes	318
		ii. Cyclic Dienes	318
	2	Co, Rh and Ir	324
	3	Cr, Mo and W	324
	4	Other Metals	327
E	Complexes Derived from Acetylenes		329
	1	Ni, Pd and Pt	329
	2	Cr, Mo and W	329
	3	Other Metals	333
F	Complexes Containing More than One Metal Atom		333
	1	Binuclear Complexes	333
	2	Polymetallic Complexes	341
		References	345

CHAPTER 14 π–Cyclopentadienyl, π–Arene and Related Complexes
By A.H. Wright

1	Introduction	354
2	Monocyclopentadienyl Complexes	354
	2.1 Titanium, Zirconium and Hafnium	354
	2.2 Vanadium, Niobium and Tantalum	355
	2.3 Chromium, Molybdenum and Tungsten	355
	2.4 Manganese and Rhenium	357
	2.5 Iron, Ruthenium and Osmium	358
	2.6 Cobalt, Rhodium and Iridium	359
	2.7 Nickel, Palladium and Platinum	361
	2.8 Lanthanides and Actinides	361
3	Sandwich Complexes	361
	3.1 Scandium and Yttrium	361
	3.2 Titanium, Zirconium and Hafnium	361
	3.3 Vanadium, Niobium and Tantalum	363
	3.4 Chromium, Molybdenum and Tungsten	365
	3.5 Manganese and Rhenium	365
	3.6 Iron, Ruthenium and Osmium	365
	3.7 Cobalt, Rhodium and Iridium	366
	3.8 Nickel	367
	3.9 Lanthanides and Actinides	367
4	Arene Complexes	367
	4.1 Chromium, Molybdenum and Tungsten	367
	4.2 Manganese and Rhenium	370
	4.3 Iron, Ruthenium and Osmium	370
	4.4 Cobalt, Rhodium and Iridium	373
	4.5 Other Arene Complexes	373
	References	374

CHAPTER 15 Homogeneous Catalysis by Transition Metal Complexes
By M. Bochmann

1	General Reviews	382
2	Hydrogenation	382
	2.1 Asymmetric Hydrogenation	385
	2.2 Heterogenized Catalysts	389
3	Hydrogen Transfer and Dehydrogenation Reactions	389

4	Hydroboration and Hydrosilylation	391
5	Catalytic Alkane Activation	393
6	Carbonylation Reactions	393
	6.1 Carbonylation of Alcohols and Esters	395
	6.2 Alcohol Homologation	396
	6.3 Carbon Monoxide Reduction and the Water Gas Shift Reaction	396
	6.4 Hydroformylations	396
7	Isomerisations	399
8	Alkene and Alkyne Metathesis	399
9	Polymerisation Reactions	402
10	Oligomerization Reactions	405
	10.1 Addition and Telomerisation Reactions	407
11	Cross-coupling Reactions	409
12	Cyclisation Reactions	411
13	Reactions with Carbon Dioxide	413
14	Transalkylations	413
15	Oxidations	413
	References	415

CHAPTER 16 Structures of Organometallic Compounds determined by Diffraction Methods
By D.R. Russell

1	Introduction	422
2	Main Table	423
3	Metals Cross Reference Table	482
	References	484

Abbreviations

Ac	acetate (MeCOO$^-$)
acac	acetylacetonate
acacen	NN'-ethylenebis(acetylacetone iminate)
Ad	adamantyl
AIBN	azoisobutyronitrile
Ar	Aryl
arphos	1-(diphenylphosphinio)-2-(diphenylarsino)ethane
ATP	adenosine triphosphate
Azb	azobenzene
9-BBN	9-borabicyclo[3.3.1]nonane
bipy	2,2'-bipyridyl
Bz	benzyl
Bzac	benzoylacetonate
cbd	cyclobutadiene
1,5,9-cdt	cyclododeca-1,5,9-triene
chd	cyclohexadiene
chpt	cycloheptatriene
[Co]	cobalamin
(Co)	cobaloxime [Co(dmg)$_2$ derivative]
cod	cyclo-octa-1,5-diene
cot	cyclo-octatraene
Cp	η^5-cyclopentadienyl
CTTM	charge transfer to metal
Cy	cyclohexyl
dab	1,4-diazabutadiene
dba	dibenzylideneacetone
DCA	9,10-dicyanoanthracene
depe	1,2-bis(diethylphosphino)ethane
depm	1,2-bis(diethylphosphino)methane
diars	o-phenylenebis(dimethyl)arsine
diarsop	{[2,2-dimethyl-1,3-dioxolan-4,5-diyl)bis(methylene)] bis-(diphenylarsine)}
dien	diethylenetriamine
diop	{[2,2-dimethyl-1,3-dioxolan-4,5-diyl)bis(methylene)] bis-(diphenylphosphine)}
diphos	1,2-bis(diphenylphosphino)ethane
DME	dimethoxyethane
DMF	NN-dimethylformamide
dmg	dimethylglyoximate
dmgH$_2$	dimethylglyoxime
dmpe	1,2-bis(dimethylphosphino)ethane
dmpm	bis(dimethylphosphino)methane
DMSO	dimethyl sulphoxide
dpa	di(2-pyridyl)amine
dpae	1,2-bis(diphenylarsino)ethane
dpam	bis(diphenylarsino)methane
dppb	1,4-bis(diphenylphosphino)butane
dppe	1,2-bis(diphenylphosphino)ethane
dppm	bis(diphenylphosphino)methane
dppp	1,3-bis(diphenylphosphino)propane
en	ethylene-1,2-diamine
EXAFS	extended X-ray absorption fine structure
F$_6$acac	hexafluoroacetylacetonate
Fc	ferrocenyl

Fp	Fe(CO)$_2$Cp
glyme	ethyleneglycol dimethyl ether
GVB	generalized valence band
HDPG	diphenylglyoximato
hfa	hexafluoroacetone
hfacac	hexafluoroacetylacetonato
hfb	hexafluorobutyne
HMPA	hexamethyl phosphoric triamide
LDA	lithium diisopropylamide
LiDBB	lithium di-t-butylbiphenyl
Me$_6$[14]dieneN$_4$	5,7,7,12,14,14-hexamethyl-1,4,8,11-tetra-azacyclotetradeca-4,11-diene
Me$_6$[14]N$_4$	5,5,7,12,12,14-hexamethyl-1,4,8,11-tetra-azacyclotetradecane
4,7-Me$_2$phen	4,7-dimethyl-1,10-phenanthroline
3,4,7,8-Me$_4$phen	3,4,7,8-tetramethyl-1,10-phenanthroline
Mes	mesityl
mcpba	metachloroperbenzoic acid
nap	1-naphthyl
nbd	norbornadiene
NBS	N-bromosuccinimide
NCS	N-chlorosuccinimide
OEP	octaethylporphyrin
Pc	phthalocyanin
PMDT	pentamethylenediethylenetetramine
PMHS	polymethylhydrosiloxane
pd	pentane-2,4-dionate
phen	1,10-phenanthroline
[PPN]$^+$	[(Ph$_3$P)$_2$N]$^+$
py	pyridine
pz	pyrazolyl
sal	salicylaldehyde
salen	NN'-bis-(salicylaldehydo)ethylenediamine
saloph	NN-bisalicylidene-o-phenylenediamine
SCF	self consistent field
TCNE	tetracyanoethylene
TCNQ	7,7,8,8-tetracyanoquinodimethane
terpy	2,2',2''-terpyridyl
TFA	trifluoroacetic acid
tfacac	trifluoroacetylacetonato
tfo	triflate, trifluoromethylsulphonate
THF	tetrahydrofuran
tht	tetrahydrothiophen
TMBD	$NNN'N'$-tetramethyl-2-butene-1,4-diamine
TMED	tetramethylethylenediamine
tmen	tetramethylethylenediamine
TMS	tetramethylsilane
tol	tolyl
TPP	*meso*-tetraphenylporphyrin
triphos	1,1,1-tris(diphenylphosphinomethyl)ethane

1
Group I: The Alkali and Coinage Metals

BY J. L. WARDELL

1 Alkali Metals

1.1 General. A timely review of organolithium structures has been published.[1] The use of ^{13}C n.m.r. spectra of ^{6}Li and ^{13}C labelled organolithiums in determining the extent of aggregation has been further illustrated[2,3] with BuLi (1), Br_2CHLi, $PhSCH_2Li$ and $Bu^tC{\equiv}CLi$.

1.2 Alkyl derivatives. A ^{1}H and ^{7}Li n.m.r. study of (1) also revealed[4] the presence in THF of dimers and tetramers, equation (1);

$$(BuLi)_4 \cdot THF + 4THF \rightleftharpoons 2[(BuLi)_2 \cdot 4THF] \qquad (1)$$

the rates of exchange of Bu groups were also obtained. Using a rapid injection n.m.r. technique at ca. -90°C, $(BuLi)_2$ was found[5] to be 10 times more reactive than $(BuLi)_4$ towards PhCHO; the reactivity of $[Bu_{4-n}Li_n(OR)_n]$ increases from n = 0 to n = 3. Butyl-sodium, obtained from BuLi and $NaOBu^t$, is insoluble in hexane but dissolves[6] on addition of TMED or THF. No evidence was found, from a ^{1}H, ^{13}C and ^{23}Na n.m.r. study, for equilibrating dimers and tetramers of BuNa in THF at low T. Interestingly, the α-CH_2 protons of (1) provide an AA'XX' system while those of BuNa are magnetically equivalent. A solid state ^{13}C n.m.r. study of $(Me^7Li)_4$ (2), $(Me^6Li)_4$ (3), $(CH_2{}^7Li_2)_n$ (4), and $(CH_2{}^6Li_2)_n$ (5) at ambient temperature has been reported.[7] In the ^{1}H and ^{6}Li decoupled ^{13}C n.m.r. spectra, singlets for (3) and (5) were at -16 and 10.5 ppm respectively. Only one type of low-binding energy carbon and lithium environment in each of (2) and (4) was indicated from an X-ray p.e. spectral study; the carbon charge was calculated to be -1.02 and -1.55 in (2) and (4) respectively.[8] The synthesis of $Me_2\overset{\frown}{C}CMe_2\overset{\frown}{C}LiR$ (6, R=Li), by pyrolysis of (6, R=H) at 170°C, its ^{1}H n.m.r. spectra and flash vapourisation mass spectra (suggesting the presence of monomers and dimers) have been reported.[9]

1.3 Functionally substituted alkyl derivatives. The core (7) of

(For references see page 10)

[Me$_2$N(CH$_2$)$_3$Li]$_4$, very similar to that of (EtLi)$_4$, consists of a distorted cube of 4C and 4Li, with each N co-ordinated to a Li atom;[10] there was n.m.r. evidence for two differently arranged tetramers in solution. Berluenga et al. have reported[11] on the synthesis of (i) MeCHMCH$_2$CH$_2$CR$_2$OM'(M=Li or K, M'=Li or MgBr); (ii) Li(CH$_2$)$_n$C(OM)RCH$_2$Li (8, n=2 or 3; M= Li, R= CH$_2$CH=CH$_2$; M= MgBr, R=H), equation (2) and (iii) (Z)-R'CH=CLiCHROM (R=Et, Pr..., R'=H or Me; M=Li or MgBr). Compound (8, M=MgBr, R=H) thermally

$$Cl(CH_2)_nC(OH)RCH_2Cl \xrightarrow{i,ii} \underset{(8)}{Li(CH_2)_nC(OM)RCH_2Li} \qquad (2)$$

Reagents: i, BuLi; ii, C$_{10}$H$_8^{-\cdot}$,Li$^+$ or i, EtMgBr; ii, Li. decomposes to Li(CH$_2$)$_n$CH=CH$_2$. In the gas phase, LiCH(SiMe$_3$)$_2$ is monomeric in contrast to the infinite polymeric solid-state structure.[12] Graphite has been used[13] as an acceptor, in the presence of a phase-transfer agent, in direct lithiation, e.g. Me$_2$C=NPr^i provided LiCH$_2$CMe=NPr^i. A MNDO calculation (3-21G, 6-31G and MP2/6-13G levels) on MCH$_2$CO$_2$M (M=Li or MgX) indicated a 1,3-1,3'-doubly bridged structure was the most stable.[14] Crystal structures of the following lithium enolates have been determined:[15] (i) [(Z)-MeCH=C(OBu^t)OLi.TMED], (ii) [Me$_2$C=C(OBu^t)OLi.TMED]$_2$, (iii) [(Z)-Bu^{t}CH=C(OMe)OLi.TMED]$_4$, (iv) [(Z)-MeCH=C(OLi)NMe$_2$.MeNHCH$_2$CH$_2$N-Me$_2$]$_2$, (v) [Bu^{t}C(OLi)=CH$_2$.MeNHCH$_2$CH$_2$NMe$_2$]$_2$ and (vi) [CH$_2$=C(OLi)-C$_6$H$_4$CH$_2$NMe$_2$-o]$_4$.

The compound, (LiCH$_2$SO$_2$Ph.TMED)$_2$, has a flattened chair-like 8-membered ring,[16] involving Li, S and O (with Ph axial and CH$_2$ groups equatorial, being outside the co-ordination sphere of Li; the short C-S bond suggests a true double bond. Also studied[17] were the dilithiated sulphones PhSO$_2$CRLi$_2$ [R=H, CH=CH$_2$ (9) and SiMe$_3$ (10)] and o-LiC$_6$H$_4$SO$_2$CHLiCH=CH$_2$ (11); compound (11) formed at low T converts to (10) at 50°C. N.m.r. spectra of (10) and PhSO$_2$CHLiSiMe$_3$ were reported.

1.4 **Aryl derivatives.** Crystal structures were reported for (i) monomeric[18] PhLi.PMDT: tetraco-ordinate Li is bound to C α of Ph; (ii) [2,4,6-Me$_3$C$_6$H$_2$Li.2THF]$_2$: centrosymmetric dimers,[19] and (iii) [(LiC$_5$H$_4$FeC$_5$H$_4$Li)$_3$.(TMED)$_2$]: three different Li environments.[20] The thermal stability of o-ClC$_6$H$_4$Li is markedly increased[21] on the addition of TMED. Benzo[b]thiophene (ArH) and Na provide[22] the 2-sodio derivative at -20°C but at -78°C the product is ArH^{2-},2Na$^+$. Ortho-lithiation was reported[23] for (i) the ylide, Ph$_3$P$^+$-$\bar{\text{C}}$H$_2$, using

Bu^{t}Li or Bu^{s}Li in ethers at -75°C, (ii) PhOH, using Bu^{t}Li in pentane and (iii) a series of m-R$_2$NC$_6$H$_4$OR' [R'=Me, MeOCH$_2$ or Et$_2$NC(O)]. The lack of o,o'-dilithiation of PhCONEt$_2$, even with excess Bu^{s}Li. TMED, was emphasised;[24] 2,6-Li$_2$C$_6$H$_4$CONEt$_2$ is available from 2,6-Br$_2$C$_6$H$_3$CONEt. The differences in the sites of lithiation of (3-MeOC$_6$H$_4$CH$_2$OH)Cr(CO)$_3$ (at C$_4$) and 3-MeOC$_6$H$_4$CH$_2$OH (at C$_2$) was explained by the relative conformation of the Cr(CO)$_3$ complex and electrostatic factors.[25]

1.5 **Benzyl derivatives.** Crystal structures have been determined for (i) [PhCH$_2$Li.Et$_2$O]$_n$: infinite chains of PhCH$_2$ and Li: Li is approximately trigonal planar;[19] (ii) [Ph$_2$CH][Li(12-crown-4)$_2$] [Ph$_3$C][Li(12-crown-4)$_2$].THF,[26a] and [(C$_6$Cl$_5$)$_3$C][K(18-crown-6)]:[26b] separate ions; planar geometry at central carbon of anions, (iii) [m-(Me$_3$SiCH)C$_6$H$_4$][Li.TMED]$_2$ (12): two Li environments (η^2 and η^3) bound to carbon moiety;[27] (iv) [PhSO$_2$CPhHLi.TMED]$_2$: symmetric dimer having 8-membered ring [(SOLiO)$_2$]: Li---C interaction in solution but not in the solid state,[28] and (v) 1,2-Ph$_2$-benzocyclobutadienediide.(LiTMED)$_2$: two modifications in which Li and TMED take up different positions w.r.t. the 4-membered ring.[29] Compound (12), was obtained in a lithiation study of m-[(Me$_3$Si)$_n$-CH$_{3-n}$C$_6$H$_4$CH$_{3-m}$(SiMe$_3$)$_m$] (n,m=0, 1 or 2).

The ^{7}Li n.m.r. spectra and a.c. conductivities of ArCH$_2$Li polyamine complexes in arenes were reported: there is a correlation between δ^7Li and pKa of ArMe.[30]

1.6 **Other derivatives.** Simultaneous deposition of Li and HC≡CH in an argon matrix at 15K gave Li(CH$_2$=CH$_2$) - a planar, lithium bridging π-system.[31] Dimetallation of PhC≡CH occurs on reaction with BuLi.KOBu^t or BuLi.TMED: the latter provided[32] isomers of LiC$_6$H$_4$C≡CLi (m+p: o = 15:85).

^{13}C n.m.r. spectral studies have been reported[33] for RR'C=CHCH$_2$M (R=R'=H, M=Li or K): M=Li; R=H, R'=Bu^{t}CH$_2$; R=Me, R'=Et or Bu^{t}CH$_2$). The compound, Bu^{t}CH$_2$CH=CHCH$_2$-^{6}Li, prepared from Bu^t-^{6}Li and CH$_2$=CHCH=CH$_2$, exists in cyclo-C$_5$H$_{10}$ in covalent (both cis- and trans-isomers) and ionic delocalised forms; the latter is the more stable and derivative in the presence of amines.[33c] Lithium 2-Li-1,1,3,3-Ph$_4$-propenide, obtained from Li and Ph$_2$C=C=CPh$_2$ in THF at low T, was the subject of two n.m.r. spectral studies[34] - with two structural conclusions, namely, (i) an ionic allylic lithium (C$_2$v symmetry) with the second Li

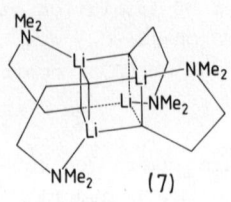

(7)

(12)

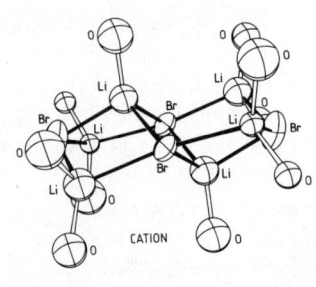

CATION

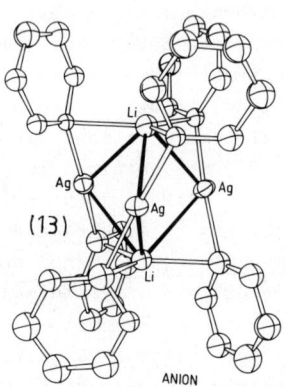

(13)

ANION

(14)

covalently bound to central carbon and (ii) another C_{2v} symmetric structure with two equivalent Li, interacting with the central C_2 carbon.

Good agreement was found between the crystal structures and MNDO calculated geometries for (i) cis-PhCHLiCH=CHCHLiPh.2TMED, obtained by dilithiation of either cis- or trans-PhCH$_2$CH=CHCH$_2$Ph [doubly-bridged (C_2) structure],[35] (ii) [Bu^{t}CLi=C=C=CLiBu^t.2THF]$_2$, produced on addition of Li to Bu^{t}C≡CC≡CBu^t in THF [dimeric cis-structure with 2 different Li environments][36] and (iii) Li$_2$-pentalenide.2DME [Li over each 5-membered ring on opposite faces of the 10π-aromatic system].[37]

A further double-bridged was suggested from ^{13}C n.m.r. spectra,[38] for Me$_2$C=CLiCLi=CMe$_2$, obtained by addition of Li to Me$_2$C=C=C=CMe$_2$. Crystal structures, mass and n.m.r. spectra were reported[39] for the monomeric [η-(Me$_3$Si)$_3$C$_5$H$_2$]Li.L (L = quinuclidine TMED or PMDT).

Coversion of H$_2$C=CH(CH$_2$)$_4$M to CH$_2$(CH$_2$)$_3$CHCH$_2$M is faster for M=Li than for R=MgX but is still 10^8-10^{10} times slower than cyclization of CH$_2$=CH(CH$_2$)$_4$. The mere observation of cyclopentylmethyl products in reactions of CH$_2$=CH(CH$_2$)$_4$X is not sufficient evidence for a radical process.[40] However the reaction of Bu^{t}Li with CH$_2$=CH(CH$_2$)$_3$CHRBr (R=H or Me) at low T does produce stable cyclized and uncyclized organolithiums as well as cyclized hydrocarbons; a radical component to the reaction was thereby indicated.[41]

2 Copper, Silver and Gold

2.1 π-Complexes. E.s.r. spectra of argon-matrix isolated MCO (M= Cu or Au) and trigonal planar Cu(CO)$_3$ were published.[42] Various [Cu diket]$_n$Lm (Hdiket = polyfluorinated β-diketone; L=CO or alkene) were obtained from Cu$_2$O, L and Hdiket.[43] Crystal structures of [Cu$_2$(COT)(CF$_3$COCHCOCF$_3$)$_2$][43] and [Cu(cyclohexene)(o-C$_5$H$_4$N)$_2$NH]$^+$ClO$_4^-$ [trigonal pyramidal Cu].[44] A photochemical study of Cu(alkene) complexes was undertaken: [Cu(H$_2$C=CH$_2$)]$^+$ in H$_2$C=CH$_2$ saturated MeOH provided[45] hexane and HCHO. Reactions of M with RCN, RNC or CH$_2$=C=CH$_2$ in a rotating cryostat at low T were investigated by e.s.r. spectroscopy.[46] Also studied were the isocyanide species [Cu(CNR)$_4$]$^+$X$^-$ (R=Me, Bu^t or C$_6$H$_3$Me$_2$; X=PF$_6$ or ClO$_4$)[47] and the carbene complexes, X$_3$Au[C(NMe$_2$)OEt] (X=Cl, Br or I) and

$X_N Au[C(NMe_2)N=CPhOMe]$ (n=1, X=Cl; n=3, X=Cl, Br or I).[48]

2.2 σ-Complexes. Photolysis of Cu in CH_4 matrices at 12K produced MeCuH, characterized by ir, uv-visible and e.s.r. spectra.[49] Thermal decomposition of MeCu at 20°C or $MeCuPR_3$ (R=cyclohexyl or Bu^t) at 100°C produced CH_4 and CH_3CH_3 as well as $CH_2=CH_2$ and C_3H_8; mechanism involved reductive elimination in a bimolecular concerted step.[50]

Crystal structures have been determined for (i) $[Li(12-crown-4)_2][CuR_2]$ (R=Me or Ph) and $[Li(12-crown-4)_2][Cu(Br)CH(SiMe_3)_2]$. PhMe: well separated cations and anions;[51] two co-ordinate Cu; (ii) trigonal bipyramidal, $Cu_4MgPh_6 \cdot Et_2O$ and $[Li(Et_2O)_4]^+[Cu_4LiPh_6]^- \cdot 2Et_2O$: Ph bridge the M_{ax}-M_{eq}; M_{eq} M_{eq} non-bonding,[52] (iii) $[Li_6Br_4(Et_2O)_{10}]^{2+}[Ag_3Li_2Ph_6]_2^-$: (13) trigonal bipyramidal anion, Li in axial sites;[53] (iv) $[Cu_2Li_2(C_6H_4CH_2NMe_2\text{-}\underline{o})_4]$: aryl groups bridge Cu---Li: Li-N co-ordination;[54] almost planar Cu_2Li_2; (v) $[\underline{o}\text{-}Me_2NCH_2C_6H_4Au.py_2]^{2+}$ $2BF_4^- \cdot Me_2CO$; square planar; Au-N co-ordination;[55] (vi) $[2,4,6\text{-}Me_3C_6H_2Cu_3(O_2CPh)_2]$, (14) obtained from $(2,4,6\text{-}Me_3C_6H_2Cu)_5$ and $(PhCO_2Cu)_4$: bridging $PhCO_2$ and mesityl groups,[56] and (vii) $\eta\text{-}MeC_5H_4Cu.PPh_3$.[57]

Fackler and co-workers have reported on further dinuclear gold ylide complexes, $[R_2PCH_2Au^ICH_2PR_2CH_2Au^ICH_2]$(15, R=Me or Ph), $[R_2PCH_2Au^{II}(X)PR_2CH_2Au^{II}(Y)CH_2]$ (16, R=Me or Ph; X=Br or I; X=Me, Me_3SiCH_2; $PhCH_2$...; X=Y=CN), $R_2PCH_2Au^{III}(X_2)CH_2PR_2CH_2Au^{III}(X_2)CH_2$ (17; R=Me or Ph; X=Cl or Br) and $XAu^ICH_2PR_2CH_2Au^{III}(Cl)CH_2PR_2CH_2$ (18, R=Me or Ph). Various crystal structures were determined including those of (15, R=Ph), (16, R=Ph; X=Br, Y=Me or CH_2Cl; X=I, Y=CF_3CH_2 or CH_3CH_2), trans,trans-(17, R=Ph, X=Br) and cis,trans (17, R=Ph, X=Cl). Compounds (16, Y=alkyl) are formed in equilibrium with (15) and alkyl halides.[58]

Other ylide complexes studied[59] were (i) $[Au\{(CH(PPh_3)CO_2R\}L]\text{-}ClO_4$ $(AuZL)ClO_4$ and $[(LAu)_2\{\mu\text{-}C(PPh_3)CO_2R\}]ClO_4$ (19, L=PPh_3 or $AsPh_3$; R=Me or Et), obtained from $(Ph_3PCH_2CO_2R)ClO_4$ and Au(acac)L, as well as AuZCl, $AuZCl_3$ $[AuZ_2]ClO_4$ and $[AuZ_2Cl_2]ClO_4$. In the solid, there is a short Au---Au contact in (19, L=PPh_3, R=Et).

Use of $(Ph_3PAu)_3O^+BF_4^-$ as an aurating agent has been illustrated[60] by the formation of various $RAu.PPh_3$ (20) from both RM (M=MgX or Li), e.g. R=$C_5H_4FeC_5H_4$, alkyl, acylcyclopropyl, benzyl, or aryl, and RH, e.g. R=$(H_3CCOC_5H_4)Mn(CO)_3$. Compound PhC≡CCu, was synthesised electrochemically by oxidation of a Cu

(15) (16) (17) (18) (19) (22) (23)

anode in a solution of PhC≡CH in Me_2CO or MeCN containing Et_4NClO_4;[61] the crystal structures were determined for (i) $[PPh_3 \cdot CuC≡CPh]_4$: tetrahedral skeleton of Cu bonded to $4PPh_3$ and to 4 μ_3 bridging PhC≡C groups,[62a] and (ii) $[Bu_4N][Au_3Cu_2(C≡CPh)_6]$: Cu apical in trigonal bipyramidal anion; Au σbonded to C≡CPh with each Cu π-bonded to 3 C≡CPh groups.[62b]

Air and moisture stable bimetallic aurate complexes, e.g. $[AuAg(C_6Cl_5)_2L]_n$ [L=THT, py, phen] and $[Ag(py)_2][Au(C_6Cl_5)_2]$ were obtained.[63] It was concluded that ^1H, ^7Li n.m.r. study at high field that in THF, and in the absence of LiX, Me_2CuLi (and higher order operates) do not exist as discrete complexes but rather in an equilibrium mixture with MeLi and Me_3Cu_2Li. However in Et_2O, Me_2CuLi behaves spectroscopically as a distinct entity.[64a] Ligand exchange occurs between lower order cuprates in ethereal solutions.[64b] Trinuclear complexes, $[X(C_6F_5)_2Au^{III}\{Ph_2PCH(Au^IL)-PPh_2\}Au^IX]$ (21) were obtained from $[(C_6F_5)_2Au(Ph_2PCHPPh_2)]$ and AuXL (X=Cl or Br; L=THT or Ph_3As; L in (21) can be replaced by Ph_3P or py. The crystal structure of (21, L=py) was also reported.[65]

Various Me_2Au^{III} complexes were studied,[66] including (i) (22) square-planar $Me_2Au[(o-C_5H_4N)_2C(OH)_2]NO_3$, obtained from $(o-C_5H_4)_2CO$ and Me_2AuNO_3 in H_2O: cis-square planar Au with N,N'-chelating ligand, (ii) $[Me_2AuL]NO_3$ [L=$(o-C_5H_4N)_n CO(N-Me-2-imidazolyl)_{2-n}$; n=0 or 1) and $(o-C_5H_4N)_n C(OH)(N-Me-2-imidazol)_{3-n}$ (n=0-3) : square planar Au with N,N'-bidentate ligand with the additional nitrogen site, if present, unco-ordinated, (iii) $Me_2AuB(pz)_4$ (23) (2-uncoordinated nitrogens), (iv) $Me_2Au(3,5-R_2-pz)$ (dimers in $CHCl_3$), (v) $[Me_2AuNMe_2]_2$: bent 4 membered Au_2N_2 ring,[67a] (vi) $(Me_2AuSEt)_2$: Au_2S_2 ring,[67b] and (vii) Me_2AuS_2X e.g. $X=PMe_2$, $P(OEt)_2$, CO_2Et S_2CPh).

3 Bibliography

<u>Organo-alkali metal</u>
P.v.R. Schleyer, D. Wilhelm, and T. Clark, <u>J. Organomet. Chem.</u>, 1985, <u>281</u>, C17. MNDO calculation on cycloheptatrienyl trilithium.
A.J. Kos, P. Stein, and P.v.R. Schleyer, <u>J. Organomet. Chem.</u>, 1985, <u>280</u>, C1. MNDO calculation on 1,4-dilithio-<u>cis</u>-w-butene.
M.L. McKee, <u>J. Am. Chem. Soc.</u>, 1985, <u>107</u>, 859. MNDO calculation on RLi-LiN(OR^1)R^2 reaction.
K.N. Houk, N.G. Rondan, P.v.R. Schleyer, E. Kaufmann, and T. Clark, <u>J. Am. Chem. Soc.</u>, 1985, <u>107</u>, 2821. 3-21G Bases set calculation on LiH and MeLi addition to $CH_2=CH_2$ and HC≡CH.
E. Kaufmann, P.v.R. Schleyer, K.N. Houk, and Y.-D. Wu, <u>J. Am. Chem. Soc.</u>, 1985, <u>107</u>, 5560. Ab initio calculation on addition of $(MeLi)_n$ (n=1 or 2) to H_2CO.

J.D. Watts and J.G. Stamper, *J. Chem. Soc. Chem. Commun.*, 1985, 5. M.O. calculation on CLi_2F_2.
V. Bellagamba, R. Ercoli, A. Gamba, and M. Simonetta, *J. Chem. Soc. Perkin Trans. 2*, 1985, 185. CNDO-2 level calculation on disociation of CH_2LiF in H_2O.
A.E. Reed and F. Weinhold, *J. Am. Chem. Soc.*, 1985, 107, 1919. SCF wave function for CLi_n^q (n=4→6; q=0, ±1, ±2).
P.A.A. Klusener, H.H. Hommes, H.D. Verkruijsee, and L.Brandsma, *J. Chem. Soc., Chem. Commun.*, 1985, 1677. Metallation of $H_2C=CMeCH=CH_2$.
J. Hooz, J.G. Calzada, and D. McMaster, *Tetrahedron Lett.*, 1985, 26, 271. Controlled lithiation of allenes.
M. Mallet and G. Queguiner, *Tetrahedron*, 1985, 41, 3433. Reactions of 3-Br-2-X-pyridines (X=F, Cl or Br): homotransmetallation.
P. Beak and C.-W. Chen, *Tetrahedron Lett.*, 1985, 26, 4979. Deprotonation of strongly acidic functions by RLi is faster than Br-Li exchange.
N.S. Narasimhan and R. Ammanamanchi, *J. Chem. Soc. Chem. Commun.*, 1985, 1368. Li-I exchange faster than OH-Li exchange.
T.N. Mitchell and W. Reimann, *J. Organomet. Chem.*, 1985, 281, 163. Formation and stabilities of $RCH=CLiSiMe_3$.
E.J. Corey, J. Kang, and K. Kyler, *Tetrahedron Lett.*, 1985, 26, 555. $Ph_3P=CHLi$.
T. Kauffmann, M. Bisling, R. König, A. Rensing, and F. Steinseifer, *Chem. Ber.* 1985, 118, 4517. Vinylcyclopentadienyllithium.
H. Yasuda, T. Nishi, S. Miyanaga, and A. Nakamura, *Organometallics*, 1985, 4, 359. $1,5-(Me_3Si)_2$-pentadienyllithium.
I.A. Garbuzova, O.G. Garkusha, B.V. Lokshin, G.K. Borisov, and T.S. Morozova, *J. Organomet. Chem.*, 1985, 279, 327. Vibrational spectra of C_5H_5M (M=Li→K).
A. Rajca and L.M. Tolbert, *J. Am. Chem. Soc.*, 1985, 107, 698. N.m.r. spectra of dipotassium diphenylmethylenemethandide.
L.M. Tolbert and A. Rajca, *J. Org. Chem.*, 1985, 50, 4805. $6,6-Me_2-2,4-Ph_2$-cyclohexadienyllithium.
K. Konishi, H. Matsumoto, K. Saito, and K. Takahashi, *Bull. Chem. Soc. Jpn.*, 1985, 58, 2294. N.m.r. spectra of o-$RR'CLiC_5H_4N$.
W. Huber, *J. Chem. Soc. Chem. Commun.*, 1985, 1630. E.s.r. spectrum of $PhH^{-\cdot}, Li^+$.
H.M. Walborsky and M. Duraisamy, *Tetrahedron Lett.*, 1985, 26, 2743. Inversion on alkylation of (S)-(+)-1-Cl-2,2-Ph_2-cyclopropane by BuLi.
P.M. Warner, S.-C. Chang, and N.J. Koszewski, *Tetrahedron Lett.*, 1985, 26, 5371. Inversion in reaction of 7-endo-Br-7-Li norcaranes with BuLi.
M. Newcomb, W.G. Williams, and E.L. Crumpacker, *Tetrahedron Lett.*, 1985, 26, 1183. Low radical extent in reaction of $BrCH_2CMe_2CH=CH_2$ and Bu^tLi in Et_2O/pentane at -23°C.
E. Juaristi, B. Gordillo, D. M. Aparicio, W.F. Bailey, and J.J. Patricia, *Tetrahedron Lett.*, 1985, 26, 1927. S_N2 mechanism for reaction of 1,3-dithianyl-Li with $XCH_2CH_2CH_2CH=CH_2$ at low T in weakly polar solvents.
M.A. Buese and T.E. Hogen-Esch, *J. Am. Chem. Soc.*, 1985, 60, 4509. Stereochemistry of methylation of $PhSOCHMeCH_2CHLiSOPh$.
H.J. Reich, N.C. Phillips, and I.L. Reich, *J. Am. Chem. Soc.*, 1985, 107, 4101. Li-I exchange reactions occur via ate complexes.
H.G. Richey, Jr., and J. Farkas, Jr., *Tetrahedron Lett.*, 1985, 26, 275. 1,4-Addition to pyridine using Et_2Mg-EtLi.
P.A. Wade, D.T. Price, J.P. McCauley, and P.J. Carroll, *J. Org. Chem.*, 1985, 50, 2804. Opposite diastereoselectivity in the addition of RLi and RMgX to 3-acylisoxazolines.
W.H. Hendrikson, Jr., W.D. McDonald, S.T. Howard, and J. Cowgado, *Tetrahedron Lett.* 1985, 26, 2939. S.e.t. in reaction of RLi with peroxides.
M.A. Al-Aseer, B.D. Allison, and S.G. Smith, *J. Org. Chem.*, 1985, 50, 2715. Mechanism of reaction of RLi and esters.
G.W. Klumpp, A.J.C. Mierop, J.J. Vrielink, A. Bringman, and M. Schakel, *J. Am. Chem. Soc.*, 1985, 107, 6740. Anti-Michael carbolithiation of $RR'C=CHCONHMe$ (R,R'=Me_3Si,Ph).
P. Beak, D.J. Kempf, and K.D. Wilson, *J. Am. Chem. Soc.*, 1985, 107, 4745. β-Lithiation of α,β-unsaturated amides.

Y. Yawakami, T. Aoki, and Y. Yamashita, Bull. Chem. Soc. Jpn., 1985, 58, 1329.
Enantioface-differentiating addition of BuLi to PhCOMe using chiral ligands,
e.g. (1S,2S,5S)-3-[(S)-2-MeO-1-R-ethylimino]-2-pinanol.
Y. Yamamoto and K. Maruyama, J. Am. Chem. Soc., 1985, 107, 6411. Enhanced
Cram selectivity using RLi.crown ethers in reactions with R'CHMeCHO.
Y. Yamamoto, T. Komatsu, and K. Maruyama, J. Organomet. Chem., 1985, 285, 31.
Control of the selectivity of $CH_2=CHCH_2M$ reactions with aldehydes.
Y. Yamamoto, T. Komatsu, and K. Maruyama, J. Org. Chem., 1985, 50, 3115.
Diasterofacial selectivity in reactions of $H_3CCH=CH=CH_2,Li^+$ with imines.
K. Tomioka, M. Sudani, Y. Shinmi,and K. Koga, Chem. Lett., 1985, 329. Enantio-
selective conjugate addition in presence of chiral ligands.
R.A. Wanat and D.B. Collum, J. Am. Chem. Soc., 1985, 107, 2078. Aggregation
effects on the stereoselectivity of alkylations of lithiated hydrazones.

Organo-copper, -silver and -gold
R. Gonzalez-Luque, M. Merchan, I. Nebot-Gil, F. Tomas,and R. Mantanana,
J. Mol. Struct., 1985, 121, 57; M. Merchan, R. Gonzalez-Luque, I. Nebot-Gil,
and F. Thomas, Chem. Phys. Lett., 1985, 114, 516. Theoretical studies on
$CH_2=CHCH_2Cu$.
B.H. Lipshutz, J.A. Kozlowski, D.A. Parker, S.L. Nguyen,and K.E. McCarthy,
J. Organomet. Chem., 1985, 285, 437. $R(2-thienyl)CuCNLi_2$.
G. Hallnemo, T. Olsson, and C. Ullenuis, J. Organomet. Chem., 1985, 282, 133.
N.m.r. investigations of reactions between Me_2CuLi and $ArCH=CHCO_2R$ (via π-complex)
E.J. Corey and N.W. Boaz, Tetrahedron Lett., 1985, 26, 6015,6019. Evidence
for d,π* cuprate enone complex in reaction of $LiCuMe_2$ with enones.
H.L. Goering, S.S. Kantner, and E.P. Seitz, Jr., J. Org. Chem., 1985, 50, 5495.
Reactivity of allylic esters with $LiCuMe_2$ (S_N2' mechanism).
C. Guo, M.L. Brownawell, and J. San Filippo,Jr., J. Am. Chem. Soc., 1985, 107,
6028. Isotope effects in reaction of $LiCuR_2$ &RI or RTos.
K.C. Dash, Indian J. Chem., Sect. A, 1985, 24, 265. Review on fluxional
organogold compounds.
M. Lan Franchi, M.A. Pellinghelli, A. Timpicchio,and F. Bonati, Acta Crystallogr.
Sect. C, 1985, 41, 52. Crystal structure of $Ph_3PAuC(OMe)=NC_6H_4Me-p$.
S. Komiya and A. Shibue, Organometallics, 1985, 4, 684. $cis-YC_6H_4AuMe_2(PAr_3)$.
J. Vicente, M.T. Chicote, M.D. Bermudez,and M. Garcia-Garcia, J. Organomet. Chem.
1985, 295, 125. $o-PhN=NC_6H_4AuCl_2$.

References

1. W. Setzer and P.v.R. Schleyer, Adv. Organomet. Chem., 1985, 24, 353.
2. J. Heinzer, J.F.M. Oth, and D. Seebach, Helv. Chim. Acta, 1985, 68, 1848.
3. G. Fraenkel and M. Slier, Am. Chem. Soc., Div. Pet. Chem., 1985, 30, 586;
   A.D. Schlüter, H. Hüber, and G. Szeimies, Angew. Chem. Int. Ed. Engl.,
   1985, 24, 404.
4. J.F.McGarrity and C.A. Ogle, J. Am. Chem. Soc., 1985, 107, 1805.
5. J.F. McGarrity, C.A. Ogle, Z. Brich,and H.-R. Loosli, J. Am. Chem. Soc.,
   1985, 107, 1810; see also, Yu.N. Baryshnikov, N.N. Koloshina,
   G.I. Vesnovskaya, and Yu.A. Kurskii, Zh. Obshch. Khim., 1984, 54, 2256.
6. C. Schade, W. Bauer,and P.v.R. Schleyer, J. Organomet. Chem., 1985, 295,
   C25.
7. J.A. Gurak, J.W. Chinn, Jr., R.J. Lagow, R.D. Kendrick,and C.S. Yannoni,
   Inorg. Chim. Acta, 1985, 96, L75.
8. G.F. Meyers, M.B. Hall, J.W. Chinn, Jr. and R.J. Lagow, J. Am. Chem. Soc.,
   1985, 107, 1413.
9. H. Kawa, B.C. Manley,and R.J. Lagow, J. Am. Chem. Soc., 1985, 107, 5313.
10. G.W. Klumpp, M. Vos, F.J.J. deKanter, C. Slob, H. Krabbendem, and
    A.L. Spek. J. Am. Chem. Soc., 1985, 107, 8292.
11. J. Barluenga, J. Florez, and M. Yus, Synthesis, 1985, 847; J. Barluenga,
    J. Fernandez and M. Yus, ibid., p.977; J. Chem. Soc. Perkin Trans. 1,
    1985, 447.

12. T. Fjeldberg, M.F. Lappert, and A.J. Thorne, J. Mol. Struct., 1985, 127, 95.
13. E.A. Mistryukov and I.K. Korshevets, Bull. Acad. Sci. USSR, Div. Chem. Sci. 1985, 34, 448.
14. J. Kaneti, P.v.R. Schleyer, and A.J. Kos, J. Chem. Soc. Chem. Commun., 1985, 1014.
15. D. Seebach, R. Amstrutz, T. Laube, W.B. Schweizer, and J.D. Dunitz, J. Am. Chem. Soc., 1985, 107, 5403; T. Laube, J.D. Dunitz, and D. Seebach, Helv. Chim. Acta, 1985, 68, 1373; J.T.B.D. Jastrzebski, G. van Koten, M.J.N. Christophersen, and C.H. Stam, J. Organomet. Chem., 1985, 292, 319.
16. H.J. Gais, H.J. Lindner, and J. Volhardt, Angew. Chem., Int. Ed. Engl., 1985, 24, 859.
17. J.J. Eisch, S.K. Dua, and M. Behrooz, J. Org. Chem., 1985, 50, 3674; J. Vollhardt, H.J. Gais, and K.L. Lukas, Angew. Chem., Int. Ed. Engl., 1985, 24, 610, 696.
18. U. Schümann, J. Kopf, and E. Weiss, Angew. Chem., Int. Ed. Engl., 1985, 24, 215.
19. M.A. Beno, H. Hope, M.M. Olmstead, and P.P. Power, Organometallics, 1985, 4, 2117.
20. I.R. Butler, W.R. Cullen, J. Ni, and S.J. Rettig, Organometallics, 1985, 4, 2197.
21. N.A.A. Al-Jabar and A.G. Massey, J. Organomet. Chem., 1985, 288, 145.
22. Y. Cohen, J. Klein, and M. Rabinovitz, J. Chem. Soc. Chem. Commun., 1985, 1033.
23. B. Schaub and M. Schlosser, Tetrahedron Lett., 1985, 26, 1623; G.H. Posner and K.A. Canella, J. Am. Chem. Soc., 1985, 107, 2571; M. Skowronska-Ptasinska, W. Verboom, and D.N. Reinhoudt, J. Org. Chem., 1985, 50, 2690.
24. R.J. Mills, R.F. Horvath, M.P. Sibi, and V. Snieckus, Tetrahedron Lett., 1985, 26, 1145.
25. M. Uemura, K. Take, K. Isobe, T. Minaru, and Y. Hayashi, Tetrahedron, 1985, 41, 5771.
26. (a) M.M. Olmstead and P.P. Power, J. Am. Chem. Soc., 1985, 107, 2174; (b) J. Veciana, J. Riera, J. Castaner, and N. Ferrer, J. Organomet. Chem., 1985, 297, 131.
27. L.M. Engelhardt, W.-P. Leung, C.L. Raston, and A.H. White, J. Chem. Soc., Dalton Trans., 1985, 337.
28. G. Boche, M. Marsch, K. Harms, and G.M. Sheldrick, Angew. Chem., Int. Ed. Engl., 1985, 24, 573.
29. G. Boche, H. Etzrodt, W. Massa, and G. Baum, Angew. Chem., Ind. Ed. Engl., 1985, 24, 863.
30. L. Assadourian, R. Faure, and G. Gau, J. Organomet. Chem., 1985, 280, 153.
31. L. Manceron and L. Andrews, J. Am. Chem. Soc., 1985, 107, 563.
32. L. Brandsma, H. Hommes, H.D. Verkpuijsee, and R.L.P. de Jong, Recl. Trav. Chim. Pays-Bas, 1985, 104, 226.
33. (a) R. Benn and A. Rufinska, Organometallics, 1985, 4, 209; (b) S. Bywater, P. Lachance and P. Black, J. Organomet. Chem., 1985, 280, 159; (c) C. Fraenkel, A.F. Halasa, V. Mochel, R. Stumpe, and D. Tate, J. Org. Chem., 1985, 50, 4563.
34. J. Bernard, C. Schneiders, and K. Müllen, J. Chem. Soc. Chem. Commun., 1985, 12; A. Rajca and L.M. Tolbert, J. Am. Chem. Soc., 1985, 107, 2969.
35. D. Wilhelm, T. Clark, P.v.R. Schleyer, H. Dietrich, and W. Mahdi, J. Organomet. Chem., 1985, 280, C6.
36. W. Neugebauer, G.A.P. Geiger, A.J. Kos, J.J. Stezowski, and P.v.R. Schleyer, Chem. Ber., 1985, 118, 1504.
37. J.J. Stezowski, H. Hoier, D. Wilhelm, T. Clark, and P.v.R. Schleyer, J. Chem. Soc., Chem. Commun., 1985, 1263.
38. A. Maercker and R. Dujardin, Angew. Chem. Int. Ed. Engl., 1985, 24, 571.
39. P. Judzi, E. Schlüter, S. Pohl, and W. Saak, Chem. Ber., 1985, 118, 1959.
40. W.F. Bailey, J.J. Patricia, V.C. Del Gobbo, R.M. Jarret, and P.J. Okarria, J. Org. Chem., 1985, 50, 1999.
41. E.C. Ashby, T.N. Pham, and B. Park, Tetrahedron Lett., 1985, 26, 4691.

42. P.H. Kasai and P.M. Jones, J. Am. Chem. Soc., 1985, 107, 813, 6385.
43. G. Doyle, K.A. Eriksen, and D. van Engen, Organometallics, 1985, 4, 830.
44. J.S. Thompson, J.C. Calabrese, and J.F. Whitney, Acta Crystallogr., Sect. C, 1985, 41, 890.
45. D. Geiger and G. Ferraudi, Inorg. Chim Acta, 1985, 101, 197.
46. J.A. Howard, R. Sutcliffe, H. Dahmane, and B. Mile, Organometallics, 1985, 4, 697; J.H.B. Chenier, J.A. Howard, and B. Mile, J. Am. Chem. Soc., 1985, 107, 4190.
47. A. Bell, R.A. Walton, D.A. Edwards, and M.A. Poulter, Inorg. Chim. Acta, 1985, 104, 171.
48. E.O. Fischer and M. Bock, J. Organomet. Chem., 1985, 287, 279.
49. J.M. Parnis, S.A. Mitchell, J. Garcia-Prieto, and G.A. Ozin, J. Am. Chem. Soc., 1985, 107, 8169.
50. S. Pasynkiewicz and J. Poplawska, J. Organomet. Chem., 1985, 282, 427; S. Pasynkiewicz, S. Pikul, and J. Poplawski, ibid., 1985, 293, 125.
51. H. Hope, M.M. Olmstead, P.P. Power, J. Sandell, and X. Xu, J. Am. Chem. Soc., 1985, 107, 4337.
52. S.I. Khan, P.G. Edwards, H.S.H. Yuan, and R. Ban, J. Am. Chem. Soc., 1985, 107, 1682.
53. M.Y. Chiang, E. Böhlen, and R. Bau, J. Am. Chem. Soc., 1985, 107, 1679.
54. G. van Koten, J.T.B.H. Jastrzebski, F. Muller, and C.H. Stam, J. Am. Chem. Soc., 1985, 107, 697.
55. J. Vicente, M.-T. Chicote, M.-D. Bermudez, P.G. Jones, and G.M. Sheldrick, J. Chem. Res. (S), 1985, 72.
56. H.L. Aalten, G. van Koten, K. Goubitz, and C.H. Stam, J. Chem. Soc., Chem. Commun., 1985, 1252.
57. T.P. Hanusa, T.A. Ulibarri, and W.J. Evans, Acta Crystallogr. Sect. C, 1985, 41, 1036.
58. J.D. Basil, H.H. Murray, J.P. Fackler, Jr., J. Tocher, A.M. Mazany, B. Trzcinska-Bancroft, H. Knackel, D.S. Dudis, J.J. Delord, and D.O. Marler, J. Am. Chem. Soc., 1985, 107, 6908; J.P. Fackler, Jr., and B. Trzcinska-Bancroft, Organometallics, 1985, 4, 1891; H.H. Murray, J.P. Fackler, Jr., and D.A. Tocher, J. Chem. Soc. Chem. Commun., 1985, 1279; H.H. Murray, A.M. Mazany, and J.P. Fackler, Jr., Organometallics, 1985, 4, 154; H.H. Murray, J.P. Fackler, Jr., ibid., p. 1633; D.S. Dudis and J.P. Fackler, Jr., Inorg. Chem., 1985, 24, 3758.
59. J. Vincente, M.T. Chicote, J.A. Cavuelas, J. Fernandez-Balza, P.G. Jones, G.M. Sheldrick, and P. Espinet, J. Chem. Soc. Dalton Trans., 1985, 1163.
60. E.G. Perevalova, M.D. Reshetova, and G.M. Kokhanyuk, Zh. Obshch. Khim., 1984, 54, 2711; E.G. Perevalova, I.G. Bolesov, Yu. T. Struchkov, I.F. Leschova, Ye. S. Kalyuzhnaya, T.I. Voyevodskaya, Yu. L. Slovokhotov, and K.I. Grandberg, J. Organomet. Chem., 1985, 286, 129; E.G. Perevalova, K.I. Grandberg, E.I. Smyslova, and E.S. Kalyuzhnaya, Bull. Acad. Sci. USSR, Div. Chem. Sci., 1985, 34, 191.
61. R. Kumar and D.G. Tuck, J. Organomet. Chem., 1985, 281, C47.
62. (a) L. Naldini, F. Demartin, M. Manassero, M. Sansoni, G. Rassu, and M.A. Zoroddu, J. Organomet. Chem., 1985, 279, C42; (b) O.M. Abu-Salah, A.-R.A. Al-Ohaly, and C.B. Knobler, J. Chem. Soc. Chem. Commun., 1985, 1502.
63. R. Uson, A. Laguna, M. Laguna, B.R. Manzano, and A. Tapia, Inorg. Chim. Acta, 1985, 101, 151.
64. (a) B.H.Lipshutz, J.A. Kozlowski, and C.M. Breneman, J. Am. Chem. Soc., 1985, 107, 3197; (b) idem., Tetrahedron Lett., 1985, 26, 5911.
65. P. Uson, A. Laguna, M. Laguna, B.R. Manzano, P.G. Jones, and G.M. Sheldrick, J. Chem. Soc., Chem. Commun., 1985, 2417.
66. P.K. Byers, A.J. Canty, L.M. Engelhardt, J.M. Patrick, and A.H. White, J. Chem. Soc. Dalton Trans., 1985, 981; P.K. Byers, A.J. Canty, N.-J. Minchin, J.M. Patrick, B.W. Skeleton, and A.H. White, ibid., p. 1183; P.K. Byers, A.J. Canty, B.W. Skeleton, and A.H. White, Aust. J. Chem., 1985, 38, 1251; P.K. Byers, A.J. Canty, K. Mills and L. Titcombe, J. Organomet. Chem., 1985, 295, 401.

67. (a) U. Grässle, W. Hiller, and J. Strähle, Z. Anorg. Allg. Chem., 1985, 529, 29; (b) H.W. Chen, C. Paparizos, and J.P. Fackler, Jr., Inorg. Chem. Acta, 1985, 96, 137.

2
Group II: The Alkaline Earths and Zinc and its Congeners

BY J. L. WARDELL

1 Beryllium

The dipolar nature of beryllocene has been confirmed in a microwave dielectric loss study. A rocking movement was suggested for the cp rings synchronous with the oscillation of Be between two equivalent positions in a η^5,σ-structure.[1] The compound, cpBeCl, has the same structure in the solid state as in the gas phase:[2] namely a monomeric η^5-structure. Successive treatment of $Me_3\overset{+}{P}\overset{-}{B}H_2P(O)Me_2$ with Bu^tLi and $BeCl_2$ produced[3] $\overline{CH_2PMe_2BH_2PMe_2XMXPMe_2BH_2PMe_2CH_2}$ (1; M=Be, X=O). Reduction of ketones by chiral cis-myrtanylberyllium compounds provides moderate optical yields of (R)-alcohols.[4]

2 Magnesium

A M.O. study was conducted on the π-complex, $H_2M(H_2C=CH_2)$ (M=Mg or Zn).[5]

A pure sample of $CH_2(MgBr)_2$ (2) has been obtained from the reaction of CH_2Br_2 and Mg/Hg in Et_2O/PhH; treatment of (2) with THF gave[6] insoluble $CH_2Mg\cdot CH_2(MgBr)_2$. The tri-Grignard reagent from cis-1,3,5-$(BrCH_2)$-cyclohexane, has been reported.[7] The products of the reaction between PhMgBr and Et_2O, during the Grignard preparation, has been further investigated.[8] Ethers can be cleaved[9] by Mg in the presence of a transition metal catalyst (obtained by reducing a chloride by Mg); e.g. THF gave $\overline{MgO(CH_2)_3CH_2}$. The compounds, RMgX.bipy (3, R=aryl or alkyl), $XMg(CH_2)_nMgX$.bipy (n=2, 4 or 5) and $[R^1_4N][RMgX_2]$.MeCN (R^1=Pr or Et) have been synthesised[10] electrochemically with a Mg electrode and RX in MeCN; (3) are less reactive than Grignards prepared by the usual route. Alkane soluble, donor-free dialkylmagnesiums, including the commercially available $BuMgBu^s$, metallate[11] various acidic hydrocarbons, RH to give R_2Mg, e.g. $R=R^1R^2C_5H_3$ ($R^1,R^2=R^1,R^2=H$, Me or $SiMe_3$), indenyl, 9-fluorenyl or RC≡C. Metallations of $PhSO_2Me$ and $PhSO_2Ph$ occur at the α- and σ-sites respectively;[11]

(For references see page 21

the ethyl-Grignard reagent metallates[12] MeRSi(C≡CH)$_2$ to give MeRSi(C≡CH)$_n$(C≡CMgBr)$_{2-n}$ (n=0 or 1).

Catalyzed (e.g. by cp$_2$ZrCl$_2$) Grignard addition occurs regioselectively to the C=C bond of various functionally substituted alkenes.[13] The intramolecular addition[14] of (E)-Me$_3$SiCH=CHCHMe-(CH$_2$)$_3$MgBr occurs in a suprafacial 5-exo-trig manner to give Me$_3$SiCHMgBrCH(CH$_2$)$_3$CHMe. The methyl-Grignard[15] and RC≡CSiMe$_3$, in the presence of a Ni^{II} and AlMe$_3$, provide (E)- and (Z)-RCMe=C(SiMe$_3$)-MgBr. Arylation of ArMgX or MgX$_2$ etherates by ArLi provides a convenient synthesis of Ar$_2$Mg.2L (4, L=Et$_2$O or THF); of interest, (4, L=Et$_2$O) but not (4, L=THF) can be decompletely desolvated. Mixed diarylmagnesiums, ArMgAr'.2THF, were shown by n.m.r. spectra, to be fluxional molecules.[16] In soluble Ph$_2$Mg is solubilized in aromatic solvents by the addition of EtOCH$_2$CH$_2$OM (M=Li→K); n.m.r. spectra suggest[17] that the resulting compounds have structures that vary with M. The bonding in allyl compounds, including cpMgCH$_2$CMe=CH$_2$ (5), was investigated[18a] by ^{1}H and ^{13}C n.m.r. spectra; it was concluded that (5), a fluxional molecule, possessed a σ-bonded allyl group. Other work on cpMg compounds, included[18b] an electron diffraction study on η-cpMgCH$_2$Bu^t and a M.O. calculation on cpMgH.

A MNDO calculation (3-21G, 6-31G, and MP2/6-31G levels) indicated the most stable structure of H$_2$CMgXCO$_2$MgX (the Ivanoff reagent) was a 1,3-1,3' doubly-bridged structure.[19] The synthesis and molecular structure of [9,10-H$_2$-9,10-(Me$_3$Si)$_2$-9,10-anthrylene]-Mg were reported: Mg bridges the 9 and 10 positions.[20] Crystal structures were also reported for (i) monomeric 1-BrMg-2-Bu^{t}CO-1,2,3,4-tetrahydroisoquinoline.3THF: octahedral Mg [bonded to C, Br, O(=C), and 3O of the THF groups];[21] (ii) [EtMg(2,2,1-cryptand)]$^+_2$[Mg$_2$(Et)$_6$]$^{2-}$ (from addition of 2,2,1-cryptand to Et$_2$Mg): anion has equivalent tetrahedral Mg with 2 bridging and 4 terminal Et groups:[22] Mg in cation is octahedral (3O, 2N and C); (iii) [Bu^{t}CH$_2$Mg(2,1,1-cryptand)]$^+$[Mg(CH$_2$Bu^t)$_3$]$^-$ (6): trigonal planar Mg in anion: pentagonal bipyramid Mg in cation[22] (4O, 2N and C): and(iv) the product of the thermal reaction of (Bu^{t}CH$_2$)$_2$Mg with 2,1,1-cryptand.[22] The ^{1}H n.m.r. spectra of (6) suggest the same ions survive in PhH.

Diethylmagnesium, in the presence of either 2,1,1-cryptand or EtLi, provides significant amounts of 1,4-addition products with pyridine.[22] A new method for assessing s.e.t. in Grignard/ketone reactions involves the comparison of the pseudo-first order rate constants for the loss of the paramagnetic intermediate with the rate of appearance of product.[23]

3 Zinc and Cadmium

Use of ultrasonics has been further illustrated by the in-situ formation and application of (i) R_2Zn, from RX (R=alkyl, aryl, vinyl or benzyl), Li and ZnX_2, and (ii) 6α-BrZn-penicillanate esters,[22] as well as in the preparation of homoallylic alcohols, from allylic halides, Zn and RCOR' in aqueous media.[24]

Reactive Cd powders,[25] obtained from $CdCl_2$ and Naph$^{-\cdot}$,Li$^+$ in glyme or THF, or Li-Cd alloy (ca. Cd_3Li, from Li and $CdCl_2$ with Naph as catalyst) have been used to obtain organocadmium compounds from RBr, $PhCH_2X$ (X=Cl or Br) or ArI. Thermal decomposition of R_2Zn (R=Me or Br) occurs via a radical chain mechanism.[26] The existence of mixed dialkyl zincs, RZnR', in mixtures of R_2Zn and R'_2Zn, has been confirmed both in a ^{13}C n.m.r. spectral study (for R=Bu^s, R'= Pr^i) and in a m.s. study (R,R'=alkyl, allyl or aryl).[27] A three co-ordinate zinc species, $[Me_3Si)_2N]PPh(=\overline{N(SiMe_3)})=NSiMe_3-\overline{Z}nPh$ (7, PhZnZ) was obtained[28] from the reaction of Ph_2Zn and $(Me_3Si)_2NP-$ $(=NSiMe_3)_2$; (7) reacts with ligands, L (L=Ph_3PO or py), to give PhZnZ.L. Ylide complexes (1, M=Zn or Cd; X=S) were formed[3] by the treatment of $LiCH_2\overset{+}{P}Me_2\overline{B}H_2P(S)Me_2$ with MS. ^{13}C N.m.r. spectra confirmed the presence of a Zn-C bond[29] in $BrZnCH_2CN$ and N$\cdots$Cd interaction[30] in $(Me_2NCH_2CH_2CH_2)_2M$ (8, M=Cd). Use of a Zn/Cu couple was made in the synthesis of $EtO_2C(CH_2)_nZnI$ (n=2 or 3) from the iodo esters;[31] also prepared[32] were the bis chelated $(Pr^iO_2CCH_2CH_2)_2Zn$ and the mono chelated $(Pr^iO_2CCH_2CH_2)_2Zn.Et_2O$ and $ClCdCH_2CH_2CO_2Pr^i$. The ate complex, $(Me_2PhSi)Bu^t_2Zn^-,Li^+$, adds[33] to both terminal and internal alkynes, in the presence of CuCN.

Preparations of trifluoromethyl compounds include (i) reaction[34] of activated metal M (M=Zn or Cd) in dry DMF with CF_2X_2 ($X_2=Br_2$, Cl_2 or BrCl) (yields >80%) and (ii) co-condensation at -196°C of Cd vapour with $CF_3\cdot$, obtained by low-energy radio frequency irradiation of $(F_3C)_2$; spectral details (e.g. ^{19}F n.m.r., m.s. and i.r.) of $(F_3C)_2Cd$ and $(F_3Si)_2Cd$, prepared similarly, were also given.[35] Complexes $(F_3C)_2M.D.$ (M=Zn or Cd; D=diglyme, glyme, py or MeCN) were reported[36] to be convenient sources of CF_2. Other perfluoroorgano cadmiums were obtained[37] by reaction in DMF of Cd powder with R_FI or C_6F_5Br. The structure of cp_2Zn in the solid state, contains infinite chains of Zn atoms with bridging cp.

Each Zn is also coordinated to a terminal cp; both σ and π interactions contributing to the cp-Zn bonds.[38] An electron diffraction study of $(C_5Me_5)_2Zn$ revealed one η^5-bound and one σ bound ring.[39]

4 Mercury

4.1 General. MNDO calculations were performed on (i) structures and fluxional behaviour of cyclopentadienylmercurials, (ii) structures, ionisation potentials, ΔH_F and dipole moments of other mercury compounds and (iii) addition of mercury ions to alkenes.[40] A p.e.s. study was carried out on RHgX (e.g. R=allyl, cyclopentadienyl, benzyl, vinyl and aryl);[41a] values of the vertical IP were determined. ^{19}F N.m.r. spectra were reported for various organomercurials, including[41b] $ArHgSC_6H_4F$-p, $ArSHgC_6H_4F$-p, $ArHgN(SO_2Ph)$-C_6H_4F-p, $RHgC\equiv CC_6H_4F$-p and $R'C\equiv CHgC_6H_4F$-p.

4.2 Methylmercury species. A useful synthesis[42a] of Me-^{203}HgCl involves the reaction of $^{203}HgCl_2$ with methylcobalamine in 0.01N HCl. Heats of solution of MeHgCl in H_2O and MeHgX (X=Cl, Br or I) in H_2O and pyridine were used to determine heats of solvation.[42b] Formation constants for MeHg^{II}-pyridine complexes decrease as the steric hindrance around N increases[43] in both CH_2Cl_2 and $MeNO_2$ but not in MeOH. Crystal structures were determined for (i) $[MeHg(\underline{o}\text{-}MeC_5H_4N)]^+$ $CF_3CO_2^-$: $CF_3CO_2^-$ is weakly bound to two Hg atoms, giving rise to loosely associated dimers,[43] (ii) $[(MeHg)_3S]^+ ClO_4^-$: trigonal-pyramidal cation:[44] ∠SHgC 172.3(14), 174.6(10) and 179.5(12)°. Hg-S 2.383(10) to 2.405(9); Hg···O 2.97(7) to 3.19(5)Å; Hg···S 3.483(9) and 3.498(11)Å, and (iii) $\overline{MeHgS\text{-}C\text{=}NC_6H_4\text{-}S\text{-}\underline{o}}$ (9): two independent molecules with secondary intramolecular interaction (Hg···C=N) and weak intermolecular (Hg···S) interactions.[45] Compound (9) and the PhHg-analogue are monomeric in solution. Other complexes to be studied include (i) MeHg^{II}-dithiol complexes (by ^1H n.m.r. spectroscopy and potentiometric titrations):[46] $HSCH_2CHSHCH_2X$ (X=OH or SO_3) but not $[HSCH(CO_2^-)_2]_2$ form chelates, (ii) methyl^{II} complexes with xanthosine (H_2Xan) (by ^1H and ^{13}C n.m.r. spectroscopy): complexes obtained[47] were MeHgHXan (two isomeric species linked via N_1-Hg or N_3-Hg) and $(MeHg)_2Xan$ (linked via N_1 and N_3) and (iii) MeHg^{II}-selenomethionine:[48] $MeHg\overset{+}{S}e(Me)CH_2CH_2CH(CO_2H)\overset{+}{N}H_3$ (at pH $\simeq$ 0.8) and $MeSeCH_2CH_2CH(CO_2H)\overset{+}{N}H_2HgMe$ (at pH$\simeq$9.0) were indicated.[4]

4.3 Other derivatives.

Complexes $[(CO)_3CrPh(CH_2)_n]HgX$ (10, RHgX: X=R; n=3 or 4), were formed[49] on reaction of $[Ph(CH_2)_n]_2$ with $Cr(CO)_6$ or $Cr(NH_3)_3(CO)_3$; also produced were (10, X=Cl). Treatment of cyclopropane with $Hg(OAc)_2$ and Bu^tOOH in 60% aqueous $HClO_4$ in CH_2Cl_2 for 3 d, followed by reaction with KBr, led[50] to the new species, $RO(CH_2)_3HgBr$ ($R=Bu^tO$, AcO or $BrHg(CH_2)_3$. Reaction of 3-MeO_2C-1,2-Ph_2-cyclopropene with $Hg(O_2CCF_3)_2$ in MeOH, followed by treatment with KCl, gave two isomeric products in a 5.5:1 ratio; the crystal structure of one (a trans-methoxymercuration adduct) was determined.[51] Syntheses were also reported for $ClHgCH_2CH(NHAc)$-$(CH_2)_8CO_2R$ (R=H or Me), via reaction[52] of $CH_2=CH(CH_2)_8CO_2R$ with MeCN and $Hg(NO_3)_2$, and trans-1-$PhSO_2$-3-cyclohexenyl-HgBr, on sulphonylmercuration[53] of 1,3-cyclohexadiene using $HgCl_2$ and $PhSO_4Na$ in aqueous acetone. Other sulphonylmercurations were also mentioned.[53] Various C-Hg bonded derivatives of 6-deoxyhex-5-enopyranosyl compounds have been obtained via oxymercuration.[54]

Partial resolution of $MeCH(HgBr)CH_2CH_2NMe_2$ (11) has been achieved[55] using dibenzoyltartaric acids; neither (11), obtained from the Grignard reagent, nor (8, M=Hg) are chelated. Further information has been given[56] on the series $(X_nMe_{3-n}SiCH_2)_2Hg$ (X=Cl or MeO; n=0-3). Treatment of $ClF(O_2N)CCO_2H$ with HgO gave $FCl(O_2N)C_2Hg$; ^{19}F n.m.r. and m.s. details were reported.[57] Dimercuration of RCH_2CHO (R=Me or Et) occurs using $HgCl_2$; the crystal structure of $EtC(HgCl)_2CHO$ was reported:[58] $\angle$C-Hg-Cl 173(2) and 172(2)°. Reaction of HgX_2 (X=Cl, Br or I) with $[Ph_2P(S)]_n[Me_2P(S)]_{3-n}CH$ (n=1 or 3) in EtOH gave $[Ph_2P(S)]_n$-$[Me_2P(S)]_{3-n}CHgX$ (12). In the crystal of (12, n=1, X=Cl), pseudo-tetrahedral Hg is co-ordinated in a tridentate fashion via 3S to the ligand. Raman and ^{31}P n.m.r. spectra of (12) and the Cd analogues were also reported.[59]

The following mercury containing radicals have been described[60] (i) the stable nitroxide, $\overline{CH_2(CH_2)_4NCH_2}\overline{CHCMe_2N(O\cdot)CMe_2CH}=CHgBr$ (13) in the crystal, there is intramolecular interaction Hg···N 2.66(2)Å, (ii) $Et_2Hg^{-\cdot}$, obtained[61] by γ-irradiation of Et_2Hg in CCl_3F and (iii) the initial product (14) of the photoreaction between R_2Hg (R=alkyl or aryl) and benzo 2,1-b; 3,4-b' dithiophen-4,5-dione.[61b] Reaction of Hg with $ArCH_2Br$ in DMF in the dark produced $ArCH_2HgX$ (13, X=Br) or (15, X=$ArCH_2$), if Br$^-$ was present;[62] ^{199}Hg and ^{13}C n.m.r. spectra were reported for (15) [Ar=$Br_nC_6H_{5-n}$ (n=0-3)]: it was concluded that there was σ-π conjugation of C-Hg bond with the ring and that there was intermolecular interaction between Hg and o-Br atoms.

The compounds, o-$(XHg)_2C_6H_4$ (16), form halide complexes $[R_4E]^+[\{o-(XHg)_2C_6H_4\}_2X]^-$ (17, $R_4E=Ph_4P$ or Bu_4N; X=Cl or Br) on reaction with R_4EX; X^- bridges two units of (16) and is strongly co-ordinated to 4 Hg atoms.[63] Decarboxylation[64] of mercury pyridine-2,3-dicarboxylate in DMSO or HMPT gave a mixture of 2-carboxylatopyridin-3-yl-HgX (major product) and 3-carboxylatopyridin-3-yl-HgX.

The crystal structure of the σ-bonded C_5Cl_5HgPh was determined; there was no evidence for secondary Cl-Hg bonding.[65] The compound is fluxional in solution. Reaction of $[ArC(O)C(=N_2)]_2Hg$ with $(F_3CSO_2)_2O$ [Tf_2O] in CH_2Cl_2 provided geometric isomers of $[ArC(OTf)=C(OTf)]_2Hg$; (17) crystal structure of (E,E)-(17, Ar=Ph) was determined.[66] Further details have been given[67] of the formation of $R^1C(SCN)=CR^2HgX$ from $R^1C\equiv CR^2$, $HgCl_2$ and SCN^-.

5 Bibliography

F. Sato, J. Organomet. Chem., 1985, 285, 53. Hydromagnesiation review.
Y. Yamamoto, T. Komatsu, and K. Maruyama, J. Org. Chem., 1985, 50, 3115. Diastereofacial selectivity in reactions of $MeCH=CHCH_2MgX$ with imines.
Y. Yamamoto and K. Maruyama, J. Am. Chem. Soc., 1985, 107, 6411. Enhanced Cram selectivity in reactions of R'CHMeCHO and RMgX.crown ether.
W. Buchowiecki, Z. Grosman-Zjawiona, and J. Zjawiony, Tetrahedron Lett., 1985, 26, 1245. Reaction of RMgX with $Ph_2P(O)\dot{N}-CH_2\dot{C}Me_2$
T. Hiyama and N. Wakasa, Tetrahedron Lett., 1985, 26, 3259. Asymmetric coupling of ArMgBr and allylic esters using $NiCl_2[(S,S)$-chiraphos] (upto 89% e.e.)
H. Brunner, W. Li, and H. Weber, J. Organomet. Chem., 1985, 288, 359. Cross-coupling of PhCHMeMgX and CH_2=CHBr in presence of chiral Ni cat (upto 32% e.e.)
A. Carpita, R. Rossi, and C.A. Veracini, Tetrahedron, 1985, 41, 1919. Ni or Pd catalysis of coupling of ArMX (M=Mg or Zn) and ArI.
E. Wenkart, J.M. Hanna, Jr., M.H. Leftin, E.L. Michelotti, K.T. Potts, and D. Usifer, J. Org. Chem., 1985, 50, 1125. Substitution of MeS in heteroarenes by H, R and Ar' via Ni catalyzed Grignard reactions.
K. Soai, H. Machida, and A. Ookawa, J. Chem. Soc. Chem. Commun., 1985, 469. Asymmetric conjugate addition of RMgX to α,β-unsaturated amides derived from (S)-2-(1-HO-1-Pri)pyrrolidine or (S)-prolinol.
J.-C. Fiaud and L. Aribi-Zvioueche, J. Organomet. Chem., 1985, 295, 383. Pd-catalysed asymmetric coupling of allylic esters and RZnX.
Yu.A. Alexander, S.A. Lebedev, and N.V. Kuznetsova, J. Organomet. Chem., 1985, 292, 39. Radicals in exchange of Cd, Zn and Hg compounds.
H.T. Kalinoski, U. Hacksell, D.F. Barofsky, E. Barofsky and G.D. Davies, Jr., J. Am. Chem. Soc., 1985, 107, 6476. Fast Atom Bombardment m.s. of RHgOAc.
M. Bassetti, B. Floris, and G. Illuminati, Organometallics, 1985, 4, 617. Reaction of $FcC\equiv CH$ with $Hg(OAc)_2$.
H.C. Brown, J.T. Kurek, M.-H. Rei, and K.L. Thompson, J. Org. Chem., 1985, 50, 1171. Solvomercuration-demercuration of alkenes in alcohols.
J. Barluenga, J.M. Martinez-Gallo, C. Najera and M.Yus, J. Chem. Res. S, 1985, 266. Chiral alcohols from oxymercuration of alkenes in presence of chiral R^*CO_2H.

J. Barluenga, F. Aznar, R. Liz and M.-P. Cabal, J. Chem. Soc., Chem. Commun., 1985, 1375. Aminomercuration of alk-3-en-1-ynes.
B. Giese, Angew Chem., Int. Ed. Engl., 1985, 24, 553. Review on Organomercurials as sources of radicals for carbon-carbon bond formation.
F.H. Gouzoules and R.A. Whitney, Tetrahedron Lett., 1985, 26, 3441. Retention in demercuration using $Hg(CH_2)_3SH$.
K.P. Butin, T.V. Magdesieva, and O.A. Reutov, Bull. Acad. Sci. USSR, Div. Chem. Sci., 1985, 34, 404; K.P. Butin, V.V. Bashilov, T.V. Magdesieva, V.I. Sokolov, and O.A. Reutov, ibid., 416. Radical mechanism for demercuration of 8-BrHgCHMe-quinoline in MeOH.
K.P. Butin and T.V. Magdesieva, J. Organomet. Chem., 1985, 292, 47. Br^- catalysed $S_E1(N)$ solvolysis of $p-O_2NC_6H_4CH(HgBr)CO_2Et$ in EtOH.
J.D. Wuest and B. Zacharie, J. Am. Chem. Soc., 1985, 107, 6121. Use of $o-(F_3CCO_2Hg)_2C_6H_4$ in reduction of thio ketones.
Y. Yamamoto and K. Maruyama, J. Organomet. Chem., 1985, 284, C45. Effect of $BF_3.Et_2O$ on the selectivity in reactions of PhCHO with $MeCH=CHCH_2MX$ (M=Cd or Hg).
S. Cacchi and G. Palmieri, J. Organomet. Chem., 1985, 282, C3. Pd catalysed reaction of ArHgX with α,β-enones.
V.V. Bashilov, E.V. Maskaeva, A.A. Musaev, V.I. Sokolov, and O.A. Reutov, Bull. Acad. Sci. USSR, Div. Chem. Sci., 1984, 33, 1466. Formation of organopalladium compounds from 8-BrHgCHR-quinoline.

References

1. S.J. Pratten, M.K. Cooper, M.J. Aroney, and S.W. Filipczink, J. Chem. Soc., Dalton Trans., 1985, 1761.
2. R. Goddard, J. Akhtar, and K.B. Starowieyski, J. Organomet. Chem., 1985, 282, 149.
3. H. Schmidbaur, E. Weiss, and W. Graf, Organometallics, 1985, 4, 1233.
4. G. Giacomelli, M. Falorni, and L. Lardicci, Gazz. Chim. Ital., 1985, 115, 289.
5. O. Gropen, A. Haaland, and D. Defrees, Acta Chem. Scand., Ser. A, 1985, 39, 367.
6. J.W. Bruin, G. Schat, O.S. Akkerman, and F. Bickelhaupt, J. Organomet. Chem., 1985, 288, 13.
7. P. Boudjouk, R. Sooriyakumaran, and C.A. Kapper, J. Organomet. Chem., 1985, 281, C21.
8. L.A. Jones, S.L. Kirby, D.M. Kean, and G.L. Campbell, J. Organomet. Chem., 1985, 284, 159.
9. E. Bartmann, J. Organomet. Chem., 1985, 284, 149.
10. P.C. Hayes, A. Osman, N. Sendeal, and D.G. Tuck, J. Organomet. Chem., 1985, 291, 1.
11. J.J. Eisch and R. Sanchez, J. Organomet. Chem., 1985, 296, C27; A.W. Duff, P.B. Hitchcock, M.R. Lappett, R.G. Taylor, and J.A. Segal, J. Organomet. Chem. 1985, 293, 271.
12. O.G. Yarosh, T.G. Ivanova, T.D. Burnashova, and M.G. Voronkov, Bull. Acad. Sci. USSR. Div. Chem. Sci., 1984, 33, 2394.
13. J.L. Fabre, M. Julia, J.-N. Verpeaux, Bull. Soc. Chim. Fr., 1985, 762, 772; U.M. Dzhemilev and O.S. Vostrikova, J. Organomet. Chem., 1985, 285, 43; U.M. Dzhemilev, O.S. Vostrikova, R.M. Sulmanov, A.G. Kukovinets, and L.M. Khalilov, Bull. Acad. Sci. USSR. Div. Chem. Sci., 1984, 33, 1873.
14. K. Utimoto, K. Imi, H. Shiragami, S. Fujikura, and H. Nozaki, Tetrahedron Lett., 1985, 26, 2101.
15. R.S.E. Conn, M. Karras, and B.B. Snider, Isr. J. Chem., 1984, 24, 108.
16. C.G. Screttas and M. Micha-Screttas, J. Organomet. Chem., 1985, 292, 325.
17. C.G. Screttas and M. Micha-Screttas, J. Organomet. Chem., 1985, 290, 1.
18. (a) R. Benn and A. Rufinska, Organometallics, 1985, 4, 209; R. Benn, H. Lehmskuhl, R. Mehler, and A. Rufinska, J. Organomet. Chem., 1985, 293, 1;

(b) R.A. Anderson, R. Blom, A. Haaland, B.E.R. Schilling, and H.V. Volden, Acta Chem. Scand., Ser. A, 1985, 39, 563.
19. J. Kaneti, P.v.R. Schleyer, and A.J. Kos, J. Chem. Soc. Chem. Commun., 1985, 1014.
20. H. Lehmkuhl, A. Shakoor, K. Mehler, C. Krüger, K. Angermund, and Y.-H. Tsay, Chem. Ber., 1985, 118, 4239.
21. D. Seebach, J. Hansen, P. Seiler, and J.M. Cromer, J. Organomet. Chem., 1985, 285, 1.
22. E.P. Squiller, R.R. Whittle, and H.G. Richey, Jnr., J. Am. Chem. Soc., 1985, 107, 432; Organometallics, 1985, 4, 1154; H.G. Richey, Jnr. and J. Farkas, Jnr., Tetrahedron Lett., 1985, 26, 275.
23. Y. Zhang, B. Wenderoth, W.-Y. Su, and E.C. Ashby, J. Organomet. Chem., 1985, 292, 29.
24. C. Petrier, J.C. de Souza Barboza, C. Dupuy, and J.-L. Luche, J. Org. Soc., 1985, 50, 5761; J.C. de Souza Barboza, C. Petrier, and J.-L. Luche, Tetrahedron Lett., 1985, 26, 829; J. Brennan and F.H.S. Hussain, Synthesis, 1985, 749; C. Petrier and J.-L. Luche, J. Org. Chem., 1985, 50, 910.
25. E.R. Burkhardt and R.D. Rieke, J. Org. Chem., 1985, 50, 416.
26. A.E. Sokolovskii and A.K. Baev, Zh. Obsch. Khim., 1984, 54, 2559; Izv. Vyssh. Uchebn. Zaved., Khim. Khim. Tekhnol., 1985, 28, 30.
27. R. Mynott, B. Gabor, H. Lehmkuhl, and I. Döring, Angew. Chem., Int. Ed. Engl., 1985, 24, 335; H. Nehl and W.R. Scheidt, J. Organomet. Chem., 1985, 289, 1.
28. V.D. Romanenko, V.V. Skopenko, and L.N. Markovskii, Zh. Obshch. Khim., 1984, 54, 2791.
29. F. Orsini, Synthesis, 1985, 500.
30. E. Langguth and K.-H. Thiele, Z. Anorg. Allg. Chem., 1985, 530, 69.
31. Y. Tamaru, H. Ochiai, T. Nakamura, K. Tsubaki, and Z.I. Yoshida, Tetrahedron Lett., 1985, 26, 5559.
32. E. Nakamura, J.-I. Shimada, and I. Kuwajima, Organometallics, 1985, 4, 641.
33. Y. Okuda, K. Wakamatsu, W. Tückmantel, K. Oshima, and H. Nozaki, Tetrahedron Lett., 1985, 26, 4629.
34. D.J. Burton and D.M. Wiemers, J. Am. Chem. Soc., 1985, 107, 5014.
35. M.A. Guerra, T.R. Bierschenk, and R.J. Lagow, J. Chem. Soc. Chem. Commun., 1985, 1550.
36. H. Lange and D. Naumann, J. Fluorine Chem., 1985, 27, 299.
37. P.L. Heinze and D.J. Burton, J. Fluorine Chem., 1985, 29, 359.
38. P.H.M. Budzelaar, J. Boersma, G.J.M. van der Kerk, A.L. Spek, and A.J.M. Duisenberg, J. Organomet. Chem., 1985, 29, 123.
39. R. Blom, A. Haaland, and J. Weidlein, J. Chem. Soc. Chem. Commun., 1985, 266.
40. M.J.S. Dewar, G.L. Grady, K.M. Merz, Jnr., and J.J.P. Stewart, Organometallics, 1985, 4, 1965; M.J.S. Dewar and K.M. Merz, Jun., p. 1967.
41. (a) V.N. Daidin, M.M. Timoshenko, Yu.V. Chizhov, Yu.A. Ustynyuk, and I.I. Kritskaya, J. Organomet. Chem., 1985, 292, 55; (b) A.S. Peregudov, U.F. Ivanov, E.I. Fedin, and D.N. Kravtsov, Bull. Acad. Sci. USSR. Div. Chem. Sci., 1985, 34, 1394; S.I. Pombrik, L.S. Golovchenko, E.V. Polunkin, A.S. Peregudov, and D.N. Kravtsov, J. Organomet. Chem., 1985, 292, 81.
42. (a) A. Naganuma, T. Urano, and N. Imura, J. Pharmacobio-Dyn., 1985, 8, 69; (b) A. Iverfeldt and I. Persson, Inorg. Chim. Acta, 1985, 103, 113.
43. R.D. Bach, H.B. Vardhan, A.F.M. Magsudar Rahman, and J.P. Oliver, Organometallics., 1985, 4, 846.
44. B. Kamenar, B. Kaitner, and S. Pocev, J. Chem. Soc. Dalton Trans., 1985, 2457.
45. J. Bravo, J.S. Casas, M.V. Castano, M. Gayoso, Y.P. Mascarenhas, A. Sanchez, C. de O.P. Santos, and J. Sordo, Inorg. Chem., 1985, 24, 3435.
46. A.P. Arnold, A.J. Canty, R.S. Reid, and D.L. Rabenstein, Can. J. Chem., 1985, 63, 2430.
47. F. Allaire, M. Simard, and A.L. Beauchamp, Can. J. Chem., 1985, 63, 2054.
48. A.A. Isab and A.P. Arnold, J. Coord. Chem., 1985, 14, 73.
49. V.S. Kaganovich and M.I. Rybinskaya, Bull. Acad. Sci. USSR. 1984, 33, 1471.
50. A.J. Bloodworth and C.J. Cooksey, J. Organomet. Chem., 1985, 295, 131.

Group II: The Alkaline Earths and Zinc and its Congeners

51. V.R. Kartashov, N.V. Gal'yanova, E.V. Skorobogatova, A.N. Chernov, and N.S. Zefirov, J. Org. Chem. USSR (Engl. Transl.), 1984, 20, 2389.
52. J. Perthuis and P. Poisson, Bull. Soc. Chim. Fr., 1985, 75.
53. O.S. Andell and J.E. Bäckvall, Tetrahedron Lett., 1985, 26, 4555.
54. R. Blattner, R.J. Ferrier, and S.R. Haines, J. Chem. Soc. Perkin Trans. 1, 1985, 2413.
55. V.V. Bashilov, E.V. Maskaeva, P.V. Petrovskii, and V.I. Sokolov, J. Organomet. Chem., 1985, 292, 89.
56. L.I. Rybin, O.A. Vyazankina, N.S. Vyazankin, T.V. Leshina, M.B. Taraban, D.V. Gendin, V.I. Mar'yasova, and M.F. Larin, Zh. Obshch. Khim., 1984, 54, 2025.
57. I.V. Martynov, V.K. Brel', L.V. Postnova, and B.I. Martynov, Bull. Acad. Sci. USSR, Div. Chem. Sci., 1984, 33, 2597.
58. B. Korpar-Colig, Z. Popovic, and M. Sikirica, Croat. Chem. Acta, 1984, 57, 689.
59. S.O. Grim, P.H. Smith, S. Nittolo, H.L. Ammon, L.C. Satek, S.A. Sangokoya, R.K. Khanna, I.J. Colquhoun, W. McFarlane, and J.R. Holden, Inorg. Chem., 1985, 24, 2889.
60. L.G. Kuzmina and Yu. T. Struchkov, Koord. Khim., 1985, 11, 118.
61. (a) J. Rideout and M.C.R. Symons, J. Chem. Soc., Chem. Commun., 1985, 129; (b) A. Alberti, F.P. Colonna, M.C. Depew, and X. Li, J. Organomet. Chem., 1985, 292, 335.
62. K.P. Butin, A.A. Ivkina, and O.A. Reutov, Bull. Acad. Sci. USSR, Div. Chem. Sci., 1985, 34, 613; Yu.K. Grishin, O.K. Sokolikova, Yu.A. Ustynnyuk, A.A. Ivkina, K.P. Butin, and O.A. Reutov, ibid., p.719.
63. J.D. Wuest and B. Zacharie, Organometallics, 1985, 4, 410.
64. G.B. Deacon and G.N. Stretton, Aust. J. Chem., 1985, 38, 419.
65. A.G. Davies, J.P. Goddard, M.B. Hursthouse, and N.P.C. Walker, J. Chem. Soc. Dalton Trans., 1985, 471.
66. G. Maas, R. Brückmann, and W. Lorenz, J. Organomet. Chem., 1985, 289, 9.
67. M. Giffard, J. Cousseau, L. Gouin, and M.R. Crahe, J. Organomet. Chem., 1985, 287, 287.

3
Boron with the Exception of the Carbaboranes

BY J. W. WILSON

1 Introduction

As in the previous two years this chapter is an attempt to give a balanced report on the significant chemistry of organoboron compounds containing at least one boron-carbon bond. It is not therefore a comprehensive review of the chemistry of organic compounds of boron.

2 Books and Reviews

Organoboron compounds feature in part three of a series published by Houben-Weyl which is now available in a fourth edition.[1] The use of organoboranes in organic synthesis has been reviewed twice[2,3] and their use in the formation of carbon-carbon and carbon-heteroatom bonds reported on extensively.[4] Reviews have also appeared on the following topics: asymmetric synthesis,[5,6,7] the application of haloboration in organic synthesis,[8] B-N perturbed metallocenes,[9] skyscraper molecules containing the prefabricated "C_3B_2"Ni unit[10] and the use of cobalt complexes containing cyclic organoboron ligands in the synthesis of pyridines.[11]

3 Uses of Organoboranes and Organoborates in Organic Synthesis

Certain hydroboration reactions which are slow under normal conditions due to their heterogeneous nature proceed much faster under the effects of ultrasound.[12] Several acyclic and cyclic trienes have been hydroborated with borane or triethylamineborane in THF and the stereochemistry of the products investigated.[13] A systematic study of the hydroboration of representative heterocyclic olefins with four different reagents has been undertaken in order to establish optimum conditions for the reactions.[14] This study has been followed by an investigation into the relative rates of reaction of 9-BBN with oxygen and sulphur containing heterocyclic olefins.[15]

The action of four hydroborating reagents on 1,4-epoxy-1,4--dihydronaphthalene has been compared. Depending on the steric

(For references see page 35

demands of the reagent the oxidation product is either the exo or the homoallylic alcohol.[16] Hydroboration of representative alkenes with disiamylborane dimer has been studied kinetically in THF and earlier results proved to be erroneous. It would appear therefore that the dissociation mechanism generally applies to hydroborations involving dialkylborane dimers.[17] An additional method for selective reduction of double bonds in the presence of reactive functional groups has been developed using catecholborane in the presence of a rhodium catalyst. Normally sluggish reactions occur without difficulty at room temperature in the presence of such catalysts which activate the C-C multiple bonds.[18]

Lithium hydro tri-sec-butylborate has been shown to be a powerful and efficient reducing agent for organic halides.[19] The conversion of glycosides and disaccharides into alditol derivatives has been achieved with ethyldiboranes in the presence of 9-methanesulphonyl-9-BBN.[20] A convenient and simple procedure for the synthesis of various optically pure lithium mono and dialkylborohydrides has been reported. These are potential asymmetric reducing agents and can be stored for extended periods without hydride loss, isomerization or racemisation. The optically pure mono or dialkylborane can be generated as required by reaction with hydrogen chloride or iodomethane.[21]

There continues to be considerable effort applied to the use of organoboranes in chiral synthesis. A convenient procedure for the synthesis of various β-chiral boronic esters of essentially 100% ee which are not available by direct asymmetric hydroboration methods has been developed and their synthetic utility described.[22] The conversion of such boronic esters into aldehydes, acids and homologated alcohols of very high enantiomeric purities is readily achieved.[23] An elegant investigation has shown that the R,R isomer of trans-2,5-dimethylborolane (1), obtained from the resolution of the racemic methoxy derivative, the first report of such a resolution, hydroborates a series of achiral olefins to give very high asymmetric inductions and is superior to any other reagent currently in use.[24] Asymmetric synthesis of the diastereomeric 1-(2-cyclohexenyl)-1-alkanals in high optical purity via B-2-cyclohexen-1-yldiisopinocampheylborane, a stereochemically stable allylic borane, has been achieved and the use of the intermediate borane in "one pot" procedures for carbon-carbon bond formation by reaction with aldehydes demonstrated.[25] Diisopinocampheylchloroborane reduces ketones at convenient rates at -25°C. The chiral induction is excellent for the reduction of aromatic prochiral ketones.[26] (+)-Diisopinocampheylborane itself has been used in the stereo-controlled total synthesis of hirsutic acid C, a fused triquinone natural product.[27] B-(3-pinanyl)-9-BBN (Alpine Borane), in concentrated solutions, has been shown to be an extremely versatile chiral reducing agent for a broad spectrum of prochiral carbonyl compounds with enantiomeric purities approaching 100%.[28] The same reagent with acyl cyanides followed by treatment with sodium borohydride and cobaltous chloride produces optically active β-amino alcohols.[29]

A method for upgrading the enantiomeric purities of (+)-longifolene and (+)-3-carene to levels approaching 100% ee has been reported.[30] This means that the corresponding dialkylboranes are now available for asymmetric hydroboration reactions.

Zinc modified cyanoborohydride has been found to be a selective and versatile reducing agent[31] and the reaction of potassium hydride with 9-BBN borinic esters gives rise to a new class of stereoselective reducing agents.[32]

9-(Phenylseleno)-9-BBN undergoes conjugate addition to a variety of α,β-unsaturated ketones to give β-seleno boron enolates. These are readily converted to unsaturated ketals.[33] Arylboronic acids obtained from directed <u>ortho</u>-metalation of benzamides and carbamates undergo efficient palladium-catalysed cross-coupling reactions with a variety of aryl halides to give unsymmetrical biaryls and heterobiaryls.[34] The use of phenylboronate as a protecting group in the stereoselective total synthesis of the complement inhibitor K-76 has demonstrated that this group serves to moderate the basicity of 1,2 diols.[35] Ethylboron groups have also been used as effective protecting groups in the synthesis of disaccharides.[36] Chlorination of ($\underline{E}$)-vinylboronic acids gives isomerically pure ($\underline{Z}$) vinly chlorides in good yields.[37] The stereochemistry of the reactions of crotylboronates with chiral α,β-dialkoxyaldehydes has been investigated and the results suggest that they should be viewed as [3,3]sigmatropic rearrangements of the intermediate allylicboronate-aldehyde complexes in which the stereochemical outcome is strongly influenced by steric effects.[38] It would appear that the high facial selectivity in the reactions of the same aldehydes with allylic boronates is the exception rather than the rule.[39]

Boron is one element that features in a study of the influence of "metal-tuning" on stereo and regio-chemical convergence in reactions of allylic carbanions with aldehydes.[40] Allylic boron compounds react with α-imino esters to give very high enantio and diastereoselective syntheses of certain amino acids.[41] The preparation and aldehyde addition reactions of a series of 2-allyl--1,3,2-dioxaborolane-4,5-dicarboxylic acid esters (2) has been described. These tartrate ester based reagents are the most highly enantioselective group of chiral allylboronate esters reported to date and as such are attractive alternatives to B-allyl diisopinocampheylborane.[42]

Organoboranes have featured in the synthesis of a variety of unsaturated systems. (Z)-β-bromo or iodo alkenyl-9-BBN undergo Michael type reactions with acyclic α,β-unsaturated ketones to give (Z)-δ-halo-γ,δ-unsaturated ketones which are stereochemically pure (>98%).[43] (Z)-1,2-dihalo-1-alkenes can be prepared stereo and regioselectively in good yields from 1-alkynes by bromoboration with BBr_3 followed by treatment with iodine monochloride or bromine chloride in the presence of sodium acetate.[44] Palladium catalysed cross-coupling reactions of haloalkenes and alkynes with 1-alkenylboranes in the presence of base proceeds stereo and regiospecifically to give conjugated alkadienes or alkenynes in high yields. The configurations of both reagents are retained during the reaction which thus provides a reliable procedure for the synthesis of stereodefined conjugated alkadienes and enynes under mild conditions.[45] Isomerically pure (E)-vinyl sulphides[46] and S-alkyl alkanethioates[47] have been synthesised via 1-iodoalk-1--enyldialkylboranes. In the synthetic procedure for the vinyl sulphides an alternative to the usual protonolysis reaction using carboxylic acids had to be devised. Aldehydes and ketones add smoothly to B-alkynyl-9-BBN derivatives to yield the corresponding propargylic alcohols in excellent yields. These boron reagents have advantages over other metal alkynyls in that they can be conveniently prepared and stored.[48] The synthesis of trimethylsilyl substituted α-allenic and β-acetylenic amines from imines and allenic or propargylic organoboranes has been investigated. The allenic to acetylenic product ratio is dramatically influenced by the structures of the reactants. The relative reactivities of representative aldehydes and ketones towards trimethylsilyl substituted propargylic boranes has been reported.[50] Lithium trialkylalkynylborates react with allylic epoxides selectively to give hexa-1,5-dien-3-ols[51] and chloromagnesium trialkylvinylborates react with carbon dioxide under pressure to give β-hydroxy-

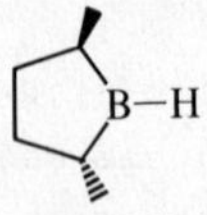
B—H

(1)

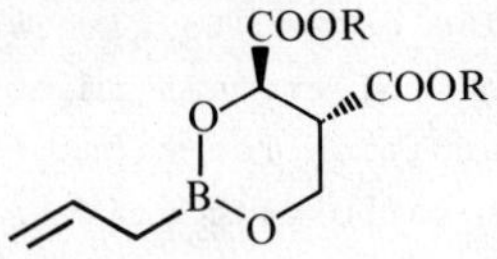
COOR
COOR
O
B
O

(2)

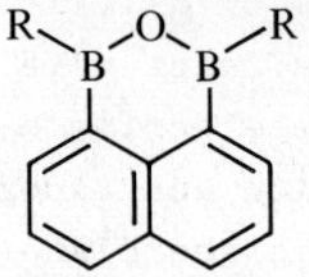
R
B
O
B
R

(3)

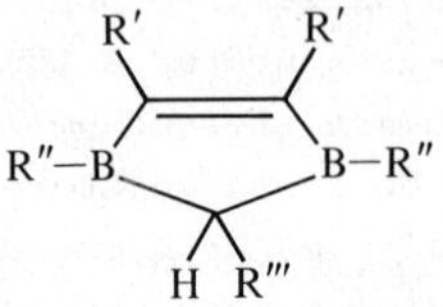
R′
R′
R″—B
B—R″
H R‴

(4)

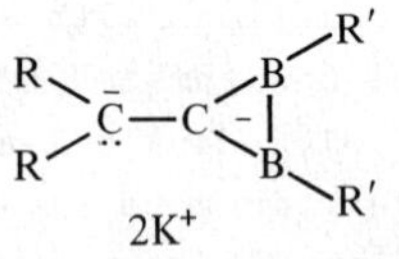
R
R
C
C
B
B
R′
R′
2K+

(5)

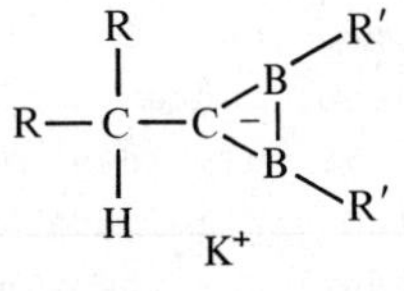
R
R—C—C
H
B
B
R′
R′
K+

(6)

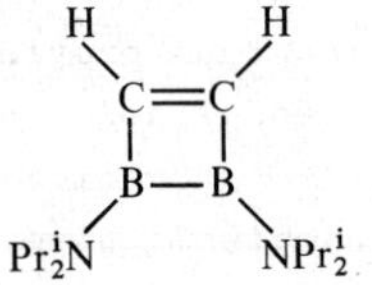
H
H
C═C
B—B
Pr$^i_2$N
NPr$^i_2$

(7)

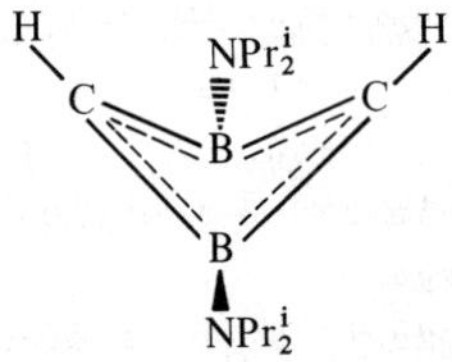
H
H
C
C
NPr$^i_2$
B
B
NPr$^i_2$

(8)

carboxylic acids in approximately 80% yields after oxidation of the intermediate.[52]

Organoboranes derived from 9-BBN react with the acetate $Ph_2C=NCH(OAc)(COOEt)$ in the presence of a hindered potassium aryloxide to yield protected higher amino acids.[53]

A general synthesis of cyclopropylboranes has been developed and its use as a stereocontrolled route to substituted cyclopropanol derivatives demonstrated.[54] The reaction of lithium dichloromethane with organoboranes has been studied systematically with a view to preparing homologated primary and secondary alcohols.[55] Low temperature oxidation of the intermediates in the reaction between arylaldehydes and boron stabilized carbanions leads to erythro-1,2-diols with high stereoselectivity. The reaction is unique to boron-Wittig reagents.[56] The same carbanions with oxiranes lead to 1,3-diols and the scope and limitations of this reaction have been investigated.[57] A similar study has been carried out on the use of lithium tris(phenylthio)methane as a reagent for the homologation of trialkylboranes.[58] Bromodimethylborane has been used under mild conditions for the interconversion of both mono-methoxy methyl and (2-methoxy ethoxy)methyl ethers into the corresponding (methylthio)methyl ethers. Dialkylacetals give O,S-acetals and the method has also been used to prepare cyanomethyl ethers.[59]

4 Preparations and Reactions of Organoboranes

Two novel routes to triethylborane have been reported. Irradiation of bromoethane and aluminium powder with ultrasound gives ethyl aluminium sesquibromide which on treatment with triethoxyborane gives triethylborane in good yields[60] and a laser initiated gas phase reaction between diborane and ethene gives yields of upto 91%.[61] Allylic boranes have been prepared from allylpotassium derivatives and chloroboranes. Hydrolysis leads to the isomerised olefin and the technique has been used to transform (+)-α-pinene into (+)-β-pinene.[62] Condensation reactions between allylboranes and acetylenes have been developed into a convenient method for the synthesis of bicyclo[3.3.1]nonane derivatives.[63] Mainly linear alkyl derivatives of 9-BBN have been synthesised from the reaction of iron carbonyls and the organoborane in a Fischer-Tropsch type reaction.[64]

A simple preparation of symmetrical and unsymmetrical borinic esters from selected boronic esters and organolithium reagents has been developed as an alternative to the hydroboration procedure.[65] Convenient and reliable methods for the synthesis of methylboronic acid and trimethylboroxin have also been published.[66] Catecholborane can now be made in yields >75% by a simple ball milling technique.[67] Methods for one carbon homologation of boronic acid and its esters using chloromethyl lithium[68] and dichloromethyl lithium[69] are now available. Pinacol[1-(trimethylstannyl)ethyl]-boronate is easily destannylated by methyl lithium at $-100°C$. The product is readily captured at this temperature by (α-haloalkyl)-boronic esters with the formation of a carbon-carbon bond.[70] The reaction between diethylborinic acid and aluminium alkyls has been studied and a two stage mechanism proposed.[71] Boronic acid derivatives of pyrimidines and 2'-deoxyribonucleosides have been synthesised and tested for anti-viral and anti-cancer activity.[72] Interaction of 1,8-naphthalenediyl bis(dimethylborane) with hydride, fluoride and hydroxide shows it to be a bidentate receptor for small anions.[73,74] In conjunction with this work a series of novel 1H,3H-naphth[1,8-cd][1,2,6]oxadiborins (3) have been synthesised and characterised.[75]

Bis(diiodoboryl)methane[76] reacts with hex-3-yne or but-2-yne to give 4,5-dialkyl-1,3-diiodo-2,3-dihydro-1,3-diboroles (1,3-diborolenes) in which the iodo functions can be readily substituted by a wide range of other groups to give (4).[77] Methods for the preparation of air stable trimethylsilyl substituted trivinylboranes[78] and 9-halo-9-borafluorenes (which are readily converted into methoxy, methylthio and diethyl amino derivatives)[79] have been published. The ability of various cuprates to transfer organo groups to 9-BBN has been investigated[80] and the reactions of aromatic organo copper compounds with organoboranes probed.[81]

The first three membered ring anions (5) and (6) have been reported and can be isolated as potassium salts. Protonation of (6) leads to the formation of a hydrogen bridged system.[82] Possible isomers of $C_2B_2H_4$ have been studied by *ab initio* M.O. methods and the analysis shows the ease with which the various structural types can isomerize.[83] The suggestion that 1,2 diboretenes should rearrange spontaneously to the more thermodynamically stable 1,3-isomers and are therefore not likely to be observed experimentally has been proved incorrect by the isolation of simple derivative[84] (7). The compound has a planar ring and isomerizes at $120°C$ to the folded 1,3-diboretene (8). Effects of first-row

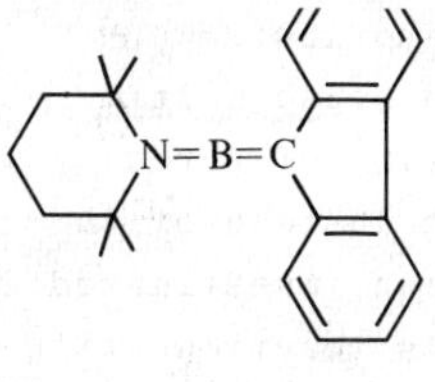

Boron with the Exception of the Carbaboranes

(9)

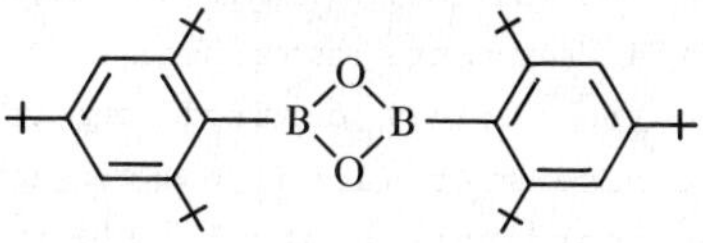

(10)

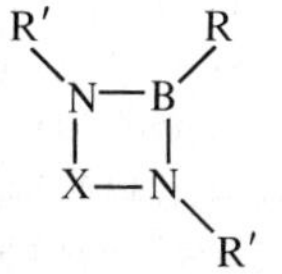

(11) X = SiMe$_2$, Me(O)P, R''(S)P, MeP

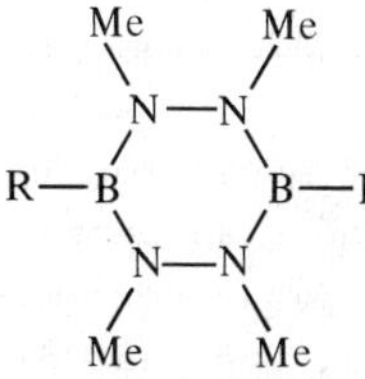

(12)

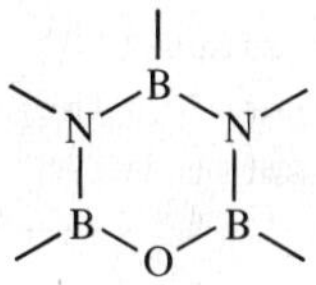

(13)

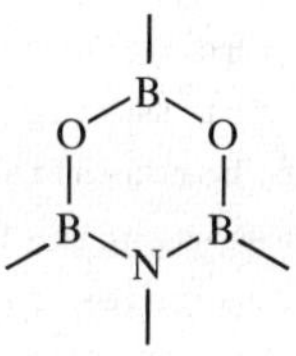

(14)

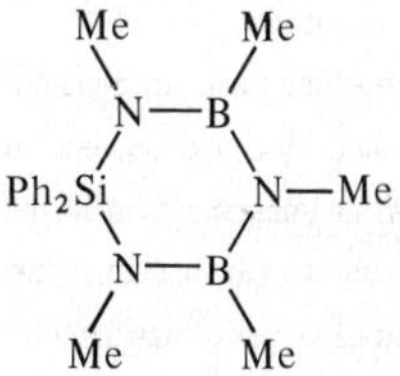

(15)

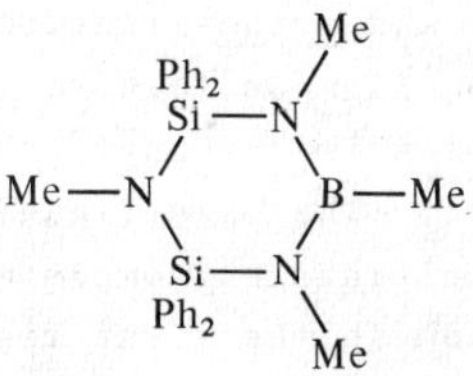

(16)

substituents on the structures and stabilities of boranes, boriranes and borirenes have been investigated by <u>ab initio</u> methods.[85] The factors controlling the formation of three coordinate, base stabilized, borenium (1+) ions instead of the simple four coordinate neutral adduct, have been investigated.[86] An allene isoster (9) has been isolated and its reactions with a range of reagents studied.[87] A kinetically stable and chemically inert iminoborane $(Me_3Si)_3CBNSiMe_3$ has been isolated.[88] Chemical and spectroscopic evidence has also been obtained for simple boraethenes $XBCH_2$ (X = OMe, NMe_2).[89]

The synthesis of boron containing heterocyclic ring systems remains an active field. A 1,3-dioxa-2,4-diboretane (10) has been synthesised in the quest for an oxoborane RBO for which it might be a precursor.[90] The diazaphospa or sila boretidines (11) have been reported[91] and tetrazadiborinanes (12) undergo [2+3] cycloaddition reactions with CO_2, COS and CS_2 to give the corresponding oxadiaza and thiadiazaborolidines.[92] A series of 1,4-dithia-2,6-diaza-3,5- -diboriranes have been subjected to a multinuclear n.m.r. spectroscopic and mass spectrometric investigation.[93] Four different routes to the boroxazines (13) and (14) have been published[94] as have the synthesis and structures of the silaborazines (15) and (16)[95] and methods for making <u>N</u>-borylated borazines.[96] The preparation of some boron-nitrogen analogues of substituted uracils,[97] of 1-aza-3-azonia-4-boratanaphthalenes[98,99] and of boron-nitrogen heterocycles from <u>N</u>'-aryl-<u>N</u>-(thiazol-2-yl)ureides and organoboranes have been reported.[100] A number of heterocyclic organoboranes containing a tertiary nitrogen atom have been prepared and their reactions with organic isocyanates discussed.[101] Cyclic chiral triarylboranes are obtained by the reaction of triphenylborane and <u>ortho</u>-lithiated aromatic amines.[102]

Organoboron ligands feature in several new metal complexes. The synthesis of a tricarbonyl cobalt 1,3-diborolyl complex has been reported.[103] The same ligand features in symmetrical and unsymmetrical multidecker complexes of platinum with iron, cobalt and nickel[104] and in the first reported hexadecker sandwich complex.[105] Thia-diborole complexes of nickel and cobalt have also been made[106] and irradiation of a 2:1 mixture of the same ligand and iron pentacarbonyl gives bis(1,2,5-thiadiborolene)monocarbonyl iron, the crystal structure of which has been determined.[107] The 1-(diisopropylamino)borolenediide ligand (17) is a versatile reagent for the formation of metal complexes. In the presence of

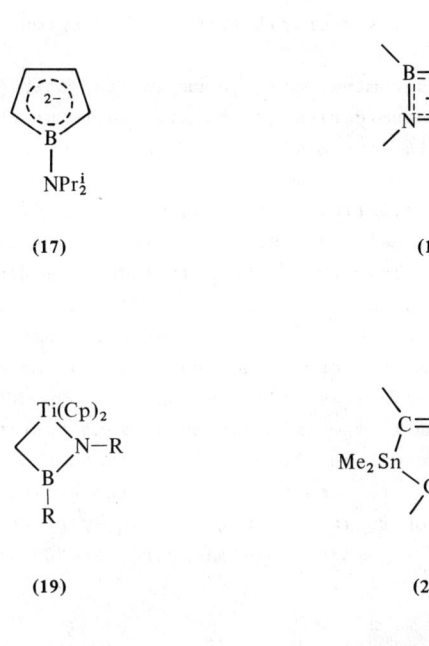

(17)   (18)

(19)   (20)

(21)   (22)

stannous chloride it is oxidized to give the Diels-Alder dimer which undergoes thermal reactions with metal carbonyls to give complexes with a range of structural types depending on the metal.[108]

Hexacarbonyls of chromium, molybdenum and tungsten react with a 1,2-azaborolinyl lithium derivative to give salts which react with MMe_3Cl (M = Ge,Sn,Pb) to form complexes containing metal-metal bonds.[109] The first main group metal complex containing this ligand, a bis-azaborolinyl tin compound which resembles other stannocenes, has been made[110] as have complexes with vanadium and titanium chlorides.[111] This report also includes the diazadiboretidine compound (18) in which the boron atoms interact with the chlorine atoms, a fact deduced from the X-ray structure. Several new borabenzene systems have been reported[112,113] as have cationic complexes of hexaalkylborazines.[114] The compound $(Bu^tBNBu^t)Co_2(CO)_6$ has been isolated in which the iminoborane acts as a bridging ligand normal to the cobalt-cobalt bond.[115] Iminoboranes react with $Cp_2Ti:CH_2$ by [2+2] cycloaddition to give the cyclobutane system[116] (19). Organoboration of suitable alkynylstannanes leads eventually to 1,2-dihydro-1,3-stannaboroles (20) in high yields.[117]

5 Physical Data

The colour of organoboron acetylacetonates has been attributed to variation in the C-O pi bond order[118] and the thermochromic effects accompanying solid state changes in a range of boron-oxygen heterocycles has been investigated in detail.[119] A new series of liquid crystalline materials based on aryl boronic esters have been developed[120] and the effect of ortho-heteroatom substituents on the rate of esterification of arylboronic acids investigated.[121] Kinetic and thermodynamic data have also been obtained for alcohol-triphenylborane and diphenylborinic ester systems.[122]

Crystal structures have been reported for $(Me_2PhSi)_3CBF(OH)$,[123] pyridine-borabenzene and pyridine-2-boranaphthalene[124] and for the chiral borane[125] (21).

Spectroscopic and theoretical studies have been carried out on 1-aza-5-boratatricycle[3.3.3.0^{1,5}]undecane (22) and related compounds[126] and the radicals $R_2\dot{C}-OBPh_2$[127] and $Bu^t\dot{N}$-9-BBN[128] observed.

References

1. Houben-Weyl Methods of Organic Chemistry, Part 3: Organoboron Compounds 4th Ed., Ed. R. Koester, Stuttgart, 1984.
2. A. Suzuki, Kagaku Zokan (Kyoto), 1985, 105, 15.
3. M. Braun, Nachr. Chem. Tech. Lab. 1985, 33, 504.
4. E. Negishi and M. J. Idacavage, Org. React. (N.Y.), 1985, 33, 1.
5. D. S. Matteson, Kagaku Zokan (Kyoto), 1985, 105, 25.
6. D. S. Matteson, K. M. Sadhu, R. Ray, P. K. Jesthi, M. L. Peterson, D. Majundar, D. J. S. Tsai, G. D. Hurst and E. Erdik, J. Organomet. Chem., 1985, 281, 15.
7. H. C. Brown and B. Singaram, Chemtech., 1985, 15, 572.
8. A. Suzuki and S. Hara, Yuki, Gosei Kagaku Kyokaishi, 1985, 43, 100.
9. G. Schmid, Comments Inorg. Chem., 1985, 4, 17.
10. W. Siebert, Angew. Chem. Int. Ed. Engl., 1985, 24, 943.
11. H. Bönnemann, Angew. Chem. Int. Ed. Engl., 1985, 24, 248.
12. H. C. Brown and U. S. Racherla, Tetrahedron Lett., 1985, 26, 2187.
13. H. C. Brown, E. Negishi and W. C. Dickason, J. Org. Chem., 1985, 50, 520.
14. H. C. Brown, J. V. N. Vara Prasad and S-H. Zee, J. Org. Chem., 1985, 50, 1582.
15. H. C. Brown, P. V. Ramachandran, J. V. N. Vara Prasad, J. Org. Chem., 1985, 50, 5583.
16. H. C. Brown, J. V. N. Vara Prasad, J. Org. Chem., 1985, 50, 3002.
17. J. Chandrasekharan and H. C. Brown, J. Org. Chem., 1985, 50, 518.
18. D. Männig and H. Nöth, Angew. Chem. Int. Ed. Engl., 1985, 24, 878.
19. S. Kim and K. Y. Yi, Bull. Chem. Soc. Jpn., 1985, 58, 789.
20. R. Koester, S. Penadés-Ullate and W. V. Dahlhoff, Angew. Chem. Int. Ed. Engl., 1985, 24, 519.
21. H. C. Brown, B. Singaram and T. E. Cole, J. Amer. Chem. Soc., 1985, 107, 460.
22. H. C. Brown, R. G. Naik, R. K. Bakshi, C. Pyun and B. Singaram, J. Org. Chem., 1985, 50, 5586.
23. H. C. Brown, T. Imai, M. C. Desai and B. Singaram, J. Amer. Chem. Soc., 1985, 107, 4980.
24. S. Masamune, B. M. Kim, J. S. Petersen, T. Sato, S. J. Veenstra and T. Imai, J. Amer. Chem. Soc., 1985, 107, 4549, 5832.
25. H. C. Brown, P. K. Jadhar and K. S. Bhat, J. Amer. Chem. Soc., 1985, 107, 2564.
26. H. C. Brown, J. Chandrasekharan and R. V. Ramachandran, J. Org. Chem., 1985, 50, 5446.
27. A. E. Greene, M-J. Luche and A. A. Serra, J. Org. Chem., 1985, 50, 3957.
28. H. C. Brown and G. G. Pai, J. Org. Chem., 1985, 50, 1384.
29. M. M. Midland and P. E. Lee, J. Org. Chem., 1985, 50, 3237.
30. P. K. Jadhav, J. V. N. Vara Prasad and H. C. Brown, J. Org. Chem., 1985, 50, 3203.
31. S. Kim, C. H. Oh, J. S. Ko, K. H. Ahn and Y. J. Kim, J. Org. Chem., 1985, 50, 1927.
32. H. C. Brown, J. S. Cha, B. Nazer and C. A. Brown, J. Org. Chem., 1985, 50, 549.
33. W. R. Leonard and T. Livingstone, J. Org. Chem., 1985, 50, 730.
34. M. J. Sharp and V. Snieckus, Tetrahedron Lett., 1985, 26, 5997.
35. J. E. McMurray and M. D. Erion, J. Amer. Chem. Soc., 1985, 107, 2712.
36. W. V. Dahlhoff and A. Geisheimer, Z. Naturforsch. B: Anorg. Chem., Org. Chem., 1985, 40B, 141.
37. S. A. Kunda, T. L. Smith, M. D. Hylarides and G. W. Kabalka, Tetrahedron Lett., 1985, 26, 277.
38. W. R. Roush, M. A. Adam and D. J. Harris, J. Org. Chem., 1985, 50, 2000, 3430.
39. W. R. Roush and A. E. Walts, Tetrahedron Lett., 1985, 26, 3427.
40. Y. Yamamoto, Y. Saito and K. Maruyama, J. Organomet. Chem., 1985, 292, 311.
41. Y. Yamamoto, W. Ito and K. Maruyama, J. Chem. Soc. Chem. Comm., 1985, 1131.

42 W. R. Roush, A. E. Walts and L. K. Hoong, J. Amer. Chem. Soc., 1985, 107, 8186.
43 Y. Satoh, H. Serizawa, S. Hara and A. Suzuki, J. Amer. Chem. Soc., 1985, 107, 5225.
44 S. Hara, T. Kato, H. Shinizu and A. Suzuki, Tetrahedron Lett., 1985, 26, 1065.
45 N. Miyawa, K. Yamada, H. Suginome and A. Suzuki, J. Amer. Chem. Soc., 1985, 107, 972.
46 M. Hoshi, Y. Masuda and A. Arase, J. Chem. Soc., Chem. Comm., 1985, 1068.
47 M. Hoshi, Y. Masuda and A. Arase, J. Chem. Soc., Chem. Comm., 1985, 714.
48 H. C. Brown, G. A. Molander, S. M. Singh and U. S. Racherla, J. Org. Chem., 1985, 50, 1577.
49 S. S. Nikam and K. K. Wang, J. Org. Chem., 1985, 50, 2193.
50 K. K. Wang and C. Liu, J. Org. Chem., 1985, 50, 2578.
51 J. M. Mas, M. Malacria and J. Goré, J. Chem. Soc., Chem. Comm., 1985, 1161.
52 M-Z. Deng, D-A. Lu and W-H. Xu, J. Chem. Soc., Chem. Comm., 1985, 1478.
53 M. J. O'Donnell and J-R. Falmagne, J. Chem. Soc., Chem. Comm., 1985, 1168.
54 R. L. Danheiser and A. C. Savoca, J. Org. Chem., 1985, 50, 2401.
55 H. C. Brown, T. Imai, P. T. Perumal and B. Singaram, J. Org. Chem., 1985, 50, 4032.
56 A. Pelter, D. Buss and A. Pitchford, Tetrahedron Lett., 1985, 26, 5093.
57 A. Pelter, G. Bugden and R. Rosser, Tetrahedron Lett., 1985, 26, 5097.
58 A. Pelter and J. M. Rao, J. Organomet. Chem., 1985, 285, 65.
59 H. E. Morton and Y. Guidon, J. Org. Chem., 1985, 50, 5379.
60 K. F. Liou, P-H. Yong and Y-T. Lin, J. Organomet. Chem., 1985, 294, 145.
61 C. Manzanares, A. Barriola and J. Carlos de Jesus, Inorg. Chim. Acta., 1985, 98, L43.
62 M. Zaidlewicz, J. Organomet. Chem., 1985, 293, 139.
63 Yu. N. Bubnov, A. I. Grandberg, M. Sh. Grigorian, V. G. Kiselev, M. I. Struchkova and B. M. Mikhailov, J. Organomet. Chem., 1985, 292, 93.
64 R. Koester and M. Yalpani, Angew. Chem. Int. Ed. Engl., 1985, 24, 572.
65 H. C. Brown, T. E. Cole and M. Srebnik, Organometallics, 1985, 4, 1788.
66 H. C. Brown and T. E. Cole, Organometallics, 1985, 4, 816.
67 D. Männig and H. Nöth, J. Chem. Soc., Dalton Trans., 1985, 1689.
68 K. M. Sadhu and D. S. Matteson, Organometallics, 1985, 4, 1687.
69 H. C. Brown, R. Naik, B. Singaram and C. Pyun, Organometallics, 1985, 4, 1925.
70 D. S. Matteson and J. W. Wilson, Organometallics, 1985, 4, 1690.
71 L. Synoradzki, M. Boleslawski and J. Lawinski, J. Organomet. Chem., 1985, 284, 1.
72 R. F. Schinazi and W. H. Prusoff, J. Org. Chem., 1985, 50, 841.
73 H. E. Katz, J. Amer. Chem. Soc., 1985, 107, 1420.
74 H. E. Katz, J. Org. Chem., 1985, 50, 5027.
75 H. E. Katz, J. Org. Chem., 1985, 50, 2575.
76 W. Siebert, U. Ender and R. Schutze, Z. Naturforsch, B: Anorg. Chem. Org. Chem., 1985, 40B, 996.
77 W. Siebert, U. Ender and W. Herter, Z. Naturforsch, B: Anorg. Chem. Org. Chem., 1985, 40B, 326.
78 N. S. Hosmane, N. N. Sirmokadam and M. N. Mollen, J. Organomet. Chem., 1985, 279, 359.
79 C. K. Narula and H. Nöth, J. Organomet. Chem., 1985, 281, 131.
80 C. G. Whiteley and I. Zwane, J. Org. Chem., 1985, 50, 1969.
81 E. Kalbarczyk and S. Pasynkiewicz, J. Organomet. Chem., 1985, 290, 257.
82 R. Wehrmann, H. Meyer, A. Berndt and K. Dimroth, Angew. Chem. Int. Ed. Engl., 1985, 24, 788.
83 P. H. M. Budzelaar, K. Krogh-Jespersen, T. Clark and P. v. R. Schleyer, J. Amer. Chem. Soc., 1985, 107, 2773.
84 W. Siebert, M. Hildenbrand and H. Pritzkow, Angew. Chem. Int. Ed. Engl., 1985, 24, 759.
85 P. H. M. Budzelaar, A. J. Kos, T. Clark and P. v. R. Schleyer, Organometallics, 1985, 4, 429.

86. C. K. Narula and H. Nöth, Inorg. Chem., 1985, 24, 2532.
87. B. Glaser and H. Nöth, Angew. Chem. Int. Ed. Engl., 1985, 24, 416.
88. M. Haase and U. Klingebiel, Angew. Chem. Int. Ed. Engl., 1985, 24, 324.
89. G. Maier, J. Henkelmann and H. P. Reisenauer, Angew. Chem. Int. Ed. Engl., 1985, 24, 1065.
90. B. Pachaly and R. West, J. Amer. Chem. Soc., 1985, 107, 2987.
91. W. Jacksties, H. Nöth and W. Storch, Chem. Ber., 1985, 118, 2030.
92. F. Kumpfmueller, D. Nölle, H. Nöth, H. Pommerening and R. Staudigl, Chem. Ber., 1985, 118, 483.
93. C. Habben, A. Meller, M. Noltemeyer and G. M. Sheldrick, J. Organomet. Chem., 1985, 288, 1.
94. R. Österle, W. Maringgelle and A. Meller, J. Organomet. Chem., 1985, 284, 281.
95. E. Hanecker and H. Nöth, Z. Naturforsch., B: Anorg. Chem. Org. Chem., 1985, 40B, 717.
96. H. Nöth, P. Otto and W. Storch, Chem. Ber., 1985, 118, 3020.
97. J. Bielawski, K. Niedenzu and J. S. Stewart, Z. Naturforsch., B: Anorg. Chem., Org. Chem., 1985, 40B, 389.
98. O. G. Boldyreva, V. A. Dorokhov and B. M. Mikhalov, Izv. Akad. Nauk SSSR, Ser. Khim., 1985, 428.
99. L. G. Vorontsova, O. S. Chizhov, O. G. Boldyreva, V. A. Dorokhov and B. M. Mikhailov, Izv. Akad. Nauk SSSR, Ser. Khim., 1985, 329.
100. V. A. Dorokhov, O. G. Boldyreva, B. M. Mikhailov, Z. A. Starikova and I. A. Teslya, Izv. Akad. Nauk SSSR, Ser. Khim., 1985, 431.
101. R. H. Cragg and T. J. Miller, J. Orgnomet. Chem., 1985, 294, 1.
102. L. Horner, V. Kaps and G. Simons, J. Organomet. Chem., 1985, 287, 1.
103. M. Bockmann, K. Geilich and W. Siebert, Chem. Ber., 1985, 118, 401.
104. H. Wadepohl, H. Pritzkow and W. Siebert, Chem. Ber., 1985, 118, 729.
105. T. Kuhlmann and W. Siebert, Z. Naturforsch., B: Anorg. Chem., Org. Chem., 1985, 40B, 167.
106. W. Siebert, M. El-Din, M. El-Essawi, R. Full and J. Heck, Z. Naturforsch. B: Anorg. Chem., Org. Chem., 1985, 40B, 458.
107. J. Edwin, W. Siebert and C. Kruegger, J. Organomet. Chem., 1985, 282, 297.
108. G. E. Herberich and H. Ohst, Chem. Ber., 1985, 118, 4303.
109. G. Schmid, F. Schmidt and R. Boese, Chem. Ber., 1985, 118, 1949.
110. G. Schmid, D. Zaika and R. Boese, Angew. Chem. Int. Ed. Engl., 1985, 24, 602.
111. G. Schmid, D. Kampmann, W. Meyer, R. Boese, P. Paetzold and K. Dalpy, Chem. Ber., 1985, 118, 2418.
112. G. E. Herberich, H. J. Becker, B. Hessner and L. Zelenka, J. Organomet. Chem., 1985, 280, 147.
113. R. Boese, N. Finke, T. Keil, P. Paetzold and G. Schmid, Z. Naturforsch. B: Anorg. Chem. Org. Chem., 1985, 40B, 1327.
114. M. SCotti, M. Valderrama, R. Ganz and H. Werner, J. Organomet.Chem., 1985, 286, 399.
115. P. Paetzold and K. Delpy, Chem. Ber., 1985, 118, 2552.
116. P. Paetzold, K. Delpy, R. P. Hughes and W. A. Herrmann, Chem. Ber., 1985, 118, 1724.
117. S. Kerschl and B. Wrackmeyer, J. Chem. Soc., Chem. Comm., 1985, 1199.
118. R. Boese, R. Koester and M. Yalpani, Chem. Ber., 1985, 118, 670.
119. M. Yalpani, W. R. Schedit and K. Seevogel, J. Amer. Chem. Soc., 1985, 107, 1684.
120. K. Seto, S. Takahashi and T. Tahara, J. Chem. Soc., Chem. Comm., 1985, 122.
121. M. Lauer, H. Boehnke, R. Grotstollen, M. Salehnia and G. Wolff, Chem. Ber., 1985, 118, 246.
122. P. J. Domaille, J. D. Druliner, L. W. Gosser, J. M. Reid Jr., E. R. Schnelzer and W. R. Stevens, J. Org. Chem., 1985, 50, 189.
123. J. L. Atwood, S. G. Bott, C. Eaborn, M. N. A. El-Khali and J. D. Smith, J. Organomet. Chem., 1985, 294, 23.
124. R. Boese, N. Finke, J. Henkelmann, G. Maier, P. Paetzold, H. P. Reisenauer and G. Schmid, Chem. Ber., 1985, 118, 1644.

125 C. S. Shiner, C. M. Garner and R. C. Haltiwanger, J. Amer. Chem. Soc., 1985 107, 7167.
126 K. J. Lee, P. D. Livant, M. L. McKee and D. S. Worley, J. Amer. Chem. Soc., 1985, 107, 5901.
127 A. Alberti and G. F. Pedulli, J. Organomet. Chem., 1985, 297, 13.
128 J. A. Baban, B. P. Roberts and A. C. H. Tsang, J. Chem. Soc., Chem. Comm., 1985, 955.

4
Carbaboranes, including their Metal Complexes

BY T. R. SPALDING

1. Introduction, Review Articles and Theoretical Aspects.

As with previous years, most of the published work appearing in 1985 has concerned the synthesis and reactions of C_2-carbaboranes or their metal derivatives.

A general review of work reported in 1984 has been published.[1] Reports concerning the synthesis of polyhedral carbaboranes,[2] and the structural chemistry of metallacarbaboranes[3] have appeared. An extensive review by Siebert of metal complexes of 2,3-dihydro-1,3-diboroles, including some compounds containing carbaboranyl ligands, has been published.[4] A detailed review of metallaboranes with up to 7 vertices is also relevant.[5] The electrochemistry of boron compounds has been reviewed.[6]

Abstracts have appeared of theses which concerned synthetic studies of ruthena-[7a] and rhodacarboranes,[7a,b] therapeutic applications of ^{10}B-labelled antibodies,[7b] chemistry of heteroboranes[7c] and pyrolysis and photolysis of small carbaboranes.[7d]

1.1 Theoretical Aspects. An analysis of the protonation of $1,6-C_2B_4H_6$ and the isoelectronic species $1,2-C_2B_4H_6$, $[CB_5H_6]^-$ and $[B_6H_6]^{2-}$ using geometry optimised *ab initio* (STO 3G) calculations showed that differences in the preferred sites of protonation could be explained in terms of the distribution of the h.o.m.o.s and the symmetry of the possible interactions.[8]

Twenty-one isomers of $C_2B_2H_4$ were studied by *ab initio* methods.[9] The basis sets used were 3-21G for geometry and 6-31G or 6-31G* with electron correlation corrections for energy calculations. No isomers with bridging H atoms were calculated. Although derivatives of a number of the isomers with local energy minima have been synthesised, neither the global energy minimum isomer nor any of its derivatives are yet known.

2. Carbaborane Synthesis, Characterisation and Reactions.

The preparation of the first 1,3-diboraallyl system[10a] with a B-H-B bridge is noteworthy in the light of the above

(For references see page 54

calculations.^{9} Reaction of $R_2C:C\{B(R')Cl\}_2$ (1) (R = Me_3Si, R' = Bu^t) with four equivalents of K yielded a diboriranide $K_2[R_2C\overline{CB(R')}B(R')]$. Further treatment with HCl gave the diboraallyl $R_2(H)C\overline{CB(R')}\mu\text{-}H\text{-}B(R')$. In contrast to this, reaction of (1) with two equivalents of K gave $R_2\overline{CB(R')}C:B(R')$ which rearranged on heating to a 1,3-diborete $(RC)_2(BR')_2$.^{10b} A 1,2-diborete has been obtained by reaction of R(Cl)BCH:CHB(Cl)R, (2) (R = NPr^i_2), with two equivalents of K.^{10b} Rearrangement to the corresponding 1,3-diborete occurred on heating (120°C).

Attempts have failed to reproduce the reported syntheses of hexaaalkyl-1,4-dibora-2,5-*cyclo*hexadienes^{11a} from R_2C_2 and the intermediate MeB:.^{11b} Instead, *nido*-2,3,4,5-tetracarbaboranes $(R_4C_4B_2R'_2)$ and various organoboron bromides were the major products.

Dimerisation of $2,3,4\text{-}Et_3\text{-}1,5\text{-}Me_2\text{-}1,5\text{-}C_2B_3$ by successive treatment with K and I_2 initially produced a C_4B_6-derivative with an adamantane type structure.^{12} On heating (160°C) this rearranged to a carbaborane with a decaborane type structure. Isolation of the initial product was unexpected since a previous reaction of the corresponding $Et_5C_2B_3$ gave the carbaborane directly. Both C_4-carbaboranes had dynamic structures in solution.

A new *arachno* carbaborane $5,6\text{-}C_2B_6H_{12}$ (3) has been synthesised by the $PtBr_2$ catalysed reaction of $1,5\text{-}C_2B_3H_5$ and B_2H_6.^{13} On heating at 65°C, $5,6\text{-}C_2B_6H_{12}$ slowly converts to $C_2B_6H_{10}$. The catalysed reaction of $1,6\text{-}C_2B_4H_6$ and diborane gave $2:1'.2'\text{-}[C_2B_4H_5][B_2H_5]$ with an *exo*polyhedral B_2H_5 group bonded *via* a B-B-B three-centre bond.

Direct reaction of B_5H_9 with $Me_3SiC:CR$ (R = Me_3Si,Me,H) gave high yields of air-stable $2\text{-}(Me_3Si)\text{-}3\text{-}R\text{-}2,3\text{-}C_2B_4H_6$ products.^{14}

All possible B-substituted mono- and di- bromo- and iodo-derivatives of $2,4\text{-}C_2B_5H_7$ have been synthesised except $1,7\text{-}Br_2\text{-}2,4\text{-}C_2B_5H_5$.^{15a} Kinetic data for the rearrangements of $5\text{-}X\text{-}2,4\text{-}C_2B_5H_6$ and $5,6\text{-}X_2\text{-}2,3\text{-}C_2B_5H_5$ isomers (X = Cl,^{15b} Br,I^{15a}) have been obtained. These studies lead to isomer stability sequences for example in the mono-derivatives, 3-> 5-> 1-X for Cl^{15b} or Br^{15a} but 5-> 3-> 1-X for I.^{15a} Possible mechanisms for the rearrangements were discussed.

B-fluoro-derivatives of $2,4\text{-}C_2B_5H_7$ were produced when $[R_4N]F$ was reacted with $5\text{-}Br\text{-}2,4\text{-}C_2B_5H_6$ or $3,5\text{-}I_2\text{-}2,4\text{-}C_2B_5H_5$.^{16a} A more general study of halide exchange showed it occured only if the incoming species had higher electronegativity.[16b] Reactions were also reported with amine adducts, e.g. $Me_3N.5,6-Br_2-2,4-C_2B_5H_5$ and $[BzEt_3N]Cl$ gave $5,6-Cl_2-2,4-C_2B_5H_5$. In a few cases rearrangement during reaction gave the more stable exchanged isomer, e.g. $3-Cl-2,4-C_2B_5H_6$ from the $5-I-2,4-C_2B_5H_6$.

An unusual addition reaction of $nido$-$Na_2[6,9-C_2B_8H_{10}]$ with HX in benzene afforded the $arachno$-$5-X-6,9-C_2B_8H_{13}$ (X = F, Cl Br, I).[17] With concentrated (96% HF), the major product was $5,5'-O-(6,9-C_2B_8H_{13})_2$ formed by the HF dehydration of $5-HO-6,9-C_2B_8H_{13}$. An improved synthesis of the parent $6,9-C_2B_8H_{14}$ carbaborane by the reaction of $5,6-C_2B_8H_{12}$ with $Na[BH_4]$ in ethanolic K[OH] has been published.[18] Halogenation under electrophilic conditions afforded $1-X-6,9-C_2B_8H_{13}$ derivatives (X = Cl, Br, I).

The synthesis and X-ray crystallographic characterisation of $Cs[nido-7(p-C_6H_4NCS)-9-I-7,8-C_2B_9H_{11}]$ have been reported.[19] This compound was prepared as part of a study of the potential use of boron compounds in neutron-capture therapy. In the same context $1-[^3H]-2-(p-C_6H_4NCS)-1,2-C_2B_{10}H_{10}$ was also prepared and the immunological properties of antibodies containing both these carbaboranes were investigated.

An improved synthesis of $9-I-1,2-C_2B_{10}H_{11}$ has been described which uses CH_2Cl_2 instead of CCl_4 as solvent.[20]

Reactions of phenyl (B-carbaboranyl) iodonium salts derived from 1,2-, 1,7-, or $1,12-C_2B_{10}H_{12}$ with various nucleophiles (L) have been reported.[21] Depending on the reagents either a nucleophilic substitution or a free radical mechanism can operate. For $L = F^-$, Cl^-, Br^-, nucleophilic substitutions gave phenyl iodide and 9-X-B-substituted products. Free radical reactions (L = OH^-, py, Ph_3P, Hg) cleaved the $I-C_6H_5$ bonds to produce $IC_2B_{10}H_{11}$ and phenyl radicals. Products from both types of reactions were observed with $L = CN^-$. New synthetic routes to 9-aryl-(1,7-carbaboranyl) halonium salts have been described.[22]

Preparations of 9-vinyl-[23,24] and 9-ethynyl-derivatives (-C:CR,R = Vy, Bu, Ph, $SiMe_3$)[23] of 1,2- and $1,7-C_2B_{10}H_{12}$ were reported. The products of the high temperature reactions of $1,7-C_2B_{10}H_{12}$ and C_6H_6 (630-750°C)[25a] or $1,2-C_2B_{10}H_{12}$ and C_6F_6 (500°C) have been analysed.[25b] The structure (X-ray) of $9-Ph-1,7-C_2B_{10}H_{11}$ was determined.[25a] Free radical initiated reactions between, $1,2-C_2B_{10}H_{12}$ and PhCN or naphthalene have been studied.[25c]

Silicon containing products have been characterised from reactions of lithium derivatives of carbaboranes and chloromethylsiloxanes[26a] or $Me_2(H)SiCl$.[26b] The synthesis and reactions of $1-(EtO)_2PCH_2-1,2-C_2B_{10}H_{11}$ were described.[27] C-bonded ($-CO_2OCBu^t$) peroxyester[28] and ($-CH_2CH(OH)CH_2O_2Bu^t$) peroxide[29] groups have been attached to C_2B_{10}-carbaboranes.

Gold (I) complexes with $1-R_2P-2-R'_2P-1,2-C_2B_{10}H_{10}$ ligands have been reported (R = Ph, R' = NMe_2).[30]

The synthesis and chemical stability of C_2B_{10}-carbaborane based polyamides,[31a] polymethyl methacrylates[31b-d] and related polymers[31e] have studied.

Monocarbon carbaboranes have received relatively little attention in 1985. New $7-L-7-CB_{10}H_{12}$ derivatives (L = Me_2S, $N_4(CH)_2)_6$, $MeCONH_2$, $HO_2CCH_2NH_2$, MeC:NH, $Et_2C:NH$, PhCH:NBz) were prepared by reactions of the H_3N-group of $7-H_3N-7-CB_{10}H_{12}$.[32] The use of $[Ph_4P]^+$ as a quantitative precipitant for $[CB_nH_{n+1}]^-$ ions (n = 9, 10, 11) was reported.[33]

2.1. Physical Properties and Uses. The Heats of Formation of $1,6-C_2B_4H_6$ and $1,2-C_2B_4H_6$ have been measured.[34]

The ^{11}B and ^1H n.m.r. spectra of CB_8H_{14} and $[CB_8H_{13}]^-$ have been recorded.[35] In the anion the *endo*-H attached to the C atom was involved in exchange with the three bridging H atoms. No such feature occurred in CB_8H_{14}.

The e.s.r. spectra of ketone adducts of boron-centered $C_2B_{10}H_{11}$ radicals were reported.[36] A study of ^{127}I n.q.r. spectra of carbaborane containing compounds with poly-coordinated I atoms has been published.[37] The molecular motion of $1,12-C_2B_{10}H_{12}$ in plastic crystalline phases was studied.[38]

The mass spectrometric fragmentation patterns of carbaborane-substituted benzene-$Cr(CO)_3$ complexes have been analysed.[39]

Among the studies of carbaborane-based products of possible commercial interest were investigations concerning fuels,[40a,b] electron transfer electrodes,[41] g.l.c. applications,[42a,b] flammability of carbaborane-siloxane polymers,[43] ceramic composites[44a,b] and related materials,[44c] and neutron-capture therapy of cancers.[19,45a-c]

3. Metallacarbaboranes Excluding Cage-bonded Compounds.

Crystallisation of $Ag[CB_{11}H_{12}]$ from benzene solution afforded $Ag[CB_{11}H_{12}].2C_6H_6$.[46] An X-ray crystallographic study showed that each $[Ag]^+$ ion was associated with two $[CB_{11}H_{12}]^-$ ions via Ag-H-B bonds and one C_6H_6 molecule was η^1-coordinated to each silver ion.

Reaction of $K[2,3-R_2-2,3-C_2B_4H_5]$ (R = H,Me) with Cu phosphine complexes afforded $[4,5-\mu-\{(R_3P)_2Cu-\}2,3-R_2-2,3C_2B_4H_5]$.[47]

Compounds containing six membered $\overline{M-C_2-Si-C_2}$ rings involving $1,2-C_2B_{10}H_{10}$ units were synthesised from $[MCl_2(bipy)]$ (M = Co, Ni, Pd, Cu) and $Li_2[(1,2-C_2B_{10}H_{10})_2SiMe_2]$.[48]

C-bonded gallium derivatives RR'GaCl (R = 1-Ph-1,2-$C_2B_{10}H_{10}$, R' = R or Me) and $RGa(O_2CCF_3)_2$ have been prepared.[49] Compounds of the type $RR'_2M(TMED)$ were obtained from reactions of R'_3M and R_2Hg (M = Al, Ga, In, R' = Me or Et).

The optimum conditions have been described for synthesis of B-bonded monothallium derivatives of 1,2-, and 1,7- $RR'C_2B_{10}H_{10}$ via carboranyl-mercury reagents.[50] A study of B-bonded $RTlX_2$ compounds (R = 1,2- or $1,7-C_2B_{10}H_{11}$, X = Cl or Br), has been reported.[51] The X-ray analysis of $9-(TlCl_2)-1,7-Me_2-1,7-C_2B_{10}H_9$ showed it contained Cl-bridged Tl_2Cl_2 units.

The electrochemical symmetrization of both C- and B-bonded carbaboranyl-mercury compounds RHgX has been investigated.[52] Free radical products of the reaction of $3,5-Bu_2^t-1,2$-benzoquinone with $(1,2-C_2B_{10}H_{11})_2Hg$ have been identified.[53]

4. Cage-Bonded Metallacarbaboranes.

Treatment of $[1-(\eta^6-C_6R_6)-2,3-Et_2-1,2,3-FeC_2B_4H_4]$ (4) (R = H, Me) with TMED gave high yields of $[1-(\eta^6-C_6R_6)-2,3-Et_2-1,2,3-FeC_2B_3H_5]^-$ species.[54] Further reaction (R = H) with K[H] followed by Fe(II)-promoted oxidative fusion gave $[2-(\eta^6-C_6H_6)-7,8,9,10-Et_4-2,7,8,9,10-FeC_4B_6H_6]$ (5). The structure of (5) was elucidated by ^{11}B 2D n.m.r. and a partially completed X-ray analysis. Treatment of (4) with naphthalene in the presence of Al and $AlCl_3$ gave several products in low yield including the unexpected B-B linked bis(ferracarbaborane) (6) which was characterised by X-ray methods.

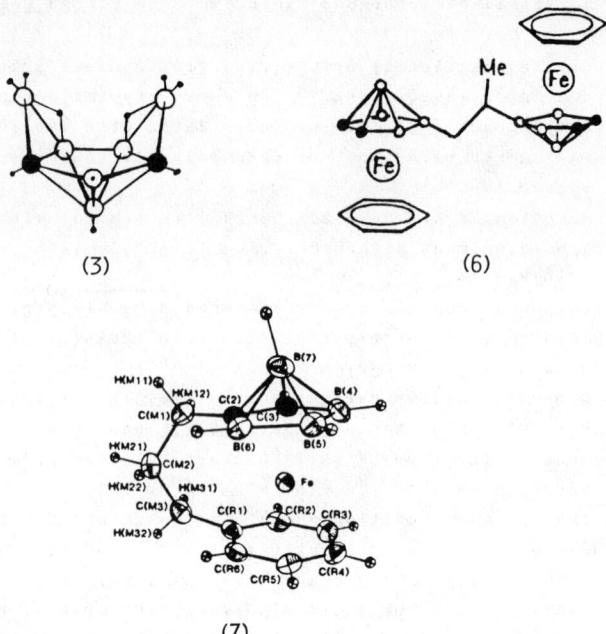

Figure 1 $[1-\{\eta^6-C_6H_5\overline{(CH_2)_3\}-1,2,3-FeC_2B_4H_5}]$
(Reproduced with permission from Organometallics, 1984, 4, 890)

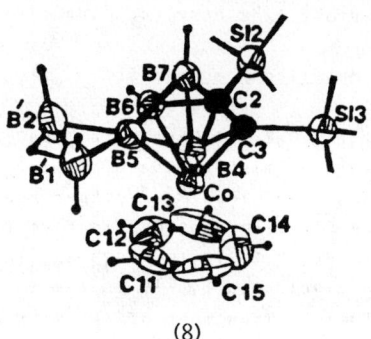

Figure 2 $5:1',2',-[1-(\eta^5-C_5H_5)-2,3-(Me_3Si)_2-1,2,3-CoC_2B_4H_3][B_2H_5]$
(Reproduced with permission from Organometallics, 1984, 4, 721)

When the C-(3-phenylpropyl)-substituted compound
[1-(η^6-C$_8$H$_{10}$)-3-Ph(CH$_2$)$_3$-1,2,3-FeC$_2$B$_4$H$_5$] was treated with
Al/AlCl$_3$, the η^6-C$_8$H$_{10}$ ligand was displaced giving (7).[55] The
structure was determined (X-ray) (Figure 1). All C$_6$-atoms were
coplanar as were the atoms in the C$_2$B$_3$-face but these planes were
mutually tilted by 7.5°. An analogous reaction with the
3-phenylethyl-substituted compound produced isomers of the
dimeric species [{η^6-C$_6$H$_5$(CH$_2$)$_2$}FeC$_2$B$_4$H$_5$]$_2$ which had each Fe atom
bound to one carbaborane unit and the C$_6$H$_5$-group attached to the
other carbaborane.

Reaction of [1-(η^6-C$_8$H$_{10}$)-2,3-Et$_2$-1,2,3-FeC$_2$B$_4$H$_4$] and
biphenyl,[54] naphthalene and phenanthrene gave
[(η^6-arene)FeEt$_2$C$_2$B$_4$H$_4$]-derivatives.[56] The biphenyl,[54]
naphthalene and one phenanthrene product[56] were characterised by
X-ray methods. A di-iron derivative of biphenyl was also
isolated.[54]

Iron and cobalt compounds [1,1-(diphos)-1-Cl-2,3-Et$_2$-
1,2,3-MC$_2$B$_4$H$_4$] were prepared by reaction of Na[Et$_2$C$_2$B$_4$H$_5$] with
MCl$_2$ and diphos.[57] Both were structurally characterised (X-ray).
Replacement of Cl by CN or I was reported for the Co compound.
Treatment of [NiBr$_2$(PPh$_3$)$_2$] with [Et$_2$C$_2$B$_4$H$_5$]$^-$ gave Ph$_3$P-B bonded
[1-Br-1,5-(PPh$_3$)$_2$-2,3-Et$_2$-1,2,3-NiC$_2$B$_4$H$_3$]. The structure of this
compound was determined (X-ray).

An unusual *exo*polyhedrally bridging B$_2$H$_5$-unit was reported
in 5:1',2'-[1-(η^5-C$_5$H$_5$)-2,3-(Me$_3$Si)$_2$-1,2,3-CoC$_2$B$_4$H$_3$][B$_2$H$_5$], (8)
(Figure 2).[58] X-ray analysis showed the B$_2$H$_5$-unit to be attached
to the one B atom opposite the two carbaborane C atoms. The
CoC$_2$B$_4$-cage structure was essentially the same as that in
[1-(η^5-C$_5$H$_5$)-2,3-Me$_2$-1,2,3-CoC$_2$B$_4$H$_4$]. Compound (8) was one of
five isolated from the reaction of thermally generated Co atoms
with B$_6$H$_{10}$, C$_5$H$_6$ and Me$_3$SiC≡CSiMe$_3$. The other products were
(C$_5$H$_5$)Co-derivatives of (Me$_3$Si)$_2$C$_2$B$_n$H$_n$ (n = 4,5 and 6 {2
isomers}). Thermolysis of (8) at 150°C gave both the above n = 4
and 5 compounds in high yields.

Degradation of [3-(η^6-C$_6$H$_6$)-3,1,2-RuC$_2$B$_9$H$_{11}$] produced
[2-(η^6-C$_6$H$_6$)-2,5,6-RuC$_2$B$_7$H$_{11}$] (9) and [1-(η^6-C$_6$H$_6$)-1,2,4-
RuC$_2$B$_8$H$_{10}$] in low yields.[59] The structure of (9) was determined
(X-ray) and showed a decaborane type arrangement with both C
atoms in the open face and H atoms bridging B$_{8,9,10}$ positions.

Relatively low temperature isomerisation of

$[3,6-(\eta^5-C_5H_5)_2-3,6,1,2-Co_2C_2B_8H_{10}]$ to the $2,4,1,7-Co_2C_2$-compound was been achieved by reaction with Na/Hg in refluxing DME or diglyme followed by oxidation.[60]

A detailed report of the reactions of isomeric 3,1,2-, 2,1,7- and 2,1,12-*closo* $[(PPh_3)_2HRh\}C_2B_9H_{11}]$ compounds with *iso*propenyl acetate and vinyl acetate, and the subsequent reactions with H_2, has been published.[61] Under mild conditions (40°C, C_6H_6) both acetates react to cleave the CO ester bond releasing alkene and forming $(PPh_3)_2Rh$-monodentate acetate or $(PPh_3)Rh$-bidentate acetate complex and PPh_3. Both systems regenerate the parent compounds on passage of H_2. These reactions are shown in Scheme 1 for $[3,3-(PPh_3)_2-3-H-3,1,2-RhC_2B_9H_{11}]$ (10). Some interesting alternative products were isolated including the Rh-alkyl complex $[2-(PPh_3)-2,2-(Me\overline{C(H)OC(Me)O})-2,1,7-RhC_2B_9H_{11}]$ from the 2,1,7-isomer and vinyl acetate, and the σ-metallated complex $[3-(PPh_3)-3,3-(PPh_2\sigma-\overline{C_6H_4})-3,1,2-RhC_2B_9H_{11}]$ (11) from the 3,1,2-isomer and either acetate reacting in refluxing benzene. With allyl acetate, the rhodacarbaboranes produced acetic acid and $[(PPh_3)(\eta^3-C_3H_5)RhC_2B_9H_{11}]$ complexes, see (12) Scheme 1. Further reaction with H_2 eliminated C_3H_6 and produced the Rh-Rh bonded dimer $[3-(PPh_3)-3,1,2-RhC_2B_9H_{11}]_2$ (13). This was also the product of the reaction of the $(PPh_3)Rh$ acetate-complex with H_2, Scheme 1. The relevance of these reactions to the catalytic hydrogenolysis of alkenyl acetate was discussed.

The synthesis and structural characterisation (X-ray) of $[2,2-(PPh_3)_2-2-H-1-Me-7-Ph-2,1,7-RhC_2B_9H_{11}]$ was reported.[62] It was notable that the *closo* and not *exo-nido* structure was adopted despite substantial steric crowding about Rh. Attempts to substitute PPh_3 by more sterically demanding phosphines $\{P(Cy)_3, P(o-tol)_3\}$ to give an *exo-nido* tautomer failed.

Low temperature (-73°C) protonation of $[3,3-(\eta^4-cod)-1,2-Me_2-3,1,2-RhC_2B_9H_9]^-$ with TFA gave an unstable intermediate with an agostic hydrogen interaction of the type $Rh-H-C_{(cod)}$.[63] At 7°C the intermediate rearranged to a remarkably stable 16 electron Rh complex, $[3-(\eta^3-C_8H_{13})-1,2-Me_2-3,1,2-RhC_2B_9H_9]$, which was structurally characterised (X-ray). The Rh atom was symmetrically bound to all five C_2B_3-atoms and the three allylic carbon atoms of the $\eta^3-C_8H_{13}$ ligand.

In a related study, protonation of $[3,3-(\eta^4$-bicyclo-

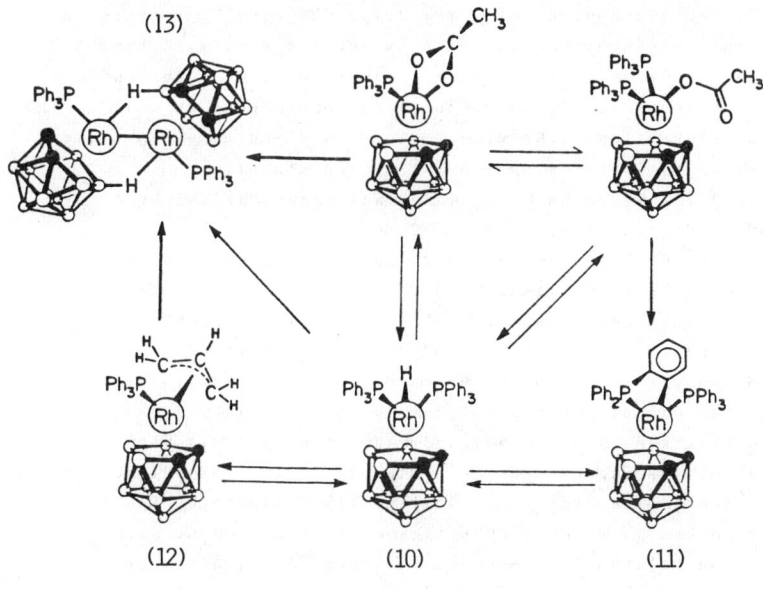

Scheme 1

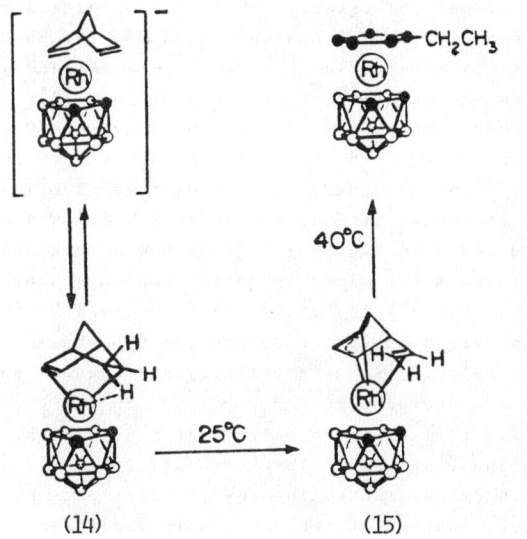

Scheme 2

[2.2.1]-hepta-2,5-diene)-1,2-Me$_2$-3,1,2-RhC$_2$B$_9$H$_9$]$^-$ produced a fluxional intermediate (14) with an agostic H atom, Scheme 2.^{64} X-ray analysis of (14), (Figure 3) showed the Rh atom bonded to the C$_7$-norbornenyl ligand by one π and one σ bond as well as the Rh-H-C interaction, d(Rh-H) = 1.9 Å. On standing at 25°C (14) isomerised to [3-(η^2-vinyl)-3-(η^4-cyclopentenyl)-1,2-Me$_2$-3,1,2-RhC$_2$B$_9$H$_9$] (15) which was characterised by X-ray methods. Heating (15) at 40°C (C$_6$H$_6$) produced [3-{η^5-C$_5$H$_4$(Et)}-1,2-Me$_2$-3,1,2-RhC$_2$B$_9$H$_9$] by a reaction which appeared to be intramolecular.

The syntheses of several anionic species [(L)$_2$RhC$_2$B$_9$H$_{11}$]$^-$ derived from 3,1,2-, 2,1,7-, and 2,1,12- isomers of [(L)$_2$HRhC$_2$B$_9$H$_{11}$] have been reported.^{65} X-ray analysis of the K[18-crown-6][3,3-(PPh$_3$)$_2$-3,1,2-RhC$_2$C$_2$B$_9$H$_{11}$].C$_4$H$_8$O.H$_2$O salt showed that the rhodacarbaborane cage structure was very similar to that of the parent hydride. However in the [Me$_4$N]$^+$ salt of 2,1,7-isomer, the cage was significantly distorted. Reactions were reported in which a PPh$_3$ ligand was replaced by either CO or C$_2$H$_4$. The related Ir compounds [3-(PPh$_3$)-3-(L)-3,1,2-IrC$_2$B$_9$H$_{11}$]$^-$ were prepared with L = PPh$_3$,CO,H$_2$.

The reaction of [RhCl(PR$_3$)cod] with [7-R-7,8-C$_2$B$_9$H$_{11}$]$^-$ appears to provide a general route to rhodacarbaborane dimers (R = H, Ph, 7'-nido-7',8'-C$_2$B$_9$H$_{11}$).^{66} One typical example is [(PPh$_3$)RhC$_2$B$_9$H$_{10}$Ph]$_2$ which has one Rh-Rh and two Rh-H-B intercage interactions. A more unusual product (16), (Figure 4) was isolated from the reaction of the anion (R = 7',8'-C$_2$B$_9$H$_{11}$) and [RhCl(PEt$_3$)cod]. There was no Rh-Rh bond but the C-C link and one Rh-H-B bridge were present. The Rh atoms were in different environments. One was bonded to a C$_2$B$_3$-face, a bridging H atom and a η^3-cyclooctenyl ligand, the other was located over a CB$_4$-face and bound to two PEt$_3$ ligands and a terminal H atom. A possible mechanism for dimer formation was suggested which involved phosphine disproportionation from mono-Rh species.

Some polymeric phosphinocarbaborane C$_2$B$_9$-based complexes of Rh have been reported to catalytically hydrogenate and isomerise unsaturated hydrocarbons.^{67}

Compounds (17) with exo-polyhedral B-H-M interactions were formed from the reaction of [W(CR)(CO)$_2$(1,2-Me$_2$-1,2-C$_2$B$_9$H$_9$)]$^-$ (R = p-tol) with [M(CO)$_2$(NCMe)$_2$(η-C$_9$H$_7$)]$^+$ salts (M = Mo,W), Scheme 3.^{68} Further reaction of the B-H-M unit and the bridging

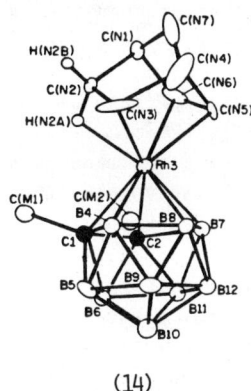

(14)

Figure 3 Structure of intermediate (14)

(Reproduced with permission from Organometallics, 1985, 4, 1692)

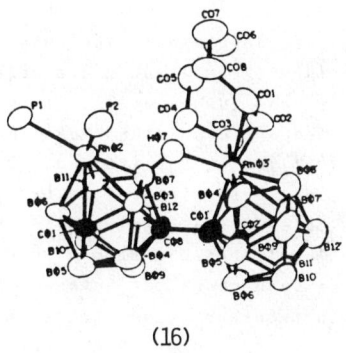

(16)

Figure 4 Structure of [RR(codH)$C_2B_9H_{10}$ Rh(H)(PEt$_3$)$_2$ $C_2B_9H_{10}$] (16)
(Reproduced with permission from J. Am. Chem. Soc., 1985, 107, 932)

alkylidyne ligand (M = Mo) occured on treatment with hex-3-yne to give (18). Both (17) and (18) were structurally characterised by X-ray methods.

The Na/Hg induced isomerisations of $[3-(\eta^5-C_5H_5)-3,1,2-MC_2B_9H_{11}]$ (M = Co, Rh) or $[3-(\eta^6-C_6H_6)-3,1,2-RuC_2B_9H_{11}]$ were reported.[60] Reaction to the 2,1,7-isomers occurred readily at room temperature for the Ru and Rh compounds. In refluxing DME or diglyme, the Co complex gave the 2,1,7- or 4,1,7- isomers respectively as the preferred products.

Reaction of $[(\eta^6-C_6Me_6)_2Co][PF_6]$ and $Tl[3,1,2-TlC_2B_9H_{11}]$ gave $[endo-H-3-(\eta^5-HC_6Me_6)-3,1,2-CoC_2B_9H_{11}]$.[59] The same carbaborane reagent reacted with $[Os(Cl)_2(NCMe)(\eta^6-C_6H_6)]$ to give the first reported arene-Os carbaborane $[3-(\eta^6-C_6H_6)-3,1,2-OsC_2B_9H_{11}]$.

The complex salt Ni.4L.[Ni(1,2-$C_2B_9H_{11}$)$_2$] was prepared by reaction of $[NiCl_2]$ and $Na[Ni(C_2B_9H_{11})_2]$ in the presence of L (L = bipy, py).[69a] The Na.2 bipy.[Ni($C_2B_9H_{11}$)$_2$] salt was also isolated. Both bipy complexes decompose at 210-230°C to give $[3-(bipy)-3,1,2-NiC_2B_9H_{11}]$ which was also prepared by thermolysis of $[Ni(C_2B_9H_{11})].bipy^{69a}$ or $[Ni(1-O_2C-1,2-C_2B_9H_{11})_2bipy]$.[69b]

Reaction of $Tl[3,1,2-TlC_2B_9H_{11}]$ and $[PdCl_2(TMED)]$ produced $[3-(TMED)-3,1,2-PdC_2B_9H_{11}]$ which was shown by X-ray methods to possess a "slipped" structure.[70] Replacement of TMED gave cod-, {P(OMe)$_3$}$_2$- and (PMe$_3$)$_2$-Pd complexes. The PMe$_3$-compound was structurally characterised (X-ray) and had a more symmetrical Pd-C$_2$B$_3$ interaction than the TMED complex. This was related to the better π-acceptor properties of the PMe$_3$ ligand.

Zwitterionic derivatives $[3,3,3-(CO)_3-3,1,2-ReC_2B_9H_{10}-1-CRR']$ (19) (R, R' = H, Me) were prepared by protonation of corresponding 1-CH$_2$OR or 1-CR:CH$_2$ complexes.[71] The X-ray analysis of (19), (R=R'=Me) showed a strong Re-CRR'-interaction and considerable double bond character between the exopolyhedral 1-C- and C(1) of the carbaborane cage. Reactions of (19) with nucleophiles (X = PR$_3$, py, CN$^-$, SCN$^-$, PhO$^-$) gave exo-polyhedral-1-CRR'X-products.

Solvent induced ^{11}B n.m.r shifts of the order of 3 ppm have been noted in the spectra of metallaboranes and - carbaboranes.[72] Clearly the possibility of solvent induced shifts cannot be ignored as they sometimes have been previously.

Thermolysis (253°C) of $[9,9-(PPh_3)_2-9,6-PtCB_8H_{12}]$ afforded the di-σ-metallated product (20), (Figure 5).[73] The carbaborane

[N(PPh$_3$)$_2$][W($\equiv$CR)(CO)$_2$(1,2-C$_2$B$_9$H$_9$Me$_2$)]

(i) $\downarrow$

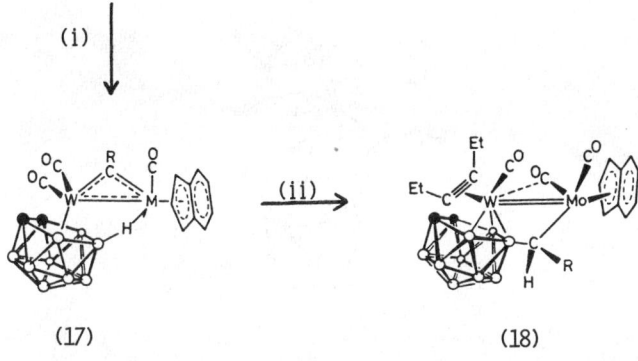

(17)   (ii) →   (18)

Reagents: i, [M(CO)$_2$(NCMe)$_2$(η-C$_9$H$_7$)][BF$_4$], (M=Mo,W); ii, EtC$_2$Et, (M=Mo)

Scheme 3

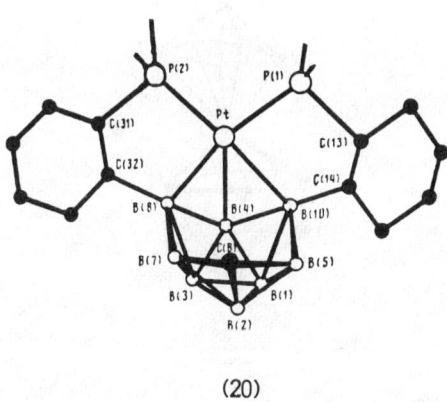

(20)

Figure 5 Structure of [9,9-($\overline{\text{Ph}_3\text{PC}_6\text{H}_4}$)$_2$-9,6-PtCB$_8H_{10}$]

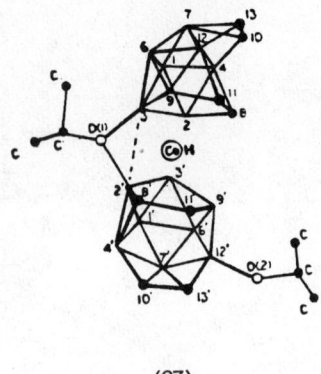

(23)

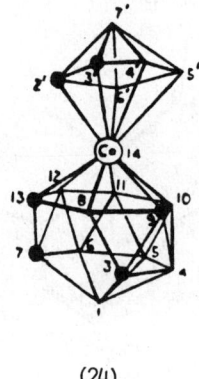

(24)

cage geometries of (20) and its precursor were essentially the same except for the somewhat shortened Pt-B(4) bond.

Several unusual C_4-carbaborane derivatives have been isolated from the reaction of [$CoCl_2$] with either [$Et_4C_4B_8H_8$]$^{2-}$ or both [B_5H_8]$^-$ and [$Et_4C_4B_8H_8$]$^{2-}$.[74] The former reaction gave [$Co_2(Et_4C_4B_8H_8)_2$] (21) and paramagnetic [$Co_2(Et_4C_4B_8H_7)_2$] (22). The structure proposed for (21) contained a pair of 14-vertex *closo* polyhedra sharing a Co-Co edge. Complex (21) and an isomer of (22) were both found as products of the reaction involving [B_5H_8]$^-$. Other products were paramagnetic [$Co(Et_4C_4B_8H_8)(Et_4C_4B_8H_7.THF)$], [$CoH(Et_4C_4B_8H_7)_2$], and [$CoH(Et_4C_4B_4H_6)(Et_4C_4B_8H_8)$]. Treatment of (21) with I_2 in acetone gave five compounds; an isomer of (22), three compounds with either Co-OH or Co-{$Et_4C_4B_8H_6(OH)$}-groups and [$CoH(Et_4C_4B_8H_7)_2(Me_2CO)_2$] (23) as the major product. The structure of (23) was determined (X-ray) and showed the cages linked *via* a B-O(acetone)-B' bridge. The second acetone was terminally B-bonded. The bridged B-B' distance was 2.259 Å indicative of weak intercage bonding. The reactions of [MCl_2] (M = Fe, Co) with [$Et_4C_4B_8H_8$]$^{2-}$ then [$Et_2C_2B_4H_5$]$^-$ were studied.[75] Products from [$CoCl_2$] included [$Co(Et_2C_2B_4H_4)(Et_4C_4B_8H_7.THF)$] (24) and [$Co(Et_2C_2B_4H_3)(Et_4C_4B_8H_6.THF)$]. The structure of (24), determined using X-ray methods, showed cobalt bonded to the C_2B_3-face of the C_2B_4-carbaborane and a C_2B_4-face in the C_4-ligand. The planes containing these faces are mutually tilted by 10.9° and the THF was bonded to B(11). The reaction with [$FeCl_2$] gave [$FeH(Et_2C_2B_3H_4)(Et_4C_4B_8H_7)$] as the major product, [$Fe(Et_2C_2B_4H_4)(Et_4C_4B_8H_8.THF)$] and [$Fe(H)_2(Et_2C_2B_4H_3)(Et_4C_4B_8H_7)$].

A number of complexes which can be related electronically to metallacarbaboranes but which contain planar π-ligands such as those derived from diboroles[4] will be mentioned here. Among the compounds reported were transition metal derivatives (Co,[76,77] Fe, Ni, Pt[76]) and a thallium complex of C_3B_2-species.[78] The X-ray determined structure of the Tl compound showed a *closo* TlC_3B_2Co-cage with the Tl atom axially opposite the Co atom. Other complexes have contained C_4B-(Cr, Mn, Fe, Co, Ni, Ru, Rh),[79] C_5B-(Co, Ni),[80] C_3BN-(Ti, V,[81] Cr, Mo, W,[82] or Sn[83]) and C_2B_2S-(Fe)[84] ligands.

References

1. T.R. Spalding, Chap. 4 in "Organometallic Chemistry", Eds. E.W. Abel and F.G.A. Stone, S.P.R., The Royal Society of Chemistry, London, 1986, Vol. 14.
2. R.B. King, Gov. Rep. Announce. Index:(US), 1985, 85, 88.
3. A.I. Yanovskii and Yu. T. Struchkov, Khim. Svyaz Str. Mol., 1984, 83.
4. W. Siebert, Angew. Chem., Int. Ed. Engl., 1985, 24, 943.
5. J.D. Kennedy, in "Progress in Inorg. Chem.", Ed. S.J. Lippard, J. Wiley & Sons, N.Y., 1984, Vol. 32.
6. J.H. Morris, H.J. Gysling, and D. Reed, Chem. Rev., 1985, 85, 51.
7. (a) P.E. Behnken, Diss. Abstr., Int. B, 1985, 45, 3227; (b) E.A. Mizusawa, Diss. Abstr. Int. B, 1985, 45, 2913; (c) A.A. Arafat, Diss. Abstr. Int. B, 1984, 45, 1461; (d) R.J. Astheimer, Diss. Abstr. Int. B, 1984, 45, 1462.
8. T. Whelan and P. Brint, J. Chem. Soc., Faraday Trans. 2, 1985, 81, 267.
9. P.H.M. Budzelaar, K. Krogh-Jespersen, T. Clark, and P.v.R. Schleyer, J. Am. Chem. Soc., 1985, 107, 2773.
10. (a) R. Wehrmann, H. Meyer and A. Berndt, Angew. Chem., Int. Ed. Engl., 1985, 24, 788; (b) M. Hildenbrand, H. Pritzkow, and W. Siebert, Angew. Chem., Int. Ed. Engl., 1985, 24, 759.
11. (a) S.M. van der Kerk, P.H.M. Budzelaar, A.L.M. van Eekeren, and G.J.M. van der Kerk, Polyhedron, 1984, 3, 271; (b) R. Schlogl and B. Wrackmeyer, Polyhedron, 1985, 4, 885.
12. R. Koster, G. Seidel, and B. Wrackmeyer, Angew. Chem., Int. Ed. Engl., 1985, 24, 326.
13. E.W. Corcoran and L.G. Sneddon, J. Am. Chem. Soc., 1985, 107, 7446.
14. N.S. Hosmane, N.N. Sirmokadam, and M.N. Mollenhauer, J. Organomet. Chem., 1985, 279, 359.
15. (a) B. Ng, T. Onak, T. Banuelos, F. Gomez, and E.W. Distefano, Inorg. Chem., 1985, 24, 4091; (b) Z.J. Abdou, M. Soltis, B. Oh, G. Siwrap, T. Banuelos, W. Nam, and T. Onak, Inorg. Chem., 1985, 24, 2363.
16. (a) B. Ng and T. Onak, J. Fluorine Chem., 1985, 27, 119; (b) B. Ng, T. Onak, and K. Fuller, Inorg. Chem., 1985, 24, 4371.
17. B. Stibr, Z. Janousek, P. Plesek, T. Jelinek, and S. Hermanek, J. Chem. Soc., Chem. Commun., 1985, 1365.
18. Z. Janousek, J. Plesek, S. Hermanek, and B. Stibr, Polyhedron, 1985, 4, 1797.
19. E.A. Mizusawa, M.R. Thompson, and M.F. Hawthorne, Inorg. Chem., 1985, 24, 1911.
20. J.S. Andrews, J. Zayas, and M. Jones, Inorg. Chem., 1985, 24, 3715.
21. V.V. Grushin, T.M. Shcherbina, and T.P. Tolstaya, J. Organomet. Chem., 1985, 292, 105.
22. I.I. Demkina, A.N. Vanchikov, V.V. Grushin, and T.P. Tolstaya, Izv. Akad. Nauk SSSR, Ser, Khim., 1985, 940.
23. L.I. Zakharkin, A.I. Kovredov, and V.A. Ol'shevskaya, Izv. Akad. Nauk SSR., Ser. Khim., 1985, 888.
24. N.M. Chistovalova, I.S. Akhrem, E.V. Reshetova, and M.E. Vol'pin, Izv. Akad. Nauk SSSR., Ser. Khim., 1984, 2343.
25. (a) Yu. A. Kabachii, A.I. Yanovskii, P.M. Valetskii, Yu. T. Struchkov, S.V. Vinogradova, and V.V. Korshak, Dokl. Akad. Nauk SSSR., 1985, 280, 1390; (b) L.I. Zakharkin, V.N. Lebedev, G.G. Zhigareva, Izv. Akad. Nauk

SSSR., Ser. Khim., 1984, 2411; (c) Yu. A. Kabachii, P.M. Valetskii, S.V. Vinogradova, and V.V. Korshak, Dokl. Akad. Nauk SSSR., 1985, 280, 1180.
26. (a) B.A. Izmailov, V.N. Kalinin, V.D. Myakuskev, A.A. Zhdanov, and L.I. Zakharkin, Dokl. Akad. Nauk SSSR., 1985, 280, 114; (b) L.M. Silva-Trivino and W.E. Hill, An. Quim., Ser. C., 1983, 79, 113.
27. A.V. Kazantsev, M.G. Meiramov and L.I. Zakharkin, Zh. Obshch. Khim., 1984, 54, 2002.
28. L.A. Churkina, T.D. Zvereva, L.I. Maleshonok, and Yu. A. Ol'dekop, Vestsi Akad. Navuk BSSR., Ser. Khim., Navuk, 1984, 105.
29. T.D. Zvereva, L.A. Churkina and Yu. A. Ol'dekop, Vestsi Akad. Navuk BSSR., Ser. Khim. Navuk, 1984, 121.
30. S. Al-Baker, W.E. Hill, and C.A. McAuliffe, J. Chem. Soc., Dalton Trans., 1985, 1387.
31. (a) V.V. Korshak, L.G. Komarova, G.A. Kats, P.V. Petrovskii, and N.I. Bekasova, Vyskomol. Soedin., Ser. B, 1985, 27, 628; (b) A.A. Askadskii, O.A. Mel'nik, G.L. Slonimskii, K.A. Bychko, T.M. Frunze, A.A. Sakharova, A.I. Solomatina, N.I. Bekasova, and V.V. Korshak, Vysokomol. Soedin., Ser. A, 1985, 27, 151; (c) S.S. Ivanchev, T.D. Korneva, I.V. Zaitseva, V.V. Konovalenko, and V.I. Stanko, Dokl. Akad. Nauk SSSR., 1985, 282, 119; (d) T.M. Frunze, A.A. Sakharova, O.A. Mel'nik, and T.N. Balykova, Deposited Doc., 1984, VINITI 2244-84; (e) N.I. Makarevich, N.I. Susho, A.I. Ivanov, L.S. Lyapina, and B.P. Parfenov, Vestsi Akad Navuk BSSR, Ser. Khim. Navuk, 1984, 63.
32. T. Jelinek, J. Plesek, S. Hermanek, and B. Stibr, Collect. Czech. Chem. Commun., 1985, 50, 1376.
33. S.F. Myasoedov, A.L. Popov, K.A. Solntsev, and N.T. Kuznetsov, Zh. Anal. Khim., 1985, 39, 2088.
34. G.L. Gal'chenko, N. Tamm, E.P. Brykina, D.B. Bekker, A.B. Petrunin, and A.F. Zhigach, Zh. Fiz. Khim., 1985, 59, 2689.
35. O.W. Howarth, M.J. Jasztal, J.G. Taylor, and M.G.H. Wallbridge, Polyhedron, 1985, 4, 1461.
36. B.L. Tumanskii, P.M. Valetskii, Yu. A. Kabachii, N.N. Bubnov, S.P. Solodovnikov, V.V. Korshak, A.I. Prokof'ev, and M.I. Kabachnik, Izv. Akad. Nauk SSSR, Ser. Khim., 1984, 2413.
37. G.K. Semin, S.I. Gushchin, S.A. Petukhov, V.V. Grushin, T.P. Tolstaya, and I.N. Lisichkina, Izv. Akad. Nauk SSSR, Ser. Khim., 1984, 2622.
38. A.L. Blyumenfel'd, S.S. Bukalov, E.I. Fedin, and L.A. Leites, Khim. Fiz., 1985, 4, 191.
39. N.I. Vasyukova, Yu. S. Nekrasov, Yu. N. Sukharev, G.K. Magomedov, and A.S. Frenkel, Izv. Akad. Nauk SSSR., Ser. Khim., 1985, 1548.
40. (a) K. Kato, A. Yamamoto, N. Suzuki, and T. Fukuda, Kogyo Kayaku, 1983, 44, 352; (b) A.A. Borisov, B.E. Gel'fand, S.A. Tsyganov, and E.I. Timofeev, Dakl. Akad. Nauk SSSR., 1985, 281, 361.
41. H.A.O. Hill, U.S.A. Pat., US 85 02861, (C.A. 104 003119).
42. (a) J.A. Yancey, J. Chromatogr. Sci., 1985, 23, 370; (b) N.S. Nikitina, Z.A. Klimak, and N.A. Nikulicheva, Neftepererab. Neftekhim., 1985, 29.
43. K.M. Gibov and T.A. Surogina, Khimiya i Tekhnol. Elementoorgan. Poluproduktiv i Polimerov, 1984, 82.
44. (a) J. Jamet, J.R. Spann, R.W. Rice, D. Lewis, and W.S. Coblenz, Ceram. Eng. Sci. Proc., 1984, 5, 677; (b) W.S. Coblenz, G.H. Wiseman, P.B. Davis, and R.W. Rice, Mater.

Sci. Res., 1984, 17, 271; (c) J. Pribyl, J. Mostecky, B.
Veruovic, V. Kubelka, and A. Dockal, Czech. Pat., Czech.,
840901 P, C.A. 103 058082.
45. (a) D. Gabel, R. Walczyna, F. Wellmann, H. Riesenberg, and
I. Hocke, Brookhaven Natl. Lab., (Rep) BNL, 1983, 225;
(b) F. Wellmann and D. Gabel, ibid, 276; (c) O. Hechter
and I.L. Schwartz, ibid, 197.
46. K. Shelly, D.C. Finster, Y.J. Lee, W.R. Scheidt, and C.A.
Reed, J. Am. Chem. Soc., 1985, 107, 5955.
47. L. Barton and P.K. Rush, Inorg. Chem., 1985, 24, 3413.
48. L.I. Zakharkin and N.F. Shemyakin, Izv. Akad. Nauk SSSR.,
Ser. Khim., 1984, 2807.
49. V.I. Bregadze, A. Ya. Usyatinskii, V. Ts. Kampel, L.M.
Golubinskaya, and N.N. Godovikov, Izv. Akad. Nauk SSSR.,
Ser. Khim., 1985, 1212.
50. V.I. Bregadze, A. Ya. Usyatinsky, and N.N. Godovikov, J.
Organomet. Chem., 1985, 292, 75.
51. A. Ya. Usyatinskii, V.I. Bregadze, N.N. Godovikov, L.E.
Vinogradova, L.A. Leites, A.I. Yanovskii, and Yu. T.
Struchkov, Izv. Akad. Nauk SSSR., Ser. Khim., 1984, 2009.
52. A. Ya. Usiatinsky, V.A. Shreider, T.M. Shcherbina, V.I.
Bregadze, N.N. Godovikov, and I.L. Knuniants, J. Organomet.
Chem., 1985, 289, 17.
53. E.S. Klimov, E.V. Gassieva, and O. Yu. Okhlobystin, Zh.
Obshch. Khim., 1985, 55, 950.
54. R.G Swisher, E. Sinn, R.J. Butcher, and R.N. Grimes,
Organometallics, 1985, 4, 882.
55. R.G. Swisher, E. Sinn, and R.N. Grimes, Organometallics,
1985, 4, 890.
56. R.G. Swisher, E. Sinn, and R.N. Grimes, Organometallics,
1985, 4, 896.
57. H.A. Boyter, R.G. Swisher, E. Sinn, and R.N. Grimes, Inorg.
Chem., 1985, 4, 3810.
58. J.J. Briguglio and L.G. Sneedon, Organometallics, 1985, 4,
721.
59. T.P. Hanusa, J.C. Huffman, T.L. Curtis, and L.J. Todd,
Inorg. Chem., 1985, 24, 787.
60. T.P. Hanusa and L.J. Todd, Polyhedron, 1985, 4, 2063.
61. R.E. King, D.C. Busby, and M.F. Hawthorne, J. Organomet.
Chem., 1985, 279, 103.
62. J.D. Hewes, M. Thompson, and M.F. Hawthorne, Organometallics,
1985, 4, 13.
63. D.M. Speckman, C.B. Knobler, and M.F. Hawthorne,
Organometallics, 1985, 4, 426.
64. D.M. Speckman, C.B. Knobler, and M.F. Hawthorne,
Organometallics, 1985, 4, 1692.
65. J.A. Walker, C.B. Knobler, and M.F. Hawthorne, Inorg. Chem.,
1985, 24, 2688.
66. P.E. Behnken, T.B. Marder, R.T. Baker, C.B. Knobler, M.R.
Thompson, and M.F. Hawthorne, J. Am. Chem. Soc., 1985,
107, 932.
67. V.N. Kalinin, O.A. Mel'nik, A.A. Sakharova, T.M. Frunze,
L.I. Zakharkin, N.V. Borunova, and V.Z. Sharf, Izv. Akad.
Nauk SSSR., Ser. Khim., 1984, 2151.
68. M. Green, J.A.K. Howard, A.P. James, A. N.de M. Jelfs,
C.M. Nunn, and F.G.A. Stone, J. Chem. Soc., Chem. Commun.,
1985, 1778.
69. (a) N.A. Maier, A.A. Erdman, Z.P. Zubreichuk, V.P.
Prokopovich, and Yu. A. Ol'dekop, J. Organomet. Chem., 1985,
292, 297; (b) V.P. Prokopovich, A.A. Erdman, N.A. Maier, and

Yu. A. Ol'dekop, Dokl. Akad. Nuak SSSR., 1984, 278, 1141.
70. H.M. Colquhoun, T.J. Greenhough, and M.G.H. Wallbridge, J. Chem. Soc., Dalton Trans., 1985, 761.
71. L.I. Zakharkin, V.V. Kobak, I.V. Pisareva, V.A. Antonovich, V.A. Ol'shevskaya, A.I. Yanovsky, and Yu. T. Struchkov, J. Organometal. Chem., 1985, 297, 77.
72. T.L. Venable, C.T. Brewer, and R.N. Grimes, Inorg. Chem., 1985, 24, 4751.
73. G.A. Kukina, V.S. Sergienko, and M.A. Porai-Koshits, Koord. Khim., 1985, 11, 400.
74. Z-T. Wang, E. Sinn, and R.N. Grimes, Inorg. Chem., 1985, 24, 826.
75. Z-T. Wang, E. Sinn, and R.N. Grimes, Inorg. Chem., 1985, 24, 834.
76. H. Wadepohl, H. Pritzkow, and W. Siebert, Chem. Ber., 1985, 118, 729.
77. M. Bochmann, K. Geilich, and W. Siebert, Chem. Ber., 1985, 118, 401.
78. K. Stumpf, H. Pritzkow, and W. Siebert, Angew. Chem., Int. Ed. Engl., 1985, 24, 71.
79. G.E. Herberich and H. Ohst, Chem. Ber., 1985, 118, 4303.
80. G.E. Herberich, H.J. Becker, B. Hessner, and L. Zelenka, J. Organomet. Chem., 1985, 280, 147.
81. G. Schmid, D. Kampmann, W. Meyer, R. Boese, P. Paetzold, and K. Delpy, Chem. Ber., 1985, 118, 2418.
82. G. Schmid, F. Schmidt, and R. Boese, Chem. Ber., 1985, 118, 1949.
83. G. Schmid, D. Zaika, and R. Boese, Angew. Chem., Int. Ed. Engl., 1985, 24, 602.
84. J. Edwin, W. Siebert, and C. Kruger, J. Organomet. Chem., 1985, 282, 297.

5
Group III: Aluminium, Gallium, Indium and Thallium

BY P. G. HARRISON

1 Structure, Spectroscopy and Bonding

The Al+C_2H_2 potential energy surface has been investigated by <u>ab initio</u> MO methods employing double-ζ + polarization basis sets and including configuration interaction in order to resolve the apparent discrepancy between existing experimental and theoretical data. Although AlCCH$_2$ is the theoretically predicted global minimum, the isomer <u>cis</u>-AlHCCH is the structure observed by e.s.r. at 4K. These calculations predict a barrier of 39.1 kcal for the isomerization of <u>cis</u>-AlHCCH to AlCCH$_2$, which would be prohibitive to isomerization under the experimental conditions at 4K. A comparison to the bare HCCH→CCH$_2$ isomerization shows that the effect of the aluminium atom is significant.[1]

As in previous years, the structures of several interesting compounds, mostly novel, have been determined. Reaction of EtAlCl$_2$ with 12-crown-4, benzo-15-crown-5, and 18-crown-6 affords cationic complexes of aluminium coordinated by 4 or 5 oxygens of the crown and two <u>cis</u> chlorines with [AlCl$_3$Et]⁻ as the gegenion.[2] Potassium phenoxide reacts cleanly with trimethylaluminium in toluene at 70° in the presence of a 1:1 ratio of dibenzo-18-crown-6 to the salt, giving [K.dibenzo-18-crown-6] [Al$_2$Me$_6$OPh] as the product. Temperatures of <u>ca</u>. 160° are necessary for complete reaction using a 1:2 ratio of crown to salt in which case the complex K[AlMe$_2$(OPh)$_2$] is formed on slow cooling, possibly as a decomposition product of the [Al$_2$Me$_6$OPh]⁻ anion. This anion has a 'Y'-shaped geometry with a Al-O-Al angle of 128.3(7)°, whilst [AlMe$_2$(OPh)$_2$]⁻ lies on a crystallographic two-fold axis.[3] Bis(diphenylphosphino)methane forms a 1:2 complex with triethylaluminium, but only 1:1 complexes with trimethylaluminium, and trimethyl-, triethyl-, and clorodimethylgallium:-

(For references see page 79

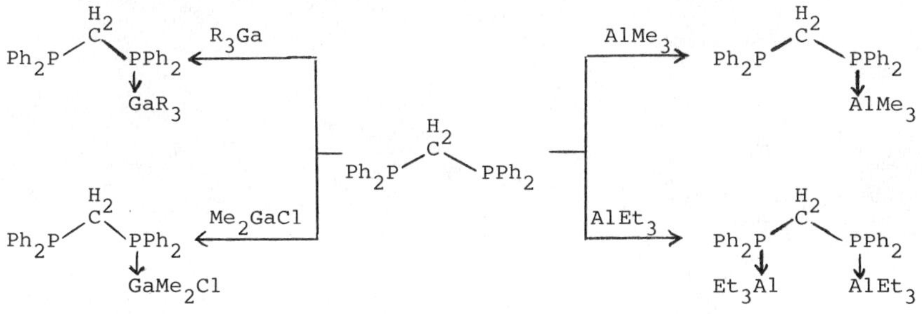

Group III: Aluminium, Gallium, Indium and Thallium

$$Ph_2P-CH_2-PPh_2 \cdot GaR_3 \xleftarrow{R_3Ga} Ph_2P-CH_2-PPh_2 \xrightarrow{AlMe_3} Ph_2P-CH_2-PPh_2 \cdot AlMe_3$$

$$Ph_2P-CH_2-PPh_2 \cdot GaMe_2Cl \xleftarrow{Me_2GaCl} \quad \xrightarrow{AlEt_3} Ph_2P-CH_2-PPh_2 \cdot Et_3Al \quad AlEt_3$$

The trialkyl complexes exhibit nmr-equivalent diphenylphosphino groups in solution in the temperature range +20 to -80°, probably because of a very rapid site exchange of the metal atoms. That only one phosphorus atom is coordinated to the metal was corroborated by crystallography for the Me_2GaCl complex, for which complex a splitting of the ^{31}P resonance into an AB quartet was observed in toluene below -60°.[4] Lithiated bis(diphenylphosphino)methane, $(Ph_2P)_2CHLi$, reacts with two equivalents of dialkylchloroalanes and -gallanes R_2MCl (R = Me, Et; M = Al, Ga), to form the compounds, $(Ph_2P)_2CH.(MR_2)_2Cl$, which exhibit unusual structures and fluxional behaviour. X-ray diffraction studies of $[Ph_2P]_2CH[Et_2Al]_2Cl$ show that these complexes may be described as the unsymmetrical adducts of the potential six-electron donor anion $[Ph_2\bar{P}-\bar{C}H-\bar{P}Ph_2]^-$ with the difunctional acceptor cation $[Et_2Al-Cl-AlEt_2]^+$. Thus, the complex possesses a central five-membered [AlClAlPC] ring formed by formal C→Al and P→Al donor-acceptor bonds, as in I (M = Al; R = Et). The heterocylic ring has an envelope conformation typical for this ring size and a centre of chirality at the central carbon atom of the anion. In toluene solution, nmr (^1H, ^{13}C and ^{31}P) data show that the acceptor portion $[R_2MClMR_2]$ of the complexes appear to undergo a rotation around the central P_2C-Al bond, thereby rendering the $[PPh_2]$ groups equivalent on the nmr time scale through breaking and reformation of the Al-P bonds, i.e.:

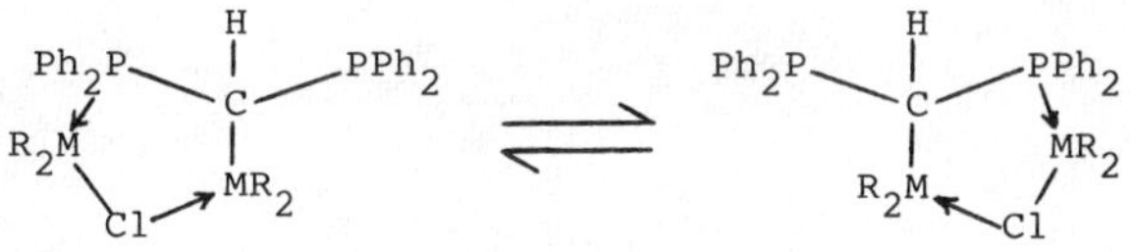

Ia ⇌ Ib

$4AlCl_3 \xrightarrow{4LiCH_2PMe_2} 2[Cl_2AlCH_2PMe_2]_2 \xrightarrow{4LiCH_2PMe_2} 2[ClAl(CH_2PMe_2)_2]_2 \xrightarrow{4LiCH_2PMe_2} 2[Al(CH_2PMe_2)_3]_2 \xrightarrow{4LiCH_2PMe_2} 4LiAl(CH_2PMe_2)_4$

$3AlCl_3 \Big| -LiCl \qquad 2AlCl_3 \Big| -2LiCl \qquad AlCl_3 \Big| -3LiCl$

$LiAl(CH_2PMe_2)_4 \qquad 2LiAl(CH_2PMe_2)_4 \qquad 3LiAl(CH_2PMe_2)_4$

$[Al(CH_2PMe_2)_3]_2 \xrightarrow{AlMe_3} \{[MeAl(CH_2PMe_2)_2]_2\} \xrightarrow{AlMe_3} [Me_2AlCH_2PMe_2]_2$

$[Al(CH_2PMe_2)_3]_2 + [Me_2AlCH_2PMe_2]_2 \rightleftharpoons 2[MeAl(CH_2PMe_2)_2]_2$

$2Me_2AlCl + 2LiCH_2PMe_2 \xrightarrow{Et_2O} [Me_2AlCH_2PMe_2]_2$

Scheme 1. Routes to (phosphinomethyl)aluminium compounds

Group III: Aluminium, Gallium, Indium and Thallium

The process is best regarded as a 'windshield wiper' motion, which leaves the Al-C bonds intact since complete dissociation of [R_2MCl] units is unlikely in view of the low activation barriers for the fluxional processes.[5] The dimeric phosphinomethyl-substituted organoaluminium compounds [(Me_2PCH_2)$_n$AlCl$_{3-n}$]$_2$ and [(Me_2PCH_2)$_n$AlMe$_{3-n}$]$_2$ (n = 1-3) have been obtained by various routes (Scheme 1), and undergo intra- and intermolecular exchange of the CH_2PMe_2, Me, and Cl substituents in solution (^{1}H, ^{13}C, ^{31}P, and ^{27}Al nmr). The central feature of these compounds is a six-membered [$Al_2C_2P_2$] ring, as in II, which was confirmed for [$Me_2AlCH_2PMe_2$]$_2$ by X-ray analysis. In the solid-state the ring in this compound adopts a chair conformation, whilst in solution donor molecules such as PMe_3 and (more readily) thf cleave the ring leading to the equilibrium:[6]

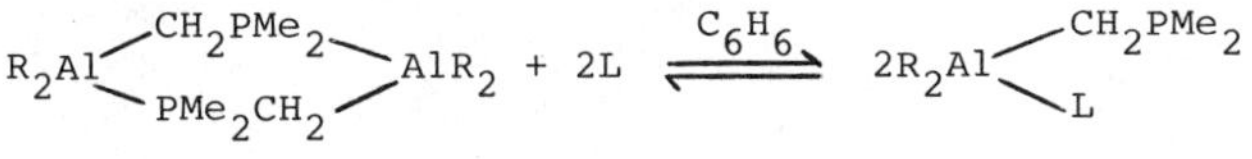

$$R_2Al\underset{PMe_2CH_2}{\overset{CH_2PMe_2}{\diagup\diagdown}}AlR_2 + 2L \underset{}{\overset{C_6H_6}{\rightleftharpoons}} 2R_2Al\underset{L}{\overset{CH_2PMe_2}{\diagup}}$$

L = PMe_3, thf

Reaction of $AlMe_3$, [$Me_2Al(CH_2PMe_2)$]$_2$, and [$Al(CH_2PMe_2)_3$]$_2$ with LiMe and $LiCH_2PMe_2$, respectively, in hydrocarbon solvents gives white precipitates, which disproportionate in diethyl ether to Li[$AlMe_4$] and polymeric [Li(Me_2PCH_2)$_2$Al(CH_2PMe_2)$_2$]$_n$ (III). In the presence of tetramethylethylenediamine and thf low-melting, toluene-soluble aluminates of the type [(tmeda)$_x$(thf)$_y$Li-(Me_2PCH_2)$_n$AlMe$_{4-n}$] can be isolated where n = 1 (x = 3/2, y = 0; x = y = 1; x = 1, y = 0) and n = 2 (x = 1, y = 0). For n = 4 [Li(Me_2PCH_2)$_2$Al(CH_2PMe_2)$_2$]$_n$ is produced as a high-melting solid, whereas for n = 3 a rapid disproportionation occurs, and [(tmeda)Li(Me_2PCH_2)$_2$AlMe$_2$] and [Li(Me_2PCH_2)$_2$Al(CH_2PMe_2)$_2$]$_n$ are produced. Similarly, [(tmeda)$_{3/2}$Li(Me_2PCH_2)Al(t-Bu)Me$_2$] slowly decomposes in solution. Reaction of [(tmeda)$_{3/2}$Li(Me_2PCH_2)AlMe$_3$] with MeI affords the ylide complex $Me_3AlCH_2PMe_3$. The observations may be rationalised by considering the nature of the (phosphino-methyl)aluminates acting as anionic phosphine ligands to Li$^+$, which is demonstrated by the X-ray analyses of [(tmeda)(thf)Li-(Me_2PCH_2)AlMe$_3$] and [(tmeda)Li(Me_2PCH_2)$_2$AlMe$_2$]. In the former compound, the lithium atom is coordinated by the thf and a chelating tmeda molecule, with distorted tetrahedral four-coordination being completed by the phosphorus donor function of

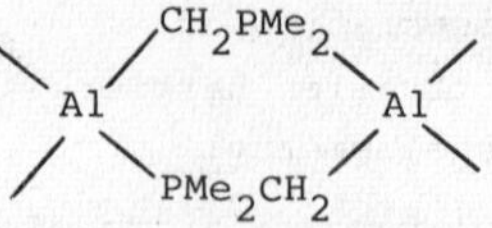

II 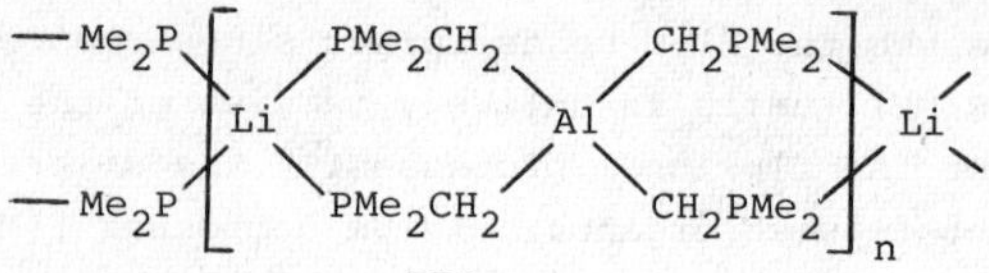 III

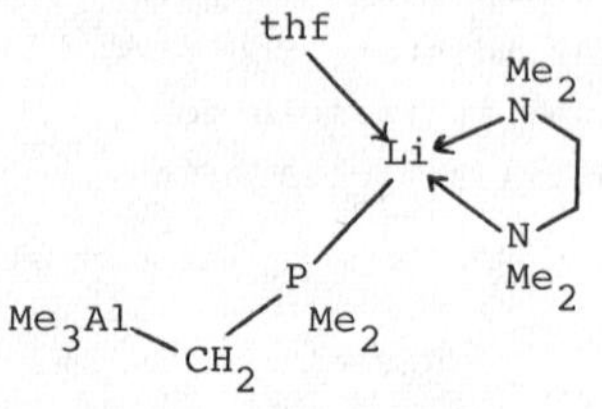

IV 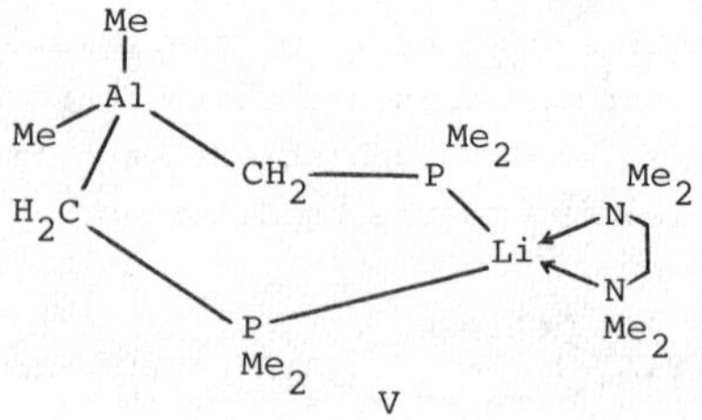 V

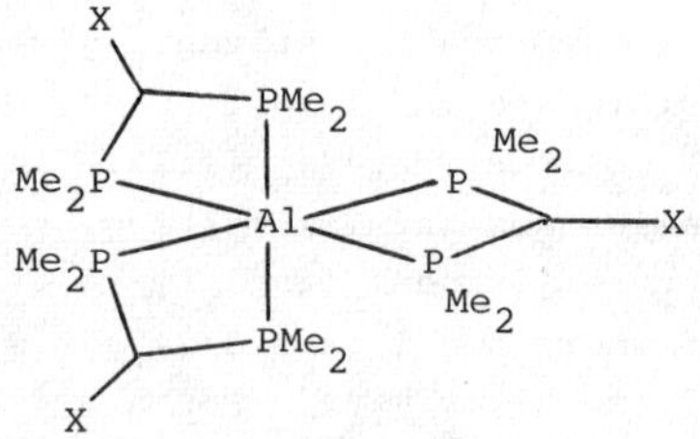

VI 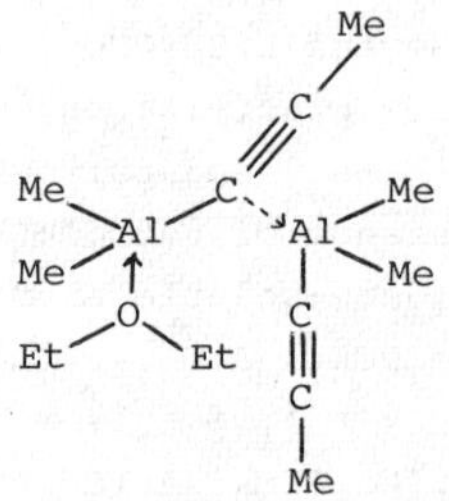 VII

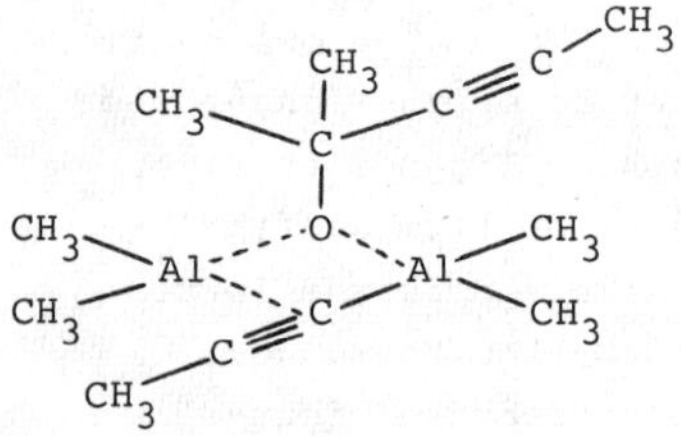

VIII 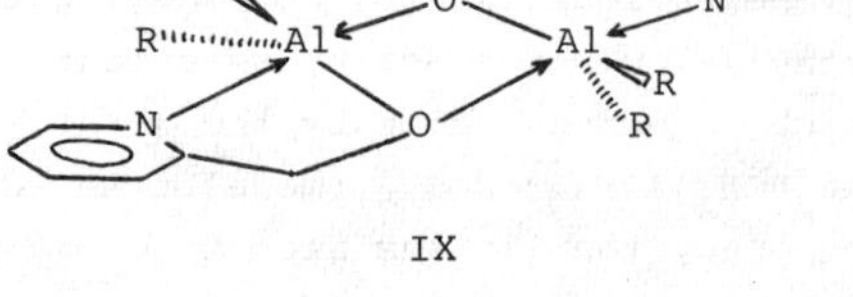 IX

the aluminate, as in IV. No other close contacts of lithium are
observed. The six-membered [LiPCAlCP] ring formed upon
complexation in [(tmeda)Li(Me$_2$PCH$_2$)$_2$AlMe$_2$], V, is in a pseudo-
envelope conformation with an almost planar Li(PC)$_2$ arrangement.
Again, the coordination at lithium is distorted tetrahedral.[7] The
neutral homoleptic phosphinomethyl aluminium compounds
Al[C(PMe$_2$)$_2$X]$_3$ (X = PMe$_2$ or SiMe$_3$), synthesised according to

$$AlCl_3 + 3Li[C(PMe_2)_2X] \xrightarrow[-3LiCl]{thf} Al[PMe_2)_2CX]_3$$

do not contain direct Al-C bonds. Rather, ^{27}Al nmr data show that
in each, the aluminium is situated in an octahedral environment
formed by the six equivalent phosphorus atoms as in VI. The
one-bond ^{27}Al-^{31}P coupling (91.6 Hz) is the first such observed.[8]
Trimethylaluminium reacts with base-free ytterbium bis[bis(trimethylsilyl)amide] in pentane to give Yb[N(SiMe$_3$)$_2$]$_2$[AlMe$_3$]$_2$. This
complex may be considered as being derived from a monomeric
{Yb[N(SiMe$_3$)$_2$]$_2$} fragment in which each lone pair of electrons on
the nitrogen atoms is coordinated to aluminium atoms so that the
coordination number of both the nitrogen and aluminium atoms is
four. Four of the methyl groups (two per aluminium) bridge the
aluminium and ytterbium atoms, with the average bridging distance
(2.019(11)Å) being significantly shorter than in Me$_6$Al$_2$
(2.125(2)Å). However, the chemistry of the ytterbium complex is
substantially different from trimethylaluminium or its
coordination complexes. In particular, the ytterbium complex
polymerizes ethylene under quite mild conditions (20° and 12
atoms).[9] Dimethylalkynylaluminium compounds form 2:1 complexes
with diethyl ether and acetone, for which the structures VII and
VIII were proposed.[10] The reactions of dialkylaluminium acetylacetonates, R$_2$Al(acac) (R = Me, Et, i-Pr) with the Lewis bases,
Et$_2$O, thf, pyridine, 2,6-dimethylpyridine, dmso, and hmpt (B)
proceeds according to:

$$3R_2Al(acac) + 2B \underset{k_2}{\overset{k_1}{\rightleftharpoons}} 2R_3Al \cdot B + Al(acac)_3$$

With strong bases, the equilibrium is shifted completely to the
right, with bases of moderate strength an equilibrium is
established, whilst with weak bases the reaction is shifted

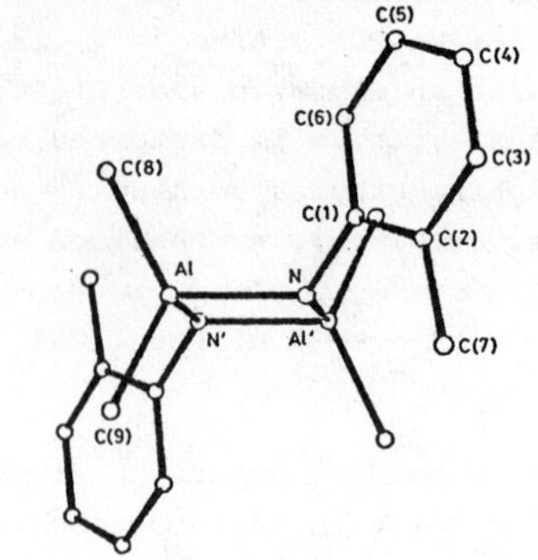

Figure 1. The structure of [IMe$_2$(NHC$_6$H$_4$Me-O)]$_2$ (reproduced from J.Chem.Soc., Dalton Trans., 1985, 1929).

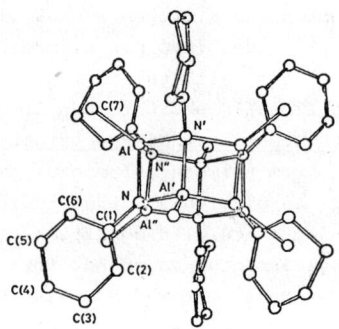

Figure 2. The structure of [AlMe(NPh)]$_6$ (reproduced from J.Chem.Soc., Dalton Trans., 1985, 1929).

completely to the left. The rate constant k_1 depends upon the nature of R and increases in the order i-Pr<Et<Me. The initial stage of the reaction is the formation of the complex $R_2Al(acac).B$, containing a five-coordinated aluminium atom, which subsequently dissociates.[11] Reacation of the alanes and chloroalanes, $R_{3-n}AlX_n$ (n = 0, 1; R = Me, Et, i-Pr, i-Bu, $PhCH_2$; X = Cl), with 2-pyridylmethanol affords dimeric $[R_{2-n}Xn\overline{AlOCH_2-2-C_5H_4N}]_2$ compounds and RH quantitatively. The principal structural features of these complexes, exemplified by an X-ray crystallographic determination of $[i-Bu_2\overline{AlOCH_2-2-C_5H_4N}]_2$, are illustrated in IX, and comprise a central four-membered $[Al_2O_2]$ ring with five-coordinate aluminium, formed by N,O-chelation of the ligand to one aluminium with oxygen bridging to the second aluminium atom. The same structural features are preserved in solution at low temperatures (^1H and ^{13}C nmr), although an equilibrium is established between this five-coordinated species IX and a four-coordinated species X involving an Al-N bond dissociation-association process with retention of the $[Al_2O_2]$ ring. This equilibrium can also be observed by ^{27}Al nmr.[12] Crystals of the arylamidoaluminium compound, $[AlMe_2(NHC_6H_4Me-\underline{O})]_2$, comprise centrosymmetrical <u>trans</u> dimers (Fig.1), in which the aromatic rings are almost perpendicular to the $(AlN)_2$ plane. Similar <u>trans</u> structures are also proposed for other homologues in the solid, but in solution mixtures of two isomers, probably <u>trans</u> dimers and <u>trans</u> trimers, appear to be present. Heating results in the formation the imido compounds, $[AlMe(NR)]_n$, with the elimination of methane. The structure of hexameric $[AlMe(NPh)]_6$ is shown in Fig.2, and exhibits a cage structure with S_6 symmetry. The hexameric nature is also preserved in the gas phase (mass spectrometry). The <u>o</u>-tolyl homologue seems to give tetrameric $[AlMe(NC_6H_4Mo-2)]_4$ molecules, whilst the corresponding <u>p</u>-tolyl derivative appears to be a mixture of tetramers and hexamers. Nmr data indicate that $[AlMe_2(NHPr-n]_3$ adopts only the <u>cis</u> configuration in benzene solution, whilst $[AlMe_2(NHBu-t)]_2$ the <u>trans</u> configuration only.[13]

Decomposition of the complex $[(\eta^5-C_5H_5)TiH_2AlH_2]_2$.tmeda in toluene affords the polycyclic complex $[(\eta^5-C_5H_5)_2Ti(\mu-H)_2Al-(\mu-H)(\eta^1:\eta^5-C_5H_4)Ti(\eta^5-C_5H_5)(\mu-H)]_2 \cdot C_6H_5CH_3$, whose structure contains two four-membered $[\overline{TiH_2Al}]$, two six-membered $[\overline{Ti(\mu-C_5H_4)Al(\mu-H)Ti(\mu-H)}]$, and one eight-membered

[$\overline{Ti(\mu-H)Ti(\mu-H)Al(\mu-H)Al(\mu-H)}$] rings. The complex promotes the hydrogenation of 1-hexene in thf, but not in hydrocarbons.[14] The interaction of $(\eta^5-C_5Me_5)_2TiH_2BH_2$ with $LiAlH_4$ in either produces [$(\eta^5-C_5Me_5)_2TiH_2AlH_2$]. The coordination polyhedron of aluminium is a distorted trigonal bipyramidal with five hydride ligands which bridge titanium and aluminium atoms. Decomposition of the complex in solution affords [$(\eta^5-C_5Me_5)_2Ti]_2AlH_5$, in which aluminium has a similar environment.[15] Dimethylamidogallane vapour contains dimeric [$Me_2NGaH_2]_2$ molecules, rather than monomers as previously suggested. Ir data indicate that the same structure is preserved both in solution and the solid state.[16] Ni(CDT) (CDT = trans,trans,trans-1,5,9-cyclododecatriene) reacts with alkali metal hydridotrialkylaluminates in ether/thf or ether/amine mixtures at temperatures below -50° to form the temperature-sensitive compounds, [$M(donor)_n$]$^+$[$R_3Al-H-N(CDT)$]$^-$ (M = Li, Na; R = Me, Et, i-Bu), which contain [Al-H-Ni] three-centre bonds.[17] Similarly, $Ni(C_2H_4)_3$ reacts both with aluminates and gallates at temperatures below -70° producing [$M(donor)_n$]$^+$[$R_3M^{III}-H-N(C_2H_4)_2$]$^-$ (M^{III} = Al, Ga). When ethene in these 16-electron compounds is replaced by CO to form an 18-electron complex; the Al-H and Ga-H bonds become so weak that the trialkylalane and -gallane are liberated, e.g.:[18]

$$[Na(THF)_6]^+[(i-Bu)_3Al-H-Ni(C_2H_4)_2]^- + 3CO \xrightarrow[-78°C]{THF}$$

$$[Na(THF)_4]^+[HNi(CO)_3]^- + 2C_2H_4 + THF.Al(i-Bu)_3$$

Molecules of the tmeda complex, $Ga(CH_2SiMe_3)_3.Me_2NC_2H_4NMe_2.-Ga(CH_2SiMe_3)_3$, obtained from $Ga(CH_2SiMe_3)_3$ and a ten-fold excess of tmeda, lie on a crystallographic inversion centre, with the tmeda ligand bridging the two gallium atoms as in XI.[19] Tris(cyclopentadienyl)gallium(III) has been synthesised from $GaCl_3$ and a slight stoichiometric excess of LiC_5H_5 at 0° in diethyl ether. Synthetic conditions must, however, be carefully controlled in order to achieve reproducibly high yields. The temperature must be maintained at or below ambient temperature at all stages. The colourless, volatile, pentane-soluble, crystalline compound decomposes slowly but irreversibly at room temperature and more rapidly at temperatures as low as 45°, as

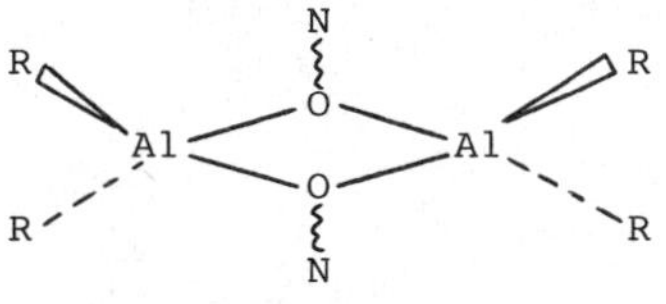

Group III: Aluminium, Gallium, Indium and Thallium

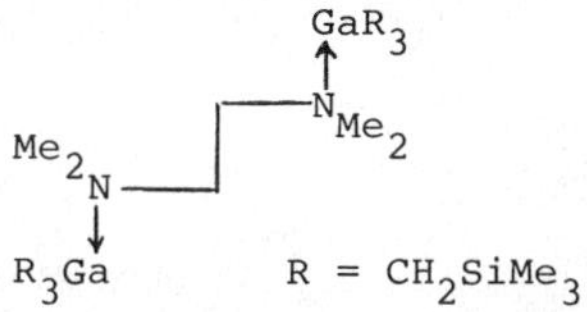

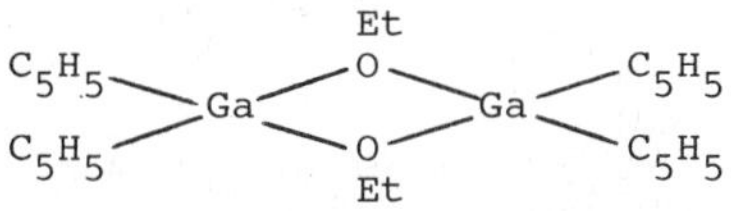

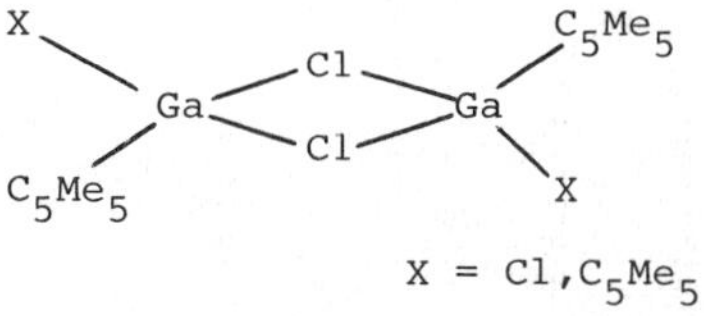

X

XI R = CH$_2$SiMe$_3$

XII

XIII X = Cl, C$_5$Me$_5$

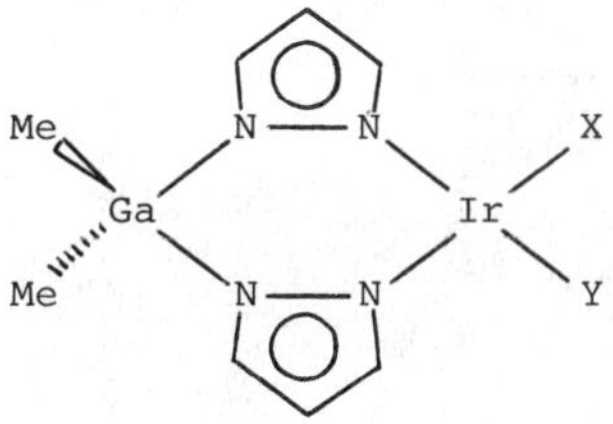

XIV

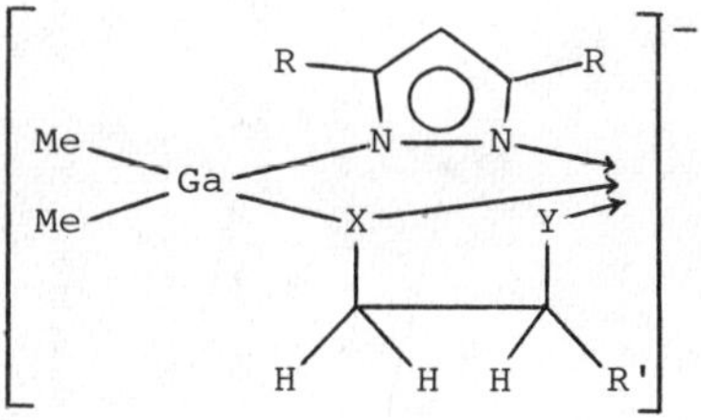

XV

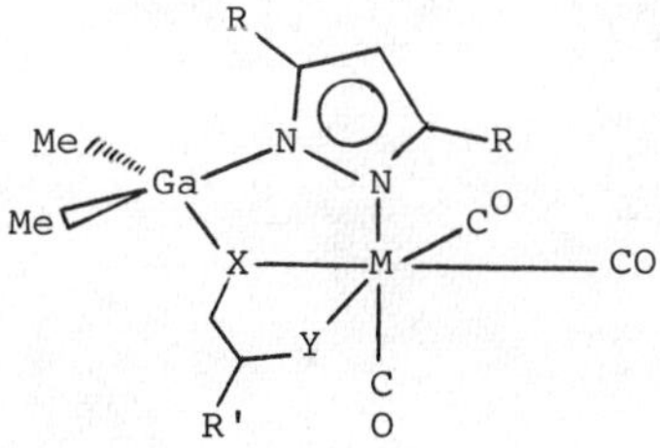

XVI

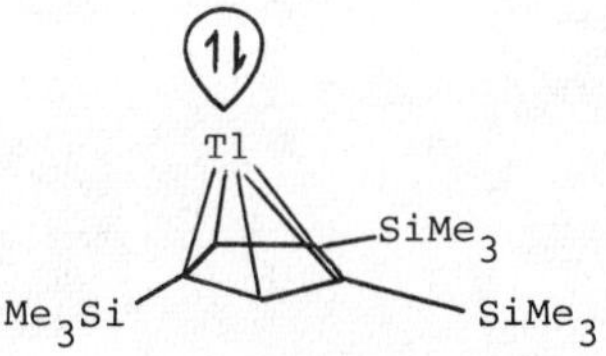

XX

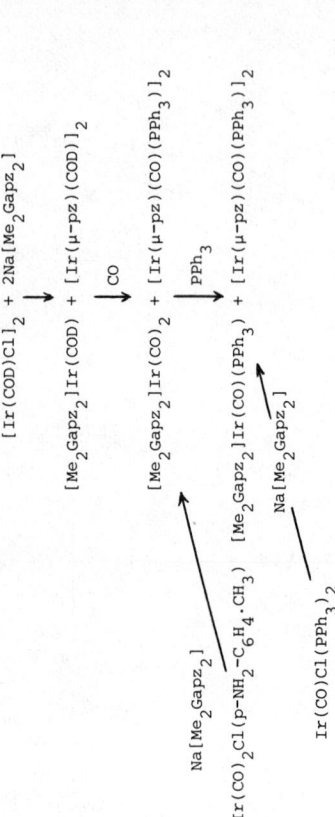

Scheme 2. Reactions of Ir(I) species with Na[Me$_2$Gapz$_2$]

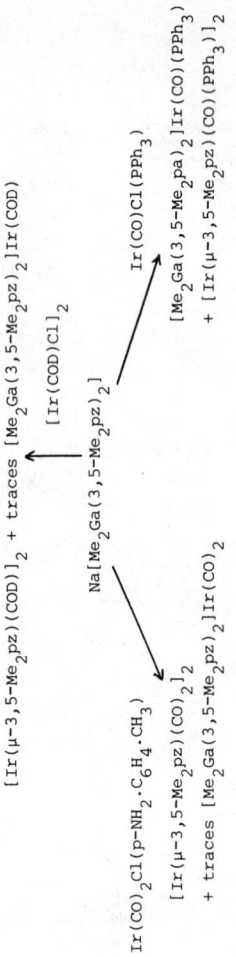

Scheme 3. Reactions of Ir(I) species with Na[Me$_2$Ga(3,5-Me$_2$pz)$_2$]

well as photolytically, to form a yellow, pentane-insoluble solid. Crystals comprise discrete isolated $Ga(C_5H_5)_3$ molecules, separated by normal van der Waal's contacts. All the cyclopentadienyl rings exhibit η^1 coordination to gallium, with the three α-carbon atoms and the gallium atom being coplanar. The compound exhibits weak Lewis acid character, with strong bases such as NMe_3 and thf affording four-coordinate complexes, but weaker bases such as diethyl ether being easily removed.[20] The nature of product of the reaction between $GaCl_3$ and LiC_5H_5 is dependent upon the temperature, at higher (ambient) temperatures, the product is the ethoxy-bridged dimer, XII, containing monohapto bonded cyclopentadienyl groups and a central planar $[Ga_2O_2]$ ring.[21] The pentamethylcyclopentadienylgallium chlorides, $Ga(C_5Me_5)_2Cl$ and $Ga(C_5Me_5)Cl_2$, are readily obtained from $GaCl_3$ and LiC_5Me_5 in diethyl ether. When the $LiC_5Me_5/GaCl_3$ stoichiometry was 3.2 or 2.2, the only gallium product was $Ga(C_5Me_5)_2Cl$, which was also isolated when thf was employed as the solvent or when the reaction mixture was refluxed. The dichloro derivative is obtained by ligand redistribution between $Ga(C_5Me_5)_2Cl$ and $GaCl_3$. Both molecules have chlorine-bridged dimeric structures (XIII), with planar, η^4-bonded pentamethylcyclopentadienyl groups. In benzene solution, the $[Ga(C_5Me_5)_2Cl]_2$ dimers dissociate, and only monomers are observed.[22]

Further chemistry of transition metal complexes containing pyrazolylgallate ligands has been published. Reactions of iridium(I) species with $Na[Me_2Gapz_2]$ and $Na[Me_2Ga(3,5-Me_2pz)_2]$ (pz = pyrazolyl; $3,5-Me_2pz$ = 3,5-dimethylpyrazolyl) are summarised in Schemes 2 and 3, respectively. The complexes, XIV (where X,Y = COD; CO,CO; CO,PPh_3), so obtained are square-planar at iridium, as exemplified by the crystal structure of $[Me_2Gapz_2]Ir(COD)$. Some of the complexes are stereochemically non-rigid in solution.[23] The structure of the rhodium(I) complex $[Me_2Ga(3,5-Me_2pz)_2]Rh(CO)PPh_3$ is similar; in both complexes, the central six-membered heterocyclic ring has the boat conformation.[24] Several complexes of the potentially terdentate ligands IV (X,Y = O,N,S) have also been synthesised. These types of ligand form octahedral the complexes XVI (M = Mn,Re) with the gallate ligand coordinated to the metal in the _fac_ mode. The X-ray structures of three of the complexes, $[Me_2Ga(pz)(OCH_2CH_2NMe_2)]Re(CO)_3$, $[Me_2Ga(pz)(OCH_2CH_2SPh)]Re(CO)_3$, and $[Me_2Ga(pz)(OCH_2CH(Et)NH_2)]Re(CO)_3$, as well as

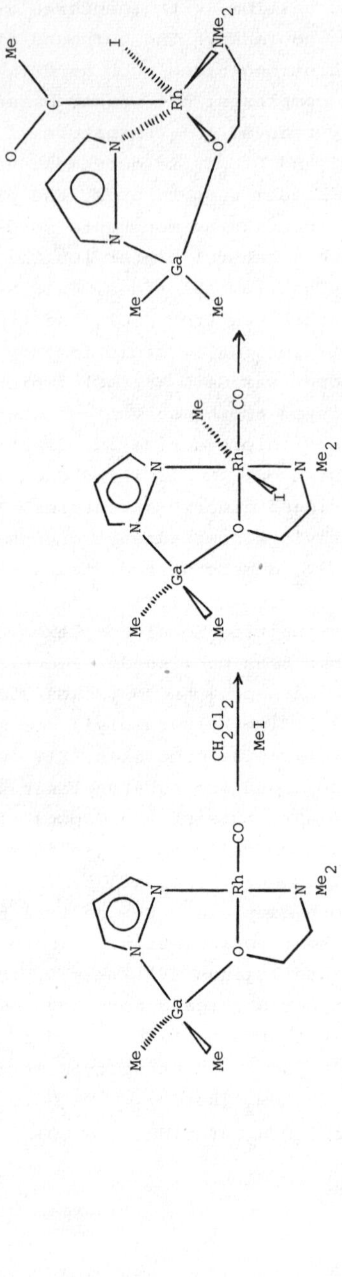

Scheme 4. Reaction of [Me$_2$Gapz(OCH$_2$CH$_2$NMe$_2$)]Rh(CO) with methyl iodide

[MeGa(pz)$_3$]Re(CO)$_3$ have been determined.[25] The rhodium(I) complex, [Me$_2$Gapz(OCH$_2$CH$_2$NH$_2$)]Rh(CO), has the square-planar structure XVII, and undergoes a facile oxidative-addition reaction with methyl iodide, which affords the five-coordinate square-pyramidal acetylrhodium(III) complex, XVIII, via the intermediate XIX.[26] The complexes, Me$_2$In(Ph$_2$pz)·½CH$_2$Cl$_2$, Me$_2$Tl(Ph$_2$pz), and Me$_2$Tl(Me$_2$pz) are formed on reaction of Me$_2$MI (M = In,Tl) with Ag(R$_2$pz). Me$_2$Tl(Me$_2$pz) is dimeric in benzene, but a monomer dimer equilibrium appears to be present for the corresponding phenyl complex, Me$_2$Tl(Ph$_2$pz).[27] The structures of two square pyramidal methylindium complexes, MeIn[MeNC(CH)$_4$N]$_2$ and MeIn{[MeN(CH$_2$)$_2$NMe]-Me$_2$In[MeN(CH$_2$)$_2$NMe]Me$_2$In}, have been described. In the latter complex, the two peripheral indium atoms are tetrahedrally coordinated.[28] Whereas the ^{1}H nmr spectra for EtInBr$_2$.tmen, EtInI$_2$.tmen (tmen = N,N,N',N'-tetramethylethylenediamine), and [Ph$_4$P][EtInI$_3$] exhibit only a broad single resonance for the ethyl protons at 60 MHz, spectra recorded at 400 MHz resolve these into a triplet + quartet for EtInBr$_2$.tmen and [Ph$_4$P][EtInI$_3$]. In the crystal both EtInX$_2$.tmen (X = Br, I) contain five-coordinated indium, whilst [Ph$_4$P][EtInI$_3$] comprises discrete [Ph$_4$P] cations and four-coordinated [EtInI$_3$] anions.[29] Carbon dioxide inserts into the In-C σ-bond of methylindium(III) porphyrins upon irradiation by visible light in dry benzene/pyridine media leading to stable acetato complexes (Scheme 5). The structure of the octa-ethylporphinatoindium acetate product was determined, confirming the chelation of the acetato group to indium.[30] 2D COSY ^{11}B-^{11}B nmr spectroscopy on nido-[Me$_2$TlB$_{10}$H$_{12}$]$^-$ enables the correlation of individual ^{11}B doublet components that are associated with particular ^{205}Tl spin-states, thereby permitting the determination of magnitudes and relative signs of various intracluster coupling constants $^n\underline{J}(^{205}$Tl-^{11}B), and hence yielding bonding information.[31]

The chemistry of arene complexes of univalent gallium, indium, and thallium, has been reviewed.[32] The structure of the mesitylene-thallium(I) complex [{1,3,5-Me$_3$H$_3$C$_6$}$_6$Tl]$_4$[GaBr$_4$]$_4$ comprises a skeletal framework of tetrameric Tl$_4$[GaBr$_4$]$_4$ which possesses crystallographic symmetry. Two types of thallium atom are present in the structure, those to which one mesitylene molecule is bonded which alternate with those thallium atoms which carry two mesitylene molecules. In all the mesitylene residues,

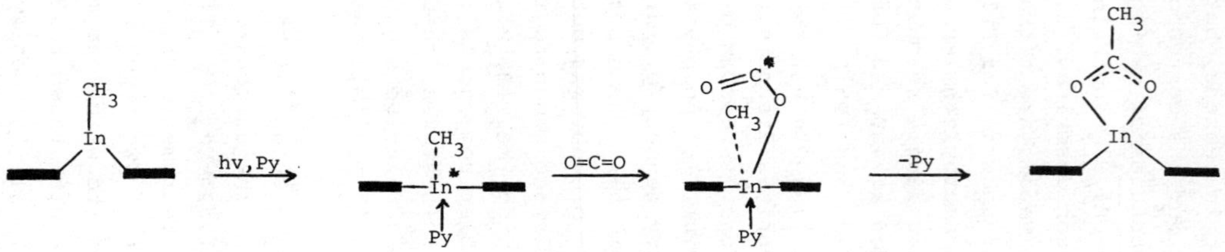

Scheme 5. Insertion of carbon dioxide into the In-C σ-bond of porphinatoindium(III) compounds

Group III: Aluminium, Gallium, Indium and Thallium

the thallium atoms lie almost symmetrically above the ring centroids (η^6 coordination). Coordination in the $[(mes)Tl]^+$ structural unit is extended by six bromine atoms to give a strongly distorted pentagonal bipyramidal arrangement. The two mesilylene molecules of the $[(mes)_2Tl]^+$ unit form an angle of 60.5° to each other, with three bromine atoms completing the coordination sphere.[33] Reaction of tris(trimethylsilyl)-cyclopentadiethyllithium with TlCl affords 1,2,4-tris-(trimethylsilyl)cyclopentadienylthallium(I) as a sublimable, air-stable, microcrystalline solid which is fairly soluble in organic solvents. From the properties, the nido structure XX was proposed.[34] The reaction of cyclopentadienylthallium(I) with (η^5-cycloentadienyl)(η^5-1,3,4,5-tetramethyl-2,3-dihydro-1H-1,3-diborolyl)cobalt in refluxing thf affords the orange brown complex XXI, crystals of which contain discrete molecules. The thallium atom does not lie over the centroid of the $[C_2B_2C]$ ring, but is displaced towards the unique carbon atom.[35]

2 Reactions and Applications in Organic Synthesis

Trimesitylaluminium, as a thf solvate, is readily obtained by treating $AlCl_3$ with either mesitylmagnesium bromide or dimesitylmagnesium in thf. Unsolvated material is produced by recrystallation from toluene. Reaction with mesityllithium yields lithium tetramesitylaluminate, whilst redistribution with $AlCl_3$ affords dimesitylaluminium chloride and mesitylaluminium dichloride, also as thf adducts.[36] Ultrasonic irradiation promotes the reaction between ethyl bromide and aluminium powder at room temperature. Reaction is complete within 10-20 minutes at room temperature. (cf. complete absence of reaction under the same conditions without irradiation). The product, ethylaluminium sesquibromide, may be used further without isolation, e.g.[37]

$$3CH_3CH_2Br + 2Al \xrightarrow{)))} (CH_3CH_2)_3Al_2Br_3 \xrightarrow{B(OCH_2CH_3)_3} (CH_3CH_2)_3B$$

The nature of organoaluminium compounds present in the $AlCl_3$-Al-C_2H_4 system has been examined. Under comparatively mild conditions (100-110°, 1 bar ethylene), both 1,1- and 1,2-ethanediylbis(dichloroaluminium) appear to be present, which give

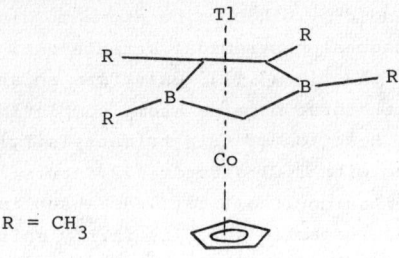

R = CH$_3$

XXI

rise to the formation of [1,1-D_2]ethane and [1,1-D_2]ethane upon treatment with D_2O:

$$>Al-CH_2CH_2-Al< + 2D_2O \rightarrow 2>AlOD + CH_2D-CH_2D$$
$$>Al-CH(CH_3)-Al< + 2D_2O \rightarrow 2>AlOD + CD_2H-CH_3$$

Although both compounds reduce $TiCl_4$, the 1,1-ethanediyl causes almost quantitative reduction to Ti^{III}, but eliminates ethylaluminium dichloride under a pressure of ethylene at 150°.[38] Di-, tri-, and tetraalkyl-2-azapyrylium salts have been obtained in high yield from the reaction of dialuminium hexahalide σ complexes of di-, tri-, and tetraalkyldobutadienes with NOCl.[39] The relative reactivities of Al-Al and Al-C bonds in tetraisobutylaluminium have been examined employing the protonic reagents H_3O^+, gaseous HCl, toluene-3,4-dithiol, and $HMn(CO)_5$. Excess H_2O at pH 1 results in 100% cleavage of both Al-Al and Al-C bonds, whereas the Al-Al bond exhibits greater reactivity towards HCl. Similarly, the dithiol also reacts more readily with the Al-Al bonds but, in contrast, the weak acid $HMn(CO)_5$ is unreactive towards $Al_2(i-Bu)_4$. With diborane, an insoluble material with the empirical stoichiometry AlB_2H_4 is obtained, for which the molecular formulation $Al_2(BH_2)_4$ with extensive intermolecular hydrogen bonding was suggested.[40]

The applications of organoaluminium compounds for performing selective reactions have been reviewed.[41] Topics covered include the organoaluminium-promoted Beckmann rearrangement of oxime sulphonates, new syntheses of polyamino macrocycles via reductive cleavage of aminals and amidines by diisobutylaluminium hydride, the diastereoselective cleavage of chiral acetals by organoaluminium compounds leading to optically active secondary alcohols, allylic alcohols, and β-substituted carbonyl compounds, and biomimetic terpene syntheses. The approximate rates and stoichiometry of the reaction of excess diisobutylaluminium hydride with a large number of organic compounds containing representative functional groups have been examined under standard conditions (toluene, 0°) in order to assess its reducing ability. Primary, secondary, and tertiary alcohols, simple phenols, and thiols evolve hydrogen rapidly and quantitatively, although only one of the active hydrogens in primary amines reacts. Aldehydes and ketones of diverse structure are reduced rapidly and

quantitatively to give the corresponding alcohols. Reduction of
norcamphor affords 7% exo- and 93% endo-norborneol. Conjugated
aldehydes and ketones such as cinnamaldehyde, methyl vinyl ketone,
and isophorone are rapidly and clearly reduced to the
corresponding allylic alcohols. Anthroquinone is reduced mainly
to 9,10-dihydro-9,10-anthracene-diol. Hexanoic acid, benzoic
acid, and crotonic acid liberate hydrogen rapidly, but only
partially, and the reduction proceeds very slowly. Acid chlorides
and esters are all reduced rapidly and quantitatively to the
corresponding alcohols. However, alkyl halides, such as n-octyl
iodide, and aromatic halides, such as p-bromotoluene, are inert.
Epoxides are reduced rapidly with an uptake of one equivalent of
hydride. Styrene oxide is reduced to give 27% 1- and 73%
2-phenylethanol. Tertiary amides are rapidly reduced (0.5h),
whereas primary amides are reduced only very slowly.
Nitrocompounds, azobenzene, and azoxybenzene are reduced
moderately. Cyclohexanone oxime liberates hydrogen rapidly,
consuming 1.2 equivalents of hydride for reduction, although
further reduction is slow. Phenyl isocyanate is rapidly reduced
to the imine stage. Pyridine reacts at a moderate rate with an
uptake of one hydride in 12h, but again further reaction is very
slow. Disulphides are rapidly reduced, whereas sulphides,
sulphones and sulphonic acid are inert under these reaction
conditions. Dimethyl sulphoxide is reduced at a moderate rate.
n-Octyl tosylate is reduced quantitatively to n-octane within
0.5h. In contrast, cyclohexyltosylate undergoes elimination
liberating one equivalent of hydrogen rapidly to give cyclohexane
in 95% yield.[42] The regioselectivity of epoxide-opening reactions
using alkylnylaluminium reagents has been examined. The factors
which govern the selectivity are often quite subtle.
Superficially innocuous changes in the amount of excess reagent
used to prepare the alkynylalanes can lead to great changes in the
resulting regioselectivity, even to the extent of reversing the
isomer ratios. Substitution of an alanate for the usual alane
reagent can sometimes be beneficial for achieving desired
regioselectivity.[43] Methylaluminium bis(2,6-di-tert-butyl-4-
alkylphenoxide), prepared in situ from trimethylaluminium and the
phenol in toluene at room temperature, is a novel reagent for
carbonyl alkylation. Reaction of this reagent with, for example,
4-tert-butylcyclohexane results in very high equatorial

selectivity (99%). This phenomenon may be ascribed to the high affinity of the oxygenophilic organoaluminium reagent for carbonyl oxygen.[44] Trialkylalanes can be cross-coupled with allyl ethers and esters, sulphides, and quaternized allylamines. The reactions proceed uncatalyzed either with mild conditions or in the presence of copper complex catalysts to give high yields of mono- and dialkenes of various structures. The alkylation of homoallyl or homopropargyl tosylates is accompanied by cyclization, which leads to alkyl-substituted cyclopropanes, cyclobutenes, and cyclopropylidenes.[45] Secondary and/or tertiary alcohols and unsymmetrical ketones have been obtained in moderate to good yields by the palladium-catalyzed carbonylative coupling of aryl halides with alkylaluminium compounds under very mild conditions (20-50°, 1 atmos of CO), e.g.:

$$PhI + CO + Et_3Al \xrightarrow[DME, 50°C]{PdCl_2(PPh_3)_2, PPh_3}$$

$$\underset{O}{Ph-\overset{\|}{C}-CH_2CH_3} + \underset{OH}{Ph-\overset{|}{CH}-CH_2CH_3} + \underset{\underset{CH_2CH_3}{|}}{Ph-\overset{OH}{\underset{|}{C}}-CH_2CH_3}$$

The type of reaction observed is dependent upon the aluminium reagent employed. While the selective formation of secondary alcohols was observed in the reacation with i-Bu$_3$Al, the use of Et$_3$Al led to the formation of a mixture of a ketone and two alcoholic products. With Et$_2$AlCl predominantly unsymmetrical ketones were produced. In all cases, the formation of directly cross-coupled products was not observed.[46] The interaction of equimolecular amounts of titanocene dichloride and MeAlCl$_2$ in chloroform with the alkyne, PhC≡C-SiMe$_3$, at -20° affords the insertion product XXII, whose structure was confirmed by X-ray crystallography:

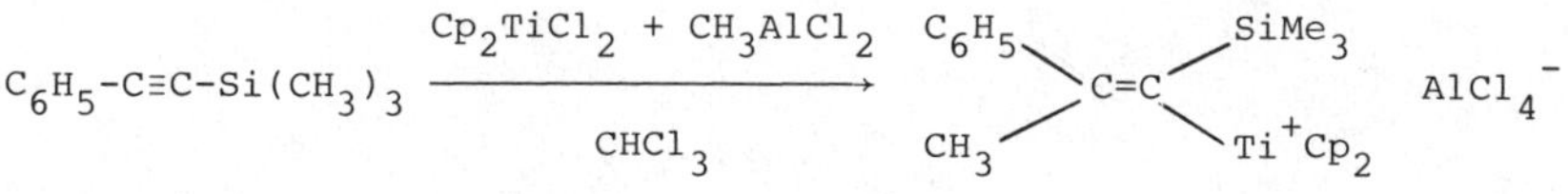

$$C_6H_5-C\equiv C-Si(CH_3)_3 \xrightarrow[CHCl_3]{Cp_2TiCl_2 + CH_3AlCl_2} \underset{CH_3}{\overset{C_6H_5}{}}C=C\underset{Ti^+Cp_2}{\overset{SiMe_3}{}} AlCl_4^-$$

XXII

From a detailed ^{1}H, ^{13}C, and ^{27}Al nmr study, it was demonstrated that XXII was the first and only insertion product formed in this catalyst system. The active component in this catalyst system was concluded to be the ion-pair species XXII, and that the crucial step in the activation of this ethylene polymerization catalyst is the isomerization of the adduct XXIV into XXIII:[47]

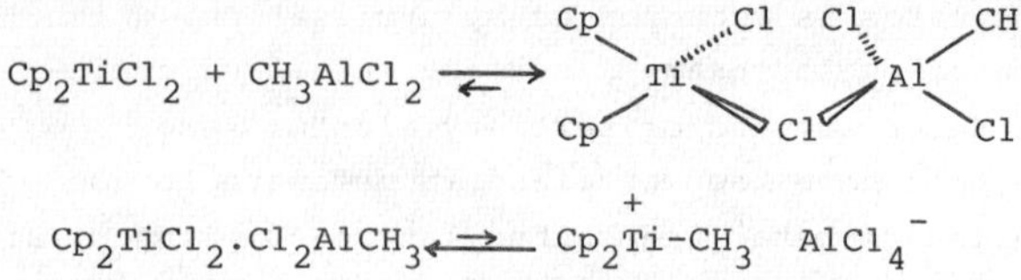

$$Cp_2TiCl_2 + CH_3AlCl_2 \rightleftharpoons \underset{Cp}{\overset{Cp}{\diagdown}}Ti\underset{Cl}{\overset{Cl}{\diagdown}}\underset{Cl}{\overset{Cl}{\diagdown}}Al\underset{Cl}{\overset{CH_3}{\diagdown}}$$

$$+$$

$$Cp_2TiCl_2 \cdot Cl_2AlCH_3 \rightleftharpoons Cp_2Ti\text{-}CH_3 \quad AlCl_4^-$$

XXIV XXIII

Highly isotactic polypropylene is obtained when chiral ethylenebis(4,5,6,7-tetrahydro-1-indenyl)-zirconium dichloride is employed in combination with methylalumoxane (MeAlO)$_n$ as the co-catalyst. This catalyst system is highly active: less than 10^{-5} mol l^{-1} suffices for the production of 7700 kg polypropylene per mole of Zr per hour at 60°. The quality of the polymer produced is also very high, with a quite small molecular weight distribution, and less than 1% solubility in hydrocarbons.[48] Trimethyl-, triethyl-, and triisobutylaluminium react with diethylhydroxyborane occurs in hydrocarbon solvents at -96° with evolution of the corresponding hydrocarbon. Reaction occurs in two stages: formation of the unstable diethylboryldialkylaluminium oxide, XXIV, which on distillation eliminates diethyl(alkyl)borane to form the oxide XXV:[49]

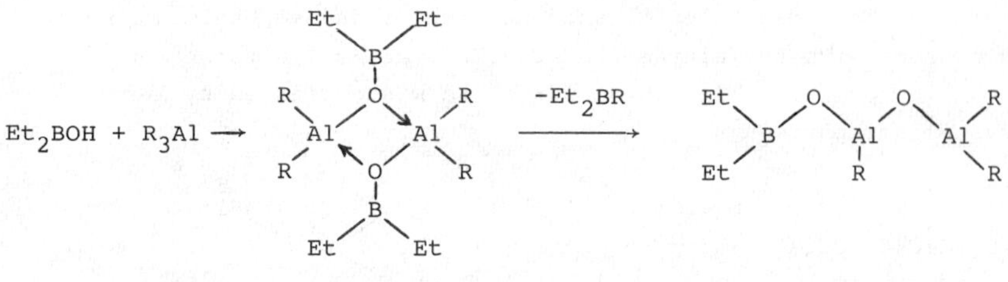

$$Et_2BOH + R_3Al \longrightarrow \text{[structure XXIV]} \xrightarrow{-Et_2BR} \text{[structure XXV]}$$

XXIV XXV

Substituted 4-alkenols undergo electrophilic cyclization with thallium(III) salts in a regio- and stereoselective manner. The organothallium intermediate was not isolated, but undergoes dethallation with concomitant 1,2-oxygen migration, leading to ring-expanded or -contracted products. The method is particulary useful as a one-step procedure for the synthesis of trans-2,5-disubstituted tetrahydrofurans.[50] Maximum yields of boron-thallated products of o- and m-carboranes are attained by consecutive treatment of the carborane with equimolecular amounts of mercuric oxide and then thallium(III) trifluoroacetate or thallium(III) acetate sesquihydrate in pure trifluoroacetic acid as the solvent.[51] Reaction of 5,6-benzobicyclo[2.2.1]hepta-2,5-diene with thallium(III) nitrate in methanol does not give rise to a ring-contracted product, but rather affords good yields of nitrate esters.[52] Thallation of vinylferrocene using thallium(III) acetate in methanol results in the formation of ferrocene derivatives of the type $(C_5H_5)Fe(C_5H_4CH(X)CH_2OCH_3)$.[53] Thallium(I) tetrafluoroborate has been employed as an electrophilic promoter for the substitution of chloride in $[Fe(\eta^5-C_5H_5)(CO)_2CH_2Cl]$ and $Mo(\eta^5-C_5H_5)(CO)_3CH_2Cl]$ by a wide range of uncharged nucleophiles.[54]

The electrochemistry, nmr spectroscopy, and UV-visible spectroscopy of gallium(III) and indium(III) porphyrins with both ionic Cl^- or σ-bonded axial organic groups, such as C_6H_5, $C_2H_2C_6H_5$, and $C_2C_6H_5(R)$. All of the compounds could be oxidized or reduced by multiple single-electron-transfer steps in which the initial step yields $[(P)M(R)]^+$ or $[(P)M(R)]^-$, where P represents the porphyrin macrocycle and M = Ga or In. In all cases, the singly reduced compounds are stable, as are the doubly reduced compounds in the case of gallium. In contrast, the singly oxidized compounds undergo a metal-carbon bond cleavage, the rate of which depends upon the electron donating properties of the axial ligand R. Physical properties of complexes containing the $C_2C_6H_5$ ligand resemble the metalloporphyrins bound by ionic ligands such as Cl.[55,56]

References

1 A.C. Scheiner and H.F. Schafer, J.Am.Chem.Soc., 1985, 107, 4451.
2 S.G. Bott, H. Elgamal and J.L. Atwood, J.Am.Chem.Soc., 1985, 107, 1796.
3 M.J. Zaworotko, C.R. Kerr and J.L. Atwood, Organometallics, 1985, 4, 238.

4   H. Schmidbaur, S. Lauteschläger and G. Müller, J.Organomet.Chem., 1985, 281, 25.
5   H. Schmidbaur, S. Lauteschläger and G. Müller, J.Organomet.Chem., 1985, 281, 33.
6   H.H. Karsch, A. Appelt, F.H. Köhler and G. Müller, Organometallics, 1985, 4, 231.
7   H.H. Karsch, A. Appelt, and G. Müller, Organometallics, 1985, 4, 1624.
8   H.H. Karsch and A. Appelt, J.Chem.Soc.,Chem.Commun., 1985, 1083.
9   J.M. Boncella and R.A. Anderson, Organometallics, 1985, 4, 205.
10 K.B. Starowieyski, A. Becalska and A. Okninski, J.Organomet.Chem., 1985, 293, 7.
11 S. Pasynkiewicz and J. Lewinski, J.Organomet.Chem., 1985, 290, 15.
12 M.R.P. van Vliet, P. Buysingh, G. van Koten, K. Vrieze, B. Kojic-Prodic and A.L. Spek, Organometallics, 1985, 4, 1701.
13 A.A.I. Al-Wassil, P.B. Hitchcock, S. Sarisaban, J.D. Smith and C.L. Wilson, J.Chem.Soc., Dalton Trans., 1985, 1929.
14 E.B. Lobkovskii, G.L. Soloveichik, A.I. Sizov and B.M. Bulychev, J.Organomet.Chem., 1985, 280, 53.
15 V.K. Bel'sky, A.I. Sizov, B.M. Bulychev and G.L. Soloveichik, J.Organomet.Chem., 1985, 280, 67.
16 P.L. Baxter, A.J. Downs, D.W.H. Rankin and H.E. Robertson, J.Chem.Soc., Dalton Trans., 1985, 807.
17 K.R. Pörschke and G. Wilke, Chem.Ber., 1985, 118, 313.
18 W. Kleimann, K.R. Pörschke and G. Wilke, Chem.Ber., 1985, 118, 323.
19 R.B. Hallock, W.E. Hunter, J.L. Atwood and O.T. Beachley, Organometallics, 1985, 4, 547.
20 O.T. Beachley, T.D. Getman, R.U. Kirss, R.B. Hallock, W.E. Hunter and J.L. Atwood, Organometallics, 1985, 4, 751.
21 A.H. Cowley, S.K. Mehrotra, J.L. Atwood and W.E. Hunter, Organometallics, 1985, 4, 1115.
22 O.T. Beachley, R.B. Hallock, H.M. Zhang and J.L. Atwood, Organometallics, 1985, 4, 1675.
23 S. Nussbaum, S.J. Rettig, A. Storr and J. Trotter, Canad.J.Chem., 1985, 63, 692.
24 B.M. Louie, S.J. Rettig, A. Storr and J. Trotter, Canad.J.Chem., 1985, 63, 503.
25 B.M. Louie, S.J. Rettig, A. Storr and J. Trotter, Canad.J.Chem., 1985, 63, 2261.
26 B.M. Louie, S.J. Rettig, A. Storr and J. Trotter, Canad.J.Chem., 1985, 63, 3019.
27 P.K. Byers, A.J. Canty, K. Mills and L. Titcombe, J.Organomet.Chem., 1985, 295, 401.
28 A.M. Arif, D.C. Bradley, D.M. Frigo, M.B. Hursthouse and B. Hussain, J.Chem.Soc., Chem.Commun., 1985, 783.
29 M.A. Khan, C. Peppe and D.G. Tuck, J.Organomet.Chem., 1985, 280, 17.
30 P. Cocolios, R. Guilard, D. Bayeul and C. Lecomte, Inorg.Chem., 1985, 24, 2058.
31 M.A. Beckett, J.D. Kennedy and O.W. Howarth, J.Chem.Soc., Chem.Commun., 1985, 855.
32 H. Schmidbaur, Angew.Chem., Int.Ed.Engl., 1985, 24, 893.
33 H. Schmidbaur, W. Bublak, J. Riede and G. Müller, Angew.Chem., Int.Ed.Engl., 1985, 24,
34 P. Jutzi and W. Leffers, J.Chem.Soc., Chem.Commun., 1985, 1735.
35 K. Stumpf, H. Pritzkow and W. Siebert, Angew.Chem., Int.Ed.Engl., 1985, 24, 71.
36 W. Seidel, Z.Anorg.Allg.Chem., 1985, 524, 101.
37 K.F. Liou, P.H. Yang and Y.T. Lin, J.Organomet.Chem., 1985, 294, 145.
38 H. Martin, H. Bretinger and F. Fürbach, Angew.Chem., Int.Ed.Engl., 1985, 24, 311.
39 Q.B. Broxterman, H. Hogeveen, R.F. Kingma and F. van Bolhuis, J.Am.Chem.Soc., 1985, 107, 5722.

40 M.A. Miller and E.P. Schram, Organometallics, 1985, 4, 1362.
41 K. Maruoka and H. Yamamoto, Angew.Chem., Int.Ed.Engl., 1985, 24, 668.
42 N.M. Yoon and Y.S. Gyoung, J.Org.Chem., 1985, 50, 2443.
43 R.S. Matthews and D.J. Erckhoff, J.Org.Chem., 1985, 50, 3923.
44 K. Maruoka, T. Itoh and H. Yamamoto, J.Am.Chem.Soc., 1985, 107, 4573.
45 G.A. Tolstikov and U.M. Dzhemiler, J.Organomet.Chem., 1985, 292, 133.
46 Y. Wakita, T. Yasunaga and M. Kojima, J.Organomet.Chem., 1985, 288, 261.
47 J.J. Eisch, A.M. Piotrowskii, S.K. Browstein, E.J. Gable and F.L. Lee, J.Am.Chem.Soc., 1985, 107, 7219.
48 W. Kaminsky, K. Külper, H.H. Brintzinger and F.R.W.P. Wild, Angew.Chem., Int.Ed.Engl., 1985, 24, 507.
49 L. Synoradzki, M. Boleslawski and J. Lewinski, J.Organomet.Chem., 1985, 284, 1.
50 J.P. Michael, P.C. Ting and P.A. Bartlett, J.Org.Chem., 1985, 50, 2416.
51 V.I. Bregadze, A.Ya Usiatinsky and N.N. Godovikov, J.Organomet.Chem., 1985, 292, 75.
52 W.J. Clayton, C.P. Brock, P.A. Crooks and S.L. Smith, J.Org.Chem., 1985, 50, 5372.
53 J.H. Youn, R. Herrmann and B. Floris, J.Organomet.Chem., 1985, 291, 355.
54 E.K. Barefield, P. McCarten and M.C. Hillhouse, Organometallics, 1985, 4, 1682.
55 K.M. Kadish, B. Boisselier-Cocolios, A. Coutsolelos, P. Mitaine and R. Guilard, Inorg.Chem., 1985, 24, 4521.
56 K.M. Kadish, B. Boisselier-Cocolios, and R. Guilard, Inorg.Chem., 1985, 24, 2139.

Bibliography

Additional references not included in the text.

'A penta-co-ordinated aluminate dimer; X-ray crystal structure', M.C. Cruickshank and L.S. Dent-Glasser, J.Chem.Soc., Chem.Commun., 1985, 84.

'The two liquid phase system NaAl(ethyl)$_4$-benzene-HMPA', J.H. Medley, N. Ahmad and M.C. Day, Canad.J.Chem., 1985, 63, 2906.

'Lithium aluminium amide, LiAl(NH$_2$)$_4$-preparation, X-ray investigation, ir spectrum, and thermal decomposition', H. Jacobs, K. Jänichen, C. Hadenfeldt and R. Juza, Z.Anorg.Allg.Chem., 1985, 531, 125.

'(i-C$_3$H$_7$SGaI$_2$)$_2$, the first "butterfly" molecule with tetracoordinated gallium', P.G. Hoffmann and C. Burschka, Angew.Chem., Int.Ed.Engl., 1985, 24, 970.

'Synthesis and characterization of ((trimethylsilyl)amino)- and (methyl(trimethylsilyl)amino)gallium dichloride', W.R. Nutt, J.A. Anderson, J.D. Odom, M.M. Williamson and B.H. Rubin, Inorg.Chem., 1985, 24, 159.

'Indium derivatives of monothio-α-diketones and the X-ray structure of tris[benzoyl(thiobenzoyl)methanato-O,S]indium(III)', C. Sreelatha, V.D. Gupta, C.K. Narula and H. Nöth, J.Chem.Soc., Dalton Trans., 1985, 2623.

'Cyclopentadienylnickel complexes', E. Hernandez and P. Royo, J.Organomet.Chem. 1985, 291, 387.

'Complexes from the trihalides of aluminium and gallium with thiols, hydrogen sulphide and selenols', G.G. Hoffmann, Chem.Ber., 1985, 118, 3320.

'Organic derivatives of aluminium. Part 1. Reactions of alkanolamines with aluminium isopropoxide', R. Duggal and R.C. Mehrotra, Inorg.Chim.Acta, 1985, 98, 121.

'Palladium- and nickel-catalysed aryl- and acyl-demetallation of organometallic compounds', N.A. Bumagin, A.B. Ponomaryor and I.P. Beletskaya, J.Organomet.Chem., 1985, 291, 129.

'The influence of (organo)metallics 'metal-tuning' on stereo- and regio-chemical convergence in reactions of allylic carbanions with aldehydes', Y. Yamamoto, Y. Saito and K. Maruyama, J.Organomet.Chem., 1985, 292, 311.

'Initiation of olefin matethesis from non-organometallic precursors', L. Bencze and A. Kraut-Vass, J.Organomet.Chem., 1985, 280, C14.

'Radicalions. LXVIII. Cyclic voltametry of organosilicon-substituted π-systems: Determination of oxidation potentials for the one-electron transfer system $AlCl_3/H_2CCl_2$', H. Bock and U. Lechner-Knoblauch, J.Organomet.Chem., 1985, 294, 295.

'Equilibrium and structural studies of silicon(IV) and aluminium(III) in aqueous solution. Part 13. A potentiometric and ^{27}Al nuclear magnetic resonance study of speciation and equilibria in the aluminium(III)-oxalic acid-hydroxide system', S. Sjöberg and L.O. Ohman, J.Chem.Soc., Dalton Trans., 1985, 2665.

'Reactivities of Lewis acid adducts of $\eta^5-C_5H_5Fe(CO)_5Y$ (Y = SPh, SO_2Ph, and PPh_2) with nucleophiles', J.J. Weers and D.P. Eyman, J.Organomet.Chem., 1985, 286, 47.

'Surface compounds of transition metals. XXIX. Reaction of surface chromiumVI/silica gel with aluminium alkyls: "Formation" of Phillips catalysts', H.L. Krauss and B. Hanke, Z.Anorg.Allg.Chem., 1985, 521, 111.

'The Hammett relationship in $CpTiCl_2OPhX$ (X = CH_3O,CH_3,Cl,NO_2) + $BuLi(Et_2AlCl)$ catalytic systems', W. Skupinski and A. Wasilewski, J.Organomet.Chem., 1985, 282, 69.

'Concerning reversal of diastereoselectivity in the BF_3 promoted addition of crotyl-organometallic compounds to aldehydes', Y. Yamamoto and K. Maruhama, J.Organomet.Chem., 1985, 284, C45.

'Dialkyl-and alkylenedithiophosphates of gallium(III)', R. Ahmad, G. Srivastara and R.C. Mehrotra, Inorg.Chim.Acta, 1985, 97, 159.

'Halogeno(phenyl)(organylthio)gallanes - synthesis, reactaivity, and properties', G.G. Hoffmann and H. Meixner, Z.Anorg.Allg.Chem., 1985, 523, 121.

'Diido(methylthio)gallane. Preparation and reactivity', G.G. Hoffmann and P. Resch', J.Organomet.Chem., 1985, 295, 137.

'Synthesis of thiogallanes via gallium halides and lead bis(thiolates)', G.G. Hoffmann, Chem.Ber., 1985, 118, 1655.

'Synthesis and properties of heteronuclear metal atom clusters $Re_4(CO)_{12}[\mu_3-GaRe(CO)_5]_4$ and $Re_2(CO)_8[\mu-GaRe(CO)_5]_2$', H.J. Haupt, P. Balsaa and B. Schwab, Z.Anorg.Allg.Chem., 1985, 521, 15.

'Kinetics and mechanisms of complex formation of gallium(III) and indium(III). The reactions with 4-(2-pyridylazo)resorcinol in water and other mixed solvents', E. Mentasti, C. Gaiocchi and L.J. Kirschenbaum', J.Chem.Soc., Dalton Trans., 1985, 2615.

'Indium(III) compounds with 2-[(dimethylamino)methyl]phenyl ligand', R.S. Steevenz, D.G. Tuck, H.A. Meinema and J.G. Noltes, Canad.J.Chem., 1985, 63, 755.

'Contributions to the chemistry of silicon-sulphur compounds. XXXV. The dimeric thalium(I)-tri-tert-butoxysilanethiolate', W. Wojnowski, K. Peters, E.M. Peters and H.G. von Schnering, Z.Anorg.Allg.Chem., 1985, 531, 147.

'Synthesis of alkylated methylene bisphosphonates via organothallium intermediates', D.W. Hutchinson and G. Semple, J.Organomet.Chem., 1985, 291, 145.

'Two ligand-bridged phthalocyanines: Crystal and molecular structure of fluoro(phthalocyaninato)gallium(III), $[Ga(Pc)F]_n$, and (μ-oxo)bis-[(phthalocyaninato)aluminium(III)], $[Al(Pc)]_2O$', K.J. Wynne, Inorg.Chem., 1985, 24, 1339.

6
Group IV: The Silicon Group

BY D. A. ARMITAGE

1 Introduction

Reactive silicon intermediates take their turn as the subject of two reviews and a book which also include silicon in organic synthesis and bioorganic silicon chemistry.[1] Reviews cover silane photochemistry and the alcoholysis of hydrosilanes,[2] protection of protic reagents by silylation, the use of organocopper reagents notably in silylation to give O-enolates, the synthetic applications of $ClCH_2SiMe_3$ and compounds with chiral silicon centres,[3] stabilisation of reactive phosphorus species through protection by large groups such as $(Me_3Si)_3C$ (Tsi), the x-ray structures of silatranes, the chemistry of organochlorosilanes, and silyl, germyl and stannyl derivatives of cyclopolyazenes $(NH)_n$ (n=3,4,5).[4] A book on carbon-functional organosilicon compounds describes (1) their use in making natural products, GLC stationary phases and in surface treatment, (2) intramolecular interactions and their influence, (3) the use of ^1H, ^{13}C and ^{29}Si NMR spectroscopy and (4) theoretical aspects of bonding.[5]

UV/PES provides information on bonding in silacyclanones and alkylgermanium halides, and X-ray/PES for structural analysis of organotin complexes of aminoacids. The roles of organotin compounds in the environment, as radical intermediates in carbon-carbon bond formation, and in regioselective manipulation of -OH groups are reviewed.[6] Applications of Mössbauer spectroscopy to Inorganic chemistry cover the structure and bonding in tin compounds, with books describing the industrial uses of tin chemicals and organotin compounds in modern technology.[7] The Gmelin Handbook of Inorganic Chemistry covers historical aspects of Inorganic silicon chemistry and Me_3Sn- and Et_3Sn-O compounds.[8]

Various aspects of the chemistry of the metal-carbon bond are described and include the silicon group elements, as does the role

(For references see page 119

of organometallic compounds in living organisms.[9] There are many brief reviews on aspects of silicon chemistry in the Japanese[10] and Russian literature.[11] This article contains about 1000 references.

2 The Carbon-Metalloid Bond

<u>Ab Initio</u> studies of the reaction of silene with formaldehyde to give silanone and ethene show it to be exothermic by ~30 kcal, with the intermediate 1,2-siloxetane even more stable, but with products involving cyclodisiloxane more stable still. The mixture of (<u>Z</u>)- and (<u>E</u>)-1-methyl-1-phenyl-2-neopentylsilene generated from vi(Ph)MeSiCl and Bu^{t}Li gives a 1,2-adduct mixture with Me$_3$SiOMe. Fluorination or reduction with LiAlH$_4$ reverses this reaction, consistent with inversion at Si. The Si=C bond length in Me$_2$Si=C(SiMe$_3$)SiMeBu$^t_2$ (170.2pm) is close to the value calculated for hindered silenes, much shorter than that found in the THF adduct (174.7pm) and in the hindered (Me$_3$Si)$_2$Si=C(OSiMe$_3$)Ad (176.4pm).[12] (Me$_3$Si)$_2$Bu^{t}Si-C(O)Ad photoisomerises to Me$_3$SiO(Me)Si=C(Ad)SiMe$_2$Bu^t which dimerises head to tail. Me$_2$Si=CH$_2$ is thought to be an intermediate amongst others in the gas-phase reaction of chloro- and chloromethyl(di)silanes with Na or K vapour.[13]

Electron energy loss spectra of Me$_4$Si are compared with other silanes and the effects of perfluoromethylcyclohexane on the photoconductivity of Me$_4$Si measured. Calculations indicate the β-stabilisation of the silyl group to be 20-30 kcal greater than that of a methyl group, in agreement with most available experimental evidence.[14] Vapour adsorption isotherms for Me$_4$Sn on graphite are determined, as are rotational tunnelling and librations of methyl groups in Me$_4$Pb.[15] Bu^{n}Li acts as a nucleophile to Me$_4$Sn, substituting Me, but Bu^{t}Li acts also as a base, giving a little Bu^{t}Me$_2$SnCH$_2$Li. Hexagonal and cubic SiC polymorphs give ^{29}Si and ^{13}C NMR spectra that indicate the number of non-equivalent lattice sites, while a photochemical cell using SiC and Pt electrodes can split water to H$_2$ with energy conversion efficiency of about 0.4% using a 500W Xe lamp.[16]

Me$_3$SiCH$_2$MgCl has now been used to prepare (2-ethoxycarbonylallyl)trimethylsilane and 2-Me$_3$SiCH$_2$-1,3-dienes, Me$_3$SiCH$_2$Cu for functionally substituted allylic silanes, and Me$_3$SiCH$_2$Li for cyclopropanols and β,δ-unsaturated ketones.[17] TMED coordinates to two moles of (Me$_3$SiCH$_2$)$_3$Ga while a series of silylmethyl complexes of Re, Y, W,

and Zr (absorbs CO) have been prepared.[18] Treating $(Me_3Sn)_4C$ with
MeLi followed by Me_3MCl gives mainly $(Me_3Sn)_3CMMe_3$, while
electrophiles (I_2,Br_2) attack $Sn-CH_2$ and $Sn-CH_3$ in $Me_3SnCH_2MMe_3$, but
acids only $Sn-CH_2$, and organometallic electrophiles only $Sn-CH_3$.[19]
Ph_3PbCH_2Li and $(Ph_3Pb)_2CHLi$ are intermediates in the preparation of
mixed R_3M-methanes, and a series of bromo- and iodomethylMPh_3
(M=Si-Pb) have been made.[20] $(Ph_2ClSn)_2CH_2$ crystallises with inter
and intramolecular chlorine bridges and the Mössbauer of
$(R_nX_{3-n}Sn)_2Y$ (X includes halogen, Y includes CH_2) determined.[21]

Me_3SiCN is used to prepare cyanosugars (from keto, epoxy and
acetal derivatives of carbohydrates), AdCN (bridgehead from halide
using $SnCl_4$), α-iminonitriles (from aldonitrones), cyanooxopentan-
oates, O-trimethylsilyl cyanohydrins and butyrolactones (from
α-bromo esters),[22] 3-siloxypropyl isocyanides (from oxetanes),
cyanocycloheptatriene-1-ol (from tropone), β-cyanoalkenylsilanes
(from ArC=CH), α-cyanated 3^y amines (from α-siloxyamines),
α-perfluoroalkyl-α-amino acid precursors, $H_2C=C(OSiMe_3)CN$ and
$R_3MCH=C(OSiMe_3)CN$ (M=Si,Ge).[23] $(CO)_5WC(OMe)Ph$ adds Me_3SiCN to give
the nitrile complex, while $(CO)_5WCPh_2$ and Bu^n_3SnCN in ether yields
$(CO)_5WNCCHPh_2$.[24]

Silylated α-diazocarbonyl compounds generate siloxyalkynes,
$(R_2N)_2PC(N_2)SiMe_3$ λ^3-phosphinocarbenes photochemically, and
$Me_3SiCH_2N_3$, triazenes and amidomethylamines (from ArCOX/KF/crown).[25]
$Ph_3P=CHNC$ results from $Me_3SiCH_2NC/Ph_3P/C_2Cl_6$, heterocycles from
silylated N-ylids and silylmethylamines, and N- or C-silylated
aziridines from Li-aziridines and R_3SiCl.[26] $N-(Me_3SiCH_2)$-amidines
and azolium salts are used to prepare N-azomethine ylids as
synthetic intermediates, while methyl 4-(Me_3Si)-3-R_2N crotonate
γ-attaches to keto electrophiles.[27] γ-Silyl and stannyl ketoximes
undergo transmetallation with $(PhCN)_2PdCl_2$ while chiral
2-$(Bu^n_3SnCH_2)$propenamides and aldehydes give α-methylene-γ-butyro-
lactones in enantiomeric excess of up to 80%.[28] A series of silylmethyl oxonium salts have been prepared from methoxy and halomethyl-
silanes with halides or ethers and used to generate dialkyloxonium
methylides $R_2O^+-{}^-CH_2$.[29] $PhCH(OMe)SiMe_3$ with $BuLi/R_2CO$ gives vinyl
ethers, some of which hydrolyse to methyl ketones. $Bu^n_3SnCH_2OR$ is a
useful precursor to $ROCH_2^-$.[30] Heteroacylsilanes result from the acyl
chloride and Me_6Si_2, and react readily with electrophiles in the
presence of CsF.[31] $(Ph_3P)_4RhH$ catalyses the efficient synthesis of
α-silyl ketones from β-silyl alcohols and γ-hydroxyalkyl silanes,

prepared stereoselectively for C(1)-C(7) segment of 6-deoxyerythronolide B, used to make keto olefins.[32] Oxidising $Pr_3SiCH_2CH_2OH$ or $Ph_3Si(CH_2)_3OH$ with NCS gave the aldehyde, Pr_3SiCH_2CHO rearranging on heating to the vinyl ether, while γ-Me$_3$Sibutyrophenone results from the rearrangement of $H_2CCH_2CHC(Ph)OSiMe_3$ formed from α-cyclopropylbenzyl alcohol, MeLi and Me_6Si_2.[33] γ-Hydroxyalkylstannanes are used in the synthesis of Brefeldin A, and γ-stannyl 3^y and benzyl alcohols used to form cyclopropanes stereospecifically with inversion at both C atoms.[34]

β-Ketosilanes have been used to prepare mono- and dicarbonyl compounds using CsF and C-electrophiles, and the rates of cleavage of the Si-C bond shown to be consistent with a cyclic intermediate. α-Branched acyltrimethylsilanes yield 1-(Me$_3$Si)allylic alkoxides from $H_2C=CHMgBr$, which rearrange through C to O migration to give the 3-siloxyallylMgBr,[35] (Me$_3$Si)$_2$acetates and RCHO provide a route to α-silyl esters while $PhCH_2SiMe_2CH_2OCOR$ photolyses to $PhCH_2CH_2SiMe_2OCOR$ through $PhCH_2$-Si cleavage to give a radical pair, and Me$_3$Si-4,4-ethylenedioxypentanoate reacts with RCHO and R_2CO to give β-hydroxyacids (or β-ethylenic ketones from ArRCO).[36] Ethyl aryl acetates result in good yield from the α-stannyl acetate and ArBr with $ZnBr_2$/Pd(II) as catalyst, and β-stannyl α,β-unsaturated esters give 3-stannylalk-3-enoates exclusively using LDA. An extensive range of addition products of $Bu^n_3SnCH_2X$ (X=CO_2Et, CN, $CONMe_2$) with RCHO have been made.[37] $Me_3SiCH_2C(O)SR$ converts acetoxyazetidinones to their thiol esters. The optically active acid $R_3Si^*CH_2CO_2H$ has been used to assess the optical purity of chiral amines using NMR spectroscopy.[38]

α-Silyl sulphonic acid and its derivatives result from Me_3SiCH_2MgCl and SO_2 or RNSO and the carbometallation of β-silyl unsaturated and saturated sulphones examined.[39] Vinylic sulphones result from carbonyl compounds and $Me_3Si(PhSO_2)CH^-$, itself being further lithiated to $PhSO_2CLi_2SiMe_3$, the first dilithio alkyl sulphone. With $(Pr^iO)_2TiCl_2$, the dititanium heterocycle results.[40] Substituted arylthiomethylsilanes $X(ArS)CHSiMe_3$ (X=MeO,Cl) are used to prepare α-ketosilanes and silylmethyl arenes, while F^- induces RCHS elimination from α-silyl disulphides which can be trapped by cyclopentadiene.[41] $CF_3CH=C(SPh)SiMe_3$ adds RCHO in the presence of F^- to give vinyl alcohols, and $H_2C=C(SPh)SiMe_3$ couples with $Me_2C=C(Li)SPh$ to give vinylcyclopropylsilanes.[42] The carbene complex $(CO)_5WC(SEt)SiPh_3$ pyrolyses to give the ketene $EtS(Ph_3Si)C=C=O$ and its $W(CO)_5$ complex, and hydrolyse to give the α-thio-α-silyl acid $Ph_3Si(EtS)CHCO_2H$. Mercaptopropylsilanised silica is used to resolve racemic arylalkyl carbinols, while $HgCl_2$ cleaves $Bu^n{}_3SnCH_2SPh$ at Sn.[43] New phosphaalkanes $Ph(Me_3Si)C=PR$ result from ROH, RNH_2, or R_2PH and $Ph(Me_3Si)C=PCl$, $Ar_3P=CHCH_2SiMe_3$ and α-methylaldehydes provide for vinyl group introduction with Cram-selectivity, while $Me_3Si(Me_2P)_2CLi$ and $AlCl_3$ gives the 6-coordinate complex $Al[(Me_2P)_2CSiMe_3]_3$ with a ^{27}Al septet.[44]

MeO⁻ substitutes Me_3SiCH_2Cl exclusively in MeOH but mainly gives the rearranged product $Me_2EtSiOMe$ in dioxan. F⁻ induced rearrangements of Me_2RSiCH_2Cl show migration for R=Ph, vinyl to be preferred, while $(Et_2N)_3S^+Me_3SiF_2^-$ catalyses the RCHO addition of α- or β-halocarbanions (generated from the haloalkylsilane).[45] Vibrational spectra of Cl_3CSiX_3 (X=F,Cl) were determined, $Bu^n{}_3SnCH_2I$ induces methylenation of sulphones and nitriles, while asymmetry parameters of $Cl(CH_2)_3SnCl_3$ support its trig. bipy. structure with equatorial C and axial chelating C-Cl.[46]

$Cp_2NdCH(SiMe_3)_2$ exhibits a very short Nd-Me contact, while diboriranides and the 1,3-diboraallyl system is stabilised by the $(Me_3Si)_2CH$ substituent. $(Me_3Si)_2CHCuBr^-$ is monomeric as a solid and has a linear, 2-coordinate geometry, while $(Me_3Si)_2CHLi$ is monomeric in the vapour state.[47] $(Me_3Si)_2C=PCl$ undergoes halogen exchange with AgF or Me_3SiX (X=Br,I), reacts with $R_5C_5M(CO)_3^-$ to give the metallaphosphaalkene, and $Na_2Fe(CO)_4$ to give the bridged complex $[\mu_2-PC(SiMe_3)_2]_2Fe_2(CO)_6$.[48] The ylid $(Me_3Si)_2C=PPh_3$ on ozonolysis gives $(Me_3Si)_2CO$ which functions as a CO^{2-} equivalent, $RPCl_2$ and $LiC(SiMe_3)_2Cl$ give $RP[=C(SiMe_3)_2]_2$, while lithiated fluorene and $[(Me_3Si)_2C=]_2PCl$ react, with further deprotonation giving the first tris(methylene)phosphate ion $[P(=C\zeta)_3]^-$. $MesP=C(SiMe_3)_2$, which results from $mesP(Cl)CH(SiMe_3)_2$, adds MeOH, and is oxidised by Me_3SiN_3 or sulphur.[49] $PhAsCl_2$ and $LiC(SiMe_3)_2Cl$ give $PhAs[=C(SiMe_3)_2]_2$ which cyclises to give the first arsirane, reducing $(Me_3Si)_2CHSbCl_2$ with $Na_2Fe(CO)_4$ gives the distibene π-complex $[(Me_3Si)_2CHSb]_2Fe(CO)_4$, while $(Me_3Si)_2CHBiCl_2$ and $Na_2W(CO)_5$ yield only Bi_2 and MeBi complexes.[50]

$(Me_3Si)_3CAs=AsC(SiMe_3)_3$ [TsiAs=AsTsi] is isostructural with its diphosphene analogue, and the solvated (with THF) lithium derivatives of $(Me_3Si)_2CHSH$ and TsiSH are dimeric. TsiSH and $Ge[N(SiMe_3)_2]_2$ in toluene give the $PhCH_2Ge(IV)$ μ-S bridged dimer as a cis-trans mixture, while $Pb[N(SiMe_3)_2]_2$ gives $[Pb(STsi)(\mu-STsi)]_2$

Group IV: The Silicon Group   89

with excess TsiSH, but $[Pb(N(SiMe_3)_2)(\mu\text{-STsi})]_2$ if 1:1.[51] $TsiTe_2^-$ reacts with Bu^t_2PCl to give the first ditellurophosphine, while TsiLi and Te yield $TsiTe_3Tsi$, the first tritelluride, which is stable to air and light.[52] $(Me_3Si)_2NBF_2$ and $(Me_3Si)_3MLi$ (M=C,Si) form $(Me_3Si)_3MBFN(SiMe_3)_2$ which decomposes to the monomeric borazyne on heating. Attempts to prepare stable Si=C and Si=N compounds through the LiX elimination using bulky substituents results in solvent attack (THF) or methanide formation at Bu^t or mes groups followed by cyclisation.[53] TsiLi and $MnCl_2$ give the anionic complex $Li(THF)_4^+[Tsi_3Mn_3Cl_4(THF)]^-$ if reacted 1:1, but Tsi_2Mn if 2:1, providing the first 2-coordinate manganese compound, m.pt. 290°C.[54]

TsiSiPh(H)OH [from TsiSiPh(H)I] reacts with ICl to give TsiSiPh(I)OH, this hydrolysing to the diol, and reacts with NaOMe/MeOH to give TsiSiPh(OMe)OH. The diol forms cyclic dimers with non-linear H-bonds, and the methoxy silanol dimers are linked by a single H-bond with no interaction at OMe. TsiSiH(OH)I gives the triol with H_2O/DMSO at 150°C. It can be readily acylated or silylated, decomposes only at its m.pt. of 285-290°C, and crystallises as hexameric H-bonded units.[55] MeO^- rapidly cleaves the acyl-O bond in $TsiSiMe_2O_2CR$ and $Bu^t_3SiO_2CCF_3$, as Si-O attack is severely hindered, to give the corresponding silanols. $TsiSiPh_2Br$ undergoes frontal nucleophilic attack by $Bu^n_4P^+Cl^-$ to give the chloride.[56] Electrophiles (ICl,Br_2,I_2,CF_3CO_2H) do not add to unsaturated substituents R in $TsiSiMe_2R$ (R=vinyl,allyl,C=CPh,Ph,$PhCH_2$) but cleave the Si-R bond. With $HNO_3/(MeCO)_2O$ and R=$PhCH_2$, ortho-para nitration occurs in the ratio 65:35, casting doubt on the predominance of o-nitration of Me_3SiCH_2Ph. Photolysis of TsiPh is free radical, giving $(Me_3Si)_2CHC_6H_4SiMe_3$-p and $(Me_3Si)_2CHPh$.[57] TsiSiMe(I)X (X=F,Cl) and AgY (Y=oxyacid anion or F^-) give rearranged products $(Me_3Si)_2C(SiMe_2X)(SiMe_2Y)$ whereas TsiSiMe(OMe)I and $TsiSi(OMe)_2I$ form unrearranged products with ICl or AgY. The alcoholysis of $(Me_3Si)_2C(SiMe_2OMe)(SiMe_2Cl)$ is established as unimolecular since it reacts more rapidly with CF_3CH_2OH than MeOH.[58]

$(PhMe_2Si)_3CLi$ reacts as expected with MeI, Me_2SiHCl and $MeSiHCl2$, but not with Me_3SiCl or other R_3SiX. $(PhMe_2Si)_3CSiMe_2H$ reacts with halogens, the iodide being more reactive than $TsiSiMe_2I$ due to added anchimeric assistance by Ph. $(PhMe_2Si)_3CB(OH)F$ results from the difluoride on hydrolysis and appears to be the first fully characterised organofluorohydroxyborane. Steric interactions prohibit intermolecular H-bonding which results in two orientations of the FB(OH) unit to the CSi_3 unit (ratio 2:3). $(Me_3Si)_4C$ with IX (X=Cl,Br) gives $(XMe_2Si)_4C$ in good yield, reduction yielding $(HMe_2Si)_4C$, which leads to other halide and oxy derivatives.[59]

Vinylsilanes mediate in polyene cyclisation, can be arylated by ArI in the presence of $Ag(I)/Pd(II)/Ph_3P$, and with OsO_4 give diols and subsequently silyl enol ethers.[60] Erythronolides result from dithiane anion addition to $Me_3SiC(CH_2)CHMeCHO$, $Me_3SiCH=CHSO_2Ph$ gives allyloxysilanes in a one-pot synthesis, while $(Me_3Si)_2C=CHCONHMe$ undergoes anti-Michael carbolithiation.[61] Dehydrochlorination of α-chloromethoxysilanes gives $(\underline{E})$-$Ph(Me_3Si)C$=CHR, while $Me_3Si(CN)C$=CH_2 results from the novel 1,4-elimination of $(Me_3Si)_2O$ from the ketenimine.[62] $Ni(cod)_2$ catalyses the cyclodimerisation of 2-(Me_3Si)buta-1,3-diene to 4 isomeric bis(silyl)substituted cyclohexenes, while silylation of trienes leads to greatly enhanced diastereoselection. The oxidation potential of 13 silyl substituted π-systems are determined.[63] Pentadienylation of aliphatic ketones or aldehydes by $Me_3Si(CH)_5SiMe_3^-$ occurs at C-3, while 1,3,5-$(Me_3Si)_3$-pentadienyllithium adds at the terminal carbon atom. RLi adds <u>trans</u> to 3-(Me_3Si)-2-propyn-1-ols giving 1-lithiovinylsilanes (20%), better yields resulting from $Me_3Si(Me_3Sn)C$=CH_2 and MeLi. Like α-Me_3Sn-vinyllithium, they undergo 1,2-addition to aldehydes and ketones to give allylic alcohols, while α,β-unsaturated ketones undergo 1,4-addition.[64]

Intramolecular iododegermylation of 1-cycloalkenylgermanes leads to cycloalkanes and alkenes, $Bu^n_3Sn(Me_3Si)C^-(CO_2Bu^t)$ and aldehydes give vinylstannanes, while ketones and ω-iodo-2-(Me_3Sn)alk-1-enes chain extend to dienes.[65] The flash pyrolysis of alkyl Bu^n_3Sn acetates gives vinylstannanes which with $EtOCH_2SnBu^n_3$ help provide an improved synthesis of ($\pm$)-AR-turmerone.[66]

Substituted allylsilanes result from the dianion of H_2C=$CMeCH_2OH$ or H_2C=$CMeCH_2NHPh$, the trianion of H_2C=$CMeCH_2SPh$, and EtOCOCOCl with Me_3SiCH_2MgCl, while <u>trans</u>-1,4-bissilylbut-2-enes are formed from $MgC_4H_6.2THF$.[67] The electrochemical reduction of allyl and benzyl, vinyl and phenyl halides in the presence of chlorosilanes affords a useful method of silylation.[68] $(allyl)_2SiMe_2$ undergoes a retroene elimination on pyrolysis to give propene and silacyclobutene, while hydroperoxidation of allylsilanes with singlet O_2 occurs with γ-migration to give the hydroperoxyvinylsilane.[69] $ClCH_2(vinyl)$- and $ClCH_2(allyl)$silanes with RO^- undergo a 1,2-alkoxy migration to give allyl and butenylmethoxysilanes. 1-(R_3Si)allylic and -propargylic alconols behave similarly, giving 3-siloxyallyl and allenyl anionic species.[70] The trialkylsilylallyl anion is alkylated by RX at the γ-position with trans-stereochemistry. These trans-vinylsilanes give (Z)-vinyl iodides which couple to give (Z)-alkenes, a procedure used to prepare insect sex pheromones.[71] Pd(II) catalyses the oxidation of allylsilanes to enones and allylic alcohols and, with Ag(I) their stereospecific arylation. (Me_3Si)allylPd complexes have been made.[72]

Allylsilanes have been used in heterocycle preparation,[73] and in the synthesis of the Prelog-Djerassi lactone, (±)-mesembrine, such hydroxylactams as (±)-isoretronecanol and (±)-epilupinine, while oxidation to give β-silyl enones is used in the preparation of (±)-desepoxy-4,5-didehydromethylenomycin A and (±)-xanthocidin.[74] Titanium promotes addition to aldehydes and oxetane,[75] $EtAlCl_2$ ring closure to hydroindanones and spiroallenes, while RCHClSR' give large ring lactones, and dienones bicyclic ring systems.[76] Fluoride introduces geometrical constraints to addition of allylsilanes to enones and readily removes Me_2PhSi from vinyl, allyl or alkoxy centres.[77] Allyl and benzylsilanes monosubstitute dicyanobenzenes photolytically, while BCl_3 promotes the conversion of nitriles to β,γ-unsaturated ketones.[78] Chloramines provide a route to silyl and silylmethylaziridines while η^4-trimethylenemethane complexes result from $Me_3SiCH_2C(CH_2)CH_2X$ (X=OAc,Cl,OSO_2Me).[79]

Tetramethallyltin results in good yield from $H_2C=CMeCH_2MgCl$ and $SnCl_4$, and gives methallyllithium with Bu^nLi. The structure of $Ph_3Snallyl$ shows no indication of π-bonding between double bond and metal, while heavy atom effects on vibrational energy transfer in $allyl_4Sn$ is compared with that in $allyl_4C$. The diastereomeric ratio of optically active organotin compounds is determined using ^{119}Sn NMR spectroscopy and $R_2Snallyl_2$ (R=$CH_2CHPhEt$) as precursor.[80] Allyltin compounds are used to make allylcarbinols from aldehydes and ketones, thioalcohols from α-methylthioaldehydes, homoallyl-amines from aldimines, and in one electron C-C bond formation.[81] They are used to prepare heterocycles such as 1,2-dihydropyridines, the pyrrolizidine alkaloid (±)-isoretronecanol and γ-butyrolactone, and couple with allylX to give 1,5-dienes.[82] α-Ethoxybutenyl$SnBu^n_3$ is intermediate in the preparation of dihomoallyl ethers, and Me_3SnCH_2 allyl ethers provide for chiral transfer in the stereospecific synthesis of hydroxylated steroid side chains. The allyl-M (M=Sn,Pb) bond inserts PhNCX (X=O,S), PhCN and CS_2.[83]

α-Allenic and β-acetylenic alcohols result from $Bu^n_3SnCH=C=CH_2$

and RCHO with $Bu^n_2SnCl_2/H_2O$, and acetylenes from alkenylsilanes and $(PhIO)_n$.[84] Allyl and alkenylsilanes add to α,β-unsaturated acylsilanes to give carbocyclic compounds, give O- and N-heterocycles and α-alkylidene-β-lactams from $ClSO_2NCO$.[85] Me_3SiCl with Li in THF silylates C_6Cl_6 at $0°C$ to give the tetrasilylallene which hydrolyses to $Me_3SiC\equiv CCH_n(SiMe_3)_{3-n}$ (n=1,2).[86] β-Silylated enones result from $PhS(Me_3Si)CHCH=CHOMe$, α-functionalised allenes, dienes and trienes from $Me_3SiCH_2C\equiv COEt$ which add to dienophiles, while the regiochemistry of the silylation of alkenyl carbamates is controlled by methyl groups.[87] Esters and amides result from silyl vinyl ketones, which react with ynamines to give bicyclo[3.1.0]hexenones, cyclohexadienones and allenic amides, while 2-aza-1,3-dienes result from N-(1-Et_3Siallyl)imines.[88] WCl_6 catalyses the metathesis of alkenylsilanes and $Re_2O_7/ SiO_2/Al_2O_3$ with R_4M (M=Sn,Pb) that of alkenes, unsaturated nitriles, and bisallylsilanes or germanes to sila or germacyclopentenes.[89]

(Z)-1-Methoxybut-1-en-3-yne has been used as a synthon for $Me_3SiC\equiv CC\equiv CSiMe_3$ while its 2-lithio-4-silyl derivative reacts with electrophiles to give the (E)-derivative. $Me_3SiC\equiv CCu$ complements 1,4-addition of acetylenic organomagnesium reagents to α,β-enones and enals, alkenylsilanes give iodonium salts with $(PhIO)_n$, propargylsilanes and α-allenic amides with acyclic N-acyliminium ions, α-allenyl ethers with acetals, while lithiated $Me_3SiCH_2C\equiv CNEt_2$ adds to ketones.[90] Ni catalyses the hydrocyanation of silylalkynes and Pd the coupling of cis-$ClCH=CHCl$ to $HC\equiv CR$ stereospecifically giving the hex-3-ene-1,5-diyne, and Cp_2ZrCl_2 the carbometallation of $Me_3SiC\equiv CC\equiv CSiMe_3$ to the but-1-en-3-yne, and $Me_3SiC\equiv C(CH_2)_nCH=CH_2$ (n=2,3) to methylenecycloalkanes.[91] These also result from Cp_2Zr and ω-vinyl-1-silyl-1-alkynes, while methylenecyclopropanes insert silylalkynes in the presence of $Ni(cod)_2$ to give cyclopentenes.[92]

Silylalkynes containing an allylic alcohol moiety cyclise to 5-membered rings when treated sequentially with $Me_3Al/Bu^i_2AlH/ZnCl_2$ and $(Ph_3P)_4Pd$, while $Me_3SiC\equiv CSiMe_3$ and silyl-1,5-diynes give benzocyclobutenes in the presence of $CpCo(CO)_2$.[93] $Me_3SiC_2Co_2(CO)_6$ induces reduction and coupling of α-substituted aryl compounds, while MePhFcSiCl (from MePhFcSiH and $PdCl_2$) couples with $LiC\equiv CR$ and $Me_2Si(C\equiv CLi)_2$.[94] Ni catalyses the ring expansion of silacyclopropenes with silaalkynes to give silacyclobutenes and disilacyclohexadienes, while $Me_3SiC\equiv CP=C(SiMe_3)_2$ adds S, Se and $H_2C=C(Me)C(Me)=CH_2$ to give heterocycles, while those from $Ph_2C^--^+N_2$ and Me_3SiN_3

decompose to $Me_3SiC\equiv CP(=CPh_2)=C(SiMe_3)_2$ and $(Me_3SiC\equiv CP[=C(SiMe_3)_2]NSiMe_3)_2$ respectively.[95]

$Me_3SiC\equiv CSiMe_3$ and CX_2 (X=S,Se) give heterobipentalenes which can be desilylated by $RXBr/F^-$ to give organic metals, $2-(Me_3SiC\equiv C)C_6H_4OH$ thermally isomerises to benzofurans through potential vinylidene precursors, while adding $Me_3SiC\equiv CLi$ to adamantanone provides a route to the vinylidene of adamantane which adds to olefins.[96] Silacycloalkynes result from $BrMgC\equiv CMgBr$ and $(ClMe_2Si)_2(CH_2)_n$ (n=0-4) with [2+2] dominating for all n, [3+3] also forming for n=0,1,2, [4+4] for n=1,2 and traces of [1+1] for n=3,4. Photoelectron spectra support a strong σ,π-interaction for [2+2] and [3+3] products (n=0).[97]

While B_5H_9 and $Me_3SiC\equiv CR$ give $(RHC=CSiMe_3)_3B$ and $2-Me_3Si-3-R-2,3-C_2B_4H_6$ quantitatively but very slowly, the vinylborane results from BH_3 and the acetylene at $0°C$, and the carborane from B_5H_9 and acetylene at $135°C$. trans-$(Me_3Si)_2C=CHCH=C(SiMe_3)_2$ results from B_5H_9 and $Me_3SiC=CSiMe_3$ at $140°C$ together with $([Me_3SiCH=C(SiMe_3)_2]_2BH)_2$ and nido-$(Me_3Si)_2C_2B_4H_6$.[98] Condensing Co atoms, B_6H_{10}, $Me_3SiC\equiv CSiMe_3$ and C_5H_6 gives silylated cobaltocarborane clusters, while propargylic and allenic organoboranes and imines yield unsaturated amines.[99]

Condensing (chloromethylene)malonic acid derivatives with tin substituted ynanilines and then $Ph_3SiC\equiv CNEt_2$ gives dien-ynamines, while $MeC\equiv CSnMe_2NEt_2$ and R_3B yield stannaborazacyclopentenes.[100] $(dppe)PdCl_2$ can be substituted by $Me_2Sn(C\equiv CR)_2$, while σ,π-interactions are extensive with alkylyltin(IV) compounds, the bond enthalpy of $SnC\equiv C$ and $SnC=C$ determined, and the vibrational spectra of $Me_3GeC\equiv CH$ and $(D_3C)_3GeC\equiv CD$ analysed.[101]

RLi desilylates $PhCH_2SiMe_3$, o-$Ph_2PC_6H_4CH_2SiMe_2H$ couples with $M_2(CO)_{10}$ (M=Mn,Re) to give the chelated complex, and the NMR and mass spectra of $(PhCH_2CH_2)_{4-n}SnR_n$ (n=0-3, R=Me,Ph,$PhCH_2$,Bu^t) analysed.[102] Similarities in the charge transfer spectra of TCNE with Ph_4M (M=C to Pb) indicate little change in the energy of the π-orbitals through size, electronegativity or π-bonding effects, while the UV spectra of $PhSiX_3$ (X=H,Me) indicate hyperconjugation to be the main effect, but d-orbital effects to dominate with X=F. The NMR spectra of phosphine complexes $ArPMe_2.BH_3$ are compared with those of $ArSiMe_3$ and indicate P to be the better π-acceptor.[103]

Electrophilic radiofluorination of $ArSiMe_3$ provides a route to ^{18}F-labelled aryl fluorides, and the mechanisms of electrophilic arylation of $C_6H_7Ru(CO)_3^+$ by $ArMMe_3$ (M=Si,Sn), and base catalysed desilylation of arylsilanes studied.[104] o-Silylated triflates provide for the first examples of the generation of aryl cations in solvolysis, while Pd complexes catalyse the alkyl(aryl)ation of imidoyl chloride giving ketenimines using R_4Sn.[105] $ArPb(OAc)_3$ result from $ArSnR_3$ and Hg(II) (or Ar_2Hg) and $Pb(OAc)_4$, and C-arylate nitroalkanes and 2-oxocyclopentane carboxylate.[106]

2,7-Bis(trimethylsilyl)-1,6-methano-[10]-annulenes can be desilylated to give the parent annulene which will then tetrasilylate 2,5,7,9 with $BuLi/Bu^tOK/TMED$. The structure shows alternate short-long bonds around the 10π-electron system, but the ^1H NMR spectra at room temperature for ring protons shows a singlet which splits into a doublet below 155K.[107] Successive halogenation (ICl) and silylethynylation of $2,3,6,7-(Me_3Si)_4$-naphthalene leads to the first linear (o-phenyleno)naphthalenes. $[Li(TMED)_2^+]_2[C_6H_4(CHSiMe_3)_2-m]^{2-}$ shows the lithium centres η^2 and η^3 to the hydrocarbyl groups.[108] Bis-(9-anthryl)dimethylsilane photoisomerises through a $[4+2]\pi$-addition giving the 9,10:1',2'-anthacene dimer. Caffeine coated silica gel provides for the separation and purification of acenaphthenylsilanes on a preparative scale.[109]

$AlCl_3$ induces cyclisation of chloromethyl(vinyl)silanes to give cyclopropylsilanes, and $1-(Me_3Si)$cyclopropanol and its silyl ether brominated to give bromoalkyl(Me_3Si)ketone. Dehalogenation of $Me_3SiCHCH_2C(Br)COCl$ with Mg/MeCN gives a stereoisomeric mixture of 4 β-keto acid chlorides, though Zn completely dehydrogenates to the cyclic dione and trione.[110] Silyl(methylene)cyclopropane on lithiation and silylation gives trisilyl syn- and anti-methylene cyclopropanes, and acidolysis of bis-silylbicyclo[n.1.0]alkanes (n=3,4,6) leads to new bis-silylmethylene cycloalkanes and alkenes. Vibrational spectra indicate the equatorial conformer of cyclobutylsilane to be the more stable.[111] Cyclic (E)-β-Bu^n_3Sn oximes with $Pb(OAc)_4$ give Δ^2-isoxazoles while linear (Z)-β-stannyl oximes are oxidised to unsaturated nitrile oxides. Oxidative fragmentation of cis- and trans-stannyl cyclohexanols gives (E)- and (Z)-enals, and is used to synthesise a mosquito pheromone.[112]

2-Me_3Si-N-methylpyrrole photolyses to the 3-silyl isomer via the 5-azabicyclo[2.1.0]pentene, the 2,5-disilyl pyrrole giving a mixture of 2,3- and 3,4-isomers. 3-Me_3Si-1-pyrazoline results from the addition of CH_2N_2 to $Me_3SiCH=CH_2$ gives 3-Me_3Si-1-pyrazoline which isomerises thermally to 1-Me_3Si-2-pyrazoline (from C_2H_4 and Me_3SiCHN_2 at 55°C). Either pyrazoline thermally eliminates N_2 to give cyclopropyl, allyl and ($\underline{E}$)- and ($\underline{Z}$)-1-propenyltrimethyldisilane via a radical pathway.[113] Mono- and bis-silylphenothiazines result from the 1,10-dilithio precursor, while 2-($Me_3SnCH_2CH_2$)oxazole can be destannylated with $TiCl_4$, the donor ligand being necessary for metallation.[114] Silylated oxazoles and thiazoles with α-asymmetric aldehydes yield O-silyl carbinols, oxygenation of 2-Me_3Si-furan gives δ-hydroxybutenolides, while epoxysilanes can be ring opened to propanols (which give alkenylsilanes) or enamines. In the case epoxides of silylcycloalkenes, hydrolysis gives α,β-dihydroxysilanes among its products.[115] Aldols condense to furans with high diastereoselectivity in the presence of $TiCl_4$, while high yields of (2-tetrahydrofuryl)silanes and germanes result from heterogeneous liquid phase hydrogenation of the 4,5-dihydro derivatives using Pd/Al_2O_3 catalyst. Base cleavage of 5-NO_2-2-(Me_3SiCH_2)thiophene gives the carbanion 5-$NO_2C_5H_2SCH_2^-$, and the rates of base cleavage of $RSiMe_n(OMe)_{3-n}$ (R=m-$ClC_6H_4CH_2$-, PhC≡C-, Cl_2CH-) with n and EtO derivatives compared.[116]

($Me_3Si)_3C_5H_2Tl$ is volatile and the ring η^5 like that in ($Me_3Si)C_5H_2LiL$ (Li-ring distance is greater for L a bidentate amine than for a monodentate one).[117] Silylcyclopentadienyl complexes of Ti, Zr, and Rh have been prepared.[118] With $CpSiMe_3$, ($THFGeCl_2$)M(CO)$_5$ gives the η^1-complex [CpGeCl(THF)]M(CO)$_5$, but η^1-$Me_5C_5SnMe_3$ η^2-complexes $Me_5C_5M'ClM(CO)_5$ (M'=Ge,Sn). $Me_3MC_5H_4Li$ (M=Si,Ge) forms $Me_3SiC_5H_4MMe_2Cl$ in hexane, but $Me_2ClMC_5H_4MMe_2Cl$ with excess Me_2MCl_2 in THF through Cl^- catalysed cleavage of Si-C.[119]

Calculations indicate $(\eta^1-C_5H_5)_2Si$ to have a $\widehat{CSiC}$ angle of 105.4° and to be a little more stable than the minimum energy form of $(\eta^5-C_5H_5)_2Si$ with D_{5d} stucture. A preliminary note indicates $(Me_5C_5)_2Si$ to have parallel rings and a very high ^{29}Si NMR signal at 577 ppm. $(Me_5C_5)_2M$ (M=Ge,Sn) show photoelectron spectra typical of a D_{5d} structure and not their bent one.[120]

Decabenzylgermanocene results from $(PhCH_2)_5C_5^-$ and GeI_2 in THF. The rings are inclined at 31° with 7 benzyl groups pointing away and 3 protecting Ge. The compound is air stable. The azaborolyl analogue of stannocene $(Bu^tNB(Me)C_3Me)_2Sn$ has rings tilted at 46.5° and occurs as a diastereoisomeric mixture (10:1) due to the prochiral character of the rings.[121] $(\eta^5-Bu^tC_5H_4)_2Sn$ is an air sensitive oil which reacts with BF_3 to give $\eta^5-Bu^tC_5H_4Sn^+BF_4^-$. $(\eta^5-C_5H_5)_2Sn \rightarrow BF_3$ has been shown to be a dimeric bridged complex in which the Sn lone pairs of electrons are coordinatively inactive. $\eta^5-C_5H_5SnCl$ gives a range of carbonyl substituted stannocenes with $RCOC_5H_4Na$.[122]

$(Cl_2CH)_2Sn.xTHF$ results from $LiCHCl_2$ and $SnCl_2$ and has been characterised as a stannylene. However $SnCl_2$ and $Me_3Si(Me_2P)_2CLi$ give not the stannylene but the non-fluxional spiro-Sn(II) complex $Me_3SiC(\overline{PMe_2})PMe_2\overline{SnPMe_2}(PMe_2)CSiMe_3$.[123] Calculations are reported on a series of insertion and cycloaddition reactions of stannylene. The germylene $[(Me_3Si)_2CH]_2Ge$ is monomeric with a $C\hat{G}eC$ angle of $107°$, while calculations show the non-planar trans-folded structure of Sn_2H_4 to be more stable than the planar one by 26 $kJmol^{-1}$, and to have a fold angle of $46°$.[124]

Me_2Ge: cycloadds 2 moles of α-substituted styrenes to give germacyclopentanes, while silacyclopentenes rearrange through intermediate alkenylsiliranes to give Me_2Si:, which also results on photolysing $Me_2Si(N_3)_2$ and $(Me_2PhSi)_2SiMe_2$.[125] Me_2Si: adds to cis,cis-hexa-2,4-diene, a 1,5-sigmatropic H-shift yielding cis-3,3-dimethyl-3-silahepta-1,4-diene, while calculations show the first transition n(Si)-3p(Si) for RSiH moves to shorter wavelength for π-donors and longer for bulky substituents, and are consistent with a λ_{max} of 450 nm for Me_2Si:. Rate constants have been determined for the reaction of Cl_2Si: with olefins and acetylenes.[126]

Calculations indicate 1,4-disilabenzene, its Dewar form, and silylene isomer to have similar thermodynamic stability in contrast to monosilabenzene and its isomers. 1,4-Disilabenzene has been isolated in an argon matrix.[127] 1,1-Di-t-butylbenzosilacyclobutene can be 4-substituted through bromination with NBS, and ring opens on heating with ROH to give 2-silatoluene (MeOH) or benzylsilane (EtOH, Pr^iOH) predominantly.[128] Vibration spectra of disilacyclobutanes and silacyclopentanes and pentenes have been analysed, and only the last has a planar equilibrium conformation. Activated $Me_2\overline{Si(CH_2)_3}$ results from $MeH\overline{Si(CH_2)_3}$ and $^1CH_2$, and decomposes to the same products as result from $Me_2\overline{Si(CH_2)_3}$ on photolysis at 254 nm.[129]

Mass spectrometry shows catalysed deuteration of silacycloalkenes to give Si-C cleavage and C_2H_3D elimination in the case of $H_2\overline{C(CHD)_2CH_2(SiMe_2)}_n$ (n=1,2), together with Me_2Si: elimination (n=2). Silenes result from 1-vinylsilacyclobutanes. Pd catalyses the extrusion of Me_2Si: from silacyclopropenes and insertion of alkenes and alkynes, while the germirene $Me_2Ge\overline{C=CCMe_2}CH_2SCH_2CMe_2$ has short Ge-C(sp^2) bonds, suggesting aromaticity.[130] Silacyclopentenes result from R_2SiCl_2 and $Mg(C_4H_6)$ together with methoxybutenylsilane (after treatment with MeOH), while thermolysis of esters of

Me$_2$SiCH=CRCR'(X)CH$_2$ (X=OC(O)NHPh,OC(S)SMe; R,R'=H,Me) gives siloles, R,R'=Me being the first methyl one stable as a monomer, as is the germole. Both undergo allyl exchange at the hetero-atom with RLi, the products giving cycloadducts with maleic anhydride and diene complexes with Fe$_2$(CO)$_9$.[131] Methoxysilanes give better yields of phenylsiloles than do chlorosilanes. Silaindene results in 85% yield on thermolysis of O-(dimethylsilyl)phenylacetylene at 800°C, the methylenesilacyclobutene also resulting at 650°C.[132] Diaza-2-silacyclohexa-3,5-dienes thermolyse to silafulvene intermediates, while diazomethyl substituted silacyclopentadiene photolyses to either silafulvene or silabenzene intermediates. C$_5$H$_5$Me$_2$SiCl and metal/Bu^{t}Li gives silafulvene dimers via the silenoid and not the silafulvene.[133]

Mg/Hg and CH$_2$Br$_2$ gives CH$_2$(MgBr)$_2$, THF addition forming CH$_2$Mg.CH$_2$(MgBr)$_2$. Both reagents give (Me$_3$Si)$_2$CH$_2$ (M=Si,Ge,Sn), while Me$_2$GeCl$_2$ gives (Me$_2$GeCH$_2$)$_n$ (n=2,3,4) and Me$_2$SnCl$_2$ the polystannacycloalkanes (Me$_2$SnCH$_2$)$_n$ (n=3,4). Small rings dominate for Ge, larger for Sn. However in liquid NH$_3$, (R$_2$NaSn)$_2$CH$_2$ and Cl(CH$_2$)$_n$Cl (n=1-4) do give distannacycloalkanes, and Me$_2$C(CH$_2$MgBr)$_2$ and Me$_2$SnCl$_2$ give the stannacyclobutane.[134] Stannacyclohexa-2,5-diene has been used to prepare boracyclodienes while GeI$_2$ and H$_2$C=C(Me)C(Me)=CH$_2$ with PhLi give 19% Ph$_2$GeCH$_2$CH=CHCH$_2$.[135] The thermal decomposition of 7-silabicyclo[2.2.1]heptadienes to silylenes is enhanced by Ph groups at the bridgehead C atom but not at Si, while carbomethoxy derivatives readily eliminate silanone via a silaenol ether intermediate. The thermal behaviour of 7-alkylidenenorbornadienes is discussed with respect to alkylidene substituents.[136] A series of endo- and exo-2-R$_3$Sn-norbornanes have been prepared, their brominolysis indicating inversion for the endo-C-Sn bond, retention for the exo-C-Sn bond, and exo-approach of the electrophile for the norbornyl system.137

The soluble tri-Grignard reagent, cis-(BrMgCH$_2$)$_3$C$_6$H$_9$ gives 44% cis-(Me$_3$SiCH$_2$)$_3$C$_6$H$_9$, and 2% silaadamantane with MeSiCl$_3$, while cis-(Me$_3$GeCH$_2$)$_3$C$_6$H$_9$ with AlCl$_3$ at 80°C gives the 1-methyl-1-germaadamantane, which can be chlorinated at Ge using HSiCl$_3$/H$_2$PtCl$_6$.[138] Solvolysis of 2-(Me$_3$Si)-2-Ad-p-nitrobenzoate shows α-silyl groups to be more stabilising than H, less than α-Me for the intermediate carbenium ion. Ph$_2$MeSn-1-Ad has distorted tetrahedral geometry about Sn with a long Sn-C bond (217.5pm), while 1-Ad$_4$Sn is thought to be associated.[139]

α-Formylpropenylsilanes act as model substances for sila-β-cyclocitral, but differ from it in fragrance.[140] Sila-β-ionone has been prepared from the silacyclohexanone $Me_2\overline{SiCOCHMe(CH_2)_3}$, while the smell of $PhCH_2PRR'$=NMe (none, later fishy) compared with isoelectronic $PhCH_2SiRR'OMe$ (honey-minty) brings the size-shape theory of Amoore into question.[141] CH_3-[Co] methylates Sn(II) and Bu^n_2Sn compounds can be detected in p.p.b. by differential pulse anode stripping voltammetry, so can be detected in river and seawater.[142] Electrophilic fluorination of ArSn compounds using CF_3OF gives ArF in 22% (Ar_4Sn) and 50% ($ArSnMe_3$) yield. The silylated perfluoro ethers $Me_nSi(C_2F_4OC_3F_7-i)_{4-n}$ and $HSi(C_2F_4OC_3F_7-i)_3$ result from chlorosilane and $i-C_3F_7OC_2F_4Li$, the rate of Si-C(F) cleavage decreasing with increasing methylation. Me_3SnCF_3 in the presence of Me_3N and BX_3 (X=Cl,Br) generates CF_2 which inserts the B-X bond, while Me_3SnF fluorinates BX_3, giving XCF_2BF_2:NMe_3 and fluoromethyl/fluoroborate anions.[143]

3 Catenation

Reducing $Bu^t_2ClSiSiClmes_2$ with alkali metals gives the disilenyl radical anion. However photolysing $(Clmes_2Si)_2$ with $Me\overline{N(CH_2)_2NMe}C=\overline{CNMe(CH_2)_2N}Me$ gives the disilyl radical with a bridging chlorine atom. The kinetics of thermal cis-trans isomerisation of hindered disilenes indicates an activation energy barrier of 25-30 kcal, in accordance with the π-bond energy.[144] Photolysing $[(2,6-Et_2C_6H_3)_2Ge]_3$ or $(2,6-Et_2C_6H_3)_2Ge(SiMe_3)_2$ gives the germylene which dimerises to the digermene, while $[(2,4,6-Pr^i_3C_6H_2)_2Sn]_2$ photolyses directly to the distannene, which retains its structural integrity in solution.[145]

Photochemical or Pd salt-catalysed cleavage of $(Bu^t_2Si)_3$ gives Bu^t_2Si: and the disilene which adds Ph_2CO to give the disilaoxetane with a very long (202.8 pm) Si-C bond. Other disilaoxetanes result from the cyclotrisilane and rearrange to disilyl enol ethers. The cyclotetrasilanes $(RR'Si)_4$ (R,R'=Bu^i, Bu^s, CH_2SiMe_3) photolyse to cyclotrisilanes (λ_{max} 300-330 nm) and disilene (λ_{max} 390-440 nm). The silapropadiene $(Me_3Si)_2\overline{C=C=Simes_2}$ adds mes_2Si: to give $(Me_3Si)_2\overline{C=CSimes_2Simes_2}$ which isomerises to the 1,2-disilacyclobutene at 170°C, this ring contracting on alcoholysis to a silacyclopropane. Similarly addition of mes_2Si: to $R(Me_3Si)C=C=Simes_2$ (R=Ph, Me_3Si) gives 2 stable disilacyclopropanes which at 170°C thermolyse quantitatively to the 5,6-benzo-1,2-disilacyclohex-3-ene (R=Ph) and the disilacyclobutene (R=Me$_3$Si). With Me$_3$N$^+$-O$^-$, the 2-oxa-1,3-disilacyclobutane results (R=Ph), whereas for R=Me$_3$Si, the oxadisilacyclopentene and silacyclopropane result, this latter possessing a long (164.3 pm) ring C-C bond.[146]

Calculations show disilene and silylsilylene isomers of Me$_4$Si$_2$ to be almost isoenergetic, the former having a very flat potential energy surface for <u>trans</u>-bending or twisting about the Si-Si bond. Si$_6$H$_6$ has about half the aromaticity of benzene, but in contrast is close in energy to its valence isomers. Methylation widens the angles at M in ions of R$_3$M and R$_6$M$_2$ (M=Al,Si,P; R=H,Me). ESR spectroscopy indicates PhMe$_2$SiSiMe$_3$$^{\cdot+}$ to be a σ-cation radical like Me$_6$Si$_2$$^{\cdot+}$ and not a perturbed benzene cation.[147] Me$_6$Si$_2$ is oxidised by atomic oxygen (^{3}P) to vibrationally excited (Me$_3$Si)$_2$O and decomposes by Si-C cleavage, while peroxidation of Me$_{10}$Si$_4$ cleaves the terminal Si-Si bonds.[148] Halogen exchange between (MeCl$_2$Si)$_2$ and (MeBr$_2$Si)$_2$ at 60°C gives equilibrium of all 6 possible products after 10 hours, while the vibrational spectra of (Me$_3$Si)$_2$SiX$_2$ and (Me$_3$Si)$_2$SiMeX (X=halogen,OMe,Ph) have been analysed.[149]

Photolysing mesitoylSi(SiMe$_3$)$_3$ gives the isomeric silene which then inserts the C-H bond of an OMe group to give a silacyclobutene or with Me$_2$Si: gives the disilacyclopropane adduct, which isomerises through ring opening to a bicyclo[4.3.0]disilacyclopentene. Halogenation of 9-fluorenylSi(SiMe$_3$)$_3$ with I$_2$ gives cleavage of only one Si-Si bond, while Br$_2$ also gives 9-Br-9-fluorenylSiBr$_2$SiMe$_3$.[150] In DME, (Me$_3$Si)$_4$Si and MeLi give [(Me$_3$Si)$_3$Si$^-$Li$^+$]$_2$DME$_3$ which possesses a 263 pm Si--Li interaction distance, while (Me$_3$Si)$_2$SiBu$^t_2$ sublimes at 280°C. (Me$_3$Si)$_3$SiLi ring substitutes Cp$_2$WCl$_2$ to give the hydride [(Me$_3$Si)$_3$SiC$_5$H$_4$]$_2$WH$_2$.[151]

(Me$_2$Si)$_6$ results in 90% yield (trace of pentamer) from Me$_2$SiCl$_2$ and Li in THF at 0°C using Ph$_3$SiSiMe$_3$ as catalyst. Photolysing in the presence of Et$_3$SiH gives the cyclopentasilane which can be readily separated 99% pure, and in CCl$_4$/CH$_2$Cl$_2$ to Cl(SiMe$_2$)$_6$Cl in 70% yield or Cl(SiMe$_2$)$_n$Cl (n=2,3) and C$_2$Cl$_6$ in the absence of CH$_2$Cl$_2$.[152] (Me$_2$Si)$_7$ adopts a twist-chair conformation with average SiSiSi angles of 116.2°, larger than found in other cyclosilanes. (RR'Si)$_x$ (R,R'=n-alkyl) show a bathochromic shift on cooling which increases with the length of the alkyl chain and is due to conformational changes. Anodic peak potentials fall as ring size decreases from 7 to 3.[153] The ESR spectra of free radical anions of cyclopolysilanes can be rationalised in terms of a singly occupied MO as a linear combination of Si-C σ^* and Si-Si σ^* orbitals.[154]

Condensing $(ClMe_2Si)_2$ with Li provides a route to cyclotetrasilanes, while ^{13}C and ^{29}Si ENDOR signals of $(Bu^tMeSi)_4^{\cdot-}$ and $(Et_2Si)_5^{\cdot-}$ have been observed for natural abundance. A small hyperfine anisotropy for ^{29}Si is consistent with Si-Si σ^* or 3d spin population, but not with π-delocalisation.[155] The tetrasilabicyclo-[1.1.0]butane $R_2SiBu^tSi-SiBu^tSiR_2$ ($R=2,6-Et_2C_6H_3$) result on reductive coupling of the 1,3-dichlorocyclotetrasilane and undergoes rapid ring inversion, while attempts to make $mes_2SiBu^tSi-SiBu^tSimes_2$ from $Bu^tCl_2SiSimes_2Cl$ and lithium naphthalenide gave only its oxidation and hydrolysis products.[156] 1,2,3-Trisilacyclopentane and hexane, silylated at the middle Si atom, eliminated silylsilylene and ring open with silyl migration on photolysis. With the 1,2,3,4-tetrasilacyclohexane and silylation at Si(2) and Si(3), only silylsilylene elimination results, whereas bis(1,1,2,3,3-pentamethyl-1,2,3-trisilacyclopentyl) only ring opens. Products of addition to Et_2MeSiH confirms these pathways. The photolytic fragmentation of trisilanes to disilane and silylene has been studied by means of the pseudopotential method and the effect of methylation shown to have little electronic effect.[157]

$PhC=CHCH=CPhGeMe_2SGeMe_2$ adds to benzyne, providing the intermediate digermabicyclooctadiene for the pyrolytic generation of $Me_2Ge=GeMe_2$ (trapped by silacyclopentadiene) and coordinates to the $Fe(CO)_3$ residue. Ge_2H_6, Me_3GeGeH_3 and Me_3SiGeH_3 react with $Co_2(CO)_8$ to give a variety of Ge-Co clusters. DBU dehydrochlorinates R_2ClGeH to give di-, tri- and tetragermanes through intermolecular condensation.[158]

α,ω-Bis(allylic acetals) undergo intramolecular reductive coupling in the presence of $Me_6Sn_2/Pd(0)$ to give a mixture of isomeric dienes, while coupling acyl chlorides with $Bu^n_6Sn_2$ in the presence of Pd(II) provides a route to unsymmetrical α-diketones via $RCOSnBu^n_3$.[159] $Bu^n_6Sn_2$ can be arylated using ArI in the presence of fluoride, while Pd catalyses the coupling of α-bromoketones with $Bu^n_6Sn_2$.[160] R_6Sn_2 is used as an anion selective ionophore and $Me_6Sn_2^{\cdot+}$ shown to be a σ(Sn-Sn) radical.[161] $(R_2ClSn)_2$ ($R=Me,Et,Bu^n$) result on electrolysis of a MeCN solution of R_2SnCl_2 and can be readily precipitated by acetate, while formic acid reduces $(R_3Sn)_2O$ (R=Bu,oct) on heating to give R_6Sn_2 and a little R_8Sn_3.[162]

cis-1,1,2,2,3,4,4,5,5,6-Decamethyl-1,2,4,5-tetrastannacyclohexane occurs with a boat conformation in contrast to the Sn-perphenyl derivative (chair), $I(Bu^t_2Sn)_4I$ has an all trans conformation, while $Me_2Sn(NEt_2)_2$ and Me_2SnH_2 in the absence of O_2, H_2O and light give $(Me_2Sn)_6$ exclusively, but in solution $(Me_2Sn)_n$ (n=5,7,8) are also formed.[163]

There is a linear correlation between electrochemical oxidation and ionisation potential of $R_3MM'R_3$ (R=Me,Et; M=M'=Si,Ge,Sn or M≠M') for the same R. HCl and CF_3CO_2H cleave both Sn-C and Sn-M bonds of Me_3SnMMe_3 (M=Si,Ge,Sn) while with Me_3SnCl, only Sn-C cleavage occurs, subsequent decomposition of Me_3MSnMe_2Cl giving Me_3SnCl and $(Me_2Sn)_n$.[164] $Me_3SnSiMe_3$ add regio- and stereospecifically to give (Z)-$Me_3SnCR=CHSiMe_3$ while $R_3SnSiR'_3$ add 1,4- to α,β-unsaturated ketones in the presence of naked CN^-. $(c-C_6H_{11})_3PbSn(c-C_6H_{11})_3$ results from $(c-C_6H_{11})_3PbLi$ and $(c-C_6H_{11})_3SnCl$, while the structure of $(c-C_6H_{11})_6Pb_2$ shows the cyclohexyl rings to be unusually flat. Mixed diplumbanes have been prepared similarly but the structure of $Ph_3(p-tolyl)_3Pb_2$ shows it to be the mixed isomer present as an equilmolar mixture of the two rotational isomers.[165]

4 Hydrogen Derivatives

2,4,6-$Bu^t_3C_6H_2SiF_3$ gives difluorides with RLi (R=Me,Bu^t,Ph) and all these fluorides can be reduced to air-stable silanes with $LiAlH_4$. This also reduces CF_3SiF_3 to CF_3SiH_3 (b.pt.$-38°C$), which hydrolyses with OH^- to CF_3H, while phase transfer catalysts provide for the $LiAlH_4$ reduction of halo- and alkoxysilanes and germanes.[166] Silane and acetylene thermolyse through silene and silylene intermediates to a variety of vinyl and ethynylsilanes with up to 3 of these groups at Si. $EtSiD_3$ decomposes by silylene and silene formation and calculations show both singlet and triplet $:CH_2$ and $:SiH_2$ to give silene. Calculations indicate that the compounds $H_2\overline{SiSiH_2MSiH_2SiH_2}$ (M=C,Si) are more stable with a spiro rather than a planar structure and are more stable than their methylene (silylene) and disilene decomposition products.[167]

With trityl perchlorate, Et_3SiH reduces RCHO to ethers and silylates ethyl diazoacetate.[168] Silanes reduce α-substituted β-ketoamides, camphor oxime to the amine and convert $H_2C=C(Me)CO_2Me$ to ketene acetals.[169] Hindered chlorosilanes result from tetrasubstituted ethers with Me_2ClSiH or $BuMeClSiH$, while $MePhSi(H)CH_2OCOX$ (X=OMe,Cl,NMe_2) pyrolyse to $PhMe_2SiOCONMe_2$ (X=NMe_2)

or $PhMe_2SiX$ (X=Cl,OMe) and CO_2.[170] Hydrosilation of dihydropyrrole and substituted ethyne derivatives are explored, while 2-(Me_2SiH)furan pyrolyses to silapyran $Me_2SiOCH=CHCH=CH$, its isomers $Me_2SiOCH=CHC=CH_2$ and $Me_2SiOC_4H_4$ (stucture unknown), while CO loss gives 4 isomers of $Me_2C_3H_4Si$ - the 3 linear unsaturated ones and the silacyclobutene.[171]

1^y silanes couple to give polysilanes with up to 10 silicon atoms in the presence of Cp_2TiR_2 (R=Me,$PhCH_2$), Cp_2MH_3 (M=Nb,Ta) and $PhMe_2SiH$ dehydrogenate to give $Cp_2MH_2SiPhMe_2$, while R'_2SiH_2 (R'=Et,Ph) decarbonylates η^5-$MeC_5H_4Mn(CO)_2PR_3$ to give complexes with 3-centre Mn-H-Si bridges.[172] $Co_2(CO)_8$ catalyses the hydrosilation of ArCN to disilylaminomethyl arenes and the ring opening of styrene oxide, oxetanes and THF,[173] rhodium complexes the reduction of α,β-unsaturated esters and enynes,[174] iridium complexes oxidatively add silanes,[175] with palladium catalysing the reduction of α,β-unsaturated ketones and aryl butadienes, the hydrosilation of vinylsilanes, and the alcoholysis of silanes.[176] (π-allyl$PdCl_2$)$_2$ with Me_3MH (M=Si,Ge,Sn) gives CH_4 and C_3H_6 for all M, and C_5H_{10}, Cl_3CSiH_3 (M=Si), Me_3GeCl and allylgermanes (M=Ge), and C_4H_8, C_6H_{12}, Me_4Sn and Me_3SnCl.[177]

Hydrogermylation has been used to make 3-Cl_3Ge propanoic acids and hydrostannation 3-stannyl propanoates using Me_3SnH or Me_2ClSnH.[178] Bu^n_3SnD reduces B_5H_8X, $Me_3SnB_4Cl_4$ to B_4H_{10} in 95% yield, and adds to $(Me_3Sn)_2C=CH_2$ giving the tristannylethane.[179] Bu^n_3SnH with chloroketones under pressure gives chloroalkoxy tins or cyclic ethers, α- and β-silyl radicals formed from the halides disobey the Baldwin-Beckwith rule, while asymmetry is induced at the α-position of $RPhMeSiCHClMe$ on reduction with Bu^n_3SnH if R=mes.[180] Bu^n_3SnH gives formate with CO_2 and $[(Me_3Si)_2CH]_2SnOs_3(\mu-H)_2(CO)_{10}$, from the stannylene and $Os_3H_2(CO)_{10}$, possesses an Os-H-Sn bridge and undergoes carbonyl insertion of Sn-C on heating.[181] The vibrational spectra of methylsilanes and the influence of conformation on Si-H vibration have been closely examined, as have torsional barriers in vinylsilanes. Sn-H vibrations help ascribe conformer composition in ethylstannanes, while the barrier to internal rotation $CH_3SnH_{3-n}D_n$ (n=1-3) is about 0.6 kcal/mol.[182]

5 Radicals and Metal Derivatives

Photolysing silanes in chlorinated solvents gives silyl radicals,

while a good linear relationship exists between the ^{29}Si coupling
constant for a Si-centred radical and $\underline{J}(^{29}\text{Si}-^{1}\text{H})$ for the
corresponding silane.[183] Aroylsilanes add phosphorus radicals,
formed under UV light, while germyl radicals result on photolysing
PhCOGeR$_3$ or aryl substituted germyl anions. Et$_3$Ge· attacks ArSO$_2$SAr'
to give sulphonyl radicals, while the rate of attack of Bu$^n{}_3$SnH by
organic radicals is faster than that of Bu$^n{}_3$GeH.[184] RS· and RSO$_2$·
radicals generate Bu$^n{}_3$Sn· from the allyl or crotylstannanes, and
Bu$^n{}_3$Sn· undergo a 1 electron transfer to nitrosugars, which lose
NO$_2{}^-$ to provide a route to C-glycosides with a 3^y C atom at C1.
While NO$_2{}^-$ cleaves the iodomethyl group from R$_3$SnCH$_2$I (R=Me,Et,Ph),
CN$^-$ gives R$_3$SnCH$_2$· and R$_3$Sn· and provides a simple route to
(R$_3$SnCH$_2$)$_2$Hg. Calculations indicate that radical cations Me$_3$SnR·$^+$
prefer a C$_{3v}$ structure.[185]

Low temperature ^{29}Si-^{6}Li coupling indicates a covalent monomeric
structure for phenylsilyl-lithium with π-polarisation of the phenyl
rings is the major cause of ^{13}C shifts. Cp$_2$MCl (M=Sm,Lu) with
Me$_3$SiLi/DME gives Li(DME)$_3{}^+$ Cp$_2$M(SiMe$_3$)$_2{}^-$, the first compound with
lanthanide-Si bonds (Sm-Si 288.0 pm). (Me$_3$Si)$_3$Al.OEt$_2$ silylates
Cp$_2$ZrCl$_2$ and CO inserts the Zr-Si and Ta-Si bonds.[186] In combination
RCu and Me$_3$SiCl accelerate and improve 1,4-addition reactions with
conjugated carbonyl compounds.[187] (PhMe$_2$Si)$_2$CuLi provides for
diastereoselectivity on the alkylation and protonation of β-silyl
enolates, preparation of 2-silylbuta-1,3-dienes from HOCH$_2$C≡CCH$_2$OH
[Bu$^n{}_3$SnCu(SMe$_2$)CuBr gives furan derivatives] and the regio
controlled preparation of allylsilanes from 2^y allylic acetates and
urethanes. Silylzincation of RC≡CH with PhMe$_2$SiZnR'$_2$Li gives an
isomeric mixture of trans and gem isomers in the presence of Cu(I),
with the former dominating for R'=Et, the latter for R=Pr^i,Bu^t.[188]
[(CF$_3$)$_3$Ge]$_2$Zn results from (CF$_3$)$_3$GeH and Ph$_2$Zn and Ph$_3$SnInR$_2$
(R=o-C$_6$H$_4$CH$_2$NMe$_2$) from R$_2$InCl and Ph$_3$SnNa.[189]

The reactions of Me$_3$SnM (M=Li,Na,K) with 1^y alkyl bromides
involve both S$_N$2 and electron-transfer pathways with radical
intermediates.[190] ^{2}H labelled 1,4-dihalobicyclo[2.2.2]octanes react
with Me$_3$SnLi to give the stannyl derivatives through a chain
mechanism involving radical anions and free radicals, and not
[2.2.2]propellane intermediacy. Pinacol (1-chloroethyl)boronate can
be converted to the α-lithio derivative through stannylation with
Me$_3$SnLi followed by cleavage with MeLi.[191] Bu$^n{}_3$SnLi converts
β-phenyl-sulphinyl β,γ-unsaturated ethers to substituted allenes,

and is used in the synthesis of α-anions of sulphides, α,α-$(Me_3Si)_2$-alkanes and nitriles.[192]

Thermal alkynes can be disilylated with Me_3SiLi or $PhMe_2SiLi$ and $MnCl_2/MeMgI$ to give alkenes as an isomeric mixture, whereas with R_3SnLi (R=Me,Bu) under the same conditions only the (Z)-isomer was formed. With Bu^n_3SnM (M=MgMe,$AlEt_2$,Zn) terminal alkynes, enol triflates or vinyl iodide give vinylstannanes in good yield, the isomer being determined by the transition metal catalyst used.[193]

6 Nitrogen Derivatives

Warming $Bu^t_2Si(Cl)N(Li)SiBu^t_3$ gives the first stable sila ketimine $Bu^t_2Si=NSiBu^t_3$ which adds to $MeOCH=CH_2$ giving the [2+2] cycloadduct, to N_2O giving the unstable [2+3] cycloadduct which decomposed to $Bu^t_3SiN_3$, and adds H_2O. Heating C-phenyl-2-Me_3Si-tetrazole gives the nitrile imine $PhC\equiv N^+$-N^-SiMe_3 which undergoes [3+2] cycloaddition to alkynes. $Me_2Bu^tSiNR'R''$ are stable to hydrolysis while $Me_2Bu^tSiNH_2$ can be distilled without deamination.[194] Aminosilanes induce carbonylation of gem-dibromocyclopropanes with $Ni(CO)_4$, while 1-Me_2Bu^tSi-pyrrole undergoes Friedel-Crafts acylation β, giving 3-pyrroloketones on NaF assisted hydrolysis. Surface-hydrated SiO_2 on treatment with Me_3SiNMe_2 at 150-250°C for 1 week gives extensive surface silylation.[195]

Both $(MeSiH_2)_2NMe$ and $(Me_2SiH)_2NMe$ are planar while calculations indicate stabilisation of the lone pair of electrons probably occurs through p_π-d_π bonding, though interaction with the σ^* Si-C orbital cannot be ruled out. Ph_2CO and $(Me_3Si)_3N$ or $(Me_3Si)_2NH/Me_3SiCl$ below 280°C give 1,1,3-triphenyl-1-H-isoindole only, but above 280°C the 2-H isomer results as well. The azaallenium cation $Ph_2C=\overset{+}{N}=CPh_2$ is thought to be the intermediate.[196]

Disilazanes condense with RCHO and R_2CO to give Schiff bases in the presence of Me_3Si triflate, and with ROCSCl to give thiourethanes $ROC(S)NR'(SiMe_3)$ and bis(thionocarboxylic acid esters) $R'N(C(S)OR)_2$. $(Me_3Si)_2NH$ silylates thiourea, the product decomposing to $(Me_3SiN)_2C$.[197] Whereas $(Me_3Si)_2NNa$ and RX (X=Br,I,p-$MeC_6H_4SO_3$) give $RN(SiMe_3)_2$, perfluoroaryl fluorides R^fF in THF give R^f_2NH.[198] The Li derivatives of Me_3SiNHR (R=Bu^t,Pr^i,Me_3Si) for which pK values are known, are used to determine the acidity of substituted pyridines. $(Me_3Si)_2NLi$ is dimeric in the vapour phase as are the trilithio derivatives of $RSi(NHSiMe_3)_3$ (R=Me,Bu^t,Ph) in the solid state. The structure comprises a trigonal antiprism array of 6 Li atoms, each bridging 3 N atoms.[199] Heptane soluble "Bu$_2$Mg" with o-Me$_3$SiNHC$_6$H$_4$NHSiMe$_3$ gives the etherated dimer with the amide ligands bridging the 2 Mg atoms to give, with ether, distorted tetrahedral bonding at Mg. The structures of Me$_2$Si(NBu^{t}AlPh$_2$)$_2$, Me$_2$Si(NHBu^t)(NBu^{t}AlCl$_2$) and [Me$_2$Si(NBu^t)$_2$InMe]$_2$ show interesting structural and bonding features, and result readily from Me$_2$Si(NHBu^t)$_2$ and R$_3$M (M=Al,Ga,In). With GaCl$_3$, (Me$_3$Si)$_2$NH gives Cl$_2$GaNHSiMe$_3$ as a dimer-trimer mixture, whereas Cl$_2$GaNMeSiMe$_3$ occurs as a <u>cis</u>,<u>trans</u> Cl-bridged dimer mixture in solution.[200]

In the vapour phase, [(Me$_3$Si)$_2$N]$_3$Sc is planar unlike the solid, while treating (Yb[N(SiMe$_3$)$_2$]$_2$)$_2$ with (Me$_3$Al)$_2$ gives (Me$_3$Al)$_2$Yb[N(SiMe$_3$)$_2$]$_2$ in which Yb interacts with 2 Si-Me and 4 Al-Me bonds, and polymerises C$_2$H$_4$. ([(Me$_3$Si)$_2$N]$_2$M(μ-SBu^t))$_2$ (M=Eu,Gd,Y) result from the chloride and Bu^{t}SLi. Mn(II) and Co(II) give 3-coordinate amides.[201] Condensing Bu^{t}N(SnMe$_3$)$_2$ with PCl$_3$ gives (Me$_3$SnNBu^t)$_2$PCl which loses Me$_4$Sn to give the cyclic zwitterion Bu^{t}N̄P$^+$NBu^{t}Sn$^-$Me$_2$Cl, (Me$_3$Si)$_2$NNa aminates and methylates SbCl$_3$ simultaneously giving MeSb[N(SiMe$_3$)$_2$]$_2$ which can be oxidised by SO$_2$Cl$_2$ to MeSbCl$_2$[N(SiMe$_3$)$_2$]$_2$ with Cl axial, while Me$_2$BiN(SiMe$_3$)$_2$ and HN$_3$ give Me$_2$BiN$_3$.[202]

A series of amino and alkoxy(amino)silanes have been prepared and complex with SnCl$_4$, Et$_2$Si(OPh)NHCH$_2$Ph giving Et$_2$Si(NHCH$_2$Ph)$_2$.2SnCl$_4$ and Et$_2$Si(OPh)$_2$.[203] Silylating N-(CF$_3$CO)amino acids with Me$_3$SiNEt$_2$ gives O-silyl derivatives.[204] O,N-Bis(trimethylsilyl)acetamidate gives cyclic derivatives with ArNCO, while vinyl compounds insert the Si-C bond of the enamine isomer $\searrow$C=CNRSiMe$_3$.[205] (ClMe$_2$Si)$_2$ and NH$_3$ gives (Me$_2$SiMe$_2$SiNH)$_n$ (n=2,3), the trimer in 3-5% yield, while Penning ionisation electron spectroscopy (PIES) help the assignment of ionisation potentials of linear and cyclosilazanes.[206]

The structures of Me$_2$Si[N(Ph)N=CMe$_2$]$_2$, (Bu^t$_2$SiNH)$_3$, (FBu^{t}SiNH)$_3$ and (Pr^i$_2$SiNBu^t)$_2$ have been determined in the solid state[207] and (Me$_2$SiNMe)$_2$ in the vapour. R$_2$Si(F)NH$_2$ (R=Pr^i,Me$_3$SiNMe) results from R$_2$SiF$_2$ and LiNH$_2$ but condenses with base to give (R$_2$SiNH)$_2$ and (Pr^i$_2$SiNH)$_n$ (n=3,4).[208] Bu^{t}NSiMe$_2$Bu^{t}NSn complexes with phosphorus ylids while Ph$_3$P=NH displaces and oxidises tin to give Me$_2$Si(NHBu^t)$_2$ and (Ph$_3$P=N)$_4$Sn which possesses the shortest Sn(IV)-N distance reported (197.5 pm). ClSn(CH$_2$CH$_2$CH$_2$)$_3$N has 3-fold symmetry with a Sn--N intramolecular distance of 237.2 pm while MeSi(CH$_2$CH$_2$CH$_2$)$_3$N methylates Me$_2$SnCl$_2$.[209] 1,2,3,4-Tetrahydro-1-trimethylsilyl-1,10-

phenanthroline has a structure with Si in a distorted trigonal bipyramid and Si-N bonds of 174.6 and 268.9 pm. The mono and dichloro derivatives are achiral and have similar 'single' Si-N bonds but much shorter donor bonds (~203 pm). The trifluoro derivative is fluxional but shows doublet/triplet structure at $-90°C$, consistent with the crystal structure.[210]

The germylene and stannylene derivatives of Pd(0) and Pt(0) $M(M'[N(SiMe_3)_2]_2]_3)$ (M=Pd,Pt;M'=Ge,Sn) have a planar structure and react with CO to give trimeric amides.[211] The stannylene mono and disubstitutes (cis- and trans-) $M(CO)_6$ (M=Mo,W) and inserts the Rh-Cl bonds of $[Rh(alkene)_2\mu-Cl]_2$.[212] The carbene analogues $[(Me_3Si)_2N]_2M$ (M=Ge,Sn,Pb) coordinate with Lewis acids, substitute binary metal carbonyls and insert M-X bonds, while silylaminogermylenes result from the dihalides, and $[(Bu^tO)_2Sn]_2$ from $[(Me_3Si)_2N]_2Sn$, which is also oxidised and then cyclised by RN_3.[213] Aminosilanes aminate $CpTiCl_3$ giving $CpTiCl_2NHR$ which decomposes on heating to give $(CpTiCl)_2(\mu-NR)_2$. With $(Me_3Si)_2NNa$, $TiCl_4$ forms titanazanes $(Cl_2TiNSiMe_3)_n$ and $[Cl_2Ti_2(NSiMe_3)_3]_n$, but $ZrCl_4$ gives $Cl_3ZrN(SiMe_3)_2$ and, with excess amide, the heterocycle $Me_3SiNSiMe_2CH_2ZrCl_2$.[214] $(R_2PCH_2SiMe_2)_2NLi$ gives monoamides with MX_4 (M=Zr,Hf), the Hf derivative giving the tris(borohydride) with $LiBH_4$ without phosphine coordination,[215] unlike $Ir(\eta^2-C_8H_{14})[N(SiMe_2CH_2PR_2)_2]$ which adds MeI with olefin displacement, this adduct adding H_2 across Ir-N. It also reacts with Me_3MCH_2Li (M=C,Si) to give the dialkyl derivative which loses Me_4C (M=C) to give the $Ir=CH_2$ complex.[216] The metallacycles $[(Me_3Si)_2N]_2MCH_2SiMe_2NSiMe_3$ (M=U,Th) insert ketones into the M-C bond and are ring opened by ROH, RC≡CH and $CpMo(CO)_3H$.[217] Me_3SiNEt_2 and $MoOCl_4$ give $(Et_2NH)MoOCl_3$ while $Me_2Si(NHBu^t)_2$ with $VO(OR)Cl_2$ (R=Et,Pr^i) forms V-N complexes with donor, single and double V-N bonds, and, from VCl_4, Me_3SiN_3 and dipy/Ph_4PCl, $Ph_4P^+(Me_3SiN\equiv VCl_4)^-$ with a V-N triple bond (Si-N 179 pm).[218]

Spectra of organotin pseudohalides are analysed and the structure of $Me_2Sn(NCS)_2\cdot 2H_2O\cdot 18$-crown-6 shown to involve trans-substitution with H_2O H-bonding to the ether. Me_3SnCl and $K_3Co(CN)_6$ form polymeric $(Me_3Sn)_3Co(CN)_6$ with a 3-D structure involving planar Me_3Sn bridges.[219] MeH_2SiX (X=pseudohalide) result from AgX and MeH_2SiBr, electron diffraction shows Me_3SiNCO to not have a linear skeleton, while Me_3SiNCS is attacked by 8-lithio-3-methyl-5,6,7,8-tetrahydroquinoline at C in low polarity solvents to give the 8-CN derivative, while intermediate protolysis gives the 8-thioamide.[220] The first organosilicon compound with a Si-N triple bond results from the pyrolysis of PhSi(N$_3$)$_3$.[221]

1,3,2-Diazasilacycloalkanes and their diones result from the transamination of Me$_2$Si(NEt$_2$)$_2$, while N,N-bis(silyl)enamines result from Me$_2$SiFe(CO)$_4$SiMe$_2$C$_6$H$_4$-o and RR'CH$_2$CN.[222] ^{15}N-^{29}Si coupling constants were determined for a wide range of Si-N compounds, Me$_2$CRCH$_2$COSiMe$_3$ (R=N$_3$,NCS) photolysed to MeN=CMeCH=C(OH)SiMe$_3$ or the pyrrolidine, while low temperature structure determinations of 1,4-(Me$_3$M)$_2$-1,4-dihydropyrazines show the rings planar, while the more crowded tetramethyl derivative has a pronounced boat conformation.[223]

Iminoboranes (Me$_3$Si)$_3$MB=NSiMe$_3$ result from (Me$_3$Si)$_3$MLi (M=C,Si) and (Me$_3$Si)$_2$NBF$_2$, while R$_3$CN=BNCMe$_2$(CH$_2$)$_3$CMe$_2$ gives silyliminium salts with Me$_3$SiX (X=CF$_3$SO$_3$,I).[224] (Me$_3$Si)$_2$NLi dehydrohalogenates (CH$_2$)$_n$NBXCR$_2$H to give (CH$_2$)$_n$N$^+$=B$^-$=CR$_2$ which with Me$_3$SiN$_3$ gives the CBN heterocycle.[225] Silaboretidines result from RBX$_2$ and Me$_2$Si(NR'Li)$_2$, diazadiboretidines were formed from Me$_3$Si(Bu^t)N-B(R)N$_3$ (X) on N$_2$ loss, while MeB=NBu^t adds Me$_3$SiN$_3$ to give the cyclotetrazole and (X)(R=Me).[226] The Cp analogue [Me$_3$SiNBMe(CH)$_3$]$^-$ forms π-complexes with TiCl$_4$ and VCl$_3$ and like (Bu^{t}NBBu^t)$_2$TiCl$_4$ have weak B--Cl interactions.[227] The cyclotriazane rings SiB$_2$N$_3$ and Si$_2$BN$_3$ phenyl substituted at Si are planar, MeBNRBMeOBMeO and MeBNRBMeNRBMeO result from MeBBr$_2$ and (Me$_3$Si)$_2$NR/(Me$_3$Si)$_2$O in 1:2 or 2:1 ratios (or from the cyclotrisiloxazane and MeBBr$_2$). ^{13}C NMR spectral data for the non-equivalent Me groups of Me$_2$N of Me$_2$N(Ph)BNHMMe$_3$ (M=C,Si) shows ΔG* greater for the silyl compound suggesting enhanced π-bonding from Me$_2$N to B.[228] XCH$_2$PX$_2$ (X=Cl,Br) and (Me$_3$Si)$_2$NNa yields the phosphaalkenes (Me$_3$Si)$_2$NP=CX$_2$, which also results (X=Cl) through elimination of HCl from (Me$_3$Si)$_2$NPH(CCl$_3$) [(Me$_3$Si)$_2$NPH$_2$ and CCl$_4$].[229] H$_2$C(PCl$_2$)$_2$ condenses 2 moles of the silyl urea Me$_3$SiN(Ph)CON(Me)SiMe$_3$ to give the bicyclic [3.3.1] derivative. (Me$_3$Si)$_2$NP(S)=CHSiMe$_3$ forms 5 membered heterocycles with diazoalkanes and Bu^{t}N$_3$, both isomerising by Me$_3$Si migration and loss of N$_2$.[230] (Me$_3$Si)$_2$NP(=NSiMe$_3$)$_2$ form chelate complexes with Al,Sn,Ti and Zn, and [(Me$_3$Si)$_2$N]$_2$Zn aminates F$_3$PNMePF$_3$NCH$_2$Cl at the ClCH$_2$ group.[231] The aminophosphene (Me$_3$Si)$_2$NP=CHSiMe$_3$ undergoes [2+2] cycloaddition with the silene Me$_3$CCH$_2$CH=SiMe$_2$ to give the 1-phospha-3-silacyclobutane and complexes with the Fe(CO)$_4$ residue more readily than

$(Me_3Si)_2NP=C(SiMe_3)_2$.[232] $(Me_3Si)_2NP=NSiMe_3$ (L) substitutes $Ni(cod)_2$ to give the tris-phospha-III-azene complex L_3Ni through P coordination. Phosphines displace L to give complexes with L η^1- and η^2-bonded to Ni while bipy oxidatively couples 2L. PtL_3 gives the trinuclear complexes $Pt_3(\mu-L)_3(CO)_3$ and $Pt_3(\mu-CNBu^t)(\mu-L)_2(CNBu^t)_3$ with CO and Bu^tNC respectively.[233] Cl_2 oxidises $(CF_3)_2AsN(SiMe_3)_2$ to the As(V) dichloride with axial Cl, excess Cl_2 inducing formation of the 4-membered cyclodiars-V-azane.[234]

Si-N intermediates are used in the preparation of $Ph_2C=NSNSO$, $(Ph_2C=NS)_2N_2S$, sulphenimines, asymmetric sulphur diimides and carbon substituted S-N rings.[235] $(Me_3SiN)_2S$ gives an amide complex with Pt(0), is protonated at N on coordinatively bridging in its adduct with $(\mu-H)_2Os_3(CO)_{10}$, extrudes S from trithiadiborinanes to give dithiadiazadiborinanes, while $Me_3SiNSNH$ exists as 2 isomers in solution and complexes with $CpW(CO)_3H$.[236] $(Me_3Sn)_3N$ and BCl_3 give $N(BCl_2)_3$, carbonylation of ArI/R_3SnX (X=OMe,NEt_2) in the presence of Pd leads to esters and amides of substituted benzoic acids, while multinuclear MR spectroscopy, in particular for ^{13}C and ^{15}N, acts as an indicator of mesomeric/inductive effects in silyl, stannyl and plumbyl pyrroles.[237]

7 Phosphorus Group Derivatives

<u>Ab Initio</u> calculations show the HOMO of $HP=SiH_2$ is π, indicating [2+2] cycloaddition reactions to be favoured. The silaphosphaalkene $mes_2Si=P(2,4,6-Bu^t_3C_6H_2)$ results from the phosphine and mes_2SiCl_2 with Bu^nLi, but is only stable in solution. It adds MeOH through protonation at P. The first germaphosphene $mes_2Ge=Pmes$ is formed as air sensitive orange crystals, m.p. 155-160°C, quantitatively from the dehydrochlorination of $mes_2GeCl-PHmes$ by Bu^tLi, and shows inequivalent mes groups at Ge, indicating restricted rotation. $[(Me_3Si)_2CH]_2SnFPH(2,4,6-Bu^t_3C_6H_2)$ is dehydrofluorinated with Bu^tLi giving the first stannaphosphene, characterised in solution but too unstable to isolate pure.[238]

Allylphosphines $H_2C=CHCH_2PHMMe_3$ (M=Si,Ge) result from the halide and phosphine in the presence of Et_3N, and new silylphosphines have been prepared by the addition of Me_3SiPH_2 to dienes and vinylphosphines, while $(allyl)_2RSiPH_2$ cyclises with AIBN to give the 1-phospha-5-silabicyclo[3.3.0]octane.[239] The Si-P bond inserts heterocumulenes and is ring opened by MeOH in the same manner as linear silylphosphines.[240] Sulphur oxidises $(Me_3Si)_3P$ and Me_3SiPPh_2 to the thiophosphate and phosphonate, but the selenium analogue of the latter disproportionates to $(Me_3Si)_2Se$ and $[Ph_2P(=Se)]_2Se$.[241]

The diphosphines $[R(Me_3Si)P]_2$ ($R=Ph, Me_3Si$) silylate $CpM(CO)_3H$ giving $CpM(CO)_3SiMe_3$, Bu^tCOCl desilylates $CpFe(CO)_2P(SiMe_3)_2$, rearrangement giving the phosphaalkenyl complex $CpFe(CO)_2P=C(Bu^t)OSiMe_3$ with P=C 170.1 pm, while the first σ-bonded diphosphene complex results from $Me_5C_5M(CO)_2P(SiMe_3)_2$ and $2,4,6-Bu^t_3C_6H_2PCl_2$.[242] $Me_3SiPRLi$ ($R=Me_3Si, Bu^t$, Ph) attacks CO in $Me_5C_5Ru(CO)_3^+$ and $CpRe(CO)_2NO^+$, the derivatives of the former losing CO if $R=Me_3Si$ or Ph to give the Ru phosphide, while for the Re compound rearrangement gives the C-bonded phosphaalkene derivative with C=P 166.2 pm.[243]

The chelate diphosphine complex $R_2P(CH_2)_2PR_2NiCl_2$ is reduced by $(Me_3Si)_2PLi$ to the disilyldiphosphene Ni(0) complex, $(Me_3Si)_3P$ gives the diphosphyne derivative $[R_2P(CH_2)_2PR_2Ni]_2P_2$ and with $(Me_3Si)_2PPh$, $NiCl_2/Ph_3P$ forms the cube cage $Ni_8(PPh)_6(PPh_3)_4$.[244] Triphos-Ni and -Co complexes [triphos=$(Ph_2PCH_2)_3CCH_3$] abstracts the P_3 unit of $P_7(SiMe_3)_3$ to give $triphosCo(P_3)$ and $[triphosNi(P_3)Nitriphos]^{2+}$. Cyclotriphosphanes $(Bu^tP)_2PX$ (X=Cl or $SnMe_3$) result from $[Me_3Sn(Bu^t)P]_2$ and PCl_3, or $(ClBu^tP)_2$ and $(Me_3Sn)_3P$.[245] Reducing $Bu^t_2SiI_2$ with excess lithium naphthalenide, followed by addition of $2,4,6-Bu^t_3C_6H_2PCl_2$ gives the diphosphene and the air stable diphosphasilirane which has long Si-P and P-P bonds, puckered phenyl rings and o-Bu^t groups bent out of the plane, while (E)- and (Z)-phenylphosphiranes result from styrene oxide and $2,4,6-Bu^t_3C_6H_2P(Li)SiMe_2Bu^t$.[246]

Oxidising $PhSb(SiMe_3)_2$ in 1,4-dioxane or benzene gives solvated $(PhSb)_6$ which possesses a chair conformation, and $(R'_2N)(R''_2N)C=PSiMe_3$ (R'=Me; R''=Me, Et) can be transsilylated readily at 20°C but $(Et_2N)_2C=PSiMe_3$ at 40°C needs 10 hr with Ph_3MCl (M=Ge, Sn) in C_6H_6.[247]

8 Oxygen Derivatives

$H_2Si=O$ is suspected as the initial product of O_3 and SiH_4, and while calculations indicate no barrier to its dimerisation, the 4-membered ring is strained to the extent of 177 kJ.[248] $Me_2Si=O$ is generated by pyrolysing 6-oxa-3,3-dimethyl-3-silabicyclo[3.1.0]hexane and the kinetics of its insertion compared with that of $Me_2Si:$. It also results on pyrolysing the ketene $Me_3Si(Me_2HSi)C=C=O$ together with

$Me_3SiC\equiv CH$ (76%). Heating $[Me_3Si(Me)SiO]_4$ gives $Me_3Si(Me)Si=O$ which adds to buta-1,3-diene as the siloxysilylene.[249] A redetermination of the structure of $Et_2Si(OH)_2$ indicates individual diol molecules connect in a chain polymer in contrast to $Bu^t_2Si(OH)_2$ which consists of H-bonded dimers linked by further H-bonds into a ladder-chain. $Bu^t_2Si(OH)F$ crystallises as cyclic tetramers with OH--OH and not OH--F H-bonds.[250] $Bu^t_2SiCl_2$ and nucleosides give sila analogues of cyclic nucleotides with a rigid <u>trans</u>-(1,2)-fused 6 to 5-membered ring system, while the GCMS of 1,2- and 1,3-diol and β-hydroxy acid derivatives gave distinctive fragmentation. $Bu^t_3SiO^-$ disubstitutes $TaCl_5$, providing an intermediate for the preparation of Ta-alkyl carbene derivatives.[251]

2^y and 3^y silanes give alkoxysilanes with 1^y or 2^y alcohols in the presence of CsF/imidazole, while adamantyl dimethylsilyl ethers result similarly using $(Ph_3P)_3RhCl$ as catalyst and are comparable in hydrolytic stability to Bu^tMe_2Si ethers.[252] Jones reagent/KF induces the conversion of silyl ethers to carbonyl compounds <u>via</u> Si-O cleavage and nickel boride allylic trimethylsilyl ethers to alkenes, while C-O/Si-O cleavage of allyl and vinyl silyl ethers is catalyst dependent. Thus with $CoH(N_2)(PPh_3)_3$ C-O cleavage gives olefins, $RhH(PPh_3)_4$ cleaves Si-O and $RuH_2(PPh_3)_4$ both. $Bu^tMe_2SiO(CH_2)_3Br$ converts terminal acetylenic alcohols into (<u>E</u>)-olefinic alcohols.[253]

Silyl enol ethers have been prepared from ketenes, α-halocarbonyl compounds, and $C_6F_5SiMe_3$ with enolisable ketones in the presence of KCN/18-crown-6.[254] They are used in the preparation of β-(PhSe)ketones, 1-silyl-3-acetyl-4-carboxyazetidine-2-one and cyanohydrins ($TiCl_4$ catalysed),[255] oxy substituted ketones,[256] diketones,[257] with ArN_2^+, α-aryl ketones and with azoesters, α-azoketones and precursors to thienamycin analogues.[258] Aldehydes give cross-aldol products, $SOCl_2$ α-oxosulphines, Hg(II) induces intramolecular cyclisation of acetylenic silyl enol ethers to 2-alkylidene-1-oxocyclopentanes,[259] while ArNO gives α-aroyl-N-phenylnitrones and β-N-phenylamino alcohols.[260]

Silyl ketene acetals dimerise to succinate esters in the presence of $TiCl_4$,[261] their thio derivatives $MeCH=C(SPh)OSiMe_3$ used to make anti aldols with RCHO, <u>syn</u>-chelated adducts with $RCHO/SnCl_4$, <u>syn</u>-α-CH_2-β-OH-δ-alkoxy esters and α-non substituted β,δ-dihydroxy-ketone and ester from α- and β-alkoxyaldehydes and β-ketoesters,[262] the α-$(PhCH_2)_2N$ derivative giving α-amino-β-hydroxy acids, N-silyl imines azetidin-2-ones, ArN_2^+ α-amino esters,[263] crotonyl and acryloyl chlorides, glutaric and γ,δ-ethylenic β-keto esters (for making (+)-turmerone) and glyceraldehydes 2-deoxy-D(and L)-riboses.[264] 2-Me$_3$SiO-buta-1,3-diene is used to make the 4-vinyl-2(5H)-furanoid unit and Danishefsky's diene [MeOCH=CH(Me$_3$SiO)C=CH$_2$] gives 4-acylcyclohexa-2,5-dienones for the first time.[265] Triphenylphosphite ozonide Ph$_3$PO$_3$ oxidises 2-Me$_3$SiO-1,3-dienes to β-hydroxy cyclic enones while Me$_3$SiO(MeO)C=CH(Me$_3$SiO)C=CH$_2$ and Ph(MeO)$_2$CCHO condense to δ-hydroxy cyclopentanone, and acacH and Ph$_2$SiCl$_2$ give a cyclic dienolate.[266] Bu^{t}Me$_2$SiOSO$_2$CF$_3$ results from CF$_3$SO$_3$H and H$_2$C=C(Me)SiMe$_3$ (66%) or Bu^{t}Me$_2$SiH (90%), Me$_3$SiOSO$_2$CF$_3$ from Me$_3$SiCCl$_3$ (88%) or (Me$_3$Si)$_2$O and (CF$_3$SO$_2$)$_2$O giving a mixture of silyl enol ethers with 2-methylcyclohexanone, while (CF$_3$SO$_2$)$_2$O substitutes Me$_3$SiOR.[267] Silyl triflates give silyl enol ethers and siloxydienes,[268] catalyse the conversion of silyl ketene acetals with Bu^{t}N=CHCN to β-ketoesters, give silylated α-diazoesters from α-diazo-phosphonates or -carboxylates.[269]

Acetolysis of XSiMe$_2$Cl (X=Cl,acetoxy,siloxy) is influenced exclusively by steric effects while carboxylates and ClCH$_2$SiMe$_2$Cl gives derivatives from Si-Cl and Si-C attack.[270] α-Hydroxy silyl esters give β-lactams with imines, Me$_3$Si-4,4-ethylenedioxypentanoate gives diketones with RCO.O.CO$_2$Et, while zinc derivatives of γ-bromoalkenyl acids add PhCHO 1 and 3, and Me$_3$SiO$_2$CC=CCO$_2$SiMe$_3$ decarboxylates to Me$_3$SiC≡CCO$_2$SiMe$_3$ with Et$_3$N.[271] Me$_3$SiOCOCCl$_3$ silylates many protic reagents, with CO$_2$ and CHCl$_3$ elimination and with alkenes and KF in a non-polar solvent gives dichlorocarbene adducts.[272]

(Me$_3$SiO)$_2$PH results from NaH$_2$PO$_2$ and Me$_3$SiCl in 3^y amine, is readily alkylated, acylated and oxidised giving R$_2$P(O)SiMe$_3$ and RC(OSiMe$_3$)[P(OSiMe$_3$)$_2$]$_2$, (Me$_3$SiO)$_3$P and ArBr gives ArP(O)(OSiMe$_3$)$_2$, and C$_3$F$_6$, CF$_3$CF=CFP(O)(OSiMe$_3$)$_2$.[273] Complexes of (Ph$_2$PO)$_2$SiMeR with Mo(0) result from Et$_3$NH$^+$[Mo(CO)$_4$(Ph$_2$PO)$_2$H]$^-$ with MeRSiCl$_2$/Et$_3$NH.[274]

Boric acid forms mixed esters with Me$_3$SiCl/(Me$_3$Si)$_2$O in Pr^{i}OH, while silylating the barium chloride silicate of composition (BaO)$_2$(BaCl$_2$)$_3$(SiO$_2$)$_2$ gives the silylated cycloheptasilicate (Me$_3$Si)$_{14}$Si$_7$O$_{21}$.[275] The silyl esters of chromyl chloride Me$_3$SiOCrO$_2$Cl and MePhSi(OCrO$_2$Cl)$_2$ notably oxidise ArCH$_3$ to ArCHO and thiols to disulphides. Silyl molybdates result from R$_3$SiOH and Mo$_2$O$_7^{2-}$, while with (Me$_3$Si)$_2$S, the mixed thiomolybdates MoO$_{3-n}$S$_n$(OSiMe$_3$)$^-$ (n=0-3) result.[276]

α-Bromo silyl enol ethers give α-silyl ketones through a 1,3-O--C silyl migration, while base catalyses the 1,3-C--O and 1,4-C--O silyl migrations in epoxy alcohols silylated at the epoxy ring.[277] TsiSiPhF$_2$ on hydrolysis with KOH gives the silanol which undergoes 1 or 2 silyl C--O migrations giving (Me$_3$Si)$_2$CHSiF(Ph)OSiMe$_3$ and Me$_3$SiCH$_2$Si(Ph)(OSiMe$_3$)$_2$, while TsiSiF$_2$R gives siloxane and silanol (R=Me), and a linear siloxane only for R=F.[278]

Kinetics of hydrolysis of chlorodi and penta-siloxanes indicates acid catalysis with the reaction first order with respect to siloxane. Polysiloxanes Me$_3$SiO(MeRSiO)$_n$SiMe$_3$ (R=vinyl,aryl) and R$_3$SiO(H$_2$SiO)$_n$SiR$_3$ (R$_3$=Me$_3$,Et$_3$,MeH$_2$,Me$_2$H) have been made.[279] NMR spectra are determined for Me$_3$SiO[SiMe(H)O]$_n$SiMe$_3$ (n=3-8 and 35), (ClMeSiO)$_4$ and (Ph$_2$SiO)$_4$.[280] A redetermination of the structure of (Ph$_2$SiO)$_3$ gives a SiÔSi angle of 132-3° and Si-O bonds of 164 pm, while Ph$_2$SiOSiMe(H)OSiPh$_2$OSiMe(H)O exists as cis- and trans-isomers, each with 2 conformations.[281] The structure of the coupled cyclotriphospha-V-azene—cyclotetrasiloxane (NP)$_3$NH(CH$_2$)$_3$(SiO)$_4$ crystallises with alternating layers of (NP)$_3$ and (SiO)$_4$ rings. Mono(β-naphthyloxymethyl)permethylcyclotetrasiloxanes and (PhMeSiO)$_n$ (n=3,4) readily form areneCr(CO)$_3$ complexes.[282]

^{19}F NMR monitoring of CF$_3$ interchange in the 10-Si-5 siliconates OC(CF$_3$)$_2$-o-C$_6$H$_4$-M-C$_6$H$_4$-o-(CF$_3$)$_2$O (M=RSi$^-$) supports a non-dissociative Berry-type pseudorotation with ΔG^* to inversion at Si for M=RSi$^-$ being less than that for M=PhP.[283] This type of intermediate provides for the inversion of the spirosilane (M=Si) in weakly nucleophilic media. While the above has an essentially trigonal bipyramidal structure with axial O, [o-OC$_6$H$_4$OSi(R)OC$_6$H$_4$O-o]$^-$ NEt$_4$$^+$ moves progressively towards a rectangular pyramid with increasing donating ability of R (1-naphthyl, Bu^n,Bu^t) and H-bonding to O introduced by Et$_3$NH$^+$.[284] Intramolecular coordination in p-RC$_6$H$_4$CO$_2$CH$_2$SiF$_3$ increase with the electron donating ability of R while the o-derivative (R=Cl,MeO) exists as a cis-trans isomer mixture.[285] Complexing energies have been calculated for X$_{ax}$(H$_3$)$_{eq}$Si--Y for X=H,F,Cl;Y=H$^-$ and X=F;Y=O=CHOH, this Y causing the weaker purterbation of FSiH$_3$. Oxidation of RSCH$_2$CH$_2$SiF$_3$ by H$_2$O$_2$ gives the sulphoxide which shows non-equivalent F due to O--Si coordination and chirality at S. At low temperature with coupling constants F$_e$-F$_e$' is less than that for F$_e$-F$_a$. ^{19}F nmr data suggests axial N--Si coordination in the trigonal bipyramidal complexes R'F$_2$SiO(CH$_2$)$_2$NR$_2$ and R'F$_2$SiN(CH$_2$CH$_2$NR$_2$)$_2$ (R'=F,Ph).[286] An extensive series of organosilicon esters of diethanolamine[287] have been prepared along with many silatranes,[288] and a few germatranes.[289] CH_2 introduction into the silatrane cage causes Si--N lengthening. The Si-C bond is susceptible to electrophilic attack by Hg(II) and the nmr spectra of a wide range of atranes interpreted from their x-ray structures.[290]

Siloxycyclopropane carboxylates expand on hydroxymethylation or with CS_2 to give THF or $\overline{(CH_2)_4S}$ derivatives, $Me_3SiO(Me_3Si)\overline{CCH_2CH_2}$ ring opens to propenes on FVP, while thermolysing siloxy(ethynyl)-cyclobutenones provides a new route to benzoquinones and cyclopentenediones.[291] Alkenoates result from $Pb(OAc)_4$ fragmentation of 1-(Me_3SiO)bicyclo[n.1.0]alkanes and o-$Me_3SiOC_6H_4SiMe_3$ is formed through o-lithiation of PhOH with Bu^tLi.[292] Metallation of 1-(Me_3SiO)-3,5-$(MeO)_2C_6H_3$ at C-4 provides for the synthesis of the trimethyl ether of sophoraflavanone A, an anti-fungal natural product, while HF and Bu^n_4NF desilylate alcoholic and phenolic silyl ethers. UV PE spectroscopy shows donor ability to decrease in the order $MeO,Me_3SiCH_2 > Me_3SiO > Me$ in siloxybenzenes.[293] Spectra of oxasilacycloalkanes are analysed.[294]

The antimuscarinic agent $Ph(c-C_6H_{11})Si(OH)CH_2CH_2\overline{N(CH_2)_4}$ is used as a reference drug in experimental pharmacology and occurs as H-bonded dimers.[295] $Me_3SiO(CF_3)_2CPH_2$ is P-halogenated by NBS or NCS and $RSi(OCH_2MMe_2)_3$ (M=P,As) displace chpt from $chptM(CO)_3$.[296] The vibrational spectra of $(Me_3Si)_2O_2$ is analysed. It oxidises phosphorus derivatives with retention of configuration at P.[297] The kinetics of phosphoryl attack of R^*_3SiCl is consistent with a siloxyphosphonium cation intermediate, while substitution of homochiral alkoxy groups in prochiral siloxanes accords with mechanistic models under retention.[298] EtCHO gives $EtCOCMe_2OSiMe_3$ while 3 of the 4 possible aldols can be made from chiral α-siloxy ketones with RCHO through appropriate choice of cation, and cis-4-(Bu^tMe_2SiO)-2-cyclopenten- 1-ol results with high stereospecificity from the epoxycyclopentan-1-ol.[299] Si-O derivatives are used in the preparation of deoxy-uridines, -aldoses and -ketoses, to protect ribonucleosides and to prepare the silyl linked analogue of thymidine dinucleotide.[300]

1,3,2-Dioxagermetanes result from Me_2GeX_2 and $Cl_3CCH(OH)_2$ and decompose on heating to generate the germanone $Me_2Ge=O$.[301] 7-Membered $Ph_8Ge_2Si_2O_3$ results from $(ClPh_2Ge)_2$ and $Ph_2Si(OH)_2$ and has a C_2 twisted chair conformation, while the bicyclic germoxane $Ph_8Ge_5O_6$ is formed from $2PhGeCl_3/3Ph_2GeCl_2$ in aqueous $AgNO_3$.[302]

With $SbCl_5$, $(R_3Sn)_2O$ gives the complex $(R_3Sn)_2OSbCl_4^+SbCl_6^-$, Ph_3SnOH dehydrates to give $(Ph_3Sn)_2O$ at 75°C and $(Ph_2SnO)_n$ at 250°C while neophyl$_3$SnOH spontaneously dehydrates.[303] The hydroxide halides $Bu^t_2Sn(OH)X$ (X=F,Cl,Br) are dimeric with OH bridges, and are further associated through O-H--X bonds. $ArOSiMe_3$ (Ar=p-tolyl) and $SnCl_4$ form the trinuclear complex $(ArO)_4Sn_3Cl_8$ with ArO bridges, and the central Sn 6-coordinate.[304] $(Me_4Sn_2Cl_2O)_2$ has a central Sn_2O_2 ring with weak Cl bridges to the outer Sn atoms, while hydrolysing Me_2SnCl_2 in the presence of Et_3N gives Et_3NH^+ $(ClMe_2Sn)_5O_3^-$ which possesses a central Sn_3O_3 ladder unit which similarly forms weak Cl bridges.[305] Pyridines assist the trifluoroacetolysis of methyltin halides, while $[PhSn(O)O_2CC_6H_{11}-c]_6$ possesses a hexagonal drum cage of tridentate O-bridges with bidentate carboxylate groups encircling it.[306] Organotin carboxylates and phenacyl bromides give esters, while Pd catalysed hydrostannolysis of α-disubstituted allyl β-keto esters give stannyl β-keto carboxylates which lose CO_2 to form stannyl enolates.[307]

$SnCl_4$ ring opens $Me_3SiO(RO)\overline{CCH_2C}H_2$ to give 3-stannyl propionates internally coordinated as are $Bu^nOCO(CH_2)_2SnCl_3$ and $PhCOCH_2CMePhGeCl_3$, while urethane formation from MeOH/PhNCO is catalysed by tetraalkyl distannoxanes.[308] π-Allyl Pd compounds are etherified by tin alkoxides and transalkoxylation of 4-chlorostann-oxy-1-alkenes and -alkynes with RCOCl gives esters.[309] $Bu^n_3SnO(CH_2)_nX$ (n=2,3;X=Cl,Br,I) gives a mixture of iminodioxalanes and oxazolidinones with RNCO, cyclic sulphites of sugars are formed from their stannylidene derivatives with $SOCl_2$, and $RP(C_6H_4OH-\underline{o})_2$ transalkoxylates $Bu^t_2Sn(OMe)_2$ to give the internally coordinated benzoxaphosphastannolin.[310] Specific NMR studies include ^{119}Sn of tetra and hexaorganodistannoxanes, ^1H, ^{13}C and ^{119}Sn of the monomer-dimer equilibrium of 1,3,2-dioxostannolanes, ^{17}O of tin carboxylates,[311] the use of ^1J ($^{117,119}Sn$,^{13}C) in predicting coordination and C$\widehat{Sn}$C angles in tin derivatives, and ^{119}Sn in establishing the presence of 3 Sn species in the $Me_2Sn(OSO_2F)_2/HSO_3F$ system.[312]

9 Sulphur, Selenium, and Tellurium Derivatives

Pyrolysis of C-silyl substituted cyclopolysulphides at 400-600°C gives viSiMe$_3$, CS_2, SiS_2, thiophene, its homologues and silylated derivatives, while fragmentation of isomeric $Me_2\overline{SiCH_2CH_2SCH_2C}H_2$ and

$Me_2\overset{\frown}{SiCH_2CH_2S}CHMe$ gives $(Me_2Si=S)^+$ as the most abundant ion.^{313} Spirobis(ethylenedithia)silane has Si-S bonds of 211.6 pm, some 3.6 pm shorter than those of $(Ph_3Si)_2S$ $(Si\hat{S}Si$ 112°). R_3SiSH ($R=Bu^tO$) gives an O-coordinated monomer with Hg(II), a dimer with Tl(I) and a tetramer with Ag(I).^{314} The silane thiols X_3SiSH (X=Me,Ph,F,Cl,Br) have been prepared from X_3SiH and S, and their spectra determined, while a series of cyclosiloxthianes result from $R_2HSiOSiHR_2$ (R=Me,Ph), $Ph_2HSiSiHPh_2$, Si_2Cl_6 and Si_3Cl_8 with sulphur compounds.^{315}

The disilathiiranes $mes(R)\overset{\frown}{SiSSi}(R)mes$ (R=mes or Bu^t) result from the disilene and sulphur, the structure of one (R=mes) showing a short Si-Si bond (228.9 pm) and Si-S 216.1 pm, 1 pm shorter than those in $(Me_2SiS)_2$. H_2S and $ClMeSi(SiMe_2)_2\overset{\frown}{SiMeCl(SiMe_2)_2}$ [$(Me_2Si)_6$ and $SbCl_5$] give $Me_2\overset{\frown}{SiMe_2SiMeSiSSiMeSiMe_2}SiMe_2$ a bicyclic silthiane.^{316} TsiH with Li/S_8 gives Tsi_2S_4 (20%) and TsiSH (60%) which on lithiation and bromination yields $(Me_3Si)_2C=S$ as a red-violet oil. Heating TsiSH gives mainly $(Me_3Si)_2CHSSiMe_3$, with Tsi_2S_4 forming $(Me_3Si)_2C=S$ and $(Me_3Si)_2S$. The thioketone adds to $H_2C=CHMeMeHC=CH_2$ to give the thiacyclohexane while soluble "Li_2S", formed from $(Me_3Si)_2S$ and MeLi couples α,ω-dibromoalkanes to thiacycloalkanes and $Br(CH_2)_nCOCl$ to thialactones.^{317}

Thiosilanes cleave halogens from $GaCl_3$ and $SnCl_2$, and $Rh_2(CO)_4Cl_2$ in the presence of Bu^t_3As to give $Rh_2(CO)_2(AsBu^t_3)_2ClSR$ [R=allyl or $CH_2CH_2Si(OMe)_3$ for silica support].^{318} With $(Me_3Si)_2S$, $(Ph_3P)_2MCl_2$ (M=Co,Ni) gives 6,7 and 8 metal atom clusters involving M, S, Ph_3P and Cl^-, whereas $(Me_3Si)_2Se$ gives 4, 6 and 9 metal atom clusters for M=Co, but $Ni_{34}Se_{22}$ for M=Ni.^{319} $Me_3SiTePh$ results in good yield from PhLi/Te or Ph_2Te_2/Na with Me_3SiCl, and ring opens lactones, epoxides and THF (cf Me_3SiI), while $(Me_3Si)_2Te$ reductively couples RCOCl to give the silylated ene-diol via $(RCO)_2Te$ and $RC(Te)OSiMe_3$.^{320}

S-Aryl thioaroates result from Me_3GeSAr and ArCOCl while $TiCl_4$ with $(Et_3Si)_2Se$ gives Cl_2TiSe.^{321} Me_2Ge: and the thioketone of adamantanone Ad=S give the digermadithiolane $\overset{\frown}{AdSGeMe_2SGeMe_2}$ and $\overset{\frown}{AdSGeMe_2GeMe_2}$ (from $Me_2Ge=GeMe_2$ and Ad=S),^{322} whereas dithiagermoles eliminate $R_2Ge=S$ above 200°C. The digermathietane $Me_2\overset{\frown}{GeSGeMe_2CH_2}$ [from $(ClMe_2Ge)_2CH_2$ and Na_2S] simultaneously dimerises on heating and decomposes to trigermathiane and dithiane through $Me_2Ge=S$ and $Me_2Ge=CH_2$ intermediates.^{323} $Cl_3CCO_2GePh_2GePh_2O_2CCCl_3$ and H_2S gives $Ph_8Ge_4S_2$ and $(Ph_2ClGe)_2/Ph_2GeCl_2$ with Na_2S or NaHSe the 5-membered ring $Ph_6Ge_3X_2$ (X=S,Se). While the dithiocarbamate complex

ClMe$_2$GeS$_2$CNMe$_2$ is 5-coordinate with a distorted trigonal bipyramidal structure (Ge-S 225.4 and 289.6 pm) the structures of Ph$_3$GeS$_2$P(OMe)$_2$ and Ph$_2$Ge[S$_2$P(OMe)$_2$]$_2$ are tetrahedral with Ge-S distances about 230 pm or more than 500 pm.^{324}

Ge, Sn and Pb esters of $\overline{(CH_2)_4NCS_2}$ have been made. Ph$_3$M (M=Sn,Pb) compounds have distorted T$_d$ structures with the non-bonded M-S distance some 70 pm longer than the bonded ones, as with Ph$_2$Bu^{n}SnS$_2$CNMe$_2$, though with PhBu^n(Cl)SnS$_2$CNEt$_2$ this difference is much less. The ^{13}C NMR spectra of Me$_2$Sn[S$_2$CN(CH$_2$)$_4$]$_2$ predicts a CŜnC angle of 138.6^o (found 137.3^o).^{325} Multinuclear NMR supports weak 5-coordination in R$_3$SnS$_2$P(OR')$_2$ and R$_2$XSnS$_2$COR, while R$_2$Sn(S$_2$COR')$_2$ are 6-coordinate. R$_3$SnCS$_2^-$ forms dithioester with RX, the first α-thiocarbonyl tin compounds, and can be isolated as R$_3$MCS$_2$M'(C$_4$H$_8$O$_2$)$_n$ (M'=Li,Na,K) which complex with metal carbonyl residues.^{326}

The structure of 13,13-dimethyl-8,13-dihydro-5H-dibenzo[d,g]-1,2-diselena-6-silonine shows the 9-membered heterocycle to have no transannular Si--Se interactions.^{327} Weak Sn-S interactions occur for the stannocane PhXSnS$\overline{(CH_2)_2S(CH_2)_2S}$ (X=Cl,Ph),^{328} while the Sn(II) compound [Sn(SC$_2$H$_4$)$_2$NBu^t]$_2$ occurs in the unit cell as 2 distinct Sn$_2$S$_2$ bridged dimers, the one with weaker Sn-S bridges (299 and 279 pm) and stronger Sn-N interactions (264 and 275 pm) than the others (Sn-S 271 pm, Sn-N 297 pm). 2D ^{119}Sn NMR spectra of CH$_2$[PhSnS(CH$_2$CH$_2$)$_2$NMe]$_2$ show 4 signals, indicating dynamic equilibrium between 3 isomers.^{329}

A series of S-stannyl 3-aminopropiolothioimidates have been prepared from aminoethynylstannanes and PhNCS, while R$_6$Sn$_2$ or R$_3$Sn$^-$ ring opens 4,5-dimethyl-1,2,3-selenadiazole, quenching with R$_3$SnCl giving MeC≡CMe and (R$_3$Sn)$_2$Se.^{330} Copper reduces (Me$_2$SnS)$_3$ to (Me$_3$Sn)$_2$S and Sn, while calculations show 2,4,6-trithia-1,3,5-tristannaadamantane to readily give the relatively stable apical radical.^{331} R$_2$SnCl$_n$(IOTG)$_{2-n}$ (n=0,1) and RSnCl$_n$(IOTG)$_{3-n}$ (n=0-2) show IOTG/Cl distribution to occur with maximum degree of ligand mixing and is related to the use of R$_2$Sn(IOTG)$_2$ in PVC stabilisation, while mercaptoester/chlorine exchange equilibria of Bu^{n}OCOCH$_2$CH$_2$SnL$_3$ (L=mercaptoester) correlate with the activity of such compounds in PVC.^{332} Toluene-3,4-dithiolate complexes of R$_2$Sn residues are associated but 4-coordinate for R$_3$Sn and 1,2-dimethyl-4,5-bis(mercaptomethyl)benzene. Ph$_2$PbS$_2$P(OCH$_2$Ph)$_2$ has trans-Ph groups (CP̂bC 165°) and anisobidentate dithiophosphate groups, and is weakly associated giving an equatorial PbS_5 arrangement. The plumbacanes $Ph_2\overline{PbS(CH_2)_2O(CH_2)_2S}$ crystallise with orthorhombic and triclinic structures, each with 3:1 chair-chair:boat-chair modifications.[333]

10 Halogen Derivatives

Aminonaphthylsilanes with CsF/imidazole discriminates 1^y and 2^y OH groups of serine n-butylamide to give fluorescent 3^y silyl ethers which can be cleaved by HF to form fluorescent silyl fluorides. Photolysing ketene in the presence of methylfluorosilanes gives products indicative of singlet and triplet methylene with the fluorosilane.[334] Cu_3Si with Cu/Zn gives 95% Me_2SiCl_2 production at 520-620K while Me_2SiCl_2, $PhSiCl_3$ and $ClMe_2SiCH_2Cl$ convert Me_4Si to Me_3SiCl in the presence of $AlCl_3$.[335] Thermodynamic data is calculated for methylhalosilanes, Bu^tMe_2SiCl with DBU used to silylate OH, NH and SH groups in high yield, which are less stable than $HMe_2CMe_2CSiMe_2$ derivatives to subsequent hydrolysis.[336] Heating trimethylsilyl-4-bromo-2-alkanoates gives Me_3SiBr in 85-90% yield.[337] Me_3SiI is used to prepare N-iodomethyl uracil, 1,2-dialkyl glycerols from benzyl glycerols, silyl enol ethers from 2-Me_3SiO-cyclopropane carboxylates, while Me_3SiCl/NaI converts trisubstituted tetrahydro-4-oxopyran-3-carboxylic esters to cyclopentenone derivatives.[338] Me_3SiI and $(Me_2N)_3P$ give the phosphonium complex in solution at $-50^\circ C$, while Me_3SiX (X=Cl, Br,I,ClO_4 and CF_3SO_3) gives 1:1 complexes with imidazole, pyridines, urea, and pyridones.[339] Acetonitrile oxide forms adducts with Me_3SiX (X=Cl,I,NCS,N_3,NEt_2, SMe) giving $MeXC=NOSiMe_3$ but not for X=F,OMe,NCO or O_2CMe, while PhSeCl cleaves Me_3SiY (Y=Br,I,CN, NEt_2,NMeCOMe) to give PhSeY.[340]

Me_4Si methylates $GeCl_4$ in the presence of $AlCl_3$ to give particular methylgermanes in good yield, depending on the ratio employed, 2-furylbromogermanes result from 2-furylLi and $GeBr_4$, while $MeGeI_3$ (x-ray) and $MeSnCl_3$ (IR-Raman) have $\underline{C}_{3v}$ structures.[341] $PhCH_2Cl/Sn$ in the absence of catalyst give $(PhCH_2)_2SnCl_2$, alkylCl chlorinates Me_4Sn in the presence of $AlCl_3$, while HCl from neoprene chlorinates Bu^n_3Sn additives.[342] While chlorine bridges in Me_2SnCl_2 have a centre of symmetry (polarised IR shows $\underline{C}_{2v}$ site in crystal), with each Cl atom interacting a different Sn neighbour, those of $(c-C_6H_{11})_2SnCl_2$ involve asymmetric chelation of both Cl atoms to one Sn atom.[343] Electrolysing $Me_2SnCl_2/PhCH_2PPh_3^+Cl^-$ with DBTTF in PhCN gives columnar DBTTF with complementing anions $Me_6Sn_3Cl_8^{2-}$ having each Sn 6-coordinate, while the anions $EtSnCl_5^{2-}$ and $Bu^nSnCl_5^{2-}$ have trans-shortened Sn-Cl bonds, and are used as dipyridylium herbicide counter ions.[344]

$Cp_2Mo(Me_2SnCl)_2$ occurs in the crystal as 2 independent molecules with different Mo-Sn bond lengths due to single weak Sn--Cl interactions (289 pm). Bu^tSnCl_3 [from $Bu^tSn(NEt_2)_3$ with Bu^tOH then Me_3SiCl] rapidly decomposes at room temperature to Bu^tCl.[345]

The enthalpies of formation of $R_3SnI.I_2$ complexes are measured, while the mechanism of attack of R_3SnI by iodine atoms shows the fragmentation of $(R_3SnI)^{\cdot+}I^-$ to be rate determining.[346] Complexes of R_3SnI and $Bu^n{}_2SnI_2$ catalyse the insertion of oxiranes by RNCO and RNCNR, and oxetanes by RNCO to give oxazin-2-ones.[347] Vibrational spectra of Ph_3SnXY^- and $Ph_2SnX_2Y^-$ are analysed (X,Y=Cl,Br,I), thermodynamic parameters measured for $MeSnCl_3/R_2SO$ showing 1:1 and 1:2 complexes coexist in solution.[348] $EtSnI_3.2Ph_2SO$ shows DPSO trans to Et and cis to each other, $Me_2SnCl_2(O=S=CPhNMePh)$ with η^1-O coordination, and $Me_2SnCl_2(H_2O)_2 \cdot 4C_5H_4N_4$ (purine) has H_2O trans-coordinated to Sn and H-bonded to purine.[349] $Me_2SnCl_2.2OP(NMe_2)_3$ also has an all trans structure while Lewis acidity of Me_3MCl increases with R.A.M. of M.[350] $Ph_2SnX(X')$ (X=NCO,NCS,N_3;X'=Cl,Br) result from Ph_3SnX and IX' with $Ph_2Sn(NCS)Br$ giving a 1:2 complex with DMSO. Ph_2SnCl_2, Ph_2SnBr_2 and $Bu^n{}_3PO$ give interconvertible isomers of $Ph_2SnCl(Br)OPBu^n{}_3$ with either Cl or Br equatorial.[351] The complexes $MeSnX_3(HOPMe_2)_2$ (X=Cl or Br) show cis-trans dynamic equilibrium in solution and $EtSnX_3.2OP(NMe_2)_3$ (X=Cl,I).[352] FAB and EI mass spectra of phenyl Ge, Sn and Pb halides are very similar except that FAB prefers halide loss, while with EI, phenyl loss is preferred. Plotting ^{13}C shift against ^1J $^{119}Sn-^{13}C$ for Me_3SnCl gives a straight line plot in 22 solvents.[353] ^{119}Sn NMR spectra of $(c-C_6H_{11})_3SnX$ (X=Cl,Br,I,OH,OAc) support monomers rather than weakly associated polymers as Mossbauer spectroscopy indicates.[354]

11 Complexes

This section lists complexes with Journal reference. Schiff base and oxine complexes of methylchlorosilanes are reported,[355] along with cyclic urea complexes of Ph_3MX (M=Sn,Pb;X=Br,I) and thiourea derivatives of Ph_2PbI_2.[356] A range of complexes with O and N ligands to $RSnCl_3$,[357] R_2SnCl_2,[358] and R_3SnCl[359] have been made, $(4-ClC_6H_4)_2$-

Group IV: The Silicon Group

SnCl$_2$(4,4'-dimethyl-2,2'-dipyridyl) occurring with both cis- and trans-SnR$_2$ configurations,[360] and their NMR spectra[361] and antitumour activity determined.[362]

Silicon-, germanium- and tin-transition metal complexes are listed by M-M' bond. Si-Zr,[363a] Si-Mn,[363b] Si-Fe,[363c] Si-Ru,[363d] Si-O-Os,[363e] Si-Co,[363f] Si-Ir,[363g] Ge-Co,[364a] Ge(Sn)-Ru(Os),[364b] Sn-Mo,[365a] Sn-W,[365b] and Sn-Re.[365c]

References

1. G.Raabe and J.Michl, Chem. Rev., 1985, **85**, 419; E.Fabry, Chem.-Ztg., 1985, **109**, 281; Organosilicon and Bioorganosilicon Chemistry, Ed. H.Sakurai, Ellis Horwood 1985.
2. J.C.Dalton, Org. Photochem., 1985, **7**, 149; E.Lukevics and M.Dzintara, J. Organomet. Chem., 1985, **295**, 265.
3. M.Lalonde and T.H.Chan, Synthesis 1985, 817; R.J.K.Taylor, ibid, 1985, 364; R.Anderson, ibid, 717; C.A.Maryanoff and B.E.Maryanoff, Asymmetric Synth., 1984, **4**, 355.
4. N.Inamoto, Yuki Gosei Kagaku Kyokaishi, 1985, **43**, 777; P.Hencsei and P.Laszlo, Kem. Kozl, 1984, **61**, 319; R.Muller, Z. Chem., 1985, **25**, 309; N.Wiberg, Adv. Organomet. Chem., Ed. R.West and F.G.A.Stone, 1985, **24**, 179, Academic Press.
5. Carbon-Functional Organosilicon Compounds, Ed. V.Chvalovsky and J.M.Bellama, Plenum 1984.
6. J.C.Maire, J. Organomet. Chem., 1985, **281**, 45; S.J.Blunden, L.A.Hobbs and P.J.Smith, Environ. Chem., 1984, **3**, 49; B.Giese, Angew. Chem. Int. Ed. Engl., 1985, **24**, 553; S.David and S.Hanessian, Tetrahedron, 1985, **41**, 643.
7. R.V.Parish from Mossbauer Spectroscopy Applied to Inorganic Chemistry, Ed. G.J.Long, Plenum Press, p.527; S.J.Blunden, P.A.Cusack and R.A.Hill, The Industrial Uses of Tin Chemicals, London RSC, 1985; C.J.Evans and S.Karpel, Organotin Compounds in Modern Technology, J. Organomet. Chem. Library, Elsevier, **16**, 1985.
8. Gmelin Handbook of Inorganic Chemistry - Si-Silicon. Part A1. History, 1984 and Sn-Organic Compounds, Part 11, 1985, Springer.
9. Chem. Met.-Carbon Bond, Ed. F.R.Hartley and S.Patai, Wiley, Vol. 2 and 3 (Chem. Abs., 1985, **103**, 215341 to 215343 and 215347 to 215349); J.S.Thayer, Organometallic Compounds and Living Organisms, Academic Press, 1984.
10. Chem. Abs., 1985, **102**, 6574, 6579, 24667, 24668, 24671, 24672, 166835, and **103**, 54111, 54114, 54116, 54117, 54119, 54120, 142043 and 215337.
11. Chem. Abs., 1985, **102**, 24659 to 24661.
12. S.M.Bachrach and A.Streitweiser, jr., J. Am. Chem. Soc., 1985, **107**, 1186; A.H.-B.Cheng, P.R.Jones, M.E.Lee, and P.Roussi, Organometallics, 1985, **4**, 581; N.Wiberg, G.Wagner and G.Muller, Angew. Chem. Int. Ed. Engl., 1985, **24**, 229; S.C.Nyburg, A.G.Brook, F.Abdesaken, G.Gutekunst and W.Wong-Ng, Acta Crystallogr., 1985, **C41**, 1632.
13. A.G.Brook, K.D.Safa, P.D.Lickiss and K.M.Baines, J. Am. Chem. Soc., 1985, **107**, 4338; L.E.Gusel'nikov, Yu.P.Polyakov, E.A.Volnina, A.V.Ivanov and N.S.Aametkin, Izv. Akad. Nauk SSSR, Ser. Khim., 1984, 2645 (Chem. Abs., 1985, **102**, 132119); L.E.Gusel'nikov, Yu.P.Polyakov, E.A.Volnina and N.S.Nametkin, J. Organomet. Chem., 1985, **292**, 189.
14. R.N.S.Sodhi S.Daniel, C.E.Brion, and G.G.B.De Souza, J. Electron Spectrosc. Relat. Phenom., 1985, **35**, 45; J.Casanovas, J.P.Guelfucci, R.L.S.Hoi, and R.Gross, J. Phys. Chem., 1985, **89**, 768; S.G.Wierschke, J.Chandrasekhar, W.L.Jorgensen, J. Am. Chem. Soc., 1985, **107**, 1496.
15. R.Brener, H.Schechter, and J.Suzanne, J. Chem. Soc., Faraday Trans. 1, 1985, 2339; M.Prager and M.Muller-Warmuth, Z. Naturforsch., Teil A, 1984, **39**, 1187 (Chem. Abs., 1985, **102**, 204059).
16. D.Farah, T.J.Karol, and H.G.Kuivila, Organometallics, 1985, **4**, 662; G.R.Finlay, J.S.Hartman, M.F.Richardson, and B.L.Williams, J. Chem. Soc.,

Chem. Commun., 1985, 159; T.Inoue and T.Yamase, Chem. Lett., 1985, 869.
17 A.Haider, Synthesis, 1985, 271; D.Djahanbini, B.Cazes, J.Gore, and F.Gobert, Tetrahedron, 1985, **41**, 867; H.Kleijn and P.Vermeer, J. Org. Chem., 1985, **50**, 5143; T.Sato, T.Kikuchi, N.Sootome, and E.Murayama, Tetrahedron Lett., 1985, **26**, 2205.
18 R.B.Hallock, W.E.Hunter, J.L.Atwood, and O.T.Beachley jr., Organometallics, 1985, **4**, 547; P.Stavropoulos, P.G.Edwards, G.Wilkinson, M.Motevalli, K.M.A.Malik, and M.B.Hursthouse, J. Chem. Soc., Dalton Trans., 1985, 2167; W.J.Evans, R.Dominguez, K.R.Levan, and R.J.Doedens, Organometallics, 1985, **4**, 1836; M.H.Chisholm, J.A.Heppert, J.C.Huffmann, and P.Thornton, J. Chem. Soc., Chem. Commun., 1985, 1466; M.F.Lappert, C.L.Raston, L.M.Engelhardt, and A.H.White, ibid, 521.
19 T.N.Mitchell and R.Wickenkamp, J. Organomet. Chem., 1985, **291**, 179; D.W.Hawker and P.R.Wells, Organometallics, 1985, **4**, 821.
20 T.Kauffmann, R.Kriegesmann, A.Rensing, R.Konig, and F.Steinseifer, Chem. Ber., 1985, **118**, 370; T.Kauffmann and A.Rensing, ibid, 380; T.Kauffmann, G.Ilchmann, R.Konig, and M.Wensing, ibid, 391.
21 J.Meunier-Piret, M.van Meerssche, K.Lurkschat, and M.Gielen, J. Organomet. Chem., 1985, **288**, 139; M.Gielen, K.Jurkschat, B.Mahieu, and D.Apers, ibid, 1985, **286**, 145.
22 F.G. De las Heras, A.S.Felix, A.Calvo-Mateo, and P.Fernandez-Resa, Tetrahedron, 1985, **41**, 3867; G.A.Olah, O.Farooq, and G.K.Surya Prakash, Synthesis, 1985, 1140; D.K.Dutta, D.Prajapati, J.S.Sandhu, and J.N.Burnah, Synth. Commun., 1985, **15**, 335; L.H.Foley, J. Org. Chem., 1985, **50**, 5204; V.H.Rawal, J.A.Rao, and M.P.Cava, Tetrahedron Lett., 1985, **26**, 4275; L.R.Krepski, L.E.Lynch, S.M.Heilmann, and J.K.Rasmussen, ibid, 981.
23 S.A.Carr and W.P.Weber, Synth. Commun., 1985, **15**, 775; P.G.Gassman and L.M.Haberman, Tetrahedron Lett., 1985, **26**, 4971; K.Saito and H.Kojima, Bull. Chem. Soc. Jpn., 1985, **58**, 1918; N.Chatani and T.Hanafusa, J. Chem. Soc., Chem. Commun., 1985, 838; N.Tokitoh and R.Okazaki, Chem. Lett., 1985, 241; Y.Yamasaki, T.Maekawa, T.Ishihara, and T.Ando, ibid, 1387; G.S.Zaitseva, O.P.Novikova, A.S.Kostyuk, and Yu.I.Baukov, Zh. Obshch. Khim., 1985, **55**, 942 (Chem. Abs., 1985, **103**, 123576).
24 H.Fischer, R.Markl, and S.Zeuner, J. Organomet. Chem., 1985, **286**, 17.
25 G.Maas and R.Bruckmann, J. Org. Chem., 1985, **50**, 2801; A.Baceiredo, G.Bertrand, and G.Sicard, J. Am. Chem. Soc., 1985, **107**, 4781; K.Nishiyama, H.Mikuni, and M.Harada, Bull. Chem. Soc. Jpn., 1985, **58**, 3381.
26 G.Zinner, W.P.Fehlhammer, and M.Schmidt, Angew. Chem. Int. Ed. Engl., 1985, **24**, 979; Y.Terao, H.Kotaki, N.Imai, and K.Achiwa, Chem. Pharm. Bull. 1985, **33**, 896; A.Padwa, Y.-Y.Chen, W.Dent, and H.Nimmesgern, J. Org. Chem., 1985, **50**, 4006; N.J.Turro, Y.Cha, I.R.Gould, A.Padwa, J.R.Gasdaski, and M.Tomas, ibid, 4415; P.F.Belloir, A.Laurent, P.Mison, R.Bartnik, and S.Lesniak, Tetrahedron Lett., 1985, **26**, 2637.
27 O.Tsuge, S.Kanemasa, and K.Matsuda, Chem. Lett., 1985, 1411; A.Padwa, U.Chiacchio, and M.K.Venkatramanan, J. Chem. Soc., Chem. Commun., 1985, 1108; G.J.Kang and T.H.Chan, Can. J. Chem., 1985, **63**, 3102.
28 H.Nishiyama, M.Matsumoto, T.Matsukura, R.Miura, and K.Itoh, Organometallics, 1985, **4**, 1911; K.Tanaka, H.Yoda, Y.Isobe, and A.Kaji, Tetrahedron Lett., 1985, **26**, 1337.
29 G.A.Olah, H.Daggweiler, J.D.Felberg, and S.Frohlich, J. Org. Chem., 1985, **50**, 4847.
30 S.Kanemasa, J.Tanaka, H.Nagahama, and O.Tsuge, Bull. Chem. Soc. Jpn., 1985, **56**, 3385; A.Duchene, D.Mouka-Mpegna, and J.-P.Quintard, Bull. Soc. Chim. Fr., 1985, 787; G.J.McGarvey and M.Kimura, J. Org. Chem., 1985, **50**, 4655.
31 A.Ricci, A.Degl'Innocenti, S.Chimichi, M.Fiorenza, G.Rossini, and H.J.Bestmann, J. Org. Chem., 1985, **50**, 130.
32 S.Sato, I.Matsuda, and Y.Izumi, Tetrahedron Lett., 1985, **26**, 4229; Y.Kobayashi, H.Uchiyama, H.Kanbara, and F.Sato, J. Am. Chem. Soc., 1985, **107**, 5541; S.R.Wilson, P.A.Zucker, C.-w.Kim, and C.A.Villa, Tetrahedron Lett., 1985, **26**, 1969.
33 L.Birkofer and W.Quittmann, Chem. Ber., 1985, **118**, 2874; J.R.Hwu, J. Chem.

Soc., Chem. Commun., 1985, 452.
34 K.Nakatani and S.Isoe, Tetrahedron Lett., 1985, **26**, 2209; I.Fleming and C.J.Urch, J. Organomet. Chem., 1985, **285**, 173.
35 M.Fiorenza, A.Mordini, S.Papaleo, S.Pastorelli, and A.Ricci, Tetrahedron Lett., 1985, **26**, 787; M.Fiorenza, A.Mordini, and A.Ricci, J. Organomet. Chem., 1985, **280**, 177; J.Enda and I.Kuwajima, J. Am. Chem. Soc., 1985, **107**, 5495.
36 R.K.Boeckman jr. and R.L.Chinn, Tetrahedron Lett., 1985, **26**, 5005; M.Kira, H.Yoshida, and H.Sakurai, J. Am. Chem. Soc., 1985, **107**, 7767; J.-L.Moreau and R.Couffignal, J. Organomet. Chem., 1985, **297**, 1.
37 M.Kosugi, Y.Negishi, M.Kameyama, and T.Migita, Bull. Chem. Soc. Jpn., 1985, **58**, 3383; E.Piers and A.V.Gavan, J. Chem. Soc., Chem. Commun., 1985, 1241; A.N.Kashin, M.L.Tulchinsky, and I.P.Beletskaya, J. Organomet. Chem., 1985, **292**, 205.
38 Y.Tajima, A.Yoshida, N.Takeda, and S.Oida, Tetrahedron Lett., 1985, **26**, 473; D.Terunuma, M.Kato, M.Kamei, H.Uchida, and H.Nohira, Chem. Lett., 1985, 13.
39 E.Wenschuh, W.Radeck, A.Porzel, A.Kolbe, and S.Edelmann, Z. Anorg. Allg. Chem., 1985, **528**, 138; J.J.Eisch, M.Behrooz, and S.K.Dua, J. Organomet. Chem., 1985, **285**, 121.
40 D.Craig, S.V.Ley, N.S.Simkins, G.H.Whitham, and M.J.Prior, J. Chem. Soc., Perkin Trans. 1, 1985, 1949; J.Vollhardt, H.-J.Gais, and K.L.Lukas, Angew. Chem. Int. Ed. Engl., 1985, **24**, 696.
41 T.Mandai, M.Yamaguchi, Y.Nakayama, J.Otera, and M.Kawada, Tetrahedron Lett., 1985, **26**, 2675; H.Ishibashi, H.Nakatani, Y.Umei, and M.Ikeda, ibid, 1985, **26**, 4373; G.A.Krafft and P.T.Meinke, ibid, 1947.
42 T.Yamazaki, K.-i.Takita, and N.Ishikawa, J. Fluorine Chem., 1985, **30**, 357; T.Cohen, J.P.Sherbine, S.A.Mendelson, and M.Myers, Tetrahedron Lett., 1985, **26**, 2965.
43 H.Hornig, E.Walther, and U.Schubert, Organometallics, 1985, **4**, 1905; C.Rosini, C.Bertucci, D.Pini, P.Altemura, and P.Salvadori, Tetrahedron Lett., 1985, **26**, 3361; D.Steinborn and U.Sedlak, Z. Chem., 1985, **25**, 376.
44 R.Appel, U.Kundgen, and F.Knoch, Chem. Ber., 1985, **118**, 1352; M.Tsukamoto, H.Iio, and T.Tokoroyama, Tetrahedron Lett., 1985, **26**, 4471; H.H.Karsch and A.Appelt, J. Chem. Soc., Chem. Commun., 1985, 1083.
45 R.L.Kreeger, P.R.Menaro, E.A.Sans, and H.Shechter, Tetrahedron Lett., 1985, **26**, 1115; R.Damrauer, V.E.Yost, S.E.Danahey, and B.K.O'Connell, Organometallics, 1985, **4**, 1779; M.Fujita and T.Hiyama, J. Am. Chem. Soc., 1985, **107**, 4085.
46 S.V.Sin'ko, Yu.A.Pentin, G.S.Gol'din, I.I.Baburina, G.M.Kuramshina, and E.M.Protasov, Vestn. Mosk. Univ., Ser.2: Khim., 1985, **26**, 131 (Chem. Abs., 1985, **103**, 142066); B.A.Pearlman, S.R.Putt, and J.A.Fleming, J. Org. Chem., 1985, **50**, 3622 and 3625; V.P.Feshin, G.V.Dolgushin, M.G.Voronkov, Ju.E.Sapozhnikov, Ja.B.Jasman, and V.I.Shirjaev, J. Organomet. Chem., 1985, **295**, 15.
47 H.Mauermann, P.N.Swepston, and T.J.Marks, Organometallics, 1985, **4**, 200; R.Wehrmann, H.Meyer, A.Berndt, and K.Dimroth, Angew. Chem. Int. Ed. Engl., 1985, **24**, 788; H.Hope, M.M.Olmstead, P.P.Power, J.Sandell, and X.Xu, J. Am. Chem. Soc., 1985, **107**, 4337; T.Fjeldberg, M.F.Lappert, and A.J.Thorne, J. Mol. Struct., 1985, **127**, 95.
48 L.N.Markovskii, V.D.Romananko, L.S.Kachkovskaya, and M.I.Povolotskii, Zh. Obshch. Khim., 1984, **54**, 2800 (Chem. Abs., 1985, **103**, 6424); D.Gudat, E.Niecke, W.Malisch, U.Hofmockel, S.Quashie, A.H.Cowley, A.M.Arif, B.Krebs, and M.Dartmann, J. Chem. Soc., Chem. Commun., 1985, 1687; A.M.Arif, A.H.Cowley, and S.Quashie, ibid, 428.
49 A.Ricci, M.Fiorenza, A.Degl'Innocenti, G.Seconi, P.Dembech, K.Witzgall, and H.J.Bestmann, Angew. Chem. Int. Ed. Engl., 1985, **24**, 1068; R.Appel, K.H.Dunker, E.Gaitzsch, and T.Gaitzsch, Z. Chem., 1984, **24**, 384; R.Appel, E.Gaitzsch, and F.Knoch, Angew. Chem. Int. Ed. Engl., 1985, **24**, 589; Z.M.Xie, P.Wisian-Neilson, and R.H.Wisian-Neilson, Organometallics, 1985, **4**, 339.
50 R.Appel, T.Gaitzsch, and K.Falk, Angew. Chem. Int. Ed. Engl., 1985, **24**, 419;

A.H.Cowley, N.C.Norman, M.Pakulski, D.L.Bricker, and D.H.Russell, J. Am. Chem. Soc., 1985, **107**, 8211; A.M.Arif, A.H.Cowley, N.C.Norman, and M.Pakulski, ibid, 1062.
51 A.H.Cowley, N.C.Norman, and M.Pakulski, J. Chem. Soc., Dalton Trans., 1985, 383; M.Aslam, R.A.Bartlett, E.Bloch, M.M.Olmstead, P.P.Power, and G.E.Sigel, J. Chem. Soc., Chem. Commun., 1985, 1674; P.B.Hitchcock, H.A.Jasim, R.E.Kelly, and M.F.Lappert, ibid, 1776.
52 F.Sladky, B.Bildstein, and D.Obendorf, J. Organomet. Chem., 1985, **295**, C1; F.Sladky, B.Bildstein, C.Rieker, A.Gieren, H.Betz, and T.Hubner, J. Chem. Soc., Chem. Commun., 1985, 1800.
53 M.Haase and U.Klingebiel, Angew. Chem. Int. Ed. Engl., 1985, **24**, 324; U.Klingebiel, S.Pohlmann, and L.Skoda, J. Organomet. Chem., 1985, **291**, 277.
54 C.Eaborn, P.B.Hitchcock, J.D.Smith, and A.C.Sullivan, J. Chem. Soc., Chem. Commun., 1985, 534; N.H.Buttrus, C.Eaborn, P.B.Hitchcock, J.D.Smith, and A.C.Sullivan, ibid, 1380.
55 Z.H.Aiube, N.H.Buttrus, C.Eaborn, P.B.Hitchcock, and J.A.Zora, J. Organomet. Chem., 1985, **292**, 177; R.I.Damja and C.Eaborn, ibid, 267; N.H.Buttrus, R.I.Damja, C.Eaborn, P.B.Hitchcock, and P.D.Lickiss, J. Chem. Soc., Chem. Commun., 1985, 1385.
56 R.I.Damja, C.Eaborn, and A.K.Saxena, J. Chem. Soc., Perkin Trans. 2, 1985, 597; R.Damrauer, P.Wheelan, H.Razavi, M.Whalen, and L.Aran, Inorg. Chim. Acta, 1985, **99**, 69.
57 R.I.Damja, C.Eaborn, and W.-C.Sham, J. Organomet. Chem., 1985, **291**, 25; H.Sakurai, H.Tshida, and M.Kira, J. Chem. Soc., Chem. Commun., 1985, 1780.
58 C.Eaborn and D.E.Reed, J. Chem. Soc., Perkin Trans. 2, 1985, 1687 and 1695; C.Eaborn, P.D.Lickiss, S.T.Najim, and M.N.Romanelli, J. Chem. Soc., Chem. Commun., 1985, 1754.
59 C.Eaborn and A.I.Mansour, J. Chem. Soc., Perkin Trans. 2, 1985, 729; J.L.Atwood, S.G.Bott, C.Eaborn, M.N.A.El-Kheli, and J.D.Smith, J. Organomet. Chem., 1985, **294**, 23; C.Eaborn and P.D.Lickiss, ibid, 305.
60 S.D.Burke, J.O.Saunders, J.A.Oplinger, and C.W.Murtiashaw, Tetrahedron Lett., 1985, **26**, 1131; K.Karabelas and A.Hallberg, ibid, 3131; P.F.Hudrlik, A.M.Hudrlik, and A.K.Kulkarni, J. Am. Chem. Soc., 1985, **107**, 4260.
61 A.K.Samaddar, T.Chiba, Y.Kobayashi, and F.Sato, J. Chem. Soc., Chem. Commun., 1985, 329; M.Ochiai, K.Sumi, and E.Fujita, Chem. Pharm. Bull. Chem. Res., 1984, **32**, 3686 (Chem. Abs., 1985, **102**, 78955); G.W.Klumpp, A.J.C.Mierop, J.J.Vrielink, A.Brugman, and M.Schakel, J. Am. Chem. Soc., 1985, **107**, 6740.
62 A.Oliva and A.Molinari, Synth. Commun., 1985, **15**, 707; Y.Kawakami, H.Hisada, and Y.Yamashita, Tetrahedron Lett., 1985, **26**, 5835.
63 T.Bartik, P.Heimbach, T.Himmler, and R.Mynott, Angew. Chem. Int. Ed. Engl., 1985, **24**, 313; R.K.Boeckman jr. and T.E.Barta, J. Org. Chem., 1985, **50**, 3421; H.Bock and U.Lechner-Knoblauch, J. Organomet. Chem., 1985, **294**, 295.
64 H.Yasuda, T.Nishi, S.Miyanaga, and A.Nakamura, Organometallics, 1985, **4**, 359; K.J.H.Kruithof, R.F.Schmitz, and G.W.Klumpp, Recl. Trav. Chim. Pays-Bas, 1985, **104**, 3; T.N.Mitchell and W.Reimann, J. Organomet. Chem., 1985, **281**, 163.
65 H.Oda, K.Oshima, and H.Nozaki, Chem. Lett., 1985, 53; A.Zapata, C.Fortoul R., and C.Acuna A., Synth. Commun., 1985, **15**, 179; E.Peirs, R.W.Friesen, and B.A.Keay, J. Chem. Soc., Chem. Commun., 1985, 809.
66 J.G.Duboudin, M.Petraud, M.Ratier, and B.Trouve, J. Organomet. Chem., 1985, **288**, C6; A.Duchene and J.-P.Quintard, Synth. Commun., 1985, **15**, 873.
67 B.M.Trost, D.M.T.Chan, and T.N.Nanninga, Org. Synth.,1984, **62**, 58 (Chem. Abs., 1985, **102**, 204017); J.Klein and L.Weiler, Isr. J. Chem., 1984, **24**, 69 (Chem. Abs., 1985, **102**, 6623); A.Haider, Synthesis, 1985, 271; W.J.Richter and B.Neugebauer, ibid, 1059.
68 J.-I.Yoshida, K.Muraki, H.Funahashi, and N.Kawabata, J. Organomet. Chem., 1985, **284**, C33; T.Shono, Y.Matsumura, S.Katoh, and N.Kise, Chem. Lett., 1985, 463.
69 N.Auner, I.M.T.Davidson, and S.Ijadi-Maghsoodi, Organometallics, 1985, **4**, 2210; J.Dubac, A.Laporterie, H.Iloughmane, J.P.Pillot, G.Deleris, and

J.Dunogues, J. Organomet. Chem., 1985, **281**, 149.
70 E.A.Sans and H.Shechter, Tetrahedron Lett., 1985, **26**, 1119; I.Kuwajima, J. Organomet. Chem., 1985, **285**, 137.
71 J.M.Muchowski, R.Neef, and M.L.Maddox, Tetrahedron Lett., 1985, **26**, 5375; T.H.Chan and K.Koumaglo, J. Organomet. Chem., 1985, **285**, 109.
72 A.Riahi, J.Cossy, J.Muzart, and J.P.Pete, Tetrahedron Lett., 1985, **26**, 839; K.Karabelas, C.Westerlund, and A.Hallberg, J. Org. Chem., 1985, **50**, 3896; T.Ohta, T.Hosokawa, S.-I.Murahashi, K.Miki, and N.Kasai, Organometallics, 1985, **4**, 2080.
73 J.-C.Gramain and R.Remuson, Tetrahedron Lett., 1985, **26**, 327; H.Hiemstra, H.P.Fortgens, and W.N.Speckamp, ibid, 3155; G.Majetich, J.Defauw, K.Hull, and T.Shawe, ibid, 4711; S.Niwayama, S.Dan, Y.Inouye, and H.Kakisawa, Chem. Lett., 1985, 957; A.Hosomi, H.Shoji, and H.Sakurai, ibid, 1049.
74 H.-F.Chow and I.Fleming, Tetrahedron Lett., 1985, **26**, 397; J.-C.Gramain and R.Remuson, Tetrahedron Lett., 1985, **26**, 4083; H.Hiemstra, M.H.A.M.Sno, R.J.Vijn, and W.N.Speckamp, J. Org. Chem., 1985, **50**, 4014; B.-W.Au-Yeung and Y.Wang, J. Chem. Soc., Chem. Commun., 1985, 825.
75 R.Imwinkelried and D.Seebach, Angew. Chem. Int. Ed. Engl., 1985, **24**, 765; S.A.Carr and W.P.Weber, J. Org. Chem., 1985, **50**, 2783.
76 D.Schinzer, S.Solyom, and M.Becker, Tetrahedron Lett., 1985, **26**, 1831; M.Wada, T.Shigehisa, K.-y.Akiba, ibid, 5191; G.Majetich, K.Hull, J.Defauw, and R.Desmond, ibid, 2747; G.Majetich, K.Hull, and R.Desmond, ibid, 2751.
77 G.Majetich, K.Hull, J.Defauw, and T.Shawe, Tetrahedron Lett., 1985, **26**, 2755; H.Oda, M.Sato, Y.Morizawa, K.Oshima, and H.Nozaki, Tetrahedron, 1985, **41**, 3257.
78 K.Mizuno, M.Ikeda, and Y.Otsuji, Tetrahedron Lett., 1985, **26**, 461; H.Hamana and T.Sugasawa, Chem. Lett., 1985, 921.
79 E.Lukevics, V.V.Dirnens, Yu.S.Goldberg, E.E.Liepinsh, M.P.Gavars, I.Y.Kalvinsh, and M.V.Shymanska, Organometallics, 1985, **4**, 1648; M.D.Jones and R.D.W.Kemmitt, J. Chem. Soc., Chem. Commun., 1985, 811.
80 D.Gampe, K.Jacob, and K.H.Theile, Z. Chem., 1985, **25**, 151; K.N.Swamy and W.L.Hase, J. Chem. Phys., 1985, **82**, 123; P.Ganis, D.Furlani, D.Marton, G.Taglivini, and G.Valle, J. Organomet. Chem., 1985, **293**, 207; J.Otera and T.Yano, Bull. Chem. Soc. Jpn., 1985, **58**, 387.
81 A.Boaretto, D.Marton, G.Taglivini, and A.Gambaro, J. Organomet. Chem., 1985, **286**, 9; J.Otero, Y.Yoshinaga, T.Yamaji, T.Yoshioka, and Y.Kawasaki, Organometallics, 1985, **4**, 1212; M.Shimagaki, H.Takubo, and T.Oishi, Tetrahedron Lett., 1985, **26**, 6235; G.E.Keck and E.J.Enholm, J. Org. Chem., 1985, **50**, 147; G.E.Keck, E.J.Enholm, J.B.Yates, and M.R.Wiley, Tetrahedron, 1985, **41**, 4079.
82 R.Yamaguchi, M.Moriyasu, M.Yoshioka, and M.Kawanisi, J. Org. Chem., 1985, **50**, 287; G.E.Keck and E.J.Enholm, Tetrahedron Lett., 1985, **26**, 3311; J.H.Simpson and J.K.Stille, J. Org. Chem., 1985, **50**, 1759; A.Hosomi, T.Imai, M.Endo, and H.Sakurai, J. Organomet. Chem., 1985, **285**, 95.
83 J.-P.Quintard, B.Elissondo, T.Hattich, and M.Pereyre, J. Organomet. Chem., 1985, **285**, 149; M.M.Midland and Y.C.Kwon, Tetrahedron Lett., 1985, **26**, 5013, 5017, and 5021; M.S.Raizada and S.N.Bhattacharya, Indian J. Chem., Sect. A, 1984, **23**, 952 (Chem. Abs., 1985, **102**, 185190).
84 A.Boaretto, D.Marton, and G.Tagliavini, J. Organomet. Chem., 1985, **297**, 149; M.Ochiai, K.Sumi, Y.Nagao, E.Fujita, M.Arimoto, and H.Yamaguchi, J. Chem. Soc., Chem. Commun., 1985, 697.
85 R.L.Danheiser and D.M.Fink, Tetrahedron Lett., 1985, **26**, 2509 and 2513; R.L.Danheiser, C.A.Kwasigroh, and Y.-M.Tsai, J. Am. Chem. Soc., 1985, **107**, 7233; J.D.Buynak, M.N.Rao, R.Y.Chandrasekaran, and E.Haley, Tetrahedron Lett., 1985, **26**, 5001.
86 B.Bennetau, D.Youhouvoulou N'Gabe, and J.Dunogues, Tetrahedron Lett., 1985, **26**, 3813.
87 T.Mandai, H.Arase, J.Otera, and M.Kawada, Tetrahedron Lett., 1985, **26**, 2677; J.Ponet, B.Khouz, and L.Miginiac, ibid, 1861; A.Hosomi, Y.Sakata, and H.Sakurai, ibid, 5175; D.Hoppe, R.Hanko, A.Bronneke, F.Lichtenberg, and E. van Hulsen, Chem. Ber., 1985, **118**, 2822.

88 K.H.Dotz, J.Muhlemeier, and B.Trenkle, J. Organomet. Chem., 1985, **289**, 257; S.-F.Chen and P.S.Mariano, Tetrahedron Lett., 1985, **26**, 47.
89 R.H.A.Bosma, G.C.N. van den Aardweg, and J.C.Mol, J. Organomet. Chem., 1985, **280**, 115; M.Berglund, C.Andersson, and R.Larsson, ibid, **292**, C15; X.Xiaoding and J.C.Mol, J. Chem. Soc., Chem. Commun., 1985, 631; V.M.Vdovin, N.V.Ushakov, E.B.Portnykh, E.Sh.Finkel'shtein, and N.P.Abashkina, Izv. Akad. Nauk SSSR, Ser. Khim., 1984, 2121 and 1892 (Chem. Abs., 1985, **102**, 113569 and 6709).
90 G.Zweifel and S.Rajagopalan, J. Am. Chem. Soc., 1985, **107**, 700; J.Drouin and G.Rousseau, J. Organomet. Chem., 1985, **289**, 223; M.Ochiai, M.Kunishima, K.Sumi, Y.Nagao, E.Fujita, M.Aromoto, and H.Yamaguchi, Tetrahedron Lett., 1985, **26**, 4501; J.Pornet, L.Miginiac, K.Jaworski, and B.Randrianoclina, Organometallics, 1985, **4**, 333; H.Hiemstra, H.P.Fortgens, S.Stegenga, and W.N.Speckamp, Tetrahedron Lett., 1985, **26**, 3151; S.I.Pennanen, Finn. Chem. Lett., 1984, 95 (Chem. Abs., 1985, **103**, 6396).
91 N.J.Fitzmaurice, W.R.Jackson, and P.Perlmutter, J. Organomet. Chem., 1985, **285**, 375; K.P.C.Vollhardt and L.S.Winn, ibid, 709; T.Kusemoto, K.Nishide, and T.Hiyama, Chem. Lett., 1985, 1409; J.A.Miller and E.-i.Negishi, Isr. J. Chem., 1984, **24**, 76 (Chem. Abs., 1985, **102**, 78942).
92 E.-i.Negishi, S.J.Holmes, J.M.Tour, and J.A.Miller, J. Am. Chem. Soc., 1985, **107**, 2568; P.Binger, Q.-H.Lii, and P.Wedemann, Angew. Chem. Int. Ed. Engl., 1985, **24**, 316.
93 S.Chatterjee and E.-i.Negishi, J. Organomet. Chem., 1985, **285**, C1; R.L.Halterman, N.H.Nguyen, and K.P.C.Vollhardt, J. Am. Chem. Soc., 1985, **107**, 1379.
94 P.Magnus and D.P.Becker, J. Chem. Soc., Chem. Commun., 1985, 640; A.El-Agamey, G.Abdel, and L.P.Asatiani, Rev. Roum. Chim., 1984, **29**, 583 (Chem. Abs., 1985, **102**, 113580).
95 M.Ishikawa, S.Matsuzawa, T.Higuchi, S.Kamitori, and K.Hirotsu, Organometallics, 1985, **4**, 2040; R.Appel and C.Casser, Chem. Ber., 1985, **118**, 3419.
96 Y.Okamoto, H.S.Lee, and S.T.Attarwala, J. Org. Chem., 1985, **50**, 2788; T.J.Barton and B.L.Groh, ibid, 159; S.Eguchi, T.Ikemoto, Y.Kobayakawa, and T.Sasaki, J. Chem. Soc., Chem. Commun., 1985, 958.
97 E.Kloster-Jensen and G.Aamdal Eliassen, Angew. Chem. Int. Ed. Engl., 1985, **24**, 565; R,Gleiter, W.Schafer, and H.Sakurai, J. Am. Chem. Soc., 1985, **107**, 3046.
98 N.S.Hosmane, N.N.Sirmokadam, and M.N.Mollenhauer, J. Organomet. Chem., 1985, **279**, 359; N.S.Hosmane, M.N.Mollenhauer, A.H.Cowley, and N.C.Norman, Organometallics, 1985, **4**, 1194.
99 S.S.Nikam and K.K.Wang, J. Org. Chem., 1985, **50**, 2193; J.L.Briguglio and L.G.Sneddon, Organometallics, 1985, **4**, 721.
100 G.Himbert and W.Brunn, Liebigs Ann. Chem., 1985, 2206; S.Kerschl and B.Wrackmeyer, Z. Naturforsch., Teil B, 1985, **40**, 845.
101 C.Stader and B.Wrackmeyer, J. Organomet. Chem., 1985, **295**, C11; C.Cauletti, C.Furlani, G.Granozzi, A.Sebald, and B.Wrackmeyer, Organometallics, 1985, **4**, 290; A.Carson, P.G.Laye, and J.A.Spencer, J. Chem. Thermodyn., 1985, **17**, 277; A.V.Belyakov, E.T.Bogoradovskii, V.S.Zavgorodnii, V.S.Nikitin, M.V.Polyakova, and I.I.Baburina, Spectrochim. Acta, Part A, 1985, **41**, 1269.
102 S.Kanemasa, J.Tanaka, H.Nagahama, and O.Tsuge, Chem. Lett., 1985, 1223; H.G.Ang and P.T.Lau, J. Organomet. Chem., 1985, **291**, 285; M.Gielen and M.Vermeulen, Bull. Soc. Chim. Belg., 1985, **94**, 19.
103 J.E.Frey, R.D.Cole, E.C.Kitchen, L.M.Suprenant, and M.S.Sylwestrzak, J. Am. Chem. Soc., 1985, **107**, 748; T.Veszpremi and J.Nagy, Kem. Kozl., 1984, **61**, 223 (Chem. Abs., 1985, **103**, 87954); J.A.Albanese, D.G.Kreider, C.D.Schaffer jr., C.H.Yoder, and M.S.Samples, J. Org. Chem., 1985, **50**, 2059.
104 M.Speranza, C.-Y.Shille, A.P.Wolf, D.S.Wilbur, and G.Angelini, J. Fluorine Chem., 1985, **30**, 97; T.I.Odiaka, J. Chem. Soc., Dalton Trans., 1985, 1049; F.Effenberger and W.Spiegler, Chem. Ber., 1985, **118**, 3872 and 3900.
105 H.J.Reich, N.H.Phillips, and I.L.Reich, J. Am. Chem. Soc., 1985, **107**, 4101; Y.Himeshima, H.Kobayashi, and T.Sonoda, ibid, 5286; T.-a.Kobayashi,

T.Sakakura, and M.Tanaka, Tetrahedron Lett., 1985, **26**, 3463.
106 R.P.Kozyrod, J.Morgan, and J.T.Pinhey, Aust. J. Chem., 1985, **38**, 1147, 713, and 1155.
107 K.Takahashi, K.Ohnishi, and K.Takase, Chem. Lett., 1985, 1079; R. Neidlein, W.Wirth, A.Gieren, V.Lamm, and T.Hubner, Angew. Chem. Int. Ed. Engl., 1985, **24**, 587.
108 H.E.Helson, K.P.C.Vollhardt, and Z.-Y.Yang, Angew. Chem. Int. Ed. Engl., 1985, **24**, 114; L.M.Engelhardt, E.-P.Leung, C.L.Raston, and A.H.White, J. Chem. Soc., Dalton Trans., 1985, 337.
109 H.Sakurai, K.Sakamoto, A.Nakamura, and M.Kira, Chem. Lett., 1985, 497; M.Daney, C.Vanucci, J.-P.Desvergne, A.Castellan, and H.Boues-Laurent, Tetrahedron Lett., 1985, **26**, 1505.
110 L.R.Robinson, G.T.Burns, and T.J.Barton, J. Am. Chem. Soc., 1985, **107**, 3935; R.F.Cunico and C.-P.Kuan, J. Org. Chem., 1985, **50**, 5410; J.-M.Wulff and H.M.R.Hoffmann, Angew. Chem. Int. Ed. Engl., 1985, **24**, 605.
111 E.Sternberg and P.Binger, Tetrahedron Lett., 1985, **26**, 301; M.Grignon-Dubois and M.Ahra, J. Organomet. Chem., 1985, **280**, 313; J.R.Durig, T.J.Geyer, T.S.Little, and M.Dakkouri, J. Phys. Chem., 1985, **89**, 4307.
112 H.Nishiyama, H.Arai, T.Ohki, and K.Itoh, J. Am. Chem. Soc., 1985, **107**, 5310; M.Ochiai, T.Uchita, Y.Nagao, and E.Fujita, J. Chem. Soc., Chem. Commun., 1985, 637.
113 T.J.Barton and G.P.Hussmann, J. Org. Chem., 1985, **50**, 5881; R.T.Conlin and Y.-W.Kwak, J. Organomet. Chem., 1985, **293**, 177.
114 A.Svensson and A.R.Martin, Heterocycles, 1985, **23**, 357; W.R.Baker, J. Org. Chem., 1985, **50**, 3942.
115 A.Dondoni, M.Fogagnola, A.Medici, and P.Pedrini, Tetrahedron Lett., 1985, **26**, 5477; S.Katsumara, K.Hori, S.Fujiwara, and S.Isoe, ibid, 4625; N.Shimizu, F.Shibata, and Y.Tsuno, Chem. Lett., 1985, 1593; P.F.Hudrlik, A.M.Hudrlik, and A.K.Kulkarni, Tetrahedron Lett., 1985, **26**, 139; G.Nagendrappa and T.J.Vidyapati, J. Organomet. Chem., 1985, **280**, 31.
116 Y.Tomo and K.Yamamoto, Tetrahedron Lett., 1985, **26**, 1061; E.Lukevics, V.N.Gevorgyan, Y.S.Goldberg, and M.V.Shymanska, J. Organomet. Chem., 1985, **294**, 163; P.Dembech, C.Eaborn, and G.Seconi, J. Chem. Soc., Chem. Commun., 1985, 1289; J.Chmielecka, J.Chojnowski, C.Eaborn, and W.A.Stanczyk, J. Chem. Soc., Perkin Trans. 2, 1985, 1779.
117 P.Jutzi and W.Leffers, J. Chem. Soc., Chem. Commun., 1985, 1735; P.Jutzi, E.Schluter, S.Pohl, and W.Saak, Chem. Ber., 1985, **118**, 1959.
118 N.Klouras, S.Voliotis, and G.Germain, Acta Crystallogr., 1984, **C40**, 1791; C.Bajgur, W.Tikkanen, and J.L.Petersen, Inorg. Chem., 1985, **24**, 2539; H.Werner, H.J.Scholz, and R.Zolk, Chem. Ber., 1985, **118**, 4531.
119 P.Jutzi, B.Hampel, K.Stroppel, C.Kruger, K.Angermund, and P.Hofmann, Chem. Ber., 1985, **118**, 2789; J.M.Rozell jr. and P.R.Jones, Organometallics, 1985, **4**, 2206.
120 C.Glidewell, J. Organomet. Chem., 1985, **286**, 289; P.Jutzi, D.Kanne, and C.Kruger, Nachr. Chem. Tech. Lab., 1985, **33**, 1049; G.Bruno, E.Ciliberto, I.L.Fragala, and P.Jutzi, J. Organomet. Chem., 1985, **289**, 263.
121 H.Schumann, C.Janiak, E.Hahn, J.Loebel, and J.J.Zuckerman, Angew. Chem. Int. Ed. Engl., 1985, **24**, 773; G.Schmid, D.Zaika, and R.Boese, ibid, 602.
122 R.Hani and R.A.Geanangel, J. Organomet. Chem., 1985, **293**, 197; T.S.Dory and J.J.Zuckerman, ibid, **281**, C1; T.S.Dory, J.J.Zuckerman, and M.D.Rausch, ibid, **281**, C8.
123 R.Hani and R.A.Geanangel, Inorg. Chim. Acta, 1985, **96**, 225; H.H.Karsch, A.Appelt, and G.Muller, Angew. Chem. Int. Ed. Engl., 1985, **24**, 402.
124 M.J.S.Dewar, J.E.Freidheim, and G.L.Grady, Organometallics, 1985, **4**, 1784; T.Fjeldberg, A.Haaland, B.E.R.Schilling, H.V.Volden, M.F.Lappert, and A.J.Thorne, J. Organomet. Chem., 1985, **280**, C43.
125 J.Kocher and W.P.Neumann, Organometallics, 1985, **4**, 400; D.Lei and P.P.Gaspar, ibid, 1471; H.Vancik, G.Raabe, M.J.Michalczyk, R.West, and J.Michl, J. Am. Chem. Soc., 1985, **107**, 4097.
126 D.Lei and P.P.Gaspar, J. Chem. Soc., Chem. Commun., 1985, 1149; Y.Apeloig and M.Karni, ibid, 1048; I.Safarnik, B.P.Ruzsicska, A.Jodhan, O.P.Strausz,

and T.N.Bell, Chem. Phys. Lett., 1985, 113, 71.
127 J.Chandrasekhar and P.von Rague Schleyer, J. Organomet. Chem., 1985, 289, 51; G.Maier, K.Schottler, and H.P.Reisenauer, Tetrahedron Lett., 1985, 26, 4079.
128 K.-T.Kang, H.-Y.Song, and H.-C.Seo, Chem. Lett., 1985, 617; K.-T.Kang, H.-C.Seo, and K.-N.Kim, Tetrahedron Lett., 1985, 26, 4761.
129 H.Schnockel, Z.Lin,N.Auer, P.Bleckmann, and M.Hinrichsen, J. Mol. Struct., 1985, 127, 1; J.Laane, ibid, 126, 99; C.A.George and R.D.Koob, J. Phys. Chem., 1985, 89, 5086.
130 A.I.Mikaya, V.G.Zaikin, N.V.Ushakov, and V.M.Vdovin, J. Organomet. Chem., 1985, 284, 5; D.Seyferth, M.L.Shannon, S.C.Vick, and T.F.O.Lim, Organometallics, 1985, 4, 57; M.P.Egorov, S.P.Kolesnikov, Yu.T.Struchkov, M.Yu.Antipin, S.V.Sereda, and O.M.Nefedov, J. Organomet. Chem., 1985, 290, C27.
131 W.J.Richter, J. Organomet. Chem., 1985, 289, 45; J.Dubac, A.Laporterie, and H.Iloughmane, ibid, 193, 295; J.Dubac, H.Iloughmane; A.Laporterie, and C.Roques, Tetrahedron Lett., 1985, 26, 1315; C.Guimon, G.Pfister-Guillouzo, J.Dubac, A.Laporterie, G.Manuel, and H.Iloughmane, Organometallics, 1985, 4, 636.
132 J.Corey, C.Guerin, B.Henner, B.Kolani, W.W.Choy, W.C.Man, and R.Corriu, C.R. Acad. Sci., Ser.2 1985, 300, 331; T.J.Barton and B.L.Groh, Organometallics, 1985, 4, 575.
133 A.Sekiguchi, H.Tanikaw, and W.Ando, Organometallics, 1985, 4, 584; P.R.Jones, J.M.Rozell jr., and B.M.Campbell, ibid, 1321.
134 J.W.Bruin, G.Schat, O.S.Akkerman, and F.Bickelhaupt, J. Organomet. Chem., 1985, 288, 13; K.Jurkschat and M.Gielen, Bull. Soc. Chim. Belg., 1985, 94,299; J.W.F.L.Seetz, B.J.J.van de Heisteeg, G.Schat, O.S.Akkerman, and F.Bickelhaupt, J. Mol. Catal., 1985, 28, 71.
135 G.Maier, J.Henkelmann, and H.P.Reisenauer, Angew. Chem. Int. Ed. Engl., 1985, 24, 1065; Y.Inoguchi, S.Okui, K.Mochida, and A.Itai, Bull. Chem. Soc. Jpn., 1985, 58, 974.
136 H.Appler, L.W.Gross, B.Mayer, and W.P.Neumann, J. Organomet. Chem., 1985, 291, 9; R.W.Hoffmann, A.Riemann, and B.Mayer, Chem. Ber., 1985, 118, 2493; H.-D.Martin, B.Mayer, R.W.Hoffmann, A.Riemann, and P.Rademacher, ibid, 2514.
137 A.Rahm, J.Grimeau, M.Petraud, and B.Barbe, J. Organomet. Chem., 1985, 286, 297; A.Rahm, J.Grimeau, and M.Pereyre, ibid, 286, 305.
138 P.Boudjouk, R.Sooriyakumaran, and C.A.Kapfer, J. Organomet. Chem., 1985, 281, C21; P.Boudjouk and C.A.Kapfer, ibid, 296, 339.
139 Y.Apeloig and A.Stanger, J. Am. Chem. Soc., 1985, 107, 2806; C.S.Frampton, R.M.G.Roberts, and J.Silver, ibid, 297, 273; C.S.Frampton, R.M.G.Roberts, J.Silver, J.F.Warmsley, and B.Yavari, J. Chem. Soc., Dalton Trans., 1985, 169.
140 R.Munstedt and U.Wannagat, Liebigs Ann. Chem., 1985, 944; U.Wannagat, R.Munstedt, and U.Harder, ibid, 950.
141 R.Munstedt and U.Wannagat, Monatsh. Chem., 1985, 116, 693 and 7.
142 P.J.Craig and S.Rapsomanikis, Inorg. Chim. Acta, 1985, 107, 39; H.Kitamura, A.Sugimae, and M.Nakamoto, Bull. Chem. Soc. Jpn., 1985, 58, 2641.
143 M.R.Bryce, R.D.Chambers, S.T.Mullins, and A.Parkin, J. Fluorine Chem., 1985, 26, 533; G.J.Chen and C.Tamborski, J. Organomet. Chem., 1985, 293, 313; H.Burger, M.Grunwald, and G.Pawelke, J. Fluorine Chem., 1985, 28, 183.
144 M.Weidenbruch, K.Kramer, K.Peters, and H.G.von Schnering, Z. Naturforsch., Teil B, 1985, 40, 601; M.Weidenbruch, K.Kramer, A.Schafer, and J.K.Blum, Chem. Ber., 1985, 118, 107; M.Michalczyk, R.West, and J.Michl, Organometallics, 1985, 4, 826.
145 S.Collins, S.Murakami, J.T.Snow, and S.Masamune, Tetrahedron Lett., 1985, 26, 1281; S.Masamune and L.R.Sita, J. Am. Chem. Soc., 1985, 107, 6390.
146 A.Schafer, M.Weidenbruch, and S.Pohl, J. Organomet. Chem., 1985, 282, 305; H.Watanabe, Y.Kougo, M.Kato, H.Kuwabara, T.Okawa, and Y.Nagai, Bull. Chem. Soc. Jpn., 1984, 59, 3019; M.Ishikawa and S.Matsuzuwa, J. Chem. Soc., Chem. Commun., 1985, 588; M.Ishikawa, S.Matsuzawa, H.Sugisawa, F.Yano, S.Kamitori, and T.Hugichi, J. Am. Chem. Soc., 1985, 107, 7706.

Group IV: The Silicon Group

147 K.Krogh-Jespersen, J. Am. Chem. Soc., 1985, **107**, 537; S.Nagase, T.Kudo, and M.Aoki, J. Chem. Soc., Chem. Commun., 1985, 1121; C.Glidewell, Inorg. Chim. Acta, 1985, **97**, 173; M.Kira, H.Nakazawa, and H.Sakurai, Chem. Lett., 1985, 1585.
148 H.Hoffmeyer, P.Potzinger, and B.Reimann, J. Phys. Chem., 1985, **89**, 4829; G.A.Razuvaev, V.V.Semenov, T.N.Brevnova, and A.N.Kornev, J. Organomet. Chem., 1985, **287**, C31.
149 H.Schmolzer and E.Hengge, Monatsh. Chem., 1984, **115**, 1125; K.Hassler, Spectrochim. Acta, Part A, 1985, **41**, 729.
150 A.G.Brook and H.-J.Wessely, Organometallics, 1985, **4**, 1487; U.Schubert, C.Steib, and S.Amberg, Chem. Ber., 1985, **118**, 4774.
151 G.Becker, H.-M.Hartmann, A.Munch, and H.Riffel, Z. Anorg. Allg. Chem., 1985, **530**, 29; U.Schubert, A.Schenkel, and J.Muller, J. Organomet. Chem., 1985, **292**, C11.
152 S.-M.Chen, A.Katti, T.A.Blinka, and R.West, Synthesis, 1985, 684; Y.Nakadaira, N.Komatsu, and H.Sakurai, Chem. Lett., 1985, 1781; R.Nakao, K.Oka, T.Dohmaru, Y.Nagata, and T.Fukumoto, J. Chem. Soc., Chem. Commun., 1985, 766.
153 F.Shafiee, J.R.Damewood jr., K.J.Haller, and R.West, J. Am. Chem. Soc., 1985, **107**, 6950; P.Trefonas III, J.R.Damewood jr., and R.West, Organometallics, 1985, **4**, 1318; H.Watanabe, K.Yoshizumi, T.Muraoka, M.Kato, Y.Nagai, and T.Sato, Chem. Lett., 1985, 1683.
154 C.L.Wadsworth, R.West, Y.Nagai, H.Watanabe, and T.Muraoka, Organometallics, 1985, **4**, 1659; C.L.Wadsworth and R.West, ibid, 1664.
155 H.Matsumoto, M.Minemura, K.Takatsuna, Y.Nagai, and M.Goto, Chem. Lett., 1985, 1005 and J. Chem. Soc., Chem. Commun., 1985, 1366; B.Kirste, R.West, H.Kurreck, J. Am. Chem. Soc., 1985, **107**, 3013.
156 S.Masumune, Y.Kabe, S.Collins, D.J.Williams, and R.Jones, J. Am. Chem. Soc., 1985, **107**, 5552; S.Collins, J.A.Duncan, Y.Kabe, S.Murakami, and S.Masamune, Tetrahedron Lett., 1985, **26**, 2837.
157 M.Ishikawa, T.Yamanaka, and M.Kumada, J. Organomet. Chem., 1985, **292**, 167; E.A.Halevi, G.Winkelhofer, M.Meisl, and R.Janoschek, ibid, **294**, 151.
158 Y.Nakadaira, H.Tobita, and H.Sakurai, New Front. Organomet. Inorg. Chem., Proc. China-Jap.-U.S.A. Trilateral Semin., 2nd 1982 (Pub. 1984) Ed. B.-K.Teo p.141; S.P.Foster, K.M.Mackay, and B.K.Nicholson, Inorg. Chem., 1985, **24**, 909; P.Riviere, A.Castel, D.Guyot, and J.Satge, J. Organomet. Chem., 1985, **290**, C15.
159 B.M.Trost and K.M.Pietrusiewicz, Tetrahedron Lett., 1985, **26**, 4039; J.-B.Verlhae, E.Chanson, B.Jousseaume, and J.-P.Quintard, ibid, 6075; N.A.Bumagin, Yu.V.Gulevich, and I.P.Beletskaya, J. Organomet. Chem., 1985, **282**, 421.
160 N.A.Bumagin, Yu.V.Gulevich, and I.P.Beletskaya, Dokl. Akad. Nauk SSSR, 1985, **280**, 633 (Chem. Abs., 1985, **103**, 71426); M.Kosugi, M.Koshiba, H.Sano, and T.Migita, Bull. Chem. Soc. Jpn., 1985, **58**, 1075.
161 U.Wuthier, H.V.Pham, E.Pretsch, D.Ammann, A.K.Beck, D.Seebach, and W.Simon, Helv. Chim. Acta, 1985, **68**, 1822; C.Glidewell, J. Organomet. Chem., 1985, **294**, 173.
162 M.Devaud, M.Engel, C.Feasson, and J.-L.Lecat, J. Organomet. Chem., 1985, **281**, 181; B.Jousseaume, E.Chanson, M.Bevilacqua, A.Saux, M.Pereyre, B.Barbe, and M.Petraud, ibid, **294**, C41.
163 H.Preut, P.Bleckmann, T.Mitchell, and B.Fabisch, Acta Crystallogr., 1984, **C40**, 370; S.Adams and M.Drager, J. Organomet. Chem., 1985, **288**, 295; B.Watta, W.P.Neumann, and J.Sauer, Organometallics, 1985, **4**, 1984.
164 K.Mochida, A.Hani, M.Yokoyama, T.Tsuchiya, S.D.Worley, and J.K.Kochi, Bull. Chem. Soc. Jpn., 1985, **58**, 2149; K.Mochida, S.D.Worley, and J.K.Kochi, ibid, 3389; M.J.Cuthbertson and P.R.Wells, J. Organomet. Chem., 1985, **287**, 25; M.J.Cuthbertson, D.W.Hawker, and P.R.Wells, ibid, 7.
165 T.N.Mitchell, H.Killing, R.Dicke, and R.Wickenkamp, J. Chem. Soc., Chem. Commun., 1985, 354; B.L.Chenard, E.D.Laganis, F.Davidson, and T.V.RajanBabu, J. Org. Chem., 1985, **50**, 3666; N.Kleiner and M.Drager, Z. Naturforsch., Teil B, 1985, **40**, 477 and J. Organomet. Chem., 1985, **293**, 323.

166 M.Weidenbruch and K.Kramer, J. Organomet. Chem., 1985, 291, 159; H.Beckers, H.Burger, and R.Eujen, J. Fluorine Chem., 1985, 27, 461; V.N.Gevorgyan, L.M.Ignatovich, and E.Lukevics, J. Organomet. Chem., 1985, 284, C31.
167 M.A.Ring, H.E.O'Neal, J.W.Erwin, and D.S.Rogers, Mater. Res. Soc. Symp. Proc. 1984, 32, 383 (Chem. Abs., 1985, 102, 6630); S.F.Rickborn, M.A.Ring, and H.E.O'Neal, Int. J. Chem. Kinet. 1984 16, 1371; K.Ohta, E.R.Davidson, and K.Morokuma, J. Am. Chem. Soc., 1985, 107, 3466; M.S.Gordon and P.Boudjouk, ibid, 1439.
168 J.-i.Kato, N.Iwasawa, and T.Mukaiyama, Chem. Lett., 1985, 743; C.S.Wilcox and R.E.Babston, Synthesis, 1985, 941.
169 M.Fujita and T.Hiyama, J. Am. Chem. Soc., 1985, 107, 8294; H.Brunner and R.Becker, Angew. Chem. Int. Ed. Engl., 1985, 24, 703; E.Yoshii and K.Takeda, Chem. Pharm. Bull. 1983, 31, 4586 (Chem. Abs., 1984, 100, 190950).
170 K.Oertle and H.Wetter, Tetrahedron Lett., 1985, 26, 5511; R.Tacke, M.Link, A.Bentlage-Felten, and H.Zilch, Z. Naturforsch., Teil B, 1985, 40, 942.
171 R.Becker, H.Brunner, S.Mohboobi, and W.Wiegrebe, Angew. Chem. Int. Ed. Engl., 1985, 24, 995; E.Lukevics, R.Ya.Sturkovich, and O.A.Pudova, J. Organomet. Chem., 1985, 292, 151; T.J.Barton and B.L.Groh, J. Am. Chem. Soc., 1985, 107, 8297.
172 C.Aitken, J.F.Harrod, and E.Samuel, J. Organomet. Chem., 1985, 279, C11; M.D.Curtis, L.G.Bell, and W.M.Butler, Organometallics, 1985, 4, 701; G.Kraft, C.Kalbas, and U.Schubert, J. Organomet. Chem., 1985, 289, 247.
173 T.Murai, T.Sakane, and S.Kato, Tetrahedron Lett., 1985, 26, 5145; K.-T.Kang and W.P.Weber, ibid, 5415 and 5753; T.Murai, S.Kato, S.Murai, Y.Hatayama, and N.Sonoda, ibid, 2683.
174 H.-J.Liu and B.Ramani, Synth. Commun., 1985, 15, 965; T.Kusumoto and T.Hiyama, Chem. Lett., 1985, 1405.
175 M.J.Auburn and S.R.Stobart, Inorg. Chem., 1985, 24, 318; A.J.Kunin, R.Farid, C.E.Johnson, and R.Eisenberg, J. Am. Chem. Soc., 1985, 107, 5315; C.E.Johnson and R.Eisenberg, ibid, 6531.
176 E.Keinan and N.Greenspoon, Tetrahedron Lett., 1985, 26, 1353; T.Hayashi and K.Kabeta, ibid, 3023; B.Marciniec, E.Mackowska, J.Gulinska, and W.Urbaniak, Z. Anorg. Allg. Chem., 1985, 529, 222; F.R.Hartley, M.D.Hodge, S.G.Murray, and A.R.Bassindale, J. Mol. Catal., 1984, 26, 313.
177 P.J.Ssebuwufu, F.Glockling, and P.Harriot, Inorg. Chim. Acta, 1985, 98, L35.
178 N.Kakimoto, Synthesis, 1985, 272; A.B.Chopa, L.C.Koll, M.C.Savini, J.C.Podesta, and W.P.Neumann, Organometallics, 1985, 4, 1036.
179 D.F.Gaines, J.C.Kunz, and M.J.Kulzick, Inorg. Chem., 1985, 24, 3336; S.L.Emery and J.A.Morrison, ibid, 1612; T.N.Mitchell, W.Reimann, and C.Nettelbeck, Organometallics, 1985, 4, 1044.
180 M.Deguiel-Castaing, B.Maillard, and A.Rahm, J. Organomet. Chem., 1985, 287, 49; J.W.Wilt, Tetrahedron, 1985, 41, 3979; G.L.Larson and E.Torres, J. Organomet. Chem., 1985, 293, 19.
181 R.J.Klingler, I.Bloom, and J.W.Rathke, Organometallics, 1985, 4, 1893; C.J.Cardin, D.J.Cardin, J.M.Power, and M.B.Hursthouse, J. Am. Chem. Soc., 1985, 107, 505.
182 R.A.Bernheim, F.W.Lampe, and J.F.O'Keefe, J. Phys. Chem., 1985, 89, 1087; D.C.McKean, A.R.Morrisson, I.Torto, and M.I.Kelly, Spectrochim. Acta, Part A, 1985, 41, 25; V.F.Kalasinsky, S.E.Rodgers, and J.A.S.Smith, ibid, 155; D.C.McKean, A.R.Morrison, and P.W.Clark, ibid, 1467; J.R.Durig, C.M.Whang, G.M.Attia, and Y.S.Li, J. Mol. Spectrosc., 1984, 108, 240.
183 J.A.Rice, J.J.Treacy, and H.W.Sidebottom, Int. J. Chem. Kinet., 1984, 16, 1505; A.Horowitz, J. Am. Chem. Soc., 1985, 107, 318; A.Hudson, R.A.Jackson, C.J.Rhodes, and A.L. Del Vecchio, J. Organomet. Chem., 1985, 280, 173.
184 A.Alberti, A.Degl'Innocenti, G.F.Pedulli, and A.Ricci, J. Am. Chem. Soc., 1985, 107, 2316; K.Mochida, K.Ichikawa, S.Okui, Y.Sakaguchi, and H.Hayashi, Chem. Lett., 1985, 1433; K.Mochida, M.Wakasa, S.-i.Ishizaka, M.Kotani, Y.Sakaguchi, and H.Hayashi, ibid, 1709; M.Kobayashi, M.Kobayashi, and M.Yoshida, Bull. Chem. Soc. Jpn., 1985, 58, 473; L.J.Johnston, J.Lusztyk, D.D.M.Wayner, A.N.Abeywickreyma, A.L.J.Beckwith, J.C.Scaiano, and K.U.Ingold, J. Am. Chem. Soc., 1985, 107, 4594.

185 G.A.Russell and L.L.Herold, J. Org. Chem., 1985, **50**, 1037; J.Dupuis, B.Giese, J.Hartung, M.Leising, H.-G.Korth, and R.Sustmann, J. Am. Chem. Soc., 1985, **107**, 4332; M.Devaud and J.-L.Lecat, Bull. Soc. Chim. Fr., 1985, **84**, 1187; M.J.S.Dewar, G.L.Grady, and D.R.Kuhn, Organometallics, 1985, **4**, 1041.
186 U.Edlund, T.Lejon, T.K.Venkatachalam, and E.Buncel, J. Am. Chem. Soc., 1985, **107**, 6408; E.Buncel, T.K.Venkatachalam, B.Eliasson, and U.Edlund, J. Am. Chem. Soc., 1985, **107**, 303; H.Schumann, S.Nickel, E.Hahn, and M.J.Heeg, Organometallics, 1985, **4**, 800; T.D.Tilley, ibid, 1452 and J. Am. Chem. Soc., 1985, **107**, 4084; J.Arnold and T.D.Tilley, ibid, 6409.
187 E.J.Corey and N.W.Boaz, Tetrahedron Lett., 1985, **26**, 6019; I.Fleming, J.H.M.Hill, D.Parker, and D.Waterson, J. Chem. Soc., Chem. Commun., 1985, 318.
188 I.Fleming and M.Taddei, Synthesis, 1985, 899 and 898; I.Fleming and A.P.Thomas, J. Chem. Soc., Chem. Commun., 1985, 411; Y.Okuda, K.Wakamatsu, T.Tuckmantel, K.Oshima, and H.Nozaki, Tetrahedron Lett., 1985, **26**, 4629.
189 N.L.Ermolaev, G.A.Vasil'eva, L.I.Vyshinskaya, M.N.Bochkarev, and G.A.Razuvaev, Zh. Obshch. Khim., 1985, **55**, 617 (Chem. Abs., 1985, **103**, 123625); R.S.Steevensz, D.G.Tuck, H.A.Meinema, and J.G.Noltes. Can. J. Chem., 1985, **63**, 755.
190 M.S.Alnajjar and H.G.Kuivila, J. Am. Chem. Soc., 1985, **107**, 416; E.C.Ashby, W.-Y.Su, and T.N.Pham, Organometallics, 1985, **4**, 1493.
191 W.Adcock, V.S.Iyer, W.Kitching, and D.Young, J. Org. Chem., 1985, **50**, 3706; D.S.Matteson and J.W.Wilson, Organometallics, 1985, **4**, 1690.
192 T.Takeda, K.Suzuki, H.Ohshima, and T.Fujiwara, Chem. Lett., 1985, 1249; T.Taneda, K.Ando, A.Mamada, and T.Fujiwara, ibid, 1149.
193 J.-i.Hibino, S.Nakatsukasa, K.Fugami, S.Matsubara, K.Oshima, and H.Nozaki, J. Am. Chem. Soc., 1985, **107**, 6416; S.Matsubara, J.-i.Hibino, Y.Morizawa, K.Oshima, and H.Nozaki, J. Organomet. Chem., 1985, **285**, 163.
194 N.Wiberg, K.Schurz, and G.Fisher, Angew. Chem. Int. Ed. Engl., 1985, **24**, 1053; C.Wentrup, S.Fisher, A.Maquestian, and R.Flammang, ibid, 56; J.R.Bowser and J.F.Bringley, Synth. React. Inorg. Metal-Org. Chem., 1985, **15**, 897.
195 T.Hirao, S.Nagata, Y.Yamana, and T.Agawa, Tetrahedron Lett., 1985, **26**, 5061; G.Simchen and M.W.Majchrzak, ibid, 5035; K.Szabo, N.Le Ha, P.Schneider, P.Zeltner, and E.sz.Kovats, Helv. Chim. Acta, 1984, **67**, 2128.
196 G.Gundersen, D.W.H.Rankin, and H.E.Robertson, J. Chem. Soc., Dalton Trans., 1985, 191; D.B.Beach and W.L.Jolly, Inorg. Chem., 1984, **23**, 4774; K.Ruhlmann, H.Schilling, H.Frey, and H.Paul, J. Organomet. Chem., 1985, **290**, 277.
197 T.Morimoto and M.Sekiya, Chem. Lett., 1985, 1371; W.Walter and M.Akram, Phosphorus Sulphur, 1985, **21**, 291; N.N.Vlasova, A.E.Pestunovich, and M.G.Voronkov, Zh. Obshch. Khim., 1985, **55**, 621 (Chem. Abs., 1985, **103**, 71376).
198 H.J.Bestmann and G.Wolfel, Chem. Ber., 1984, **117**, 1250; G.G.Furin, A.O.Miller, and G.G.Yakobson, Izv. Sib. Otd. Akad. Nauk SSSR, Ser. Khim. Nauk, 1985, 127 (Chem. Abs., 1985, **103**, 87947).
199 R.R.Fraser, T.S.Mansour, and S.Savard, J. Org. Chem., 1985, **50**, 3232; T.Fjeldberg, M.F.Lappert, and A.J.Thorne, J. Mol. Struct., 1984, **125**, 265; D.J.Brauer, H.Burger, G.R.Liewald, and J.Wilke, J. Organomet. Chem., 1985, **287**, 305.
200 A.W.Duff, P.B.Hitchcock, M.F.Lappert, R.G.Taylor, and J.A.Segal, J. Organomet. Chem., 1985, **293**, 271; M.Veith, H.Lange, O.Recktenwald, and W.Frank, ibid, **294**, 273; M.Veith, H.Lange, A.Belo, and O.Recktenwald, Chem. Ber., 1985, **118**, 1600; W.R.Nutt, J.A.Anderson, J.D.Odom, M.M.Williamson, and B.H.Rubin, Inorg. Chem., 1985, **24**, 159.
201 T.Fjeldberg and R.A.Andersen, J. Mol. Struct., 1985, **128**, 49; J.M.Boncella and R.A.Andersen, Organometallics, 1985, **4**, 205; H.C.Aspinall, D.C.Bradley, M.B.Hursthouse, K.D.Sales, and N.P.C.Walker, J. Chem. Soc., Chem. Commun., 1985, 1585; B.D.Murray and P.P.Power, Inorg. Chem., 1984, **23**, 4584.
202 M.Burklin, E.Hanecker, H.Noth, and W.Storch, Angew. Chem. Int. Ed. Engl.,

1985, **24**, 999; W.Kolonbra, W.Schwarz, and J.Weidlein, Z. Naturforsch., Teil B, 1985, **40**, 872; J.Muller, U.Muller, A.Loss, J.Lorberth, H.Donath, and W.Massa, ibid, 1320.
203 J.Pikies and W.Wojnowski, Z. Anorg. Allg. Chem., 1985, **521**, 173; S.P.Narula and N.Kapur, Inorg. Chim. Acta, 1984, **86**, 37.
204 G.Michael, Z. Chem., 1985, **25**, 371 and 19; K.Kawashiro, S.Morimoto, and H.Yoshida, Bull. Chem. Soc. Jpn., 1985, **58**, 1903.
205 W.Kantlehner, E.Haug, P.Speh, and H.-J.Brauner, Liebigs Ann. Chem., 1985, 65; M.Fortinon, B.de Jeso, and J.-C.Pommier, J. Organomet. Chem., 1985, **289**, 239.
206 U.Wannagat and T.Blumenthal, Monatsh. Chem., 1985, **116**, 557; T.Veszpremi, L.Bihatsi, Y.Harada, K.Ohno, and H.Mutoh, J. Organomet. Chem., 1985, **280**, 39.
207 W.Clegg, Acta Crystallogr., 1984, **C40**, 529; W.Clegg, G.M.Sheldrick, and D.Stalke, ibid, 433 and 816, W.Clegg, M.Haase, G.M.Sheldrick, and N.Vater, ibid, 871.
208 E.Gergo, G.Schultz, and I.Hargittai, J. Organomet. Chem., 1985, **292**, 343; U.Kliebisch, U.Klingebiel, and N.Vater, Chem. Ber., 1985, **118**, 4561.
209 M.Veith and V.Huch, J. Organomet. Chem., 1985, **293**, 161; K.Jurkschat, A.Tzschach, J.Meunier-Piret, and M.van Meerssche, ibid, **290**, 285; K.Jurkschat, C.Mugge, J.Schmidt, and A.Tzschach, ibid, **287**, C1.
210 G.Klebe, J. Organomet. Chem., 1985, **293**, 147; G.Klebe, J.W.Bats, and K.Hensen, J. Chem. Soc., Dalton Trans., 1985, 1; G.Klebe and K.Hensen, ibid, 5.
211 T.A.K.Al-Allaf, C.Eaborn, P.B.Hitchcock, M.F.Lappert, and A.Pidcock, J. Chem. Soc., Chem. Commun., 1985, 548; P.B.Hitchcock, M.F.Lappert, and M.C.Misra, ibid, 863; G.K.Campbell, P.B.Hitchcock, M.F.Lappert, and M.C.Misra, J. Organomet. Chem., 1985, **289**, C1.
212 J.E.Shade, B.V.Johnson, D.H.Gibson, W.-L.Hsu, and C.D.Schaeffer jr., Inorg. Chim. Acta, 1985, **99**, 99; S.M.Hawkins, P.B.Hitchcock, and M.F.Lappert, J. Chem. Soc., Chem. Commun., 1985, 1592.
213 M.F.Lappert and P.P.Power, J. Chem. Soc., Dalton Trans., 1985, 51; A.Meller and C.-P.Grabe, Chem. Ber., 1985, **118**, 2020; T.Fjeldberg, P.B.Hitchcock, M.F.Lappert, S.J.Smith, and A.J.Thorne, J. Chem. Soc., Chem. Commun., 1985, 939; A.M.Khmaruk and A.M.Pinchuk, Zh. Org. Khim., 1984, **20**, 1805 (Chem. Abs., 1985, **102**, 95813).
214 C.T.Jekel-Vroegop and J.H.Teuben, J. Organomet. Chem., 1985, **286**, 309; S.K.Vasisht, R.Sharma, and V.Goyal, Indian J. Chem., Sect. A, 1985, **24**, 37 (Chem. Abs., 1985, **103**, 215446).
215 M.D.Fryzuk, A.Carter, and A.Westerhaus, Inorg. Chem., 1985, **24**, 642; M.D.Fryzuk, S.J.Rettig, A.Westerhaus, and H.D.Williams, ibid, 4316.
216 M.D.Fryzuk, P.A.MacNeil, and S.J.Rettig, Organometallics, 1985, **4**, 1145 and J. Am. Chem. Soc., 1985, **107**, 6708.
217 A.Dormond, A.El Bouadili, A.Aaliti, and C.Moise, J. Organomet. Chem., 1985, **288**, C1; A.Dormond, A.El Bouadili, and C.Moise, J. Chem. Soc., Chem. Commun., 1985, 914.
218 S.K.Vasisht and G.Singh, Z. Anorg. Allg. Chem., 1985, **526**, 161; F.Preuss, E.Fuchslocher, and W.S.Sheldrick, Z. Naturforsch., Teil B, 1985, **40**, 1040; E.Schweda, K.D.Scherfise, and K.Dehnicke, Z. Anorg. Allg. Chem., 1985, **528**, 117.
219 P.C.Srivastava and S.K.Srivastava, Spectrochim. Acta, Part A, 1985, **41**, 687; R.Barbieri, A.Silvestri, G.Ruisi, and G.Alonzo, Inorg. Chim. Acta, 1985, **97**, 113; G.Valle, G.Ruisi, and U.Russo, ibid, **99**, L21; K.Yunlu, N.Hock, and R.D.Fisher, Angew. Chem. Int. Ed. Engl., 1985, **24**, 879.
220 J.M.Bellama and S.K.Tandon, Inorg. Chim. Acta, 1985, **102**, 23; S.Cradock, C.M.Huntley, and J.D.Durig, J. Mol. Struct., 1985, **127**, 319; R.Crossley and R.G.Shepherd, J. Chem. Soc., Perkin Trans. 1, 1985, 1917.
221 H.Bock and R.Dammel, Angew. Chem. Int. Ed. Engl., 1985, **24**, 111.
222 R.H.Cragg and R.D.Lane, J. Organomet. Chem., 1985, **294**, 7; W.Maringgele, Z. Naturforsch., Teil B, 1985, **40**, 277; R.J.P.Corriu, J.J.E.Moreau, and M.Pataud-Sat, Organometallics, 1985, **4**, 623.

M.Pataud-Sat, Organometallics, 1985, **4**, 623.
223 E Kupce, E.Liepins, and E.Lukevics, Angew. Chem. Int. Ed. Engl., 1985, **24**, 568; D.I.Gasking and G.H.Whitham, J. Chem. Soc., Perkin Trans. 1, 1985, 409; H.D.Hausen, O.Mundt, and W.Kaim, J. Organomet. Chem., 1985, **296**, 321.
224 M.Haase and U.Klingebiel, Angew. Chem. Int. Ed. Engl., 1985, **24**, 324; H.Noth and S.Weber, Chem. Ber., 1985, **118**, 2144.
225 B.Glaser and H.Noth, Angew. Chem. Int. Ed. Engl., 1985, **24**, 416.
226 W.Jacksties, H.Noth, and W.Storch, Chem. Ber., 1985, **118**, 2030; P.Paetzold, E.Schroder, G.Schmid, and R.Boese, ibid, 3205; K.Delpy, H.-U.Meier, P.Paetzold, and C.von Plotho, Z. Naturforsch., Teil B, 1984, **39**, 1696.
227 G.Schmid, D.Kampmann, W.Meyer, R.Boese, P.Paetzold, and K.Delpy, Chem. Ber., 1985, **118**, 2418.
228 E.Hanecker and H.Noth, Z. Naturforsch., Teil B, 1985, **40**, 717; R.Oesterie, W.Maringgele, and A.Meller, J. Organomet. Chem., 1985, **284**, 281; R.H.Cragg, T.J.Miller, and D.O'N Smith, ibid, **291**, 273.
229 A.A.Prishchenko, A.V.Gromov, Yu.N.Luzikov, A.A.Borisenko, E.I.Lazhko, K.Klaus, and I.F.Lutsenko, Zh. Obshch. Khim., 1984, **54**, 1520 (Chem. Abs., 1985, **102**, 6660); E.Niecke, W.Gueth, and M.Lysek, Z. Naturforsch., Teil B, 1985, **40**, 331.
230 R.Neidlein, H.-J.Degener, A.Gieren, G.Weber, and T.Hubner, Z. Naturforsch., Teil B, 1985, **40**, 1532; E.Niecke, J.Boeske, B.Krebs, and M.Dartmann, Chem. Ber., 1985, **118**, 3227.
231 V.D.Romanenko, V.F.Shul'gin, V.V.Skopenko, and L.N.Markovskii, Zh. Obshch. Khim., 1985, **55**, 538 and 1984, **54**, 2791 (Chem. Abs., 1985, **103**, 142085 and **102**, 185203); H.Hahn, W.Meindl, and K.Utvary, Monatsh. Chem., 1985, **116**, 157.
232 R.R.Ford, B.L.Li, R.H.Neilson, and R.J.Thoma, Inorg. Chem., 1985, **24**, 1993.
233 O.J.Scherer, R.Walter, and W.S.Sheldrick, Angew. Chem. Int. Ed. Engl., 1985, **24**, 525; O.J.Scherer, R.Konrad, E.Guggolz, and M.L.Zeigler, Chem. Ber., 1985, **118**, 1.
234 R.Bohra and H.W.Roesky, J. Fluorine Chem., 1984, **25**, 145.
235 T.Chivers, R.T.Oakley, R.Pieters, and J.F.Richardson, Can. J. Chem., 1985, **63**, 1063; T.Morimoto, Y.Nezu, K.Achiwa, and M.Sekiya, J. Chem. Soc., Chem. Commun., 1985, 1584; A.Schwobel and G.Kresze, Liebigs Ann. Chem., 1985, 453; R.T.Boere, A.W.Cordes, and R.T.Oakley, J. Chem. Soc., Chem. Commun., 1985, 929; H.W.Roesky, A.Theil, M.Noltemeyer, and G.M.Sheldrick, Chem. Ber., 1985, **118**, 2811.
236 N.P.C.Walker, M.B.Hursthouse, C.P.Warrens, and J.D.Woollins, J. Chem. Soc., Chem. Commun., 1985, 227; G.Suss-Fink, W.Buhlmyer, M.Herberhold, A.Gieren, and T.Hubner, J. Organomet. Chem., 1985, **280**, 129; C.Habben, A.Meller, M.Noltemeyer, and G.M.Sheldrick, ibid, **288**, 1; M.Herberhold, W.Jellen, W.Buhlmyer, W.Ehrenreich, and J.Reiner, Z. Naturforsch., Teil B, 1985, **40**, 1229; M.Herberhold, W.Jellen, W.Buhlmyer, and W.Ehrenreich, ibid, 1233.
237 R.Lang, H.Noth, P.Otto, and W.Storch, Chem. Ber., 1985, **118**, 86; N.A.Bumagin, Yu.V.Gulevich, and I.P.Beletskaya, J. Organomet. Chem., 1985, **285**, 415; B.Wrackmeyer, ibid, **297**, 265.
238 J.-G.Lee, J.E.Boggs, and A.H.Cowley, J. Chem. Soc., Chem. Commun., 1985, 773; C.N.Smit, F.M.Lock, and F.Bickelhaupt, Tetrahedron Lett., 1984, **25**, 3011; J.Escudie, C.Couret, J.Satge, M.Andrianarison, and J.D.Andriamizaka, J. Am. Chem. Soc., 1985, **107**, 3378; C.Couret, J.Escudie, J.Satge, A.Raharinirina, and J.D.Andriamizika, ibid, 8280.
239 D.M.Schubert and A.D.Norman, Inorg. Chem., 1985, **24**, 1107 and 4130; K.Issleib, U.Kuhne, and F.Krech, Phosphorus Sulphur, 1985, **21**, 367.
240 G.Becker, J.Harer, G.Uhl, and H.J.Wessely, Z. Anorg. Allg. Chem., 1985, **520**, 120; G.Becker, W.Massa, R.E.Schmidt, and G.Uhl, ibid, 139.
241 H.-G.Horn and H.-J.Lindner, Chem.-Ztg., 1985, **109**, 77.
242 J.Grobe and R.Haubold, Z. Anorg. Allg. Chem., 1985, **526**, 145; L.Weber, K.Reizig, R.Boese, and M.Polk, Angew. Chem. Int. Ed. Engl., 1985, **24**, 604; L.Weber and K.Reizig, Angew. Chem. Int. Ed. Engl., 1985, **24**, 865.
243 L.Weber, K.Reizig, and R.Boese, Organometallics, 1985, **4**, 2097; L.Weber and K.Reizig, Angew. Chem. Int. Ed. Engl., 1985, **24**, 53.

244 H.Schafer, D.Binder, and D.Fenske, Angew. Chem. Int. Ed. Engl., 1985, 24, 522; D.Fenske, J.Hachgenei, and F.Rogel, ibid, 982.
245 M.Peruzzini and P.Stoppioni, J. Organomet. Chem., 1985, 288, C44; M.Baudler and B.Makowka, Z. Anorg. Allg. Chem., 1985, 528, 7.
246 M.Weidenbruch, M.Herrndorf, A.Schafer, K.Peters, and H.G.von Schnering, J. Organomet. Chem., 1985, 295, 7; M.Yoshifuji, K.Toyota, and N.Inamoto, Chem. Lett., 1985, 441.
247 H.J.Breunig, K.Haberle, M.Drager, and T.Severengiz, Angew. Chem. Int. Ed. Engl., 1985, 24, 72; L.N.Markovskii, V.D.Romanenko, T.V.Saria-Pidvarko, and M.I.Povolotskii Zh. Obshch. Khim., 1985, 55, 221 (Chem. Abs., 1985, 103, 6462).
248 R.J.Glinski, J.L.Gole, and D.A.Dixon, J. Am. Chem. Soc., 1985, 107, 5891; T.Kudo and S.Nagase, ibid, 2589; M.O'Keefe and G.V.Gibbs, J. Phys. Chem., 1985, 89, 4574.
249 I.M.T.Davidson, A.Fenton, G.Manuel, and G.Bertrand, Organometallics, 1985, 4, 1324; I.M.T.Davidson and A.Fenton, ibid, 2040 and 2060; T.J.Barton and B.L.Groh, J. Am. Chem. Soc., 1985, 107, 7221; T.J.Barton and G.P.Hussmann, ibid, 7581.
250 P.E.Tomlins, J.E.Lydon, D.Akrigg, and B.Sheldrick, Acta Crystallogr., 1985, C41, 941; N.H.Buttrus, C.Eaborn, P.B.Hitchcock, and A.K.Saxena, J. Organomet. Chem., 1985, 284, 291 and 287, 157.
251 K.Furusawa and T.Katsura, Tetrahedron Lett., 1985, 26, 887; C.J.W.Brooks, W.J.Cole, and G.M.Barrett, J. Chromatogr., 1984, 315, 119; R.E.Lapointe, P.T.Wolczanski, and G.D.van Duyne, Organometallics, 1985, 4, 1810.
252 L.Horner and J.Mathias, J. Organomet. Chem., 1985, 282, 155; E.Wanek, Y.-M.Pai, and W.P.Weber, Synth. Commun., 1985, 15, 185.
253 H.-J.Liu and I.-S.Han, Synth. Commun., 1985, 15, 759; D.N.Sharma and R.P.Sharma, Tetrahedron Lett., 1985, 26, 371; S.Komiya, R.S.Srivastava, A.Yamamoto, and T.Yamamoto, Organometallics, 1985, 4, 1504; J.W.Patterson, Synthesis, 1985, 337.
254 L.M.Baigrie, H.N.Seikaly, and T.T.Thomas, J. Am. Chem. Soc., 1985, 107, 5391; L.M.Baigrie, D.Lenoir, H.R.Seikaly, and T.T.Tidwell, J. Org. Chem., 1985, 50, 2105; J.M.Poirier and L.Hennequin, Synth. Commun., 1985, 15, 217; L.Duhamel, F.Tombret, and J.M.Poirier, Org. Prep. Proced. Int., 1985, 17, 99; M.Marsi, K.C.Brinkman, C.A.Lisensky, G.D.Vaughn, and J.A.Gladysz, J. Org. Chem., 1985, 50, 3396; O.A.Vyazankina, B.A.Gostevskii, and N.S.Vyazankin, J. Organomet. Chem., 1985, 292, 145.
255 C.C.Silveira, J.V.Comasseto, and V.Catani, Synth. Commun., 1985, 15, 931; G.Cainello, M.Contento, A.Drusiani, M.Panunzio, and L.Plessi, J. Chem. Soc., Chem. Commun., 1985, 240; M.T.Reetz, K.Kesseler, and A.Jung, Angew. Chem. Int. Ed. Engl., 1985, 24, 989.
256 M.A.Ibragimov, M.I.Lazareva, and W.A.Smit, Synthesis, 1985, 880; E.Akgun and U.Pindur, Liebigs Ann. Chem., 1985, 2472; C.Rochin, O.Babot, and F.Duboudin, J. Organomet. Chem., 1985, 281, C24; R.M.Moriarty, O.Prakash, and M.P.Duncan, 1985, 943.
257 I.Stahl, Chem. Ber., 1985, 118, 1798; R.M.Moriarty, O.Prakash, and M.P.Duncan, J. Chem. Soc., Chem. Commun., 1985, 420; K.Saigo, T.Yamashita, A.Hongu, and M.Hasegawa, Synth. Commun., 1985, 15, 715; J.W.Huffman, S.M.Potnis, and A.V.Satish, J. Org. Chem., 1985, 50, 4266.
258 T.Sakakura, M.Hara, and M.Tanaka, J. Chem. Soc., Chem. Commun., 1985, 1545; R.M.Moriarty and I.Prakash, Synth. Commun., 1985, 15, 649; Y.Ueda, G.Roberge, and V.Vinet, Can. J. Chem., 1984, 62, 2936.
259 T.Mukaiyama, S.Kobayashi, and M.Murakami, Chem. Lett., 1985, 447; B.G.Lenz, H.Regeling, H.L.M.van Rosendaal, and B.Zwanenburg, J. Org. Chem., 1985, 50, 2930; J.Drouin, M.-A.Boaventura, and J.-M.Conia, J. Am. Chem. Soc., 1985, 107, 1726.
260 T.Sasaki, K.Mori, and M.Ohno, Synthesis, 1985, 279 and 280.
261 G.E.Totten, G.Wenke, and Y.E.Rhodes, Synth. Commun., 1985, 15, 291; E.N.Jacobssen, G.E.Totten, G.Wenke, A.C.Karydas, and Y.E.Rhodes, ibid, 301.
262 C.Gennari, A.Bernardi, S.Cardani, and C.Scolastico, Tetrahedron Lett., 1985, 26, 797; C.Gennari, A.Bernardi, G.Poli, and C.Scolastico, ibid, 2373;

A.Bernardi, S.Cardani, C.Gennari, G.Poli, and C.Scolastico, ibid, 6509;
J.-i.Uenishi, H.Tomozane, and M.Yamato, ibid, 3467; I.Stahl, Chem. Ber.,
1985, **118**, 3159.
263 G.Guanti, L.Banfi, E.Narisano, and C.Scolastico, Tetrahedron Lett., 1985,
26, 3517; E.W.Colvin and D.G.McGarry, J. Chem. Soc., Chem. Commun., 1985,
539; T.Sakakura and M.Tanaka, J. Chem. Soc., Chem. Commun., 1985, 1309.
264 G.Rousseau and L.Blanco, Tetrahedron Lett., 1985, **26**, 4191 and 4195; Y.Kita,
H.Yasuda, O.Tamura, F.Itoh, Y.Y.Ke, and Y.Tamura, Tetrahedron Lett., 1985,
26, 5777.
265 S.I.Pennanen, Synth. Commun., 1985, **15**, 865; L.B.Jackson and A.J.Waring, J.
Chem. Soc., Chem. Commun., 1985, 857.
266 C.Iwata, Y.Takemoto, A.Nakamura, and T.Imanishi, Tetrahedron Lett., 1985,
26, 3227; T.H.Chan and M.A.Brook, ibid, 2943;K.M.Taba and W.V.Dahlhoff, J.
Organomet. Chem., 1985, **280**, 27.
267 P.F.Hudrlik and A.K.Kulkarni, Tetrahedron Lett., 1985, **26**, 1389;
J.M.Aizpurua and C.Palomo, ibid, 6113 and Synthesis, 1985, 206; C.Aubert
and J.-P.Begue, ibid, 759.
268 L.N.Mander and S.P.Sethi, Tetrahedron Lett., 1985, **26**, 5953; I.Matsuda,
S.Sato, M.Hattori, and Y.Izumi, ibid, 3215; K.Krageloh, G.Simchen, and
K.Schweiker, Liebigs Ann. Chem., 1985, 2352.
269 K.Okano, T.Morimoto, and M.Sekiya, J. Chem. Soc., Chem. Commun., 1985, 119;
T.Allspach, H.Gumbel, and M.Regitz, J. Organomet. Chem., 1985, **290**, 33.
270 U.Scheim, H.Grosse-Ruyken, K.Ruhlmann, and A.Porzel, J. Organomet. Chem.,
1985, **293**, 29; C.H.Yoder, S.L.Tesno, S.M.Heaney, and C.Bohan, Synth. React.
Inorg. Metal-Org. Chem., 1985, **15**, 321.
271 F.P.Cossio and C.Palomo, Tetrahedron Lett., 1985, **26**, 4239; J.-L.Moreau and
R.Couffignal, J. Organomet. Chem., 1985, **294**, 139; M.Bellassoued,
M.Gaudemar, A.El Borgi, and B.Baccar, ibid, **280**, 165; G.Simchen and
H.H.Hergott, Chimia, 1985, **39**, 53 (Chem. Abs., 1985, **103**, 105038).
272 J.M.Renga and P.-C.Wang, Tetrahedron Lett., 1985, **26**, 1175; E.V.Dehmlow and
W.Leffers, J. Organomet. Chem., 1985, **288**, C41.
273 S.A.Shaidulin, O.A.Litvinov, V.A.Naumov, T.Ya.Stepanova, and G.V.Romanov,
Zh. Strukt. Khim., 1985, **26**, 175; K.Issleib, W.Mogelin, A.Balszuweit, Z.
Chem., 1985, **25**, 370 and Z. Anorg. Allg. Chem., 1985, **530**, 16; K.Issleib,
A.Balszuweit, J.Kotz, St.Richter, and R.Lentloff, ibid, **529**, 151; U.von
Allworden and G.-V.Roschenthaler, Chem.-Ztg., 1985, **109**, 81.
274 G.M.Gray and K.A.Redmill, J. Organomet. Chem., 1985, **280**, 105 and Inorg.
Chem., 1985, **24**, 1279.
275 H.Wada, S.Araki, K.Kuroda, and C.Kato, Polyhedron, 1985, **4**, 653; D.Hoebbel,
K.Ujszaszy, T.Reiher, G.Engelhardt, and A.Winkler, Z. Anorg. Allg. Chem.,
1985, **524**, 51.
276 J.M.Aizpurua, M.Juaristi, B.Lecea, and C.Palomo, Tetrahedron, 1985, **41**,
2903; W.G.Klemperer, V.V.Mainz, R.-C.Wang, and W.Shum, Inorg. Chem., 1985,
24, 1968; Y.Do, E.D.Simhon, and R.H.Holm, ibid, 1831.
277 P.Sampson and D.F.Wiemer, J. Chem. Soc., Chem. Commun., 1985, 1746;
K.Yamamo, T.Kimura, and Y.Tomo, Tetrahedron Lett., 1985, **26**, 4505.
278 U.Klingebiel, S.Pohlmann, L.Skoda, C.Lensch, and G.M.Sheldrick, Z.
Naturforsch., Teil B, 1985, **40**, 1023.
279 K.Ruhlmann, J.Brumme, U.Scheim, and H.Grosse-Ruyken, J. Organomet. Chem.,
1985, **291**, 165; B.J.Brisdon and A.M.Watts, J. Chem. Soc., Dalton Trans.,
1985, 2191; D.Seyferth and C.C.Prud'homme, Inorg. Chem., 1984, **23**, 4412.
280 Y.-M.Pai, W.P.Weber, and K.L.Servis, J. Organomet. Chem., 1985, **288**, 269;
H.Jancke and A.Porzel, Z. Chem., 1985, **25**, 251; T.C.Moore, W.D.McCormick,
and C.G.Wade, J. Phys. Chem., 1985, **89**, 3936.
281 P.E.Tomlins, J.E.Lydon, D.Akrigg, and B.Sheldrick, Acta Crystallogr., 1985,
C41, 292; Yu.E.Ovchinnikov, V.E.Shklover, Yu.T.Struchkov, A.A.Remizova,
B.D.Lavrukhin, T.V.Astapova, and A.A.Zhdanov, Z. Anorg. Allg. Chem., 1985,
524, 75.
282 Yu.E.Ovchinnikov, V.E.Shklover, Yu.T.Struchkov, A.A.Remizova, V.M.Kopylov,
V.A.Kovyazin, and V.V.Kireev, Z. Anorg. Allg. Chem., 1985, **523**, 14;
C.G.Francis, P.D.Morand, and N.J.Spare, Organometallics, 1985, **4**, 1958;

J.F.Harrod, A.Shaver, and A.Tucka, ibid, 2166.
283 W.H.Stevenson III, S.Wilson, J.C.Martin, and W.B.Farnham, J. Am. Chem. Soc., 1985, **107**, 6340; W.H.Stevenson III and J.C.Martin, ibid, 6352.
284 R.R.Holmes, R.O.Day, V.Chandrasekhar, and J.M.Holmes, Inorg. Chem., 1985, **24**, 2009; R.R.Holmes, R.O.Day, V.Chandrasekhar, J.J.Harland, and J.M.Holmes, ibid, 2016.
285 Yu.L.Frolov, N.N.Chipanina, G.A.Gavrilova, E.E.Shestakov, L.I.Gubanova, and M.G.Voronkov, Dokl. Akad. Nauk SSSR, 1984, **277**, 647 (Chem. Abs., 1985, **102**, 62312); G.A.Gavrilova, N.N.Chipanina, Yu.L.Frolov, L.I.Gubanova, V.M.D'yakov,, and M.G.Voronkov, Izv. Akad. Nauk SSSR, Ser. Khim., 1984, 2251 and 1985, 103 (Chem. Abs., 1985, **102**, 149322 and **103**, 6401).
286 Yu.L.Frolov, S.G.Shevchenko, and M.G.Voronkov, J. Organomet. Chem., 1985, **292**, 159, V.A.Pestunovich, M.F.Larin, M.S.Sorokin, A.I.Albanov, and M.G.Voronkov, ibid, **280**, C17; R.Krebs, D.Schomburg, and R.Schmulzler, Z. Naturforsch., Teil B, 1985, **40**, 282.
287 I.Urtane, G.Zelchans, and E.Lukevics, Z. Anorg. Allg. Chem., 1985, **520**, 179.
288 Chem. Abs., 1985, **102**, 62316, 62317, 95710, 113571, 113579, 132118, 204016, 220932 and **103**, 87960, 87961, 160563, 178317.
289 Chem. Abs., 1985, **102**, 149409 and **103**, 22683, 178349.
290 P.Henscei, I.Kovacs, and L.Parkanyi, J. Organomet. Chem., 1985, **293**, 185; J.D.Nies, J.M.Bellama, and N.Ben-Zvi, ibid, **296**, 315; E.Kupce, E.Liepins, A.Lapsina, I.Urtane, G.Zelcans, and E.Lukevics, ibid, **279**, 343.
291 C.Bruckner and H.-U.Reissig, J. Chem. Soc., Chem. Commun., 1985, 1512 and Angew. Chem. Int. Ed. Engl., 1985, **24**, 588; R.F.Cunico, Organometallics, 1985, **4**, 2176; J.O.Karlsson, V.N.Nghi, L.D.Foland, and H.W.Moore, J. Am. Chem. Soc., 1985, **107**, 3392.
292 G.M.Rubottom, E.C.Beedle, C.-W.Kim, and R.C.Mott, J. Am. Chem. Soc., 1985, **107**, 4230; G.H.Posner and K.A.Canella, ibid, 2571.
293 B.M.Trost and M.G.Saulnier, Tetrahedron Lett., 1985, **26**, 123; E.W.Collington, H.Finch, and I.J.Smith, ibid, 681; W.Kaim, J. Organomet. Chem., 1985, **282**, 1.
294 R.H.Cragg and R.D.Lane, J. Organomet. Chem., 1985, **289**, 23 and **291**, 153; V.Chvalovsky and W.S.El-Hamouly, Org. Mass Spectrom., 1985, **20**, 73 (Chem. Abs., 1985, **103**, 6394).
295 R.Tacke, H.Linoh, H.Zilch, J.Wess, U.Moser, E.Mutschler, and G.Lambrecht, Liebigs Ann. Chem., 1985, 2223; W.S.Sheldrick, H.Linoh, R.Tacke, G.Lambrecht, U.Moser, and E.Mutschler, J. Chem. Soc., Dalton Trans., 1985, 1743.
296 H.Kischkel and G.-V.Roschenthaler, Chem. Ber., 1985, **118**, 4842; J.Grobe, N.Krummen, and Duc Le Van, Z. Naturforsch., Teil B, 1984, **39**, 1711.
297 I.S.Ignat'ev, A.V.Ganyushkin, and A.N.Lazarev, Zh. Strukt. Khim., 1985, **26**, 67 (Chem. Abs., 1985, **103**, 142070); L.Wozniak, J.Kowalski, and J.Chojnowski, Tetrahedron Lett., 1985, **26**, 4965.
298 J.Chojnowski, M.Cypryk, J.Michalski, and L.Wozniak, J. Organomet. Chem., 1985, **288**, 275; W.J.Richter, ibid, **286**, 1.
299 S.D.Young, C.T.Buse, and C.H.Heathcock, Org. Synth., 1985, **63**, 79; C.H.Heathcock and S.Arseniyadis, Tetrahedron Lett., 1985, **26**, 6009; M.Asami, ibid, 5803.
300 H.Tanaka, H.Hayakama, S.Iijima, K.Haraguchi, T.Miyasaka, Tetrahedron, 1985, **41**, 861; K.B.G.Torssell, A.C.Hazell, and R.G.Hazell, ibid, 5569; L.W.McLaughlin, N.Piel, and T.Hellmenn, Synthesis, 1985, 322; K.K.Ogilvie and J.F.Cormier, Tetrahedron Lett., 1985, **26**, 4159.
301 J.Barrau, G.Rima, M.El Amine, and J.Satge, J. Chem. Res. (S), 1985, 30; C.Guimon, G.Pfister-Guillouzo, G.Rima, M.El Amine, and J.Barrau, Spectrosc. Lett., 1985, **18**, 7.
302 H.Puff, T.R.Kok, P.Nauroth, and W.Schuh, J. Organomet. Chem., 1985, **281**, 141; K.Haberle and M.Drager, Z. Naturforsch., Teil B, 1984, **39**, 1541.
303 S.P.Narula, N.Kapur, and R.K.Sharma, Inorg. Chim. Acta, 1985, **85**, 37; J.D.Donaldson, S.M.Grimes, A.F.L.Holding, and M.Hornby, Polyhedron, 1985, **4**, 1293; T.P.Lockhart, J. Organomet. Chem., 1985, **287**, 179.
304 H.Puff, H.Hevendehl, K.Hofer, H.Reuter, and W.Schuh, J. Organomet. Chem.,

1985, **287**, 163; H.Jolibois, F.Theobald, R.Mercier, and C.Devin, Inorg. Chim. Acta, 1985, **97**, 119.
305 D.Dakternieks, R.W.Gable, and B.F.Hoskins, Inorg. Chim. Acta, 1984, **85**, L43; N.W.Alcock, M.Pennington, and G.R.Willey, J. Chem. Soc., Dalton Trans., 1985, 2683.
306 J.J.Bonire, Polyhedron, 1985, **4**, 1707; V.Chandrasekhar, R.O.Day, and R.R.Holmes, Inorg. Chem., 1985, **24**, 1970.
307 S.T.Vijayaraghavan and T.R.Balasubramanian, J. Organomet. Chem., 1985, **282**, 17; F.Guibe, Y.T.Xian, A.M.Zigna, and G.Balavoine, Tetrahedron Lett., 1985, **26**, 3559.
308 E.Nakamura, J.-i.Shimada, and I.Kuwajima, Organometallics, 1985, **4**, 641; M.V.Garad, Inorg. Chim. Acta, 1984, **87**, 79; S.N.Gurkova, A.I.Gusev, N.V.Alekseev, T.K.Gar, and N.A.Viktorov, Zh. Strukt. Khim., 1984, **25**, 174 (Chem. Abs., 1985, **102**, 166882); J.Otera, T.Yano, and R.Okawara, Chem. Lett., 1985, 901.
309 E.Keinan, M.Sahai, Z.Roth, A.Nudelman, and J.Herzig, J. Org. Chem., 1985, **50**, 3558; A.Boaretto, D.Marton, and G.Tagliavini, J. Organomet. Chem., 1985, **288**, 283.
310 A.Baba, H.Kishiki, I.Shibata, and H.Matsuda, Organometallics, 1985, **4**, 1329; A.Guiller, C.H.Gagnieu, and H.Pacheco, Tetrahedron Lett., 1985, **26**, 6067; A.Tzschach, E.Nietzschmann, and C.Mugge, Z. Anorg. Allg. Chem., 1985, **523**, 21.
311 T.Yano, K.Nakashima, J.Otera, and R.Okawara, Organometallics, 1985, **4**, 1501; T.P.Lockhart, W.F.Manders, and F.E.Brinckman, J. Organomet. Chem., 1985, **286**, 153; S.Roelens and M.Taddei, J. Chem. Soc., Perkin Trans. 2, 1985, 799; A.Lycka and J.Holecek, J. Organomet. Chem., 1985, **294**, 179.
312 T.P.Lockhart, W.F.Manders, and J.J.Zuckerman, J. Am. Chem. Soc., 1985, **107**, 4546; T.P.Lockhart and W.F.Manders, ibid, 5863 and J. Organomet. Chem. 1985, **297**, 143; T.Birchall and V.Manivannan, Can. J. Chem., 1985, **63**, 2211.
313 E.A.Chernyshev, O.V.Kuz'min, A.V.Lebedev, V.G.Zaikin, and A.I.Mikaya, J. Organomet. Chem., 1985, **289**, 231; N.A.Tarasenko, V.V.Volkova, V.G.Zaikin, A.I.Mikaya, A.A.Tishenkov, V.G.Avakyan, L.E.Gusel'nikov, M.G.Voronkov, S.V.Kirpichenko, and E.N.Suslova, ibid, **288**, 27.
314 W.Wojnowski, K.Peters, M.C.Boehm, and H.G.von Schnering, Z. Anorg. Allg. Chem., 1985, **523**, 169; W.Wojnowski, K.Peters, E.-M.Peters, and H.G.von Schnering, ibid, **525**, 121 and **531**, 153, 147, and **530**, 79.
315 H.-G.Horn and M.Hemeke, Chem.-Ztg., 1985, **109**, 1 and 409.
316 R.West, D.J.De Young, and K.J.Haller, J. Am. Chem. Soc., 1985, **107**, 4942; W.Wojnowski, B.Dreczewski, A.Herman, K.Peters, E.-M.Peters, and H.G.von Schnering, Angew. Chem. Int. Ed. Engl., 1985, **24**, 992.
317 A.Ricci, A.Degl'Innocenti, M.Fiorenza, P.Dembech, N.Ramodan, G.Seconi, and D.R.M.Walton, Tetrahedron Lett., 1985, **26**, 1091; E.Bloch and M.Aslam, ibid, 2259; K.Steliou, P.Salama, and J.Corriveau, J. Org. Chem., 1985, **50**, 4969.
318 G.G.Hoffmann, Z. Naturforsch., Teil B, 1984, **39**, 1216, J. Organomet. Chem., 1984, **273**, 187, and Z. Anorg. Allg. Chem., 1985, **524**, 185; G.G.Hoffmann and H.Meixner, ibid, **53**, 121; W.-W.du Mont and M.Grenz, Chem. Ber., 1985, **118**, 1045; H.Schumann, S.Jurgis, E.Hahn, J.Pickardt, J.Blum, and M.Eisen, Chem. Ber., 1985, **118**, 2738.
319 D.Fenske, J.Hachgenei, and J.Ohmer, Angew. Chem. Int. Ed. Engl., 1985, **24**, 706 and 993.
320 K.Sasaki, Y.Aso, T.Ostubo, and F.Ogura, Tetrahedron Lett., 1985, **26**, 453; T.Severengiz, W.-W.du Mont, D,Lenoir, and H.Voss, Angew. Chem. Int. Ed. Engl., 1985, **24**, 1041.
321 S.Kozuka, S.Tamura, T.Yamazaki, S.Yamaguchi, and W.Tagaki, Bull. Chem. Soc. Jpn., 1985, **58**, 3277; A.I.Charov, Khim. Elementoorg. Soedin. 1983, 42 (Chem. Abs., 1985, **102**, 6732).
322 W.Ando, T.Tsumuraya, and A.Sekiguchi, Tetrahedron Lett., 1985, **26**, 4523.
323 H.Lavayssiere and G.Dousse, J. Organomet. Chem., 1985, **297**, C17; J.Barrau, N.Ben Hamida, and J.Satge, ibid, **282**, 315.
324 M.Drager and K.Haberle, J. Organomet. Chem., 1985, **280**, 183; R.K.Chadha, J.E.Drake, and A.B.Sarkar, Inorg. Chem., 1984, **23**, 4769 and 1985, **24**, 3156.

325 E.M.Holt, F.A.K.Nasser, A.Wilson jr., and J.J.Zuckerman, Organometallics, 1985, **4**, 2073; V.G.K.Das, C.Wei, and E.Sinn, J. Organomet. Chem., 1985, **290**, 291; T.P.Lockhart, W.F.Manders, and E.O.Schlemper, J. Am. Chem. Soc., 1985, **107**, 7451.
326 H.C.Clark, V.K.Jain, R.C.Mehrotra, B.P.Singh, G.Srivastava, and T.Birchell, J. Organomet. Chem., 1985, **279**, 385; D.Dakternicks, B.F.Hoskins, P.A.Jackson, E.R.T.Tiekink, and G.Winter, Inorg. Chim. Acta, 1985, **101**, 203; D.Dakternieks, B.F.Hoskins, E.R.T.Tiekink, and G.Winter, ibid, 1984, **85**, 215; U.R,Kunze, J. Organomet. Chem., 1985, **281**, 79.
327 Yu.E.Ovchinnokov, V.E.Shklover, Yu.T.Struchkov, V.A.Palyulin, V.I.Rokitskaya, M.Yu.Aismont, and V.F.Traven, J. Organomet. Chem., 1985, **290**, 25.
328 M.Drager, Z. Anorg. Allg. Chem., 1985, **527**, 169 and Z. Naturforsch., Teil B, 1985, **40**, 1511.
329 K.Jurkschat, M.Scheer, A.Tzschach, J.Meunier-Piret, and M.van Meerssche, J. Organomet. Chem., 1985, **281**, 173; C.Wynants, G.van Binst, C.Muegge, K.Jurkschat, A.Tzschach, H.Pepermans, M.Gielen, and R.Willem, Organometallics, 1985, **4**, 1906.
330 G.Himbert, Liebigs Ann. Chem., 1985, 1669; R.Arad-Yellin and F.Wudl, J. Organomet. Chem., 1985, **280**, 197.
331 V.I.Shcherbakov, I.K.Grigor'eva, O.S.D'yachkovskaya, and G.A.Razuvaev, Zh. Obshch. Khim., 1985, **55**, 107 (Chem. Abs., 1985, **102**, 220964); M.J.S.Dewar and G.L.Grady, Organometallics, 1985, **4**, 1327.
332 S.J.Blunden, L.A.Hobbs, P.J.Smith, A.G.Davies, and S.B.Teo, J. Organomet. Chem., 1985, **282**, 9; J.W.Burley, R.E.Hutton, and R.D.Dworkin, ibid, **284**, 171.
333 K.Gratz, F.Huber, A.Silvestri, G.Alonzo, and R.Barbieri, J. Organomet. Chem., 1985, **290**, 41; M.G.Begley, C.Gaffney, P.G.Harrison, and A.Steel, ibid, **289**, 281; M.Drager and N.Kleiner, Z. Anorg. Allg. Chem., 1985, **522**, 48.
334 L.Horner and J.Mathias, J. Organomet. Chem., 1985, **282**, 175; T.N.Bell, A.G.Sherwood, and G.Soto-Garrido, J. Phys. Chem., 1985, **89**, 1155.
335 T.C.Frank, K.B.Kester, and J.L.Falconer, J. Catal. 1985, **91**, 44; M.Bordeau, S.M.Djamei, R.Calas, and J.Duogues, J. Organomet. Chem., 1985, **288**, 131.
336 H.Aleman, S.C.L.Lau, and J.Lielmezs, Thermochim. Acta, 1985, **84**, 57 (Chem. Abs., 1985, **103**, 123572); J.M.Aizpurna and C.Palomo, Tetrahedron Lett., 1985, **26**, 475; H.Wetter and K.Oerlte, ibid, 5515.
337 M.Bellassoued, A.El Borgi, and M.Gaudemar, Synth. Commun., 1985, **15**, 973.
338 N.G.Kundu and S.G.Khatri, Synthesis, 1985, 323; H.C.Berk, K.E.Zwickelmaier, and J.E.Franz, Synth. Commun., 1985, **15**, 57; H.-U.Reissig, Tetrahedron Lett., 1985, **26**, 3943; T.Sakai, K.Mujata, and A.Takeda, Chem. Lett., 1985, 1137.
339 M.Cypryk, J.Chojnowski, and J.Michalski, Tetrahedron, 1985, **41**, 2471; A.R.Bassindale and T.Stout, Tetrahedron Lett., 1985, **26**, 3403.
340 R.F.Cunico and L.Bedell, J. Organomet. Chem., 1985, **281**, 135; S.Tomoda, Y.Takeuchi, and Y.Nomura, Synthesis, 1985, 212.
341 M.Bordeau, S.M.Djamei, and J.Dunogues, Organometallics, 1985, **4**, 1087; E.Lukevics, L.M.Ignatovich, J.Popelis, S.Rozite, and I.Mazeika, Latv. PSR Zinat. Akad. Vestis, Kim. Ser. 1985, 73 (Chem. Abs., 1985, **103**, 6465); R.K.Chadha, J.E.Drake, and M.K.H.Neo, J. Cryst. Spectrosc. Res., 1985, **15**, 39 (Chem. Abs., 1985, **102**, 88010); M.S.Soliman, M.A.Khattab, and A.G.El-Kourashy, Bull. Soc. Chim. Belg., 1985, **94**, 87.
342 B.Mahieu, C.Mestdagh-Peters, and M.de Laet, Bull. Soc. Chim. Belg., 1985, **94**, 797; M.Bordeau, S.M.Djamei, R.Calas, and J.Dunogues, Bull. Soc. Chim. Fr., 1985, **84**, 488; D.W.Allen, S.Bailey, J.S.Brooks, and B.F.Taylor, Chem. Ind. (London), 1985, 826.
343 J.M.Landry, J.E,Katon, and J.M.Hughes, Spectrochim. Acta, Part A, 1985, **41**, 291; K.C.Molloy, K.Quill, and I.W.Howell, J. Organomet. Chem., 1985, **289**, 271.
344 G.Matsubayashi, K.Ueyama, and T.Tanaka, J. Chem. Soc., Dalton Trans., 1985, 465; G.Matsubayashi, R.Shimizu, and T.Tanaka, Chem. Lett., 1985, 973;

K.A.Paseshnitchenko, L.A.Aslanov, A.V.Jatsenko, and S.V.Medvedev, J. Organomet. Chem., 1985, **287**, 187; N.Burke, K.C.Molloy, T.G.Purcell, and M.P.Singh, Inorg. Chim. Acta, 1985, **106**, 129.
345 V.K.Bel'skii, A.N.Protskii, B.M.Bulychev, and G.L.Soloveichik, J. Organomet. Chem., 1985, **280**, 45; D.Hanssgen, H.Puff, and N.Beckermann, ibid, **293**, 191.
346 S.Hoste, G.G.Herman, W.Lippens, L.Verdonck, and G.P.van der Kelen, Spectrochim. Acta, Part A, 1985, **41**, 925; P.H.de Ryck, L.Verdonck, and G.P.van der Kelen, Bull. Soc. Chim. Belg., 1985, **94**, 621; L.Verdonck, P.H.de Ryck, S.Hoste, and G.P.van der Kelen, J. Organomet. Chem., 1985, **288**, 289.
347 A.Baba, I.Shibita, K.Masuda, and H.Matsuda, Synthesis, 1985, 1144; A.Baba, I.Shibata, M.Fujiwara, and H.Matsuda, Tetrahedron Lett., 1985, **26**, 5167.
348 I.Wharf, R.Cuenca, and M.Onyszchuk, Can. J. Spectrosc. 1984, **29**, 31; H.Fujiwara, F.Sakai, A.Kawamura, N.Shimizu, and Y.Sasaki, Bull. Chem. Soc. Jpn., 1985, **58**, 2331.
349 A.V.Jatsenko, S.V.Medvedev, K.A.Paseshnitchenko, and L.A.Aslanov, J. Organomet. Chem., 1985, **284**, 181; R.J.F.Jans, G.van Koten, K.Vrieze, B.Kojic-Prodic, A.L.Spek, and J.L.de Boer, Inorg. Chim. Acta, 1985, **105**, 193; G.Valle, G.Plazzogna, and R.Ettorre, J. Chem. Soc., Dalton Trans., 1985, 1271.
350 A.L.Rheingold, S.W.Ng, and J.J.Zuckerman, Inorg. Chim. Acta, 1984, **86**, 179; J.N.Spencer, S.W.Barton, B.M.Cader, C.D.Corsico, L.E.Harrison,, M.E.Mankuta, and C.H.Yoder, Organometallics, 1985, **4**, 394.
351 P.C.Srivastava, S.K.Srivastava, and S.B.Sharma, Can. J. Chem., 1985, **63**, 329; R.Colton and D.Dakternieks, Inorg. Chim. Acta, 1985, **102**, L17.
352 A.A.Muratova, E.G.Yarkova, N.P.Morozova, N.R.Safiullina, and A.N.Pudovik, Zh. Obshch. Khim., 1984, **54**, 1745 (Chem. Abs., 1985, **102**, 6730); A.I.Tursina, L.A.Aslanov, S.V.Medvedev, and A.V.Yatsenko, Koord. Khim. 1985, **11**, 417 (Chem. Abs., 1985, **103**, 88006); L.A.Aslanov, A.I.Tursina, V.V.Chernyshev, S.V.Medvedev, and A.V.Yatsenko, ibid, 277 (Chem. Abs., 1985, **103**, 54174).
353 J.M.Miller and A.Fulcher, Can. J. Chem., 1985, **63**, 2308; R.M.Davidson, H.G.Grant, and D.O'Smith, Spectrochim. Acta, Part A, 1985, **41**, 581.
354 S.J.Blunden and R.Hill, Inorg. Chim. Acta, 1985, **98**, L7; K.C.Molloy and K.Quill, J. Chem. Soc., Dalton Trans., 1985, 1417.
355 Acta Crystallogr., 1984, **C40**, 476; Inorg. Chim. Acta, 1984, **82**, 211 and 1985, **107**, 231.
356 J. Organomet. Chem., 1985, **295**, 149; J. Indian Chem. Soc., 1985, **62**, 60.
357 Synth. React. Inorg. Metal-Org. Chem., 1985, **15**, 779, 907; Inorg. Chim. Acta, 1984, **87**, 47 and 1985, **99**, 87; J. Org. Chem., 1985, **50**, 2086; J. Chem. Soc., Chem. Commun., 1985, 231.
358 J. Chem. Soc., Dalton Trans., 1985, 321, 333, 487; Inorg. Chim. Acta, 1984, **83**, 79 and 1985, **105**, 9 and **107**, 57; Indian J. Chem., Sect. A. 1984, **23**, 878 and 917; Synth. React. Inorg. Metal-Org. Chem., 1985, **15**, 133; Acta Crystallogr., 1985, **C41**, 1027; J. Organomet. Chem., 1985, **279**, 373; Polyhedron, 1985, **4**, 747; Z. Naturforsch., Teil B, 1985, **40**, 1173.
359 Inorg. Chim. Acta, 1985, **108**, 141; J. Chem. Soc., Chem. Commun., 1985, 894; Polyhedron, 1985, **4**, 447.
360 J. Organomet. Chem., 1985, **291**, C17.
361 J. Organomet. Chem., 1985, **280**, 323; Polyhedron, 1985, **4**, 2089.
362 Inorg. Chim. Acta, 1985, **93**, 179.
363 (a) Z. Naturforsch., Teil B, 1985, **40**, 447 (b) Z. Anorg. Allg. Chem., 1985, **523**, 199 (c) Organometallics, 1985, **4**, 1509 and Inorg. Chem., 1985, **24**, 2976 (d) J. Organomet. Chem., 1985, **286**, 225 (e) Nouv. J. Chim., 1985, **9**, 155 (f) Chem. Ber., 1985, **118**, 1143 and J. Organomet. Chem., 1985, **289**, 157 (g) J. Am. Chem. Soc., 1985, **107**, 6391 and 266.
364 (a) Chem. Ber., 1985, **118**, 1746, 1758, and 1770 (b) J. Organomet. Chem., 1985, **295**, C3.
365 (a) J. Organomet. Chem., 1985, **287**, 221 (b) J. Chem. Soc., Dalton Trans., 1985, 1331 (c) J. Organomet. Chem., 1985, **293**, 69.

7
Group V: Arsenic, Antimony, and Bismuth

BY J. L. WARDELL

1 Tervalent Compounds

1.1 <u>Metal-Metal Bonded Species</u>.- The diarsene, $(Me_3Si)_3CAs=$ $AsC(SiMe_3)_3$, is isostructural with the phosphorus analogue; it adopts a trans-planar configuration in the solid state [As=As 2.242(1) and 2.245(1) Å in two independent molecules].[1] Octa-t-butyltricyclo[6.4.0.0^{2,6}]dodeca-arsane, isolated from the reaction between Bu^tAsCl_2, $AsCl_3$ and Mg in refluxing THF, is stable at RT in the absence of light and air, but is less stable in solution.[2] Controlled aerial oxidation of a mixture of $PhSb(SiMe_3)_2$ and dioxane provided crystalline cyclo-$(SbPh)_6$.dioxane (1); in the solid state, the six membered Sb ring adopts a chair conformation with the Ph group in equatorial positions.[3] The synthesis, Scheme (1), crystal structure and spectra of 4-Me-1,2,6-tri-

$$MeC(CH_2Br)_3 \xrightarrow{i} MeC(CH_2SbPh_2)_3 \xrightarrow{ii} MeC(CH_2SbCl_2)_3 \xrightarrow{iii}$$

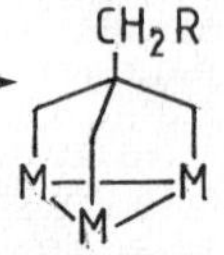

Reagents: i, $Ph_2SbNa, liqNH_3$; ii, HCl; iii, Na, THF

(2, R=H; M=Sb)

Scheme 1

stibatricyclo[2.2.1.0^{2,6}]heptane (2, M=Sb; R=H) were reported.[4] The Sb-Sb bond lengths in (1) and (2, M=Sb; R=H) are 2.836(1)-2.839(1) and 2.796(3)-2.817(3) Å respectively. The p.e. spectra and MNDO calculations on (2, M=As, Sb or Bi; R=H) were also reported. An analogous procedure to Scheme (1), employing $C(CH_2Cl)_4$, provided[5] (2, M=As; R=Cl). The arsirane, $PhAsC(SiMe_3)_2\overset{.}{C}(SiMe_3)_2$ (3) was obtained[6] from the reaction of $PhAsCl_2$ and $LiCCl(SiMe_3)_2$ <u>via</u> the intermediacy of $PhAs[=C(SiMe_3)_2]_2$; the crystal structure of (3) was also determined.

1.2 <u>Triorganometallic Species</u>.- A series of dimetallo-triptycenes, $(\underline{o}\text{-}C_6F_4)_3MM'$ (4, MM'=As_2, Sb_2, Bi_2, PAs, PSb or $AsSb$) have been synthesised;[7] <u>e.g.</u> (MM'=As_2) was generated from $(\underline{o}\text{-}BrC_6F_4)_3As$ on successive treatment with BuLi and $AsCl_3$. Vibrational and mass

(For references see page 142

spectra were reported for (4) and for $(\underline{o}\text{-}C_6H_4)_3Sb_2$, prepared[8] from Sb and $(\underline{o}\text{-}C_6H_4Hg)_3$ at 260°C. Routes to R_2BiR' (5, R,R=alkyl or aryl) include[9] reactions of R_2BiM (M=Li or Na) with R'X and of Ph_2BiCl with R'Li. Compounds (5, R=alkyl) are thermally labile, e.g. Me_2BiAr disproportionates to Me_3Bi and $MeBiAr_2$ at 90°C. Reaction of PhLi with $(Ph_2Bi)_2CH_2$ provided[10] the synthetically useful Ph_2BiCH_2Li. Interestingly the reactivity towards electrophiles of Ph_2AsCH_2Li, obtained from RLi and either Ph_2AsCH_2I or $Ph_2AsCH_2SnPh_3$, depends[11] on the precursor. Resolution[12] of $H_2NCH_2CH_2AsRPh$ (R=Me or Bu) was achieved using chiral Pd^{II} complexes.

Among other triorgano-derivatives studied were (i) $PhAs(COBu^t)_2$ (6), prepared from[13] $PhAs(SiMe_3)_2$ and Bu^tCOCl, (ii) $Ph\overline{MCH=CHCH=CH}$ (M=As or Bi),[14] (iii) $(\underline{E}\text{-}Ph_2AsCH_2CH=CHCH=CH_2$ and [15]$(\underline{Z},\underline{Z})\text{-}Ph_2AsCH=CHCH_2CH=CHSnBu_3$, (iv) $(H_2C=CHCH_2CH_2)_nAs(CH_2CH_2CH_2\text{-}AsMe_2)_{3-n}$ (n=1 or 2),[16] (v) $Ph_2As(CH_2)_nC_5H_5$ (n=2 or 4), $Ph_2As(CH_2)_3AsPh(CH_2)_2C_5H_5$, $PhAs[(CH_2)_2C_5H_5]_2$ and[17] $1\text{-}(Ph_2AsCH_2CH_2)\text{-}$indene, and (vi) $1,3\text{-}(Ph_2AsCH_2CH_2)$indene.[18] In addition, various polydentate, heterocyclic compounds were obtained,[19] having 3, 4 or 6 As atoms and As, O and S atoms eg (7)/(8). Crystal structures of (6)[13] and 9-R-9-stibafluorene (R=Ph or Cl)[20] were determined.

1.3 <u>Other Compounds</u>.- The crystal structure of pyramidal $(Me_3Si)_2CHSbCl_2$ was determined.[21] The reaction of 2,4,6-$Bu^t_3C_6H_2Li$ with $SbCl_3$ produces[21] the heterocycle, 4,6-$Bu^t_2\overline{C_6H_2SbClCH_2}CMe_2$-2. Compounds, $\underline{m}$ and $\underline{p}$-$(Cl_2Sb)_2C_6H_4$,[22] and the chelated[23] $Cl_nBi(CH_2CH_2CO_2Pr^i)_{3-n}$ (n=1 or 2) were synthesized. Also reported were the syntheses, spectral properties and crystal structures of (i) $Ph\overline{MSCH_2CH_2XCH_2CH_2}S$ (9, M=Sb, or Bi; X=O or S): strong intramolecular M...X interactions were indicated in the solid state;[24] (ii) $PhM[S_2P(OPr^i)_2]_2$ (M=As or Sb): octahedral geometry with lone pair <u>trans</u> to Ph in axial sites;[25] (iii) $MeSb(S_2CNEt_2)_2$;[26] and (iv) Me_2MN_3 (M=Sb or Bi): distorted trigonal bipyramidal structures with M linked to α-N of the N_3 unit so forming endless zig-zag chains.[27]

2 <u>Quinquevalent Compounds</u>

Reactions of $(Ph_2As)_2CH_2$ with MeI and $MeOSO_2F$ gave $Ph_2M\overset{+}{e}AsCH_2AsPh_2$, I^-(10) and $(Ph_2\overset{+}{As}Me)_2CH_2$, $2FSO_3^-$(11) respectively;[28] (10) and (11) on base treatment provided $MePh_2As=CHAsPh_2$ and $MeP\overset{+}{h}_2As=C=$

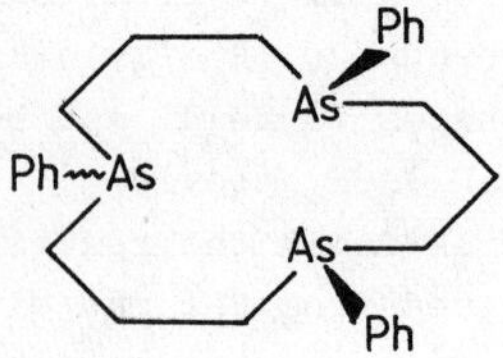
Ph
As
Ph—As
As
Ph

(7)

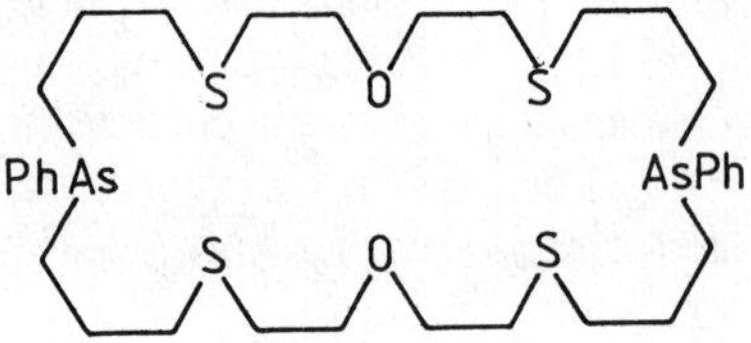
S
O
S
PhAs
AsPh
S
O
S

(8)

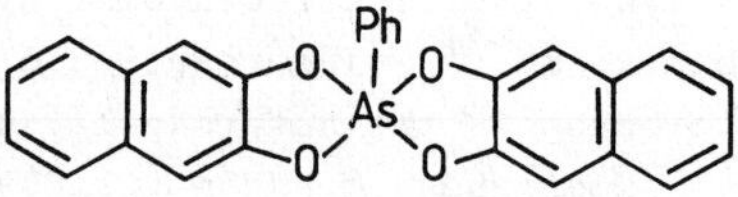
Ph
O
O
As
O
O

(12)

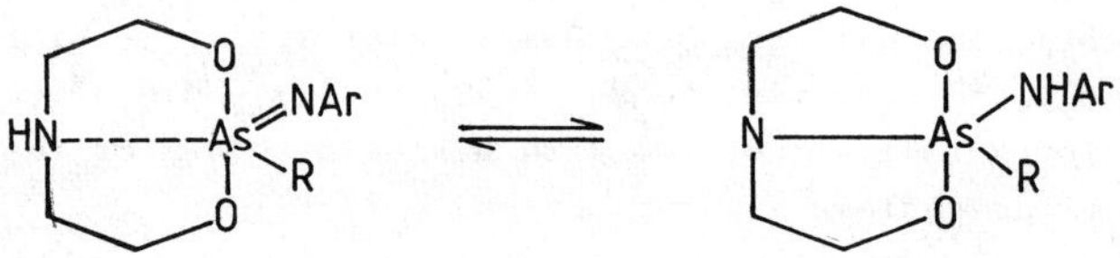
O
NAr
HN
As
R
O
O
NHAr
N
As
R
O

(13)

AsPh$_2$Me. The stibonium salts, Ph$_3$ArSb$^+$X$^-$ (X=BF$_4$ or ClO$_4$), were obtained[29] by electrochemical oxidation of Ph$_3$Sb in MeCN solution containing ArH and NaClO$_4$ or Et$_4$NBF$_4$. Crystal structures of the following compounds were determined: (i) trigonal bipyramidal Ph$_4$SbOSO$_2$Ph.H$_2$O (obtained from Ph$_4$SbOH and PhSO$_3$H):[30] long Sb-O(SO$_2$Ph) bond, (ii) trigonal bipyramidal (HOCH$_2$CH$_2$SO$_3$SbPh$_3$)$_2$O (produced[31] from Ph$_3$SbO and HOCH$_2$CH$_2$SO$_3$H): O apical with linear SbOSb; (iii) (Ph$_2$SbBrO)$_2$ from Ph$_2$SbBr and Bu^{t}OOH: planar Sb$_2$O$_2$ ring,[32] and (iv) various spirocyclic arsoranes[33] e.g. trigonal bipyramidal, MeAs(OCMe$_2$CMe$_2$O)$_2$ and rectangular pyramidal (12). The formation was reported of pentacoordinate arsinimines, containing N-As coordination, e.g. (13): these compounds dimerize to diarsadiazocyclobutanes.[34]

3 Uses in Organic Synthesis

Ketones are selectively reduced to alcohols using Ph$_2$SbH and AlCl$_3$; esters, acid chlorides, alkyl halides or alkenes are unaffected.[35] An indirect oxidation of ArCH$_2$OH to ArCHO involves[36] the formation of Ph$_2$SbOCH$_2$Ar and subsequent reaction with Br$_2$. Akiba et al. have reported[37] on the oxidation of α-hydroxyketones to α-diketones using Ph$_3$SbBr$_2$ in the presence of a base and on the transformation RCH$_2$Br to RCH$_3$ (R=ArCO or Ar) using Bu$_3$Sb. Barton et al.[38] have published full details of phenylations of phenols, enols and enolate anions by pentavalent phenylbismuth species.

4 Bibliography

W. Wolfsberger, Chem. Ztg., 1985, 109, 53. Review on addition of As-H and Sb-H bonds to alkenes and alkynes.

F.D. Yambushev, G.I. Kokorev, and F.G. Khalitov, Zh. Obshch. Khim., 1984, 84, 2005. EtArAr'As.

N.A.A. Al-Jabar and A.G. Massey, J. Organomet. Chem., 1985, 288, 145. (2-ClC$_6$H$_4$)$_3$M (M=As, Sb or Bi).

M. Gielen, P. Bras and J. Wolters, Org. Mass. Spectrom., 1984, 19, 647. M.s. of ArAr'Ar"Bi.

C. Srinivasen and K. Pitchumani, Can. J. Chem., 1985, 63, 2285. Oxidation of Ph$_3$M (M=As or Sb) using peroxydisulphate.

W.-W. du Mont, Z. Naturforsch., Teil B, 1985, 40, 1453. Reactions of Bu$^t_3$As with Se.

V.A. Dodonov, T.I. Starostina, A.N. Grishin and N.A. Zelenova, Zh. Obshch. Khim., 1985, 55, 586. Reactions of Bu$_3$Bi with CCl$_4$ or HCCl$_3$.

N.M. Zaripov, R.M. Galiakberov, A.V. Golubinskii, L.V. Vilkov and N.A. Chadaeva, J. Struct. Chem., 1984, 25, 528. E.d. of $MeAsSCH_2CH_2S$ and $ClAsCH_2CH_2CH_2CH_2$.

M. Wieber, I. Fetzer-Kremling, D. Wirth and H.G. Rüdling, Z. Anorg. Allg. Chem., 1985, 520, 59. $RM(SR')_2$ (M=Sb or Bi).

P. Raj, A.K. Saxena, K. Singhal and A. Ranjan, Polyhedron., 1985, 4, 251. Reactions of $(C_6F_5)_3Sb$ leading to $(C_6F_5)_3SbXY$.

V.A. Dodonov, A.V. Gushchin and T.G. Brilkina, Zh. Obshch. Khim., 1985, 55, 73. $Ph_3Bi(O_2CR)_2$.

N.K. Jha and D.M. Joshi, Polyhedron., 1985, 4, 2093. $R_3Sb(L)OMe$. LH=Hacac, 8-HOquinoline, etc.

R. Nomura, M. Kori and H. Matsuda, Chem. Lett., 1985, 579. $[MeNHCH_2CH_2OSb(OH)Ph_2]_2O$.

Y. Huang, L. Shi, and J. Yang, Tetrahedron Lett., 1985, 26, 6447. Reaction of $Ph_3\overset{+}{As}CH_2CHO$, Br^- with RCHO.

W. Tao, F. Kuang, H. Pu, J. Liu and T. Lu, Wuhan Daxue Xuebao Ziran Kexueban, 1984, 63; Chem. Abstr., 1985, 102, 6719.   Reaction of 2-($Ph_3As=CHCO$)thiophene and RR'CO.

W. Tao and H. Pu, Huazue Xuebao, 1985, 43, 403; Chem. Abstr., 1985, 103, 195952. Reactions of $R_3\overset{+}{As}CHCOR'$ and ketene dimer.

X. Huang, Z. Chen, W. Dai, L. Zhou and J. Shi, Youji Huaxue, 1985, 144; Chem. Abstr., 1985, 103, 195962. Reactions of $Ph_3As=CHCO_2Me$ with o-HO-aryl-COR.

W. Tao and H. Pu, Youji Huaxue, 1985, 137; Chem. Abstr., 1985, 103, 196175. $Ph_3As=C(COR)CONHPh$.

References

1. A.H. Cowley, N.C. Norman and M. Pakulski, J. Chem. Soc. Dalton Trans., 1985, 383.
2. M. Baudler and S. Wietfeldt-Haltenhoff, Angew. Chem., Int. Ed. Engl., 1985, 24, 991.
3. H.J. Breunig, K. Häberle, M. Dräger and T. Severengiz, Angew. Chem., Int. Ed. Engl., 1985, 24, 72.
4. R. Gleiter, H. Köppel, P. Hofmann, H.R. Schmidt, and J. Ellermann, Inorg. Chem., 1985, 24, 4020; J. Ellermann, E. Köck and H. Burzlaff, Acta Crystallogr., Sect. C, 1985, 41, 1437; J. Ellermann and A. Veit, J. Organomet. Chem., 1985, 290, 307.
5. J. Ellermann and L. Brehm, Chem. Ber., 1985, 118, 4794.
6. R. Appel, T. Gaitzsch and F. Knoch, Angew. Chem., Int. Ed. Engl., 1985, 24, 419.
7. N.A.A. Al-Jabar and A.G. Massey, J. Organomet. Chem., 1985, 287, 57.
8. N.A.A. Al-Jabar, D. Bowen and A.G. Massey, J. Organomet. Chem., 1985, 295, 29.
9. M. Wieber and I. Sauer, Z. Naturforsch., Teil B, 1985, 40, 1476; T. Kauffmann and F. Steinseifer, Chem. Ber., 1985, 118, 1031.
10. T. Kauffmann, F. Steinseifer and N. Klas, Chem. Ber., 1985, 118, 1039.
11. T. Kauffmann, B. Altepeter, N. Klas and R. Kriegesmann, Chem. Ber., 1985, 118, 2353.
12. Y. Shigetomi, M. Kojima and J. Fujita, Bull. Chem. Soc. Jpn., 1985, 58, 258.
13. G. Becker, B. Becker, M. Birkham, O. Mundt and R.E. Schmidt, Z. Anorg. Allg. Chem., 1985, 529, 97.
14. A.J. Ashe and F.J. Drone, Organometallics, 1985, 4, 1478.
15. T. Kauffmann and K.-R. Gaydoul, Tetrahedron Lett., 1985, 26, 4067, 4071.
16. F.M. Ashmawy, F.R. Benn, C.A. McAuliffe and D.G. Watson, J. Organomet. Chem., 1985, 287, 65.

17. T. Kauffmann, K. Berghus and J. Ennen, Chem. Ber., 1985, 118, 3724; T. Kauffmann, K. Berghus, A. Rensing and J. Ennen, p.3737.
18. T. Kauffmann, K. Berghus and A. Rensing, Chem. Ber., 1985, 118, 4507.
19. K. Kauffmann and J. Ennen, Chem. Ber., 1985, 118, 2692, 2703, 2714.
20. V.K. Belskii, J. Struct. Chem., 1984, 25, 965.
21. A.H. Cowley, N.C. Norman, M. Pakulski, D.L. Bricker and D.H. Russell, J. Am. Chem. Soc., 1985, 107, 8211.
22. R. Stricker-Lennartz and H.P. Latscha, Z. Naturforsch., Teil B, 1985, 40, 1045.
23. E. Nakamura, J.-I. Shimoa and I. Kuwajima, Organometallics, 1985, 4, 641.
24. H.M. Hoffmann and M. Dräger, J. Organomet. Chem., 1985, 295, 33; M. Dräger and B.M. Schmidt, J. Organomet. Chem., 1985, 290, 133.
25. R.K. Gupta, A.K. Rai, R.C. Mehrotra, V.K. Jain, B.F. Hoskins, and E.R.T. Tiekink, Inorg. Chem., 1985, 24, 3280.
26. M. Wieber, D. Wirth, J. Metter and C. Burschka, Z. Anorg. Allg. Chem., 1985, 520, 65.
27. J. Müller, U. Müller, A. Loss, J. Lorberth, H. Donath and W. Massa, Z. Naturforsch., Teil B, 1985, 40, 1320.
28. H. Schmidbaur and P. Nusstein, Organometallics, 1985, 4, 344.
29. E.V. Nikitin, O.V. Parakin and Yu.M. Kargin, Zh. Obshch. Khim., 1984, 54, 1789.
30. R. Rüther, F. Huber and H. Preut, J. Organomet. Chem., 1985, 295, 21.
31. H. Preut, R. Rüther, and F. Huber, Acta Crystallogr., Sect. C, 1985, 41, 358.
32. D.M. Wesolek, D.B. Sowerby and M.J. Begley, J. Organomet. Chem., 1985, 293, C5.
33. C.A. Poutasse, R.O. Day, J.J. Holmes and R.R. Holmes, Organometallics, 1985, 4, 708; R.R. Holmes, R.O. Day and A.C. Sau, p.714.
34. P. Maroni, Y. Madule, T. Seminario and J.-G. Wolf, Can. J. Chem., 1985, 63, 636.
35. Y. Huang, Y. Shen and C. Chen, Tetrahedron Lett., 1985, 26, 5171.
36. Y. Haung, Y. Shen and C. Chen, Synthesis, 1985, 651.
37. K.Y. Akiba, H. Ohnari and K. Ohkata, Chem. Lett., 1985, 1577; K.Y. Akiba, A. Shimuzu, H. Ohnari and K. Ohkata, Tetrahedron Lett., 1985, 26, 3211.
38. D.H.R. Barton, N.Y. Bhatnagar, J.-C. Blazejewski, B. Charpiot, J.-P. Finet, D.J. Lester, W.B. Motherwell, M.T.B. Paroula and S.P. Stanforth, J. Chem. Soc. Perkin Trans. 1, 1985, 2657; D.H.R. Barton, J.-C. Blazejewski, B. Charpiot, J.-P. Finet, W.B. Motherwell, M.T.B. Papoula and S.P. Stanforth, p.2667.

8
Metal Carbonyls

BY B. J. BRISDON

1 Introduction

The format of this report is similar to that used last year with relevant reviews described below and papers of a more specific nature considered in subsequent sections. The continued expansion in the pertinent literature has necessitated some selectivity, and particular emphasis has been placed on new species and novel reactions.

Included in the published account of the plenary lectures presented at the International Conference on Chemical Thermodynamics held during 1986 is an interesting article on metal-ligand bond energies in organometallic compounds, which includes data on metal carbonyls.[1] Relevant n.m.r. data are to be found in two different sources,[2,3] and ion-pairing effects on the structures and reactivities of metal carbonyl anions have been described.[4] A timely review of the photochemistry of M-M bonds deals almost exclusively with metal carbonyl derivatives;[5] these also feature in articles on transition metal-hydrogen bonds.[6] The associative substitution reactions of metal carbonyls,[7] and the preparations, bonding and reactivity of thiocarbonyl complexes[8] were also reviewed in 1985. As befits current interest, the largest number of reviews centre on metal carbonyl cluster compounds. General topics covered included electron transfer reactions,[9] ligand and cluster transformations,[10] and the chemistry of metal clusters containing nitrosyl and nitrido ligands.[11] More specific topics reviewed include sulphido-osmium carbonyl cluster compounds,[12] and homonuclear platinum clusters.[13] The preparations of $[Nb(CO)_6]^-$, $[M_2(CO)_{10}(\mu-H)]^-$ (M = Cr or W), $Mn_2(CO)_8X_2$ and $Re(CO)_5X$ (X = Cl, Br or I) are described in Volume 23 of Inorganic Syntheses.[14]

2 General and Theoretical Studies

Several interesting and significant papers concerning ligated CO appeared in 1985. Quantitative aspects of the reaction of OH^- and OMe^- with $M(CO)_5$ and $M_3(CO)_{12}$ (M = Fe, Ru or Os) have been reported and the implications with regard to the water gas shift reaction were noted.[15,16] A full account of a method for computing the most favourable initial site for nucleophilic attack on a transition metal complex has been described and applied to reactions of cationic iron carbonyl derivatives.[17] Nucleophilic attack by H_2O, OH^- and H^- on dicationic Ir^{III} carbonyl complexes yields cationic hydroxycarbonyl and formyl

cations. The latter was protonated to afford electrophilic dicationic hydroxycarbene complexes.[18] Reversible C-C bond formation by interconversion of CO and
μ-acyl ligands on the face of an osmium cluster has been observed,[19] and the
reactivity of η^2-acyl complexes has also been analysed theoretically.[20] Reversible insertion of CO into the Zr-Si bond of $ZrCl(cp)_2SiMe_3$ has been recorded.[21]
However, only the C atom of CO inserts into a μ-CSiMe₃ ligand in $W_2(\mu\text{-CSiMe}_3)_2$-
$(OPr)_4$ with cleavage of the CO triple bond and formation of a σ,π-CCSiMe₃
ligand.[22] It has been shown that formyl formation *via* intermolecular hydride
transfer from one Ru complex to a CO co-ordinated to a second Ru is possible.
This lends support to the suggestion that formyl formation *via* intermolecular
hydride transfer may occur during the catalytic hydrogenation of CO.[23] Theoretical studies of various CO insertion processes have been published,[24] as have
calculations on the protonation energies of some d^8 and d^{10} metal carbonyl
species,[25] and the M-H homolysis of $Co(CO)_4H$.[26]

A fascinating coupling reaction involving CO occurs on treatment of $[Mo(CO)_3$-
$(2,4\text{-Me}_2pd)]^-$ with MeI at -78° C. A CO ligand is converted into an O-bound
alkoxide ligand concomitant with conversion of the pentadienyl group to a cyclo-
hexenyl ligand (1).[27] Systematics of the formation of π-CO ligands in four-iron
cluster have been described in three important papers.[28-30] The formation of
π-CO ligands is induced in $[Fe_4(CO)_{13}]^{2-}$ by the attachment of electron-acceptor
ligands, which also have a marked effect on details of the M-M bonding within
the cluster.[29] Solution equilibria between two isomeric forms of the products
of such reactions have been reported, and the selective cleavage of C- and O-bonded
CO described.[30] Reductive homologation of CO to a ketenecarboxylate has been
observed.[31] This ligand is formally dervied from three CO molecules by two one-
electron reductions plus the appropriate couplings. Further examples of metal(IV)
carbonyl complexes have been prepared,[32] and a linear Ti-O-C chain has been found
in $[Ti(\eta\text{-Me}_5C_5)]_2(\mu\text{-OC})_2[MoCO)_2cp]_2$.[33]

Much consideration has been given to the interactions between CO and transition metal surfaces.[34-37] An analysis has been presented of the similarities
and differences in the binding of CO to sites on Ti, Cr, Fe, Co and Ni,[34] and an
explanation given for the very low CO stretching frequency for CO chemisorbed
on Fe.[35] Factors determining the CO binding site preferences on Pd and Pt
sites have also been studied.[36]

In the area of polynuclear clusters, the stabilisation of triangular transition metal frameworks of the type M_3L_6 has been analysed theoretically,[37] and
reasons sought for the stability of *closo*-M_4E_2 clusters (M = metal atom in an ML_3
local co-ordination and E = S, Te) which have 8 skeletal electron pairs instead
of the expected 7.[38] A theoretical analysis has been used to show how the MOs
of three-connected polyhedral molecules differ from those of deltahedra,[39] and a
systematic investigation of the electronic requirements for structures based on

four-connected polyhedra has also been presented.[40] The inter-relationships between the polyhedral skeletal electron pair theory and the newly developed topological electron counting procedure have also been explored in successive papers.[41,42]

3 Chemistry of Metal Carbonyls

3.1 Mononuclear Carbonyl Derivatives.-In a series of papers on the structure and bonding in metal carbonyl and nitrosyl derivatives, gas phase electron diffraction studies of $Cr(NO)_4$, $Mn(CO)(NO)_3$, $Fe(CO)_2(NO)_2$ and $Co(CO)_3NO$ have been reported.[43,44] Changes in bond lengths of the type $\Delta r(M-C)$ and $\Delta r(M-N)$ were found to correspond to generally increasing bond orders as the atomic number of M increases, while at the same time the total bond order of the M bonds changes very little. The structure of the volatile, red pyridinium salt of $[V(CO)_6]^-$ has also been determined and the significance of the $\overset{+}{N}-H--OC$ interactions discussed in relation to the structure of $[V(CO)_6H]^-$.[45] Further structural studies on the encrypted sodium salts of $[M(CO)_5]^{2-}$ (M = Mo or W) have been reported,[46] and ^{13}C and high resolution solid-state ^{17}O n.m.r. spectra have been recorded on several other early transition metal carbonyls and carbonylate anions.[47] The bonding and isomerism of $Fe(CO)_5$ has been investigated,[48] and the relative binding energies of this compound to a series of gaseous anions determined.[49]

"Naked" $Cr(CO)_5$ is produced by laser photolysis at 351 nm of $Cr(CO)_6$ in the gas phase,[50] whereas in the absence of collisions $Cr(CO)_4$ is the principal product formed on pulsed laser excitation at 249 nm.[51] Flash photolysis of $Cr(CO)_6$ in cyclohexane yields $Cr(CO)_5$.solvent. The $Cr(CO)_5$ moiety adopts a square pyramidal geometry at room temperature.[52] Complex formation between $M(CO)_5$ fragments (M = Cr or W) and CO_2 or N_2O has also been demonstrated in matrix photochemical studies. Such association activates these molecules, so allowing photoreduction to CO or N_2 to occur with concomitant oxidation of the carbonyls to products which include *trans*-dioxotungsten tetracarbonyl.[53] Electron momentum distributions have been measured for $Cr(CO)_6$ at several different bonding energies and the results compared with theoretical calculations,[54] and kinetic studies of the $M(CO)_6$ (M = Cr or W) catalysed water gas shift reactions have been reported.[55] The fragmentation characteristics of $[Cr(CO)_6]^+$ have been studied,[56] and cationic transition metal carbonyl cluster fragments of the type $[M_x(CO)_y]^+$ have been prepared by dissociative ionisation of the parent carbonyls $Cr(CO)_6$ and $Fe(CO)_5$.[57] A novel application of metal carbonyl fragments as a class of markers in molecular biology is also worthy of note.[58] Papers on the identification, magnetic properties and substitution behaviour of several metal carbonyl radicals have been published.[59-65] E.s.r. measurements on matrix isolated carbonyls $M(CO)_n$ (M = Cu or Ag, n = 1 or 3 and M = Au, n = 1) have been used to infer structural and bonding data.[62-65] Spectra of $Ag(CO)_3$ recorded under differing conditions have been interpreted in terms of two different ground states. Further examples of Cu^I[66] and Pd^{II} carbonyl derivatives[67,68] have been reported, including the first mononuclear dicarbonylpalladium(II) complexes.[68]

3.2 Binuclear Carbonyl Derivatives.-An improved synthesis of $Re_2(CO)_{10}$ has been reported. Carbonylation of NH_4ReO_4 at 250-290° C and high pressure in toluene yields the product in $ca.$ 70% yield.[69] Laser photolysis and pulse radiolysis studies of this carbonyl have also been described.[70,71] The relative rate constants for ion-molecule reactions in $Mn_2(CO)_{10}$ and $MnRe(CO)_{10}$ have been determined; cluster ions with up to 8 metal atoms were identified.[72] A complete assignment of the u.v. photoelectron spectrum of $MnRe(CO)_{10}$ has been made, and the properties of ions and radicals from $M_2(CO)_{10}$ (M = Mn or Re) by electron attachment and protonation in the gas phase have been described.[74]

Reaction of $[Ru(CO)_4]^{2-}$ with CO_2 yields the new dianion $[Ru_2(CO)_8]^{2-}$, which has a novel configuration (2) in which trigonal bipyramidal and pyramidal Ru(CO) units are linked via a Ru-Ru bond (2.936 Å).[75] The binuclear species $[Ni_2(CO)_8]^+$, $[Ni_2(CO)_6]^+$ and $[Ni_2(CO)_7]^-$ have been identified in freshly irradiated samples of $Ni(CO)_4$ in krypton. Both cations apparently contain two bridging CO ligands, whereas the anion has only a single CO bridge.[76]

3.3 Polynuclear Carbonyl Derivatives.-Two major problems in cluster chemistry were addressed in 1985. As cluster carbonyl species have become larger, the description of the spatial distribution of metals and ligands have become a nontrivial problem. A relatively simple system has been proposed which introduces new notation to define these parameters unequivocally.[77] The largest molecular clusters now being synthesised have diameters only some 10 Å less than the smallest metal crystallites produced by physical means, and there is much interest in this borderline area. A theoretical paper has appeared in which "magic numbers" corresponding to the commonest polygons and polyhedra are derived and applied to structural aspects of large cluster chemistry.[78] Several aspects of $M_3(CO)_{12}$ (M = Fe, Ru or Os) chemistry have been reported, including kinetic measurements on substitution reactions of the Fe and Os analogues,[79,80] their photolysis (M = Fe or Ru),[81,82] reactivity,[83] spectroscopic properties on oxide surfaces,[84] and electrochemistry[85] (M = Ru). The solid-state structure of the $[Fe_4(CO)_{13}]^{2-}$ anion has been shown to differ in its PPN^+ and $[Fepy_6]^{2+}$ salts.[86] Crystallographic studies on the tetranuclear species $Ru_2Co_2(CO)_{13}$ and $[RuIr_3(CO)_{12}]^-$ have been published, and the solution structure of $Co_4(CO)_{12}$ has been reinvestigated using ^{13}C and ^{59}Co n.m.r. techniques.[87] A re-evaluation of the thermodynamic parameters relating to the equilibrium between $Rh_2(CO)_8$ and $Rh_4(CO)_{12}$ under CO has appeared.[88] Further details of the preparation of $[Rh_6(CO)_{14}]^{4-}$ from $Rh_6(CO)_{16}$ and aqueous alkali have been published. This anion is highly reactive towards electrophiles and readily undergoes redox condensations with other rhodium carbonyl cluster derivatives.[89] The structural rearrangement of the

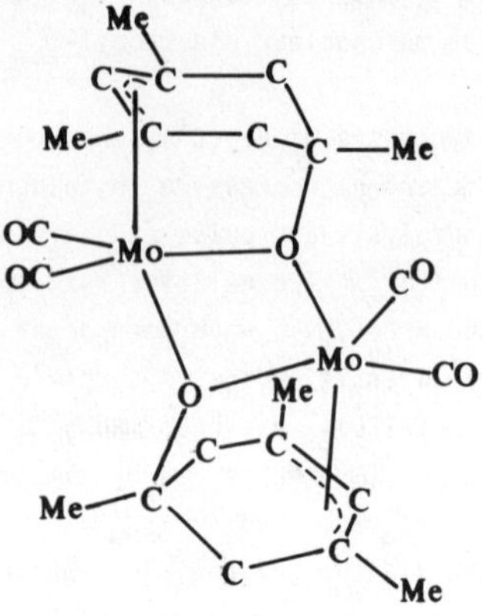

(1)

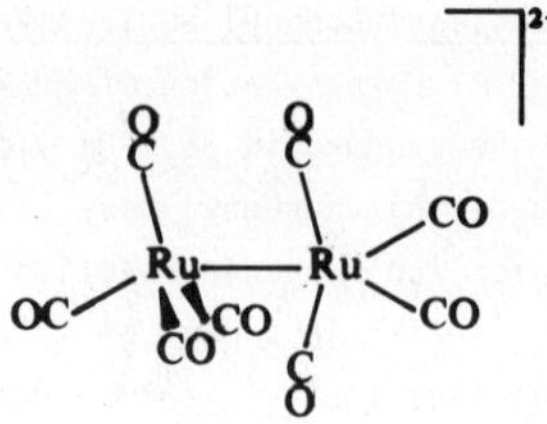

(2)

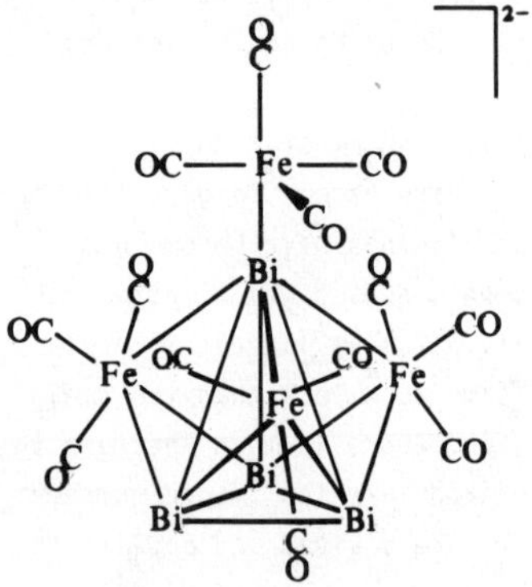

(3)

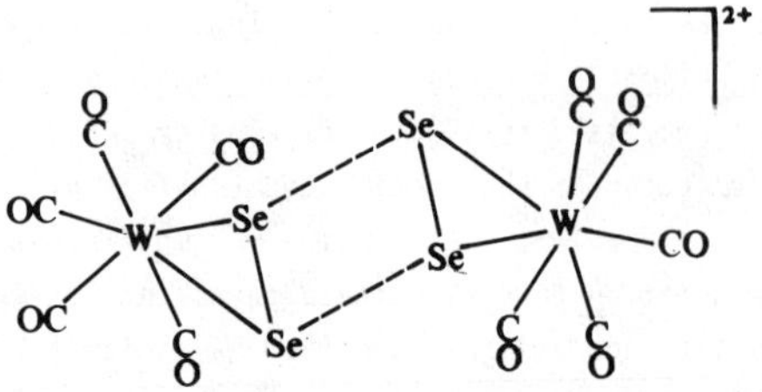

(4)

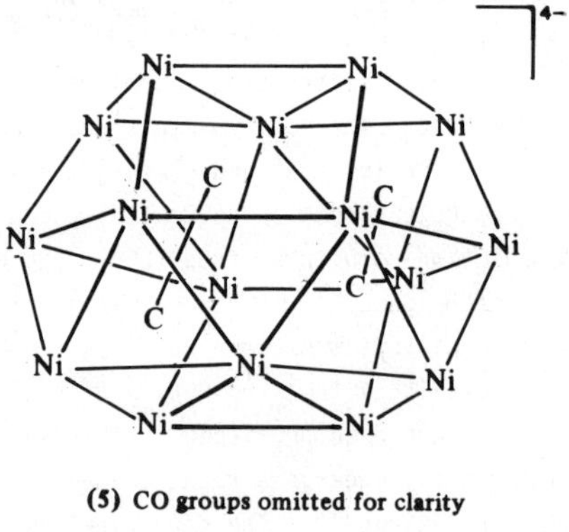

(5) CO groups omitted for clarity

bicapped tetrahedral core of $Os_6(CO)_{18}$ during formation of $[Os_6(CO)_{18}]^{2-}$ has been monitored,[90] and the reduction of CO_2 by anionic ruthenium carbonyl clusters has been briefly reported.[91]

Several new heteronuclear carbonyl clusters containing Cu have been characterised.[92,93] Reaction of $[Co(CO)_4]^-$ with CuCl yields $[CuCo(CO)_4]_4$ and polymeric $[CuCo(CO)_4]_n$. The former contains an almost linear environment for the Cu atoms, whereas the latter has an infinite zig-zag metal atom chain.[92] Interesting structures were also found for $[Cu(PPh_3)_2]_2Fe(CO)_4$ and $[Cu(diphos)_2]_2$-$Cu_6Fe_4(CO)_{16}$.[93] The synthesis and structure of the first hybrid Zintl-metal carbonyl $(NEt_4)_2[Bi_4Fe_4(CO)_{13}]$ has been described.[94] The four Bi atoms in the anion define a tetrahedron in which three faces are capped by $Fe(CO)_3$ groups while the fourth is bare (3). In $(\mu_3\text{-}Bi)_2Fe_3(CO)_9$ the Bi atoms occupy apical sites of a trigonal bipyramidal Bi_2Fe_3 core.[95] A series of complexes $[M(CO)_5]_3(\mu_3\text{-}M')$ (where M = Cr, W or Mn and M' = As, Sb, Ge or Sn) have been prepared by reaction of various carbonyl metallate anions with either $[M(CO)_5]_2$-M'Cl or M'Cl$_4$. The crystal structures of five of these complexes have been determined.[96] The reaction of $[W(CO)_5]_3Sn$ with H_2Te yields $[W(CO)_5]_3Te_2$ which contains a Te_2 unit σ-bonded to two $W(CO)_5$ moieties and π-bonded to the third. However, the $[Se_4W_2(CO)_{10}]^{2+}$ cation, formed in the reaction of $W(CO)_6$ with $Se_4(Sb_2F_{11})_2$ in SO_2, contains a 6-membered W_2Se_4 ring formed by dimerisation of two $[Se_2W(CO)_5]^+$ moieties (4).[97] All Ge-Ge and Ge-H bonds are cleaved in the reactions of polygermanes with $Co_2(CO)_8$. Products include $Ge_2Co_6(CO)_{20}$ which loses CO above 50° C to yield $Ge_2Co_6(CO)_{19}$ whose structure was reported.[98] The structures of several other heteronuclear cluster carbonyls containing Co and Zn, Hg or Au have also been determined.[99-101]

4 Cluster Carbonyls Containing C, N, or S

Electrochemical and X-ray investigations of the isoelectronic carbido- and nitrido-carbonyl clusters $Fe_5C(CO)_{15}$, $[Fe_5N(CO)_{14}]^-$ and $[Fe_5C(CO)_{14}]^{2-}$ have been described.[102] The Fe skeleton of these 74-electron systems has a similar distorted square-based pyramidal geometry and obeys Wade's rules. The electrochemical reduction of $Fe_5C(CO)_{15}$ leads to $[Fe_5C(CO)_{14}]^{2-}$ via a 2-electron transfer followed by loss of a CO ligand. New nitrosyl containing Fe clusters have been obtained by reaction of $Fe_6C(CO)_{16}$ with $NOBF_4$ in CH_2Cl_2. The initial product $Fe_6C(CO)_{15}(NO)$ yields $Fe_6C(CO)_{11}(NO)_4$ on treatment with excess reagent.[103] Successive degradation of $closo$-octahedral metal carbido carbonyl clusters and addition of different metal carbonyl fragments to the resultant $nido$-tetragonal-pyramidal clusters have been used to generate $[Fe_5CoC(CO)_{16}]^-$, $[Fe_4CoC(CO)_{14}]^-$, $Fe_4CoRhC(CO)_{16}$ and $[Fe_4CoPdC(CO)_{15}]^-$.[104] A reactive butterfly carbide complex $[Fe_3RhC(CO)_{12}]^-$ is formed in a redox condensation reaction of $[Fe_3(CO)_9(CCO)]^{2-}$ with $Rh_2(CO)_4Cl_2$. Monoprotonation of this product yields the butterfly methylidene derivative $Fe_3Rh(CH)(CO)_{12}$.[105] The anions $[Fe_3MnC(CO)_{13}]^-$,

[Fe$_3$Ni$_3$C(CO)$_{14}$]$^-$ and [Fe$_4$CoC(CO)$_{14}$]$^-$ were also prepared from the same three-iron ketenylidene precursor. Two isomeric forms of Ru$_5$C(CO)$_{13}$(NO)(AuPEt$_3$) have been structurally characterised. Both have a square pyramidal Ru$_5$ core, but different bonding modes for the AuPEt$_3$ ligand.[106] More complete preparative and structural details have been reported for [Co$_6$C(CO)$_{15}$]$^{2-}$ amd [Co$_8$C(CO)$_{18}$]$^{2-}$,[107] and the new carbido species containing Co$_{13}$, Rh$_{12}$ and Os$_{10}$ cores characterised for the first time.[108-110] Both [Os$_{10}$C(CO)$_{24}$(μ-I)]$^-$ and Os$_{10}$C(CO)$_{24}$(μ-I)$_2$ are formed by electrophilic addition to [Os$_{10}$C(CO)$_{24}$]$^{2-}$ and show enhanced reactivity towards nucleophiles such as PR$_3$.[110] Halogen-containing carbido species were also isolated during studies of the reactivity of [Re$_7$C(CO)$_{21}$]$^{3-}$.[111,112] One such product, [Re$_4$C(CO)$_{15}$I]$^-$, contains an exposed carbide atom at the centre of a tetrahedrally distorted square of Re atoms.[111] Three preliminary reports on the first nickel carbidocarbonyl clusters have appeared. Reaction of [Ni$_6$-(CO)$_{12}$]$^{2-}$ with CCl$_4$, C$_2$Cl$_4$ or C$_2$Cl$_6$ resulted in the isolation of [Ni$_{10}$C(CO)$_{18}$]$^{2-}$, [Ni$_9$C(CO)$_{17}$]$^{2-}$ and [Ni$_8$C(CO)$_{16}$]$^{2-}$. The [Ni$_{10}$C(CO)$_{18}$]$^{2-}$ dianion may be converted into the tetra-anion [Ni$_{16}$C(CO)$_{23}$]$^{4-}$ by reaction with PPh$_3$. This species has a unique structure with a large cavity within the metallic polyhedron which is occupied by 4 interstitial carbide atoms linked in pairs. Within each pair the C-C separation (average 1.38 Å) is the shortest found in related species (5).[114] Reaction of [Ni$_9$C(CO)$_{17}$]$^{2-}$ with Co$_3$CCl(CO)$_9$ yields [Co$_3$Ni$_7$C$_2$(CO)$_{16}$]$^{2-}$, which contains a 3,4,3 stack of metal atoms of C$_{2h}$ ideal symmetry.[115]

Several new nitrido complexes of Co, Rh and Ru have been prepared.[116-118] Tetranuclear nitrido Ru complexes are formed by CO reduction of the μ$_2$-NO ligand in Ru$_3$(CO)$_{10}$(μ-NO)$_2$.[118] The products Ru$_4$N(CO)$_{12}$(μ$_2$-NO) and Ru$_4$N(CO)$_{12}$-(μ$_2$-NCO) both adopt a 64-electron butterfly structure in which the "hinge" M-M vector is long and supports a μ$_2$-NO or μ$_2$-NCO bridging group.[118] Bi- and tri-nuclear iron carbonyl sulphido species Fe$_2$S$_2$(CO)$_6$, Fe$_3$S$_2$(CO)$_9$, Fe$_3$S(CO)$_{10}$ and Fe$_3$S(CO)$_9$(SO) are formed in reaction of [FeH(CO)$_4$]$^-$ or [Fe$_3$H(CO)$_{11}$]$^-$ with either sulphite or polyslphide.[119] The structure of Fe$_2$CoS(CO)$_8$(NO), prepared by nitrosation of HFe$_2$CoS(CO)$_9$, has been reported,[120] and the mixed metal clusters [Fe$_3$(CO)$_9$SM(CO)$_5$]$^{2-}$ (M = Cr or W)[121] and Os$_3$(CO)$_9$(μ$_3$-S)(μ$_4$-S)W(CO)$_5$[122] have been characterised.

5 Metal Carbonyl Hydrides

Three independent reports have been published citing evidence for the formation of Cr(CO)$_5$(H$_2$) from Cr(CO)$_6$ and H$_2$ under differing reaction conditions.[122-125] Both cis-Cr(CO)$_4$(H$_2$)$_2$ and Cr(CO)$_4$(H$_2$) were also tentatively identified in a photolysis study.[125] A neutron diffraction study on Ph$_4$P[W$_2$H(CO)$_{10}$] has revealed W-H and W-W distances of 1.897(5) and 3.340(5) Å respectively, and a W-H-W angle of 123.4°.[126] Both the molybdenum anion [Mo$_2$H(CO)$_{10}$]$^-$ and the mononuclear anions [MH(CO)$_5$]$^-$ (M = Cr, Mo or W) have been used for the selective reduction of organic carbonyl compounds.[127,128] The mixed heterobimetallic complexes PPN[FeMH(CO)$_9$] (M = Cr, Mo or W) are olefin isomerisation catalysts.[129] A further report on the relative acidity of some metal carbonyl hydrides has been published,[130] and e.s.r. measurements on radical anions have been interpreted.[131] Detailed n.m.r. measurements have been published on a series of trinuclear derivatives,[132] and the heat of formation of the iron-formyl anion [Fe(CO)$_4$(CHO)]$^-$ has been derived using new data for the hydride affinity of [FeH(CO)$_4$]$^-$.[133] A very low activation energy has been determined for the carbonyl rearrangement in FeH$_2$(CO)$_4$.[134] The chemistry of [Ru$_3$H(CO)$_{11}$]$^-$ has been explored with respect to the catalysis of the water gas shift reaction,[135] and the synthesis and characterisation of HRuRh$_3$(CO)$_{12}$ has been described.[136] This cluster belongs to the class of hydride clusters HMM'$_3$(CO)$_{12}$ (M = Fe, Ru or Os and M' = Co, Rh or Ir) of which only three examples are known. The tetranuclear clusters Os$_4$H$_2$(CO)$_{15}$ and Os$_4$H$_3$X(CO)$_{13}$ (X = Cl, Br or I) have been synthesised by the reaction of OsH$_2$(CO)$_4$ with Os$_3$ cluster carbonyl derivatives.[137] Deprotonation of the butterfly cluster Os$_4$H$_3$(CO)$_{12}$I with PPN(NO$_2$) yields PPN[Os$_4$H$_2$(CO)$_{12}$-(µ-I)]$^-$, which contains a tetrahedral Os$_4$ core with a terminal iodide ligand.[138] A low yield of Os$_4$H$_3$(CO)$_{11}$(NO) also results from this reaction.[139] A new Rh$_{12}$ metallic array of idealised D$_{2d}$ symmetry was found in [Rh$_{12}$H$_2$(µ$_3$-CO)(µ-CO)$_9$-(CO)$_{13}$], which results from the treatment of [Rh$_{14}$(CO)$_{25}$]$^{4-}$ with CF$_3$COOH,[140] and [Ni$_{38}$Pt$_6$(CO)$_{48}$H$_{6-n}$]$^{n-}$ (n = 5 or 4) ions have been prepared and characterised for use as molecular models for "cherry" crystallites.[141]

6 Metal Carbonyl Halides

The [Nb$_2$X$_3$(CO)$_8$]$^-$ (X = Cl, Br or I) anions have been obtained by halogen oxidation of Nb(CO)$_6^-$ in the presence of HX. The chloro-bridged anion has idealised C$_{2v}$ symmetry and a Nb-Nb separation of 3.631(1) Å.[142] Halo bridges are also present in both W$_2$Br$_4$(CO)$_8$ and W$_2$Br$_4$(CO)$_7$.[143] The latter complex is thought to be formed by CO loss from the former and it contains three bridging and one terminal Br ligand. The new [Re$_2$(CO)$_9$I]$^-$ anion was formed in the reaction of [Re(CO)$_5$]$^-$ with MeC(CH$_2$I)$_3$,[144] and Re(CO)$_5$Me has been used to prepare [Re$_2$(CO)$_{10}$F]$^+$, which yields Re(CO)$_5$F in MeCN.[145] Co-condensation of Re, Ru, Rh, Ir and Pt atoms with oxalyl chloride gives metal chlorocarbonyl derivatives used as precursors to Re$_2$(CO)$_8$Cl$_2$, α-[Ru$_2$(CO)$_6$Cl$_2$(µ-Cl)$_2$], Rh$_2$(CO)$_4$(µ-Cl)$_2$ and cis-Pt(CO)$_2$Cl$_2$.[146] The role of rhodium carbonyl iodides in the catalytic reactions of MeOH with CO has also received attention.[147]

References

1   H.A. Skinner and J.A. Conner, Pure Appl. Chem., 1985, 57, 79.
2   J.J. Dechter, Prog. Inorg. Chem., 1985, 33, 393.
3   M. Minelli, J.H. Enemark, R.T.C. Browlee, M.J. O'Conner, and A.G. Wedd, Coord. Chem. Rev., 1985, 68, 169.
4   M.Y. Darensbourg, Prog. Inorg. Chem., 1985, 33, 221.
5   T.J. Meyer and J.V. Casper, Chem. Rev., 1985, 85, 187.

6   R.G. Pearson, ibid, 1985, 85, 41.
7   F. Basolo, Inorg. Chim. Acta, 1985, 100, 33.
8   P.V. Broadhurst, Polyhedron, 1985, 4, 1801.
9   W. Geiger and N.G. Connelly, Adv. Organomet. Chem., 1985, 24, 1.
10 R.D. Adams and I.T. Horvath, Prog. Inorg. Chem., 1985, 33, 127.
11 W.L. Gladfelter, Adv. Organomet. Chem., 1985, 24, 41.
12 R.D. Adams, Polyhedron, 1985, 4, 2003.
13 D.M.P. Mingos and R.W.M. Wardle, Transition Met. Chem., 1985, 10, 441.
14 F. Calderazzo and G. Pampaloni, Inorg. Synth., 1985, 23, 34; M.D. Gillone, ibid., p.27; F. Calderazzo, R. Poliand and D. Vitali, ibid., p.32; S.P. Schmidt, W.C. Trogler and F. Basolo, ibid., p.41.
15 R.J. Trautman, D.C. Gross and P.C. Ford, J. Am. Chem. Soc., 1985, 107, 2355.
16 D.C. Gross and P.C. Ford, J. Am. Chem. Soc., 1985, 107, 585.
17 D.A. Brown, N.J. Fitzpatrick and M.A. McGinn, J. Organomet. Chem., 1985, 293, 235.
18 M.A. Lilga and J.A. Ibers, Organometallics, 1985, 4, 590.
19 A.A. Koridze, O.A. Kizas, N.E. Kolobova and P.V. Petrovskii, J. Organomet. Chem., 1985, 292, C1.
20 K. Tatsumi, A. Nakamura, P. Hofmann, P. Stauffert and R. Hoffmann, J. Am. Chem. Soc., 1985, 107, 4440.
21 T.D. Tilley, ibid., 1985, 107, 4084.
22 M.M. Chisholm, J.A. Heppert, J.C. Huffman and W.E. Streib, J. Chem. Soc., Chem. Commun., 1985, 1771.
23 D.S. Barrett and D.J. Cole-Hamilton, J. Chem. Soc., Chem. Commun., 1985, 1557.
24 G. Blyholder, K-M. Zhao and M. Lawless, Organometallics, 1985, 4, 1371 and 2170; N. Kogan and K. Morokuma, J. Am. Chem. Soc., 1985, 107, 7230.
25 T. Ziegler, Organometallics, 1985, 4, 675.
26 C. Daniel, I. Hyla-Kryspin, J. Demuynck and A. Veillard, Nouv. J. Chem., 1985, 9, 581.
27 M.S. Kralik, J.P. Hutchinson and R.D. Ernst, J. Am. Chem. Soc., 1985, 107, 8296.
28 P.G. Rasmussen, J.E. Anderson, O.H. Bailey, M. Tamres and J.C. Bayon, ibid., 1985, 107, 281.
29 C.P. Horwitz, E.M. Holt, C.P. Brock and D.F. Shriver, ibid., 1985, 107, 8136.
30 C.P. Horwitz and D.F. Shriver, ibid., 1985, 107, 8147.
31 W.J. Evans, J.W. Gate, L.A. Hughes, H. Zhang and J.L. Atwood, ibid., 1985, 107, 3729.
32 M.M. Millar, T. O'Sullivan, N. de Vries and S.A. Koch, ibid., 1985, 107, 3714.
33 E.J.M. de Boer, J. de With and A.G. Orpen, J. Chem. Soc., Chem. Commun., 1985, 1666.
34 S-S. Sung and R. Hoffmann, J. Am. Chem. Soc., 1985, 107, 578.
35 D.W. Moon, S.L. Bernasek, D.J. Dwyer and J.L. Gland, ibid., 1985, 107, 4363.
36 A.B. Anderson and Md. K. Awad, ibid., 1985, 107, 7854.
37 C. Mealli, ibid., 1985, 107, 2245.
38 J-F. Halet, R. Hoffmann and J-Y. Sailland, Inorg. Chem., 1985, 24, 1695.
39 R.L. Johnston and D.M.P. Mingos, J. Organomet. Chem., 1985, 280, 407.
40 R.L. Johnston and D.M.P. Mingos, ibdi., 1985, 280, 419.
41 D.M.P. Mingos, Inorg. Chem., 1985, 24, 114.
42 B.K. Ten, ibid., 1985, 24, 115,
43 L. Hedberg, K. Hedberg, S.K. Satija and B.I. Swanson, ibid, 1985, 24, 2766.
44 K. Hedberg, L. Hedberg, K. Hagen, R.R. Ryan and L.H. Jones, ibid., 1985, 24, 2771.
45 F. Calderazzo, G. Pampaloni, M. Lanfranchi and G. Pelizzi, J. Organomet. Chem., 1985, 296, 1.

46 J.M. Maher, R.P. Beatty and N.J. Cooper, Organometallics, 1985, 4, 1354.
47 E. Oldfield, M.A. Keniry, S. Shimoda, S. Schramm, T.L. Brown and H.S. Gutowsky, J. Chem. Soc., Chem. Commun., 1985, 791; K. Ihmels, D. Rehder and V. Pank, Inorg. Chim. Acta, 1985, 96, L69.
48 G. Blyholder and J. Springs, Inorg. Chem., 1985, 24, 224.
49 K.R. Lane, L. Sallans and R.R. Squires, J. Am. Chem. Soc., 1985, 107, 5369.
50 T.A. Seder, S.P. Church, A.J. Ouderkirk and E. Weitz, ibid., 1985, 107, 1432.
51 T.R. Fletcher and R.N. Rosenfeld, ibid., 1985, 107, 2203.
52 S.P. Church, F-W. Grevels, H. Hermann and K. Schaffner, Inorg. Chem., 1985, 24, 418.
53 M.J. Almond, A.J. Downs and R.N. Perutz, ibid., 1985, 24, 275.
54 D.J. Chornay, M.A. Coplan, J.A. Tossell, J.H. Moore, E.J. Baerends and A. Rozendaal, ibid., 1985, 24, 877.
55 B.H. Weiller, J-P. Liu and E.R. Grant, J. Am. Chem. Soc., 1985, 107, 1595.
56 P.R. Das, T. Nishimura and G.G. Meisels, J. Phys. Chem., 1985, 89, 2808.
57 D.J.A. Fredeen and D.H. Russel, J. Am. Chem. Soc., 1985, 107, 3762.
58 G. Jaouen, A. Vessières, S. Top, A.A. Ismail and I.S. Butler, ibid., 1985, 107, 4778.
59 S.P. Church, F.W. Grevels, H. Hermann, J.M. Kelly, W.E. Klotzbücher and K. Schaffner, J. Chem. Soc., Chem. Commun., 1985, 594.
60 J.M. McCall, J.R. Morton and K.F. Preston, J. Magn. Res., 1985, 64, 414.
61 T.R. Herrinton and T.L. Brown, J. Am. Chem. Soc., 1985, 107, 5700.
62 P.H. Kasai and P.M. Jones, ibid., 1985, 107, 813.
63 P.H. Kasai and P.M. Jones, J. Phys. Chem., 1985, 89, 1147.
64 C.A. Hampson, J.A. Howard and B. Mite, ibid., 1985, 966.
65 P.H. Kasai and P.M. Jones, J. Am. Chem. Soc., 1985, 107, 6385.
66 J.S. Thompson and R.M. Swiatek, Inorg. Chem., 1985, 24, 110.
67 R.D. Feltham, G. Elbaze, R. Ortega, C. Eck and J. Dubrawski, ibid., 1985, 24, 1503.
68 R. Uson, J. Fornies, M. Tomás and B. Menjón, Organometallics, 1985, 4, 1913.
69 F. Calderazzo and R. Poli, Gazz. Chim. Ital., 1985, 115, 573.
70 K. Yasufuku, H. Noda, J-I. Iwai, H. Ohtani, M. Hoshino and T. Kobayashi, Organometallics, 1985, 4, 2174.
71 W.K. Meckstroth, D.T. Reed and A. Wojcicki, Inorg. Chim. Acta, 1985, 105, 147.
72 W.K. Meckstroth, R.B. Freas, W.D. Reents Jr. and D.P. Ridge, Inorg. Chem., 1985, 24, 3139.
73 P.R. Andréa, A. Terpstra, D.J. Stufkens and A. Oskam, Inorg. Chim. Acta, 1985, 96, L57.
74 W.K. Meckstroth and D.P. Ridge, J. Am. Chem. Soc., 1985, 107, 2281.
75 L-Y Hsu, N. Bhattacharyya and S.G. Shore, Organometallics, 1985, 4, 1483.
76 J.R. Morton and K.F. Preston, Inorg. Chem., 1985, 24, 3317.
77 T-Y Luh, H.N.C. Wong and B.F.G. Johnson, Angew. Chem. Int. Ed. Engl., 1985, 24, 45.
78 K.T. Boon and N.J.A. Sloane, Inorg. Chem., 1985, 24, 4545.
79 A. Shojaie and J.D. Atwood, Organometallics, 1985, 4, 187.
80 A. Poc and V.C. Sekhar, Inorg. Chem., 1985, 24, 4376.
81 V.V. Kane, J.R.C. Light and M.C. Whiting, Polyhedron, 1985, 4, 533.
82 F.C. Descrosiers, D.A. Wink and P.C. Ford, Inorg. Chem., 1985, 24, 1.
83 G.A. Foulds, B.F.G. Johnson and J. Lewis, J. Organomet. Chem., 1985, 296, 147.
84 S. Dobos, I. Boszormenyi, V. Silberer, L. Guczi and J. Mink, Inorg. Chim. Acta, 1985, 96, L13.
85 J.C. Cyr, J.A. DeGray, D.K. Gosser, E.S. Lee and P.H. Rieger, Organometallics, 1985, 4, 950.
86 G. van Buskirk, C.B. Knobler and H.D. Kaesz, ibid., 1985, 4, 149.

87 E. Roland and H. Vahrenkamp, Chem. Ber., 1985, 118, 1133; S. Aime, R. Gobetto, D. Osella, L. Milone, G.E. Hawkes and E.W. Randall, J. Magn. Res., 1985, 65, 308.
88 A. Fumagalli, F. Demartin and A. Sironi, J. Organomet. Chem., 1985, 279, C33; F. Oldani and G. Bor, ibid., 1985, 279, 459.
89 S. Martinengo, A. Fumagalli and P. Chini, ibid., 1985, 284, 275.
90 B. Tulyathan and W.E. Geiger, J. Am. Chem. Soc., 1985, 107, 5960.
91 B-H Chang, J. Organomet. Chem., 1985, 291, C31.
92 P. Klufers, Angew. Chem. Int. Ed. Engl., 1985, 24, 70.
93 G. Doyle, K.A. Eriksen and P. Van Engen, J. Am. Chem. Soc., 1985, 107, 7914.
94 K.H. Whitmire, M.R. Churchill and J.C. Fettinger, ibid., 1985, 107, 1056.
95 M.R. Churchill, J.C. Fettinger and K.H. Whitmire, J. Organomet. Chem., 1985, 284, 13.
96 G. Huttner, U. Weber, B. Sigwarth, O. Scheidsteger, H. Lang and L. Zsolnai, ibid., 1985, 282, 331.
97 O. Scheidsteger, G. Huttner, K. Dehnicke and J. Pebler, Angew. Chem. Int. Ed. Engl., 1985, 24, 428; C. Belin, T. Makani and J. Rozière, J. Chem. Soc., Chem. Commun., 1985, 118.
98 S.P. Foster, K.M. Mackay and B.K. Nicholson, Inorg. Chem., 1985, 24, 909.
99 J.M. Burlitch, S.E. Hayes and J.T. Lemley, Organometallics, 1985, 4, 167.
100 P. Braunstein, J. Rosé, A. Tiripicchio and M. Tirpicchio-Camellini, Angew. Chem. Int. Ed. Engl., 1985, 24, 767.
101 J.M. Ragosta and J.M. Burlitch, J. Chem. Soc., Chem. Commun., 1985, 1187.
102 A. Gourdon and Y. Jeannin, J. Organomet. Chem., 1985, 290, 199.
103 A. Gourdon and Y. Jeannin, ibid., 1985, 282, C39.
104 V.E. Lopatin, S.P. Gubin, N.M. Mikova, M.Ts. Tsybenov, Yu.L. Slovokhotov and Yu.T. Struchkov, ibid., 1985, 292, 275.
105 J.A. Hriljac, P.N. Swepston and D.F. Shriver, Organometallics, 1985, 4, 158.
106 K. Henrick, B.F.G. Johnson, J. Lewis, J. Mace, M. McPartlin and J. Morris, J. Chem. Soc., Chem. Commun., 1985, 1617.
107 S. Martinengo, D. Strumolo, P. Chini, V.G. Albano and D. Braga, J. Chem. Soc., Dalton Trans., 1985, 35.
108 V.G. Albano, D. Braga, A. Fumagalli and S. Martinengo, ibid., 1985, 1137.
109 V.G. Albano, D. Braga, D. Strumolo, C. Seregni and S. Martinengo, ibid., 1985, 1309.
110 R.J. Goudsmit, P.F. Jackson, B.F.G. Johnson, J. Lewis, W.J.H. Nelson, J. Puga, M.D. Vargas, D. Braga, K. Henrick, M. McPartlin and A. Sironi, ibid., 1985, 1795.
111 T. Beringhelli, G. Ciani, G. D'Alfonso, A. Sironi and M. Freni, J. Chem. Soc., Chem. Commun., 1985, 978.
112 T. Beringhelli, G. D'Alfonso, M. Freni, G. Cianni and A. Sironi, J. Organomet. Chem., 1985, 295, C7; A. Ceriotti, G. Longoni, M. Manassero, N. Masciocchi, L. Resconi and M. Sansoni, J. Chem. Soc., Chem. Commun., 1985, 181.
113 A. Ceriotti, G. Longoni, M. Mannassero, M. Perego and M. Sansoni, Inorg. Chem., 1985, 24, 117.
114 A. Ceriotti, G. Longoni, M. Manassero, N. Masciocchi, G. Piro, L. Resconi and M. Sansoni, J. Chem. Soc., Chem. Commun., 1985, 1402.
115 A. Arrigoni, A. Ceriotti, R.D. Rergola, G. Longoni, M. Manassero and M. Sansoni, J. Organomet. Chem., 1985, 296, 243.
116 R.E. Stevens, P.C.C. Liu and W.L. Gladfelter, ibid., 1985, 287, 133.
117 M.L. Blohm and W.L. Gladfelter, Organometallics, 1985, 4, 45.
118 J.P. Attard, B.F.G. Johnson, J. Lewis, J.M. Mace and P.R. Raithby, J. Chem. Soc., Chem. Commun., 1985, 1526.
119 C. Glidewell, J. Organomet. Chem., 1985, 295, 73.
120 K. Fischer, W. Deck, M. Schwarz and H. Vahrenkamp, Chem. Ber., 1985, 118, 4946.
121 J. Tukács and L. Markó, Transition Met. Chem., 1985, 10, 21.
122 R.D. Adams, I.T. Horváth and S. Wang, Inorg. Chem., 1985, 24, 1728.

123 R.K. Upmacis, G.E. Gadd, M. Poliakoff, M.B. Simpson, J.J. Turner,
     R. Whyman and A.F. Simpson, J. Chem. Soc., Chem. Commun., 1985, 27.
124 S.P. Church, F.W. Grevels, H. Hermann and K. Schaffner, ibid., 1985, 30.
125 R.L. Sweany, J. Am. Chem. Soc., 1985, 107, 2374.
126 D.W. Hart, R. Bau and T.F. Koetzle, Organometallics, 1985, 4, 1590.
127 D.H. Gibson and Y.S. El-Omrani, ibid., 1985, 4, 1473.
128 P.L. Gans, S.C. Kao, K. Youngdahl and M.Y. Darensbourg, J. Am. Chem. Soc., 1985, 107, 2428.
129 P.A. Tooley, L.W. Arndt and M.Y. Darensbourg, ibid., 1985, 107, 2422.
130 S.C. Kao, C.T. Spillett, C. Ash, R. Lusk, Y.K. Park and M.Y. Darensbourg, Organometallics, 1985, 4, 83.
131 T. Sowa, T. Kawamura and T. Yonezawa, J. Organomet. Chem., 1985, 284, 337; T. Sowa, T. Kawamura, T. Yamabe and T. Yonezawa, J. Am. Chem. Soc., 1985, 107, 6471.
132 T. Beringhelli, H. Moluiari and A. Pastore, J. Chem. Soc., Dalton Trans., 1985, 1899; T. Beringhelli, G. D'Alfonso and H. Moluiari, J. Organomet. Chem., 1985, 295, C35; E. Rosenberg, C.B. Thorsen, L. Milone and S. Aime, Inorg. Chem., 1985, 24, 231; G.E. Hawkes, L.Y. Lian, E.W. Randall, K.D. Sales and S. Aime, J. Chem. Soc., Dalton Trans., 1985, 225.
133 K.R. Laue, L. Sallons and R.R. Squires, Organometallics, 1985, 4, 408.
134 K.H. Whitmire and T.R. Lee, J. Organomet. Chem., 1985, 282, 95.
135 J.C. Bricker, C.C. Nagel, A.A. Bhattacharyya and S.G. Shore, J. Am.Chem. Soc., 1985, 107, 377.
136 J. Pursiainen, T.A. Pakkanen and J. Jääskeläinen, J. Organomet. Chem., 1985, 290, 85.
137 E.J. Ditzel, B.F.G. Johnson, J. Lewis, P.R. Raithby and M.J. Taylor, J. Chem. Soc., Dalton Trans., 1985, 555.
138 J. Puga, R. Sánchez-Delgado, A. Andriollo, J. Ascanio and D. Braga, Organometallics, 1985, 4, 2064.
139 J. Puga, R. Sánchez-Delgado and D. Braga, Inorg. Chem., 1985, 24, 3971.
140 G. Ciani, A. Sironi and S. Martinengo, J. Chem. Soc., Chem. Commun., 1985, 1757.
141 A. Ceriotti, F. Demartin, G. Longoni, M. Manassero, M. Marchionna, G. Pira and M. Sansoni, Angew. Chem., Int. Ed. Engl., 1985, 24, 697.
142 F. Calderazzo, M. Castellani, G. Pampaloni and P.F. Zanazzi, J. Chem. Soc., Dalton Trans., 1985, 1989.
143 F.A. Cotton, L.R. Falvello and J.H. Meadows, Inorg. Chem., 1985, 24, 514.
144 R. Poli, G. Wilkinson, M. Motevalli and M.B. Hursthouse, J. Chem. Soc., Dalton Trans., 1985, 931.
145 K. Raab and W. Beck, Chem. Ber., 1985, 118, 3830.
146 P.R. Brown, F.G.N. Cloke and M.L.H. Green, Polyhedron, 1985, 4, 869.
147 A.G. Kent, B.E. Mann and G.A. Manuel, J. Chem. Soc., Chem. Commun., 1985, 728.

9
Organometallic Compounds containing Metal–Metal Bonds

BY W. E. LINDSELL

1 Introduction

The general format of this chapter remains the same as in previous years.

1.1 Reviews. – The photochemistry of binuclear metal-metal bonded carbonyl complexes is presented in a general review[1] and in an article on photochemical disproportionation reactions.[2] Both bridged-bimetallic and cluster compounds are included in a survey of electron transfer reactions of polynuclear organo-transition metal complexes.[3] Reactions of triply bonded complexes $[M_2(OR)_6]$ (M = Mo,W) with molecules containing C≡C, C≡N or C≡O functions are also reviewed.[4]

An account of the reactivity of metal carbonyl clusters has appeared.[5] Other reviews of polynuclear systems cover selective metal-ligand interactions in heterometallic transition metal clusters,[6] metal exchange reactions in clusters[7] and the chemistry of phosphine carbonyl clusters of palladium and platinum[8a] or of platinum alone.[8b] Transition metal clusters incorporating specific ligand groupings are the subject of reviews on alkyne substituted clusters,[9] clusters with nitrosyl or nitrido ligands[10] and sulphido-osmium carbonyl clusters.[11] Metal-metal bonded systems, including clusters, are also discussed in accounts of α-diimine complexes,[12] of the cleavage of ligand P-C bonds[13] and of complexes containing multiply bonded As, Sb or Bi ligands.[14] Heteronuclear metal-metal bonded binuclear[15a] and cluster[15b] compounds reported in 1982/83 have been surveyed.

1.2 Theoretical Studies. – Molecular mechanics have been applied to anions $[M_2Me_8]^{4-}$ and when M = Cr a δ contribution to the quadruple bond of 11 kcal/mol is estimated.[16] Xα-SW calculations on $[W_2(O_2CH)_4Me_2]$ suggest the presence of a quadruple bond in spite of strong axial ligation by alkyl groups.[17] Fenske-Hall calculations on $[V_2\{2,6-C_6H_3(OMe)_2\}_4]$ indicate a $\sigma^2\pi^2\delta^2$ triple bond (V≡V).[18] Qualitative MO studies on binuclear Au(I) compounds of the type $[Au_2\{(CH_2)_2PR_2\}_2]$ are reported.[19]

The interrelationships between and relative merits of the Topological Electron Counting (TEC) and Polyhedral Skeletal Electron Pair (PSEP) theories are discussed;[20] the TEC theory has been generalised to include 1D, 2D and

(For references see page 183)

mixed metal/non-metal clusters[20c] and justified in terms of MO theory.[21] Stone's Tensor Surface Harmonic (TSH) theory has been applied to 3- and 4-connected polyhedral molecules[22] and TSH theory has been used in both simplified[23a] and extended[23b] forms to study various clusters. The metal cluster-borane analogy has been analysed.[24] Electron counting rules for high nuclearity clusters have been derived[25] and also atom counting rules for preferred nuclearities (magic numbers) of polygonal and polyhedral clusters.[26]

EHMO calculations are reported for heterometallic alkyne clusters M_3C_2, indicating the preferred orientation of the alkyne;[27] hydrogen migrations in 1-alkyne bi- and tri-nuclear complexes have also been examined.[28] Electronic structures of square bipyramidal clusters M_4E_2 (E = PR,GeR,S,Te) have been analysed with M in ML_3 local co-ordination, and related to ML_2 containing species.[29] UV-PES and theoretical studies are reported for ferracyclopentadienyl complexes[30a] and for μ_3-allenyl Ru_3 or Os_3 clusters.[30b] The association of triangular Pt_3 or Ni_3 clusters into chains has been analysed by EHMO calculations.[31] EHMO calculations on penta- and hepta-metallic, homo- and hetero-nuclear phosphine-gold clusters are also reported.[32] Steric effects in clusters $[Fe_4(\mu_4-CR)(CO)_{12}]^-$ can be probed using interactive molecular graphics.[33]

1.3 Physical Studies. - H-H couplings, observed in ^1H-^{187}Os satellite n.m.r. spectra have potential applications in structural assignments for osmium clusters.[34] Quadruplar effects of ^{59}Co in n.m.r. spectra of some cobalt clusters have been monitored[35] and ^{63}Cu n.m.r. spectroscopy has been applied to copper containing clusters.[36] 2D ^{13}C n.m.r. experiments on $[Os_3(\mu-H)_2(CO)_{10}]$ illustrate applications to shift assignments and to slow exchange processes.[37] Detailed ^{13}C n.m.r. studies on anions $[Re_3(\mu-H)_{5-n}(CO)_{12}]^{n-}$ (n = 1,2)[38] and, also, e.s.r. studies of $[Mn_3(\mu-H)_3(CO)_{12}]^-$ [39a] and $[Re_2(\mu-H)_2(CO)_6L_2]^-$ {L_2 = $(CO)_2$ or dppm},[39b] have been reported. Tin complexes investigated by multinuclear n.m.r. include examples containing V-Sn,[40a] Ir-Sn[40b] and Pt-Sn[40c] bonds. Mössbauer effect data for trinuclear reduced carbido-iron clusters can be correlated with structure and reactivity.[41]

Magnetic susceptibilities of osmium clusters show the evolution of "metallic" behaviour for high nuclearity species.[42] Electrochemical measurements have been made on several cluster systems, including $[Os_6(CO)_{18}]$[43a] and $M_2^1M_2^2$ species (M^1 = Pd,Pt; M^2 = Cr,Mo,W).[43b]

Measured heats of hydrogenation of metal-metal bonded complexes $[M_2(CO)_6Cp_2]$ are -3.3, +6.3 and -1.5 kcal/mol, for M = Cr, Mo and W, respectively.[44] Further studies of reactions of gas phase transition metal cluster ions with hydrocarbons have been reported.[45]

1.4 Surface Bound Species - There have been a number of reports involving surface bound clusters in catalysis; only results concerning structural aspects of the cluster are briefly considered here. An X-Ray structure determination of $[Os_3(\mu-H)(\mu-OSiEt_3)(CO)_{10}]$ provides a characterised molecular analogue of a surface bound entity on silica.[46] EXAFS results are reported for pyrolysis products of osmium clusters anchored by S or P ligands to silica[47] and for iron clusters on dehydroxylated alumina.[48] Spectroscopic studies on supported catalysts include, FT-IR/ photoacoustic spectra and ^{13}C-CPMAS n.m.r. spectra of $[Mo_2(C_3H_5)_4]$ on alumina or silica,[49] IR spectra of intact $[H_3Os_4(CO)_{12}]^-$ on alumina,[50] and IR, UV and ^{1}H n.m.r. spectra of supported $[Re_4(\mu-OH)_4(CO)_{12}]$ from reaction of $[Re_4H_4(CO)_{12}]$ with silica.[51] Metal segregation occurs on treating a RhOs$_3$ cluster on γ-Al$_2$O$_3$ with CO and H$_2$ at 200°C.[52] $[M_3H(NPh)(CO)_9]^-$ (M = Fe,Os) is an intermediate in the catalytic phenylisocyanate formation from nitrobenzene over polymer supported $[M_3H(CO)_{11}]^-$.[53] Phosphine (L) anchored derivatives $[M_3(CO)_9(dppm)L]$ (M = Ru,Os) or $[M_4(CO)_8\{HC(PPh_2)_3\}L]$ (M = Co,Rh,Ir), attached to silica or polymer, are described[54] and the intercalation of metal cluster cations of Ru, Os or Ir into smectite clay (hectorite) has been studied.[55]

2 Compounds with Homonuclear Transition Metal Bonds

2.1 Titanium, Zirconium and Hafnium. - Preparation and reactions of $[M_2(\mu-Cl)_2Cp_4]$ (M = Zr,Hf) are described, including conversion into diamagnetic $[Zr_2Me_2Cp_4]$ using LiMe.[56] Tetrameric zirconocene $[ZrCp_2]_4$ is formed by reduction of $[ZrH_2Cp_2]$ and may contain Zr-Zr bonds.[57] Paramagnetic $[Ti_5(\mu_3-S)_6Cp_5]$ contains a distorted trigonal bipyramidal Ti$_5$ core and the delocalised bonding comprises an element of direct equatorial Ti-Ti interaction.[58]

2.2 Niobium. - Fulvalene bridged derivatives $[Nb_2(\mu-H)(\mu-X)(\mu-\eta^5,\eta^5-C_{10}H_8)Cp_2]$ (X = H, Halogen, OR) possess 3-centre Nb-H-Nb bonds.[59] Niobium (II) complexes $[Nb_2(\mu-Cl)_2(CO)_4Cp_2']$ and $[Nb_2(\mu-Cl)_2(\eta-C_2Ar_2)_2Cp_2']$ ($Cp' = \eta^5$-C$_5$H$_4$Me) have similar single Nb-Nb bonds but the former possess a folded 'butterfly' Nb$_2$Cl$_2$ core and the latter a planar Nb$_2$Cl$_2$ unit.[60]

2.3 Chromium, Molybdenum and Tungsten. - Structurally characterised $[Mo_2(\mu-TePh)_2(CO)_8]$ is reported[61] and studies of substitution of $[M_2(\mu-PEt_2)_2(CO)_8]$ (M = Cr,W) indicate a dissociative mechanism for the process.[62] Stibinidene intermediates $[W_2(\mu-SbR)(CO)_{10}]$ react with W(CO)$_5$ to form $[W_2(\mu-SbR\{W(CO)_5\})(CO)_{10}]$ containing one W-W bond.[63]

Ring alkylated cyclopentadienyl compexes $[M_2(CO)_6(\eta-C_5H_4R)_2]$ (M = Cr, R = Me; M = Mo,W, R = CHAr$_2$)[64] have been characterised and studied.

The mechanisms for photochemical disproportionation of $[Mo_2(CO)_6Cp'_2]$ by halide ion have been investigated in various solvents[65] and, in a related photochemical reaction with NO_2^-, reduction to co-ordinated NO occurs:[66] the product of the latter reaction $[Mo(CO)_2(NO)Cp]$ is also formed thermally from $[Mo_2(CO)_4Cp_2]$ with NO_2^- or NO_3^-.[66]

Disproportionation of fulvalene complex $[Mo_2(\mu\text{-}\eta^5,\eta^5\text{-}C_{10}H_8)(CO)_6]$ with phosphine (L) gives a charge localised, zwitterion $[(OC)_3Mo^-(\mu\text{-}C_{10}H_8)\text{-}Mo^+(CO)_2L_2]$.[67] A Fischer-type carbene fulvalene complex $[Mo_2(\mu\text{-}\eta^5,\eta^5\text{-}C_{10}H_8)(=\overline{COCH_2CH_2CH_2})(CO)_5]$ is fluxional and liberates propene on degradation.[68]

The new triply bonded complexes $[Mo_2(CO)_4(\eta^5\text{-indenyl})_2]$[69] and $[Cr_2(\mu\text{-}CO)_3(\eta^6\text{-}C_6H_6)_2]$[70] have been structurally characterised. Studies employing cross-over experiments of the reversible CO addition to $[Mo_2(CO)_4Cp_2]$ (Mo≡Mo) appear to rule out a mechanism involving facile CO dissociation from $[Mo(CO)_3Cp]$ formed by Mo-Mo homolysis.[71] Phosphorus ligands will displace one CO in $[Mo_2(CO)_4Cp_2^*]$ ($Cp^* \equiv \eta^5\text{-}C_5Me_5$) in spite of steric crowding.[72] Two CO groups of $[Mo_2(CO)_4Cp_2]$ become bridging and act as linear 0-donors to Ti on reaction with $[Ti(\underline{neo}\text{-}C_5H_{11})Cp_2^*]$ giving complex (1).[73]

Photochemical reaction of ethyne with $[Mo_2(CO)_4Cp_2^*]$ generates a side-on bridging vinylidene ligand in complex (2);[74] thermal isomerisation of (2) forms $[Mo_2(\mu\text{-}HCCH)(CO)_4Cp_2^*]$, reaction with CH_2N_2 gives $[Mo_2(\mu\text{-}CH_2CCH_2)(CO)_4Cp_2^*]$ and mono- or di-protonation with CF_3COOH gives complex (3) or (4), respectively.[74] Studies of η^2-alkyne bridged binuclear carbonyl complexes cover fluxional properties of $[M_2(\mu\text{-}RCCR')(CO)_4Cp'_2]$ (M = Mo,W),[75a] formation of stabilised cation $[Mo_2(\mu\text{-}CH_2C_2CH_2)(CO)_4Cp_2]^{2+}$ [75b] and an X-ray diffraction structure determination of a bridged fulvalene species.[75c]

Protonation of carbyne complexes $[W(CR)(CO)_2Cp]$ (R = p-tol,Me) by HBF_4 affords cationic $[W_2(\mu\text{-}H)(\mu\text{-}RCCR)(CO)_4Cp_2]^+$ which can be reversibly deprotonated;[76] other transformations of the same carbyne precursors include the formation of alkyne-di-tungsten complexes incorporating a ($\eta^5\text{-}C_5R_2^1R_2^2R$) ligand, by coupling with $[W(CO)(R^1C_2R^2)_3]$,[77] and the formation of $[W_2(\mu\text{-}CHR)(CO)_4Cp_2]$ by reaction with $[WH(CO)_5]^-$.[78] Methylisocyanide insertion into a W-H-W bridge of unsaturated $[W_2(\mu\text{-}H)_2(CO)_4Cp_2]$ yields a dimetallacyclopropene complex.[79]

The bridging phosphaalkyne ligand of $[Mo_2(\mu\text{-}PCBu^t)(CO)_4Cp_2]$ co-ordinates via phosphorus to Rh(I), Pd(II) or Pt(II) centres.[80] Full details have been published of several derivatives containing a bridging As_2 group acting as a 4-, 6- or 8- electron donor; e.g. the As_2 group in $[M_2(\mu\text{-}As_2)(CO)_4Cp_2]$ (M = Mo,W) may be further co-ordinated to one or two metal centres.[81] Oxo complex $[M_2(\mu\text{-}O)_2(O)_2Cp_2^*]$ (M = Cr) has a centrosymmetric trans-structure[82a] in contrast to the cis-structure of the molybdenum analogue which may be formed by oxidation with $[NiO_2(CNBu^t)_2]$.[82b] Reactions of dithioester with $[Mo_2(CO)_4(\eta^5-C_5H_{5-n}Me_n)_2]$ produce the structurally characterised complexes $[Mo_2\{\mu-MeC(S)SEt\}(CO)_4Cp_2]$ and $[Mo_2(\mu-SCMe)(\mu-SEt)(CO)_2Cp_2']$.[83] Other chalcogen ligands bridging two M-M bonded $M(CO)_2Cp$ units include η^2-XCR_2 (X = S,Se; R = H,Me; M = Cr),[84a] $\overline{S(CH_2)_n\dot{C}H_2}$ (n = 3-5; M = Mo)[84b] and two SSO_2 ligands (M = W).[84c]

1,2-Dimetallacyclo-1,2-hexyne compounds $[\overline{M_2\{(CH_2)_3\dot{C}H_2\}}(NMe_2)_4]$ (M = Mo,W) are formed from 1,4-dilithiobutane and $1,2-[M_2Cl_2(NMe_2)_4]$.[85] Phosphines and $1,2-[Mo_2Br_2(CH_2SiMe_3)_4]$ afford terminal bis(carbene) complexes, e.g. $[Mo_2Br_2(=CHSiMe_3)_2(PMe_3)_4]$ (Mo≡Mo 2.276(1)Å).[86] Reactions of alkynes with d^3-d^3 (W≡W) complexes are reported in several papers: $[W_2(\mu-NMe_2)_2(\mu-C_2Me_2)Cl_4(py)_2]$ is the product of ligand redistribution from reaction of 2-butyne and $[W_2Cl_2(NMe_2)_4]$ in pyridine;[87a] ethyne and $[W_2Cl_3(NMe_2)_3(PMe_2Ph)_2]$ form $[W_2(\mu-Cl)(\mu-NMe_2)(\mu-C_2H_2)Cl_2(NMe_2)_2(PMe_2Ph)_2]$ at 5°C but at room temperature complex (5) containing σ,η^2-vinyl and σ,η^2-CH_2NMe bridges is produced;[87b] $[W_2(CH_2Ph)_2(OPr^i)_4]$ and 2-butyne initially form $[W_2(CH_2Ph)_2(OPr^i)_4(\eta^2-C_2Me_2)_2]$ which eliminates toluene at 60°C to give $[W_2H(\mu-C_4Me_4)(\mu-CPh)(OPr^i)_4]$.[87c] Trans-$[Pt(C≡CH)_2(PMe_2Ph)_2]$ reacts with $[W_2(OBu^t)_6]$ by successive eliminations of Bu^tOH to form heterometallic products linked by dicarbido(4-) ligands, as in $[Pt\{C_2W_2(OBu^t)_5\}_2(PMe_2Ph)_2]$.[88] Both CO^{89} and di-p-tolylcarbodiimide[90] add across the triple bond of complexes $[W_2(OR)_6]$.

Complexes $[W_2(\mu-CSiMe_3)_2X_4]$ (X = CH_2SiMe_3 or OR) with planar $W_2(\mu-CR)_2$ cores undergo insertion into one of the bridging alkylidyne ligands on reaction with alkynes or allenes;[91] with isocyanides or CO, cleavage of C-N or C-O bond respectively, occurs and a μ-$CCSiMe_3$ bridge is formed.[92] Halogens add to $W_2(\mu-CR)_2$ centres to form $[W_2(\mu-CSiMe_3)_2X_2(CH_2SiMe_3)_4]$ with δ-type HOMO and temperature dependent paramagnetism.[93] A tetranuclear species $[W_4(\mu-CSiMe_3)_2(OEt)_{14}]$ is formed on ethanolysis and contains two binuclear units coupled by OEt bridges.[94]

Reported trinuclear μ_3-alkylidyne bridged derivatives include $[W_3(\mu_3-CMe)(\mu-OPr^i)_3(OPr^i)_6]$,[94] $[Mo_3(\mu_3-CMe)(\mu-Br)_3(\mu-OAc)_3(H_2O)_3]^+$ [95] and paramagnetic $[W_3(\mu_3-O)(\mu_3-CMe)(OAc)_6(H_2O)_3]^{2+}$ which may be reduced to a +1 cation.[96] The dimer of trimers $[W_3(\mu-O)(\mu-CC_3H_7)(OBu^t)_5O]_2$ is also structurally characterised.[97]

The trinuclear carbonyl cluster $[Mo_3(\mu_3-As)(CO)_6Cp_3]$ forms adducts with other metals and gives a spiro-cluster on reaction with a Wittig reagent.[98] The phosphinidene-bridged cluster $[Cr_3(\mu_3-PBu^t)(\mu-Bu^tPPBu^t)(\mu-CO)(CO)_9]$ also contains a Z-diphosphene moiety.[99]

2.4 Manganese and Rhenium.

Binuclear dioxycarbene complexes $[M_2(=\overline{COCH_2CH_2O})_n(CO)_{10-n}]$ (M = Mn,Re; n = 1 or 2) have been prepared from oxirane.[100] Substituted products [Re(CO)$_{10-n}$L$_n$] (L = CNR, PR$_3$) are formed by using heterogeneous palladium catalysts and structures have been determined for species with L = CNR, n = 1, 2 or 3.[101] Crystal structures are also reported for [Mn$_2$(CO)$_8${P(OMe)$_3$}$_2$][102] and [Mn(μ-dppm)$_2$(CO)$_6$][103] and for terminal and bridged tetrahydrothiophene complexes [Mn$_2$(CO)$_9$(SC$_4$H$_8$)] and [Mn$_2$(μ-SC$_4$H$_8$)(CO)$_8$].[104] Photochemical reaction of [Re$_2$(CO)$_{10}$] with P(OPh)$_3$ affords highly substituted products including [Re$_2$(CO)$_4${P(OPh)$_3$}$_6$].[105] The complex <u>mer</u>,<u>fac</u>-[Mn$_2$(μ-dmpm)$_2$(CO)$_6$] (dmpm = Me$_2$PCH$_2$PMe$_2$) is readily oxidised to a mono-cation and is a potent one-electron photoreductant in MeCN solution.[106]

Mechanisms involving spontaneous or induced Mn-Mn bond homolysis of [Mn$_2$(CO)$_8$(PPh$_3$)$_2$] account for its thermal reactions with C$_2$H$_4$Cl$_2$, C$_{16}$H$_{33}$I, O$_2$, NO, P(OEt)$_3$ or PBu$_3^n$.[107] Photochemical reactions of [Mn$_2$(CO)$_8$L$_2$] (L = PR$_3$) with organic halides involve 17e mononuclear intermediates, however in flash photolytic experiments the only products are halides [MnX(CO)$_4$L][108] whereas in some continuous irradiation experiments equimolar amounts of halide and alkyl complexes are formed.[109]

Photochemical reactions of alkenes with phosphine complexes [Re$_2$(CO)$_9$L] or 1,2-[Re$_2$(CO)$_8$L$_2$] (L = monodentate or L$_2$ = μ-bidentate) yield products [Re$_2$(μ-H)(μ-CH=CHR)(CO)$_{8-n}$L$_n$]; monodentate species reacting <u>via</u> M-M bond cleavage and bidentate <u>via</u> photodissociation of CO.[110] Related photochemical reactions of 1-alkynes yield several products including [Re$_2$(μ-H)-(μ-L͡L)(C≡CR)(CO)$_7$], [Re$_2$(μ-H)(μ-σ,η^2-C≡CR)(μ-L͡L)(CO)$_6$] and [Re$_2$(μ-σ,η^2-CR=CH$_2$)(μ-σ,η^2-C≡CR)(CO)$_5$(L͡L)] (L͡L = dppm or dmpm).[111] Complexes [Re$_2$(μ-L͡L)$_2$(CO)$_6$] (L͡L = dppm and/or dmpm) are less reactive but do react at high temperatures with alkynes and with H$_2$O, MeOH or a hydrogen donor.[112]

Protonation with HBF$_4$.Et$_2$O of [Mn$_2$(μ-η^2-CO)(μ-dppm)$_2$(CO)$_4$] gives [Mn$_2$H(μ-η^2-CO)(μ-dppm)$_2$(CO)$_4$]$^+$ which decomposes above -60°C to [Mn$_2$H(μ-dppm)$_2$(CO)$_6$]$^+$.[113] Mono-protonation of complexes [Mn$_2$H(μ-dppm)$_2$(CN)-(CO)$_n$] occurs; when n = 5 a terminal CNH ligand is formed, but when n = 4 the bridging η^2-CN is transformed into bridging η^2-CNH.[114] The synthesis and some reactions of [Mn$_2$(μ-H)$_2${μ-(EtO)$_2$POP(OEt)$_2$}(CO)$_6$] are described.[115]

Binuclear mono- and di-carbonyl adducts are formed by [Re$_2$(μ-dppm)$_2$Cl$_4$] and the structurally characterised dicarbonyl species [Re$_2$(μ-Cl)(μ-CO)(μ-dppm)$_2$Cl$_3$(CO)] undergoes chloride substitution by nitriles or isonitriles in the presence of TlPF$_6$.[116] Non-cleavage products of [Re$_2$(μ-dppm)$_2$Cl$_4$] are formed in reactions with isonitriles, including the unusual paramagnetic complex [Re$_2$(μ-Cl)(μ-C=NHBu^t)(μ-dppm)$_2$Cl$_2$(CNBu^t)$_2$]PF$_6$ formed in a complex redox-protonation process.[117]

Photolyses at a MLCT band of various complexes [(OC)$_5$MM'(CO)$_3$(α-diimine)] {M,M' = Mn,Re; α-diimine = various ligands, but especially R-dab (RN=CHCH=NR)} have been studied in inert gas matrices, in PVC films and in Me-thf solutions

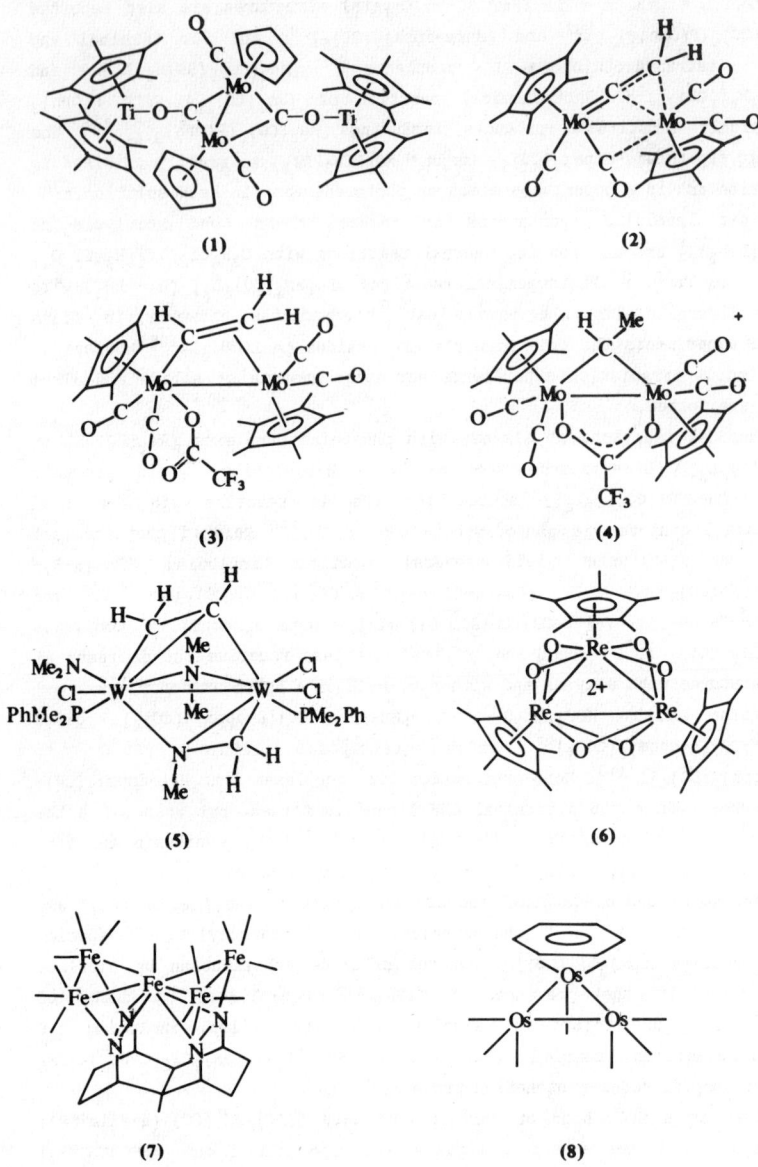

over a range of temperatures: various reactions are observed including a change in R-dab co-ordination, carbonyl substitution and photocatalytic formation of $[Mn(CO)_3(\alpha\text{-diimine})(PBu_3^n)]^+[Mn(CO)_5]^-$ in the presence of PBu_3^n.[118] Eight electron bonded Bu^t-dab is present in $[Mn_2(\mu\text{-}Bu^t\text{-dab})(CO)_6]$ and C-C coupling of Bu^t-dab generates the unsymmetrically bridging ligand in $[Mn_2(\mu\text{-}\{Bu^tN=CHC(H)(NR)\}_2)(CO)_6]$.[119]

The structure of $[Mn_2\{\mu\text{-}C(Fc)CO\}(CO)_6Cp']$, containing a bridging ferrocenyl-ketenyl ligand, has been determined.[120]

Deoxygenation of $[ReO_3Cp^*]$ in the presence of varying amounts of air produces $[Re_2(\mu\text{-}O)_2O_2Cp_2^*]$, $[Re_2(\mu\text{-}O)_2(O)(OReO_3)_2Cp_2^*]$ or the cation $[Re_3(\mu\text{-}O)_6Cp_3^*]^{2+}$ (6).[121] Reaction of $[Re_2(\mu\text{-}O)_2(O)_2Cp_2^*]$ with o-quinones affords both mono- and bi-nuclear oxo-rhenium products.[122] Phenols react with $[Re_3(\mu\text{-}H)_4(CO)_{10}]^-$ to form $[Re_3(\mu\text{-}H)_3(\mu\text{-}OC_6R_5)(CO)_{10}]^-$ (R = H,F).[123] Fragmentation of $[Re_7C(CO)_{21}]^{3-}$ by I_2 produces $[Re_4C(CO)_{15}I]^-$ which contains a carbide atom in the centre of a tetrahedrally distorted Re_4 square.[124] Square pyramidal $[Re_5(\mu_4\text{-}PMe)(\mu_3\text{-}P\{Re(CO)_5\})(\mu\text{-}PMe_2)(CO)_{14}]$ and trigonal prismatic $[Re_6(\mu_4\text{-}PMe)_3(CO)_{18}]$ are reported.[125]

2.5 Iron. - The structure of $[Fe_2\{\mu\text{-}CH(Ph)C_6H_4\}(CO)_6]$ has been determined by neutron diffraction.[126] Complexes $[Fe_2(\mu\text{-}COEt)(\mu\text{-}\sigma,\eta^2\text{-}CRCR'H)(CO)_6]$ [127a] react with activated alkynes to give insertion products and structurally characterised species are $[Fe_2\{\mu\text{-}C(OEt)C(COOMe)C(COOMe)\}(\mu\text{-}CPhCPhH)(CO)_5]$ and $[Fe_2\{\mu\text{-}C(CF_3)C(CF_3)CHCHCH(OMe)\}(CO)_6]$.[127b] Reactions of electrophiles with $[Fe_2\{\mu\text{-}CPhCPhC(CF_3)H\}(CO)_6]$ cause F^- abstraction from a CF_3 group and subsequent ethanolysis affords $[Fe_2\{\mu\text{-}CPhCPhC(CF_3)CHC(OEt)_2\}(CO)_6]$.[128]

The structures of $[Fe_2(\mu\text{-}CO)(\mu\text{-}dmpm)_2(CO)_4]$[129] and $[Fe_2(\mu\text{-}CO)\{\mu\text{-}Me_2PCH_2P(Me)CH_2PMe_2Fe(CO)_4\}(CO)_6]$[130] are reported. Photochemical rearrangement of asymmetric $[Fe_2(\mu\text{-}dad)(CO)_5\{P(OR)_3\}]$ (dad = 1,4-diaza-1,3-diene) results in the apparent migration of $P(OR)_3$ between inequivalent Fe atoms.[131]

The structure of diphosphido-complex $[Fe_2(\mu\text{-}PhP(CH_2)_3PPh)(CO)_6]$ has been determined.[132] Neutral diphosphides $[Fe_2(\mu\text{-}PR_2)_2(CO)_6]^z$ (z = 0) can be reduced by Na or $LiAlH_4$ to dianions (z = -2) and reactions of the former with $[BEt_3H]^-$ or protonations of the latter complexes afford $[Fe_2(\mu\text{-}PR_2)(\mu\text{-}CO)(CO)_5(PR_2H)]$;[133] intermediates in these processes have been identified and include formyl- and hydrido- iron complexes. Alkylations of the dianions give acyl derivatives $[Fe_2(\mu\text{-}PR_2)_2(COR)(CO_5)]^-$ and alkyl-iron intermediates are observed in some cases.[133] $[Fe_2(\mu\text{-}\{P(NPr_2^i)\}_2CO)(CO)_6]$ is formed from $PCl_2(NPr_2^i)$ and $[Fe(CO)_4]^{2-}$ in a process involving carbonyl migration from Fe to P; a product of hydride reduction of this diphosphido-complex is $[Fe_2\{\mu\text{-}PH(NPr_2^i)CHP(NPr_2^i)\}(CO)_6]$.[134] The bridging phosphaalkenyl complex $[Fe_2\{\mu\text{-}P=C(SiMe_3)_2\}_2$-

$(CO)_6$] has been characterised.[135] New co-ordination modes for iminophosphines include 6-electron donation in [$Fe_2(\mu-\eta^2-Bu^tP=NBu^t)(CO)_6$] and coupling with carbon monoxide in [$Fe_2\{\mu-Bu^tPN(Bu^t)CO\}(CO)_7$].[136] The structurally characterised complexes [$Fe_2(\mu-Bu^tPCRCR')(CO)_6$] may be formulated as 1-phospha-2-ferracyclobutadiene complexes of $Fe(CO)_3$.[137]

Thioketone complex [$Fe_2(\mu-\eta^2-SC\overline{CCMe_2(CH_2)_3C}Me_2)(CO)_6$] forms asymmetric adducts with CO, phosphines and phosphoranes and also reacts with hydride to form a thioacyl anion which yields a half-opened thioaldehyde-complex on protonation.[138] [Et_3NH]$^+$[$Fe_2(\mu-SR)(\mu-CO)(CO)_6$]$^-$ reacts with halogen compounds to generate binuclear complexes [$Fe_2(\mu-L)(\mu-SR)(CO)_6$] with L = RCO, PPh_2, C_3H_6, η^1,η^2-allenyl.[139] The monoanionic complexes [$Fe_2(\mu-S)(\mu-SR)(CO)_6$]$^-$ or dianionic [$Fe_2(\mu-S)_2(CO)_6$]$^{2-}$ formed from [$Fe_2(\mu-S_2)(CO)_6$] by S-S bond cleavage, are precursors to thio-alkyl, thio-acyl or S-S coupled tetranuclear complexes;[140] the dianion has also been utilised in the synthesis of [$MoFe_3S_6(CO)_6$]$^{2-}$ which on oxidative decarbonylation forms double-$MoFe_3S_4$ cubanes of biological relevance.[141] Other studies on disulphido-bridged species of the type [$Fe_2(\mu-SR)_2(CO)_6$] are also described,[142] including their application in catalysis of carbonyl ligand substitution.[142a] Disulphide bridges are present in [$Fe_2\{\mu-S_2P(C_6H_4OMe-p)Fe(CO)_4\}(CO)_6$][143] and diselenide bridges in metallapropellane [$Fe_2(\mu-Se_2C_{13}H_{16}O)(CO)_6$].[144]

A thiocarbonyl group of <u>trans</u> or <u>cis</u> [$Fe_2(\mu-CS)_2(CO)_2Cp_2$] can be alkylated or can form an adduct with $HgCl_2$.[145] The formally 18-electron complex [$Fe_2(\mu-CO)_3Cp_2^*$] has a triplet ground state with a doubly occupied, 2-fold degenerate ($\pi^*\pi^*$) HOMO.[146] The mechanism for interconversion of complexes [$Fe_2(\mu-L)(\mu-CO)(CO)_2Cp_2$]$^+$ (L = CCH_2R or σ,η^2-CH=CHR) has been studied by V.T. n.m.r. spectroscopy.[147] Syntheses, structure and reactions of vinylcarbyne complexes [$Fe_2(\mu-CCH=CHR)(\mu-CO)(CO)_2Cp_2$]$^+$ are described.[148] Ethenylidene complex [$Fe_2(\mu-L)(\mu-CO)(CO)_2Cp_2$]$^+$ (L = $C=CH_2$) reacts on photolysis with PhC≡CPh to give a metallacyclopentenone and, subsequently a μ-σ,η^3-allyl complex;[149a] cross coupling of complexes with L = $C=CH_2$ and CH forms a tetranuclear species [$\{Fe_2(\mu-CO)(CO)_2Cp_2\}_2\mu-C_3H_3$]$^+$.[149b]

Deprotonation of [$Fe_3(\mu_3-CH)(\mu-H)_3(CO)_9$] gives [$Fe_3(\mu_3-H-CH)(\mu-H)(CO)_9$]$^-$ containing one agostic Fe-H-C interaction and reprotonation at low temperatures affords an intermediate with a second agostic link which then rearranges to starting μ_3-CH cluster.[150] [$Fe_3(\mu_3-CCO)(CO)_9$]$^{2-}$ can be acylated at ketenylidene O-atom to produce the $\mu_3\eta^2$-CCOC(O)Me containing cluster.[151] Bis-alkylidyne complex [$Fe_2(\mu_3-CF)_2(CO)_9$] may be formed from $CFBr_3$.[152] A reversible alkylidyne-alkylidyne coupling of [$Fe_3(\mu_3-CMe)(\mu_3-COEt)(CO)_9$] to form μ_3-η^2-(MeC≡COEt) ligand is induced by addition of CO;[153a] alkynes may also induce coupling but are incorporated into the ligand to give ferracyclopentadiene rings.[153b]

Formyl, acyl and carbene derivatives of clusters $[Fe_3(\mu_3-EPh)_2(CO)_9]$ (E = N or P) are described, but further coupling of carbene ligand with μ-EPh yields free (E = N) or co-ordinated (E = P) PhE=C(OEt)Ph.[154] The synthesis and structures of $[Fe_3(\mu-PPh_2)_2(\mu-H)_2(CO)_8]$ and related Ru_3 and Os_3 derivatives are reported.[155] The reversible opening and closing of Fe-Fe bonds in $[Fe_3(\mu_3-PR)(CO)_{10-n}L_n]$ on successive addition and elimination of one or two ligands is likened to a "breathing" process and is the mechanism for ligand substitution.[156] Thermal reactions of diphosphines $R_{2-n}P(H_n)CH_2PR_{2-m}(H_m)$ with iron carbonyls give various clusters containing μ_2-PRCH$_2$PR$_2$ or μ_2-PRCH$_2$PR and/or fragmentation ligands PR$_2$Me, μ_2-PRMe, μ_3-PR, μ_4-PH or μ_4-P, e.g. in clusters $[Fe_3(\mu-H)(\mu-PRCH_2PR_2)(CO)_9]$, $[Fe_4(\mu_4-PH)(\mu-H)(\mu-PRMe)(\mu-PRCH_2PR)-(CO)_{10}]$ and spiro-$[Fe_4(\mu_4-P)(\mu-PRCH_2PR_2)(CO)_{13}]$.[157] 1,2,3-Triphenyl-triphosphaindan ($\overline{C_6H_4PPhPPh}$Ph) reacts with $Fe_2(CO)_9$ or $Fe_3(CO)_{12}$ to form products including three characterised Fe_3P_3 clusters.[158]

The sulphur atoms in $[Fe_3(\mu_3-S)(\mu-H)(CO)_9]^-$ and $[Fe_3(\mu-S)(CO)_9]^{2-}$ act as donors to $M(CO)_5$ (M = Cr,W) or may be alkylated.[159]

Ethyne initially forms $[Fe_4(\mu_3-CMe)(CO)_{12}]^-$, by reaction with $[Fe_4H(CO)_{13}]^-$, but a second molecule produces structurally characterised $[Fe_4(\mu_4-\eta^3-CMeCHCH)(\mu-CO)_2(CO)_9]^-$.[160] Structural and electrochemical studies of $[Fe_5C(CO)_{15}]$, $[Fe_5N(CO)_{14}]^-$ and $[Fe_5C(CO)_{14}]^{2-}$ [161a] and syntheses and structures of $[Fe_6C(CO)_{15}(NO)]$ and $[Fe_6C(CO)_{11}(NO)_4]$ are reported.[161b] A syn-bis-diazene ligand stabilises the pentametallic unit in $[Fe_5(\mu_5-\eta^4-C_{11}H_{16}N_4)(CO)_{13}]$, (7).[162]

2.6 - Ruthenium and Osmium. - An unbridged Ru-Ru bond {3.003(1)Å} is present in $[Ru_2(CN-xylyl)_{10}]^{2+}$.[163]

Complexes $[Ru_2(\mu-R-dab)(CO)_n]$ (n = 6, R-dab = RN=CHCH=NR) react with carbodiimides or thiofluorenone-S-oxide ($C_{12}H_8C=S=O$) by C-C coupling to form new bridging ligands;[164a] reactions of the same R-dab complexes (n = 5 or 6) with H_2 afford oxidative-addition products and when R = neo-C_5H_{11} a linear tetranuclear complex is obtained, $[H_2Ru_4(\mu-R-dab)_2(\mu-CO)_2(CO)_6]$.[164b] Structures are reported for binuclear complexes $[Ru_2(\mu-PPh_2)(\mu-OCPh)(CO)_5(PPh_3)]$,[165] $[Ru_2(\mu-H)(\mu-PhPC_6H_4CO)(\mu-Ph_2PCHPPh_2)(CO)_4]$ and $[Ru_2(\mu-Cl)(\mu-PPh_2)(\mu-dppm)-(CO)_4]$,[166] all formed by thermal degradation of co-ordinated phosphine ligand.

Carboxylato bridged binuclear species reported include the catalytically active $[Ru_2(\mu-O_2CC_6H_4F-p)_2(CO)_5(H_2O)]$[167] and $[Ru_2(\mu-O_2CEt)_2(CO)_4L_2]$ (L = solvent) which in the absence of L associates via Ru-O interactions.[168] Complexes $[Ru_2(\mu-RCONH)_2\{\mu-Ar_2POC(R)N\}_2Ar_2]$ (Ar = Ph, p-tol) are produced by aryl group transfer from PAr$_3$ to Ru in $[Ru_2(\mu-RCONH)_4Cl]$.[169]

Eighteen μ-alkylidene complexes $[Ru_2(\mu-CRR')(NO)_2Cp_2]$ have been synthesised from $[Ru_2(\mu-NO)_2Cp_2]$ and their isomerisation, protonation and thermolysis investigated.[170] The alkylidene ligand of $[Ru_2(\mu-L)(\mu-CO)(dppm)Cp_2]$ (L = CH_2 or CHMe) is deprotonated on 2e oxidation to form complexes with L = CH^+ or $CHCH_2^+$, respectively;[171] these cationic products react with nucleophiles and, when L = CH^+, with dihydrogen to give the complex with L = CH_3^+. Reactions between complexes $[Ru_2(\mu-L)(\mu-CO)(CO)_2Cp_2]$ with L = CMe^+ and $C=CH_2$ yield $[\{Ru_2(\mu-CO)(CO)_2Cp_2\}_2(\mu-CMeCHCH)]^+$ and $[\{Ru_3(\mu-CO)_3Cp_3\}(\mu-CCH_2CHC)\{Ru_2(\mu-CO)(CO)_2Cp_2\}]$.[172]

Spectroscopically characterised $[Ru_3(\mu-CH_2)(\mu-CO)(CO)_{10}]$ exhibits fluxionality via bridge/terminal methylene interchange, rearranges thermally to μ_3-ketenylidene complex and forms a μ-CCH_2 bridge on reaction with CH_2N_2.[173] 50-Electron clusters $[Os_3(\mu-X)(\mu-CH_2)(CO)_{10}]^-$ (X = Cl,Br,I,NCO,NO_2) have been characterised and readily insert CO to form $[Os_3(\mu-X)(\mu-CH_2CO)(CO)_{10}]^-$ thus illustrating the halide-ion promotion of such reactions on carbonyl clusters.[174] Synthetic and mechanistic aspects of ligand substitution on $[Ru_3(\mu-H)(\mu-CX)(CO)_{10}]$ (X = OMe,NMe_2) and of the related reversible H_2 addition to form $[Ru_3(\mu_3-CX)(\mu-H)_3(CO)_9]$ (X = OMe) have been investigated; a CO dissociative mechanism is implicated for both processes.[175]

Pyrolysis of structurally characterised $[Os_3(\mu_3-CPh)(\mu-H)(CO)_{10}]$ produces the ortho-metallated product $[Os_3(\mu_3-\eta^3-CHC_6H_4)(\mu-H)(CO)_9]$.[176] The mixed carbene-carbyne cluster $[Os_3(\mu_3-CPh)(\mu-H)\{\eta^1-C(OMe)_2\}(CO)_9]$ is reported.[177] $[Ru_3(\mu_3-CMe)(\mu-CO)_3Cp_3]$ may be oxidised to +1 and +2 cations and the latter species deprotonates to form $[Ru_3(\mu_3-CCH_2)(\mu-CO)_3Cp_3]^+$ which generates a μ_3-CEt bridge on reaction with LiMe.[178] $[Ru_3(\mu_3-CCO)(\mu-CO)_3(CO)_6]^{2-}$ may be alkylated by $MeOSO_2CF_3$ at the ketenylidene bridge but protonated at the metal framework.[179]

Coupling reactions of methylidyne ligands of $[Ru_3(\mu_3-CX)(\mu-H)_3(CO)_9]$ (X = OMe,Me,Ph,$CH_2CH_2CMe_3$) with alkynes RCCR' give regioisomeric products $[Ru_3(\mu_3-\eta^3-XCCRCR')(CO)_9]$ accompanied by cis-alkene and subsequent hydrogenation (X = OMe) yields $[Ru_3(\mu_3-CCHRCH_2R')(\mu-H)_3(CO)_9]$;[180] when R and R' = H, a second hydrogenation product is $[Ru_3(\mu_3-\eta^2-MeCCOMe)(\mu-H)_2(CO)_9]$.[181] Cluster $[Ru_3(\mu_3-CSEt)(\mu-H)_3(CO)_9]$ also reacts with alkynes C_2R_2 to form $[Ru_3(\mu_3-\eta^2-EtSCCRCR)(CO)_9]$ and $[Ru_3(\mu_3-CSEt)(\mu-cis-CRCHR)(CO)_9]$.[182]

Benzene, formed by dehydrogenation of cyclohexadiene, is co-ordinated in a face capping mode in $[Os_3(\mu_3-\eta^2,\eta^2,\eta^2-C_6H_6)(CO)_9]$ (8);[183] a similar co-ordination occurs in $[Ru_6(\mu_6-C)(\mu_3-\eta^2,\eta^2,\eta^2-C_6H_6)(CO)_{11}(\eta^6-C_6H_6)]$.

Oxidative addition of furan to $[Os_3(CO)_{10}(NCMe)_2]$ produces $[Os_3(\mu-H)(\mu-\eta^2-C_4H_3O)(CO)_{10}]$.[184] Competitive oxidative addition reactions of $\alpha\beta$-unsaturated aldehydes form clusters $[Os_3(\mu-H)(\mu-L)(CO)_{10}]$ with L = RC=CHCHO or RCH=CHCO.[185] The C-C acyl bond of $[\overline{Os_3\{\mu_3-\eta^2-CC(Ph)C(CO)=CPhRe(CO)_4\}}(\mu-H)(CO)_9]$ is reversibly cleaved by PMe_2Ph giving $[\overline{Re(CO)_4(PMe_2Ph)}]^+[Os_3(\mu-H)\{\mu-CC(Ph)C\equiv CPh\}(CO)_{10}]^-$.[186]

Cleavage of N-O bonds of $[Ru_3(\mu\text{-H})(\mu\text{-NO})(CO)_{10}]$ or $[Ru_3(\mu\text{-NOMe})(CO)_{10}]$ by molecular hydrogen affords nitrogen containing clusters $[Ru_3(\mu_3\text{-NH})(\mu\text{-H})_2\text{-}(CO)_9]$, $[Ru_3(\mu\text{-H})(\mu\text{-NH}_2)(CO)_{10}]$ and traces of $[Ru_4(\mu\text{-H})_3(\mu\text{-NH}_2)(CO)_{12}]$;[187a] related Os_3 species are formed by similar reactions, accompanied by $[Os_3(\mu_3\text{-NH})(\mu\text{-H})_4(CO)_8]$.[187b] $[Ru_3(CO)_{10}(bipy)]$ has been structurally characterised.[188] The useful synthetic precursors $[Ru_3(CO)_{12-n}(NCMe)_n]$ (n = 1,2) are described;[189a] the acetonitrile ligands are readily displaced by $P(OMe)_3$, PPh_3, C_2Ph_2 and by bipy but reactions with pyridines and related nitrogen heterocycles also give <u>ortho</u>-metallated derivatives.[189] Other cluster products of metallated nitrogen heterocycles of relevance to hydrodenitrogenation catalysis, produced by reactions with $M_3(CO)_{12}$ (M = Ru,Os), are also described, including X-ray structures of quinoline derivatives.[190] 2-Ethenylpyridine forms open trinuclear clusters $[\{HOs(CO)_3L\}\text{-}Os_2(\mu\text{-NC}_5H_4CH\text{=}CH)(CO)_6]$ (L = CO,PMe$_2$Ph) from closed Os_3 precursors.[191]

Cluster products of reactions between $[Os_3(CO)_{10}(NCMe)_2]$, $[Os_3(\mu\text{-H})_2(CO)_{10}]$ or, at higher temperatures, $[Os_3(CO)_{12}]$ and amides,[192] dimethylcyanamide (Me_2NCN),[193] or various sulphur diimides $\{S(NX)_2; X = SiMe_3,PBu_2,AsBu_2\}$[194] have been studied. Reversible CO addition to $[Ru_3(\mu\text{-R-dab})(CO)_n]$ (n = 8) gives 50-electron species (n = 9) without Ru-Ru bond rupture[195a] whereas reaction of the same cluster (n = 8) with CH_2N_2 affords $[Ru_3(\mu\text{-CH}_2)(\mu\text{-R-dab})(CO)_8]$ with cleavage of a single Ru-Ru bond.[195b]

Kinetics of substitution on $[Ru_3(CO)_{12-2n}(dppm)_n]$ (n = 0,1) by dppm[196a] and of substitution or fragmentation of $[M_3(CO)_{12-n}(PBu_3)_n]$ (M = Ru,Os; n = 0,1,2) by PBu_3[196b] have been investigated; fragmentation is a bimolecular F_N2 process. New isomers of $[Os_3(CO)_{12-n}(PMe_2Ph)_n]$ (n = 2,3) are described[197] and other reported phosphorus-ligand substituted trinuclear clusters are $[Ru_3(\mu\text{-CO})_2(CO)_6\{PPh(OMe)_2\}_4]$,[198] unsaturated $[Os_3(\mu\text{-H})_2(\mu\text{-dppm})(CO)_8]$[199] and $[Ru_3\{\mu_3\text{-}(PPh_2)_3CH\}(CO)_9]$.[166] N.m.r. studies of stereochemical non-rigidity of clusters containing phosphorus(III) ligands, 1,2-$[Os_3(CO)_{10}(PMe_2Ph)_2]$ and 1,2,3-$[Os_3(CO)_9\{P(OMe)_3\}L_2]$ (L = PMe_2Ph,PPh_3),[200] $[Os_3(CO)_7\{P(OMe)_3\}_5]$[201] and $[Ru_3(\mu_3\text{-C}_2Bu^t)(\mu\text{-H})(CO)_8\{PMe(Bz)Ph\}]$[202] show localised PR_3 exchange at a single metal centre and support a C_3 rotational mechanism which also interchanges axial and equatorial CO ligands.

Phosphido clusters $[Os_3(\mu\text{-H})(\mu\text{-PR}^1R^2)(CO)_{10}]$ have been systematically synthesised and characterised.[203] Other trinuclear phosphido derivatives are formed by P-C bond cleavage on thermolysis or thermal hydrogenolysis of co-ordinated ditertiary[204] or tritertiary[166] phosphine clusters; e.g. from dppm products include $[Ru_3(\mu_3\text{-}\eta^2\text{-PPhCH}_2PPh_2)(\mu\text{-H})(CO)_9]$ and $[Ru_3\{\mu_3\text{-}\eta^3\text{-PPhCH}_2PPh(C_6H_4)\}(CO)_9]$.[204] The anion $[Ru_3(\mu_3\text{-PPhCH}_2PPh_2)(CO)_9]^-$ reacts with allyl chloride to form a symmetrical $\mu\text{-}\eta^3\text{-C}_3H_5$ bridge.[205]

Products $[Os_3(\mu\text{-H})(\mu\text{-X})(CO)_{10}]$ (X = Cl,Br,I,CF$_3$CO$_2$,OH,S$_2$CNEt$_2$,S$_2$COEt) result from combined protonation and anion addition to $[Os_3(\mu-H)(\mu-OCH=CH_2)(CO)_{10}]$ with liberation of acetaldehyde via identifiable intermediates.[206] Crystal structures are reported for $[Os_3(\mu-H)(\mu-O_2CMe)(CO)_{10}]^{207}$ and $[Ru_3(\mu-H)(\mu-SEt)(CO)_{10}]$.[208] A sulphide ligand in $[Os_3(\mu_3-S)_2(CO)_9]$ forms a donor bond to $W(CO)_5^{209}$ and $[Os_3(\mu_3-S)(\mu_3-CO)(CO)_9]$ reacts with NMe_3 at 125°C to form an amino-carbene cluster $[Os_3(\mu_3-S)(\mu-H)_2\{CH(NMe_2)\}(CO)_8]$ which is a catalyst for transalkylation of tertiary amines.[210] $[Os_4(\mu_3-S)(CO)_{14}]$ exhibits two site reactivity with PhC_2H yielding non-interconvertible, structurally characterised products $[Os_4\{\mu_4-CC(Ph)H\}(\mu_3-S)(CO)_{12}]$ and $[Os_4(\mu_4-\eta^2-SCPhCH)-(CO)_{12}]$.[211]

There is an unbridged Os→Os donor-acceptor bond {2.940Å (mean)} in $[Os_3\{Os(CO)_4(PMe_3)\}(CO)_{11}]$.[212] Tetranuclear clusters can be synthesised by reaction of $[OsH_2(CO)_4]$ with trinuclear systems, e.g. $[Os_4(\mu-H)_2(CO)_{14}(NCMe)]$ from $[Os_3(CO)_{10}(NCMe)_2]$.[213]

Thermal decomposition of the isocyanato ligand in $[Ru_4(\mu-NCO)(CO)_{13}]^-$ produces new nitrido clusters $[Ru_4(\mu_4-N)(CO)_{12}]^-$, $[Ru_5(\mu_5-N)(\mu-CO)(CO)_{13}]^-$ and $[Ru_6(\mu_6-N)(\mu-CO)_2(CO)_{14}]^-$.[214] Reduction of the nitrosyl group in $[Ru_3(\mu-NO)_2(CO)_{10}]$ by CO gives 64-electron, "butterfly" clusters $[Ru_4(\mu_4-N)(\mu-L)(CO)_{12}]$ (L = NO or NCO).[215] The crystal structure of $[Os_4(\mu-H)_3(CO)_{11}(NO)]$, formed using $[PPN]NO_2$,[216] and the synthesis and auration of $[Ru_5(\mu_5-C)(CO)_{13}(NO)]^-$ [217] are described.

The phosphinidene cluster $[Ru_4(\mu_3-PPh)(CO)_{13}]$ reacts with PhC≡CPh to give structurally characterised products $[Ru_4(\mu_3-PPhCPhCPh)(\mu-CO)_2(CO)_{10}]$ and $[Ru_4(\mu_4-\eta^2-PhCCPh)(\mu_3-PPh)(\mu-CO)_2(CO)_9]$ which may be considered to contain nido-Ru_4PC_2 and closo-Ru_4PC_2 frameworks, respectively.[218] Bis-(diphenylphosphino)ethyne reacts with $[Ru_3(CO)_{12}]$ in boiling thf to form $[Ru_4(\mu_4-PPh)(\mu_4-PhC_2PPh_2)(CO)_{10})]$, a flattened "butterfly" cluster, accompanied by a linked binuclear species $[\{Ru_2(C_2PPh_2)(PPh)(CO)_5\}_2]$;[219] the same reactants also afford $[\{Ru_3(CO)_{11}\}_2(\mu-Ph_2PC_2PPh_2)]^{220a}$ and subsequent thermolysis gives $[Ru_5(\mu_5-\eta^2-C_2PPh_2)(\mu-PPh_2)(CO)_{13}]$ an open "swallow" cluster with a core consisting of three edge-fused Ru_3 triangles.[220] This latter pentanuclear cluster adds two molecules of CO to form two structural isomers of $[Ru_5(\mu_5-C_2PPh_2)(\mu-PPh_2)(CO)_{15}]$, one of which incorporates a Ru→Ru donor-acceptor bond;[221a] hydrogenation of $[Ru_5(\mu_5-\eta^2-C_2PPh_2)(\mu-PPh_2)(CO)_{13}]$ at 10 atm. affords structurally characterised $[Ru_4(\mu_4-\eta^2-HC_2PPh_2)(\mu-H)_3(\mu-PPh_2)-(CO)_{10}]^{221b}$ but at 1 atm. successive addition of three molecules of H_2 yields pentanuclear clusters containing μ_5-vinylidene, -methylidyne and -carbide ligands, respectively:[221c] the overall reaction converts μ-C_2PPh_2 into C and $PMePh_2$ in $[Ru_5(\mu_5-C)(\mu-H)_3(\mu-PPh_2)(CO)_{11}(PMePh_2)]$.

$[Os_6(CO)_{17}(NCMe)]$ reacts with alkynes under mild conditions to produce structurally characterised clusters $[Os_6(\mu_4-\eta^2-C_2Et)(\mu-H)(CO)_{17}]$,

$[Os_6(\mu_4-\eta^2-HCCEt)(CO)_{17}]$ and $[Os_6(\mu_3-\eta^2-MeCCEt)(CO)_{16}]$.[222]

Halide, nitrosyl and phosphite derivatives of $[Os_{10}(\mu_6-C)(CO)_{24}]^{2-}$ are obtained by reactions with electrophilic and nucleophilic reagents; products characterised by X-ray diffraction are $[Os_{10}(\mu_6-C)(\mu-I)(CO)_{24}]^-$, $[Os_{10}(\mu_6-C)-(\mu-I)_2(CO)_{24}]$, $[Os_{10}(\mu_6-C)I(CO)_{22}(NO)]^{2-}$ $[Os_{10}(\mu_6-C)(\mu-I)_2(CO)_{23}\{P(OMe)_3\}]$ and $[Os_{10}(\mu_6-C)(\mu_n-CO)(CO)_{20}\{P(OMe)_3\}_4]$ (two isomers n = 2 or 3).[223]

2.7 Cobalt.

Alkyne complex $[Co_2(\mu-\eta-HCCSiMe_3)(CO)_6]$ is readily deprotonated by $LiNR_2$ and the resulting lithiated derivative reacts with various electrophiles to give products characteristic of SET reactions, including dimer $[\{Co_2(CO)_6\}_2(\mu-Me_3SiC\equiv C-C\equiv CSiMe_3)]$ formed by C-C coupling; a related trimer is also structurally characterised and extended coupling forms a polyacetylene chain fully η^2-co-ordinated to $Co_2(CO)_6$ units.[224] An iminoborane acts like the isoelectronic alkyne in forming $[Co_2(\mu-\eta-Bu^tNBBu^t)(CO)_6]$.[225] Reversible one-electron reduction of "flyover" complexes $[Co_2(\mu-C_6R_6)(CO)_4]$ occurs most readily for R = CF_3, and the radical anion is stable in thf, decomposes to form $C_6(CF_3)_6$ in acetone or CH_2Cl_2 and undergoes ETC catalysed substitution by MeCN.[226] There is a full report of the synthesis of $[Co_2(\mu-CCH_2)(CO)_2Cp_2]$ from anion $[Co_2(CO)_2Cp_2]^-$ and of its protonation to $\mu-CCH_3^+$, reduction to $\mu-CHCH_3$ and transformation to heteronuclear clusters.[227] The related vinylidene complex $[Co_2(\mu-CCMe_2)(CO)_2Cp_2^*]$ is produced by reaction of N_2CCMe_2 precursors with $[Co_2(CO)_2Cp_2^*]$.[228]

Carbonyl complex $[Co_2\{\mu-MeN(PF_2)_2\}_3(CO)_2]$ has been structurally characterised;[229a] it is reversibly reduced by two one-electron steps and the paramagnetic monoanion has been studied spectroscopically.[229b]

Phosphinidene species $[Co_2(\mu-PAr)(CO)_6]$ (Ar = 2,4,6-$Bu_3^tC_6H_2$) has a trigonal planar geometry at the P-atom and this is consistent with multiple Co-P bonds.[230] Crystal structures have been determined for $[Co_2(\mu-PH\{CH(SiMe_3)_2\})-(CO)_4Cp]$[231] and for metallaphosphazenes $[M_2\{\mu-\overline{P(NPCl_2)_2N}\}(CO)_2Cp_2]$ (M = Co, Rh).[232]

The non metal-metal bonded $[Co_2(\mu-PPh_2)_2(CO)_n(PPh_2H)_2]$ (n = 4) loses two CO molecules on heating to form product (n = 2) for which spectroscopic data imply a Co=Co double bond.[233] Cobaltocene and secondary phosphines give complexes $[Co_2(\mu-PR_2)_2Cp_2]$ and related linked biphosphido species;[234] the Co-Co bond of these derivatives is cleaved by insertion of SO_2 and is protonated to form a 3-centre, 2-electron closed CoHCo interaction.[234] The bridging CoHCo link in $[Co_2(\mu-H)(\mu-PMe_2)_2Cp_2]^+$ is cleaved by addition of ligands and also by insertion of MeNC to form a μ-formimidoyl derivative.[235] The Co-Co bonded $[Co_2(\mu-CH_2PMe_2)(\mu-PMe_2)Cp_2]$ is obtained by reductive dehalogenation of a product of reaction between CH_2Br_2 and $[Co_2(\mu-PMe_2)_2Cp_2]$.[236]

$[Co_2(\mu-P_2)(CO)_5L]$ (L = CO, PR_3) forms adducts by donation from phosphorus to one or two metal fragments $M(CO)_5$ (M = Cr,Mo,W) or $Fe(CO)_4$.[237] The binuclear unit $[Co_2(\mu\text{-}SR)_2Cp_2]$ becomes linked through sulphur to a third Co-atom in a metallacyclopentadiene ring after reaction with $CF_3C\equiv CCF_3$.[238]

Crystal structures of $[Co_3(\mu_3\text{-}C\overline{CHCH_2CH_2})(CO)_9]$[239] and of hydroformylation catalyst $[Co_3(\mu_3\text{-}CMe)(\mu\text{-}dppm)(CO)_7]$[240] are reported. Structurally characterised clusters $[Co_3(\mu_3\text{-}NO)_2Cp_3]$ (Cp = $\eta\text{-}C_5H_5$ and $\eta\text{-}C_5H_4Me$) are reversibly oxidised and reduced to form respective mono-ions.[241]

Reactions of mononuclear complexes $L_nM(PH_3)$ with $Co_2(CO)_8$ yield clusters $[Co_3(\mu_3\text{-}PML_n)(CO)_9]$ (M = Cr,Mo,W,Mn);[242] primary phosphine complexes yield open trinuclear species. Crystal structures of $[Co_3(\mu\text{-}PPh_2)_3(CO)_6]$[243] and $[Co_3\{\mu_3\text{-}PCH(SiMe_3)_2\}(CO)_9]$[231] have been determined.

Structurally characterised triptycene cluster $[Co_4(CO)_9(\eta^6\text{-}C_{20}H_{14})]$ is described[244] and both the crystal structure and vibrational analysis of $[Co_4(\mu_4\text{-}\eta\text{-}C_2H_2)(\mu\text{-}CO)_2(CO)_8]$ are reported.[245] Ethylene in $[Co(C_2H_4)_2Cp]$ rearranges on heating to μ_3-hydrido and μ_3-ethylidyne ligands in forming tetrahedral $[Co_4(\mu_3\text{-}H)(\mu_3\text{-}CMe)Cp_4]$; $[Co(C_2H_4)_2Cp]$ and formaldehyde give $[Co_4(\mu_3\text{-}CO)_4Cp_4]$.[246] Diphosphazane substituted tetranuclear carbonyl clusters have been synthesised, including structurally characterised $[Co_4(\mu\text{-}CO)_3\{\mu\text{-}\{(PhO)_2P\}_2NEt\}(CO)_7]$.[247] The substitution reactions of $[Co_4\{\mu_3\text{-}HC(PPh_2)_3\}(CO)_9]$ have been studied mechanistically[248a] and electrochemical investigations also establish that electrocatalysed substitution involves the relatively stabilised radical anion.[248b]

Bicapped μ_4-phosphinidene clusters $[Co_4(\mu_4\text{-}PR)_2(\mu\text{-}CO)_2(CO)_8]$ have been structurally characterised[231,249] and undergo limited substitution with simple phosphine ligands but, interestingly, $MeN(PF_2)_2$ forms $[Co_4(\mu_4\text{-}PPh)_2\{\mu\text{-}MeN\text{-}(PF_2)_2\}_4(CO)_3]$ with an opened Co-Co bond.[249]

$[PPN]NO_2$ reacts with $Co_4(CO)_{12}$ to form $[Co_6(\mu_6\text{-}N)(\mu\text{-}CO)_9(CO)_6]^-$.[250] Carbido clusters $[Co_{13}(\mu_6\text{-}C)_2(\mu\text{-}CO)_{12}(CO)_{12}]^{3-}$ [251a] $[Co_6(\mu_6\text{-}C)(\mu\text{-}CO)_9(CO)_6]^{2-}$, $[Co_6(\mu_6\text{-}C)(\mu\text{-}CO)_6(CO)_8]^-$ and $[Co_8(\mu_8\text{-}C)(\mu\text{-}CO)_{10}(CO)_8]^{2-}$ [251b] have been synthesised and the former two characterised by X-ray diffraction.

2.8 Rhodium and Iridium.

Complex $[Ir_2(Tcbiim)_2(CO)_2(MeCN)_2\{P(OEt)_3\}_2]$ ($H_2Tcbiim$ = tetracyanobiimidazole) contains an unbridged Ir-Ir bond {2.826(2)Å}.[252] Vibrational and electronic spectra of isocyanide bridged Rh(II) and Ir(II) complexes $[M_2(CNNC)_4L_2]^{n+}$ have been analysed.[253]

Non-carbonyl containing binuclear complex $[Ir_2(\mu\text{-}I)_2I_2(cod)_2]$ {Ir-Ir = 2.914(1)Å} has been structurally characterised[254] and an Ir-Ir interaction is also probably present in $[Ir_2(\mu\text{-}H)(\mu\text{-}Cl)H_2(cod)(PPh_3)_2]$.[255] The monohydrido-bridged $[Ir_2(\mu\text{-}H)H_2(PMe_3)_2Cp_2^*]^+$ is reported,[256a] and a unique bridging borohydride ligand is found in $[Ir_2(\mu\text{-}H)(\mu\text{-}H_2BH_2)H_2Cp_2^*]$.[256b]

Crystal structures are reported for $[Rh_2(\mu\text{-}CO)\{\mu\text{-}\eta^5,\eta^5\text{-}(C_5H_4)_2CH_2\}(CO)_2]$

containing a methylene-linked bicyclopentadienyl bridge,[257] and for [Rh$_2$(μ-H)(μ-CO)(μ-dppm)Cp$_2$]$^+$ formed by protonation of [Rh$_2$(μ-CO)(μ-dppm)-Cp$_2$].[258] Fulvalene bridged [Rh$_2$(μ-η^5,η^5-C$_{10}$H$_8$)(CO)$_2$(PPh$_3$)$_2$]$^{n+}$ (n = 2) possesses a Rh-Rh bond but reduction to neutral species (n = 0) causes cleavage of this bond and the adoption of a <u>trans</u>-geometry.[259]

Alkali metal reduction of [Rh$_2$(μ-CO)$_2$Cp$_2^*$]$^{n-}$ (n = 0) generates anions with n = 1 or 2 and the crystalline dianion is associated in pairs of Rh$_2$ units joined by aggregates of four K$^+$ ions;[260] appropriate alkylations of these neutral or anionic dirhodium complexes afford [Rh$_2$(μ-CO)$_2$(R)Cp$_2^*$]$^-$ and [Rh$_2$(μ-CO)$_2$R$_2$Cp$_2^*$]. Reactions of diazoalkenes with [Rh$_2$(μ-CO)$_2$Cp$_2^*$] produce μ-vinylidene complexes, including [Rh$_2$(μ-CCMe$_2$)(CO)$_2$Cp$_2^*$] characterised by <u>X</u>-ray diffraction.[228] Full details are given for reactions of cyclopropenes with [Rh$_2$(μ-CO)$_2$Cp$_2^*$], or with related dicobalt or heteronuclear complexes; C=C double bond cleavage occurs in forming 1:1 adducts, as in structurally characterised [Rh$_2$(μ-CO)(μ-COCHCMe$_2$CH)Cp$_2^*$] for which protolysis reactions are described.[261] Additions of various alkynes to [Rh$_2$(μ-CO)$_2$Cp$_2^*$] give [Rh$_2$(μ-σ,σ-RC$_2$R')(μ-CO)$_2$Cp$_2^*$] and/or [Rh$_2$(μ-η^2,η^2-COCRCR')(μ-CO)Cp$_2^*$] as main product(s).[262]

Additions to [Rh$_2$(μ-CF$_3$C$_2$CF$_3$)(μ-CO)Cp$_2$] of carbene or nitrene units from various organic precursors have been studied;[263] initially formed μ-CR$_2$ bridges subsequently migrate to the alkyne ligand with C-C (R = H) or C-O (R = COOEt) coupling[263a] and nitrenes become incorporated in a {μ-η^3-C(CF$_3$)C(CF$_3$)CONR} bridge.[263b] Binuclear Rh(IV) complexes [Rh$_2$(μ-CH$_2$)$_2${μ-CH$_2$CR(CH$_2$CR=CH$_2$)CH$_2$}Cp$_2^*$] and [Rh$_2$(μ-CH$_2$)$_2$(μ-P̂P̂)Cp$_2^*$]$^{2+}$ (P̂P̂ = dppm,dppe) are reported and have ^{103}Rh-^{103}Rh n.m.r. coupling constants in the range 13.5 - 11.9 Hz.[264]

Additions of activated alkynes to [Ir$_2$(μ-pz)$_2$(cod)$_2$] afford "parallel" (μ-σ,σ-RCCR) bridged products.[265] [Rh$_2$(μ-pz)$_2$(CO)$_2${PPh$_2$(o-BrC$_6$F$_4$)}$_2$] undergoes a thermal <u>ortho</u>-metallation to form [Rh$_2$(μ-pz)$_2$(μ-Ph$_2$PC$_6$F$_4$)Br(CO)-{PPh$_2$(<u>o</u>-C$_6$F$_4$Br)}];[266] <u>ortho</u>-metallation of PPh$_3$ has also yielded [Rh$_2$(μ-OAc)$_2$(μ-Ph$_2$PC$_6$H$_4$)$_2$].2L.[267] Complexes containing Rh$_2$ units bridged by deprotonated 1<u>H</u>-pyrrolo[2,3-<u>b</u>]pyridine (HL) are described, e.g. [Rh$_4$(μ-Cl)$_2$(μ-L)$_2$(μ-CO)$_2$(nbd)$_2$].[268]

Complex [Rh$_2$(μ-dppm)$_2$(CO)$_3$] possesses an asymmetric, non-A-frame structure[269a] and reacts with PhC≡CH by competitive C-H activation or C≡C co-ordination to form a μ-CCHPh or μ-η^2-PhCCH bridge, respectively.[269b] Insertion of alkynes into two Ir-H bonds of a binuclear precursor affords complex [Ir$_2$(μ-dppm)$_2$Cl$_2$(MeO$_2$CC$_2$CHCO$_2$Me)$_2$(CO)$_2$].[270] Studies on transformations of several dirhodium A-frame derivatives have involved characterised species with Rh-Rh bonds: e.g. [Rh$_2$(μ-CO)(μ-dppm)$_2$Cl$_2$],[271a] [Rh$_2$(μ-CO){μ-(PPh$_2$)$_2$NMe}X$_2$] (X = halogen, pseudohalogen),[271b] [Rh$_2$(μ-pz)$_2$(μ-dppm)I$_2$(CO)$_2$],[271c] [Rh$_2$(μ-3,5-Me$_2$pz)(μ-dppm)$_2$I$_2$(CO)$_2$]$^+$ [271d] and

$[Rh_2(\mu-O_2CCF_3)(\mu-CO)(\mu-dppm)_2(CO)_2]^+$ 271e

The geometric isomerisations occurring for $[Rh(\mu-PBu^t_2)_2(CO)_4]$ as a neutral entity and during its reversible reduction to structurally characterised dianion have been investigated both experimentally[272a] and theoretically.[272b]

Structures are reported for open trirhodium complexes bridged by triphosphine, i.e. $[Rh_3\{\mu_3-PhP(CH_2PPh_2)_2\}_2(\mu-I)_3I(CO)]^+$, [273a] $[Rh_3\{\mu_3-PhP(CH_2PPh_2)_2\}_2(\mu-Cl)(\mu-CO)Cl(CO)]^+$ [273b] and its product of H_2 addition $[Rh_3\{\mu_3-PhP(CH_2PPh_2)_2\}_2(\mu-Cl)_2(H)_2(CO)_2]^+$. [273c] The properties of 46-electron closed cluster $[Rh_3(\mu-CO)_3(\mu-dppm)_2(CO)_3]^+$ are described.[274] Triangular cluster $[Rh_3(\mu_3-CO)_2Cp^*_3]$ reacts with molecular hydrogen to form $[Rh_3(\mu_3-CO)(\mu-H)_2(\mu-CO)Cp^*_3]$ and both have been structurally characterised;[275] related mixed-metal species are also reported.[275] Addition of rhodium or cobalt nucleophiles to $[Rh_2(\mu-CO)(\mu-CF_3C_2CF_3)Cp_2]$ produces several trinuclear clusters including the structurally characterised open species $[Rh_3\{\mu_3-COC(CF_3)C(CF_3)\}(\mu-CO)(CO)Cp_2Cp^*]$.[276] The crystal structure of non-carbonyl cluster $[Ir_3(\mu_3-O)_2(\mu-I)(cod)_3]$ is reported.[254]

<u>Ortho</u>-metallation of co-ordinated $Ph_2PCH=CHPPh_2$ on a tetrahedral Ir_4 cluster affords $[Ir_4(\mu_3-Ph_2PCH=CHPhC_6H_4)(\mu-H)(CO)_7(Ph_2PCH=CHPPh_2)]$.[277] The structure and fluxional behaviour of $[Ir_4(\mu-CO)_2(\mu-SO_2)(CO)_9]$ is described.[278] Electrochemical reduction of $[Rh_4\{\mu_3-HC(PPh_2)_3\}(CO)_9]$ to the monoanion is described[248b] and clusters $[Rh_4(CO)_{10}\{(-)-diop\}]$ and $[Rh_6(CO)_{10}\{(-)-diop\}_3]$ have been utilised in catalytic asymmetric hydrogenations.[279]

$[Rh_6(\mu_4-AsBu^t)(\mu-AsBu^t_2)_2(\mu-CO)_2(CO)_9]$ contains a pentagonal-pyramidal Rh_6 core.[280] $[PPN][NO_2]$ is a good reagent for synthesis of $[Rh_6N(CO)_{15}]^-$ and of new clusters $[Ir_4(\mu-CO)_3(CO)_8(NCO)]^-$ and $[Ir_6(\mu-CO)_4(CO)_{11}(NO)]^-$. [250] New carbido cluster $[Rh_{12}(\mu_6-C)_2(\mu-CO)_{10}(CO)_{13}]^{4-}$ has been characterised.[281]

2.9 Nickel. - Reaction of unbridged $[Ni_2(CNMe)_8]^{2+}$ cation with dppm gives the asymmetrically bridged complex $[Ni_2(\mu-CNMe)(\mu-dppm)_2(CNMe)_3]^{2+}$ [282] Bridged, dimeric cyclopentadienylnickel complexes $[Ni_2(\mu-CNCF_3)_2Cp_2]$[283] and $[Ni_2(\mu-n-PCBu^t)Cp_2]$[284] are described. Symmetrical allyl bridges are present in derivatives $[Ni_2(\mu-X)(\mu-C_3H_5)(PRR'_2)_2]$ (X = Cl, Br, C_3H_5).[285] New, structurally characterised nickel carbide clusters are $[Ni_8(\mu_8-C)(\mu-CO)_8(CO)_8]^{2-}$ [286a] $[Ni_9(\mu_8-C)(\mu_3-CO)_4(\mu-CO)_4(CO)_9]^{2-}$ [286a] $[Ni_{10}(\mu_{10}-C_2)(\mu-CO)_{10}(CO)_6]^{2-}$ [286b] and $[Ni_{16}(\mu_9-C_2)_2(\mu_3-CO)(\mu-CO)_{10}(CO)_{12}]^{4-}$. [286c]

2.10 Palladium and Platinum. - Photolysis of $[M_2(CNMe)_6]^{2+}$ (M = Pd, Pt) primarily forms 15-electron radicals by M-M bond cleavage.[287] Complex $[PdCl_2(CO)(PEt_2Ph)_3]$ possesses a semi-bridging carbonyl ligand.[288] Addition of small molecules to $[Pd_2(\mu-Br)(\mu-Cp)(PR_3)_2]$ generates new bridged, Pd-Pd bonded species, including structurally characterised $[Pd_2(\mu-SO_2)Br(PEt_3)_2Cp]$.[289]

The synthesis and structure of $[Pt_2(\mu-H)(\mu-CHBz)(dppe)_2]^+$ are described.[290] Mononuclear $[Pt(\eta^2-ECS)(diphos)]$ (E = S,Se) is transformed into $[Pt_2(\mu-E)(CS)(diphos)(PPh_3)]$ by reaction with $[Pt(PPh_3)_2L]$ and subsequent reaction with diphosphine affords $[Pt_2(\mu-ECS)(diphos)_2]$.[291] Studies involving bis-diphosphine bridged complexes $[Pt_2(\mu-dppm)_2R(\eta^1-dppm)]$ (R = Et,Pr,Bu)[292a] and $[Pt_2(\mu-dppm)_2(C\equiv CAr)_2]$[292b] are reported.

Triangular clusters $[Pt_3(\mu-CO)_3L_3]$ (L = bulky phosphine) undergo facile phosphine exchange and, when L = PBu_2^tPh, are cleaved to binuclear products by CS_2, OCS or SO_2.[293] $[Pt_3(\mu-CO)_3(PCy_3)_3]$ reacts with chelating phosphines to give 44-electron clusters $[Pt_3(\mu-CO)_3(PCy_3)_2\{PPh_2(CH_2)_nPPh_2\}]$ (n = 2,3)[294a] and with isocyanide to give structurally characterised species $[Pt_3(\mu-CO)-(\mu-CNR)_2(CNR)(PCy_3)_2]$ and $[Pt_3(\mu-CNR)_3(CNR)_2(PCy_3)]$ (R = xylyl).[294b] Structures are also reported for $[Pt_3(\mu-L)_3(CO)_3]$ and $[Pt_3(\mu-CNBu^t)(\mu-L)_2(CNBu^t)_3]$ (L = $Bu^t(Me_3Si)NP=NBu^t$).[295]

$[Pd_3(\mu_3-X)(\mu_3-CO)(\mu-dppm)_3]^+$ (X = Cl) has been structurally characterised and related species with X = CF_3CO_2, Br or Cl,[296a] or with X = H,[296b] are reported. Isomeric forms of $[Pt_3(\mu-PPh_2)_3Ph(PPh_3)_2]$, containing either a closed Pt_3 triangle or a bent Pt_3 chain, crystallise from different solvents.[297] Cluster $[Pd_4(\mu-CO)_5(PPh_3)_4]$ has a distorted tetrahedral core with one non Pd-Pd bonded edge.[288]

<u>2.11 Copper and Gold</u>. - Structures of trinuclear $[Cu_3(\mu-O_2CPh)_2(\mu-mesityl)]$ with a 3-centre, 2-electron Cu-C-Cu interaction,[298] and of tetranuclear $[Cu_4(\mu_3-C\equiv CPh)_4(PPh_3)_4]$[299] are reported.

Studies of oxidative-addition of organic halides to $[Au_2\{\mu-(CH_2)_2PPh_2\}_2]$ have involved structurally characterised Au-Au bonded species $[Au_2\{\mu-(CH_2)_2PPh_2\}_2X(R)]$ (X = Br, R = Me;[300a] X = I, R = Et,CH_2CF_3[300b]) and, also a species (X = Br, X = CH_2Cl)[300c] in which a secondary Au..Cl interaction is consistent with further reaction to form μ-CH_2 derivative, as found in Au(III) species $[Au_2(\mu-CH_2)\{\mu-(CH_2)_2PPh_2\}_2(CN)_2]$ without an Au-Au bond.[300d] Isomerisation of $[Au_2\{\mu-(CH_2)_2PPh_2\}_2Cl_2]$ occurs to form a non Au-Au bonded mixed-valence Au(I)-Au(III) species.[301] Complex $[Au_2(\mu-C(PPh_3)CO_2Et)(PPh_3)_2]^+$ has a short Au-Au separation (2.892(2)Å).[302]

3 Compounds with Heteronuclear Transition Metal Bonds

Tables 1 and 2 list, respectively, the bi- and poly-nuclear complexes containing hetero-transition metal bonds that have been structurally characterised by <u>X</u>-ray diffraction: complexes are entered under the metal of earliest periodic group and arranged in order of (i) increasing group number of other metal(s), (ii) increasing nuclearity (Table 2) and (iii) decreasing

Table 1   X-Ray determined structures of heterobinuclear metal-metal bonded complexes

Thorium

[ThRuI(CO)$_2$CpCp*$_2$]^{303}
[ThNi(μ-PPh$_2$)(CO)$_2$Cp*$_2$]^{304}

Titanium and Zirconium

[TiW{μ-C(tol-p)=CH$_2$}(μ-CO)(CO)Cp$_3$]^{305}
[TiRu(CO)$_2$(NMe$_2$)$_3$Cp]^{306}
[TiCu(μ-SCH$_2$CH$_2$PPh$_2$)$_2$Cp$_2$]$^+$ ^{307}
[ZrMo(μ-PPh$_2$)$_2$(CO)$_4$Cp$_2$]^{308}
[ZrW(μ-PPh$_2$)$_2$(CO)$_4$Cp$_2$]^{309}
[ZrRu(μ-σ,η^5-C$_5$H$_4$)(CO)$_3$Cp$_2$]^{310}
[ZrRu(μ-σ,η^5-C$_5$H$_4$)(μ-CO)(CO)(PMe$_3$)Cp$_2$]^{310}

Chromium, Molybdenum and Tungsten

[CrFe{μ-PH(NPr$^i_2$)}(μ-CO)(CO)$_6$Cp]^{311}
[MoW{μ-C$_2$(tol-p)$_2$}(CO)$_4$Cp(η^5-C$_9$H$_6$COMe)]^{312}
[MoW{μ-C(tol-p)}(CO)$_3$(η^5-C$_9$H$_7$)-
(η^5-C$_2$B$_9$H$_9$Me$_2$)]^{313}
[MoW{μ-σ,η^3-CH(tol-p)}(C$_2$B$_9$H$_8$Me$_2$)}(CO)$_3$-
(C$_2$Et$_2$)(η^5-C$_9$H$_7$)]^{313}
[MoMn(μ-H)(μ-PPh$_2$)(CO)$_6$Cp]^{314}
[MoMn{μ-σ,η^3-CH$_2$CMeCMe}(μ-PPh$_2$)(CO)$_5$Cp]^{314}
[MoMn{μ-σ,η^4-CHCHCH$_2$CHPPh$_2$}(CO)$_4$Cp]^{314}
[MoFe(μ-η^2-Te$_2$Br)(CO)$_5$Cp]^{315}
[MoFe{μ-η^2-Te$_2$SC(S)NEt$_2$}(CO)$_5$Cp']^{315}
[MoFe(μ-H)$_2$I$_2$(NCMe)Cp$_2$]^{316}
[MoRu(μ-CO)(μ-dppm)$_2$(CO)$_5$]^{317}

Molybdenum and Tungsten (continued)

[MoCo(μ-H)$_2$Br$_2$(NCMe)Cp$_2$]^{316}
[MoNi{μ-η^2,η^2-CMeCMeCO}(CO)$_2$CpCp']^{318}
[MoCu(μ-CO)$_2$(CO)(TMED)Cp]^{319}
[MoAu(CO)$_3$(PPh$_3$)(η^5-C$_5$H$_4$CHO)]^{320}
[WMn(μ-CCHCOOMe)(CO)$_6$Cp]^{321}
[WRe(μ-PPh$_2$)(CO)$_9$]^{322}
[WRe(μ-PPh$_2$)$_2$Me(CO)$_7$]^{323}
[WFe{μ-PH(NPr$^i_2$)}(μ-CO)(CO)$_5$Cp]^{311}
[WRu(μ-Cl)(μ-CMe)Cl(CO)$_2$(PPh$_3$)$_2$Cp]^{324}
[WIr(CO)$_7$Cp*]^{325}
[WNi(μ-CO)$_2$(CO)(PPh$_3$)$_2$Cp]^{326}
[WPt(μ-PPh$_2$)(CO)$_3$(PPh$_3$)Cp]^{327}
[WPt{μ-C(tol-p)=CH$_2$}(CO)$_2$(PMe$_3$)$_2$Cp]^{305}
[WCu(μ-Cl)(μ-CO)(μ-dppm)$_2$(CO)$_2$]^{328}

Manganese and Rhenium

[MnRe(CO)$_8$(Pr^{i}N=CHCH=NPr^i)]^{329}
[MnFe(μ-H){μ-N(tol-p)CH$_2$CH$_2$N(tol-p)}(CO)$_6$]^{330}
[MnFe(μ-H)(μ-C$_5$H$_4$NCH=NBu^t)(CO)$_6$]^{331}
[MnRh(μ-CCH$_2$)(μ-CO)(CO)(PPr$^i_3$)Cp$_2$]^{332}
[MnFe(μ-H)(μ-CO)(CO)$_4$(PEt$_3$)$_2$]^{333}
[MnPt(H)(μ-dppm)$_2$Br(CO)$_3$]^{334}
[RePt{μ-H)(μ-PPh$_2$)(NO)(PPh$_3$)$_2$Cp]$^+$ ^{335}
[ReAu(N$_2$C$_6$H$_4$OMe-p)(CO)(PPh$_3$)Cp]^{336}

Iron, Ruthenium and Osmium

[FeRu(μ-CO)$_2$(CO)$_2$Cp$_2$]^{337}
[FeCo{μ-P(NPCl$_2$)$_2$N}(CO)$_5$Cp]^{232}
[FeRh(μ-CCH$_2$)(μ-CO)(CO)$_3$(PPr$^i_3$)-
Cp$_2$·$\frac{1}{2}$CH$_2$Cl$_2$]^{332}
[FeNi(μ-σ,σ,η^4-C$_4$Ph$_4$)Cp$_2$]^{338}
[RuRh(μ-H)(μ-PhPCH$_2$PPh$_2$)Ph(dppm)-
(cod)]^{339}
[OsRh(μ-CO)$_2$(μ-dppm)$_2$BrCl$_2$]^{340}

Cobalt and Rhodium

[CoCu(μ-CO)(CO)$_3$(TMED)]^{341}
[CoAu(CO)$_3$(PPh$_3$)$_2$]^{342}
[RhPt(μ-Ph$_2$AsCH$_2$PPh$_2$)$_2$(CO)X$_3$]
(X = Cl,I)^{343} ^{343}

Platinum

[PtAg(C$_6$F$_5$)$_3$(PPh$_3$)(SC$_4$H$_8$)]^{344}
[PtAg(μ-Cl)Cl(C$_6$F$_5$)$_2$(PPh$_3$)]$^-$ ^{345}

number of carbonyl groups. The abbreviations Cp* ≡ η^5-C_5Me_5 and Cp' ≡ η^5-C_5H_4Me are used.

3.1 Binuclear Complexes. - Hetero-binuclear derivatives studied in 1985, which are neither listed in Table 1 nor are simple ligand modified derivatives of these listed complexes, contain bonding interactions between the following pairs of metals: Ti and W or Fe;[309] Zr and Fe;[309,403] Nb and Fe or Co;[404] Cr and W,[78] Re,[353,405] Fe,[356] Rh[332] or Cu;[319,328] Mo and Mn,[321] Re,[353] Ru,[406] Ni,[326] Pt[327] or Cu;[328] W and Mn,[405] Re,[353,405] Fe,[356] Os[407] or Cu;[319] Mn and Re,[118,405] Fe,[311,408] Ru,[406] Co[242] or Pt;[409] Re and Os;[405] Fe and Co;[366] Os and Cu;[410] Co and Rh,[261,275,361] Ir,[261,275,394] or Ni;[284] Rh and Ir;[261,275] Pd and Pt.[287,291]

An unbridged metal-metal bond linking "early-late" transition metals is present in the novel actinide complex [ThRuI(CO)$_2$CpCp$_2^*$][303] and in [TiRu(CO)$_2$(NMe$_3$)$_3$Cp].[306] Stereochemically non-rigid complexes [Nb(ML$_n$)(CO)Cp(η^5-C_5H_4CHMePh)] {ML$_n$ = Co(CO)$_4$, FeH(CO)$_4$} probably contain labile Nb-M bonds.[404] The unbridged species [WIr(CO)$_7$Cp*], with a dative Ir→W bond, undergoes a fluxional carbonyl exchange via a postulated bridged intermediate.[325] [WFeH(CO)$_9$]$^-$, for which a hydrido bridged structure was suggested (see last year's review), and the related [CrFeH(CO)$_9$]$^-$, for which very preliminary structural data is given,[356] are now believed to possess unbridged Fe-M bonds with terminal cis-Fe-H groups.[356] [FeCo(CO)$_8$]$^-$ is formed in a radical chain process from [Co$_2$(CO)$_8$] and [Fe$_2$(CO)$_8$]$^{2-}$.[366] and [PdPt(CNMe)$_6$]$^{2+}$ is a product of photolysis of a mixture of homo-binuclear species.[287] A useful synthesis of some heterobinuclear carbonyl complexes involves intermolecular reductive elimination of C-H bond from L$_n$MR and L$_m'$M'H.[405]

Binuclear Au-M carbonyl complexes (M = Mo,[320] Re[336] or Co[342]) have non-bridged metal-metal bonds whereas copper species Cu-M (M = Mo,[319] W,[328] Co[341]) and related derivatives possess asymmetric carbonyl bridges. There is an unbridged Pt-Ag bond in [(C$_6$F$_5$)$_3$(C$_4$H$_8$S)PtAg(PPh$_3$)][344] but a chlorine bridge is found in [(C$_6$F$_5$)$_2$ClPt(μ-Cl)Ag(PPh$_3$)]$^-$.[345]

Two hydride bridges are present in complexes [Cp$_2$Mo(μ-H)$_2$MX$_2$(NCMe)] (M = Fe,Co).[316] [MnPt(μ-H)(μ-CO)(CO)$_4$(PEt$_3$)$_2$] undergoes a fluxional process which appears to involve rotation of the PtL$_2$ unit about an Mn-H bond.[333] Photochemical reactions of dienes or alkynes with the μ-H group in [CpMoMn(μ-H)(μ-PPh$_2$)(CO)$_4$] afford novel organic bridges with C-H-Mn interactions as in [CpMoMn(μ-σ,η^3-H-CHCMeCHMe)(μ-PPh$_2$)(CO)$_4$].[314]

Full details of the structure and cis-trans isomerisation of [FeRu(μ-CO)$_2$(CO)$_2$Cp$_2$] are published;[337] its transformation into complexes bridged by various organic ligands is also described. Paramagnetic

[WNi(μ-CO)$_2$(CO)(PPh$_3$)$_2$Cp] contains two asymmetric carbonyl bridges;[326] [MoNi(μ-CO)$_2$(CO)$_2$CpCp'] reacts with alkynes to form [MoNi(μ-CRCR'CO)(CO)$_2$CpCp'] as well as [MoNi(μ-RCCR')(CO)$_2$CpCp'].[318] Addition of cyclopropenes[261] or metal species[275,361,394] to unsaturated complexes [MM'(μ-CO)$_2$Cp$_2$] (M,M' = Co,Rh,Ir) are reported.

Alkylidyne precursor [W(CR)(CO)$_2$Cp] has again, directly or indirectly, provided interesting organic bridged binuclear complexes: species described (R = p-tol) include [CrW(μ-CHR)(CO)$_6$LCp],[78] [MoW(μ-C$_2$R$_2$)(CO)$_4$Cp(η^5-C$_9$H$_6$COMe)] in which acetyl migration to indenyl ring has occurred,[312] and alkenyl bridged complexes [TiW(μ-CR=CH$_2$)(μ-CO)(CO)Cp$_3$] and [PtW(μ-CR=CH$_2$)(CO)$_2$(PMe$_3$)$_2$Cp], now described in a full paper.[305] [WRu(μ-Cl)(μ-CMe)Cl(CO)$_2$(PPh$_3$)$_2$Cp] is obtained from the alklidyne precursor (R = Me) and [RuCl$_2$(PPh$_3$)$_3$].[324] The synthesis and reactivity of methoxyalkylidyne complex [FeMn(μ-COMe)(μ-CO)(CO)$_2$CpCp'] are reported.[408] Vinylidene bridges are formed by reactions of mononuclear vinylidene derivatives (Mn, Rh or Os) with a substitutionally labile complex giving [CpMMn(μ-CCHCOOMe)(CO)$_6$] (M = Mo,W),[321] [CpMnPt(μ-CCHPh)(CO)$_2$L$_2$],[409] [CpRhM(μ-CCH$_2$)(μ-CO)(PPr$_3^i$)L$_n$] (M = Cr,Mn,Fe)[332] and [(η^6-C$_6$H$_6$)OsCu(μ-CCHPh)Cl(PPr$_3^i$)].[410]

Ligand induced cleavage of one Zr-Ru bond of [Cp$_2$Zr{Ru(CO)$_2$Cp}$_2$] affords a binuclear species and [RuH(CO)$_2$Cp]; thus CO reversibly produces [Cp$_2$(OC)Zr(μ-σ,η^5-C$_5$H$_4$)Ru(CO)$_2$] which undergoes facile CO substitution by PMe$_3$.[310] The structure and properties of ferrocene analogues [CpFe{Ni(CRCR^{1}CR^{2}CR^3)Cp}] are described.[338]

Structurally characterised complexes with ligands derived from α-diimines include simple chelated [MnRe(CO)$_8$(dab)], for which spectral and photolytic studies are also reported,[118,329] [MnFe(μ-H)(μ,μ'-RNCH$_2$CH$_2$NR)(CO)$_6$] with symmetric diamido bridge,[330] and asymmetrically bridged [MnFe(μ-H)-(μ-2-pyCH=NBu^t)(CO)$_6$].[331] Several species bridged by dppm are reported: [MoRu(μ-CO)(μ-dppm)$_2$(CO)$_5$] exists in three readily interconvertible forms;[317] [MnPt(μ-dppm)$_2$X(CO)$_3$] (X = Cl,Br) is reversibly protonated;[334] [RuRh(μ-H)(μ-Cl)(μ-dppm)H(dppm)(cod)] is reactive and undergoes P-C bond cleavage with LiMe to form [RuRh(μ-H)(μ-PhPCH$_2$PPh$_2$)Ph(dppm)(cod)].[339] "Early-late" heterobimetallic complexes with bifunctional, phosphine-containing bridges are [Cp$_2$Ti(μ-SCH$_2$CH$_2$PPh$_2$)$_2$Cu]$^+$ [307] and [Cp(Bu^{t}O)Zr(μ-η^5-C$_5$H$_4$PPh$_2$)ML$_n$] (M = Fe,Co).[403]

Two phosphido bridges link early metals, Th,[304] Ti[309] or Zr[308,309] to later transition metals in species which may have some direct metal-metal interaction. The W-Re bond in [WRe(μ-PPh$_2$)(CO)$_9$] is displaced on addition of ligands to Re[322] and if PPh$_2$H is added, subsequent thermolysis affords dibridged species [WRe(μ-PPh$_2$)$_2$H(CO)$_7$] which can be transformed via deprotonation to alkyl derivatives;[323] related WOs derivatives are also described.[407] Protonation of [CpMPt(μ-PPh$_2$)(CO)$_3$(PPh$_3$)] forms a hydrogen bridge when M = Mo but a terminal W-H bond when M = W which very slowly isomerises to become bridging;[327] related terminal-bridging hydrogen isomerisation occurs for [CpRePtH(μ-PCy$_2$)(NO)(PPh$_3$)$_2$]$^+$.[335]

Chalcogen bridges are present in [PdPt(μ-Se)(CS)(dppe)(PPh$_3$)] and [PdPt(μ-SeCS)(dppe)$_2$][291] and in [MoFe(μ-Te$_2$X)(CO)$_5$Cp], containing a hypervalent Te centre.[315]

3.2 Tri- and Higher- Nuclearity Complexes.

Clusters structurally characterised by X-ray diffraction are listed in Table 2. Limitations of space for this article prevent a full discussion of all these heteronuclear systems.

Syntheses and structural characterisations are reported for heteronuclear binary carbonyl or hydrido-carbonyl clusters containing Fe[366] or Ru[366,375,377,383] and also for phosphine substituted species [RuM$_3$H(CO)$_{10}$(PPh$_3$)$_2$] (M = Co,Rh).[377] The radical anion of [Co$_2$Rh$_2${μ_3-(PPh$_2$)$_3$CH}(CO)$_9$] is stabilised by the tripod ligand.[248b] Several carbido-carbonyl-clusters have been formed from Fe$_3$,[368] Fe$_4$[368] or Fe$_6$[367] precursors: reported species contain Fe$_3$MC (M = Mn,Rh),[368] Fe$_3$Ni$_3$C,[368] Fe$_4$CoC,[367,368] Fe$_5$CoC[367] and Fe$_4$CoMC (M = Rh,Pd)[367] cores and chemical oxidation of Fe$_4$MC systems (M = Co,Rh) affords new clusters with two heteroatoms, M.[411]

Trinuclear clusters incorporating cyclopentadienyl ligands include unsaturated [MRh$_2$(μ_3-CO)$_2$Cp$_3^*$] (M = Co or Ir), which adds H$_2$ to form [MRh$_2$(μ_3-CO)(μ-H)$_2$(μ-CO)Cp$_3^*$],[275] and also [CoRh$_2$(μ-COCCF$_3$CCF$_3$)-(μ-CO)(CO)Cp$_2$Cp*].[276] Interestingly, 46-electron [Cp*IrCo$_2$(μ-CO)$_2$Cp$_2$] has a paramagnetic, triplet ground state.[394] [RuRh$_2$(μ_3-CO)(μ-CO)(CO)$_2$(η^4-cod)Cp$_2^*$] is also characterised[380] and several tetranuclear Ru$_{4-n}$Rh$_n$ species (n = 1,[381] 2,[281] 3[380,382]) with cyclopentadienyl ligands are reported: thus [Ru$_3$Rh(μ-H)$_2$(μ-CO)(CO)$_9$Cp] has isomeric structures when Cp = η^5-C$_5$H$_5$ or η^5-C$_5$Me$_5$ and is formed reversibly from [Ru$_3$(CO)$_{12}$] and respective complex [CpRh(CO)$_2$].[381] Other tetranuclear clusters studied are [CpNiOs$_3$(μ-H)$_3$(CO)$_8$(PPh$_2$C≡CPr^i)], which will co-ordinate from the unsaturated phosphine ligand to Co$_2$(CO)$_6$,[386] and structurally characterised [Co$_2$Rh$_2$(μ_3-CO)$_2$(μ-CO)(CO)$_4$Cp$_2^*$].[393]

Fluxionality has been studied in chiral alkyne complexes [CpNiFeM(μ_3-PhCCCO$_2$Pr^i)(CO)$_3$] (M = NiCp,Co(CO)$_3$,Mo(CO)$_2$Cp).[350] The 62-electron alkyne cluster [Cp$_2$Ni$_2$Fe$_2$(μ_4-C$_2$Ph$_2$)(CO)$_6$] possesses a distorted square-planar arrangement of metal atoms and can be expanded to pentametallic species.[369] Alkyne cleavage occurs in [Cp$_2$W$_2$Os(μ_3-C$_2$R$_2$)(CO)$_7$] (R = p-tol) on thermal decarbonylation to form [Cp$_2$W$_2$Os(μ_3-CR)(μ-CR)(CO)$_5$]; this process is reversed

Table 2. X-Ray determined structures of heterometallic tri- and higher- nuclearity complexes

Chromium

$[Cr_2Co_2(\mu_3-S)_4\{P(OMe)_3\}_2Cp_2^*]^{346}$

Molybdenum

$[MoMn_2(\mu-CCHCOOMe)_2(CO)_6Cp_2]^{321}$
$[Mo_2Fe(\mu_3-Te)_2(CO)_7Cp_2]^{347}$
$[Mo_2Fe(\mu_3-S)_2(\mu-S_2)(CO)_2Cp_2^*]^{348}$
$[Mo_2Fe(\mu_3-S)_2\{\mu-SC(O)S\}(CO)_2Cp_2]^{348}$
$[Mo_2Fe_2(\mu_3-Te)_2(\mu_3-CO)(\mu-CO)(CO)_5Cp_2]^{347}$
$[Mo_2Fe_2(\mu_3-S)_4(CO)_4Cp_2^*]^{348}$
$[MoFeCo_2(\mu-CCO_2Pr^i)(CO)_5Cp_3]^{349}$
$[MoFeNi(\mu_3-PhC_2COOPr^i)(CO)_5Cp_2]^{350}$
$[MoCo_2(\mu_3-CMe)(\mu-CO)(CO)_2Cp_3]^{227}$
$[Mo_2Co(\mu_3-S)_2(\mu-\eta^2-AsS)(CO)_2Cp_2^*]^{351}$
$[Mo_2Co_2(\mu_3-S)_4(S_2CNEt_2)_2(CO)_2(NCMe)_2]^{352}$

Tungsten

$[WReCo_2\{\mu_3-C(tol-\underline{p})\}(CO)_{11}]^{353}$
$[WFe_2(\mu_3-SCMe)(CO)_8Cp]^{354}$
$[WFe_2(\mu_3-S)\{\mu-C(tol-\underline{p})\}(CO)_7Cp]^{354}$
$[WFe_2\{\mu-OCCH_2(tol-\underline{p})\}(\mu-PPh_2)_2(CO)_6-(PPh_2H)Cp]^{355}$
$[WFe_2\{\mu_3-OCCH_2(tol-\underline{p})\}(\mu-PPh_2)_2(CO)_5Cp]^{355}$
$[WFeAu(CO)_9(PPh_3)]^{356}$
$[W_2Ru(\mu_3-C_2Me_2)(CO)_7Cp_2]^{324}$
$[WOs_3(\mu-H)(\mu-O)\{\mu-CCH(tol-\underline{p})\}-(CO)_9Cp]^{357}$

Tungsten (continued)

$[WOs_3(\mu-H)(\mu-O)\{\mu-CHCH_2(tol-\underline{p})\}(CO)_9Cp]^{358}$
$[WOs_3\{\mu_3-CCH_2(tol-\underline{p})\}(\mu-O)(CO)_9Cp]^{359}$
$[WCo_2(\mu_3-CMe)(\mu-H)(\mu-PPh_2)(CO)_6Cp]^{360}$
$[WRh_2(\mu_3-CMe)(\mu-CO)(CO)_2CpCp_2^*]^{361}$
$[W_2Ir_2\{\mu_3-OC(OEt)CH\}\{\mu-CH(COOEt)\}(CO)_7-Cp_2]^{362}$

Manganese

$[MnCo_2(\mu_3-CMe)(\mu-CO)_2(CO)_3Cp_2]^{227}$
$[Mn_4Cu_2(\mu_3-C_2H)_2(\mu-H)_4\{\mu-(EtO)_2POP(OEt)_2\}_2-(CO)_{12}]^{115}$

Rhenium

$[ReCo_2\{\mu_3-C(tol-\underline{p})\}(CO)_{10}]^{353}$
$[\{Re_7Ag(\mu_6-C)(CO)_{21}\}_2(\mu-Br)]^{5-}$ 363

Iron

$[FeCo_2\{\mu_3-P(NPCl_2)_2N\}(CO)_9]^{232}$
$[Fe_2Co(\mu_3-S)(CO)_8(NO)]^{364}$
$[Fe_2Co(\mu_3-CSMe)(\mu-CO)_2(CO)_3Cp_2]^{365}$
$[Fe_2Co(\mu_3-CSMe)(\mu-CO)_2(CO)_2(PPh_3)Cp_2]^{365}$
$[Fe_3Co(\mu-CO)_3(CO)_{10}]^{-}$ 366
$[Fe_4Co(\mu_5-C)(\mu-CO)(CO)_{13}]^{-}$ 367, 368
$[Fe_4CoRh(\mu_6-C)(\mu-CO)_2(CO)_{14}]^{367}$
$[FeCoNi(\mu_3-PhC_2CO_2Pr^i)(CO)_6Cp]^{350}$
$[Fe_3Rh(\mu_4-C)(CO)_{12}]^{-}$ 368

Iron (continued)

$[FeNi_2(\mu_3-PhC_2CO_2Pr^i)(CO)_3Cp_2]^{350}$
$[Fe_2Ni_2(\mu_4-\eta^2-C_2Ph_2)(CO)_6Cp_2]^{369}$
$[Fe_2Ni_2(\mu_4-\eta^2-C_2Pb)(\mu-PPh_2)(CO)_5Cp_2]^{370}$
$[Fe_2Pt(CO)_6(NO)_2(NCPh)_2]^{371}$
$[FeCu_2(CO)_4(PPh_3)_4]^{372}$
$[Fe_3Cu(\mu_3-CMe)(CO)_{10}(PPh_3)]^{373}$
$[Fe_4Cu(\mu-CO)(CO)_{12}(PPh_3)]^{-}$ 374
$[Fe_4Cu_6(CO)_{16}]^{2-}$ 372
$[Fe_2Au\{\mu-SC(C_{11}H_{19})\}(CO)_6(PPh_3)]^{138b}$
$[Fe_4Au(\mu-\eta^2-CO)(CO)_{12}(PEt_3)]^{-}$ 374
$[Fe_4Au(\mu_4-\eta^2-COMe)(CO)_{12}(PEt_3)]^{374}$
$[Fe_3Au_2(\mu_3-S)(CO)_9(PPh_3)_2]^{364}$

Ruthenium

$[Ru Co_2(\mu-CO)(CO)_{10}]^{375}$
$[RuCo_2\{\mu_3-CCH(R)\}(CO)_9]$ (R = Ph, Bu^t) 376
$[RuCo_2(\mu_3-PhCCPh)(\mu-CO)(CO)_8]^{376}$
$[RuCo_2(\mu_3-HCCBu^t)(\mu-CO)(CO)_8]^{376}$
$[Ru_2Co_2(\mu-CO)_4(CO)_9]^{375}$
$[RuCo_3(\mu_3-H)(\mu-CO)_3(CO)_7(PPh_3)_2]^{377}$
$[RuCo_2Rh(\mu_4-PPh)(CO)_{10}Cp(PPh_3)_2]^{378}$
$[RuCo_nRh_{3-n}Au(\mu-CO)_3(CO)_9(PPh_3)]^{379}$
$[RuRh_2(\mu_3-CO)(\mu-CO)(CO)_2(\eta-C_8H_{10})Cp_2^*]^{380}$
$[Ru_3Rh(\mu-H)_2(\mu-CO)(CO)_9Cp]^{381}$
$[Ru_3Rh(\mu-H)_2(\mu-CO)(CO)_9Cp^*]^{381}$

Table 2 (continued)

Ruthenium (continued)

[Ru$_3$Rh(μ-H)$_4$(CO)$_9$Cp*]^{381}

[RuRh$_3$(μ-H)(μ-CO)$_3$(CO)$_9$]^{377}

[RuRh$_3$(μ-H)(μ-CO)$_3$(CO)$_7$(PPh$_3$)$_2$]^{377}

[RuRh$_3$(μ_3-CO)$_2$(CO)$_3$Cp$_3$*]^{380,382}

[RuIr(μ-CO)$_3$(CO)$_9$]$^-$ ^{383}

[Ru$_3$Au(μ_3-S)(μ-H)(CO)$_9$(PPh$_3$)]^{384}

[Ru$_3$Au(μ_3-SBu^t)(CO)$_9$(PPh$_3$)]^{384}

[Ru$_3$Au$_2$(μ_3-S)(CO)$_9$(PPh$_3$)$_2$]^{384}

[Ru$_3$Au$_2$(μ_3-CCHBu^t)(CO)$_9$(PPh$_3$)$_2$]^{385}

[Ru$_5$Au(μ_5-C)(μ-CO)(CO)$_{12}$(NO)(PEt$_3$)]^{217}

Osmium

[Os$_3$Ni(μ-H)$_3$(CO)$_8$(Ph$_2$PC$_2$Pr^i)Cp]^{386}

[Os$_3$Ni(μ-H)$_3$(CO)$_8$(Ph$_2$PC$_2$Pr^i(Co$_2$(CO)$_6$))Cp]^{386}

[Os$_3$Pt(μ-H)$_2$(CO)$_{11}$(PCy$_3$)]^{387}

[Os$_3$Pt(μ-H)$_4$(CO)$_{10}$(PCy$_3$)]^{387}

[Os$_3$Pt(μ-H)$_2$(μ-CH$_2$)(CO)$_{10}$(PCy$_3$)]^{387}

Osmium (continued)

[Os$_3$Pt(μ_4-C)(μ-H)$_2$(CO)$_{10}$(PCy$_3$)]^{388}

[Os$_3$Pt(μ_3-S)$_2$(CO)$_9$(PMe$_2$Ph)$_2$]^{389}

[Os$_3$Pt$_2$(μ_5-C)(μ-H)$_2$(μ-CO)(CO)$_9$(PCy$_3$)$_2$]^{388}

[Os$_3$Pt$_2$(μ_5-C)(μ-H)(μ-OMe)(μ-CO)(CO)$_9$-(PCy$_3$)$_2$]^{388}

[Os$_6$Pt$_2$(CO)$_{16}$(cod)$_2$]^{390}

[Os$_3$Au(μ-CHCHAr)(CO)$_{10}$(PPh$_3$)]
(Ar = Ph, C$_6$F$_5$) ^{391}

[Os$_4$Au$_2$(μ-CO) (CO)$_{12}$(PEt$_3$)$_2$]^{392}

[Os$_4$Au$_2$(CO)$_{12}$(PMePh$_2$)$_2$]^{392}

Cobalt

[Co$_2$Rh$_2$(μ_3-CO)$_2$(μ-CO)(CO)$_4$Cp$_2$*]^{393}

[Co$_2$Ir(μ-CO)$_2$Cp$_2$Cp*]^{394}

[Co$_3$Ni(μ_{10}-C$_2$)(μ-CO)$_{10}$(CO)$_6$]$^{2-}$ ^{395}

[CoCu(CO)$_4$]$_n$ ^{396}

Rhodium

[Rh$_2$Pd{μ_3-(Ph$_2$PCH$_2$)$_2$AsPh}$_2$(μ-Cl)Cl$_2$(CO)$_2$]^{397}

[Rh$_2$Ag(μ-CO)(μ-dppm)(OPF$_2$O)Cp$_2$]^{398}

[Rh$_2$Au(μ-CO)(μ-dppm)(PPh$_3$)Cp$_2$]$^+$ ^{258}

Nickel

[Ni$_{38}$Pt$_6$H(μ_3-CO)$_{18}$(μ-CO)$_{12}$(CO)$_{18}$]$^{5-}$ ^{399}

[Ni$_{38}$Pt$_6$H$_2$(CO)$_{48}$]$^{4-}$ ^{399}

Palladium

[Pd$_3$Pt$_3$(μ_4-PCBu^t)$_3$(PPh$_3$)$_5$]^{400}

Platinum

[Pt$_2$Ag(μ-C$_6$F$_5$)$_2$(C$_6$F$_5$)$_4$(OEt$_2$)]$^-$ ^{344}

[Pt$_2$Ag$_2$(μ-Cl)$_4$(C$_6$F$_5$)$_4$]$^{2-}$ ^{345}

[Pt$_6$Ag(μ-CO)$_6$(PPr$^i_3$)$_6$]$^+$ ^{401}

Copper

[Cu$_2$Au$_3$(μ-σ,η^2-C$_2$Ph)$_6$]$^-$ ^{402}

on exposure to CO.[412] 1-Alkyne ligands in [Co$_2$Ru(μ_3-RC$_2$H)(CO)$_9$] are thermally transformed into vinylidene ligands (μ_3-C=CHR)[376] and the vinylidene bridge in [Mn$_2$Mo(μ-CCHCO$_2$Me)(CO)$_6$Cp$_2$] can originate from mononuclear alkyne or vinylidene Mn precursors.[321] Clusters [Fe$_3$Rh(μ_4-η^2-CCHR)(μ-H)(CO)$_{11}$] have been synthesised and studied for catalytic activity.[413] Addition of the M-H bond from species [HMn(CO)$_5$] or [HMo(CO)$_2$LCp] to the vinylidene bridge of [Cp$_2$Co$_2$(μ-CCH$_2$)(CO)$_2$] affords a trinuclear herocluster with a μ_3-CMe bridge.[227] Binuclear precursors combine to form acetylide cluster [Cp$_2$Ni$_2$Fe$_2$(μ_4-η^2-C$_2$Ph)(μ-PPh$_2$)(CO)$_5$] with a spiked-triangular metal framework.[370]

Syntheses are reported for several heteronuclear alkylidyne bridged clusters from [Co$_3$(μ_3-CR)(CO)$_9$], including the "butterfly" species [Cp$_3$Co$_2$MoFe(μ_4-CCO$_2$Pr^i)(CO)$_5$].[349] Terminal alkylidyne ligands have been converted, by condensation with binuclear complexes, into μ_3-bridged clusters [MCo$_2$Re(μ_3-CR)(CO)$_{15}$] (M = Cr,Mo,W), containing a MCo$_2$ triangle and a single M-Re bond,[353] [Co$_2$Re(μ_3-CR)(CO)$_{10}$] (R = p-tol)[353] and also, when R = Me, [WRhM(μ_3-CMe)(μ-CO)(CO)$_2$CpCp$_2^*$] (M = Co,Rh) and a related compound with μ_3-C(O)CMe ligand (M = Rh).[361] The alkylidyne bridge in clusters [Cp$_2$Fe$_2$Co(μ_3-CSMe)(μ-CO)$_2$(CO)$_2$L] is bonded to the three metal atoms by the C-atom but also to Co by S.[365] Compounds [CpWFe$_2$(μ_3-CR)(μ-CO)(CO)$_8$] react with sulphur or selenium to give μ_3-E (E = S,Se) and (when R = Me) μ_3-SCMe bridged clusters;[354] reactions of the same precursor (R = p-tol) with PPh$_2$H can produce μ_3-acyl derivatives, e.g. [CpWFe$_2$(μ_3-OCCH$_2$R)(μ-PPh$_2$)$_2$(CO)$_5$],[355] or μ-phosphido-μ-hydrido-μ_3-alkylidyne clusters.[360]

Crystal structures are reported for alkylidyne species [CpWOs$_3$-(μ_3-CCH$_2$tol-p)(μ-O)(CO)$_9$][359] and for alkylidene species [CpWOs$_3$(μ-H)(μ-O)(μ-CHCH$_2$tol-p)(CO)$_9$],[358] formed by its hydrogenation. Cluster [Cp$_2$W$_2$Ir$_2${μ_3-OC(OEt)CH}{μ-CH(COOEt)}(CO)$_7$] contains two differently bonded ethylcarboxylatocarbene bridges; both span a W-Ir bond but one is also co-ordinated to the second W-atom via its carbonyl O-atom.[362]

New structurally characterised sulphido-bridged clusters include cubanes [Cp$_2^*$Cr$_2$Co$_2$(μ_3-S)$_4$(PR$_3$)$_2$],[346] [Cp$_2^*$Mo$_2$Fe$_2$(μ_3-S)$_4$(CO)$_4$],[348] and [Mo$_2$Co$_2$(μ_3-S)$_4$-(S$_2$CNEt$_2$)$_2$(CO)$_2$(NCMe)$_2$],[352] and trinuclear clusters [Cp$_2^*$Mo$_2$Fe(μ_3-S)$_2$(μ-L)-(CO)$_2$] (L = S$_2$,SCOS)[348] and [Cp$_2^*$Mo$_2$Co(μ_3-S)$_2$(μ-AsS)(CO)$_2$].[351] Metal exchange and nitrosation reactions have been investigated on [FeCo$_2$(μ_3-S)(CO)$_9$] and [Fe$_2$Co(μ_3-S)(μ-H)(CO)$_9$].[364] Thermolysis of [Cp$_2$Mo$_2$Fe(μ_3-Te)$_2$(CO)$_7$] affords three clusters, including structurally characterised tetrahedral species [Cp$_2$Mo$_2$Fe$_2$(μ_3-Te)$_2$(CO)$_7$] with Te atoms capping each Mo$_2$Fe face.[347]

Phosphido systems [M^{1}M^{2}M^3(μ_3-PR)(CO)$_9$] (M^1,M^2,M^3 = variously Fe,Ru,Co) add CpRh(CO) fragments giving products with M^{1}M^{2}M^{3}Rh(μ_4-PR) and M^{1}M^{2}M^{3}Rh$_2$(μ_4-PR) cores.[378] The properties of double clusters M^{1}M$_2^2$(μ_3-P)-R-(μ_3-P)M^{1}M$_2^2$

(M^1,M^2 = Fe,Co) have been studied.[414] Kinetic studies of reversible Mn-Fe bond opening in [CpMnFe$_2$(μ_3-PR)(CO)$_8$] give $\Delta H^{\neq} \simeq$ 90 kJ/mol,[415] and reactivity comparisons of clusters [Co$_2$Fe(μ_3-EMe)(CO)$_9$] with E = P or As indicate that the larger μ_3-AsMe group enhances the metal-metal bond lability.[416]

The platinum species [Pt{Fe(CO)$_3$NO}$_2$(NCPh)$_2$] contains a linear Fe-Pt-Fe chain;[371] the triangular cluster [FePdPt(μ-dppm)$_2$(CO)$_4$] isomerises by CO migration between Pd and Pt.[417] Addition of CO, H$_2$ or CH$_2$ (from CH$_2$N$_2$) to unsaturated cluster [Os$_3$Pt(μ-H)$_2$(CO)$_{10}$(PCy$_3$)] gives products characterised by X-ray analysis.[387] Reactions between [Pt(C$_2$H$_4$)$_2$(PCy$_3$)] and alkylidyne clusters [Os$_3$H(CX)(CO)$_{10}$] (X = H,OMe) afford carbido-clusters [Os$_3$Pt(μ_4-C)-(μ-H)$_2$(CO)$_{10}$(PCy$_3$)] and [Os$_3$Pt$_2$(μ_5-C)(μ-H)(μ-X)(μ-CO)(CO)$_9$(PCy$_3$)$_2$].[388] Skeletal rearrangement of [Os$_3$Pt(μ_3-S)$_2$(CO)$_9$(PMe$_2$Ph)$_2$] has been studied,[389] and phosphaalkyne bridged [Pd$_2$Pt$_3$(μ_4-PCBu^t)$_3$(PPh$_3$)$_5$] may be considered to contain a <u>closo</u>-hexagonal bipyramidal Pd(Pt$_3$P$_3$)Pd core.[400] In [Pt$_6$Ag(μ-CO)$_6$(PPr$^i_3$)$_6$]$^+$ an Ag$^+$ ion is sandwiched between staggered Pt$_3$ triangular units.[401]

A linear metal chain is found in [Fe{Cu(PPh$_3$)$_2$}$_2$(CO)$_4$],[372] a polymeric chain in [CoCu(CO)$_4$]$_n$ [396] and a bent chain in [(Ph$_3$P)AuFeW(CO)$_9$]$^-$.[356] There is a central Cu-Cu bond in [Mn$_4$Cu$_2$(μ-H)$_6$(μ-{(EtO)$_2$P}$_2$O)$_2$(CO)$_{12}$] within the planar Mn$_4$Cu$_2$ skeleton.[115] [Cu$_2$Au$_3$(μ-C$_2$Ph)$_6$]$^-$ contains a trigonal bipyramidal Cu$_2$Au$_3$ core with Cu-Au bonds but no bonds between equatorial Au$_3$ atoms.[402] Ag$^+$ or (Ph$_3$P)Au$^+$ ions forms bridged adducts with binuclear Mn$_2$[115] or Rh$_2$[258,398,418] complexes.

Reactions of anionic or hydrido-clusters with [R$_3$PMCl] (M = Cu,Ag,Au), [M(NCMe)$_4$]$^+$ (M = Cu,Ag), [(R$_3$P)Au]$^+$ or [O{Au(PPh$_3$)}$_3$]$^+$ have produced more heteronuclear systems containing Group 1B metals (see Table 2). Clusters [Fe$_4$(MPR$_3$)(CO)$_{13}$]$^-$ (M = Cu,Au) can have tetrahedral Fe$_4$ units, face-capped by MPR$_3$, or "butterfly" Fe$_4$ units, "hinge"-bridged by MPR$_3$ with a π-CO ligand;[374,419] [Fe$_3$Cu(μ_3-CMe)(CO)$_{10}$(PPh$_3$)] contains a butterfly Fe$_3$Cu core.[373] Isomerism between μ_3-AuPR$_3$ and μ_2-AuPR$_3$ geometries is observed in the two forms of [Ru$_5$Au(μ_5-C)(μ-CO)(CO)$_{12}$(NO)(PEt$_3$)] found in a single crystal.[217] Several other gold-ruthenium clusters are reported; of interest are di-gold systems with trigonal bipyramidal Ru$_3$Au$_2$ cores incorporating direct Au-Au bonds, <u>i.e.</u> [Ru$_3$Au$_2$(μ_3-L)(CO)$_9$(PPh$_3$)$_2$] (L = S,[384] CCHBu^t [385]). Group 1B metal exchange (M,M$'$ = Cu,Ag,Au; L = PPh$_3$) and displacement of ligands (L = NCMe) by mono- or bi-phosphines (M,M$'$ = Cu,Ag) occur for clusters [Ru$_4$(ML)(M$'$L)(μ_3-H)$_2$(CO)$_{12}$].[420] Acetylides [(Ph$_3$P)MC$_2$Ar] (M = Cu,Ag,Au) react with [Os$_3$H$_2$(CO)$_{10}$] to give several products including [MOs$_3$(μ-CHCHAr)(CO)$_{10}$(PPh$_3$)].[391] The reactivity and structure are described for unsaturated [Os$_4$Au$_2$(CO)$_n$(PR$_3$)$_2$] (n = 12) prepared by decarbonylation of the cluster with n = 13.[392]

Structurally characterised higher nuclearity clusters include [Cu$_6$Fe$_4$(CO)$_{16}$]$^{2-}$ with tetracapped Cu$_6$ octahedral core,[372] [{Re$_7$Ag(μ_6-C)-

$(CO)_{21}\}_2Br]^{5-}$ with linked trans-bicapped octahedral units,[363] $[Os_6Pt_2(CO)_{16}(cod)_2]$ with two edge-fused Os_4 tetrahedra, one being bi-capped by Pt atoms,[390] and paramagnetic $[Co_3Ni_7(\mu_{10}\text{-}C_2)(\mu\text{-}CO)_{10}(CO)_6]^{2-}$ containing a 3,4,3-C_{2h} stack of metal atoms.[395] The highest nuclearity cluster to be fully characterised by X-ray analysis is $[Ni_{38}Pt_6H(\mu_3\text{-}CO)_{18}(\mu\text{-}CO)_{12}(CO)_{18}]^{5-}$ containing a Pt_6 octahedron encapsulated in an outer ν_3 Ni_{38} octahedron, resulting from a c.c.p. arrangement of metal atoms.[399]

4 Compounds containing Bonds between Transition and Main Group Metals.

4.1 Lithium. - Crystal structures are reported for cluster anions $[Cu_4LiPh_6]^-$ [421a] and $[Ag_3Li_2Ph]^-$ [421b] with trigonal bipyramidal cores.

4.2 Group II. - Cluster $[Cu_4MgPh_6]$ has trigonal bipyramidal geometry.[421a] Zinc cluster $[Co_4Zn_2(\mu\text{-}CO)(CO)_{14}]$ incorporates two $ZnCo(CO)_4$ groups bridging a Co-Co bond.[422] A full report of $[Ni_2Zn_4Cp_6]$ is published[423a] and the structure of $[Cp_2Zn_2\{\mu\text{-}Ni(PPh_3)Cp\}(\mu\text{-}C_5H_5)]$ is described.[423b]

Structurally characterised mercury derivatives include linearly bonded $[Cp'Mo(HgCl)(CO)_3]$[424] and $[(TPP)SnMn(CO)_4HgMn(CO)_5]$,[425] capped tetrahedral cluster $[Fe_4(\mu_3\text{-}HgMe)(CO)_{13}]^-,$ [374b] and $[Hg\{Co_3Ru(\mu\text{-}CO)_3(CO)_9\}_2]$ in which Hg links two Co_3Rh tetrahedra by their Co_3 faces.[426] The skeleton of $[Hg_9Co_6(CO)_{18}]$ is an approximate trigonal prism with a Co atom at each vertex and an Hg atom at the midpoint of each edge.[427]

4.3 Gallium. - Spectroscopically characterised clusters $[Re_4\{\mu_3\text{-}GaRe(CO)_5\}_4$-$(CO)_{12}]$ and $[Re_2\{\mu\text{-}GaRe(CO)_5\}_2(CO)_8]$ are reported, with preliminary structural data for the former species.[428]

4.4 Group IV. - Structurally characterised M-Si bonded species include $[Cp_2Sm(SiMe_3)_2]^-$ [429] and $[Cp_2Zr(SiMe_3)X]$ (X = S_2CNEt_2);[430a] a related Zr species with X = Cl undergoes CO insertion to form $[Cp_2Zr(\eta^2\text{-}OCSiMe_3)Cl]$.[430b] Complexes $[Cp_2MH_2(SiMe_2Ph)]$ with M = Nb and Ta (structurally characterised) have been investigated[431] and $[Cp^*Ta(SiMe_3)Cl_3]$ undergoes a CO insertion process into the Ta-Si bond which also involves subsequent ether cleavage.[432] $[H_2Si\{MnH(CO)_2Cp^*\}_2]$ has a bent Mn-Si-Mn chain $\{124.4(3)°\}$.[433] A study is reported of the intramolecular migration of silyl group from Fe to Cp ligand in [FpSiMe_2R] when reacted with strong base.[434] Iridium(I) complex $[\overline{Ir\{Si(Me)_2CH_2CH_2}PPh_2\}(PPh_3)(CO)_2]$ has been structurally characterised.[435]

Various transformations of clusters $[Co_3(\mu_3\text{-}EX)(CO)_9]$ (E = Si,Ge; X = R,ML_n) are described, including the formation of heterometallic species;[436]

tetranuclear $[Co_4(\mu_4\text{-GeR})_2(CO)_{11}]$ is also reported[436b] and structures have been determined for $[Co_3(\mu_3\text{-GeFp})(CO)_9]$,[436b] $[CpNiCo_3(\mu_4\text{-GeBu}^t)_2(\mu\text{-CO})(CO)_7]$,[436b] and $[Co_4\{\mu_4\text{-GeCo}(CO)_4\}_2(\mu\text{-CO})(CO)_{10}]$.[437]

Digermane complexes $[Fe(GeH_2GeH_3)X(CO)_4]$ (X = GeH_3, Ge_2H_5) have been prepared.[438] Structures are reported for two conformers of $[Ge\{Mn(CO)_2Cp^*\}_2]$ with linear multiple Mn-Ge-Mn bonding.[439] Trigonal-planar co-ordination of Ge or Sn occurs in $[W_2\{\mu\text{-E=W}(CO)_5\}(CO)_{10}]$ (E = Ge,Sn).[440] Clusters $[M_3(\mu\text{-H})(ER_3)(CO)_{10}L]$ (M = Ru,Os; E = Ge,Sn) are readily synthesised.[441]

Crystal structures are reported for the following M-Sn species: $[Cp_2Mo(SnMe_2Cl)_2]$,[442a] $[Cp_2Re(SnMe_{3-n}Cl_n)]$ (n = 1-3),[442b] $[CpW(SnPh_3)\{CH(tol-\underline{p})\}(CO)_2]$,[78] $[Rh\{SnCl(NR_2)_2\}(\eta^2\text{-}C_8H_{14})(\eta^6\text{-tol})]$ (R = $SiMe_3$)[443] and $[CpIrH_3(SnPh_3)]$.[256b] Clusters $[M_3\{\mu\text{-Sn}(NR_2)_2\}_3(CO)_3]$ (M = Pd,Pt; R = $SiMe_3$) with planar M_3Sn_3 rings reversibly form monoanions on reduction.[444] $[Os_3(SnR_2)(\mu\text{-H})_2(CO)_{10}]$ thermally transforms into $[Os_3(SnR)(\mu\text{-H})_2(\mu\text{-OCR})(CO)_9]$ and reacts with $C_2(COOMe)_2$ to form $[Os_3(SnR_2)\{\mu\text{-C(COOMe)CH}_2(COOMe)\}(CO)_9]$.[445]

A linear multiple-bonded Mn-Pb-Mn chain is present in $[Pb\{Mn(CO)_2Cp\}_2]$.[446]

4,5 Bismuth. - Crystal structures are reported for $[Fp(Bi\{SC(S)NEt_2\}_2)]$,[447] $[W_2(\mu\text{-}\eta^2\text{-}Bi_2)(\mu\text{-Bi}\{W(CO)_5\}Me)(CO)_8]$,[448] trigonal bipyramidal $[Fe_3(\mu_3\text{-Bi})_2(CO)_9]$[449] and $[Bi_4Fe_4(CO)_{13}]^{2-}$ containing a Bi_4 tetrahedron tricapped by $Fe(CO)_3$ units and linked by a single bond from apical Bi to $Fe(CO)_4$.[450]

References

1   T.J.Meyer and J.V.Casper, Chem. Rev., 1985, **85**, 187.
2   A.E.Stiegman and D.R.Tyler, Coord. Chem. Rev., 1985, **63**, 217.
3   W.E.Geiger and N.G.Connelly, Adv. Organomet. Chem., 1985, **24**, 87.
4   M.H.Chisholm, D.M.Hoffman and J.C.Huffman, Chem. Soc. Rev., 1985, **14**, 69.
5   R.D.Adams and I.T.Horvath, Prog. Inorg. Chem., 1985, **33**, 127.
6   E.Sappa, A.Tiripicchio and P.Braunstein, Coord. Chem. Rev., 1985, **64**, 219.
7   H.Vahrenkamp, Comments Inorg. Chem., 1985, **4**, 253.
8   (a)N.K.Eremenko, E.G.Mednikov and S.S.Kurasov, Russ. Chem. Rev., 1985, **54**, 394; (b) D.M.P.Mingos and R.W.M.Wardle, Transition Met. Chem., (Weinheim, Ger.), 1985, **10**, 441.
9   P.R.Raithby and M.J.Rosales, Adv. Inorg. Chem. Radiochem., 1985, **29**, 170.
10 W.L.Gladfelter, Adv. Organomet. Chem., 1985, **24**, 41.
11 R.D.Adams, Polyhedron, 1985, **4**, 2003.
12 K.Krieze and G.van Koten, Inorg. Chim. Acta., 1985, **100**, 79.
13 P.E.Garrou, Chem. Rev., 1985, **85**, 171.
14 O.J.Scherer, Angew. Chem., Int. Ed. Engl., 1985, **24**, 924.
15 (a) M.I.Bruce, J. Organomet. Chem., 1985, **283**, 339; (b) idem, J. Organomet. Chem., Library, 1985, **17**, 399.
16 J.C.A.Boeyens, Inorg. Chem., 1985, **24**, 4149.
17 M.D.Braydich, B.E.Bursten, M.H.Chisholm and D.L.Clark, J. Am. Chem. Soc., 1985, **107**, 4459.
18 F.A.Cotton, M.P.Diebold and I.Shim, Inorg. Chem., 1985, **24**, 1510.
19 Y.Hang, S.Alvarez and R.Hoffmann, Inorg. Chem., 1985, **24**, 749.
20 (a)D.M.P.Mingos, Inorg. Chem., 1985, **24**, 114; (b) B.K.Teo, ibid, p.115; (c) idem, ibid, p.4209.
21 B.K.Teo, Inorg. Chem., 1985, **24**, 1627.

22 R.L.Johnston and D.M.P.Mingos, J. Organomet. Chem., 1985, 280, 407 and 419.
23 (a) A.Ceulemans and P.W.Fowler, Inorg. Chim. Acta., 1985, 105, 75; (b) P.W.Fowler and W.W.Porterfield, Inorg. Chem., 1985, 24, 3511.
24 R.G.Wooley, Inorg. Chem., 1985, 24, 3519 and 3525.
25 D.M.P.Mingos, J. Chem. Soc., Chem. Commun., 1985, 1352.
26 B.K.Teo and N.J.A.Sloane, Inorg. Chem., 1985, 24, 4545.
27 J-F.Halet, J-Y.Saillard, R.Lissillour, M.J.McGlinchey and G.Jaouen, Inorg. Chem., 1985, 24, 218.
28 J.Silvestre and R.Hoffmann, Helv. Chim. Acta, 1985, 68, 1461.
29 J-F.Halet, R.Hoffmann and J-Y.Saillard, Inorg. Chem., 1985, 24, 1695.
30 (a) M.Casarin D.Ajo, G.Granozzi, E.Tondello and S.Aime, Inorg. Chem., 1985, 24, 1241; (b) G.Granozzi, E.Tondello, R.Bertoncello, S.Aime and D.Osella, ibid, 1985, 24, 570.
31 D.J.Underwood, R.Hoffmann, K.Tatsumi, A.Nakamura and Y.Yamamoto, J. Am. Chem. Soc., 1985, 107, 5968.
32 D.G.Evans and D.M.P.Mingos, J. Organomet. Chem., 1985, 295, 389.
33 J.M.Newsam and J.S.Bradley, J. Chem. Soc., Chem. Commun., 1985, 759.
34 J.S.Holmgren, J.R.Shapley and P.A.Belmonte, J. Organomet. Chem., 1985, 284, C5.
35 T.Saito and S.Sawada, Bull. Chem. Soc. Jpn., 1985, 58, 459.
36 G.Doyle, B.T.Heaton and E.Occhiello, Organometallics, 1985, 4, 1224.
37 G.E.Hawkes, L.Y.Lian, E.W.Randall, K.D.Sales and S.Aime, J. Chem. Soc., Dalton Trans., 1985, 225.
38 T.Beringhelli, G.Ciani, G.D'Alfonso, H.Molinari and A.Sironi, Inorg. Chem., 1985, 24, 2666; T.Beringhelli, H.Molinari and A.Pastore, J. Chem. Soc., Dalton Trans., 1985, 1899; T.Beringhelli, G.D'Alfonso and H.Molinari, J. Organomet. Chem., 1985, 295, C35.
39 (a) T.Sowa, T.Kawamura and T.Yonezawa, J. Organomet. Chem., 1985, 284, 337; (b) T.Sowa, T.Kawamura, T.Yamabe and T.Yonezawa, J. Am. Chem. Soc., 1985, 107, 6471.
40 (a) M.Hoch and D.Rehder, J. Organomet. Chem., 1985, 288, C25; (b) M.Kretschmer, P.S.Pregosin, A.Albinati and A.Togni, ibid, 1985, 281, 365; (c) A.Albinati, U.von Gunten, P.S.Pregosin and H.J.Ruegg, ibid, 1985, 295, 239.
41 C.G.Benson, G.J.Long, J.W.Kolis and D.F.Shriver, J. Am. Chem. Soc., 1985, 107, 5297.
42 D.C.Johnson, R.E.Benfield, P.P.Edwards, W.J.H.Nelson and M.D.Vargas, Nature, 1985, 314, 231.
43 (a) B.Tulyathan and W.E.Geiger, J. Am. Chem. Soc., 1985, 107, 5960; (b) R.Jund, P.Lemoine, M.Gross, R.Bender and P.Braunstein, J. Chem. Soc., Dalton Trans., 1985, 711.
44 J.T.Landrum and C.D.Hoff, J. Organomet. Chem., 1985, 282, 215.
45 D.B.Jacobson and B.S.Freiser, J. Am. Chem. Soc., 1985, 107, 1581; M.B.Wise, D.B.Jacobson and B.S.Freiser, ibid, p.1590; R.L.Hettich and B.S.Freiser, ibid, p.6222; D.J.Trevor, R.L.Whetten, D.M.Cox and A.Kaldor, ibid, p.518.
46 L.D'Ornelas, A.Choplin, J.M.Basset, L-Y.Hsu and S.Shore, Nouv. J. Chim., 1985, 9, 155.
47 N.Binsted, S.L.Cook, J.Evans and G.N.Greaves, J. Chem. Soc., Chem. Commun., 1985, 1103.
48 M.A.Drezdzon, C.Tessier-Youngs, C.Woodcock, P.M.Blonsky, O.Leal, B-K.Teo, R.L.Burwell and D.F.Shriver, Inorg. Chem., 1985, 24, 2349.
49 W.P. McKenna and E.M.Eyring, J. Mol. Catal., 1985, 29, 363.
50 T.R.Krause, M.E.Davies, J.Lieto and B.C.Gates, J. Catal., 1985, 94, 195.
51 P.S.Kirlin and B.C.Gates, Inorg. Chem., 1985, 24, 3914.
52 J.R.Budge, B.F.Lucke, B.C.Gates and J.Toran, J. Catal., 1985, 91, 272.
53 H.Marrakchi, J-B.Nguini Effa, M.Haimeur, J.Lieto and J-P.Aune, Bull. Soc. Chim. Fr., 1985, 390.
54 D.F.Foster, J.Harrison, B.S.Nicholls and A.K.Smith, J. Organomet. Chem., 1985, 295, 99.
55 E.P.Giannelis and T.J.Pinnavaia, Inorg. Chem., 1985, 24, 2115 and 3602.
56 T.Cuenca and P.Royo, J. Organomet. Chem., 1985, 293, 61; idem, ibid, 1985, 295, 159.
57 H.Kopf and T.Klapotke, Z. Naturforsch., B., 1985, 40, 447.

58 F.Bottomley, G.O.Egharevba and P.S.White, J. Am. Chem. Soc., 1985, 107, 4353.
59 D.A.Lemenovskii, I.F.Urazowski, Yu.K.Grishin and V.A.Roznyatovsky, J. Organomet. Chem., 1985, 290, 301; E.G.Perevalova, I.F.Urazowski, D.A.Lemenovskii, Yu.L.Slovokhotov and Yu.T.Struchkov, ibid, 1985, 289, 319; D.A.Lemenovskii, I.F.Urazowski, I.E.Nifant'ev and E.G.Perevalova, ibid, 1985, 292, 217.
60 M.D.Curtis and J.Real, Organometallics, 1985, 4, 940.
61 T.Vogt and J.Strahle, Z. Naturforsch., B., 1985, 40, 1599.
62 M.Basato, J. Chem. Soc., Dalton Trans., 1985, 91.
63 U.Weber, G.Huttner, O.Scheidsteger and L.Zsolnai, J. Organomet. Chem., 1985, 289, 357.
64 (a) R.M.Medina, A.Alvarez-Valdes and J.R.Masaguer, J. Organomet. Chem., 1985, 294, 209; (b) R.Drews and U.Behrens, Chem. Ber., 1985, 118, 888.
65 A.E.Stiegman and D.R.Tyler, J. Am. Chem. Soc., 1985, 107, 967.
66 K.Isobe, S.Kimura and Y.Nakamura, J. Chem. Soc., Chem. Commun., 1985, 378.
67 M.Tilset and K.P.C.Vollhardt, Organometallics, 1985, 4, 2230.
68 J.S.Drage and K.P.C.Vollhardt, Organometallics, 1985, 4, 191.
69 M.A.Greaney, J.S.Merola and T.R.Halbert, Organometallics, 1985, 4, 2057.
70 P.Klufers, L.Knoll, C.Reiners and K.Reiss, Chem. Ber., 1985, 118, 1825.
71 N.N.Turaki and J.M.Huggins, Organometallics, 1985, 4, 1766.
72 J.G.Riess, U.Klement and J.Wachter, J. Organomet. Chem., 1985, 280, 215.
73 E.J.M.de Boer, J.de With and A.G.Orpen, J. Chem. Soc., Chem. Commun., 1985, 1666.
74 N.M.Doherty, C.Elschenbroich, H-J.Kneuper and S.A.R.Knox, J. Chem. Soc., Chem. Commun., 1985, 170.
75 (a) P.Bougeard, S.Peng, M.Mlekuz and M.J.McGlinchey, J. Organomet. Chem., 1985, 296, 383; (b) O.A.Reutov, I.V.Barinov, V.A.Chertkov and V.I.Sokolov, ibid, 1985, 297, C25; (c)J.Bularzik, K.Kourtakis, J.Nitschke and W.Totsche, Acta Crystallogr., Sec. C., 1985, 41, 1426.
76 J.A.K.Howard, J.C.Jeffery, J.C.V.Laurie, I.Moore, F.G.A.Stone and A.Stringer, Inorg. Chim. Acta, 1985, 100, 23.
77 G.A.Carriedo, J.A.K.Howard, D.B.Lewis, G.E.Lewis and F.G.A.Stone, J. Chem. Soc., Dalton Trans., 1985, 905.
78 D.Hodgson, J.A.K.Howard, F.G.A.Stone and M.J.Went, J. Chem. Soc., Dalton Trans., 1985, 1331.
79 H.G.Alt and T.Frister, J. Organomet. Chem., 1985, 293, C7.
80 M.F.Meidine, C.J.Meir, S.Morton and J.F.Nixon, J. Organomet. Chem., 1985, 297, 255.
81 G.Huttner, B.Sigwarth, O.Scheidsteger, L.Zsolnai and O.Orama, Organometallics, 1985, 4, 326.
82 (a) M.Herberhold, W.Kremnitz, A.Razavi, H.Schollhorn and U.Thewalt, Angew. Chem., Int. Ed. Engl., 1985, 24, 601; (b) H.Arzoumanian, A.Baldy, M.Pierrot and J.F.Petrignani, J. Organomet. Chem., 1985, 294, 327.
83 H.Alper, F.W.B.Einstein, F.W.Hartstock and A.C.Willis, J. Am. Chem. Soc., 1985, 107, 173.
84 (a) W.A.Herrmann, J.Rohrmann, H.Noth, C.K.Nanila, I.Bernal and M.Draux, J. Organomet. Chem., 1985, 284, 189; (b) K.Blechschmitt, E.Guggolz and M.L.Ziegler, Z. Naturforsch. B., 1985, 40, 85; (c) G.J.Kubas, H.J.Wasserman and R.R.Ryan, Organometallics, 1985, 4, 419.
85 M.J.Chetcuti, M.H.Chisholm, H.T.Chiu and J.C.Huffman, Polyhedron., 1985, 4, 1213.
86 K.J.Ahmed, M.H.Chisholm and J.C.Huffman, Organometallics, 1985, 4, 1168.
87 (a) K.J.Ahmed, M.H.Chisholm, and J.C.Huffman, Organometallics, 1985, 4, 1312; (b) K.J.Ahmed, M.H.Chisholm, K.Folting and J.C.Huffman, J. Chem. Soc., Chem. Commun., 1985, 152; (c) M.H.Chisholm, B.W.Eichhorn and J.C.Huffman, ibid, 1985, 861.
88 R.J.Blau, M.H.Chisholm, K.Folting and R.J.Wang, J. Chem. Soc., Chem. Commun., 1985, 1582.
89 M.H.Chisholm, D.M.Hoffman and J.C.Huffman, Organometallics, 1985, 986.
90 F.A.Cotton, W.Schwotzer and E.S.Shamshoum, Organometallics, 1985, 4, 461.
91 (a) M.H.Chisholm, J.C.Huffman and J.A.Heppert, J. Am. Chem. Soc, 1985, 107,

5116; (b) M.H.Chisholm, K.Folting, J.A.Heppert and W.E.Streib, J. Chem. Soc., Chem. Commn., 1985, 1755.
92 M.H.Chisholm, J.A.Heppert, J.C.Huffman and W.E.Streib, J. Chem. Soc., Chem. Commun., 1985, 1771.
93 M.H.Chisholm, J.A.Heppert, J.C.Huffman and P.Thornton, J. Chem. Soc., Chem. Commun., 1985, 1466.
94 M.H.Chisholm, K.Folting, J.A.Heppert, D.M.Hoffman and J.C.Huffman, J. Am. Chem. Soc., 1985, 107, 1234.
95 A.Birnbaum, F.A.Cotton, Z.Dori, M.Kapon, D.Marler, G.M.Reisner and W.Schwotzer, J. Am. Chem. Soc., 1985, 107, 2405.
96 F.A.Cotton, Z.Dori, M.Kapon, D.O.Marler, G.M.Reisner, W.Schwotzer and M.Shaia, Inorg. Chem., 1985, 24, 4381.
97 F.A.Cotton, W.Schwotzer and E.S.Shamshoum, J. Organomet. Chem., 1985, 296, 55.
98 (a) K.Blechschmitt, H.Pfisterer, T.Zahn and M.L.Ziegler, Angew. Chem., Int. Ed. Engl., 1985, 24, 66; (b) K.Blechschmitt, T.Zahn and M.L.Ziegler, ibid, 1985, 24, 702.
99 J.Borm, G.Huttner and L.Zsolnai, Angew. Chem., Int. Ed. Engl., 1985, 24, 1069.
100 M.M.Singh and R.J.Angelici, Inorg. Chim. Acta., 1985, 100, 57.
101 G.W.Harris and N.J.Colville, Organometallics, 1985, 4, 908; G.W.Harris, J.C.A.Boeyens and N.J.Colville, J. Chem. Soc., Dalton Trans., 1985, 2277; idem, Organometallics, 1985, 4, 914.
102 H.Masuda, T.Taga, T.Sowa, T.Kawamura and T.Yonezawa, Inorg. Chim. Acta, 1985, 101, 45.
103 F.A.Cotton and K.B.Shiu, Gazz. Chim. Ital., 1985, 115, 705.
104 E.Guggolz, K.Layer, F.Oberdorfer and M.Ziegler, Z. Naturforsch, B., 1985, 40, 77.
105 C.S.Young, S.W.Lee and C.P.Cheng, J. Organomet. Chem., 1985, 282, 85.
106 F.R.Lemke and C.P.Kubiak, J. Chem. Soc., Chem. Commun., 1985, 1729.
107 A.Poe and C.V.Sekhar, J. Am. Chem. Soc., 1985, 107, 4874.
108 R.S.Herrick, T.R.Herrington, H.W.Walker and T.L.Brown, Organometallics, 1985, 4, 42.
109 M.A.Biddulph, R.Davis, C.H.J.Wells and F.I.C.Wilson, J. Chem. Soc., Chem. Commun., 1985, 1287.
110 K.W.Lee and T.L.Brown, Organometallics, 1985, 4, 1030.
111 K.W.Lee, W.T.Pennington, A.W.Cordes and T.L.Brown, J. Am. Chem. Soc., 1985, 107, 631.
112 K.W.Lee and T.L.Brown, Organometallics, 1985, 4, 1025.
113 A.J.Deeming and S.Donovan-Mtunzi, Organometallics, 1985, 4, 693; H.C.Aspinall and A.J.Deeming, J. Chem. Soc., Dalton Trans., 1985, 743.
114 A.J.Deeming and S.Donovan-Mtunzi, J. Chem. Soc., Dalton Trans., 1985, 1609.
115 V.Riera, M.A.Ruiz, A.Tiripicchio and M.T.Camellini, J. Chem. Soc., Chem. Commun., 1985, 1505.
116 F.A.Cotton, L.M.Daniels, K.R.Dunbar, L.R.Falvello, S.M.Tetrick and R.A.Walton, J. Am. Chem. Soc., 1985, 107, 3524; F.A.Cotton, K.R.Dunbar, L.R.Falvello and R.A.Walton, Inorg. Chem., 1985, 24, 4180.
117 L.B.Anderson, T.J.Barder and R.A.Walton, Inorg. Chem., 1985, 24, 1421; T.J.Barder, D.Powell and R.A.Walton, J. Chem. Soc., Chem. Commun., 1985, 550.
118 M.W.Kokkes, D.J.Stufkens and A.Oskam, Inorg. Chem., 1985, 24, 2934 and 4411; M.W.Kokkes, W.G.J.de Lange, D.J.Stufkens and A.Oskam, J. Organomet Chem., 1985, 294, 59.
119 J.Keijsper, G.van Koten, K.Vrieze, M.Zoutberg and C.H.Stam, Organometallics, 1985, 4, 1306.
120 E.O.Fischer, J.K.R.Wanner, G.Muller and J.Riede, Chem. Ber., 1985, 118, 3311.
121 W.A.Herrmann, R.Serrano, U.Kusthardt, E.Guggolz, B.Nuber and M.L.Ziegler, J. Organomet. Chem., 1985, 287, 329; W.A.Herrmann, R.Serrano, M.L.Ziegler, H.Pfisterer and B.Nuber, Angew. Chem. Int. Ed. Engl., 1985, 24, 50.
122 W.A.Herrmann, U.Kusthardt and E.Herdtweck, J. Organomet. Chem., 1985, 294, C33.
123 T.Beringhelli, G.Ciani, G.D'Alfonso, A.Sironi and M.Freni, J. Chem. Soc., Dalton Trans., 1985, 1507.
124 T.Beringhelli, G.Ciani, G.D'Alfonso, A.Sironi and M.Freni, J. Chem. Soc., Chem.

125 Commun., 1985, 978.
 N.J.Taylor, J. Chem. Soc., Chem. Commun., 1985, 478.
126 I.Bkouche-Waksman, J.S.Rici, T.F.Koetzle, J.Weichmann and W.A.Herrmann, Inorg. Chem., 1985, 24, 1492.
127 (a)X.Solans, M.Font-Altaba, J.Ros, R.Yanez and R.Mathieu, Acta Crystallogr., Sec. C., 1985, 41, 1186; (b)J.Ros, G.Commenges, R.Mathieu, X.Solans and M.Font-Altaba, J. Chem. Soc., Dalton Trans., 1985, 1087.
128 J. Ros, X.Solans, C.Miravitlles and R.Mathieu, J. Chem. Soc., Dalton Trans., 1985, 1981.
129 W.K.Wong, K.W.Chiu, G.Wilkinson, M.Motevalli and M.B.Hurthouse, Polyhedron, 1985, 4, 1231.
130 D.J.Brauer, S.Hietkamp, H.Sommer, O.Stelzer, G.Muller, M.J.Romao and C.Kruger, J. Organomet Chem., 1985, 296, 411.
131 H.W.Fruhauf, D.Meyer and J.Breuer, J. Organomet. Chem., 1985, 297, 211.
132 A.L.Rheingold, Acta Crystallogr., Sect. C., 1985, 41, 1043.
133 A.Wojcicki, Inorg. Chim. Acta., 1985, 100, 125; S.G.Shyu and A.Wojcicki, Organometallics, 1985, 4, 1457.
134 R.B.King, F.J.Wu, N.D.Sadanani and E.M.Holt, Inorg. Chem., 1985, 24, 4449.
135 A.M.Arif, A.H.Cowley and S.Quashie, J. Chem. Soc., Chem. Commun., 1985, 428.
136 A.M.Arif, A.H.Cowley and M.Pakulski, J. Am. Chem. Soc., 1985, 107, 2553.
137 H.Lang, L.Zsolnai and G.Huttner, Chem. Ber., 1985, 118, 4426.
138 (a) H.Umland, D.Wormsbacher and U.Behrens, J. Organomet Chem., 1985, 284, 353; (b)H.Umland and U.Behrens, ibid, 1985, 287, 109.
139 D.Seyferth, G.B.Womack and J.C.Dewan, Organometallics, 1985, 4, 398.
140 D.Seyferth, R.S.Henderson, L.C.Song and G.B.Womack, J. Organomet. Chem., 1985, 292, 9; D.Seyferth and A.M.Kiwan, ibid, 1985, 286, 219; D.Seyferth, A.M.Kiwan and E.Sinn, ibid, 1985, 281, 111.
141 J.A.Kovacs, J.K.Bashkin and R.H.Holm, J. Am. Chem. Soc., 1985, 107, 1784.
142 (a)S.Aime, M.Botta, R.Gobetto and D.Osella, Organometallics, 1985, 4, 1475; (b)J.Messelhauser, I.P.Lorenz, K.Haug and W.Hiller, Z. Naturforsch., B., 1985, 40, 1064; (c)A.I.Nekhaev, S.D.Alekseeva, N.S.Nametkin, V.D.Tyurin, B.I.Kolobkov, G.G.Aleksandrov, N.A.Parpiev, M.T.Tashev and H.B.Dustov, J. Organomet. Chem., 1985, 297, C33.
143 G.J.Kruger, S.Lotz, L.Linford, M.van Dyk and H.G.Raubenheimer, J. Organomet. Chem., 1985, 280, 241.
144 A.J.Mayr, H.S.Lien, K.H.Pannell and L.Parkanyi, Organometallics., 1985, 4, 1580.
145 R.J.Angelici and J.W.Dunker, Inorg. Chem., 1985, 24, 2209.
146 J.P.Blaha, B.E.Bursten, J.C.Dewan, R.B.Frankel, C.L.Randolph, B.A.Wilson and M.S.Wrighton, J. Am. Chem. Soc., 1985, 107, 4561.
147 C.P.Casey, S.R.Marder and B.R.Adams, J. Am. Chem. Soc., 1985, 107, 7700.
148 C.P.Casey and S.R.Marder, Organometallics., 1985, 4, 411; C.P.Casey, M.S.Konings, R.E.Palermo and R.E.Colborn, J. Am. Chem. Soc., 1985, 107, 5296.
149 (a)C.P.Casey, W.H.Miles, P.J.Fagan and K.J.Haller, Organometallics., 1985, 4, 559; (b)C.P.Casey, S.R.Marder and A.L.Rheingold, ibid, 1985, 4, 762.
150 J.C.Vites, G.Jacobsen, T.K.Dutta and T.P.Fehlner, J. Am. Chem. Soc., 1985, 107, 5563.
151 J.A.Hriljac and D.F.Shriver, Organometallics., 1985, 4, 2225.
152 D.Lentz, I.Brudgam and H.Hartl, Angew. Chem., Int. Ed.Engl., 1985, 24, 119.
153 (a)D.Nuel, F.Dahan and R.Mathieu, Organometallics., 1985, 4, 1436; (b) idem, J. J. Am. Chem. Soc., 1985, 107, 1658.
154 G.D.Williams, G.L.Geoffroy and R.R.Whittle, J. Am. Chem. Soc., 1985, 107, 729.
155 V.D.Patel, A.A.Cherkas, D.Nucciarone, N.J.Taylor and A.J.Carty, Organometallics., 1985, 4, 1792.
156 K.Knoll, G.Huttner, L.Zsolnai, I.Jibril and M.Wasiucionek, J. Organomet. Chem., 1985, 294, 91.
157 D.J.Brauer, S.Hietkamp, H.Sommer and O.Stelzer, J. Organomet. Chem., 1985, 281, 187; idem, Z. Naturforsch, B, 1985, 40, 1677; D.J.Brauer, G.Hasselkuss, S.Hietkamp, H.Sommer and O.Stelzer, ibid, 1985, 40, 961.
158 E.P.Kyba, K.L.Hassett, B.Sheikh, J.S.McKennis, R.B.King and R.E.Davis, Organometallics., 1985, 4, 994.
159 J.Takacs and L.Marko, Transition Met. Chem., (Weinheim Ger.), 1985, 10, 21.

160 M.K.Alami, F.Dahan and R.Mathieu, Organometallics., 1985, 4, 2122.
161 (a)A.Gourdon and Y.Jeannin, J. Organomet. Chem., 1985, 290, 199; (b)idem, ibid, 1985, 282, C39.
162 G.Fischer, G.Sedelmeier, H.Prinzbach, K.Knoll, P.Wilharm, G.Huttner and I.Jibril, J. Organomet. Chem., 1985, 297, 307.
163 A.A.Chalmers, D.C.Liles, E.Meintjies, H.E.Oosthuizen, J.A.Pretorius and E.Singleton, J. Chem. Soc., Chem. Commun., 1985, 1340.
164 (a)J.Keijsper, L.Polm, G.van Koten, K.Vrieze, C.H.Stam and J.D.Schagen, Inorg. Chim. Acta., 1985, 103, 137; (b)J. Keijsper, L.H.Polm, G.van Koten, K.Vrieze, E.Nielsen and C.H.Stam, Organometallics., 1985, 4, 2006.
165 A.Basu, S.Bhaduri, H.Khwaja, P.G.Jones, T.Schroeder and G.M.Sheldrick, J. Organomet. Chem., 1985, 290, C19.
166 J.A.Clucas, M.M.Harding, B.S.Nicholls and A.K.Smith, J. Chem. Soc., Dalton Trans. 1985, 1835.
167 M.Rotem, I.Goldberg and Y.Shvo, Inorg. Chim. Acta., 1985, 97, L27.
168 G.Suss-Fink, G.Herrmann, P.Morys, J.Ellermann and A.Veit, J. Organomet. Chem., 1985, 284, 263.
169 A.R.Chakravarty and F.A.Cotton, Inorg. Chem., 1985, 24, 3584.
170 W.A.Herrmann, M.Floel, C.Weber, J.L.Hubbard and A.Schafer, J. Organomet. Chem. 1985, 286, 369.
171 N.G.Connelly, N.J.Forrow, B.P.Gracey, S.A.R.Knox and A.G.Orpen, J. Chem. Soc., Chem. Commun., 1985, 14.
172 D.L.Davies, J.A.K.Howard, S.A.R.Knox, K.Marsden, K.A.Mead, M.J.Morris and M.C.Rendle, J. Organomet. Chem., 1985, 279, C37.
173 J.S.Holmgren and J.R.Shapley, Organometallics., 1985, 4, 793.
174 E.D.Morrison, G.L.Geoffroy and A.L.Rheingold, J. Am. Chem. Soc., 1985, 107, 254; E.D.Morrison, G.L.Geoffroy, A.L.Rheingold and W.C.Fultz, Organometallics., 1985, 4, 1413; E.D.Morrison and G.L.Geoffroy, J. Am. Chem. Soc., 1985, 107, 3541.
175 D.M.Dalton, D.J.Barnett, T.P.Duggan, J.B.Keister, P.T.Malik, S.P.Modi, M.R.Shaffer and S.A.Smesko, Organometallics, 1985, 4, 1854; L.M.Bavaro and J.B.Keister, J. Organomet. Chem., 1985, 287, 357.
176 W.Y.Yeh, J.R.Shapley, Y.Li and M.R.Churchill, Organometallics, 1985, 4, 767.
177 J.R.Shapley, W.Y.Yeh, M.R.Churchill and Y.Li, Organometallics, 1985, 4, 1898.
178 N.G.Connelly, N.J.Forrow, S.A.R.Knox, K.A.Macpherson and A.G.Orpen, J. Chem. Soc., Chem. Commun., 1985, 16.
179 M.J.Sailor and D.F.Shriver, Organometallics, 1985, 4, 1476.
180 L.R.Beanan and J.B.Keister, Organometallics, 1985, 4, 1713.
181 M.R.Churchill, J.C.Fettinger, J.B.Keister, R.F.See and J.W.Ziller, Organometallics, 1985, 4, 2112.
182 D.M.Dalton and J.B.Keister, J. Organomet. Chem., 1985, 290, C37.
183 M.P.Gomez-Sal, B.F.G.Johnson, J.Lewis, P.R.Raithby and A.H.Wright, J. Chem. Soc., Chem. Commun., 1985, 1682.
184 D.Himmelreich and G.Muller, J. Organomet. Chem., 1985, 297, 341.
185 A.J.Arce, Y.de Sanctis and A.J.Deeming, J. Organomet. Chem., 1985, 295, 365.
186 A.A.Koridze, O.A.Kizas, N.E.Kolobova and P.V.Petrovskii, J. Organomet. Chem., 1985, 292, C1.
187 (a)J.A.Smieja, R.E.Stevens, D.E.Fjare and W.L.Gladfelter, Inorg. Chem., 1985, 24, 3206; (b)J.A.Smieja and W.L.Gladfelter, J. Organomet. Chem., 1985, 297, 349.
188 T.Venalainen, J.Pursiainen and T.A.Pakkanen, J. Chem. Soc., Chem. Commun., 1985, 1348.
189 (a)G.A.Foulds, B.F.G.Johnson and J.Lewis, J. Organomet. Chem., 1985, 296, 147; (b)idem, ibid, 1985, 294, 123.
190 A.Eisenstadt, C.M.Giandomenico, M.F.Frederick and R.M.Laine, Organometallics, 1985, 4, 2033.
191 K.Burgess, H.D.Holden, B.F.G.Johnson, J.Lewis, M.B.Hursthouse, N.P.C.Walker, A.J.Deeming, P.J.Manning and R.Peters, J. Chem. Soc., Dalton. Trans., 1985, 85.
192 T.I.Odiaka, J. Organomet. Chem., 1985, 284, 95.
193 J.Banford, M.J.Mays and P.R.Raithby, J. Chem. Soc., Dalton Trans., 1985, 1355.
194 (a)G.Suss-Fink, W.Buhlmeyer, M.Herberhold, A.Gieren and T.Hubner, J. Organomet. Chem., 1985, 280, 129; (b)W.Ehrenreich, M.Herberhold, G.Herrmann, G.Suss-Fink, A.Gieren and T.Hubner, ibid, 1985, 294, 183; (c)G.Suss-Fink, K.Guldner,

M.Herberhold, A.Gieren and T.Hubner, ibid, 1985, 279, 447.
195 (a)J.Keijsper, L.H.Polm, G.van Koten, K.Vrieze, P.F.A.B.Seignette and C.H.Stam, Inorg. Chem., 1985, 24, 518; (b)J.Keijsper, L.H.Polm, G.van Koten, K.Vrieze, K.Goubitz and C.H.Stam, Organometallics, 1985, 4, 1876.
196 (a)B.Ambwani, S.Chawla and A.Poe, Inorg. Chem., 1985, 24, 2635; (b)N.Brodie, A.Poe and V.Sekhar, J. Chem. Soc., Chem. Commun., 1985, 1090; A.Poe and V.C.Sekhar, Inorg. Chem., 1985, 24, 4376.
197 A.J.Deeming, S.Donovan-Mtunzi, S.E.Kabir and P.J.Manning, J. Chem. Soc., Dalton Trans., 1985, 1037.
198 M.I.Bruce, J.G.Matisons, J.M.Patrick, A.H.White and A.C.Willis, J. Chem. Soc., Dalton Trans., 1985, 1223.
199 J.A.Clucas, M.M.Harding and A.K.Smith, J. Chem. Soc., Chem. Commun., 1985, 1280.
200 A.J.Deeming, S.Donovan-Mtunzi and S.E.Kabir, J. Organomet. Chem., 1985, 281, C43.
201 R.F.Alex and R.K.Pomeroy, J. Organomet. Chem., 1985, 284, 379.
202 E.Rosenberg, C.B.Thorsen, L.Milone and S.Aime, Inorg. Chem., 1985, 24, 231.
203 S.B.Colbran, B.F.G.Johnson, J.Lewis and R.M.Sorrell, J. Organomet. Chem., 1985, 296, C1.
204 (a)N.Lugan, J.-J.Bonnet and J.A.Ibers, J. Am. Chem. Soc., 1985, 107, 4484; (b)M.I.Bruce, O.B.Shawkataly and M.L.Williams, J. Organomet. Chem., 1985, 287, 127; M.I.Bruce, P.A.Humphrey, B.W.Skelton, A.H.White and M.L.Williams, Aust. J. Chem., 1985, 38, 1301.
205 M.I.Bruce and M.L.Williams, J. Organomet. Chem., 1985, 288, C55.
206 A.J.Arce, A.J.Deeming, S.Donovan-Mtunzi and S.E.Kabir, J. Chem. Soc., Dalton Trans., 1985, 2479.
207 P.M.Lausarot, G.A.Vaglio, M.Valle, A.Tiripicchio, M.T.Camellini and P.Gariboldi, J. Organomet. Chem., 1985, 291, 221.
208 M.R.Churchill, J.W.Ziller and J.B.Keister, J. Organomet. Chem., 1985, 297, 93.
209 R.D.Adams, I.T.Horvath and S.Wang, Inorg. Chem., 1985, 24, 1728.
210 R.D.Adams, H.S.Kim and S.Wang, J. Am. Chem. Soc., 1985, 107, 6107.
211 R.D.Adams and S.Wang, Organometallics, 1985, 4, 1902.
212 F.W.B.Einstein, L.R.Martin, R.K.Pomeroy and P.Rushman, J. Chem. Soc., Chem. Commun., 1985, 345.
213 E.J.Ditzel, B.F.G.Johnson, J.Lewis, P.R.Raithby and M.J.Taylor, J. Chem. Soc., Dalton Trans., 1985, 555.
214 M.L.Blohm and W.L.Gladfelter, Organometallics, 1985, 4, 45.
215 J.P.Attard, B.F.G.Johnson, J.Lewis, J.M.Mace and P.R.Raithby, J. Chem. Soc., Chem. Commun., 1985, 1526.
216 J.Puga, R.Sanchez-Delgado and D.Braga, Inorg. Chem., 1985, 24, 3971.
217 K.Henrick, B.F.G.Johnson, J.Lewis, J.Mace, M.McPartlin and J.Morris, J. Chem. Soc., Chem. Commun., 1985, 1617.
218 J.Lunniss, S.A.MacLaughlin, N.J.Taylor, A.J.Carty and E.Sappa, Organometallics, 1985, 4, 2066.
219 J.C.Daran, Y.Jeannin and O.Kristiansson, Organometallics, 1985, 4, 1882.
220 (a)M.I.Bruce, M.L.Williams, J.M.Patrick and A.H.White, J. Chem. Soc., Dalton Trans. 1985, 1229; (b)J.C.Daran, O.Kristiansson and Y.Jeannin, C.R.Seances Acad. Sci., 1985, 300, 943.
221 (a)M.I.Bruce and M.L.Williams, J. Organomet. Chem., 1985, 282, C11; (b)M.I.Bruce, M.L.Williams, B.W.Skelton and A.H.White, ibid, 1985, 282, C53. (c)M.I.Bruce, B.W.Skelton, A.H.White and M.L.Williams, J. Chem. Soc., Chem. Commun., 1985, 744.
222 M.P.Gomez-Sal, B.F.G.Johnson, R.A.Kamarudin, J.Lewis and P.R.Raithby, J. Chem. Soc., Chem. Commun., 1985, 1622.
223 R.J.Goudsmit, P.F.Jackson, B.F.G.Johnson, J.Lewis, W.J.H.Nelson, J.Puga, M.D.Vargas, D.Braga, K.Henrick, M.McPartlin and A.Sironi, J. Chem. Soc., Dalton Trans., 1985, 1795.
224 P.Magnus and D.P.Becker, J. Chem. Soc., Chem. Commun., 1985, 640.
225 P.Paetzold and K.Delpy, Chem. Ber., 1985, 118, 2552.
226 C.M.Arewgoda, A.M.Bond, R.S.Dickson, T.F.Mann, J.E.Moir, P.H.Rieger, B.H.Robinson and J.Simpson, Organometallics, 1985, 4, 1077.

227 E.N.Jacobsen and R.G.Bergman, J. Am. Chem. Soc., 1985, 107, 2023.
228 W.A.Herrmann, C.Weber, M.L.Ziegler and O.Serhadli, J. Organomet. Chem., 1985, 297, 245.
229 (a)R.B.King, M.Chang and M.G.Newton, J. Organomet. Chem., 1985, 296, 15; (b)F.Babonneau, M.Henry, R.B.King and N.El Murr, Inorg. Chem., 1985, 24, 1946.
230 A.M.Arif, A.H.Cowley and M.Pakulski, J. Chem. Soc., Chem. Commun., 1985, 1707.
231 A.M.Arif, A.H.Cowley, M.Pakulski, M.B.Hursthouse and A.Karauloz, Organometallics, 1985, 4, 2227.
232 H.R.Allcock, P.R.Suszko, L.J.Wagner, R.R.Whittle and B.Boso, Organometallics, 1985, 4, 446.
233 G.L.Geoffroy, W.C.Mercer, R.R.Whittle, L.Marko and S.Vastag, Inorg. Chem., 1985, 24, 3771.
234 (a)L.Chen, D.J.Kountz and D.W.Meek, Organometallics, 1985, 4, 598; (b)H.Werner, W.Hofmann, R.Zolk, L.F.Dahl, J.Kocal and A.Kuhn, J. Organomet. Chem., 1985, 289, 173.
235 R.Zolk and H.Werner, Angew. Chem., Int. Ed. Engl., 1985, 24, 577.
236 H.Werner and R.Zolk, Organometallics, 1985, 4, 601.
237 A.Visi-Orosz, G.Palyi, L.Marko, R.Boese and G.Schmid, J. Organomet. Chem., 1985, 288, 179.
238 F.Y.Petillon, J.L.Le Quere, F.Le Floch-Perennou, J.E.Guerchais and P.L'Haridon, J. Organomet. Chem., 1985, 281, 305.
239 P.F.Seidler, S.T.McKenna, M.A.Kulzick and T.M.Gilbert, Acta Crystallogr., Sect. C., 1985, 41, 352.
240 G.Balavoine, J.Collin, J.J.Bonnet and G.Lavigne, J. Organomet. Chem., 1985, 280, 429.
241 K.A.Kubat-Martin, A.D.Rae and L.F.Dahl, Organometallics, 1985, 4, 2221.
242 R.L.De and H.Vahrenkamp, Z.Naturforsch, B, 1985, 40, 1250.
243 M.K.Markiewicz, M.E.Frazer, R.K.Ungar, S.Fortier and M.C.Baird, Acta Crystallogr., Sect. C., 1985, 41, 336.
244 R.A.Gancarz, J.F.Blount and K.Mislow, Organometallics, 1985, 4, 2028.
245 G.Gervasio, R.Rossetti and P.L.Stanghellini, Organometallics, 1985, 4, 1612.
246 S.Gambarotta, C.Floriani, A.Chiesi-Villa and C.Guastini, J. Organomet. Chem., 1985, 296, C6.
247 N.J.Bailey, G.de.Leeuw, J.S.Field, R.J.Haines, I.C.D.Stuckenberg and R.B.English, S. Afr. J. Chem., 1985, 38, 139.
248 (a)D.J.Darensbourg and D.J.Zalewski, Organometallics, 1985, 4, 92; (b)J.Rimmelin, P.Lemoine, M.Gross, A.A.Bahsoun and J.A.Osborn, Nouv. J. Chim., 1985, 9, 181.
249 M.G.Richmond, J.D.Korp and J.K.Kochi, J. Chem. Soc., Chem. Commun., 1985, 1102.
250 R.E.Stevens, P.C.C.Liu and W.L.Gladfelter, J. Organomet. Chem., 1985, 287, 133.
251 (a)V.G.Albano, D.Braga, A.Fumagalli and S.Martinengo, J. Chem. Soc., Dalton Trans., 1985, 1137; (b)S.Martinengo, D.Strumolo, P.Chini, V.G.Albano and D.Braga, ibid, 1985, 35.
252 P.G.Rasmussen, J.E.Anderson, O.H.Bailey, M.Tamres and J.C.Bayon, J. Am. Chem. Soc., 1985, 107, 279.
253 V.M.Miskowski, T.P.Smith, T.M.Loehr and H.B.Gray, J. Am. Chem. Soc., 1985, 107, 7925.
254 F.A.Cotton, P.Lahuerta, M.Sanau and W.Schwotzer, J. Am. Chem. Soc., 1985, 107, 8284.
255 G.G.Hlatky, B.F.G.Johnson, J.Lewis and P.R.Raithby, J. Chem. Soc., Dalton Trans., 1985, 1277.
256 (a)T.M.Gilbert and R.G.Bergman, J. Am. Chem. Soc., 1985, 107, 3502; (b)T.M.Gilbert, F.J.Hollander and R.G.Bergman, ibid, 1985, 107, 3508.
257 H.Werner, H.J.Scholz and R.Zolk, Chem. Ber., 1985, 118, 4531.
258 S.L.Schiavo, G.Bruno, F.Nicolo, P.Piraino and F.Faraone, Organometallics, 1985, 4, 2091.
259 M.J.Freeman, A.G.Orpen, N.G.Connelly, I.Manners and S.J.Raven, J. Chem. Soc., Dalton Trans., 1985, 2283.
260 M.J.Krause and R.G.Bergman, J. Am. Chem. Soc., 1985, 107, 2972.
261 M.Green, A.G.Orpen, C.J.Schavarien and I.D.Williams, J. Chem. Soc., Dalton

262 R.S.Dickson, G.S.Evans and G.D.Fallon, Trans. 1985, 2483.
263 (a)R.S.Dickson, G.D.Fallon, R.J.Nesbit and G.N.Pain, Organometallics, 1985, 4, 355; (b)R.S.Dickson, R.J.Nesbit, H.Pateras and W.Baimbridge, ibid, 1985, 4, 2128.
264 B.E.Mann, N.J.Meanwell, C.M.Spencer, B.F.Taylor and P.M.Maitlis, J. Chem. Soc., Dalton Trans., 1985, 1555.
265 G.W.Bushnell, M.J.Decker, D.T.Eadie, S.R.Stobart, R.Vefghi, J.L.Atwood and M.J.Zaworotko, Organometallics, 1985, 4, 2106.
266 F.Barcelo, P.Lahuerta, M.A.Ubeda, C.Foces-Foces, F.H.Cano and M.Martinez-Ripoll, J. Chem. Soc., Chem. Commun., 1985, 43.
267 A.R.Chakravarty, F.A.Cotton, D.A.Tocher and J.H.Tocher, Organometallics, 1985, 4, 8.
268 L.A.Oro, M.A.Ciriano, B.E.Villarroya, A.Tiripicchio and F.J.Lahoz, J. Chem. Soc., Dalton Trans., 1985, 1891.
269 (a)C.Woodcock and R.Eisenberg, Inorg. Chem., 1985, 24, 1285; (b)D.H.Berry and R.Eisenberg, J. Am. Chem. Soc., 1985, 107, 7181.
270 B.R.Sutherland and M.Cowie, Organometallics, 1985, 4, 1801.
271 (a)L.Gelmini, D.W.Stephan and S.J.Loeb, Inorg. Chim. Acta., 1985, 98, L3; (b)J.Ellermann and G.Szucsanyi, Chem. Ber., 1985, 118, 1868; (c)L.A.Oro, M.T.Pinillos, A.Tiripicchio and M.Tiripicchio-Camellini, Inorg. Chim. Acta., 1985, 99, L13; (d)L.A.Oro, D.Carmona, P.L.Perez, M.Esteban, A.Tiripicchio and M.Tiripicchio-Camellini, J. Chem. Soc., Dalton Trans., 1985, 973; (e)S.P.Deraniyagala and K.R.Grundy, Inorg. Chim. Acta, 1985, 101, 103.
272 (a)J.G.Gaudiello, T.C.Wright, R.A.Jones and A.J.Bard, J. Am. Chem. Soc., 1985, 107, 888; (b)S.K.Kang, T.A.Albright, T.C.Wright and R.A.Jones, Organometallics, 1985, 4, 666.
273 (a)A.L.Balch and M.M.Olmstead, Isr. J. Chem., 1985, 25, 189; (b)A.L.Balch, L.A.Fossett, R.R.Guimerans and M.M.Olmstead, Organometallics, 1985, 4, 781; (c)A.L.Balch, J.C.Linehan and M.M.Olmstead, Inorg. Chem., 1985, 24, 3975.
274 S.P.Deraniyagala and K.R.Grundy, J. Chem. Soc., Dalton Trans., 1985, 1577.
275 A.C.Bray, M.Green, D.R.Hankey, J.A.K.Howard, O.Johnson and F.G.A.Stone, J. Organomet. Chem., 1985, 281, C12.
276 R.S.Dickson, G.S.Evans, G.D.Fallon and G.N.Pain, J. Organomet. Chem., 1985, 295, 109.
277 V.G.Albano, D.Braga, R.Ros and A.Scrivanti, J. Chem. Soc., Chem. Commun., 1985, 866.
278 D.Braga, R.Ros and R.Roulet, J. Organomet. Chem., 1985, 286, C8.
279 R.Mutin, W.Abboud, J.M.Basset and D.Sinou, J. Mol. Catal., 1985, 33, 47.
280 R.A.Jones and B.R.Whittlesey, J. Am. Chem. Soc., 1985, 107, 1078.
281 V.G.Albano, D.Braga, D.Strumolo, C.Seregni and S.Martinengo, J. Chem. Soc., Dalton Trans., 1985, 1309.
282 D.L.DeLaet, D.R.Powell and C.P.Kubiak, Organometallics, 1985, 4, 954.
283 D.Lentz, Chem. Ber., 1985, 118, 560.
284 R.Bartsch, J.F.Nixon and N.Sarjudeen, J. Organomet. Chem., 1985, 294, 267.
285 R.Hanko, Angew. Chem., Int. Ed. Engl., 1985, 24, 704.
286 (a)A.Ceriotti, G.Longoni, M.Manassero, M.Perego and M.Sansoni, Inorg. Chem., 1985, 24, 117; (b)A.Ceriotti, G.Longoni, M.Manassero, N.Masciocchi, L.Resconi and M.Sansoni, J. Chem. Soc., Chem. Commun., 1985, 181; (c)A.Ceriotti, G.Longoni, M.Manassero, N.Masciocchi, G.Piro, L.Resconi and M.Sansoni, ibid, 1985, 1402.
287 M.K.Reinking, M.L.Kullberg, A.R.Cutler and C.P.Kubiak, J. Am. Chem. Soc., 1985, 107, 3517.
288 R.D.Feltham, G.Elbaze, R.Ortega, C.Eck and J.Dubrawski, Inorg. Chem., 1985, 24, 1503.
289 P.Thometzek, K.Zenkert and H.Werner, Angew. Chem., Int. Ed. Engl., 1985, 24, 516.
290 G.Minghetti, A.Albinati, A.L.Bandini and G.Banditelli, Angew. Chem., Int. Ed. Engl., 1985, 24, 120.
291 M.Ebner, H.Otto and H.Werner, Angew. Chem., Int. Ed. Engl., 1985, 24, 518.
292 (a)M.P.Brown, A.Yavari, R.H.Hill and R.J.Puddephatt, J. Chem. Soc., Dalton Trans., 1985, 2421; (b) C.E.Langrick, P.G.Pringle and B.L.Shaw, ibid, 1985, 1015.
293 C.S.Browning, D.H.Farrar, R.R.Gukathasan and S.A.Morris, Organometallics, 1985, 4,

1750.
294 (a)M.F.Hallam, N.D.Howells, D.M.P.Mingos and R.W.M.Wardle, J. Chem. Soc., Dalton Trans., 1985, 845; (b)C.E.Briant, D.I.Gilmour, D.M.P.Mingos and R.W.M.Wardle, ibid, 1985, 1693.
295 O.J.Scherer, R.Konrad, E.Guggolz and M.L.Ziegler, Chem. Ber., 1985, 118, 1.
296 (a)L.Manojlovic-Muir, K.W.Muir, B.R.Lloyd and R.J.Puddephatt, J. Chem. Soc., Chem. Commun., 1985, 536; (b)B.R.Lloyd and R.J.Puddephatt, J. Am. Chem. Soc., 1985, 107, 7785.
297 R.Bender, P.Braunstein, A.Tiripicchio and M.T.Camellini, Angew. Chem., Int. Ed. Engl., 1985, 24, 861.
298 H.L.Aalten, G.van Koten, K.Goubitz and C.H.Stam, J. Chem. Soc., Chem. Commun., 1985, 1252.
299 L.Naldini, F.Demartin, M.Manassero, M.Sansoni, G.Rassu and M.A.Zoroddu, J. Organomet. Chem., 1985, 279, C42.
300 (a)J.D.Basil, H.H.Murray, J.P.Fackler, J.Tocher, A.M.Mazany, B.Trzcinska-Bancroft, H.Knachel, D.Dudis, T.J.Delord and D.O.Marler, J. Am. Chem. Soc., 1985, 107, 6908; (b)H.H.Murray, J.P.Fackler and B.Trzcinska-Bancroft, Organometallics, 1985, 4, 1633; (c)H.H.Murray, J.P.Fackler and D.A.Tocher, J. Chem. Soc., Chem. Commun., 1985, 1278; (d)H.H.Murray, A.M.Mazany and J.P.Fackler, Organometallics, 1985, 4, 154.
301 J.P.Fackler and B.Trzcinska-Bancroft, Organometallics, 1985, 4, 1891.
302 J.Vicente, M.T.Chicote, J.A.Cayuelas, J.Fernandez-Baeza, P.G.Jones, G.M.Sheldrick and P.Espinet, J. Chem. Soc., Dalton Trans., 1985, 1163.
303 R.S.Sternal, C.P.Brock and T.J.Marks, J. Am. Chem. Soc., 1985, 107, 8270.
304 J.M.Ritchey, A.J.Zozulin, D.A.Wrobleski, R.R.Ryan, H.J.Wasserman, D.C.Moody and R.T.Paine, J. Am. Chem. Soc., 1985, 107, 501.
305 M.R.Awang, R.D.Barr, M.Green, J.A.K.Howard, T.B.Marder and F.G.A.Stone, J. Chem. Soc., Dalton Trans., 1985, 2009.
306 W.J.Sartain and J.P.Selegue, J. Am. Chem. Soc., 1985, 107, 5818.
307 G.S.White and D.W.Stephan, Inorg. Chem., 1985, 24, 1499.
308 L.Gelmini, L.C.Matassa and D.W.Stephan, Inorg. Chem., 1985, 24, 2585.
309 T.S.Targos, R.P.Rosen, R.R.Whittle and G.L.Geoffroy, Inorg. Chem., 1985, 24, 1375.
310 C.P.Casey, R.E.Palermo and R.F.Jordan, J. Am. Chem. Soc., 1985, 107, 4597.
311 R.B.King, W-K.Fu and E.M.Holt, Inorg. Chem., 1985, 24, 3094.
312 J.C.Jeffery, J.C.V.Laurie and F.G.A.Stone, Polyhedron, 1985, 4, 1135.
313 M.Green, J.A.K.Howard, A.P.James, A.N.de M.Jelfs, C.M.Nunn and F.G.A.Stone, J. Chem. Soc., Chem. Commun., 1985, 1778.
314 A.D.Horton, M.J.Mays and P.R.Raithby, J. Chem. Soc., Chem. Commun., 1985, 247.
315 L.E.Bogan, T.B.Rauchfuss and A.L.Rheingold, Inorg. Chem., 1985, 24, 3720.
316 V.K.Bel'skii, B.M.Bulychev and A.V.Aripovskii, Koord. Khim., 1985, 11, 540.
317 B.Chaudret, F.Dahan and S.Sabo, Organometallics, 1985, 4, 1490.
318 M.C.Azar, M.J.Chetcuti, C.Eigenbrot and K.A.Green, J. Am. Chem. Soc., 1985, 107, 7209.
319 G.Doyle, K.A.Eriksen and D.van Engen, Organometallics, 1985, 4, 2201.
320 B.N.Strunin, K.I.Grandberg, V.G.Andrianov, V.N.Setkina, E.G.Perevalova, Yu.T.Struchkov and D.N.Kursanov, Dokl. Akad. Nauk SSSR, 1985, 281, 599.
321 N.E.Kolobova, L.L.Ivanov, O.S.Zhvanko, A.S.Batsanov and Yu.T.Struchkov, J. Organomet. Chem., 1985, 279, 419.
322 W.C.Mercer, R.R.Whittle, E.W.Burkhardt and G.L.Geoffroy, Organometallics, 1985, 4, 68.
323 W.C.Mercer, G.L.Geoffroy and A.L.Rheingold, Organometallics, 1985, 4, 1418.
324 J.A.K.Howard, J.C.V.Laurie, O.Johnson and F.G.A.Stone, J. Chem. Soc., Dalton Trans., 1985, 2017.
325 F.W.B.Einstein, R.K.Pomeroy, P.Rushman and A.C.Willis, Organometallics, 1985, 4, 250.
326 L.Carlton, W.E.Lindsell, K.J.McCullough and P.N.Preston, Organometallics, 1985, 4, 1138.
327 J.Powell, J.F.Sawyer and S.J.Smith, J. Chem. Soc., Chem. Commun., 1985, 1312.
328 A.Blagg, A.T.Hutton, B.L.Shaw and M.Thornton-Pett, Inorg. Chim. Acta, 1985, 100, L33.

329 M.W.Kokkes, T.L.Snoeck, D.J.Stufkens, A.Oskam, M.Christophersen and C.H.Stam, J. Mol. Struct., 1985, 131, 11.
330 J.Keijsper, P.Grimberg, G.van Koten, K.Vrieze, M.Christophersen and C.H.Stam, Inorg. Chim. Acta, 1985, 102, 29.
331 J.Keijsper, P.Grimberg, G.van Koten, K.Vrieze, B.Kojic-Prodic and A.L.Spek, Organometallics, 1985, 4, 438.
332 H.Werner, F.J.G.Alonso, H.Otto, K.Peters and H.G.von Schnering, J. Organomet. Chem., 1985, 289, C5.
333 P.Braunstein, G.L.Geoffroy and B.Metz, Nouv. J.Chim., 1985, 9, 221.
334 S.W.Carr, B.L.Shaw and M.Thornton-Pett, J. Chem. Soc., Dalton Trans., 1985, 2131.
335 J.Powell, J.F.Sawyer and M.V.R.Stainer, J. Chem. Soc., Chem. Commun., 1985, 1314.
336 C.F.Barrientos-Penna, F.W.B.Einstein, T.Jones and D.Sutton, Inorg. Chem., 1985, 24, 632.
337 B.P.Gracey, S.A.R.Knox, K.A.Macpherson, A.G.Orpen and S.R.Stobart, J. Chem. Soc, Dalton Trans., 1985, 1935.
338 S.B.Colbran, B.H.Robinson and J.Simpson, Organometallics, 1985, 4, 1594.
339 B.Delavaux, B.Chaudret, F.Dahan and R.Poilblanc, Organometallics, 1985, 4, 935.
340 J.A.Iggo, D.P.Markham, B.L.Shaw and M.Thornton-Pett, J. Chem. Soc., Chem. Commun., 1985, 432.
341 G.Doyle, K.A.Eriksen and D.Van Engen, Organometallics, 1985, 4, 877.
342 J.Bashkin, C.E.Briant, D.M.P.Mingos and R.W.M.Wardle, Transition Met. Chem. (Weinheim, Ger), 1985, 10, 113.
343 A.L.Balch, R.R.Guimerans, J.Linehan, M.M.Olmstead and D.E.Oram, Organometallics, 1985, 4, 1445.
344 R.Uson, J.Fornies, M.Tomas and J.M.Casas, J. Am. Chem. Soc., 1985, 107, 2556.
345 R.Uson, J.Fornies, B.Menjon, F.A.Cotton, L.R.Falvello and M.Tomas, Inorg. Chem., 1985, 4, 4651.
346 H.Brunner, W.Meier, J.Wachter, H.Pfisterer and M.L.Ziegler, Z. Naturforsch. B., 1985, 40, 923.
347 L.E.Bogan, T.B.Rauchfuss and A.L.Rheingold, J. Am. Chem. Soc., 1985, 107, 3843.
348 H.Brunner, N.Janietz, J.Wachter, T.Zahn and M.L.Ziegler, Angew. Chem., Int. Ed. Engl., 1985, 24, 133.
349 M.Mlekuz, P.Bougeard, B.G.Sayer, R.Faggiani, C.J.L.Lock, M.J.McGlinchey and G.Jaouen, Organometallics, 1985, 4, 2046.
350 M.Mlekuz, P.Bougeard, B.G.Sayer, S.Peng, M.J.McGlinchey, A.Marinetti, J-Y.Saillard, J.B.Naceur, B.Mentzen and G.Jaouen, Organometallics, 1985, 4, 1123.
351 H.Brunner, H.Kauermann, U.Klement, J.Wachter, T.Zahn and M.L.Ziegler, Angew. Chem., Int. Ed. Engl., 1985, 24, 132.
352 T.R.Halbert, S.A.Cohen and E.I.Stiefel, Organometallics, 1985, 4, 1689.
353 J.C.Jeffery, D.B.Lewis, G.E.Lewis and F.G.A.Stone, J. Chem. Soc., Dalton Trans., 1985, 2001.
354 E.Delgado, A.T.Emo, J.C.Jeffery, N.D.Simmons and F.G.A.Stone, J. Chem. Soc., Dalton Trans., 1985, 1323.
355 J.C.Jeffery and J.G.Lawrence-Smith, J. Chem. Soc., Chem. Commun., 1985, 275.
356 L.W.Arndt, M.Y.Darensbourg, J.P.Fackler, R.J.Lusk, D.O.Marler and K.A.Youngdahl, J. Am. Chem. Soc., 1985, 107, 7218.
357 M.R.Churchill and Y-J.Li, J. Organomet. Chem., 1985, 294, 367.
358 M.R.Churchill and Y-J.Li, J. Organomet. Chem., 1985, 291, 61.
359 M.R.Churchill, J.W.Ziller and L.R.Beanan, J. Organomet. Chem., 1985, 287, 235.
360 J.C.Jeffery and J.G.Lawrence-Smith, J. Organomet Chem., 1985, 280, C34.
361 J.C.Jeffery, C.Marsden and F.G.A.Stone, J. Chem. Soc., Dalton Trans., 1985, 1315.
362 M.R.Churchill, L.V.Biondi, J.R.Shapley and C.H.McAteer, J. Organomet. Chem., 1985, 280, C63.
363 T.Beringhelli, G.D'Alfonso, M.Freni, G.Ciani and A.Sironi, J. Organomet. Chem., 1985, 295, C7.
364 K.Fischer, W.Deck, M.Schwarz and H.Vahrenkamp, Chem. Ber., 1985, 118, 4946.
365 N.C.Schroeder, J.W.Richardson, S-L.Wang, R.A.Jacobson and R.J.Angelici, Organometallics, 1985, 4, 1226.
366 C.P.Horwitz, E.M.Holt and D.F.Shriver, Organometallics, 1985, 4, 1117.
367 V.E.Lopatin, S.P.Gubin, N.M.Mikova, M.Ts.Tsybenov, Yu.L.Slovokhotov and

Yu.T.Struchkov, J. Organomet. Chem., 1985, 292, 275.
368 J.A.Hriljac, P.N.Swepston and D.F.Shriver, Organometallics, 1985, 4, 158.
369 E.Sappa, A.M.M.Lanfredi, G.Predieri, A.Tiripicchio and A.J.Carty, J. Organomet. Chem., 1985, 288, 365.
370 C.Weatherell, N.J.Taylor, A.J.Carty, E.Sappa and A.Tiripicchio, J. Organomet. Chem., 1985, 291, C9.
371 P.Braunstein, G.Predieri, F.J.Lahoz and A.Tiripicchio, J. Organomet. Chem., 1985, 288, C13.
372 G.Doyle, K.A.Eriksen and D.Van Engen, J. Am. Chem. Soc., 1985, 107, 7914.
373 S.Attali, F.Dahan and R.Mathieu, J. Chem. Soc., Dalton Trans., 1985, 2521.
374 (a)C.P.Horwitz, E.M.Holt and D.F.Shriver, J. Am. Chem. Soc., 1985, 107, 281; (b)C.P.Horwitz, E.M.Holt, C.P.Brock and D.F.Shriver, ibid, 1985, 107, 8136.
375 E.Roland and H.Vahrenkamp, Chem. Ber., 1985, 118, 1133.
376 E.Roland, W.Berhardt and H.Vahrenkamp, Chem. Ber., 1985, 118, 2858.
377 J.Pursiainen, T.A.Pakkanen and J.Jaaskelainen, J. Organomet. Chem., 1985, 290, 85.
378 D.Mani and H.Vahrenkamp, Angew Chem., Int. Ed. Engl., 1985, 24, 424.
379 J.Pursiainen, M.Ahlgren and T.A.Pakkanen, J. Organomet. Chem., 1985, 297, 391.
380 L.J.Farrugia, J.C.Jeffery, C.Marsden and F.G.A.Stone, J. Chem. Soc., Dalton Trans., 1985, 645.
381 W.E.Lindsell, C.B.Knobler and H.D.Kaesz, J. Organomet. Chem., 1985, 296, 209; idem, J. Chem. Soc., Chem. Commun., 1985, 1171.
382 R.P.Hughes, A.L.Rheingold, W.A.Herrmann and J.L.Hubbard, J. Organomet. Chem., 1985, 286, 361.
383 A.Fumagalli, F.Demartin and A.Sironi, J. Organomet. Chem., 1985, 279, C33.
384 M.I.Bruce, O.B.Shawkataly and B.K.Nicholson, J. Organomet. Chem., 1985, 286, 427.
385 M.I.Bruce, E.Horn, O.B.Shawkataly and M.R.Snow, J. Organomet. Chem., 1985, 280, 289.
386 E.Sappa, G.Predieri, A.Tiripicchio and M.T.Camellini, J. Organomet. Chem., 1985, 297, 103.
387 L.J.Farrugia, M.Green, D.R.Hankey, M.Murray, A.G.Orpen and F.G.A.Stone, J. Chem. Soc., Dalton Trans., 1985, 177.
388 L.J.Farrugia, A.D.Miles and F.G.A.Stone, J. Chem. Soc., Dalton Trans., 1985, 2437.
389 R.D.Adams and S.Wang, Inorg. Chem., 1985, 24, 4447.
390 C.Couture and D.H.Farrar, J. Chem. Soc., Chem. Commun., 1985, 197.
391 M.I.Bruce, E.Horn, J.G.Matisons and M.R.Snow, J. Organomet. Chem., 1985, 286, 271.
392 C.M.Hay, B.F.G.Johnson, J. Lewis, R.C.S.McQueen, P.R.Raithby, R.M.Sorrell and M.J.Taylor, Organometallics, 1985, 4, 202.
393 P.Braunstein, H.Lehner, D.Matt, A.Tiripicchio and M.T.Camellini, Nouv. J. Chim., 1985, 9, 597.
394 W.A.Herrmann, C.E.Barnes, T.Zahn and M.L.Ziegler, Organometallics, 1985, 4, 172.
395 A.Arrigoni, A.Ceriotti, R.D.Pergola, G.Longoni, M.Manassero and M.Sansoni, J. Organomet. Chem., 1985, 296, 243.
396 P.Klufers, Angew. Chem., Int. Ed. Engl., 1985, 24, 70.
397 A.L.Balch, L.A.Fossett, M.M.Olmstead, D.E.Oram and P.E.Reedy, J. Am. Chem. Soc., 1985, 107, 5272.
398 G.Bruno, S.L.Shiavo, P.Piraino and F.Faraone, Organometallics, 1985, 4, 1098.
399 A.Ceriotti, F.Demartin, G.Longoni, M.Manassero, M.Marchionna, G.Piva and M.Sansoni, Angew. Chem., Int. Ed. Engl., 1985, 24, 697.
400 S.I.Al-Resayes, P.B.Hitchcock, J.F.Nixon and D.M.P.Mingos, J. Chem. Soc., Chem. Commun., 1985, 365.
401 A.Albinati, K-H.Dahmen, A.Togni and L.M.Venanzi, Angew. Chem., Int. Ed. Engl., 1985, 24, 766.
402 O.M.Abu-Salah, A-R.A.Al-Ohaly and C.B.Knobler, J. Chem. Soc., Chem. Commun., 1985, 1502.
403 C.P.Casey and F.Nief, Organometallics, 1985, 4, 1218.
404 J-F.Reynoud, J-C.Leblanc and C.Moise, Transition Met. Chem., (Weinheim, Ger.), 1985, 10, 291.
405 K.E.Warner and J.R.Norton, Organometallics, 1985, 4, 2150.

406 S.Sabo, B.Chaudret and D.Gervais, J. Organomet. Chem., 1985, 292, 411.
407 S.Rosenberg, G.L.Geoffroy and A.L.Rheingold, Organometallics, 1985, 4, 1184.
408 R.H.Fong and W.H.Hersch, Organometallics, 1985, 4, 1468.
409 A.B.Antonova, S.V.Kovalenko, E.D.Korniyets, P.V.Petrovsky, G.R.Gulbis and A.A.Johansson, Inorg. Chim. Acta., 1985, 96, 1; A.B.Antonova, S.V.Kovalenko, E.D.Korniyets, P.V.Petrovsky, A.A.Johansson and N.A.Deykhina, ibid, 1985, 105, 153.
410 R.Weinland and H.Werner, J. Chem. Soc., Chem. Commun., 1985, 1145.
411 V.E.Lopatin, V.A.Ershova, M.Ts.Tsybenov and S.P.Gubin, Bull. Acad. Sci. USSR, Div. Chem. Sci., 1984, 33, 2429.
412 Y.Chi and J.R.Shapley, Organometallics, 1985, 4, 1900.
413 S.Attali and R.Mathieu, J. Organomet. Chem., 1985, 291, 205.
414 U.Honrath, L.Shu-Tang and H.Vahrenkamp, Chem. Ber., 1985, 118, 132.
415 J.Schneider, M.Minelli and G.Huttner, J. Organomet. Chem., 1985, 294, 75.
416 W-R.Dietrich and H.Vahrenkamp, J. Chem. Res.(S), 1985, 200.
417 P.Braunstein, J.Kervennal and J-L.Richert, Angew. Chem., Int. Ed. Engl., 1985, 24, 768.
418 W.A.Herrmann and W.Kalcher, Chem. Ber., 1985, 118, 3861.
419 C.P.Horwitz and D.F.Shriver, J. Am. Chem. Soc., 1985, 107, 8147.
420 (a)I.D.Salter, J. Organomet. Chem., 1985, 295, C17; (b)S.S.D.Brown, I.D.Salter and B.M.Smith, J. Chem. Soc., Chem. Commun., 1985, 1439.
421 (a)S.I.Khan, P.G.Edwards, H.S.H.Yuan and R.Bau, J. Am. Chem. Soc., 1985, 107, 1682; (b)M.Y.Chiang, E.Bohlen and R.Bau, ibid, 1985, 107, 1679.
422 J.M.Burlitch, S.E.Hayes and J.T.Lemley, Organometallics, 1985, 4, 167.
423 (a)P.H.M.Budzelaar, J.Boersma, G.J.M.van der Kerk, A.L.Spek and A.J.M.Duisenberg, Organometallics, 1985, 4, 680; (b) idem, J. Organomet. Chem., 1985, 287, C13.
424 M.Cano, R.Criado, E.Gutierrez-Puebla, A.Monge and M.P.Pardo, J. Organomet. Chem., 1985, 292, 375.
425 S.Onaka, Y.Kondo, M.Yamashita, Y.Tatematsu, Y.Kato, M.Goto and T.Ito, Inorg. Chem., 1985, 24, 1070.
426 P.Braunstein, J.Rose, A.Tiripicchio and M.T.Camellini, Angew. Chem., Int. Ed. Engl., 1985, 24, 767.
427 J.M.Ragosta and J.M.Burlitch, J. Chem. Soc., Chem. Commun., 1985, 1187.
428 H-J.Haupt, P.Balsaa and B.Schwab, Z. Anorg. Allg. Chem., 1985, 521, 15.
429 H.Schumann, S.Nickel, E.Hahn and M.J.Heeg, Organometallics, 1985, 4, 800.
430 (a)T.D.Tilley, Organometallics, 1985, 4, 1452; (b)idem, J. Am. Chem. Soc., 1985, 107, 4084.
431 M.D.Curtis, L.G.Bell and W.M.Butler, Organometallics, 1985, 4, 701.
432 J.Arnold and T.D.Tilley, J. Am. Chem. Soc., 1985, 107, 6409.
433 W.A.Herrmann, E.Voss, E.Guggolz and M.L.Ziegler, J. Organomet. Chem., 1985, 284, 47.
434 S.R.Berryhill, G.L.Clevenger and F.Y.Burdurlu, Organometallics, 1985, 4, 1509.
435 M.J.Auburn, S.L.Grundy, S.R.Stobert and M.J.Zaworoko, J. Am. Chem. Soc., 1985, 107, 266.
436 (a)P.Gusbeth and H.Vahrenkamp, Chem. Ber., 1985, 118, 1143, 1758 and 1770; (b)idem, ibid, 1985, 118, 1746.
437 S.P.Foster, K.M.Mackay and B.K.Nicholson, Inorg. Chem., 1985, 24, 909.
438 A.Bonny, K.M.Mackay and F.S.Wong, J. Chem. Res.,(S), 1985, 40; (M), 1985, 558.
439 J.D.Korp, I.Bernal, R.Horlein, R.Serrano and W.A.Herrmann, Chem. Ber., 1985, 118, 340.
440 G.Huttner, U.Weber, B.Sigwarth, O.Scheidsteger, H.Lang and L.Zsolnai, J. Organomet. Chem., 1985, 282, 331.
441 K.Burgess, C.Guerin, B.F.G.Johnson and J.Lewis, J. Organomet. Chem., 1985, 295, C3.
442 (a)V.K.Bel'skii, A.N.Protskii, B.M.Bulychev and G.L.Soloveichik, J. Organomet. Chem., 1985, 280, 45; (b)V.K.Belsky, A.N.Protsky, I.V.Molodnitskaya, B.M.Bulychev and G.L.Soloveichik, ibid, 1985, 293, 69.
443 S.M.Hawkins, P.B.Hitchcock and M.F.Lappert, J. Chem. Soc., Chem. Commun., 1985, 1592.
444 G.K.Campbell, P.B.Hitchcock, M.F.Lappert and M.C.Misra, J. Organomet. Chem.,

1985, 289, Cl.
445 C.J.Cardin, D.J.Cardin, J.M.Power and M.B.Hursthouse, J. Am. Chem. Soc., 1985, 107, 505.
446 W.A.Herrmann, H-J.Kneuper and E.Herdtweck, Angew. Chem., Int. Ed. Engl., 1985, 24, 1062.
447 M.Wieber, D.Wirth and C.Burschka, Z. Naturforsch., B, 1985, 40, 258.
448 A.M.Arif, A.H.Cowley, N.C.Norman and M.Pakulski, J. Am. Chem. Soc., 1985, 107, 1062.
449 M.R.Churchill, J.C.Fettinger and K.H.Whitmire, J. Organomet. Chem., 1985, 284, 13.
450 K.H.Whitmire, M.R.Churchill and J.C.Fettinger, J. Am. Chem. Soc., 1985, 107, 1056.

10
Ligand Substitution Reactions of Metal and Organometal Carbonyls with Group V and VI Donor Ligands

BY D. A. EDWARDS

1 Introduction and Reviews

As in previous Volumes of this Series, this Chapter surveys substitution reactions of mononuclear metal carbonyls in some depth, but space limitations permit only a selective coverage of such reactions involving metal-metal bonded and organometal carbonyls. These compounds are reviewed more extensively elsewhere in this Volume.

A special issue of *Inorganica Chimica Acta* commemorating the 100th volume contains reviews on substitution reactions of metal carbonyls known to proceed by associative mechanisms,[1] and on metal carbonyl complexes containing $RN=C(R')(R'')C=NR$ or $2-RN=C(R')C_5H_4N$ ligands.[2] The first review since 1977 of transition metal thiocarbonyls has been published.[3] Two reviews of the photochemistry of metal-metal bonded carbonyl dimers have appeared.[4] The ability of multiply-bonded species e.g. E_2 (E = P, As, Sb, or Bi), PN, phosphorins, phospha-alkenes and diphosphenes to act as complex ligands has been reviewed.[5] Other reviews have been concerned with PF_3 complexes,[6] metal carbonyls containing ligands of biological significance e.g. porphyrins, nucleosides and amino acids,[7] and carbonyl phosphine clusters of palladium and platinum.[8]

2 Papers of General Interest

The ability of $[Fe_2(CO)_6(\mu-SMe)_2]$ to act as a catalyst in substitution reactions of metal carbonyls has been noted. Important features of the catalytic cycles are a change in the coordination mode of the thiolate ligand and the formation of a 17-electron metal centre.[9] Electron transfer catalysed substitutions of metal carbonyls, e.g. $Fe(CO)_5$ and $Os_3(CO)_{12}$, using traces of $[CpFe(CO)_2]_2$ as initiator have been studied. The reactions involve generation of 19-electron substitution-labile intermediates.[10] One carbonyl ligand of $[M(CO)_4(bipy)]$ (M = Cr,

(For references see page 222)

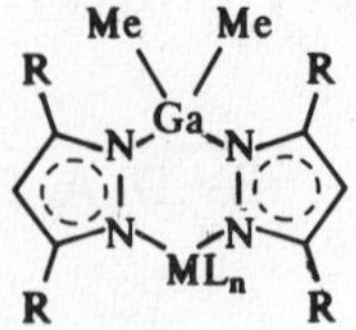

(1) ML_n = $Re(CO)_3(PPh_3)$, R = H;
ML_n = $Rh(CO)(PPh_3)$, R = Me;
ML_n = $Ir(CO)_2$ or $Ir(CO)(PPh_3)$,
R = H or Me

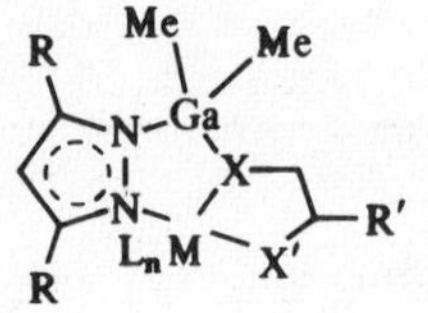

(2) ML_n = $Mn(CO)_3$ or $Re(CO)_3$, R = H or Me
R' = H, Me or Et, X = O or S,
X' = NH_2, NMe_2, SPh or SEt;
ML_n = Rh(CO), R = H, R' = H or Me,
X = O, X' = NH_2, NMe_2, SPh or SMe

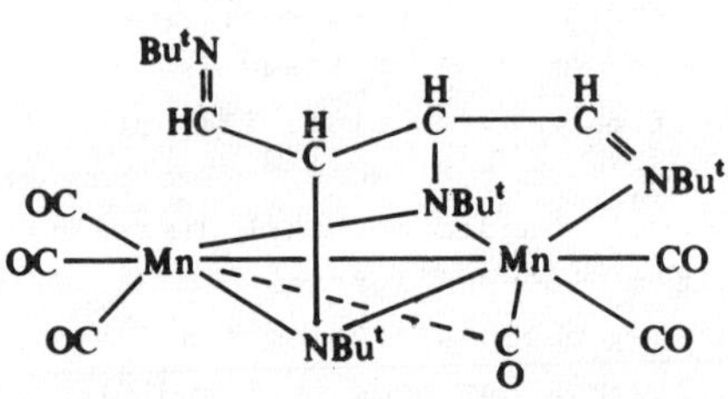

(3)

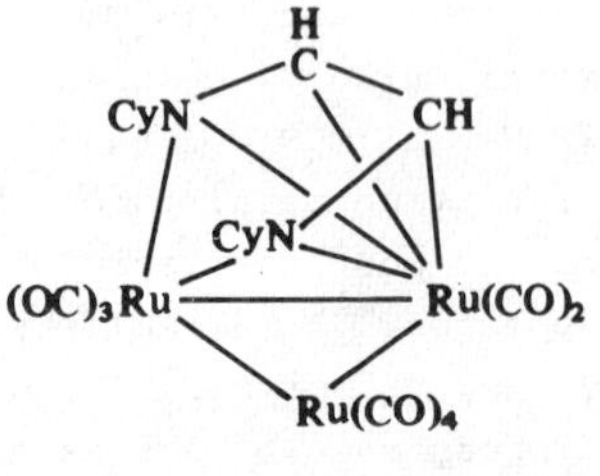

(4)

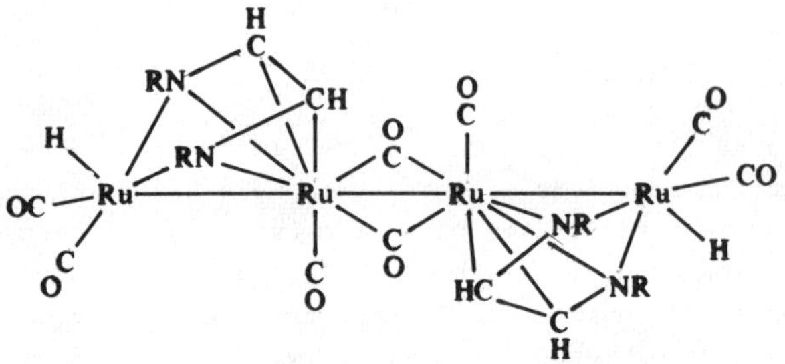

(5) R = neo-pentyl

Mo, or W) can be replaced by PPh_3 in an electrode-catalysed process. Elimination of CO following the electrode reduction to $[M(CO)_4(bipy)]^-$ radical anions is the rate-determining step.[11] The $[Cp_2Fe]^+$ oxidation of $[CpFe(CO)_2]_2$, $Co_2(CO)_8$ and $[CpNi(CO)]_2$ in the presence of PPh_3 results in the formation of mononuclear $[CpFe(CO)_2(PPh_3)]^+$, <u>trans</u>-$[Co(CO)_3(PPh_3)_2]^+$ and $[CpNi(PPh_3)_2]^+$, respectively.[12]

2.1 <u>Nitrogen Donor Ligands</u>.— The coordination properties of several uninegative chelate pyrazolylgallate ligands have been examined and complexes of the types (1) and (2) characterized.[13] Metal carbonyl - DAB complexes, [DAB = RN=C(R')(R')C=NR; R = alkyl or aryl, R' = H or alkyl], have been extensively studied. Complexes prepared or studied by spectroscopic or structural methods include: (i) $[M(CO)_5(\eta^1\text{-DAB})]$ (M = Cr, Mo, or W) and their conversion into $[M(CO)_4(DAB)]$ chelate complexes[14]; (ii) two isomers of $[Fe_2(CO)_5(DAB)P(OR)_3]$ (R = Me or Pr^i)[15]; (iii) $[(\eta\text{-}C_5R_5)Co(DAB)]$ (R = H or Me)[16]; (iv) $[MoBr(\eta\text{-}C_3H_5)(CO)_2(DAB)]$ (DAB = CyN=CHCH=NCy)[17]; (v) He(I) and He(II) P.E.S. of $[ML_n(DAB)]$, $[ML_n = Cr(CO)_4, Mo(CO)_4, W(CO)_4, ReCl(CO)_3, Fe(CO)_3, Ru(CO)_3,$ and $Ni(CO)_2]$[18]; (vi) $[(OC)_5MM'(CO)_3(DAB)]$ (M,M' = Mn and Re) - IR, resonance Raman, PES and photochemistry.[19] The reaction between $[MnBr(CO)_3(Bu^tN=CHCH=NBu^t)]$ and $[CpFe(CO)_2]^-$ leads to two manganese products, $[Mn_2(CO)_6(Bu^tN=CHCH=NBu^t)]$, containing 8-electron (σ-N, σ-N', η-C=N, η-C=N') donor DAB, and (3) in which the bridging 8-electron donor ligand has been formed by reductive C-C coupling of two DAB ligands.[20] Reaction of $[MnBr(CO)_3(\underline{p}\text{-tolN=CHCH=Ntol-}\underline{p})]$ with $[HFe(CO)_4]^-$ followed by protonation gives $[(OC)_3Mn(\mu\text{-}H)(\mu,\mu'\text{-}\underline{p}\text{-tolNCH}_2CH_2Ntol\text{-}\underline{p})Fe(CO)_3]$ in which the σ-N,σ-N' chelate DAB ligand has been reduced to a formal 8-electron donor substituted 1,2-diaminoethane dianion. Reactions between $[ReBr(CO)_3\{RN=CH(R')C=NR\}]$ (R = Pr^i or Cy, R' = H; R = Pr^i, R' = Me) and $[HFe(CO)_4]^-$ follow a different course producing the metal-metal bonded complexes $[(OC)_3Re(\mu\text{-}H)\{RN=CH(R')C=NR\}Fe(CO)_3]$ in which the DAB ligand acts as a 6-electron donor (σ-N, σ-N' bonded to Re and η^2-C=N' bonded to Fe).[21] The ligands RN=CHCH=NR (R = Pr^i, Cy, neo-pentyl, Bu^i and $\underline{p}$-tol) react with $Ru_3(CO)_{12}$ to give $[Ru(CO)_3(DAB)]$ intermediates which can be converted into the known $[Ru_2(CO)_n(DAB)]$ (n = 5 or 6) dimers. The novel trinuclear complex (4) containing an 8-electron donor DAB ligand, results from the reaction of

[$Ru_2(CO)_6(CyN=CHCH=NCy)$] with $Ru_3(CO)_{12}$. The novel addition products [$Ru_3(CO)_8(\mu-CH_2)(RN=CHCH=NR)$] (R = Bu^i or neo-pentyl) also contain 8-electron donor DAB but only two Ru-Ru bonds. Whereas [$Ru_2(CO)_5(RN=CHCH=NR)$] (R = Pr^i, Cy, or $CyCH_2$) reacts with H_2 to give [$H_2Ru_2(CO)_5(RN=CHCH=NR)$] oxidative addition products containing 6-electron donor DAB, when R = neo-pentyl the product is the novel complex (5) with a Ru_4 linear chain and 8-electron donor DAB ligands.

Carbodiimides RN=C=NR (R = p-tol, Pr^i or Cy) react with [$Ru_2(CO)_6(Bu^tN=CHCH=NBu^t)$] to give coupled products of type (6).[22]

2.2 Phosphorus and the Heavier Group V Donor Ligands.

A number of studies of phosphinidene complexes have been reported. The anions [$M_2(CO)_{10}$]$^{2-}$ (M = Cr, Mo, or W) react with PX_3 (X = Cl, Br, or I) or $RPCl_2$ [R = mesityl or $CH(SiMe_3)_2$] to give the species [$\{M(CO)_5\}_2(\mu-PE)$] (E = X or R) as well as the diphosphene complexes [$\{M(CO)_5\}_n(RP=PR)$] (n = 1 or 2). Where E = X conversion into [$\{M(CO)_5\}_2(\mu-PL)$] chelate complexes can be achieved by reaction with HL (HL = tropolone or 8-hydroxyquinoline).[23] A variety of phosphinidene complexes can be isolated from reactions of $Fe_2(CO)_9$ with either $RPCl_2$ or the methylenebis(phosphines) $HRPCH_2PRH$ and $R_2PCH_2PR_2$ (R = alkyl). Products formed by cleavage of P-H or P-Cl bonds and/or the P-C-P skeleton include [$Fe_3(CO)_8(\mu-H)_2(PR_2Me)(\mu_3-PR)$], [$Fe_3(CO)_9(PR_2Me)(\mu_3-PR)$], [$Fe_3(CO)_9(\mu_3-PR)_2$], [$Fe_3(CO)_9(\mu-H)(\mu-PRMe)(\mu_3-PR)$], [$Fe_3(CO)_{10}(\mu_3-PR)$], [$Fe_4(CO)_{11}(\mu_4-PR)_2$] and [$Fe_4(CO)_{10}(\mu-H)(\mu-Pr^iPCH_2PPr^i)(\mu-PPr^iMe)(\mu_4-PH)$], the latter complex possessing an unsubstituted phosphinidene capping a planar trapezoid Fe_4 moiety.[24] Phosphinidene complexes of cobalt that have been isolated include [$\{CpCo(CO)\}_2(\mu-PR)$], containing an 'open' structure, [$\{Co(CO)_3\}_2(\mu-PR)$] (R = 2,4,6-$Bu^t_3C_6H_2$) containing a novel Co_2P three-membered ring system with Co-P multiple bonding, and [$Co_4(CO)_8(\mu-CO)_2\{\mu_4-PCH(SiMe_3)_2\}(\mu_4-PCH_2SiMe_3)$], a cluster with two different capping phosphinidenes. Carbonyl substitution reactions of [$Co_4(CO)_{10}(\mu_4-PPh)_2$] using $P(OMe)_3$, dppm, dppe and $MeN(PF_2)_2$ have also been studied.[25]

Arsine decomposes in the presence of [$(\eta-C_5Me_5)Mn(CO)_2(THF)$] to give an arsinidene fragment which acts as a symmetrical bridging ligand in [$\{(\eta-C_5Me_5)Mn(CO)_2\}_2(\mu-AsH)$].[26] The arsinidene and stibinidene complexes [$\{M(CO)_5\}_2(\mu-ER)$] (M = Cr, Mo, or W; E = As or Sb; R = Cl, Me, Bu^t or Ph), can be stabilized by the formation of [{M(CO)$_5$}$_2$(μ-ER)L] mono-adducts with Lewis bases [L = PPh$_3$, PBu$^n_3$ or SC(NHMe)$_2$]. The tungsten stibinidene complexes also react with the Lewis acid fragment W(CO)$_5$ giving [{W(CO)$_5$}$_3$(μ$_3$-SbR)]. These observations may be rationalised by the valence tautomerism [(OC)$_5$W⋯Sb(R)⋯W(CO)$_5$] ⇌ [(OC)$_5$W-Sb(R)-W(CO)$_5$] the former acting as a Lewis acid, the latter as a Lewis base.[27] Other compounds isolated include [Fe$_2$(CO)$_8$(μ-SbR)] [R = CH(SiMe$_3$)$_2$] and the first coordinated bismuthinidene complex (7).[28]

Transition metal-diphosphene complexes continue to be explored. The complexes [M(CO)$_5$(RP=PR')] (M = Cr, Mo, or W; R = Ph or 2,4,6-Bu$^t_3$C$_6$H$_2$) may be prepared by reactions of the diphosphenes with [M(CO)$_5$(THF)] or RPCl$_2$ with [M(CO)$_5$(RPH$_2$)] in the presence of a base. E - to - Z-photoisomerization of a number of these complexes has been noted.[29] In [Cr$_3$(CO)$_9$(μ-CO)(μ$_3$-PBu^t)(μ$_2$,η-Bu^{t}P=PBu^t)], the product of pyrolysis of [{Cr(CO)$_5$}$_2$(μ-PBu^t)], the Z-diphosphene ligand is σ-P, σ-P'-bonded to two metals and η^2-P=P bonded to the third metal in the cluster. Products of the type [{Cr(CO)$_5$}$_2$(μ-L)] [L = PhP(Br)P(Br)Ph, PhP(H)P(X)Ph and PhPYPPh; X = OH, OMe, NHBu^n, or Cl; Y = S or CH$_2$] arise from reactions of [{Cr(CO)$_5$}$_2$(μ-PhP=PPh)].[30] Elimination of Me$_3$SiCl on reaction of [(η-C$_5$Me$_5$)M(CO)$_2$P(SiMe$_3$)$_2$] with RPCl$_2$ (M = Fe or Ru; R = 2,4,6-Bu$^t_3$C$_6$H$_2$) affords the first diphosphenyl complexes [(η-C$_5$Me$_5$)M(CO)$_2$(P=PR)].[31] The first bridging phospha-alkenyl complex [Fe$_2$(CO)$_6${μ$_2$-P=C(SiMe$_3$)$_2$}$_2$] has been reported.[32] The compound RP=CH$_2$ (R = 2,4,6-Bu$^t_3$C$_6$H$_2$) is a versatile ligand, being P-bonded in [Ni(CO)$_3$(RP=CH$_2$)] and in one form of [Fe(CO)$_4$(RP=CH$_2$)], η^2-bonded in a second form of the latter complex and η^1,η^2-bonded in [{Fe(CO)$_4$}$_2$(RP=CH$_2$)].[33]

Reactions of metal carbonyls with the cyclotetraphosph(III)-azane [MePNMe]$_4$ yield [M(CO)$_3$(MePNMe)$_4$] (M = Cr, Mo, W or Mn$^+$) and [Ni$_2$(CO)$_2$(μ-CO)(MePNMe)$_4$] complexes.[34] Routes to several P-bonded metallocyclotriphosph(V)azene complexes e.g. [CpCr(CO)$_3$(N$_3$P$_3$Cl$_5$)], [CpM(CO)$_3$(N$_3$P$_3$Cl$_4$η^1-C$_5$H$_5$)] (M = Mo or W), [Cp$_2$M$_2$(CO)$_2$(N$_3$P$_3$Cl$_4$)] (M = Co or Rh) and [(OC)$_4$FeCo(CO)Cp(N$_3$P$_3$Cl$_4$)] have been described.[35] 2-Amino-1,3,2-dioxaphosphorinanes act as unidentate P-donor ligands in the complexes [RhCl(CO)L$_2$], [RhCl(CO)L]$_2$, (L = RPOCH$_2$CH$_2$O; R = Me$_2$N, Et$_2$N, Ph$_2$N, morpholino, or piperidino), [RhCl(CO)L'$_2$], [RhCl(CO)L']$_2$, Mo(CO)$_5$L' and [Co(CO)$_3$L']$_2$ (L' = Et$_2$NPOCH$_2$CH$_2$CMe$_2$O).[36] A variety of P,P'-chelate 1,2-bis(diphenylphosphino)o-carborane complexes [L$_n$M{Ph$_2$PC(B$_{10}$H$_{10}$)CPPh$_2$}] [L$_n$M = Ni(CO)$_2$, Fe(CO)$_3$,

$Cr(CO)_4$, $Mo(CO)_4$, $W(CO)_4$ and $CpMn(CO)$] and some arsino-analogues have been characterised.[37]

2.3 Group VI Donor Ligands. — The anions $[CpCo\{P(O)R_2\}_3]^-$ (R = OMe or OEt), formally isoelectronic with $C_5H_5^-$, act as tridentate oxygen donor ligands in forming $[Rh(CO)_2(ligand)]$, $[Rh_2(\mu-CO)_3(ligand)_2]$ and $[M(CO)_3(ligand)]$ (M = Mn or Re) complexes.[38] The tetrathiosquarate anion $C_4S_4^{2-}$ displays several modes of co-ordination, acting as a bis(chelate) ligand in $[L_nM(C_4S_4)ML_n]^{x-}$ [x = 0, ML_n = $Mn(CO)_4$ or $Rh(CO)_2$; x = 2, ML_n = $Cr(CO)_4$, $Mo(CO)_4$, $W(CO)_4$, or $Re(CO)_3Br$], as a bis(unidentate) ligand in $[L_nM(C_4S_4)ML_n]$ [ML_n = $Re(CO)_5$, $CpMo(CO)_3$, or $CpW(CO)_3$], and as a tetrakis(unidentate) ligand in $[(L_nM)_2(C_4S_4)(ML_n)_2]^{2+}$ [ML_n = $Re(CO)_5$ or $CpW(CO)_3$].[39] Stannanedithioesters behave as unidentate ligands in $[L_nM\{S=C(SR)SnPh_3\}]$ complexes [L_nM = $W(CO)_5$, $CpMn(CO)_2$, or $Re_2(CO)_9$]. Unidentate stannanedithiocarboxylate complexes e.g. $[L_nM\{SC(=S)SnPh_3\}]$ [L_nM = $Re(CO)_5$, $Re(CO)_4(PPh_3)$, $Mn(CO)_3(bipy)$, $Mn(CO)_3(dppe)$, or $CpFe(CO)_2$] and bidentate chelate analogues e.g. $[L_nM(S_2CSnPh_3)]$ [L_nM = $Mn(CO)_4$, $Mn(CO)_3(PPh_3)$, $Re(CO)_4$, $CpFe(CO)$, $CpMo(CO)_2$, or $CpW(CO)_2$] have also been described.[40] The ferrocenedithiocarboxylate anion similarly can act as a unidentate or a bidentate chelating ligand in bonding to various metal carbonyl and cyclopentadienyl metal carbonyl moieties.[41]

3 Groups IV and V

The complexes $[Cp_2M(CO)(PMe_3)]$ (M = Ti, Zr, or Hf) result from the photolysis of pentane solutions of $[Cp_2M(CO)_2]$ in the presence of PMe_3.[42]

Many complexes of the type $Et_4N[V(CO)_5L']$ have been prepared by ligand exchange reactions using $Et_4N[V(CO)_5L]$ as starting materials (L = DMSO, THF or CO; L' = phosphines, including potentially multidentate phosphines, phosphites, nitriles, amines, pyridines, sulphoxides, acetone, η^1-SO_2, η^1-N_2, η^2-CS_2, ethers, thio-, seleno-, and telluroethers). The ^{51}V n.m.r. δ values of many of these complexes have been correlated with ligand strength as assessed by Graham's σ and π parameters.[43] Reactions of $Na[V(CO)_4(dppe)]$ with H_2S, Na_2SeO_3 and Na_2TeO_3 lead to the chalcogen-bridged species $[\{V(CO)_3(dppe)\}_2(\mu-E)]$ (E = S, Se, or Te) containing V-E multiple bonds.[44]

The seven-coordinate complexes [TaX(CO)$_3$(PMe$_3$)$_3$] (X = Cl, Br, or I) result from halogen oxidation of [Ta(CO)$_6$]$^-$ in the presence of PMe$_3$ or carbonylation of [TaClH$_2$(PMe$_3$)$_4$]. The η^1-BH$_4$ complex [Ta(BH$_4$)(CO)$_3$(PMe$_3$)$_3$] has also been isolated and found to slowly convert into the 1:1 electrolyte [Ta(CO)$_3$(PMe$_3$)$_4$][Ta(CO)$_5$(PMe$_3$)].^{45}

A range of complexes of the types [CpV(CO)$_3$L] (L = thio-, seleno-, and telluroethers, thioureas, DMSO, amines, pyridines, nitriles or imidazoles) and cis-[CpV(CO)$_2$(L-L)] (L-L = bipy or phen) can be prepared by substitution reactions of [CpV(CO)$_3$(THF)].^{46} Photolysis of [CpM(CO)$_4$] in the presence of EMe$_2$ affords [CpM(CO)$_3$(EMe$_2$)] complexes (M = V, E = S, Se, or Te; M = Nb, E = S).^{47} The kinetics of substitution of [(η^5-C$_9$H$_7$)V(CO)$_4$] by PR$_3$ (R = Bu^n or OEt) occurs by a dissociative S$_{N1}$ process to give [(η^5-C$_9$H$_7$)V(CO)$_3$(PR$_3$)].^{48}

4 Group VI

4.1 Carbonyl Complexes of Cr^0, Mo^0, and W^0.— The stable imine complexes [W(CO)$_5$(HN=CRR')] (R = H, R' = Me, Ph, or p-MeOC$_6$H$_4$; R = Me, R' = Me or Ph) have been isolated from reactions of [W(CO)$_5$(THF)] with aldoximes and ketoximes.^{49} Photolysis of W(CO)$_6$ in the presence of aminoacid derivatives affords [W(CO)$_5$L] (L = ethyl glycinate or glycylglycine).^{50} Substitution reactions of cis-[Mo(CO)$_4$(py)$_2$], [Mo(CO)$_5$(py)] and [Mo(CO)$_5$(4-Mepy)] with bipy, phen and MeC(=NPh)C(=NPh)Me have been studied as a function of temperature and pressure and mechanisms proposed.^{51} The ligands Ph(R)C=N(CH$_2$)$_2$N=C(R)Ph (R = H, Me, or Ph) form [Cr(CO)$_4$(chelate)] complexes on reaction with Cr(CO)$_6$ but ligand-bridged binuclear [{Cr(CO)$_4$(EPh$_3$)}$_2${μ-Ph(R)C=N(CH$_2$)$_2$N=C(R)Ph}] complexes on reaction with [Cr(CO)$_5$(EPh$_3$)] (E = P or As).^{52} Photolysis of a variety of [W(CO)$_4$(α-diimine)] complexes in the presence of PR$_3$ (R = alkyl, Ph, or MeO) effects substitution of a carbonyl ligand. Quantum yields are low and depend on the basicity, cone angle and concentration of the PR$_3$ indicating an associative mechanism.^{53} The O$_2$N$_2$- and O$_2$N$_3$-donor macrocycles L^1 and L^2 [structure (8)] react with Mo(CO)$_6$ and [(η-PhMe)Mo(CO)$_3$] to afford [Mo(CO)$_4$L^1] and [Mo(CO)$_3$L^2], respectively. Only the nitrogen atoms of the macrocycles are coordinated to the metal.^{54} Some complexes containing 4,4'-bipyridyl (4,4'-bipy), 1,2-bis(2-pyridyl)ethene (2-bpe) and 1,2-bis(4-pyridyl)ethene (4-bpe) have been characterised. Examples include polymeric

[Mo(CO)$_4$(4,4'-bipy)]$_n$, mononuclear cis-[Mo(CO)$_4$(2-bpe)$_2$] and [Mo(CO)$_5$(4,4'-bipy)] and binuclear [Mo(CO)$_4$(μ-4-bpe)]$_2$ and [{Mo(CO)$_3$(bipy)}$_2$(μ-4,4'-bipy)].[55] The photophysical properties of some [M(CO)$_4$(chelate)] complexes have been explored [M = Mo or W; chelate = substituted bipy or phen,[56a] 2,2'-bi(4H-5,6-dihydrothiazine),[56b] biacetylbis(phenylimine),[56c] or 4,4'-bipyrimidine[56d]]. In the course of these studies some bis-chelating ligands have been used to generate complexes of type (9) as well as analogous complexes containing azo-2,2'-bipyridine or 3,6-bis(2-pyridyl)-1,2,4,5-tetrazine ligands.[57]

The relative sizes of 12 PR$_3$ ligands have been assessed by determining the cis:trans ratios of [W(CO)$_4$(PR$_3$)L] complexes obtained from reactions of [W(CO)$_4$(py)L], [L = PPhMe$_2$, PPh$_2$Et, or P(tol-p)$_3$], with PR$_3$. Generally, decreasing cis:trans ratios correlated with increasing PR$_3$ cone angle. The dissociative loss of pyridine is the rate-determining step in these reactions.[58] Other complexes containing unidentate ligands that have been characterised include [M(CO)$_5$L], [M = Cr or W; L = R$_2$NPH$_2$, R$_2$N = Cy$_2$N, Pr$^i_2$N, or 2,2,6,6-Me$_4$-piperidino[59a]; M = Mo, L = C$_5$H$_5$P[59b]; M = Cr, Mo, or W, L = the cyclotristibine MeC(CH$_2$Sb)$_3$[59c]], [Cr(CO)$_{6-n}$L$_n$], [L = RPS(CH$_2$)$_2$S or RPS(o-C$_6$H$_4$)S; R = Me or CF$_3$; n = 1 or 2][60] and [Mo(dppe)(PF$_3$)$_4$] and [Mo(F$_2$PNMe$_2$)$_6$] both prepared by photochemical reactions of [Mo(CO)$_4$(dppe)] with the appropriate ligand.[61]

The ligand dmpm is chelating in cis-[Cr(CO)$_2$(dmpm)$_2$], but bridging in [Mo$_2$(CO)$_6$(μ-dmpm)$_3$] and [(OC)$_3$Cr(μ-dppm)$_2$Fe(CO)$_3$]. The latter complex involves square pyramidal Cr(CO)$_3$P$_2$ and trigonal bipyramidal Fe(CO)$_3$P$_2$ moieties with little, if any, direct Cr---Fe interaction.[62] Other [M(CO)$_4$(P,P'-chelate)] complexes that have been characterised contain the ligands (Bu$^t_2$P)$_2$Te (M = Cr, Mo, or W),[63a] (Ph$_2$PCH$_2$)$_2$PPh (M = Cr, Mo, or W),[63b] (Ph$_2$P)$_2$NPPh$_2$ (M = Cr),[63c] [Ph(Pr^{i}NH)P]$_2$NPr^i (M = Mo),[63d] and (Ph$_2$PO)$_2$Si(Me)R or (Ph$_2$PO)$_2$P(O)R (M = Mo; R = alkyl, haloalkyl, or aryl).[63e]

The complexes [M(CO)$_4${Ph$_2$PC(S)NHR}].THF (M = Cr or Mo, R = Me or Ph), prepared by photolysis of [M(CO)$_5$(THF)] with phosphinothioformamides, contain planar P,S-chelating ligands with additional weak N-H---OC$_4$H$_8$ hydrogen bonds.[64]

The ligands Me$_2$EP(CF$_3$)$_2$ (E = P or As) are unidentate in [M(CO)$_4${Me$_2$EP(CF$_3$)$_2$}$_2$] but bridging in [{M(CO)$_5$}$_2${μ-Me$_2$EP(CF$_3$)$_2$}] and [M(CO)$_4${μ-Me$_2$EP(CF$_3$)$_2$}]$_2$, (M = Cr, Mo, or W).[65] Aspects of

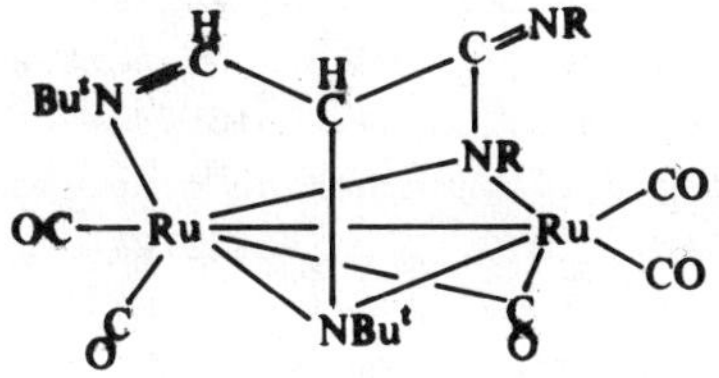

(6) R = p-tol, Pr^i, or Cy

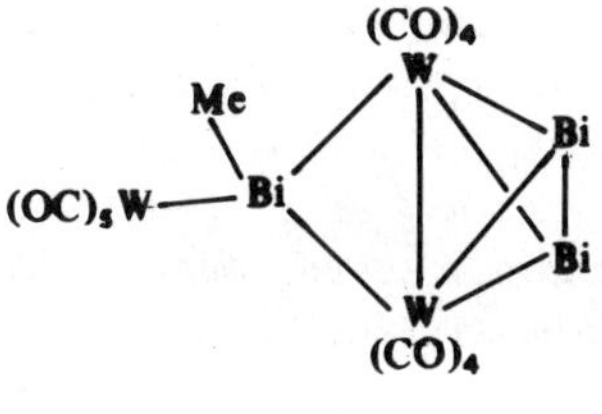

(7)

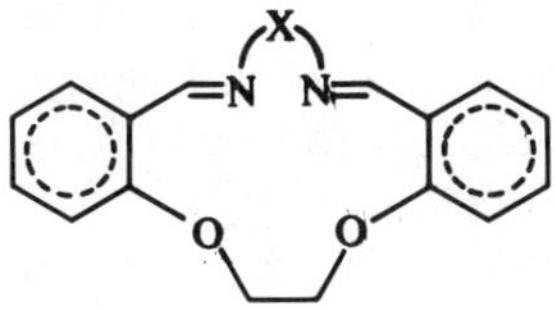

(8) L^1, X = $-(CH_2)_3-$;
L^2, X = $-(CH_2)_3N(Me)(CH_2)_3-$

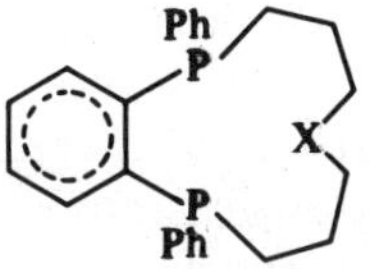

(10) X = PH, PPh, or NMe

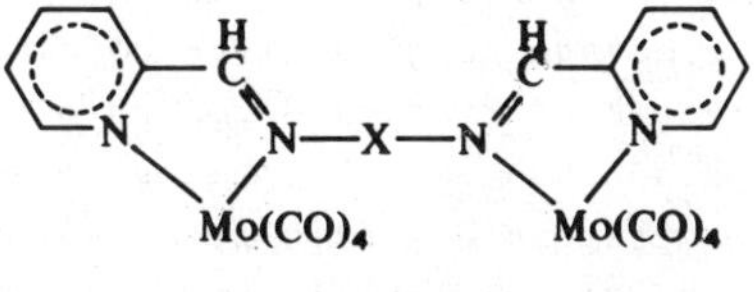

(9) X = 1,4-C$_6$H$_4$, 1,4-C$_6$H$_4$C$_6$H$_4$, 1,4-CH$_2$C$_6$H$_4$CH$_2$ or is absent

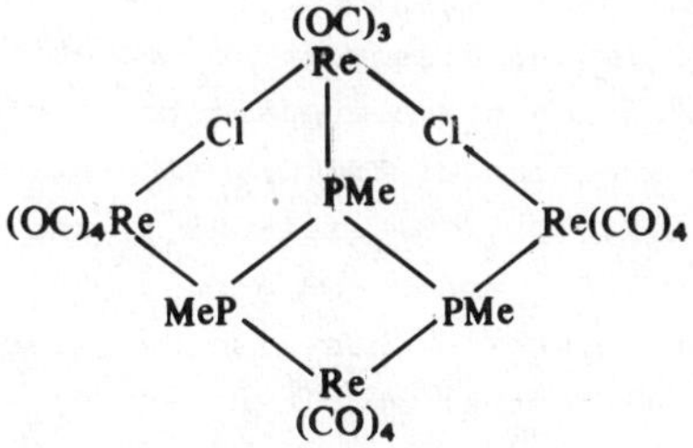

(11)

the chemistry e.g.(As-N bond cleavage) of metal carbonyl-amino-arsine complexes containing ligands such as $(Me_2As)_2NH$, Me_2AsNH_2, Me_2AsNMe_2 and $MeAs(NMe_2)_2$ have been explored.[66] Reactions between $[M(CO)_5(PPh_2H)]$ (M = Cr, Mo, or W) and the sulphonyl chlorides RSO_2Cl (R = Me or p-tol) give the bridging diphosphoxane complexes $[\{M(CO)_5\}_2(\mu-Ph_2POPPh_2)]$ via non-isolable $[M(CO)_5\{Ph_2POS(O)R\}]$ intermediates. Base hydrolysis of these products using Et_3N leads to the formation of the complexes $[Et_3NH][M(CO)_5(Ph_2PO)]$ stabilized by very strong N-H---O-P cation - anion interactions.[67] The tridentate macrocycles (10) react with $M(CO)_6$ to form fac-$[M(CO)_3(macrocycle)]$ (M = Mo, X = PH or NMe; M = W, X = PPh) complexes.[68] The complexes mer-$[Cr(CO)_3(L-L)_2]$ (L-L = dppm, or $(Ph_2P)_2NR$; R = H or Me), synthesised from $[Cr(CO)_3(\eta-chpt)]$, contain both chelating and unidentate L-L ligands. The electro-chemical and chemical oxidation of these complexes has been investigated.[69] The ten-membered ring complexes $[(OC)_4M(\mu-dppe)_2M'(CO)_4]$ result from reactions of cis-$[M(CO)_4(Ph_2PH)_2]$ with cis-$[M'(CO)_4(Ph_2PCH=CH_2)_2]$ (M, M' = Cr, Mo, or W) in the presence of $KOBu^t$ as catalyst.[70] The complexes $[M(CO)_4\{(Ph_2P)_2C=CH_2\}]$ undergo Michael-type additions with a range of amines, hydrazines, acetylides and dichlorocarbenes producing functionalized diphosphine complexes.[71]

Pulsed-laser induced photochemical reactions of $[W(CO)_4\{Bu^tS(CH_2)_nSBu^t\}]$ (n = 2 or 3) with phosphines and phosphites proceed by W-S bond cleavage to give five-coordinate ring-opened intermediates. The rates of ring closure and of reaction of the intermediates with phosphines and phosphites have been determined. Pulsed-laser flash photolysis has also been used to cleave the M-N bond of various $[M(CO)_4(chelate)]$ complexes [M = Cr, Mo, or W; chelate = $R_2N(CH_2)_nPPh_2$; R = Me or Et, n = 2; R = Me, n = 3]. The rate of subsequent ring closure for the square pyramidal intermediates depends on the identity of M and R and ring size.[72] The reaction of $[(OC)_4\overline{M(\mu-PEt_2)_2M}(CO)_4]$ (M = Cr or W) with PBu^n_3 occurs by a carbonyl ligand dissociative mechanism.[73]

Rather less attention has been paid to complexes with Group VI donor ligands. The reaction of $Cr(CO)_6$ with Et_4NOH in aqueous THF solution affords $[Et_4N]_4[Cr_4(CO)_{12}(\mu_3-OH)_4]$, the cubane-like anion having alternating $Cr(CO)_3$ and μ_3-OH groups and no metal-metal bonds. The analogous phenoxide has also been isolated.[74] The bridging thio- and selenoether complexes $[M(CO)_4(\mu-ER_2)]_2$ (M = Cr or W, $ER_2 = SMe_2$; M = Cr or Mo, $ER_2 = SEt_2$; M = Cr, $ER_2 = SeEt_2$)

are formed by irradiation of solutions of [M(CO)$_5$(ER$_2$)] at -20°C.[75]
The syntheses and electrochemical behaviour of [Et$_4$N]$_2$[Mo(CO)$_4$(SR)$_2$]
(R = Ph or Bz) and [Et$_4$N]$_2$[W$_2$(CO)$_8$(SR)$_2$] (R = H, Ph or Bu^t) have
been reported.[76] Various anionic chelating ligands such as
xanthates and dithiocarbamates react with Mo(CO)$_6$ and
[Mo(CO)$_4$(bipy)] to yield [Mo(CO)$_4$(chelate)]$^-$ anions isolated as
Ph$_4$P$^+$ salts.[77] The structure and some reactions of
[Et$_4$N]$_2$[Cr(CO)$_3$(S$_2$C$_6$H$_4$)], which has a trigonal bipyramidal 16-
electron anion, have been reported. A number of other thioether
and thiolate complexes have been characterised e.g.
[Mo(CO)$_3${S(CH$_2$)$_2$S(CH$_2$)$_2$S}]$^{2-}$, [Mo(CO)$_3$(ttcn)] (ttcn = 1,4,7-
trithiacyclononane), [(OC)$_4$MoSC$_6$H$_4$S(CH$_2$)$_2$SC$_6$H$_4$SMo(CO)$_4$]$^{2-}$ and
[(OC)$_2$Mo(MeSC$_6$H$_4$S)$_2$Mo(CO)$_4$].[78] The S-bonded thione complexes
[M(CO)$_5$L] [M = Cr, Mo, or W; L = MeN(CH$_2$)$_2$N(Me)C=S] prepared from
M(CO)$_6$ using thermal or photochemical methods, react with
Et$_2$NC(S)SSC(S)NEt$_2$, (Et$_4$dts), to give [M(CO)$_4$(Et$_4$dts)] complexes.[79]
The complexes [M(CO)$_5$(Me$_3$SiCH$_2$EECH$_2$SiMe$_3$)] and
[M(CO)$_5$(Me$_2$CCH$_2$EECH$_2$)] (M = Cr, Mo, or W; E = S or Se) have been
prepared. The variable-temperature ^{1}H n.m.r. spectra indicate the
presence of two dynamic processes, pyramidal chalcogen inversion
and a 1,2-metal shift between adjacent chalcogen atoms. Pyramidal
inversion of the coordinated sulphur atoms in cis and trans
configurational isomers of [Cr(CO)$_4${(MeS)$_2$CHCH(SMe)$_2$}] has also
been examined.[80] The first stable phosphine telluride complexes
[M(CO)$_5$(Te=PBu$^t_3$)] (M = Cr, Mo, or W) result from photolysis of
M(CO)$_6$ in the presence of Bu$^t_3$P=Te. The synthesis and structure
of [Mo(CO)$_4$(TePh)]$_2$ has also been reported.[81]

4.2 Carbonyl Complexes of Mo^{II} and W^{II}. — Complexes of the type
[MoBr$_2$(CO)$_x$L] containing the new phosphines Ph$_2$PCH$_2$Si(Me)$_2$CH=CH$_2$
(x = 3 and 4) and (Ph$_2$PCH$_2$)$_2$SiMe$_2$ (x = 3) have been prepared.[82]
Improved preparations of [M(CO)$_x$(S$_2$CNR$_2$)$_2$] (x = 2 or 3; R = Me or
Et) utilise [M(CO)$_5$I]$^-$ or [M(CO)$_4$I$_3$]$^-$ as starting materials.
Addition of F$^-$ or N$_3^-$ to [M(CO)$_2$(S$_2$CNR$_2$)$_2$] affords the seven-
coordinate anions [M(CO)$_2$(S$_2$CNR$_2$)$_2$X]$^-$ (X = F or N$_3$).[83] A number
of Mo^{II} dithiolate complexes e.g. [Mo(CO)$_3$(PMe$_3$)$_2$(S$_2$C$_6$H$_4$)] and
[Mo(CO)$_2$(PMe$_3$){SC$_6$H$_4$S(CH$_2$)$_3$SC$_6$H$_4$S}] have been prepared from
[MoCl$_2$(CO)$_2$(PMe$_3$)$_3$].[84] The five-coordinate anions [M(CO)$_2$(SR)$_3$]$^-$
(M = Mo or W; R = 2,4,6-Pr$^i_3$C$_6$H$_2$, 2,4,6-Me$_3$C$_6$H$_2$, 2,6-Pr$^i_2$C$_6$H$_3$ or
C$_6$F$_5$) result from reactions of [MBr$_2$(CO)$_4$] with sterically hindered

thiolate anions. However, the reaction between $[MoBr_2(CO)_4]$ and $[2,6-Ph_2C_6H_3S]^-$ leads to an alternative 16-electron thiolate product $[Mo(CO)(SC_6H_3Ph_2)\{(\eta^6-Ph)PhC_6H_3S\}]$ containing a chelate η^6-arene-thiolate anion.[85]

4.3 Cyclopentadienyl and Arene Complexes.

Reactions of $[(\eta-C_5H_4Me)Cr(CO)_3]_2$ with PR_3 (R = Ph, OMe, or OEt) give $[(\eta-C_5H_4Me)Cr(CO)_2(PR_3)]_2$ dimers as products.[86] The ligand-bridged dimers $[\{CpMo(CO)_2\}_2(\mu-H)(\mu-ER_2)]$ ($ER_2 = Bu^t_2P$ or Me_2As) arise from reactions of $[CpMo(CO)_3]_2$ with Bu^t_2PH or Me_2AsH.[87] The binuclear complexes $[(\eta-C_5Me_5)_2Mo_2(CO)_3L]$, [L = $P(OMe)_3$ or $\overline{PhPO(CH_2)_2NH(CH_2)_2O}$ - the open tautomeric form of the bicyclic aminophosphorane $\overline{PhP(H)(OCH_2CH_2)_2N}$], are the first examples of pentamethylcyclopentadienyl molybdenum dimers containing phosphorus donor ligands, as well as being the first unsymmetrically substituted derivatives of $[(\eta-C_5Me_5)Mo(CO)_2]_2$. Their isolation contrasts with the previously reported inertness of this metal-metal bonded carbonyl towards phosphorus donor ligands.[88] The bicyclic $\overline{P[OC(Me)_2CH_2]_2N}$ reacts with $[CpMo(CO)_3R]$ (R = H, Cl, or Me) to give P-bonded substitution or insertion products of the type $[R(Cp)(CO)_2Mo\{\overline{P[OC(R')_2CH_2]_2N}\}]$ (R' = H, Cl, or COMe). Some acyclic analogues e.g. $[CpMoCl(CO)_{3-x}L_x]$ [L = $(EtO)_2PNMe_2$; x = 1 or 2] have also been characterised.[89] Dimeric $[CpMo(CO)_2(\mu-EPh)]_2$, formed in reactions of $[CpMo(CO)_2]_2$ with Ph_2E_2 (E = S, Se, or Te), do not contain metal-metal bonds. Spectroscopic measurements reveal the presence of <u>cis</u> and <u>trans</u> isomers each with <u>syn</u> or <u>anti</u> orientations of the phenyl groups. Reactions of $[CpMo(CO)_3Cl]$ with Ph_2E_2 yield $[\{CpMo(CO)(SPh)\}_2(\mu-Cl)(\mu-SPh)]$ and $[CpMo(CO)Cl(\mu-EPh)]_2$ (E = Se or Te).[90] The S,S'-chelate complexes $[CpM(CO)_2\{S_2P(OR)_2\}]$ (M = Mo or W; R = Me, Et, or Pr^i) result from reactions of $[CpM(CO)_3Cl]$ with O,O'-dialkyldithiophosphoric acids $HSP(S)(OR)_2$.[91] The weakly coordinated fluoro-ligands X in $[CpM(CO)_2LX]$, [M = Mo or W; L = CO, PPh_3 or $P(OPh)_3$; X = FBF_3, FPF_5, $FAsF_5$ or $FSbF_5$] are easily replaced by O- or S-donor ligands (e.g. L' = H_2O, H_2S, Me_2CO, EtOH, Et_2O, Ph_2S, or Ph_2SO) giving the cationic complexes $[CpM(CO)_2LL']X$.[92] The reactions between $[CpM(CO)_3H]$ and R_2EER_2 or RE'E'R (M = Cr, Mo, or W; R = Me or CF_3; E = P or As; E' = S or Se) yield the complexes $[CpM(CO)_3(ER_2)]$ and $[CpM(CO)_3(E'R)]$. Reactions of $[CpM(CO)_3H]$ with the diphosphines Ph_2PPPh_2, $Ph(CN)PP(CN)Ph$ and $(CN)_2PP(CN)_2$

proceed in an analogous manner yielding [CpM(CO)$_3$(PRR')] and HPRR' (R,R' = Ph or CN) products. The tendency to form binuclear [CpM(CO)$_2$(μ-X)]$_2$ (X = ER$_2$, E'R or PRR') complexes in these reactions has also been assessed.[93]

Photolysis of a variety of [(η^6-arene)Cr(CO)$_3$] complexes in the presence of MeCN or p-CH$_3$C$_6$H$_4$CN affords [(η^6-arene)Cr(CO)$_2$(NCR)] products. Solutions of the MeCN complexes yield, after warming or u.v. photolysis, the novel products [(η^6-arene)$_2$Cr$_2$(μ-CO)$_3$] containing a formal Cr≡Cr triple bond bridged by 3 carbonyl ligands.[94] The reaction between [(η^6-PhCO$_2$Me)Cr(CO)$_2$(CS)] and excess P(OMe)$_3$ in toluene unexpectedly produces mer-[Cr(CO)$_2$(CS){P(OMe)$_3$}$_3$] with two mutually trans carbonyl ligands.[95] A kinetic study of the formation of fac-[Cr(CO)$_3${P(OMe)$_3$}$_3$] from [(η^6-naphthalene)Cr(CO)$_3$] and P(OMe)$_3$ has been published.[96] The heats of reaction of [(η^6-PhMe)Mo(CO)$_3$] with phosphines and phosphites have been measured by solution calorimetry.[97] One carbonyl ligand of [(η^6-arene)Cr(CO)$_3$], (arene = PhMe or PhCO$_2$Me), is displaced by photolysis in the presence of the thioethers (PhCH$_2$)$_2$S or 1,3-dithiane. In contrast, substitution of the η^6-arene ring occurs when thioethers are replaced by the dithioester PhC(S)SMe or the trithiocarbonate S(CH$_2$)$_2$SC=S, the products being of the type [Cr(CO)$_5$L].[98]

A kinetic study of the reaction of [(η-C$_7$H$_7$)W(CO)$_3$]$^+$ with PPh$_3$ suggests that an initial rapid pre-equilibrium formation of a π-complex is followed by a rate-determining attack by a second PPh$_3$ at the metal. Subsequent attack of a third PPh$_3$ at the metal leads to displacement of the tropylium ring and the formation of [W(CO)$_3$(PPh$_3$)$_3$].[99]

5 Group VII

5.1 Carbonyl, Carbonyl Halide and Related Complexes.—
The photochemistry of M$_2$(CO)$_{10}$ (M = Mn or Re) has been investigated by laser flash photolysis. Two primary photoprocesses are observed; metal-metal bond cleavage producing ·M(CO)$_5$ radicals, and carbonyl ligand dissociation producing M$_2$(CO)$_9$ species. The second process is more dominant for Re$_2$(CO)$_{10}$ than for Mn$_2$(CO)$_{10}$.[100] Substitution of ·Mn(CO)$_5$ radicals by PPh$_3$ or AsPh$_3$ occurs by an associative process.[101] The kinetics of the thermal reactions of [Mn$_2$(CO)$_8$(PPh$_3$)$_2$] with nucleophiles such as O$_2$, NO, P(OR)$_3$, CO and

PBu$^n_3$ (R = Et or Ph) have been explored.^{102} The palladium-catalysed reactions between Re$_2$(CO)$_{10}$ and a selection of phosphines or P(OMe)$_3$ in xylene afford [Re$_2$(CO)$_{10-n}$(PR$_3$)$_n$] (n = 1 or 2) as the major products. These complexes can also be prepared from Re$_2$(CO)$_{10}$ and PR$_3$ in refluxing dichloromethane in the presence of Me$_3$NO but substantial amounts of fac-[ReCl(CO)$_3$(PR$_3$)$_2$] are also formed.^{103} The photolysis of hexane solutions of Re$_2$(CO)$_{10}$ in the presence of P(OPh)$_3$ under vacuum gives four new products, 1-ax-2',4'-di-eq-[Re$_2$(CO)$_7$L$_3$], 1,1'-di-ax-2,2'-di-eq- and 2,2',4,4'-tetra-eq-[Re$_2$(CO)$_6$L$_4$] and 1,1'-di-ax,2,2',4,4'-tetra-eq-[Re$_2$(CO)$_4$L$_6$] as well as mer-[ReH(CO)$_3$L$_2$].^{104} The reaction between Mn$_2$(CO)$_{10}$ and dmpm gives two products, separable by Soxhlet extraction. They are symmetrical mer,mer-[Mn$_2$(CO)$_6$(dmpm)$_2$] and the novel mer,fac-isomer which is the first example of a cis,trans-bis(dmpm)-bridged complex.^{105} In [Mn$_2$(CO)$_6$(dppm)$_2$] two transoid dppm ligands bridge the metal atoms.^{106} Thermal, photochemical and Me$_3$NO-induced reactions of [Re$_2$(CO)$_8$(μ-L-L)] (L-L = dppm or dmpm) with MeCN or additional L-L have been studied. Products isolated include [Re$_2$(CO)$_7$(μ-L-L)(NCMe)] and [Re$_2$(CO)$_6$(μ-L-L)$_2$]. Subsequent reactions with alkynes lead to complexes such as [Re$_2$(CO)$_4$(μ-H)(μ-C≡CPh)(μ-L-L)$_2$]. The photochemical reactions of [Re$_2$(CO)$_9$L], 1,2-[Re$_2$(CO)$_8$L$_2$] (L = PMe$_3$ or PPh$_3$) and [Re$_2$(CO)$_8$(μ-L-L)] (L-L = dmpm, dppe, or dmpe) with alkenes have also been explored. Products of the type [Re$_2$(CO)$_{8-n}$(μ-H)(μ-CH=CH$_2$)L$_n$] (n = 0-2) are most commonly formed.^{107} The tetrahydrothiophene complexes [M$_2$(CO)$_9$(C$_4$H$_8$S)] and [M$_2$(CO)$_8$(C$_4$H$_8$S)] (M = Mn and Re) have been characterized, the ligand symmetrically bridging two metal-metal bonded M(CO)$_4$ units in the octacarbonyls.^{108} Thermal reactions between M$_2$(CO)$_{10}$ (M = Mn or Re) and o-Ph$_2$PC$_6$H$_4$CH$_2$Si(H)Me$_2$ produce the mononuclear chelate complexes [M(CO)$_4${o-Ph$_2$PC$_6$H$_4$CH$_2$SiMe$_2$}]. However, a u.v.-induced reaction merely leads to the formation of binuclear [Mn$_2$(CO)$_9${o-Ph$_2$PC$_6$H$_4$CH$_2$Si(H)Me$_2$}].^{109} The pyrolysis of a 1:5 mixture of Ni(PMeCl$_2$)$_4$ and Re$_2$(CO)$_{10}$ at 175° under argon affords the novel complex (11) in which the 8-electron donor (MeP)$_3$ ligand forms a four-membered ReP$_3$ ring. More extensive pyrolysis of (11) in the presence of Re$_2$(CO)$_{10}$ gives two new cluster compounds, [Re$_6$(CO)$_{18}$(μ$_4$-PMe)$_3$] containing a trigonal prismatic Re$_6$ array with three capping methylphosphinidene ligands and the unique [Re$_5$(CO)$_{14}$(μ$_4$-PMe)(μ-PMe$_2$){μ$_3$-P[Re(CO)$_5$]}].^{110}

The reaction between $Na[Mn(CO)_5]$ and the cyclodiphosphazane $ClP(NBu^t)P(NBu^t)Cl$ proceeds via P-P bond forming and P-N bond breaking and re-forming processes to give the coupled ligand $P_4(NBu^t)_4$ which is coordinated as shown in structure (12).[111] The favoured products of the reactions of $Me_2P(CH_2)_2SiX_3$ (X = Cl, F, or OMe) with $[Mn(CO)_5]^-$ are the chelate complexes $[Mn(CO)_4\{Me_2P(CH_2)_2SiX_2\}]$ because of the stability of the Mn-Si bond.[112] Elimination of Me_3SiCl on reaction of $Mn(CO)_5Cl$ with Me_3SiPH_2 allows formation of $[Mn(CO)_4(\mu-PH_2)]_2$ and $[Mn(CO)_4(\mu-PH_2)]_3$ containing four- and six-membered rings, respectively.[113] The complexes $[MBr(CO)_4\{Ph_2P(CH_2)_2Cl\}]$ (M = Mn or Re; n = 5 or 6), prepared from $MBr(CO)_5$ and $Ph_2P(CH_2)_nCl$, have been subjected to reductive cyclo-elimination by reaction with sodium amalgam. Only the seven-membered ring compounds $[M(CO)_4\{Ph_2P(CH_2)_4CH_2\}]$ could be isolated in reasonable yields.[114] The metallophosphorane $[\underline{o}-\overline{OC_6H_4OP}Mn(CO)_5]$ reacts with PR_3 ligands to give the products $[\underline{o}-\overline{OC_6H_4OP}Mn(CO)_4(PR_3)]$ (R = Ph, cis and trans isomers; R = OMe or OPh cis-isomers only). There is no indication of insertion of CO into the P-Mn bond.[115] Routes to the chelating phosphino-alkyl and phosphino-acyl ligand complexes $[\overline{Mn(CO)_2(Ph_2PCH_2CH_2)}(dppm)]$ and $[Mn(CO)_2\{Ph_2PCH_2CH_2C(O)\}(dppm)]$ have been published.[116] Reaction of $ReBr(CO)_5$ with $Li[W(CO)_5(PPh_2)]$ gives $[(OC)_5\overline{W(\mu-PPh_2)Re}(CO)_5]$ but decarbonylation on work-up leaves $[(OC)_5\overline{W(\mu-PPh_2)Re}(CO)_4]$ containing a donor-acceptor W-Re bond. The latter complex reacts with ligands to give $[(OC)_5W(\mu-PPh_2)Re(CO)_4L]$ species (L = MeCN, $PMePh_2$ or PPh_2H). Thermolysis of the PPh_2H derivative regenerates the W-Re bond by the formation of $[(OC)_4\overline{W(\mu-PPh_2)_2Re}(CO)_3H]$.[117]

The complexes $[MBr(CO)_{5-n}(ER_2)_n]$ (M = Mn or Re; E = S, Se, or Te; R = alkyl; n = 1 or 2) have been spectroscopically detected in solution and in some cases isolated. The kinetics of displacement of EMe_2 from $[ReBr(CO)_3(EMe_2)_2]$ by CO have been investigated.[118] The dinuclear complexes $[Et_4N][Mn_2(CO)_6(\mu-SR)_3]$ (R = Ph, Me, or Bu^t) have been prepared from $Mn_2(CO)_{10}$, $[Mn(CO)_4Br_2]^-$, $[Mn_2(CO)_6(\mu-Br)_3]^-$ or $[Mn(CO)_4(\mu-Br)]_2$. Subsequent reactions with $[Me_3O][BF_4]$ afford tetranuclear $[Mn(CO)_3(\mu_3-SR)]_4$ or if in the presence of CO, $[Mn(CO)_4(\mu-SR)]_2$. Reactions with $[Me_3O][BF_4]$ in the presence of PMe_3 lead to cis-$[Mn(CO)_3(PMe_3)(\mu-SR)]_2$ and $[Mn_2(CO)_4(\mu-CO)(\mu-SR)_2(PMe_3)_2]$. Electrochemical and chemical oxidation of $[Et_4N][Mn_2(CO)_6(\mu-SPh)_3]$ occurs by two sequential one-electron steps. In the intermediate

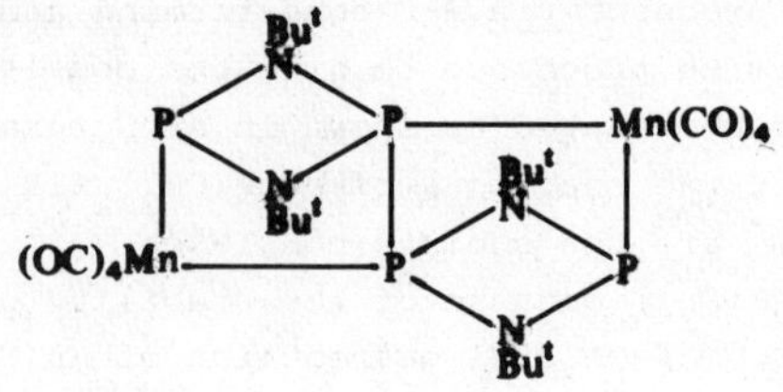

(12)

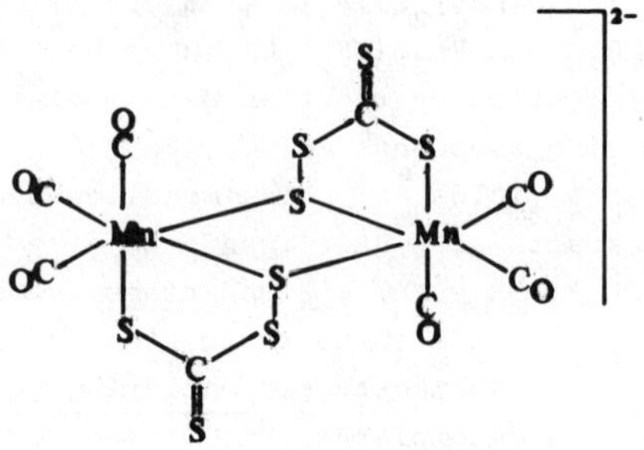

(13) [BzNEt$_3$]$^+$ salt

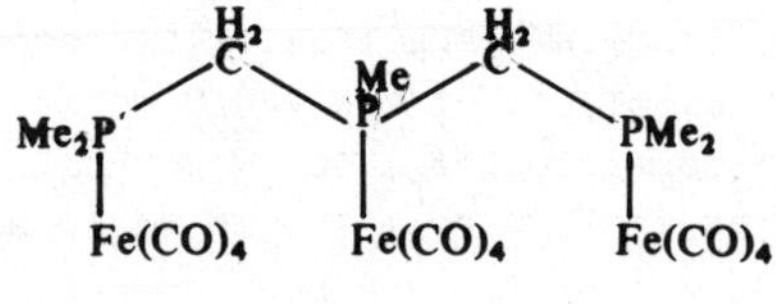

(14)

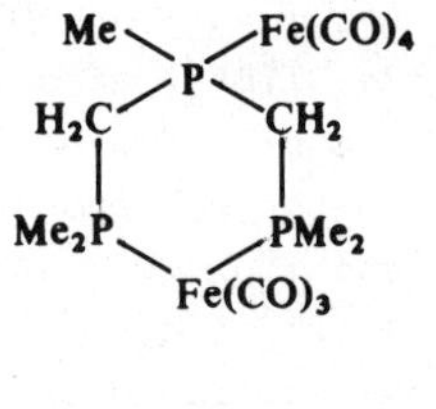

(15)

[$Mn_2(CO)_6(\mu-SPh)_3$] the unpaired electron is completely delocalized over the two metal centres.[119] The ($\underline{E}$)- and ($\underline{Z}$)-forms of [$\overline{ReBr(CO)_3\{Ph_2PC(S)NRR'\}}$] (R = R' = Me; R = Me, R' = Ph) and [$\overline{ReBr(CO)_3\{Ph_2PC(NR)SMe\}}$] (R = Me or Ph) are the products of reactions of $ReBr(CO)_5$ with chelating phosphinothioformamides and -thioformimidate ligands. The phosphinothioformimidato-complex [$\overline{Mn(CO)_4\{SC(NMe)PPh_2\}}$] reacts with $P(OMe)_3$, $P(OPh)_3$, PEt_3 or $PMePh_2$ to give $\underline{fac}$-[$\overline{Mn(CO)_3\{SC(NMe)PPh_2\}(PR_3)}$] species as kinetic products which rearrange on warming into thermodynamically more stable $\underline{mer}$-forms. The isomerisation is controlled by steric factors because reactions with PPh_3 or PCy_3 give exclusively $\underline{mer}$-isomers whereas those with Ph_2PCl or Ph_2PH give $\underline{fac}$-isomers. Sterically less crowded rhenium analogues also prefer the $\underline{fac}$-configuration.[120] The perthiocarbonate complex (13) is the unexpected product of a phase-transfer catalysed reaction between $MnBr(CO)_5$ and elemental sulphur using [$BzNEt_3$]Cl as the phase-transfer agent.[121]

Syntheses of the fluorosulphate and triflate complexes [$M(CO)_5X$] (M = Mn or Re; X = SO_3F or O_3SCF_3) have been reported. The triflates are attacked by nucleophiles yielding [$Mn(CO)_5L$][O_3SCF_3] or [$Mn(CO)_3L_3$][O_3SCF_3] products.[122] The monocarbonyl cations $\underline{trans}$-[$Mn(CO)(chelate)_2L$]X [chelate = dppm or dppe; L = RCN, $NC(CH_2)_nCN$, or $\underline{o}$-$C_6H_4(CN)_2$; R = Me, Et, Ph, or Bz; n = 1 or 2; X = ClO_4 or PF_6] and $\underline{trans}$-[{$Mn(CO)(chelate)_2$}$_2${μ-$NC(CH_2)_nCN$}]X_2 have been prepared from $\underline{trans}$-[$Mn(CO)(chelate)_2Br$].[123] The anions $P_3O_9^{3-}$ and $\underline{cis}$-[$Nb_2W_4O_{19}$]$^{4-}$ act as tridentate oxygen donor ligands on reaction with $\underline{fac}$-[$M(CO)_3(NCMe)_3$]$^+$ cations (M = Mn or Re). The products are [$M(CO)_3(P_3O_9)$]$^{2-}$, with a structure derived from that of P_4O_{10} by replacing one P=O unit by a $M(CO)_3$ group, and [$M(CO)_3(Nb_2W_4O_{19})$]$^{3-}$ in which the heteropolyanion coordinates by three adjacent bridging oxygens.[124]

5.2 Cyclopentadienyl and Other Complexes.— The three products [(η-C_5H_4Me)$Mn(CO)_2(NC_5H_4CN)$], [(η-C_5H_4Me)$Mn(CO)_2(NCC_5H_4N)$] and [{(η-C_5H_4Me)$Mn(CO)_2$}$_2$(μ-NCC_5H_4N)], separable by low temperature chromatography, result from the reaction of [(η-C_5H_4Me)$Mn(CO)_2(THF)$] with 4-cyanopyridine.[125] The deep blue paramagnetic complexes [$CpMn(CO)_2(NHR)$]$^{\cdot}$ (R = Ph, $\underline{o}$-, $\underline{m}$-, or $\underline{p}$-tol, 2,6-, 3,5-, 3,4- or 2,4-xylyl, or $\underline{m}$-XC_6H_4; X = Cl, Br, or I) have been regarded by Sellmann $\underline{et\ al.}$ as aminyl-radical species. However, Kaim $\underline{et\ al.}$ using e.s.r. evidence formulate such complexes

with R = p-Me$_2$NC$_6$H$_4$, p-tol, or 4-pyridyl as well as
[CpMn(CO)$_2$(SBu^t)] and [CpMn(CO)$_2$(SePh)] as low-spin d^5 Mn(II)
complexes rather than stabilized radical species. These workers
suggest that of the compounds examined only [CpMn(CO)$_2$(4-NCC$_5$H$_4$N)]$^{\overline{\cdot}}$
and [{CpMn(CO)$_2$}$_2$(μ-L-L)]$^{\overline{\cdot}}$ [L-L = 4,4'-bipy or 1,4-(NC)$_2$C$_6$H$_4$] can be
regarded as a combination of a diamagnetic d^6 manganese(I) fragment
with a radical anion ligand.   The complexes [(η-C$_5$Me$_5$)Mn(CO)$_2$(SR)]$^{\cdot}$
(R = Bu^t or 2-adamantyl) and [(η-C$_5$Me$_5$)Mn(CO)$_2$(SePh)]$^{\cdot}$ have also
been formulated as radical species.[126]

The very unstable sulphur monoxide, generated by thermolysis
of thiirane $\underline{S}$-oxide, has been trapped in [{CpMn(CO)$_2$}$_2$(μ-SO)].[127]
The complexes [(η-C$_5$H$_4$Me)Mn(CO)$_2$(EMe$_2$)] (E = S or Se) have been
isolated and their structures determined. In solution the Me$_2$S
complex is in equilibrium with [{(η-C$_5$H$_4$Me)Mn(CO)$_2$}$_2$(μ-SMe$_2$)] and
free Me$_2$S.[118]

Thermolysis of the rhenacyclopentane [CpRe(CO)$_2${(CH$_2$)$_3$CH$_2$}] in
the presence of PR$_3$ (R = Me or Ph) affords methylcyclopropane and
[CpRe(CO)$_2$(PR$_3$)], the reactions being first-order in rhenacyclopentane and independent of phosphine concentration. It is
suggested that a $\eta^5 \rightleftharpoons \eta^3$ cyclopentadienyl ring slip occurs to
generate a vacant coordination site.[128] The η^5-indenyl complex
[(η^5-C$_9$H$_7$)Re(CO)$_3$] reacts with ligands yielding η^1-indenyl
derivatives $\underline{fac}$-[(η^1-C$_9$H$_7$)Re(CO)$_3$L$_2$] (L = PMe$_3$ or PBu^n$_3$; L$_2$ =
bipy).[129]   A new synthesis of [(η^5-C$_4$H$_4$N)Mn(CO)$_3$] and the first
preparations of [(η^5-C$_8$H$_6$N)Mn(CO)$_3$] (C$_8$H$_6$N = indolyl or 1-
pyrindinyl) have been reported.   Kinetic studies of their
reactions with PR$_3$ (R = Ph, Cy, Bu^n, or OEt) are indicative of
second-order processes.[130] The Me$_3$NO-induced substitution of a
carbonyl ligand of [(η-C$_4$H$_4$N)Mn(CO)$_3$] by PPh$_3$ has been reported.
The pyrrolyl ring nitrogen of the product possesses considerable
σ-donor capacity as illustrated by the formation of
[{(η-C$_4$H$_4$N)Mn(CO)$_2$(PPh$_3$)}$_2$PdCl$_2$].[131] The cyclohexadienyl complexes
[(η^5-C$_6$H$_6$Ph)Mn(CO)$_{3-n}$L$_n$] (L = PPh$_3$, n = 1; L$_2$ = dppe; n = 2) have
been prepared by photolysis of the tricarbonyl in the presence of a
phosphine.[132]   Kinetic data have been reported for carbonyl ligand
substitution reactions of [(η^3-C$_3$H$_4$X)Mn(CO)$_4$] using a variety of
phosphines (X = substituent in the 1- or 2-position of the allyl
e.g. H, Me, Ph, Bu^t or Cl). The substitutions occur by a
dissociative process, the rates being first-order in substrate
concentration and zero-order in entering nucleophile
concentration.[133]

6 Group VIII : Iron, Ruthenium, and Osmium

6.1 Iron, Ruthenium, and Osmium Carbonyl Complexes.— A new route to [Fe(CO)$_{5-n}${P(OR)$_3$}$_n$] complexes (R = alkyl, n = 2 or 3) is the atmospheric pressure and room temperature carbonylation of anhydrous FeCl$_2$ using manganese as reducing agent.[134] Many of the trigonal bipyramidal [M(CO)$_4$L] complexes, [M = Fe, Ru, or Os, L = PPh$_3$, AsPh$_3$, SbPh$_3$, PMe$_3$, or P(OCH$_2$)$_3$CMe; M = Ru or Os, L = SbMe$_3$] exhibit axial-equatorial isomerism in solution. The tendency to give the less-common equatorial isomer follows the orders Ru > Os > Fe, Sb > As > P, Ph > Me and P(OCH$_2$)$_3$CMe > PMe$_3$, PPh$_3$. In the solid state, [Ru(CO)$_4$(AsPh$_3$)] and [Ru(CO)$_4$(SbMe$_3$)] form axial isomers whereas in [Os(CO)$_4$(SbPh$_3$)] the stibine takes up an equatorial position.[135] The complex [Fe(CO)$_4$(SbBu$^t_3$)], prepared from Fe(CO)$_5$ and cyclo-Sb$_4$Bu$^t_4$, is the axial isomer.[136] Preparations of trigonal bipyramidal cis-[Fe(CO)$_2$(η^1-dmpm)(η^2-dmpm)], square pyramidal [Fe(CO)(η^2-dmpm)$_2$], and binuclear [Fe$_2$(CO)$_{8-2n}$(μ-CO)(μ-dmpm)$_n$] (n = 1 or 2) have been published.[62,137] The S-bonded isomers [(C$_3$F$_7$)Fe(CO)$_3$I(S=PHR$_2$)] (R = alkyl) have been obtained by carbonyl ligand substitution in [(C$_3$F$_7$)Fe(CO)$_4$I]. Elimination of HI from these products gives the η^2-thiophosphinito-complexes [(C$_3$F$_7$)Fe(CO)$_3$(S=PR$_2$)].[138] The aminomethylene phosphine (Me$_3$Si)$_2$NP=CHSiMe$_3$ reacts with Fe$_2$(CO)$_9$ to afford a P-bonded [Fe(CO)$_4$L] complex.[139] The novel compounds [Fe$_2$(CO)$_6$(Bu^{t}N=PBu^t)], containing a distorted tetrahedral Fe$_2$PN core and a six-electron donor iminophosphine ligand, and [Fe$_2$(CO)$_7${Bu^{t}PN(Bu^t)CO}], containing both Fe$_2$P and FePNC rings, result from the reaction of the iminophosphine with Fe$_2$(CO)$_9$.[140] The triphosphine MeP(CH$_2$PMe$_2$)$_2$ reacts with Fe$_2$(CO)$_9$ to give complexes (14)-(17).[141] The dialkylaminodichlorophosphines R$_2$NPCl$_2$ (R = Pr^i or Cy; R$_2$N = 2,2,6,6-tetramethylpiperidino) react with Na$_2$Fe(CO)$_4$,1.5C$_4$H$_8$O$_2$ to give the unusual compounds [Fe$_2$(CO)$_6${R$_2$NPC(=O)PNR$_2$}] in which a carbonyl group has migrated from iron to phosphorus.[142] Several products arise from the reaction of Fe$_2$(CO)$_9$ with Lawesson's reagent, (p-MeOC$_6$H$_4$)P(S)SP(S)(C$_6$H$_4$OMe-p)S, including [Fe$_2$(CO)$_6${μ-S$_2$P(C$_6$H$_4$OMe-p)Fe(CO)$_4$}] which has a S-P(R)-S unit both S,S'-bridging the iron-iron bond of a Fe$_2$(CO)$_6$ moiety and P-bonded to a Fe(CO)$_4$ group.[143] The reactions of 1,2,3-triphenyltriphosphaindane, o-C$_6$H$_4$P(Ph)P(Ph)PPh, with Fe$_2$(CO)$_9$ and Fe$_3$(CO)$_{12}$ afford at least five compounds, two of which have previously been reported. The three new products each contain P$_3$Fe$_3$ clusters.[144]

(16)

(17)

(18)

The kinetics of the substitution reactions of $M_3(CO)_{12}$, (M = Fe, Ru, or Os) with monodentate phosphines and phosphites, particularly PBu^n_3, have been studied. Substitution, leading to $[M_3(CO)_{12-n}L_n]$ (n = 1-3) products, and cluster fragmentation processes are observed. Whereas the former process is ligand-independent, the latter process follows a bimolecular pathway assigned the mechanistic designation $F_N 2$.[145] The final product of the reaction between $Ru_3(CO)_{12}$ and dppm is $[Ru_3(CO)_8(\mu\text{-dppm})_2]$, with $[Ru_3(CO)_{10}(\mu\text{-dppm})]$ and $[Ru_3(CO)_9(\mu\text{-dppm})(\eta^1\text{-dppm})]$ as intermediates.[146] The flash photolysis of $Ru_3(CO)_{12}$ in the presence of $P(OMe)_3$ or PPh_3 initiates competing substitution and fragmentation reactions. The intermediate $[(OC)_4\overline{Ru(\mu\text{-CO})Ru(CO)_3}Ru(CO)_4]$, an isomer of $Ru_3(CO)_{12}$ with a vacant coordination site at one metal centre, may be generated on photolysis.[147]

Among the nitrogen donor ligand complexes that have been isolated from reactions with $Ru_3(CO)_{12}$ or $[Ru_3(CO)_{12-n}(NCMe)_n]$ (n = 1 or 2) are $[Ru_3(CO)_{10}L_2]$ (L_2 = bipy or pyridazine) and the ortho-metallated clusters $[HRu_3(CO)_{10}L]$ where LH = pyridine, 2-, 3-, or 4-methylpyridine, quinoline, isoquinoline, pyrimidine, or pyrazine.[148]

Isomeric mixtures of $[Os_3(CO)_{12-n}(PMe_2Ph)_n]$ (n = 2 or 3) result from the reactions of $[Os_3(CO)_{10}L_2]$ with the phosphine (L_2 = η^4-s-cis-C_4H_6, μ-s-trans-C_4H_6, or 2 MeCN). The bis-phosphine product can be separated into non-interconvertible 1,1- and 1,2-isomers and the tris-phosphine product is separable into non-interconvertible 1,2,3- and 1,1,2-isomers. Intramolecular phosphine exchange mechanisms in the 1,2-isomer of $[Os_3(CO)_{10}(PMe_2Ph)_2]$ and the 1,2,3-isomers of $[Os_3(CO)_9\{P(OMe)_3\}L_2]$ (L = PMe_2Ph or PPh_3) have also been studied.[149] The complex $[Os_3(CO)_7\{P(OMe)_3\}_5]$ is the most highly phosphite-substituted derivative of $Os_3(CO)_{12}$ yet reported.[150] The complex $[Ru_3(CO)_6(\mu\text{-CO})_2\{PPh(OMe)_2\}_4]$ is the first $Ru_3(CO)_{12}$ derivative shown to possess an $Fe_3(CO)_{12}$-type structure.[151] The reaction between $Ru_3(CO)_{12}$ and $HC(PPh_2)_3$ is complex, yielding the ligand-capped product $[Ru_3(CO)_9\{HC(PPh_2)_3\}]$ and five other complexes each containing various fragments of the triphosphine viz., $[Ru_3(CO)_9\{Ph_2PCHP(Ph)C_6H_4PPh\}]$, $[Ru_2(CO)_4(\mu\text{-CO})(\mu\text{-Ph})\text{-}\{(Ph_2P)_2C(H)PPh\}]$, $[Ru_2(CO)_4(\mu\text{-Cl})(\mu\text{-PPh}_2)(\mu\text{-dppm})]$, $[Ru_2(CO)_4(\mu\text{-H})(Ph_2PCHPPh_2)(PhPC_6H_4)]$ and $[Ru_2(CO)_4(\mu\text{-H})(Ph_2PCHPPh_2)\text{-}\{PhPC_6H_4C(O)\}]$. In the latter two complexes the $Ph_2PCHPPh_2$ moieties bridge Ru-Ru bonds and the central carbon is also bonded to one ruthenium generating a RuPC ring.[152] The complexes [$M_3(CO)_{10}$(dppm)] (M' = Ru or Os) undergo selective substitution by phosphines to give [$M_3(CO)_9$(dppm)L], [L = PPh_3 or $Ph_2P(CH_2)_2Si(OEt)_3$], in which L is coordinated to the previously unsubstituted $M(CO)_4$ group.[153] Preparations of the isostructural compounds [$M_3(CO)_8(\mu-H)_2(\mu-PPh_2)_2$] (M = Fe, Ru, or Os) have been described. By-products such as [$Fe(CO)_4(PPh_2H)$], [$Ru_2(CO)_6(\mu-PPh_2)_2$], and [$Os_3(CO)_{10}(\mu-H)(\mu-PPh_2)$] are also formed.[154] The radical-ion initiated reaction between $Ru_3(CO)_{12}$ and $Ph_2PC\equiv CPPh_2$ produces [{$Ru_3(CO)_{11}$}$_2(\mu-Ph_2PC\equiv CPPh_2)$] which on heating is converted into [$Ru_5(CO)_{13}(\mu-PPh_2)(\mu_5-\eta^2-C\equiv CPPh_2)$]. The seven-electron donor $-C_2PPh_2$ moiety interacts with all five metal atoms of an open Ru_5 cluster built from three edge-fused triangular Ru_3 units.[155] The dithiolato-complex [{$Fe(CO)_3$}$_2(\mu-SCH_2CH_2S)$] has been obtained by thermal fragmentation of thiirane S-oxide in the presence of $Fe_3(CO)_{12}$ and by photolysis of a C_2H_4-[$Fe(CO)_3S$]$_2$ mixture.[156]

6.2 Cyclopentadienyl Complexes.

The cations [$CpFe(CO)_{3-n}L_n$]$^+$, [n = 1-3; L = $P(OR)_3$; R = alkyl or Ph], [$CpFe(CO)_2(\eta^1-REER)$]$^+$ and [{$CpFe(CO)_2$}$_2(\mu-REER)$]$^{2+}$, (E = S, Se, or Te; R = alkyl or Ph) and the Se-bonded [$CpFe(CO)_2(Me_3Sb=Se)$]$^+$ have been prepared from [$CpFe(CO)_2(THF)$]$^+$.[157] Reaction of the disulphides $(RO)_2P(S)SSP(S)(OR)_2$ (R = Et or Pr^i) with [$Cp^*Fe(CO)_2$]$_2$ (Cp^* = $\eta-C_5H_5$, $\eta-C_5H_4Me$, or $\eta-C_5Me_5$) affords the monodentate dithiophosphate complexes [$Cp^*Fe(CO)_2\{\eta^1-SP(S)(OR)_2\}$].[158] When photolysed in the presence of the appropriate ligand [$CpFe(CO)_2(CS)$]$^+$ is converted into [$CpFe(CS)L_2$]$^+$ (L = MeCN or Me_2S). The carbonyl ligand of [$CpFe(CO)(CS)I$] can be substituted by ligands yielding [$CpFe(CS)(L)I$] complexes, (L = PEt_3, PPh_3, $AsPh_3$, $SbPh_3$, $P(OMe)_3$, or $P(OPh)_3$].[159] The heterobimetallic complexes [$CpFe(CO)_2\{PH(NPr_2^i)\}ML_n$], [$ML_n$ = $M(CO)_5$ or $CpMn(CO)_2$; M = Cr, Mo, or W] arise from reactions of Na[$CpFe(CO)_2$] with [$M(CO)_5\{PH(X)NPr_2^i\}$] (X = halide). Photolysis of these products promotes decarbonylation and metal-metal bond formation yielding complexes of the type [$CpFe(CO)(\mu-CO)\{\mu-PH(NPr_2^i)\}ML_{n-1}$].[160] The labilizing effect of the hydride ligand of [$CpRu(CO)_2H$] permits both carbonyl ligands to be replaced by dppe, dppb or (R)-$Ph_2PCH_2CH(Me)PPh_2$ to give [$CpRu$(chelate)H] products.[161]

7 Group VIII : Cobalt, Rhodium, and Iridium

7.1 Carbonyl Complexes.— The cluster complexes $[M_4(CO)_9\{HC(PPh_2)_3\}]$ (M = Co, Rh, or Ir) undergo stereoselective substitution at the apical metal atom to give $[M_4(CO)_8\{HC(PPh_2)_3\}L]$ products [M = Co, Rh, or Ir, L = PPh_3 or $Ph_2PCH_2CH_2Si(OEt)_3$; M = Co, L = $P(OMe)_3$, PMe_3, or PBu^n_3]. When L is a phosphinated silica or a phosphinated polystyrene-divinylbenzene, surface-anchored clusters of catalytic potential are produced. The electrochemical behaviour of $[M_4(CO)_9\{HC(PPh_2)_3\}]$ (M_4 = Co_4, Co_2Rh_2, or Rh_4) has been studied and the radical anion $[Co_4(CO)_9\{HC(PPh_2)_3\}]^{\bar{}}$ found to be more substitution labile than its neutral precursor.[153,162] The diphosphazane complexes $[Co_4(CO)_{12-2n}\{\mu-(RO)_2PN(Et)P(OR)_2\}_n]$ (n = 1 or 2; R = Me, Pr^i, or Ph) have been prepared from $Co_4(CO)_{12}$.[163]

Several rhodium complexes containing bridging pyrazolato-anions have been isolated including $[Rh(\mu-pz)(CO)_2]_2$, $[\{Rh(CO)_2\}_2(\mu-Cl)(\mu-pz)]$ (a rare example of a binuclear complex displaying one-dimensional stacking in the crystal lattice), $[Rh_2(CO)_4I_2(\mu-pz)_2]$, $[Rh_2(CO)_4I_4(\mu-pz)_2]$ and the metal-metal bonded rhodium(II) complex $[Rh_2(CO)_2I_2(\mu-dppm)(\mu-pz)_2]$.[164] A number of benzotriazole (btzH) and benzotriazolato (btz) complexes have also been reported viz. $[Rh(CO)_2Cl(btzH)]$, $[Rh(CO)(PPh_3)(\mu-btz)]_2$, $[Rh_2(CO)_4(\mu-btz)(\mu-N_3)]_x$, and tetranuclear $[(OC)_2ClRh\{Rh(CO)_2(\mu-btz)\}RhCl(CO)_2]$.[165] Many $[Rh(CO)L(chelate)]$ and $[RhL_2(chelate)]$ compounds have been prepared from $[Rh(CO)_2(chelate)]$ complexes. Examples include those with chelate = various β-diketonates, N-o-tolylsalicylaldiminato, picolinato, or 1-phenyl-3-methyl-4-benzoylpyrazolonato and L = PPh_3 or $P(OPh)_3$.[166]

The nature of the products from reactions between $[Rh(CO)_2Cl]_2$ and N,N-bis(diphenylphosphino)alkyl- and -arylamines depend on temperature, solvent and reactant ratios. Complexes isolated include the cationic chelates $[Rh\{(Ph_2P)_2NR\}_2][Rh(CO)_2Cl_2]$ and $[Rh(CO)\{(Ph_2P)_2NR\}_2][Rh(CO)_2Cl_2]$, the A-frame complexes $[Rh_2(\mu-CO)\{\mu-(Ph_2P)_2NR\}_2X_2]$ and $[Rh_2(CO)_2(\mu-CO)(\mu-Cl)\{\mu-(Ph_2P)_2NR\}_2]$ -$[Rh(CO)_2Cl_2]$, (R = Me, Et, Ph, or p-tol; X = halide or pseudohalide).[167] Some heterobimetallic complexes containing bridging dppm ligands have been prepared from $[Rh(CO)_2Cl]_2$ including $[XRh(\mu-CO)_2(\mu-dppm)_2OsX'_2]$ (X = Cl or Br; X' = halide or pseudohalide), $[Rh(CO)Cl(\mu-dppm)_2M(CN)_2]$ (M = Pd or Pt) and

[Rh(CO)(μ-dppm)$_2$(σ,η-C$_2$Me)Pt(C$_2$Me)] in which the σ,η-C$_2$Me ligand is σ-bonded to platinum and unsymmetrically side-on bonded to rhodium.[168] The reaction of Ph$_2$AsCH$_2$PPh$_2$, dapm, with [Rh(CO)$_2$Cl]$_2$ leads sequentially to trans-[Rh(CO)Cl(η^1-dapm)$_2$] and then to two isomers of [Rh$_2$(CO)$_2$Cl$_2$(μ-dapm)$_2$].[169] An equilibrium mixture of three isomers of [Rh$_2$(CO)$_2$Cl$_2${μ-(Ph$_2$P)$_2$py}$_2$] results from the reaction of [Rh(CO)$_2$Cl]$_2$ with 2,6-bis(diphenylphosphino)pyridine. Further reaction with SnCl$_2$ unexpectedly gives [Rh(CO)Cl$_2${μ-(Ph$_2$P)$_2$py}$_2$(μ-SnCl)Rh(CO)(SnCl$_3$)] containing a central tin atom bonded to two pyridyl nitrogen atoms, two rhodium atoms, and one chlorine atom.[170] The potentially tridentate ligand (Ph$_2$PCH$_2$)$_2$AsPh, dpma, reacts with [Rh(CO)$_2$Cl]$_2$ or with [Ir(CO)$_2$Cl(p-MeC$_6$H$_4$NH$_2$)] to give the P,P-bonded macrocycles [M$_2$(CO)$_2$Cl$_2$(μ-dpma)$_2$] (M = Rh or Ir). A third metal can be coordinated to the donor arsenic atoms of these products leading to trinuclear species such as [M$_2$Pd(CO)$_2$Cl$_3$(μ-dpma)$_2$][BPh$_4$].[171]

The very air-sensitive anion [Rh(CO)$_4$]$^-$ is generated in THF solution by sodium amalgam reduction of [Rh(CO)$_2$Cl]$_2$ or [Rh(CO)Cl(PPh$_3$)$_2$] in the presence of carbon monoxide. Such solutions may be used to produce other tetrahedral Rh(-1) anions of the types [Rh(CO)$_3$L]$^-$ (L = various phosphines) and [Rh(CO)$_2$L$_2$]$^-$ [L = P(OPh)$_3$; L$_2$ = dppe].[172] The bis(phosphino)pyrazole 3,5-(Ph$_2$PCH$_2$)$_2$C$_3$H$_2$N$_2$ loses the -NH proton on reaction with [Rh(CO)$_2$Cl]$_2$, the resulting bis(phosphino)pyrazolato anion acting as a planar quadridentate ligand in forming [Rh$_2$(CO)$_2$(μ-Cl){μ-3,5-(Ph$_2$PCH$_2$)$_2$-C$_3$HN$_2$}].[173] The reaction between [Rh(CO)$_2$Cl]$_2$ and 8-quinolinyl phenyl ketone produces the novel 1,3-dimetallacyclobutane derivative (18). Carbon dioxide is also liberated as a result of transfer of a ketone carbonyl oxygen to a carbonyl ligand of [Rh(CO)$_2$Cl]$_2$.[174] The complex [Rh(CO)Cl(dpcb)] is the product of the reaction between [Rh(CO)$_2$Cl]$_2$ and 1,2-bis(diphenylphosphino)-1,2-dicarbadodecaborane.[175] The complexes [Ir$_2$(CO)$_2$L$_2$(NCMe)$_2$-(tcbiim)$_2$], [L = CO or P(OEt)$_3$; H$_2$tcbiim = tetracyanobiimidazole], are the first examples of Ir(II) complexes with metal-metal bonds but no bridging ligands.[176]

Carbonylation of the A-frame complexes [Rh$_2$(CO)$_2$(μ-OR)(μ-dppm)$_2$][ClO$_4$] (R = H, Me, or Et) gives the 46-electron cluster [Rh$_3$(CO)$_3$(μ-CO)$_3$(μ-dppm)$_2$][ClO$_4$] whose structure resembles that of an A-frame complex with a bridging Rh(CO)(μ-CO)$_2$ unit. The unique terminal carbonyl ligand of this moiety can readily be replaced by phosphine ligands.[177] Mild heating of solutions of $[Rh_3(CO)_3X_2(\mu\text{-dpmp})_2][BPh_4]$, [X = halide; dpmp = $(Ph_2PCH_2)_2PPh$] effects conversion into $[Rh_3(CO)(\mu\text{-CO})X(\mu\text{-X})(\mu\text{-dpmp})_2][BPh_4]$ containing bent Rh_3 chains with trans triphosphine ligands.[178] Reaction of $Ir_4(CO)_{12}$ with cis-$Ph_2PCH=CHPPh_2$ (dpp) in refluxing THF leads to the stepwise formation of $[Ir_4(CO)_{10}(dpp)]$, $[Ir_4(CO)_8(dpp)_2]$ and finally ortho-metallated $[Ir_4H(CO)_7(dpp)(PhC_6H_4PCH=CHPPh_2)]$.[179]

The thioethers Me_3SiSR [R = Bu^t or $(CH_2)_nSi(OMe)_3$; n = 2 or 3] react with $[Rh(CO)_2Cl]_2$ in the presence of $Bu^t{}_3As$ to give binuclear $[\{Rh(CO)(AsBu^t{}_3)\}_2(\mu\text{-Cl})(\mu\text{-SR})]$ products.[180] The cluster complex $[Rh_6(CO)_9(\mu\text{-CO})_2(\mu\text{-AsBu}^t{}_2)_2(\mu_4\text{-AsBu}^t)]$, which arises from the reaction of $[Rh(CO)_2Cl]_2$ with $LiAsBu^t{}_2$, contains the first example of a pentagonal pyramidal Rh_6 framework.[181]

7.2. Cyclopentadienyl Complexes.

The stereochemistry and absolute configuration of the optically active chelate complex $[CpCoI\{NC_5H_4C(Me)=NCH(Me)Ph\}]I$ has been elucidated. It is the product of the reaction of $[CpCo(CO)_2I_2]$ with the Schiff base.[182] The major reactions of $[CpCo(CO)X(SnX_3)]$ (X = halide) with $P(OPh)_3$ and PPh_3 involve substitution affording $[CpCo(PR_3)X(SnX_3)]$ species. A limited amount of reductive elimination may also occur producing $[CpCo(CO)_2]$ and $[CpCo(CO)(PR_3)]$.[183]

8 Group VIII : Nickel, Palladium, and Platinum

As part of a study of the conformational behaviour of dimethyl-2,2'-bipyridyl ligands, $[Ni(CO)_2(4,6\text{-Me}_2\text{bipy})]$ has been prepared from $Ni(CO)_4$ and its structure elucidated.[184] Improved syntheses of $[Ni(CO)_3L]$, [L = Me_2NPF_2, $(Me_2N)_2PF$, or $MeN(CH_2)_2N(Me)PF$] have been reported. Further complexes of this type with L = $(Me_2N)_2PH$ or $MeN(CH_2)_2N(Me)PH$ have been prepared by reduction of the corresponding fluorophosphine complexes.[185] The reaction between $Ni(CO)_4$ and cyclo-$P_3Pr_3{}^i$ yields $[Ni_2(CO)_4(\mu\text{-}P_3Pr_3{}^i)_2]$ containing bidentate bridging $P_3Pr_3{}^i$ ligands within the six-membered Ni_2P_4 ring. Mononuclear $[Ni(CO)_3(\eta^1\text{-}P_3Pr_3{}^i)]$ has been identified as an intermediate.[186] The product of the reaction between $Ni(CO)_4$ and $ClP=C(SiMe_3)_2$ is $[Ni(CO)\{\eta^2\text{-}ClP=C(SiMe_3)_2\}_2]$ whereas use of $Ph_2PP=C(SiMe_3)_2$ leads to the formation of $[Ni(CO)_3\{\eta^1\text{-}Ph_2PP=C(SiMe_3)_2\}]$ which subsequently dimerises at room temperature to afford $[Ni(CO)\{Ph_2PP=C(SiMe_3)_2\}]_2$. In this final product the ligands are σ-PPh$_2$ bonded to one nickel atom and η^2-P=C bonded to the other.[187] Trinuclear [{Ni(CO)$_3$}$_3${(Me$_2$PCH$_2$)$_2$-PMe}] and binuclear [Ni(CO)$_2${(Me$_2$PCH$_2$)$_2$PMe}Ni(CO)$_3$] of structures related to (14) and (15) have been isolated from the reaction of Ni(CO)$_4$ with the triphosphine.[141]

The metal-metal bonded complex [Pd$_2$Cl$_2$(μ-dmpm)$_2$] is the product of the reaction between [Pd(CO)Cl]$_n$ and dmpm. Substitution of hydroxide for chloride occurs in water leading to the stable species [Pd$_2$(OH)$_2$(μ-dmpm)$_2$].[188] The preparations of the complexes [Pd(CO)(PPh$_3$)$_3$], [Pd$_3$(CO)$_6$(PPh$_3$)$_4$], [Pd$_4$(CO)$_5$(PPh$_3$)$_4$] and [Pd$_{10}$(CO)$_{12}$(PPh$_3$)$_6$] and their interconversions have been described.[189] The complexes [Pd$_4$(CO)$_5$(PR$_3$)$_4$] (R$_3$ = Ph$_3$, Et$_3$, MePh$_2$ or Me$_2$Ph) result from carbonylation of [Pd(NO$_2$)$_2$(PR$_3$)$_2$]. However, carbonylation of [Pd(NO$_2$)$_2$(PEt$_2$Ph)$_2$] affords the novel palladium(I) dimer [Pd$_2$(CO)Cl$_2$(PEt$_2$Ph)$_3$], the first example of a palladium complex with a semi-bridging carbonyl ligand.[190] The complexes [Pt$_3$(CO)$_3$L$_3$] undergo facile phosphine ligand exchange on reaction with three equivalents of L' (L = PBu$^t_3$, L' = PBu$^t_2$Ph; L = PBu$^t_2$Ph, L' = PBu^{t}Ph$_2$, PCy$_3$, or PPr$^i_3$; L = PCy$_3$, L' = PPr$^i_3$). The products, [Pt$_3$(CO)$_3$L'$_3$], are formed by an associative process with the triangular Pt$_3$ array remaining intact throughout.[191]

References

1. F. Basolo, Inorg. Chim. Acta, 1985, 100, 33.
2. K. Vrieze and G. van Koten, Inorg. Chim. Acta, 1985, 100, 79.
3. P.V. Broadhurst, Polyhedron, 1985, 4, 1801.
4. T.J. Meyer and J.V. Caspar, Chem. Rev., 1985, 85, 187; A.E. Stiegman and D.R. Tyler, Coord. Chem. Rev., 1985, 63, 217.
5. O.J. Scherer, Angew. Chem., Int. Ed. Engl., 1985, 24, 924.
6. J.F. Nixon, Adv. Inorg. Chem. Radiochem., 1985, 29, 41.
7. A.A. Ioganson, Russ. Chem. Rev. (Engl. Transl.), 1985, 54, 277.
8. D.M.P. Mingos and R.W.M. Wardle, Transition Met. Chem., 1985, 10, 441; N.K. Eremenko, E.G. Mednikov, and S.S. Kurasov, Russ. Chem. Rev. (Engl.Transl) 1985, 54, 394.
9. S. Aime, M. Botta, R. Gobetto, and D. Osella, Organometallics, 1985, 4, 1475.
10. A.S. Goldman and D.R. Tyler, Inorg. Chim. Acta, 1985, 98, L47.
11. D. Miholova and A.A. Vlcek, J. Organomet. Chem., 1985, 279, 317.
12. H. Schumann, J. Organomet. Chem., 1985, 290, C34.
13. B.M. Louie, S.J. Rettig, A. Storr, and J. Trotter, Can. J. Chem., 1985, 63, 503, 703, 2261 and 3019; S. Nussbaum, S.J. Rettig, A. Storr, and J. Trotter, ibid., p. 692.
14. M.J. Schadt, N.J. Gresalfi, and A.J. Lees, Inorg. Chem., 1985, 24, 2942.
15. H.-W. Fraühauf, D. Meyer, and J. Breuer, J. Organomet. Chem., 1985, 297, 211.
16. H. tom Dieck and M. Haarich, J. Organomet. Chem., 1985, 291, 71.
17. A.J. Graham, D. Akrigg, and B. Sheldrick, Acta Crystallogr., Sect. C, 1985, 41, 995.

18. R.R. Andréa, J.N. Louwen, M.W. Kokkes, D.J. Stufkens, and A. Oskam, J. Organomet. Chem., 1985, 281, 273.
19. M.W. Kokkes, T.L. Snoeck, D.J. Stufkens, A. Oskam, M. Christophersen, and C.H. Stam, J. Mol. Struct., 1985, 131, 11; M.W. Kokkes, D.J. Stufkens, and A. Oskam, Inorg. Chem., 1985, 24, 2934 and 4411; R.R. Andréa, D.J. Stufkens and A. Oskam, J. Organomet. Chem., 1985, 290, 63; M.W. Kokkes, W.G.J. de Lange, D.J. Stufkens, and A. Oskam, J. Organomet. Chem., 1985, 294, 59.
20. J. Keijsper, G. van Koten, K. Vrieze, M. Zoutberg, and C.H. Stam, Organometallics, 1985, 4, 1306.
21. J. Keijsper, P. Grimberg, G. van Koten, K. Vrieze, M. Christophersen, and C.H. Stam, Inorg. Chim. Acta, 1985, 102, 29; J. Keijsper, P. Grimberg, G. van Koten, K. Vrieze, B. Kojic-Prodic, and A.L Spek, Organometallics, 1985, 4, 438.
22. J. Keijsper, L.H. Polm, G. van Koten, K. Vrieze, P.F.A.B. Seignette, and C.H. Stam, Inorg. Chem., 1985, 24, 518; J. Keijsper, L.H. Polm, G. van Koten, K. Vrieze, K. Goubitz, and C.H. Stam, Organometallics, 1985, 4, 1876; J. Keijsper, L.H. Polm, G. van Koten, K. Vrieze, E. Nielsen, and C.H. Stam, ibid., p. 2006; J. Keijsper, L.H. Polm, G. van Koten, K. Vrieze, C.H. Stam, and J.-D. Schagen, Inorg. Chim. Acta, 1985, 103, 137.
23. H. Lang, L. Zsolnai, and G. Huttner, Z. Naturforsch., Teil B, 1985, 40, 500; H. Lang, O. Orama, and G. Huttner, J. Organomet. Chem., 1985, 291, 293.
24. D.J. Brauer, S. Hietkamp, H. Sommer, and O. Stelzer, J. Organomet. Chem., 1985, 281, 187; Z. Naturforsch., Teil B, 1985, 40, 1677; D.J. Brauer, S. Hietkamp, H. Sommer, O. Stelzer, G. Müller, and C. Krüger, ibid., 288, 35; D.J. Brauer, G. Hasselkuss, S. Hietkamp, H. Sommer, and O. Stelzer, Z. Naturforsch., Teil B, 1985, 40, 961; H. Lang, L. Zsolnai and G. Huttner, J. Organomet. Chem., 1985, 282, 23.
25. A.M. Arif, A.H. Cowley, N.C. Norman, A.G. Orpen, and M. Pakulski, J. Chem. Soc., Chem. Commun., 1985, 1267; A.M. Arif, A.H. Cowley, and M. Pakulski, ibid., p. 1707; A.M. Arif, A.H. Cowley, M. Pakulski, M.B. Hursthouse, and A. Karauloz, Organometallics, 1985, 4, 2227; M.G. Richmond, J.D. Korp, and J.K. Kochi, J. Chem. Soc., Chem. Commun., 1985, 1102.
26. W.A. Herrmann, B. Koumbouris, A. Schäfer, T. Zahn, and M.L. Ziegler, Chem. Ber., 1985, 118, 2472.
27. B. Sigwarth, U. Weber, L. Zsolnai, and G. Huttner, Chem. Ber., 1985, 118, 3114; U. Weber, G. Huttner, O. Scheidsteger, and L. Zsolnai, J. Organomet. Chem., 1985, 289, 357.
28. A.H. Cowley, N.C. Norman, M. Pakulski, D.L. Bricker, and D.H. Russell, J. Am. Chem. Soc., 1985, 107, 8211; A.M. Arif, A.H. Cowley, N.C. Norman, and M. Pakulski, ibid., p. 1062.
29. M. Yoshifuji, T. Hashida, K. Shibayama, and N. Inamoto, Chem. Lett., 1985, 287; M. Yoshifuji, T. Hashida, N. Inamoto, K. Hirotsu, T. Horiuchi, T. Higuchi, K. Ito, and S. Nagase, Angew. Chem., Int. Ed. Engl., 1985, 24, 211.
30. J. Borm, G. Huttner, and L. Zsolnai, Angew. Chem., Int. Ed. Engl., 1985, 24, 1069; J. Borm, G. Huttner, O. Orama, and L. Zsolnai, J. Organomet. Chem., 1985, 282, 53.
31. L. Weber and K. Reizig, Angew. Chem., Int. Ed. Engl., 1985, 24, 865.
32. A.M. Arif, A.H. Cowley, and S. Quashie, J. Chem. Soc., Chem. Commun., 1985, 428.
33. R. Appel, C. Casser, and F. Knoch, J. Organomet. Chem., 1985, 293, 213.
34. K.D. Gallicano and N.L. Paddock, Can. J. Chem., 1985, 63, 314.
35. H.R. Allcock, G.H. Riding, and R.R. Whittle, J. Am. Chem. Soc., 1984, 106, 5561; H.R. Allcock, P.R. Suszko, L.J. Wagner, R.R. Whittle, and B. Boso, Organometallics, 1985, 4, 446.
36. E.E. Nifantyev, A.T. Teleshev, G.M. Grishina, and A.A. Borisenko, Phosphorus Sulfur, 1985, 24, 333; E.E. Nifantyev, Yu. I. Blokhin, A.T. Teleshev, Yu. G. Chikishev, B.V. Rozynov, and G.A. Vakhtberg, Zh. Obshch. Khim., 1985, 55, 1274.
37. L.I. Zakharkin, A.V. Kazantsev, and M.G. Meiramov, Zh. Obshch. Khim., 1984, 54, 1536; Bull. Acad. Sci. USSR, Div. Chem. Sci., 1984, 33, 1505.

38. W. Kläui, M. Scotti, M. Valderrama, S. Rojas, G.M. Sheldrick, P.J. Jones, and T. Schroeder, Angew. Chem., Int. Ed. Engl., 1985, 24, 683; W. Kläui, J. Okuda, M. Scotti, and M. Valderrama, J. Organomet. Chem., 1985, 280, C26.
39. F. Götzfried, R. Grenz, G. Urban, and W. Beck, Chem. Ber., 1985, 118, 4179; K. Raab and W. Beck, ibid., p. 3830.
40. U.R. Kunze, J. Organomet. Chem., 1985, 281, 79.
41. M. Herberhold, J. Ott, and L. Haumaier, Chem. Ber., 1985, 118, 3143.
42. L.B. Kool, M.D. Rausch, H.G. Alt, M. Herberhold, B. Wolf, and U. Thewalt, J. Organomet. Chem., 1985, 297, 159.
43. K. Ihmels and D. Rehder, Organometallics, 1985, 4, 1334 and 1340; Chem. Ber., 1985, 118, 895; D. Rehder, K. Ihmels, D. Wenke, and P. Oltmanns, Inorg. Chim. Acta, 1985, 100, L11.
44. N. Albrecht, P. Hübener, U. Behrens, and E. Weiss, Chem. Ber., 1985, 118, 4059.
45. M.L. Luetkens, D.J. Santure, J.C. Huffman, and A.P. Sattelberger, J. Chem. Soc., Chem. Commun., 1985, 552; M.L. Luetkens, J.C. Huffman, and A.P. Sattelberger, J. Am. Chem. Soc., 1985, 107, 3361.
46. M. Hoch and D. Rehder, J. Organomet. Chem., 1985, 288, C25.
47. A. Belforte, F. Calderazzo, and P.F. Zanazzi, Gazz. Chim. Ital., 1985, 115, 71.
48. R.M. Kowaleski, D.O. Kipp, K.J. Stauffer, P.N. Swepston, and F. Basolo, Inorg. Chem., 1985, 24, 3750.
49. D. Czarkie and Y. Shvo, J. Organomet. Chem., 1985, 280, 123.
50. A.A. Ioganson, Yu. G. Kovalev, and L.L. Milovanova, Bull. Acad. Sci. USSR, Div. Chem. Sci., 1984, 33, 1539.
51. H.-T. Macholdt and R. van Eldik, Transition Met. Chem., 1985, 10, 323.
52. S.C. Tripathi, S.C. Srivastava, A.K. Shrimal, and O.P. Singh, Inorg. Chim. Acta, 1985, 98, 19.
53. H.K. van Dijk, P.C. Servaas, D.J. Stufkens, and A. Oskam, Inorg. Chim. Acta, 1985, 104, 179.
54. P. Leoni, E. Grilli, M. Pasquali, and M. Tomassini, J. Chem. Soc., Dalton Trans., 1985, 2561.
55. J.A. Connor, E.J. James, and C. Overton, Polyhedron, 1985, 4, 69.
56. (a) P.C. Servaas, H.K. van Dijk, T.L. Snoeck, D.J. Stufkens, and A. Oskam, Inorg. Chem., 1985, 24, 4494; (b) M.J. Blandamer, J. Burgess, and T. Digman, Transition Met. Chem., 1985, 10, 274; (c) H.-T. Macholdt, R. van Eldik, H. Kelm, and H. Elias, Inorg. Chim. Acta, 1985, 104, 115; S. Ernst and W. Kaim, Angew. Chem., Int. Ed. Engl., 1985, 24, 430.
57. M.-A. Haga and K. Koizumi, Inorg. Chim. Acta, 1985, 104, 47; S. Kohlmann, S. Ernst, and W. Kaim, Angew. Chem., Int. Ed. Engl., 1985, 24, 684.
58. M.L. Boyles, D.V. Brown, D.A. Drake, C.K. Hostetler, C.K. Maves, and J.A. Mosbo, Inorg. Chem., 1985, 24, 3126.
59. (a) R.B. King and N.D. Sadanani, Inorg. Chem., 1985, 24, 3136; (b) A.J. Ashe, W. Butler, J.C. Colburn, and S. Abu-Orabi, J. Organomet. Chem., 1985, 282, 233; (c) J. Ellermann and A. Veit, ibid., 290, 307.
60. J. Grobe, D. Le Van, and J. Szameitat, J. Organomet. Chem., 1985, 289, 341.
61. G. Andolfatto, R. Granger, and L.K. Peterson, Inorg. Chim. Acta, 1985, 99, L1.
62. W.K. Wong, K.W. Chiu, G. Wilkinson, M. Motevalli, and M.B. Hursthouse, Polyhedron, 1985, 4, 1231.
63. (a) R. Hensel, W.-W. du Mont, R. Boese, D. Wewers, and L. Weber, Chem. Ber., 1985, 118, 1580; (b) A.L. Balch, R.R. Guimerans, and J. Linehan, Inorg. Chem., 1985, 24, 290; (c) J. Ellermann and W. Wend, J. Organomet. Chem., 1985, 281, C29; (d) T.G. Hill, R.C. Haltiwanger, and A.D. Norman, Inorg. Chem., 1985, 24, 3499; (e) G.M. Gray and K.A. Redmill, Inorg. Chem., 1985, 24, 1279 and J. Organomet. Chem., 1985, 280, 105.
64. U. Kunze, H. Jawad, W. Hiller, and R. Naumer, Z. Naturforsch., Teil B, 1985, 40, 512.
65. J. Grobe, M. Kühne-Wächter, and D. Le Van, J. Organomet. Chem., 1985, 280, 331.

66. F. Kober and M. Kerber, Z. Anorg. Allg. Chem., 1985, 522, 65 and 76.
67. C. Zeiher, J. Mohyla, I.-P. Lorenz, and W. Hiller, J. Organomet. Chem., 1985, 286, 159; C. Zeiher, W. Hiller, and I.-P. Lorenz, Chem. Ber., 1985, 118, 3127.
68. E.P. Kyba and S.-T. Liu, Inorg. Chem., 1985, 24, 1613; E.P. Kyba, R.E. Davis, S.-T. Liu, K.A. Hassett, and S.B. Larson, ibid., p. 4629.
69. A. Blagg, S.W. Carr, G.R. Cooper, I.D. Dobson, J.B. Gill, D.C. Goodall, B.L. Shaw, N. Taylor, and T. Boddington, J. Chem. Soc., Dalton Trans., 1985, 1213.
70. J.A. Iggo and B.L. Shaw, J. Chem. Soc., Dalton Trans., 1985, 1009.
71. G.R. Cooper, F. Hassan, B.L. Shaw, and M. Thornton-Pett, J. Chem. Soc., Chem. Commun., 1985, 614.
72. G.R. Dobson, C.B. Dobson and S.E. Mansour, Inorg. Chem., 1985, 24, 2179; G.R. Dobson, S.S. Basson, and C.B. Dobson, Inorg. Chim. Acta, 1985, 105, L17; G.R. Dobson, C.B. Dobson and S.E. Mansour, ibid., 100, L7; G.R.Dobson, I. Bernal, G.M. Reisner, C.B. Dobson, and S.E. Mansour, J. Am. Chem. Soc., 1985, 107, 525.
73. M. Basato, J. Chem. Soc., Dalton Trans., 1985, 91.
74. T.J. McNeese, T.E. Mueller, D.A. Wierda, D.J. Darensbourg, and T.J. Delord, Inorg. Chem., 1985, 24, 3465.
75. G. Bremer, P. Klüfers, and T. Kruck, Chem. Ber., 1985, 118, 4224.
76. L.D. Rosenhein, W.E. Newton, and J.W. McDonald, J. Organomet. Chem., 1985, 288, C17; B. Zhuang, J.W. McDonald, F.A. Schultz, and W.E. Newton, Inorg. Chim. Acta, 1985, 99, L29.
77. A.E. Sanchez-Pelaez, M.F. Perpinan, and A. Santos, J. Organomet. Chem., 1985, 296, 367.
78. D. Sellmann, W. Ludwig, G. Huttner, and L. Zsolnai, J. Organomet. Chem., 1985, 294, 199; D. Sellmann and L. Zapf, ibid., 289, 57; M.T. Ashby and D.L. Lichtenberger, Inorg. Chem., 1985, 24, 636.
79. K.S. Jasim and C. Chieh, Inorg. Chim. Acta, 1985, 99, 25; T.C.W. Mak, K.S. Jasim and C. Chieh, ibid., p. 31.
80. E.W. Abel, S.K. Bhargava, P.K. Mittal, K.G. Orrell, and V. Sik, J. Chem. Soc., Dalton Trans., 1985,1561; E.W. Abel, P.K. Mittall, K.G. Orrell, and V. Sik, ibid., p. 1569; E.W. Abel, K.M. Higgins, K.G. Orrell, V. Sik, E.H. Curzon, and O.W. Howarth, ibid., p. 2195.
81. N. Kuhn, H. Schumann, and G. Wolmershäuser, J. Chem. Soc., Chem. Commun., 1985, 1595; T. Vogt and J. Strähle, Z. Naturforsch., Teil B., 1985, 40, 1599.
82. E.C. Alyea, R.P. Shakya, and A.E. Vougioukas, Transition Met. Chem., 1985, 10, 435.
83. S.J.N. Burgmayer and J.L. Templeton, Inorg. Chem., 1985, 24, 2224.
84. D. Sellmann and W. Reisser, J. Organomet. Chem., 1985, 294, 333.
85. P.J. Blower, J.R. Dilworth, J. Hutchinson, T. Nicholson, and J.A. Zubieta, J. Chem. Soc., Dalton Trans., 1985, 2639; P.T. Bishop, J.R. Dilworth, and J.A. Zubieta, J. Chem. Soc., Chem. Commun., 1985, 257.
86. R.M. Medina, A. Alvarez-Valdes, and J.R. Masaguer, J. Organomet. Chem., 1985, 294, 209.
87. R.A. Jones, S.T. Schwab, A.L. Stuart, B.R. Whittlesey, and T.C. Wright, Polyhedron, 1985, 4, 1689.
88. J.G. Riess, U. Klement, and J. Wachter, J. Organomet. Chem., 1985, 280, 215.
89. J. Febvay, F. Casabianca, and J.G. Riess, Inorg. Chem., 1985, 24, 3235.
90. P. Jaitner and W. Wohlgenannt, Inorg. Chim. Acta, 1985, 101, L43; P. Jaitner, W. Wohlgenannt, A. Gieren, H. Betz, and T. Hübner, J. Organomet. Chem., 1985, 297, 281.
91. W. De Oliveira, J.-L. Migot, M. B.G. De Lima, J. Sala-Pala, J.-E. Guerchais, and J.-Y. Le Gall, J. Organomet. Chem., 1985, 284, 313.
92. K. Sünkel, G. Urban, and W. Beck, J. Organomet. Chem., 1985, 290, 231; G. Urban, K. Sünkel, and W. Beck, ibid., p. 329.
93. J. Grobe and R. Haubold, Z. Anorg. Allg. Chem., 1985, 522, 159; ibid., 526, 145.
94. P. Klüfers, L. Knoll, C. Reiners, and K. Reiss, Chem. Ber., 1985, 118, 1825.

95. P.H. Bird, A.A. Ismail, and I.S. Butler, Inorg. Chem., 1985, 24, 2911.
96. J.A.S. Howell, D.T. Dixon, J.C. Kola, and N.F. Ashford, J. Organomet. Chem., 1985, 294, C1.
97. S.P. Nolan and C.D. Hoff, J. Organomet. Chem., 1985, 290, 365.
98. S. Lotz, M. Schindehutte, M.M. van Dyk, J.L.M. Dillen and P.H. van Rooyen, J. Organomet. Chem., 1985, 295, 51.
99. T.I. Odiaka, Inorg. Chim. Acta, 1985, 103, 9.
100. K. Yasufuku, H. Noda, J. Iwai, H. Ohtani, M. Hoshino, and T. Kobayashi, Organometallics, 1985, 4, 2174; T. Kobayashi, K. Yasufuku, J. Iwai, H. Yesaka, H. Noda, and H. Ohtani, Coord. Chem. Rev., 1985, 64, 1.
101. T.R. Herrinton and T.L. Brown, J. Am. Chem. Soc., 1985, 107, 5700.
102. A. Poë and C.V. Sekhar, J. Am. Chem. Soc., 1985, 107, 4874.
103. G.W. Harris, J.C.A. Boeyens, and N.J. Coville, J. Chem. Soc., Dalton Trans., 1985, 2277.
104. C.S. Young, S.W. Lee, and C.P. Cheng, J. Organomet. Chem., 1985, 282, 85.
105. F.R. Lemke and C.P. Kubiak, J. Chem. Soc., Chem. Commun., 1985, 1729.
106. F.A. Cotton and K.-B. Shiu, Gazz. Chim. Ital., 1985, 115, 705.
107. K.-W. Lee, W.T. Pennington, A.W. Cordes, and T.L. Brown, J. Am. Chem. Soc., 1985, 107, 631; K.-W. Lee and T.L. Brown, Organometallics, 1985, 4, 1025 and 1030.
108. E. Guggolz, K. Layer, F. Oberdorfer, and M. Ziegler, Z. Naturforsch., Teil B, 1985, 40, 77.
109. H.G. Ang and P.T. Lau, J. Organomet. Chem., 1985, 291, 285.
110. N.J. Taylor, J. Chem. Soc., Chem. Commun., 1985, 476 and 478.
111. D.A. DuBois, E.N. Duesler, and R.T. Paine, Inorg. Chem., 1985, 24, 3.
112. J. Grobe and R. Martin, Z. Anorg. Allg. Chem., 1985, 523, 199.
113. H. Schäfer, J. Zipfel, B. Gutekunst, and U. Lemmert, Z. Anorg. Allg. Chem., 1985, 529, 157.
114. E. Lindner, F. Zinsser, W. Hiller, and R. Fawzi, J. Organomet. Chem., 1985, 288, 317.
115. S.K. Chopra, S.S.C. Chu, P. de Meester, D.E. Geyer, M. Lattman, and S.A. Morse, J. Organomet. Chem., 1985, 294, 347.
116. G.A. Carriedo, J.B. ParraSoto, V. Riera, X. Solans, and C. Miravitlles, J. Organomet. Chem., 1985, 297, 193.
117. W.C. Mercer, G.L. Geoffroy, and A.L. Rheingold, Organometallics, 1985, 4, 1418; W.C. Mercer, R.R. Whittle, E.W. Burkhardt, and G.L. Geoffroy, ibid., p. 68.
118. A. Belforte, F. Calderazzo, D. Vitali, and P.F. Zanazzi, Gazz. Chim. Ital., 1985, 115, 125.
119. P.M. Treichel and M.H. Tegen, J. Organomet. Chem., 1985, 292, 385; J.W. McDonald, Inorg. Chem., 1985, 24, 1734.
120. U. Kunze and A. Bruns, Z. Naturforsch., Teil B, 1985, 40, 127; J. Organomet. Chem., 1985, 292, 349; D. Rehder, V. Pank, U. Kunze, and A. Bruns, Inorg. Chim. Acta, 1985, 99, L35.
121. H. Alper, F. Sibtain, F.W.B. Einstein, and A.C. Willis, Organometallics, 1985, 4, 604.
122. S.P. Mallela and F. Aubke, Inorg. Chem., 1985, 24, 2969; J. Nitschke, S.P. Schmidt, and W.C. Trogler, ibid., p. 1972.
123. F.J. Garcia-Alonso, V. Riera, and M.J. Misas, Transition Met. Chem., 1985, 10, 19.
124. C.J. Besecker, V.W. Day, and W.G. Klemperer, Organometallics, 1985, 4, 564; C.J. Besecker, V.W. Day, W.G. Klemperer, and M.R. Thompson, Inorg. Chem., 1985, 24, 44.
125. R. Gross and W. Kaim, J. Organomet. Chem., 1985, 292, C21.
126. D. Sellmann and J. Müller, J. Organomet. Chem., 1985, 281, 249; R. Gross and W. Kaim, Angew. Chem., Int. Ed. Engl., 1985, 24, 856; A. Winter, G. Huttner, M. Gottlieb, and I. Jibril, J. Organomet. Chem., 1985, 286, 317.
127. I.-P. Lorenz, J. Messelhäuser, W. Hiller, and K. Haug, Angew. Chem., Int. Ed. Engl., 1985, 24, 228.

128. G.K. Yang and R.G. Bergman, Organometallics, 1985, 4, 129.
129. C.P. Casey and J.M. O'Connor, Organometallics, 1985, 4, 384.
130. L.-N. Ji, D.L. Kershner, M.E. Rerek, and F. Basolo, J. Organomet. Chem., 1985, 296, 83.
131. N.I. Pyshnograeva, A.S. Batsanov, Yu. T. Struchkov, A.G. Ginzburg, and V.N. Setkina, J. Organomet. Chem., 1985, 297, 69.
132. N.G. Connolly, M.J. Freeman, A.G. Orpen, A.R. Sheehan, J.B. Sheridan, and D.A. Sweigart, J. Chem. Soc., Dalton Trans., 1985, 1019.
133. G.T. Palmer and F. Basolo, J. Am. Chem. Soc., 1985, 107, 3122.
134. L.P. Battaglia, T. Boselli, G.P. Chiusoli, M. Nardelli, C. Pelizzi, and G. Predieri, Gazz. Chim. Ital., 1985, 115, 395.
135. L.R. Martin, F.W.B. Einstein, and R.K. Pomeroy, Inorg. Chem., 1985, 24, 2777.
136. A.L. Rheingold and M.E. Fountain, Acta Crystallogr., Sect. C, 1985, 41, 1162.
137. W. K. Wong, K.W. Chiu, G. Wilkinson, A.J. Howes, M. Motevalli, and M.B. Hursthouse, Polyhedron, 1985, 4, 603.
138. E. Lindner, C.-P. Krieg, W. Hiller, and R. Fawzi, Chem. Ber., 1985, 118, 1398.
139. R.R. Ford, B.-L. Li, R.H. Neilson, and R.J. Thoma, Inorg. Chem., 1985, 24, 1993.
140. A.M. Arif, A.H. Cowley, and M. Pakulski, J. Am. Chem. Soc., 1985, 107, 2553.
141. D.J. Brauer, S. Hietkamp, H. Sommer, O. Stelzer, G. Müller, M.J. Romao, and C. Krüger, J. Organomet. Chem., 1985, 296, 411.
142. R.B. King, F.-J. Wu, N.D. Sadanani, and E.M. Holt, Inorg. Chem., 1985, 24, 4449.
143. G.J. Kruger, S. Lotz, L. Linford, M. van Dyk, and H.G. Raubenheimer, J. Organomet. Chem., 1985, 280, 241.
144. E.P. Kyba, K.L. Hassett, B. Sheikh, J.S. McKennis, R.B. King, and R.E. Davis, Organometallics, 1985, 4, 994.
145. A. Shojaie and J.D. Atwood, Organometallics, 1985, 4, 187; N. Brodie, A. Poë, and V.C. Sekhar, J. Chem. Soc., Chem. Commun., 1985, 1090; A. Poë and V.C. Sekhar, Inorg. Chem., 1985, 24, 4376.
146. B. Ambwani, S. Chawla, and A. Poë, Inorg. Chem., 1985, 24, 2635.
147. M.F. Desrosiers, D.A. Wink, and P.C. Ford, Inorg. Chem., 1985, 24, 1.
148. T. Venäläinen, J. Pursiainen, and T.A. Pakkanen, J. Chem. Soc., Chem. Commun., 1985, 1348; G.A. Foulds, B.F.G. Johnson, and J. Lewis, J. Organomet. Chem., 1985, 294, 123; 296, 147.
149. A.J. Deeming, S. Donovan-Mtunzi, S.E. Kabir, and P.J. Manning, J. Chem. Soc., Dalton Trans., 1985, 1037. A.J. Deeming, S. Donovan-Mtunzi, and S.E. Kabir, J. Organomet. Chem., 1985, 281, C43.
150. R.F. Alex and R.K. Pomeroy, J. Organomet. Chem., 1985, 284, 379.
151. M.I. Bruce, J.G. Matisons, J.M. Patrick, A.H. White, and A.C. Willis, J. Chem. Soc., Dalton Trans., 1985, 1223.
152. J.A. Clucas, M.M. Harding, B.S. Nicholls, and A.K. Smith, J. Chem. Soc., Dalton Trans., 1985, 1835.
153. D.F. Foster, J. Harrison, B.S. Nicholls, and A.K. Smith, J. Organomet. Chem., 1985, 295, 99.
154. V.D. Patel, A.A. Cherkas, D. Nucciarone, N.J. Taylor, and A.J. Carty, Organometallics, 1985, 4, 1792.
155. M.I. Bruce, M.L. Williams, J.M. Patrick, and A.H. White, J. Chem. Soc., Dalton Trans., 1985, 1229; J.-C. Daran, O. Kristiansson, and Y. Jeannin, C.R. Hebd. Seances Acad. Sci., Ser. II, 1985, 300, 943.
156. J. Messelhäuser, I.-P. Lorenz, K. Haug, and W. Hiller, Z. Naturforsch., Teil B, 1985, 40, 1064.
157. H. Schumann, J. Organomet. Chem., 1985, 293, 75; N. Kuhn and H. Schumann, ibid., 287, 345 and 288, C51.
158. M. Moran and I. Cuadrado, J. Organomet. Chem., 1985, 295, 353.
159. R.J. Angelici and J.W. Dunker, Inorg. Chem., 1985, 24, 2209.
160. R.B. King, W.-K. Fu, and E.M. Holt, Inorg. Chem., 1985, 24, 3094.

161. C. White and E. Cesarotti, *J. Organomet. Chem.*, 1985, **287**, 123.
162. D.J. Darensbourg and D.J. Zalewski, *Organometallics*, 1985, **4**, 92; J. Rimmelin, P. Lemoine, M. Gross, A.A. Bahsoun, and J.A. Osborn, *Nouv. J. Chim.*, 1985, **9**, 181.
163. N.J. Bailey, G. de Leeuw, J.S. Field, R.J. Haines, I.C.D. Stuckenberg, and R.B. English, *S. Afr. J. Chem.*, 1985, **38**, 139.
164. B.M. Louie, S.J. Rettig, A. Storr, and J. Trotter, *Can. J. Chem.*, 1985, **63**, 688; L.A. Oro, M.T. Pinillos, A. Tiripicchio, and M. Tiripicchio-Camellini, *Inorg. Chim. Acta*, 1985, **99**, L13.
165. L.A. Oro, M.T. Pinillos, and C. Tejel, *J. Organomet. Chem.*, 1985, **280**, 261.
166. G.J. Lamprecht, J.G. Leipoldt, and G.J. van Zyl, *Inorg. Chim. Acta*, 1985, **97**, 31; J.G. Leipoldt, S.S. Basson, E.C. Grobler, and A. Roodt, *ibid.*, **99**, 13; J.G. Leipoldt, G.J. Lamprecht, and D.E. Graham, *ibid.*, **101**, 123; A.M.Trzeciak and J.J.Ziolkowski, *ibid.*, **96**, 15; R. van Eldik, S.Aygen, H.Kelm, A.M.Trzeciak, and J.J.Ziolkowski, *Transition Met. Chem.*, 1985, **10**, 167; F. Bonati, L.A.Oro, and M.T.Pinillos, *Polyhedron*, 1985, **4**, 357.
167. J.Ellermann and G.Szucsanyi, *Z. Anorg. Allg. Chem.*, 1985, **520**, 113; *Chem.Ber,* 1985, **118**, 1868; J.Ellermann, G.Szucsanyi and E.Wilhelm, *ibid.*, 1588.
168. J.A. Iggo, D.P. Markham, B.L.Shaw, and M.Thornton-Pett, *J. Chem. Soc., Chem. Commun.*, 1985, 432; F.S.M. Hassan, D.P.Markham, P.G.Pringle, and B.L.Shaw, *J. Chem. Soc., Dalton Trans.*, 1985, 279; A.T.Hutton, C.R.Langrick, D.M.McEwan, P.G.Pringle, and B.L.Shaw, *ibid.*, p.2121.
169. A.L.Balch, R.R.Guimerans, J.Linehan and F.E.Wood, *Inorg. Chem.*, 1985, **24**, 2021.
170. A.L.Balch, H.Hope, and F.E.Wood, *J. Am. Chem. Soc.*, 1985, **107**, 6936.
171. A.L.Balch, L.A.Fossett, M.M.Olmstead, D.E.Oram, and P.E.Reedy, *J.Am. Chem. Soc.*, 1985, **107**, 5272.
172. A.S.C.Chan, H.-S.Shieh, and J.R.Hill, *J. Organomet. Chem.*, 1985, **279**, 171.
173. T.G.Schenck, J.M.Downes, C.R.C.Milne, P.B.MacKenzie, H.Boucher, J.Whelan, and B.Bosnich, *Inorg. Chem.*, 1985, **24**, 2334; T.G.Schenck, C.R.C.Milne, J.F.Sawyer, and B.Bosnich, *ibid.*, p.2338.
174. J.W.Suggs, M.J.Wovkulich and K.S.Lee, *J.Am. Chem. Soc.*, 1985, **107**, 5546.
175. F.A.Hart and D.W.Owen, *Inorg. Chim. Acta*, 1985, **103**, L1.
176. P.G.Rasmussen, J.E.Anderson, O.H.Bailey, M.Tamres, and J.C.Bayon, *J.Am. Chem. Soc.*, 1985, **107**, 279.
177. S.P.Deraniyagala and K.R.Grundy, *J.Chem. Soc., Dalton Trans.*, 1985, 1577.
178. A.L.Balch, L.A.Fossett, R.R.Guimerans, and M.M.Olmstead, *Organometallics,* 1985, **4**, 781.
179. V.G.Albano, D.Braga, R.Ros, and A.Scrivanti, *J.Chem. Soc., Chem. Commun.*, 1985, 866.
180. H.Schumann, S.Jurgis, E.Hahn, J.Pickardt, J.Blum, and M.Eisen, *Chem. Ber.*, 1985, **118**, 2738.
181. R.A.Jones and B.R.Whittlesey, *J. Am. Chem. Soc.*, 1985, **107**, 1078.
182. I.Bernal, G.M.Reisner, H.Brunner,and G.Riepl, *Inorg. Chim. Acta*, 1985, **103**, 179.
183. P.T.Murray and A.R.Manning, *J. Organomet. Chem.*, 1985, **288**, 219.
184. J. Sieler, N.-N.Than, R.Benedix, E.Dinjus, and D.Walther, *Z.Anorg. Allg. Chem.*, 1985, **522**, 131.
185. S.S.Snow, D.-X. Jiang and R.W. Parry, *Inorg. Chem.*, 1985, **24**, 1460.
186. M.Baudler, F.Salzer, J.Hahn and E.Dürr, *Angew. Chem., Int. Ed. Engl.*, 1985, **24**, 415.
187. R.Appel, C.Casser, and F.Knoch, *J. Organomet. Chem.*, 1985, **297**, 21.
188. M.C.Kullberg, F.R.Lemke, D.R.Powell,and C.P.Kubiak, *Inorg. Chem.*, 1985, **24**, 3589.
189. E.G.Mednikov and N.K.Eremenko, *Bull. Acad. Sci. USSR, Div. Chem. Sci.*, 1984, **33**, 2547.
190. R.D.Feltham, G.Elbaze, R.Ortega, C.Eck and J.Dubrawski, *Inorg. Chem.*, 1985, **24**, 1503.
191. C.S.Browning, D.H.Farrar, R.R.Gukathasan, and S.A.Morris, *Organometallics,* 1985, **4**, 1750.

11
Complexes Containing Metal–Carbon σ-Bonds of the Groups Scandium to Manganese, including Carbenes and Carbynes

BY M. J. WINTER

1. Introduction

In this chapter Cp´ represents C_5H_4Me and Cp* indicates C_5Me_5.

There is material of general relevance in a number of reviews including that in articles on transition metal complexes in organic synthesis[1], complexes with heteronuclear metal-metal bonds[2], C-H activation[3,4], the photochemistry of metal alkyls, alkylidenes and alkylidynes[5], u.v. photoelectron spectroscopy of metal alkyls[6], reactive intermediates derived from metallocenes[7], and M-C bond energies for organometallics[8]. Some other material appears in a series of "literature highlights"[9,10,11] and conference abstracts on alkene metathesis.[12]

2. Group 3 (Sc, Y, and La), Lanthanides, and Actinides.

Scandium chemistry is reviewed[13] while other reviews concern organolanthanides in catalysis[14], the organometallic chemistry of the rare earths[15], organolanthanide chemistry[16], and compounds of the type $U(=CHPR_3)Cp_3$.[17]

Nuclear magnetic resonance studies show that $ScCp_3$ is a fluxional dimer for which there is exchange between σ and π bonded Cp rings[18] while the use of ^{89}Y n.m.r. spectroscopy allows analysis of alkylyttrium complexes containing the $YCp´_2$ unit.[19]

The Th-R disruption energies in $ThRCp_3$ are larger than those for ThR_2Cp* and lie in the range 77.3-86.1 kcal mol^{-1}.[20] Reaction of $URCp_3$ (R = Me, Et) with BH_3L (L = BH_3, THF, SMe_2) gives $U(BH_4)Cp_3$ via the primary characterised monoborane insertion product $U(H_3BR)Cp_3$.[21]

Reaction of RLi (R = Bu^t, Bu^s) with $M(THF)Cp_3$ (M = Nd, Lu) gives $MR(THF)Cp_2$, the Bu^s derivatives of which decompose at $^-30^\circ$C to butene and $[MH(THF)Cp_2]_2$. The Bu^t derivatives react with H_2 at ambient temperature to give the same dimers.[22] The methyl species $TbMe(THF)Cp´_2$ photoluminesces differently from $[TbMeCp´_2]_2$. Since

(For references see page 257)

monomer/dimer ratios depend on the solvent, this gives a probe for complex solvation.[23] The trichlorides MCl_3 (M = La, Lu) react with MeLi, Cp*[-] and TMED to form $\overline{M[MeLi(TMED)Me]Cp*_2}$ while use of $LiCH_2SiMe_3$ in place of MeLi gives $[Lu(CH_2SiMe_3)Cp*_2]^-$. The corresponding reactions of MCl_3 (M = Yb, Lu) (different stoicheiometries) result in $[MMe_3Cp*]^-$.[24] Treatment of $MClCp_2$ (M = Er, Lu) with MeLi and TMED/THF gives $\overline{M[MeLi(TMED)Me]Cp_2}$, but $\overline{M[MeLi(THF)_2Me]Cp_2}$ on substituting DME for TMED.[25] The alkyls $\overline{Lu[MeLi(THF)_2Me]Cp*_2}$ or $[LuMe_3Cp*]^-$ react with Bu^tOH or Bu^tSH to form various Bu^tO or Bu^tS complexes and methane.[26]

The anions $[Li(ether)_2][MCl_2Cp*_2]$ (M = La, Nd, Nd, Lu) react with $LiCH(SiMe_3)_2$ to give $M[CH(SiMe_3)_2]Cp*_2$,[27,28] while $[Li(ether)_2][LuCl_2\{\eta^5,\eta^{5'}-C_5Me_4Si(Me)_2C_5Me_4\}]$ and $LiCH(SiMe_3)_2$ provide $Lu[CH(SiMe_3)_2]\{\eta^5,\eta^{5'}-C_5Me_4Si(Me_2)C_5Me_4\}$,[29] the latter undergoes reaction with Bu^tLi to give the anion $[\overline{Lu\{CH_2Si(Me)_2CHSiMe_3\}}\{\eta^5,\eta^{5'}-C_5Me_4SiMe_2C_5Me_4\}]^-$. The dimer $[YClCp_2]_2$ reacts with $LiCH_2SiMe_3$ to give $[Y(CH_2SiMe_3)_2Cp_2]^-$ and a dimeric complex of similar stoicheiometry.[30] Treatment of $LuClCp_2$ with $LiCH_2PMe_2$ leads to $\overline{Lu[CH_2PMe_2]Cp_2}$, which in turn reacts with $PPh_3=CH_2$ to form the metallacycle $\overline{Lu[C_6H_4P(Ph)CH_2]Cp_2}$.[31]

The alkyls MMe_2Cp*_2 (M = Th, U) undergo consecutive insertions with CO_2 resulting in acetates $MMe(Ac)Cp*_2$ and $M(Ac)_2Cp*_2$, the latter is also available from NaAc and MCl_2Cp*_2.[32] The complexes MMe_2Cp_2 (M = U, Th) adsorb onto dehydroxylated alumina supports to give materials which are catalytic for propene hydrogenation, butene isomerisation and ethene polymerisation.[33] These active species are tractable to high resolution solid state ^{13}C n.m.r. analysis.[34] The metallacycle $Hf\{\overline{CH_2(CH_2)CH_2}\}Cp_2$ reacts with $PPh_3=CH_2$ to form $Hf(Et)(CH=PPh_3)Cp_2$, $Hf[\overline{CH_2CH_2CH_2}]Cp_2$, $H_2C=CH_2$, and PPh_3. The latter complex reacts with PhCHO to form the metallacycle (1).[35]

Isonitriles, RNC (R = Cy or xylyl), give iminoalkylamido insertion products $U(NEt_2)[\eta^2-C(NEt_2)=NR]Cp_2$ or $U[\eta^2-C(NEt_2)=NR]_2Cp_2$ upon reaction with $U(NEt_2)_2Cp_2$, depending on the stoicheiometry. In related reactions $UCl(NEt_2)Cp*_2$ reacts with CyNC to form $UCl[\eta^2-C(NEt_2)=N(Cy)]Cp*_2$ and with MeLi to form $UMe(NEt_2)Cp*_2$, the methyl of which is more prone to insertion than the NEt_2 group.[36] The compound $U(THF)Cp'_3$ reacts with PhNCO to give the complex $U[\eta^2-C(OUCp'_3)=NPh]Cp'_3$,[37] while $UMeCp_3$ undergoes insertion with CyNC to form $U[\eta^2-C(Me)=NCy]Cp_3$.[38]

Tetrachlorides MCl_4 (M = U, Th) give 95% yields of metallacycles

(2) on addition of NaN(SiMe$_3$)$_4$. Ketones insert into the M-CH$_2$ bond to form larger metallacycles.^{39} Compounds (2) also react with other reagents, these include PhC≡CH which results in the alkynide M(CCPh)[N(SiMe$_3$)$_2$]$_3$, py which affords M(η^2-C$_5$H$_4$N)[N(SiMe$_3$)$_2$]$_3$, and CpH which gives $\overline{\text{M[CH}_2\text{Si(Me)}_2\text{N}}$(SiMe$_3$)]Cp$_2$.^{40}

Various MCl$_3$ (M = La, Pr, Er) react with RLi [R = (CH$_2$)$_3$NMe$_2$] to form Li$_3$[MCl$_3$R$_3$] while the analogous Ce reaction gives Li$_3$[CeR$_6$].^{41} The metals M (M = Yb, Eu, Sm, Ce) react with C$_6$F$_5$Br to form "MBr(C$_6$F$_5$)" which are reactive towards Ph$_3$SnCl and water while the corresponding iodides are accessible from the reaction of MIPh with C$_6$F$_5$H. The same metals react with α-iodothiophene to form $\overline{\text{MI(C=CHCH=SC}}$H).^{42} Ethyllanthanoid iodides ("MEtI"; M = Pr, Nd, Sm), prepared from the metal and EtI, are precursors of catalysts for Tishchenko condensation of aldehydes to esters.^{43} Organocerium compounds CeCl$_2$R react with α,β-unsaturated carbonyls R^{1}R^{2}C=CHCOR^3 to form 1,2-addition products in high regioselectivity by a polar pathway.^{44} Excess "YbIPh" reacts with PhHC=CHCOPh to give PhCH$_2$CH=CPh$_2$, trans-PhHC=CHPh (C-C cleavage product), and PhCH$_2$OH.^{45}

Amalgams M/Hg (M = Sm, Eu, Yb) react with HgFc$_2$ or Hg[(C$_5$H$_4$)Mn(CO)$_3$]$_2$ to give the alkyls MFc$_2$ or M[(C$_5$H$_5$)Mn(CO)$_3$]$_2$^{46} while the mercury derivatives Hg[η^6-C$_6$H$_5$Cr(CO)$_3$]$_2$ react with M filings (M = Sm, Eu, Yb) to form M[η^6-C$_6$H$_5$Cr(CO)$_3$]$_2$ as THF solvates.^{47}

Base free Yb$_2$[N(SiMe$_3$)$_2$]$_4$ reacts with AlMe$_3$ to form Yb[N(SiMe$_3$)$_2$]$_2$[AlMe$_3$]$_2$, which has four Yb-Me-Al interactions. It polymerises C$_2$H$_4$ under mild conditions when AlMe$_3$ will not.^{48}

The aryls M(o-C$_6$H$_4$CH$_2$NMe$_2$)(THF)(η-cot) and M(o-C$_6$H$_4$CH$_2$NMe$_2$)$_3$ (M = Er, Yb, Lu) react with sterically demanding RC≡CH forming alkynides but do not react with C$_2$H$_4$ or H$_2$.^{49} Hydrogen reacts with HfMe$_2$[=P(Bu^t)$_2$]Cp* to form a dimeric methyl complex which undergoes reaction with C$_2$H$_4$ to form HfMe(Et)[=P(Bu^t)$_2$]Cp*.^{50}

3. Group 4 (Ti, Zr, and Hf)

A review concerning M(η^4-diene)Cp$_2$ (M = Zr, Hf) is relevant.^{51}

The anions [ZrMe$_2$Cp$_2$]$^-$ form on variable temperature electrochemical reduction of ZrMe$_2$Cp$_2$.^{52} Methylation (AlMe$_3$ or MeLi) of [ZrClCp$_2$]$_2$ gives [ZrMeCp$_2$]$_2$ which reacts with CO or CNR to form [ZrMe(L)Cp$_2$]$_2$ (L = CO, CNR). These equilibrate with their respective η^2-acyls. The bis-zirconaethene (3) forms in the

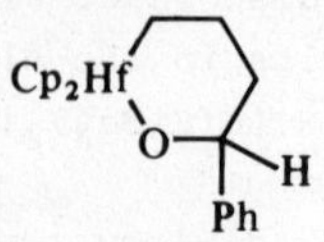
Cp₂Hf
O
H
Ph

(1)

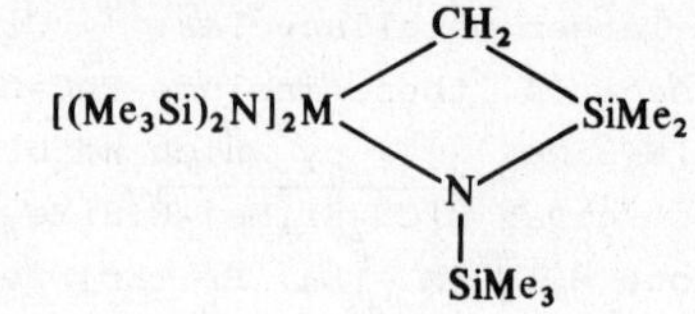
CH₂
[(Me₃Si)₂N]₂M
SiMe₂
N
SiMe₃

(2)

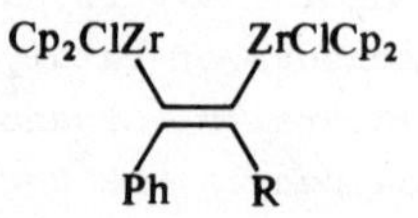
Cp₂ClZr
ZrClCp₂
Ph
R

(3)

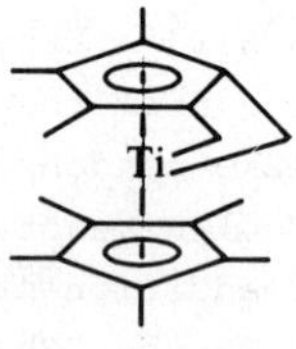
Ti

(4)

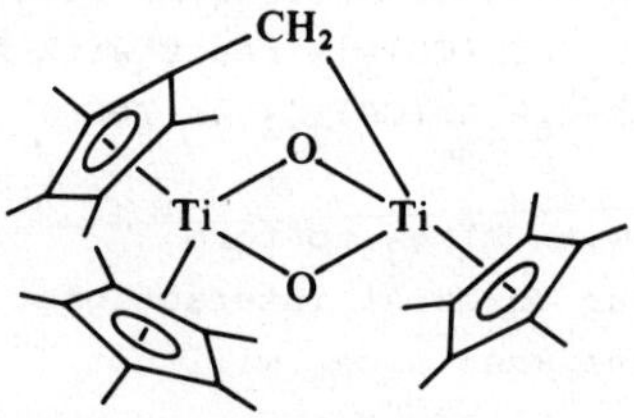
CH₂
O
Ti
Ti
O

(5)

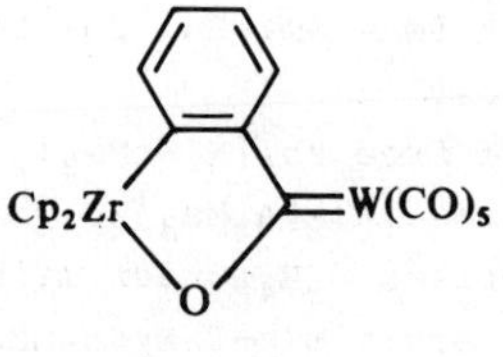
Cp₂Zr
W(CO)₅
O

(6)

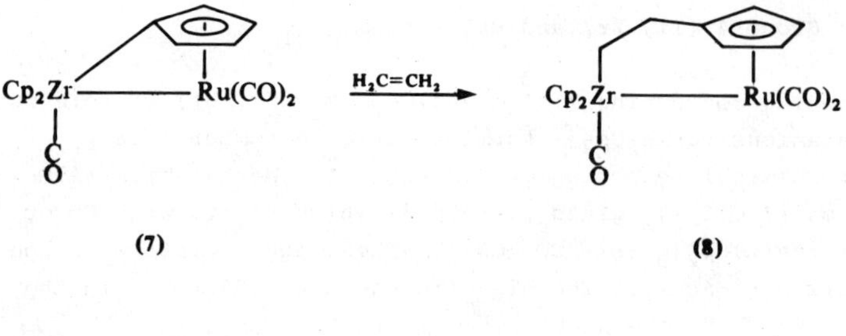
Cp₂Zr
Ru(CO)₂
C
O
H₂C=CH₂
Cp₂Zr
Ru(CO)₂
C
O

(7) (8)

reaction of $[ZrClCp_2]_2$ with $PhC\equiv CR$ (R = H, Ph).[53] Treatment of $ZrCl_2Cp_2$ with $LiCH_2PMe_2$ gives $ZrCl(CH_2PMe_2)Cp_2$ while $LiCH_2PPh_2$ reacts with $ZrCl_2Cp*_2$ to form the related $ZrCl(CH_2PPh_2)Cp*_2$.[54] Addition of $LiCH_2SPh$ to $TiCl_2Cp_2$ provides $Ti(CH_2SPh)_2Cp_2$.[55] Norbornyl lithium (norLi) reacts with $TiCl_2Cp_2$ to form $TiCl(nor)Cp_2$, $Ti(nor)_2Cp_2$, and $[Ti(nor)_3Cp_2]^-$, depending on the stoicheiometry, the last decomposes below room temperature to $Ti(nor)_3Cp$.[56] Aryl derivatives $Zr(ar)(N=CHR)Cp_2$ are formed from ArLi and $ZrCl(N=CHR)Cp_2$.[57] The chlorides MCl_2Cp*_2 (M = Zr, Hf) metathesise with $LiC\equiv CBu^t$ to the alkynides $M(CCBu^t)Cp*_2$.[58] Reduction of $ZrCl(CH_2PPh_2)Cp_2$ with $LiAlH(OBu^t)_3$ provides $ZrH(CH_2PPh_2)Cp_2$ which upon thermolysis or treatment with BuLi gives metallacycle $\overline{Zr(CH_2PPh_2)}Cp_2$ and $ZrHCp_2$. On the other hand, reaction with phosphines gives $ZrH(PR_3)Cp_2$ and the metallated derivatives $\overline{Zr(C_6H_4PR_2)}Cp_2$ in various proportions.[59] Reaction of MCl_2Cp_2 with $LiCH_2C_6H_4ER_2\text{-}\underline{o}$ leads to $MCl(CH_2C_6H_4ER_2\text{-}\underline{o})Cp_2$ or $M(CH_2C_6H_4ER_2\text{-}\underline{o})_2Cp_2$ (M = Zr, Hf; ER_2 = NMe_2, PPh_2) according to the reaction conditions.[60]

Various alkynes $R^1C\equiv CR^2$ react with MH_2Cp*_2 (M = Zr, Hf) to give $MH(CR^1=CR^2H)Cp*_2$ or $M(CR^1=CR^2H)Cp*_2$, some of the former in the presence of excess alkyne proceed to zirconacyclopentenes.[58] The dihydride $Hf(H)_2Cp*_2$ reacts with PhLi to form $HfH(Ph)Cp*_2$ and inserts with alkenes $R^1R^2C=CH_2$ to give the alkyls $HfH(CH_2CHR^1R^2)Cp*_2$. The latter gives $\overline{Hf[CH_2(CH_2)_2CH_2]}Cp*_2$ and $MeCHR^1R^2$ with excess C_2H_4. The alkyne $Bu^tC\equiv CH$ reacts with $Hf(H)_2Cp*_2$ to form the vinyl $HfH(CH=CBu^tH)Cp*_2$ which in turn inserts with C_2H_4 to form $HfEt(CH=CBu^tH)Cp*_2$ and ultimately $\overline{Hf[CH_2(CH_2)_2CH_2]}Cp*_2$.[61]

The zirconium species $ZrMe_2Cp_2$ inserts with $Ph_2C=C=O$ to give $ZrMe[OC(Me)=CPh_2]Cp_2$ initially while ArNCO forms $\overline{ZrMe[OC(Me)NAr]}Cp_2$, and $R^1N=C=NR^1$ leads to $\overline{Zr[N(R)C(Me)NR]}Cp_2$.[62] The same $ZrMe_2Cp_2$ reacts with $MoH(CO)_3Cp$ by evolution of CH_4 and formation of the isocarbonyl $ZrMe[OCMo(CO)_2Cp]Cp_2$.[63] The ylide $PPh_3=CH_2$ reacts with $[Zr(H)ClCp_2]_x$ to give $ZrClMeCp_2$, with $ZrPh_2Cp_2$ to form $ZrPh(CH=PPh_3)Cp_2$ while $ZrMe_2Cp_2$ is unreactive up to $170°C$.[64] Photolysis of $ZrPh_2Cp_2$ in the presence of hexadienes gives biphenyl and $Zr(diene)Cp_2$ complexes.[65] The rates of ligand redistribution in ZrR_2Cp_2, $ZrR(X)Cp_2$, and ZrX_2Cp_2 varies as a function of the ligand in the order F,I > Cl,Br and Me > Ph.[66]

The methyl groups of one Cp* ligand are activated in the conversion of $TiRCp*_2$ to $Ti(\eta^5\text{-}C_5Me_4CH_2)Cp*$ by the elimination of

RH (R = Me, Et, Pr). This species subsequently loses H_2 to form (4).[67] Another activation occurs in the reaction of $TiCp^*_2$ with N_2O which forms the unusual compound (5) while eliminating Cp^*H and N_2.[68]

The aryl $ZrPh_2Cp_2$ reacts with $W(CO)_6$ to form (6), a reaction that proceeeds via an unisolated $Zr(benzyne)Cp_2$ complex.[69] Thermolysis of $ZrPh[CH(SiMe_3)SMe]Cp_2$ gives $Zr(SMe)[CH(SiMe_3)Ph]Cp_2$ while further heating provides $PhCH_2SiMe_3$. The corresponding Cl, Me, and Bz derivatives are unreactive.[70]

Alpha hydrogen activation is not a significant factor in either the rate or stereochemistry of alkene insertion into the Ziegler polymerisation system $TiCl[(CH_2)_nCH=CH_2]Cp_2/AlCl_2Et$.[71] Insertion of $PhC\equiv CSiMe_3$ into the Zr-C bond of the soluble Ziegler system $TiCl_2Cp_2/AlCl_2Me$ to give $[Ti\{C(SiMe_3)=C(Ph)Me\}Cp_2][AlCl_4]$ is a direct observation of an unsaturated hydrocarbon insertion for this system.[72] The dehydrogenative coupling of primary silanes to form linear polysilanes is catalysed by TiR_2Cp_2 (R = Me, Bz).[73]

The structures of the fulvene complex $Ti(\eta^5-C_5H_4CPh_2)_2$ and its CO adduct are described.[74] The heterobimetal complex $Zr[Ru(CO)_2Cp]_2Cp_2$ and CO equilibrate with (7). Complex (7) reacts with C_2H_4 by insertion to give (8), which in turn reacts with Bu^tOH to form $Zr[(CH_2CH_2C_5H_4RuH)(CO)_2](OBu^t)Cp_2$.[75] The compounds $O[M(CH_2PPh_2)Cp_2]_2$ (M = Ti, Zr) are suitable synthons for heterobimetallic complexes.[76]

The X-ray structures of $M(diene)Cp_2$ (M = Zr, Hf) suggest they are best regarded as σ^2,π complexes but the σ/π ratio is less for M = Zr and acyclic dienes. These structural differences affect the reactivity.[77] The bis-phosphine complex $Ti(PMe_3)_2Cp_2$ reacts with $HC\equiv CH$ to give the titanacycle $\overline{Ti(CH=CHCH=CH)}Cp_2$ and polyacetylene via $Ti(HCCH)(PMe_3)Cp_2$. Substituted alkynes lead to various isomeric metallacycles.[78] Zirconacycles $\overline{Zr(CPh=CRCR=CPh)}Cp'_2$ arise in the reaction of $[ZrH(\mu-H)Cp'_2]_2$ with $PhC\equiv CR$ (R = H, Ph).[79] The Grignard type complex $Ti(CH_2MgBr)_2Cp_2$ reacts with MCl_2Cp_2 to form the 1,3-metallatitanacyclobutanes $\overline{Cp_2MCH_2TiCH_2}Cp_2$ (M = Ti, Zr, Hf).[80] The titanacycle $\overline{Ti[CH_2C(Me)Pr^iCH_2]}Cp_2$ decomposes at 0°C to give the short lived $Ti(=CH_2)Cp_2$ which is trappable through its reaction with $RB\equiv NR'$ which affords $\overline{Ti(CH_2BR=NR')}Cp_2$.[81] There is evidence for radical mechanisms in the reactions of $Ti(=CH_2)Cp_2$ (from $\overline{Ti(CH_2CH_2CH_2)}Cp_2$) with RX. Thus reaction of $\overline{Ti(CH_2CH_2CH_2)}Cp_2$ and BzCl leads to $TiCl(CH_2Bz)Cp_2$ and Bz_2.[82] Styrene reacts with $Ti[CH_2Al(Me)_2Cl]Cp_2$ to form $\overline{Ti[CH_2CH(Ph)CH_2]}Cp_2$ and

Ti($CH_2CH_2\overline{C}HPh$)Cp_2, the former of which isomerises to the latter, apparently by an intermolecular process involving readdition of styrene to Ti(=CH_2)Cp_2.[83]

Alkynes, nitriles, CO_2, and MeCHO add to the coordinated ethene of Ti(C_2H_4)$Cp*_2$ to form metallacyclopentenes, metallacycloimines, metallacyclolactones, and oxymetallacyclopentanes respectively. Many of these products are reactive in their own right.[84] Methylene cyclopropanes $CH_2CH(R)C=CH_2$ react with Ti(C_2H_4)$Cp*_2$ to form titanaspiroheptanes (**9**) (R = H, Ph).[85] The alkyne compound Ti(PhCCPh)Cp_2 reacts with Me_2CO or PhCHO to form oxymetallacyclopentenes $\overline{Ti[OC(R^1)}(R^2)\overline{CPh=CPh})Cp_2$ ($R^1 = R^2 = Me$; $R^1 = Ph$, $R^2 = H$).[86] The enynes $R_3SiC≡C(CH_2)_nCH=CH_2$ react in the presence of $ZrCl_2Cp_2$ and $Mg/HgCl_2$ to form (**10**) (n = 3, 4), these give α-silylcyclopentenones upon carbonylation.[87]

Carbonylaton of ZrCl($SiMe_3$)Cp_2 reversibly gives the η^2-acyl ZrCl(η^2-COMe)Cp_2.[88] The η^2-acyls ZrX(η^2-COMe)Cp_2 (X = Me, Cl; R = Me, Ph) undergo initial hydrogen transfer with MH_2Cp_2 (M = Mo, W), $ReHCp_2$ or ReH_4(PMe_2H)Cp to form ZrX[OCH(R)ML_n]Cp_2, the fate of which depends upon the particular example.[89] The electronic and geometric structures of various η^2-acyls MX(η^2-COR)Cp_2 correlate with their reactivity.[90,91] Treatment of the η^2-acyl (**11**) with $ZrMe_2Cp_2$ leads to a trinuclear η^2-acetone complex (**12**).[92]

The Grignard CH_2=CHCH$_2$MgCl reacts with $ZrCl_3Cp$ to form Zr(allyl)$_3$Cp, which subsequently reacts with more $ZrCl_3Cp$ to give Zr(allyl)Cl_2Cp. Subsequent treatment of the monoallyl with RLi gives Zr(allyl)R_2Cp (R = tol, Me).[93] Reduction of ZrR_2Cp with RLi/OEt_2 affords Zr(Bz)Cp.OEt_2 (R = Bz) or Zr(Ph)Cp.$3OEt_2$ (R = Ph).[94] Addition of the Grignards RMgBr (R = Me, Et) to TiCl(dmpe)(chpt) provides TiR(dmpe)(chpt), the methyl complex is also available for the P(Pr^i)$_2CH_2CH_2$P(Pr^i)$_2$ case.[95]

The alkyls $TiMe_nX_{(4-n)}$ (n = 1, 2, 3) form during reaction of $TiMe_4$ and TiX_4. Thermolysis of $TiMeX_3$ gives TiX_3 while thermolysis of $TiMe_2X_2$ gives Ti(II)X_2 in a very reactive form.[96]

The results of calculations on =CH_2, =CHX, and =CX_2 (X = H, OH, and SH) derivatives of H_2Ti, Cl_2Ti, and Cp_2Ti fragments are linked to their geometries and rotational barriers.[97] Calculations on the isomerisation of $\overline{Ti(CH_2CH_2CH_2})Cl_2$ to Ti(=CH_2)(C_2H_4)Cl_2 suggest the process is a formal 2π + 2π process with no activation barrier and that the metallacycle is more stable than the alkene-alkylidene complex by 49 kJ mol^{-1}.[98]

The spectroscopic properties of M(1-norbornyl)$_4$ (M = Ti, Zr, Hf)

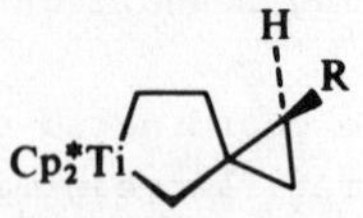

(9)

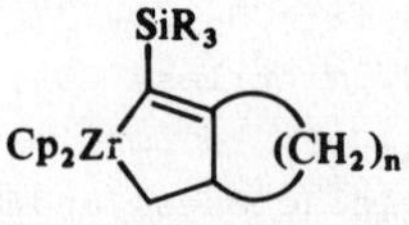

(10)

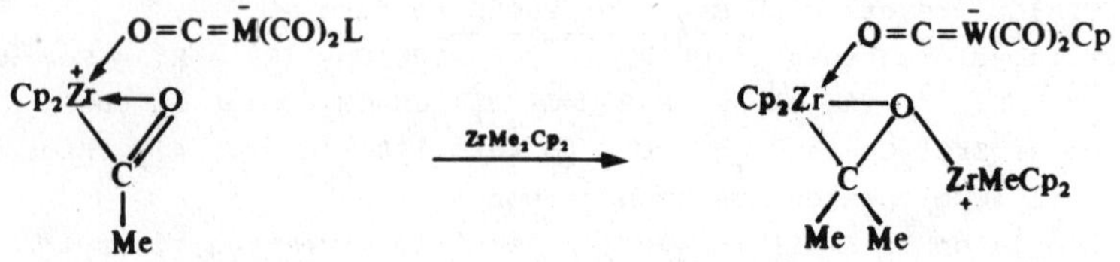

(11) L = Cp or Cp′;
M = Mo or W

(12)

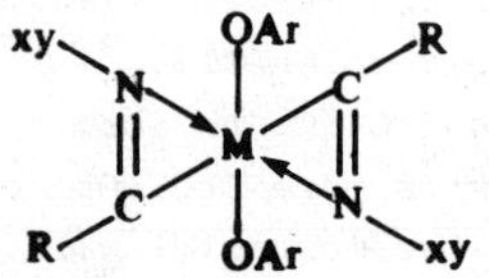

(13) R = Me, Bz; M = Ti, Zr

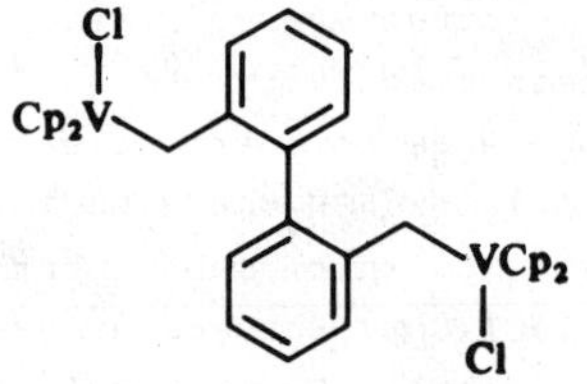

(14)

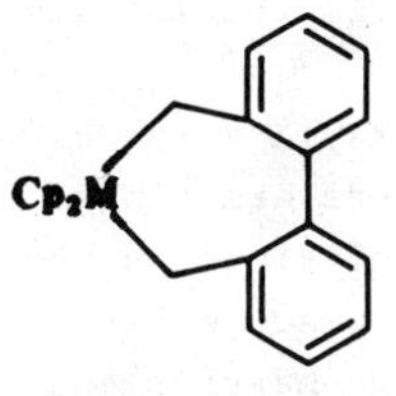

(15) M = Nb, Ta

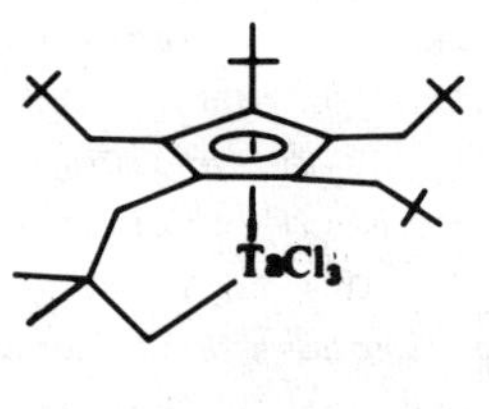

(16)

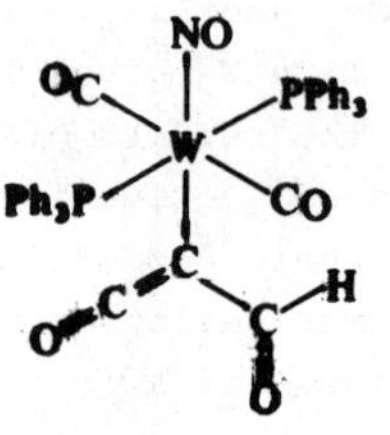

(17)

and the 2-norbornyl derivatives of Ti and Zr are reported.[99]

The sterically demanding alcohol 2,6-di-Bu^t-phenol (Ar^1OH) gives the substitution products $Zr(OAr^1)(Bz)_3$ and $Zr(OAr^1)_2(Bz)_2$ on reaction with $Zr(Bz)_4$.[100] Thermolysis of $M(OAr^1)_2(Bz)_2$ (M = Ti, Zr) results in formation of $M(OAr^1)(Bz)[OC_6H_3(Bu^t)C(Me)_2CH_2]$ and loss of toluene, an example of intramolecular C-H bond activation by M(IV).[101] The isonitrile xyNC (xy = xylyl) reacts with $MR_2(OAr)_2$ (M = Ti, Zr; R = Me, Bz) to form the η^2-iminoacyls (13), via the monoinsertion products. These undergo intramolecular cis coupling during thermolysis to form chelating enediamides complexes.[102,103]

Electrophiles such as HCl react with $Zr[CH(R)Al(Bu^i)_2Cl$ by either C-Al cleavage (HMPA present) to give $ZrCl(CH_2R)Cp_2$ or C-Zr cleavage (no HMPA) resulting in $ZrCl_2Cp_2$.[104]

4. Group 5 (V, Nb, and Ta)

The results of EHMO calculations on various diene derivatives of TaCp and TaCp* suggest they are best regarded as bent metallacyclopent-3-enes.[105] The bonding and behaviour of these complexes is the subject of review.[106]

Methyl lithium reacts with $VClCp*_2$ to form $VMeCp*_2$. This compound undergoes hydrogenolysis to the hydride $VHCp*_2$ and inserts with CO (1 atm.) to form $V(COMe)(CO)Cp*_2$ via the two intermediates $VMe(CO)Cp*_2$ and $V(COMe)Cp*_2$.[107] Interaction of VCl_2Cp_2 with $(\underline{o}-[(THF)_nClMgCH_2]C_6H_4)_2$ gives the alkyl (14) while the corresponding Nb and Ta reactions lead to (15).[108] A new synthesis of bis-niobocene involves the reaction of $NbCl_2Cp_2$ with NaH.[109]

Strong π-acceptors such as CO, CO_2, or CNxy, but not alkynes or nitriles, react with $V(\eta^1-C_3H_5)Cp_2$ to form respectively the dicarbonyl $V(CO)_2[\eta^4-C_5H_5(C_3H_5)]Cp$, the carboxylate $V(OCOC_3H_5)Cp_2$, and the iminoacyl $V[\eta^1-C(C_3H_5)=Nxy]Cp_2$ (xy = xylyl).[110] Photolysis of VMe_2Cp_2 produces methane and ethane, the yield of ethane is dependent on solvent, reaction time, and temperature. Carbonylation in the presence of Na results in $V(CO)_4Cp$.[111]

Hydride to alkene migration of $NbH(CH_2=CHR)Cp*_2$ in the presence of L (L = CNMe, CO) leads to the alkyls $Nb(CH_2CH_2R)LCp*_2$, the reaction kinetics are interpretable in terms of concerted four centre cyclic transition states.[112] Alkynes with electron withdrawing substituents insert into the M-H bond of $MH(CO)Cp_2$ resulting in the formation of the alkynyls $M(CR=CHR)(CO)Cp_2$ (M =

Nb, Ta) in regio- and stereoselective reactions.[113]

The dimeric alkylidynes $Zn(\mu-Cl)_2[Ta(Cl)_2(DME)(\mu-CBu^t)]_2$ undergo complex reactions with alkynes $R^1C\equiv CR^2$ forming products in which two alkynes and one CBu^t group combine to form cyclopentadienyl rings. However with dineopentylethyne, intramolecular metallation of one or two neopentyl groups occurs in the reaction of dineopentylacetylene, forming (16) and a similar, but dimetallated complex.[114] The reaction of sodium sand, $TaCl_5$, and PMe_3 gives $Ta(PMe_3)_3(\eta^2-CH_2PMe_2)(\eta^2-CHPMe_2)$.[115]

5. Group 6 (Cr, Mo, and W)

Reviews on C-H activation[116] and molybdenum and tungsten chemistry[117] contain some relevant material.

Hydrolysis of the M-C bond of α-hydroxyalkyl Cr(III) complexes proceeds by a mechanism in which the electrophile is a solvent H_2O rather than a cis aqua ligand.[118] Reaction of Cr(II)(aq) with the $CH(OH)_2$ radical proceeds via a transient species to the hydrated formyl $[Cr(CH(OH)_2)(OH_2)_5]^{2+}$ which decomposes to the formyl $[Cr(CHO)(OH_2)_5]^{2+}$. This formyl reacts with $H_2C=O$ with formation of MeOH, CO, and $[Cr(OH_2)_6]^{3+}$.[119] The alkyl $[Cr(CH_2CH_2OEt)(OH_2)_5]^{2+}$ decomposes by an acid catalysed process in aqueous solution to $[Cr(OH_2)_6]^{2+}$, ethene, and ethanol.[120]

The porphyrin $Mo(CO)_2(TPP)$ undergoes reaction with chlorinated benzene impurities to form the σ-phenyl $Mo(Cl)(\eta^1-Ph)(TPP)$.[121] The protonations of $MoO(S_2CNMe_2)_2(R^1CCR^2)$ with CF_3COOH which form Mo(VI) vinyl oxo complexes are used as models for the nitrogenase enzyme catalysed formation of ethene from ethyne.[122]

Methyl iodide reacts with $W(CO)_3(P\{Pr^i\}_3)[\eta^3-MeC(S)SMe]$ to form $WI[CMe(SMe)-SMe](CO)_3(P\{Pr^i\}_3)$.[123]

The kinetics of CO_2 insertion into anionic alkyl and aryl complexes cis-$[WR(CO)_4\{P(OMe)_3\}]^-$ to give primarily $[W(CO_2Me)(CO)_4\{P(OMe)_3\}]^-$ are comparable with those of the corresponding CO insertion reactions.[124] The reaction proceeds with retention of configuration at the α-C atom.[125] The reaction of $[W(CO)_5]^{2-}$ with $PPh_2[CH_2]_3Cl$ leads to the cyclic acyl $[\overline{W\{C(O)(CH_2)_3PPh_2\}}(CO)_4]^-$. This acyl decarbonylates on heating to the anion $[\overline{W\{(CH_2)_3PPh_2\}}(CO)_4]^-$ which is reactive towards CO_2.[126] The metallaketene (17) forms in the reaction of $WH(NO)(CO)_2(PPh_3)_2$ with carbon suboxide.[127] The insertion compounds $\overline{M[P(Ph)-CR=CR-C(O)]}(CO)_5$ (M = Cr, Mo, W; R = Ph, Et) are in

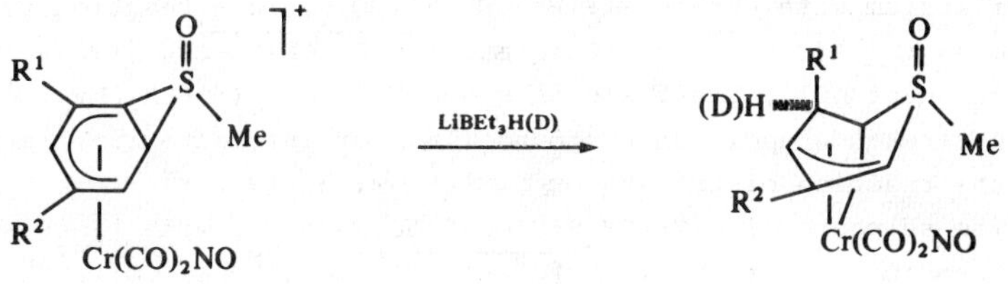

(19) R^1, R^2 = alkyl, Ph   (18)

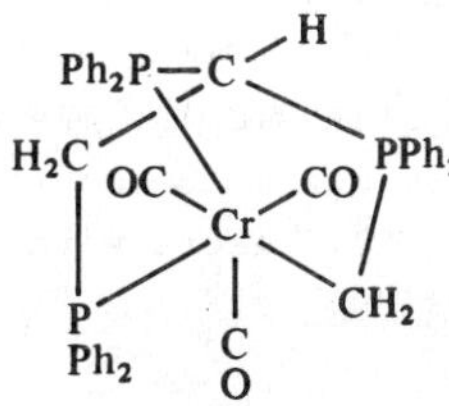

(20)

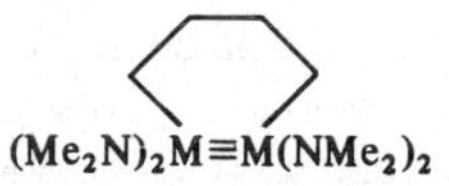

(21) M = Mo, W

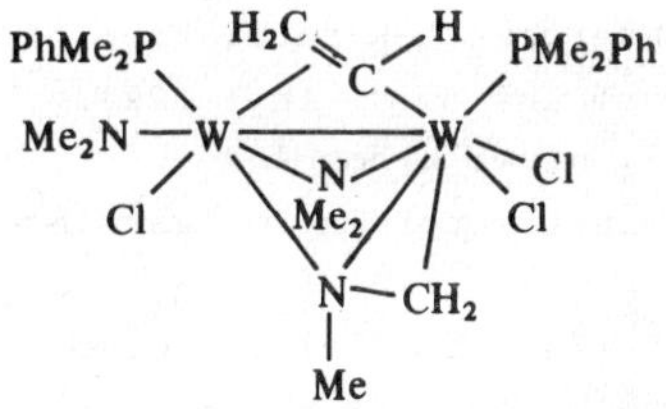

(22)

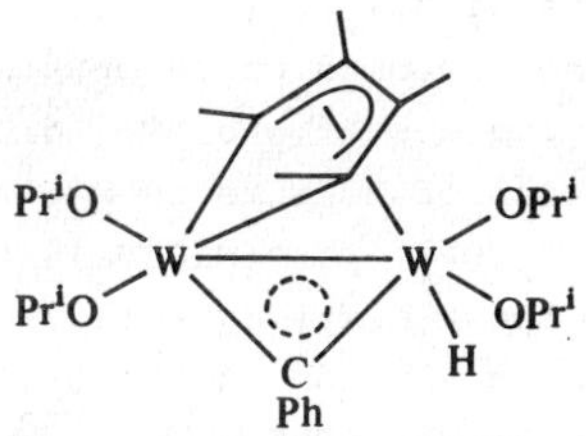

(23)

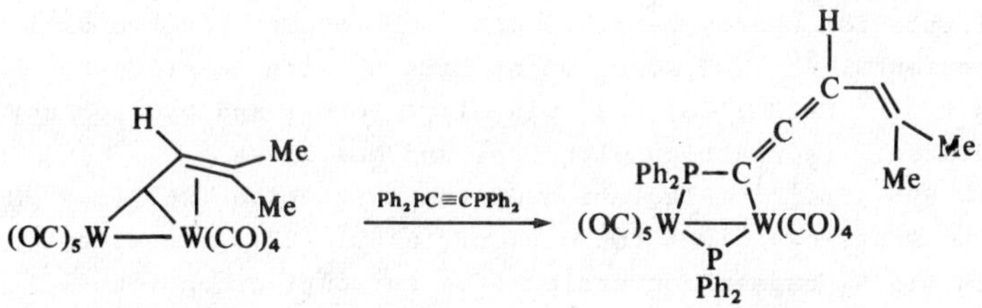

(24)   (25)

equilibrium with CO and $M[\overline{P(Ph)CR=CR}](CO)_5$, in a reaction that may proceed via $\overline{M(CR=CR-PPh_2)}(CO)_5$ and $\overline{M[C(O)-CR=CR-PPh_2]}(CO)_5$.[128] Reaction of WCl_6, PMe_3, and Na gives $WH(PMe_3)_4(\eta^2-CH_2PMe_2)$.[115]

A rare endo addition compound (**18**), or its deuteride, is formed in the reaction of $LiEt_3BH$ or $LiEt_3BD$ with (**19**).[129]

The sulphur ylide $Cr[CH_2S(O)Me_2](CO)_5$ reacts with $(Ph_2P)_2C=CH_2$ to form the electrophilic $\overline{Cr[CH_2P(Ph)_2C(=CH_2)PPh_2]}(CO)_4$.[130] In related reactions with $(R_2P)(R_nE)C=PR_3$ the analogous complexes $\overline{Cr[CH_2PPh_2C(=PR_3)PPh_2]}(CO)_4$ arise ($ER_n = PR_2$ or SPh; R various combinations of Me, Ph).[131] The same ylide reacts with $(Ph_2P)_2CHCH_2PPh_2$ to form the tripod compound (**20**),[132] while its reactions with $(R_2As)_2X$ (X = $C=PPh_3$, NMe, O) result in the formation of the ylide chelates $\overline{Cr[CH_2As(R)_2XAsR_2]}(CO)_4$. Related reactions with $R_2ECH_2CH_2E_2R_2$ (E = P, As) give six membered ring chelates.[133] The bis-ylide $Cr[CH_2S(O)Me_2]_2(CO)_4$ reacts with $Te(PR_2)_2$ or $(PBu^t{}_2)_2CH_2$ to form $Cr(PR_2TePR_2)(CO)_4$ or $\overline{Cr[P(Bu^t)_2CH_2P(Bu^t)_2]}(CO)_4$ respectively.[134]

Salt elimination in the reaction of $Li(CH_2)_4Li$ with $1,2-M_2Cl_2(NMe_2)_4$ (M = Mo, W) gives (**21**) containing a MM triple bond bridged by a polymethylene chain.[135] One equivalent of $HC\equiv CH$ reacts with $W_2Cl_3(NMe_2)_3(PMe_2Ph)$ to form a transversely bridged alkyne compound which on standing undergoes C-H activation of a methyl group of one NMe_2 group to form the σ,η^2-vinyl (**22**).[136] Addition of $MeC\equiv CMe$ to $W_2(Bz)_2(OPr^i)_4$ gives a complex $W_2(Bz)_2(OPr^i)_4(\eta^2-MeCCMe)_2$ that eliminates toluene on warming to form the μ-alkylidyne (**23**).[137]

Carbonylation of the μ-alkylidene (**24**) and alcohol solvolysis leads to $Me_2C=CHCH_2CO_2R$.[138] The same W_2 complex reacts with $Ph_2PC\equiv CPPh_2$ with rupture of a P-C bond and formation of (**25**).[139]

The "isolated" C-H methyl stretches are reported for $M(CHD_2)(CO)_3Cp$ (M = Cr, Mo, W). The C-H bond strength decreases in the order Cr>Mo>W, reflecting the increase in M-C bond strength (55-70% from Cr-W). The methyl group is subject to appreciable internal rotation energy barriers and each rotamer has two distinct CH environments.[140] The reaction of $LiEt_3BH$ with $MoMe(CO)_3Cp$ provides $[Mo(\eta^2-MeCHO)(CO)_2Cp]^-$ via alkyl formyl and hydrido acyl intermediates, its reaction with PPh_3 and MeI gives $MoMe(CO)_2(PPh_3)Cp$.[141] Alkylation of the zwitterion (**26**) (L = PMe_3 or $L_2 = Me_2PCH_2PMe_2$) gives the cationic alkyl (**27**). Treatment of (**27**) with $LiAlH_4$ causes conversion of a carbonyl group into a methyl, so forming (**28**).[142] Disubstituted alkynes $R^1C\equiv CR^2$ react with the alkyls $WR(CO)_3L$ (L = Cp, Cp´, Cp*) to form acyls $W(COR)(CO)(R^1CCR^2)L$ which thermally dismute to $W(CR^1=CR^2-CR=O)(CO)_2L$ and $WR(CO)(R^1CCR^2)L$.[143,144] Treatment of $W(CH=CH-CMe=O)(CO)_2Cp$ with NO at $-78°C$ gives $WMe(=O)(\eta^2-HCCH)Cp$ and $W(COMe)(=O)(\eta^2-HCCH)Cp$ in a mechanistically obscure reaction. The former is formed exclusively in the reaction of $WMe(CO)(\eta^2-HCCH)Cp$ and NO.[145] Dehydration of $Mo[CH_2C=CC(R^1)(OH)R^2](CO)_3Cp$ with HBF_4 gives cationic η^2-butatrienes $[Mo(CO)_3(H_2C=C=C=CR^1R^2)Cp]^+$.[146] Treatment of $[M(CO)_3Cp]^-$ (M = Mo, W) with $ClCH_2COR$ (R = OEt, Me) gives the alkyls $M(CH_2COR)(CO)_3Cp$ which photochemically decarbonylate to the η^3-oxaallyls $M(CO)_2[\eta^3-CH_2C(R)O]Cp$.[147] The tungsten complex reacts with PPh_3 by cleavage of the M-O bond to form $W(CH_2COR)(CO)_2(PPh_3)Cp$.[148] The cation $[Mo(CO)_3Cp]^+$ undergoes reaction with $Mo(CH_2OMe)(CO)_3Cp$ resulting in $MoMe(CO)_3Cp$ and the cation $[Mo(CO)_3(=CH\{OMe\})Cp]^+$, while its reaction with $MoEt(CO)_3Cp$ gives the dinuclear cation $[Mo_2(CO)_4(\mu,\eta^2-EtCO)Cp_2]^+$.[149]

Oxidative decarbonylation of $Mo(\eta^1-C_3H_5)(CO)_3Cp$ with Me_3NO gives $Mo(\eta^3-C_3H_5)(CO)_2Cp$.[150] Photolysis of $M(\eta^1-Bz)(CO)_3Cp$ (M = Mo, W) gives toluene and $[Mo(CO)_3Cp]_2$ or $[W(CO)_3Cp]_2$ and $W(\eta^3-Bz)(CO)_2Cp$ according to the metal.[151] The benzyls $Mo(Bz)(CO)_3L$ (L = Cp, Cp´, Cp*) form acyls in their reactions with PR_3 and RNC, the kinetics depending on L. Only small cone angle PR_3 ligands react with the Cp* system.[152] Infrared monitoring of the photolysis of $W(\eta^1-C_3H_5)(CO)_3Cp$ in matrices suggests that the $\eta^1-\eta^3$ allyl reaction proceeds via the 16-electron $W(\eta^1-C_3H_5)(CO)_2Cp$.[153] The alkynide $W(C=CPh)(CO)_3Cp$ reacts with $C_2(CN)_4$ to form the σ-butadienyl $W[\eta^1-C=CPhC(CN)_2C(CN)_2](CO)_3Cp$ which photochemically decarbonylates to $W(CO)_2[\eta^3-C(CN)_2C(Ph)C(CN)_2]Cp$.[154]

Treatment of $[MI_2(NO)Cp]_2$ with RMgCl in Et_2O leads to $Mg[MR_2(NO)Cp]I_2(OEt_2)$, which hydrolyse to stable 16-electron alkyl complexes $MR_2(NO)Cp$ (M = Mo, W; R = CH_2Bu^t, CH_2SiMe_3).[155] Excess Grignard Me_3SiCH_2MgX gives $W(CH_2SiMe_3)_3(O)Cp$ as a byproduct. This oxide undergoes thermolysis to form the alkyl carbene $W(CH_2SiMe_3)(=CHSiMe_3)(O)Cp$ while the $W(CH_2SiMe_3)_2(NO)Cp$ is oxidised by O_2 to $W(CH_2SiMe_3)(O)_2Cp$.[156] Sulphur inserts into the W-C bonds of $W(CH_2SiMe_3)_2(NO)Cp$ to form $W(SR)_2(NO)Cp$ via two characterised intermediates.[157]

Molybdenum η^2-vinyl compounds arise in the reactions of $LiBHBu^s_3$, $LiCuAr_2$, or $LiCuMe_2$ reagents with various alkyne complexes. THese are described as either η^2, 3 electron vinyls or metallacyclopropenes.[158] Insertion of alkynes into the M=C bond of

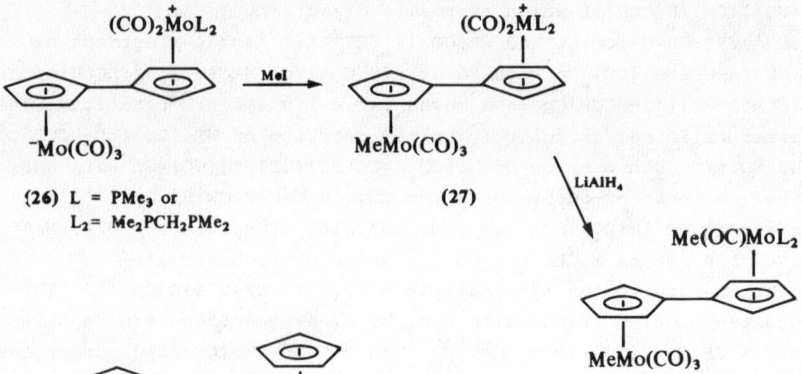

(26) L = PMe$_3$ or
L$_2$ = Me$_2$PCH$_2$PMe$_2$

(27)

(28)

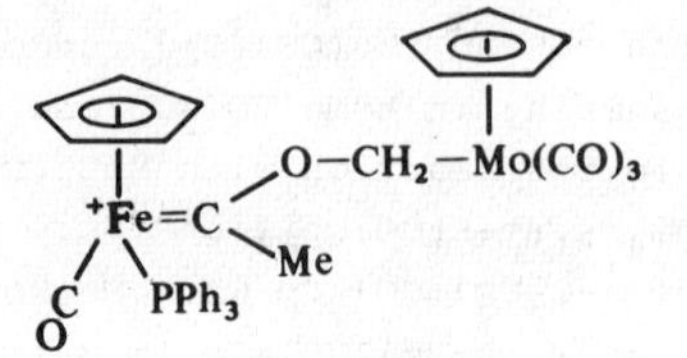

(29)

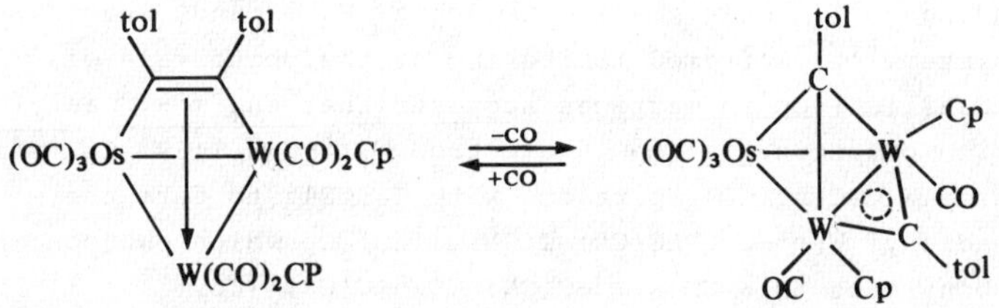

(30)

(31)

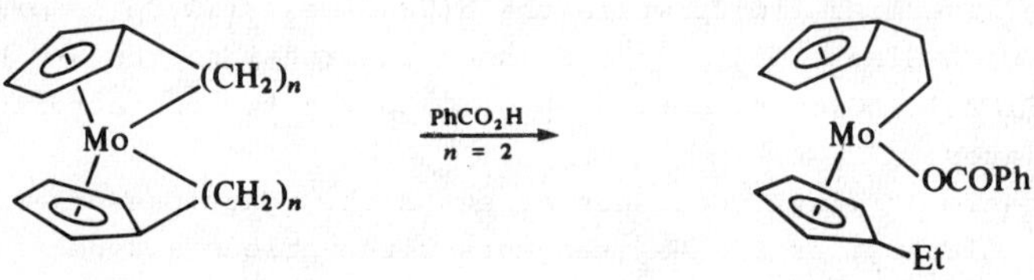

(32) n = 2,4

(33)

the η^2-vinyl $\overline{W(F_3CCCCF_3)[\eta^2\text{-}C(CF_3)C(CF_3)S(R)]}$Cp give isomeric products and shows that η^2-vinyl compounds are plausible intermediates in metal promoted alkyne oligomerisation.[159]

The cationic methylidene [Mo(=CH$_2$)(CO)$_3$Cp]$^+$ alkylates the acyl oxygen of Fe(COMe)(CO)(PPh$_3$)Cp forming (**29**).[160]

The CC triple bond of the bridging alkyne of cluster (**30**) is cleaved in a reversible decarbonylation to form the bis-alkylidyne (**31**).[161] Two molecules of HCCH are incorporated into a C$_4$ ligand in a photochemical reaction of Cp(OC)$_2$Mo(μ-PPh$_2$)(μ-H)Mn(CO)$_4$.[162] A rare μ_3,η^2-methylenearsanediyl complex is formed in the reaction of PPh$_3$=CH$_2$ with the μ_3-As cluster As[Mo(CO)$_2$Cp]$_3$.[163]

The vinyls MH(η^1-CR^1=CR^2)Cp$_2$ (M = Mo, W) react with HX to form MX(η^1-CR^1=CR^2)Cp$_2$[164,165] The electronic properties of these are tractable to study by n.m.r. methods.[166] Under sufficiently dilute conditions the hydrido alkyl W(H)MeCp$_2$ undergoes intramolecular elimination of CH$_4$. At higher concentrations attack upon the C-H of another WHMeCp$_2$ results in H/Me hydrogen exchange which competes with CH$_4$ elimination.[167]
Formaldehyde reacts with Mo[CH(Ph)CH(Ph)]Cp$_2$ by elimination of cis-PhHC=CHPh and formation of Mo(OCH$_2$)Cp$_2$, the same compound arises in the reaction of MoH$_2$Cp$_2$ with formaldehyde.[168] Two mechanisms are proposed for the interconversion of [WH[CH$_2$P(Me)$_2$Ph]Cp$_2$]$^+$ and [WMe[P(Me)$_2$Ph]Cp$_2$]$^+$, one of which involves an equilibrium between [WH(CH$_2$)Cp$_2$]$^+$ and [WMeCp$_2$]$^+$.[169]

Molybdenum atoms react with spiro-[4.4]-nona-1,3-diene forming (**32**; n = 4), this complex further reacts with I$_2$ or protons. Compound (**32**; n = 2) reacts with benzoic acid to form (**33**), which is reactive towards Me$_3$SiCl.[170] The dihydride WH$_2$Cp*$_2$ undergoes intramolecular C-H activation during photolysis forming the "tucked in" compound (**34**) which in turn leads to a doubly "tucked in" compound.[171]

The reaction of CrCp$_2$ with Me$_3$NO yields Cr$_4$O$_4$Cp$_4$ and Cr$_4$O$_3$(μ_3,η^2-C$_5$H$_4$)Cp$_2$.[172] The halide MoX(dppe)(chpt) (X = Cl, I) undergoes X for Me exchange giving MoMe(dppe)(chpt) in its reaction with MeLi.[173]

Deprotonation of [W(S$_2$CNEt$_2$)$_2$(η^4-C$_4$Ph$_4$H)]$^+$ with NEt$_3$/H$_2$O/CH$_2$Cl$_2$ gives the metallacyclopentadiene (**35**).[174] Cleavage of the CX (X = NR, O) triple bond occurs in their reactions with [W(OPr^i)$_2$(μ-CSiMe$_3$)]$_2$, which generate compounds (**36**) as the carbon atom adds to the μ-CSiMe$_3$ ligand.[175]

Reaction of the σ-mesityl complex Mo(mes)$_2$(O)$_2$ with Bu$^n_3$P=CH$_2$

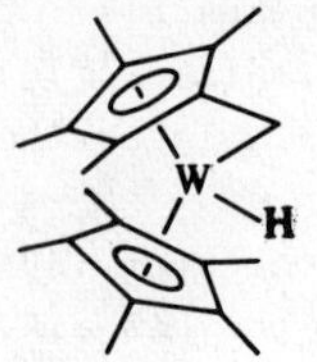

(34)

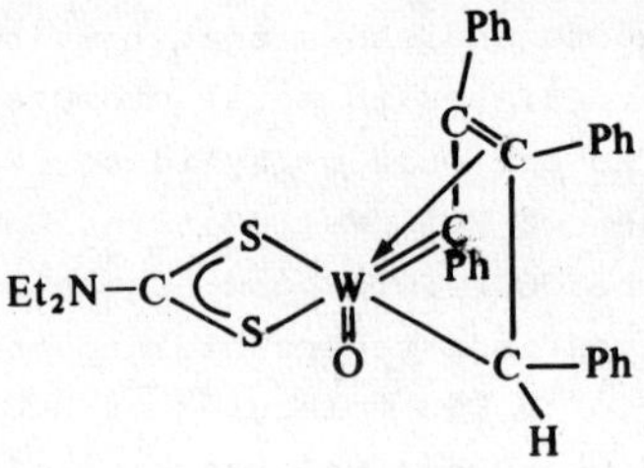

(35)

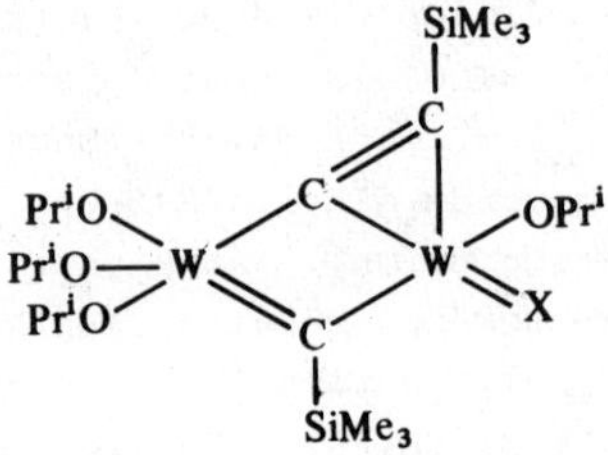

(36)

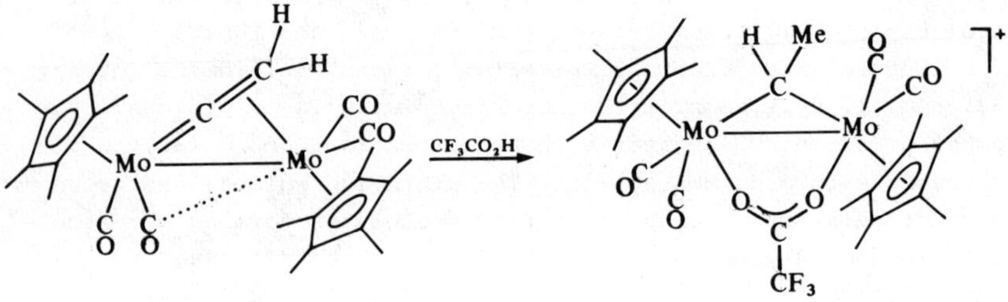

(37) →[CF_3CO_2H] (38)

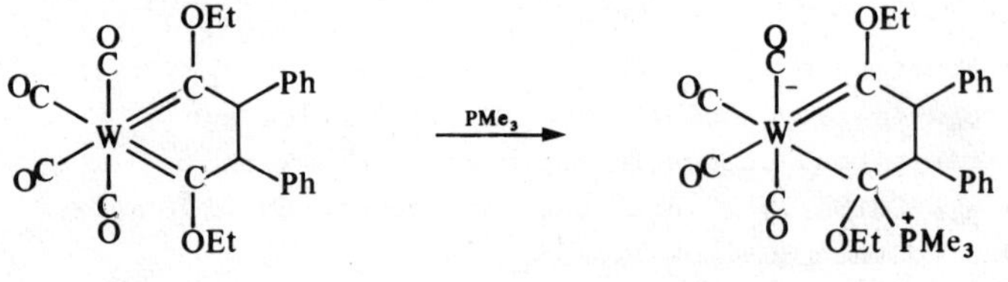

(40) →[PMe_3] (39)

provides the alkylidene $Mo(mes)(O)_2[C(PBu^n_3)(mes)]$.[176] A general synthesis of pentacoordinate W(IV) alkylidenes $WX_2(OCH_2Bu^t)_2(=CHBu^t)$ (X = Br, I, CF_3SO_3, OCH_2Bu^t) is achieved by X for halide ligand exchange. Some of the arising products are metathesis catalysts while others show Wittig type reactivity with esters and lactones.[177] The alkylidene $WBr_2(OCH_2Bu^t)3(=CHBu^t)$ reacts with $GaBr_3$ and various norbornadienes to give a carbene complex used to make polymers.[178] The complex $W(OAr)_2(Cl)(CH_2R^1)(=CHR)(OR^2_2)$ is a well defined, Lewis acid free, homogeneous metathesis catalyst whose activity varies with OAr.[179]

The bis-alkyne complex $M(X)(CF_3CCCF_3)_2Cp$ (X = CF_3, Cl, SR) reacts with Bu^tNC initially to form the η^2-vinyl $Mo(SR)(CF_3CCCF_3)[\eta^2-C(CF_3)C(CF_3)(CNBu^t)]$ which converts to a complex metallacycle.[180]

The structure of a hydrido-alkylidene WOs3 cluster is reported.[181,182]. Photolysis of $[Mo(CO)_2Cp^*]_2$ in the presence of HC≡CH gives the $\mu-\eta^1,\eta^2-C=CH_2$ complex (**37**). This isomerises thermally and reacts with CH_2N_2 or CF_3COOH. The last reaction results in a neutral bridging vinyl which itself reacts further with protons to form the μ-alkylidene cation (**38**).[183]

The reaction of $M(CO)_6$ (M = Cr, W) with LiC(OMe)=CH(OMe) followed by H_2SO_4 or Me_3O^+ results in stereospecific formation of the carbenes $M(CO)_5[=C(OR)(COMe=CHOMe)]$ (R = H, Me).[184] Aldehydes RCHO react with $W(CO)_5[N(Cy)=C=C(OEt)Ph]$ to form the carbenes $W(CO)_5[=CC(OEt)(Ph)OC(H)(R)N(Cy)]$.[185] Alkoxide catalysed reaction of $W(CO)_5[=C(OMe)(tol)]$ with 4-hydroxybut-1-ene gives the carbene $W(CO)_5[=C(OCH_2CH_2CH=CH_2)(tol)]$. Gentle warming gives a cyclopropane derivative via an alkene-carbene complex while reaction with $NHMe_2$ gives $W(CO)_5[=C(NMe_2)(tol)]$.[186]

Treatment of $M(CO)_5[C(OMe)(C≡CPh)]$ (M = Cr, W) with CpH modifies the carbene by Diels-Alder type addition of the alkyne across CpH.[187] Carbenes $Cr(CO)_5$(carbene) react with 1-alkynols to form hydroquinone and/or vinyl lactone derivatives via vinyl ketone intermediates.[188] Annulation reactions of Cr carbene complexes with 3-carbomethoxy-5-hexynoate and CO form part of a new regiospecific route relevant to the synthesis of 4-demethoxydaunomycinone.[189]

Treatment of $W(CO)_6$ with $ZrPh_2Cp_2$ at 80°C gives $W[=CC_6H_4Zr(Cp_2)O](CO)_5$ via an unisolated benzyne intermediate.[69] Alhough the reaction of $Cr(CO)_5[=C(OMe)R]$ with hept-1-en-6-yne gives isomeric bicyclo-[3.2.0]-heptan-6-ones there is no reaction

with the W analogues.[190] The acidic protons α to the carbene in
$Cr(CO)_5[=C(OMe)Me]$ undergo H/D exchange to give statistical $CH_{3-n}D_n$
complexes.[191] Treatment of $Cr(CO)_5[=C(OEt)NR_2]$ with BCl_3 gives
cationic carbynes $[Cr(CO)_5(CNR_2)]^+$ that exchange ^{13}CO for CO with
no site preference and react with Cl^- to form the carbenes
$Cr(CO)_5[=C(Cl)NR_2]$. Thermolysis of the last species results in the
carbyne $CrCl(CO)_4(=CNR_2)$, perhaps by an intramolecular process.[192]
Treatment of the carbene anions $[Cr\{=C(OMe)=CHR\}(CO)_5]^-$ with
$R^1R^2C=O$ and $BF_3 \cdot OEt_2$ or $TiCl_4$ followed by hydrolysis gives
α-hydroxy carbene complexes $M(CO)_5[=C(OMe)CHRC(OH)R^1R^2]$ that have
considerable synthetic potential.[193]

Isonitriles R^2NC insert into the Cr=C bond of $Cr(CO)_5[=C(OMe)R^1]$
to form $Cr(CO)_5[NR^2=C=C(OMe)R^1]$ (R^2 = Me, Cy, Bu^t).[194] Thiocarbene
complexes are formed in the reactions of lithio-thioacetals with
Fischer carbenes $M(CO)_5[=C(OMe)Ph]$.[195] Treatment of
$Cr(CO)_5[=C(OEt)NMe_2]$ with $HAuCl_4$ gives gold carbenes
$AuCl_3[=C(OEt)NMe_2]$.[196] Both the compounds $W(CO)_4ReR(CO)_3(\mu-PPh_2)_2$
(R = Me, Et) form isomeric Fischer-type tungsten carbene complexes
upon treatment with R'Li followed by $EtOSO_2F$.[197] Thermolysis of
$W(CO)_5[=C(SEt)SiPh_3]$ gives $W(CO)_5[S(Et)C(=C=O)SiPh_3]$ and
$(EtS)(SiPh_3)C=C=O$.[198] The reaction of $[Cr(COMe)(CO)_5]^-$ with
toluene sulphonyl chloride followed by arCH=NMe reagents gives
azetidinylidene complexes which are oxidisable to β-lactams.[199]

Nitriles R^1R^2NCN (R^1, R^2 = alkyl) react with $M(CO)_5[=C(Ph)R^3]$ (M
= Cr, Mo, W; R^3 = OMe, Ph) by second order kinetics to form
$M(CO)_5[C(NR^1R^2)(N=CPhR^3)]$.[200] Irradiation of $W(CO)_6$ in the presence
of RCCH gives $W(CO)_4(=C=CHR)(\eta^2-RCCH)$ via initial alkyne and
vinylidene species which are catalytically active in the
photoassisted polymerisation of alkynes.[201] Treatment of $M(CO)_6$ (M
= Cr, W) with $N=CAr_2^-$ followed by Et_3O+ forms
$M(CO)_5[=C(OEt)N=CAr_2]$.[202] Removal of OEt by BF_3 gives cationic
2-azaallenylidenes $[M(CO)_5(=C=N=CAr_2)]^+$ which on photolysis generate
$Ar_2C=CAr_2$.[203] The trans CO of $[W(CO)_5(=C=NAr_2)][AlBr_4]$ is replaced by
Br^- from the counter anion resulting in trans-$WBr(CO)_4(=C=N=CAr_2)$,
the CNC fragment is bent in the product but linear in the starting
material.[204] Reductive dimerisation of the cation
$[Cr(CO)_5(=C=N=CPh_2)]^+$ in THF forms $Ph_2C[Cr(CO)_5(CN)]$ by elimination
of a CPh_2 group.[205]

The carbenes $M(CO)_5(=CR^1R^2)$ (M = Cr, Mo, W) react with BH_3CN^- or
other nitriles to form nitrile compounds; the new species
correspond formally to the products of NC insertion of RCN into the

M=C bond and the carbene into the C-R (of RCN) bond.[206] The alkylidenes $M(CO)_5(=CPh_2)$ (M = Cr, W) react with Me_3SiN_3 to form $M(NH=CPh_2)(CO)_5$ and with Bu^nN_3 (W only) to give $W(NBu^n=CPh_2)(CO)_5$.[207] THe carbene ylide (**39**) is the product of reaction between the bis-carbene (**40**) and PMe_3.[208] The same bis-carbene undergoes either nucleophilic substitution of the OEt groups by NR^1R^2 or abstraction of the carbene α-protons to give 1,4-diethoxy-2,3-phenylbutadiene and cis-$W(NHR^1R^2)_2(CO)_4$ in reactions with amines.[209] The carbene alkene complex $\overline{W(CO)_4[=C(OMe)CH_2CH_2CH=CH_2]}$ reacts with EtC≡CEt to give $W(CO)_6$ and a bicyclo-[4.1.0]-heptane derivative.[210] The anion $[Cr(CO)_5(CPhNH)]^-$ reacts with CS_2 followed by Et_3O^+ to give the carbene $Cr(CO)_5[=C(SEt)Ph]$ while the anion $[Cr(CO)_5\{C(OEt)CH_2\}]^-$ gives related carbenes with the same reagents.[211] Reaction of the ferrocenyl anion $Fe(C_5H_4Li)(C_5H_4PPh_2)$ with $M(CO)_6$ followed by Me_3O^+ gives the zwitterion (**41**) and the chelated carbene (**42**).[212]

Treatment of the anion $[Mo(CNMe)(CO)_2Cp]^-$ with $I(CH_2)_3I$ leads to cis-$\overline{MoI(CO)_2[=C(CH_2)_3NMe]Cp}$ rather than the anticipated trans species.[213] This compound and other related carbene complexes undergo reduction with $NaC_{10}H_8$ resulting in the carbene anions $[M(CO)_2(carbene)Cp]^-$.[214] The fulvene dianion $[(C_{10}H_8)\{Mo(CO)_3\}_2]^{2-}$ reacts with $I[CH_2]_3I$ to form the fluxional carbene (**43**) and with MeI to form the bis-methyl $(C_{10}H_8)[MoMe(CO)_3]_2$. Thermolyis or photolysis results in evolution of propene.[215]

Protonation or alkylation of trans-$M(CNBu^t)_2(dppe)_2$ gives a mixture of cis and trans $[M(CNRBu^t)(CNBu^t)(dppe)_2]^+$ (R = H, Me, Et) and other products.[216] Oxidation of the μ-alkylidyne by X_2 $[W(CH_2SiMe_3)_2(μ-CSiMe_3)]_2$ gives $[WX(CH_2SiMe_3)_2(μ-CSiMe_3)]_2$.[217] There is an extensive chemistry of the same and closely related compounds with alkynes, alcohols[218], and allenes.[219] Other work concerns the thermolysis of alkylidenes.[220]

Treatment of tungstena-cbd $\overline{W(OBu^t)[OC(Me_2)C(Me_2)O](CEtCEtCEt)}$ with 3,7-decadiyne proceeds with insertion of each alkyne bond into the metallacycle and resulting formation of $[W(OBu^t)(O)_2(\eta-C_5Et_4CH_2-)]_2$ containing linked Cp groups.[221] The tungstena-cbd $\overline{W[C(Bu^t)CHC(Bu^t)][OCH(CF_3)_2]_3}$[224] and the deprotonated $\overline{WCl[(CBu^t)CC(Bu^t)]Cp}$[223] have planar tungstena-cbd rings while the compound $\overline{WCl_2[C(Ph)C(Bu^t)C(Ph)]Cp}$ has a nonplanar ring.[224] This molecule reacts with PMe_3 to form $WCl_2(PMe_3)[C_3(Bu^t)_2Me]Cp$ containing a tetrahedral C_3W core.[225]

Interaction of the carbyne anion $[WCl_4(\equiv CBu^t)]^-$ with S_4N_4 gives

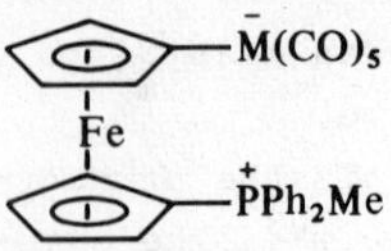

(41) M = Cr, W

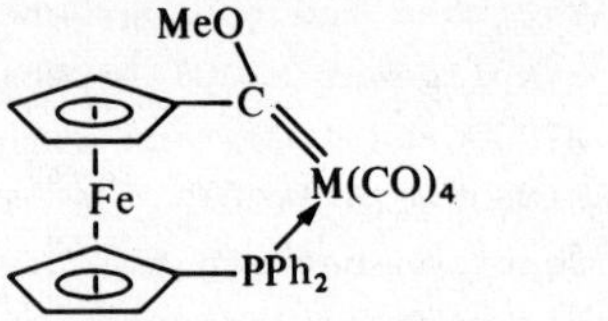

(42) M = Cr, W

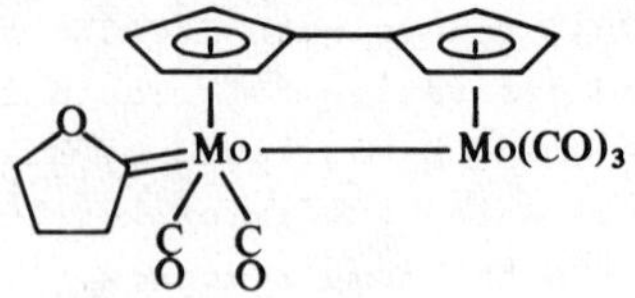

(43)

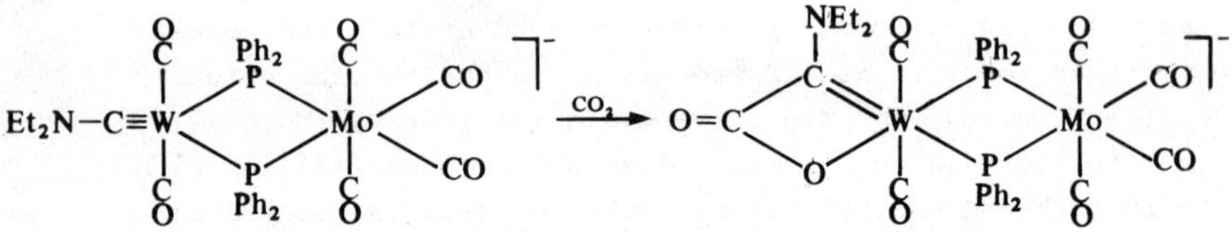

(44) → (45) (CO₂)

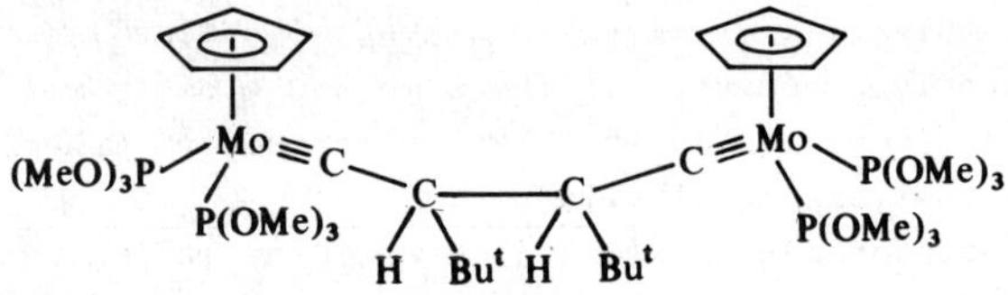

(46)

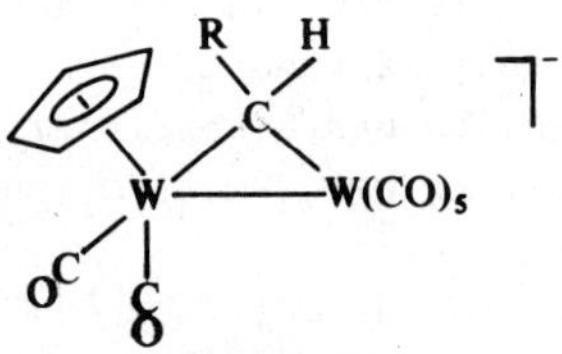

(47) R = Me, tol

[WCl$_3$(O)(OS$_2$N$_2$)]$^-$.[226] The geometry of the compound [WCl$_3$(PMe$_3$)$_3$(=CBu^t)] is monocapped octahedral with capping neopentylidene.[227] The Mo≡Mo complex 1,2-Mo$_2$Br$_2$(CH$_2$SiMe$_3$)$_4$ reacts with PMe$_3$ to form Mo$_2$Br$_2$(=CHSiMe$_3$)$_2$(PMe$_3$)$_4$. This reacts further to give Mo$_2$Br$_4$(PMe$_3$)$_4$ and MoBr(PMe$_3$)$_4$(≡CSiMe$_3$).[228] Treatment of MoCl$_2$(O)$_2$ with Bu^{t}CH$_2$MgCl leads to Mo(CH$_2$Bu^t)$_3$(≡CBu^t), this in turn reacts with HX (X = Cl, Br) to form MoX$_3$(≡CBu^t). These are precursors for the respective alkoxides, some of which react with alkynes to generate molybdena-cbd compounds.[229] The complex Mo[OCH(CF$_3$)$_2$]$_3$(dme)(≡CBu^t) reacts with Bu^{t}C≡CH to form alkylidene Mo(IV) complexes.[230] Synthetic routes to the alkylidynes W(OBu^t)$_3$(≡CR^1) include the reactions of R^{1}C≡CR^1 or R^{1}C≡CR^2 with W$_2$(OBu^t)$_6$ or W(OBu^t)$_3$(CMe) (R^1 = alkyl, Ph, vinyl, CH$_2$NR^2 etc.; R^2 = Me, Et). The W≡C bond in these compounds behaves as if polarised W(+)≡C(-).[231] Treatment of W(OBu^t)$_3$(≡CBu^t) with two equivalents of HX leads to the neopentylidenes WX$_2$(OBu^t)$_2$(=CHBu^t) (X = Cl, Br, RCO$_2$, OAr) in reactions reversed with LiOBu^t but six coordinate compounds on addition of py or protonation with [pyrH]X.[232] A minor product in the reaction of trans-Pt(C≡CH)$_2$(PMe$_2$Ph) and W$_2$(OBu^t)$_6$ is the bis-carbyne (Bu^{t}O)$_3$W≡C-C≡W(OBu^t)$_3$.[233] The reaction of W$_2$(OBu^t)$_6$ with PhC≡CPh gives high yields of W$_2$(OBu^t)$_4$(μ-CPh)$_2$ while its reaction with PhC≡CMe gives W(OBu^t)$_3$(≡CPh).[234]

The anions [M(CO$_2$R)(CO)$_5$]$^-$ react with reagents such as C$_2$O$_2$Cl$_2$, C$_2$O$_2$Br$_2$, COCl$_2$, (CF$_3$CO)$_2$O or CO(OCCl$_3$)Cl followed by L$_2$ (L$_2$ = 2 py or TMED) resulting in high yields of the carbynes MX(CO)$_2$(L$_2$)(=CR). (X = Cl, Br, O$_2$CF$_3$)[235] The carbyne WX(CO)$_4$(≡CNEt$_2$) undergoes carbonyl substitution on reaction with L$_2$ (L$_2$ = bipy, phen) resulting in WX(L$_2$)$_2$(≡CNEt$_2$)$_2$. These react with cis-Mo(CO)$_4$(PPh$_2$K)$_2$ forming the anionic carbyne (**44**), which in turn reacts with CO$_2$ to form (**45**).[236] The dianionic carbyne fac-[W(SCN)$_3$(CO)$_2$(≡CNEt$_2$)]$^{2-}$ is formed by ligand substitutions in the reaction of W(S-tol)(CO)$_4$(≡CNEt$_2$) with SCN$^-$.[237] Treatment of WBr(CO)$_2$(≡CNEt$_2$)(L$_2$) (L$_2$ = bipy or phen) with CN$^-$ affords the fluxional complex WCN(CO)$_2$(L$_2$)(≡CNEt$_2$).[238]

The carbynes WX(CO)$_2$(L)$_2$(≡CPh) (X = halide, L = 2 electron donor) luminesce in solution on excitation with visible radiation.[239] Excess alkenes (maleic anhydride or fumaronitrile) substitute a carbonyl of the carbyne WCl(CO)$_2$(py)$_2$(≡CR) to form stable alkene carbyne complexes W(Cl)(CO)(alkene)(py)$_2$(≡CR).[240] Reactions of the carbynes (OC)$_5$ReM(CO)$_4$(≡CMe) (M = Cr, Mo, W) with

$Co_2(CO)_8$ result in tetranuclear clusters containing μ_3-C(tol) ligands.[241]

Treatment of the carbyne MoL($\equiv$CCH$_2$Bu^t)Cp [L = P(OMe)$_3$] with BuLi gives an equilibrium mixture of the carbyne anion [MoL$_2$($\equiv$CCHBu^t)Cp]$^-$ and the vinylidene anion [MoL$_2$(=C=CHBu^t)Cp]$^-$. One electron oxidation (CuI or Fc$^+$) gives meso and R,S bis-carbynes (46) and they react with reagents such as MeI, CH$_2$CH$_2$O, ClCH$_2$OEt etc. by attack of an R group at the α-carbon resulting in carbynes Mo(L)$_2$($\equiv$CCHRBu^t)Cp.[242] Treatment of the carbyne W(CO)$_2$($\equiv$Ctol)Cp with Cr(Ph)$_3$(THF)$_3$ or ZnPh$_2$ results in the substituted benzyl W(CO)$_2$[η^3-CH(Ph)tol]Cp.[243] The same carbyne or W(CO)$_2$($\equiv$CMe)Cp reacts with $\overline{\text{Ti[Cl(AlMe}_2\text{)CH}_2\text{]}}Cp_2$ to form μ-vinyl complexes[244] or with [WH(CO)$_5$]$^-$ to form the μ-alkylidene anions (47). The tol derivative reacts with Ph$_3$SnCl by W-W bond cleavage and formation of the alkylidene W(SnPh$_3$)(CO)$_2$[=C(H)tol]Cp.[245] Protonation of the carbynes W(CO)$_2$($\equiv$CR)Cp (R = Me, tol) with HBF$_4$.OEt$_2$ results in an easily deprotonated ditungsten μ-alkyne cation [W$_2$H(CO)$_4$(μ-RCCR)Cp$_2$]$^+$ arising from linkage of carbyne units while treatment with aqueous HI provides the iodo alkylidenes WI(CO)$_2$(=CHR)Cp. The tol derivative reacts with BH(CHMeEt)$_3^-$ to form the η^3-benzyl W(CO)$_2$(CH$_2$tol)Cp.[246]

The Ru(II) complex RuCl$_2$(PPh$_3$)$_3$ reacts with W(CO)$_2$($\equiv$CMe)Cp to form an μ-alkylidyne species (48) which is reactive towards MeC=CMe.[247] The tris-alkyne W(CO)(RCCR)$_3$ (R = Ar, Et) and W(CO)$_2$($\equiv$Ctol)Cp interact to form several W$_2$ complexes containing μ-alkynes and Cp ligands constructed from alkyne and alkylidyne fragments.[248] A complex reaction between MoMe(CO)$_3$(indenyl) and W(CO)$_2$($\equiv$CR)Cp (R = tol) results in the μ-alkyne heterobimetallic compound MoW(μ-RCCR)(CO)$_4$(η^5-C$_5$H$_4$COMe).[249] Treatment of W(CO)$_2$($\equiv$CMe)Cp with Rh$_2$(CO)$_2$Cp*$_2$ results in the formation of a Rh$_2$W cluster which loses CO irreversibly to form a (μ_3-CR)-Rh$_2$W cluster.[250]

The cation [M(CO)$_2$(CNMe)$_2$(indenyl)]$^+$ reacts with the carbyne W(CO)$_2$($\equiv$CR)(η-1,2-C$_2$B$_9$H$_9$Me$_2$) to form dimetallic μ-carbyne complexes that are reactive towards EtC$\equiv$CEt.[251] The clusters Fe$_2$W(μ-CO)(μ_3-CR)(CO)$_8$Cp (R = tol, Me) react with sulphur and selenium.[252] A mixture of Mo(CO)$_6$, acetic acid, acetic anhydride, and NaBr react to give a Mo$_3$ cluster containing a μ_3-CMe group.[253]

Reaction of CO$_2$ with the acetylide anion [W(C$\equiv$CR)(CO)$_3$(dppe)]$^-$ results in formation of the vinylidene anion [W[=C=C(R)CO$_2$](CO)$_3$(dppe)]$^-$ which reacts with Me$_3$O$^+$ to form the vinylidene $W[=C=C(R)(CO_2Me)](CO)_3(dppe)$.[254]

6. Group 7 (Mn, Tc, and Re)

The reaction of $[Li(THF)_4(Li\{C[SiMe_3]_3\}_2)]$ with $MnCl_2$ gives $Mn[C(SiMe_3)_3]_2$, a two coordinate Mn complex.[255] Gas phase electron diffraction results on $Mn(CH_2Bu^t)_2$ are consistent with a monomeric structure with a CMnC angle of 180° and Mn-C distances of 2.104(6) Å.[256] Various vinyl or allyl Li reagents react with $MnCl_2$ to form manganates $[Mn(vinyl)_4]^{2-}$, $[Mn(allyl)_4]^{2-}$, and $[Mn(allyl)_5]^{3-}$.[257] The reaction of $MnCl_2$ with $(Me_3Si)_3Cl$ forms the anion $Mn_3(Cl)_4(THF)[C(SiMe_3)_3]^{3-}$ (49) which is a model for reagents generated in the reactions of $MnCl_2$ with RMgBr or RLi.[258] Rhenium oxides react with Grignard reagents $MeMgX$ or Me_3SiCH_2MgX to form $(ReMe_4O)_2Mg(THF)_4$ or $[Re(CH_2SiMe_3)_4O]_2Mg(THF)_2$, these oxidise to ReR_4O (R = Me, CH_2SiMe_3) and $Re_2(CH_2SiMe_3)_6O_3$.[259]

Photolysis of a mixture of $R^1C\equiv CH$ and $Re(CO)_4(PR_2CH_2PR_2)Re(CO)_4$ results in η^1-alkynides, μ-η^1,η^2-alkynides, and μ-η^1,η^2-vinyl complexes.[260] Treatment of $Re_2(CO)_6(\mu$-dppm$)_2$ or related compounds with $PhC\equiv CH$ leads to μ-η^1,η^2-alkynide species.[261]

Ion cyclotron resonance and photoelectron studies concerning ionisation kinetics and thermochemical properties of $Mn(Bz)(CO)_5$ are related to the Mn-Mn bond strength of $Mn_2(CO)_{10}$.[262] Photolysis of $MnMe(CO)_5$ in C_6H_6 results in the evolution of CH_3D and CH_4, while photolysis of $MnPh(CO)_5$ gives C_6H_6 but no biphenyl. Photolysis of $MnBz(CO)_5$ gives toluene and bibenzyl.[151] The anion $[Re(CO)_5]^-$ reacts with $ClCH_2COR$ (R = Me, OEt) to form $Re(CH_2COR)(CO)_5$, which undergo CO substitution with PPh_3.[148]

Reaction of alkyls such as $MnMe(CO)_5$, $ReEt(CO)_5$, FpMe, and $OsMe_2(CO)_4$ with hydrides such as $ReH(CO)_5$, $OsH_2(CO)_4$, $MnH(CO)_5$, and $WH(CO)_3Cp$ in coordinating solvents gives dinuclear products with a vacant coordination site at the metal originally bound to the alkyl. The organic products are generally aldehydes or, on occasion alkanes.[263] The reaction of $Mn(CH_2C_6H_4OMe$-$\underline{p})(L)(CO)_4$ with $MnH(L)(CO)_4$ [L = CO, $P(C_6H_4OMe$-$\underline{p})_3$] proceeds by diverse mechanisms. In benzene the products include MeC_6H_4OMe-$\underline{p}$ but in Me_2CO or MeCN the aldehyde $MeOC_6H_4(CH_2CHO)$-$\underline{p}$ is produced.[264] Studies on the decarbonylations of $Mn(\eta^1$-$C_3H_5)(L)(CO)_4$ (L = CO, PPh_3) rule out an η^3-η^1-η^3 mechanism for CO substitution reactions of $Mn(CO)_4(\eta^3$-$C_3H_5)$.[265] A complex reaction of $MnR(CO)_{5-n}(L)_n$ with $HgPh_2$ give cyclic compounds (50) (L = PPh_3, $AsPh_3$, $SbPh_3$).[266]

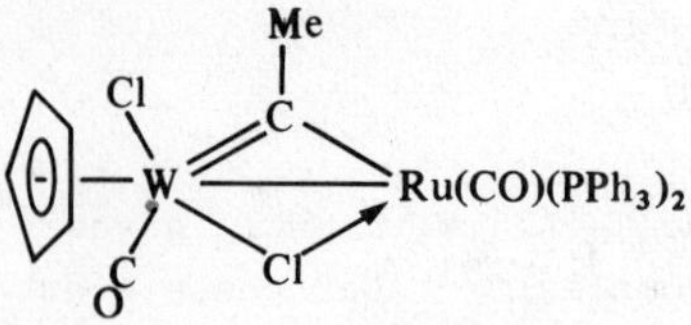

(48)

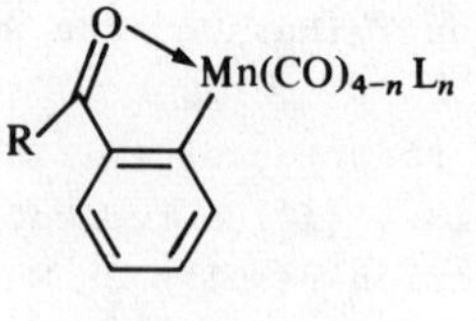

(50)

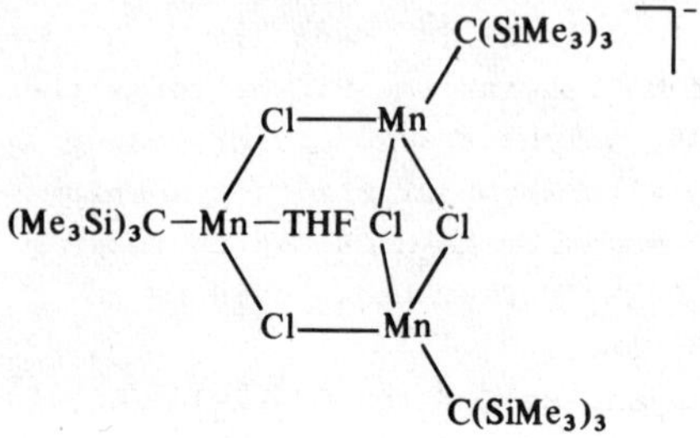

(49)

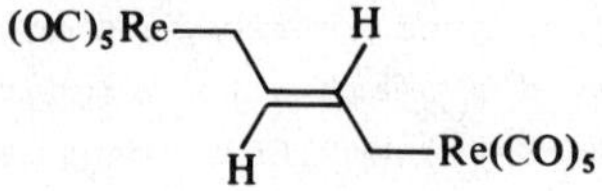

(51)

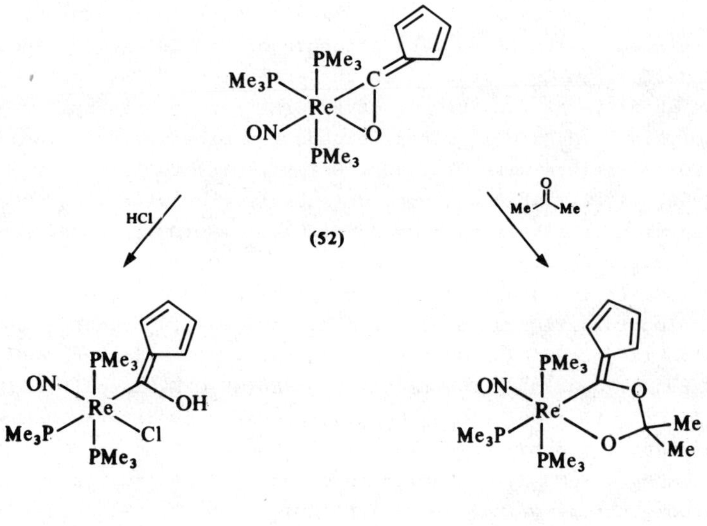

(52)

(53)

(54)

Reduction of $MBr(CO)_4[P(Ph)_2(CH_2)_5Cl]$ (M = Mn, Re) with Na/Hg leads to the cycle $\overline{M[PPh_2(CH_2)_4CH_2]}(CO)_4$ which undergo insertion at the metal alkyl bond with CO to form an acyl.[267] Similar reductions of $MnBr(CO)_2(Ph_2PCH_2CH_2Cl)(dppm)$ and $MnBr(CO)_3(L)(Ph_2POCH_2CH_2Cl)$ lead to $\overline{Mn(CH_2CH_2PPh_2)}(CO)_2(dppm)$[268] and $\overline{Mn(CH_2CH_2OPPh_2)}(CO)_3(L)$ (L = CO, PPh_3)[269] respectively.

Addition of silver or copper alkynides to $MnBr(CO)_n(L)_{5-n}$ results in replacement of the halide by C≡CR. The same alkynides form in the reaction of $Mn(CO)_n(L)_{5-n}(OClO_3)$ with RC≡CH in the presence of base.[270]

Photolysis of $Mn_2(CO)_8(PR_3)_2$ (R = Et, Bu^n, OEt) and R^1X causes oxidative addition of R^1X across the Mn-Mn bond forming equimolar quantities of $MnR^1(CO)_4(PR_3)$ and $MnX(CO)_4(PR_3)$ via mononuclear intermediates.[271] Photolysis of $Re_2(CO)_9L$ (L = PMe_3, PPh_3) and ethene gives a complex mixture of bridging hydride complexes containing σ-π-vinyl groups, these undergo fluxional processes in which the μ-alkenyl σ and π bonds interchange.[272]

Deprotonation of $(OC)_4W(\mu-PPh_2)_2ReH(CO)_3$ with Bu^nLi results in an anion whose nucleophilic behaviour is consistent with the negative charge being rhenium localised. It reacts with MeI, $EtOSO_2CF_3$, $COCl_2$, and PhCOBr to form the corresponding Me, Et, Cl and Br derivatives.[197]

The reaction of $MeC(CH_2X)_3$ (X = Br, I) with $[M(CO)_5]^-$ (M = Mn, Re) results in the alkyl $\overline{Re[CH_2C(Me)CH_2CH_2]}(CO)_5$ or the acyl $\overline{Mn(COCH_2CMeCH_2CH_2)}(CO)_5$.[273] The reaction of $[M(CO)_5]^-$ (M = Mn, Re) with oxalyl chloride results in $(OC)_5M-C(=O)-C(=O)-M(CO)_5$ which contains a strictly planar M-CO-CO-M unit.[274] Reaction of $[Mn(CO)_5]^-$ with $ClC(O)[CH_2]_4C(O)Cl$ gives the disubstituted product $(OC)_5MnC(=O)[CH_2]_4C(=O)Mn(CO)_5$ which thermally decarbonylates to the dimetallabutane $(OC)_5Mn[CH_2]_4Mn(CO)_5$.[275] Treatment of the butadiene cation $[Re(CO)_5(C_4H_6)]^+$ with $[Re(CO)_5]^-$ at $-40°C$ gives an intermediate whose nature is not clear but undergoes rearrangement at $20°C$ to form dimetallabutene (51).[276] Ketenimines $Ph_2C=C=NR$ react with $[Re(CO)_5]^-$ with regio- and site-specificity by a formal [2+2]-cycloaddition resulting in $[(OC)_4Re\{C=CPh_2)N(R)C=O\}]^-$.[277]

The reactions of $M_2(CO)_{10}$ (M = Mn, Re) with $\overline{CH_2CH_2O}$ in the presence of catalytic Br^- give carbenes $(OC)_5M-M(CO)_4[=\overline{CO(CH_2)_2O}]$ in high yield. These undergo M-M cleavage on reaction with Br_2 forming $MBr(CO)_5$ and $MBr(CO)_4[=\overline{CO(CH_2)_2O}]$.[278]

Treatment of the η^1-Cp $Re(\eta^1-C_5H_5)(Me)(CO)(NO)(PMe_3)_2$ with PMe_3 results in formation of the ketene (52), which reacts with HCl or

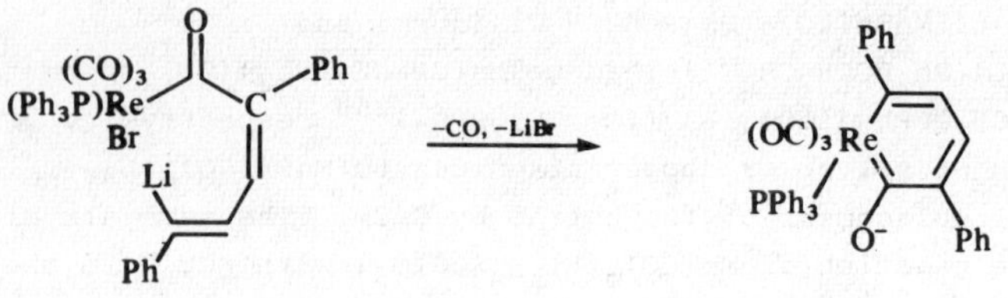

(55) → (56)

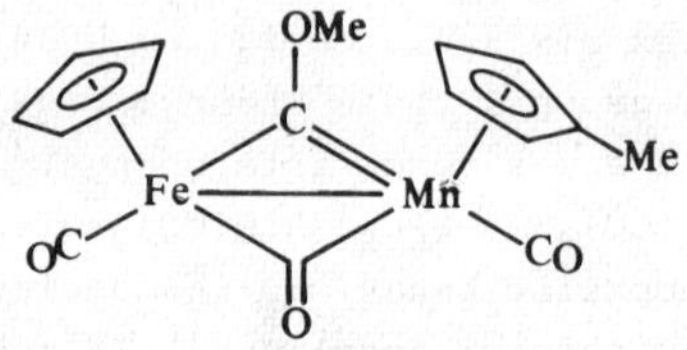

−CO, −LiBr

(57)

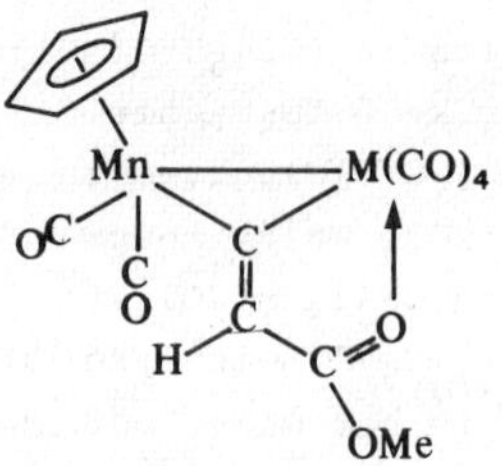

(58) M = Mo or W

acetone to form (53) or (54).[279]

The reaction of dilithiodiphenylbutadiene with $ReBr(CO)_4(PPh_3)$ results in the acyl (55) which loses LiBr to form the alkyl $Re[CPh=CH-CH=C(Ph)Li](CO)_4(PPh_3)$. This last complex cyclises to form an inferred "rhenabenzene" (56).[280]

Inter- and intramolecular oxidative addition of C-H bonds into $Re(PMe_3)_2Cp$ or $Re(L)(PMe_3)Cp*$ (L = CO, PMe_3) occurs on photolysis of an appropriate complex. Primary, cyclopropyl, methane, aromatic, and vinyl C-H bonds are attacked but not secondary centres.[281]

The conformational reactivity relationships for $ReR(NO)(PPh_3)Cp$ (R = alkyl or aryl) are described.[282] Electrochemical oxidation of $ReR(PPh_3)(NO)Cp$ has implications for hydride abstraction by Ph_3C^+.[283] Treatment of $Re(CH_2CN)(PPh_3)(NO)Cp$ successively with Bu^nLi and $MeOSO_2CF_3$ leads to (SR,RS)-$Re(CHMeCN)(PPh_3)(NO)Cp$ via $[Re(CNCN)(PPh_3)(NO)Cp]^-$ suggesting stereospecific alkylation.[284] Among other reactions, the anion $[Re(PPh_3)(NO)Cp]^-$ reacts with MeI, $ClCH_2CH=CH_2$, $ClCH_2CN$ and $(PhCO)O$ to form $ReR(PPh_3)(NO)Cp$ (R = Me, $CH_2CH=CH_2$, CH_2CN, and $C(O)Ph$).[285] Addition of the phosphide $LiPR(SiMe_3)$ (R = Bu^t, $SiMe_3$, Ph) to $[Re(CO)_2(NO)Cp*]^+$ results in eventual formation of $Re[C(OSiMe_3)=PR](CO)(NO)Cp$, initially as the (E) form, but which isomerises to the (Z) form.[286]

Stoicheiometric quantities of aqueous OH- add to the cation $[Re(N_2Ar)(CO)_2Cp]^+$ to form the stable carboxylic acid derivatives $Re(COOH)(CO)(N_2Ar)$. These decarboxylate to the hydride on treatment with solid NaOH and CH_2Cl_2 but forms isolable carboxylates with excess aqueous NaOH.[287] Nucleophiles such as $MeNH_2$, Me_2NH, NH_3, and OMe- add to $[Re(CO)_2(N_2Ar)L]^+$ (L = Cp, Cp*) to form $Re(COX)(CO)(N_2Ar)L$ (X = NHMe, NMe_2, NH_2, OMe). Similar but less stable Mn complexes are also available.[288] Benzyl bromide adds with high stereoselectivity to $Re(COMe=CHMe)(PPh_3)(NO)Cp$ with elimination of MeBr to form the acyl $Re[C(O)CH(Me)Bz](PPh_3)(NO)Cp$ while MeI adds in the opposite direction.[289]

The reaction of the dihydride $ReH_2(CO)_2Cp$ with $I[CH_2]_4I$ in the presence of a non-coordinating base leads to the metallacycle $\overline{Re([CH_2]_3CH_2)}(CO)_2Cp$. Thermolysis in the presence of PR_3 (R = Me, Ph) results in methylcyclopropane and $Re(CO)_2(PR_3)Cp$.[290]

Trimethylphosphine and $ReMe(PMe_3)(NO)Cp$ equilibrate with the η^1-Cp $ReMe(\eta^1-C_5H_5)(PMe_3)_3(NO)$. Thermolysis of the equilibrium mixture (high PMe_3 concentration) results in the cation $[ReMe(PMe_3)_4(NO)][C_5H_5]$ and a 1:1 mixture of both isomers of the substituted cyclopentadienes ReMe(η^1-C_5H_5)(PMe$_3$)$_3$(NO). Thermolysis of the salt in the absence of PMe$_3$ causes reversion to ReMe(PMe$_3$)(NO)Cp.[291]

Treatment of Re(CO)$_3$(η^5-C_9H_7) with PMe$_3$ results in the η^1-indenyl fac-Re(η^1-C_9H_7)(CO)$_3$(PMe$_3$)$_2$ in higher yield at a faster rate than the corresponding Cp derivative.[292]

Reduction of ReCp*$_2$ with a potassium mirror leads to [ReCp*$_2$]$^-$ which alkylates with MeI to form ReMeCp*$_2$.[293]

Boron trichloride treatment of Re(=CPhOMe)(CO)$_2$Cp´ leads to the cationic carbyne [Mn(=CPh)(CO)$_2$Cp´]$^+$ which reacts with NH$_2$CH$_2$CH=CH$_2$ to form Mn[=C(Ph)NHCH$_2$CH=CH$_2$](CO)$_2$Cp´. This compound undergoes thermal decarbonylation to form Mn[=C(Ph)NHCH$_2$CH=CH$_2$](CO)Cp´.[294] Thermolysis of the carbene Mn[=C(OUCp$_3$)CHPMe$_2$Ph](CO)$_2$Cp gives the alkynide Mn(C≡CPMe$_2$Ph)(CO)$_2$Cp and U(OH)Cp$_3$.[295] Addition of BCl$_3$ to Mn[=C(Fc)(OMe)](CO)$_2$Cp´ leads to the carbyne cation [Mn(CO)$_2$(=CFc)Cp´]$^+$, which reacts with LiXPh (X = S, Se, Te) to form the carbene Mn(CO)$_2$[C(Fc)(XPh)]Cp´.[296] Addition of MeOSO$_2$CF$_3$ to [Mn(CO)Cp´(μ-CO)$_2$Fe(CO)Cp]$^-$ gives the heterobimetallic carbyne (57) as cis and trans isomers.[297]

The reaction of Mn(=C=CPhH)(CO)$_2$Cp with sources of Pt(PPh$_3$)$_2$ leads to the μ-vinylidene Mn(CO)$_2$Cp(μ-C=CPhH)Pt(PPh$_3$)$_2$[298] which undergoes substitution reactions with PR$_3$ or CO.[299] Addition of M(CO)$_5$(THF) (M = Mo, W) to Mn(CO)$_2$[=C=CH(CO$_2$Me)]Cp gives the bridging vinylidene (58).[300] Bridging vinylidenes are also formed in the reaction of Mn(CO)$_2$(THF)Cp or the photolytic reaction of Cr(CO)$_3$(η-C_6H_6) with Rh(=C=CH$_2$)(PPr$^i_3$)Cp.[301]

Methyl lithium or LiEt$_3$BH react with PPh$_2$[W(CO)$_5$][Re(CO)$_4$(PMePh$_2$)] in the presence of phosphine to form acyl or formyl compounds PPh$_2$[W(CO)$_5$][Re(CRO)(CO)$_3$(PMePh$_2$)]$^-$ (R = Me or H). In the absence of phosphine, two unstable formyls arise which decompose to hydride compounds over $^-$20°C.[302]

References

1. L.S. Hegedus, J. Organomet. Chem., 1985, **283**, 1.
2. M.I. Bruce, J. Organomet. Chem., 1985, **283**, 339.
3. M.L.H. Green, and D. O'Hare, Pure. Appl. Chem., 1985, **57**, 1897.
4. I.P. Rothwell, Polyhedron, 1985, **4**, 177.
5. D.P. Pourreau and G.L. Geoffroy, Adv. Organomet. Chem., 1985, **24**, 249.
6. H. van Dam and A. Oskam, Transition Met. Chem., 1985, **9**, 125, Eds., G.A. Melson and B.N. Figgs, M. Dekker Inc.
7. K. Jonas, Angew. Chem.,Int. Ed. Engl., 1985, **24**, 295.
8. H.A. Skinner and J.A. Connor, Pure. Appl. Chem., 1985, **57**, 79.
9. J.A. McCleverty, Transition Met. Chem., 1985, **10**, 118.
10. M. Green and M. Green, Transition Met. Chem., 1985, **10**, 196.
11. M. Green and M. Green, Transition Met. Chem., 1985, **10**, 31.
12. Abstr. Fifth Intnl. Symp. Olefin Metathesis, Graz, Austria, 1983, J. Mol. Cat., 1985, **28**, 1.
13. E.C. Constable, Coord. Chem. Rev., 1985, **62**, 131.
14. P.L. Watson and G.W. Parshall, Acc. Chem. Res., 1985, **18**, 51.
15. H. Schumann, J. Organomet. Chem., 1985, **281**, 95.
16. W.J. Evans, Adv. Organomet. Chem., 1985, **24**, 131.
17. J.W. Gilje, R.E. Cramer, M.A. Bruck, K.T. Higa, and K. Panchanetheswaran, Inorg. Chim. Acta., 1985, **110**, 139.
18. P. Bougeard, M. Marcini, B.G. Sayer, and M.J. McGlinchey, Inorg. Chem., 1985, **24**, 93.
19. W.J. Evans, J.H. Meadows, A.G. Kostka, and G.L. Closs, Organometallics, 1985, **4**, 324.
20. D.C. Sonnenberger, L.R. Morss, and T.J. Marks, Organometallics, 1985, **4**, 352.
21. G. Rossetto, M. Porchia, F. Ossola, P. Zanella, and R.D. Fischer J. Chem. Soc., Chem. Commun., 1985, 1460.
22. H. Schumann and G. Jeske, Angew. Chem.,Int. Ed. Engl., 1985, **24**, 225.
23. H.G. Brittain, J.H. Meadows, and W.J. Evans., Organometallics, 1985, **4**, 1585.
24. I. Albrecht, E. Hahn, J. Pickardt, and H. Schumann, Inorg. Chim. Acta., 1985, **110**, 145.
25. H. Schumann, H. Lauke, E. Hahn, M.J. Heeg, and D. van der Helm., Organometallics, 1985, **4**, 321.
26. H. Schumann, I. Albrecht, and E. Hahn, Angew. Chem.,Int. Ed. Engl., 1985, **24**, 985.
27. G. Jeske, H. Lauke, H. Mauermann, P.N. Swepston, H. Scumann, and T.J. Marks, J. Am. Chem. Soc., 1985, **107**, 8091.
28. H. Mauermann, P.N. Swepston, and T.J. Marks, Organometallics, 1985, **4**, 200.
29. G. Jeske, L.E. Schock, P.N. Swepston, H. Schumann, and T.J. Marks, J. Am. Chem. Soc., 1985, **107**, 8103.
30. W.J. Evans, R. Dominguez, K.R. Levan, and R.J. Doedens, Organometallics, 1985, **4**, 1836.
31. H. Schumann, F.-W. Reier, and E. Palamidio, J. Organomet. Chem., 1985, **297**, C30.
32. K.G. Moloy and T.J. Marks, Inorg. Chim. Acta, 1985, **110**, 127.
33. M.-Y. He, G. Xiong, P.J. Toscano, R.L. Burwell, Jr., and T.J. Marks, J. Am. Chem. Soc., 1985, **107**, 641.
34. P.J. Toscano and T.J. Marks, J. Am. Chem. Soc., 1985, **107**, 653.
35. G. Erker, P. Czisch, C. Krüger, and J.M. Wallis, Organometallics, 1985, **4**, 2059.
36. A. Dormond, A. Aaliti, and C. Moïse, J. Chem. Soc., Chem. Commun., 1985, 1231.
37. J.G. Brennan and R.A. Andersen, J. Am. Chem. Soc., 1985, **107**, 514.
38. P. Zanella, G. Paolucci, G. Rosetto, F. Benetollo, A. Polo, R.D. Fischer, and G. Bombieri, J. Chem. Soc., Chem. Commun., 1985, 96.
39. A. Dormond, A. el Bouadili, A. Aciliti, and C. Moïse, J. Organomet. Chem., 1985, **288**, C1.

40. A. Dormond, A.A. el Bouadili, and C. Moïse, J. Chem. Soc., Chem. Commun., 1985, 914.
41. A. Shakoor, K. Jacob, and K.-H. Thièle, Z. Anorg. Allg. Chem., 1985, **521**, 57.
42. O.P. Syutkina, L.F. Rybakova, E.S. Petrov, and I.P. Beletskaya, J. Organomet. Chem., 1985, **280**, C67.
43. K. Yokoo, N. Mine, H. Taniguchi, and Y. Fujiwaru, J. Organomet. Chem., 1985, **279**, C19.
44. T. Imamoto, Y. Sugiura, J. Organomet. Chem., 1985, **285**, C21.
45. Z. Hou, N. Mine, Y. Fujiwara, and H. Taniguchi, J. Chem. Soc., Chem. Commun., 1985, 1700.
46. G.Z. Suleimanov, Y.S. Bogatchev, L.T. Abdullaeva, I.L. Zhuravleva, K.S. Khalilov, L.F. Rybakova, and I.P. Beletskanya, Polyhedron, 1985, **4**, 29.
47. G. M.Z. Suleimanov, V.N. Khandozhko, P.V. Petrovskii, R. Yu. Mekhdiev, N.E. Kolobova, and I.P. Beletskaya, J. Chem. Soc., Chem. Commun., 1985, 596.
48. J.M. Boncella and R.A. Andersen, Organometallics, 1985, **4**, 205.
49. A.L. Wayda, R.D. Rogers, Organometallics, 1985, **4**, 1440.
50. D.M. Roddick, B.D. Santarsiero, and J.E. Bercaw, J. Am. Chem. Soc., 1985, **107**, 4670.
51. G. Erker, C. Krüger, and C. Müller, Adv. Organomet. Chem., 1985, **24**, 1.
52. E. Samuel, D. Guery, J. Vedel, and F. Basile, Organometallics, 1985, **4**, 1073.
53. T. Cuenca and P. Royo, J. Organomet. Chem., 1985, **295**, 159.
54. S.J. Young, M.M. Olmstead, M.J. Knudsen, and N.E. Schore, Organometallics, 1985, **4**, 1432.
55. D. Steinborn and R. Taube, J. Organomet. Chem., 1985, **284**, 395.
56. V. Dimitrov, J. Organomet. Chem., 1985, **282**, 321.
57. W. Frömberg and G. Erker, J. Organomet. Chem., 1985, **280**, 343.
58. C. McDade and J.E. Bercaw, J. Organomet. Chem., 1985, **279**, 281.
59. R. Choukroun and D. Gervais, J. Chem. Soc., Chem. Commun., 1985, 225.
60. J.J. Koh, P.H. Reiger, and W.R. Riser, Jr., Inorg. Chem., 1985, **24**, 2312.
61. D.M. Roddick, M.D. Fryzuk, P.F. Seidler, G.L. Hillhouse, and J.E. Bercaw, Organometallics, 1985, **4**, 97.
62. S. Gambarotta, S. Strologo, C. Floriani, A. Chiesi-Villa, and C. Guastini, Inorg. Chem., 1985, **24**, 654.
63. B. Longato, B.D. Martin, J.R. Norton, and O.P. Anderson, Inorg. Chem., 1985, **24**, 1389.
64. G. Erker, P. Czisch, R. Mynott, Y.-H. Tsay, and C. Krüger, Organometallics, 1985, **4**, 1310.
65. G. Erker, K. Engel, U. Korek, P. Czisch, H. Berke, P. Coubere, and R. Vanderesse, Organometallics, 1985, **4**, 1531.
66. R.F. Jordan, J. Organomet. Chem., 1985, **294**, 321.
67. J.W. Pattiasina, C.E. Hissink, J.L. de Boer, A. Meetsma, and J.H. Teuben, J. Am. Chem. Soc., 1985, **107**, 7758.
68. F. Bottomley, G.O. Egharevba, I.J.B. Lin, and P.S. White, Organometallics, 1985, **4**, 550.
69. G. Erker, U. Dork, R. Mynott, Y.-H. Tsay, and C. Krüger, Angew. Chem.,Int. Ed. Engl., 1985, **24**, 584.
70. E.A. Mintz, A.S. Ward, and D.S. Tice, Organometallics, 1985, **4**, 1308.
71. L. Clawson, J. oto, S.L. Buchwold, M.L. Steigerwald, and R.H. Grubbs, J. Am. Chem. Soc., 1985, **107**, 3377.
72. J.J. Eisch, A.M. Piotrowski, S.K. Brownstein, E.J. Gab, and F.L. Lee, J. Am. Chem. Soc., 1985, **107**, 7219.
73. C. Aitken, J.F. Harrod, and E. Samuel, J. Organomet. Chem., 1985, **279**, C11.
74. J.A. Bandy, V.S.B. Mtetwa, K. Prout, J.C. Green, C.E. Davies, M.L.H. Green N.J. Hazel, A. Izquierdo, and J.J. Martin-Polo, J. Chem. Soc., Dalton Trans., 1985, 2037.
75. C.P. Casey, R.E. Palermo, R.F. Jordan, and A.L. Rheingold, J. Am. Chem. Soc., 1985, **107**, 4597.
76. F. Senocq, M. Basso-Bert, R. Choukroun, and D. Gervais, J. Organomet. Chem., 1985, **297**, 155.

77. C. Krüger, G, Müller, G. Erker, U. Dorf, and K. Engel, Organometallics, 1985, **4**, 215.
78. H.G. Alt, H.E. Engelhardt, M.D. Rausch, and L.B. Kool, J. Am. Chem. Soc., 1985, **107**, 3717.
79. S.B. Jones and J.L. Petersen, Organometallics, 1985, **4**, 966.
80. B.J.J. van de Heisteeg, G. Schat, O.S. Akkerman, and F. Bickelhaupt, Organometallics, 1985, **4**, 1141.
81. P. Paetzold, K. Delpy, R.R. Hughes, and W.A. Herrmann, Chem. Ber., 1985, **118**, 1724.
82. S.L. Buchwald, E.V. Anslyn, and R.H. Grubbs, J. Am. Chem. Soc., 1985, **107**, 1766.
83. T. Ikariya, S.C.H. Ho, and R.H. Grubbs, Organometallics, 1985, **4**, 199.
84. S.A. Cohen and J.E. Bercaw, Organometallics, 1985, **4**, 1006.
85. K. Mashima and H. Takaya, Organometallics, 1985, **4**, 1464.
86. V.B. Shur, V.V. Burlakov, A.I. Yanovsky, P.V. Petrovsky, Yu.T. Struchov, and M.E. Vol´pin, J. Organomet. Chem., 1985, **297**, 51.
87. E.-i. Negishi, S.J. Holmes, J.M. Tour, and J.A. Miller, J. Am. Chem. Soc., 1985, **107**, 2568.
88. T.D. Tilley, J. Am. Chem. Soc., 1985, **107**, 4084.
89. J.A. Marsella, J.C. Huffman, K. Fotting, and K.G. Caulton, Inorg. Chim Acta, 1985, **96**, 161.
90. P. Hofmann, P. Stauffert, K. Tatsumi, A. Nakamura, and R. Hoffmann, Organometallics, 1985, **4**, 404.
91. K. Tatsumi, A. Nakamura, P.Hofmann, P. Stauffert, and R. Hoffmann, J. Am. Chem. Soc., 1985, **107**, 4440.
92. B.D. Martin, S.A. Matchett, J.R. Norton, and O.P. Anderson, J. Am. Chem. Soc., 1985, **107**, 7952.
93. G. Erker, K. Berg, R. Benn, and G. Schroth, Chem. Ber., 1985, **118**, 1383.
94. K.-H. Thiele and A. Krüger, Z. Anorg. Allg. Chem., 1985, **527**, 95.
95. C.E. Davies, I.M. Gardiner, J.C. Green, M.L.H. Green, N.J. Hazel, P.D. Grebenik, V.S.B. Mtetwa, and K. Prout, J. Chem. Soc., Dalton Trans., 1985, 669.
96. M. Schlegel, and K.-H. Thiele, Z. Anorg. Allg. Chem., 1985, **526**, 43.
97. A.R. Gregory and E.A. Mintz, J. Am. Chem. Soc., 1985, **107**, 2179.
98. T.H. Upton and A.K. Rappé, J. Am. Chem. Soc., 1985, **107**, 1206.
99. V. Dimitrov, K.-H. Thiele, and D. Schenke, Z. Anorg. Allg. Chem., 1985, **527**, 85.
100. S.L. Latesky, A.K. McMullen, G.P. Niccolai, I.P. Rothwell, and J.C. Huffman, Organometallics, 1985, **4**, 902.
101. S.L. Latesky, A.K. McMullen, I.P. Rothwell, and J.C. Huffman, J. Am. Chem. Soc., 1985, **107**, 5981.
102. S.L. Latesky, A.K. McMullen, G.P. Niccolai, I.P. Rothwell, and J.C. Huffman, Organometallics, 1985, **4**, 1896.
103. A.K. McMullen, I.P. Rothwell, and J.C. Huffman, J. Am. Chem. Soc., 1985, **107**, 1072.
104. S.M. Clift and J. Schwartz, J. Organomet. Chem., 1985, **285**, C5.
105. H. Yasuda, K. Tatsumi, T. Okamoto, K. Mashima, K. Lee, A. Nakamura, Y. Kai, N. Kanehisa, and N. Kasai, J. Am. Chem. Soc., 1985, **107**, 2410.
106. H. Yasuda, K. Tatsumi, and A. Nakamura, Acc. Chem. Res., 1985, **18**, 120.
107. C.J. Curtis, J.C. Smart, and J.L. Robbins, Organometallics, 1985, **4**, 1283.
108. S.I. Bailey, L.M. Engelhardt, W.-P. Leung, I.M. Ritchie, and A.H. White, J. Chem. Soc., Dalton Trans., 1985, 1747.
109. D.A. Lemonovskii, I.F. Urazowski, I.F. Nifantev, and E.G. Perevalova, J. Organomet. Chem., 1985, **292**, 217.
110. J. Nieman and J.H. Teuben, J. Organomet. Chem., 1985, **287**, 207.
111. D.F. Foust and M.D. Rausch, J. Organomet. Chem., 1985, **287**, 195.
112. N.M. Doherty and J.E. Bercaw, J. Am. Chem. Soc., 1985, **107**, 2670.
113. J. Amadrut, J.-C. Leblanc, C. Moise, and J. Sala-Pala, J. Organomet. Chem., 1985, **295**, 167.
114. H. Van der Heijden, A.W. Gal, P. Pasman, and A.G. Orpen, Organometallics,

1985, **4**, 1847.
115. V.C. Gibson, C.E. Graimann, P.M. Hare, M.L.H. Green, J.R. Bandy, P.D. Grebenik, and K. Prout, J. Chem. Soc., Dalton Trans., 1985, 2025.
116. R.H. Crabtree, Chem. Rev., 1985, **85**, 245.
117. R. Colton, Coord. Chem. Rev., 1985, **62**, 145.
118. A. Rotman, H. Cohen, and D. Meyerstein, Inorg. Chem., 1985, **24**, 4158.
119. H. Cohen, D. Meyerstein, A.J. Shusterman and M. Weiss, J. Chem. Soc., Chem. Commun., 1985, 424.
120. H. Cohen and D. Meyerstein, Angew. Chem.,Int. Ed. Engl., 1985, **24**, 779.
121. J. Colin and B. Chevrier, Organometallics, 1985, **4**, 1090.
122. G.J.-J. Chen, J.W. MacDonald, and W.E. Newton, Organometallics, 1985, **4**, 422.
123. W.A. Schenk, D. Rüb, and C. Burschka, Angew. Chem.,Int. Ed. Engl., 1985, **24**, 971.
124. D.J. Darensbourg, R.K. Hanckel, C.G. Bauch, M. Pala, D. Simmons, and J.N. White, J. Am. Chem. Soc., 1985, **107**, 7463.
125. D.J. Darensbourg and G. Grotsch, J. Am. Chem. Soc., 1985, **107**, 7473.
126. D.J. Darensbourg, R. Kudaroski, and T. Delord, Organometallics, 1985, **4**, 1094.
127. G.L. Hillhouse, J. Am. Chem. Soc., 1985, **107**, 7772.
128. A. Mannetti, J. Fischer, and F. Mathey, J. Am. Chem. Soc., 1985, **107**, 5001.
129. L. Weber and R. Boese, Chem. Ber., 1985, **118**, 1545.
130. L. Weber and D. Wewars, Chem. Ber., 1985, **118**, 3560.
131. L. Weber and D. Wewars, Chem. Ber., 1985, **118**, 541.
132. L. Weber, D. Wewars, and R. Boese, Chem. Ber., 1985, **118**, 3570.
133. L. Weber and D. Wewars, Organometallics, 1985, **4**, 841.
134. R. Heusel, W.-W. duMont, R. Boese, D. Wewars, and L. Weber, Chem. Ber., 1985, **118**, 1580.
135. M.J. Chetcuti, M.H. Chisholm, H.T. Chiu, and J.C. Huffman, Polyhedron, 1985, **4**, 1213.
136. K.J. Ahmed, M.H. Chisholm, K. Folting, and J.C. Huffman, J. Chem. Soc., Chem. Commun., 1985, 152.
137. M.H. Chisholm, B.W. Eichorn, and J.C. Huffman, J. Chem. Soc., Chem. Commun., 1985, 861.
138. D. Navane, F. Rose-Munch, and H. Rudler, J. Organomet. Chem., 1985, **284**, C15.
139. J. Levisalles, F. Rose-Munch, H. Rudler, J.C. Daran, and Y. Jeannin, J. Organomet. Chem., 1985, **279**, 413.
140. D.C. McKean, G.P. McQuillan, A.R. Morrisson, and I. Torto, J. Chem. Soc., Dalton Trans., 1985, 1207.
141. J.T. Gauntlett, B.F. Taylor, and M.J. Winter, J. Chem. Soc., Dalton Trans., 1985, 1815.
142. M. Tilset and K.P.C. Vollhardt, Organometallics, 1985, **4**, 2230.
143. H.G. Alt, J. Organomet. Chem., 1985, **288**, 149.
144. H.G. Alt, H.E. Engelhardt, U. Thewalt, J. Riede, J. Organomet. Chem., 1985, **288**, 165.
145. H.G. Alt and H.I. Hayen, Angew. Chem.,Int. Ed. Engl., 1985, **24**, 497.
146. F. Giuliéri and J. Benaim, Nouv. J. Chim., 1985, **9**, 335.
147. J.L. Dorey, R.G. Bergman and C.H. Heathcock, J. Am. Chem. Soc., 1985, **107**, 3724.
148. C.H. Heathcock, J.C. Dorey, and R.G. Bergman, Pure Appl. Chem., 1985, **57**, 1789.
149. J. Markham, W. Tolman, K. Menard, and A. Cutler, J. Organomet. Chem., 1985, **294**, 45.
150. T.-Y. Luh, and C.S. Wong, J. Organomet. Chem., 1985, **287**, 231.
151. T.E. Gismondi and M.D. Rausch, J. Organomet. Chem., 1985, **284**, 59.
152. J.D. Cotton and H.A. Kimlin, J. Organomet. Chem., 1985, **294**, 213.
153. R.B. Hittam, K.A. Mahmoud, and A.J. Rest, J. Organomet. Chem., 1985, **291**, 321.
154. M.I. Bruce, T.W. Hambley, M.R. Snow, and A.G. Swincer, Organometallics, 1985, **4**, 494.

155. P. Legdzins, S.J. Rettig, L. Sánchez, B.E. Burston, and M.G. Gatter, J. Am. Chem. Soc., 1985, **107**, 1411.
156. P. Legdzins, S.J. Rettig, and L. Sánchez., Organometallics, 1985, **4**, 1470.
157. P. Legdzins and L. Sanchez, J. Am. Chem. Soc., 1985, **107**, 5525.
158. S.R. Allen, R.G. Beevor, M. Green, N.C. Norman, A.G. Orpen, and I.D. Williams, J. Chem. Soc., Dalton Trans., 1985, 435.
159. L. Carlton, J.L. Davidson, P. Ewing, L. Manojlović-Muir, and K.W. Muir, J. Chem. Soc., Chem. Commun., 1985, 1474.
160. T.W. Bodnar and A.R. Cutler, Organometallics, 1985, **4**, 1558.
161. Y. Chi and J.R. Shapley, Organometallics, 1985, **4**, 1900.
162. A.D. Horton, M.J. Mays, and P.R. Raithby, J. Chem. Soc., Chem. Commun., 1985, 247.
163. K. Blechschmitt, T. Zahn, and M.L. Ziegler, Angew. Chem.,Int. Ed. Engl., 1985, **24**, 702.
164. M. Cariou, M.M. Kubicki, R. Kergoat, L.C.G. de Lima, H. Scordia, and J.E. Guerchais, Inorg. Chim. Acta, 1985, **104**, 185.
165. M.M. Kubicki, R. Kergoat, L.C.G. de Lima, M. Cariou, H. Scordia, J.E. Guerchais, and P. L´Haridan, Inorg. Chim. Acta, 1985, **104**, 191.
166. L.C.G. de Lima, M. Cariou, H. Scordia, R. Kergoat, M.M. Kubicki, and J.E. Guerchais, J. Organomet. Chem., 1985, **290**, 321.
167. R.M. Bullock, C.E.L. Headford, S.E. Kegley, and J.R. Norton, J. Am. Chem. Soc., 1985, **107**, 727.
168. G.E. Herberich and J. Okuda, Angew. Chem.,Int. Ed. Engl., 1985, **24**, 402.
169. J.C. Green, M.L.H. Green and C.P. Morley, Organometallics, 1985, **4**, 1303.
170. M.L.H. Green and D. O´Hare, J. Chem. Soc., Dalton Trans., 1985, 1585.
171. F.G.N. Cloke, J.C. Green, M.L.H. Green and C.P. Morley, J. Chem. Soc., Chem. Commun., 1985, 945.
172. F. Bottomley, D.E. Paez, L. Sutin, and P.S. White, J. Chem. Soc., Chem. Commun., 1985, 598.
173. M.L.H. Green and R.B.A. Pardy, Polyhedron, 1985, **4**, 1035.
174. J.R. Morrow, T.L. Tonker, and J.L. Templeton J. Am. Chem. Soc., 1985, **107**, 5004.
175. M.H. Chisholm, J.A. Heppert, J.C. Huffman, and, W.E. Streib, J. Chem. Soc., Chem. Commun., 1985, 1771.
176. H. Arzoumanian, A. Baldy, R. Lau, J. Metzger, M.-L. Nkeng-Peh, M. Pierrot, J. Chem. Soc., Chem. Commun., 1985, 1151.
177. A. Aguero, J. Kreiss, and J.A. Osborn, J. Chem. Soc., Chem. Commun., 1985, 793.
178. J. Kreiss, J.A. Osborn, R.M.E. Greene, K.J. Ivin, and J.J. Rooney, J. Chem. Soc., Chem. Commun., 1985, 874.
179. F. Quignard, M. Leconte, and J.-M. Basset, J. Chem. Soc., Chem. Commun., 1985, 1816.
180. J.L. Davidson, W.F. Wilson, and K.W. Muir, J. Chem. Soc., Chem. Commun., 1985, 460.
181. M.R. Churchill and Y.-J. Li, J. Organomet. Chem., 1985, **291**, 61.
182. M.R. Churchill and Y.-J. Li, J. Organomet. Chem., 1985, **294**, 367.
183. N.M. Doherty, C. Elschenbroich, H.-J. Kneuper, and S.A.R. Knox, J. Chem. Soc., Chem. Commun., 1985, 170.
184. K.H. Dotz, W. Kuhn, and U. Thewalt, Chem. Ber., 1985, **118**, 1126.
185. R. Aumann, E. Kuckert, and H. Heinen, Angew. Chem.,Int. Ed. Engl., 1985, **24**, 978.
186. C.P. Casey and A.J. Shusterman, Organometallics, 1985, **4**, 736.
187. K.-H. Dötz and W. Kuhn, J. Organomet. Chem., 1985, **286**, C23.
188. K.-H. Dötz and W. Sturm, J. Organomet. Chem., 1985, **285**, 205.
189. K.-H. Dötz and M. Popall, J. Organomet. Chem., 1985, **291**, C1.
190. W.D. Wulff and R.W. Kaeslar, Organometallics, 1985, **4**, 1461.
191. N.Q. Dao, M. Jouan, G.P. Fonseca, N.H.T. Huy, and E.O. Fischer, J. Organomet. Chem., 1985, **287**, 215.
192. H. Fischer, A. Motsch, R. Märkl, and K. Ackermann, Organometallics, 1985, **4**, 726.
193. W.D. Wulff and S.R. Gilbertson, J. Am. Chem. Soc., 1985, **107**, 503.

194. R. Aumann and H. Heinen, Chem. Ber., 1985, 118, 952.
195. H.G. Raubenheimer, G.J. Kruger, A. Van A. Lombard, L. Linford, and J.C. Viljoen, Organometallics, 1985, 4, 275.
196. E.O. Fischer and M. Böck, J. Organomet. Chem., 1985, 287, 279.
197. W.C. Mercer, G.L. Geoffroy, and A.L. Rheingold, Organometallics, 1985, 4, 1418.
198. H. Hörnig, E. Walther, and U. Schubert, Organometallics, 1985, 4, 1905.
199. A.G.M. Bassett, C.P. Brock, and M.A. Sturgess, Organometallics, 1985, 4, 1903.
200. H. Fischer and R. Märkl, Chem. Ber., 1985, 118, 3683.
201. S.L. Landon, P.M. Shulman, and G.L. Geoffroy, J. Am. Chem. Soc., 1985, 107, 6739.
202. F. Seitz, H. Fischer, and J. Riede, J. Organomet. Chem., 1985, 287, 87.
203. H. Fischer, F. Seitz, J. Riede, and J. Vogel, Angew. Chem.,Int. Ed. Engl., 1985, 24, 121.
204. H. Fischer, F. Seitz, and J. Riede, J. Chem. Soc., Chem. Commun., 1985, 537.
205. F. Seitz and H. Fischer, J. Organomet. Chem., 1985, 290, C31.
206. H. Fischer, R. Märkl, and S. Zeuner, J. Organomet. Chem., 1985, 286, 17.
207. H. Fischer and S. Zeuner, J. Organomet. Chem., 1985, 286, 201.
208. N.H.T. Huy, E.O. Fischer, H.G. Alt, and K.H. Dotz., J. Organomet. Chem., 1985, 284, C9.
209. N.H.T. Huy and E.O. Fischer, Nouv. J. Chim., 1985, 9, 257.
210. A. Parlier, H. Rudler, N. Platzer, M. Fontanille, and A. Soum., J. Organomet. Chem., 1985, 287, C8.
211. H.G. Raubenheimer, G.J. Kruger, and H.W. Viljoen, J. Chem. Soc., Dalton Trans., 1985, 1963.
212. I.R. Butler, W.R. Cullen, F.W.B. Einstein, and A.C. Willis, Organometallics, 1985, 4, 603.
213. H. Adams, N.A. Bailey, V.A. Osborn, and M.J. Winter, J. Organomet. Chem., 1985, 284, C1.
214. V.A. Osborn and M.J. Winter, J. Chem. Soc., Chem. Commun., 1985, 1744.
215. J.S. Drage and K.P.C. Vollhardt, Organometallics, 1985, 4, 191.
216. M.F.N.N. Carvalho, C.M.C. Laranjeira, A.T.Z. Nobre, A.J.L. Pombeiro, A.C.A.M. Viegas, and R.L. Richards, Transition Met. Chem., 1985, 10, 427.
217. M.H. Chisholm, J.A. Heppert, J.C. Huffman, and P. Thornton, J. Chem. Soc., Chem. Commun., 1985, 1466.
218. M.H. Chisholm, J.C. Huffman, and J.A. Heppert, J. Am. Chem. Soc., 1985, 107, 5116.
219. M.H. Chisholm, K. Folting, J.A. Heppert, and W.E. Streib, J. Chem. Soc., Chem. Commun., 1985, 1755.
220. A. Schäfer and W.A. Herrman, J. Organomet. Chem., 1985, 297, 229.
221. S.A. MacLaughlin, R.C. Murray, J.C. Dewar, and R.R. Schrock, Organometallics, 1985, 4, 796.
222. M.R. Churchill and J.W. Ziller, J. Organomet. Chem., 1985, 286, 27.
223. M.R. Churchill and J.W. Ziller, J. Organomet. Chem., 1985, 281, 237.
224. M.R. Churchill and J.W. Ziller, J. Organomet. Chem., 1985, 279, 403.
225. M.R. Churchill and J.C. Fettinger, J. Organomet. Chem., 1985, 290, 375.
226. J. AnHaus, Z.A. Siddiqui, H.W. Roesky, and J.W. Bats, J. Chem. Soc., Dalton Trans., 1985, 2453.
227. M.R. Churchill and Y.-J. Li, J. Organomet. Chem., 1985, 282, 239.
228. K.J. Ahmed, M.H. Chisholm, and J.C. Huffman, Organometallics, 1985, 4, 1168.
229. L.G. McCullogh, R.R. Schrock, J.C. Dewar, and J.C. Murdzek, J. Am. Chem. Soc., 1985, 107, 5987.
230. H. Strutz, J.C. Dewar, and R.R. Schrock, J. Am. Chem. Soc., 1985, 107, 5999.
231. M.L. Listemann and R.R. Schrock, Organometallics, 1985, 4, 74.
232. J.H. Freudenberger and R.R. Schrock, Organometallics, 1985, 4, 1937.
233. R.J. Bau, M.H. Chisholm, K. Folting, and R.J. Wang, J. Chem. Soc., Chem. Commun., 1985, 1582.

234. F.A. Cotton, W. Schwotzer, and E.S. Shamshoum, J. Organomet. Chem., 1985, **296**, 55.
235. A. Mayr, G.A. McDermott, and A.M. Dorries, Organometallics, 1985, **4**, 608.
236. E.O. Fischer, A.C. Filippou, H.G. Alt, and U. Thewalt, Angew. Chem.,Int. Ed. Engl., 1985, **24**, 203.
237. E.O. Fischer and D. Wittmann, J. Organomet. Chem., 1985, **292**, 245.
238. E.O. Fischer, A.C. Filippou, and H.G. Alt, J. Organomet. Chem., 1985, **296**, 69.
239. A.B. Bocalsky, R.E. Cameron, H.-D. Rubin, G.A. McDermott, C.R. Wolff, and A. Mayr, Inorg. Chem., 1985, **24**, 3976.
240. A. Mayr, A.M. Dorries, G.A. McDermott, S.J. Geib, A.L. Rheingold, J. Am. Chem. Soc., 1985, **107**, 7775.
241. J.C. Jeffrey, D.B. Lewis, G.E. Lewis, and F.G.A. Stone, J. Chem. Soc., Dalton Trans., 1985, 2001.
242. R.G. Beever, M.J. Freeman, M. Green, C.E. Morton, and A.G. Orpen, J. Chem. Soc., Chem. Commun., 1985, 68.
243. J.C. Jeffrey, A.L. Ratermain, and F.G.A. Stone, J. Organomet. Chem., 1985, **289**, 367.
244. M.R. Awang, R.D. Barr, M.Green, J.A.K. Howard, T.B. Marder, and F.G.A. Stone, J. Chem. Soc., Dalton Trans., 1985, 2009.
245. D. Hodgson, J.A.K. Howard, and M.J. Went, J. Chem. Soc., Dalton Trans., 1985, 1331.
246. J.A.K. Howard, J.C. Jeffrey, J.C.V. Laurie, I. Moore, F.G.A. Stone, and A.J. Stringer, Inorg. Chim. Acta, 1985, **100**, 23.
247. J.A.K. Howard, J.C.V. Laurie, O. Johnson, and F.G.A. Stone, J. Chem. Soc., Dalton Trans., 1985, 2017.
248. G.A. Carriedo, J.A.K. Howard, D.B. Lewis, G.E. Lewis, and F.G.A. Stone, J. Chem. Soc., Dalton Trans., 1985, 905.
249. J.C. Jeffrey, J.C.V. Laurie, and F.G.A. Stone, Polyhedron, 1985, **4**, 1135.
250. J.C. Jeffrey, C. Marsden, and F.G.A. Stone, J. Chem. Soc., Dalton Trans., 1985, 1315.
251. M. Green, J.A.K. Howard, A.P. James, A.N. de M. Jelfs, C.M. Nunn, and F.G.A. Stone, J. Chem. Soc., Chem. Commun., 1985, 1778.
252. E. Delgado, A.T. Emo, J.C. Jeffrey, N.D. Simmons, and F.G.A. Stone, J. Chem. Soc., Dalton Trans., 1985, 1323.
253. A. Birnbaum, F.A. Cotton, Z. Dori, M. Kapon, D. Marler, G.M. Reisner, and W. Schwotzer, J. Am. Chem. Soc., 1985, **107**, 2405.
254. K.R. Birdwhistell and J.L. Templeton, Organometallics, 1985, **4**, 2063.
255. N.H. Buttrus, C. Eaborn, P.B. Hitchcock, J.D. Smith, and A.C. Sullivan, J. Chem. Soc., Chem. Commun., 1985, 1380.
256. R.A. Anderson, A. Haaland, K. Rypdal, and H.C. Volden, J. Chem. Soc., Chem. Commun., 1985, 1807.
257. D. Gampe, K. Jacob, and K.-H. Thiele, Z. Anorg. Allg. Chem., 1985, **526**, 36.
258. C. Eaborn, P.B. Hitchcock, J.D. Smith, and A.C. Sullivan, J. Chem. Soc., Chem. Commun., 1985, 535.
259. P. Stauvopoulos, P.G. Edwards, G. Wilkinson, M. Motevalli, K.M.A. Malik, M.B. Hursthouse, J. Chem. Soc., Dalton Trans., 1985, 2167.
260. K.-W. Lee, W.T. Pennington, A.W. Cordes, and T.L. Brown, J. Am. Chem. Soc., 1985, **107**, 631.
261. K.-W. Lee and T.L. Brown, Organometallics, 1985, **4**, 1025.
262. J.A.M. Simões, J.C. Schultz, and J.L. Beauchamp, Organometallics, 1985, **4**, 1238.
263. K.E. Warner and J.L. Norton, Organometallics, 1985, **4**, 2150.
264. M.J. Nappa, R. Santi, and J. Halpern, Organometallics, 1985, **4**, 34.
265. G.T. Palmer and F. Basolo, J. Am. Chem. Soc., 1985, **107**, 3122.
266. H.-J. Haupt, G. Lohmann, and U. Flörke, Z. Anorg. Allg. Chem., 1985, **526**, 103.
267. F. Lindner, F. Zinsser, W. Hiller, and R. Fawzi, J. Organomet. Chem., 1985, **288**, 317.
268. G.A. Carriedo, J.B.P. Soto, V. Riera, X. Solans, and M. Miravitlles, J. Organomet. Chem., 1985, **297**, 193.

269. E. Lindner and A. Brösamle, Chem. Ber., 1985, **118**, 2134.
270. D. Miguel and V. Riera, J. Organomet. Chem., 1985, **293**, 379.
271. M.A. Biddulph, R. Davies, C.H.J. Wells and F.I.C. Wilson, J. Chem. Soc., Chem. Commun., 1985, 1287.
272. K.-W. Lee and T.L. Brown, Organometallics, 1985, **4**, 1030.
273. R. Poli, G. Wilkinson, M. Motevalli, and M.B. Hursthouse, J. Chem. Soc., Dalton Trans., 1985, 931.
274. E.J.M. de Boer, J. de With, N. Meijboom, and A.G. Orpen, Organometallics, 1985, **4**, 259.
275. S.F. Mapolie, J.R. Moss, and L.G. Scott, J. Organomet. Chem., 1985, **297**, C1.
276. W. Beck, K. Raab, U. Nagel, and W. Sacher, Angew. Chem.,Int. Ed. Engl., 1985, **24**, 505.
277. W.P. Fehlhammer, P. Hirschmann, and A. Volkl, J. Organomet. Chem., 1985, **294**, 251.
278. M.M. Singh and R.J. Angelici, Inorg. Chim. Acta, 1985, **100**, 57.
279. C.P. Casey, J.M. O´Connor, and K.J. Haller, J. Am. Chem. Soc., 1985, **107**, 3172.
280. R. Ferede, J.F. Hinton, W. Korfmacher, J.P. Freeman, and N.T. Allson, Organometallics, 1985, **4**, 614.
281. R.G. Bergman, P.F. Seidler, and T.J. Wenzel, J. Am. Chem. Soc., 1985, **107**, 4358.
282. J.I. Seeman and S.G. Davies, J. Am. Chem. Soc., 1985, **107**, 6522.
283. M.F. Asaro, G.S. Bodner, J.A. Gladysz, S.R. Cooper, and N.J. Cooper, Organometallics, 1985, **4**, 1020.
284. G.L. Crocco and J.A. Gladysz, J. Am. Chem. Soc., 1985, **107**, 4103.
285. G.L. Crocco and J.A. Gladysz, J. Chem. Soc., Chem. Commun., 1985, 283.
286. L. Weber and K. Reizig, Angew. Chem.,Int. Ed. Engl., 1985, **24**, 53.
287. C.F. Barrientos-Penna, A.B. Gilchrist, A.H. Klahn-Oliva, A.J. Hanlan, and D.Sutton, Organometallics, 1985, **4**, 478.
288. C.F. Barrientos-Penna, A.H. Klahn-Oliva, and D. Sutton, Organometallics, 1985, **4**, 367.
289. D.E. Smith and J.A. Gladysz, Organometallics, 1985, **4**, 1480.
290. G.K. Yang and R.G. Bergman, Organometallics, 1985, **4**, 129.
291. C.P. Casey, J.M. O´Connor, and K.J. Haller, J. Am. Chem. Soc., 1985, **107**, 1241.
292. C.P. Casey and J.M. O´Connor, Organometallics, 1985, **4**, 384.
293. F.G.N. Cloke and J.P. Day, J. Chem. Soc., Chem. Commun., 1985, 967.
294. M.J. McGeary, J.L. Tonker, and J.L. Templeton, Organometallics, 1985, **4**, 2102.
295. R. Cramer, K.T. Higa, and J.W. Gilje, Organometallics, 1985, **4**, 1140.
296. E.O. Fischer and J.K.R. Wanner, Chem. Ber., 1985, **118**, 2489.
297. R.H. Fong and W.H. Hersh, Organometallics, 1985, **4**, 1468.
298. A.B. Antonova, S.V. Kovalenko, E.D. Korniyets, P.V. Petrovsky, G.R. Gulbis, and A.A, Johansson, Inorg. Chim. Acta, 1985, **96**, 1.
299. A.B. Antanova, S.V. Kovalenko, E.D. Korniyets, P.V. Petrovsky, A.A. Johansson, and N.A. Deykhina, Inorg. Chim. Acta, 1985, **105**, 153.
300. N.E. Kolobova, L.L. Ivanov, O.S. Zhvanko, A.S. Batsanov, and Yu.T. Struchov, J. Organomet. Chem., 1985, **279**, 419.
301. H. Werner, F.J.G. Alonso, H. Otto, K. Peters, and H.G. von Schnering, J. Organomet. Chem., 1985, **289**, C5.
302. W.C. Mercer, R.R. Whittle, E.W. Burkhardt, and G.L. Geoffroy, Organometallics, 1985, **4**, 68.

12
Complexes Containing Metal–Carbon σ-Bonds of the Groups Iron, Cobalt, and Nickel

BY A. K. SMITH

1. Introduction

As in previous volumes, this chapter is divided into the following four sections: reviews and articles of general interest; metal-carbon σ-bonds of the Group VIII triads in the sequence iron, cobalt, and nickel; carbene and carbyne complexes including complexes containing μ-CH_2 and μ-CH ligands; and a short bibliography giving details of some papers that have not been included in the main body of the text due to space limitations.

2. Reviews and Articles of General Interest

The second volume of the series 'The Chemistry of the Metal-Carbon Bond', sub-titled 'The Nature and Cleavage of Metal Carbon Bonds', has been published.[1] The activation of C-H bonds[2,3] and the organometallic chemistry of alkanes[4] have been the subjects of reviews in 1985, as well as ligand substitution reactions of square planar molecules,[5] and the photochemistry of alkyl, alkylidene, and alkylidyne complexes.[6] Articles of general interest include a discussion of the current state of organocobalt(IV) and organorhodium(IV) chemistry,[7] a review of the reactions of alkylcobalamins with platinum complexes,[8] the chemistry of organopalladium and nickel complexes relevant to catalysis - including reductive elimination, CO insertion, and nucleophilic attack on coordinated ligands,[9] and a review of Pd-ligand and Pt-ligand bond energies which includes data on Pd-C and Pt-C bonds.[10]

3. Metal-Carbon σ-Bonds involving Group VIII Metals

3.1 The Iron Triad. - The gas phase reactions of $FeCH_3^+$ and $CoCH_3^+$ with a variety of alkenes and alkynes have been described,[11] and the reactions of $CoFe^+$ with a number of hydrocarbons have also been

(For references see page 293

reported.[12]

The dialkyl-iron complexes cis-[FeMe$_2$(dmpe)$_2$],[13] cis-[FeMe$_2$(dmpm)$_2$],[14] trans-[Fe(Bz)$_2$(dmpm)$_2$],[14] and [C$_6$H$_4$(CH$_2$)$_2$Fe-(dmpm)$_2$][14] [dmpm = bis(dimethylphosphino)methane] have been prepared. The di-iodoiron complex [Fe(C$_5$Me$_5$)(NO)I$_2$] may be alkylated with RLi or RMgX to give the dialkyl complexes [Fe(C$_5$Me$_5$)(NO)R$_2$] (R=Me, Bz, or CH$_2$SPh).[15] The addition of TCNE to [Fe(η^3-allyl)$_2$-(CO)$_2$] leads to the formation of the 16-electron complex [Fe(η^3-allyl){$\overline{CHCH_2C(CN)_2C(CN)_2CH_2}$}(CO)$_2$] which adds PPh$_3$ to give the 18-electron complex [Fe(η^3-allyl){$\overline{CHCH_2C(CN)_2C(CN)_2CH_2}$}(CO)$_2$-(PPh$_3$)].[16] The photolysis reactions of [M(Cp)(CO)$_2$R] (M=Fe, Ru; R=Me or Et) have been studied by a combination of solution (-30°C) and matrix-isolation (12K) techniques.[17] Near u.v. irradiation of [Fe(C$_5$R$_5$)(CO)$_2$(Bz)] (R=H or Me) yields both loss of CO and Bz radicals, and the relative importance of these two processes has been investigated.[18] The methyliron complexes [Fe(CO)$_2$(I)(Me)(pdmp)] [pdmp = o-phenylenebis(dimethylphosphine)] and fac-[Fe(CO)$_3$Me(CO)$_3$]$^+$ have been prepared from the acyl complex [Fe(CO)$_2${C(O)Me}(I)(pdmp)] by thermal decarbonylation and by CO migration in the presence of halide acceptors, respectively.[19] The cationic acyl complex [Fe(CO)$_2${C(O)Me}(diars)(PMe$_3$)]$^+$ undergoes a stereospecific decarbonylation in the solid state to give [Fe(CO)$_2$(Me)(diars)(PMe$_3$)]$^+$.[20] A number of organoiron complexes of the type [Fe(Cp)(LL)$_2$R] (LL=dppe or dppb; R=Me, Et, Bu, cyclopropyl, vinyl, 2-butenyl) have been synthesised.[21] A number of C$_2$ molecules and ligands, such as acetic acid ester, η^2-ethene, and η^1-ethyl, have been generated from the carbalkoxymethyl ligand in [Fe(Cp)(CO)(CH$_2$CO$_2$R){P(OMe)$_3$}] (R=Me or Et).[22] The indenyliron complexes [Fe(η^5-indenyl)(CO)(L)R] (L=CO, PPh$_3$; R=Me, COMe) have been synthesised.[23] The substitution of chloride in [Fe(Cp)(CO)$_2$(CH$_2$Cl)] by a wide range of nucleophiles to give the cationic complexes [Fe(Cp)(CO)$_2$(CH$_2$X)]$^+$ (X=PMe$_2$Ph, NEt$_3$, SMe$_2$, etc) is promoted by TlBF$_4$.[24] The synthesis and some reactions of 1-acetoxyalkyl derivatives of Fe(II) have been reported.[25]

The carbonylation of [Fe(CO)$_2$(I)(Me)(PMe$_3$)$_2$] at 250K gives cis-[Fe(CO)$_2$(I)(MeCO)(PMe$_3$)$_2$] which isomerises at 298K to give the trans isomer.[26] Labelling and ^{13}C n.m.r. studies have shown that the mechanism involves a formal methyl migration and a square planar intermediate.[26] In complexes of the type [Fe(Cp)(CO)$_2$(Me)] it has been found that carbonylation to give the corresponding acetyl complex is favoured by replacing Cp with indenyl and CO with

PPh$_3$.^{27} Reaction conditions, such as solvent and the presence of an acid catalyst, are also important.^{27} A conformational analysis based on extended Hückel calculations for the complex [Fe(Cp)(CO)(PPh$_3$)R] (R=alkyl and aryl) has been published, and the results used to explain the high stereospecificities that have been observed in the reactions of this complex.^{28} The conformationally labile complex [Fe(Cp)(CO)(PPh$_3$)(CH$_2$SiMe$_3$)] has been studied using NOE difference n.m.r. spectroscopy.^{29}

The reactions of a variety of alkylmetal carbonyl complexes, including [Fe(Cp)(CO)$_2$(Me)] and [Os(CO)$_4$(Me)$_2$], with various metal hydrides (such as [ReH(CO)$_5$], [OsH$_2$(CO)$_4$], [MnH(CO)$_5$], and [WH(CO)$_3$(Cp)]) have been shown to lead to dinuclear complexes or polynuclear hydrides with the organic products eliminated usually being aldehydes although alkane elimination is also seen.^{30}

The alkenyliron complexes [Fe(Cp)(CO){η^1-C(CH$_2$Nuc)=CH$_2$}-{P(OPh)$_3$}] (Nuc=H,Me,Ph, CH=CH$_2$, C≡CMe,SPh) have been prepared by treating [Fe(Cp)(CO)(η^2-CH$_2$=C=CH$_2$)][BF$_4$] with the appropriate anionic nucleophile.^{31} These alkenyliron complexes undergo a 1,3-hydrogen shift isomerisation reaction on chromatography on alumina to give [Fe(Cp)(CO){η^1-C(Me)=C(H)Nuc}{P(OPh)$_3$}]. The isomerisation of the alkenyliron complexes has been investigated using D-labelling and protonation studies.^{31} The alkenyliron complexes [Fe(Cp)(CO)$_2$(η^1-diene)] (diene=1,3-butadiene, 3-methyl--1,3-butadiene) have been found to form Diels-Alder adducts with a variety of electrophilic dienophiles.^{32} The new compound [Fe(η^5-C$_5$Me$_5$)(CO)$_2$(η^1-C$_5$H$_5$)] has been prepared, and its reactions with dimethyl fumarate to give a [3+2] cycloadduct and with bis-(trifluoromethyl)ketene to give 2:1 adducts have been investigated and compared to the reactivity of the analogous complex [Fe(η^5-C$_5$H$_5$)(CO)$_2$(η^1-C$_5$H$_5$)].^{33} Treatment of both [Fe(Cp)(CO)$_2$-(endo-norbornyl)] and [Fe(Cp)(CO)$_2$(exo-norbornyl)] with Ph$_3$C$^+$BF$_4^-$ gives a mixture of the migratory CO insertion products, [Fe(Cp)(CO)$_2$(endo-2-norbornylcarbonyl)] and [Fe(Cp)(CO)$_2$-(exo-2-norbornylcarbonyl)], respectively, and the product of β-elimination, [Fe(Cp)(CO)$_2$(η^2-exo-norbornene)][BF$_4$].^{34} A new class of σ-propenyl complexes of the type [Fe(Cp)(CO)$_2${C=C(Ph)-C(Ph)R}] (R=H,OMe,CN) has been obtained by the reaction of [Fe(Cp)(CO)$_2${$\overline{\text{C}^+\text{C(Ph)C(Ph)}}$}][BF$_4^-$] with nucleophiles.^{35} NaCN reacts with the cationic di-iron complex [(Cp)(CO)$_2$FeC≃CPh≃C{Fe(CO)$_2$(Cp)}-CHPh]$^+$BF$_4^-$ to give the cyclobutene complex (1), which on heating

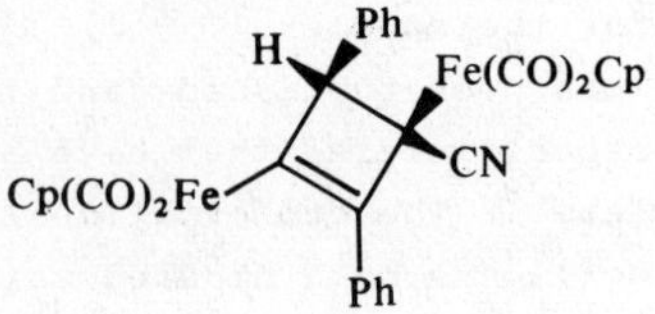

(1)

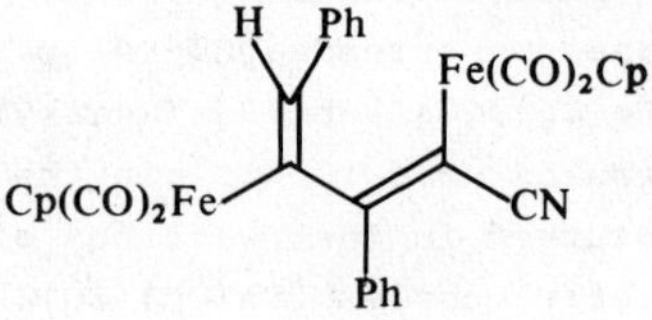

(2)

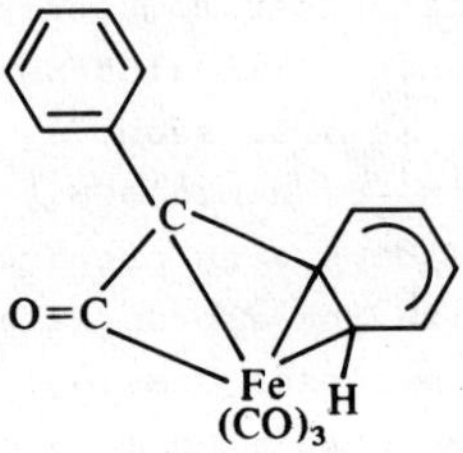

(3)

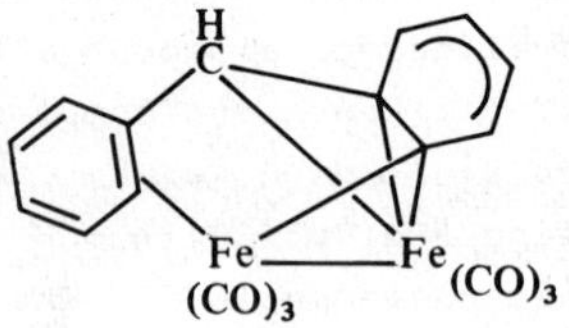

(4)

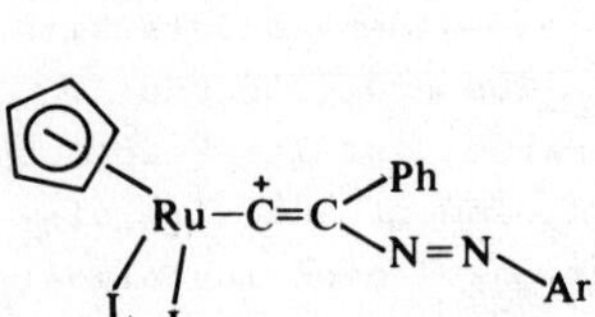

(5)

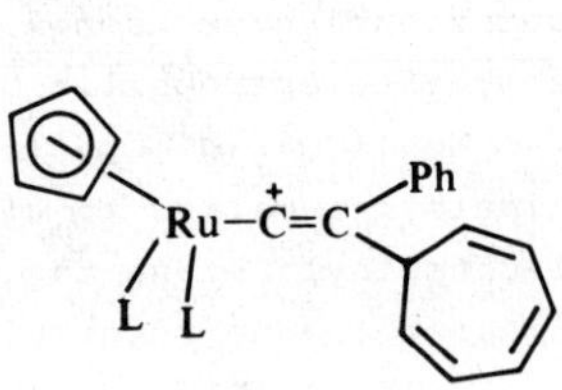

(6)

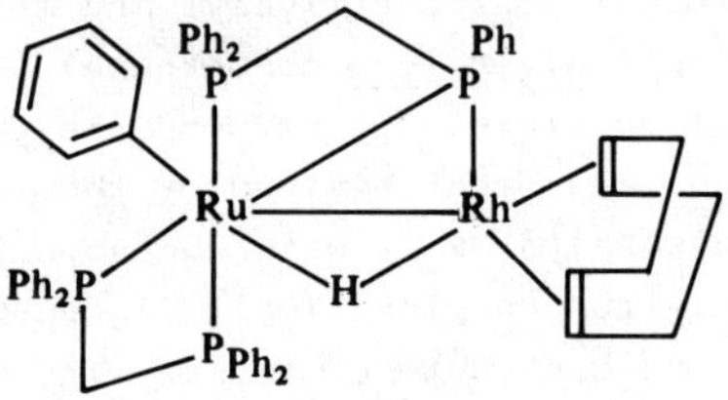

(7)

in solution undergoes a ring-opening reaction to give the butadiene derivative (2).[36]

Ferracyclopentanes of the type $[(CO)_3(L)\overline{FeCH_2CX_2CX_2CH_2}]$ and ferrahydrindanes of general formula $[(CO)_3(L)\overline{FeCH_2CH(CH_2)_4CHCH_2}]$ [X=H,D, or Me; L=CO,PPh$_3$ or P(OMe)$_3$] have been synthesised.[37] The μ-(1,n)-alkanediyliron complexes $[\{(Cp)(CO)_2Fe\}_2\{\mu-(CH_2)_n\}]$ (n=7-12) have been prepared.[38] These complexes together with those for which n=3-6 have been studied using mass spectrometry and differential scanning calorimetry.[38] Their reaction with Ph$_3$CPF$_6$ to give the cationic di-iron complexes $[\{(Cp)(CO)_2Fe\}_2\{\mu-(C_nH_{2n-1})\}]^+$ (n=3-6) has been reported.[39] The $\eta^1:\eta^3$-diphenylketene complex (3) is obtained upon photolysis of Fe(CO)$_5$ in the presence of diphenylketene.[40] Treatment of (3) with [Fe$_2$(CO)$_9$] gives (4).[40] The very high stability of the metallacyclopentadienyl ring in complexes of the type (5) and (6) has been ascribed to the presence of a very strong σ-M-C interaction, following a UV photoelectron spectroscopic and ab initio study.[41]

The synthesis of a number of 6-co-ordinate nitrosyl alkyl- and aryl-iron porphyrin complexes, and their electrochemical characterisation, has been reported.[42] An account of the synthesis and electrochemistry of several 5-coordinate aryliron porphyrins has also been published.[43]

Syntheses and X-ray crystal structures of a number of carbido cluster complexes containing iron have been reported. Such complexes include [Fe$_4$CoC(CO)$_{14}$]$^-$,[44] [Fe$_4$CoRhC(CO)$_{16}$],[44] [Fe$_5$C(CO)$_{15}$],[45] [NBu$_4$]$_2$[Fe$_5$C(CO)$_{14}$],[45] [(Ph$_3$P)$_2$N][Fe$_6$C(CO)$_{15}$NO],[46] and [Fe$_6$C(CO)$_{11}$(NO)$_4$].[46]

The complexes [Ru(C$_6$Me$_6$)(CO)(Me)(PPh$_3$)][BF$_4$] and [Ru(C$_6$Me$_6$)(C$_2$H$_4$)(Me)(PPh$_3$)][BF$_4$] have been prepared by the reaction of [Ru(C$_6$Me$_6$)(Me)$_2$(PPh$_3$)] with HBF$_4$ in the presence of CO or C$_2$H$_4$, respectively.[47] The ethylene complex undergoes a CH$_4$ elimination reaction at room temperature to form the 2-styryldiphenylphosphine complex, [Ru(C$_6$Me$_6$)(H)(PPh$_2$C$_6$H$_4$CH=CH$_2$)][BF$_4$].[47] The complexes [Ru(C$_6$H$_6$)(C$_2$H$_4$R)(PMe$_3$)$_2$][PF$_6$] undergo hydride abstraction reactions with CPh$_3$PF$_6$ to give arene(olefin) Ru(II) dicationic complexes.[48] The dicationic complex with R=H reacts with PR'$_3$ to give [Ru(C$_6$H$_6$)(C$_2$H$_4$PR'$_3$)(PMe$_3$)$_2$][PF$_6$]$_2$.[48]

The ruthenium(II) complexes [Ru(η^5-C$_5$R$_5$)L$_2$X] (R=H, Me; X=Br, Cl) are formed by thermal decomposition of [Ru(η^5-C$_5$R$_5$)(η^3-C$_5$H$_5$)(Me)X] in the presence of neutral ligands L.[49] SO$_2$ insertion into the Ru-C bonds in the complexes [Ru(Cp)LL'(R)] (L,L'=CO, PPh$_3$; R=Me, Bz) occurs to give the corresponding S-sulphinato complexes, except when L=L'=CO and R=Me.[50] [RuCl(CO)H(PPh$_3$)$_3$] reacts with methyl acetate and N,N,-dimethylacrylamide to give the corresponding insertion products [Ru{CH$_2$CH$_2$C(O)OMe}Cl(CO)-(PPh$_3$)$_2$] and [Ru{CH$_2$CH$_2$C(O)NMe$_2$}Cl(CO)(PPh$_3$)$_2$], respectively.[51] The insertion products of 2-vinylpyridine and 5-ethyl-2-vinyl-pyridine have also been prepared in this way.[51]

The acetylide complexes [Ru(Cp)(L)$_2$(C≡CPh)] react with halogens to give halovinylidene complexes of the type [Ru(Cp)(L)$_2${C=C(Br)(C$_6$H$_4$Br-4)}]$^+$Br$_3^-$ (L=PPh$_3$)[52] and with arene-diazonium or tropylium salts to give the cationic aryldiazovinyl-idene complex(5) and the cycloheptatrienylvinylidene complex(6) (L=PPh$_3$ or L$_2$=dppe), respectively.[53]

The chiral complexes [Ru(CO)(Me)(L)(triphos)]X [L=^{t}BuNC, ^{s}BuNC, P(OMe)$_3$, and PMe$_2$Ph; X=halide] have been resolved and the absolute configurations of the (+)-enantiomer have been determined.[54]

The treatment of [Ru$_2$Cl{3,5-(OMe)$_2$C$_6$H$_3$CONH}$_4$] with PPh$_3$ or P(p-tolyl)$_3$ in the presence of [NBu$_4$]ClO$_4$ results in aryl group transfer and the formation of the arylruthenium complexes [Ru$_2$(Ar)$_2${3,5-(OMe)$_2$C$_6$H$_3$CONH}$_2${Ar$_2$POC(3,5-(OMe)$_2$C$_6$H$_3$}N)$_2$] (Ar=Ph or tol).[55] Aryl group transfer also occurs on treatment of [RuRhH$_2$Cl(cod)(dppm)$_2$] with MeLi, when the phenylruthenium complex (7) is formed.[56] The reduction of [MCl$_2$(PMe$_3$)$_2${P(OMe)$_3$}$_2$] (M=Ru,Os) with Na/Hg in benzene or toluene gives the aryl(hydrido) metal complexes cis, trans, cis-[MH(Ph)(PMe$_3$)$_2${P(OMe)$_3$}$_2$] (M=Ru, Os) and [OsH(C$_6$H$_4$Me)(PMe$_3$)$_2${P(OMe)$_3$}$_2$], while the reduction of trans-[RuCl$_2$(PMe$_3$)$_4$] with Na/Hg in the presence of PPh$_3$ affords the ortho-metallated complex fac-[RuH(C$_6$H$_4$PPh$_2$)(PMe$_3$)$_3$].[57] The reduction of trans-[RuCl$_2$(PMe$_3$)$_4$] with Na/Hg and of trans-[OsCl$_2$(PMe$_3$)$_4$] with Na/naphthalene gives the complexes [MH(η^2-CH$_2$PMe$_2$)(PMe$_3$)$_3$].[58]

A number of ruthenium complexes containing ortho-metallated ligands have been reported. Such complexes include [Ru(CO) -{C$_6$H$_3$Y'C(O)C$_6$H$_4$Y}XL$_2$] (Y=Y'=Me, L=PMe$_2$Ph or AsMe$_2$Ph; Y=Y'=Cl, L=PMe$_2$Ph; Y=Me,Y'=Cl, L=PMe$_2$Ph; X=Cl or I),[59] [Ru$_3$H(CO)$_{10}$L] (where L is an ortho-metallated methylpyridine, quinoline, isoquinoline, or diazine),[60] and [Ru(bipy)L]$^+$ [where L is cyclometallated

2-(3-nitrophenyl)pyridine, phenylpyridine, benzo[h]quinoline, azobenzene, or p-(dimethylamino)azobenzene].[61] A two-dimensional ^{1}H n.m.r. study of the complex [Ru(bipy)L]$^+$ [L=cyclometallated 2-(4-nitrophenyl)pyridine] has been reported.[62] The reaction of [RuH(Cl)(PPh$_3$)$_3$] with alkynes produces the ortho-metallated complex [RuCl{P(C$_6$H$_4$)Ph$_2$}(PPh$_3$)$_2$] which is present during catalytic alkyne hydrogenation reactions.[63] One of the complexes isolated and structurally characterised from the reaction between [Ru$_3$(CO)$_{12}$] and HC(PPh$_2$)$_3$ is the dppm-metallated complex (8).[64]

A number of organoruthenium porphyrin complexes have been prepared from reactions involving cleavage of the dimeric complex [Ru(por)]$_2$ (por=OEP, TTP).[65]

The structures of two isomeric carbido-cluster complexes [Ru$_5$C(CO)$_{13}$(NO)(μ_3-AuPEt$_3$)] and [Ru$_5$C(CO)$_{13}$(NO)(μ_2-AuPEt$_3$)], present in a single crystal, have been reported.[66]

The air-stable, tetrahedral osmium(IV) compounds, [OsR$_4$] (R=cyclohexyl, o-methylphenyl) have been synthesised by treatment of [Os$_2$(O$_2$CMe)$_4$Cl$_2$] and OsO$_4$ with the appropriate Grignard reagent.[67] A mechanistic study of the elimination reactions of [Os(CO)$_4$(H)R] and [Os(CO)$_4$R$_2$] has been carried out.[68] Alkane elimination from cis-[Os(CO)$_4$(H)R] proceeds via a rate determining isomerisation (probably to an acyl hydride); this isomeric intermediate can then react with unisomerised complex to give intermolecular R-H elimination or with another nucleophile L to give intramolecular R-H elimination. Simple intramolecular reductive elimination from any complex of the type cis-[Os(CO)$_4$RR'] has not been observed. Rather, the formation of CH$_4$ from cis-[Os(CO)$_4$(Me)$_2$] proceeds via an intermediate (probably a methyl radical) which shows high H/D selectivity in attacking alkane solvents.[68]

The arene-osmium complex [Os(p-cymene)Cl$_2$L] (L=CO, CNBu^t, Me$_2$SO, or PMe$_3$), on treatment with Al$_2$Me$_6$, gives [Os(p-cymene)Cl(Me)L].[69] When L=Me$_2$SO, this methylosmium complex reacts with L'[L'=PPh$_3$ or P(OPh)$_3$] to give [Os(p-cymene)Cl(Me)L'] which reacts with Al$_2$Me$_6$ to afford the ortho-metallated complexes [Os(p-cymene){PPh$_2$(o-C$_6$H$_4$)}X] (X=Cl,Me) and [Os(p-cymene){P(OPh)-(OC$_6$H$_4$-o)$_2$}] as major products, although when L=PPh$_3$, the complexes [Os(p-cymene)(PPh$_3$)(Me)X] (X=Cl or Me) are also obtained.[69] The complex [Os(p-cymene)(C$_2$H$_4$)(PMe$_3$)] reacts with excess PMe$_3$ in benzene to give the inter-molecular C-H addition product [OsH(Ph)(PMe$_3$)$_4$]. This is in contrast to the analogous reaction with

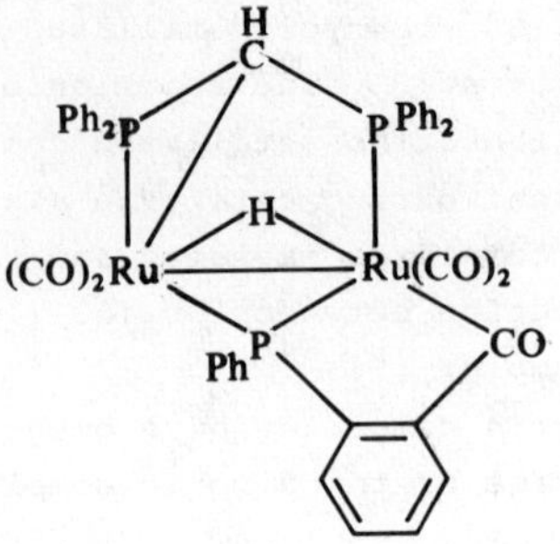
H
C
Ph₂P
PPh₂
H
(CO)₂Ru
Ru(CO)₂
Ph
P
CO

(8)

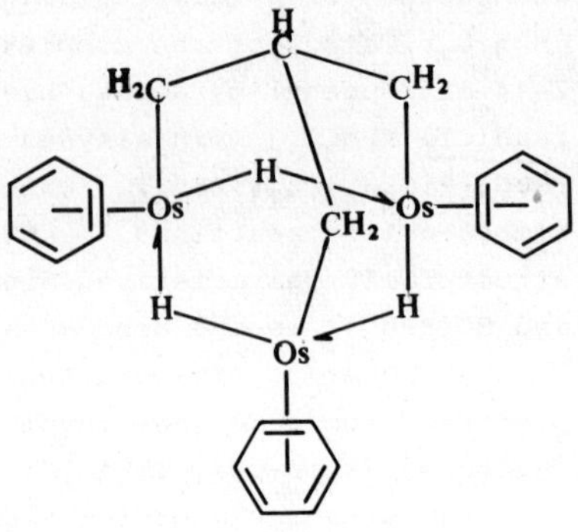
H
C
H₂C
CH₂
H
Os
CH₂
Os
H
H
Os

(9)

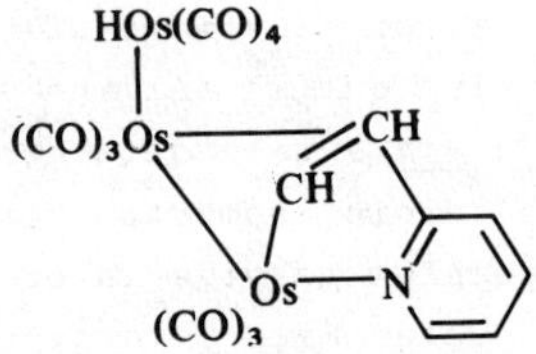
HOs(CO)₄
(CO)₃Os
CH
CH
Os
N
(CO)₃

(10)

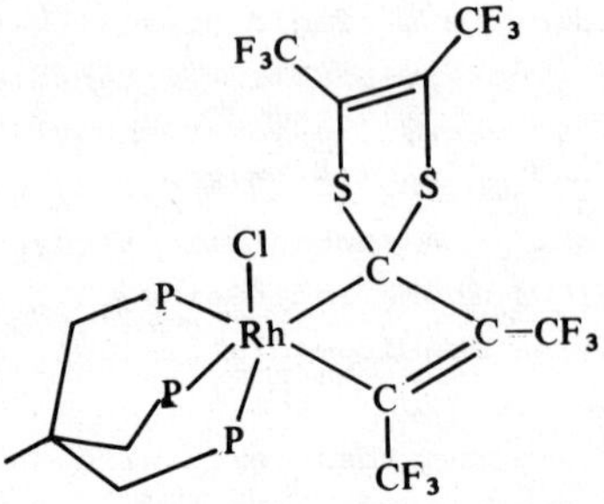
CF₃
F₃C
S
S
Cl
C
P
Rh
C—CF₃
P
C
P
CF₃

(11)

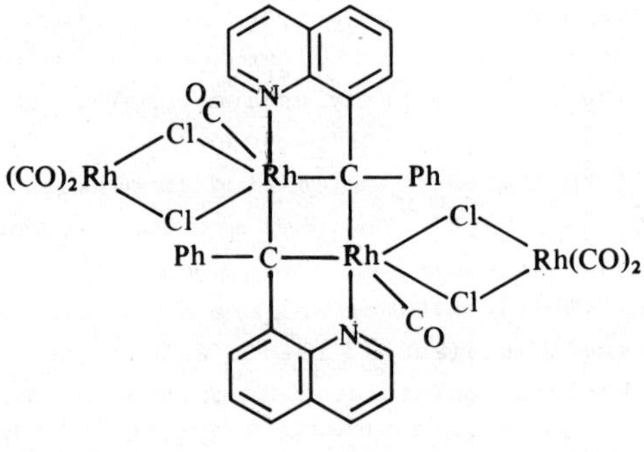
N
OC
Cl
(CO)₂Rh
Rh—C—Ph
Cl
Cl
Ph—C—Rh
Rh(CO)₂
Cl
CO
N

(12)

[Os(C_6H_6)(C_2H_4)(PMe$_3$)] which undergoes an intramolecular C-H addition reaction.[70] The reaction of [Os(C_6H_6)I$_2$(PMe$_3$)] with NaC$_{10}$H$_8$ in the presence of C_6H_6 or C_6D_6 gives the inter-molecular C-H addition products [Os(C_6H_6)H(C_6H_5)(PMe$_3$)] and [Os(C_6H_6)D-(C_6D_5)(PMe$_3$)], respectively.[71] The vinylidene-osmium complex [Os(C_6H_6)(=C=CHPh)(PPr$_3^i$)] has been prepared from [Os(C_6H_6)I$_2$]$_2$ via alkynyl(iodo)-, vinyl(hydrido)- and vinyl(halogeno)- osmium(arene) complexes.[72] The co-condensation of Os atoms and a benzene: 2-methylpropane mixture gives the Os$_3$ complex [Os(η-C_6H_6)(μ-H)]$_3${μ_3-(CH$_2$)$_3$CH}] (9), in which the μ-H ligands have been shown to come from the 2-methylpropane.[73]

Metallation of 2-ethenylpyridine occurs on reaction with [Os$_3$H$_2$(CO)$_{10}$] or [Os$_3$(CO)$_{10}$(MeCN)$_2$], giving the open Os$_3$ complex [Os$_3$H(CO)$_{10}$(NC$_5$H$_4$CH=CH)] (10).[74] Furan oxidatively adds to the complex [Os$_3$(CO)$_{10}$(MeCN)$_2$] to give the μ,η^2-furyl compound [HOs$_3$(CO)$_{10}$(μ-η^2-C_4H_3O)].[75]

The carbido-clusters [Os$_3$Pt(μ-H)$_2$(μ_4-C)(CO)$_{10}$(PCy$_3$)], [Os$_3$Pt$_2$(μ-H)$_2$(μ_5-C)(μ-CO)(CO)$_9$(PCy$_3$)$_2$], and [Os$_3$Pt$_2$(μ-H)(μ_5-C)-(μ-OMe)(μ-CO)(CO)$_9$(PCy$_3$)$_2$] have been prepared and structurally characterised.[76]

3.2 The Cobalt Triad. - Oxidative addition reactions between hydrocarbon radicals and complexed cobalt(II) centres in the gas phase have been studied using negative chemical ionisation mass spectrometry.[77] The gas-phase chemistry of Co$^+$ with a series of 1-chloro-n-alkanes and alcohols has been investigated, and it has been found that for RCl, as the chain length increases, Co$^+$ insertion into internal C-C bonds becomes preferred to insertion into the C-Cl bond.[78]

^{59}Co n.m.r. parameters for a range of organo-cobalt(I) and -cobalt(III) compounds have been determined, and have also been related to the chemical properties of the compounds.[79] E.s.r. spectroscopy has been used to determine the ground state configuration of the paramagnetic complex [Co(η^6-toluene)(C_6F_5)$_2$].[80]

The cis-dialkylcobalt(III) complexes, cis-[CoR$_2$(bipy)$_2$](ClO$_4$) (R=Me,Et, and Bz) undergo photo-induced Co-C cleavage which induces selective C-C bond formation from benzyl and allyl bromides to yield 1,2-diphenylethane and 1,5-hexadiene, respectively.[81] One-electron oxidation of cis-[CoR$_2$(bipy)$_2$][ClO$_4$] (R=Me or Et) by organic oxidants results in Co-R bond cleavage to yield R-R exclusively.[82] A study of the reactions of the complex [Co(Cp)(H)(C$_2$H$_4$){P(OMe)$_3$}], which contains an agostic hydride, has shown that the same factors that favour a bridging agostic over a terminal hydride structure will facilitate alkyl migration. Thus, [Co(Cp)(H)(C$_2$H$_4$){P(OMe)$_3$}] reacts with P(OMe)$_3$ by displacement of the C-H bridge and formation of [Co(Cp)(CH$_2$CH$_3$){P(OMe)$_3$}$_2$].[83] The complexes [Co(Cp)(CO)(PMe$_3$)] or [Co(Cp)(C$_2$H$_4$)(PMe$_3$)] react with CH$_2$ClI to give mixtures of [Co(Cp)(CH$_2$Cl)(PMe$_3$)I] and [Co(Cp)Cl(PMe$_3$)I].[84] The same reaction in the presence of L gives the carbenoidcobalt(III) compounds [Co(Cp)(CH$_2$Cl)(PMe$_3$)L]PF$_6$ [L=PMe$_3$, P(OMe)$_3$, and CNMe] or the ylide complexes [Co(Cp)(CH$_2$PMe$_3$)-(PMe$_3$)X]PF$_6$ (L=NEt$_3$, X=Br,I).[84] The dinuclear CH$_2$PMe$_2$-bridged complexes, [(Cp)XCo(μ-CH$_2$PMe$_2$)$_2$CoX(Cp)] (X=Br,I) are formed by treating [Co(Cp)(μ-PMe$_2$)]$_2$ with CH$_2$Br$_2$ or CH$_2$I$_2$.[85] These complexes react with Na/Hg to give [(Cp)Co(μ-PMe$_2$)(μ-CH$_2$PMe$_2$)Co(Cp)].[85]

The regioselectivity of the reaction between [Co(Cp)(PPh$_3$)-(η^2-R^{1}C≡CR^2)] and R^{3}NC to form the iminocobaltacyclobutenes [(Cp)(PPh$_3$)CoC(R^1)=C(R^2)C(=NR^3)] (R^1=H,Ph,Me; R^2=Ph,CO$_2$Me,CN; R^3=Ph,2,6-Me$_2$C$_6$H$_3$,p-MeC$_6$H$_4$) has been shown to be due to the electronic effects of the substituents on the alkyne ligand.[86] The structures of the complexes trans-[Co(CH$_2$CF$_3$)(L)(DH)$_2$] [DH=dimethylglyoximato; L=4-cyanopyridine, PPh$_3$, and P(OMe)$_3$] have been determined, and the geometry of the CH$_2$CF$_3$ group, which departs significantly from that expected for sp^3 hybridisation at the C having the F substituents, used to provide information concerning the electron donation of the alkyl group to the metal.[87]

Reports on organocobalt B$_{12}$ models include the syntheses and structures of [Co(dmgH)$_2${P(OR)$_3$}(1-adamantyl)] (R=Me,Ph),[88] [Co(gH)$_2$(py)R] (gH=glyoximato, R=Me,Et,Pr^i),[89] and [Co(saloph)R] (R=Me,Pr^i).[90] A review of the effect of axial ligands on the structural and coordination chemistry of cobaloximes has been published.[91] A number of dinuclear dialkylcobalt(III) complexes of general formula [R(chel)CoLCo(chel)R] [R=Me,CH$_2$Cl; L=pyrazine, bipy,dppe; chel=dmgH, cyclohexanedionedioxime, bis(acetylacetone)-ethylenediimine, or salen) have been synthesised.[92] The preparation of a new series of σ-alkylcobalt(III) complexes of the type [Co(R)(7-Me-salen)L$_2$]$^+$ (L$_2$=monodentate or bidentate Lewis base) has been reported,[93] and the kinetics of their acid-induced decomposition (R=Me,Et, and Bu^n; L$_2$=en) have been studied.[94] The cobalt(I) complex, [Co(salen)]$^-$ reacts with

3,3,3-trifluoroprop-1-yne to give [Co(C_2CF_3)(salen)(py)], and [{Co(salen)}$_2O_2$] or [Co(salen)(acac)] react with PhC≡CH to give [Co(C_2Ph)(salen)(py)].[95]

The carbido-clusters [$Co_3Ni_7C_2(CO)_{16}$]$^{2-}$,[96] [$Co_6C(CO)_{14}$]$^-$,[97] [$Co_6C(CO)_{15}$]$^{2-}$,[97] [$Co_8C(CO)_{18}$]$^{2-}$,[97] and [$Co_{13}C_2(CO)_{24}$]$^{3-}$,[98] have been synthesised.

The syntheses and fluxional behaviour of the complexes [M(CO)(R)(triphos)] (M=Rh, R=CH_2SiMe_3; M=Ir, R=Me,CH_2CMe_3, CH_2SiMe_3) have been reported.[99] The complex [RhCl(η^2-CS_2)-(triphos)] underoges a reaction with F_3CC≡CCF_3 to give the rhodacyclobutene complex (11).[100]

From a study of the equilibration of [Rh(C_5Me_5)(PMe$_2$Bz)-(Ph)H] with its cyclometallated analogue [Rh(C_5Me_5)(PMe$_2CH_2C_6H_4$)H] and by equilibrating [Rh(C_5Me_5)PMe$_2CH_2CH_2CH_2$)H] with benzene, it has been found that kinetic selectivity favours inter-molecular over intra-molecular C-H bond activation when neat solvent is the competing reactant and that there is a thermodynamic preference for intra-molecular C-H bond activation.[101] The reaction of aldehydes RCHO with [M(C_5Me_5(Me)$_2$(Me$_2$SO)] (M=Rh,Ir) gives the complexes [M(C_5Me_5)(Me)(R)(CO)] which are also formed by the carbonylation of [M(C_5Me_5)(Me)R(Me$_2$SO)].[102] Treatment of the rhodium(I) complexes, [Rh(C_5Me_5)(CO)$_2$] or [Rh(C_5Me_5)(CO)L].[L=PMe$_3$,PMe$_2$Ph, P(OMe)$_3$], with CH_2X_2 and CHX$_3$ (X=Cl,Br,I,CN) gives [Rh(C_5Me_5)(L)(CH_2X)X] and [Rh(C_5Me_5)(L)-(CHX$_2$)X], respectively.[103] Some reactions of these complexes, involving substitution and Rh-C bond cleavage, have been reported.[103] Hydride abstraction using [CPh$_3$]BF$_4$ occurs from the β-carbon atom of the ethyl group in [Rh(Cp)(Et)(PMe$_3$)$_2$]$^+$ to give the ethylene complex [Rh(Cp)(C_2H_4)(PMe$_3$)$_2$][BF$_4$]$_2$.[104]

The effect of iodide and acetate on the oxidative addition of MeI to [Rh(CO)$_2$I]$^-$ in acetic acid has been investigated.[105] A study has been made of the oxidative addition and reductive elimination reactions of MeI and acetyl chloride with homobimetallic complexes of Rh(I) and Ir(I) containing the binucleating ligand bis(diphenylphosphino)pyrazole.[106] It is found that oxidative addition on one metal leads to deactivation of the other metal despite the absence of a M-M bond.[106] The 1,3-dimetallacycle (12) is formed on treatment of [Rh(CO)$_2$Cl]$_2$ with 8-quinolinylphenylketone.[107] With 8-quinolinecarboxaldehyde, followed by pyridine, [Rh(C_2H_4)$_2$Cl]$_2$ reacts to give the ethyl complex (13).[108]

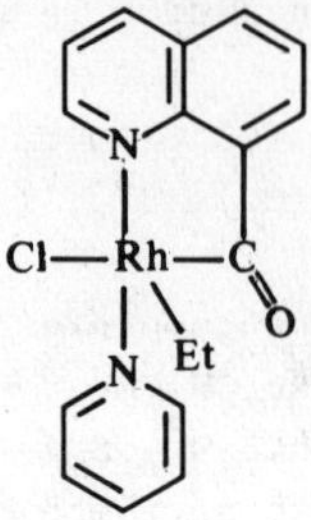
N
Cl—Rh—C
O
Et
N

(13)

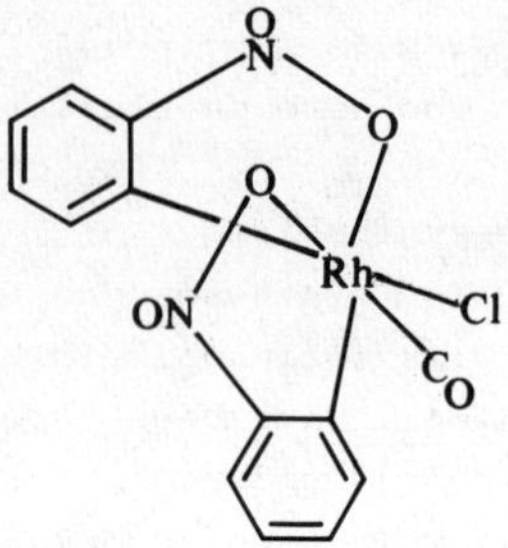
O
N
O
O
Rh
Cl
ON
CO

(14)

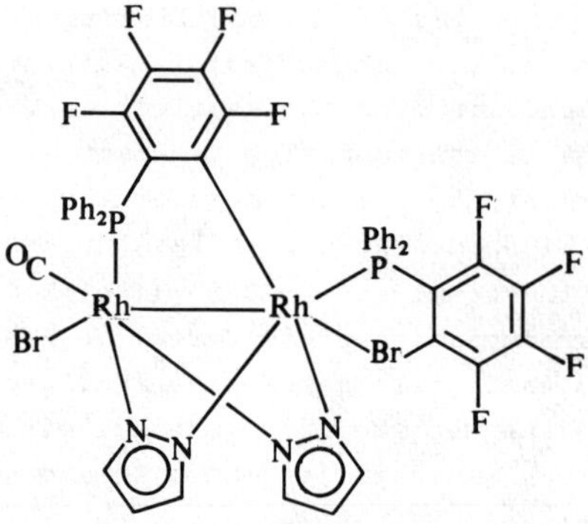
F F
F F
Ph₂P
OC
Rh
Rh
Br
Ph₂
P
F
F
F
F
Br
N N
N N

(15)

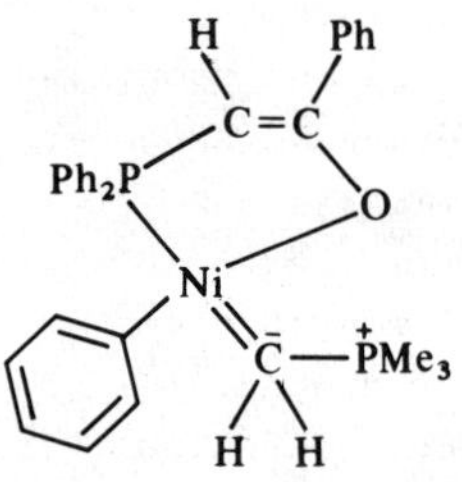
H Ph
C=C
Ph₂P
O
Ni
C̄—P̄Me₃
H H

(16)

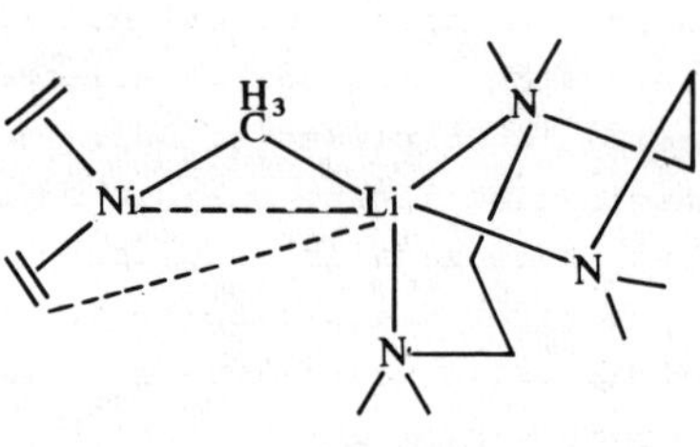
H₃
C
Ni
Li
N
N
N

(17)

The anionic rhodium complexes [Rh(C_5Me_5)(CO)]⁻ or [Rh(C_5Me_5)(CO)]²⁻ react with 1,3-bis(p-tolylsulphonyloxy)propane to give the dimetallacycle [{Rh(C_5Me_5)(CO)}$_2$(CH_2)$_3$], or with MeI to give cis- and trans-isomers of [Rh(C_5Me_5)(CO)(Me)]$_2$.[109] Treatment of the neutral dimer [Rh(C_5Me_5)(CO)]$_2$ with RLi followed by alkyltosylates, leads to addition of RLi across the Rh=Rh double bond to give the complexes [Rh(C_5Me_5)(CO)(R)]$_2$ (R=Me,Et).[109]

A cyclometallated ortho-nitrophenyl complex, (14), is obtained on reaction of [RhCl(CO)$_2$]$_2$ with [Hg(2-$C_6H_4NO_2$)$_2$].[110] The oxidative addition of one C-Cl bond in the complex [Rh(CO)(Cl)-{PR_2(CH_2)$_3$Cl}$_2$] (R=Ph,Cy) occurs in refluxing benzene to give the rhodacyclopentane derivative [(Cl)$_2${PR_2(CH_2)$_3$Cl}(CO)Rh$PR_2CH_2CH_2$-CH_2],[111] which reacts in turn with PMe_3 to give the cyclic acyl complex [(Cl)$_2$(Me$_3$P)$_2$Rh$PR_2CH_2CH_2CH_2$C(O)] (R=Ph).[111] Ortho-metallation of the complex [Rh(μ-pz)(CO)L] [L=PPh_2(o-BrC_6F_4)] occurs in refluxing toluene to give the Rh-Rh bonded complex (15).[112]

A number of studies involving rhodium-porphyrin complexes containing Rh-C σ-bonds have been published. These studies include the formation of Rh-benzyl derivatives from the reaction of [Rh(OEP)]$_2$ with methyl-substituted aromatics;[113] the kinetics of the insertion reactions of styrene into the Rh-Rh and Rh-H bonds of [Rh(OEP)]$_2$ and [Rh(OEP)H], respectively;[114] the synthesis and reactivity of [Rh(OEP)(CH_2OH)];[115] and the α-metallation of ketones on reaction with [Rh(OEP)](ClO$_4$) to give [Rh(OEP){CH(R)C-(O)Me}] (R=H, COMe, CO_2Et).[116]

The rhodium carbido-cluster, [NPr$_4$]$_4$[Rh$_{12}$C$_2$(CO)$_{23}$] has been synthesised and characterised by an X-ray crystallographic study.[117]

The crystal structure and a number of addition reactions of the complex trans-[Ir(CO)(Me)(PPh$_3$)$_2$] have been reported.[118] The reactions of the complexes [Ir(CO)(R)(PPh$_3$)$_2$] (R=Me,CH_2CMe_3, CH_2SiMe_3,Ph,2-MeC_6H_4, and 2,4,6-Me$_3C_6H_2$) with H_2 produce RH and [IrH$_3$(CO)(PPh$_3$)$_2$].[119] The thermal decomposition of [Ir(CO)(R)(L)$_2$] (R=n-alkyl) has been investigated.[120] The complexes [Ir(CO)(R)-(PPh$_3$)$_2$] (R=Me,CH_2CN,Ph) undergo electrochemical one-electron oxidation in the presence of PPh$_3$ to give [Ir(CO)(R)(PPh$_3$)$_3$]⁺, and further one-electron oxidation to give highly reactive d^6-cations which, upon addition of chloride ions, yield [IrCl$_2$(CO)(R)-(PPh$_3$)$_2$].[121] The complexes [Ir(Me)I{N(SiMe$_2$CH$_2$PR$_2$)$_2$}] (R=Ph,Pr^i), formed by the oxidative addition of MeI to [Ir (η^2-C_8H_{14})-

{N(SiMe$_2$CH$_2$PR$_2$)$_2$}], react with H$_2$ to give [Ir(Me)I(H){NH-(SiMe$_2$CH$_2$PR$_2$)$_2$}],[122] and with R^{1}Li (R^1=CH$_2$CMe$_3$ or CH$_2$SiMe$_3$) to give the dialkyl iridium complexes [Ir(Me)(R^1){N(SiMe$_2$CH$_2$PPh$_2$)$_2$}].[123] Photolysis of the neopentyl derivative yields an iridium dihydride complex and the square-planar iridium methylidene complex [Ir(=CH$_2$){N(SiMe$_2$CH$_2$PPh$_2$)$_2$}].[123]

The molecular hydrogen complex [IrH(H$_2$)(PPh$_3$)$_2$(C$_{13}$H$_8$N)]$^+$ (where C$_{13}$H$_8$N=metallated 7,8-benzoquinoline) has been synthesised.[124] Structural data on a number of iridium complexes such as [IrH$_2$(PPh$_3$)$_2$L]$^+$ (L=8-methylquinoline, caffeine), containing a C-H---Ir bridge have been used to construct a C-H+M⟶C-M-H reaction trajectory.[125] On the basis of this trajectory, conformational and steric effects are proposed to play an important role in deciding whether cyclometallation or attack on an external substrate such as an alkane will take place.[125] Heating the complex [Ir(C$_5$Me$_5$)(PMe$_3$)(Cy)H] under ethylene (20 atm.) gives cyclohexane, [Ir(C$_5$Me$_5$)(PMe$_3$)(CH=CH$_2$)H] and [Ir(C$_5$Me$_5$)(PMe$_3$)-(CH$_2$=CH$_2$)].[126] The ethylene π-complex is not an intermediate in the formation of the hydrido-vinyl complex, but is formed concurrently.[126] Treatment of the η^1-allyl complex [Ir(C$_5$Me$_5$)(Cl)-(η^1-C$_3$H$_5$)(PMe$_3$)] with LiEt$_3$BH gives the iridacyclobutane complex [(C$_5$Me$_5$)(PMe$_3$)IrCH$_2$CH$_2$CH$_2$].[127] The η^3-allyl complex [Ir(C$_5$Me$_5$)(H)(η^3-C$_3$H$_5$)] reacts with PMe$_3$ in benzene to give [Ir(C$_5$Me$_5$)(PMe$_3$)$_2$] and [Ir(C$_5$Me$_5$)(PMe$_3$)(Ph)(Pr)].[127] Deprotonation followed by alkylation of [Ir(C$_5$Me$_5$)(H)$_2$(PMe$_3$)] provides a route to hydrido(alkyl)iridium complexes of the type [Ir(C$_5$Me$_5$)-(R)(H)(PMe$_3$)] (R=Me, n-pentyl, CH$_2$CMe$_3$, SiMe$_3$).[128]

One of the products of the reaction between [Ir$_2$Cl$_2$-(C$_8$H$_{14}$)$_4$] and PBu^{t}CH$_2$CMe$_3$ is the metallacyclic complex [(Bu$_2^t$PCH$_2$CMe$_3$)Ir(CH$_2$CMe$_2$CH$_2$PBu$_2^t$)(Cl)(H)].[129] Similar reactions with PBu$_{3-n}^t$(CH$_2$CMe$_3$)$_n$ (n=3,2,1) give the cyclometallated hydrides [(C$_8$H$_{14}$)$_2$Ir(μ-Cl)$_2$Ir{CH$_2$CMe$_2$CH$_2$P(CH$_2$CMe$_3$)$_2$}{P(CH$_2$CMe$_3$)$_3$}(H)], [{Bu^t(CH$_2$CMe$_3$)$_2$}$_2$(H)$_2$Ir(μ-Cl)$_2$Ir(H){CH$_2$CMe$_2$CH$_2$PBu^t(CH$_2$CMe$_3$)}-{PBu^t(CH$_2$CMe$_3$)$_2$}], and [Ir(CH$_2$CMe$_2$CH$_2$PBu$_2^t$)(H)(Cl)]$_2$.[130] Complexes of the type fac-[(chel-P$_3$)Ir(H){C(O)C$_6$H$_{4-n}$Me$_n$CH$_2$}] [chel-P$_3$=PhP-{(CH$_2$)$_3$PPh$_2$}$_2$, MeP{(CH$_2$)$_3$PPh$_2$}$_3$, PhP{(CH$_2$)$_2$PPh$_2$}$_2$, MeC(CH$_2$PPh$_2$)$_3$] have been prepared.[131] The di-iridacyclobutene complexes, [Ir(cod)(μ-pz)(μ-alkyne)]$_2$ (alkyne = CF$_3$C≡CCF$_3$, MeO$_2$CC≡CCO$_2$Me, or HC≡CCO$_2$Me) are formed on treatment of [Ir(cod)(μ-pz)]$_2$ with the appropriate alkyne.[132] Treatment of [Ir$_2$(H)$_2$(Cl)$_2$(CO)$_2$(dppm)$_2$] with MeO$_2$CC≡CCO$_2$Me gives the Ir-C σ-bonded complex

[$Ir_2H(Cl)_2(MeO_2CC=CHCO_2Me)(CO)_2(dppm)_2$], while a similar reaction of [$Ir_2(H)_4Cl(CO)_2(dppm)_2$][$BF_4$] involves alkyne insertion into two Ir-H bonds to afford [$Ir_2(Cl)_2(MeO_2CC=CHCO_2Me)_2(CO)_2(dppm)_2$].[133] The complexes [($C_5Me_5$)M($C_6H_4CO_2$)($Me_2SO$)] (M=Rh,Ir) and [(p-cymene)Os($C_6H_4CO_2$)(Me_2SO)], containing cyclo-metallated benzoic acid have been prepared, and the crystal structure of the iridium derivative has been determined.[134] [Ir(C_5Me_5)(Me)$_2$(Me_2SO)] reacts with arenes (C_6H_5X) to give methane and [Ir(C_5Me_5)(Me)(C_6H_4X)-(Me_2SO)] (X=Cl,Br,I,CF_3,CN,NO_2,COMe,CO_2Me,NH_2).[135] Further reaction occurs at higher temperatures to give [Ir(C_5Me_5)(C_6H_4X)$_2$-(Me_2SO)] (X=H,Cl,I).[135] A mechanism for this reaction, involving an iridium(V) intermediate, has been proposed.[135]

3.3 The Nickel Triad.- An extensive series of compounds of the type trans-[NiR(R^1)L_2] (L=PMe_2Ph,PEt_3; R=C_2Cl_3,C_6H_4Cl, or $C_6H_2Me_3$-2,4,6; R^1=Ph or substituted aryl group) has been prepared and some of their reactions investigated.[136] An investigation of the d-d excitation of the complexes trans-[Ni{2,6-(MeO)$_2C_6H_3$}$_2L_2$] (L=PMe_3 or PMe_2Ph) has shown that reversible phosphine dissociation and/or organic group rotation about the Ni-C bond occurs.[137] An arene replacement reaction occurs on treatment of [Ni(arene)(C_6F_5)$_2$] with TlCp to give [Ni(Cp)(C_6F_5)$_2$]$^-$.[138] A convenient synthetic route to [Ni(arene)R_2] (R=C_6F_5,C_6F_4H,CCl_5, various halosilanes) complexes has been reported, involving the treatment of [Ni(CF_3CO_2)$_2$] with RLi followed by an arene/BF_3 mixture.[139] The structures of [Ni(η^4-nbd)(C_6F_5)$_2$] and [Ni(η^4-cod)(C_6F_5)$_2$] have been determined.[140]

The thermolysis of compounds of general formula [Ni(Cp)-(η^2-propene)(R)] (R=Me,CD_3,Et,Pr, and CH_2SiMe_3) has been studied.[141] The complex [Ni(C_2H_4)(Et)(CN){P(O-o-tol)$_3$}] is formed quantitatively on treatment of [Ni(C_2H_4){P(O-o-tol)$_3$}] with ethylene and HCN.[142] The reaction of the ethyl(cyano)nickel complex with P(O-o-tol)$_3$ leads to the reductive elimination of propionitrile.[142] The ease of oxidative addition of various nickel(0) complexes to the central bond of biphenylene has been examined as a function of the ligands attached to nickel.[143] Various reactions of the dibenzonickelole products have also been investigated, and a unified mechanistic scheme for the nickel-catalysed Reppe trimerisation and tetramerisation of alkynes has been proposed.[143] The formation of the new o-xylylnickel derivatives [NiCl(η^1-CH$_2$C$_6$H$_4$Me)(PMe$_3$)$_2$], [NiCl(η^3-CH$_2$C$_6$H$_4$Me)(PMe$_3$)], and [Ni$_3$(η^1-CH$_2$C$_6$H$_4$Me)$_4$(PMe$_3$)$_2$(μ_3-OH)$_2$], has been reported.[144]

Synthetic routes to complexes of the type [Ni(Cp)(PPh$_3$)(CH$_2$SPh)][145] and [Ni(Cp)(L)(CH$_2$SMe)] [L=PPh$_3$,PPhMe$_2$,P(OMe)$_3$][146] have been reported. Treatment of [Ni(cod)$_2$] with Me$_3$P=CH$_2$ and Ph$_3$P=CHC(O)Ph yields the complex (16), which is a highly active catalyst for the polymerisation of ethylene.[147] [Ni(1,5,9-cdt)] and [Ni(C$_2$H$_4$)$_3$] react with R$_3$P=CH$_2$ (R=Me or Ph) to afford [Ni(1,5,9-cdt)(CH$_2$PR$_3$)] (R=Me) and [Ni(C$_2$H$_4$)$_2$(CH$_2$PR$_3$)] (R=Me,Ph).[148] The η^2-(C-C)ketene complex, [Ni(CH$_2$=C=O)(PPh$_3$)$_2$], is formed by carbonylation of nickelacyclobutane complexes or by carbonylation of the product from the reaction of [Ni(PPh$_3$)$_4$] with CH$_2$Br$_2$ in the presence of zinc.[149] The carbanionic methyl group of the LiMe moiety forms a σ-bond to nickel in complexes of general formula [Ni(LiMe)(L)$_m$(π-ligand)$_n$] (L=chelating amine or THF; π-ligand=1,5,9-cdt, ethene,or CO), as shown by the crystal structure of the [Ni(LiMe)(PMDT)(C$_2$H$_4$)$_2$] (17) derivative.[150]

The reactions of a number of substituted allenes with [Ni(PPh$_3$)$_3$] have been investigated and the factors affecting the regio- and stereochemistry of π-complex formation and coupling to form nickelacyclopentane complexes have been analysed.[151] The nickelacyclic complex [(Et$_3$P)Ni{C(Ph)=C(Ph)C(Ph)=CPh}] has been synthesised from [NiBr$_2$(PEt$_3$)$_2$] and (E-E)-1,4-dilithio-1,2,3,4-tetraphenyl-1,3-butadiene.[152] The nickelacyclopentane (18) is obtained by the oxidative addition of Me$_2$C=C=C=CMe$_2$ to [Ni(bipy)(cod)].[153] The oxidative addition of 3,3-dimethylcyclopropene to [Ni(PPh$_3$)$_3$] or [Ni(PPh$_3$)$_2$(C$_2$H$_4$)] gives the nickelacyclic complex (19).[154]

The nickel carbido-clusters [AsPh$_4$]$_2$[Ni$_{10}$(CO)$_{16}$C$_2$][155] and [Ni$_{16}$(CO)$_{23}$C$_4$]$^{4-}$ [156] have also been synthesised and structurally characterised.

An ab initio M.O. study has determined the optimised transition state structure for the carbonyl insertion reaction of [Pd(Me)(H)(CO)(PH$_3$)] and has shown that the reaction proceeds via methyl migration.[157] A molecular orbital study of β-elimination from [Pd(Et)(H)(PH$_3$)] suggests that an agostic CH···M interaction may be responsible for facile β-elimination.[158] A theoretical study of the ring-opening reaction of Pd and cyclopropane has been carried out.[159]

[$Pd_2Cl(Me)(\mu$-dppm$)_2$], prepared by treatment of [Pd_2Cl_2-(μ-dppm$)_2$] with Al_2Me_6, reacts with ethanol at low temperature to give [$Pd_2H(Me)(\mu$-Cl)(μ-dppm$)_2$]$^+$, which on warming to $-20°C$ eliminates CH_4 in an intramolecular reaction.[160]

The synthesis and study of the reactivity of a series of complexes [Pd(Me)$_2$(L)] and [Pd(Me)I(L)] [L=various poly(pyrazol-1-yl)methane ligands] has been published.[161] The dimethyl-palladium complex [Pd(Me)$_2$(PEt$_2$Ph)$_2$] undergoes facile <u>trans-cis</u> isomerisation in THF containing Me_2Mg or MeMgBr.[162] Kinetic and isotopic studies of this reaction suggest that the isomerisation occurs through an inter-molecular methyl exchange process between MePd and MeMg compounds.[162] The thermolysis of <u>cis</u>-[Pd(Et)$_2$(PR$_3$)$_2$] in the presence of added tertiary phosphine yields ethane and ethene <u>via</u> a β-elimination process rather than the reductive elimination of butane.[163] The η^1-allyl complexes [Pd(Ar)(η^1-CH$_2$CH=CHR)(dppe)] (Ar=C_6F_5, C_6HCl_4; R=H,Me) undergo selective Pd-allyl bond <u>cleavage</u> on reaction with electrophiles EX [EX=HCl, Br$_2$, $\overline{BrNC(O)CH_2CH_2C(O)}$] to produce [Pd(Ar)X(dppe)] and CH_2=CHCHR(E).[164] The mechanism of insertion of isocyanides into the Pd-C bond of [PdCl(dppe)$\{C_5H_3N(Cl-6)-\underline{C^2}\}$] to give [PdCl(dppe)$\{C(=NR)C_5H_3N(Cl-6)-\underline{C^2}\}$] has been studied.[165] The oxidative addition of PhX (X=Cl,Br,I) to palladium complexes and the reactivity of <u>trans</u>-[Pd(Ph)(I)(PMePh$_2$)$_2$] and <u>trans</u>-[Pd(COPh)(I)(PMePh$_2$)$_2$] towards amines and CO have been examined as part of a study on the Pd catalysed carbonylation of aryl halides to give α-ketoamines.[166] The carbonylation of <u>cis</u>-[M(C$_6$X$_5$)$_2$(THF)$_2$] (M=Pd,Pt; X=F,Cl) leads to the formation of <u>cis</u>-[M(C$_6$X$_5$)$_2$(CO)$_2$].[167] The pentafluorophenyl complexes [Pd(C$_6$F$_5$)(L)$_2$(APPY)]$^+$ (L=PPh$_3$, PBu$_3^n$; L$_2$=bipy; APPY=Ph$_3$PCHCOMe) have been prepared in which the APPY ligand is co-ordinated <u>via</u> the O atom.[168] The σ-vinylpalladium complexes [Pd(σ-CH=CHCOOR)-(PPh$_3$)$_2$X] (X=Cl,I; R=Me,Et) have been prepared and shown to <u>rearrange on heating</u> to give the isomeric η^2-olefin-ylide complexes [Pd$^-\{$CH(COOR)CHPPh$_3^+\}$(PPh$_3$)X].[169] Palladium acetate reacts with acrylonitrile to give [Pd(MeCO$_2$)(CH$_2$=CHCN)]$_4$ in which the cyanoethylacetate moiety acts as a bidentate bridging ligand with a Pd-C σ-bond.[170]

The alkynyl complex [Pd(C≡CR)(dppe)(SnMe$_2$Cl)] (R=H,Me,Bu^t) is one of the products of the reaction between

[PdCl$_2$(dppe)] and bis(alkynyl)stannanes.[171] The μ-alkynediyl complexes [Cl(PMe$_3$)$_2$MC≡CM(PMe$_3$)$_2$Cl] (M=Pd,Pt) are formed when [M(C≡CH)$_2$(PMe$_3$)$_2$] is treated with [MCl$_2$(PMe$_3$)$_2$] in the presence of CuCl in diethylamine.[172]

A review of the application of cyclopalladated compounds in synthesis has been published.[173] Among the ligands involved in cyclopalladated complexes are the N-methyl-2,2'-bipyridylium ion,[174] 2,4'-bipyridine,[175] N,N-dimethylneopentylamine,[176] primary and secondary benzylamines,[177] N,N-dimethylbenzylamines,[178] benzophenone phenylhydrazone,[179] and 3,3'-substituted bipyridines.[180] Reactions involving cyclopalladated ligands include the reaction of [Pd(μ-Cl)(P-C)]$_2$ (P-C=Bu$_2^t$PCMe$_2$CH$_2$) with bidentate ligands to produce [(P-C)ClPd(dppm)PdCl(P-C)] and [Pd(P-C)(Cl)-(L$_2$)] (L$_2$=Ph$_2$AsCH$_2$CH$_2$PPh$_2$, diars, or bipy),[181] the protonation of cyclometallated 2-pyrimidyl or 2-pyrazyl complexes,[182] electrophilic and nucleophilic attack on bis(1-methoxynaphthalene-8-C,O) palladium(II),[183] and the reaction of [M(o-C$_6$H$_4$CH$_2$ER$_2$)Cl]$_2$ (M=Pd,Pt; E=N,P) with o-LiC$_6$H$_4$CH$_2$E'R$_2$ (E'=N,P,As) to form the asymmetric metallacycles [M(o-C$_6$H$_4$CH$_2$ER$_2$)(o-C$_6$H$_4$CH$_2$E'R$_2$)].[184]

A wide range of complexes of general formula cis- and trans-[Pt(PBu$_3$)$_2$(C$_6$H$_4$X)$_2$] have been synthesised.[185,186] The treatment of [PtCl$_2$L$_2$] with barium polyfluorobenzenesulphinates leads to SO$_2$ elimination and the formation of [Pt(R)$_2$L$_2$] (R=C$_6$F$_5$ or p-HC$_6$F$_4$; L$_2$=trans-py$_2$; R=C$_6$F$_5$; L$_2$=bipy).[187] The reaction of [Pt(cod)X$_2$] (X=Cl,Br) with excess diazomethane gives the halomethyl complexes [Pt(CH$_2$X)$_2$(cod)] in which the cod ligand may be subsequently replaced by phosphine ligands.[188]

The oxidative addition of a number of alkyl and aryl halides to Pt atoms has been studied and it has been shown that the Pt atoms react less efficiently than Ni or Pd atoms.[189] The photochemically initiated oxidative addition of Pr^{i}I to [Pt(Me)$_2$(phen)] has been investigated and shown to involve overall iodine atom abstraction from the Pr^{i}I.[190] Photo-induced addition of aryl bromides or iodides to the pyrophosphite complex [Pt$_2$(μ-P$_2$O$_5$H$_2$)$_4$] gives the axially disubstituted complexes [Pt$_2$(μ-P$_2$O$_5$H$_2$)$_4$(Ar)-X]$^{4-}$.[191] The oxidative addition of MeI to cis-[Pt(Me)$_2$(SMe$_2$)$_2$] gives fac-[PtI(Me)$_3$(SMe$_2$)$_2$] via an intermediate characterised as fac-[Pt(Me)$_3$(SMe$_2$)$_2$(CD$_3$CN)]I.[192] In a slower reaction, the dimer [Pt$_2$(Me)$_2$(μ-SMe$_2$)$_2$] reacts with MeI to give a mixture of fac-[PtI(Me)$_3$(SMe$_2$)$_2$] and [PtI(Me)$_3$]$_4$.[192] The reaction of a series of

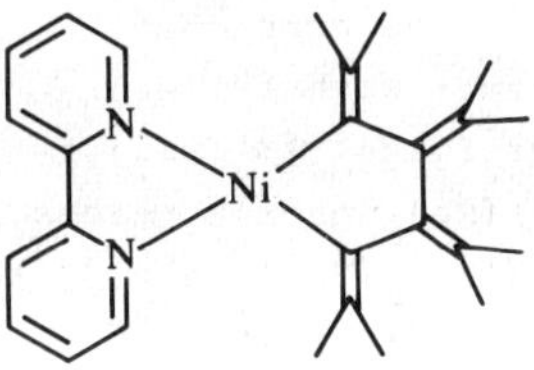

(18) 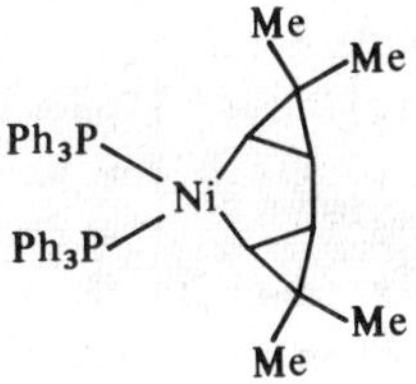 (19)

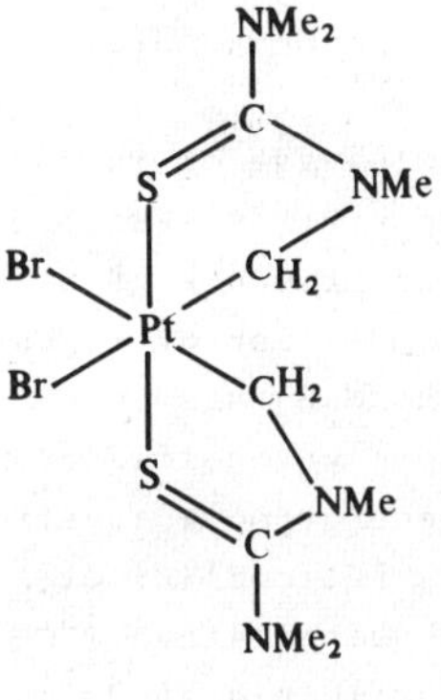

(20) 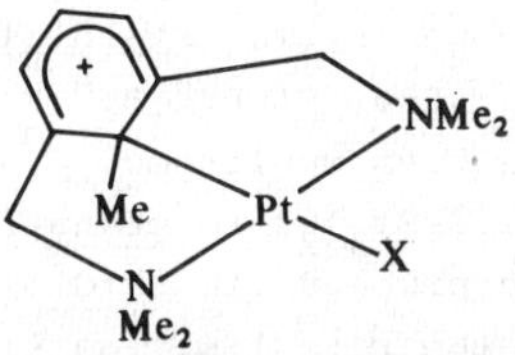 (21)

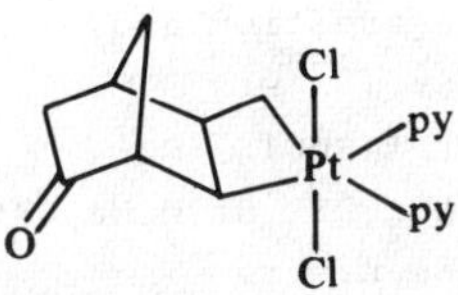

(22) 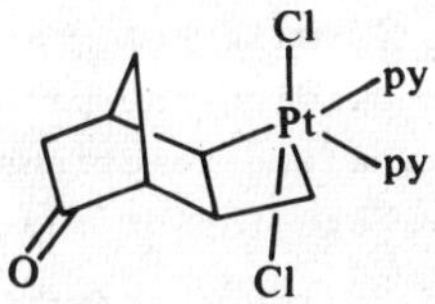 (23)

organic halides with the complex [Pt(Me)$_2$(phen)] gives the platinum(IV) complexes [PtX(Me)$_2$R(phen)] [R=Me,Et,Pr^n,Bu^n, CH$_2$C$_6$H$_4$CH$_2$Br,(CH$_2$)$_n$X(n=0,1,3,5); X=Br,I].[193] The complexes [PtI(Me)$_2${(CH$_2$)$_n$I}(phen)] (n=3,5) react further with [Pt(Me)$_2$-(phen)] to yield the binuclear complexes [Pt$_2$I$_2$(Me)$_4${μ-(CH$_2$)$_n$}-(phen)$_2$].[193] In the presence of oxygen, [Pt(Me)$_2$(phen)] reacts with Pr^{i}X (X=Br,I) to give a mixture of [PtX(Me)$_2$(Pr^i)(phen)], the alkylperoxoplatinum(IV) complex [PtI(Me)$_2$(OOPr^i)(phen)], and [Pt(I)$_2$(Me)$_2$(phen)].[194] Treatment of [Pt(Me)$_4$(N-N)] (N-N=bipy or phen) with HX yields <u>fac</u>-[Pt(Me)$_3$X(N-N)] (X=Cl,O$_2$CCF$_3$,O$_2$CMe, OPh,SPh,OMe), with HgCl$_2$ yields <u>fac</u>-[Pt(Me)$_3$Cl(N-N)], with SO$_2$ gives <u>fac</u>-[Pt(Me)$_3$(SO$_2$Me)(N-N)], and with C$_2$F$_4$ gives [{Pt(Me)$_3$(N-N)}$_2$(μ-C$_2$F$_4$)].[195] The oxidative addition of tetramethylthiourea (tmtu) to [Pt(tmtu)Br$_2$]$_2$ gives the octahedral platinum(IV) complex (20).[196]

A mechanistic study of the electrophilic cleavage of the Pt-C σ-bond in <u>cis</u>-[Pt(Ph)$_2$(PEtPh$_2$)$_2$] and <u>trans</u>-[PtX(R)(PEt$_3$)$_2$] (X=Cl,Br; R=Me,Et,Pr^n,Bu^n,Bz) by protons in the presence of halide ions has been published.[197] The thermal decomposition of <u>trans</u>-[PtCl(Et)(PEt$_3$)$_2$] to <u>trans</u>-[PtCl(H)(PEt$_3$)$_2$] and C$_2$H$_4$ has been shown to proceed via [PtCl(H)(C$_2$H$_4$)(PEt$_3$)$_2$] from which loss of C$_2$H$_4$ is the rate limiting step.[198] The free energy of activation for β-hydride elimination from [Pt(cod)(Et)$_2$], [Pt(cod)(Et)Cl], and [Pt(cod)(Et)I] has been determined from a quantitative ^{13}C n.m.r. study.[199] A differential scanning calorimetric study of β-hydride elimination from the complexes <u>cis</u>- and <u>trans</u>-[Pt(PEt$_3$)$_2$(n-alkyl)Cl] (n-alkyl=Et,Pr,Bu) has been reported.[200]

The ionic complexes (21) (X=Cl,Br,I), prepared from the reaction of [PtX{C$_6$H$_3$(CH$_2$NMe$_2$)$_2$-<u>o</u>,<u>o</u>'}] with methyl triflate, and (21) (X=tolyl), formed from the reaction of [Pt(tol){C$_6$H$_3$-(CH$_2$NMe$_2$)$_2$-<u>o</u>,<u>o</u>'}] with MeI, have been characterised, and provide new evidence regarding the pathways of oxidative addition and reductive elimination reactions.[201] Multinuclear (^{195}Pt, ^{119}Sn and ^{31}P) n.m.r. studies of the complexes <u>trans</u>-[Pt(SnCl$_3$)(R)-(PEt$_3$)$_2$] {R=Me,Bz,COPh,C$_6$Cl$_5$,C$_6$H$_4$Y; Y=<u>m</u>- and <u>p</u>-NO$_2$,CF$_3$,Br,H,Me,OMe, or [Pt(SnCl$_3$)(PEt$_3$)$_2$]} have been reported and the n.m.r. parameters have been rationalised in terms of the different σ- and π-bonding abilities of the ligand R.[202] The synthesis, structure, and dynamic behaviour in solution of the complexes [PtX(Me)$_3$(MeECH=CHEMe)] (E=S or Se; X=Cl,Br,I) have been published.[203]

An n.m.r. study of several organoplatinum compounds has shown that ^{195}Pt chemical shifts may be used to characterise metal-diene systems as metallacyclopentene or η^2-bonded diene.[204] Among the platinacyclic compounds to be reported are the platinacyclobutan-3-one complexes [Pt(CHRCOCHR)L$_2$] (R=COMe, L=PPh$_3$ or AsPh$_3$),[205] the benzalazine derivatives [PtCl{p-ClC$_6$H$_3$CH=NN=CH(p-ClC$_6$H$_4$)}]$_2$ and [PtCl(p-ClC$_6$H$_3$CH=N)]$_n$,[206] the isomeric platinacyclobutane complexes (22) and (23),[207] and the cycloheptatriene derived platinacyclopentane complex(24).[208] The reaction of [PtL$_4$] (L=PR$_3$ or Cl) with CH$_2$Br and CO in the presence of zinc, or the direct reaction of [PtL$_4$] with ketene, yields the η^2-(C,C)ketene complexes [Pt(CH$_2$=C=O)L$_2$] and [PtCl$_2$(CH$_2$=C=O)(PMe$_2$Ph)$_2$].[209] The thermal decomposition of 3,3,4,4-tetramethylplatinacyclopentane complexes yields 2,2,3,3-tetramethylbutane, via a heterogeneous process catalysed by Pt(O), and 1-methyl-1-tert-butylcyclopropane, via a homogeneous process, as major products.[210] The heterogeneous process is suppressed by the addition of mercury(O).[210]

Two synthetic routes to cis-[Pt(C≡CPh)$_2$(PMePh$_2$)$_2$] have been reported, and isomerisation to the trans isomer has been found to be catalysed by Hg(II) halides.[211] Treatment of [Pt(C$_2$O$_4$)-(dppe)] with PhC≡CH produces [Pt(C≡CPh)$_2$(dppe)], while similar treatment of [Pt(C$_2$O$_4$)(dppm)] gives [Pt(C≡CPh)$_2$(dppm)] which rearranges to cis,cis-[Pt$_2$(C≡CPh)$_4$(μ-dppm)$_2$].[212] [Pt(C≡CPh)$_2$-(dppm)] is also formed by the reaction of [Pt(Cl)$_2$(dppm)] with PhC≡CH/KOH/18-crown-6 or with [SnMe$_3$(C≡CPh)].[212] In a similar reaction, the complexes [Pt(C≡CR)$_2$(dppm)] (R=H,Me,Bu^t,Ph) have been prepared by treatment of [Pt(Cl)$_2$(dppm)] with [SnMe$_2$(C≡CR)$_2$].[213] The monomeric bis(acetylide) complexes readily convert to the dimers trans,trans-[Pt$_2$(C≡CR)$_4$(μ-dppm)$_2$].[212-214] Other acetylide complexes to be synthesised include [Pt(C≡CMe)$_2$(L-L)] (L-L=dppm, Et$_2$PCH$_2$PEt$_2$,Pr$_2^i$PCH$_2$PPr$_2^i$),[214] [(RC≡C)Pt(μ-dppm)$_2$(μ-H)Pt(C≡CR)]Cl,[215] [(RC≡C)Pt(μ-dppm)$_2$Pt(C≡CR)] (R=Ph or p-tolyl),[215] and the hetero-binuclear complexes [(MeC≡C)Pt(μ-dppm)$_2$(σ,η-C≡CMe)Rh(CO)]Cl and [(RC≡C)Pt(μ-dppm)$_2$(σ,η-C CR)RhCl] (R=Ph or p-tolyl).[216]

The carbonylation reactions of a series of complexes of the type [PtCl(Ph)(P-Y)] (P-Y=dppe, arphos, dppp, Ph$_2$PCH$_2$CH$_2$SMe, Ph$_2$PCH$_2$CH$_2$NMe$_2$) have been investigated and shown to depend on the ligand (P-Y).[217] Complexes the type [PtR$_2$(dppm)] (R=Me,CH$_2$CMe$_3$,

Et,Bz,Ph,C_6H_4Me-p,C_6H_4OMe-2,$C_6H_2Me_3$, or C_6H_4Me-o) have been prepared.[218] Treatment of a number of these dialkyl or diaryl complexes with MeI gives the <u>trans</u> addition product, [$PtR_2Me(I)$-(dppm)].[218] Other bidentate phosphine complexes of platinum to be studied include the reaction of MeI with [$Pt_2Me_4(\mu$-$R_2PCH_2PR_2)_2$] (R=Me,Et,Ph) which was found to depend on the bulk of the substituents R;[219] the reaction of [$Pt_2Me_4(\mu$-dmpm$)_2$] with halogens to give [$Me_3Pt(\mu$-X$)(\mu$-dmpm$)_2$PtMe]X (X=I,Br,Cl) and, for X=I, [$Me_3Pt(\mu$-I$)(\mu$-dmpm$)_2$PtMe]I_3 and [$Me_3Pt(\mu$-I$)(\mu$-dmpm$)_2$PtI]I_3;[220] the reaction of [$Pt_2H(L)(\mu$-dppm$)_2$][PF_6] with CH_2N_2 to yield [$Pt_2Me(\mu$-$CH_2)(L)(\mu$-dppm$)_2$][PF_6] (L=CO,PMe_2Ph) or [$Pt_2Me(\mu$-CH_2)($CH_2PPh_3)(\mu$-dppm$)_2$][PF_6] (L=PPh_3);[221] and the reaction of [$Pt_2(\mu$-dppm$)_3$] with RI followed by the addition of KPF_6 to give the first alkyldiplatinum(I) complexes containing β-hydrogen atoms, [$Pt_2R(\mu$-dppm$)_2$(dppm-p)][PF_6] (R=Et,CD_3CH_2,Pr,or Bu).[222] The photochemical and thermal decomposition of the ethyl derivative was investigated.[222] A number of heterobimetallic and unsymmetrical diplatinum complexes have been prepared from <u>cis</u>-[PtR_2(dppm-p$)_2$] (R=Me,1-naphthyl, or C_6H_4Me-o).[223] Such complexes include [(1-naphthyl)$_2$Pt(μ-dppm$)_2$RhCl(CO)] and <u>cis</u>,<u>cis</u>-[(C_6H_4Me-o)$_2$Pt(μ-dppm$)_2$PtMe$_2$].[223]

The complex [$Pt_3(\mu$-$PPh_2)_3$(Ph)($PPh_3)_2$], formed by heating an acetone solution of [$Pt(C_2H_4)(PPh_3)_2$], undergoes skeletal isomerisation upon recrystallisation in various solvents, the two isomers differing in their Pt-Pt distances and Pt-P-Pt angles.[224] The dicarbido complexes <u>trans</u>-[Pt(C≡CH){C_2W_2(OBu$^t)_5$}($PMe_2Ph)_2$] and <u>trans</u>-[Pt{C_2W_2(OBu$^t)_5$}$_2$($PMe_2Ph)_2$] (25) are formed on reaction of <u>trans</u>-[Pt(C≡CH)$_2$($PMe_2Ph)_2$] with [W(OBu$^t)_6$].[225]

4 Carbene and Carbyne Complexes of the Group VIII Metals

4.1 The Iron Triad. - Treatment of the complexes [Fe(C_5Me_5)(CO)(L)(CH_2OMe)] (L=CO,PPh_3) with acid gives the cationic methylene complexes [Fe(C_5Me_5)(CO)(L)(=CH_2)]$^+$.[226] A similar reaction of [Fe(Cp)(CO)(L)(CH_2OMe)] with HBF_4[227] or [Mo(Cp)(CO)$_3$][PF_6][228] generates the carbene complexes [Fe(Cp)(CO)(L)(=CH_2)]$^+$ and [Fe(Cp)(CO)(L){=CH(OMe)}]$^+$. The methylidene complex [Fe(Cp)(CO)$_2$(=CH_2)]$^+$ has been used as an alkylating agent towards vinyl and allyl ligands.[229] Treatment of either [Fe(Cp)(CO)(L){C(Me)=CH_2}]

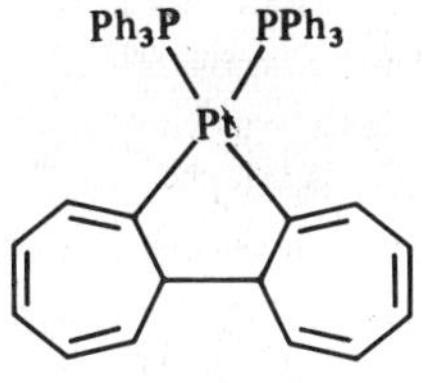
Ph3P
PPh3
Pt

(24)

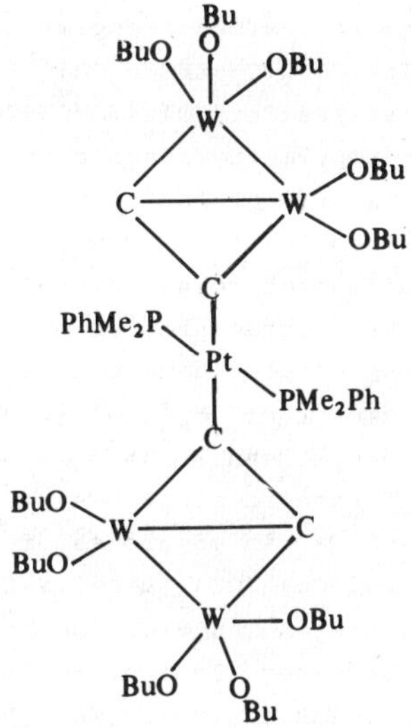
Bu
O
BuO
OBu
W
C
W
OBu
OBu
C
PhMe2P
Pt
PMe2Ph
C
BuO
W
C
BuO
W
OBu
BuO
O
Bu

(25)

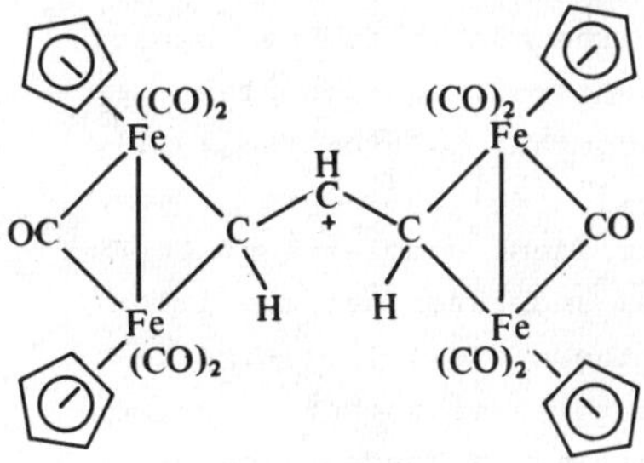
(CO)2
Fe
OC
C
H
C
+
H
C
H
(CO)2
Fe
CO
Fe
(CO)2
Fe
(CO)2

(26)

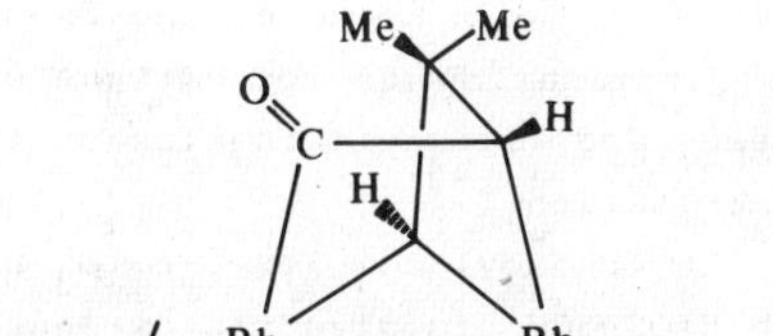
Me
Me
O
C
H
H
Rh
Rh
C
O

(27)

(L=CO,PPh$_3$) or [Fe(Cp)(CO)$_2${C(OMe)Me$_2$}] with HBF$_4$ at -23°C yields the carbene complex [Fe(Cp)(CO)(L)(=CMe$_2$)]$^+$ which for L=CO, rearranges at -11°C to [Fe(Cp)(CO)$_2$(CH$_2$=CHMe)]$^+$, while when L=PPh$_3$ the carbene complex is stable as a solid.[230] Reaction of [Fe(Cp)(CO)$_2$(CH=CHCMe$_2$OH)] with HBF$_4$ gives the vinyl carbene complex [Fe(Cp)(CO)$_2$(=CHCH=CMe$_2$)]$^+$.[230] The synthesis of the cyclopropylmethylidene complex [Fe(Cp)(CO)$_2${=CH(c-C$_3$H$_5$)}]$^+$ and the transfer of the carbene moiety to alkenes has been described.[231] Halide abstraction from [Fe(Cp)(CO)(PPh$_3$)(CF$_3$)] using BF$_3$ produces the difluorocarbene complex [Fe(Cp)(CO)(PPh$_3$)(=CF$_2$)][BF$_4$].[232] The X-ray structures of this complex and the analogous dichlorocarbene complex have been determined.[232] The ethyoxyarylcarbene complexes [Fe(C$_4$H$_6$)(CO)$_2${C(OEt)Ar}] (Ar=Ph, substituted Ph, 1-naphthyl, and 2-thienyl) have been synthesised.[233]

A kinetic study has revealed that the rearrangement of μ-CR to μ-CH=CR' in complexes of the type [Fe$_2$(Cp)$_2$(CO)$_2$(μ-CO)$_2$(μ-CR)] and [Fe$_2$(Cp)$_2$(CO)$_2$(μ-CO)(μ-CR)]$^+$ is accelerated by alkyl substituents on the β-carbon atom of the alkylidyne group, R.[234] The complexes [Fe$_2$(Cp)$_2$(CO)$_2$(μ-CO)(μ-CR)]$^+$ (R=Me,H) react with aldehydes RCHO (R=C$_6$H$_4$Me-p,CH=CHPh,Bu^t) to produce μ-vinylcarbene complexes.[235] The vinylcarbene complex [Fe$_2$(Cp)$_2$(CO)$_2$(μ-CO) (μ-CCH=CHCH$_2$Me)]$^+$ is formed on reaction of [Fe$_2$(Cp)$_2$(CO)$_2$(μ-CO) {μ-C=CH(CH$_2$)$_2$Me}] with Ph$_3$C$^+$PF$_6^-$.[236] Some reactions of this vinylcarbene complex are reported.[236] The μ-methylidyne complex [Fe$_2$(Cp)$_2$(CO)$_2$(μ-CO)(μ-CH)]$^+$ and the μ-ethenylidene complex [Fe$_2$(Cp)$_2$(CO)$_2$(μ-CO)(μ-C=CH$_2$)] react together to from the μ-allyl complex [{Fe$_2$(Cp)$_2$(CO)$_2$(μ-CO)}$_2$(μ-C$_3$H$_3$)]$^+$ (26).[237] The thermolyses of a large number of μ-alkylidene complexes of Fe,Ru, Os,Co,Rh, and Ir have been investigated and three decomposition patterns have been found to dominate; these are (i) elimination of the carbene bridge as an alkane of the same number of carbon atoms with two H atoms coming from ancillary π-bonded ligands, (ii) dimerisation of the carbene bridge to form the corresponding alkene, and (iii) elimination of the carbene with intramolecular isomerisation.[238]

The benzoyl complex [Fe$_3$(μ_3-NPh)$_2$(COPh)(CO)$_8$] may be alkylated with EtOSO$_2$CF$_3$ to give the carbene-nitrene complex [Fe$_3$(μ_3-NPh)$_2$ (CO)$_8${C(OEt)Ph}].[239] Carbene-nitrene coupling to give the imidate PhN=C(OEt)Ph occurs on exposure of this complex to CO over several days.[239] The μ_3-carbyne cluster complex [Fe$_3$(CO)$_9$(μ_3-CF)$_2$] is

formed on treatment of [Fe(CO)$_5$] or [Fe$_2$(CO)$_9$] with CFBr$_3$.^{240}
Alkylidyne-alkylidyne coupling occurs on treatment of
[Fe$_3$(CO)$_9$(μ_3-CMe)(μ_3-COEt)] with CO to give [Fe$_3$(CO)$_{10}$(μ_3-η^2-CH$_3$C≡COEt)]. This reaction is reversed in solution or the solid
state at room temperature.^{241} Alkylidyne-alkylidyne coupling is
also induced by alkynes. Thus, treatment of [Fe$_3$(CO)$_9$(μ_3-CMe)-(μ_3-COEt)] with RC≡CR' produces complexes such as
[Fe$_3$(CO)$_8${μ_3-η^4-C(OEt)C(Me)C(R)CR'}] (R=R'=Ph or SiMe$_3$).^{242} The
mixed-metal butterfly clusters [MFe$_3$(μ_3-CMe)(CO)$_{10}$(PPh$_3$)]
(M=Cu or Au) are formed on reaction of [Fe$_3$(μ_3-CMe)(CO)$_{10}$]$^-$ with
[MCl(PPh$_3$)].^{243}

The stereochemistry of the reaction of [Ru(Cp)(Ph$_2$PCHRCHR'PPh$_2$)-{C(OMe)CH$_2$Ph}] with LiAlH$_4$ to give the corresponding 2-phenylethyl
complexes has been investigated and shown to take place with
retention of configuration at the Ru atom.^{244} Deprotonation of
[Ru(Cp)(L)(PPh$_3$){C(OR')CH$_2$R}] (R=R'=Me; R=Ph,R'=Me,Et) by NaOMe
gives the corresponding vinylether derivatives. This reaction is
reversed by the addition of HPF$_6$.^{245}

A wide range of μ-alkylidene complexes [(Cp)(NO)Ru(μ-CRR')-Ru(Cp)(NO)] have been synthesised and their isomerisation and
protonation reactions have been studied.^{246} The μ-CH$^+$ complex
[Ru$_2$(Cp)$_2$(μ-CH)(μ-CO)(μ-dppm)]$^+$ has been synthesised by oxidation
of [Ru$_2$(Cp)$_2$(μ-CH$_2$)(μ-CO)(μ-dppm)] to the dication followed by
deprotonation with 2,6-dimethylpyridine.^{247} This monocationic
complex reacts with nucleophiles to generate derivatives of the
original μ-methylene complex, and with H$_2$ under pressure to give
the μ-methyl complex [Ru$_2$(Cp)$_2$(μ-CH$_3$)(μ-CO)(μ-dppm)].$^+$ ^{247}
Ethylidyne-vinylidene linking occurs on reaction of
[Ru$_2$(CO)$_2$(μ-CO)(μ-CMe)(Cp)$_2$]$^+$ with [Ru$_2$(CO)$_2$(μ-CO)(μ-CCH$_2$)(Cp)$_2$]
to give [{Ru$_2$(CO)$_3$(Cp)$_2$}$_2$(μ-CMeCHCH)]$^+$ and [{Ru$_3$(CO)$_3$(Cp)$_3$}-(μ-CCH$_2$CHC){Ru$_2$(CO)$_3$(Cp)$_2$}].^{248} The synthesis of the heterobi-metallic complex [RuW(μ-Cl)(μ-CMe)Cl(CO)$_2$(PPh$_3$)$_2$(Cp)] and its
reaction with MeC≡CMe have been reported.^{249}

The complex [Ru$_3$(CO)$_{10}$(μ-CO)(μ-CH$_2$)] has been isolated as the
initial product of the reaction between [Ru$_3$(CO)$_{12}$] and CH$_2$N$_2$.
This μ-methylene complex undergoes rearrangement above 45°C to
[Ru$_3$(H)$_2$(CO)$_9$(CCO)] and yields the C-C coupling product
[Ru$_3$(H)$_2$(CO)$_9$(CCH$_2$)] on further reaction with CH$_2$N$_2$.^{250}
[Ru$_3$(CO)$_8$(R-dab)] (R=Bu^i or neo-pentyl) reacts with CH$_2$N$_2$ to give

the μ-methylene cluster [Ru$_3$(CO)$_8$(μ-CH$_2$)(R-dab)].^{251} The reaction of alkynes RC≡CR with [Ru$_3$H$_3$(μ_3-CX)(CO)$_9$] (X=OMe,Me,Ph, or CH$_2$CH$_2$CMe$_3$) or with [Ru$_3$H (μ-CX)(CO)$_{10}$] leads to the alkylidyne-alkyne coupling products [Ru$_3$H(μ_3-η^3-XCCRCR)(CO)$_9$] (R=H,Me,Bu^n,Ph, CO$_2$Me).^{252} Deprotonation of [Ru$_3$(C$_5$Me$_5$)$_3$(μ_3-CMe)(μ-CO)$_3$]$^{2+}$ gives the μ_3-vinylidene cation [Ru$_3$(C$_5$Me$_5$)$_3$(μ_3-CCH$_2$)(μ-CO)$_3$]$^+$ which, on treatment with MeLi, gives [Ru$_3$(C$_5$Me$_5$)$_3$(μ_3-CEt)(μ-CO)$_3$].^{253}

The addition of trimethylamine to the sulphido-cluster [Os$_3$(CO)$_9$(μ_3-S)(μ_3-CO)] gives the carbene cluster [Os$_3$(CO)$_8$-{C(H)NMe$_2$}(μ-H)$_2$(μ_3-S)] which promotes transalkylation reactions of tertiary amines.^{254} Sequential treatment of [Os$_3$(μ-H)(μ-COMe)-(CO)$_{10}$] with PhLi and MeOSO$_2$CF$_3$ at 0°C gives the carbene-carbyne cluster [Os$_3$(μ-H)(CO)$_9${η^1-C(OMe)$_2$}(μ_3-CPh)],^{255} which on extended treatment with MeOSO$_2$CF$_3$ gives [Os$_3$(μ-H)(μ_3-CPh)(CO)$_{10}$].^{256} The μ-methylene clusters [Os$_3$(CO)$_{10}$(μ-X)(μ-CH$_2$)]$^-$ (X=Cl,Br,I,NCO) react with CO, very much faster than [Os$_3$(CO)$_{11}$(μ-CH$_2$)], to give the μ-ketene clusters [Os$_3$(CO)$_{10}$(μ-X)(μ-CH$_2$CO)]$^-$, illustrating a halide-promoting effect for this reaction.^{257,258} Treatment of the unsaturated cluster [Os$_3$Pt(μ-H)$_2$(CO)$_{10}$(PCy$_3$)] with CH$_2$N$_2$ produces the μ-methylene cluster [Os$_3$Pt(μ-H)$_2$(CO)$_{10}$(μ-CH$_2$)-(PCy$_3$)].^{259}

4.2 The Cobalt Triad. - Treatment of the metal chlorides CoCl$_2$ or RhCl$_3$.3H$_2$O with CNCH$_2$CHROH (R=H or Me) yields the homoleptic carbene complexes [M{CN(H) CH$_2$CHRO}$_6$]Cl$_3$ (M=Co,Rh).^{260} The mononuclear carbene complex [Co(Cp)(CO){=CN(Me)CH$_2$CH$_2$NMe}] has been prepared and some reactions with electrophilic reagents have been investigated.^{261} The preparation of a number of carbene complexes of Co and Rh from optically active electron-rich alkenes such as (S)-proline, (R)-(-)-trans-N-N'-dimethyl-1,2-diamino-cyclohexane, (S)-leucine, and (S)-alanine, has been reported.^{262} The aminocarbyne complex [(Cp)Co(μ-PMe$_2$)$_2${μ-CNH(Me)}Co(Cp)][PF$_6$] has been prepared by methyl isocyanide insertion into the Co-H-Co bridge in the complex [(Cp)Co(μ-PMe$_2$)$_2$(μ-H)Co(Cp)][PF$_6$].^{263} The μ-vinylidene species [{Co(Cp)(CO)}$_2$(μ-CCH$_2$)] results from the reaction of [Co(Cp)(CO)]$_2$Na with 1,1-dibromoethane.^{264} Protonation of the μ-vinylidene complex give the μ-ethylidyne cation [{Co(Cp)(CO)}$_2$(μ-CMe)]$^+$ which decomposes in solution at room temperature to give [Co$_3$(Cp)$_2$(CO)$_2${μ_3-CMe}]$^+$.^{264}

Several other reactions of the μ-vinylidene and μ-ethylidyne complexes are described, including reactions with metal hydrides.[264] The structure of $[Co_3(CO)_9(\mu_3\text{-}C\text{-}\underline{c}\text{-}C_3H_5)]$ has been reported.[265] The reaction of $[MRh(\mu\text{-}CO)(C_5Me_5)_2]$ (M=Co or Rh) with $[W(\equiv CMe)(CO)_2(Cp)]$ gives the carbyne clusters $[MRhW(\mu\text{-}CO)(\mu_3\text{-}CMe)\text{-}(CO)_2(Cp)(C_5Me_5)_2]$.[266]

Studies of the reactivity of $RhCH_2^+$ in the gas phase have been described, and the bond strength $D(Rh^+\text{-}CH_2)$ was found to be 395 ± 21 kJ mol^{-1}.[267] The square-planar vinylidene complexes $[M\{=C=CH(R)\}Cl(PPr_3^i)]$ (R=H,Me,Ph; M=Rh,Ir) have been prepared by elimination of pyridine from the alkynyl(hydrido)compound $[Rh(C\equiv CR)(H)Cl(py)(PPr_3^i)]$ or from $[Ir(\eta^2\text{-}RC\equiv CH)Cl(PPr_3^i)]$ <u>via</u> the intermediate $[Ir(C\equiv CR)(H)Cl(PPr_3^i)]$.[268] Heterobinuclear μ-vinylidene complexes have been prepared by treating $[Rh(=C=CH_2)(PPr_3^i)(Cp)]$ with $[Cr(C_6H_6)(CO)_3]$, $[Fe_2(CO)_9]$, or $[Mn(Cp)(CO)_2(THF)]$.[269] μ-Vinylidene complexes of the type $[\{Rh(C_5Me_5)(CO)\}_2(\mu\text{-}C=CR_2)]$ have been prepared by treating $[Rh(C_5Me_5)(CO)]_2$ with $N_2=C=CR_2$.[270,271] The reaction of PhC≡CH at 28°C with $[(CO)Rh(\mu\text{-}dppm)_2Rh(CO)_2]$ gives the μ-phenylvinylidene complex $[(CO)Rh(\mu\text{-}dppm)_2(\mu\text{-}C=CHPh)Rh(CO)]$; at higher temperatures the η^2-alkyne complex $[(CO)Rh(\mu\text{-}dppm)_2\{\mu\text{-}\eta^2\text{-}C(H)C(Ph)\}Rh(CO)]$ is also formed.[272] The μ-alkylidene complexes $[Rh_2(Cp)_2(\mu\text{-}CO)\text{-}(\mu\text{-}CR^1R^2)(\mu\text{-}CF_3C_2CF_3)]$ ($R^1=R^2$=H or CO_2Et) have been prepared and shown to undergo alkylidene migration in solution at room temperature to give $[Rh_2(Cp)_2(CO)\{\mu\text{-}C(CF_3)C(CF_3)CH_2\}]$ and $[Rh_2(Cp)_2\{\mu\text{-}C(CF_3)=C(CF_3)OC(OEt)=C(CO_2Et)\}]$.[273] The complexes $[Rh_2(C_5Me_5)_2(\mu\text{-}CH_2)_2\{\mu\text{-}CH_2CR(CH_2CR=CH_2)CH_2\}]$ (R=H or Me) and $[Rh_2(C_5Me_5)_2(\mu\text{-}CH_2)_2\{\mu\text{-}Ph_2P(CH_2)_nPPh_2\}]^{2+}$ (n=1 or 2) have been synthesised and characterised largely by n.m.r. spectroscopy.[274] The reaction of 3,3-dimethylcyclopropene with $[MM'(\mu\text{-}CO)_2(C_5Me_5)_2]$ (M=M'=Rh or Co; M=Rh,M'=Co; M=Ir, M'=Co; M=Rh, M'=Ir) leads to C=C bond cleavage and the formation of μ-carbene complexes of the type (27).[275]

The formyl complexes <u>trans</u>-$[IrX(CHO)(dppe)_2]^+$ (X=H,Cl) are protonated by strong acids to yield the dicationic hydroxycarbene complexes <u>trans</u>-$[IrX(CHOH)(dppe)_2]^{2+}$.[276] The dichlorocarbene complex $[IrCl_2(=CCl_2)(CO)(PPh_3)_2]^+$ is formed when $[IrCl_2(CF_3)(CO)(PPh_3)_2]$ is treated with BCl_3.[277] The dichlorocarbene ligand attacks a phenyl ring on a PPh_3 ligand to give

[IrCl$_2$($\overline{\text{o-PPh}_2\text{C}_6\text{H}_4}CCl_2$)(CO)(PPh$_3$)]. The hydrolysis and further reaction of this metallacycle with BCl$_3$ have been investigated.[277] The μ-methylene complexes [{IrX(μ-Bu^{t}S)(CO)L}$_2$(μ-CH$_2$)] [X=I, L=CO, P(OMe)$_3$,PPh$_3$,PPh$_2$Me,PMe$_3$; X=Br,L=PPh$_3$] have been prepared by reacting [Ir(μ-Bu^{t}S)(CO)L]$_2$ with dihalomethane.[278].

4.3 The Nickel Triad

The crystal structure of the methyl-(methoxy)carbenenickel complex trans-[Ni(C$_6$Cl$_5$)(PMe$_2$Ph)$_2${C(OMe)-Me}][BF$_4$] has been determined.[279] The reaction of CF$_3$C≡CCF$_3$ with the η^2-CS$_2$ complexes [Ni(η^2-CS$_2$)(L$_3$)] [L$_3$=triphos or tris(diphenyl-phosphinoethyl)amine] affords the carbene complexes [Ni(L$_3$)-{=$\overline{\text{CSC(CF}_3\text{)=C(CF}_3\text{)S}}$}].[280] The carbene complex [Ni{=C(OEt)(NPr$_2^i$)}-(CO)$_2$(PPh$_3$)] has been prepared and reacted with BCl$_3$ to give the cationic carbynenickel complex [Ni(≡CNPr$_2^i$)(CO)$_2$(PPh$_3$)]$^+$.[281]

Carbene complexes of the type trans-[PtH(carbene)(PPh$_3$)$_2$]$^+$ [carbene=C(OMe)$_2$, $\overline{\text{CO(CH}_2\text{)}_n\text{O}}$, $\overline{\text{CS(CH}_2\text{)}_2\text{S}}$] have been prepared.[282] The synthesis and structure of the μ-alkylidene(μ-hydrido) diplatinum complex [Pt$_2$(μ-H)(μ-CHCH$_2$Ph)(dppe)$_2$]$^+$ has been reported.[283] The complex [PtW{μ-C(C$_6$H$_4$Me-4)=CH$_2$}(CO)$_2$(PMe$_3$)$_2$(Cp)] is one of the products formed by treatment of [PtW{μ-C(C$_6$H$_4$Me-4)}(CO)$_2$(PMe$_3$)$_2$(Cp)] with [Ti$\overline{\text{{Cl(AlMe}_2\text{)CH}_2\text{}}}(Cp)_2$].[284]

5 Bibliography

L. Weber, K. Reizig, and G. Meine, Z. Naturforsch., Teil B, 1985, 40, 1698. Phospha-alkenyliron and -ruthenium complexes with M=C co-ordination.

H$_a$ Lang, L. Zsolnai, and G. Huttner, Chem. Ber., 1985, 118, 4426. η^4-1-phospha-2-ferracyclobutadiene complexes.

G. Sundararajan and J. San Filippo, Organometallics, 1985, 4, 606. Evidence for intramolecular pathways in C-C bond-forming reactions proceeding from binuclear iron complexes.

R.L. Cerny, B.P. Sullivan, M.M. Bursey, and T.J. Meyer, Inorg. Chem., 1985, 24, 397. FAB and field desorption mass spectrometry of organometallic derivatives of Ru(II) and Os(II).

S. Gopinathan, I.R. Unny, and C. Gopinathan, Polyhedron, 1985, 4, 1569. Synthesis and characterisation of chelated (1-cyanoethyl)ruthenium(II) carbonyl complexes.

C.D. Hoff, F. Ungváry, R.B. King, and L. Markó, J. Am. Chem. Soc., 1985, 107, 666. A kinetic study of the reactions of [Co(CO)$_4$(CH$_2$COOEt)] with ^{13}CO, PPh$_3$, [HCo(CO)$_4$], and H$_2$.

A.M. Trzeciak and J.J. Ziolkowski, Transition Met. Chem., 1985, 10, 385. Reactions of the ortho-metallated complex [Rh{P(OC$_6$H$_4$)(OPh)$_2$}{P(OPh)$_3$}$_3$] with HX.

K.A. King, P.J. Spellane, and R.J. Watts, J. Am. Chem. Soc., 1985, 107, 1431. Excited-state properties of a triply ortho-metallated iridium(III) complex.

W. Seidel, Z. Chem., 1985, 25, 411. Synthesis of (mesityl)nickel complexes.

A. Albinati, P.S. Pregosin, and R. Rüedi, Helv. Chim. Acta,1985, 68, 2046. Reactions of cyclopalladated benzylidene-aniline Schiff's base complexes.

H. Nishiyama, M. Matsumoto, T. Matsukura, R. Miura, and K. Itoh, Organometallics, 1985, 4, 1911. A new cyclopalladation reaction involving chelated transmetallation: selective cleavage of C-Sn and C-Si bonds of stannyl and silyl ketoximes.

G.C. Dash and B.K. Mohapatra, J. Indian Chem. Soc., 1985, 62, 14. Hetero β-diketone C-3-bonded platinum(II) complexes.

J.R. Lisko and W.M. Jones, Organometallics, 1985, 4, 944. Photoinduced ring expansion of a cyclopropyliron σ-complex to a metallacyclopentenone: an example of alkyl group rearrangement from C to Fe to generate a carbene complex.

M.M. Singh and R.J. Angelici, Inorg. Chim. Acta,1985, 100, 57. Dioxy carbene complexes from reactions of metal carbonyls with oxirane.

C.P. Casey, W.H. Miles, P.J. Fagan, and K.J. Haller, Organometallics, 1985, 4, 559. Photochemical reaction of a μ-ethenylidene iron complex with alkynes.

B.P. Gracey, S.A.R. Knox, K.A. MacPherson, A.G. Orpen, and S.R. Stobart, J. Chem. Soc., Dalton Trans., 1985, 1935. Organo-iron-ruthenium chemistry from the precursor complex [FeRu(CO)$_2(\mu$-CO)$_2$(Cp)$_2$].

References

1 "The Chemistry of the Metal-Carbon Bond. Vol. 2. The Nature and Cleavage of Metal Carbon Bonds", Eds. F.R. Hartley and S. Patai, J. Wiley & Sons, New York, 1985.
2 M.L.H. Green and D. O'Hare, Pure Appl. Chem., 1985, 57, 1897.
3 J. Halpern, Inorg. Chim. Acta, 1985, 100, 41
4 R.H. Crabtree, Chem. Rev., 1985, 85, 245.
5 R.J. Cross, Chem. Soc. Rev., 1985, 14, 197.
6 D.B. Pourreau and G.L. Geoffroy, Adv. Organomet. Chem., 1985, 24, 249.
7 M.E. Vol'pin, I.Y. Levitin, A.L. Sigan, and A.T. Nikitaev, J. Organomet. Chem., 1985, 279, 263.
8 Y-T.Fanchiang, Coord. Chem. Rev., 1985, 68, 131.
9 A. Yamamoto, T. Yamamoto, and F. Ozawa, Pure Appl. Chem., 1985, 57, 1799.
10 C.T. Mortimer, Rev. Inorg. Chem., 1984, 6, 233.
11 D.B. Jacobson and B.S. Freiser, J. Am. Chem. Soc., 1985, 107, 5876.
12 D.B. Jacobson and B.S. Freiser, J. Am. Chem. Soc., 1985, 107, 1581.
13 G.S. Girolami, G. Wilkinson, A.M.R. Galas, M. Thornton-Pett, and M.B. Hursthouse, J. Chem. Soc., Dalton Trans., 1985, 1339.
14 W.K. Wong, K.W. Chiu, G. Wilkinson, A.J. Howes, M. Motevalli, and M.B. Hursthouse, Polyhedron, 1985, 4, 603.
15 B.N. Diel, J. Organomet. Chem., 1985, 284, 257.
16 R. Bertani, A. Scrivanti, and G. Carturan, Inorg. Chim. Acta, 1985, 98, L9.
17 K. A. Mahmoud, A.J. Rest, and H.G. Alt, J. Chem. Soc., Dalton Trans., 1985, 1365.
18 J.P. Blaha and M.S. Wrighton, J. Am. Chem. Soc., 1985, 107, 2694.
19 C.R. Jablonski and Y-P.Wang, Organometallics, 1985, 4, 465.
20 C.R. Jablonski, Y-P.Wang, and N.J. Taylor, Inorg. Chim. Acta, 1985, 96, L17.
21 H. Lehmkuhl and G. Mehler, Chem. Ber., 1985, 118, 2407.
22 E.J. Crawford, C. Lambert, K.P. Menard, and A.R. Cutler, J. Am. Chem. Soc., 1985, 107, 3130.
23 T.C. Forschner and A.R. Cutler, Inorg. Chim. Acta, 1985, 102, 113.
24 E.K. Barefield, P. McCarten, and M.C. Hillhouse, Organometallics, 1985, 4, 1682.
25 S.E. Himmel, G.B. Young, D.C.M. Fung, and C. Hollinshead, Polyhedron, 1985, 4, 349.

26 S.C. Wright and M.C. Baird, J. Am. Chem. Soc., 1985, 107, 6899.
27 T.C. Forschner and A.R. Cutler, Organometallics, 1985, 4, 1247.
28 J.I. Seeman and S.G. Davies, J. Am. Chem. Soc., 1985, 107, 6522.
29 B.K. Hunter and M.C. Baird, Organometallics, 1985, 4, 1481.
30 K.E. Warner and J.R. Norton, Organometallics, 1985, 4, 2150.
31 D.L. Reger and K.A. Belmore, Organometallics, 1985, 4, 305.
32 P.S.Waterman, J.E. Belmonte, T.E. Bauch, P.A. Belmonte, and W.P. Giering, J. Organomet. Chem., 1985, 294, 235.
33 M.E. Wright, G.O. Nelson and R.S. Glass, Organometallics, 1985, 4, 245.
34 R.S. Bly, G.S. Silverman, and R.K. Bly, Organometallics, 1985, 4, 374.
35 R. Gompper and E. Bartmann, Angew. Chem., Int. Ed. Engl., 1985, 24, 209.
36 N.E. Kolobova, T.V. Rozantseva, Y.T. Struchkov, A.S. Batsanov, and V.I. Bakhmutov, J. Organomet. Chem., 1985, 292, 247.
37 E. Lindner, E. Schauss, W. Hiller,and R. Fawzi, Chem. Ber., 1985, 118, 3915.
38 J.R. Moss, L.G. Scott, M.E. Brown, and K.J. Hindson, J. Organomet. Chem., 1985, 282, 255.
39 J.W. Johnson and J.R. Moss, Polyhedron, 1985, 4, 563.
40 I. Bkouche-Waksman, J.S. Ricci, T.F. Koetzle, J. Weichmann, and W.A. Hermann, Inorg. Chem., 1985, 24, 1492.
41 M. Casarin, D. Ajò, G. Granozzi, E. Tondello, and S. Aime, Inorg. Chem., 1985, 24, 1241.
42 R. Guilard, G. Lagrange, A. Tabard, D. Lançon, and K.M. Kadish, Inorg. Chem, 1985, 24, 3649.
43 R. Guilard, B. Boisselier-Cocolois,A. Tabard, P. Cocolios, B. Simonet, and K.M. Kadish, Inorg. Chem., 1985, 24, 2509.
44 V.E. Lopatin, S.P. Gubin, N.M. Mikova, M.T. Tsybenov, Y.L. Slovokhotov, and Y.T. Struchkov, J. Organomet. Chem., 1985, 292, 275.
45 A. Gourdon and Y. Jeannin, J. Organomet. Chem., 1985, 290, 199.
46 A. Gourdon and Y.Jeannin, J. Organomet. Chem., 1985, 282, C39.
47 H. Kletzin and H. Werner, J. Organomet. Chem., 1985, 291, 213.
48 H. Werner and R. Werner, Chem. Ber., 1985, 118, 4543.
49 H. Nagashima, K. Yamaguchi, K. Mukai, and K. Itoh, J. Organomet. Chem., 1985, 291, C20.
50 M.F. Joseph and M.C. Baird, Inorg. Chim. Acta, 1985, 96, 229.
51 K. Hiraki, N. Ochi, Y. Sasada, H. Hayashida, Y. Fuchita, and S. Yamanaka, J. Chem. Soc., Dalton Trans., 1985, 873.
52 M.I. Bruce, M.G. Humphrey, G.A. Koutsantonis, and B.K. Nicholson, J. Organomet. Chem., 1985, 296, C47.
53 M.I. Bruce, C. Dean, D.N. Duffy, M.G. Humphrey, and G.A. Koutsantonis, J. Organomet. Chem., 1985, 295, C40.
54 S.I. Hommeltoft, A.D. Cameron, T.A. Shackleton, M.E. Fraser, S. Fortier, and M.C. Baird, J. Organomet. Chem., 1985, 282, C17.
55 A.R. Chakravarty and F.A. Cotton, Inorg. Chem., 1985, 24, 3584.
56 B. Delavaux, B. Chaudret, F. Dahan, and R. Poilblanc, Organometallics, 1985, 4, 935.
57 H. Werner and J. Gotzig, J. Organomet. Chem., 1985, 284, 73.
58 J. Gotzig, R. Werner, and H. Werner, J. Organomet. Chem., 1985, 285, 99.
59 Z. Dauter, R.J. Mawby, C.D. Reynolds, and D.R. Saunders, J. Chem. Soc., Dalton Trans., 1985, 1235.
60 G.A. Foulds, B.F.G. Johnson, and J. Lewis, J. Organomet. Chem., 1985, 294, 123.
61 P. Reveco, R.H. Schmehl, W.R. Cherry, F.R. Fronczek, and J. Selbin, Inorg. Chem., 1985, 24, 4078.
62 P. Reveco, J.H. Medley, A.R. Garber, N.S. Bhacca, and J. Selbin, Inorg. Chem., 1985, 24, 1096.
63 A.M. Stolzenberg and E.L. Muetterties, Organometallics, 1985, 4, 1739.
64 J.A. Clucas, M.M. Harding, B.S. Nicholls,and A.K. Smith, J. Chem. Soc., Dalton Trans., 1985, 1835.

65 J.P. Collman, P.J. Brothers, L. McElwee-White, E. Rose, and L.J. Wright, J. Am. Chem. Soc., 1985, 107, 4570.
66 K. Henrick, B.F.G. Johnson, J. Lewis, J. Mace, M. McPartlin, and J. Morris, J. Chem. Soc., Chem. Commun., 1985, 1617.
67 R.P. Tooze, P. Stavropoulos, M. Motevalli, M.B. Hursthouse, and G. Wilkinson, J. Chem. Soc., Chem. Commun., 1985, 1139.
68 W.J. Carter, S.J. Okrasinski, and J.R. Norton, Organometallics, 1985, 4, 1376.
69. J.A. Cabeza and P.M. Maitlis, J. Chem. Soc., Dalton Trans., 1985, 573.
70 H. Werner and K. Zenkert, J. Chem. Soc., Chem. Commun., 1985, 1607.
71 H. Werner and K. Roder, J. Organomet. Chem., 1985, 281, C38.
72 R. Weinand and H. werner, J. Chem. Soc., Chem. Commun., 1985, 1145.
73 M.L.H. Green and D. O'Hare, J. Chem. Soc., Chem. Commun., 1985, 355.
74 K. Burgess, H.D. Holden, B.F.G. Johnson, J. Lewis, M.B. Hursthouse, N.P.C. Walker, A.J. Deeming, P.J. Manning, and R. Peters, J. Chem. Soc., Dalton Trans., 1985, 85.
75 D. Himmelreich and G. Müller, J. Organomet. Chem., 1985, 297, 341.
76 L.J. Farrugia, A.D. Miles, and F.G.A. Stone, J. Chem. Soc., Dalton Trans., 1985, 2437.
77 G.W. Dillow, I.K. Gregor, and M. Guilhaus, J. Organomet. Chem., 1985, 294, 131..
78 A. Tsarbopoulos and J. Allison, J. Am. Chem. Soc., 1985, 107, 5085.
79 R. Benn, K. Cibura, P. Hofmann, K. Jonas, and A. Rufińska, Organometallics 1985, 4, 2214.
80 J.H. Ammeter, C. Elschenbroich, T.J. Groshens, K.J. Klabunde, R.O. Kühne, and R. Möckel, Inorg. Chem., 1985, 24, 3307.
81 S. Fukuzumi, K. Ishikawa, and T. Tanaka, Chem. Lett., 1985, 1355.
82 S. Fukuzumi, K. Ishikawa, and T. Tanaka, J. Chem. Soc., Dalton Trans., 1985, 899.
83 G.F. Schmidt and M. Brookhart, J. Am. Chem. Soc., 1985, 107, 1443.
84 L. Hofmann and H. Werner, J. Organomet. Chem., 1985, 289, 141.
85 H. Werner and R. Zolk, Organometallics, 1985, 4, 601.
86 Y. Wakatsuki, S. Mija, S. Ikuta, and H. Yamazaki, J. Chem. Soc., Chem. Commun., 1985, 35.
87 N. Bresciani-Pahor, M. Calligaris, L. Randaccio, L.G. Marzilli, M.F. Summers, P.J. Toscano, J. Grossman, and D. Liotta, Organometallics, 1985, 4, 630.
88 N. Bresciani-Pahor, L. Randaccio, E. Zangrando, M.F. Summers, J.H. Ramsden, P.A. Marzilli, and L.G. Marzilli, Organometallics, 1985, 4, 2086.
89 N. Bresciani-Pahor, L. Randaccio, E. Zangrando, and P.J. Toscano, Inorg. Chim. Acta, 1985, 96, 193.
90 L.G. Marzilli, M.F. Summers, N. Bresciani-Pahor, E. Zangrando, J-P. Charland, L. Randaccio, J. Am. Chem. Soc., 1985, 107, 6880.
91 N. Bresciani-Pahor, M. Forcolin, L.G. Marzilli, L. Randaccio, M.F. Summers, and P.J. Toscano, Coord. Chem. Rev., 1985, 63, 1.
92 A.W. Herlinger and K. Ramakrishna, Polyhedron, 1985, 4, 551.
93 I.Y. Levitin, A.N. Kitaigorodskii, A.T. Nikitaev, V.I. Bakhmutov, A.L. Sigan, and M.E. Vol'pin, Inorg. Chim. Acta, 1985, 100, 65.
94 A.D. Ryabov, I.Y. Levitin, A.T. Nikitaev, A.N. Kitaigorodskii, V.I. Bakhmutov, I.Y. Gromov, A.K. Yatsimirsky, and M.E. Vol'pin, J. Organomet. Chem., 1985, 292, C4.
95 A.M. Van den Bergen, R.L. Elliot, C.J. Lyons, K.P. MacKinnon, and B.O. West, J. Organomet. Chem., 1985, 297, 361.
96 A. Arrigoni, A. Ceriotti, R. Della Pergola, G. Longoni, M. Manassero, and M. Sansoni, J. Organomet. Chem., 1985, 296, 243.
97 S. Martinengo, D. Strumolo, P. Chini, V.G. Albano, and D. Braga, J. Chem. Soc., Dalton Trans., 1985, 35.
98 V.G. Albano, D. Braga, A. Fumagalli, and S. Martinengo, J. Chem. Soc., Dalton Trans., 1985, 1137.
99 L. Dahlenburg and F. Mirzaei, Inorg. Chim. Acta, 1985, 97, L1.

100 C. Bianchini, C. Meall, A. Mell, and M. Sabat, Organometallics, 1985, 4, 421.
101 W.D. Jones and F.J. Feher, J. Am. Chem. Soc., 1985, 107, 620.
102 M. Gómez, J.M. Kisenyi, G.J. Sunley, and P.M. Maitlis, J. Organomet. Chem., 1985, 296, 197.
103 W. Paul and H. Werner, Chem. Ber., 1985, 118, 3032.
104 H. Werner, R. Feser, and L. Hofmann, J. Organomet. Chem., 1985, 292, 361.
105 M.A. Murphy, B.L. Smith, G.P. Torrence, and A. Aguiló, Inorg. Chim. Acta, 1985, 101, L47.
106 T.G. Schenk, C.R.C. Milne, J.F. Sawyer, and B. Bosnich, Inorg. Chem., 1985, 24, 2338.
107 J.W. Suggs, M.J. Wovkulich, and K.S. Lee, J. Am. Chem. Soc., 1985, 107, 5546.
108 J.W. Suggs, M.J. Wovkulich, and S.D. Cox, Organometallics, 1985, 4, 1101.
109 M.J. Krause and R.G. Bergman, J. Am. Chem. Soc., 1985, 107, 2972.
110 J. Vicente, J. Martín, M-T. Chicote, X. Solans, and C. Miravitlles, J. Chem. Soc., Chem. Commun., 1985, 1004.
111 E. Lindner, R. Fawzi, and H.A. Mayer, Z. Naturforsch., Teil B, 1985, 40, 1333.
112 F. Barceló, P. Lahuerta, M.A. Ubeda, C. Foces-Foces, F.H. Cano, and M. M. Martinez-Ripoll, J. Chem. Soc., Chem. Commun., 1985, 43.
113 K.J. Del Rossi and B.B. Wayland, J. Am. Chem. Soc., 1985, 107, 7941.
114 R.S. Panonessa, N.C. Thomas, and J. Halpern, J. Am. Chem. Soc., 1985, 107, 4333.
115 S.L. Van Voorhees and B.B. Wayland, Organometallics, 1985, 4, 1887.
116 Y. Aoyama, T. Yoshida, and H. Ogoshi, Tetrahedron Lett., 1985, 26, 6107.
117 V.G. Albano, D. Braga, D. Strumolo, C. Seregni, and S. Martinengo, J. Chem. Soc., Dalton Trans., 1985, 1309.
118 W.M. Rees, M.R. Churchill, Y-J. Li, and J.D. Atwood, Organometallics, 1985, 4, 1162.
119 L. Dahlenburg, F. Mirzaei, and B. Pietsch, Inorg. Chim. Acta, 1985, 97, L5.
120 A.A. Vitale and J. SanFilippo, J. Organomet. Chem., 1985, 286, 91.
121 S. Zecchin, G. Zotti, and G. Pilloni, J. Organomet. Chem., 1985, 294, 379.
122 M.D. Fryzuk, P.A. MacNeil, and S.J. Rettig, Organometallics, 1985, 4, 1145.
123 M.D. Fryzuk, P.A. MacNeil, and S.J. Rettig, J. Am. Chem. Soc., 1985, 107, 6708.
124 R.H. Crabtree and M. Lavin, J. Chem. Soc., Chem. Commun., 1985, 794.
125 R.H. Crabtree, E.M. Holt, M. Lavin, and S.M. Morehouse, Inorg. Chem., 1985, 24, 1986.
126 P.O. Stoutland and R.G. Bergman, J. Am. Chem. Soc., 1985, 107, 4581.
127 W.D. McGhee and R.G. Bergman, J. Am. Chem. Soc., 1985, 107, 3388.
128 T.M. Gilbert and R.G. Bergman, J. Am. Chem. Soc., 1985, 107, 3503.
129 L. Dahlenburg and N. Höck, Inorg. Chim. Acta, 1985, 104, L29.
130 L. Dahlenburg and A. Yardimcioglu, J. Organomet. Chem., 1985, 291, 371.
131 E. Arpac and L: Dahlenburg, Chem. Ber., 1985, 118, 3188.
132 G.W. Bushnell, M.J. Decker, D.T. Eadie, S.R. Stobart, R. Vefghi, J.L. Atwood, and M.J. Zaworotko, Organometallics, 1985, 4, 2106.
133 B.R. Sutherland and M. Cowie, Organometallics, 1985, 4, 1801.
134 J.M. Kisenyi, J.A. Cabeza, A.J. Smith, H. Adams, G.J. Sunley, N.J. S. Salt, and P.M. Maitlis, J. Chem. Soc., Chem. Commun., 1985, 770.
135 M. Gómez, P.I.W. Yarrow, D.J. Robinson, and P.M. Maitlis, J. Organomet. Chem., 1985, 279, 115.
136 J.M. Coronas, G. Muller, M. Rocamora, C. Miravitlles, and X. Solans, J. Chem. Soc., Dalton Trans., 1985, 2333.
137 M. Wada and M. Kumazoe, J. Chem. Soc., Chem. Commun., 1985, 1204.
138 M.M. Brezinski, K.J. Klabunde, S.K. Janikowski, and L.J. Radonovich, Inorg. Chem., 1985, 24, 3305.
139 S-T. Lin, R.N. Narske, and K.J. Klabunde, Organometallics, 1985, 4, 571.
140 M.W. Eyring and L.J. Radonovich, Organometallics, 1985, 4, 1841.
141 S. Pasynkiewicz and H. Lehmkuhl, J. Organomet. Chem., 1985, 289, 189.

142 R.J. McKinney and D.C. Roe, J. Am. Chem. Soc., 1985, 107, 261.
143 J.J. Eisch, A.M. Piotrowski, K.I. Han, C. Kruger, and T.H. Tsay, Organometallics, 1985, 4, 224.
144 E. Carmona, J.M. Marín, P. Palma, M. Paneque, and M.L. Poveda, Organometallics, 1985, 4, 2053.
145 R. Taube, D. Steinborn, and W. Höbold, J. Organomet. Chem., 1985, 284, 385.
146 J.G. Davidson, E.K. Barefield, and D.G. Van Derveer, Organometallics, 1985, 4, 1178.
147 K.A. Ostoja Starzewski and J.Witte, Angew. Chem., Int. Ed. Engl., 1985, 24, 599.
148 K-R. Pörschke, G. Wilke, and R. Mynott, Chem. Ber., 1985, 118, 298.
149 A. Miyashita, H. Shitara, and H. Nohira, J. Chem. Soc., Chem. Commun., 1985, 850.
150 K-R. Pörschke, K. Jonas, G. Wilke, R. Benn, R. Mynott, R. Goddard, and C. Krüger, Chem. Ber., 1985, 118, 275.
151 D.J. Pasto, N-Z.Huang, and C.W. Eigenbrot, J. Am. Chem. Soc., 1985, 107, 3160.
152 J.J. Eisch, A.M. Piotrowski, A.A. Aradi, C. Krüger, and M.J. Romão, Z. Naturforsch., Teil B, 1985, 40, 624.
153 L. Stehling and G. Wilke, Angew. Chem., Int. Ed. Engl., 1985, 24, 496.
154 T.A. Peganova, P.V. Petrovskii, L.S. Isaeva, D.N. Kravtsov, D.B. Furman, A.V. Kudryashev, A.O. Ivanov, S.V. Zotova, and O.V. Bragin, J. Organomet. Chem., 1985, 282, 283.
155 A. Ceriotti, G. Longoni, M. Manassero, N. Masciocchi, L. Resconi, and M. Sansoni, J. Chem. Soc., Chem. Commun., 1985, 181.
156 A. Ceriotti, G. Longoni, M. Manassero, N. Masciocchi, G. Piro, L. Resconi, and M. Sansoni, J. Chem. Soc., Chem. Commun., 1985, 1402.
157 N. Koga and K. Morokuma, J. Am. Chem. Soc., 1985, 107, 7230.
158 N. Koga, S. Obara, K. Kitaura and K. Morokuma, J. Am. Chem. Soc., 1985, 107, 7109.
159 J-E. Bäckvall, E.E. Björkman, L. Pettersson, P. Siegbahn, and A. Strich, J. Am. Chem. Soc., 1985, 107, 7408.
160 B. Kellenberger, S.J. Young, and J.K. Stille, J. Am. Chem. Soc., 1985, 107, 6105.
161 P.K. Byers and A.J. Canty, Inorg. Chim. Acta, 1985, 104, L13.
162 F. Ozawa, K. Kurihara, T. Yamamoto, and A.Y Yamamoto, J. Organomet. Chem., 1985, 279, 233.
163 F. Ozawa, K. Kurihara, T. Yamamoto, and A. Yamamoto, Bull. Chem. Soc. Jpn., 1985, 58, 399.
164 H. Kurosawa and A. Urabe, Chem. Lett., 1985, 1839.
165 A. Campagnaro, A. Mantovani, and P. Uguagliati, Inorg. Chim. Acta., 1985, 99, L15.
166 F. Ozawa, H. Soyama, H. Yanagihara, I. Aoyama, H. Takino, K. Izawa, T. Yamamoto, and A. Yamamoto, J. Am. Chem. Soc., 1985, 107, 3235.
167 R. Usón, J. Forniés, M. Tomás, and B. Menjón, Organometallics, 1985, 4, 1912.
168 R. Usón, J. Forniés, R. Navarro, P. Espinet, and C. Mendívil, J. Organomet. Chem., 1985, 290, 125.
169 L.V. Rybin, E.A. Petrovskaya, M.I. Rubinskaya, L.G. Kuz'mina, Y.T. Struchkov, V.V. Kaverin, and N.Y. Koneva, J. Organomet. Chem., 1985, 288, 119.
170 M. Lenarda, G. Nardin, G. Pellizer, E. Braye, and M. Graziani, J. Chem. Soc., Chem. Commun., 1985, 1536.
171 C. Stader and B. Wrackmeyer, J. Organomet. Chem., 1985, 295, C11.
172 H. Ogawa, T. Joh, S. Takahashi, and K. Sonogashira, J. Chem. Soc., Chem. Commun., 1985, 1220.
173 A.D. Ryabov, Russ. Chem. Rev., 1985, 54, 153.
174 F.L. Wimmer and S. Wimmer, Polyhedron, 1985, 4, 1665.
175 E.C. Constable, J. Chem. Soc., Dalton Trans., 1985, 1719.

176 Y. Fuchita, K. Hiraki, and Y. Matsumoto, J. Organomet. Chem., 1985, 280, C51.
177 P.W. Clark and S.F. Dyke, J. Organomet. Chem., 1985, 281, 389.
178 A.D. Ryabov, I.K. Sakodinskaya, and A.K. Yatsimirsky, J. Chem. Soc., Dalton Trans., 1985, 2629.
179 P.W. Clark, S.F. Dyke, G. Smith, C.H.L. Kennard, and A.H. White, Acta Cryst., Sect. C, 1985, 41, 1742.
180 G.R. Newkome, W.E. Puckett, G.E. Kiefer, V.K. Gupta, F.R. Fronczek, D.C. Pantaleo, G.L. McClure, J.B. Simpson, and W.A. Deutsch, Inorg. Chem., 1985, 24, 811.
181 A.B. Goel and S. Goel, Inorg. Chim. Acta, 1985, 98, 67.
182 B. Crociani, F. Dibianca, A. Giovenco, and A. Scrivanti, J. Organomet. Chem., 1985, 291, 259.
183 H. Ossor and M. Pfeffer, J. Chem. Soc., Chem. Commun., 1985, 1540.
184 H.-P. Abicht and K. Issleib, J. Organomet. Chem., 1985, 289, 201.
185 W-D. Müller and H.A. Brune, Chem. Ber., 1985, 118, 4347.
186 W-D. Müller, G. Schmidtberg, and H.A. Brune, Chem. Ber., 1985, 118, 4653.
187 G.B. Deacon and I.L. Grayson, J. Organomet. Chem., 1985, 292, 1.
188 R. McCrindle, G.J. Arsenault, and R. Farwaha, J. Organomet. Chem., 1985, 296, C51.
189 S.T. Lin and K.J. Klabunde, Inorg. Chem., 1985, 24, 1961.,
190 R.H. Hill and R.J. Puddephatt, J. Am. Chem. Soc., 1985, 107, 1218.
191 D.M. Roundhill, J. Am. Chem. Soc., 1985, 107, 4354.
192 R.J. Puddephatt and J.D. Scott, Organometallics, 1985, 4, 1221.
193 P.K. Monaghan and R.J. Puddephatt, Organometallics, 1985, 4, 1406.
194 G. Ferguson, P.K. Monaghan, M. Parvez, and R.J. Puddephatt, Organometallics, 1985, 4, 1669.
195 J.E. Hux and R.J. Puddephatt, Inorg. Chim. Acta, 1985, 100, 1.
196 P. Castan, J. Jaud, N.P. Johnson, and R. Soules, J. Am. Chem. Soc., 1985, 107, 5011.
197 G. Alibrandi, D. Minniti, R. Romeo, P. Uguagliati, L. Calligaro, U. Belluco, and B. Crociani, Inorg. Chim. Acta, 1985, 100, 107.
198 R.L. Brainard and G.M. Whitesides, Organometallics, 1985, 4, 1550.
199 H.E. Bryndza, J. Chem. Soc., Chem. Commun., 1985, 1696.
200 G. Alibrandi, D. Minniti, R. Romeo, G. Cum, and R. Gallo, J. Organomet. Chem., 1985, 291, 133.
201 J. Terheijden, G. Van Koten, I.C. Vinke, and A.L. Spek, J. Am. Chem. Soc., 1985, 107, 2891.
202 A. Albinati, U. Von Gunten, P.S. Pregosin, and H.J. Ruegg, J. Organomet. Chem., 1985, 295, 239.
203 E.W. Abel, S.K. Bhargava, K.G. Orrell, A.W.G. Platt, V. Šik, and T.S. Cameron, J. Chem. Soc., Dalton Trans., 1985, 345.
204 R. Benn, R-D. Reinhardt, and A. Rufińska, J. Organomet. Chem., 1985, 282, 291.
205 A. Imran, R.D.W. Kemmitt, A.J.W. Markwick, P. McKenna, D.R. Russell, and L.J.S. Sherry, J. Chem. Soc., Dalton Trans., 1985, 549.
206 R.M. Ceder and J. Sales, J. Organomet. Chem., 1985, 294, 389.
207 R.A. Ekeland and P.W. Jennings, J. Organomet. Chem., 1985, 281, 397.
208 W.R. Winchester, M. Gawron, G.J. Palenik and W.M. Jones, Organometallics, 1985, 4, 1894.
209 A. Miyashita, H. Shitara, and H. Nohira, Organometallics, 1985, 4, 1463.
210 G.M. Whitesides, M. Hackett, R.L. Brainard, J-P.P.M. Lavallege, A.F. Sowinski, A.N. Izumi, S.S. Moore, D.W. Brown, and E.M. Staudt, Organometallics, 1985, 4, 1819.
211 R.J. Cross and M.F. Davidson, Inorg. Chim. Acta, 1985, 97, L35.
212 G.K. Anderson and G.J. Lumetta, J. Organomet. Chem., 1985, 295, 257.
213 A. Sebald and B. Wrackmeyer, Z. Naturforsch, Teil B, 1985, 40, 1481.
214 A.J. McLennan and R.J. Puddephatt, Organometallics, 1985, 4, 485.
215 C.R. Langrick, P.G. Pringle, and B.L. Shaw, J. Chem. Soc., Dalton Trans., 1985, 1015.

216 A.T. Hutton, C.R. Langrick, D.M. McEwan, P.G. Pringle, and B.L Shaw, J. Chem. Soc., Dalton Trans., 1985, 2121.
217 G.K. Anderson and G.L. Lumetta, Organometallics, 1985, 4, 1542.
218 F.S.M. Hassan, D.M. McEwan, P.G. Pringle, and B.L. Shaw, J. Chem. Soc., Dalton Trans., 1985, 1501.
219 S.S.M. Ling, I.R. Jobe, L. Manojlović-Muir, K.W. Muir, and R.J. Puddephatt, Organometallics, 1985, 4, 1198.
220 S.S.M. Ling, N.C. Payne, and R.J. Puddephatt, Organometallics, 1985, 4, 1546.
221 K.A. Azam, A.A. Frew, B.R. Lloyd, L. Manojlović-Muir, K.W. Muir, and R.J. Puddephatt, Organometallics, 1985, 4, 1400.
222 M.P. Brown, A. Yavari, R.H. Hill, and R.J. Puddephatt, J. Chem. Soc., Dalton Trans.,1985, 2421.
223 A.T. Hutton, P.G. Pringle, and B.L. Shaw, J. Chem. Soc., Dalton Trans., 1985, 1677.
224 R. Bender, P. Braunstein, A. Tiripicchio, M. Tiripicchio Camellini, Angew. Chem., Int. Ed. Engl., 1985, 24, 861.
225 R.J. Blau, M.H. Chisholm, K. Folting, and R.J. Wang, J. Chem. Soc., Chem. Commun., 1985, 1582.
226 V. Guerchais and D. Astruc, J. Chem. Soc., Chem. Commun., 1985, 835.
227 S.G. Davies and T.R. Maberly, J. Organomet. Chem., 1985, 296, C37.
228 J. Markham, W. Tolman, K. Menard, and A.R. Cutler, J. Organomet. Chem., 1985, 294, 45.
229 T.W. Bodnar and A.R. Cutler, Organometallics, 1985, 4, 1558.
230 C.P. Casey, W.H. Miles, and H. Tukada, J. Am. Chem. Soc., 1985, 107, 2924.
231 M. Brookhart,and W.B. Studabaker, Organometallics, 1985, 4, 943.
232 A.M. Crespi and D.F. Shriver, Organometallics, 1985, 4, 1830.
233 C. Jiabi, L. Guixin, X. Weihua, J. Xianglin, S. Meicheng, and T. Youqi, J. Organomet. Chem., 1985, 286, 55.
234 C.P. Casey, S.R. Marder, and B.R. Adams, J. Am. Chem. Soc., 1985, 107, 7700.
235 C.P. Casey, M.S. Konings, R.E. Palermo, and R.E. Colborn, J. Am. Chem. Soc., 1985, 107, 5296.
236 C.P. Casey and S.R. Marder, Organometallics, 1985, 4, 411.
237 C.P. Casey, S.R. Marder, and A.L Rheingold, Organometallics, 1985, 4, 762.
238 A. Schäfer and W.A. Herrmann, J. Organomet. Chem., 1985, 297, 229.
239 G.D. Williams, G.L. Geoffroy, R.R. Whittle, and A.L. Rheingold, J. Am. Chem. Soc., 1985, 107, 729.
240 D. Lentz, I. Brüdgam, and H. Hartl, Angew. Chem., Int. Ed. Engl., 1985, 24, 119.
241 D. Nuel, F. Dahan, and R. Mathieu, Organometallics, 1985, 4, 1436.
242 D. Nuel, F. Dahan, and R. Mathieu, J. Am. Chem. Soc., 1985, 107, 1658.
243 S. Attali, F. Dahan, and R. Mathieu, J. Chem. Soc., Dalton Trans., 1985, 2521.
244 G. Consiglio, F. Bangerter, and F. Morandini, J. Organomet. Chem., 1985, 293, C29.
245 M.I. Bruce, D.N. Duffy, M.G. Humphrey, and A.G. Swincer, J. Organomet. Chem., 1985, 282, 383.
246 W.A. Herrmann, M. Flöel, C. Weber, J.L. Hubbard, and A. Schäfer, J. Organomet. Chem., 1985, 286, 369.
247 N.G. Connelly, N.J. Forrow, B.P. Gracey, S.A.R. Knox,and A.G. Orpen, J. Chem. Soc., Chem. Commun., 1985, 14.
248 D.L. Davies, J.A.K. Howard, S.A.R. Knox, K. Marsden, K.A. Mead, M.J. Morris,and M.C. Rendle, J. Organomet. Chem., 1985, 279, C37.
249 J.A.K. Howard, J.C.V. Laurie, O. Johnson, and F.G.A. Stone, J. Chem. Soc., Dalton Trans., 1985, 2017.
250 J.S. Holmgren and J.R. Shapley, Organometallics, 1985, 4, 793.
251 J. Keijsper, L.H. Polm, G. Van Koten, K. Vrieze, K. Goubitz, and C.H. Stam, Organometallics, 1985, 4, 1876.

252　L.R. Beanan and J.B. Keister, Organometallics, 1985, 4, 1713.
253　N.G. Connelly, N.J. Forrow, S.A.R. Knox, K.A. MacPherson, and A.G. Orpen, J. Chem. Soc., Chem. Commun., 1985, 16.
254　R.D. Adams, H-S.Kim, and S. Wang, J. Am. Chem. Soc., 1985, 107, 6107.
255　J.R. Shapley, W-Y. Yeh, M.R. Churchill,and Y-J.Li, Organometallics, 1985, 4, 1898.
256　W-Y. Yeh, J.R. Shapley, Y. Li, and M.R. Churchill, Organometallics, 1985, 4, 767.
257　E.D. Morrison, G.L. Geoffroy, and A.L. Rheingold, J. Am. Chem. Soc., 1985, 107, 254.
258　E.D. Morrison and G.L. Geoffroy, J. Am. Chem. Soc., 1985, 107, 3541.
259　L.J. Farrugia, M. Green, D.R. Hankey, M. Murray, A.G. Orpen, and F.G.A. Stone, J. Chem. Soc., Dalton Trans., 1985, 177.
260　U. Plaia, H. Stolzenberg, and W.P. Fehlhammer, J. Am. Chem. Soc., 1985, 107, 2171.
261　D.W, Macomber and R. D. Rogers, Organometallics, 1985, 4, 1485.
262　A.W. Coleman, P.B. Hitchcock, M.F. Lappert, R.K. Maskell, and J.H. Müller, J. Organomet. Chem., 1985, 296, 173.
263　R. Zolk and H. Werner, Angew. Chem., Int. Ed. Engl.,1985, 24, 577.
264　E.N. Jacobsen and R.G. Bergman, J. Am. Chem. Soc., 1985, 107, 2023.
265　P.F. Seidler, S.T. McKenna, M.A. Kulzick,and T.M. Gilbert, Acta Cryst., Section C, 1985, 41, 352.
266　J.C. Jeffery, C. Marsden, and F.G.A. Stone, J. Chem. Soc., Dalton Trans., 1985, 1315.
267　D.B. Jacobson and B.S. Freiser, J. Am. Chem. Soc., 1985, 107, 5870.
268　F.J. Garcia Alonso, A. Höhn, J. Wolf, H. Otto, and H. Werner, Angew. Chem., Int. Ed. Engl., 1985, 24, 406.
269　H. Werner, F.J. Garcia Alonso, H. Otto, K. Peters, and H.G. Von Schnering, J. Organomet. Chem., 1985, 289, C5.
270　W.A. Herrmann and C. Weber, J. Organomet Chem., 1985, 282, C31.
271　W.A. Herrmann, C. Weber, M.L. Zeigler, and O. Serhadli, J. Organomet. Chem., 1985, 297, 245.
272　D.H. Berry and R. Eisenberg, J. Am.Chem. Soc., 1985, 107, 7181.
273　R.S. Dickson, G.D. Fallon, R.J. Nesbit, and G.N. Pain, Organometallics, 1985, 4, 355.
274　B.E. Mann, N.J. Meanwell, C.M. Spencer, B.F. Taylor, and P.M. Maitlis, J. Chem. Soc., Dalton Trans., 1985, 1555.
275　M. Green, A.G. Orpen, C.J. Schaverien, and I.D. Williams, J. Chem. Soc., Dalton Trans., 1985, 2483.
276　M.A. Lilga and J.A. Ibers, Organometallics, 1985, 4, 590.
277　G.R. Clark, T.R. Greene, and W.R. Roper, J. Organomet. Chem., 1985, 293, C25.
278　M. El Amane, A. Maisonnat, F. Dahan, R. Pince, and R. Poilblanc, Organometallics, 1985, 4, 773.
279　D. Xu, K. Miki, Y. Kai, N. Kasai, and M. Wada, J. Organomet. Chem., 1985, 287, 265.
280　C. Bianchini, A. Meli, and G. Scapacci, Organometallics, 1985, 4, 264.
281　E.O. Fischer and J.R. Schneider, J. Organomet. Chem., 1985, 295, C29.
282　R.A. Michelin, G. Facchin, and R. Ros, J. Organomet. Chem., 1985, 279, C25.
283　G. Minghetti, A. Albinati, A.L. Bandini, and G. Banditelli, Angew. Chem., Int. Ed. Engl., 1985, 24, 120.
284　M.R. Awang, R.D. Barr, M. Green, J.A.K. Howard, T.B. Marder, and F.G.A. Stone, J. Chem. Soc., Dalton Trans., 1985, 2009.

13
Metal–Hydrocarbon π-Complexes, other than π-Cyclopentadienyl and π-Arene Complexes

BY J. A. S. HOWELL

A. Reviews

Specific reviews have been published on nickel olefin complexes,[1] on nickel[2,3] and palladium[4] allyl complexes, on zirconium diene complexes[5], and on open chain pentadienyl complexes.[6] Additionally, reviews on metal borole complexes,[7] on metal heterodiene complexes,[8] on oxidation[9] and electron transfer reactions[10] of organometallics, on the reactions of organic compounds with Os_3 clusters[11] and C-C and C-H activation in clusters,[12] and on electron density determinations in organometallics[13] contain material of relevance to this chapter. General articles on the use of transition metal complexes in organic synthesis have appeared,[14,15] and a new text on organometallic chemistry contains material on the same topic.[16]

B. Allyl Complexes and Complexes Derived from Monoolefins
1. Ni, Pd and Pt

Calculations confirm a singlet ground state for $Ni(C_2H_4)$ with a C-C bond length of 1.45 Å, a bend-back angle of 21° and a binding energy of about 80 kJ mol^{-1}.[17] Calculations on $(PH_3)_2ML$ (M=Ni,Pd,Pt) and $(PH_3)_4ML^+$ (M=Co,Rh,Ir) complexes (L=O_2, C_2H_2, C_2H_4) indicate the greater importance of back-bonding; binding energies follow the order $O_2 > C_2H_2 \approx C_2H_4$ and Ni>Pd>Pt.[18]

Reaction of MeLi/TMED with $Ni(C_2H_4)_3$ yields the $[CH_3Ni(C_2H_4)_2]^-$ ion as its $Li(TMED)_2^+$ salt; C_2H_4 may be displaced by CO to give $[CH_3Ni(CO)_3]^-$. Use of PMDT yields a 1:1 complex of structure (1) containing a covalent Li-Ni interaction.[19] Treatment of $Ni(C_2H_4)_3$ with $R_3P=CH_2$ yields $R_3PCH_2Ni(C_2H_4)_2$ from which ethylene may be displaced with CO and which decompose thermally (R=Me) to give $(Me_3P)Ni(C_2H_4)_2$.[20] Reaction of $Ni(C_2H_4)_3$ with M_2EHR_3 (M=Li, Na, E=Al,Ga, R=alkyl) at -70 °C generates $[R_3E-H-Ni(C_2H_4)_2]^-$ which undergoes elimination of ER_3 on reaction with CO to give $[HNi(CO)_3]^-$.[21] Reactions of Ni(1,5,9-cdt) parallel those of $Ni(C_2H_4)_3$, though reactivities indicate an order of acceptor strength for the metal fragment of $Ni(CO)_3 > Ni(1,5,9-cdt) > Ni(C_2H_4)_2$.[22]

(For references see page 345)

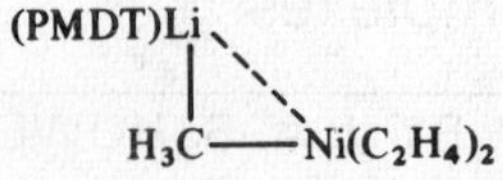

(1) (PMDT)Li⋯H₃C—Ni(C₂H₄)₂

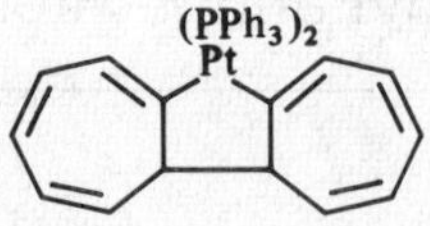

(2) Pt(PPh₃)₂ dibenzo structure

(3) bicyclic alkene

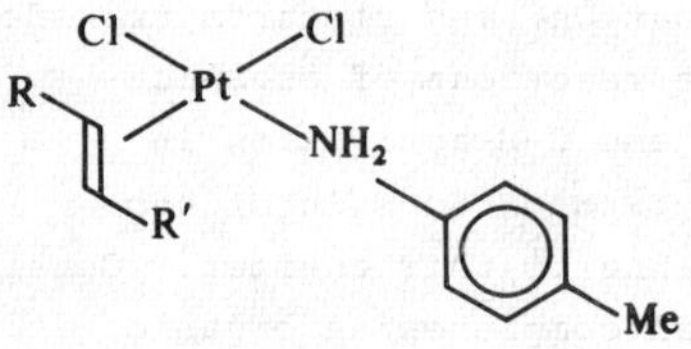

(4) R = Et, neopentyl; R' = H
(5) R = H; R' = alkoxy

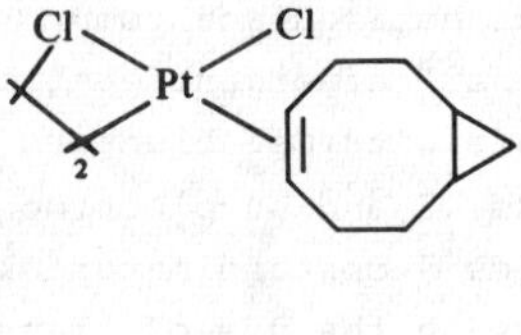

(6)

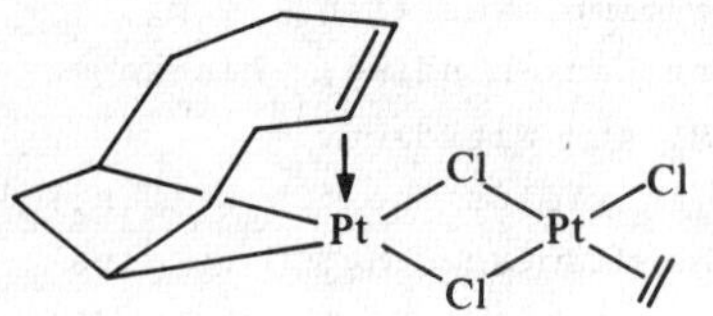

(7)

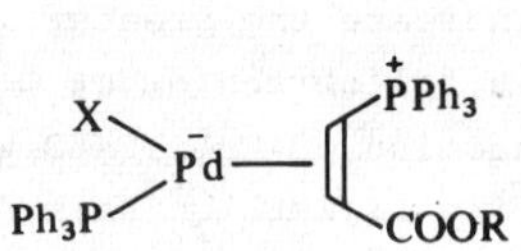

(8)

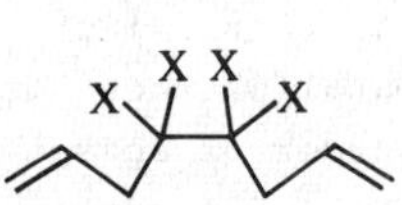

(9) X = CN

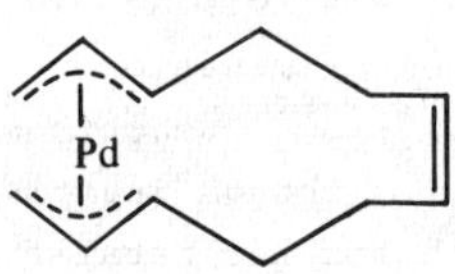

(10)

Two reports confirm that the phosphaalkene complex $(PPh_3)_2Pt[Ph_2PC=P(mesityl)]$ adopts an η^1-mode in the solid state, but exists predominantly in solution as an η^2-complex.[23,24] The phosphalkene complex $[ClP=C(SiMe_3)_2]_2NiCO$ exhibits an η^2-coordination in the solid state.[25]

Substituted allenes react with $Ni(PPh_3)_3$ at -70 °C to give mono- and cis- and trans-bis(allene)$Ni(PPh_3)_n$ (n=2 or 1 respectively) derivatives; complexation involves predominantly the unsubstituted bond. On warming, the cis-bis(allene) complexes undergo coupling to give nickelacyclopentane derivatives which above -15 °C undergo hydrogen migration and reductive elimination. Rates of coupling increase with substituent in the order Bu^t<Et<1,1-Me$_2$<<MeO<Ph<CN.[26,27] Cycloheptatetraene and its benzo- and dibenzo-derivatives may be trapped as their $Pt(PPh_3)_2$ complexes in which the metal is η^2-bound to one of the allenic double bonds; reaction with a second mole of cycloheptatetraene results in coupling to give (2).[28,29] The anti-Bredt olefin (3) is stabilized by complexation to Pt(0) or Pt(II); data indicate strong back-donation.[30]

A crystal structure determination of $\{(bipy)Ni[Et(H)C=C(Me)-COH]\}_2 \cdot H_2O$ shows some interaction of the metal with the aldehydic carbon; the bonding is intermediate between olefinic and allylic.[31] Reaction of $M(PPh_3)_2(C_2H_4)$ or $PPh_3/M(cod)_2$ with the chelating (2-vinylphenyl)PPh_2 yields $(PPh_3)_2ML$ complexes; the C=C bond is coordinated in the Pt and Pd derivatives, but not in the case of Ni.[32]

Cis-Cl_2Pt(p-toluidine)(olefin) complexes adopt exclusively stereoisomer (4) where R=Et or neopentyl but exclusively (5) where R=alkoxy due to hydrogen bonding between oxygen and NH_2; this also greatly increases the barrier to olefin rotation.[33,34] Addition of diimines (RN=CRCR=NR) to Cl_2Pt(olefin) fragments produces either five-coordinate chelated complexes or square planar complexes containing monodentate or bridging diimine; chelation is favoured by increasing σ-donor character of the diimine, by alkyl substitution of the imine carbons, and by increasing π-acceptor character of the olefin.[35]

Thermolysis of (6) in the presence of excess Zeise's dimer yields (7) containing a platinocyclobutane ring and a formal Pt(IV)-olefin bond.[36] Reactions of $Pd(PPh_3)_4$ with $XCH=CHCO_2R$ (X=halogen, R=Me,Et) yield $Pd(\sigma-CH=CHCOOR)(PPh_3)_2X$ complexes which on heating rearrange to the ylid structure (8).[37]

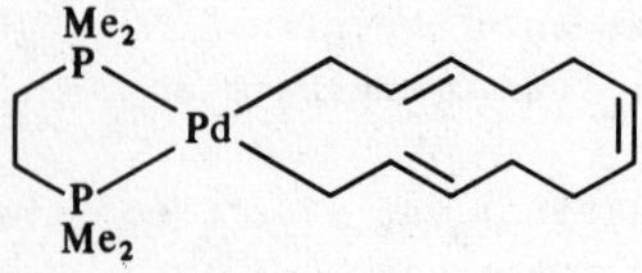

(11)

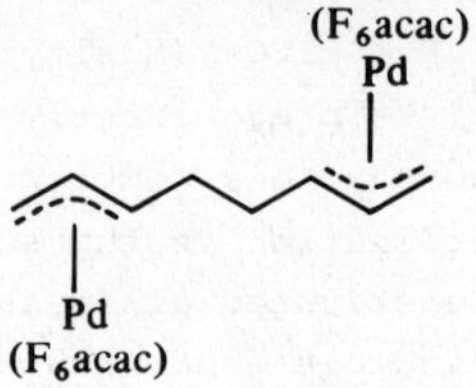

(12)

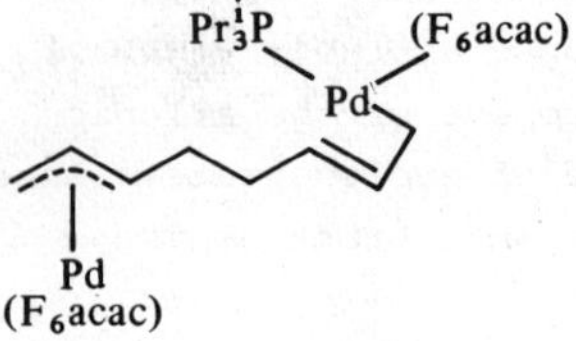

(13)

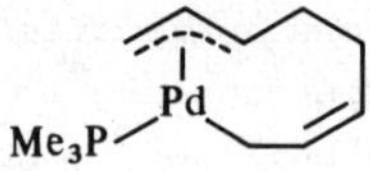

(14)

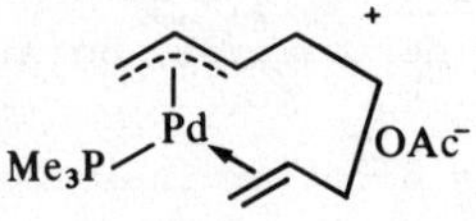

(15)

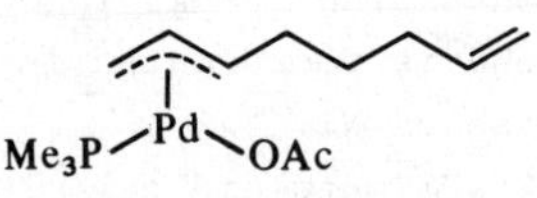

(16)

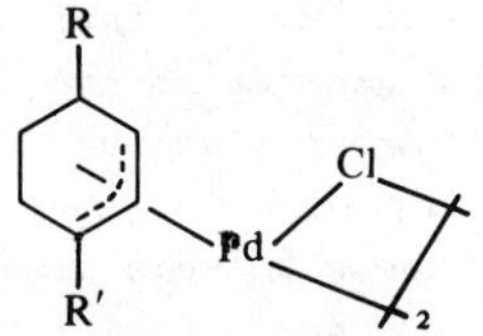

(17) R = Me
R' = Pr^i, Bu^t

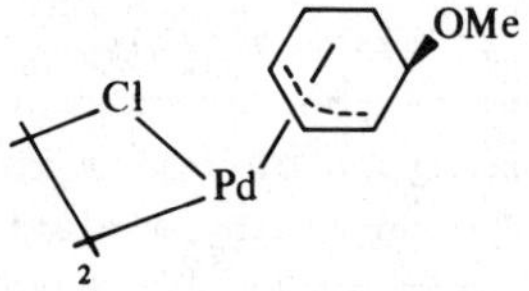

(18)

(η^3-allyl)$_2$Pt reacts with TCNE to give an adduct of formula (η^3-allyl)$_2$Pt(TCNE) and with PPh$_3$ to give (η^3-allyl)Pt(PPh$_3$)-(η^1-allyl); reaction of the latter with TCNE or the former with PPh$_3$ results in elimination of Pt metal and linear addition of TCNE to give (9).[38] Elimination of alkene on reaction of (η^3-2-methylallyl)$_2$Ni with HSiR$_3$ in the presence of toluene provides a route to (η^6-toluene)Ni(SiR$_3$)$_2$ complexes (R$_3$=F$_3$, Cl$_3$, Cl$_2$Me); reaction with SiCl$_4$ or GeCl$_4$ yields instead [(η^3-methylallyl)NiCl]$_2$.[39]

SCF calculations on <u>trans</u>-Ni(allyl)$_2$ show that the energy minimum involves a bending of the <u>syn</u> and C-2 hydrogens towards the metal (6.7 and 13.4°) and a bending away of the <u>anti</u> hydrogens (31.4°), in agreement with a neutron diffraction study; reduction of steric interaction between Ni and anti-hydrogens is the determining factor.[40] Nmr studies of (allyl)$_2$M (M=Ni,Pd,Pt) and several other allyl complexes show that ^{1}J(^{13}C,^{13}C) values can be used as a guide to relative donor/acceptor character of the allyl ligand, and that J(^{13}C,H) and J(H,H) values are sensitive to deviations of the allyl hydrogens from the C$_3$ plane.[41] Incorporation of paramagnetic phosphine in complexes such as (2-methylallyl)PdLCl also provides a structural probe for allyl geometry, including non-planarity of the terminal hydrogen atoms.[42]

Butadiene reacts with Pd(dba)$_2$ to give (10), which reacts at -40 °C with dmpe to give the metallocycle (11); on heating above -20 °C or on reaction with PPh$_3$, a mixture of linear dodecatetraenes are liberated.[43] The 1,4-polymerization of butadiene using {(allyl)Ni[P(OR)$_3$]$_2$}$^+$ as catalyst proceeds <u>via</u> butadiene insertion in a <u>syn</u>-(RCHCHCH$_2$)NiP(OR)$_3$(η^2-butadiene) intermediate to give an <u>anti</u>-allyl structure; the cycle is completed by rate-dermining isomerization to the <u>syn</u>-geometry.[44] Butadiene reacts with Pd(F$_6$acac)$_2$ to generate (12); treatment with one mole of PPr$_3^i$ gives (13) which undergoes insertion of CO$_2$ into the Pd-C bond to give, after acidolysis, pelargonic acid.[45] Reaction of butadiene/PMe$_3$ with Pd(2-methylallyl)$_2$ yields (14); protonation with HOAc yields initially (15) which undergoes displacement of olefin to give (16).[46] Maleic anhydride induces reductive elimination from (allyl)$_2$Pd complexes; this has been used to produce phosphine-free Pd(0) which catalyses dimerization of butadiene to 4-vinylcyclohexene.[47] This has also been applied to the catalytic coupling of allyl halides and allylSnR$_3$ to yield 1,5-dienes using [(allyl)PdCl]$_2$/maleic anhydride as catalyst.[48]

Reaction of 1,4-disubstituted cyclohexa-1,3-dienes with $PdCl_4^{2-}$/MeOH at 60 °C yields allyl complexes of structure (17) in which the bulkier R group is incorporated into the allyl ligand.[49] At lower temperature, methoxypalladation of chd occurs to give (18). This complex undergoes either cis- or trans-addition of OAc$^-$ to the allyl depending on reaction conditions; trans-attack is observed using HOAc/p-benzoquinone, whereas reaction with AgOAc to displace chloride, followed by treatment with p-benzoquinone yields the product of cis-addition.[50]

Complexes (19) and (20) exist in equilibrium in solution, but treatment with Ag^+ in MeCN generates exclusively (21), characterized crystallographically. The C=C bond may be displaced by MeNC to regenerate the (19) ⇌ (20) equilibrium in the analogous $[(allyl)Pd(MeCN)_2]^+$ cation.[51,52] Asymmetric allylic alkylation may be accomplished using Pd complexes of S,S-chiraphos as catalyst; enantiomeric excess is determined by the major diastereoisomeric $[(\eta^3\text{-allyl})Pd(chiraphos)]^+$ cation formed on oxidative addition. In the 1,1-bis(3,5-dimethylphenyl)-3--phenylallyl complex, a crystal structure shows that the most stable diastereoisomer results from minimization of interactions of allyl substituents with the fixed chiraphos chelate ring.[53] Chelated $[(2\text{-methylallyl})Pd(Ph_2PCH_2CH_2SR)]^+$ complexes (R=Me, Et, Ph) have been reported.[54]

2. Co, Rh and Ir.

Olefin may be displaced from $Co(propene)(PMe_3)_3$ and $HCo(butene)(PMe_3)_3$ by anthracene to give $Co(\eta^2\text{-anthracene})(PMe_3)_3$ and $HCo(\eta^2\text{-anthracene})(PMe_3)_3$ respectively; reaction of $Ph_3SnCo(PMe_3)_3$ with anthracene yields $Ph_3SnCo(\eta^4\text{-anthracene})(PMe_3)_2$, characterized crystallographically.[55] Matrix photolysis of $CpRh(C_2H_4)_2$ yields $CpRh(C_2H_4)$ which undergoes oxidative addition with Et_3SiH.[56] For the Cp* complex, the equilibria $Cp*Rh(C_2H_4)_2 + HSiEt_3 \rightleftharpoons Cp*Rh(C_2H_4)(SiEt_3)H + C_2H_4$ and $Cp*Rh(C_2H_4)(SiEt_3)H + HSiEt_3 \rightleftharpoons Cp*Rh(SiEt_3)_2H_2 + C_2H_4$ may be induced either thermally or photochemically starting from either side.[57] Thermolysis of $CpCo(C_2H_4)_2$ yields $Cp_4Co_4(\mu_3\text{-H})(\mu_3\text{-CMe})$ through rearrangement of C_2H_4 to H and CMe fragments.[58]

$Cp*Ir(\eta^3\text{-allyl})X$ (X=H,Cl) have been prepared; though the chloride reacts with PMe_3 to give $Cp*Ir(\eta^1\text{-allyl})(PMe_3)Cl$, the hydride yields $Cp*Ir(PMe_3)(C_3H_7)Ph$ on reaction with PMe_3 in benzene. Formation of coordinated propene, followed by oxidative addition of benzene, is postulated as the initial

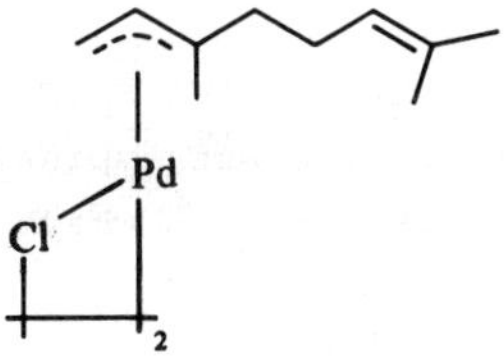

(19)

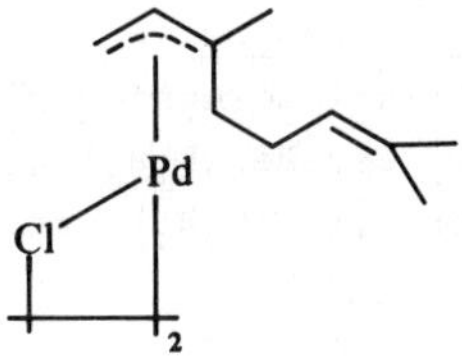

(20)

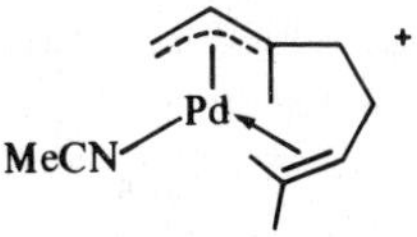

(21)

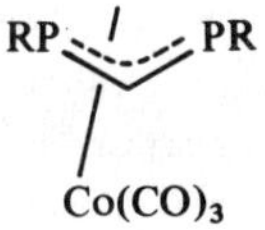

(22) R = 2,4,6-Bu$_3^t$C$_6$H$_2$

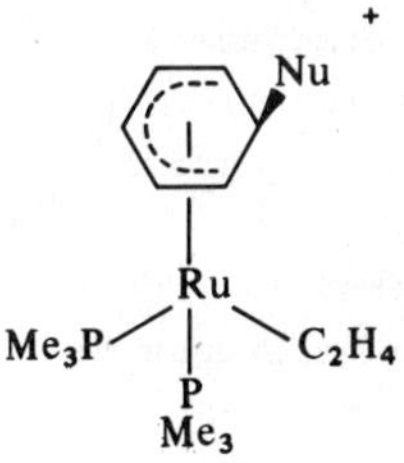

(23)

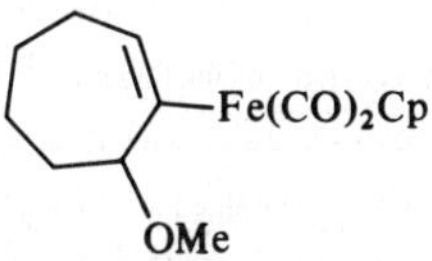

(24)

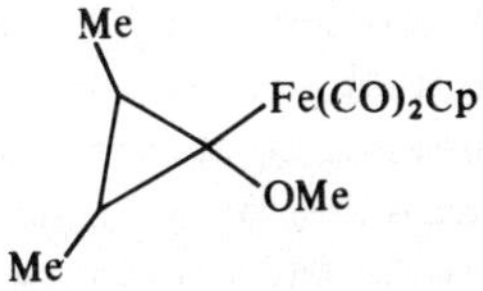

(25)

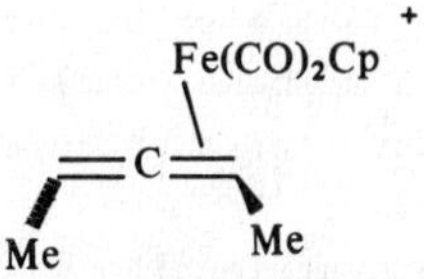

(26)

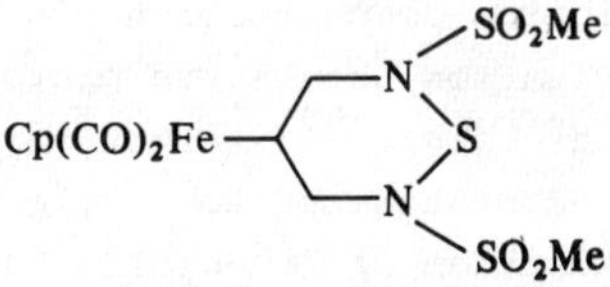

(27)

step.[59] The 1,3-diphosphaallyl complex (22) has been prepared and characterized crystallographically.[60]

3. Fe, Ru and Os.

Matrix interaction of Fe or Fe_2 with C_2H_4 yields $Fe(C_2H_4)$, $Fe(C_2H_4)_2$ and $Fe_2(C_2H_4)_2$. Though the latter two are typical π-complexes, $Fe(C_2H_4)$ exists as two isomeric forms in which iron is bound via the hydrogen atoms; photoexcitation results in reversible photoinsertion to give $HFeC_2H_3$. A similar structure and insertion is observed for $Fe(C_2H_2)$ formed in matrices at low concentrations of Fe and acetylene.[61,62] Photolysis of tetramethylethylene and $Fe(CO)_5$ in pentane at -130 °C yields the radicals $HFe_2(CO)_8$ and $(\eta^3$-1,1,2-trimethylallyl$)Fe(CO)_3$; the presumed intermediates are (olefin)$Fe(CO)_n$ (n=3,4) and (allyl)$Fe(CO)_3H$. Unsymmetrical olefins yield isomeric mixtures of (allyl)$Fe(CO)_3$ radicals whose proportions are kinetically controlled at low temperature.[63] The crystal structure of $(C_2H_4)Ru(PMe_3)_4$ shows a long C-C bond consistent with a metallocyclopropane description.[64]

$[(\eta^6$-Benzene$)M(PMe_3)_2C_2H_5]^+$ (M=Ru,Os) reacts with Ph_3C^+ to give $[(\eta^6$-benzene$)M(PMe_3)_2(\eta^2$-$C_2H_4)]^{2+}$. Treatment with soft nucleophiles [PR_3, $P(OR)_3$] results in addition to ethylene, whereas hard nucleophiles (OMe, Me) yield cyclohexadienyl complexes of structure (23).[65] The $[CpFe(CO)_2(\eta^2$-1,2-cycloheptadiene$)]^+$ cation has been prepared by protonation of (24); the iron migrates between the two double bonds.[66] Methoxide extraction from (25) yields the allene cation (26), presumably via the intermediacy of a cyclopropylidene complex.[67] Reversible protonation of (27) and a related derivative occurs at nitrogen to generate the olefin complex $\{CpFe(CO)_2[\eta^2$-CH_2=$CHCH_2N(SO_2Me)$-$SNH(SO_2Me)]\}^+$, characterized crystallographically.[68] Reaction of $[CpFe(CO)(L)(\eta^2$-allene$)]^+$ [L=$P(OPh)_3$] with a variety of nucleophiles yields $CpFe(CO)(L)[\eta^1$-$C(CH_2Nu)$=$CH_2]$ complexes (Nu=H,Me,Ph, CH=CH_2, C≡CMe, SPh) which isomerize on chromatography to $CpFe(CO)(L)(\eta^1$-E-C(Me)=CHNu), though where Nu=Me, an equilibrium concentration of the Z-isomer may also be detected. Isomerization occurs via protonation to a carbene complex of structure (28).[69] Alkoxycarbene complexes of structure (29) may be deprotonated to vinylether complexes.[70]

Reaction of $(\eta^4$-cod$)(\eta^6$-1,3,5-cyclooctatriene$)Ru$ with 3-butenoic acid and PPh_3 yields (30).[71] Treatment of $CpRu(\eta^3$-allyl$)X_2$ (X=Cl,Br) with MeLi gives $CpRu(\eta^3$-allyl$)(X)Me$

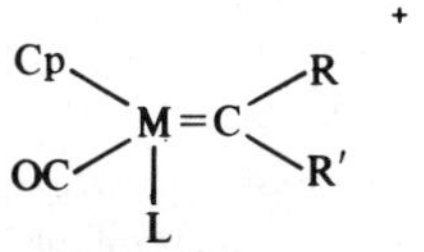

(28) M = Fe; L = P(OPh)$_3$;
R = Me; R' = CH$_2$Nu
(29) M = Ru; L = PPh$_3$;
R = OMe; R' = CH$_2$Ph, CH$_2$Me

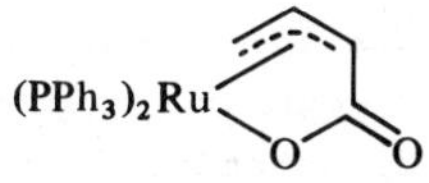

(30)

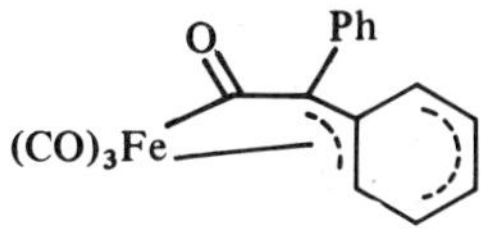

(31)

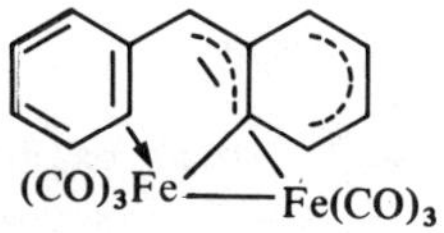

(32)

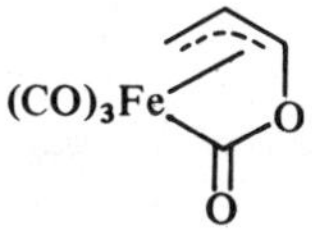

(33)

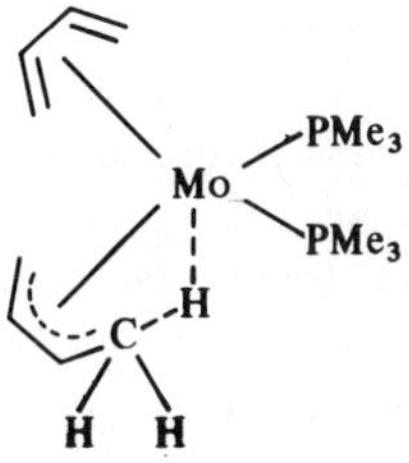

(34)

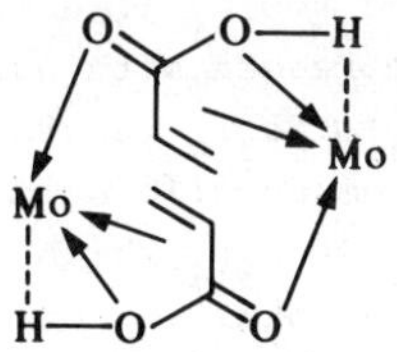

(35)

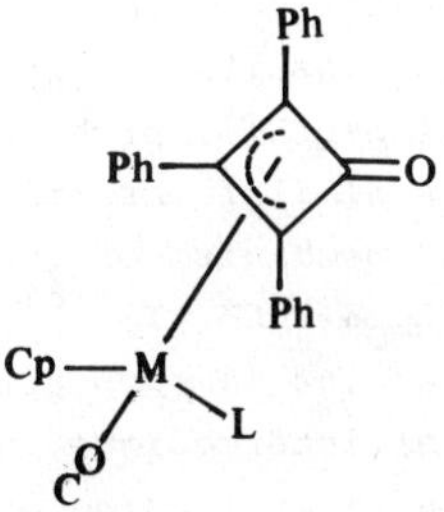

(37) Cp = C$_5$H$_5$; M = Mo;
L = PMe$_2$Ph
(38) Cp = Me$_5$C$_5$; M = Cr, Mo, W;
L = CO

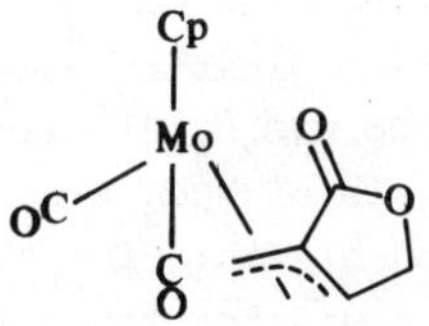

(36)

which on thermolysis in the presence of L, yields butene and $CpRuL_2X$ (L=CO, Bu^tNC, PPh_3, $P(OPh)_3$; L_2=cod, nbd, dppe).[72] Reaction of $(C_6H_6)Ru(\eta^3$-allyl)Cl with PhC≡CPh in the presence of Ag^+ results in coupling to give $[(C_6H_6)Ru(1,2-Ph_2Cp)]^+$; the reaction can be extended to substituted allyls, and use of $CpRh(\eta^3$-allyl)Cl provides a route to substituted rhodicenium salts.[73] Acidolysis of $OsH(\eta^3$-2-methylallyl)$(PPh_3)_2$ in the presence of CO yields $[OsH(CO)_2(OH_2)(PPh_3)_2]^+$, characterized crystallographically.[74]

Photolysis of $Fe(CO)_5$ and diphenylketene yields (31) which on treatment with $Fe_2(CO)_9$ yields (32).[75] Vinyl epoxide ring opening with $Fe_2(CO)_9$ to give lactone derivatives of structure (33) has been used in the synthesis of several naturally occurring δ-lactones.[76]

4. Cr, Mo and W

Condensation of Mo with PMe_3 yields $Mo(PMe_3)_6$ which exists in solution in equilibrium with $HMo(PMe_3)_4(\eta^2-CH_2=PMe_2)$ and free PMe_3. Reactions with C_2H_4 and butadiene yield <u>trans</u>-$Mo(C_2H_4)_2$-$(PMe_3)_4$ and <u>cis</u>-$Mo(butadiene)_2(PMe_3)_2$ respectively. Protonation of the butadiene complex with HBF_4 yields (34), in which the agostic interaction scrambles hydrogen between the four terminal carbons of the two C_4 ligands.[77] $HW(PMe_3)_4(CH_2=PMe_2)$ has been prepared, and reacts with butadiene in the same way.[78] <u>Trans</u>-$Mo(C_2H_4)_2(PMe_3)_4$ reacts with CO_2 <u>via</u> formal insertion into C-H to give the acrylic acid complex (35); hydrogenation yields $MoH(OOCCH_2CH_3)(PMe_3)_4$ which on treatment with $BuLi/C_2H_4$ regenerates the starting ethylene complex.[79]

Reduction of $[(C_6H_6)Mo(dmpe)(\eta^3$-allyl)$]^+$ in the presence of C_2H_4 yields $(C_6H_6)Mo(dmpe)(C_2H_4)$ which protonates to give $[(C_6H_6)Mo(dmpe)(C_2H_4)H]^+$. Treatment with PMe_3 results in transfer of C_2H_5 to benzene to give $[(C_6H_5Et)Mo(dmpe)(PMe_3)H]^+$, probably <u>via</u> initial insertion of C_2H_4 into Mo-H.[80] A similar insertion, followed by α-elimination seems the probable mechanism of the reaction of $[Cp_2W(C_2H_4)H]^+$ with I_2 to give $[Cp_2W(CHCH_3)I]^+$.[81]

Photolysis of $W(CO)_5$(olefin) complexes results in competitive loss of CO and olefin; for the $W(CO)_4$(olefin) intermediate, no evidence was found for formation of $HW(CO)_4(\eta^3$-allyl) using olefins containing allylic hydrogens.[82] <u>Trans</u>-L^1L^2mer$(CO)_3W$(olefin) complexes [$L^1=L^2=P(OMe)_3$; $L^1=P(OMe)_3$, $L^2=P(OEt)_3$; olefin=C_2H_4, methylmaleate, -acrylate, -fumarate] exhibit olefin rotational barriers in the range 28-60 kJ mol^{-1}; the ground state structure has the C=C axis parallel to the P-W-P axis.[83,84] A crystal structure of the chelate complex $(CO)_4Mo(Me_2POCHPhC=CH_2)$ shows the olefinic bond parallel to the P-Mo-CO axis.[85]

$CpMo(CO)_3(\eta^1$-allyl) is converted to $CpMo(CO)_2(\eta^3$-allyl) on reaction with Me_3NO.[86] Reaction of non-carbon nucleophiles (H^-, OMe^-, SR^-) with $[CpMo(CO)(NO)(\eta^3$-allyl)$]^+$ (allyl=C_3H_5, cyclooctenyl) occurs stereospecifically cis to NO via the exo conformation only to give neutral $CpMo(CO)(NO)(\eta^2$-olefin) complexes.[87] Allyl complexes of structure (36) may also be converted to $[CpMo(CO)(NO)(\eta^3$-allyl)$]^+$ cations; reaction with nucleophiles occurs cis to NO, but via the endo conformation.[88] Protonation of $CpMo(CO)_2(\eta^3$-allyl) with HBF_4 generates $CpMo(CO)_2(\eta^2$-propene)FBF_3 from which both olefin and anion can be displaced by PPh_3 to give $[CpMo(CO)_2(PPh_3)_2]^+$.[89] Reaction of $Cp_2Mo_2(CO)_4$ with 2-substituted azirines gives $CpMo(CO)_2$-(η^3-HN-CR-CH_2) (R=Ph, p-tolyl) which rearrange in solution via a 1,3-hydrogen shift to give the $Cp(CO)_2Mo\overset{+}{=}N=C(R)Me$ isomer.[90]

Dissolution of $MoCl(\eta^3$-allyl)$(CO)_2(NCMe)_2$ in methanol or water generates $[Mo(\eta^3$-allyl)$(CO)_2(MeOH)_3]Cl$ and $[Mo(\eta^3$-allyl)$(CO)_2(H_2O)_x]Cl$ cations respectively; water may be displaced by bidentate ligands to generate complexes such as $MoCl(\eta^3$-allyl)$(CO)_2$(bipy).[91] A crystal structure determination of the related $MoBr(\eta^3$-allyl)$(CO)_2$(N,N'-dicyclohexylethamediimine) has been reported.[92]

Addition of $C_3Ph_3BF_4$ to $Mo(CO)_3(MeCN)_3$ yields $[Mo(\eta^3-C_3Ph_3)(CO)_2(MeCN)_3]BF_4$ which on treatment with TlCp gives $CpMo(CO)_2(\eta^3-C_3Ph_3)$. Thermal reaction of the latter with PMe_2Ph results in ring expansion to give (37).[93] Reaction of $[Cp*M(CO)_3]^-$ (M=Cr, Mo, W) with $C_3Ph_3^+$ yields directly the oxocyclobutenyl complex (38).[94] Reaction of $[Cp*Mo(CO)_3]^+$ with sodium 2,3-diphenyl-2-cyclopropene-1-carboxylate yields (39), containing a chelated cyclopropene.[95] $Cp(CO)_2W(\equiv CC_6H_4Me-4)$ reacts with $CrPh_3(thf)_3$ and ZnR_2 (R=Ph, Pr^i, $CH_2C_6H_4Me$-4) to give η^3-benzyl complexes of structure (40). With $ZnEt_2$, $Zn\{CpW(CO)_2$-(η^2-MeCH=CHC_6H_4Me-4)$\}_2$ is formed initially, but undergoes hydrolysis to (40) (R=Et).[96]

5. Other Metals

Cyclic and acyclic $[CpMn(CO)_2(\eta^3$-allyl)$]^+$ complexes may be prepared by protonation of the free double bond of $CpMn(CO)_2(\eta^2$-cyclohexadiene) or protonation of $CpMn(CO)_2(\eta^2$-allyl-alcohol) complexes; addition of PR_3 occurs at the allyl ligand to

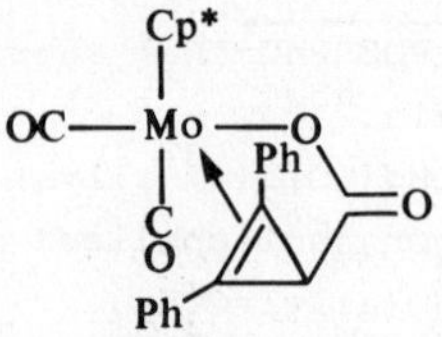

(39)

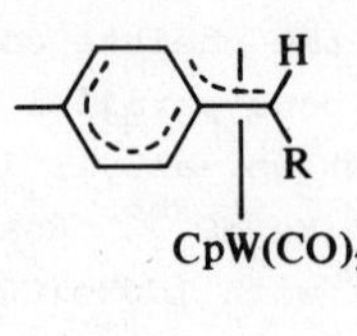

(40)

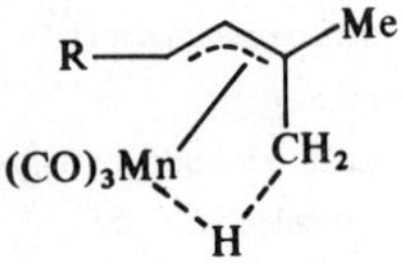

(41) R = H
(43) R = Me

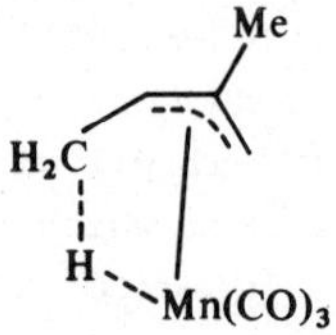

(42)

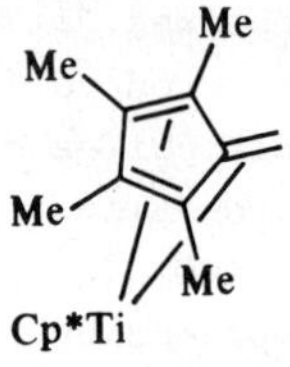

(44)

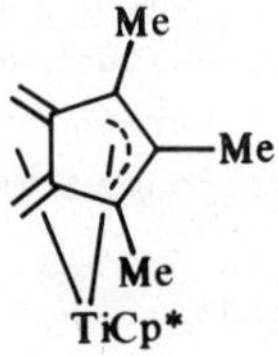

(45)

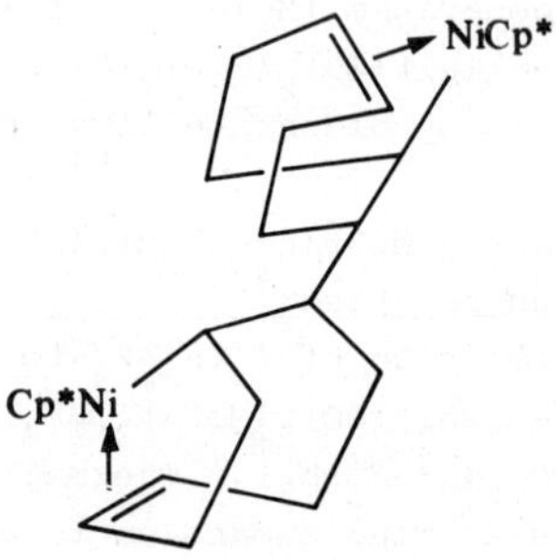

(46)

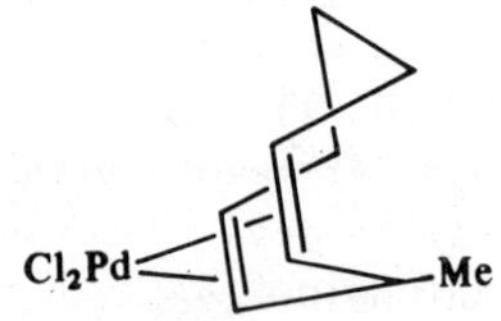

(47)

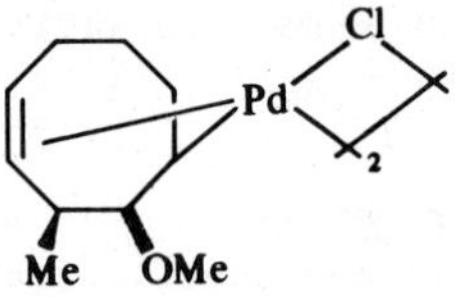

(48)

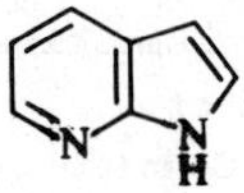

(49)

generate phosphonium salts.[97] Matrix photolysis of
$Mn(CO)_5(\eta^1\text{-allyl})$ yields directly $Mn(CO)_4(\eta^3\text{-allyl})$; with
$CpW(CO)_3(\eta^1\text{-allyl})$, the 16e$^-$ intermediate $CpW(CO)_2(\eta^1\text{-allyl})$ may
be detected before closure to the η^3-complex.[98] Thermolysis of
$Mn(CO)_4(\eta^3\text{-1,1-dimethylallyl})$ yields (41) and (42), containing
agostic M-H-C interactions; both equilibration of (41) and (42)
$\Delta G^* = 72$ kJ mol^{-1}) and exchange of the three protons of the
bridging methyl ($\Delta G^* = 37$ kJ mol^{-1}) can be observed. Deprotonation
yields $[(\eta^4\text{-isoprene})Mn(CO)_3)]^-$ which on methylation results in
C-C bond formation to give a mixture of isomers such as (43) in
which the agostic interaction is regenerated.[99] (η^3-Pentadienyl)-
$Mn(dmpe)_2$ has been prepared, and undergoes elimination of
1,3-pentadiene on reaction with $NH_4PF_6/P(OMe)_3$ to give
$\{\underline{cis}\text{-}Mn(dmpe)_2[P(OMe)_3]_2\}PF_6$.[100]

Propene reacts thermally with $(PPh_3)_2ReH_7$ to give
$(PPh_3)_2ReH_3(\eta^4\text{-}CH_2=CHMeCH=CHMe)$, though photolysis yields
$(PPh_3)_2ReH(\eta^3\text{-allyl})_2$. The diene complex may also be prepared
from $(PPh_3)_2ReH_7$ and the free diene.[101] A similar reaction with
2,3-dimethylbutadiene gives $(PPh_3)_2ReH_3(\eta^4\text{-2,3-dimethylbutadiene})$,
though use of the chelated derivative $(dppe)ReH_7$ yields instead
$(dppe)ReH_4(\eta^3\text{-2-isopropylallyl})$ due to steric constraints imposed
by the P_2 chelate.[102]

Cp_2V reacts with activated olefins such as acrylonitrile to
give $Cp_2V(\eta^2\text{-olefin})$ adducts which are best regarded as
metallocycles.[103] Photolysis of $V(CO)_6^-$ in the presence of olefins
or acetylenes gives $[V(CO)_5(\eta^2\text{-olefin/acetylene})]^-$ complexes which
decompose above -50 °C.[104] Thermolysis of $Cp_2^*TiCH_3$ results in
sequential loss of CH_4 and H_2 to give (44) and (45);[105]
similar compounds are formed on sequential photolytic loss of H_2
from $Cp_2^*WH_2$.[106] Calculations show that conversion of the
metallocycle $Cp_2\overline{TiCH_2CH_2CH_2}$ to $Cp_2Ti(CH_2)(\eta^2\text{-}C_2H_4)$ has no
activation barrier, with the metallocycle more stable by 50 kJ
mol^{-1}.[106] $Cp_2^*Ti(\eta^2\text{-}C_2H_4)$ undergoes coupling on reaction with
acetylenes to give the metallocyclopentenes $Cp_2^*\overline{TiC(R)=CMeCH_2CH_2}$
(R=Me, H); similar coupling occurs on reaction with nitriles,
acetaldehyde and CO_2.[107] $(Cp_2^*MH)_2$ (M=La, Nd) react with propene
to give propane and $Cp_2^*M(\eta^3\text{-allyl})$; reaction with butadiene
gives $Cp_2^*La(\eta^3\text{-syn-1-methylallyl})$.[108]

A variety of $[L_2Cu(\eta^2\text{-olefin}]ClO_4$ complexes have been
prepared (L_2=TMED, bipy, phen, di-2-pyridylamine; olefin=C_2H_4,
propene, cyclohexene).[109,110] A crystal structure of

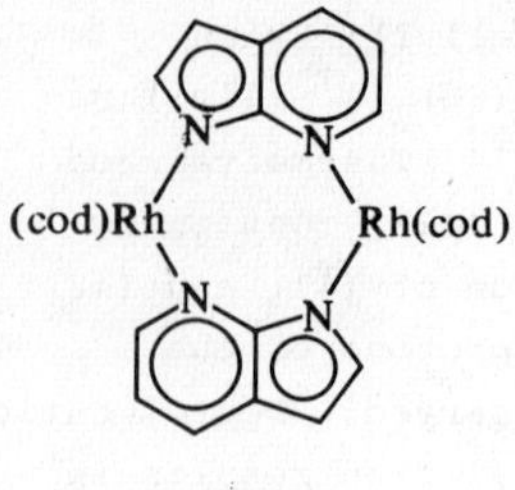

(50)

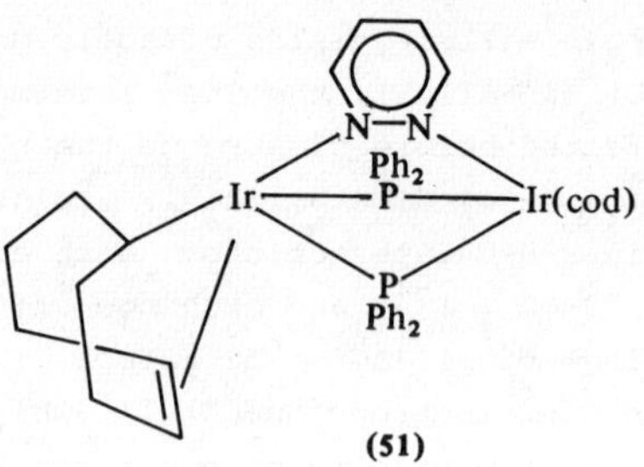

(51)

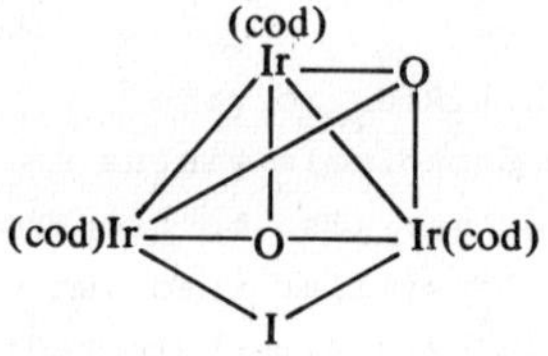

(52)

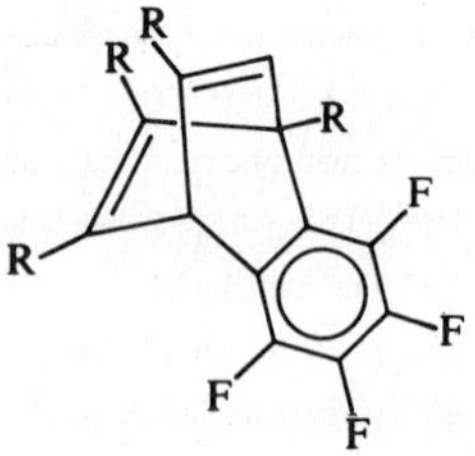

(53) a; R = H
b; R = Me

Ru(PMe$_2$Ph)$_3$

(54)

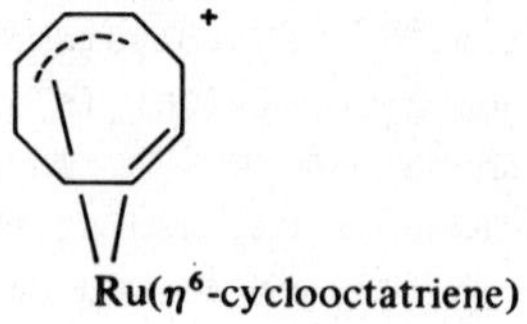

Ru(η^6-cyclooctatriene)

(55)

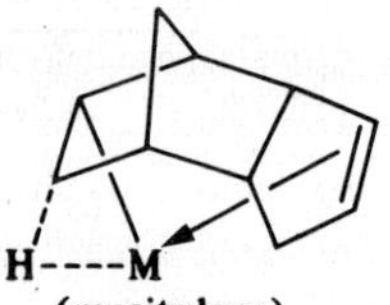

(mesitylene)

(56) M = Ru, Os

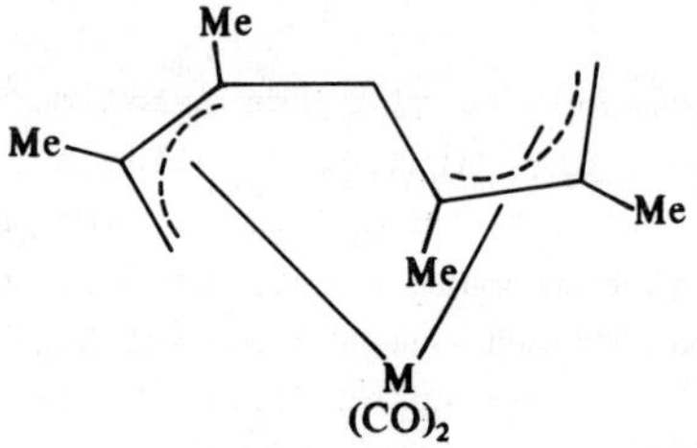

(57)

CuOSO$_2$CF$_3$(η^2-cyclohexene) reveals a basic tetrameric unit bridged by triflate.[111]

C. Complexes Derived from Unconjugated Dienes

1. Ni, Pd and Pt

Reaction of Ni(cod)$_2$ with Cp*$_2$Zn results in coupling to give (46).[112] Crystal structures of (η^4-diene)Ni(C$_6$F$_5$)$_2$ (diene = nbd, cod) reveal square planar coordination.[113] A new low temperature diffraction study of (cod)PtCl$_2$ indicates a symmetrical bonding of the C=C bonds.[114] Nbd is more easily displaced than cod from the (diene)PtMe$_2$ complex by nitrogen donors to give cis-L$_2$PtMe$_2$ derivatives.[115] (3-methyl-1,4-cyclooctadiene)-PdCl$_2$ adopts the boat-chair conformation (47) in the solid state, unlike the unsubstituted diene which is boat-boat; (47) rearranges thermally to (eq-3-methyl-1,5-cod)PdCl$_2$ which thermally equilibrates with its axial isomer. Reaction of (47) with OMe$^-$ yields (48).[116]

2. Co, Rh and Ir

Reaction of [Rh(diene)$_2$]$^+$ or [RhCl(diene)]$_2$ (diene = nbd, cod and others) with (49) yields monodentate [Rh(diene)L$_2$]$^+$ or RhCl(diene)L complexes; CO easily displaces the diene, and treatment of the chloro-complex with KOH yields the bridged derivative (50), characterized crystallographically.[117] Analogous Ir complexes may also be prepared, and the related complex (cod)Ir(μ-pz)$_2$Ir(cod) (Hpz=pyrazole) undergoes addition of Ph$_2$PH with hydrogenation of cod and elimination of Hpz to give (51).[118] Reaction of [(cod)IrCl]$_2$ with PPh$_3$/SnCl$_2$ yields Ir(SnCl$_3$)(cod)-(PPh$_3$)$_2$; in the presence of H$_2$ or HCl, IrH$_2$(SnCl$_3$)(cod)(PPh$_3$) and IrH(SnCl$_3$)$_2$(cod)(PPh$_3$) are obtained respectively.[119] Reaction of IrH$_5$(PPh$_3$)$_2$ with [M(cod)Cl]$_2$ (M = Rh, Ir) yields (cod)M(μ-H)(μ-Cl)IrH$_2$(PPh$_3$)$_2$, characterized crystallographically.[120] Treatment of [(cod)Ir(OMe)]$_2$ with I$_2$ yields the Ir(III) dimer (cod)$_2$Ir$_2$I$_2$(μ-I)$_2$ which on treatment with AgOAc gives the bioxo-capped cluster (52).[121]

Crystal structures of several [(η^6-arene)Rh(diene)]$^+$ cations [diene = (53a,b)] show that the C=C bonds generally eclipse the unsubstituted arene carbons.[122]

3. Other Metals

Reaction of WBr$_2$(CO)$_4$ with nbd yields 16e$^-$ WBr(CO)$_2$(nbd), containing trans-carbonyls.[123] In (nbd)$_2$Mo(CO)$_2$, the two nbd ligands are bound at 90° with respect to one another, consistent with the geometry of the dimer isolated on thermolysis.[124]

7-Butoxynbd also yields a (diene)$_2$Mo(CO)$_2$ complex as two isomers; one is isostructural with the above, but the second contains one nbd bound in an <u>exo</u> fashion <u>via</u> one C=C bond and the oxygen.^{125}

[OsCl$_2$(cod)]$_x$ reacts with hydrazines to give [OsCl(cod)-(NH$_2$NR$_2$)$_3$]$^+$ (R = H,Me) and [Os(cod)(NH$_2$NH$_2$)$_4$]$^{2+}$. Treatment of the former (R=H) with donor ligands in acetone yields the hyrazone complexes <u>trans</u>-Os(NH$_2$N=CMe$_2$)$_2$L$_4$ [L=2,6-xylylNC, P(OMe)$_3$, P(OMe)$_2$Ph], though with Bu^{t}NC, the complex Ox(cod)(CNBu^t)$_2$-(NH$_2$N=CMe$_2$)$_2$ may be isolated.^{126}

An agostic M-H-C interaction with the <u>endo</u>-hydrogen adjacent to the allyl group is present in [(η^3-cyclooctenyl)RuL$_3$]$^+$ (L=P(OMe)Ph$_2$), obtained from thermal isomerization of [(cod)RuHL$_3$]$^+$.^{127} Reaction of either [(cod)RuHL$_3$]$^+$ or [(η^3-cyclooctenyl)RuL$_3$]$^+$ [L=PMe$_3$, PMe$_2$Ph, P(OMe)$_2$Ph, P(OMe)Ph$_2$] with cyclic or acyclic dienes results in elimination of 1,3-cyclooctadiene and formation of analogous [(η^3-enyl)RuL$_3$]$^+$ complexes which, in the case of the cyclic complexes, show 1,2-metal migration around the ring.^{128,129} Reaction of [(cod)RuH(PMe$_2$Ph)$_3$]$^+$ with cot yields the bicyclic 18e$^-$ cation (54).^{130} Treatment of [(cod)RuH(NH$_2$NMe$_2$)$_3$]$^+$ with excess cot yields (55); isocyanides displace the η^6-cyclooctatriene ligand, but reaction with phosphines or phosphites results in isomerization to give [(η^5-cyclooctadienyl)RuL$_3$]$^+$ [L=PMe$_2$Ph, P(OMe)$_3$].^{131}

Protonation of (cod)Ru(η^6-cyclooctatriene) in the presence of arenes gives [(η^6-arene)Ru(η^5-cyclooctadienyl)]$^+$ salts,^{132} while protonation of (η^6-mesitylene)Os(η^4-<u>endo</u>-dicyclopentadiene) (M=Ru,Os) yields the 16e$^-$ cation (56) containing an agostic interaction.^{133} Four fluxional processes have been characterized for (cod)Os(η^6-cot); metal migration takes place by both η^4-1,2,5,6- and η^4-1-4 intermediates, the latter being of higher energy.^{134}

Hydride abstraction from (cod)Ru(η^6-cyclooctatriene) yields [(η^6-cyclooctatriene)Ru(1-3,5,6-η^5-cyclooctadienyl)]$^+$, from which cyclooctatriene may be displaced by CO or P(OMe)$_3$.^{135} Hydrogenation of (cod)Ru(η^6-cyclooctatriene) in the presence of PCy$_3$ gives RuH$_6$(PCy$_3$)$_2$.^{136} Cod may easily be displaced from CpRu(cod)Cl to give CpRuL$_2$Cl (L$_2$ = butadiene, dppe, 2L=RNC); cot yields the non-fluxional [CpRu(η^6-cot)]$^+$ cation.^{137}

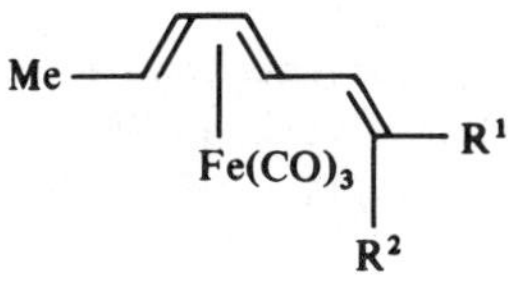

(58) R^1 = H; R^2 = CN, COMe, CO_2Et
R^1 = R^2 = CN, CO_2Et

(59)

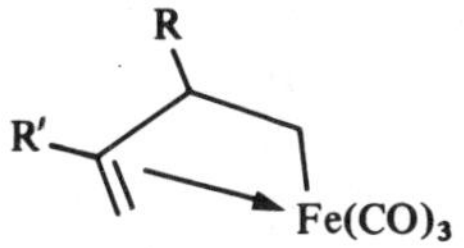

(60) R = $CHPh_2$
R' = Me, Et, OMe

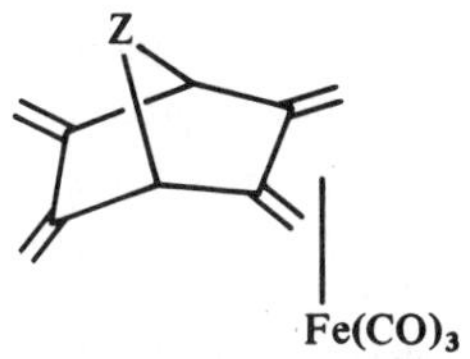

(61) *endo*
a; Z = CH=CH   d; Z = O
b; Z = CH_2CH_2 e; Z = CO
c; Z = CH_2 f; Z = $C(O)CH_2$

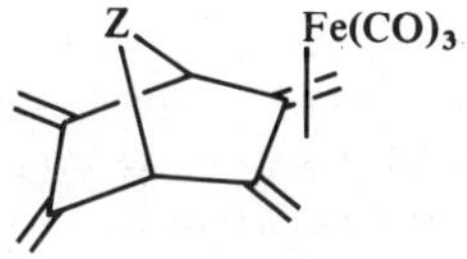

(62) *exo*

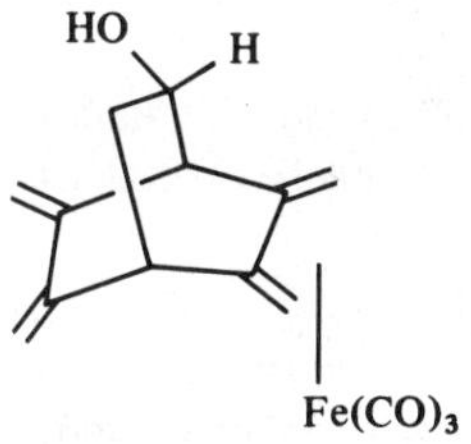

(63)

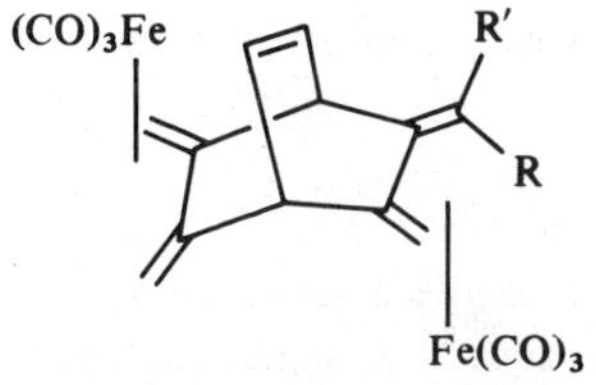

(64) R = R' = H
(65) R = H; R' = COMe
(66) R = COMe; R' = H

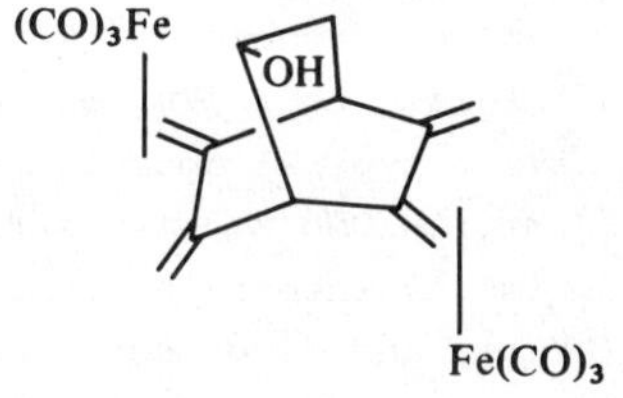

(67)

D. Complexes Derived from Conjugated Dienes
1. Fe, Ru and Os
   i. Acyclic Dienes

Spectral data on a variety of (η^4-benzylideneacetone)Fe(CO)$_2$L complexes [L=PR$_3$, P(OR)$_3$] are consistent with a strong π-acceptor character for the heterodiene ligand.[138]

Thermolysis of (2,3-dimethylbutadiene)$_2$M(CO) (M=Fe, Ru) in the presence of CO results in coupling to give (57).[139] Reaction of allylic halides or phosphates [e.g. Me$_2$C=CHCH$_2$OP(O)(OEt)$_2$] provides a general, though poorly yielding, route to (diene)Fe(CO)$_3$ complexes [e.g. (isoprene)Fe(CO)$_3$].[140] Base-catalysed reaction of active hydrogen compounds R^{1}R^{2}CH$_2$ with (sorbaldehyde)Fe(CO)$_3$ yields complexes of structure (58), together, where R^1=H, with the isomer (59) formed by base-catalysed metal migration. The chain may be extended by reduction of (59; R^2=CN) to the aldehyde, followed by further reaction with carbanion.[141] Incorporation of deuterium into the liberated olefin on reaction of (60) with D$^+$ is consistent with scrambling through a ferracyclobutane intermediate.[142]

Photoelectron spectra of the exo and endo isomers (61) and (62) show substantial differences.[143] Reduction of the endo complex (61f) occurs under electronic control to give mainly (63);[144] hydroboration of (64) occurs stereospecifically to give (67).[145]

Oxidation of (64) or (66) results in selective loss of the endo Fe(CO)$_3$ moiety, whereas the exo Fe(CO)$_3$ group is lost from (65).[146] The delocalized quinodimethane cations (68) and (69) have been prepared; nucleophile addition (OH$^-$, OMe$^-$) to (68) occurs at C-1.[147]

   ii. Cyclic Dienes

(Trimethylenemethane)Fe(PMe$_3$)$_3$ has been prepared, and may easily be oxidized to a 17e$^-$ cation which a crystal structure shows to have a typical umbrella shaped ligand.[148] Reaction of CH$_2$=C(CH$_2$SiMe$_3$)CH$_2$OSO$_2$Me with metal complexes provides a general route to trimethylenemethane complexes; thus, Os(CO)$_3$(PPh$_3$)$_2$, OsCl(NO)(PPh$_3$)$_3$ and [(cyclooctene)$_2$IrCl]$_2$ yield (C$_4$H$_6$)Os(CO)$_2$PPh$_3$, (C$_4$H$_6$)OsCl(PPh$_3$)(NO) and (C$_4$H$_6$)IrCl(PPh$_3$)$_2$ respectively.[149] Photoelectron studies on (70a-c) show a strong back-donation which confers some aromatic character on the ring.[150]

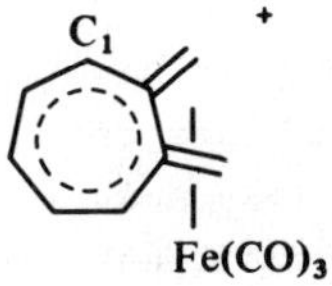

(68)

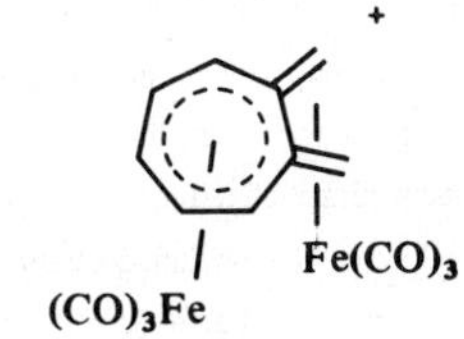

(69)

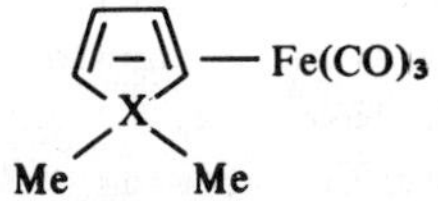

(70) a; X = C
b; X = Si
c; X = Ge

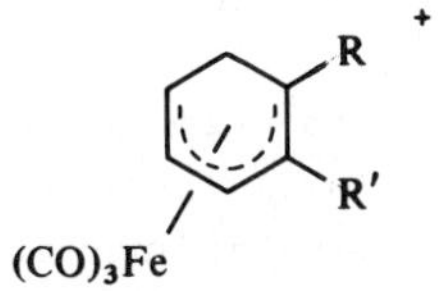

(71) R = H; R' = Me
(76) R = CO$_2$Me; R' = H

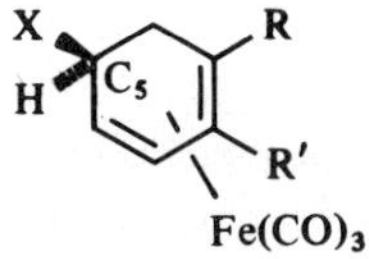

(72) R = H; R' = Me; X = SO$_2$Ar
(75) R = H; R' = Me; X = D
(77) R = H; R' = OMe; X = $\overset{+}{P}$Ph$_3$
(79) R = R' = H; X = $\overset{+}{P}$Ph$_2$(neomenthyl)

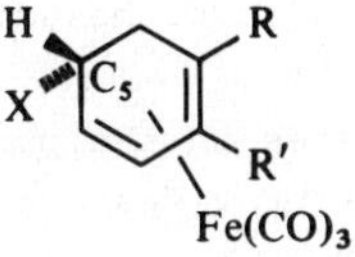

(73) R = H; R' = Me; X = SO$_2$Ar
(74) R = H; R' = Me; X = D
(78) R = H; R' = OMe; X = $\overset{+}{P}$Ph$_3$
(80) R = R' = H; X = D

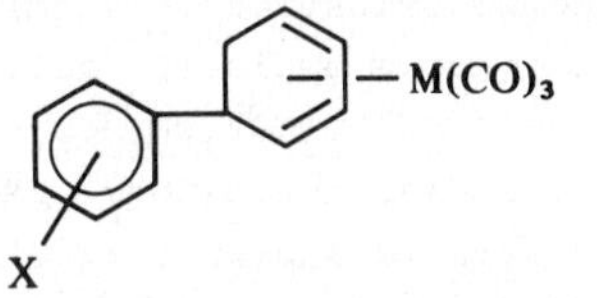

(82) M = Ru; X = Me, OMe, NMe$_2$

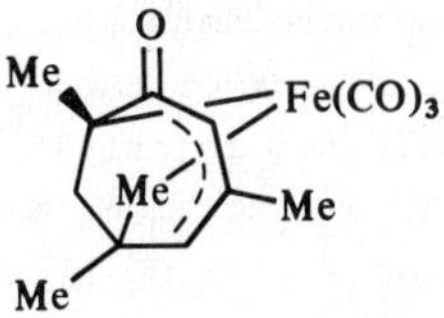

(81)

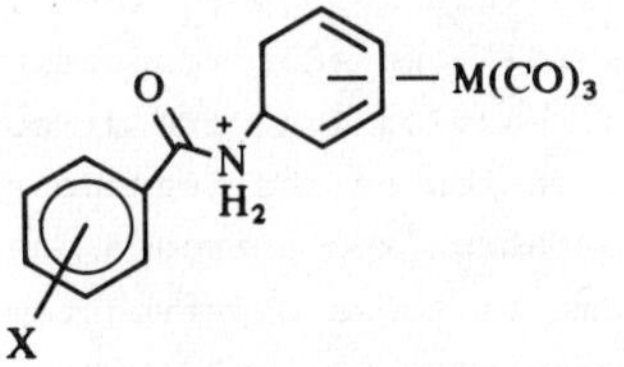

(83) X = OMe, Cl

Principles governing the use of $(chd)Fe(CO)_3$ and $[(cyclohexadienyl)Fe(CO)_3]^+$ complexes in organic synthesis have been reviewed,[151] and specific application to the syntheses of spiro[4,5]decanes,[152] tetrahydrocarbazolones,[153] and (+)- and (−)-gabaculine [using fully resolved (1-carboxycyclohexadiene)-$Fe(CO)_3$][154] have been reported. The course of protonation and demethoxylation of (1- and (2-methoxycyclohexadiene)$Fe(CO)_3$ complexes has been investigated in detail,[155] and may be used to prepare isomerically pure $[(2,4-dimethylcyclohexadienyl)Fe(CO)_3]^+$ from an isomeric mixture of $(chd)Fe(CO)_3$ complexes.[156] Reaction of (71) with $ArSO_2^-$ yields the <u>exo</u>-sulphinate (72) which may be converted to the <u>endo</u> complex (73) by deprotonation/protonation at low temperature. Reaction with BD_4^- results in S_N2 displacement of sulphinate to give (74) and (75) respectively; use of a resolved salt such as (+)-(76), followed by decomplexation, produces enantiomers differing only in the CHD configuration at C-5.[157] Both <u>exo</u> and <u>endo</u> isomers (77) and (78) may be similarly prepared, and the resolved diastereoisomer (79) may be converted into enantiomerically pure (80) by reaction with $LiAlD_4$.[158] Reaction of $(chd)Fe(CO)_3$ complexes with $AlCl_3/CO$ occurs with high stereospecificity; thus, (1,3,5,5-tetramethylcyclohexadiene)$Fe(CO)_3$ yields exclusively the ketone (81).[159]

Me_3NO oxidation of $(chd)Fe(CO)_3$ complexes in the presence of PR_3 (including optically active phosphines) provides a general route to $(chd)Fe(CO)_2PR_3$ complexes.[160] [(Cyclohexa- and [(cycloheptadienyl)$M(CO)_2(CNMe)]^+$ complexes (M=Fe, Ru) adopt a square pyramidal structure with CNMe in a basal position.[161]

Calculations suggest that particularly for hard nucleophiles, the metal atom or CO is the favored site for nucleophilic attack in $[(cyclohexadienyl)Fe(CO)_3]^+$, with metal attack favored in solvents of lower polarity.[162] Reaction of [(cyclohexadienyl)-$M(CO)_3]^+$ with $XC_6H_4EMe_3$ (E=Si,Sn; M=Ru) and $XC_6H_4CONH_2$ (M=Fe) yields (82) and (83) respectively; kinetic measurements for the former reaction indicate direct electrophilic attack of the dienyl cation on the benzene ring.[163,164] Relative reactivities of phosphorus and nitrogen donors towards free carbocations are the same as those observed towards $[(cyclohexadienyl)M(CO)_3]^+$.[165]

Treatment of $[CpRu(NCMe)_2CO]^+$ with chd or acyclic dienes yields $[CpRu(\eta^4-diene)CO]^+$ salts; hydride reduction occurs in an exo fashion to give $CpRu(CO)(\eta^3-allyl)$ complexes. Reaction with cot yields $[CpRu(CO)(\eta^4-cot)]^+$ which on photolysis gives

[CpRu(η^6-cot)]$^+$.^{166} Nucleophilic attack by MeCOCH$_2^-$ on [(XC$_6$H$_5$)FeCp]$^+$ where X is electron withdrawing (NO$_2$, Cl, COPh, CN) occurs in an exo fashion and regiospecifically ortho to X.^{167} Attack on [(η^6-chpt)FeCp]$^+$ by PR$_3$, H$^-$ or Me$^-$ is also exo, yielding rather unstable CpFe(cycloheptadienyl) complexes; the activating effect of CpFe$^+$ is much less than Mn(CO)$_2$L$^+$ (L=CO, PPh$_3$).^{168,169} Attack on [(arene)Fe(cyclohexadienyl)]$^+$ occurs at the dienyl terminus to give, after decomplexation, 5-substituted 1,3-cyclohexadienes.^{170}

(η^2-methylacrylate)$_2$Fe(CO)$_3$ couples thermally with chpt to give (84), which on irradiation in the presence of CO rearranges to (85); adducts similar to (84) are obtained with cot and 1,3,5-cyclooctatriene. Chd and 1,3-cycloheptadiene yield (86) and (87) respectively.^{171} Reduction of (cycloheptadienyl)Fe(CO)$_2$I yields [(cycloheptadienyl)Fe(CO)$_2$]$^-$ which on treatment with MeCOCl/CO gives the endo complex (88); prior treatment with CO, followed by addition of MeCOCl yields the exo-complex (89).^{172} A facile synthesis of (η^4-chpt)Fe(CO)$_2$P(OPh)$_3$ and its protonation to [(cycloheptadienyl)Fe(CO)$_2$P(OPh)$_3$]$^+$ have been reported.^{173}

Kinetic studies of the 3+2 cycloaddition of TCNE to (η^4-chpt)Fe(CO)$_3$ are consistent with a concerted mechanism.^{174} (Carbomethoxy)maleic anhydride undergoes concerted 4+2 addition to give (90).^{175} Reinvestigation of the reaction of TCNE with (η^4-tropone)Fe(CO)$_3$ shows initial concerted formation of 3+2 and 4+2 adducts in the ratio 96:4; the 3+2 adduct undergoes rearrangement to (91), the product of formal 5+2 addition.^{176} (η^4-tropone)Fe(CO)$_3$ has been resolved into its enantiomers by HPLC; racemization by 1,3-shift occurs ($\Delta G^* = 107$ kJ mol^{-1}).^{177}

Both solution and matrix photolysis of (η^4-cot)Fe(CO)$_3$ result in competitive CO and polyene dissociation.^{178} Ring whizzing in (η^4-cot)Fe(CO)$_2$(CNPr^i) occurs by a [1,5]-suprafacial sigmatropic shift, consistent with Woodward-Hoffmann rules.^{179} C$_3$Ph$_3^+$ undergoes cycloaddition to (η^4-cot)M(CO)$_3$ (M=Fe, Ru) to give (92); hydride addition occurs at C-8.^{180} Complex (93), obtained by oxidative dimerization of (η^4-cot)Fe(CO)$_3$, reacts with BH$_4^-$ to give (94), but is reduced by [BH(CHMeEt)$_3$]$^-$ to give (95). Oxidation of (94) results in regioselective ring opening to give (96), whereas oxidation of (95) gives (97).^{181,182}

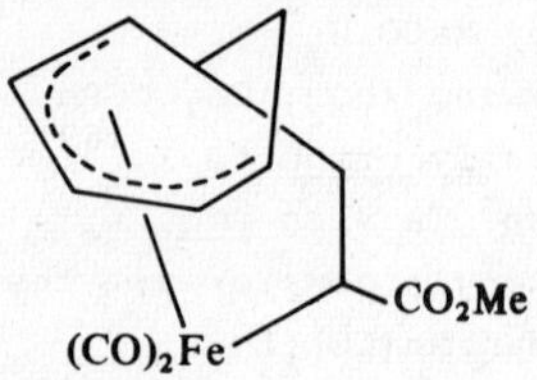

(84)

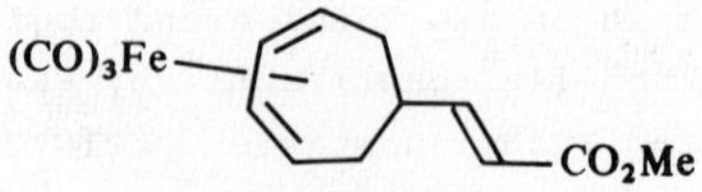

(85)

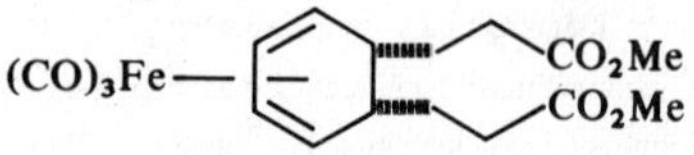

(86)

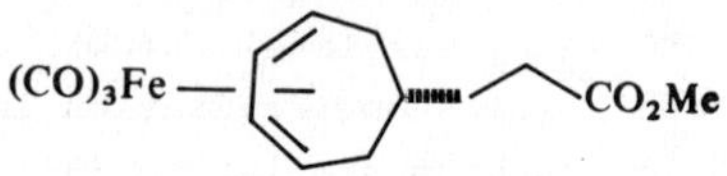

(87)

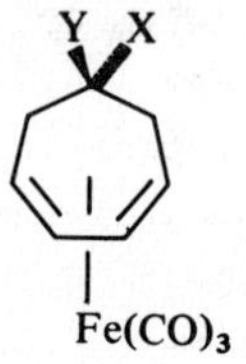

(88) X = COMe; Y = H
(89) X = H; Y = COMe

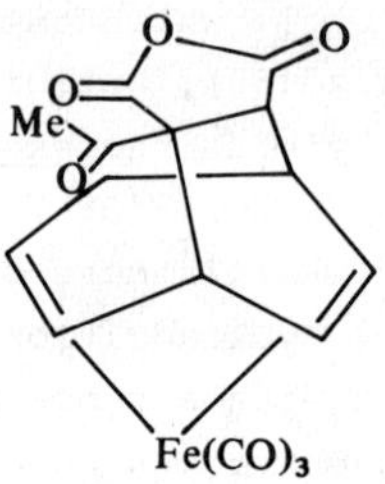

(90)

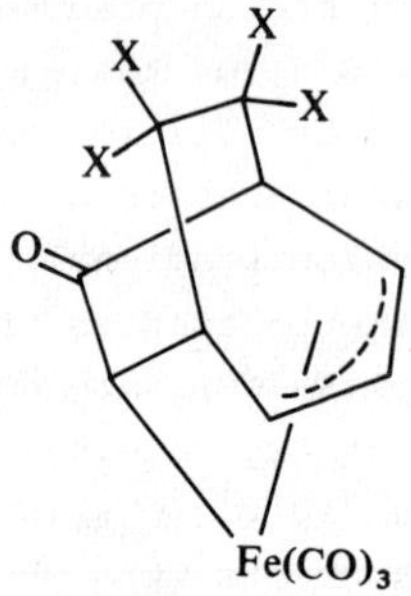

(91). X = CN

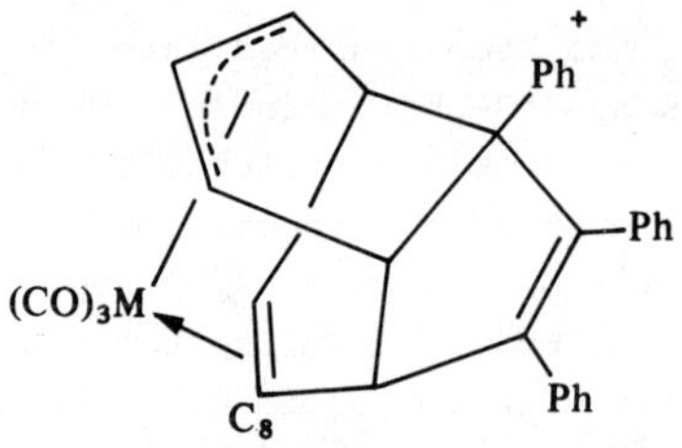

(92)

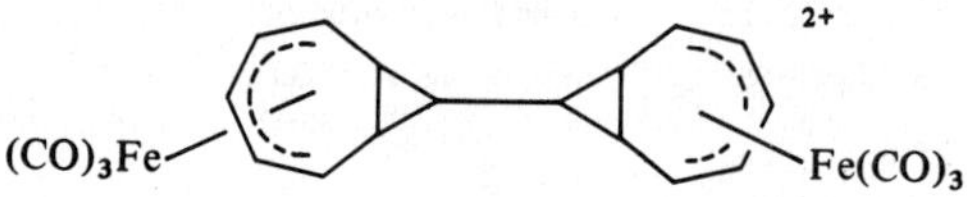

(93)

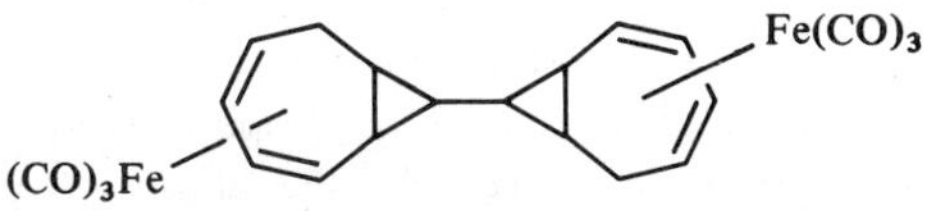

(94)

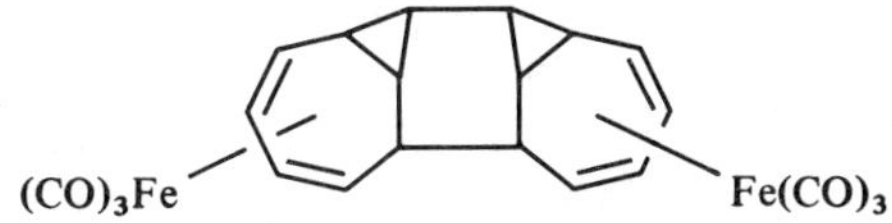

(95)

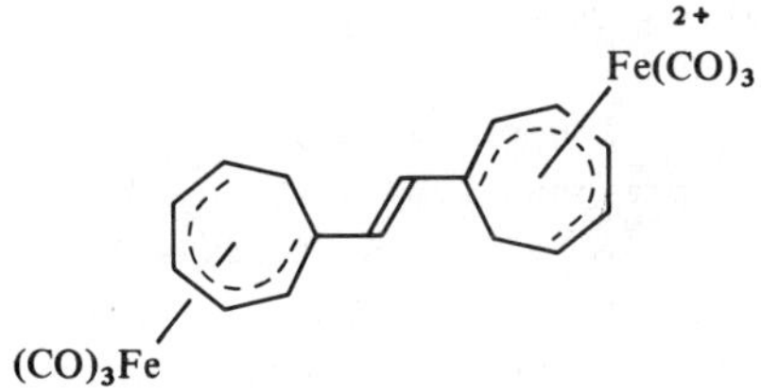

(96) (CO)$_3$Fe ... Fe(CO)$_3$ $^{2+}$

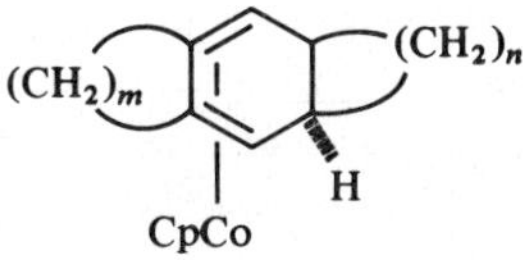

(98) $m = 4$; $n = 3,4$

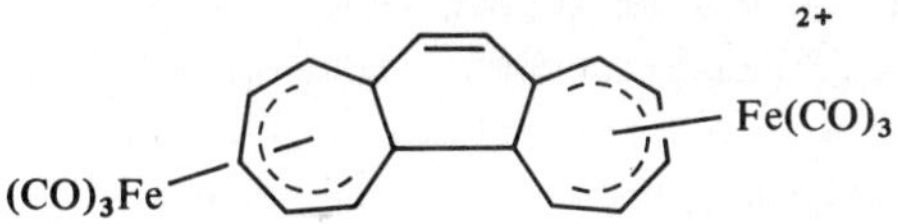

(97) (CO)$_3$Fe ... Fe(CO)$_3$ $^{2+}$

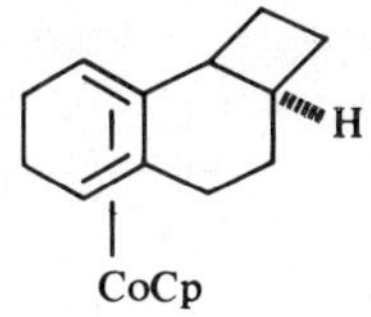

(99)

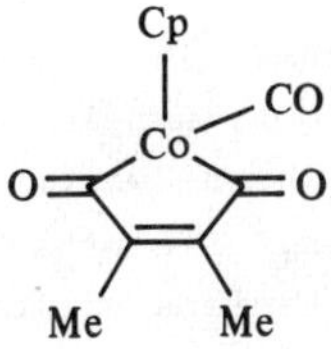

(100)

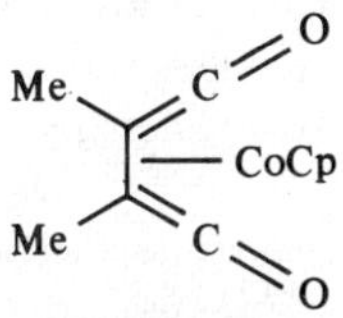

(101)

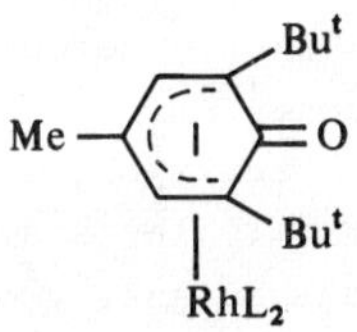

(102) L_2 = cod
L = PPh$_3$, C$_2$H$_4$

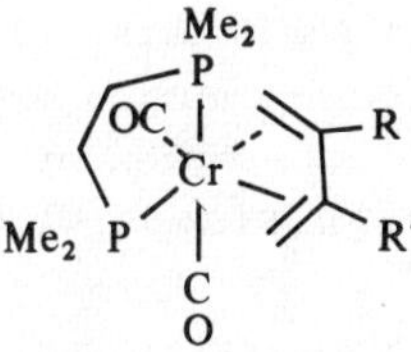

(103)

2. Co, Rh and Ir

Reaction of $CoBr(PMe_3)_3$ with acyclic dienes (butadiene, isoprene) yields square pyramidal $[(\eta^4\text{-diene})Co(PMe_3)_3]^+$ cations.[183] Details of the application in organic synthesis of the $CpCo(CO)_2$ catalysed [2+2+2] cycloaddition of α,δ,ω-enediynes to give complexes of structure (98) have appeared.[184-186] Attempts to prepare the (n=2, m=4) complex result in double bond migration to give (99).[187] Insertion of $CpCo(CO)_2$ into dimethylcyclobutenedione gives (100). Photolysis in the presence of acetylenes yields $CpCo(\eta^4\text{-benzoquinone})$ complexes from which the ligand may be easily liberated; in the absence of acetylenes, photolysis gives the ketene complex (101).[188,189] Flash vacuum pyrolysis of $(\eta^4\text{-thiophene-1,1-dioxide})CoCp$ complexes gives the (cbd)CoCp derivative via SO_2 extrusion.[190]

Reaction of $[IrH_2(PPh_3)_2(Me_2CO)_2]^+$ with chd gives $[(\text{cyclohexadienyl})IrH(PPh_3)_2]^+$ which undergoes loss of H_2 on thermolysis to give $[(C_6H_6)Ir(PPh_3)_2]^+$.[191] Reaction of several chloro-rhodium complexes with $2,6\text{-Bu}^t_2\text{-4-MeC}_6H_2O^-$ yields complexes of structure (102) in which the ligand is bound in an oxocyclohexadienyl fashion.[192]

3. Cr, Mo and W

Treatment of $W(PMe_3)_5H_2$ with chd yields $(\text{chd})W(PMe_3)_3H_2$, a complex which is also formed on cocondensation of W atoms with PMe_3/cyclohexene or PMe_3/cyclohexane. Cocondensation of W atoms with chd or cyclohexene yields $(\text{chd})W(\text{benzene})H_2$.[193,194] Arene may be displaced from $(\text{toluene})_2M$ (M = Mo, W) or $(\text{toluene})_2Ti$ by dimethyl- or diphenylfulvene to give (toluene)M(fulvene) and $Ti(\text{fulvene})_2$ complexes; crystal structures show strong bending of the exocyclic carbon towards the metal.[195] $(\text{Fulvene})Mo(CO)_3$ complexes undergo nucleophilic attack by PR_3 at the exocyclic carbon.[196]

Matrix photolysis of $(\text{butadiene})Cr(CO)_4$ and derivatives results primarily in CO dissociation.[197] Photolysis of $Cr(CO)_4(\text{dmpe})$ with acyclic dienes yields complexes of structure (103) which, where $R \neq R^1$, exist as interconverting isomers in solution.[198] Photolysis of $Cr(CO)_5L$ [L=CO, $P(OMe)_3$, PMe_3] with 1,3-cycloheptadiene yields (104), in which the agostic interaction exchanges between C-5 and C-7 ($\Delta G^* \approx 28$ kJ mol^{-1}).[199] A complex of similar structure is formed on photolysis of $Cr(CO)_5P(OMe)_3$ with 2,4-dimethyl-1,3--pentadiene.[200]

Though the effect is small, rotational barriers ($\sim$40 kJ mol^{-1})

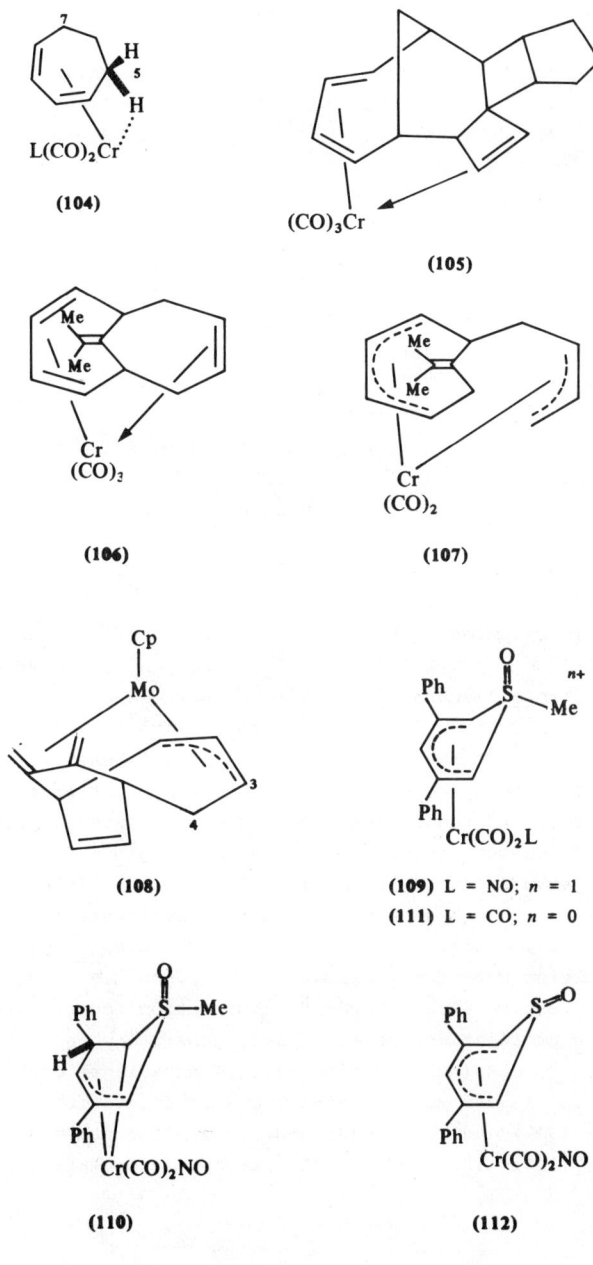

in (η^6-chpt)Cr(CO)$_3$ decrease with electron withdrawing 7-exo substituents and increase with sterically demanding 7-exo substituents.[201] Thermochemical measurements show an order of stability for the Mo(CO)$_3$ complex of chpt > hexamethylbenzene > mesitylene > cot > toluene > benzene.[202] Photolysis of (chpt)Cr(CO)$_3$ with tricyclo[6.3.0.0^{2,7}]undeca-3,5-diene results in 4+6 addition to give (105).[203] Depending on diene substituent, acyclic dienes undergo photolytic cycloaddition to (6,6-dimethylheptafulvene)Cr(CO)$_3$ to give either (106) or (107); treatment of (107) with CO results in ring closure to give (106).[204]

Attack of OMe$^-$ on [(C$_7$H$_7$)M(CO)$_3$]$^+$ (M=Mo, W) occurs initially at the metal, in agreement with calculation; dissociative rearrangement yields (C$_7$H$_7$)M(CO)$_2$CO$_2$Me and eventually (7-exo--methoxychpt)M(CO)$_3$.[205,206] Reaction between [(C$_7$H$_7$)(CO)$_3$]$^+$ and aniline and PPh$_3$ occurs via initial rapid formation of a π-complex; with aniline, the product of exo-ring addition is isolated, whereas PPh$_3$ displaces C$_7$H$_7^+$ to give W(CO)$_3$(PPh$_3$)$_3$.[207,208] A crystal structure of [(C$_7$H$_6$OMe)Cr(CO)$_3$]$^+$ shows a C-OMe bond length intermediate between single and double.[209]

Regio- and stereospecific nucleophilic addition to [(indenyl)Mo(CO)$_2$(chd)]$^+$ complexes, followed by regiospecific and stereospecific reaction with Ph$_3$C$^+$ may be used to convert 1-methoxychd into 1-alkyl-1,3-cyclohexadienes and to convert 1-substituted into 2-substituted 1,3-cyclohexadienes. The diene may be liberated from the salt by treatment with CO.[210] Iodination of the intermediate CpMo(CO)$_2$(η^3-cyclohexenyl) complexes yields lactones where the C$_6$ ring has a carboxy substituent, whereas others yield iodocyclohexenes of use in natural product synthesis.[211] Unlike the Cp complex, acyclic[Cp*Mo(CO)$_2$(diene)]$^+$ salts do not show cis/trans isomerism of diene substituents in the 1-position; this has been applied to a synthesis of (E,E)-undeca-1,3,5-triene.[212] [CpMo(NCMe(MeC≡CMe)$_2$]$^+$ undergoes 6+2 cycloaddition with cot, followed by hydrogen migration and proton loss, to give (108); hydride abstraction occurs reversibly at C-4, while protonation occurs at C-3, generating an agostic interaction.[212] CpMo(NO)(2,5-dimethyl-2,4-hexadiene) contains the diene bound in the unusual η^4-s-trans configuration.[213]

Complex (109) undergoes endo hydride addition to give (110).[214] Treatment of (111) with LiBHEt$_3$ followed by nitrosylation yields (112).[215]

4. Other Metals

Crystal structures of [(6-exo-phenylcyclohexadienyl)Mn(CO)-(dppe)]$^{n+}$ (n=0,1) show mainly a lengthening of the Mn-P bonds in the 17e$^-$ cation.[216] Phosphorus and nitrogen nucleophiles add to [(cyclohexadienyl)Mn(CO)$_2$NO]$^+$ in the usual exo fashion, though reaction with hydride is endo. Starting with [(arene)Mn(CO)$_3$]$^+$, the results provide a general method for double nucleophile addition to coordinated arenes.[217]

Cocondensation of Re atoms with benzene/chpt gives (benzene)Re(η^5-cycloheptadienyl) and (benzene)Re(η^5-cycloheptatrienyl).[218] Photolysis of Re$_2$(CO)$_{10}$ and chpt yields the analogous Re(CO)$_3$ complexes, together with mainly the dimer (113).[219] Photolysis of ReH$_5$(PMe$_2$Ph)$_3$ with cot yields (η^4-cyclooctatriene)-ReH$_3$(PMe$_2$Ph)$_2$ which thermally rearranges to (η^5-cyclooctadienyl)-ReH$_2$(PMe$_2$Ph)$_2$; transfer of hydrogen is specifically endo.[220] Photoelectron spectra of open chain (pentadienyl)Mn(CO)$_3$ complexes show little difference in metal ionization relative to CpMn(CO)$_3$.[221] Reaction of MCl$_4$(PEt$_3$)$_2$ (M=Zr, Nb, Mo) with 2,4-dimethylpentadienyl anion yields (2,4-dimethylpentadienyl)-M(PEt$_3$). The isostructural Zr and Nb complexes adopt an eclipsed syn(U) pentadienyl conformation, but the Mo complex has one pentadienyl in an anti(S) configuration; C-1 deviates substantially from planarity.[222]

Structure determinations of Cp$_2$M(s-cis-η^4-diene) complexes [M=Zr, Hf; diene-2,3-dimethylbutadiene, 1,2-bis(methylene)-cyclohexane] confirm the metallocyclopentene nature of the bonding, with Hf best approaching this limiting structure.[223] Cp$_2$Zr[2,3-bis(methylene)bicyclo[2.2.2]octane] also adopts the s-cis configuration, and shows the characteristic flipping between diene faces ($\Delta G^* = 29.7$ kJ mol^{-1}).[224] Cp$_2$Zr(s-trans-Z,Z-2,4-hexadiene) undergoes sequential thermal isomerization to Cp$_2$Zr(s-trans-Z,E-2,4-hexadiene) and Cp$_2$Zr(s-trans/s-cis-E,E-2,4-hexadiene); slippage to an η^3-allyl intermediate appears to be the most likely pathway.[225] The complex (Bu^{t}C$_5$H$_4$)$_2$Zr-(s-cis-butadiene) additionally shows evidence for hindered Cp rotation at low temperature.[226]

Treatment of Cp- or Cp*TaCl$_4$ with 1 or 2 equivalents of magnesium-butadiene and its substituted derivaties yields Cp*TaCl$_2$(η^4-butadiene) and Cp*Ta(η^4-butadiene)$_2$

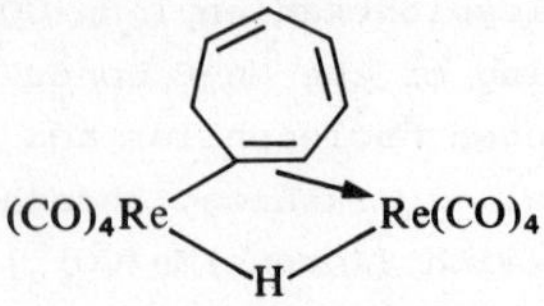

(113)

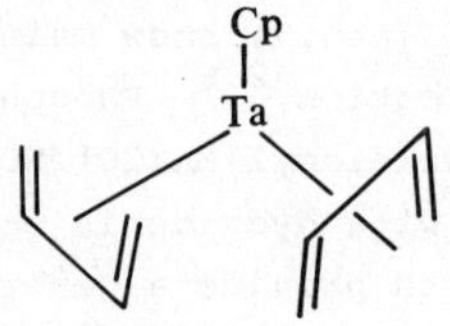

(114)

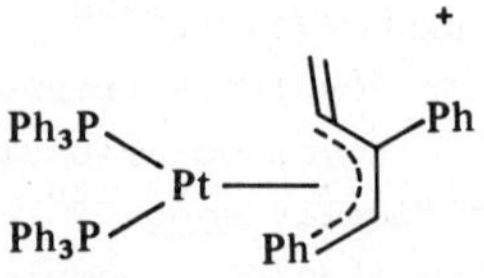

(115)

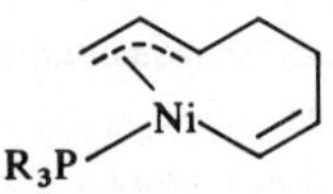

(116)

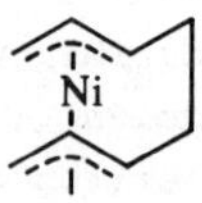

(117)

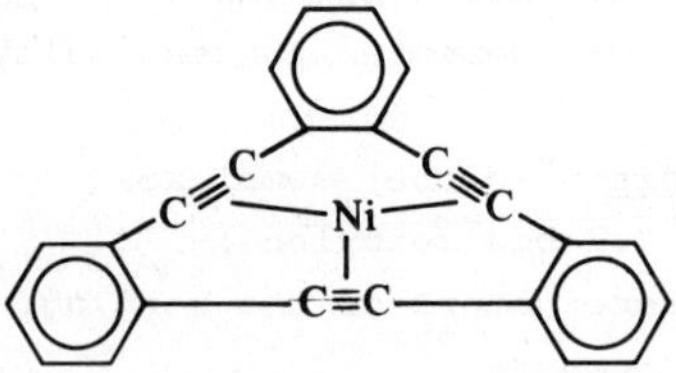

(118)

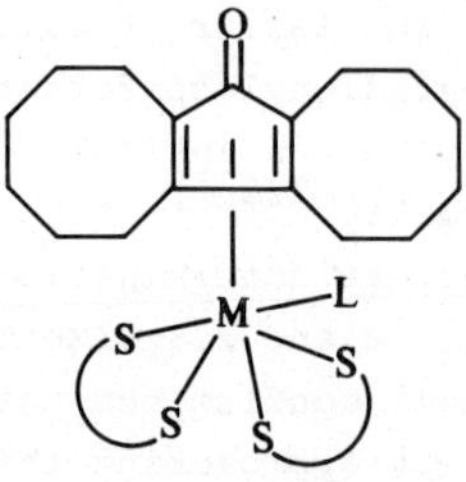

(119) S—S = S_2CNMe_2
a; L = CO
b; L = cyclooctyne

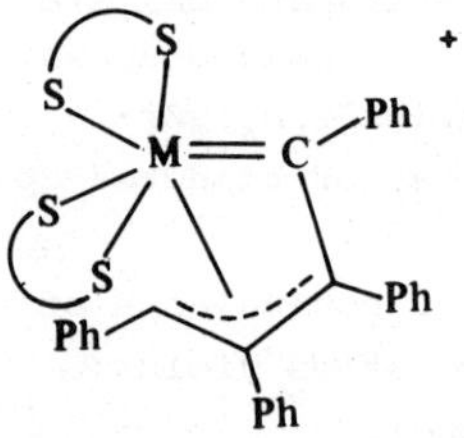

(120) S—S = S_2CNR_2

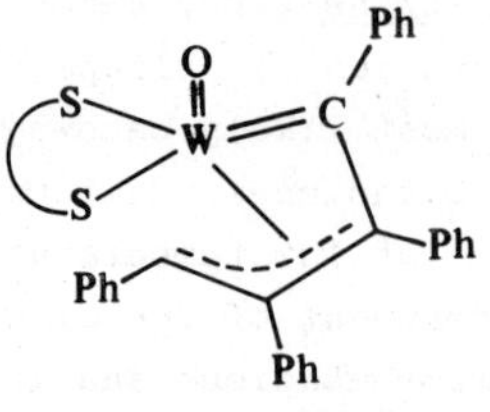

(121) S—S = S_2CNR_2

complexes. Structural studies show that the former are best regarded as metallocyclopentenes; the latter adopt the unusual prone-supine geometry (114), in agreement with calculation.[227]

Reduction of Cp*MCl$_3$ in the presence of chpt or cot yields Cp*M(η^7-C$_7$H$_7$) (M=Ti, Zr, Hf) and Cp*M(η^8-cot) (M=Ti, Zr).[228] Ti(toluene)$_2$ reacts with chpt/AlEtCl$_2$ in thf to give (η^7-C$_7$H$_7$)Ti(η^5-C$_7$H$_9$) and [(C$_7$H$_7$)Ti(thf)(μ-Cl)]$_2$; the latter reacts with dmpe to give (C$_7$H$_7$)Ti(dmpe)Cl which on treatment with MeMgBr yields (C$_7$H$_7$)Ti(dmpe)Me. Though 16e$^-$, no agostic interaction is observed.[229]

(Tetraphenylcyclobutadiene)Ni(PEt$_3$)$_2$ has been prepared; a variety of organic compounds are liberated on carbonylation, oxidation, reduction or acidolysis.[230] Reaction of (PPh$_3$)$_2$Pt(C$_2$H$_4$) with MePh$_2$C$_3^+$ results in ring opening and hydrogen shift to give (115).[231] Minimum energy pathways for haptotropic rearrangements in d^8 and d^{10} (polyene)ML$_2$ complexes (polyene = pentadienyl, benzyl, fulvene, cyclobutadiene, benzene, Cp) have been calculated.[232]

Cp$_2$Ni reacts with vinylLi to give CpNi(CH=CH$_2$) which undergoes reductive coupling to give Cp$_2$Ni$_2$(μ-butadiene) or reacts with C$_2$H$_4$ to give CpNiCH$_2$CH$_2$CH=CH$_2$; the latter undergoes isomerization to CpNi(η^3-1-methylallyl).[233] Full details of the stoichiometric reactions of butadiene and derivatives with Ni(1,5,9-cdt)(PR$_3$) or Ni(cod)$_2$/PR$_3$ to give sequentially complexes of structure (116) and (117) have appeared.[234]

E. Complexes Derived from Acetylenes

1. Ni, Pd and Pt

Reactions of (Me$_3$P)$_2$Ni(C$_2$H$_4$) or (dmpe)$_2$Ni$_2$(C$_2$H$_4$)$_3$ with acetylenes give (Me$_3$P)Ni(RC≡CR), (dmpe)Ni(RC≡CR) and (dmpe)$_2$Ni$_2$(RC≡CR) complexes (R=H, Me, Ph); both (dmpe)Ni(PhC≡CPh) and (Ph$_3$P)$_2$Ni(HC≡CH) have been characterized crystallographically.[235,236] Reaction of the free ligand with Ni(cod)$_2$ gives (118); the structure is nearly planar.[237]

2. Cr, Mo and W

Cp*W(CO)$_2$Me undergo photoreaction with acetylenes to give Cp*W(CO)(COMe)(RC≡CR) complexes which undergo thermal dismutation to Cp*W(CO)(RC≡CR)Me and Cp*(CO)$_2$WC(R)=C(R)COMe; the former undergo facile addition of 2e$^-$ ligands to give Cp*W(COMe)(RC≡CR)L complexes.[238,239] Reaction of Cr(CO)$_2$(RC≡CR)$_2$ (R=SiMe$_3$) with PhC≡CPh yields 16e$^-$ (tetraphenylcyclobutadiene)Cr(CO)$_2$(RC≡CR) which may be reduced sequentially to the 18e$^-$ dianion.[240]

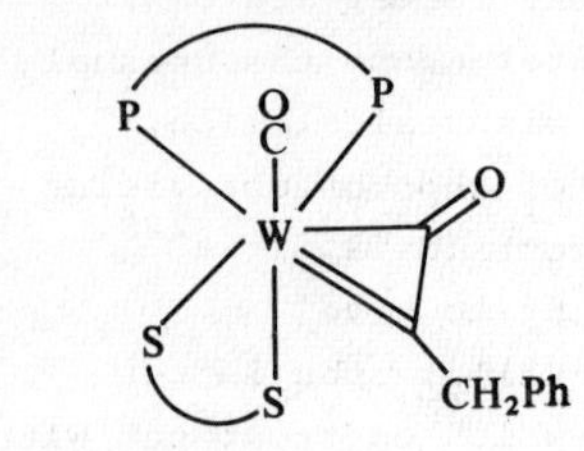

(122) S—S = S_2CNMe_2; P—P = dppe

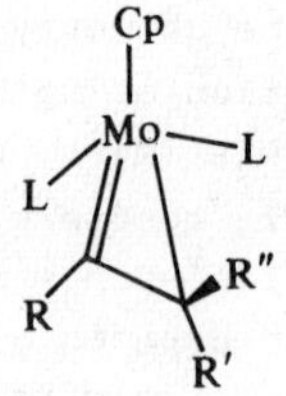

(123) R = R' = Ph; R" = H
(124) R = Me; R' = R" = Ph

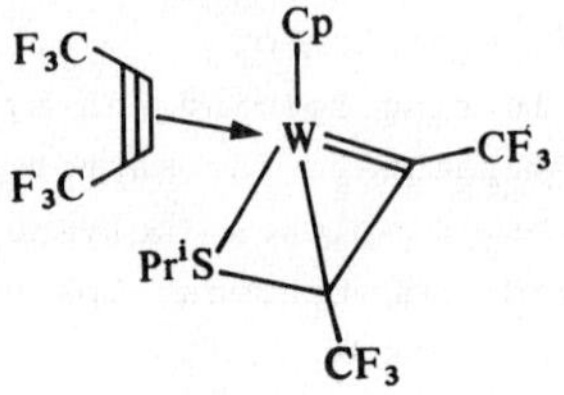

(125)

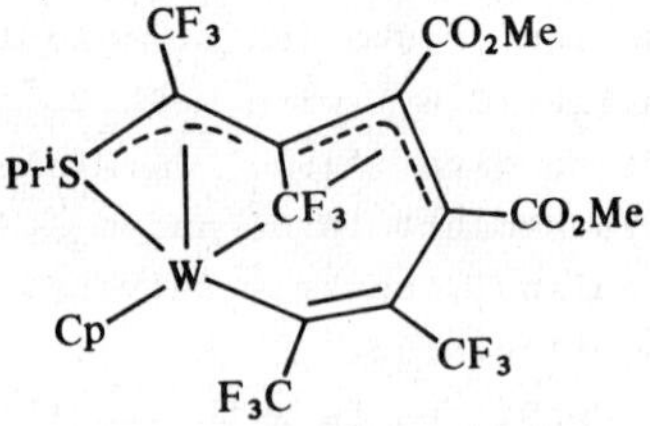

(126)

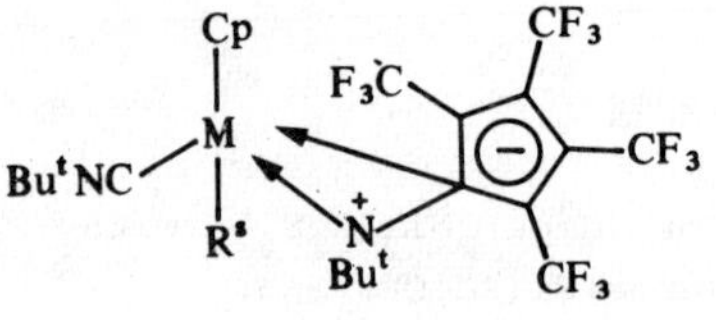

(127)

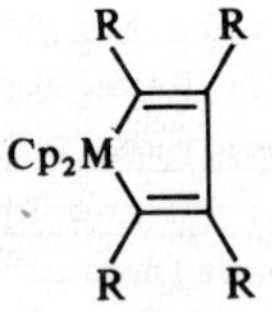

(128) M = Zr, Hf; R = Et, Ph
(129) M = Ti; R = H

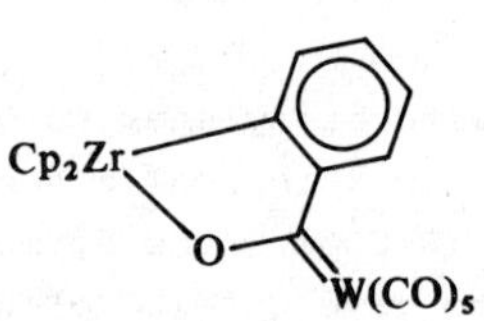

(130)

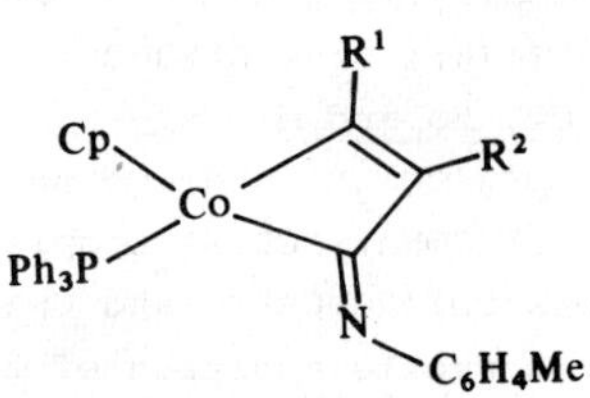

(131) R^1 = Ph; R^2 = CO_2Me
(132) R^1 = CO_2Me; R^2 = Ph

Cyclooctyne reacts in a 1:1 ratio with $M(S_2CNMe_2)_2(CO)_2L$ (M=Mo, L=PPh$_3$; M=W, L=CO, PPh$_3$) to give $M(S_2CNMe_2)_2(CO)$(cyclooctyne) complexes which may be oxidized by X_2 (X=Br, I) to $MX_2(S_2CNMe_2)_2$(cyclooctyne). In the presence of excess cyclooctyne, coupling occurs to give (119a,b).[241,242] $M(CO)(S_2CNR_2)_2$(acetylene) complexes (M=Mo, W) react with electron poor olefins (TCNE, maleic anhydride) to give $M(S_2CNR_2)_2$(acetylene)(olefin) complexes. Olefin and acetylene are <u>cis</u> and parallel; nucleophilic addition by PR$_3$ occurs at a terminal acetylene carbon.[243] Protonation of $M(S_2CNR_2)_2(PhC\equiv CPh)_2$ results in coupling to give (120), which on treatment with NEt$_3$/H$_2$O (M=W), yields (121).[244] Protonation of $OMo(S_2CNR_2)_2(MeO_2CC\equiv CCO_2Me)$ with CF$_3$COOH gives $OMo(S_2CNR_2)_2$-$(CF_3CO_2)(\eta^1$-C(CO$_2$Me)=C(H)CO$_2$Me).[245]

[WBr$_2$(CO)$_4$]$_2$ reacts with RC≡CR (R=Me, Et, Ph) to give [(RC≡CR)$_2$W(CO)(Br)(μ-Br)]$_2$ complexes which may be cleaved by 2e$^-$ donors (CNBu^t, P(OMe)$_3$, PPh$_3$).[246] Reaction of <u>trans</u>-Cr(N$_2$)$_2$-(dmpe)$_2$ with PhC≡CPh gives (dmpe)Cr(PhC≡CPh)$_2$.[247]

Reaction of [(C$_7$H$_7$)Mo(NCMe)(dppe)]$^+$ with PhC≡CH yields the vinylidene [(C$_7$H$_7$)Mo(dppe)(=C=CHPh)]$^+$ which may be deprotonated to (C$_7$H$_7$)Mo(dppe)C≡CPh.[248] Similar rearrangements occur on reaction of PhC≡CH with <u>trans</u>-ReCl(N$_2$)(dppe)$_2$[249] and <u>fac</u>-W(CO)$_3$(dppe)-(acetone)[250] to give <u>trans</u>-ReCl(=C=CHPh)(dppe)$_2$ and <u>mer</u>-W(CO)$_3$-(dppe)(=C=CHPh) respectively. The latter may be protonated to give [<u>mer</u>-W(CO)$_3$(dppe)(≡CCH$_2$Ph)]$^+$. Loss of CO, followed by treatment with S$_2$CNMe$_2^-$ results in coupling to give the ketenyl complex (122) which undergoes alkylation to give [(S$_2$CNMe$_2$)(dppe)-W(CO)(η^2-ROC≡CCH$_2$Ph)]$^+$. <u>Trans</u>-MCl(RC≡CH)(PPr$_3^i$)$_2$ complexes (M=Rh, Ir; R=Me, H, Ph) undergo transformation to <u>trans</u>-MCl(=C=CHR)(PPr$_3^i$)$_2$ <u>via</u> initial oxidative addition to give HMCl(C≡CR)(PPr$_3^i$)$_2$.[251]

[CpW(PMe$_3$)$_2$(RC≡COR^1)]$^+$ complexes (R=Me, <u>p</u>-tolyl, R^1=Me, Et) have been prepared and structurally characterized.[252] Treatment of [CpMo(RC≡CR^1)L$_2$]$^+$ [L=P(OMe)$_3$; R=R^1=Ph] with KBHBu$_3^s$ yields (123); alkylation with LiCuPh$_2$ (R=Me, M^1=Ph) yields (124) of similar structure, which rearranges slowly in solution to CpMoL$_2$(η^3-1,1-diphenylallyl). In some cases, direct metal attack is observed, such as the reaction of LiCuPh$_2$ with [CpMoL$_2$-(MeC≡CBu^t)]$^+$ to give CpMoL(MeC≡CBu^t)Ph.[253] The related σ-vinyl complex (125) undergoes insertion on reaction with RC≡CR (R=CO$_2$Me, Me, CF$_3$) to give complexes such as (126),[254] while the reaction of CpM(SR)(CF$_3$C≡CCF$_3$)$_2$ (M=Mo, W) with Bu^{t}NC

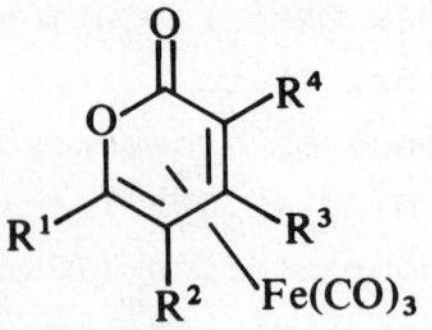

(133) R^1 = EtO; R^2 = R^3 = Me; R^4 = Ph
(134) R^1 = EtO; R^2 = Ph; R^3 = R^4 = Me

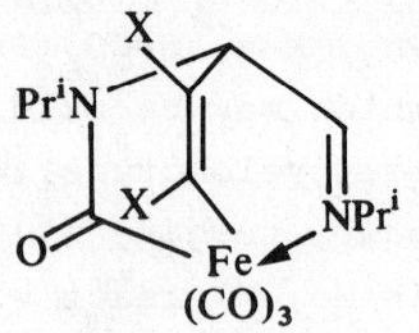

(135) X = CO_2Me

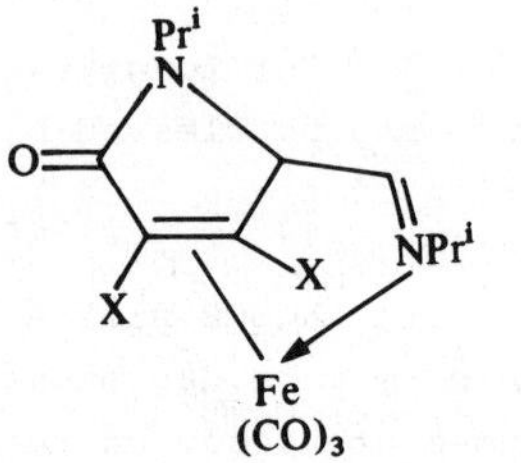

(136) X = CO_2Me

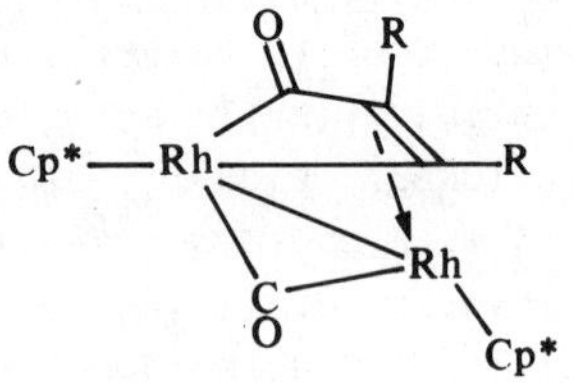

(137)

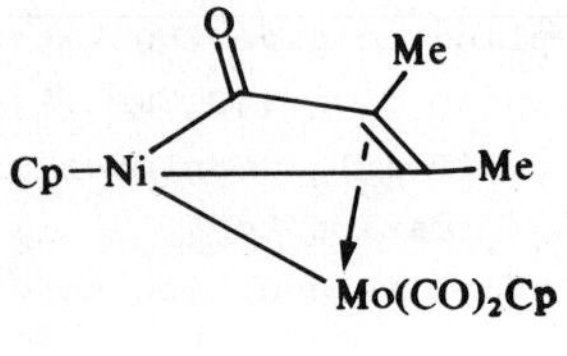

(138)

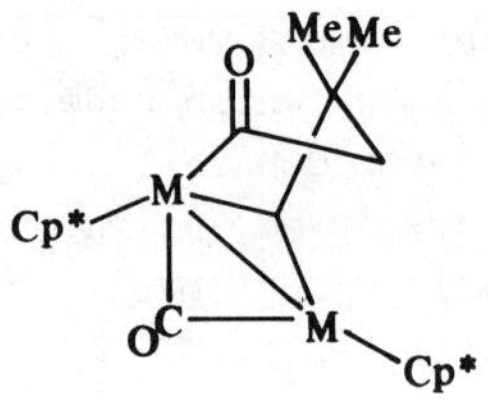

(139)

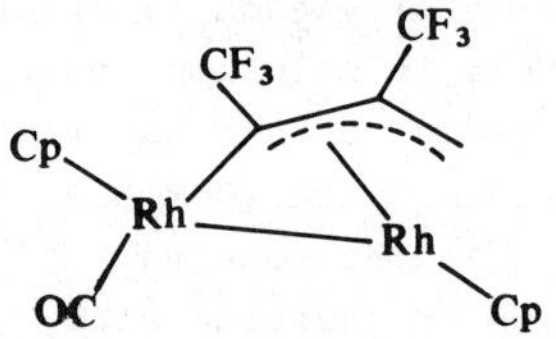

(140)

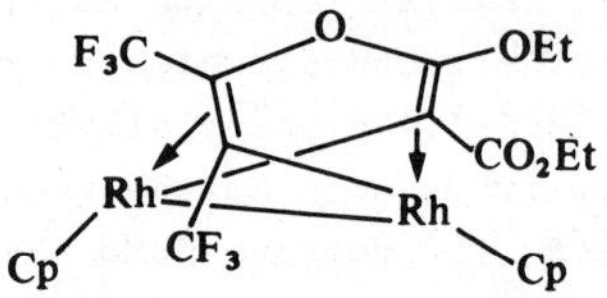

(141)

to give (127) proceeds via initial attack on acetylene to give a
σ-vinyl intermediate.[255]

3. Other Metals

Photoreaction of $CpV(CO)_4$ with acetylenes gives $CpV(CO)_2$-
(RC≡CR) complexes; rotational barriers are in the range
44-68 kJ mol^{-1}.[256] $(MeC_5H_4)NbCl_4$ is reduced by Al in the presence
of RC≡CR (R=p-tolyl) to give $(MeC_5H_4)NbCl_2(RC≡CR)$ which may be
further reduced with Na/Hg to $[(MeC_5H_4)Nb(RC≡CR)(\mu-Cl)]_2$.[257]

Photolysis of $Cp_2M(CO)_2$ (M=Zr, Hf) with RC≡CR (R=Et, Ph)
results in coupling to give (128).[258] A complex of the same
structure is obtained on reaction of $[(MeC_5H_4)ZrH(\mu-H)]_2$ with
PhC≡CPh, together with products of acetylene hydrogenation.[259]
The unsubstituted complex (129) is obtained by sequential
displacement of PMe_3 from $Cp_2Ti(PMe_3)_2$.[260] $Cp_2Zr(\eta^2$-benzyne) may
be generated in situ by photolysis of Cp_2ZrPh_2 and reacts with
$W(CO)_6$ to give (130).[261]

The complexes $CpRe(CO)_2(\eta^2$-RC≡CR^1) (R=R^1=Ph; R=Ph, R^1=Me)
have been structurally characterized.[262] $(CO)_5MnFBF_3$ reacts with
MeC≡CMe to give $[(tetramethylcyclobutadiene)Mn(CO)_4]^+$.[263]
Reaction of $ReOX_3(AsPh_3)_2$ (X=Cl, Br, I) with acetylenes gives
tetrahedral ReOX(acetylene)$_2$; treatment with Ag$^+$/L yields
$[ReO(acetylene)_2L]^+$ cations (L=py, PPh$_3$).[264]

$CpCo(PPh_3)(PhC≡CCO_2Me)$ reacts with MeC_6H_4NC under electronic
control to give (131) which isomerizes thermally to the
thermodynamic product (132).[265]

Reaction of $(CO)_4FeC(OEt)Ph$ with acetylenes under CO
stereospecifically yields (pyrone)$Fe(CO)_3$ complexes via
ferracyclobutene intermediates; thus, MeC≡CMe yields (133) which
isomerizes thermally to an equilibrium with (134).[266]
(Diazadiene)$Fe(CO)_3$ complexes react with RC≡CR (R=CO_2Me) under
CO give an initial adduct (135) which isomerizes easily to the
1,5-dihydropyrrol-2-one complex (136).[267]

F. Complexes Containing More Than One Metal Atom

1. Binuclear Complexes

Kinetic studies of the reaction of nbd with $Co_2(CO)_8$ to
give $Co_2(CO)_6(nbd)$ indicate a facile dissociation to
$Co_2(CO)_7$.[268] Kinetic studies of the hydrogenation of
$Co_2(CO)_6(\mu-HC≡CH)$ to give C_2H_4 indicate both H_2 coordination and
CO dissociation in the rate determining step.[269] Reaction of
$[Co(CO)_3L]^-$ (L=PPh$_3$, CO) with octafluorocot yields
$Co_2(CO)_4L_2(\mu-C_8F_6)$; the ring is puckered in the CO complex, but

planar where L = PPh$_3$.^{270} Co$_2$(CO)$_6$(μ-HC≡CC(Me)=CH$_2$) may be acetylated to give stable [Co$_2$(CO)$_6$(μ-HC≡CC(Me)CH$_2$COR)]$^+$ salts which undergo nucleophilic attack to give Co$_2$(CO)$_6$(μ-HC≡CC(Me)-(Nu)CH$_2$COR).^{271} Co$_2$(CO)$_6$(μ-LiC≡CR)(R=SiMe$_3$), formed from deprotonation of Co$_2$(CO)$_6$(μ-HC≡CR), undergoes reaction with electrophiles by electron transfer to give [Co$_2$(CO)$_6$]$_n$-(RC≡C-C≡CR)(n=1,2); where n=2, solution polymerization occurs to yield a complexed polyacetylene.^{272}

Reaction of Cp*$_2$Rh$_2$(CO)$_2$ with acetylenes yields Cp*$_2$Rh$_2$(CO)$_2$(μ-RC≡CR) and the complex of structure (137) as an equilibrium solution mixture. Proportions are dependent on acetylene substituent; where R=C$_6$F$_5$, only the μ-RC≡CR complex is observed, whereas where R=CF$_3$, the equilibrium lies almost entirely towards (137).^{273} Reaction of Cp$_2$NiMo(CO)$_4$ with MeC≡CMe yields (138) of similar structure.^{274} Reaction of Cp$_2^*$M$_2$(CO)$_2$ (M=Rh, Co) with 3,3-dimethylcyclopropene results in ring opening to give (139).^{275}

Treatment of the coordinatively unsaturated Cp$_2$Rh$_2$(μ-CO)-(μ-CF$_3$C≡CCF$_3$) with diazoalkanes gives Cp$_2$Rh$_2$(μ-CO)(μ-CR$_2$)-(μ-CF$_3$C≡CCF$_3$) (R=H, CO$_2$Et) which convert in solution to (140) and (141) respectively.^{276} Reaction of Cp$_2$Rh$_2$(μ-CO)(μ-CF$_3$C≡CCF$_3$) with PhN$_3$, RNCNR (R=Pr^i, Cy) or RNCO (R=Me, Bu^t) yields the bridging acrylamide complex (142).^{277}

CO and SO$_2$ insert into the metal-metal bond of Rh$_2$X$_2$-(μ-CF$_3$C≡CCF$_3$)(dppm)$_2$ (X=Cl, I), whereas MeNC substitutes in a terminal fashion to yield eventually [Rh$_2$(MeNC)$_4$(μ-CF$_3$C≡CCF$_3$)-(dppm)$_2$]$^{2+}$.^{278} Thermolysis of Ir$_2$Cl$_2$(CO)$_2$(μ-RC≡CR)(dppm)$_2$ (R=CF$_3$, CO$_2$Me) yields Ir$_2$Cl$_2$(CO)(μ-RC≡CR(dppm)$_2$, whereas halide abstraction yields [Ir$_2$Cl(CO)$_2$(μ-RC≡CR)(dppm)$_2$]$^+$ in which halide exchanges between the two metal atoms. The cation reacts reversibly with CO to give [Ir$_2$Cl(CO)$_3$(μ-RC≡CR)(dppm)$_2$]$^+$ and with excess Bu^{t}NC to give [Ir$_2$(Bu^{t}NC)$_4$(μ-RC≡CR)(dppm)$_2$]$^{2+}$ (R=CO$_2$Me).^{279,280} Reaction of RC≡CR (R=CF$_3$, CO$_2$Me) with (cod)$_2$Ir$_2$(μ-pz)$_2$ (Hpz=pyrazole) yields parallel-bonded (cod)$_2$Ir$_2$(μ-pz)$_2$(μ-RC≡CR) complexes.^{281} Treatment of Rh$_2$(CO)$_3$(dppm)$_2$ with PhC≡CH yields either Rh$_2$(CO)$_2$(μ-HC≡CPh)-(dppm)$_2$ or the vinylidene Rh$_2$(CO)$_2$(μ-C=CHPh)(dppm)$_2$, depending on reaction conditions; though the former slowly isomerizes to the latter, kinetic studies indicate a direct reaction path to the vinylidene complex.^{282}

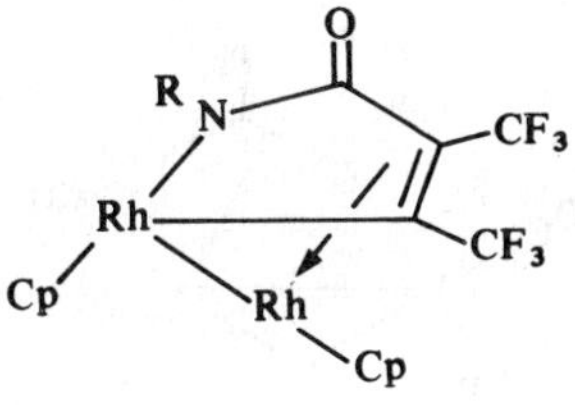

(142)

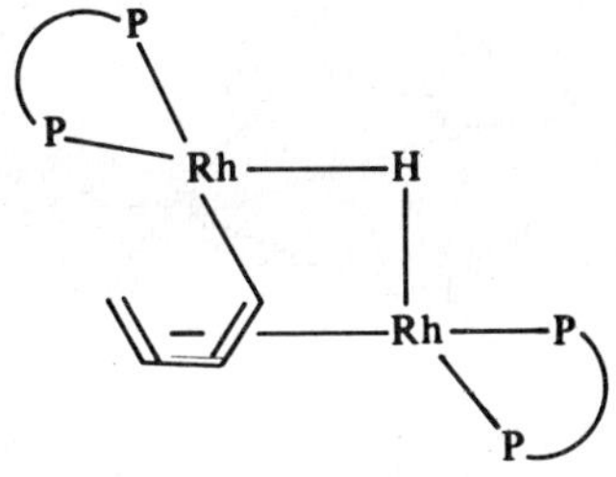

(143)

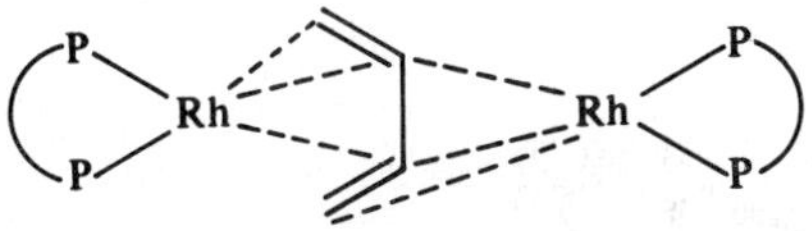

(144)

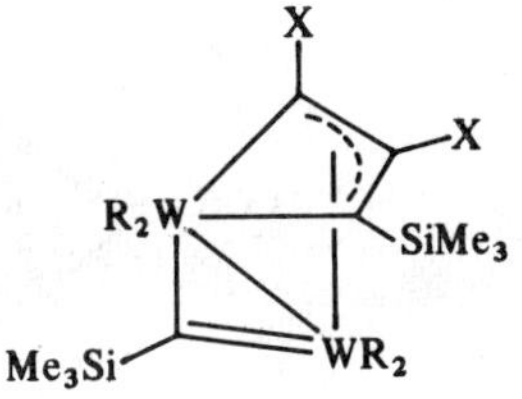

(145)

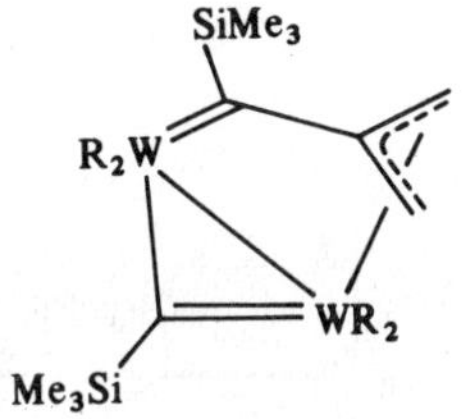

(146)

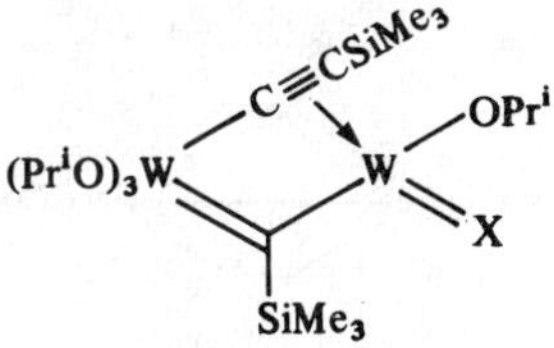

(147) X = O, NR

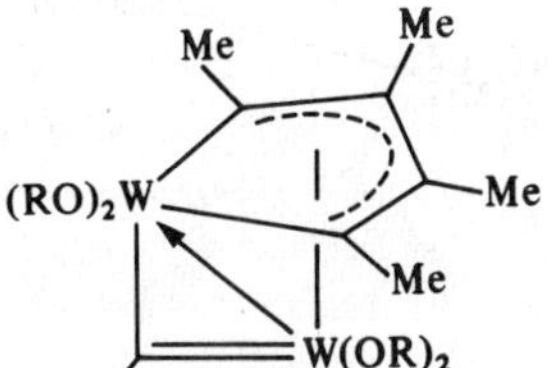

(148)

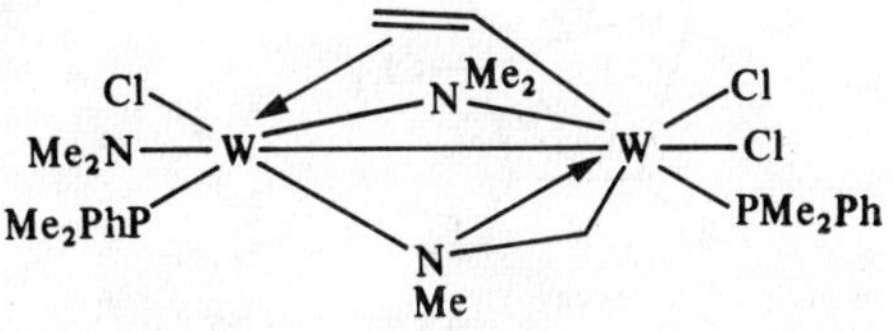

(149)

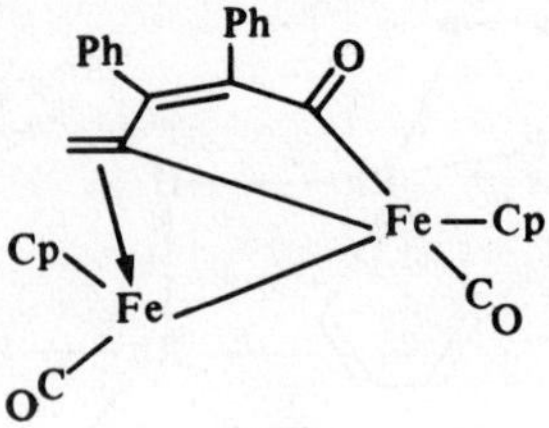

(150)

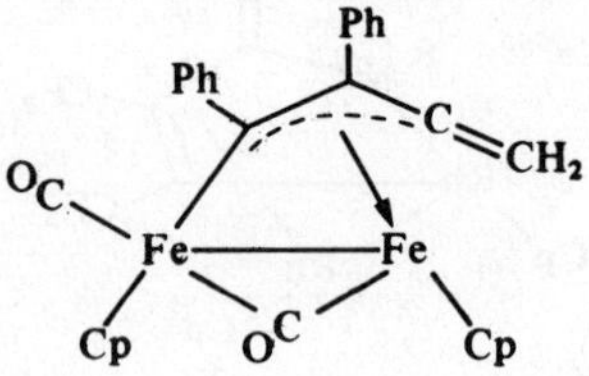

(151)

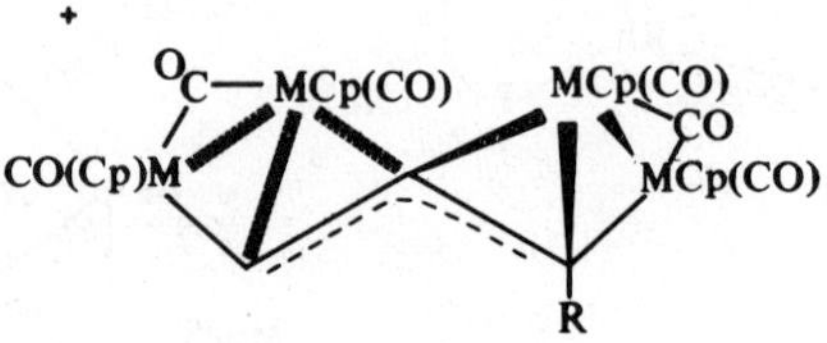

(152) M = Ru; R = Me
M = Fe; R = H

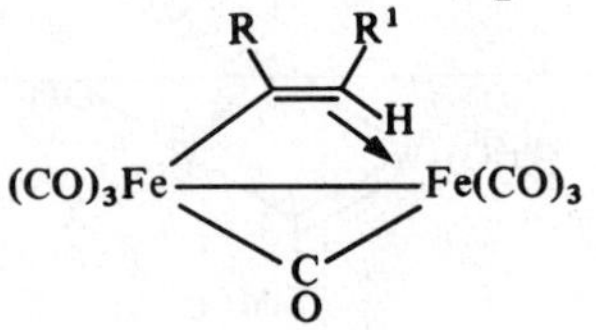

(153)

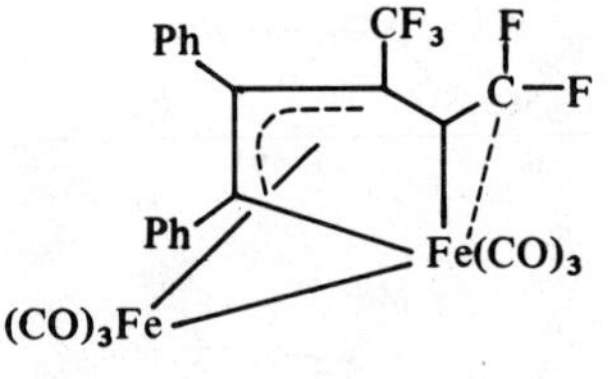

(154)

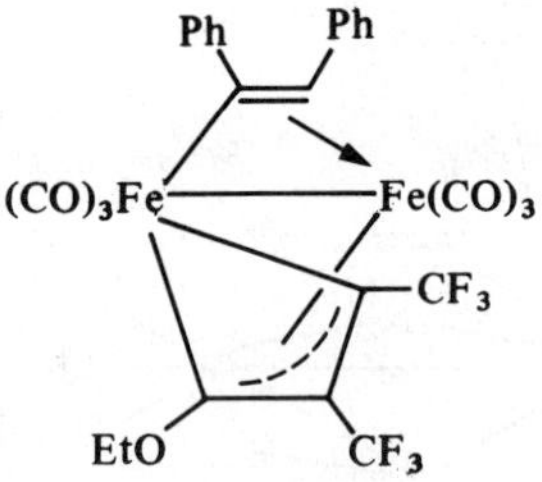

(155)

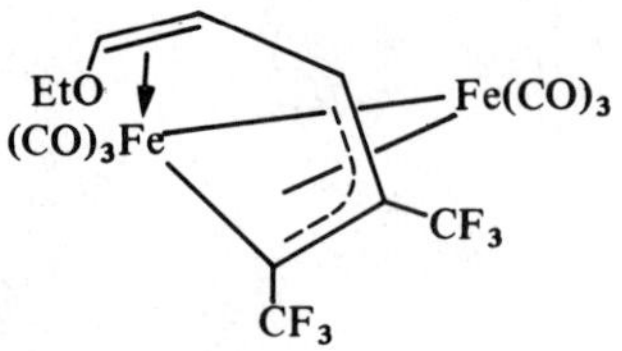

(156)

Though $[(Pr^i_2P(CH_2)_nPPr^i_2)_2Rh]_2(\mu-H)_2$ (n=2) reacts with butadiene to give (143), the n=3 complex gives (144) containing an unusual $\eta^3:\eta^3$-coordination.[283]

Reaction of $[R_2W(\mu-CSiMe_3)]_2$ (R=CH_2SiMe_3, OPr^i) with acetylenes yields initial η^2-adducts which undergo insertion to give (145); in some cases (e.g. X=H), thermal isomerization of $SiMe_3$ to the thermodynamically favored internal carbon occurs.[284] Reaction of $[R_2W(\mu-CSiMe_3)]_2$ with allene also yields an initial η^2-adduct which undergoes insertion to give (146).[285] Reaction of CO or CNR with $[(Pr^iO)_2W(\mu-CSiMe_3)]_2$ results in CO or CN bond cleavage to give (147).[286]

$1,2$-$W(CH_2Ph)_2(OPr^i)_4$ reacts with MeC≡CMe to give a 2:1 η^2-adduct which on thermolysis undergoes loss of toluene to give (148).[287] $W_2Cl_3(NMe_2)_3(PMe_2Ph)_2$ reacts with HC≡CH to give the transversely bonded $[Cl(NMe_2)(PMe_2Ph)W]_2(\mu-Cl)(\mu-NMe_2)(\mu-HC\equiv CH)$ which undergoes thermal rearrangement to give (149).[288]

Thermolysis of $CpMo(CO)_3C\equiv CPh$ yields $Cp_2Mo_2(CO)_4$-$(\mu-PhC\equiv C-C\equiv CPh)$.[289] $W(CO)(RC\equiv CR)_3$ (R=Ph, Et) reacts with $Cp(CO)_2W(\equiv CC_6H_4Me)$ to give $(C_5Ph_4C_6H_4Me)W(CO)(\mu-CO)_2WCp(\eta^2-PhC\equiv CPh)$ and $Cp(CO)W(\mu-EtC\equiv CEt)W(CO)(C_5Et_4C_6H_4Me)$ respectively.[290] Terminal-semibridging carbonyl exchange in $M_2(CO)_4(\mu$-acetylene) complexes has a much lower barrier where M=W.[291]

Photoreaction of $[CpFe(CO)]_2(\mu-CO)(\mu-C=CH_2)$ with PhC≡CPh yields the metallocyclopentenone (150) which undergoes photolytic loss of CO to give (151).[292] Reaction of $[CpM(CO)]_2(\mu-CO)$-$(\mu-C=CH_2)$ with $\{[CpM(CO)]_2(\mu-CO)(\mu-CR)\}^+$ (M=Fe, R=H; M=Ru, R=Me) yields a complex best represented as (152).[293,294] The anion (153) may be alkylated at the bridging CO.[295] Reaction of (153) with $CF_3C\equiv CCF_3$ results in insertion into the C-H bond. Attempted alkylation of the product results in F^- abstraction to give (154).[296] Ethylation of (153) (R=R^1=H) followed by reaction with $CF_3C\equiv CCF_3$ results in insertion into the carbyne bridge to give (155), whereas where R=R^1=H, insertion into the vinyl bridge occurs to give (156).[297] Reaction of $Fe_3(CO)_{10}(\mu_3-PC_6H_4OMe)$ with RC≡CR (R=H, Ph) yields the novel η^4-complex (157) which undergoes CO loss on thermolysis to give (158).[298] The complex $(CO)_3FeRh(CO)_2(\mu-C_7H_7)$ undergoes reversal of the η^3-Fe, η^4-Rh bonding mode on reaction with PPh_3 to give (159).[299]

Photolysis of $Re_2(CO)_8(\mu-R_2PCH_2PR_2)$ (R=Me, Ph) with terminal acetylenes $R^1C\equiv CH$ (R^1=H, Ph) yields initially (160), which is converted on further photolysis into (161); reaction with a further

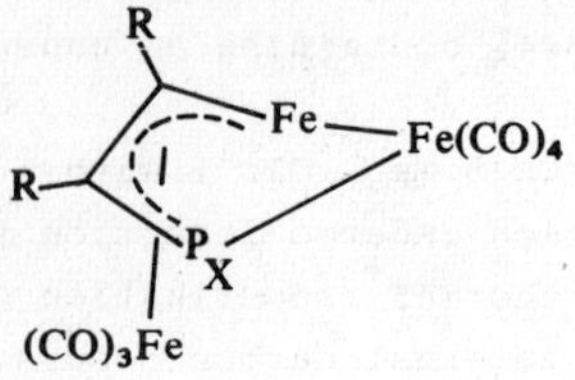

(157)

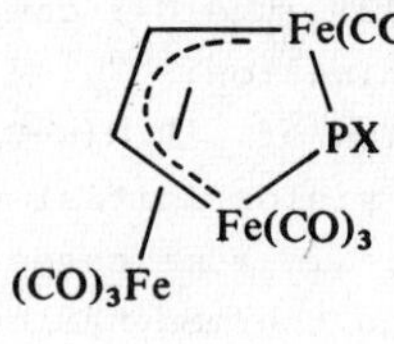

(158)

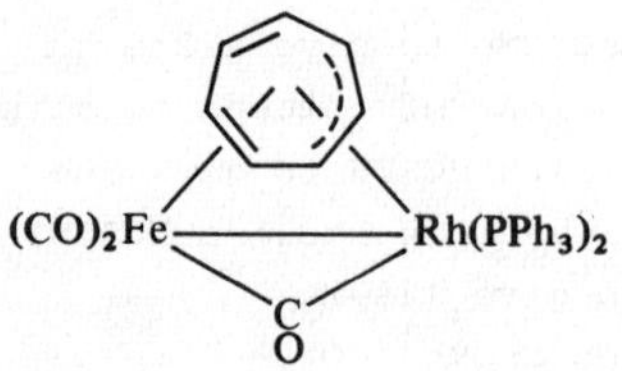

(159)

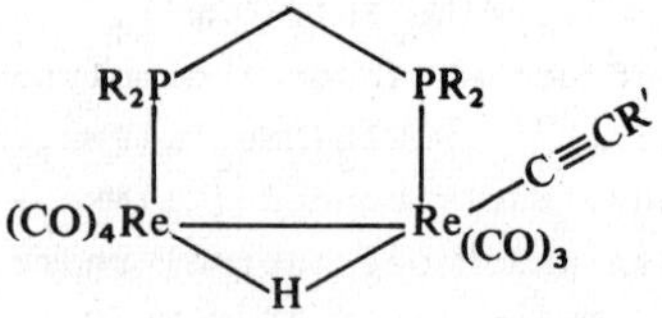

(160)

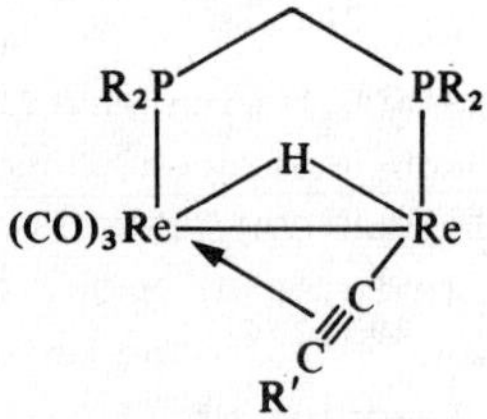

(161)

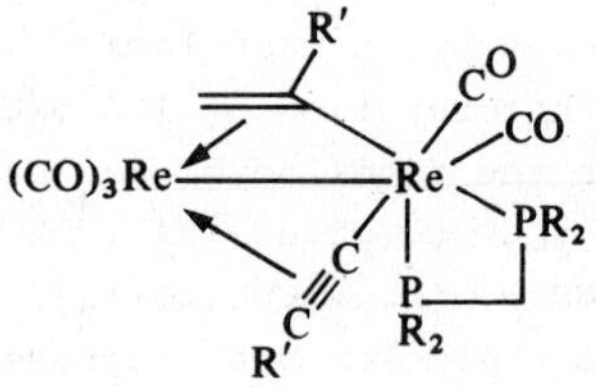

(162)

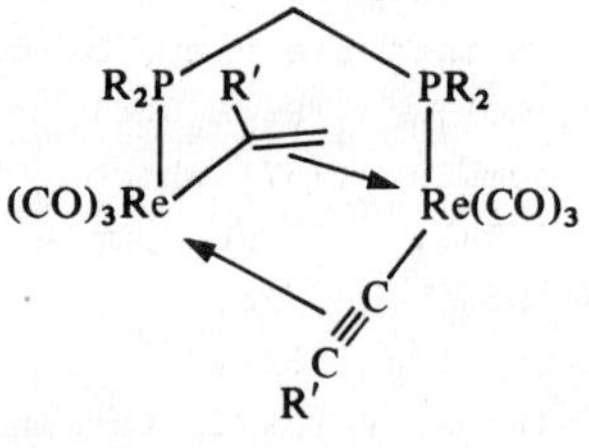

(163)

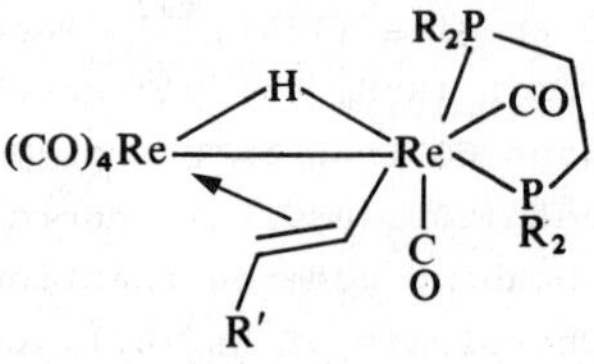

(164)

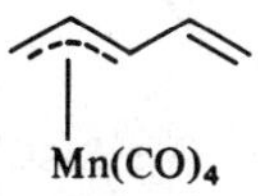

(165)

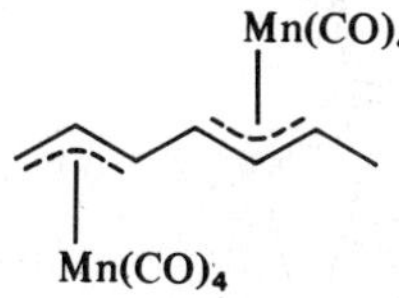

(166)

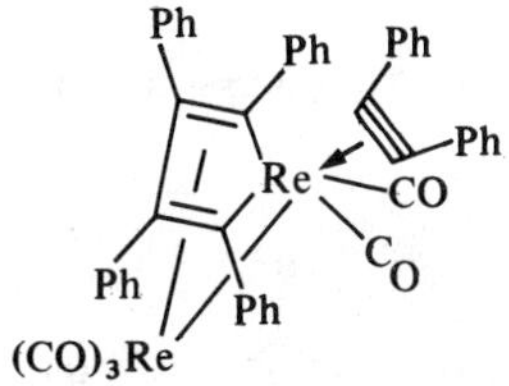

(167)

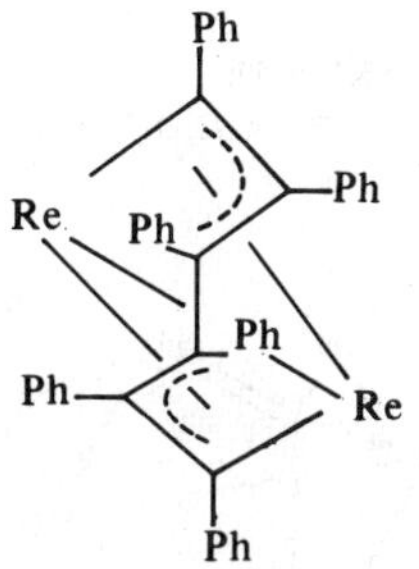

(168)

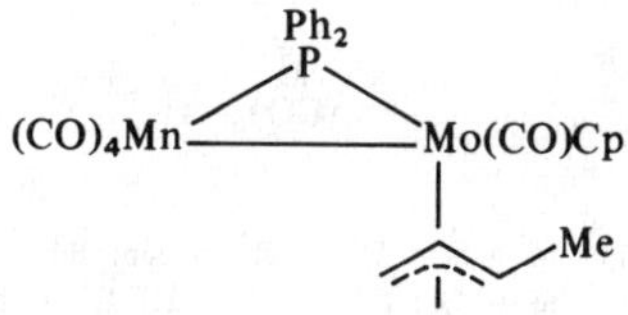

(169)

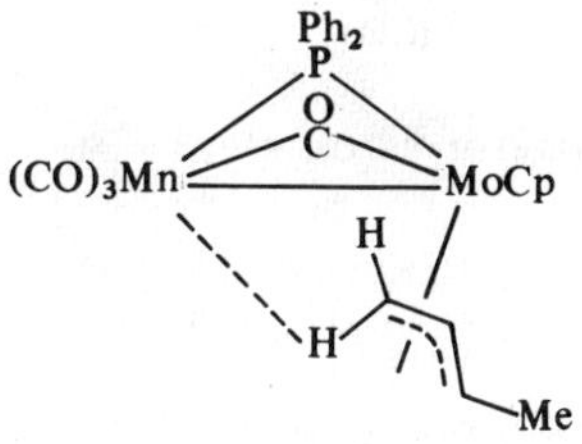

(170)

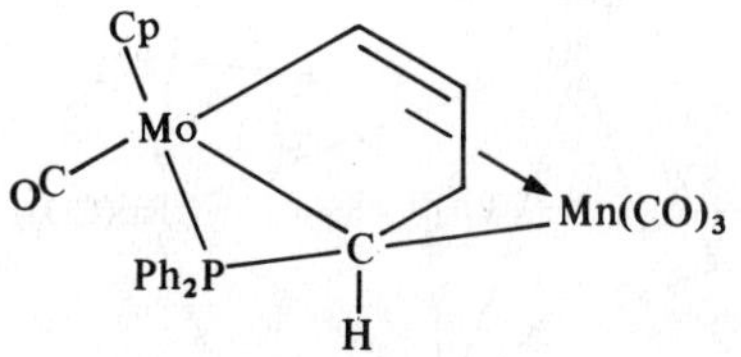

(171)

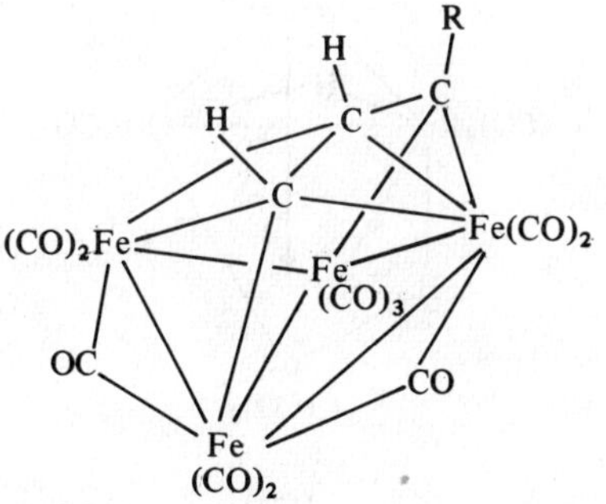

(172)

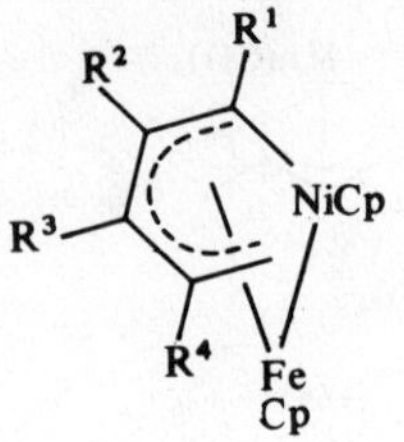

(173) R¹-R⁴ = Ph
(174) R¹ = R⁴ = H; R² = R³ = Ph

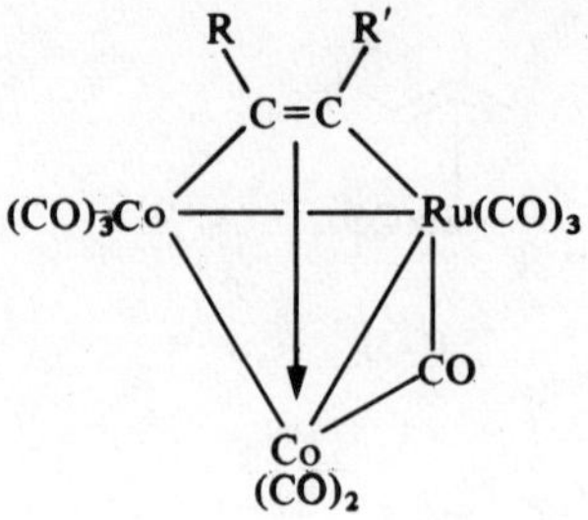

(175) R = Me, Bu^t, Ph; R' = H

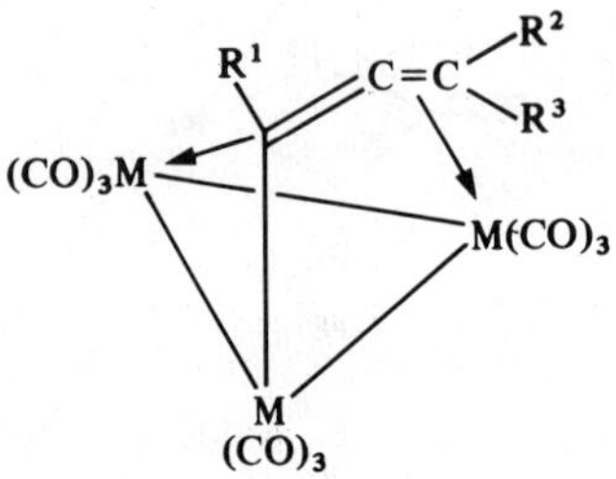

(176) M = Ru, Os; R¹-R³ = Me

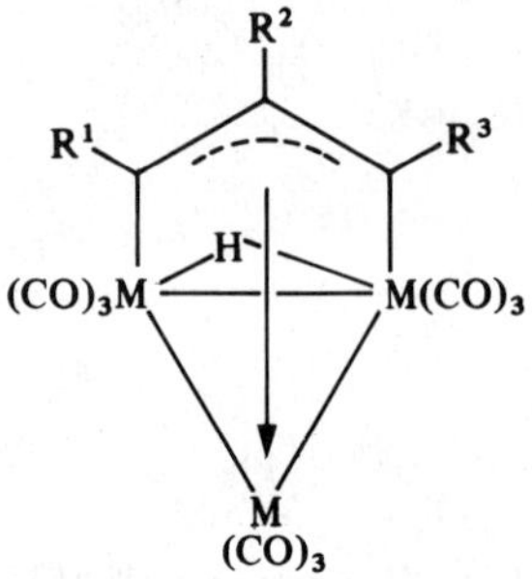

(177) M = Ru; R¹ = R³ = Me; R² = H
     M = Os; R¹ = R² = H; R³ = Me
(178) M = Ru; R¹ = R² = H; R³ = OMe
(180) M = Ru; R¹ = R² = Me, Ph; R³ = SEt

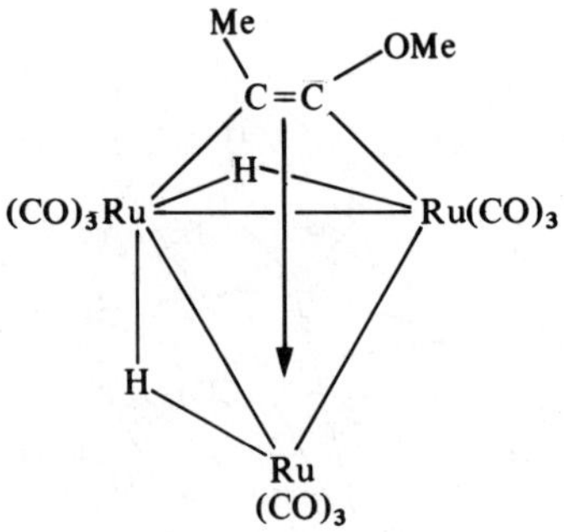

(179)

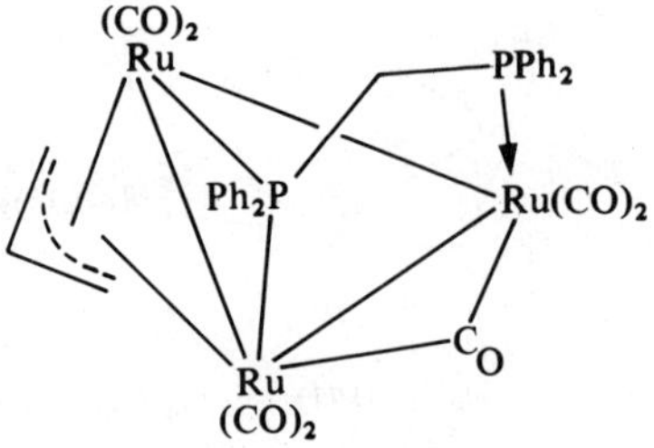

(181)

mole of acetylene yields both (162) and (163).[300] Photolysis of $Re_2(CO)_8(\mu-Me_2PCH_2PMe_2)$ with $H_2C=CHR^1$ yields alkenyl analogues of (161), whereas with $Re_2(CO)_8(\mu-R_2P(CH_2)_2PR_2)$ (R=Me, Ph), rearrangement to the chelate (164) occurs.[301] Though similar products are obtained using $Re_2(CO)_9L$ or $Re_2(CO)_8L_2$ (L=PMe$_3$, PPh$_3$), M-M cleavage rather than CO dissociation is the primary photolysis pathway. $Mn_2(CO)_{10}$ undergoes photolysis with cis- or trans-1,3-pentadiene to give (165) which thermally loses CO to give the (η^5-pentadienyl)Mn(CO)$_3$ complex; similar products are obtained with other substituted butadienes.[302] Photolysis of $Mn_2(CO)_{10}$ with 1,3,5-hexatriene yields (166), together with its (EZ) isomer.[303] Complexes (167) and (168) have been structurally characterized as products from the photoreaction of $H_3Re_3(CO)_{12}$ with PhC≡CPh.[304] Photolysis of $CpMo(CO)_2(\mu-H)(\mu-PPh_2)Mn(CO)_4$ with butadiene yields (169) which reversibly loses CO to give (170); reaction with acetylene results in coupling to give (171).[305]

2. Polymetallic Complexes

Fe$_3$(CO)$_9$(μ_3-CMe)(μ_3-COEt) reacts reversibly with CO to couple the alkylidyne ligands, giving $Fe_3(CO)_{10}(\mu_3-\eta^2-MeC\equiv COEt)$; alkylidyne coupling is also observed on reaction with RC≡CR (R=Ph, SiMe$_3$) to give ferracyclopentadiene derivaties.[306,307] [HFe$_4$(CO)$_{13}$]$^-$ reacts with HC≡CH to give [Fe$_4$(μ_3-CMe)(CO)$_{12}$]$^-$; coupling with a second molecule of acetylene yields (172). Monosubstituted acetylenes RC≡CH yield either an analogue of (172) when R is electron withdrawing, or [Fe$_4$(μ_3-C≡CR)(CO)$_9$]$^-$ when R is electron donating.[308]

Reaction of Cp$_2$Fe$_2$(CO)$_4$ with Cp$_2$Ni and PhC≡CPh gives (173); with HC≡CPh, an interconverting isomeric mixture is obtained, with (174) being the major isomer.[309] The acetylene complexes (175) undergo rotation (ΔG^*=65 kJ mol^{-1}); thermolysis results in isomerization to a vinylidene isomer.[310] Acetylene rotation in the heterometallic clusters (RC≡CR^1)MM^{1}Fe(CO)$_3$ (M=CpNi, M^1=CpMo(CO)$_2$, Co(CO)$_3$, CpNi) also has barriers in the range 60-70 kJ mol^{-1}.[311] Orientation of the acetylene on the M$_3$ face is in agreement with calculation.[312]

Photoelectron spectra of (176) and its thermodynamically more stable isomer (177) have been reported.[313] Reaction of $H_3Ru_3(\mu_3-COMe)(CO)_9$ or $HRu_3(\mu-COMe)(CO)_{10}$ with acetylenes results in coupling to yield complexes such as (178); with unsymmetrical acetylenes, regioisomers are obtained. Hydrogenation of (178)

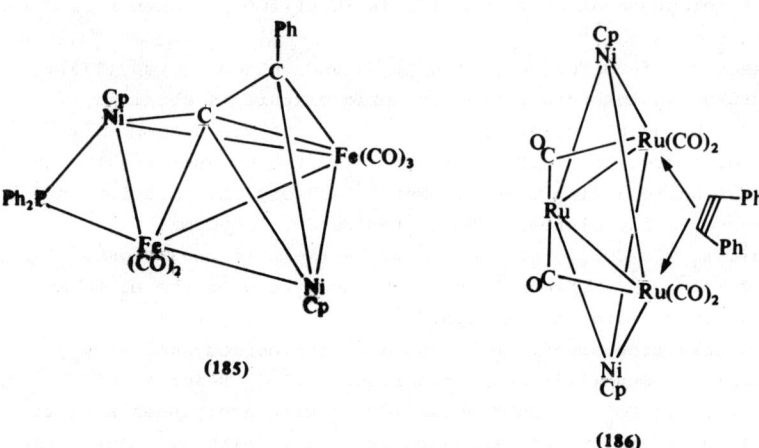

(182)

(183)

(184)

(185)

(186)

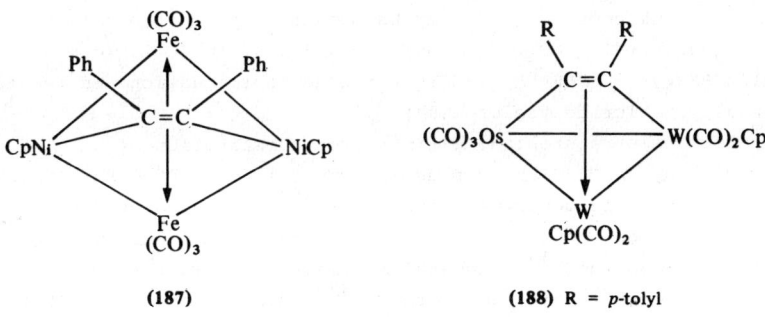

(187)

(188) R = *p*-tolyl

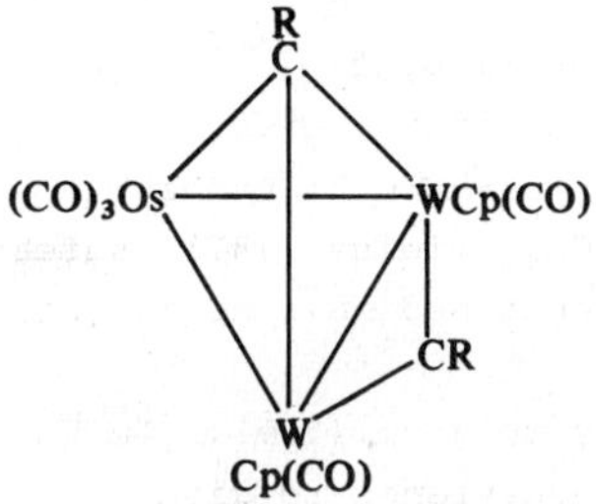

(189)

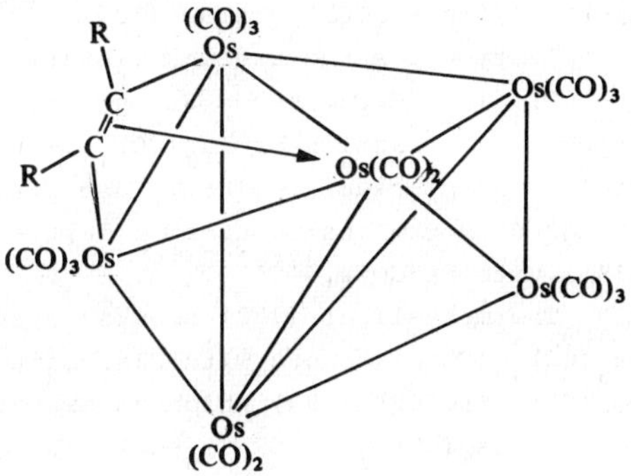

(190)

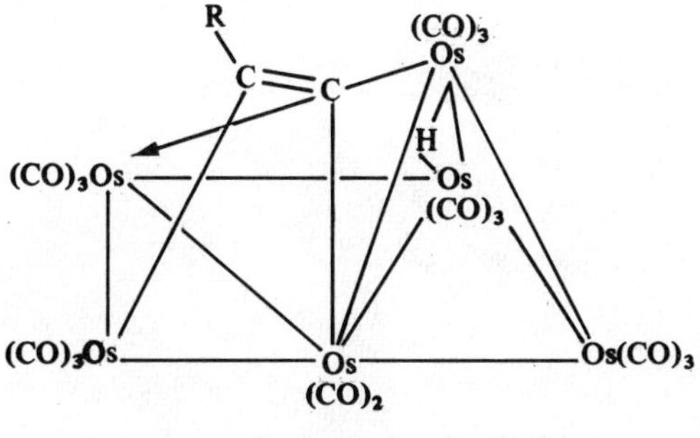

(191)

yields mainly $H_3Ru_3(\mu_3\text{-CEt})(CO)_9$, together with (179).[314,315] The analogous complex (180) may be obtained by reaction of RC≡CR with $H_3Ru_3(\mu_3\text{-CSEt})(CO)_9$.[316] Hydrogenation of $Ru_3(CO)_{10}(dpm)$ yields $HRu_3(\mu_3\text{-PhPCH}_2PPh_2)(CO)_9$ which on deprotonation and reaction with allyl chloride yields (181).[317]

Room temperature radical-initiated substitution of $Ru_3(CO)_{12}$ by $Ph_2PC\equiv CPPh_2$ yields $[Ru_3(CO)_{11}]_2(\mu\text{-Ph}_2PC\equiv CPPh_2)$ which on thermolysis is converted into (182).[318] The cluster (183) is isolated as one product of the thermal reaction of $Ru_3(CO)_{12}$ with the same phosphine.[319] Sequential cleavage of two Ru-Ru bonds occurs on reaction of (182) with CO.[320] Hydrogenation of (182) at 1 atm results in sequential gain of three moles of H_2 to give $(\mu\text{-H})_3Ru_5C(CO)_{11}(PMePh_2)(\mu\text{-PPh}_2)$ resulting from formal conversion of $\mu\text{-C}_2PPh_2$ to carbon and $PMePh_2$; under 10 atm H_2, complex (184) is obtained instead.[321] Reaction of $Cp_2Ni_2(CO)_2$ with $Fe_2(CO)_6$-$(\mu\text{-PPh}_2)(\mu\text{-}\eta^2\text{-C}\equiv CPh)$ yields (185).[322]

A crystal structure and vibrational analysis of $Co_4(CO)_{10}(\mu_4\text{-HC}\equiv CH)$ shows strong back-bonding from the Co_4 moiety.[323] Reaction of $Ru_3(CO)_{12}$ with $Cp_2Ni_2(PhC\equiv CPh)$ under H_2 yields the μ_4-cluster (186); use of $Fe_3(CO)_{12}$ yields (187) instead, though clusters analogous to (186) can be obtained on reaction of (187) with $Fe(CO)_5$.[324,325]

Thermolysis of (188) occurs reversibly to give (189).[326] $Os_6(CO)_{17}(NCMe)$ reacts with disubstituted acetylenes to give $Os_6(CO)_{16}(RC\equiv CR)$ (190) which undergoes C-C rupture on thermolysis to give $Os_6(CO)_{16}(\mu_4\text{-CR})(\mu_3\text{-CR})$. Reaction with terminal acetylenes yield an initial $Os_6(CO)_{17}(\mu_4\text{-RC}\equiv CH)$ adduct which on thermolysis undergoes C-H cleavage to give (191).[327]

References

1. K. Jonas, Angew. Chem. Int. Ed. Engl., 1985, 24, 295.
2. D.C. Billington, Chem. Soc. Rev., 1985, 14, 93.
3. G. Wilke, Pure Appl. Chem., 1984, 56, 1635.
4. P.W. Jolly, Angew. Chem. Int. Ed. Engl., 1985, 24, 283.
5. G. Erker, C. Kruger and G. Muller, Adv. Organomet. Chem., 1985, 24, 1.
6. R.D. Ernst, Accts. Chem. Res., 1985, 18, 56.
7. W. Siebert, Angew. Chem. Int. Ed. Engl., 1985, 24, 943.
8. K. Vrieze and G. van Koten, Inorg. Chim. Acta, 1985, 100, 79.
9. J. Halpern, Accts. Chem. Res., 1985, 24, 274.
10. W.E. Geiger and N.G. Connelly, Adv. Organomet. Chem., 1985, 24, 87.
11. K. Burgess, Polyhedron, 1984, 3, 1175.
12. R.D. Adams and I.T. Horvath, Prog. Inorg. Chem., 1985, 33, 127.
13. K. Angermund, K.H. Claus, R. Goddard and C. Kruger, Angew. Chem. Int. Ed. Engl., 1985, 24, 237.
14. D.M. Hollinshead and S.V. Ley, Gen. Synth. Methods, 1985, 7, 233.
15. Ed. F.R. Hartley and S. Patai, "Chemistry of the Metal-Carbon Bond", volumes 2 and 3, J. Wiley and Sons, 1985.
16. A.J. Pearson, "Metallo-Organic Chemistry", J. Wiley and Sons, 1985.
17. P.O. Widmark, B.O. Roos and P.E.M. Siegbahn, J. Phys. Chem., 1985, 24, 1547.
18. T. Ziegler, Inorg. Chem., 1985, 24, 1547.
19. K.R. Porschke, K. Jonas, G. Wilke, R. Benn, R. Mynott, R. Goddard and C. Kruger, Chem. Ber., 1985, 118, 275.
20. K.R. Porschke, G. Wilke and R. Mynott, Chem. Ber., 1985, 118, 298.
21. K.R. Porschke and G. Wilke, Chem. Ber., 1985, 118, 313.
22. W. Kleimann, K.R. Porschke and G. Wilke, Chem. Ber., 1985, 118, 323.
23. T.A. van der Knapp, F. Bickelhaupt, J.G. Kraaykamp, G. van Koten, J.P.C. Bernards, H.I. Edzes, W.S. Veeman, E. de Boer and E.J. Baerends, Organometallics, 1984, 3, 1804.
24. H.W. Kroto, S.I. Klein, M.R. Meidine, J.F. Nixon, R.K. Harris, K.J. Packer and P. Reams, J. Organomet. Chem., 1985, 280, 281.
25. R. Appel, C. Casser and F. Knoch, J. Organomet. Chem., 1985, 297, 21.
26. D.J. Pasto and N.Z. Huang, Organometallics, 1985, 4, 1386.
27. D.J. Pasto, N.Z. Huang and C.W. Eigenbrot, J. Am. Chem. Soc., 1985, 107, 3160.
28. W.R. Winchester and W.M. Jones, Organometallics, 1985, 4, 2228.
29. W.R. Winchester, M. Gawron, G.J. Palenik and W.M. Jones, Organometallics, 1985, 4, 1894.
30. S. Godleski, K.B. Gundlach and R.S. Valpey, Organometallics, 1985, 4, 296.
31. E. Dinjus, H. Gorls, E. Uhlig, D. Walter, J. Sieler, O. Linqvist and L. Andersen, J. Organomet. Chem., 1984, 276, 257.
32. M.A. Bennett and C. Chiraratvatana, J. Organomet. Chem., 1985, 296, 255.
33. R. Lazzaroni, G. Ucello-Baretta, S. Bertozzi and P. Salvadori, J. Organomet. Chem., 1984, 275, 145.
34. R. Lazzaroni, G. Ucello-Baretta, S. Bertozzi and P. Salvadori, J. Organomet. Chem., 1985, 297, 117.
35. V.G. Albano, F. Demartin, A. de Renzi, G. Morelli and A. Saporito, Inorg. Chem., 1985, 24, 2032.
36. E.J. Parsons, R.D. Larsen and P.W. Jennings, J. Am. Chem. Soc., 1985, 107, 1793.
37. L.V. Rybin, E.A. Petrovkaya, M.I. Rybinskaya, L.G. Kuzmina, Y.T. Struchkov, V.V. Kaverin and N.Y. Koneva, J. Organomet. Chem., 1985, 288, 119.
38. R. Bertani, G. Carturan, A. Scrivanti and A. Wojcicki, Organometallics, 1985, 4, 2139.
39. S.T. Lin, R.N. Narske and K.J. Klabunde, Organometallics, 1985, 4, 571.
40. R. Goddard, C. Krugar, F. Mark, R. Stansfield and K. Zhang, Organometallics, 1985, 4, 285.
41. R. Benn and A. Rufinska, Organometallics, 1985, 4, 209.

42. J.W. Faller, C. Blankenship, B. Whitmore and S. Sena, Inorg Chem., 1985, 24, 4483.
43. R. Benn, P.W. Jolly, R. Mynott and G. Schenker, Organometallics, 1985, 4, 1136.
44. R. Taube, J.P. Gehrke and R. Radeglia, J. Organomet. Chem., 1985, 291, 101.
45. A. Behr and G. von Ilseman, J. Organomet. Chem., 1984, 276, C77.
46. R. Benn, P.W. Jolly, R. Mynott, B. Raspel, G. Schenker, K.P. Schick and G. Schroth, Organometallics, 1985, 4, 1945.
47. A. Goliaszewska and J. Schwartz, Organometallics, 1985, 4, 415.
48. A. Goliaszewski and J. Schwartz, Organometallics, 1985, 4, 417.
49. S. Imaizumi, T. Matsuhisa and Y. Senda, J. Organomet Chem., 1985, 280, 441.
50. J.E. Backvall, R.E. Nordberg and D. Wilhelm, J. Am. Chem. Soc., 1985, 107, 6892.
51. B. Akermark and A. Vitagliano, Organometallics, 1985, 4, 1275.
52. R. Ciajolo, M.A. Jama, A. Tuzi and A. Vitagliano, J. Organomet. Chem., 1985, 295, 233.
53. D.H. Farrar and N.C. Payne, J. Amer. Chem. Soc., 1985, 107, 2054.
54. M. Bressan, L. DalMas, F. Morandini and B. Prozzo, J. Organomet. Chem., 1985, 287, 275.
55. H.F. Klein, K. Ellrich, S. Lamac, G. Lull, L. Zsolnai and G. Huttner, Z. Naturforsch., Teil B, 1985, 40, 1377.
56. D.M. Heddleton and R.N. Perutz, J. Chem. Soc. Chem. Comm., 1985, 1372.
57. P.O. Bentz, J. Ruiz, B.E. Mann, C.M. Spencer and P.M. Maitlis, J. Chem. Soc. Chem. Comm., 1985, 1374.
58. S. Gambarotta, C. Floriani, A. Chiesa-Villa and C. Gaustini, J. Organomet. Chem., 1985, 296, C6.
59. W.D. McGhee and R.G. Bergman, J. Am. Chem. Soc., 1985, 107, 3308.
60. R. Appel, W. Schuhn and F. Knoch, Angew. Chem., Int. Ed. Engl., 1985, 24, 420.
61. Z.H. Kafafi, R.H. Hauge and J.L. Margrave, J. Am. Chem. Soc., 1985, 107, 7550.
62. E.S. Kline, Z.H. Kafafi, R.H. Hauge and J.L. Margrave, J. Am. Chem. Soc., 1985, 107, 7559.
63. P.J. Krusic, R. Briere and P. Rey, Organometallics, 1985, 4, 801.
64. W.W. Wong, K.W. Chiu, J.A. Statler, G. Wilkinson, M. Motevalli and M.B. Hursthouse, Polyhedron, 1984, 3, 1255.
65. H. Werner and R. Werner, Chem. Ber., 1985, 118, 4543.
66. F.J. Manganiello, S.M. Oon, M.D. Ratcliffe and W.M. Jones, Organometallics, 1985, 4, 1069.
67. J.R. Lisko and W.M. Jones, Organometallics, 1985, 4, 612.
68. Y.R. Hu, T.W. Leung, S.C.H. Su, A. Wojcicki, M. Calligaris and G. Nardin, Organometallics, 1985, 4, 1001.
69. D.L. Reger and K.L. Belmore, Organometallics, 1985, 4, 305.
70. M.I. Bruce, D.N. Duffy, M.G. Humphrey and A.G. Swincer, J. Organomet. Chem., 1985, 282, 383.
71. K. Sano, T. Yamamoto and A. Yamamoto, Z. Naturforsch., Teil B, 1985, 40, 210.
72. H. Nagashima, K. Yamaguchi, K. Mukai and K. Itoh, J. Organomet. Chem., 1985, 291, C20.
73. Z.L. Lutsenko, G.G. Aleksandrov, P.V. Petrovskii, E.S. Shubina, V.G. Andrianov, Y.T. Struchkov and A.Z. Rubezhov, J. Organomet. Chem., 1985, 281, 349.
74. C. Conway, R.D.W. Kemmitt, A.W.G. Platt, D.R. Russell and L.J.S. Sherry, J. Organomet. Chem., 1985, 292, 419.
75. I. Bkouche-Waksman, I.S. Ricci, T.F. Koetzle, J. Weichmann and W.A. Herrmann, Inorg. Chem., 1985, 24, 1492.
76. A.M. Horton and S.V. Ley, J. Organomet. Chem., 1985, 285, C17.
77. M. Brookhart, K. Cox, F.G.N. Cloke, J.C. Green, J. Bashkin, A.E. Derome and P.D. Grebenik, J. Chem. Soc., Dalton Trans., 1985, 423.
78. V.C. Gibson, C.E. Gramain, P.M. Hare, M.L.H. Green, J.A. Bandy, P. Grebenik and K. Prout, J. Chem. Soc., Dalton Trans., 1985, 2025.

79. R. Alvarez, E. Carmora, D.J. Cole-Hamilton, A. Galindo, E. Gutierrez-Puebla, A. Morge, M.L. Poveda and C. Ruiz, J. Am. Chem. Soc., 1985, 107, 5529.
80. M. Canestrari, M.L.H. Green and A. Izquieredo, J. Chem. Soc., Dalton Trans., 1984, 2795.
81. G.A. Miller and N.J. Cooper, J. Am. Chem. Soc., 1985, 107, 709.
82. K.R. Pope and M.S. Wrighton, Inorg. Chem., 1985, 24, 2792.
83. H. Berke and C. Sontag, Z. Naturforsch., Teil B, 1985, 40, 794.
84. H. Berke, G. Huttner, C. Sontag and L. Zsolnai, Z. Naturforsch., Teil B, 1985, 40, 799.
85. R.T. de Pue, D.B. Collum, J.W. Ziller and M.R. Churchill, J. Am. Chem. Soc., 1985, 107, 2131.
86. T.Y. Luh and C.S. Wong, J. Organomet. Chem., 1985, 287, 231.
87. W. Vanarsdale, R.E.K. Winter and J.K. Kochi, J. Organomet. Chem., 1985, 296, 31.
88. C. Benamon and J. Benaim, J. Organomet. Chem., 1985, 280, 377.
89. J. Markham, A. Cutler and K. Menard, Inorg. Chem., 1985, 24, 1581.
90. M. Green, R.J. Mercer, C.E. Morton and A.G. Orpen, Angew. Chem., Int. Ed. Engl., 1985, 24, 422.
91. B.J. Brisdon, M. Cartwright and A.G. Hodson, J. Organomet. Chem., 1984, 277, 85.
92. A.J. Graham, D. Akrigg and B. Sheldrick, Acta Crystallogr., Section C., 1985, 41, 995.
93. R.P. Hughes, J.W. Reisch and A.L. Rheingold, Organometallics, 1985, 4, 1754.
94. R.P. Hughes, W. Klaui, J.W. Reisch and A. Muller, Organometallics, 1985, 4, 1761.
95. R.P. Hughes, J.W. Reisch and A.L. Rheingold, Organometallics, 1985, 4, 241.
96. J.C. Jeffery, A.L. Paterman and F.G.A. Stone, J. Organomet. Chem., 1985, 289, 367.
97. B. Buchmann and A. Salzer, J. Organomet. Chem., 1985, 295, 63.
98. R.B. Hitam, K.A. Mahmoud and A.J. Rest, J. Organomet. Chem., 1985, 291, 321.
99. M. Timmers and M. Brookhart, Organometallics, 1985, 4, 1365.
100. J.R. Bleeke and J.J. Kotyk, Organometallics, 1985, 4, 194.
101. D. Baudry, J.M. Cormier, M. Ephretikhine and H. Felkin, J. Organomet. Chem., 1984, 277, 99.
102. D. Baudry, P. Baydell, M. Ephretikhine, H. Felkin, J. Guilhem, C. Pascard and E.T.H. Dau, J. Chem. Soc., Chem. Comm., 1985, 670.
103. M. Moran, J.J. Santos-Garcia, J.R. Masaguer and V. Fernandez, J. Organomet. Chem., 1985, 295, 327.
104. K. Ihmels and D. Rehder, Chem. Ber., 1985, 118, 895.
105. J.W. Pattsina, C.E. Hissink, J.L. de Beor, A. Meetsma and J.H. Teuben, J. Am. Chem. Soc., 1985, 107, 7758.
106. F.G.N. Cloke, J.C. Green, M.L.H. Green and C.P. Morley, J. Chem. Soc. Chem. Comm., 1985, 945.
107. S.A. Cohen and J.E. Bercaw, Organometallics, 1985, 4, 1006.
108. G. Jeske, H. Lauke, H. Mauermann, P.N. Swepston, H. Schumann and T.J. Marks, J. Am. Chem. Soc., 1985, 107, 8091.
109. J.S. Thompson and R.M. Swiatek, Inorg. Chem., 1985, 24, 110.
110. J.S. Thompson, J.C. Calabrese and J.F. Whitney, Acta Crystallogr., Sect. C, 1985, 41, 891.
111. P.J.J.A. Timmermans, A. Mackor, A.L. Spek and B. Kojic-Prodic, J. Organomet. Chem., 1984, 276, 287.
112. B. Fischer, J. Boersma, B. Kojik-Prodic and A.L. Spek, J. Chem. Soc. Chem. Comm., 1985, 1237.
113. M.W. Eyring and L.J. Radonovich, Organometallics, 1985, 4, 1841.
114. A. Syed, E.D. Stevens and S.G. Cruz, Inorg. Chem., 1984, 23, 3673.
115. T.G. Appleton, J.R. Hall, D.W. Neale and M.A. Williams, J. Organomet. Chem., 1984, 276, C73.

116. G.R. Wiger, S.S. Tomita, M.F. Rettig and R.M. Wing, Organometallics, 1985, 4, 1157.
117. L.A. Oro, M.A. Ciriano, B.E. Villarroya, A. Tiripicchio and F.J. Lahoz, J. Chem. Soc. Dalton Trans., 1985, 1891.
118. G.W. Bushnell, S.R. Stobart, R. Vefghi and M.J. Zaworotko, J. Chem. Soc. Chem. Comm., 1984, 282.
119. M. Kretschmer, P.S. Pregosin, A. Albinati and A. Togni, J. Organomet. Chem., 1985, 281, 365.
120. G.G. Hlatky, B.F.G. Johnson, J. Lewis and P.R. Raithby, J. Chem. Soc. Dalton Trans., 1985, 1277.
121. F.A. Cotton, P. Lahuerta, M. Samau and W. Schwotzer, J. Am. Chem. Soc., 1985, 107, 8284.
122. F.H. Cano, C. Foces-Foces and L.A. Oro, J. Organomet. Chem., 1985, 288, 225.
123. F.A. Cotton and J.H. Meadows, Inorg. Chem., 1984, 23, 4688.
124. T.J. Chow, M.Y. Wu and L.K. Liu, J. Organomet. Chem., 1985, 281, C33.
125. T.J. Chow and Y.S. Chao, J. Organomet. Chem., 1985, 296, C23.
126. H.E. Oosthuizen, E. Singleton, J.S. Field and G.C. van Niekerk, J. Organomet. Chem., 1985, 279, 433.
127. T.V. Ashworth, D.C. Liles and E. Singleton, Organometallics, 1984, 3, 1851.
128. T.V. Ashworth, A.A. Chalmers, E. Meintjies, H. Oosthuizen and E. Singleton, J. Organomet. Chem., 1985, 286, 237.
129. T.V. Ashworth, A.A. Chalmers, E. Meintjies, H. Oosthuizen and E. Singleton, J. Organomet. Chem., 1984, 276, C19.
130. T.V. Ashworth, A.A. Chalmers, D.C. Liles, E. Meintjies, H.E. Oosthuizen and E. Singleton, J. Organomet. Chem., 1984, 276, C49.
131. T.V. Ashworth, A.A. Chalmers, D.C. Liles, E. Meintjies, H.E. Oosthuizen and E. Singleton, J. Organomet. Chem., 1985, 284, C19.
132. G. Vitulli, P. Pertici and C. Bigelli, Gazz., 1985, 115, 79.
133. M.A. Bennett, I.J. McMahon, S. Pelling, B.G. Robertson and W.A. Wickramasinghe, Organometallics, 1985, 4, 754.
134. M. Grassi, B.E. Mann and C.M. Spencer, J. Chem. Soc. Chem. Comm., 1985, 1169.
135. F. Bonachir, B. Chaudret, D. Neibecker and I. Tkatchenko, Angew. Chem., Int. Ed. Engl., 1985, 24, 347.
136. B. Chaudret and R. Poilblanc, Organometallics, 1985, 4, 1722.
137. M.O. Albers, H.E. Oosthuizen, D.J. Robinson, A. Shaver and E. Singleton, J. Organomet. Chem., 1985, 282, C49.
138. E.J.S. Vichi, F.Y. Fujiwara and E. Stein, Inorg. Chem., 1985, 24, 286.
139. D.N. Cox, R. Roulet and G. Chapuis, Organometallics, 1985, 4, 2001.
140. Y. Butsugan, A. Yamashita and S. Araki, J. Organomet. Chem., 1985, 287, 103.
141. A. Hafner, W. von Phillipsborn and A. Salzer, Angew. Chem., Int. Ed. Engl., 1985, 24, 126.
142. M.F. Semmelhack and H.T.M. Le, J. Am. Chem. Soc., 1985, 107, 1455.
143. G. Granozzi, E. Lorenzoni, R. Roulet, J.P. Daudey and D. Ajo, Organometallics, 1985, 4, 836.
144. J.C. Zwick and P. Vogel, Angew. Chem., Int. Ed. Engl., 1985, 24, 787.
145. E. Tagliaferri, P. Campiche, R. Roulet, R. Gabioud, P. Vogel and G. Chapuis, Helv. Chim. Acta, 1985, 68, 126.
146. E. Tagliaferri, U. Hamisch, R. Roulet, P. Vogel and K.J. Schenk, Helv. Chim. Acta, 1985, 68, 1362.
147. N. Morita, M. Oda and T. Asao, Tet. Letters, 1984, 25, 5419.
148. J.M. Grosselin, H. Le Bozec, C. Moinet, L. Toupet and P.H. Dixneuf, J. Am. Chem. Soc., 1985, 107, 2809.
149. M.D. Jones and R.D.W. Kemmitt, J. Chem. Soc. Chem. Comm., 1985, 811.
150. C. Guimon, G. Pfister-Guillouzo, J. Dubac, A. Laporterie, G. Manuel and H. Iloughmane, Organometallics, 1985, 4, 636.

151. A.J. Birch and L.F. Kelly, J. Organomet. Chem., 1985, 285, 267.
152. A.J. Pearson and T.R. Perrior, J. Organomet. Chem., 1985, 285, 253.
153. G.R. Stephenson, J. Organomet. Chem., 1985, 286, C41.
154. B.M.R. Bandara, A.J. Birch and L.F. Kelly, J. Org. Chem., 1984, 49, 2496.
155. A.J. Birch, B. Chauncy, L.F. Kelly and D.T. Thompson, J. Organomet. Chem., 1985, 286, 37.
156. H. Curtis, B.F.G. Johnson and G.R. Stephenson, J. Chem. Soc., Dalton Trans., 1985, 1723.
157. L.F. Kelly and A.J. Birch, Tet. Letters, 1984, 25, 6065.
158. J.A.S. Howell and M.J. Thomas, Organometallics, 1985, 4, 1054.
159. P. Eilbracht, R. Jelitte and L. Walz, Chem. Ber., 1984, 117, 3473.
160. A.J. Birch and L.F. Kelly, J. Organomet. Chem., 1985, 286, C5.
161. M. Moll, H. Behrens, W. Popp and P. Wurstl, Z. Anorg. Allg. Chem., 1984, 516, 127.
162. D.A. Brown, N.J. Fitzpatrick and M.A. McGinn, J. Organomet. Chem., 1985, 293, 235.
163. T.I. Odiaka, J. Chem. Soc., Dalton Trans., 1985, 1049.
164. T.I. Odiaka and J.I. Okogun, J. Organomet. Chem., 1985, 288, C30.
165. J. Alavosus and D.A. Sweigart, J. Am. Chem. Soc., 1985, 107, 985.
166. M. Crocker, M. Green, C.E. Morton, K.R. Nagle and A.G. Orpen, J. Chem. Soc., Dalton Trans., 1985, 2145.
167. R.G. Sutherland, R.L. Choudbury, A. Piorko and C.C. Lee, J. Chem. Soc., Chem. Comm., 1985, 1296.
168. C.A. Camaioni and D.A. Sweigart, J. Organomet. Chem., 1985, 282, 107.
169. E.D. Honig, M.Q. Jin, W.T. Robinson, P.G. Willard and D.A. Sweigart, Organometallics, 1985, 4, 871.
170. D. Astruc, P. Michaud, A.M. Madonik, J.Y. Saillard and A. Hoffmann, Nouv. J. Chim., 1985, 9, 41.
171. R. Goddard, F.W. Grevels and R. Schrader, Angew. Chem., Int. Ed. Engl., 1985, 24, 353.
172. G.M. Williams and D.E. Rudisill, J. Am. Chem. Soc., 1985, 107, 3357.
173. A.J. Pearson and B. Chen, J. Org. Chem., 1985, 50, 2587.
174. P. McArdle and S. Kavanagh, J. Organomet. Chem., 1985, 282, C1.
175. Z. Goldschmidt, S. Antebi, H. Gottlieb and D. Cohen, J. Organomet. Chem., 282, 369.
176. Z. Goldschmidt, H.E. Goffieb and D. Cohen, J. Organomet. Chem., 1985, 294, 219.
177. A. Tajiri, N. Morita, T. Asao and M. Hatano, Angew. Chem., Int. Ed. Engl., 1985, 24, 329.
178. T.H. Chang and J.I. Zink, Inorg. Chem., 1985, 24, 4016.
179. M.J. Hails, B.E. Mann and C.M. Spencer, J. Chem. Soc., Dalton Trans., 1985, 693.
180. K. Broadley, N.G. Connelly, P.G. Graham, J.A.K. Howard, W. Risse and M.W. Whiteley, J. Chem. Soc., Dalton Trans., 1985, 777.
181. N.G. Connelly, M.J. Freeman, A.G. Orpen, J.B. Sheridan, A.N.D. Symonds and M.W. Whiteley, J. Chem. Soc., Dalton Trans., 1985, 1027.
182. N.G. Connelly, A.R. Lucy, R.M. Mills, J.B. Sheridan and P. Woodward, J. Chem. Soc., Dalton Trans., 1985, 699.
183. L.C.A. de Carvalho, Y. Peres, M. Dartiguenave, Y. Dartiguenave and A.L. Beauchamp, Organometallics, 1985, 4, 2021.
184. E.D. Sternberg and K.P.C. Vollhardt, J. Org. Chem., 1984, 49, 1564.
185. E.D. Sternberg and K.P.C. Vollhardt, J. Org. Chem., 1984, 49, 1574.
186. J.M. Boncella, F. Okino, C. Orvig, R.P. Planalp and G.L. Rosenthal, Acta Crystallogr., Sect. C, 1985, 41, 833.
187. E. Dunach, R.L. Haltermann and K.P.C. Vollhardt, J. Am. Chem. Soc., 1985, 107, 1664.
188. L.S. Liebeskind and C.F. Jewell, J. Organomet. Chem., 1985, 285, 305.
189. C.F. Jewell, L.S. Liebeskind and M. Williamson, J. Am. Chem. Soc., 1985, 107, 6715.
190. J.S. Drage and K.P.C. Vollhardt, Organometallics, 1985, 4, 389.
191. R.H. Crabtree and C.P. Parnell, Organometallics, 1985, 4, 519.

192. L. Dahlenberg and N. Hock, J. Organomet. Chem., 1985, 284, 129.
193. M.L.H. Green and G. Parkin, J. Chem. Soc., Chem. Comm., 1985, 771.
194. M.L.H. Green, D. O'Hare and G. Parkin, J. Chem. Soc., Chem. Comm., 1985, 356.
195. J.A. Bandy, V.S.B. Mtetwa, K. Prout, J.C. Green, C.E. Davies, M.L.H. Green, N.J. Hazel, A. Izquiredo and J.J. Martin-Polo, J. Chem. Soc., Dalton Trans., 1985, 2037.
196. R. Drews and U. Behrens, Chem. Ber., 1985, 118, 888.
197. W. Gerhartz, F.W. Grevels, W.E. Klotzbucher, E.A.K. von Gustorf and R.N. Perutz, Z. Naturforsch., Teil B, 1985, 40, 518.
198. C.G. Kreiter and M. Kotzian, J. Organomet. Chem., 1985, 289, 295.
199. G. Michael, J. Kaub and C.G. Kreiter, Chem. Ber., 1985, 118, 3944.
200. G. Michael, J. Kaub and C.G. Kreiter, Angew. Chem., Int. Ed. Engl., 1985, 24, 502.
201. S.D. Reynolds and T.A. Albright, Organometallics, 1985, 4, 980.
202. C.D. Hoff, J. Organomet. Chem., 1985, 282, 201.
203. S. Ozkar and C.G. Kreiter, J. Organomet. Chem., 1985, 283, 229.
204. E. Michels, W.S. Sheldrick and C.G. Kreiter, Chem. Ber., 1985, 118, 964.
205. D.A. Brown, M.J. Fitzpatrick, W.K. Glass and T.H. Taylor, J. Organomet. Chem., 1984, 275, C9.
206. D.A. Brown, N.J. Fitzpatrick and M.A. McGinn, J. Organomet. Chem., 1984, 275, C5.
207. T.I. Odiaka and L.A.P. Kane-Maguire, J. Organomet. Chem., 1985, 284, 35.
208. T.I. Odiaka, Inorg. Chim. Acta, 1985, 103, 9.
209. U. Behrens, J. Kopf, K. Lal and W.E. Watts, J. Organomet. Chem., 1984, 276, 193.
210. M. Green, S. Greenfield and M. Kersting, J. Chem. Soc., Chem. Comm., 1985, 18.
211. A.J. Pearson, M.N.I. Khan, J.C. Clardy and H. Cun-Leng, J. Am. Chem. Soc., 1985, 107, 2748.
212. D.C. Bourner, L. Brammer, M. Green, G. Moran, A.G. Orpen, C. Reeve and C.J. Schaverien, J. Chem. Soc., Chem. Comm., 1985, 1409.
213. A.D. Hunter, P. Legzdins and C.R. Nurse, J. Am. Chem. Soc., 1985, 107, 1791.
214. L. Weber and R. Boese, Chem. Ber., 1985, 118, 1545.
215. L. Weber, Z. Naturforsch., Teil B, 1985, 40, 373.
216. N.G. Connelly, M.J. Freeman, A.G. Orpen, A.R. Sheehan, J.B. Sheridan and D.A. Sweigart, J. Chem. Soc., Dalton Trans., 1985, 1019.
217. Y.K. Chung, D.A. Sweigart, N.G. Connelly and J.B. Sheridan, J. Am. Chem. Soc., 1985, 107, 2388.
218. M.L.H. Green and D.O'Hare, J. Chem. Soc., Chem. Comm., 1985, 332.
219. C.G. Kreiter, K.H. Franzreb, W. Michaels, U. Schubert and K. Ackermann, Z. Naturforsch., Teil B, 1985, 40, 1188.
220. M.C.L. Trimarchi, M.A. Green, J.C. Huffmann and K.G. Caulton, Organometallics, 1985, 4, 514.
221. J.C. Green, M. Paz-Sandoval and P. Powell, J. Chem. Soc., Dalton Trans., 1985, 2677.
222. L. Stahl, J.P. Hutchinson, D.R. Wilson and R.D. Ernst, J. Am. Chem. Soc., 1985, 107, 5016.
223. C. Kruger, G. Muller, G. Erker, U. Dorf and K. Engel, Organometallics, 1985, 4, 215.
224. G. Erker, C. Kruger, Y.H. Tsay, E. Samuel and P. Vogel, Z. Naturforsch., Teil B, 1985, 40, 150.
225. G. Erker, K. Engel, U. Korek, P. Czisch, H. Berke, P. Caubere and R. Vanderesse, Organometallics, 1985, 4, 1531.
226. G. Erker, T. Muhlenbernd, R. Benn, A. Rufinska, Y.H. Tsay and C. Kruger, Angew. Chem., Int. Ed. Engl., 1985, 24, 321.
227. H. Yasuda, K. Tatsumi, T. Okamoto, K. Mashima, K. Lee, A. Nakamura, Y. Kai, N. Kanehisa and N. Kasai, J. Am. Chem. Soc., 1985, 107, 2410.
228. J. Blenkers, P. Bruin and J.H. Teuben, J. Organomet. Chem., 1985, 297, 61.
229. C.E. Davies, I.M. Gardiner, J.C. Green, M.L.H. Green, N.J. Hazel, P.D. Grebenik, V.S.B. Mtetwa and K. Prout, J. Chem. Soc., Dalton Trans., 1985, 669.

230. J.J. Eisch, A.M. Piotrowski, A.A. Aradi, C. Kruger and M.J. Romao, Z. Naturforsch., Teil B, 1985, **40**, 624.
231. R.P. Hughes, J.M.J. Lambert and A.L. Rheingold, Organometallics, 1985, **4**, 2055.
232. J. Silvestre and T.A. Albright, J. Am. Chem. Soc., 1985, **107**, 6829.
233. H. Lehmkuhl and C. Naydowski, J. Organomet. Chem., 1984, **277**, C18.
234. R. Benn, B. Bussemeier, S. Holle, P.W. Jolly, R. Mynott, I. Tkatchenko and G. Wilke, J. Organomet. Chem., 1985, **279**, 63.
235. K.R. Porschke, Y.H. Tsay and C. Kruger, Angew. Chem., Int. Ed. Engl., 1985, **24**, 323.
236. K.R. Porschke, R. Mynott, K. Angermund and C. Kruger, Z. Naturforsch., Teil B, 1985, **40**, 199.
237. J.D. Ferrara, C. Tessier-Youngs and W.J. Youngs, J. Am. Chem. Soc., 1985, **107**, 6719.
238. H.G. Alt, J. Organomet. Chem., 1985, **288**, 149.
239. H.G. Alt, H.E. Engelhardt, U. Thewalt and J. Riede, J. Organomet. Chem., 1985, **288**, 165.
240. D.J. Wink, J.R. Fox and N.J. Cooper, J. Am. Chem. Soc., 1985, **107**, 5012.
241. M.A. Bennett and I.W. Boyd, J. Organomet. Chem., 1985, **290**, 165.
242. M.A. Bennett, I.W. Boyd, G.B. Robertson and W.A. Wickramasinghe, J. Organomet. Chem., 1985, **290**, 181.
243. J.R. Morrow, T.L. Tonker and J.L. Templeton, J. Am. Chem. Soc., 1985, **107**, 6956.
244. J.R. Morrow, T.L. Tonker and J.L. Templeton, J. Am. Chem. Soc., 1985, **107**, 5004.
245. G.J.J. Chen, J.W. McDonald and W.E. Newton, Organometallics, 1985, **4**, 422.
246. J.L. Davidson and G. Vasapollo, J. Chem. Soc., Dalton Trans., 1985, 2239.
247. J.E. Salt, G.S. Girolami, G. Wilkinson, M. Motevalli, M. Thornton-Pett and M.B. Hursthouse, J. Chem. Soc., Dalton Trans., 1985, 685.
248. J.S. Adams, M. Cunningham and M.W. Whiteley, J. Organomet. Chem., 1985, **293**, C13.
249. A.J.L. Pombiero, J.C. Jeffery, C.J. Pickett and R.L. Richards, J. Organomet. Chem., 1984, **277**, C10.
250. K.R. Birdwhistell, T.L. Tonker and J.L. Templeton, J. Am. Chem. Soc., 1985, **107**, 4474.
251. F.J.G. Alonso, A. Hohn, J. Wolf, H. Otto and H. Werner, Angew. Chem., Int. Ed. Engl., 1985, **24**, 406.
252. F.R. Kreissl, W.J. Sieber, P. Hoffmann, J. Riede and M. Wolfgruber, Organometallics, 1985, **4**, 788.
253. S.R. Allen, R.G. Beevor, M. Green, N.C. Norman, A.G. Orpen and I.D. Williams, J. Chem. Soc., Dalton Trans., 1985, 435.
254. L. Carlton, J.L. Davidson, P. Ewing, L. Manojlovic-Muir and K.W. Muir, J. Chem. Soc., Chem. Comm., 1985, 1474.
255. J.L. Davidson, W.F. Wilson and K.W. Muir, J. Chem. Soc., Chem. Comm., 1985, 460.
256. H.G. Alt and H.E. Engelhardt, Z. Naturforsch., Teil B., 1985, **40**, 1134.
257. M.D. Curtis and J. Real, Organometallics, 1985, **4**, 940.
258. D.J. Sikora and M.D. Rausch, J. Organomet. Chem., 1984, **276**, 21.
259. S.B. Jones and J.L. Petersen, Organometallics, 1985, **4**, 966.
260. H.G. Alt, H.E. Engelhardt, M.D. Rausch and L.B. Kool, J. Am. Chem. Soc., 1985, **107**, 3717.
261. G. Erker, U. Dorf, R. Mynott, Y.H. Tsay and C. Kruger, Angew. Chem., Int. Ed. Engl., 1985, **24**, 584.
262. F.W.B. Einstein, K.G. Tyers and D. Sutton, Organometallics, 1985, **4**, 489.
263. K. Raab, M. Appel and W. Beck, J. Organomet. Chem., 1985, **291**, C28.
264. J.M. Mayer, D.L. Thorn and T.H. Tulip, J. Am. Chem. Soc., 1985, **107**, 7454.
265. Y. Wakatsuki, S. Miya, S. Ikuta and H. Yamazaki, J. Chem. Soc., Chem. Comm., 1985, 35.
266. M.F. Semmelhack, R. Tamura, W. Schnatter and J. Springer, J. Am. Chem. Soc., 1984, **106**, 5363.
267. H.W. Fruhauf, F. Sells, R.J. Goddard and M.J. Romao, Organometallics, 1985, **4**, 948.

268. F. Ungvary, J. Shamshool and L. Marko, J. Organomet. Chem., 1985, 296, 155.
269. D. Osella, S. Aime, D. Boccardo, M. Castiglioni and L. Milone, Inorg. Chim. Acta, 1985, 100, 97.
270. S.J. Doig, R.P. Hughes, R.E. Davis, S.M. Gadol and K.D. Holland, Organometallics, 1984, 3, 1921.
271. W.A. Smit, A.A. Shchegolev, A.S. Gubin, G.S. Mikaelyan and R. Caple, Synthesis, 1984, 887.
272. P.D. Magnus and D.P. Becker, J. Chem. Soc., Chem. Comm., 1985, 640.
273. R.S. Dickson, G.S. Evans and G.D. Falton, Aust. J. Chem., 1985, 38, 273.
274. M.C. Azar, M.J. Chetcuti, C. Eigenbrot and K.A. Green, J. Am. Chem. Soc., 1985, 107, 7209.
275. M. Green, A.G. Orpen, C.J. Schaverien and I.D. Williams, J. Chem. Soc., Dalton Trans., 1985, 2483.
276. R.S. Dickson, G.D. Fallon, R.J. Nesbit and G.N. Pain, Organometallics, 1985, 4, 355.
277. R.S. Dickson, R.J. Nesbit, H. Pateras, W. Bainbridge, J.M. Patrick and A.H. White, Organometallics, 1985, 4, 2128.
278. M. Cowie, R.S. Dickson and B.W. Hames, Organometallics, 1984, 3, 1879.
279. B.R. Sutherland and M. Cowie, Organometallics, 1984, 3, 1869.
280. J.T. Mague, C.L. Klein, R.J. Majeste and E.D. Stevens, Organometallics, 1984, 3, 1860.
281. G.W. Bushnell, M.J. Decker, D.T. Eadie, S.R. Stobart, R. Vefghi, J.L. Atwood and M.J. Zaworotko, Organometallics, 1985, 4, 2106.
282. D.H. Berry and R. Eisenberg, J. Am. Chem. Soc., 1985, 107, 7181.
283. M.D. Fryzuk, W.E. Piers and S.J. Rettig, J. Am. Chem. Soc., 1985, 107, 8259.
284. M.H. Chisholm, J.C. Huffmann and J.A. Heppert, J. Am. Chem. Soc., 1985, 107, 5116.
285. M.H. Chisholm, K. Folting, J.A. Heppert and W.E. Streib, J. Chem. Soc., Chem. Comm., 1985, 1755.
286. M.H. Chisholm, J.A. Heppert, J.C. Huffmann and W.E. Streib, J. Chem. Soc., Chem. Comm., 1985, 1771.
287. M.H. Chisholm, B.W. Eichorn and J.C. Huffmann, J. Chem. Soc., Chem. Comm., 1985, 861.
288. K.J. Ahmed, M.H. Chisholm, K. Folting and J.C. Huffmann, J. Chem. Soc., Chem. Comm., 1985, 152.
289. N.A. Ustynuk, V.N. Vinogradova, V.N. Korneva, D.N. Kravtsov, V.G. Andrianov and Y.T. Struchkov, J. Organomet. Chem., 1984, 277, 285.
290. G.A. Garriedo, J.A.K. Howard, D.B. Lewis, G.E. Lewis and F.G.A. Stone, J. Chem. Soc., Dalton Trans., 1985, 905.
291. P. Bougeard, S. Peng, M. Mlekuz and M.J. McGlinchey, J. Organomet. Chem., 1985, 296, 383.
292. C.P. Casey, W.H. Miles, P.J. Fagan and K.J. Haller, Organometallics, 1985, 4, 559.
293. C.P. Casey, S.R. Marder and A.L. Rheingold, Organometallics, 1985, 4, 762.
294. D.L. Davies, J.A.K. Howard, S.A.R. Knox, K. Marsden, K.A. Mead, M.J. Morris and M.C. Rendle, J. Organomet. Chem., 1985, 279, C37.
295. X. Solans, M. Font-Alba, J. Ros, R. Yanez and R. Mathieu, Acta Crystallogr., Sect. C, 1985, 41, 1186.
296. J. Ros, X. Solans, C. Miravitlles and R. Mathieu, J. Chem. Soc., Dalton Trans., 1985, 1981.
297. J. Ros, G. Commenges, R. Mathieu, X. Solans and M. Font-Alba, J. Chem. Soc., Dalton Trans., 1985, 1087.
298. K. Knoll, O. Oroma and G. Huttner, Angew. Chem., Int. Ed. Engl., 1984, 23, 976.
299. G.Y. Lin and J. Takats, J. Organomet. Chem., 1984, 269, C4.
300. K.W. Lee, W.T. Pennington, A.W. Cordes and T.L. Brown, J. Am. Chem. Soc., 1985, 107, 631.
301. K.W. Lee and T.L. Brown, Organometallics, 1985, 4, 1030.

302. C.G. Kreiter and M. Leyendecker, J. Organomet. Chem., 1985, 280, 225.
303. C.G. Kreiter and M. Leyendecker, J. Organomet. Chem., 1985, 292, C18.
304. D.B. Pourreau, R.R. Whittle and G.L. Geoffroy, J. Organomet. Chem., 1984, 273, 333.
305. A.D. Horton, M.J. Mays and P.R. Raithby, J. Chem. Soc., Chem. Comm., 1985, 247.
306. D. Nuel, F. Dahan and R. Mathieu, Organometallics, 1985, 4, 1436.
307. D. Nuel, F. Dahan and R. Mathieu, J. Am. Chem. Soc., 1985, 107, 1658.
308. M.K. Alami, F. Dahan and R. Mathieu, Organometallics, 1985, 4, 2122.
309. S.B. Colbran, B.H. Robinson and J. Simpson, Organometallics, 1985, 4, 1594.
310. E. Roland, W. Bernhardt and H. Vahrenkamp, Chem. Ber., 1985, 118, 2858.
311. M. Mlekuz, P. Bougeard, B.G. Sayer, S. Peng, M.J. McGlinchey, A. Marinetti, J.Y. Saillard, J.B. Naceur, B. Mentzen and G. Jaouen, Organometallics, 1985, 4, 1123.
312. J.F. Halet, J.Y. Saillard, R. Lissilour, M.J. McGlinchey and G. Jaouen, Inorg. Chem., 1985, 24, 218.
313. G. Granozzi, F. Tondello, R. Bertonello, S. Aime and D. Osella, Inorg. Chem., 1985, 24, 570.
314. M.R. Churchill, J.C. Fettinger, J.B. Keister, R.F. See and J.W. Ziller, Organometallics, 1985, 4, 2112.
315. L.A. Beanan and J.B. Keister, Organometallics, 1985, 4, 1713.
316. D.M. Dalton and J.B. Keister, J. Organomet. Chem., 1985, 290, C37.
317. M.I. Bruce and M.L. Williams, J. Organomet. Chem., 1985, 288, C55.
318. M.I. Bruce, M.L. Williams, J.M. Patrick and A.H. White, J. Chem. Soc., Dalton Trans., 1985, 1229.
319. J.C. Daran, Y. Jeannin and O. Kristiansson, Organometallics, 1985, 4, 1882.
320. M.I. Bruce and M.L. Williams, J. Organomet. Chem., 1985, 282, C11.
321. M.I. Bruce, B.W. Shelton and A.H. White, J. Organomet. Chem., 1985, 282, C53.
322. C. Weatherell, N.J. Taylor, A.J. Carty, E. Sappa and A. Tiripicchio, J. Organomet. Chem., 1985, 291, C9.
323. G. Gervasio, R. Rosetti and P.L. Stanghellini, Organometallics, 1985, 4, 1612.
324. A. Tiripicchio, M.T. Camellini and E. Sappa, J. Chem. Soc., Dalton Trans., 1984, 627.
325. E. Sappa, A.M.M. Lanfredi, G. Prediari, A. Tiripicchio and A.J. Carty, J. Organomet. Chem., 1985, 288, 365.
326. Y. Chi and J.R. Shapley, Organometallics, 1985, 4, 1900.
327. M.P. Gomez-Sal, B.F.G. Johnson, R.A. Kamarudin, J. Lewis and P.R. Raithby, J. Chem. Soc., Chem. Comm., 1985, 1622.

14
π-Cyclopentadienyl, π-Arene, and Related Complexes*

BY A. H. WRIGHT

1 Introduction

The π-bound benzene ligand and the related cyclopentadienyl continue to fascinate the chemist. Investigations into the bonding in arene complexes have been reviewed for the major class of (arene)Cr(CO)$_3$ molecules.[1] Within the field dealing with sandwich complexes (those containing two cyclic polyolefinic ligands) the most noticeable trend is the continued increase in reports dealing with early transition metal sandwich complexes. A slightly different aspect, multilayered sandwich compounds, has been reviewed.[2] The synthetic application of the degradation of metallocenes has also been reviewed.[3] In other cases the aromatic ring is important for its influence on the metal,[4,5] or as a unit for elaboration in organic synthesis. This latter aspect has also been reviewed.[6]

Emphasis in this report is placed on chemistry that involves the aromatic ligand ultimately. Reactions that involve replacement or elaboration of other ligands present are generally not included. Similarly, a very selective approach is taken to Cp or arene complexes that contain metal-metal bonds, carbene or carbyne, or hydrocarbyl ligands (see Chapters 9-13). Only those metalloborane and metallocarborane complexes that incorporate a Cp or arene ligand are described and crystal structure determinations are included only where results of particular significance or solutions to structural problems are provided.

2 Monocyclopentadienyl Complexes

2.1 Titanium, Zirconium and Hafnium.- A mild transfer reagent for the Cp ligand, Cp$_2$Mg has been reacted with ZrCl$_4$ to give CpZrCl$_3$.[7] Functionalised Cp rings, C$_5$H$_4$X (X = halogen) have been introduced

---

* Throughout this review the abbreviations Cp, Cp', Cp* and hmb explicitly denote (η^5-C$_5$H$_5$), (η^5-C$_5$H$_4$Me), (η^5-C$_5$Me$_5$) and (η^6-C$_6$Me$_6$) respectively.

into titanium complexes.[8] The cluster $Cp_5Ti_5S_6$ is formed in the reaction of H_2S with $Cp_2Ti(CO)_2$.[10] The use of $CpTiCl_2(OC_6H_4X)$ for olefin hydrogenation and polymerisation has been tested.[11] Heterocycles can be used with $CpTiCl_3$ to substitute the chlorides.[12]

2.2 Vanadium, Niobium and Tantalum.- A highly substituted cyclopentadienyl ligand is generated by cyclisation of two acetylenes with a carbyne ligand in the complex (1)[13] when $Cp*TaCl_3(SiMe_3)$ is treated with CO, a product involving insertion into the Ta-Si bond can be identified (2).[14] Reduction of $CpMCl_4$ (M = Nb, Ta) with Al/CO gives chlorobridged dimeric products.[15]

2.3 Chromium, Molybdenum and Tungsten.- The half sandwich complex anions $[CpM(CO)_3]^-$ (M = Cr, Mo, W) can be synthesised directly from C_5H_6 and $LiEt_3BH$.[16] While the direct reaction of indene with $Mo(CO)_6$ is reported to give the dimer $(indenyl)_2Mo_2(CO)_4$.[17] The reaction of fulvene with both Mo and W carbonyl complexes gives fulvene products and also dimeric complexes with Cp ligands.[18]

The addition of cyclopentadienyl ligands to electron rich complexes $WH_{(12-n)}(PMe_3)_{n/2}$ has been studied and products include hydrogenation of the Cp ring.[19] A Cp ring is formed from the combination of two acetylene and an isocyanide ligand in (3).[20] The unusual bistungsten complex (4) results from the coupling of 3,7-decadiyne and $W(C_3Et_3)(OCMe_2CMe_2O)(OCMe_3)$.[21]

The lability of Cp ligands has been utilised in the oxidative clusterfication of Cp_2Cr with Me_3NO to give $Cp_4Cr_4O_3-\eta^2-C_5H_4$).[22] Ring slippage $\eta^5-\eta^3$ remains a subject of study and the direction of ring slippage has been investigated in $(indenyl)Mo(CO)_2(\eta^3-C_3H_5)$ and $(indenyl)IrH(PPh_3)_2$.[23] The 16-electron complexes $CpM(NO)R_2$, (M = Mo, W and R = CH_2CMe_3, CH_2SiMe_3), can be made from the dimers $[CpM(NO)I_2]_2$.[24] The cationic 16-electron intermediate $[CpW(CO)_3]^+$ can be utilised synthetically starting with $CpW(CO)_3(BF_4)$.[25,26]

High valent complexes including $Cp*WCl_4$ are produced when $Cp*W(CO)_3Me$ is reacted with PCl_5.[27] Other high valent chemistry includes that of $CpWO_2R$[28] complexes. Oxidation of $[CpMo(CO)_3]^-$ gives the carbonate complex $[CpMo(CO)_2(CO_3)]^-$.[29] By contrast, oxidation of $Cp*Cr(CO)_2(NO)$ leads to $Cp_2Cr_2O_4$.[30]

(1), (2), (3), (4), (5), (6), (7), (8)

The reactions of $CpM(CO)_x$ (M = Cr, Mo, W) with other Group VIII compounds have been reported.[31-36] The different conformations of $Cp_2Mo_2S_4$ have been explored using interactive molecular graphics.[37]

A report of $CpW(CO)_3H$ reacting with SO_2 to give dimeric products has appeared.[38] Cleavage of dimeric complexes has also been examined. The hydrolysis of $Cp*_2Cr_2(CO)_4$ gives $Cp*Cr(CO)_3H$ and $[Cp*_4Cr_4(OH)_6]^{2+}$.[39] $Cp'_2Mo_2(\mu-S)_2(\mu-SH)_2$ has been tested as a catalyst for the reduction of SO_2 with H_2.[40]

Chemistry of the dimeric complexes $Cp_2M_2(CO)_{6 \text{ or } 4}$ (M = Cr, Mo or W) includes the addition of H_2, which is reported to be exothermic for Cr and W and endothermic for Mo.[41] When the dimers are reacted with nitrite, nitrosyl complexes are produced.[42] Photodisproportionation of the Cp' dimers of Mo has been studied with halide ions.[43]

Reactions involving unsaturated organic substrates include the regioselective and stereoselective manipulation of 1,3-dienes using indenyl Mo complexes.[44] $CpMoMn(CO)_6(\mu-H)-(\mu-PPh_2)$ has been photolysed in the presence of dienes and acetylenes.[45] Addition of 1,3-diodopropane to the carbonyl ligand of the dimer (5) has been reported.[46]

2.4 Manganese and Rhenium.-

Reports of reactions in which the Cp ligand is implicated in the reactions is particularly common for Re. Lithiation of the ring is the initial product when $CpRe(PPh_3)(NO)H$ is treated with ^nBuLi. Subsequent rearrangement gives $Li[CpRe(PPh_3)(NO)]$.[47]

Cleavage of the Cp ring from $CpRe(CO)(NO)Me$ can be effected using PMe_3.[48] An initial product containing η^1-Cp (6) can be identified. Further PMe_3 then substitutes the ligand. Alternatively, treatment of (6) with CO leads to multiple ligand migration reactions to give (7).[49] In a comparable reaction (indenyl)$Re(CO)_3$ has been treated with PMe_3 to give $(\eta^1\text{-indenyl})Re(CO)_3(PMe_3)_2$.[50] The η^5-η^6 equilibrium for the system $\underline{1}$:

$$(\eta^5\text{-indenyl})Mn(CO)_3 \underset{-H^+}{\overset{H^+}{\longrightarrow}} [(\eta^6\text{-indene})Mn(CO)_3]^+ \quad (1)$$

has been examined.[51]

In the thermolytic reaction of $CpRe(CO)_2H$ with phosphines, kinetic studies indicate the initial product results from ring slippage rather than CO loss.[52] When $(\eta^4-C_5H_6)Re(PPh_3)_2H_3$ is heated, $CpRe(PPh_3)_2H_2$ is formed.[53]

Reports of the use of $CpM(CO)_2$ (M = Mn and Re) fragments include the synthesis of amine complexes[54] and acetylene complexes.[55] The $Cp*Re(CO)(PMe_3)$ fragment has been reported to activate C-H bonds.[56]

High oxidation state Cp* complexes have been made by treating $Cp*Re(CO)_3$ with H_2O_2[57] and the chemistry the resulting $Cp*ReO_3$ is unfolding.[58,59] Air oxidation of $Cp*Mn(CO)_2S_2$ leads to oxidation of the S_2 ligand giving $Cp*Mn(CO)_2(S_2O)$.[60]

Electrochemical investigations of $CpMn(CO)_2(ER)$[61] (E = S, Se) and $CpRe(PPh_3)(NO)R$[62] have been made. Polynuclear complexes involving Ge[63] and As[64] with $CpMn(CO)_2$ fragments have been characterised.

2.5 Iron, Ruthenium and Osmium.- The range of indenyl complexes has been increased with the synthesis of $(indenyl)Ru(PPh_3)_2Cl$ from $Ru(PPh_3)_3Cl_2$.[65] A more conventional approach to Cp complexes has succeeded with the reaction of $Ru_3(CO)_{12}$ with C_5H_6 and chelating phosphines to give complexes of the form $CpRuH(LL)$.[66] The direct reaction of dicyclopentadiene with $Fe(CO)_4(CNAr)$ generates isocyanide-substituted dimers including $Cp_2Fe_2(CO)_3(CNAr)$.[67]

When $CpFe(CO)_2(CH_2Ph)$ is photolysed in the presence of PPh_3, $CpFe(CO)_2(PPh_3)$ and (8) are formed.[68] The scope of the lithiation/migration reaction of $CpFe(CO)_2(SiMe_3)$ to give $Li[(C_5H_4SiMe_3)Fe(CO)_2]$ has been investigated.[69] The investigation of Cp-rich-sulphur systems continues with the report of the sulphur-rich complex (9).[70]

The use of unsaturated Cp complexes has been extended with the use of $CpRu(COD)Cl$[71,72] (COD = cyclooctadiene). For iron the photochemical dissociation of CO in $CpFe(CO)_2R$ has been examined[73,74]. The related cation $[CpFe(CO)_2(THF)]^+$ has also been investigated.[75] Low temperature photodissociation of xylene from $[CpFe(xylene)]^+$ gives the intermediate $[CpFe(NCMe)_3]$.[76]

The new dimeric complex, $Cp*_2Fe_2(\mu-CO)_3$, has been structurally characterised and reported to exist in a triplet ground state.[77] Anionic bimetallic dimers result from the reaction of $[CpFe(CO)_2]^-$ with $Cp'Mn(CO)_2(NCMe)$.[78]

Spectroscopic results indicate an increasing stability of the complexes $[CpFe(CO)_n(P(OR)_3)_{3-n}]^+$ as the degree of phosphite substitution increases.[79]

Oxygen exchange in the CO_2 complexes of $[CpFe(CO)_2]^-$ has been investigated.[80] Thionitrosyl ligands have been generated by reacting $[NS][AsF_6]$ with $[CpFe(CO)_2(SO_2)]^+$.[81] The dimeric complex (10) contains a linear CN bridge.[82] Reactions of the Fe-P multiple bond in $CpFe(CO)_2(PPh_2)$ have been investigated.[83]

The reaction of $CpRu(PPh_3)_2Cl$ with metal anions has been examined,[84] as has the solvolysis reactions of $CpRu(PMe_3)X$ (X = halogen).[85] Various $Cp*Fe(CO)_2X$ (X = halogen) complexes can be made by cleaving the parent dimeric complex.[86]

2.6 Cobalt, Rhodium and Iridium.-

A simple reaction of $Rh_2(COD)_2Cl_2$ with NaC_5H_5 and phosphines gives $CpRh(PR_3)_2$.[87] New functionalised Cp ligands have been introduced in, for example, the cobalt complex $(carbomethoxyCp)Co(CO)_2$.[88] A number of Cp*Ir (polyhydrides) including $Cp*IrH_4$ have been reported.[89] The synthesis of dimeric complexes such as (11) has allowed a study of reactivity including the tendency to form metal-metal bonds.[90]

An investigation of ^{59}Co n.q.r. spectra of $CpCoL_n$ and $Cp*CoL_n$ complexes has shown that the Cp* inductively donates more electron density to the metal than Cp, but the change is relatively small (less than 10%).[92] The photoelectron spectra of complexes $CpM(PMe_3)X$ (M = Co, Rh and X = CO, CS) have been interpreted in terms of a stronger interaction of Rh with orbitals in the Cp ring.[93]

A photochemical study of C-H activation of benzene by $Cp*Ir(PPh_3)H(OEt)$ has lead to a proposed intermediate derived from either ring slippage or alkoxide migration to the ring.[94] The activation of C-H bonds is also possible with $Cp*Ir(\eta^3-C_3H_5)H$.[95] Related complexes such as $CpRh(C_2H_4)_2$[96] and $CpM(DMSO)Me_2$[97,98] (M = Ir, Rh and DMSO = dimethyl- sulphoxide). In the case of $CpCo(C_2H_4)_2$, clusterification results from heating, giving $Cp_4Co_4H(CCH_3)$.[99]

Catalytic applications include cyclotrimerisation reactions using various $CpRhL_2$ complexes.[100] Using CpRh(COD) the catalytic reaction of S_8 with acetylenedicarboxylates has been reported.[102] $Cp*Co(PR_3)(olefin)H$ has been used for alkene polymerisation.[103]

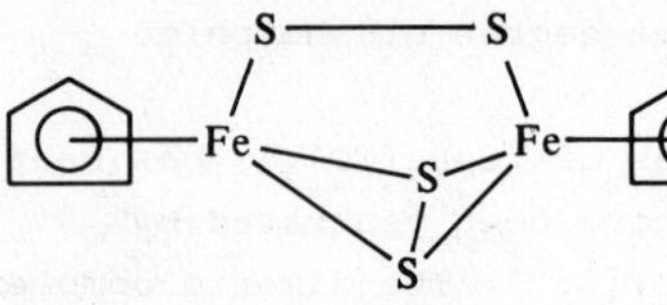

(9)

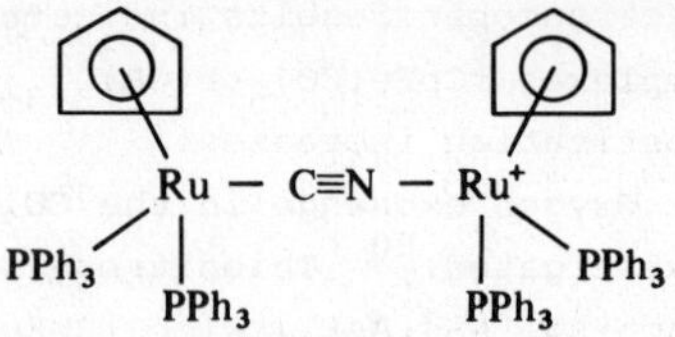

(10)

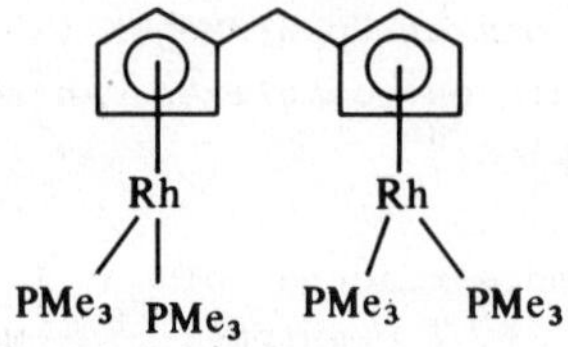

(11)

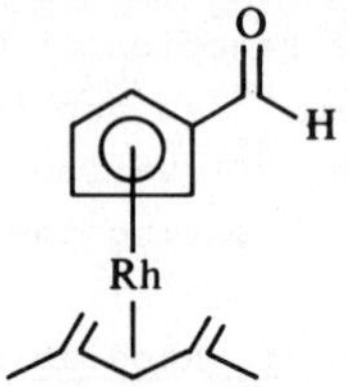

(12)

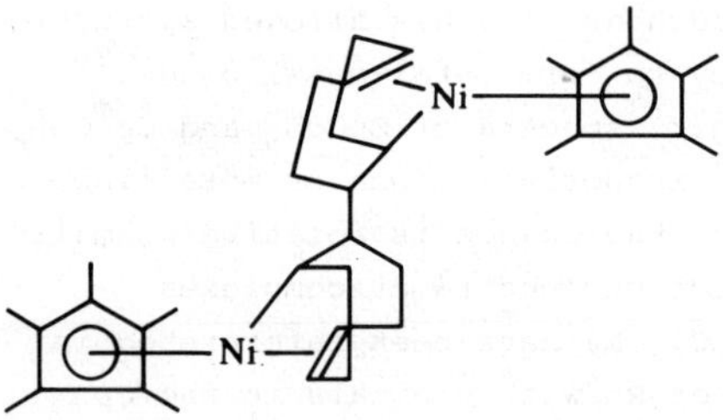

(13)

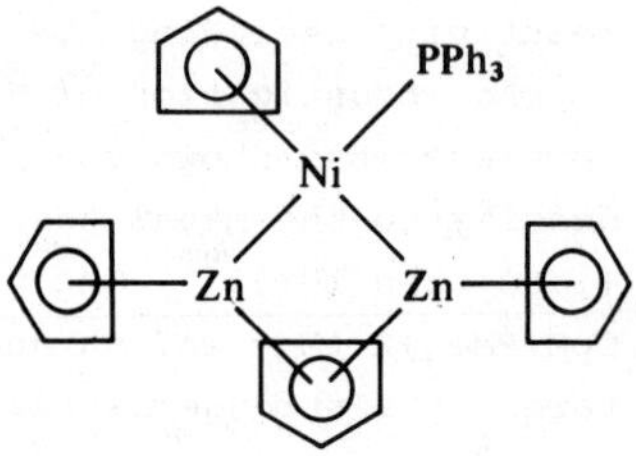

(15)

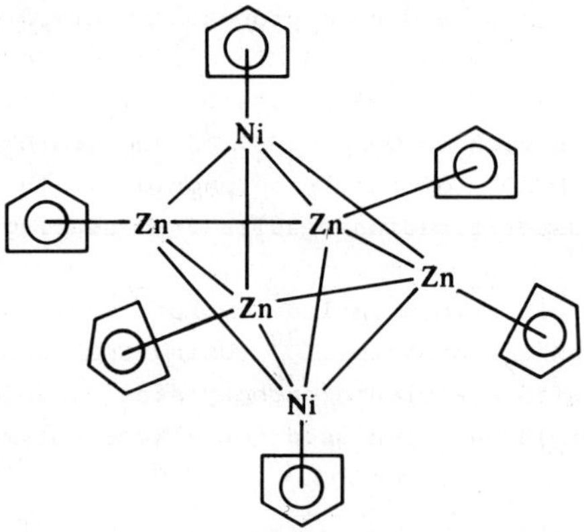

(14)

Olefin hydrosylation has been investigated using Cp*RhH$_2$(SiEt$_3$)$_2$.[104] CpCo(COT) (COT = cyclooctatetrene) has been used to catalyse the conversion of nitriles and acetylenes into pyridines.[105]

The structure of the Co(II) complex Cp*Co(py)Cl has been determined.[106] The structure of the rhodium complex (12) shows that the C$_3$, C$_4$ bond lengths to Rh are longer than the others. This has been interpreted as a concentration of electron density in the C$_3$-C$_4$ bond.[107]

2.7 Nickel, Palladium and Platinum.- An alternative Cp transfer reagent that is being extensively exploited is Cp*$_2$Zn. Products include (13)[108] from the reaction with Ni(COD)$_2$; (14)[109] and (15)[110] result when Cp$_2$Zn is used. A variety of CpNi(PPh$_3$)X (X = halogen) complexes can be generated from reaction of Ni(PPh$_3$)$_2$X$_2$ with TlCp.[111] The bridging Cp in (16) migrates to a terminal position when the complex is treated with CO.[112]

2.8 Lanthanides and Actinides.- A volatile uranium complex CpU(BH$_4$)$_3$[113] has been prepared.

3 Sandwich Complexes

Complexes containing two cyclic polyolefinic ligands will be included in this section if one of the rings is cyclopentadienyl.

3.1 Scandium and Yttrium.- Fluoroscandium complexes have been identified such as [Cp$_2$ScF]$_3$ which contains bridging fluoride ligands.[114] ^{89}Y n.m.r. spectra have been reported for bis cyclopentadienyl complexes using a wide bore magnet and 300 second pulse delays.[187] Dimeric complexes such as Cp*$_2$Y(μ-Cl)YClCp*$_2$ have been structurally characterised.[191]

3.2 Titanium, Zirconium and Hafnium.- The preparation and structures of the asymmetric sandwich complexes of the type [Cp*Ti(COT)]$^+$ have been described.[115] Tying the rings together has been achieved with (17). The Zr complex was also made and reduction chemistry investigated.[116] A variant on this type of molecule can be made where the tie is an olefinic[117] or ethyl[118] linkage.

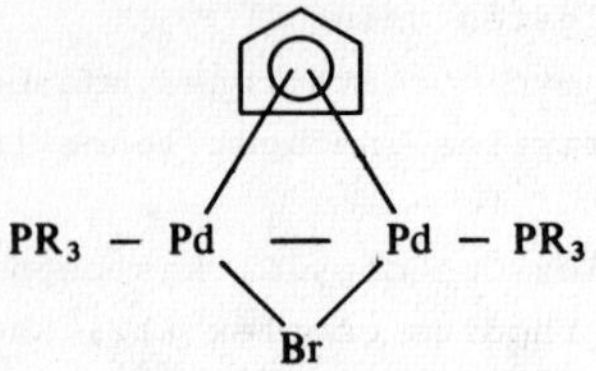

(16)

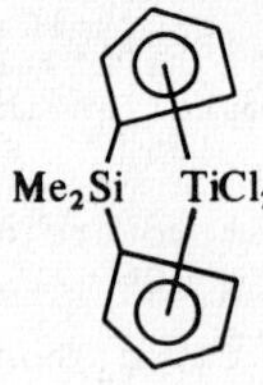

(17)

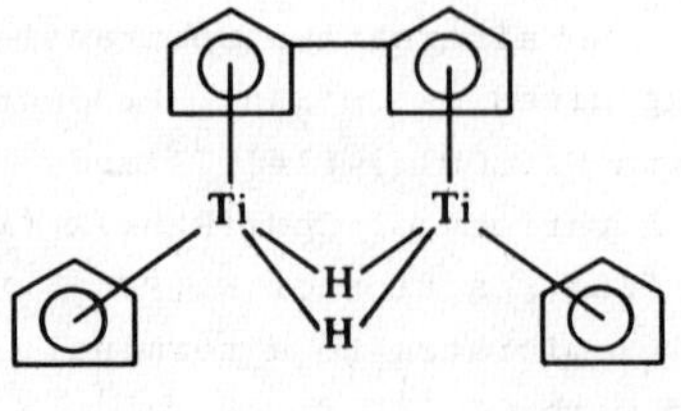

(18)

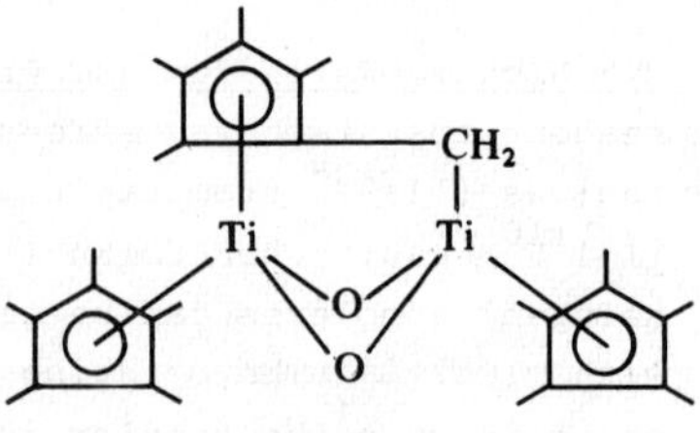

(19)

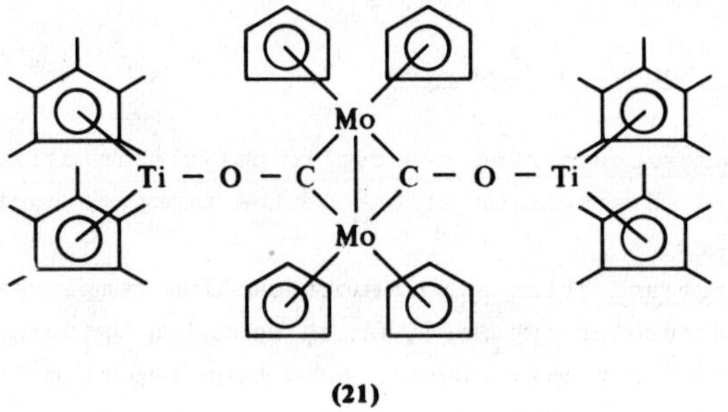

(21)

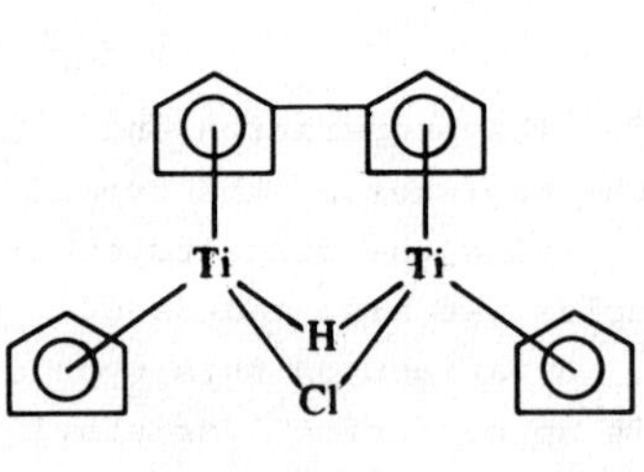

(20)

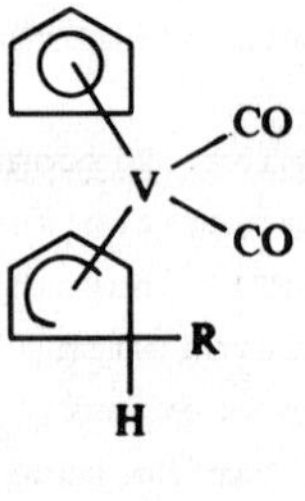

(22)

A comparison of the tied and untied complexes, $Me_2C_2(C_5H_4)_2TiCl_2$, has been reported using electrochemical techniques.[119] A different type of tie leads to the dimeric molecule (18) which has been shown to be paramagnetic at elevated temperatures.[120] Dimeric species $Cp_4M_2Cl_2$ (M = Zr and Hf) result from Na/Hg reduction of Cp_2MCl_2 complexes. Diamagnetism of the products has been used to infer a metal-metal interaction.[121]

Structural studies of molecules $Cp*_2ZrX_2$ (X = OH, Cl) show intramolecular interactions between the methyl H's and the coordinated oxygens.[122] Oxidation of $Cp*_2Ti$ with N_2O leads to C-H activation and the bridged dimer (19).[123] Reductive dimerisation has also been reported (20)[124] when Cp_2TiCl_2 is treated with potassium. Deuteration of the Cp rings in Cp_2Zr complexes has been reported.[125] However, when Cp_2TiCl_2 is treated with Li norbornyl[126] one of the Cp rings is lost.

Reports of the application of sandwich complexes as catalysts include the use of polymer supported Cp_2Zr and Hf complexes in olefin hydrogenation.[127] The catalytic activity of the decomposition products of the dimers $[Cp_2Ti(AlH_4)]_2$ has been examined.[128,129]

The reactive intermediate complex $[Cp_2Ti(THF)]^+$ has been proposed from e.s.r. and ENDOR spectral studies of $[Cp_2TiX]_2ZnX_2$ (X = Cl, Br) systems in THF.[130] A photochemical study of Cp_2TiI_2 has led to the identification of both $[Cp_2TiI]_2$ and Cp_2Ti.[131] Considerable use of Cp_2Ti can be made from $Cp_2Ti(PMe_3)_2$ since both PMe_3 ligands are labile.[132] The $Cp*_2Ti$ fragment can be used as a Lewis acid in complexes such as (21).[133]

Mixed ring complexes $Cp*CpMCl_2$ (M = Hf, Ti) have been structurally investigated,[134] and the related molecules Cp_2MCl_2 have been studied for antitumor properties.[135] When the rings are phosphine-substituted, bimetallic species are readily generated.[136]

3.3 <u>Vanadium, Niobium and Tantalum</u>.- A number of reports of reactions involving the cyclopentadienyl ring have been made. Reactions involving migration of groups to the Cp ring include Cp_2VR reacting with CO to give (22).[137] Loss of a Cp ring occurs when Cp_2V is treated with Me_3NO, leading to clusterification and products such as $Cp_{11}V_{13}O_{18}(NMe_2)_2$.[138]

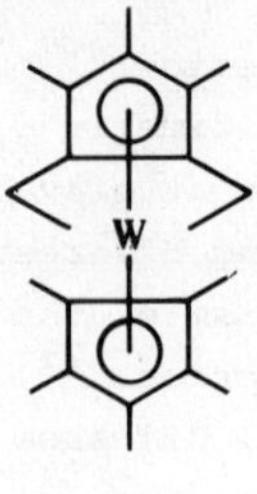

(23)

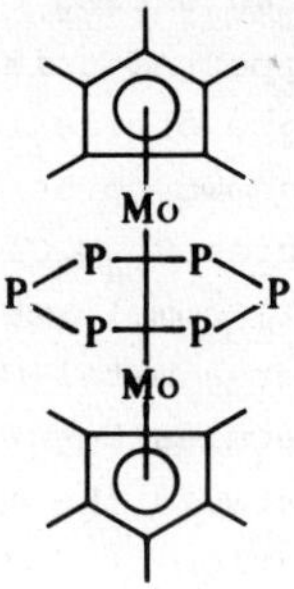

(24)

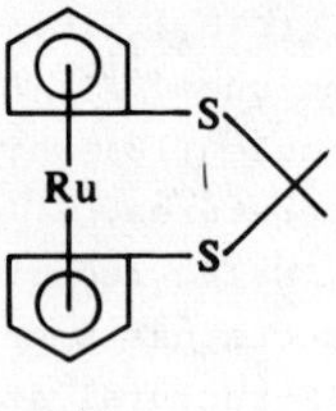

(25)

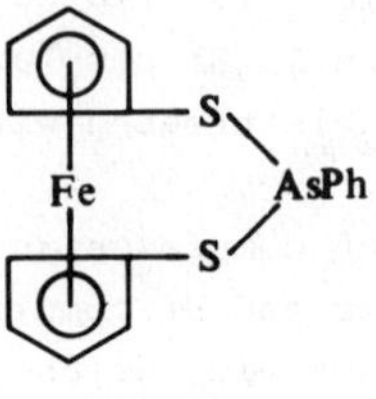

(26)

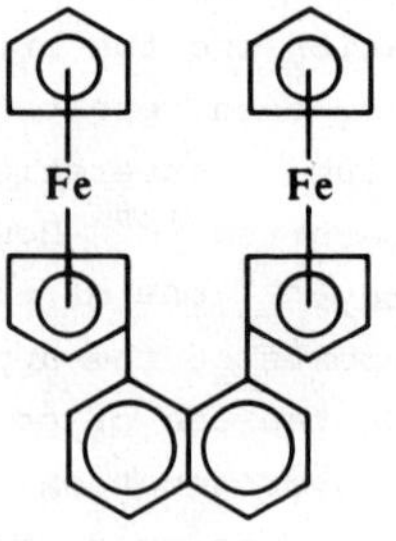

(27)

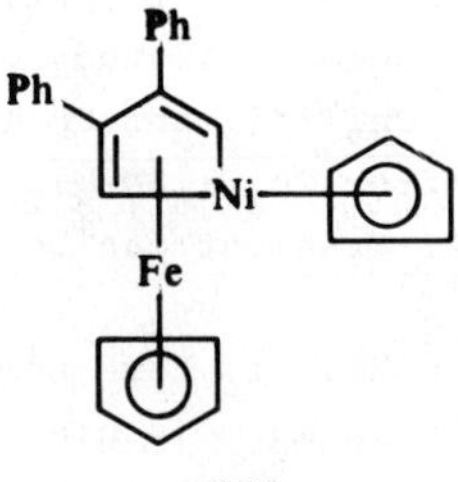

(28)

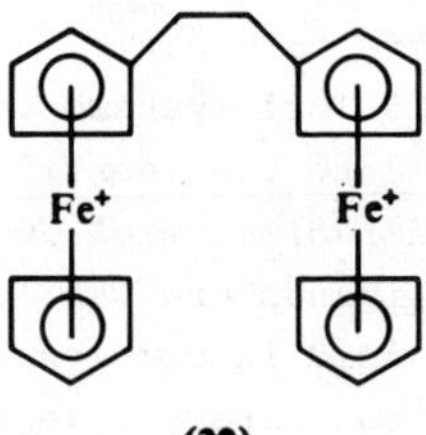

(29)

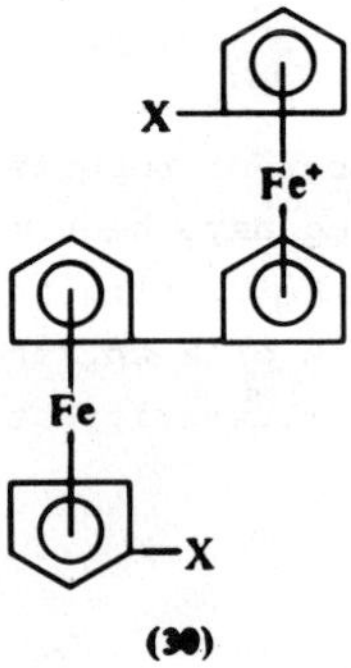

(30)

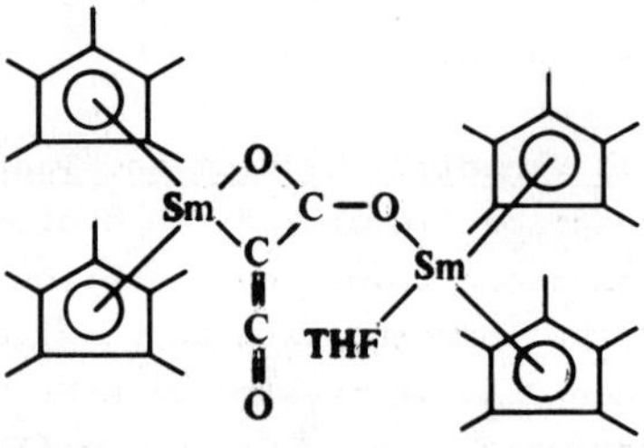

(31)

The photochemistry of complexes such as Cp_2MH_3 (M = Ta, Nb) has been examined.[139] Ethane is an identifiable product from the photolysis of Cp_2VMe_2.

A study of insertion/β-elimination reactions has been made using $Cp*_2NbH(olefin)$.[141] The first linear M-C-O-M system has been reported for $Cp_2V-O-C-V(CO)_5$.[142] Nitrene complexes have been generated from the reaction of $Cp*_2V$ with azides.[143] The chemistry of tied complexes of Nb corresponding to (18) has been investigated.[144]

3.4 Chromium, Molybdenum and Tungsten.-
The complex $Cp*_2WH_2$ has been generated using a metal vapour synthesis.[145] Subsequent photolysis leads to double metallation of one ring to give (23). The complex (24) results from the treatment of Mo vapour with spiro[4.4]nona-1,3-diene.[146]

The reaction of $Cp_2W(Me)(OCOPh)$ with hydride and deuteride sources has demonstrated that initial nucleophilic attack takes place at the Cp ring followed by endo transfer of the hydride to W.[147] Substitution at the ring takes place when Cp_2WCl_2 is reacted with $Li[Si(SiMe_3)_3]$[148].

The triple-decker sandwich complex (24) contains the first P_6 ring.[149]

3.5 Manganese and Rhenium.-
A very short Re-Sn bond is generated in the product of the reaction of Cp_2ReH with methyl tin compounds.[150]

3.6 Iron, Ruthenium and Osmium.-
The preparation and characterisation of Ru and Os metallocenes includes the structure of Cp_2*Ru which contains eclipsed rings.[151] Different Cp rings have been introduced to both Fe and Ni metallocenes using LiCp* on Fe or $NiBr_2.DME$ (DME = 1,2-dimethoxyethane).[152] Interest in metallocenophanes has led to the synthesis of molecules such as (25)[153] and (26).[186] A variety of bridging atoms can be used including -SeGeSe-.[154] A step on the way to columnar organometallics has been made with tied metallocenes such as (27).[155] When a ferrocenophane with three ties between rings is oxidised, the structure reveals longer Fe-C bonds than in the parent neutral complex.[156]

Electron distributions of substituted ferrocenes with substituents such as the styryl group, have been estimated,[157] and mass-analysed ion kinetic energy spectra of monosubstituted ferrocenes reported.[158] Mono- and diketone-substituted ferrocenophanes have been subjected to electrochemical, Mossbauer and ^{13}C n.m.r. studies.[159,160]

Hydride abstraction from the permethyl ring of Cp*$_2$Ru gives the tetramethylfulvene complex.[161] An unusual dimetallocene (28) is formed in a reaction of Cp$_2$Fe$_2$(CO)$_4$ with Cp$_2$Ni in the presence of phenylacetylene.[162] Coupling of ferrocenylcarbenium cations gives the paramagnetic complex (29).[163]

The catalytic generation of H$_2$ by ferrocenophanes in strong acid media has been investigated.[164] Another catalytic application is the use of chiral ferrocenes as ligands for rhodium complexes that can then be used as asymmetric hydrogenation catalysts.[165]

Dithiaferrocenophanes have been studied electrochemically and by ^{13}C n.m.r. to investigate the interaction of the S with the Fe.[166] Metals have been used in a number of cases to provide the link between the rings of ferrocenophanes.[167,168] In other cases aromatic rings have been used.[169,170] Ruthenocenophanes, appropriately substituted, have been used to extract metal ions and have been found more efficient than the analogous ferrocenophanes.[171,172]

Charge transport through films of electropolymerised 1,1'-bischloromethylferrocene has been studied.[173] The degree of charge localisation in the dimeric complexes (30) has been found to depend on the solid state structures.[174]

Triple decker sandwich complexes have been reported with thiadiborolene as the middle ring.[175] Ferrocenecarbodithiolates have been used to stabilise high oxidation state oxomolybdenum complexes.[176] The interaction of I$_2$ with ferrocene to give ferrocenum polyiodide complexes has also been investigated.[177]

3.7 Cobalt, Rhodium and Iridium.- The decomposition of [Cp$_2$Co]$^+$ has been studied using mass spectroscopy.[178] When Cp$_2$Co is reacted with secondary phosphines, phosphido-bridged dimeric complexes can be isolated.[179]

3.8 Nickel

Nickelocene has been reacted with dithiols to give polymeric products.[180] Dimeric compounds such as $Cp_3Ni_2(\mu\text{-}CNCF_3)_2$ result when Cp_2Ni is reacted with $Ni(CNCF_3)_4$.[181]

3.9 Lanthanides and Actinides

Convenient solution preparations of $Cp^*_2Sm(THF)_2$ and related complexes have been reported,[182] as well as a range of preparations of related complexes using metal vapour syntheses.[183] The photoluminescence spectra of $Cp^*_2Eu(OEt)_2$ and quenching by $Cp^*_2Yb(OEt)_2$ have been examined.[185] Luminescence spectra have been used to distinguish between monomers and dimers of Cp'_2TbX (X = Cl or Me).

A study of the bond energies for $Cp_3Th\text{-}R$ bonds gives an estimate of around 80 kcalmol^{-1}; a figure greater than that estimated for the Cp^* analogues.[188]

Hexenes can be hydrogenated using $[Cp^*_2NdH]_2$. The catalyst can also be used to polymerise ethylenes.[189] CO homologation results from the reaction of $Cp^*_2Sm(THF)_2$ with CO, (31).[190] Cyclopentadienyl uranium complexes react with TiW_5O_{19} anions to give clusters.[192,193]

The CO of $CpMn(CO)_3$ can be deoxygenated by reaction with $Cp_3U=CHPMe_2Ph$.[194] A multiple U-P bond can be generated in the reaction of $Cp_3U(NEt_2)_3$ with $PHPh_2$.[195] Dimeric complexes result from the reaction of $Cp^*_2Th(PPh_2)_2$ with $Ni(COD)_2$.[196] The reaction of $[Cp_2SmH]_2$ with CO has been investigated[197] and the lithium adduct of $Cp_2Er(Me)_2$, (32) has been structurally characterised.[198]

4 Arene Complexes

The development of π-arene chemistry includes two new ways in which benzene can bind to a metal, η^1 and $\eta^3:\eta^2:\eta^2:\eta^2$. Besides new syntheses, diverse reports include thermochemical data and application of π-arene chemistry to organic synthesis.

4.1 Chromium, Molybdenum and Tungsten

The use of metal vapour syntheses continues to dominate the reports of new arene complexes including new complexes of Cr, W and Re with benzene and PMe_3.[199,200] Variants of the conventional method include synthesis with the solid ligand at 77K, (33).[201] More conventional syntheses via $Cr(CO)_6$ or $Cr(CO)_3L_3$ led to the $(\eta^6\text{-triptycene})Cr(CO)_3$ compound.[202]

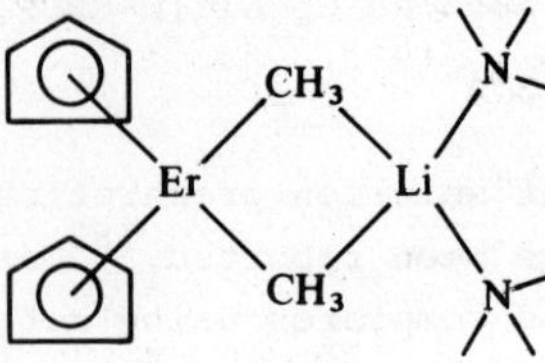

(32)

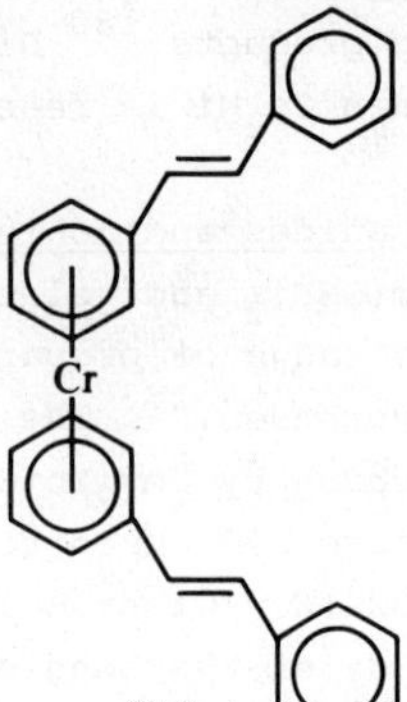

(33)

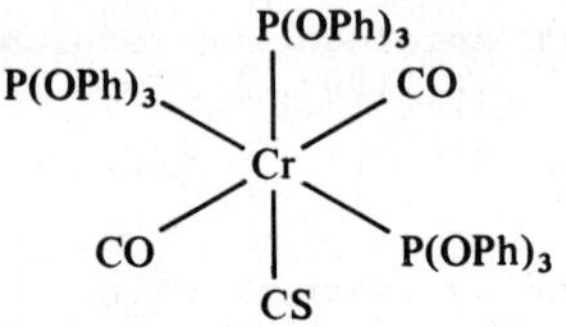

(34)

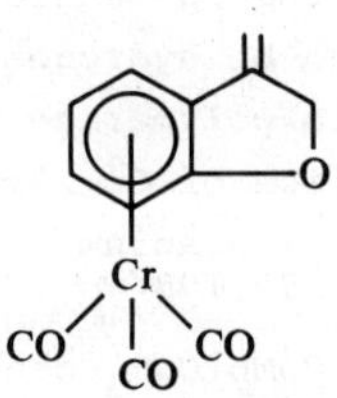

(35)

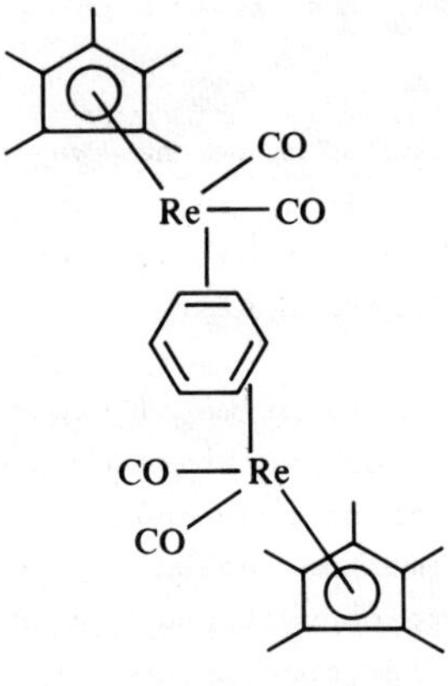

(36)

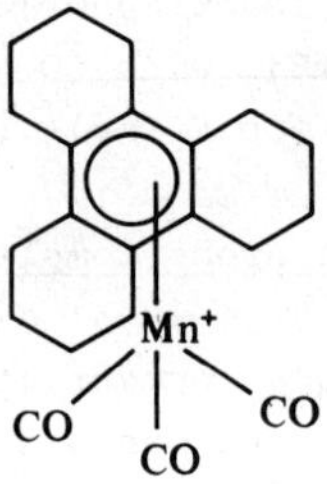

(37)

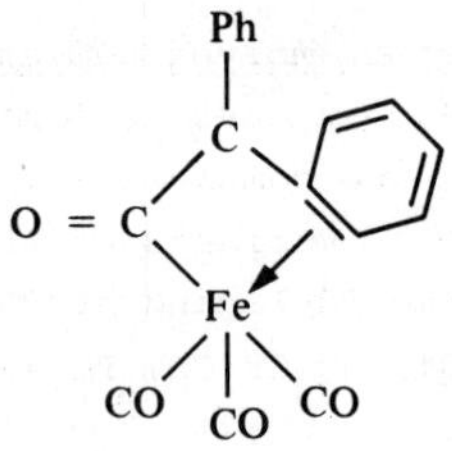

(39)

The exploitation of the synthetic potential of carbene complexes has led to a range of highly substituted η^6-arenes.[203] The use of bisnaphthalene chromium and (naphthalene)Cr(CO)$_3$ complexes as reagents for further arene complexes has been examined.[204] The (naphthalene)Cr(CO)$_3$ substitutes $10^3/10^4$ times faster than monocyclic arenes.

A molybdenum(II) arene complex results from the reaction of MoBr$_2$(CO)$_4$ with sodium-2,6-diphenylbenzenethiolate. The arene ring dissociates reversibly when CO is added.[205]

This contrasts with the nucleophilic substitution reactions reported for Mo(η^6-PhPR$_2$)(PPhR$_2$)(DPPE) (DPPE = PPh$_2$CH$_2$CHPPh$_2$) in which the σ-bound phosphine is substituted rather than the π-bound ligand.[206] Subtle effects can determine whether or not the arene complex is isolated. For 3-benzoylpyrrole, the η^6-arene complex has been characterised but for the 2-benzoylpyrrole case the σ-chelate of the Cr(III) has been reported.[207]

The electronic spectra of arene chromium complexes have been reexamined and the transitions reassigned in terms of Rydberg transitions.[208] Thermochemical data has been collected for the arene displacement reaction by pyridine of (arene)Mo(CO)$_3$ complexes. The resulting order of stability was found to be; benzene<toluene<mesitylene<hmb<cyclopentadienyl/hydride<tris pyridine. The reported difference between the strength of the Mo-C$_6$H$_6$ and Mo-hmb bonds is 6 kcalmol^{-1}.[209]

The (toluene)Mo(CO)$_3$ complex has been used similarly to generate a number of other reactions that give thermochemical data including the strength of the C$_5$H$_5$-M bonds[210] and M-Pr$_3$ bonds.[212]

The loss of the arene ring is a reaction that has been used to generate a unexpected isomer of (34).[210] The isolation of this particular isomer is attributed to the electronic effect of the thiocarbonyl ligand. The arene ring is also lost in a gas phase reaction of (benzene)Cr(CO)$_3$ with deuterium to give [HCr(CO)$_4$]$^-$.[214]

The structure of (1,2-dimethoxybenzene)Cr(CO)$_3$ shows that the metal is displaced away from the methoxy groups.[215] The migration of M(CO)$_3$ groups from one ring to an adjacent ring continues to fascinate the chemist. The most common example, η^6-η^5 rearrangements[216,217] can be triggered by base deprotonation of the cyclopentadiene ring[218]

The photochemical reaction of (arene)M(CO)$_3$ (M = Cr, Mo and W) with acetylenes has been reported.[219] Alternatively the arene complex can be activated by electrochemical reduction prior to addition of an electrophile.[220]

Studies of catalytic applications of (arene)Cr(CO)$_3$ complexes have appeared including the addition of CCl$_4$ to alkenes.[221] The arene complex can also be a substrate for a catalytic reaction as in the case of Pd(PPh$_3$)$_4$ catalysing the acetylation of the arene group in (arene)Cr(CO)$_3$.[222] (Arene)Cr(CO)$_3$ can be used to hydrogenate alkynes to Z-alkenes.[223]

Reports of the applications of arene complexes in organic synthesis include the synthesis of benzofurans from the reaction of Cr(CO)$_3$ complexes with oximes,[224] heterocycle elaboration,[225] making dibenzocrownethers,[226] cyclophane synthesis,[227] the use of anions made by deprotonation of benzylic carbons,[228,229,230] and the use of (trimethylsilylarene)Cr(CO)$_3$ complexes.[231,232] Cr(CO)$_3$ complexes have also been used with hexestrol in the detection of oestradiol receptor sites.[233] The use of the Cr(CO)^3 group to give stereochemical control has been the subject of several reports[230,235,236] as has the use of the group to stabilise otherwise unstable isomers of (35).[237]

4.2 Manganese and Rhenium.- New arene rhenium complexes, made using metal vapour syntheses, have been characterised using 2-D n.m.r.[238] An unusual example of a double η^2 benzene complex (36) has been reported from the photochemical reaction of Cp*Re(CO)$_3$ in benzene.[239]

The first X-ray crystal structure of an [(arene)Mn(CO)$_3$]$^+$ complex has been carried out (37).[240] Double nucleophilic, stereospecific addition to related complexes has been achieved.[241]

4.3 Iron, Ruthenium and Osmium.- The osmium benzene complex (38) has been made using a metal vapour synthesis.[242] Matrix isolation of bis arene iron complexes has led to the report that while (hmb)$_2$Fe is stable at room temperature, (C$_6$H$_6$)$_2$Fe decomposes above 200K.[242] Bis naphthalene iron was identified in two forms, η^4-η^4 at less than 100K and η^6-η^6 at greater than 150K.[243] The η^2-complex (39) has been isolated from the reaction of Fe(CO)$_5$ with diphenylketene.[244]

The anthracene complex (40) has been identified as an intermediate in the catalytic hydrogenation of anthracene.[245] Other arene ruthenium complexes have been identified in the reactions of $CpRu(PPh_3)_2Cl$, such as $[(\eta^6-C_6H_5BPh_2)CpRu]^+$.[246] Complexes of the type(41) can be made in five different combinations of oxidation state.[247] New bimetallic iron carboranes have been made using biphenyl[248,249] or polyaromatics.[250]

A face-capping benzene ligand has been found in the osmium cluster (42),[251] while an η^6-benzene undergoes slippage to the face-capping geometry in the synthesis of $Ru_6C(CO)_{11}(C_6H_6)_2$. An improved synthesis of $[(arene)RuCp]^+$ complexes has been made from ruthenocene with $AlCl_3/Al$.[252] Functionalised arenes have been used in the sandwich complexes $CpFe(C_6H_5X)$, to coordinate to rhodium complexes.[253] Catalytic activity of related complexes has also been tested.[254]

M.O. calculations have been used to rationalise the frontier-orbital control of nucleophilic attack on $[(C_6H_6)Fe(C_6H_7)]^+$ to give cyclohexadienes.[255] The $[(hmb)FeCp]^+$ complex can be treated with TCNQ to give electrically conducting single crystals with a one dimensional stacks of cations and anions.[256] Protonation of the related [(arene)MCp]+ (M = Ru, Os) complexes has also been investigated.[257] Reactions can be carried out on the carborane skeleton of $(C_6H_6)Ru(C_2B_9H_{11})$.[258]

The structure of the metacyclophane complex (43) has been compared with that of the free ligand.[259] Related structures have also been determined with xanthene and 2-methylthiaxanthene.[260] Substitution of the arene ring in $[(arene)FeCp]^+$ complexes can be achieved both thermally and photochemically.[261] While in the reaction of $[(toluene)FeCp]^+$ with ethylene oxide, both the Cp and arene ligands are displaced and two crown ether ligands are generated.

Two-dimensional n.m.r. has been applied to elucidate the products of the reactions of $[(cymene)OsCl_2]_2$ complexes.[263] The reduction and substitution reactions of $(arene)Os(PR_3)I_2$ complexes have also been studied.[264] With complexes such as (44) aromatic C-H activation has been reported together with loss of the arene ligand.[266] Nucleophilic addition to related dicationic complex $[(C_6H_6)Ru(PMe_3)_2(C_2H_4)]^{2+}$ leads to attack at the ethylene with phosphines and at the benzene with harder nucleophiles such as NEt_3 of CH_3^-.[267]

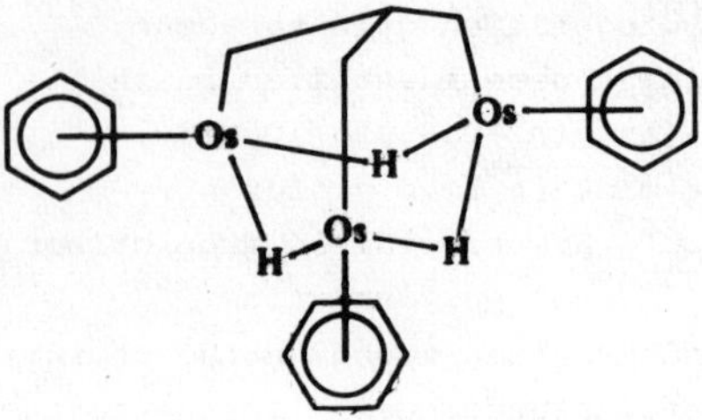

(38)

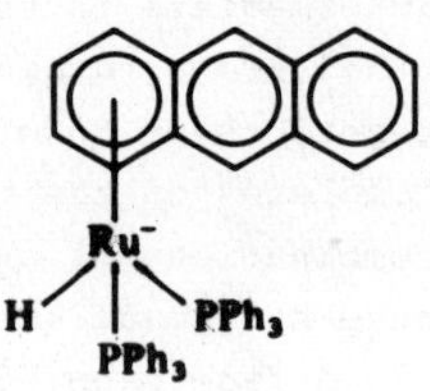

(40)

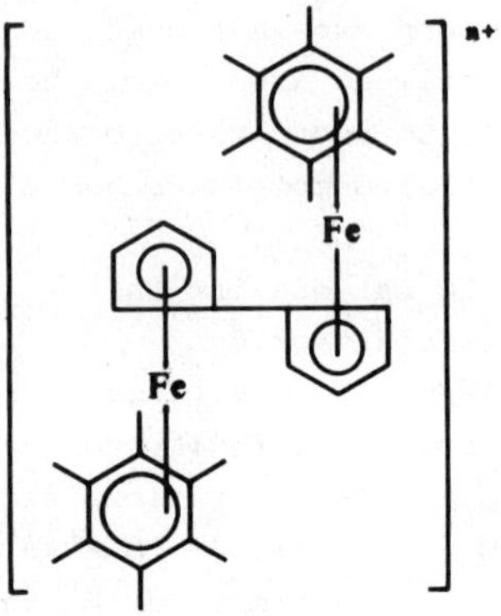

(41)

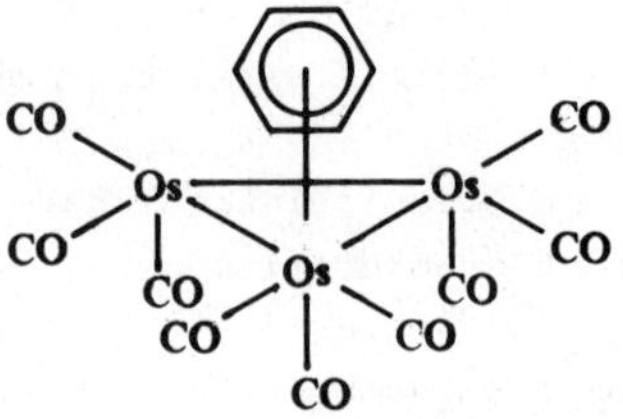

(42)

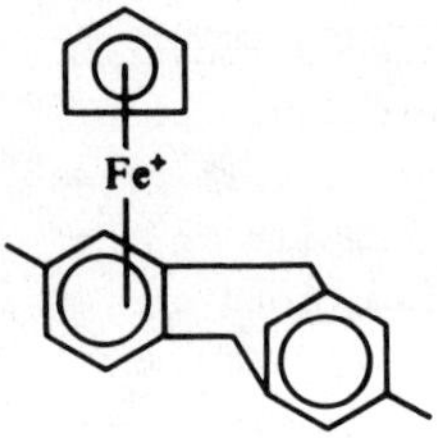

(43)

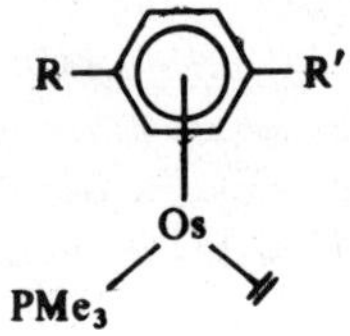

(44)

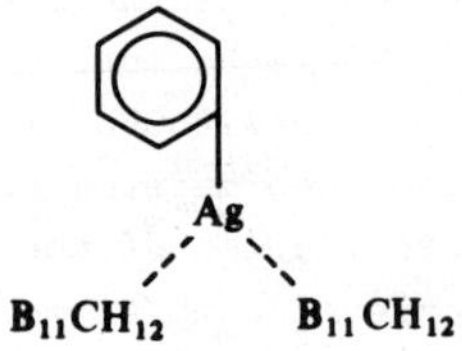

(45)

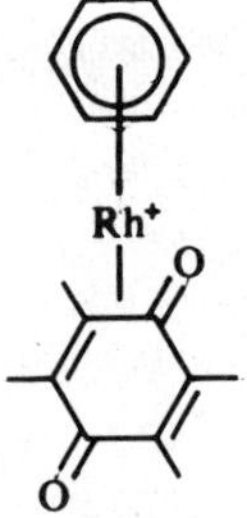

(46)

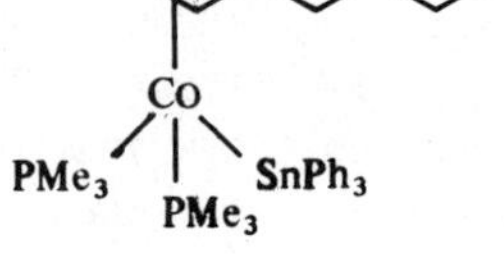

(47)

4.4 Cobalt, Rhodium and Iridium.

Arene complexes of iridium [(arene)Ir(PPh$_3$)$_2$]$^+$ contain labile arene ligands and are active catalysts for the dehydrogenation of cyclohexane.[269] The loss of arene ligands has also been examined for a number of (arene)ML$_2$ (M = Rh[270] and Ni[271,273]) complexes.

The crystal structure of (46) shows the toluene to be boat shaped.[273] The structure of the related mesitylene complex has also been determined.[274] Naphthalene rhodium complexes have also been prepared,[275] as have the first neutral complexes (toluene)Rh(C$_8$H$_{14}$)Sn(NR$_2$)$_2$.[276] With (toluene)Co(C$_6$F$_5$)$_2$, e.s.r. studies show that the unpaired electron is predominantly d-orbital based.[277]

Two new ways in which anthracene can be bound to metals (47) and (48) have been structurally characterised.[265] In the case of (47) the η^4-coordinated anthracene contains a naphthalene-like planar section which folds away from the Co atom. In (48) the anthracene is best considered to be η^2-coordinated to Co(0).

4.5 Other Arene Complexes.

The new η^1-benzene ligand has been identified in the silver complex (45).[268] The first uranium(IV) complex of hmb has been reported (49).[278]

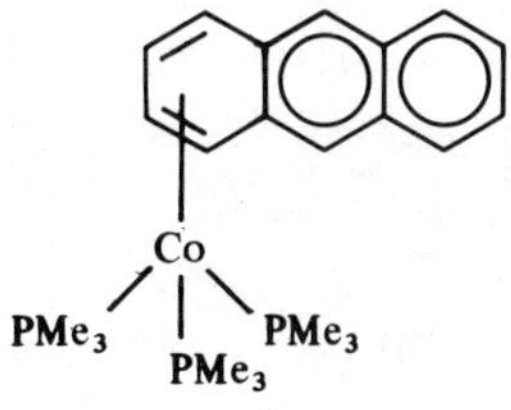

(48)

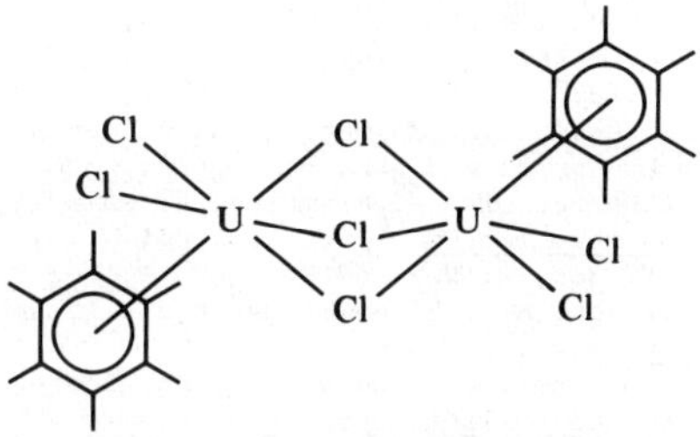

(49)

1. A. Solladie-Cavallo, Polyhedron, 1985, 4, 901.
2. W. Siebert, Ang. Chem. Int. Ed, 1985, 24, 943.
3. K. Jonas, Ang. Chem. Int. Ed, 1985, 24, 295.
4. M. Draganjac and T.B. Rauchfuss, Ang. Chem. Int. Ed, 1985, 24, 742.
5   K. Angermund, K.H. Claus, R. Goddard and C. Kruger, Ang. Chem. Int. Ed, 1985, 24, 237.
6. H. Bonneman, Ang. Chem. Int. Ed, 1985, 24, 248.
7. A.W. Duff, P.B. Hitchcock, M.F. Lappert, R.G. Taylor and J.A. Segal, J.Organomet.Chem., 1985, 293, 271.
8. B.G. Conway and M.D. Rausch, Organometallics, 1985, 4, 688.
9. R.D. Rogers, M.A. Benning, L.K. Kurihara, K.J. Moriarty and M.P. Rausch, J.Organomet.Chem., 1985, 293, 51.
10. F. Bottomley, G.O. Egharevba and P.S. White, J.Amer.Chem. Soc., 1985, 107, 4353.
11. W. Skupinski and A. Wasilewski, J.Chem.Soc., Chem.Commun., 1985, 282, 69.
12. O.P Pandey, S.K. Sengupta and S.C. Tripathi, Bull. Chem. Soc. Jpn, 1985, 58, 2395.
13. H. Heijden, A.W. Gal, P. Pasman, A.G. Orpen, Organometallics, 1985, 4, 1847.
14. J. Arnold and T.D. Tilley, J.Amer.Chem.Soc., 1985, 107, 6409.
15. M.D. Curtis and J. Real, Organometallics, 1985,4, 940.
16. M.A. El-Hinnawi and A.K. El-Qasser, J.Organomet.Chem., 1985, 281, 119.
17. M.A. Greaney, J.S. Merola and T.R. Halbert, Organometallics, 1985, 4, 2057.
18. R. Drews and U. Behrens, Chem.Ber., 1985, 118, 888.
19. M.L.H. Green and G. Parkin, J.Chem.Soc., Chem.Commun., 1985, 1383.
20. J.L. Davidson, W.F. Wilson and K.W. Muir, J.Chem.Soc., Chem. Commun., 1985, 460.
21. S.A. MacLaughlin, R.C. Murray, J.C. Dewan and R.R. Schrock, Organometallics, 1985, 4, 796.
22. F. Bottomley, D.E. Paez, L. Sutin and P.S. White, J.Chem. Soc., Chem.Commun., 1985, 597.
23. J.W. Faller, R.H. Crabtree and A. Habib, Organometallics, 1985, 4, 929.
24. P. Legzdins, S.J. Rettig, L. Sanchez, B.E. Bursten and M.G. Gatter, J.Amer.Chem.Soc., 1985, 107, 1411.
25. K. Sunkel, G. Urban and W. Beck, J.Organomet.Chem., 1985, 290, 231.
26. G. Urban, K. Sunkel and W. Beck, J.Organomet.Chem., 1985, 290, 329.
27. R.C. Murray, L. Blum, A.H. Liu and R.R. Schrock, Organometallics, 1985, 4, 953.
28. P. Legzdins, S.J. Rettig and L. Sanchez, Organometallics, 1985, 4, 1470.
29. M.D. Curtis and K.R. Han, Inorg.Chem., 1985, 24, 378.
30. M. Herberhold, W. Kremnitz, A. Razavi, H. Schollharn and W. Thewalt, Ang.Chem.Int.Ed., 1985, 24, 601.
31. L.Y. Goh, C. Wei and E. Sinn, J.Chem.Soc., Chem.Commun., 1985, 462.
32. L.E. Bogan, T.B. Rauchfuss and A.L. Rheingold, J.Amer.Chem. Soc., 1985, 107, 3843.
33. G. Huttner, B. Sigwarth, O. Scheidsteger, L. Zsolnai and O. Orama, Organometallics, 1985, 4, 326.
34. M.A. El-Hinnawi and A.K. El-Quaseer, J.Organomet.Chem., 1985, 296, 393.
35. P. Jaitner, W. Wohlgenannt, A. Gieren, H. Betz and T. Hubner, J.Organomet.Chem., 1985, 295, 281.
36. K. Blechschmitt, H. Pfisterer, T. Zahn and M.L. Ziegler, Ang.Chem.Int.Ed., 1985, 24, 66.
37. J.M. Newsam and T.R. Halbert, Inorg.Chem., 1985, 24, 491.
38. G.J. Kubas, H.J. Wasserman and R.R. Ryan, Organometallics, 1985, 4, 419.
39. D. Warmsbacher, K.M. Nicholas and A.L. Rheingold, J.Chem. Soc., Chem.Commun., 1985, 721.
40. G.J. Kubas and R.R. Ryan, J.Amer.Chem.Soc., 1985, 107, 6138.
41. J.T. Landrum and C.D. Hoff, J.Organomet.Chem., 1985, 282, 215.

42. K. Isobe, S. Kimura and Y. Nakamura, J.Chem.Soc., Chem. Commun., 1985, 378.
43. A.E. Stiegman and D.R. Tyler, J.Amer.Chem.Soc., 1985, 107, 967.
44. M. Green, S. Greenfield and M. Kersting, J.Chem.Soc., Chem. Commun., 1985, 18.
45. A.D. Horton, M.J. Mays and P.R. Raithby, J.Chem.Soc., Chem. Commun., 1985, 247.
46. J.S. Drage and K.P.C. Vollhardt, Organometallics, 1985, 4, 191.
47. G.L. Crocco and J.A. Gladysz, J.Chem.Soc., Chem.Commun., 1985, 283.
48. C.P. Casey, J.M. O'Connor and K.J. Maller, J.Amer.Chem.Soc., 1985, 107, 1241.
49. C.P. Casey, J.M. O'Connor and K.J. HallerJ.Amer.Chem.Soc., 1985, 107, 3172.
50. C.P. Casey and J.M. O'Connor,Organometallics, 1985, 4, 384.
51. M.G. Yezernitskaya, B.V. Lokshin, V.I. Zelanovich, I.A. Lobanova and N.E. Kolobova, J.Organomet.Chem., 1985, 282, 363.
52. G.K. Yang and R.G. Bergman, Organometallics, 1985, 4, 129.
53. W.D. Jones and J.A. Mcguire, Organometallics, 1985, 4, 951.
54. D. Sellmann and J. Muller, J.Organomet.Chem., 1985, 281, 249.
55. F.W.B. Einstein, K.G. Tyers and D. Sutton, Organometallics, 1985, 4, 489.
56. R.G. Bergman, P.F. Seidler and T.T. Wenzel, J.Amer.Chem. Soc., 1985, 107, 4358.
57. W.A. Herrmann, E. Voss and M. Floel, J.Organomet.Chem., 1985, 297, C5.
58. W.A. Herrmann, R. Serrano, U. Kusthardt, E. Guggolz, B. Nuber and M.L. Ziegler, J.Organomet.Chem., 1985, 287, 329.
59. W.A. Herrmann, R. Serrano, M.L. Ziegler, H. Pfisterer and B. Nuber, Ang.Chem.Int.Ed., 1985, 24, 50.
60. M. Herberhold, B. Schmidkonz, M.L. Ziegler and T. Zahn, Ang.Chem.Int.Ed., 1985, 24, 515.
61. A. Winter, G. Huttner, M. Gottlieb and I. Jibril, J.Organomet.Chem., 1985, 286, 317.
62. M.F. Asaro, G.S. Bodner, J.A. Gladysz, S.R. Cooper and N.J. Cooper, Organometallics, 1985, 4, 1020.
63. J.D. Korp, I. Bernal, R. Horlein, R. Serrano, W.A. Herrmann, Chem.Ber., 1985, 118, 340.
64. W.A. Herrmann, B. Koumbouis, A. Schafer, T. Kahn and M.L. Ziegler, Chem.Ber., 1985, 118, 2472.
65. L.A. Oro, M.A. Ciriano, M. Campo, C. Foces-Foces and F.H. Cano, J.Organomet.Chem., 1985, 289, 117.
66. C. White and E. Cesarotti, J.Organomet.Chem., 1985, 287, 123.
67. A.R. Manning, G. McNally and P. Soye, J.Organomet.Chem., 1985, 288, C53.
68. J.P. Blaha and M.S. Wrighton, J.Amer.Chem.Soc., 1985, 107, 2694.
69. S.R. Berryhill, G.L. Clevenger and F.Y. Burdurlu, Organometallics, 1985, 4, 1509.
70. R. Weberg, R.C. Haltiwanger and M.R. DuBois, Organometallics, 1985, 4, 1315.
71. M.O. Albers, H.E. Oosthuizen, D.J. Robinson, A. Shaver and E. Singleton, J.Organomet.Chem., 1985, 282, C49.
72. M.O. Albers, H.E. Oosthuizen, D.J. Robinson, A. Shaver and E. Singleton, J.Organomet.Chem., 1985, 282, C49.
73. G.L. Trainor,J.Organomet.Chem., 1985, 282, C43.
74. K.A. Mahmoud, A.J. Rest and H.G. Alt, J.Chem.Soc., Dalton Trans., 1985, 1365.
75. N. Khun and H. Schumann, J.Organomet.Chem., 1985, 288, C51.
76. D.C. Boyd, D.A. Bohling and K.R. Mann, J.Amer.Chem.Soc., 1985, 107, 1641.
77. J.P. Blaha, B.E. Bursten, J.C. Dewan, R.B. Frankel, C.L. Randolph, B.A. Wilson and M.S. Wrighton,J.Amer.Chem.Soc., 1985, 107, 4561.
78. R.H. Fong and W.H. Hersh, Organometallics, 1985, 4, 1468.
79. H. Schumann, J.Organomet.Chem., 1985, 293, 75.
80. G.R. Lee and N.J. Cooper, Organometallics, 1985, 4, 794.

81. G. Hartmann and R. Mews, Ang.Chem.Int.Ed., 1985, 24, 202.
82. G.J. Baird, S.G. Davies, S.D. Moon, S.J. Simpson and R.H. Jones, J.Chem.Soc., Dalton Trans., 1985, 1479.
83. W. Angerer, W.S. Sheldrick and W. Malisch, Chem.Ber., 1985, 118, 1261.
84. G. Doyle and D. van Engen, J.Organomet.Chem., 1985, 280, 253.
85. P.M. Treichel and P.J. Vincenti, Inorg.Chem., 1985, 24, 228.
86. V. Guerchais and D. Astruc, J.Chem.Soc., Chem.Commun., 1985, 835.
87. H. Werner, L.H. Hofmann, R. Feser and W. Paul, J.Organomet.Chem., 1985, 281, 317.
88. W.P. Hart, D. Shihua and M.D. Rausch, J.Chem.Soc., Chem. Commun., 1985, 282, 111.
89. T.M. Gilbert, F.J. Hollander and R.G. Bergman, J.Amer.Chem. Soc., 1985, 107, 3508.
90. H. Werner, H.J. Scholz and R. Zolk, Chem.Ber., 1985, 118, 4531.
91. D. Villemin and E. Schigeko, J.Organomet.Chem., 1985, 293, C10.
92. E.J. Miller, S.J. Landon and T.B. Brill, Organometallics, 1985, 4, 533.
93. C. Cauletti, C. Furlani, C. Puliti and H. Werner, J.Organomet.Chem., 1985, 289, 417.
94. L.J. Newman and R.G. Bergman, J.Amer.Chem.Soc., 1985, 107, 5314.
95. W.D. McGhee and R.G. Bergman, J.Amer.Chem.Soc., 1985, 107, 3388.
96. D.M. Haddleton and R.N. Perutz, J.Chem.Soc., Chem.Commun., 1985, 1372.
97. M. Gomez, J.M. Kisenyi, G.J. Sunley and P.M. Maitlis, J.Organomet.Chem., 1985, 296, 197.
98. M. Gomez, P.I.W. Yarrow, D.J. Robinson and P.M. Maitlis, J.Organomet.Chem., 1985, 279, 115.
99. S. Gambarotta, C. Floriani, A. Chiesi-Villa and C. Guastini, J.Organomet.Chem., 1985, 296, C6.
100. K. Abdulla, B.L. Booth and C. Stacey, J.Organomet.Chem., 1985, 293, 103.
101. P. Zanella, G. Paulucci, G. Rossetoo, F. Benetollo, A. Polo, R.D. Fischer and G. Bombieri, J.Chem.Soc., Chem.Commun., 1985, 96.
102. M. Kajitani, T. Suetsugo, R. Wakabayashi, A. Igarashi, T. Akiyama and A. Sugimori, J.Organomet.Chem., 1985, 293, C15.
103. G.F. Schmidt and M. Brookhart, J.Amer.Chem.Soc., 1985, 107, 1443.
104. P.O. Bentz, J. Ruiz, B.E. Mann, C.M. Spencer and P.M. Maitlis, J.Chem.Soc., Chem.Commun., 1985, 1374.
105. Y. Wakatsuki and H. Yamazaki, Bull. Chem. Soc. Jpn, 1985, 58, 2715.
106. E. Raabe and U. Koelle, J.Organomet.Chem., 1985, 279, C29.
107. M. Arthurs, H. Karodia, M. Sedgwick, D.A. Morton-Blake, C.J. Cardin and H. Paige, J.Organomet.Chem., 1985, 291, 231.
108. B. Fischer, J. Boersma, B. Kojic-Prodic and A.L. Spek, J.Chem.Soc., Chem.Commun., 1985, 1237.
109. P.H.M. Budzelaar, J. Boersma, G.J.M. van der Kerk, A.L. Spek and A.J.M. Duisenberg, Organometallics, 1985, 4, 680.
110. P.H.M. Budzelaar, J. Boersma, G.J.M. van der Kerk, A.L. Spek and A.J.M. Duisenberg, J.Organomet.Chem., 1985, 287, C13.
111. E. Hernandez and P. Royo, J.Organomet.Chem., 1985, 291, 387.
112. P. Thometzek, K. Zenkert and H. Werner, Ang.Chem.Int.Ed., 1985, 24, 516.
113. D. Baudry, P. Charpin, M. Ephritikhine, G. Folcher, J. Lambard, M. Lance, M. Nierlich and J. Vigner, J.Chem.Soc., Chem.Commun., 1985, 1553.
114. F. Bottomley, D.E. Paez and P.S. White, J.Organomet.Chem., 1985, 291, 35.
115. L.B. Kool, M.D. Rausch and R.D. Rogers, 1985, 297, 289.
116. C.S. Bajgur, W.R. Tikkanen and J.L. Peterson, Inorg.Chem., 1985, 24, 2539.
117. F.R. Wild, M. Waslucionek, G. Huttner, H.H. Brintzinger, J.Organomet.Chem., 1985, 288, 63.
118. F. Wochner, L. Zsolnai, G. Huttner and H.H. Brintzinger, J.Organomet.Chem., 1985, 288, 69.
119. H. Schwemlein, W. Tritschler, H. Kiesele and H.H. Brintzinger, J.Organomet.Chem., 1985, 293, 353.
120. D.A. Lemenovskii, I.F. Urazowski, Y.K. Grishin and V.A. Roznyatovsky, J.Organomet.Chem., 1985, 290, 301.

121. T. Cuenca and P. Royo, J.Organomet.Chem., 1985, 293, 61.
122. R. Bortolin, V. Patel, I. Munday, N.J. Taylor and A.J. Carty, J.Chem.Soc., Chem.Commun., 1985, 456.
123. F. Bottomley, G.O. Egharevba, I.J.B. Lin and P.S. White, Organometallics, 1985, 4, 550.
124. E.G. Perevalova, I.F. Urazowski, D.A. Lemenovskii, Y.L. Slovokhotov and Y.U.T. Struchkov, J.Organomet.Chem., 1985, 289, 319.
125. C.S. Bajgur, S.B. Jones and J.L. Petersen, Organometallics, 1985, 4, 1929.
126. V. Dimitrov, J.Organomet.Chem., 1985, 282, 321.
127. B-H. Chang, R.H. Grubbs and C.H. Brubaker, J.Organomet. Chem., 1985, 280, 365.
128. E.B. Lobkovskii, G.L. Soloveichik, A.I. Sizou and B.M. Bulychev,J.Organomet.Chem., 1985, 280, 53.
129. V.K. Belsky, A.I. Sizou, B.M. Bulychev and G.L. Soloveichik, J.Organomet.Chem., 1985, 280, 67.
130. D. Gonrier, D. Vivieu and E. Samuel, J.Amer.Chem.Soc., 1985, 107, 7418.
131. M.R.M. Bruce and D.R. Tyler, Organometallics, 1985, 4, 528.
132. L.B. Kool, M.D. Rausch, H.G. Alt, M. Herberhold, U. Thewalt and B. Wolf, Ang.Chem.Int.Ed., 1985, 24, 394.
133. E.J.M. de Boer, J. de With and A.G. Orpen, J.Chem.Soc., Chem.Commun., 1985, 1666.
134. R.D. Rogers, M.W. Benning, L.K. Kurihara, K.J. Moriarty and M.D. Rausch, J.Organomet.Chem., 1985, 293, 51.
135. J.H. Toney and T.J. Marks, J.Amer.Chem.Soc., 1985, 107 947.
136. C.P. Casey and F. Nief, Organometallics, 1985, 4, 1218.
137. R. Chonkronin and D. Gervais, J.Chem.Soc., Chem.Commun., 1985, 224.
138. F. Bottomley, D.E. Paez and P.S. White, J.Amer.Chem.Soc., 1985, 107, 7226.
139. R.F.G. Baynham, J. Chetwynd-Talbot, P. Grebenik, R.N. Perutz and M.H.A. Powell, J.Organomet.Chem., 1985, 284, 229.
140. D.F. Foust and M.D. Rausch, J.Organomet.Chem., 1985, 287, 195.
141. N.G. Doherty and J.E. Bercaw, J.Amer.Chem.Soc., 1985, 107, 2670.
142. J.H. Osborne, A.L. Rheingold and W.C. Trogler, J.Amer.Chem. Soc., 1985, 107, 6292.
143. J.H. Osborne, A.L. Rheingold and W.C. Troqler, J.Amer.Chem. Soc., 1985, 107, 7945.
144. D.A. Lemenovskii, I.F. Urazowski, I.E. Nifantev and E.G. Perevalova, J.Organomet.Chem., 1985, 292, 217.
145. F.G.N. Cloke, J.C. Green, M.L.H. Green and C.P. Morley, J.Chem.Soc., Chem.Commun., 1985, 945.
146. M.L.H. Green and D. O'Hare, J.Chem.Soc., Dalton Trans., 1985, 1585.
147. R.M. Bullock, C.E.L. Headford, S.E. Kegley and J.R. Norton, J.Amer.Chem.Soc., 1985, 107, 727.
148. U. Schubert, A. Schenkel and J. Muller, J.Organomet.Chem., 1985, 292, C11.
149. G.E. Herberich and J. Okuda, Ang.Chem.Int.Ed., 1985, 24, 402.
150. V.K. Belsky, A.N. Protsky, I.V. Molodnitskaya, B.M. Bulychev and G.L. Solovechik, J.Organomet.Chem., 1985, 293, 69.
151. D.C. Liles, A. Shaver, E. Singleton and M.B. Wiege, J.Organomet.Chem., 1985, 288, C33.
152. U. Kolle, B. Fuss, F. Khouzami and J. Gersdorf, J.Organomet. Chem., 1985, 290, 77.
153. S. Akabori, T. Kumagai and M. Sato, J.Organomet.Chem., 1985, 286, 69.
154. A.G. Osborne, A.J. Blake, R.E. Hollands, R.F. Bryan and S. Lockhart, J.Organomet.Chem., 1985, 287, 39.
155. M. Lee, B.M. Foxman and M. Rosenblum, Organometallics, 1985, 4, 539.
156. M. Hillman and A. Larson, J.Organomet.Chem., 1985, 280, 389.
157. A.G. Nagy and S. Toma, J.Organomet.Chem., 1985, 282, 267.
158. N. Filipovic-Marinic, V. Rapic, V. Kramer and B. Kralj, J.Organomet.Chem., 1985, 296, 405.
159. S. Toma, E. Solcaniova and A.G. Nagy, J.Organomet.Chem., 1985, 288, 331.

160. B. Lukas, C.W. Patterson, R.M.G. Roberts and J. Silver, J.Organomet.Chem., 1985, 286, 209.
161. U. Kolle and J. Grub, J.Organomet.Chem., 1985, 189, 133.
162. S.B. Colbran, B.H. Robinson and J. Simpson, Organometallics, 1985, 4, 1594.
163. E.I. Fedin, A.L. Blumenfeld, P.V. Petrovskii, A.Z. Kreindlin, S.S. Fadeeva and M.I. Rybinskaya, J.Organomet.Chem., 1985, 292, 257.
164. M. Hillman, S. Michaile, S.W. Feldberg and J.J. Eisch, Organometallics, 1985, 4, 1258.
165. T.D. Appleton, W.R. Cullen, S.V. Evans, T-J. Kim and J. Trotter, J.Organomet.Chem., 1985, 279, 5.
166. M. Sato, S. Tanaka, S. Ebine, K. Morinaga and S. Akabon, J.Organomet.Chem., 1985, 282, 247.
167. I.R. Butler, W.R. Cullen, T. Kim, S.J. Rettig and J. Trotter, Organometallics, 1985, 4, 972.
168. B. McCulloch, D.L. Ward, J.D. Woolins and C.H.Brubaker, Organometallics, 1985, 4, 1425.
169. M. Sato, S. Tanaka, S. Ebine, K. Morinaga and S. Akabon, J.Organomet.Chem., 1985, 289, 91.
170. A. Kasahara, T. Izumi, I. Shimizu, T. Oikawa, H. Ymezawa, M. Murakami and O. Watanabe, Bull. Chem. Soc. Jpn, 1985, 58, 1560.
171. Y. Habata, S. Akabari and M. Sato, Bull.Chem.Soc.Jpn., 1985, 58, 3540.
172. S. Akabari, H. Munegumi, Y. Habata, S. Sato, K. Kawazoe, C. Tamura and M. Sato, Bull.Chem.Soc.Jpn., 1985, 58, 2185.
173. M. Nishihara and K. Aramaki, J.Chem.Soc., Chem.Commun., 1985, 709.
174. T-Y. Dang, M.J. Cohn, D.N. Hendrickson and C.G. Pierpont, J.Amer.Chem.Soc., 1985, 107, 4777.
175. J. Edwin, W. Siebert and C. Kruger, J.Organomet.Chem., 1985, 282, 297.
176. M. Nakamoto, K. Tanaka and T. Tanaka, Bull. Chem. Soc. Jpn, 1985, 58, 1816.
177. E.W. Neuse and M.S. Loonat, J.Organomet.Chem., 1985, 286, 329.
178. D.V. Zagorevskii, Y.S. Nekrasov, A.V. Malkov and E.V. Leonova, J.Organomet.Chem., 1985, 284, 100.
179. L. Chen, D.J. Kountz and D.W. Meek, Organometallics, 1985, 4, 598.
180. W. Douglas, J.Organomet.Chem., 1985, 287, C39.
181. D. Lentz, Chem.Ber., 1985, 118, 560.
182. W.J. Evans, J.W. Grate, H.W. Choi, I. Bloom, W.E. Hunter and J.L. Atwood, J.Amer.Chem.Soc., 1985, 107, 941.
183. W.J. Evans, I. Bloom, W.E. Hunter and J.L. Atwood, Organometallics, 1985, 4, 112.
184. D.M. Ruddick, M.D. Fryzuk, P.F. Seidler, G.L. Hillhouse and J.E. Bercaw, Organometallics, 1985, 4, 97.
185. A.C. Thomas and A.B. Ellis, Organometallics, 1985, 4, 2223.
186. A.G. Osborne, R.E. Hollands, R.F. Bryan and S. Lockhart, J.Organomet.Chem., 1985, 288, 207.
187. W.J. Evans, J.H. Meadows, A.G. Kostka and G.L. Closs, Organometallics, 1985, 4, 324.
188. D.C. Sonnenberger, L.R. Morss and T.J. Marks Organometallics, 1985, 4, 352.
189. H. Manermann, P.N. Swepston and T.J. Marks, Organometallics, 1985, 4, 200.
190. W.J. Evans, J.W. Grate, L.A. Hughes, H. Zhang and J.L. Atwood, J.Amer.Chem.Soc., 1985, 107, 3728.
191. W.J. Evans, T.T. Peterson, M.D. Rausch, W.E. Hunter, H. Zhang and J.L. Atwood, Organometallics, 1985, 4, 554.
192. V.W. Day, C.W. Earley, W.G. Klemperer and D.J. Maltbie, J.Amer.Chem.Soc., 1985, 107, 8261.
193. V.W. Day, W.G. Klemperer and D.J. Malbie, Organometallics, 1985, 4, 104.
194. R.E. Cramer, K.T. Higa and J.W. Gilje, Organometallics, 1985, 4, 1140.
195. G. Paolucci, G. Rossetto, P. Zanella and R.D. Fischer, J.Organomet.Chem., 1985, 284, 213.

196. J.M. Ritchley, A.J. Zozulin, D.A. Wrobleski, R.R. Ryan, H.J. Wasserman, D.C. Moody and R.T. Paine, J.Amer.Chem.Soc., 1985, 107, 501.
197. W.J. Evans, J.W. Grate and R.J. Doedens, J.Amer.Chem.Soc., 1985, 107, 1671
198. H. Schumann, H. Lauke, E. Hahn, M.J. Heeg and D. Helm, Organometallics, 1985, 4, 321.
199. M.L.H. Green, D.O'Hare and G. Parkin, J.Chem.Soc., Chem. Commun., 1985, 356.
200. M.L.H. Green and G. Parkin, J.Chem.Soc., Chem.Commun., 1985, 771.
201. R.J. Markle, T.M. Pettijohn and J.J. Lagowski, Organometallics, 1985, 4, 1529.
202. R.A. Gancarz, J.F. Blount and K. Mislow, Organometallics, 1985, 4, 2028.
203. K.H. Dotz and M. Popall, J.Organomet.Chem., 1985, 291, C1.
204. E.P. Kundig, C. Perret, S. Spichiger and G. Bernardinelli, J.Organomet.Chem., 1985, 286, 183.
205. P.T. Bishop, J.R. Dilworth and J.A. Zubieta, J.Chem.Soc., Chem.Commun., 1985, 257.
206. J.J. Frizzell, R.L. Luck, R.H. Morris and S.H. Peng, J.Organomet.Chem., 1985, 284, 243.
207. N.J. Gogan, J. Doull and J. Evans, Can.J.Chem., 1985, 63, 3147.
208. G.T.W. Wittman, G.W. Krynauw, S.Lotz and W. Ludwig, J. Organomet.Chem., 1985, 293, C33.
209. C.D. Hoff, J.Organomet.Chem., 1985, 282, 201.
210. S.P. Nolan, C.D. Hoff and J.T. Landrum, J.Organomet.Chem., 1985, 282, 357.
211. Tatsumi, A. Nakamura, P. Hofmann, P. Stauffert and R. Hoffmann, J.Amer.Chem.Soc., 1985, 107, 4440.
212. S.P. Nolan and C.D. Hoff, J.Organomet.Chem., 1985, 290, 365.
213. P.H. Bird, A.A. Ismail and I.S. Butter, Inorg.Chem., 1985, 24, 2911.
214. K.R. Lane and R.R. Squires, J.Amer.Chem.Soc., 1985, 107, 6403.
215. J-C. Boutonnet, J. Levisalles, F. Rose-Munch and E. Rose, J.Organomet.Chem., 1985, 290, 153.
216. A. Ceccon, A. Gambaro and A. Venzo, J.Organomet.Chem., 1985, 281, 221.
217. A. Ceccon, A. Gambaro and A. Venzo, J.Chem.Soc., Chem. Commun., 1985, 540.
218. L.N. Novikova, N.A. Ustynyuk, V.E. Zvorykin, L.S. Dneprovskaya and Y.A. Ustynyuk, J.Organomet.Chem., 1985, 292, 237.
219. P.G. Sennikov, M.V. Panicheva, A.N. Egorochkin, N.I. Sirotkin and Y.Y. Zhiltsova,J.Organomet.Chem., 1985, 280, 81.
220. C. Degrand, F. Gasquez and P-L. Compagnon, J.Organomet. Chem., 1985, 280, 87.
221. W.J. Bland, R. Davis and J.L.A. Durrant, J.Organomet.Chem., 1985, 280, 95.
222. D. Villemin and E. Schigeko, J.Organomet.Chem., 1985, 293, C10.
223. M. Sodeoka and M. Shibasaki, J.Org.Chem., 1985, 50, 1147.
224. A. Alemagna, C. Baldoli, P.D. Buttero, E. Licandro and S. Maiorana, J.Chem.Soc., Chem.Commun., 1985, 417.
225. J. Blagg, S.G. Davies and B.E. Mobbs, J.Chem.Soc., Chem. Commun., 1985, 619.
226. C. Baldoli, P.D. Buttero, S. Maiorana and A. Papagni, J.Chem. Soc., Chem.Commun., 1985, 1181.
227. R.H. Mitchell, T.K. Vinod and G.W. Bushnell, J.Amer.Chem. Soc., 1985, 107, 3340.
228. M.C. Senechal-Tocquer, D. Senechal, J.L. Bihan, D. Gentric and B. Caro, J.Organomet.Chem., 1985, 291, C5.
229. J-C. Boutonnet, F. Rose-Munch, E. Rose and G. Precigoux, J.Organomet.Chem., 1985, 284, C25.
230. M. Uemura, K. Isobe and Y. Hayashi, Tet.Lett., 1985, 26, 767.
231. J-C. Boutonnet, F. Rose-Munch, E. Rose, Y. Jeannin and F. Robert, J.Organomet.Chem., 1985, 297, 185.
232. F. Effenberger and K. Schollkopf, Chem.Ber., 1985, 118, 4356.
233. S. Tondu, Siden Top, A. Vessieres and G. Jaouen, J.Chem.Soc., Chem.Commun., 1985, 326.

234. P.H.M. Budzelaar, J. Boersma, G.J.M. van der Kerk, A.L. Spek and A.J.M. Duisenberg, J.Organomet.Chem., 1985, 281, 123.
235. A. Solladie-Cavallo and J. Suffert, Tet.Lett., 1985, 26, 429.
236. A. Solladie-Cavallo and J. Suffert, Synthesis, 1985, 659.
237. P.J. Beswick and D.A. Widdowson, Synthesis, 1985, 492.
238. M.L.H. Green and D.O'Hare, J.Chem.Soc., Chem.Commun., 1985, 332.
239. H. van der Heijden, A.G. Orpen and P. Pasman, J.Chem.Soc., Chem.Commun., 1985, 1576.
240. L.H.P. Gommans, L. Main and B.K. Nicholson, J.Organomet. Chem., 1985, 284, 345.
241. Y.K. Chung, D.A. Swegart, N.G. Connelly and J.B. Sheridan, J.Amer.Chem.Soc., 1985, 107, 2388.
242. M. .H. Green and D.O'Hare, J.Chem.Soc., Chem.Commun., 1985, 355.
243. P.D. Morand and C.G. Francis, Organometallics, 1985, 4, 1653.
244. I. Bkouche-Waksman, J.S. Ricci, T.F. Koetzle, J. Weichmann and W.A. Herrmann, Inorg.Chem., 1985, 24, 1492.
245. W.A. Fordyce, R. Wilczynski and J. Halpern, J.Organomet. Chem., 1985, 296, 115.
246. T. Wilczewski, J.Organomet.Chem., 1985, 295, 331.
247. M-H. Desbois, D. Astruc, J. Guillin, J-P. Mariot and F. Variet, J.Amer.Chem.Soc., 1985, 107, 5280.
248. R.G. Swisher, E. Sinn, R.J. Butcher and R.N. Grimes, Organometallics, 1985, 4, 882.
249. R.G. Swisher, E. Sinn and R.N. Grimes, Organometallics, 1985, 4, 890.
250. R.G. Swisher, E. Sinn and R.N. Grimes, Organometallics, 1985, 4, 896.
251. M.P. Gomez-Sal, B.F.G. Johnson, J. Lewis, P.R. Raithby, A.H. Wright, J.Chem.Soc., Chem.Commun., 1985, 1682.
252. N.A. Vol'Kenau, I.N. Bolesova, L.S. Shul'pina, A.N. Kitagorodskii and D.N. Krautsov, J.Organomet.Chem., 1985, 288, 341.
253. E. Roman, V. Castro and M. Camus, J.Organomet.Chem., 1985, 293, 93.
254. B-H. Chang, C-P. Lau, R.H. Grubbs, C.H. Brubaker, J.Organomet.Chem., 1985, 281, 213.
255. D. Astruc, P. Michaud, A.M. Madonik, J.Y. Saillard and R. Hoffmann, Nouveau.J.Chem., 1985, 1, 41.
256. R. Lequan, M. Lequan, G. Jeouen, L. Ouahab, P. Batail, J. Padion and R.G. Southerland, J.Chem.Soc., Chem.Commun., 1985, 116.
257. M.A. Bennett, I.J. McMahon, S. Pelling, G.B. Robertson and W.A. Wickramasinghe, Organometallics, 1985, 4, 754.
258. T.P. Hanusa, J.C. Huffman, T.L. Curtis and L.J. Todd, Inorg. Chem., 1985, 24, 787.
259. A.R. Koray, T. Zahn and M.L. Ziegler, J.Organomet.Chem., 1985, 291, 53.
260. S.H. Simonsen, V.M. Lynch, R.G. Sutherland and A. Piorko, J.Organomet.Chem., 1985, 290, 387.
261. C.C. Lee, M. Iqbal, U.S. Gill and R.G. Sutherland, J.Organomet.Chem., 1985, 288, 89.
262. K. Meier and G. Rihs, Ang.Chem.Int.Ed., 1985, 24, 858.
263. J.A. Cabeza, B.E. Mann, C. Brevard and P.M. Maitlis, J.Chem.Soc., Chem.Commun., 1985, 65.
264. H. Werner and K. Roder, J.Organomet.Chem., 1985, 281, C38.
265. H-F. Klein, K. Ellrich, S. Lamac, G. Lull, L. Zsolnai and G. Huttner, Z. Naturforsch., 1985, 40B, 1377
266. H. Werner and K. Zenkert, J.Chem.Soc., Chem.Commun., 1985, 1607.
267. H. Werner and R. Werner, Chem.Ber., 1985, 118, 4543.
268. K. Shelly, D.C. Finster, Y.J. Lee, W.R. Scheidt and C.A. Reed, J.Amer.Chem.Soc., 1985, 107, 5955.
269. R.H. Crabtree and C.P. Parnell, Organometallics, 1985, 4, 519.
270. E. Bittersmann, K. Hildenbrand, A. Cervilla and P. Lahuerta, J.Organomet.Chem., 1985, 287, 255.
271. M.M. Brezinski, K.J. Klabunde, S.K. Janikowski and L.J. Radonovich, Inorg.Chem., 1985, 24, 3305.

272. M.... Eyring and L.J. Radonovich, Organometallics, 1985, 4, 1841.
273. M. Valderrama, M. Scotti, R. Ganz, L.A. Oro, F.J. Lahoz, C. Foces-Foces and F.H. Cano, J.Organomet.Chem., 1985, 288, 97.
274. F.H. Cano, C. Foces-Foces and L.A. Oro, J.Organomet.Chem., 1985, 288, 225.
275. M. Scotti, M. Valderrama, R. Ganz and H. Werner, J.Organomet. Chem., 1985, 286, 399.
276. S M. Hawkins, P.B. Hitchcock and M.F. Lappert, J.Chem.Soc., Chem.Commun., 1985, 1592.
277. J.H. Ammeter, C. Elschenbroich, T.J. Groshens, K.J. Klabunde, R.O. Kuhne and R. Mockel, Inorg.Chem., 1985, 24, 3307.
278. F.A. Cotton and W. Schwotzer, Organometallics, 1985, 4, 942.

15
Homogeneous Catalysis by Transition-metal Complexes

BY M. BOCHMANN

1 General Reviews

Diene telomerisation, synthesis of pyridines, iodide promoted rhodium catalysts, alcohol homologation and water gas shift are the subjects of a recent addition to a well-known series.[1] Another book concentrates on homogeneous rhodium and iridium catalysts[2] and complements an earlier book on the organometallic chemistry of these elements.[3] A monograph on organo-f element chemistry includes a section on lanthanides in catalysis,[4] a subject also dealt with in a review.[5] Other reviews highlight new chemical processes (mainly carbonylations)[6] and the use of transition metals in organic synthesis during 1983, covering oligomerisations, metathesis, isomerisations, carbonylations[7] and hydroformylation, water gas shift, hydrogenation and hydrocarbon oxidation.[8] Evidence has been collected to point out the close relationship between transformations of organometallic complexes and the active sites on metal surfaces.[9] Transition metal mediated P-C bond cleavage and its relevance to homogeneous catalyst deactivation has been discussed,[10] and a brief review highlights various reactions including cytochrome P-450 model oxidation and asymmetric C-C bond formation.[11] A new ESCA technique takes advantage of the steric control of diffusion of solution species into porous polymer films: catalytically active species carrying long-chain substituents are left on the surface where they can be studied. Examples include decarbonylations by Rh catalysts and the ring-opening polymerisation of cyclopentene.[12] A systematic study of heterometallic clusters as catalysts in various reactions shows a dependence on cluster structure, though activities and selectivities were modest in all cases.[13] Many interesting contributions have appeared on the occasion of the 60th birthdays of J. Halpern[14] and G. Wilke.[15]

2. Hydrogenation

Low temperature NMR studies indicate that the hydrido cation $[IrH_6L_2]^+$ contains two short H-H distances and should be formulated $[IrH_2(H_2)_2L_2]^+$ (L = PCy_3); there is exchange on warming.[16] The molecular structures

(For references see page 415)

of cationic Rh hydrogenation catalysts containing the ferrocenyl phosphine (1) have been determined; bulky groups (R = Bu^t, R' = Bu^t, Ph) enhance hydrogenation rates.[17] The lanthanide hydrides [(C$_5$Me$_5$)$_2$MH]$_2$ are highly active as hydrogenation catalysts for terminal olefines; the activity decreases in the order M = Lu > Sm > Nd > La.[18] Treatment of magnesium anthracene with CrCl$_3$ or TiCl$_4$ and H$_2$ gives hydrogenation catalysts; high surface area MgH$_2$ is produced in this way which is a suitable medium for reversible hydrogen storage.[19] The hydrogenation of terminal, internal and cyclic olefins, alkynes, aldehydes and ketones at 145°C under pressure is catalysed by the cyclopentadienone complex [(η^4-Ph$_4$C$_4$CO)Ru(CO)$_2$]$_2$. The cyclopentadienone ligand is resistant to hydrogenation and is retained in the catalyst.[20] The rhodacarborane (2) has been structurally characterised and catalyses the hydrogenation of the hindered olefin (3); the catalyst was recovered unchanged.[21] Of carbonyl clusters with M$_3$E structure, only those with E = PR and at least one Ru atom exhibit high hydrogenation activity and can be recovered unaltered, though none of the clusters reached the catalytic activity of comparable mononuclear complexes.[22] Cp$_2$Ti(C$_2$Ph$_2$)(PMe$_3$) hydrogenates alkynes to alkanes under ambient conditions.[23] Whereas RuH$_2$(dppb)$_2$ is inactive in alkyne hydrogenation, [RuH(dppb)$_2$]PF$_6$ in the presence of excess dppb catalyses the selective reduction of 1-heptyne to 1-heptene.[24] Selective cis-hydrogenation of internal alkynes to alkenes is achieved using [RuH(PMe$_2$Ph)$_5$]PF$_6$ as catalyst; no isomerisation is observed, and the rates are enhanced by excess PMe$_2$Ph or PPh$_3$.[25] A mechanistic study of alkyne hydrogenation in the presence of RuHCl(PPh$_3$)$_3$ has revealed the presence of the ortho-metallated complex RuCl[P(C$_6$H$_4$)Ph$_2$])(PPh$_3$)$_2$ in the catalytic cycle; it reacts with H$_2$ to regenerate the Ru hydride complex.[26]

The thermal decomposition of tris(allyl)chromium at 358 K in heptane gives a catalyst for the hydrogenation and isomerisation of 1,5-hexadiene.[27] Conjugated dienes are selectively reduced to mono-olefins using Co(NCS)(PPh$_3$)$_3$ as catalyst precursor (eqn.1),[28] whereas 1,4-dienes are reduced to a mixture of cis-mono-olefin isomers by a (naphthalene)chromium tricarbonyl catalyst.[29] Cationic iridium catalysts effect the stereoselective reduction of exocyclic C=C bonds, e.g. endo-(3) is reduced to endo-exo-(4) exclusively (eqn.2).[30] The stereochemistry of this reduction has been determined by deuteration studies.[31] The stereochemistry of the olefin hydrogenation catalysed by a related complex, [Ir(COD)py(PCy$_3$)]PF$_6$, is controlled by functional (carboxamide and carbalkoxy) group of the substrate.[32] Hydroxy groups direct the hydrogenation of (5); RhCl(PPh$_3$)$_3$ gives 95% selectivity to (6), whereas [Rh(NBD)(dppb)]BF$_4$ provides pure (7) (eqn.3).[33]

Ruthenium hydrido anions with two PPh$_3$ ligands are responsible for the unusual reduction of anthracene to 1,2,3,4-tetrahydroanthracene, and several

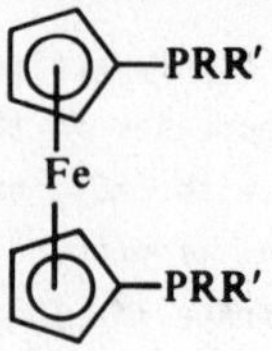

(1)

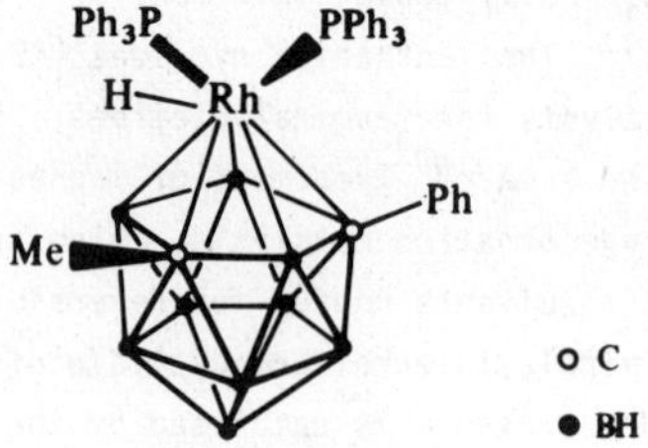

(2)

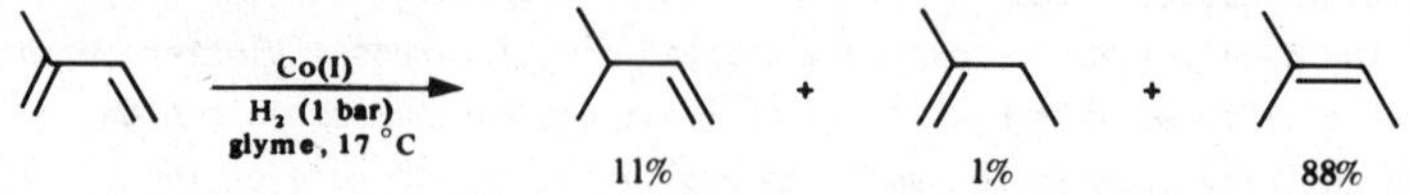

○ C
● BH

$$\text{isoprene} \xrightarrow[\text{glyme, 17 °C}]{\text{Co(I)} \atop \text{H}_2 \text{ (1 bar)}} \quad 11\% \quad + \quad 1\% \quad + \quad 88\% \tag{1}$$

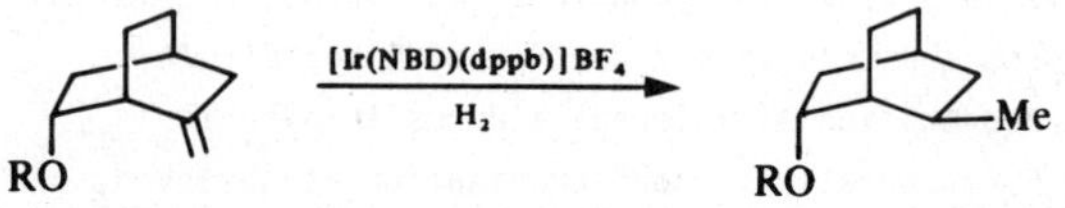

$$endo\text{-}(3) \xrightarrow[\text{H}_2]{[\text{Ir(NBD)(dppb)}]\text{BF}_4} endo\text{=}exo\text{-}(4) \tag{2}$$

$$(5) \xrightarrow{\text{H}_2} (6) \quad + \quad (7) \tag{3}$$

$$\text{quinoline} \xrightarrow[\text{150 psi D}_2]{\text{RuDCl(PPh}_3)_3} \text{ product} \tag{4}$$

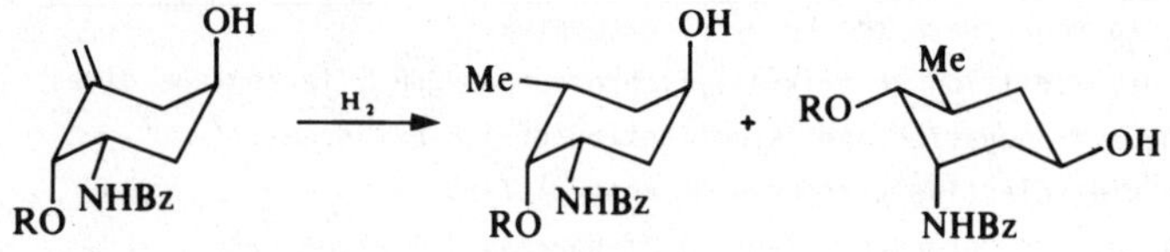

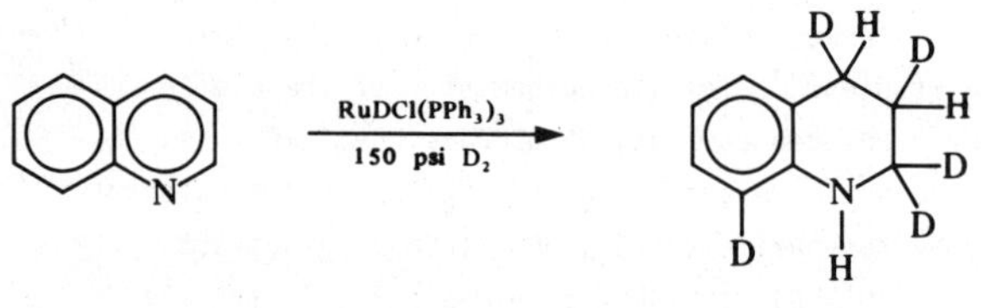

such complexes have been characterised (Scheme 1).[34] N and S polycyclic aromatics as models for coal liquefaction are hydrogenated by $RuHCl(PPh_3)_3$ to give compounds with saturated heteocycles; a mechanism based on deuteration studies is proposed (eqn.4).[35]

The kinetics of the hydrogenation of Schiff bases to secondary amines with $Co_2(CO)_8$ and with $HCo(CO)_4$ are identical. The mechanism is supported by stoichiometric reactions (Scheme 2).[36] The hydrogenation of olefins, aldehydes and ketones with Ru complexes of N,P polydentate ligands is likely to involve cationic intermediates. Olefin hydroformylation with these catalysts gives good selectivities for linear aldehydes.[37] Aldehydes are reduced by formic acid in the presence of $RuCl_2(PPh_3)^3$; other reducible groups are not affected.[38] $RuHCl(PPh_3)_3$ is preferable to $RuH_2(PPh_3)_4$ in the selective reduction of (8) (eqn.5).[39] Ketones are reduced by Rh bipyridyl complexes in basic methanol[40] and by Rh cations with chelating ligands.[41] Di- and tetranuclear Ru carbonyl acetates reduce dimethyl oxalate to $HOCH_2COOMe$,[42] whereas $Ru(CO)_2(OAc)_2(PBu^n_3)_2$ generates 1,2-dimethoxyethane which is slowly converted to ethylene glycol.[43]

Photochemical olefin hydrogenations catalysed by $Fe(CO)_5$ follows a radical pathway, according to high-pressure IR and UV studies.[44] Irradiation with visible light of the system triethanolamine/$Ru(bipy)_3Cl_2$/DMF induces the reduction of CO_2 to formate.[45] A similar system with a Ru(III) complex as electron acceptor cleaves acetylene photolytically ($\lambda > 400$ nm) to methane (eqn.6).[46]

2.1 Asymmetric hydrogenation The kinetics and mechanism of H_2 addition to $[Rh(dppp)_2]^+$ and to $[Rh(9)_2]^+$ have been compared; the latter is suitable for the hydrogenation of olefinic acids, _e.g._ (10).[47] Ir analogues of Rh asymmetric hydrogenation catalysts have been prepared, and the crystal structure of $[Ir(10, R = Me)_2]BF_4$ was determined.[48] The asymmetric hydrogenation of (10) in the presence of tetra- and hexanuclear Rh clusters containing the phosphines (9), (12) and (13) proceeds after cluster degradation to mononuclear hydrides.[49] Several new asymmetric ligands have been prepared, and complexes $[Rh(L)(L^*)]X$, [L = NBD, COD; L* = (11),[50] (14),[51] (15),[52] and (16),[53] X = ClO_4, BF_4] are highly reactive in the reduction of (10). The latter catalyst is suitable for hydrogenations in aqueous media (up to 60% e.e.).[53] $RuHCl[(-)-(17)]_2$ gives (S)-enantiomers of aminoacids (up to 92% e.e.) while Rh catalysts give the (R) forms with equal selectivity.[54] Rh complexes of (17) under H_2 pressure also selectively reduce the C2 double bond in geraniol and nerol (18), in optical yields up to 66%.[55] Cobalt complexes, _e.g._ $Co_2(CO)_6[PPh_2(neomenthyl)]_2$, reduce the C=C bonds of α,β-unsaturated ketones with modest optical yields.[56] Rh catalysts supported on silica

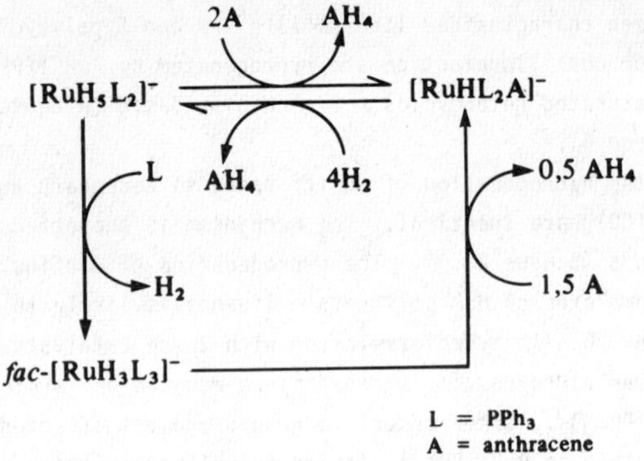

Scheme 1

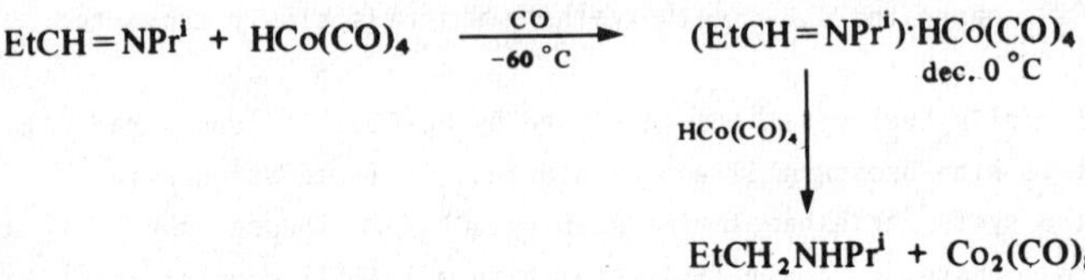

$$[RuH_5L_2]^- \rightleftharpoons [RuHL_2A]^-$$

(with 2A → AH$_4$ forward; L, AH$_4$, 4H$_2$ reverse)

$[RuH_5L_2]^-$ → (−L, −H$_2$) → fac-$[RuH_3L_3]^-$

fac-$[RuH_3L_3]^-$ → (−0.5 AH$_4$, −1.5 A) → $[RuHL_2A]^-$

L = PPh$_3$
A = anthracene

Scheme 2

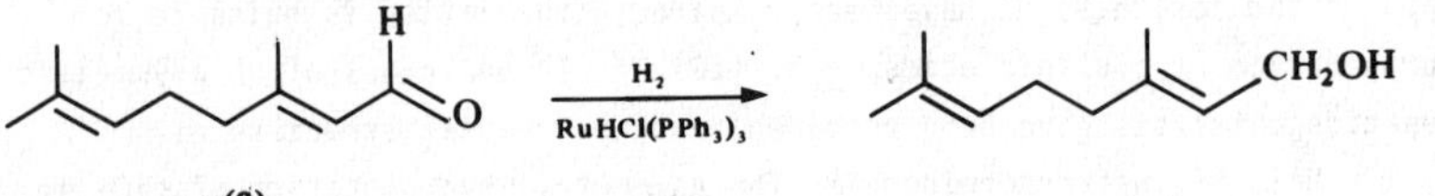

$$EtCH=NPr^i + HCo(CO)_4 \xrightarrow[-60\,°C]{CO} (EtCH=NPr^i)\cdot HCo(CO)_4$$
dec. 0 °C

↓ HCo(CO)$_4$

$$EtCH_2NHPr^i + Co_2(CO)_8$$

$$\text{(8)} \xrightarrow[RuHCl(PPh_3)_3]{H_2} \text{geraniol} \quad (5)$$

$$HC\equiv CH + 6H^+ + 6e^- \xrightarrow[Ru(II)/Ru(III)]{h\nu} 2\,CH_4 \quad (6)$$

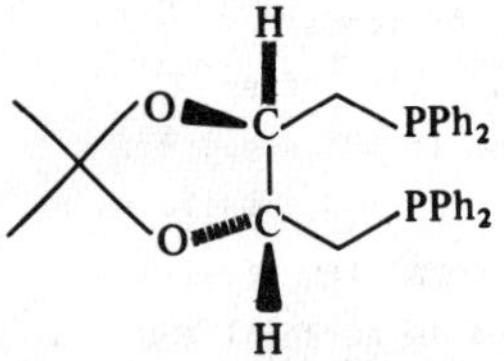

(R,R)-DIOP
(9)

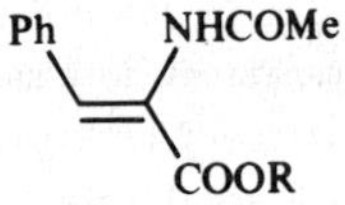

(10)

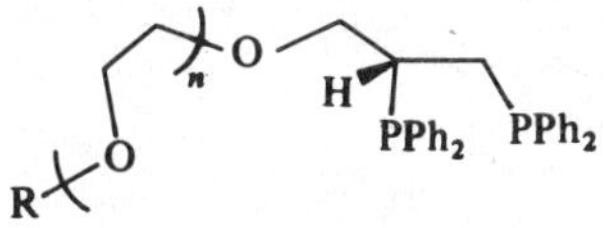

Homogeneous Catalysis by Transition-metal Complexes 387

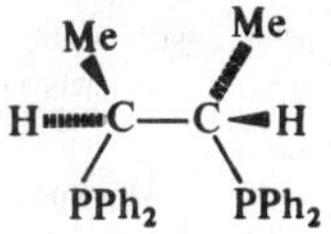

(11) $n = 1-5$; $R = Me, Bu^n, Ph, CH_2CH_2NEt_2$

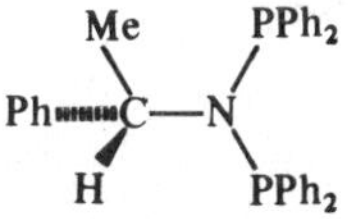

(12) S,S-CHIRAPHOS

(13) (S)-PEAP

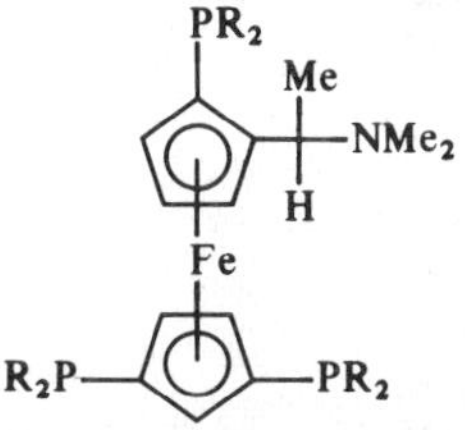

(14) $R = Bu^t$

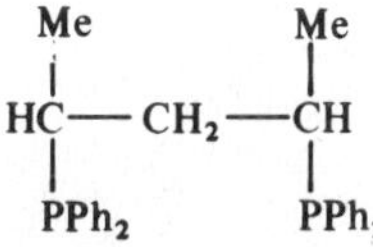

(15)

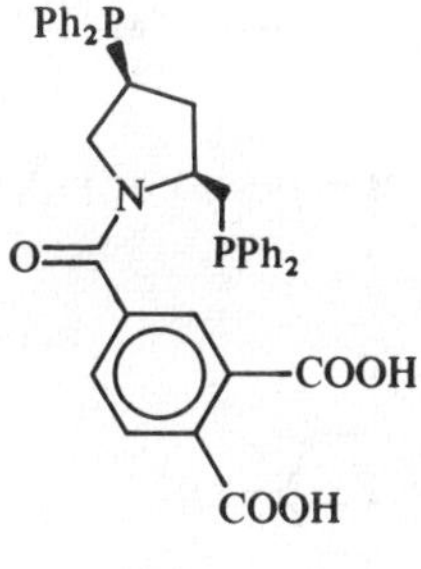

(16)

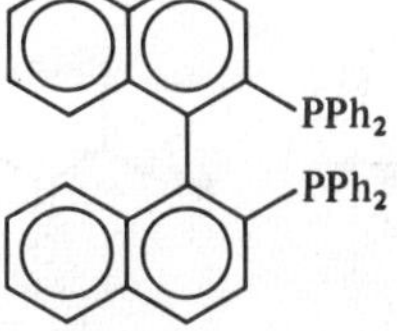

(17) BINAP

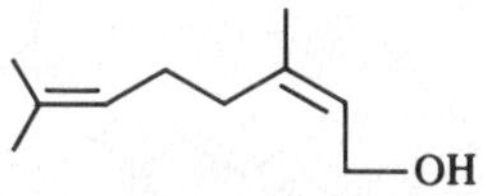

(18)

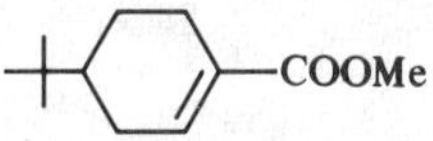

(19)

$$PhCH=CHCHO \rightarrow PhCH_2CH_2CHO + PhCH_2CH_2CH_2OH \quad (7)$$
$$(20) \qquad (21)$$
$$+ \; PhCH=CHCH_2OH$$
$$(22)$$

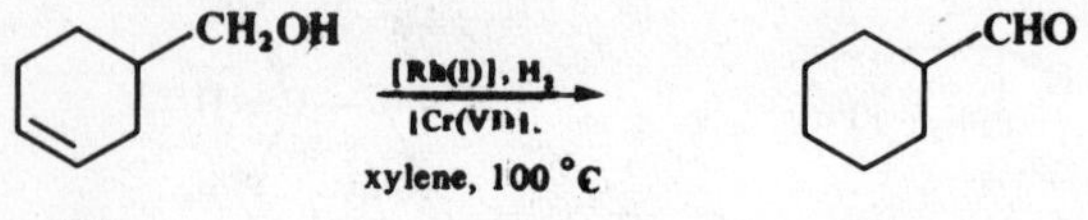

$$\text{(23)} \xrightarrow[\text{xylene, 100 °C}]{[Rh(I)], H_2, [Cr(VI)]} \text{CHO} \quad 80\% \qquad (8)$$

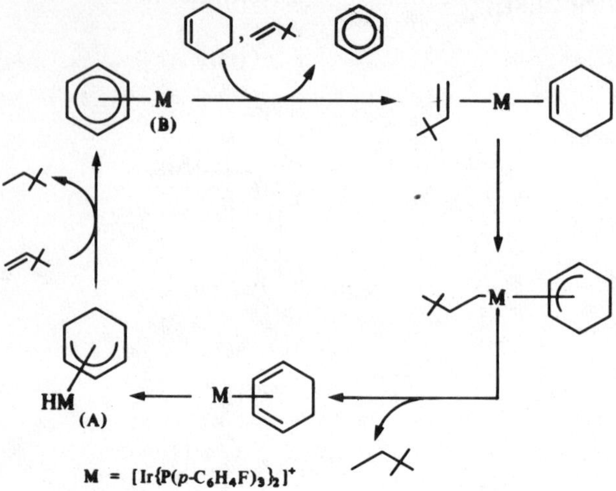

M = $[Ir\{P(p-C_6H_4F)_3\}_2]^+$

Scheme 2A

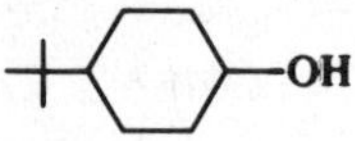

(24)

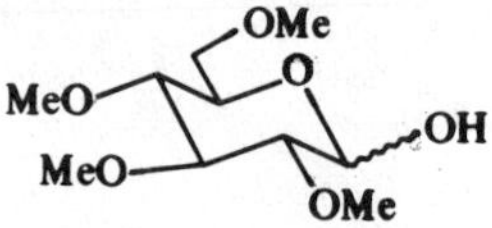

(25)

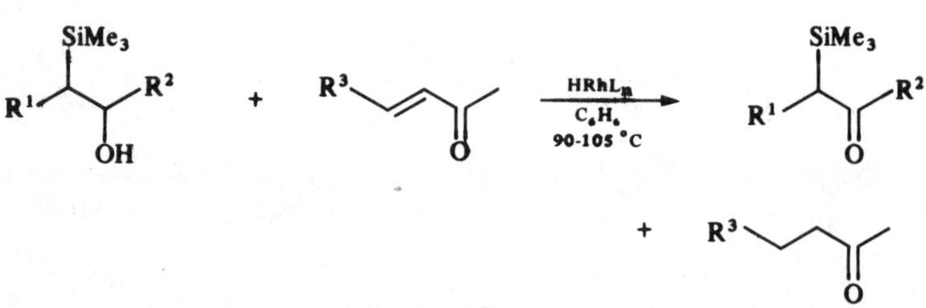

$$\underset{\text{OH}}{\overset{\text{SiMe}_3}{R^1\text{-}R^2}} + R^3\diagdown\diagdown\text{=O} \xrightarrow[\substack{C_6H_6 \\ 90\text{-}105\,°C}]{HRhL_n} \underset{O}{\overset{\text{SiMe}_3}{R^1\text{-}R^2}} + R^3\diagdown\diagdown\text{=O} \qquad (9)$$

carrying -(CH$_2$)$_n$-P(menthyl)$_2$ chains are more selective than comparable homogeneous systems and achieve up to 87% e.e. in the reduction of (10).[57]

2.2 Heterogenized catalysts Rhodium metal gives cis, RhCl(PPh$_3$)$_3$ trans-hydrogenation products with t-butylcyclohexenes, e.g. (19).[58] Rh and Pt phosphido complexes attached to resin are slow catalysts for the stereospecific deuteration of benzene to cis-d$_6$-cyclohexane.[59] The activity of [RhCl(CO)$_2$]$_2$ and RuH$_2$(CO)(PPh$_3$)$_3$ as homogeneous catalysts and attached to silica via amine side-chains has been compared in the reduction of cinnamaldehyde (eqn.7). Rh catalysts give mainly (20), Ru in solution leads to (21), whereas heterogenized Ru catalysts give predominantly (22).[60] Polymer-supported Rh catalysts give 10-20 times higher rates than RhCl(PPh$_3$)$_3$ in the selective reduction of polynuclear N and S heteroaromatics.[61] The same complex dissolved in the stationary liquid phase of a gas chromatography column is suitable for the gas-phase deuteration of olefins.[62] The preparative procedure of attaching Ru and Rh complexes to phosphinated polymeric supports influences their performance. Bi-anchoring and absence of uncoordinated PPh$_2$ groups are necessary for high activity in 1-hexene hydrogenation.[63] Palladium attached to phosphinated cellulose gives only low optical yields in attempted asymmetric hydrogenations.[64] A different approach to heterogenized catalysis consists of entrapping the metal in a porous polymer (HDPE) while an incompatible reaction is carried out simultaneously, e.g. the reduction/oxidation of (23) (eqn.8).[65] Several papers explore the catalytic activity of electropositive metals. Polymer-attached Cp$_2$ZrCl$_2$, CpZrCl$_3$, Cp$_3$ZrCl and Hf analogues, activated by Bu^{n}Li, are considerably more active than solution species in the stepwise reduction of diphenylacetylene to dibenzyl and in allylbenzene isomerisation,[66] reactions which are also used to test supported CpNb and Ta complexes. Catalysts derived from CpTaCl$_4$ are also able to dimerise ethylene to 1-butene.[67] (C$_5$Me$_5$)$_2$MMe$_2$ (M = Th,U) on Al$_2$O$_3$ are considerably better propene hydrogenation catalysts than Cp$_3$M-R derivatives; they are converted to hydrides. Ethylene is polymerised.[68]

3. Hydrogen Transfer and Dehydrogenation Reactions

Cyclohexane and neohexene are converted to benzene and neohexane in the presence of [IrH$_2$(Me$_2$CO)$_2$(PR$_3$)$_2$]SbF$_6$. Without neohexene, benzene and two equivalents cyclohexane are produced. The mechanism (Scheme 2A)is based on the isolated complexes (A) and (B).[69] 1,5-cyclooctadiene is slowly converted to cyclooctene by Rh$_6$(CO)$_{16}$ in isopropanol, probably via isomersiation to 1,4- and 1,3-cyclooctadiene.[70] The oxidation of cis-(24) by cyclohexenone in the presence of RhH(PPh$_3$)$_4$ is faster than of trans-(24) due to steric interactions

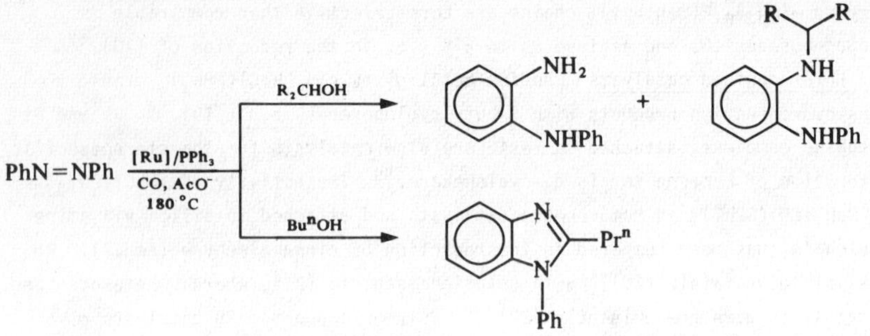

$PhN=NPh \xrightarrow[\substack{CO, AcO^- \\ 180\ ^\circ C}]{[Ru]/PPh_3}$ ...

Scheme 3

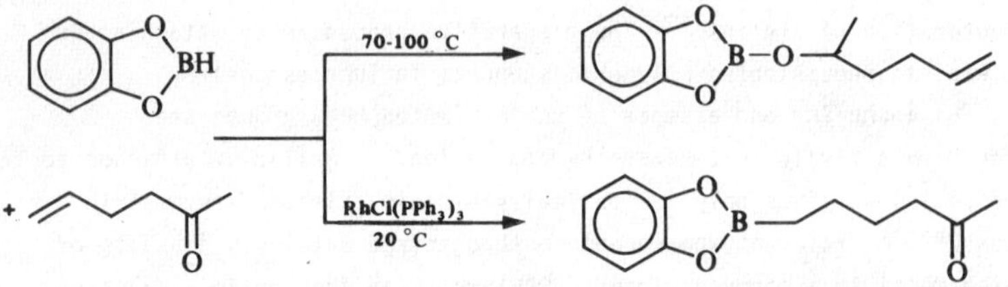

Scheme 4

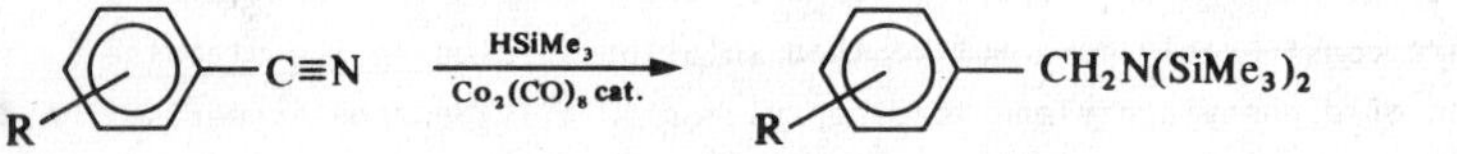

$$R\!-\!\!\!\bigcirc\!\!\!-C\equiv N \xrightarrow[Co_2(CO)_8 \text{ cat.}]{HSiMe_3} R\!-\!\!\!\bigcirc\!\!\!-CH_2N(SiMe_3)_2 \qquad (10)$$

$$R-\equiv-\equiv-R + R'_3SiH \xrightarrow{\substack{H_2PtCl_6 \\ 80\text{-}90\ ^\circ C}}$$

R = SiMe$_3$

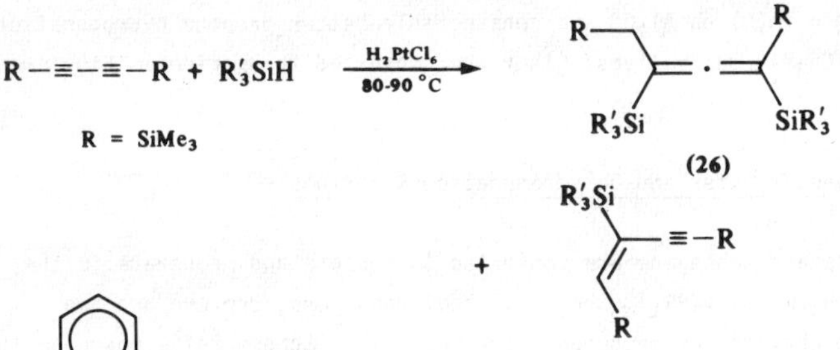

(26)

+

(27)   (11)

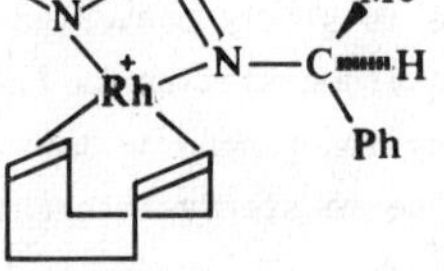

(28)

in the intermediate Rh complex, cis/trans isomerisation is slow.[71] Sugar derivatives, e.g. (25), are similarly oxidised by benzalacetone.[72] A mixture of [IrCl(COD)]$_2$, base, P(OMė)$_3$ and R-(−)-mandelic acid as cheap asymmetric ligand catalyses the asymmetric reduction of acetophenone by isopropanol in modest (<12%) optical yields;[73] a similar system with a bidentate phosphinite as asymmetric ligand works slightly better (17.9% e.e.).[74] The reduction of cinnamaldehyde (eqn.7) by isopropanol in the presence of an [IrCl(COD)]$_2$/ tris(o-methoxyphenyl)phosphine catalyst and KOH gives up to 100% (22). The reaction is strongly influenced by the phosphine ligand.[75]

Ketones are reduced to alcohols in the phase-transfer catalysed H-transfer reduction with isopropanol, or better PhCH$_2$CH$_2$OH, in the presence of Fe$_3$(CO)$_{12}$. Fe(CO)$_5$ is far less active. Mono-, di- and trinuclear iron hydride carbonyl anions are generated in situ.[76] α-Trimethylsilylketones can be prepared via Rh catalysed oxidations with butenones (eqn.9).[77] Azobenzene is isomerised and reduced to o-phenylenediamines by a RuCl$_3$/PPh$_3$/CO/LiOAc system in secondary alcohols.[78] By contrast, n-butanol leads to formation of benzimidazoles[79] (Scheme 3).

The photolytic dehydrogenation of methanol in the presence of Rh catalysts is sensitized by acetone; products are CH$_2$O, H$_2$, CH$_2$(OMė)$_2$ and those derived from CH$_2$OH radicals.[80] Binuclear Rh and Pd dppm complexes are also effective. Addition of acetone to the methanol medium changes the gas-phase composition from pure H$_2$ to methane and CO.[81] The first thermal methanol dehydrogenation uses Ru$_2$(OAċ)$_4$Cl/PEtPh$_2$ as catalyst precursor.[82]

4. Hydroboration and Hydrosilylation

Acetylenes are reduced to cis-olefins by NaBH$_4$ and Fe(II)/lipoamide and Fe(II)/dithiol catalysts in ethanol,[83] and further to alkanes by NaBH$_4$/-polyethyleneglycol/PdCl$_2$.[84] Allylic tosylates are selectively desulfonylated with LiBHEt$_3$ in the presence of PdCl$_2$(dppṗ) without isomerisation.[85] RhCl(PPh$_3$)$_3$ changes the direction of the hydroboration of hexenone from C=O to C=C addition (Scheme 4).[86]

[IrCl(cyclooctenė)$_2$]$_2$ catalyses the hydrosilylation of 1,3-dienes and of cyclohexenone to give mainly 1,4- and 1,2-addition products, respectively.[87] The sterically hindered isocyanide complexes (2,6-RC$_6$H$_3$-NC)M(CO)$_5$, M = Cr, Mo, W, R = Me, Pr^i, catalyse the hydrosilylation of 2,3-dimethylbutadiene.[88] Vinylsilanes are hydrosilylated by Pd phosphine complexes.[89] and aromatic nitriles are converted to benzyl amines by Co$_2$(CO)$_8$ (eqn.10).[90] The hydrosilylation of dialkynes is silane dependent; Et$_3$SiH gives 100% (26), Pr^{i}SiH up to 92% (27) (eqn.11).[91]

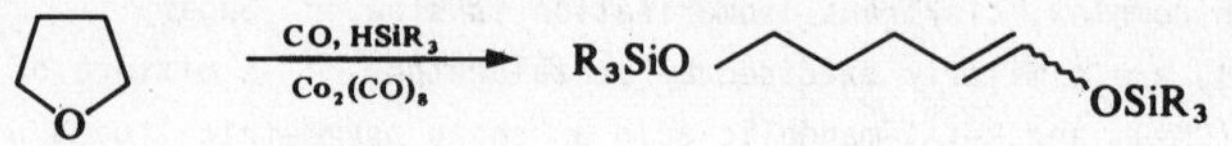

$$\text{THF} \xrightarrow[\text{Co}_2(\text{CO})_8]{\text{CO, HSiR}_3} R_3\text{SiO}\diagup\diagdown\diagup\diagdown\text{OSiR}_3 \qquad (12)$$

$$R-\equiv-R + \text{Ni(CO)}_4 \xrightarrow[\text{MeOH}]{\text{HCl}} \text{[dimeric Ni complex with Cl bridges]} \qquad (13)$$

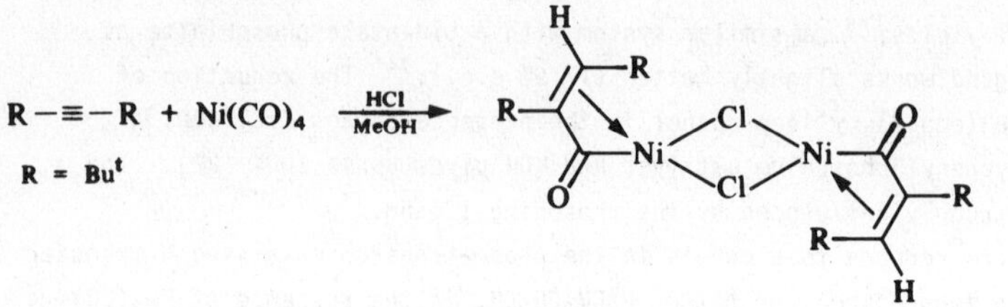

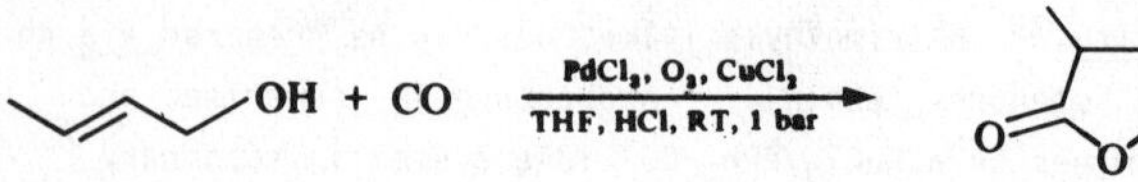

R = Bu^t

$$\diagup\diagdown\diagup\text{OH} + \text{CO} \xrightarrow[\text{THF, HCl, RT, 1 bar}]{\text{PdCl}_2, \text{O}_2, \text{CuCl}_2} \text{[butyrolactone]} \qquad (14)$$

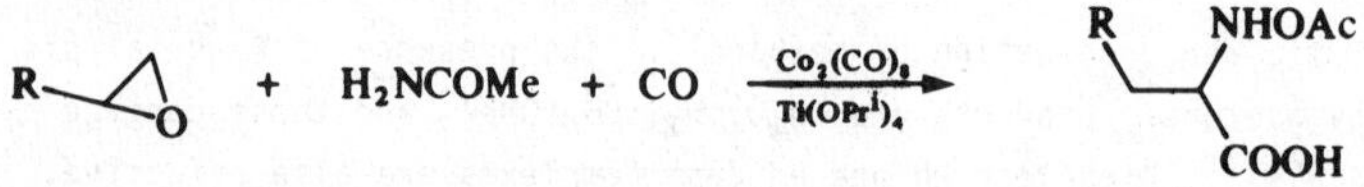

$$R\text{-epoxide} + \text{H}_2\text{NCOMe} + \text{CO} \xrightarrow[\text{Ti(OPr}^i)_4]{\text{Co}_2(\text{CO})_8} \begin{array}{c} R\quad\text{NHOAc}\\ \diagdown\diagup\\ \text{COOH}\end{array} \qquad (15)$$

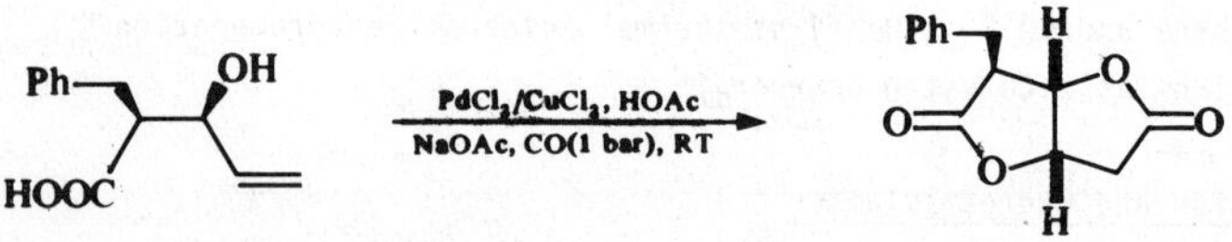

$$\begin{array}{c}\text{Ph}\diagdown\quad\text{OH}\\ \diagup\diagdown\diagup\\ \text{HOOC}\end{array} \xrightarrow[\text{NaOAc, CO(1 bar), RT}]{\text{PdCl}_2/\text{CuCl}_2, \text{HOAc}} \text{[bicyclic bis-lactone]} \qquad (16)$$

$$\begin{array}{c}\text{Ph}\diagdown\diagup\text{R}\\ \diagdown\text{O}\diagup\end{array} + \text{CO} \xrightarrow[\text{NaOH, C}_6\text{H}_6, \text{R}_4\text{N}^+\text{Br}^-]{\text{Co}_2(\text{CO})_8, \text{MeI}} \text{[tetronic acid derivative]} \qquad (17)$$

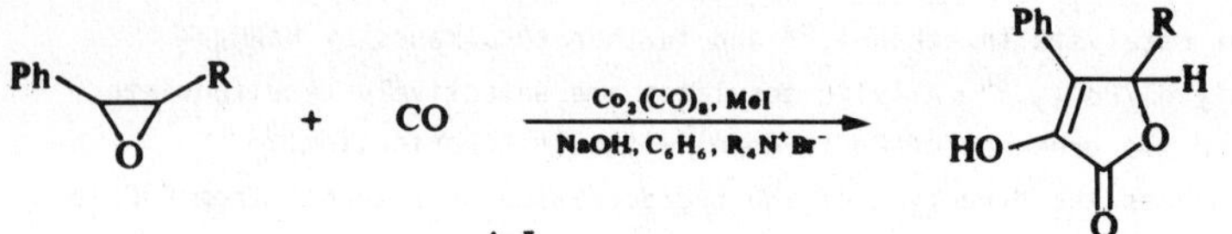

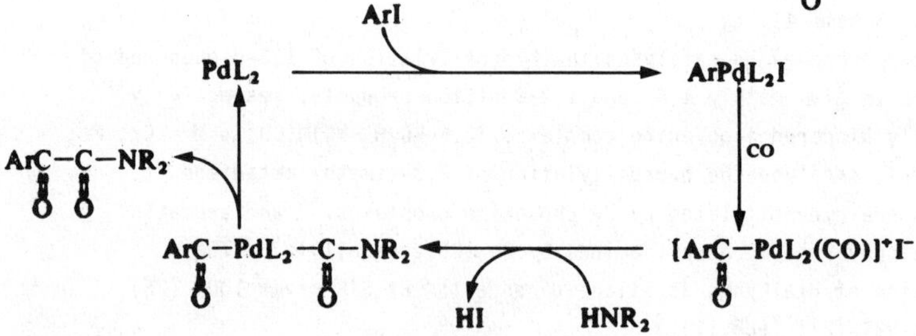

Scheme 5

The asymmetric hydrosilylation of ketones is catalysed by Rh complexes of chelating N,P ligands and gives optical yields of up to 19.6%.[92] The solid and solution structures of cationic Rh hydrosilylation catalysts with optically active ligands, e.g. (28), have been determined.[93] The stereoselective splitting of a C-C bond is observed in the RhCl(PPh$_3$)$_3$ catalysed reaction of (-)-(1R,4R)-campheroxime with Ph$_2$SiH$_2$.[94]

The Co$_2$(CO)$_8$ catalysed silylcarbonylation of THF[95] and oxetane[96] with HSiR$_3$ and CO gives enol silyl ethers and disilyl ethers (eqn.12). With styrene oxide no CO incorporation takes place.[97]

5. Catalytic Alkane Activation

The organometallic chemistry of alkanes[98] and the alkane activation with oxide bound complexes derived from Rh(allyl)$_3$[99] have been reviewed. IrH$_2$(CF$_3$COO)(PCy$_3$)$_2$ catalyses the photochemical dehydrogenation of cyclooctane to cyclooctene and H$_2$ without the need for a hydrogen acceptor.[100] H/D exchange between C$_6$D$_6$ and Bu^{t}CH=CH$_2$ in the presence of IrH$_5$(PPr$^i_3$)$_2$ demonstrates that the activation of vinylic C-H bonds occurs more readily than the addition-elimination sequence of a metal hydride across the double bond.[101]

6. Carbonylation Reactions

An intermediate in the carbonylation of alkynes to esters can be isolated in a stoichiometric reaction with Ni(CO)$_4$ (eqn.13).[102] The mechanism of 1-octene hydroesterification with Co$_2$(CO)$_8$ in methanol has been re-investigated by high-pressure IR and UV spectroscopy.[103] Polar solvents and SnCl$_2$ as additive to PdCl$_2$(PPh$_3$)$_2$ improve the selectivity to diethylglutarate in the carbonylation of CH$_2$=CHCH$_2$COOEt.[104] Ni(CN)$_2$ reacts with CO and base to give [Ni(CO)$_3$CN]$^-$ which catalyses the carbonylation of allyl halides to carboxylic acids.[105] Carbonylation of allylic alcohols under oxidative conditions generates cyclic lactones (eqn.14).[106] Ether solvents participate in the carbonylation of benzyl bromide with [RhCl(1,5-hexadiene)]$_2$/KI to give esters.[107] Unsaturated amides are prepared from allylic phosphates, CO and secondary amines in the presence of Rh$_6$(CO)$_{16}$ under pressure.[108]

Mixed-metal catalysts based on Co$_2$(CO)$_8$ have proved effective in the synthesis of N-acyl aminoacids, starting with either allylic alcohols, oxiranes (eqn.15) or trifluoropropene.[109] The oxidative carbonylation of organomercury compounds is subject to solvent effects.[110] The Pd catalysed carbonylative cross-coupling of aryl iodides with triisobutylaluminium gives secondary benzyl alcohols; a variety of functional groups are tolerated.[111] Carbamate esters

$$CH_3C^*H_2-[Rh] \rightleftharpoons H-[Rh](C^*H_2CH_2) \rightleftharpoons C^*H_3CH_2-[Rh] \quad (18)$$
$$[Rh] = Rh(CO)I_3^-$$

$$MeCOOMe \xrightarrow[Pd/Rh]{CO/H_2} MeCH(OAc)_2 + (Ac)_2O + AcOH \quad (19)$$

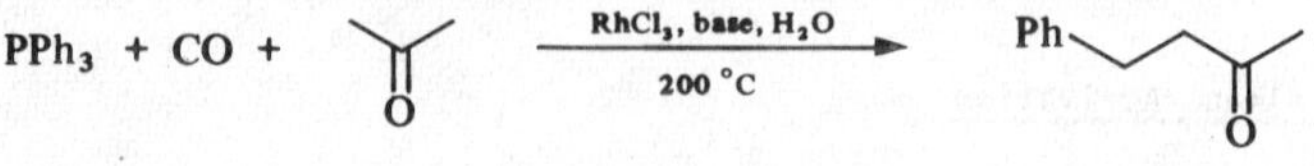

$$PPh_3 + CO + \underset{O}{\text{Me-CO-Me}} \xrightarrow[200\ °C]{RhCl_3,\ base,\ H_2O} Ph\text{-CH}_2\text{-CH}_2\text{-CO-Me} \quad (20)$$
(29)

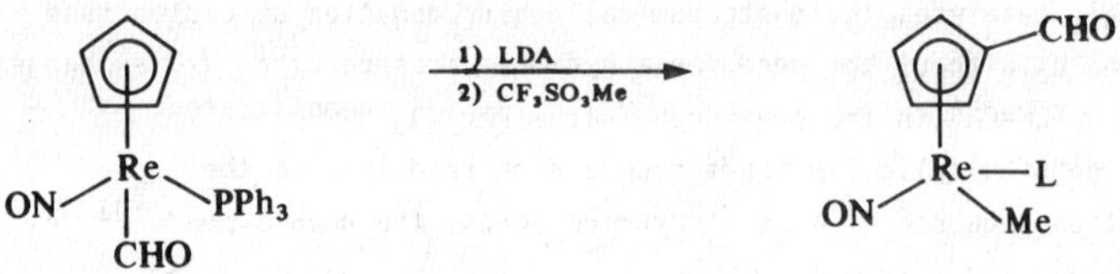

(21) — Re complex with Cp, ON, PPh$_3$, CHO ligands → via 1) LDA, 2) CF$_3$SO$_3$Me → Re complex with Cp-CHO, ON, L, Me

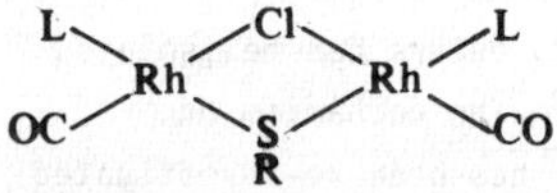

(30), L = PPh$_3$, AsBu$_3^t$

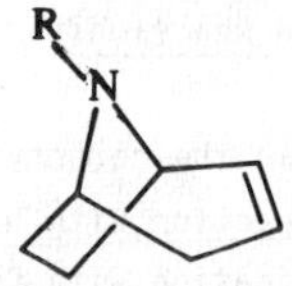

(31)

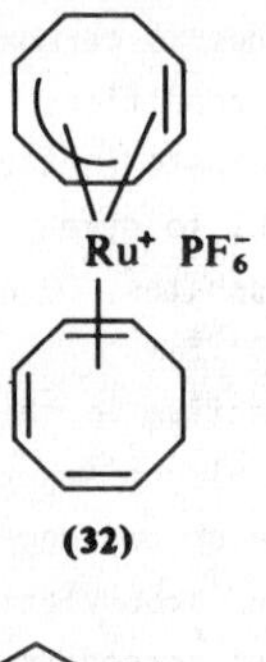

(32) Ru$^+$ PF$_6^-$ (bis-cyclooctadiene/cyclooctatriene sandwich)

$$\underset{(33)}{\text{MeCH=CH-CH}_2\text{-CN}} \xrightleftharpoons[]{Ni[P(O\text{-}p\text{-tol})_3]_4/H^+} \underset{\text{kinetic product}}{CH_2=CH\text{-CH}_2\text{-CH}_2\text{-CN}} + \underset{\text{thermodynamic product}}{MeCH=CH\text{-CH}_2\text{-CN}} \quad (22)$$

are synthesized under mild oxidative conditions ($PdCl_2/CuCl_2$, O_2, RT) from aniline, CO and alcohols.[112] Under reducing conditions in the presence of a $PtCl_2(PPh_3)_2/SnCl_4$ catalyst, aniline, CO and nitrobenzene derivatives are combined to (un-)symmetric arylureas.[113] The stereoselective formation of bicyclic bis-lactones is possible by Pd catalysed oxidative carbonylation (eqn.16),[114] and stereoselective carbonylation routes to (E)-alkylidene-succinates have been described.[115]

Several papers deal with the products of (formally) "double carbonylation" reactions. The carbonylation product of styrene oxide in the presence of a $Co_2(CO)_8$/MeI catalyst under phase-transfer conditions contains two CO derived carbons (eqn.17).[116] Two groups report the synthesis of α-keto esters from aryl iodides, CO, alcohols and tertiary amines; the selectivity depends on ROH.[117,118] Replacement of alcohol by water leads to α-keto acids.[119] α-Keto-amides are prepared similarly from secondary amines. In this case the mechanism has been studied in detail; oxidative addition of the aryl halide is the rate-determining step (Scheme 5).[120]

6.1 Carbonylation of Alcohols and Esters.

The mechanism of the Rh/I⁻ catalysed alcohol carbonylation has been studied in detail. Rates decrease sharply from methanol to n-propanol. Formation of isobutyric acid as a by-product points to a β-H elimination-reinsertion sequence.[121] This sequence has also been demonstrated for ethanol carbonylation by selective ^{13}C labelling (eqn.18).[122] The reductive carbonylation of methanol in the presence of CoI_2 and PPh_3 generates acetaldehyde, ethanol and methyl acetate. Only diphenylether and alkanes as solvents did not decompose under the reaction conditions (170°C, 5000 psi).[123] The carbonylation of methanol to methyl formate is catalysed by $W(CO)_6$/OMe⁻ or $[W_2(CO)_{10}H]^-$; CO_2 inhibits the reaction.[124] The cobalt carbonyl catalysed-desulfurization of benzyl thiols gives COS and hydrocarbons; CO insertion does not take place.[125] Synthesis gas reacts with ethyl orthoformate under Ru/I⁻ catalysis to a mixture of hydrogenolysis and carbonylation products.[126] A similar catalyst under more severe conditions converts methyl formate into acetic acid and methylacetate; CO_2 and methane are by-products.[127] The effectiveness of iodide promoters in the Ru catalysed carbonylation of methyl acetate decreases in the order $PPh^+_4I^-$ > KI > NaI   LiI > CsI due to ion-pairing effects with the active species. The crystal structure of $[K(18-crown-6)]^+[\underline{fac}-Ru(CO)_3I_3]^-$ reveals weak interactions between K and the iodide ligands.[128] Mixed Pd/Rh/MeI/PBu^n_3 catalysts convert methyl acetate mainly to ethylidene diacetate (eqn.19).[129]

6.2 Alcohol Homologation Solvent and promoter effects on the cobalt carbonyl catalysed methanol homologation have been studied under synthesis gas pressure.[130] The main product in a methanol/hydrocarbon two-phase system is 1,1-dimethoxyethane (ca. 70% selectivity).[131] Using similar iodide promoted cobalt catalysts, $R_2C(OMe)_2$ and dimethylcarbonate are converted to acetaldehyde with up to 87% selectivity.[132] Ruthenium in the presence of Co, I_2 and dppe improves the ethanol selectivity in the homologation of dimethylether. Best results are achieved in inert solvents with high dielectric constants, e.g. sulfolane (ϵ = 44), and with BF_3 as activator.[133]

6.3 Carbon Monoxide Reduction and the Water Gas Shift Reaction A review of the role of metal cluster catalysts in solution and attached to supports for CO hydrogenation is optimistic about the use of cluster-derived heterogeneous catalysts on industrial processes.[134] Selective reduction of CO to methanol occurs on Pt cathodes coated with $K_2Fe_2(CN)_6$ in methanol solutions containing Fe(III) or Cr(III) complexes.[135,136]

A review describes the application of the homogeneously catalysed water gas shift reaction to the hydrogenation of olefins and carbonyl compounds, hydroformylation and hydrocarboxylation.[137] The activation energy for the photochemical $Cr(CO)_6$ catalysed WGS reaction is 30 kJ mol^{-1}, compared to 120 kJ mol^{-1} for the thermal process. In-situ UV and IR studies under pressure have confirmed $[Cr(CO)_5(O_2CH)]^-$ as the first detectable intermediate in the catalytic cycle.[138] The thermodynamic parameters of the pulsed laser initiated WGS reaction catalysed by Cr and W hexacarbonyls in basic aqueous methanol solution have been determined. Large H_2O concentrations are inhibitive.[139] The cluster $[Ru_3H(CO)_{11}]^-$ is an active catalyst for WGS, whereas $[Ru_4H_3(CO)_{12}]^-$ is not. The paper underlines the importance of trinuclear species as catalysts.[140] Metal clusters in the presence of N-chelates catalyse the reduction of nitrobenzene to aniline with CO/H_2O; $Ru_3(CO)_{12}$ is superior to clusters of Rh, Ir and Os.[141] Phenyl groups derived from PPh_3 are incorporated into organic products under WGS conditions. Introduction of H_2 increases the yield of (29) up to 40% (eqn.20).[142]

6.4 Hydroformylations. A new computer program able to propose mechanistic schemes of metal catalysed reactions has been applied to the hydroformylation of ethylene.[143] The hydroformylation of formaldehyde (160°C, 186 bar) in the presence of alcohols gives glycol monoethers; catalyst precursors are $Co_2(CO)_8$ with alkyls of O, S, Se, Sn and Ge.[144] Under milder conditions (70°C) $RhCl(CO)(PPh_3)_2$ in γ-picoline as solvent converts formaldehyde into glycol monoaldehyde in up to 90% selectivity.[145] Methyl and dimethyl formamide are synthesized directly from CO, H_2 and NH_3 using Ru catalysts in molten $PBu_4^+I^-$, though activity is low.[146]

The first example of a formylation by a metal-bound formyl group has been reported (eqn.21).[147] There is evidence for radical intermediates in the $HCo(CO)_4$ catalysed hydroformylation of styrene.[148] A mechanism based on intact Ru_3 clusters accounts for the isotope distribution in the deuteroformylation of C_2H_4 in the presence of $[Ru_3H(CO)_{11}]^-$.[149] In-situ Raman spectroscopic studies of the hydroformylation of propene by $Co_2(CO)_6(PPh_3)_2$ reveal the formation of Co hydrides both in solution and with catalysts attached to SiO_2, but not Al_2O_3 which gives inferior activity.[150]

Binuclear Rh complexes (30), attached to silica via S, are active for olefin hydroformylation, hydrogenation and isomerisation.[151,152] Kinetic data of the hydroformylation of 1-hexene by $[Rh(SBu^t)(CO)\{P(OMe)_3\}]_2$ suggest catalysis by a mononuclear complex.[153] The efficiency of pyrazolato bridged Rh complexes in this reaction increases in the order L = PPh_3 < $P(OMe)_3$- < $P(OPh)_3$.[154]

The Rh/PPh_3 catalysed hydroformylation of vinyl ethers gives mainly branched alkoxyaldehydes.[155] A complex mixture of linear, cyclic and ring-contracted (mainly saturated) monoaldehydes are formed in the hydroformylation of cyclo-octatetraene.[156] The regio- and stereoelectivities in the hydroformylation of heterocycles, e.g. (31), have been studied.[157] Butadiene is converted to pentanal in the presence of Ru-diphosphine catalysts; dppe is most effective.[158]

Supported $[Os_3H(CO)_{11}]^-$ converts 1-hexene to n-heptanal in high selectivity; the Fe analogue breaks up.[159] The activities and selectivities of zeolite-supported Rh hydroformylation catalysts is generally inferior to homogeneous systems; Rh attached to the outside zeolite surface leaches into the solution.[160] The activity of Rh SLP (supported liquid phase) catalysts, prepared by impregnating a highly porous support with solutions of Rh complexes in PPh_3, depends on the degree of pore-filling.[161] The kinetics of 1-butene hydroformylation[162] and the thermal deactivation of these catalysts have been reported.[163,164]

In the asymmetric hydroformylation of mono- and disubstituted olefins, Rh complexes of (12) gave higher (up to 45% e.e.) optical yields than catalysts containing (9).[165] The results of the asymmetric hydroformylation of 1-butene by $[PtCl(SnCl)_3(9)]$ may be falsified by concomitant isomerisation to 2-butenes; optical purity is therefore temperature dependent.[166]

at 25 °C: > 90% *B*
at −78 °C: 94% *A*

7. Isomerisations

The clusters $Ru_4H_3(AuL)(CO)_{12}$ and $Ru_4H_2(AuL)(CO)_{12}$ (L = PPh_3) show greater reactivity than $Ru_4H_4(CO)_{12}$ in the isomerisation of 1-pentene.[167] Allylbenzene is converted to cis- and trans-propenylbenzene by $[FeMH(CO)_9]^-$ (M = Cr,Mo,W) and their $P(OMe)_3$ derivatives.[168] The isomerisation of 1,5- to 1,3-cyclooctadiene is catalysed by $Ru_2(O_2CCF_3)_2(\mu-O_2CCF_3)_2(\mu-H_2O)(COD)_2$.[169] The cationic Ru allyl complex (32) isomerises 1-hexene to 2-hexene under an H_2 atmosphere; it is also active in the dimerisation of ethylene, the ring-opening polymerisation of norbornene and the polymerisation of butadiene mainly to the trans-1,4-polymer.[170] The rates of 1-butene isomerisation by $Os_3(CO)_{12}$ attached to a resin through OH groups compare unfavourably with the activity of the same cluster on silica or alumina supports.[171] The isomerisation of (33), important for the synthesis of adiponitrile via Ni catalysed hydrocyanation of butadiene, is kinetically controlled (eqn.22); the ratio between the kinetic and the thermodynamic product can be as high as 70:1 and is ligand influenced.[172] The optically active Rh complexes $[Rh(17)L]ClO_4$ [L = COD, (17)] isomerises (34) into the enamine in near-quantitative enantiomeric excess (eqn.23); both (R) and (S) isomers can be prepared. The crystal structure of $[Rh(17)_2]ClO_4$ has been determined.[173]

Ruthenium isomerises allylic alcohols to ketones; the regiochemistry was elucidated by deuteration studies.[174] The Pd catalysed allylic rearrangement of allyl tosylates can be applied to the synthesis of α,β-unsaturated ketones.[175] The 3-aza Cope rearrangement of N-allyl enamines is mediated by $Pd(PPh_3)_4$/-CF_3COOH,[176] while Pd(0), Pd(II), Rh(I) and Ir(I) complexes effect the closely related conversion of allyl imidates into allylic amides (eqn.24).[177] The isomerisation of allylic alcohols under oxidative conditions is catalysed by $V(O)(acac)_2$ or $MoO_2(acac)_2$ and can be temperature dependent (eqn.25).[178]

8. Alkene and Alkyne Metathesis

Papers presented on the 5th International Symposium on Olefin Metathesis (Graz, August, 1983) are collected in a special issue;[179] some of these are reviewed here in more detail. The tungstenacyclobutadiene complex $WC(Et)C(Et)C(Et)[(CF_3)_2CHO]_3$ metathesises 3-heptyne at 25°C; the structure of (35) is described.[180] A cyclisation (eqn.26) lends support to the mechanism of metal carbene catalysed alkyne polymerisation.[181] Vinylidene complexes which initiate the polymerisation of terminal and internal alkynes can be generated in situ by photolysis of $W(CO)_6$ and RC≡CH; preformed catalysts are not necessary.[182] The polymer formed by the WCl_6/PhC≡CH initiated ring-opening polymerisation of cyclopentene has a block of poly(phenylacetylene) attached at the start of the chain, indicating that the propagation involves initially phenylacetylene derived carbene species which "switch over" to the olefinic substrate in the absence of free alkyne. Traces of diphenylacetylenequench the metathesis.[183] The polymerisation of functionalised acetylenes HC≡C-(CH$_2$)$_3$X (X = Cl, CN, COOMe) is catalysed by WCl$_6$/SnMe$_4$.[184] A different route to polyacetylene-type polymers is the ring-opening polymerisation of cyclooctatetraene.[185]

Titanacyclobutane complexes exhibit moderate activity in olefin metathesis.[186] A "living" propagating W carbene complex has been detected by NMR in the ring-opening polymerisation of norbornene with W(CHBu^t)(OCH$_2$-Bu$^t_2$)$_2$Br$_2$/GaBr$_3$.[187] A ^{13}C analysis of metathetically generated norbornene polymers suggest the presence of two types of metal carbene intermediates which give rise to "cis after cis" or "trans after trans" double bond stereoselectivity.[188]

The ether complexes WCl(OAr)(CH$_2$Bu^t)(CHBu^t)(OR$'_2$) are highly active olefin metathesis catalysts in the absence of a Lewis acid component.[189] Another one-component catalyst is WCl$_2$(CO)$_3$(AsPh$_3$)$_2$[190] although the activation of this complex by AlCl$_3$ is also reported.[191] Replacement of AsPh$_3$ by PPh$_3$ makes the catalyst ineffective, as does removal of CO as in WCl$_2$(NO)$_2$-(AsPh$_3$)$_2$.[190] It is suggested that the active carbene species is formed via a 1,2-H shift of coordinated olefin.[192]

WCl$_2$(OAr)$_4$/AlEtCl$_2$ catalysts are more stereoselective than WCl$_4$(OAr)$_2$.[193] The stereochemistry of the olefinic products is influenced by the steric demand of the aryloxo ligands.[194] Molybdenum nitrosyl complexes, e.g. MoCl$_2$(NO)$_2$(MeCN)$_2$, in the presence of AlEtCl$_2$ are probably converted into catalytically active Mo nitrides; MoNCl$_3$L$_3$/AlEtCl$_2$ is active even at -40°C (L = OPPh$_3$, bipy, MeCN).[195] The kinetics of olefin metathesis with WCl$_6$/SnMe$_4$ catalysts have been measured,[196] and EPR evidence suggests the reduction by the olefin of W(VI) to W(V) and W(III) as catalyst precursors.[197] WCl$_6$/SnMe$_4$ also catalyses the ring-opening polymerisation of functionalised olefins (eqn.27),[198] the self- and co-metathesis of unsaturated nitriles[199] and the polymerisation of 1,2-dichlorocyclobut-3-ene.[200] Chromium carbene complexes (CO)$_5$Cr=CPhR (E = Ph,OMe) catalyse the metathesis of vinyl ethers and the polymerisation of 2,3-dihydrofurans.[201] Tungsten carbene complexes have been applied to the synthesis of (36), an insect phenomenon.[202] Unsaturated polymers can be degraded by metathesis for GC-MS analysis.[203] Impregnation of silica with Re$_4$H$_4$(CO)$_{12}$ in the presence of water generates supported [Re(CO)$_3$OH]$_4$ which is a stable catalyst for propene metathesis. 93% of the cluster can be desorbed from the support after use.[204]

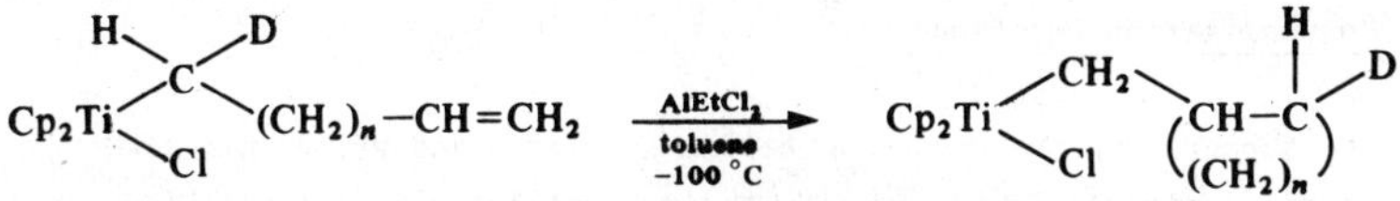

$$\text{Cp}_2\text{Ti}\begin{array}{c}\text{H}\diagdown\ \diagup\text{D}\\ \text{C}\\ \diagup\ \diagdown(\text{CH}_2)_n-\text{CH}=\text{CH}_2\\ \text{Cl}\end{array} \xrightarrow[-100\,°\text{C}]{\text{AlEtCl}_2 \atop \text{toluene}} \text{Cp}_2\text{Ti}\begin{array}{c}\diagup\text{CH}_2\\ \diagdown\text{Cl}\end{array}\begin{pmatrix}\text{CH}-\overset{\text{H}}{\underset{(\text{CH}_2)_n}{\text{C}}}-\text{D}\end{pmatrix} \quad (28)$$

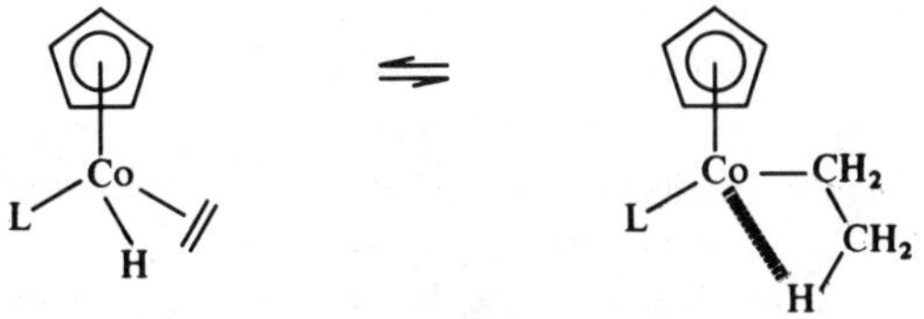

(37) L = P(OMe)$_3$

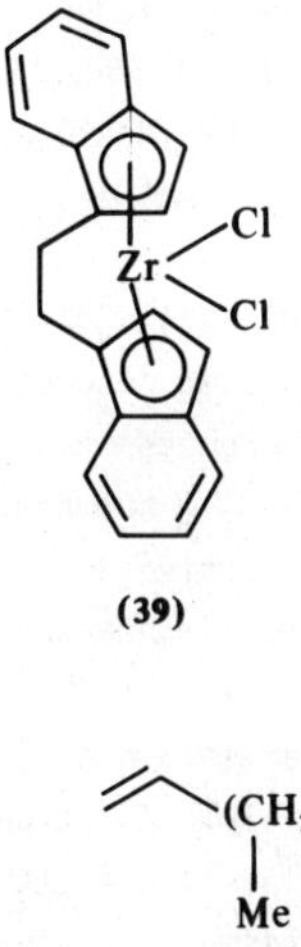

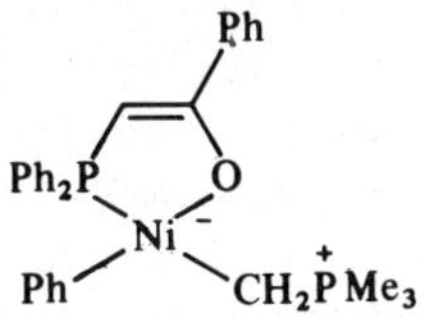

(38)

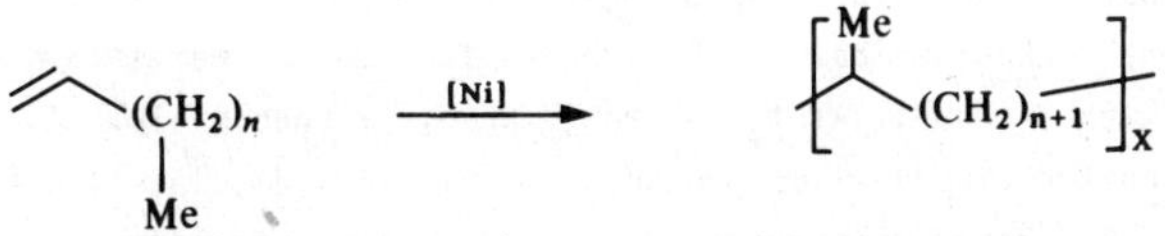

(39)

$$\diagup\!\!\!\diagdown\!(\text{CH}_2)_n\!\!-\!\!\underset{\text{Me}}{|} \xrightarrow{[\text{Ni}]} \left[\underset{\phantom{(\text{CH}_2)}}{\overset{\text{Me}}{|}}\!\!-\!\!(\text{CH}_2)_{n+1}\right]_x \quad (29)$$

PhC*≡C*H in PhC≡CH

$\xrightarrow[\text{toluene}]{\text{Ti}(\text{OBu}^n)_4/\text{AlEt}_3}$

$\xrightarrow[\text{toluene}]{\text{MoCl}_5/\text{SnPh}_4}$

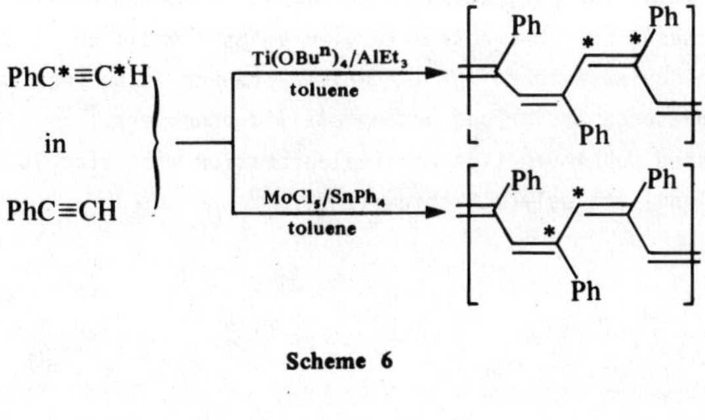

Scheme 6

9. Polymerisation Reactions

Further support for the Cossée mechanism of olefin insertion into Ti-C bonds in Ziegler catalyst is provided by an intramolecular cyclisation (eqn.28); no breaking of α-C-H(D) bonds is observed.[205] In another attempt to model such a stoichiometric insertion, PhC_2SiMe_3 was reacted with Cp_2TiCl_2 and $MeAlCl_2$; the structure of the resulting cationic Ti vinyl complex was determined.[206] The assumption that systems which undergo very facile H-migration can also act as polymerisation catalysts was verified by using (37) to polymerise ethylene.[207] Complexes of V(II), e.g. $(THF)_4V(\mu-Cl)_2ZnCl_2$, are better ethylene polymerisation catalysts than V(III), especially if "activated" by halocarbons.[208] $[(\eta^7-C_7H_7)Ti(THF)(\mu-Cl)]_2$ also polymerises ethylene in the presence of aluminium alkyls.[209] Lanthanide complexes, e.g. $(C_5Me_5)_2MCH(SiMe_3)_2$, are excellent ethylene polymerisation catalysts, while higher olefins are not polymerised. The activity decreases for M = La > Nd ≫ Lu.[210] The Ni ylid complex (38) polymerises $0.5.10^5$ mol C_2H_4/Ni (10 bar, 90°C); the activity is influenced by the basicity of the ylid.[211] The chiral complex (39), activated by $(MeAlO)_n$, produces several thousand kg of highly isotactic polypropylene per mol Zr, whereas $Cp_2ZrMe_2/(MeAlO)_n$ only leads to an atactic product.[212] The polymerisation of C_3-C_{20} α-olefins with a $Ni(O)/(Me_3Si)_2N-P$ (=$NSiMe_3$)$_2$ catalyst does not follow the usual 1,2-mode, but gives 2,ω-connected products (eqn.29); a mechanism involving Ni migrations is proposed.[213] Homogeneous catalysts prepared from Cr(III) stearate and $AlEt_2Cl$ polymerise C_2H_4 but not propene;[214] a heterogeneous system containing $MgCl_2$ is more effective.[215]

$CpTiCl_3$ attached to aluminosilicate gels polymerises isoprene; homogeneous analogues, e.g. $CpTiCl_2(OAr)$, are inactive.[216]

Two different mechanisms have been revealed for the polymerisation of phenylacetylene with $Ti(OBu^n)_4/AlEt_3$ and $MoCl_5/SnPh_4$, respectively. Addition of doubly ^{13}C labelled alkyne gives products in which the ^{13}C labels are either separated by a double bond, as expected for an insertion mechanism, or a single bond as predicted for a metal carbene pathway (Scheme 6).[217] Metal alkyl activated $MoCl_5$ polymerises 1-chloro-1-octyne to give a functionalised polyacetylene of high molecular weight.[218] 1-Phenyl-1-propyne is polymerised by $TaCl_5$ and $NbCl_5$ in the presence of various organometallic promoters.[219]

The Ni or Pd catalysed dehalogenation of dihaloaromatics with zinc is a route to poly(1,4-phenylene) and poly(2,5-thienylene).[220]

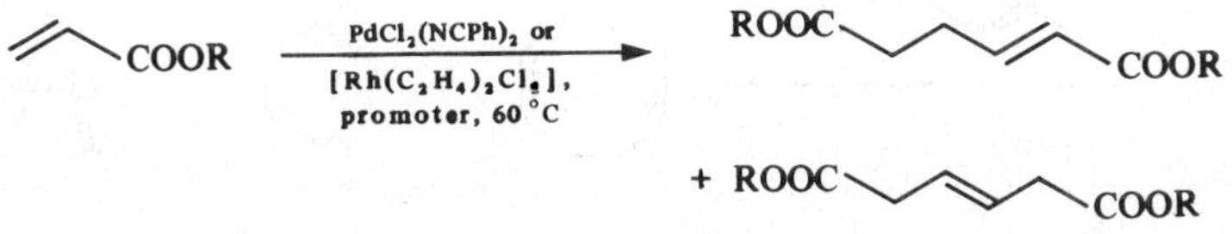

$$\text{CH}_2=\text{CH-COOR} \xrightarrow[\text{promoter, 60 °C}]{\text{PdCl}_2(\text{NCPh})_2 \text{ or } [\text{Rh}(\text{C}_2\text{H}_4)_2\text{Cl}_2],} \text{ROOC-CH}_2\text{-CH=CH-CH}_2\text{-COOR} + \text{ROOC-CH}_2\text{-CH=CH-CH}_2\text{-COOR} \quad (30)$$

$$\text{C}_2\text{H}_4 + \text{CO} \xrightarrow[\text{ROH}]{[\text{Rh}]} \text{H-(CH}_2\text{CH}_2\overset{\text{O}}{\underset{\|}{\text{C}}}\text{-})_n\text{OR} + \text{H-(CH}_2\text{CH}_2\overset{\text{O}}{\underset{\|}{\text{C}}}\text{-})_n\text{CH}_2\text{CH}_3 \quad (31)$$

$n = 1-4$

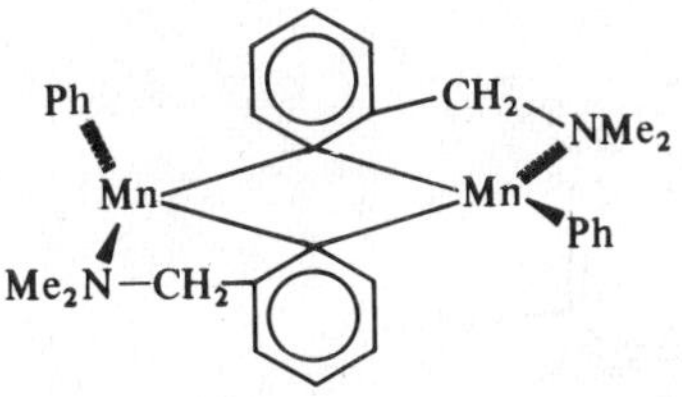

(40) — dimanganese complex with Ph, NMe₂, CH₂ bridges

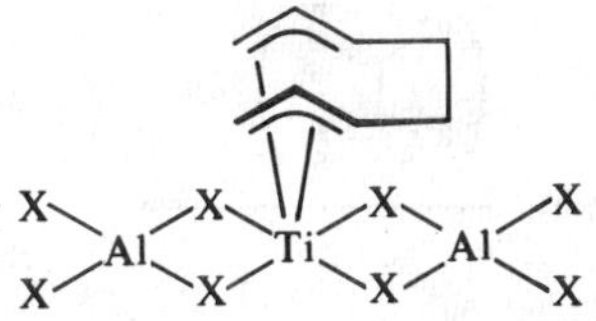

(41) — Ti(COD) with two AlX₄ groups

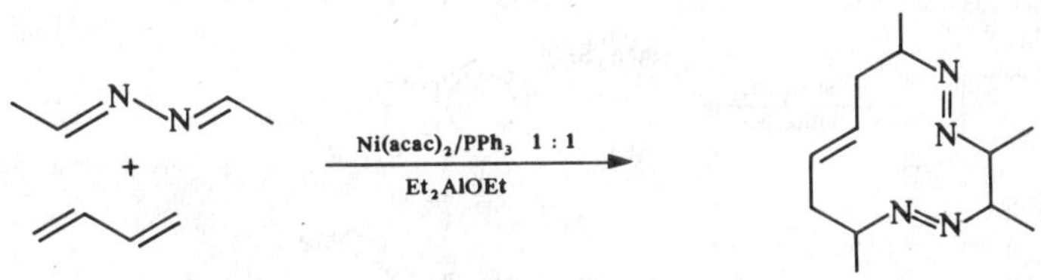

$$\text{CH}_3\text{CH=N-N=CHCH}_3 + \text{CH}_2=\text{CH-CH=CH}_2 \xrightarrow[\text{Et}_2\text{AlOEt}]{\text{Ni(acac)}_2/\text{PPh}_3 \ 1:1} \text{macrocyclic bis-azo product} \quad (32)$$

$$\text{CH}_2=\text{CH-CH=CH-COOMe} \xrightarrow{\text{Ni(COD)}_2/\text{PCy}_3} \text{cyclooctadiene-(COOMe)}_2 + \text{cyclohexene-(COOMe)}_2 \quad (33)$$

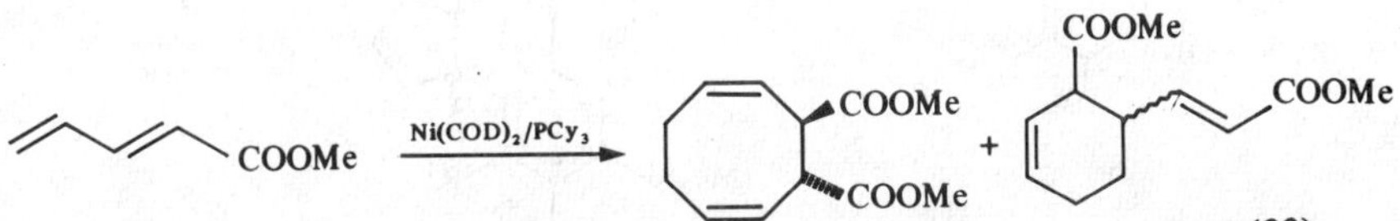

Selectivity; with Et₂AlOEt: 81% 19%
without Et₂AlOEt: 2.5% 93.5%

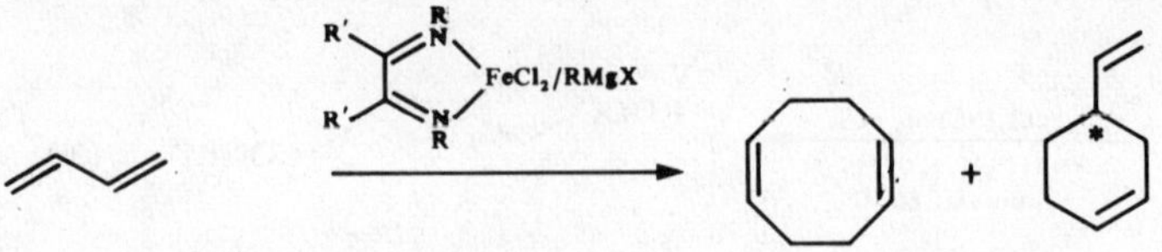

$$\text{CH}_2=\text{CH-CH}=\text{CH}_2 \xrightarrow{\text{FeCl}_2/\text{RMgX}} \text{cycloocta-1,5-diene} + \text{4-vinylcyclohexene} \qquad (34)$$

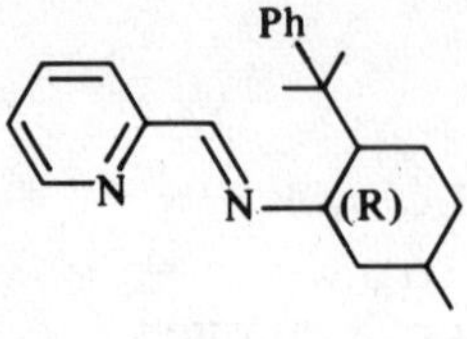

(42)

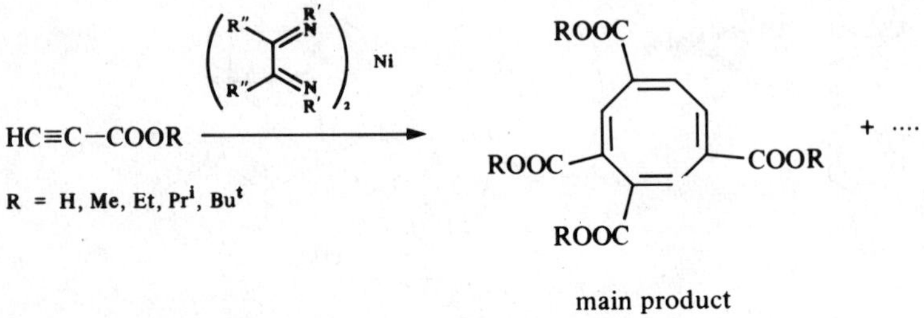

$$\text{HC}\equiv\text{C-COOR} \xrightarrow{\text{Ni}} \text{tetrasubstituted cyclooctatetraene (ROOC)}_4 + \cdots \qquad (35)$$

R = H, Me, Et, Pr^i, Bu^t

main product

R^1 = SiMe$_3$, H

R^2 = $-\text{CH}_2-\text{C(=CH}_2)\text{-OMe}$

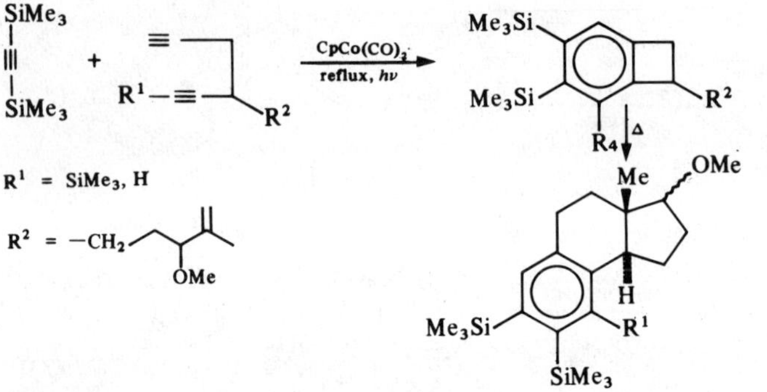

Scheme 7

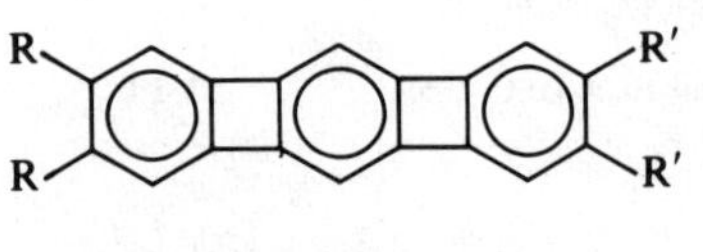

(43) R, R' = H, SiMe$_3$

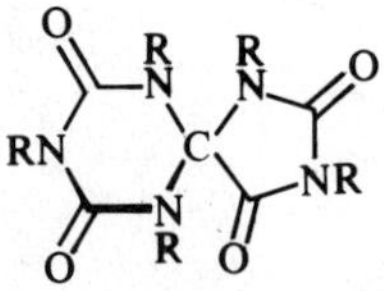

(44)

10. Oligomerisation Reactions

Electrochemically generated Ni(I) in the presence of a sacrificial Zn, Al or Cd anode in propylene carbonate catalyses the dimerisation of ethylene to 1- and 2- butene.[221] The linear dimerisation of acrylates is catalysed by $PdCl_2(NCPh)_2$ or $[Rh(C_2H_4)_2Cl]_2$ promoted by Lewis acids and by proton sources ($AlCl_3 \ll SnCl_4 < FeCl_3$; $HCl < MeOH$) (eqn.30).[222] A kinetic study of the dimerisation of acrylonitrile by Ru catalysts in the presence of N-methyl pyrrolidine supports a mechanism based on Ru hydride species which retain one nitrogen base ligand throughout the cycle.[223] Ethylene is co-oligomerised with CO by $RhH(CO)(PPh_3)_3$ and $RhCl(CO)(PPh_3)_2$ via a single stepwise mode of chain propagation (eqn.31).[224]

A new catalyst for the cyclotrimerisation of butadiene to 1,5,9-cyclododecatriene is the Mn(II) complex (40).[225] The cyclotrimerisation of butadiene by $(C_6H_6)Ti(\mu-X)_4Al_2X_4$ (X = Cl,Br) is second order in butadiene, probably leading to (41) as intermediate.[226] The reaction is also catalysed by Ni supported on poly(vinylpyridine/styrene) resin; all three cyclododecatrienes (Z,Z,E, Z,E,E, and E,E,E isomers) are produced, and the catalyst can be recycled after reactivation with Et_2AlOEt.[227] Sterically hindered 2-Bu^t and 2-SiMe$_3$ substituted butadienes are dimerised by $Ni(COD)_2$ to vinylcyclohexene derivatives. Whereas PPh_3 activates the dimerisation of butadiene, it has a deactivating effect in this system.[228] An intermediate in the Pd catalysed butadiene trimerisation, (η^3,η^3-dodecatrienyl)palladium, has been prepared stoichiometrically.[229] In the presence of maleic anhydride, phosphine-free Pd catalysts dimerise butadiene to vinylcyclohexene.[230] Nitrogen heterocycles are prepared by co-trimerising butadiene with heterodienes (eqn.32).[231] The direction of cyclodimerisation of 2,4-pentanedienoic methyl ester is strongly affected by Et_2AlOEt (eqn.33).[232] New trimers of cyclopentadiene are formed by Pd catalysts under acidic conditions.[233] Iron(0) 1,4-diaza-1,3-dienes complexes catalyse the dimerisation of butadiene to 1,5-COD and vinylcyclohexene. Optically active ligands transfer their chirality to vinylcyclohexene in up to 62% enantiomeric excess with (42) (eqn.34).[234]

Benzyl chloride is dehydrochlorinated by $M(CO)_5Cl$ catalysts (M = Mn,Re); branched oligomers (n = 40-50) are formed.[235]

η^2-Vinyl complexes may play a role as intermediates in alkyne oligomerisations, as evidenced by the stoichiometric reactions of η^2-vinyl tungsten complexes.[236] The rates of alkyne trimerisations to benzene derivatives, catalysed by $(\eta^5-C_5R_5)RhL_2$ (L = $^1/_2$ COD $C_2H_4 >$ CO > PF_3), are dependent on the π-acceptor strengths of the Cp ligand and on L; a mechanism is proposed in which one L remains coordinated at all stages.[237] Rh fluorenyl and indenyl

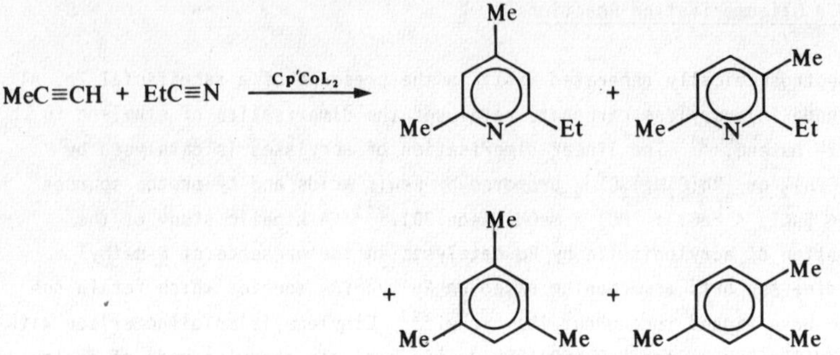

$$\text{MeC} \equiv \text{CH} + \text{EtC} \equiv \text{N} \xrightarrow{\text{Cp'CoL}_2}$$

[products: 4-Me-2-Me-6-Et pyridine + 3-Me-2-Me-6-Et pyridine + 1,3,5-trimethylbenzene + 1,2,4-trimethylbenzene]

(36)

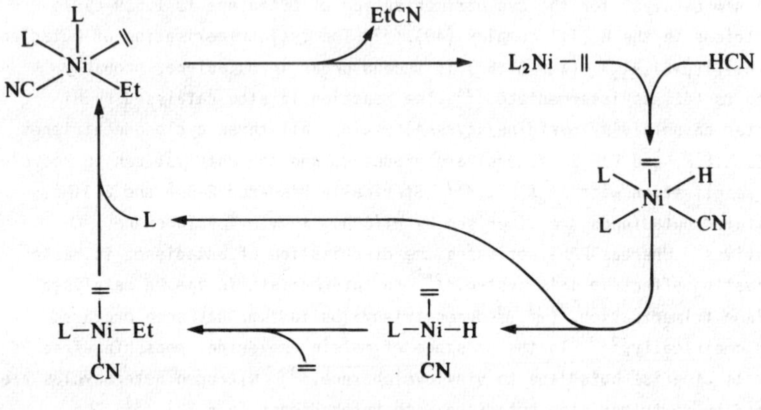

L = tris(*o*-tolyl)phosphite

Scheme 8

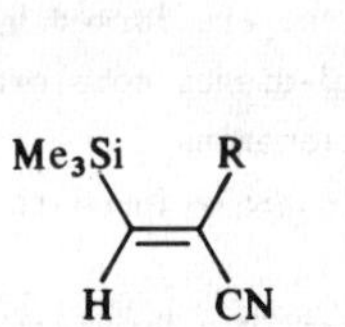

(45) R = H, Me, Ph

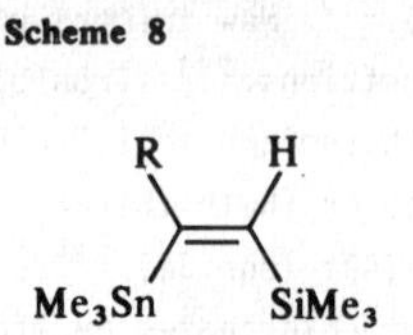

(46) R = Bu^n, Ph, CH$_2$Ph, CH$_2$NMe$_2$, CH$_2$OH, CMe$_2$OH, *etc.*

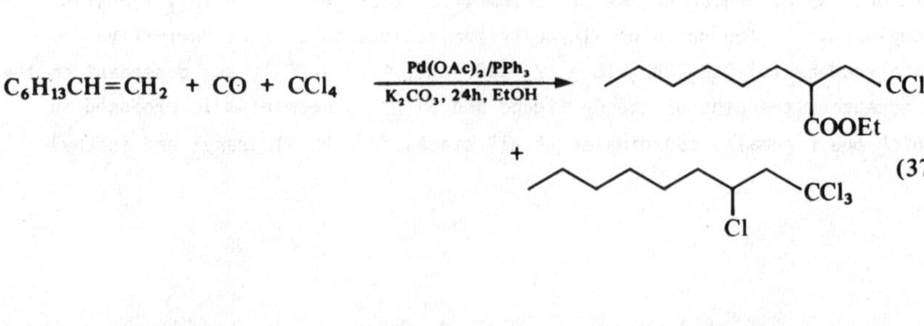

$$\text{C}_6\text{H}_{13}\text{CH}=\text{CH}_2 + \text{CO} + \text{CCl}_4 \xrightarrow[\text{K}_2\text{CO}_3,\ 24\text{h},\ \text{EtOH}]{\text{Pd(OAc)}_2/\text{PPh}_3}$$

[product with COOEt and CCl$_3$ groups + product with Cl and CCl$_3$ groups]

(37)

complexes also trimerise terminal and internal alkynes. With diphenylacetylene, the complexes (Ind)Rh(PhC CPh)$_2$ and (Ind)Rh(η^4-C$_4$Ph$_4$) (Ind = indenyl) were isolated and found to be inactive as catalysts.[238] Phenyl acetylene is trimerised by Ni(CO)$_3$[P(OPr^i)$_3$] to 1,2,4-triphenylbenzene in >95% selectivity.[239] A large number of Ni diazadiene complexes were examined for the tetramerisation of terminal alkynes (eqn.35).[240] Thiophenes are obtained from alkynes and sulphur in the presence of CpRh(COD).[241] Synthetic applications of the alkyne trimerisation reaction include the synthesis of silylated arenes under photolytic conditions (Scheme 7),[242] and the construction of biphenylenes (43).[243]

The co-trimerisation of alkynes with nitriles (eqn.36) is catalysed by Cp'Co(COD) complexes and has been described comprehensively.[244] The catalytic activity increases with the electron-withdrawing character of the Cp' ligands and gives a linear correlation with the ^{59}Co NMR chemical shift.[244,245] The reaction also proceeds in the presence of (MeOOC-C$_5$H$_4$)Co(COT).[246] The co-trimerisation of trimethylstannyl acetylenes with acetonitrile has been employed for the synthesis of vitamin B$_6$.[247]

The reductive tetramerisation of isocyanates is catalysed by [Ru$_3$H(SiEt$_3$)$_2$-(CO)$_{10}$]$^-$ and leads to the spirocycle (44).[248]

10.1 Addition and Telomerisation Reactions

The rate of hydrocyanation of ethylene is independent of the concentrations of C$_2$H$_4$ and HCN, but second order in the L$_2$Ni(C$_2$H$_4$) catalyst since the rate-determining steps are HCN addition to Ni(0) and the regeneration of L$_2$Ni(C$_2$H$_4$) (Scheme 8).[249] The hydrocyanation of butadiene can be catalysed by CuBr/Cl$_3$CCOOH.[250] The direction of the Ni[P(OPh)$_3$]$_4$ catalysed hydrocyanation of alkyl trimethylsilyl acetylenes is dependent on the alkyl substituent; the major product is (45). The reaction also works with acetone cyanohydrin instead of HCN.[251] Olefins of a complementary stereochemistry can be made by PdCl$_2$/pyridine catalysed cis-addition of Me$_3$SiCN to aryl acetylenes.[252] Terminal and aryl acetylenes and those carrying functional groups are regiospecifically silyl-stannylated by Pd(PPh$_3$)$_4$ to give (46).[253] Internal alkynes can be disilylated by a R$_3$SiLi/-MeMgI/MnCl$_2$ reagent.[254]

Radicals are involved in the addition of CCl$_4$ to olefins in the presence of (naphthalene)Cr(CO)$_3$.[255] The system Pd(OAc)$_2$/PPh$_3$/K$_2$CO$_3$/CO also adds CCl$_4$ to 1-octene;[256] CO may be incorporated at 20-40 bar pressure (eqn.37).[257]

A state-of-the-art account has been given on the role of η^3-allyl Pd complexes in diene oligomerisations and (mainly) telomerisations.[258] The 1,4-acetoxychlorination of conjugated dienes is catalysed by Pd(OAc)$_2$/LiCl/LiOAc/HOAc in the presence of benzoquinone and proceeds stereoselectively <u>cis</u>

(38)

(47)

(48)

(39)

R = Ph, C_nH_{2n+1} (n = 4–8), $PhCH_2CH_2$, $Me_2C{=}CHCH_2CH_2$

(40)

(49)

(41)

E = COOMe

+ OAc^-

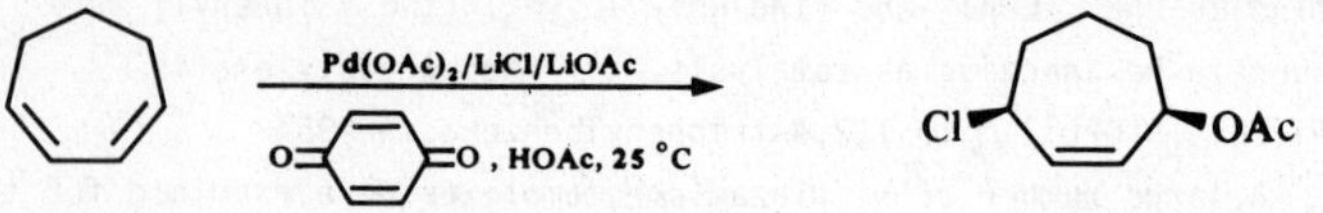

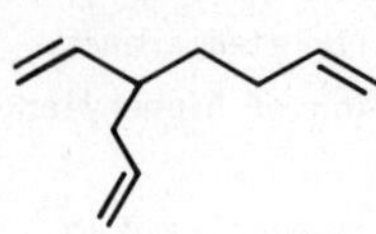

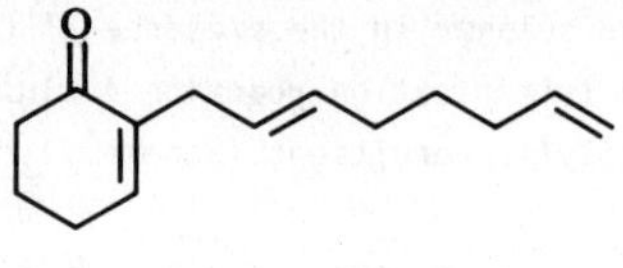

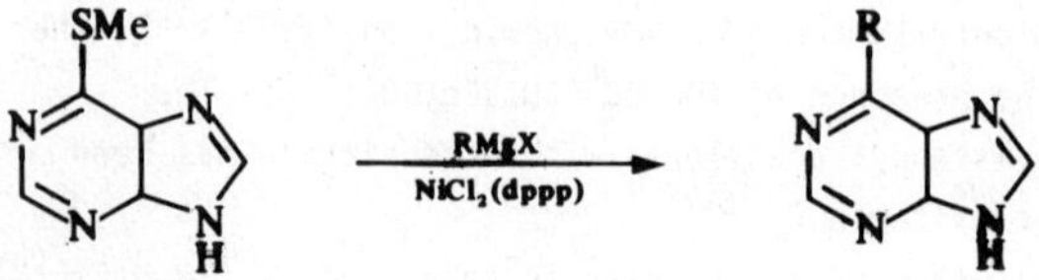

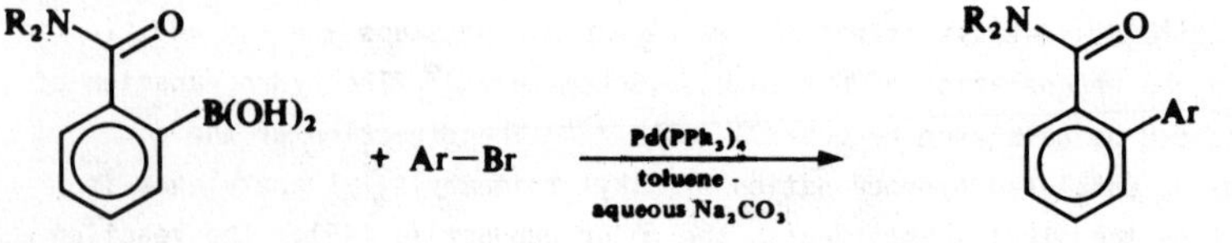

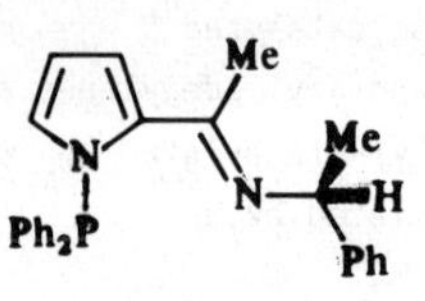

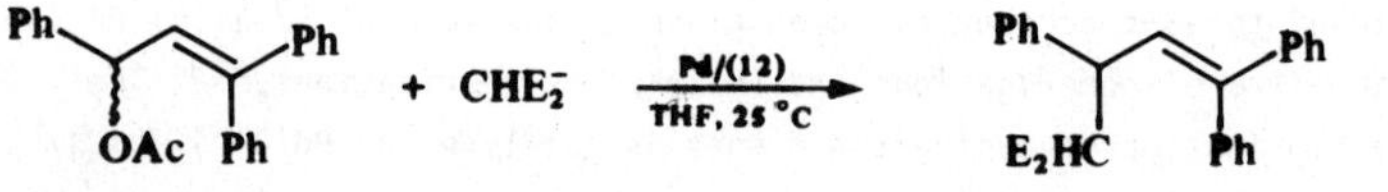

(eqn.38).[259] Aryl acetates react with cyclopentene-epoxide in the presence of Pd(PPh$_3$)$_4$ to give cis-cyclopentene-1,4-diol derivatives.[260]
N-C bonds are broken by Pd catalysts in the telomerisation of butadiene with allyl diethylamine to give mainly (47).[261] The reaction of isoprene with HNEt$_2$ in the presence of cationic Pd complexes gives mainly (50-70%) head-to-head telomers.[262] Alkane- and arene sulphonamides react with butadiene like primary amines.[263] Isoprene reacts with water (Pd catalyst, 90°C) to give a mixture of terpenols; a catalyst supported on a phosphinated resin is prone to leaching.[264] Butadiene reacts with cyclohexenone to give (48) in moderate yield.[265]

11. Cross-Coupling Reactions

Numerous examples describe the formation of carbon-carbon bonds via coupling reactions involving activated organic halides. Almost without exception, Ni or Pd catalysts are employed.

(E)-1-alkenyl ethers are produced from EtOCH=CHOEt and Grignard reagents in the presence of NiCl$_2$(dppp) at 0°C.[266] Organozinc reagents are prepared from alkyl halides, lithium metal and zinc dihalides with the help of ultrasound; they give Ni catalysed 1,4-addition with α,β-unsaturated ketones.[267] Allenyl bromides can be coupled in high yield with isobutyl Al, Mg or Zn reagents.[268] 6-(Methylthio)purines can be converted to the 6-alkyl derivatives (eqn.39).[269]

Allylic bromides and chlorides couple with allyl tin reagents in the presence of a [(allyl)PdCl]$_2$/maleic anhydride catalyst to give 1,5-dienes without allylic rearrangements.[270] Tin analogues of the Reformatsky reagent are coupled with aryl halides by PdCl$_2$(PR$_3$)$_2$ in DMF or HMPA. The yield is improved by Zn halides, while other Lewis acids act as poisons.[271] Alkenyl boranes couple with vinyl bromides and bromoalkynes to give conjugated dienes or alkenynes in high isomeric purity.[272] The same catalyst - Pd(PPh$_3$)$_4$ - couples aryl bromides with aryl boronic acids (eqn.40).[273] Instead of the usual coupling products, aryl iodides react with vinyltrimethylsilane in the presence of AgNO$_3$ and a Pd catalyst to give (E)-trimethylsilylstyrenes.[274] Copper acetylides prepared in situ couple with (chlorobenzene)chromium tricarbonyl to give (alkynylbenzene)Cr(CO)$_3$ complexes in high yield (Pd/PPh$_3$ catalyst).[275] Unsymmetrical diaryl selenides are made from ArSeNa and aryl halides using a NiBr$_2$(bipy) catalyst.[276]

Nickel catalysts with optically active ligands, e.g. (49), couple optically active Grignard reagents with vinyl halides in up to 32.9% enantiomeric excess.[277] A Pd/neomenthyldiphenylphosphine catalyst achieves only a modest 12% e.e. in the allylation of PhZnCl with allylic acetates.[278] Chelating asymmetric

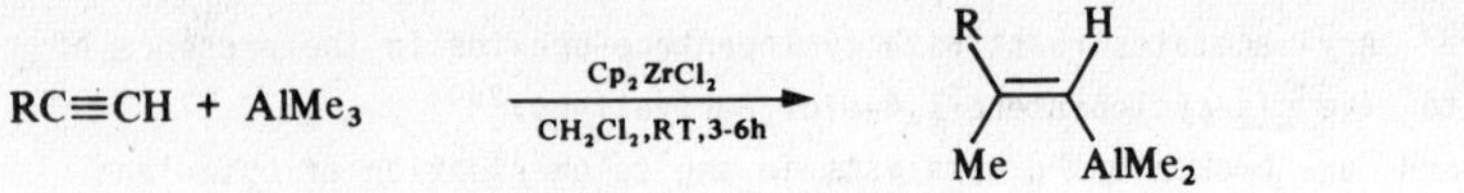

$$RC\equiv CH + AlMe_3 \xrightarrow[CH_2Cl_2, RT, 3-6h]{Cp_2ZrCl_2} \begin{array}{c}RH\\ \diagdown\diagup\\ C=C\\ \diagup\diagdown\\ MeAlMe_2\end{array} \quad (42)$$

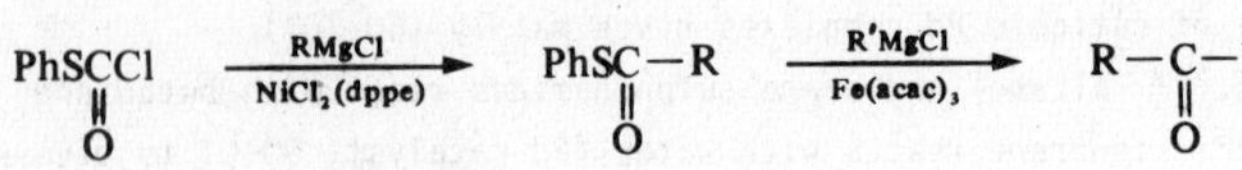

$$\underset{\underset{O}{\|}}{PhSCCl} \xrightarrow[NiCl_2(dppe)]{RMgCl} \underset{\underset{O}{\|}}{PhSC-R} \xrightarrow[Fe(acac)_3]{R'MgCl} \underset{\underset{O}{\|}}{R-C-R'} \quad (43)$$

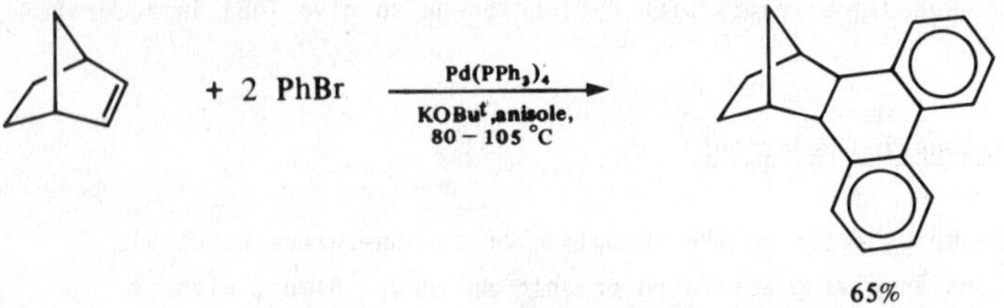

(44) Norbornadiene + 2 PhBr → Pd(PPh₃)₄, KOBuᵗ, anisole, 80–105 °C → product, 65%

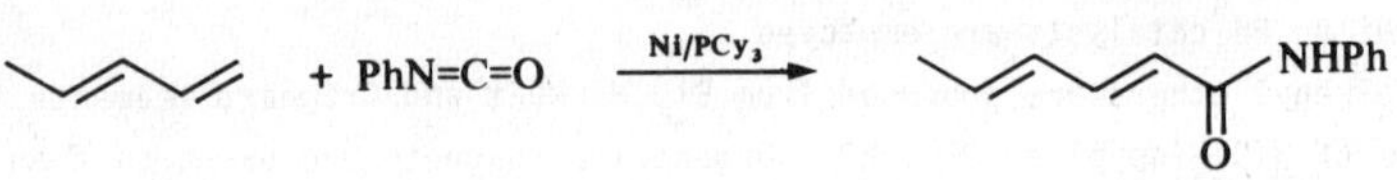

(45) diene + PhN=C=O → Ni/PCy₃ → dienamide-NHPh

$$\underset{\underset{O}{\|}}{R^1}\diagup\!\!\!\diagdown + \underset{\underset{O}{\|}}{H}\diagdown\!\!\!R^2 \xrightarrow{RhH(PPh_3)_4} \underset{\underset{O}{\|}OH}{R^1R^2}$$

(50), selectivity 97% (46)

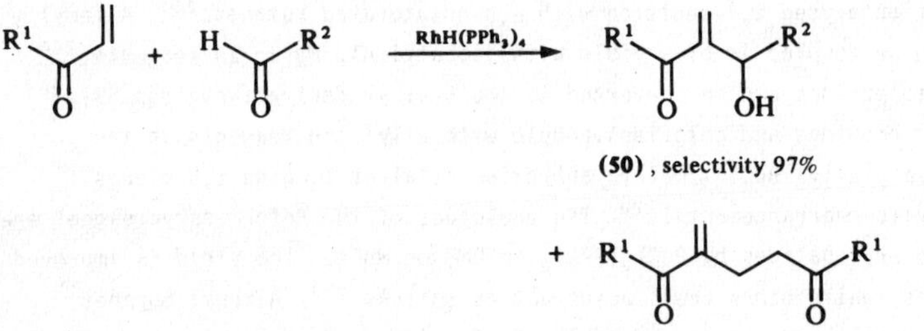

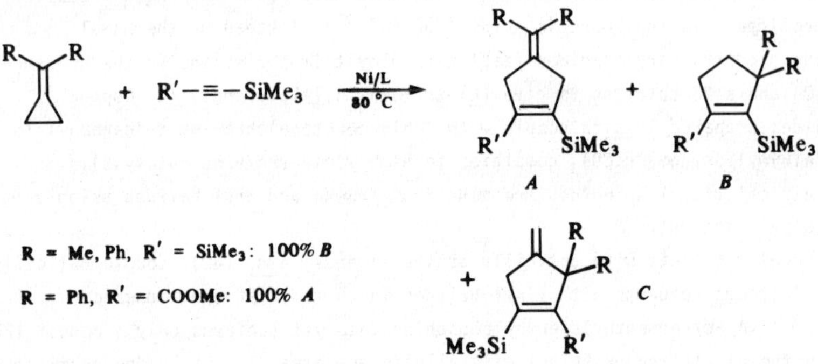

$$+ \underset{\underset{O}{\|}}{R^1}\diagdown\diagup\diagdown\underset{\underset{O}{\|}}{\diagup R^1}$$

(47)

R = Me, Ph, R′ = SiMe₃ : 100% B
R = Ph, R′ = COOMe : 100% A

Products A, B, C as shown (cyclopentene derivatives with SiMe₃ and R, R′ substituents).

phosphines, e.g. (12), give more effective Pd allylation catalysts (eqn.41);[279] the molecular structure of such a catalyst, [(η^3-C(xylyl)$_2$CHCHPh)Pd-(S,S-chiraphos)]BPh$_4$, has been determined.[280]

Terminal and internal alkynes are carbometallated by AlMe$_3$ in the presence of a Cp$_2$ZrCl$_2$ catalyst with high stereo- and regiospecificity (eqn.42).[281] Norbornenes are similarly metallated by magnesium reagents.[282] Palladium phosphine complexes catalyse the synthesis of unsymmetrical ketones from acyl chlorides and alkyl or alkynyl aluminium reagents,[283] from acyl chlorides and alkyl iodides with the aid of a Cu/Zn couple,[284] or from aryl iodides, triaryl aluminium and CO (1 bar).[285] Imidoyl chlorides and organotin compounds generate ketimines.[286] Alternatively, Ni and Fe catalysts can be used in succession (eqn.43).[287] Zinc in the presence of NiBr$_2$(dppe)/NEt$_4$I in THF reductively couples α-haloketones[288] and aryl bromides to give polycyclic compounds.[289]

The formation of C-C bonds does not necessarily require main group organometallics. A review deals with vitamin B$_{12}$ in organic synthesis.[290] Palladium N-allylates amides with 2-allyl isourea.[291] Aryl iodides, thiourea and NaBH$_3$CN in the presence of Ni complexes gives aryl thiols; Me, OMe, NH$_2$, Cl or Br as para-substituents are tolerated.[292] The reaction products of norbornene with bromobenzene (eqn.44)[293] and with dibromobenzene[294] indicate the intermediacy of ortho-metallated Pd complexes. Propargylic alcohols react with aryl iodides in a Pd-catalysed reduction with formic acid to give γ,γ-diaryl allyclic alcohols,[295] while Ni/PCy$_3$ catalysts provide a simple route to sorbanilides (eqn.45).[296] The coupling of vinyl ketones with aldehydes and RhH(PPh$_3$)$_4$ as catalyst leads to (50) in high selectivity (eqn.46).[297]

12. Cyclisation Reactions

The cyclopropanation of alkenes by ethyl diazoacetate is catalysed by W(CO)$_5$ complexes derived from (CO)$_5$W=C(OMe)Ph.[298] Fumarate esters of the chiral alcohols (-)-dimenthol and (-)-borneol are cyclopropanated by CH$_2$X$_2$/Zn in the presence of Co(0) catalysts in acetonitrile in up to 71% optical yield.[299]

The product selectivity in the reaction of methylene cyclopropanes with alkynes is strongly substituent-dependent; the reaction is catalysed by Ni(0)/tris(o-tolyl)phosphite (eqn.47).[300] The intramolecular hydroacylation of 4-pentenal is catalysed by low valent Co/PPh$_3$ complexes and produces mainly cyclopentanone.[301] Cyclopentenones and cyclohexenones are the products of the intramolecular carbonylation of vinyl iodides (eqn.48).[302] A large variety of heterocycles are accessible via the intramolecular Pd catalysed cyclisation of unsaturated α-haloamides.[303] Treatment with base and Pd(OAc)$_2$/PPh$_3$ converts 2,6-dibromo-1,6-dienes into cyclopentanes with exocyclic double

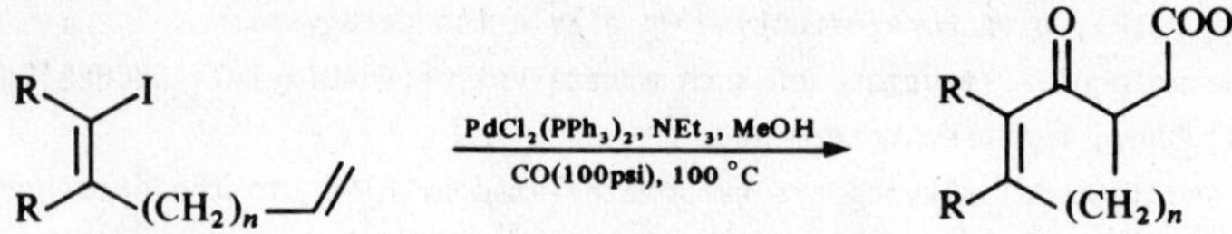

E = electron withdrawing group

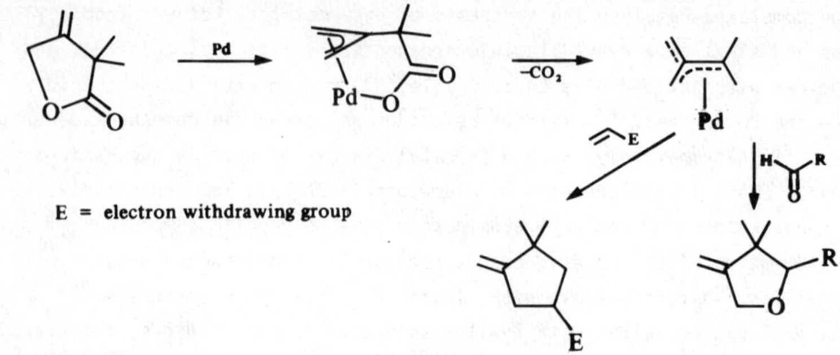

Scheme 9

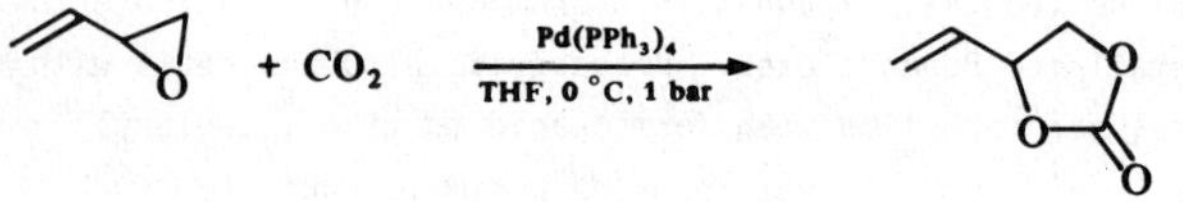

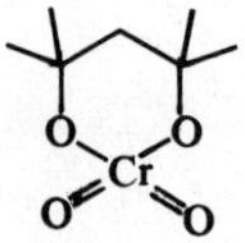

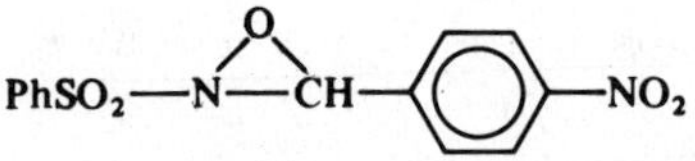

(51) (52)

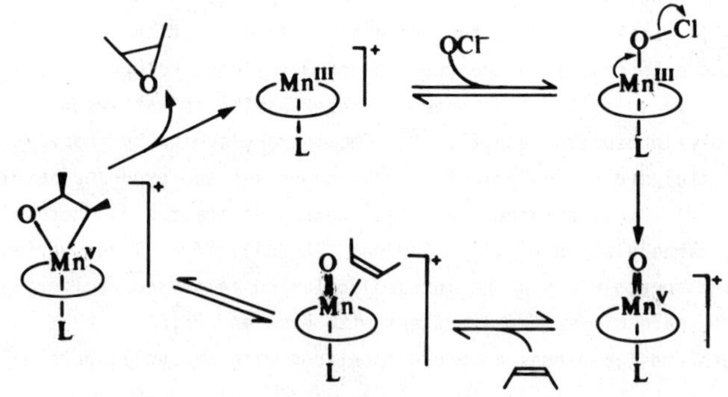

Scheme 10

bonds.[304] β-Methylene lactones are converted to hydrofurans or methylenecyclopentane derivatives via a (trimethylenemethane)Pd intermediate (Scheme 9).[305] Furan derivatives are accessible via intramolecular cyclisation of 3-alkyn-1,2-diols with Pd complexes.[306] 1,3-propanediols react with anilines and nitrobenzene as H-acceptor and a $RuCl_3/PBu_3$ catalyst at 180°C to give quinolines in up to 50% yield.[307]

13. Reactions with Carbon Dioxide

The synthesis of organic chemicals by catalytic reactions of CO_2 has been reviewed.[308] Benzyl and allyl halides are carboxylated electrolytically in the presence of Co(salen) catalysts.[309] Palladium(0) catalysts smoothly carboxylate butadiene epoxide to the carbonate; a mechanism via π-allyl Pd complexes is proposed (eqn.49).[310]

14. Transalkylations

The cluster $Os_3(\mu_3\text{-}S)(\mu\text{-}CO)(CO)_9$ reacts with NMe_3 at 125°C to give the aminocarbene hydride $Os_3(\mu\text{-}H)(\mu_3\text{-}S)[\mu\text{-}CHNMe_2](CO)_8$ which catalyses the transalkylation of NEt_3 and NPr_3 to give $NEtPr_2$ and NEt_2Pr. Four Os_3 clusters with different aminocarbene ligands were recovered from the reaction mixture.[311] The kinetics of the same reaction catalysed by $Ru_3(CO)_{12}$ under a CO atmosphere has been studied; the proposed mechanism invokes α-C-H oxidative addition to the cluster.[312]

15. Oxidations

Reviews discuss the advantages of heterogeneous and homogeneous catalytic oxidations for organic synthesis,[313] and catalytic oxygen atom transfer reactions.[314] The new Cr(VI) reagent (51) catalyses the oxidation of secondary alcohols to ketones with peroxyacetic acid.[315] The tetra(diperoxotungsto)phosphate(3-) anion has been structurally characterised; it catalyses stereospecific olefin epoxidations.[316]

The mechanism of olefin oxidation by O_2 in the presence of $PdCl(NO_2)$-$(MeCN)_2$/HOAc to glycol monoacetates and ketones has been studied by ^{18}O labelling.[317] Resin-supported $Pd(OAc)_2$ catalyses the selective oxidation of alkenes to alkynes (predominantly) and ketones.[318] Carbon tetrachloride in the presence of $Pd(II)/PPh_3/K_2CO_3$ oxidises primary alcohols to aldehydes and esters; allylic alcohols add a CCl_3 group and form ketones.[319] Substituted bromobenzenes are oxidatively coupled to norbornene by $Pd(PPh_3)_4$ if NaOPh is used as a base.[320]

The conversion of terminal olefins to ketones by O_2 is catalysed by a
$[Rh(H_2O)_6]^{3+}/Cu^{2+}/LiCl$ system; 1,5-COD gives 4-octenone.[321] Cationic Rh
complexes of chelating phosphines work similarly; the second oxygen of O_2 is
used to oxidise the alcohol solvent.[322]

The oxidation of hydrocarbons by PhIO is strongly influenced by the size
and position of the substituents in the Fe porphyrin catalyst. Steric hindrance
near the Fe centre improves the yield by preventing the destructive dimerisation
of the catalyst.[323] The highly hindered $Mn^{III}(TTPPP)(OAc)$ gives good
selectivity for primary alcohols in the oxidation of long-chain alkanes
[TTPPP = 5,10,15,20-tetrakis(2,4,6-triphenylphenyl)porphyrinate].[324] Related
Mn(III) complexes catalyse the oxidation of cycloalkanes by aqueous NaOCl under
phase-transfer conditions; tert.-C-H reacts 10-25 times faster than sec.C-H.[325]
The mechanism proposed for olefin epoxidations by Mn(III) cytochrome P-450
models includes formation of a Mn(V) oxo-olefin complex which is in equilibrium
with a Mn(V) alkoxo-alkyl species (Scheme 10).[326] The kinetics of cyclohexene
epoxidation by NaOCl/Mn(III) under phase-transfer conditions show that the Mn(V)
oxo complex is formed in a pyridine catalysed step.[327] Imidazole is essential
for the epoxidation of cyclic and terminal olefins by H_2O_2 catalysed by
MnCl[tetrakis-(2,6-dichlorophenyl)porphyrin].[328] Similar catalysts are also
effective in the olefin epoxidation by (52).[329] The kinetics of olefin
epoxidations by C_6F_5IO catalysed by Fe porphyrin chloride show that the
rate-determining step is the O transfer from Fe to the olefin rather than from
iodosylbenzene to Fe.[330] Dioxo(tetramesitylporphyrinato)Ru(VI) catalyses the
aerobic epoxidation of olefins at ambient conditions.[331]

The mechanism of olefin epoxidation with iodosylbenzene in the presence of
Cr(III) Schiff base complexes has been studied.[332] The same reaction is
catalysed by $VO(acac)_2$, probably via free radicals.[333] Trans-stilbene is
epoxidised by $NaIO_4$ in the presence of $RuCl_3$ and substituted phenanthroline
ligands.[334] Vanadium(V) supported on a functionalised polystyrene resin is a
good catalyst for the epoxidation of allylic alcohols by Bu^tOOH; a similar
Mo(VI) catalyst is more suitable for cyclohexene epoxidation.[335]

REFERENCES

1. Aspects of Homogeneous Catalysis, Vol. 5, R .Ugo (Ed.), D. Reidel Publ., Dordrecht 1984.
2. Homogeneous Catalysis with Compounds of Rhodium and Iridium. R.S.Dickson, D. Reidel Publ., Dordrecht 1985.
3. Organometallic Chemistry of Rhodium and Iridium. R.S. Dickson, Academic Press, New York 1983.
4. Fundamental and Technological Aspects of Organo-f Element Chemistry. T.J. Marks, I.L. Fragala (Eds.), D. Reidel Publ., Dordrecht 1985.
5. P.L. Watson, and G.W. Parshall, Acc.Chem.Res. 1985, 18, 51.
6. F.J. Waller, J.Mol.Catal. 1985, 31, 123.
7. L.S. Hegedus, J.Organomet.Chem. 1985, 283, 1.
8. L. Markó, J.Organomet.Chem. 1985, 283, 221.
9. J.J. Rooney, J.Mol.Catal. 1985, 31, 147.
10. P.E. Garrou, Chem.Rev. 1985, 85, 171.
11. B.F.G. Johnson, and M.V. Twigg, Transition Met. Chem. 1985, 10, 278.
12. P.G. Gassman, D.W. Macomber, and S.M. Willging, J.Am.Chem.Soc. 1985, 107, 2380.
13. C.U. Pittman, W.Honnick, M. Absi-Halabi, M.G. Richmond, R. Bender, and P. Braunstein, J.Mol.Catal. 1985, 32, 177.
14. J.Organomet.Chem. 1985, 279, No. 1-2.
15. Angew.Chem. 1985, 97, No.4.
16. R.H. Crabtree, and M. Lavin, J.Chem.Soc.,Chem.Comm. 1985, 1661.
17. W.R. Cullen, T.J. Kim, F.W.B. Einstein, and T. Jones, Organometallics, 1985, 4, 346.
18. G. Jeske, H. Lauke, H. Mauermann, H. Schumann, and T.J. Marks, J.Am.Chem.Soc. 1985, 107, 8111.
19. B. Bogdanovic, Angew.Chem. 1985, 97, 253.
20. Y. Blum, D. Czarkie, Y. Rahamim, and Y. Shvo, Organometallics, 1985, 4, 1459.
21. J.D. Hewes, M. Thompson, and M.F. Hawthorne, Organometallics 1985, 4, 13.
22. D. Mani, and H. Vahrenkamp, J.Mol.Catal. 1985, 29, 305.
23. B. Demerseman, P. Le Coupanec, P.H. Dixneuf, J.Organomet.Chem. 1985, 287, C35.
24. M.O. Albers, E. Singleton, and M.M. Viney, J.Mol.Catal. 1985, 33, 77.
25. M.O. Albers, E. Singleton, and M.M. Viney, J.Mol.Catal. 1985, 30, 213.
26. A.M. Stolzenberg, and E.L. Muetterties, Organometallics 1985, 4, 1739.
27. W. Kramarz, and S.S. Kurek, J.Mol.Catal. 1985, 33, 305.
28. T. Nakayama, and H. Kanai, Bull.Chem.Soc.Jpn. 1985, 58, 16.
29. J.R. Tucker, and D.P. Riley, J.Organomet.Chem. 1985, 279, 49.
30. J.M. Brown, and S.A. Hall, Tetrahedron 1985, 41, 4639.
31. J.M. Brown, A.E. Derome, and S.A.Hall, Tetrahedron 1985, 41, 4647.
32. A.G. Schultz, and P.J. McCloskey,J.Org.Chem. 1985, 50, 5905.
33. A.S. Machado, A. Olesker, S. Castillon, and G. Lukacs, J.Chem.Soc. Chem.Comm. 1985, 330.
34. W.A. Fordyce, R. Wilczynski, and J. Halpern, J.Organomet.Chem. 1985, 296, 115.
35. R.H. Fish, J.L.Tan, and A.D. Thormodsen, Organometallics 1985, 4 1743.
36. A. Baranyai, F. Ungváry, and L. Markó, J.Mol.Catal. 1985, 32, 343.
37. T. Suarez, and B. Fontal, J.Mol.Catal. 1985, 32, 191.
38. B.T. Khai, and A. Arcelli, Tetrahedron Lett. 1985, 26, 3365.
39. K. Hotta, J.Mol.Catal. 1985, 29, 105.
40. H. Pasternak, E. Lancman, and F. Pruchnik, J.Mol.Catal. 1985, 29, 13.
41. K. Tani, E. Tanigawa, Y. Tatsuno, and S. Otsuka, J.Organomet.Chem. 1985, 279, 87.
42. U. Matteoli, M. Bianchi, G. Menchi, P. Frediani, and F. Piacenti, J.Mol.Catal. 1985, 29, 269.
43. U. Matteoli, G. Menchi, M. Bianchi, and F. Piacenti, J.Organomet.Chem. 1986, 299, 233.
44. H. Nagorski, and M.J.Mirbach, J.Organomet.Chem. 1985, 291, 199.
45. J. Hawecker, J.M. Lehn, and R. Ziessel, J.Chem.Soc., Chem.Comm. 1985, 56.
46. Y. Degani, and I. Willner, J.Chem.Soc., Chem.Comm. 1985, 648.

47 B.R. James, and D. Mahajan, J.Organomet.Chem. 1985, 279, 31.
48 N.W. Alcock, J.M. Brown, A.E. Derome, and A.R. Lucy, J.Chem.Soc.,Chem.Comm. 1985, 575.
49 R. Mutin, W. Abboud, J.M. Basset, and D. Sinou, J.Mol.Catal. 1985, 33, 47.
50 Y. Amrani, D. Lafont, and D. Sinou, J.Mol.Catal. 1985, 32, 333.
51 T.D. Appleton, W.R. Cullen, S.V. Evans, T.J. Kim, and J. Trotter, J.Organomet.Chem. 1985, 279, 5.
52 J. Bakos, I. Tóth, B. Heil, and L. Markó, J.Organomet.Chem. 1985, 279, 23.
53 R. Benhamza, Y. Amrani, and D. Sinou, J.Organomet.Chem. 1985, 288, C37.
54 T. Ikariya, Y. Ishii, H. Kawano, T. Arai, M. Saburi, S. Yoshikawa, S. Akutagawa, J.Chem.Soc.,Chem.Comm. 1985, 922.
55 S. Inoue, M. Osada, K. Koyano, and H. Takaya, Chem. Lett. 1985, 1007.
56 P. Le Maux, V. Massonneau, and G. Simonneaux, J.Organomet.Chem. 1985, 284, 101.
57 A. Kinting, H. Krause, and M. Capka, J.Mol.Catal. 1985, 33, 215.
58 A. Yanagawa, Y. Suzuki, I. Anazawa, Y. Takagi, and S. Yada, J.Mol.Catal. 1985, 29, 41.
59 R.A. Jones, and M.H. Seeberger, J.Chem.Soc.,Chem.Comm. 1985, 373.
60 Z. Brouckova, M. Czakova, and M. Capka, L.Mol.Catal. 1985, 30, 241.
61 R.H. Fish, A.D. Thormodsen, and H. Heinemann, J.Mol.Catal. 1985, 31, 191.
62 A.I. Mikaya, V.G. Zaikin, and V.M. Vdovin, J.Mol.Catal. 1985, 32, 353.
63 S. Torroni, G. Innorta, A. Foffani, S. Scagnolari, A. Modelli, J.Mol.Catal. 1985, 33, 37.
64 K. Kaneda, H. Yamamoto, T. Imanaka, and S. Teranishi, J.Mol.Catal. 1985, 29, 99.
65 D.E. Bergbreiter, and R. Chaudran, J.Am.Chem.Soc. 1985, 107, 4792.
66 B.H. Chang, R.H. Grubbs, and C.H. Brubaker, J.Organomet.Chem. 1985, 280, 365.
67 B.H. Chang, C.P. Lau, R.H. Grubbs, and C.H. Brubaker, J.Organomet.Chem. 1985, 281, 213.
68 M.Y. He, G. Xiong, P.J. Toscano, R.L. Burwell, and T.J. Marks, J.Am.Chem.Soc. 1985, 107, 641.
69 R.H. Crabtree, and C.P. Parnell, Organometallics 1985, 4, 519.
70 J. Kaspar, R. Spogliarich, and M. Graziani, J.Organomet.Chem. 1985, 281, 299.
71 M. Massoui, D. Beaupère, G. Goethals, and R. Uzan, J.Mol.Catal. 1985, 33, 209.
72 ibid. 1985, 29, 7.
73 B. Heil, P. Kvintovics, L. Tarszabó, and B.R. James, J.Mol.Catal. 1985, 33, 71.
74 P.K. Kvintovics, J. Bakos, and B. Heil, J.Mol.Catal. 1985, 32, 111.
75 M. Visintin, R. Spogliarich, J. Kaspar, and M. Graziani, J.Mol.Catal. 1985, 32, 349.
76 K. Jothimony, S. Vancheesan, and J.C. Kuriacose, J.Mol.Catal. 1985, 32, 11.
77 S. Sato, I. Matsuda, and Y. Izumi, Tetrahedron Lett. 1985, 26, 4229.
78 A. Spencer, J.Organomet.Chem. 1985, 295, 91.
79 ibid. 1985, 294, 357, and 295, 79.
80 T. Takahashi, S. Shinoda, and Y. Saito, J.Mol.Catal. 1985, 31, 301.
81 H. Yamamoto, S. Shinoda, and Y. Saito, J.Mol.Catal. 1985, 30, 259.
82 S. Shinoda, H. Itagaki, and Y. Saito, J.Chem.Soc.Chem.Comm. 1985, 860.
83 M. Kijima, Y. Nambu, and T. Endo, Chem.Lett. 1985, 1851.
84 N. Suzuki, Y. Kaneko, T. Tsukanaka, T. Nomoto, Y. Ayaguchi, and Y. Izawa, Tetrahedron 1985, 41, 2387.
85 M. Mohri, H. Kinoshita, K. Inomata, H. Kotake, Chem.Lett. 1985, 451.
86 D. Männig, and H. Nöth, Angew.Chem. 1985, 97, 854.
87 D.C. Apple, K.A. Brady, J.M. Chance, N.E. Heard, and T.A. Nile, J.Mol.Catal. 1985, 29, 55.
88 K.P.Adams, J.A. Joyce, T.A. Nile, A.I. Patel, C.D. Reid, and J.M. Walters, J.Mol.Catal. 1985, 29, 201.
89 B. Marciniec, E. Mackowska, J. Gulinski, and W. Urbaniak, Z.anorg.allg.Chem. 1985, 529, 222.
90 T. Murai, T. Sakane, and S. Kato, Tetrahedron Lett. 1985, 26, 5145.
91 T. Kusumoto, and T. Hiyama, Chem.Lett. 1985, 1405.
92 H. Brunner, and H. Weber, Chem.Ber. 1985, 118, 3380.
93 H. Brunner, P. Beier, G. Riepl, I. Bernal, G.M. Reisner, R. Benn, and A. Rufinska, Organometallics 1985, 4, 1732.

94 H. Brunner, and R. Becker, Angew.Chem. 1985, 97, 713.
95 T. Murai, S. Kato, S. Murai, Y. Hatayama, and N. Sonoda, Tetrahedron Lett. 1985, 26, 2683.
96 K.T. Kang, and W.P. Weber, Tetrahedron Lett. 1985, 26, 5753.
97 ibid., 1985, 26, 5415.
98 R.H. Crabtree, Chem.Rev. 1985, 85, 245.
99 J. Schwartz, Acc.Chem.Res. 1985, 18, 302.
100 M.J. Burk, R.H. Crabtree, and D.V. McGrath, J.Chem.Soc.Chem.Comm. 1985, 1829.
101 J.W. Faller, and H. Felkin, Organometallics 1985, 4, 1488.
102 B. Fell, H.U. Hög, and C. Krüger, J.Mol.Catal. 1985, 30, 57.
103 M.F. Mirbach, and M.J. Mirbach, J.Mol.Catal. 1985, 32, 59.
104 G. Cavinato, L. Toniolo, and C. Botteghi, J.Mol.Catal. 1985, 32, 211.
105 F. Joó, and H. Alper, Organometallics 1985, 4, 1775.
106 H. Alper, and D. Leonard, Tetrahedron Lett. 1985, 26, 5639.
107 C. Buchan, N. Hamel, J.B. Woell, and H. Alper, Tetrahedron Lett. 1985, 26, 5743.
108 S. Murahashi, and Y. Imada, Chem.Lett. 1985, 1477.
109 I. Ojima, K. Hirai, M. Fujita, and T. Fuchikami, J.Organomet.Chem. 1985, 279, 203.
110 A. Chiesa, and R. Ugo, J.Organomet.Chem. 1985, 279, 215.
111 Y. Wakita, T. Yasunaga, and M. Kojima, J.Organomet.Chem. 1985, 288, 261.
112 H. Alper, F.W. Hartstock, J.Chem.Soc.Chem.Comm. 1985, 1141.
113 Y. Tsuji, R. Takeuchi, and Y. Watanabe, J.Organomet.Chem. 1985, 290, 249.
114 Y. Tamaru, H. Higashimura, K. Naka, M. Hojo, and Z. Yoshida, Angew.Chem. 1985, 97, 1070.
115 Y. Yamamoto, R. Degushi, and J. Tsuji, Bull.Chem.Soc.Jpn. 1985, 58, 3397.
116 H. Alper, H. Arzoumanian, J.F. Petrignani, and M. Saldana-Maldonado, J.Chem.Soc.Chem.Comm. 1985, 340.
117 M. Tanaka, T.A. Kobayashi, T. Sakakura, H. Itatani, S. Danno, and K. Zushi, J.Mol.Catal. 1985, 32, 115.
118 F. Ozawa, N. Kawasaki, T. Yamamoto, and A. Yamamoto, Chem.Lett. 1985, 567.
119 M. Tanaka, T. Kobayashi, and T. Sakakura, J.Chem.Soc.Chem.Comm. 1985, 837.
120 F. Ozawa, H. Soyama, H. Yanagihara, I. Aoyama, H. Takino, K. Izawa, T. Yamamoto, and A. Yamamoto, J.Am.Chem.Soc. 1985, 107, 3235.
121 T.W. Dekleva, and D. Forster, J.Am.Chem.Soc. 1985, 107, 3565.
122 T.W. Dekleva, and D. Forster, J.Mol.Catal. 1985, 33, 269.
123 R.W. Wegman, and D.C. Busby, J.Mol.Catal. 1985, 32, 125.
124 D.J. Darensbourg, R.L. Gray, C. Ovalles, and M. Pala, J.Mol.Catal. 1985, 29, 285.
125 S.C. Shim, S. Antebi, and H.Alper, Tettrahedron Lett. 1985, 26, 1935.
126 G. Braca, A.M. Raspolli Galetti, S. Sbrana, and R. Lazzaroni, J.Organomet.Chem. 1985, 289, 107.
127 M. Lütgendorf, B.O. Elveroll, and M. Röper, J.Organomet.Chem. 1985, 289, 97.
128 A.M. Raspolli Galetti, G. Braca, S. Sbrana, and F. Marchetti, J.Mol.Catal. 1985, 32, 291.
129 K. Kudo, S. Mori, and N. Sugita, Chem.Lett. 1985, 265.
130 M.F. Mirbach, and M.J. Mirbach, J.Mol.Catal. 1985, 33, 23.
131 G. Jenner, and Y.Chreim, J.Mol.Catal. 1985, 33, 259.
132 R.W. Wegman, and J.B. Letts, J.Mol.Catal. 1985, 33, 357.
133 R. Bartek, M.M. Habib, and W.R. Pretzer, J.Mol.Catal. 1985, 33, 245.
134 J. Zwart, and R. Snel, J.Mol.Catal. 1985, 30, 305.
135 K. Ogura, and M. Kaneko, J.Mol.Catal. 1985, 31, 49.
136 K. Ogura, and S. Yamasaki, J.Chem.Soc.Dalton Trans. 1985, 2499.
137 P. Escaffre, A. Thorez, and P. Kalck, J.Mol.Catal. 1985, 33, 87.
138 H. Nagorski, M.J. Mirbach, and M.F. Mirbach, J.Organomet.Chem. 1985, 297, 171.
139 B.H. Weiller, J.P.Liu, and E.R. Grant, J.Am.Chem.Soc. 1985, 107, 1595.
140 J.C. Bricker, C.C. Nagel, A.A. Bhattacharyya, and S.G. Shore, J.Am.Chem.Soc. 1985, 107, 377.
141 E. Alessio, G. Clauti, and G. Mestroni, J.Mol.Catal. 1985, 29, 77.

142 W.F. Brill, J.Mol.Catal. 1985, 32, 17.
143 I. Theodosiou, R. Barone, and M. Channon, J.Mol.Catal. 1985, 32, 27.
144 J.F. Knifton, J.Mol.Catal. 1985, 30, 281.
145 T. Okano, M. Makino, H. Konishi, and J. Kiji, Chem.Lett. 1985, 1793.
146 J.F. Knifton. J.Chem.Soc.Chem.Comm. 1985, 1412.
147 P.C. Heah, and J.A. Gladysz, J.Mol.Catal. 1985, 31, 207.
148 T.M. Bockman, J.F. Garst, R.B. King, L. Markó, and F. Ungváry, J.Organomet.Chem. 1985, 279, 165.
149 G. Süss-Fink, and G. Herrmann, J.Chem.Soc.Chem.Comm. 1985, 735.
150 S.I. Woo, and C.G. Hill, J.Mol.Catal. 1985, 29, 231.
151 M. Eisen, J. Blum, H. Schumann, and S.Jurgis, J.Mol.Catal. 1985, 31, 317.
152 H. Schumann, S. Jurgis, E. Hahn, J. Pickardt, J.Blum, and M. Eisen, Chem.Ber. 1985, 118, 2738.
153 T.G. Southern, J.Mol.Catal. 1985, 30, 267.
154 P. Kalck, A. Thorez, M.T. Pinillos, and L.A, Oro, J.Mol.Catal. 1985, 31, 311.
155 R. Lazzaroni, S. Bertozzi, P.Pocai, F.Troiani, and P. Salvadori, J.Organomet.Chem. 1985, 295, 371.
156 B. Falk, B. Fell, and W. Meltzow, J.Mol.Catal. 1985, 31, 93.
157 D. Banach, G.O. Evans, D.G. McIntyre, T. Predmore, M.G. Richmond, J.H. Supple, and R.P.Stewart, J.Mol.Catal. 1985, 31, 15.
158 P.W.N.M. van Leeuwen, and C.F. Roobeek, J.Mol.Catal. 1985, 31, 345.
159 H. Marrakchi, J.B. Nguini Effa, M. Haimeur, J. Lieto, and J.P. Aune, J.Mol.Catal. 1985, 30, 101.
160 M.E. Davies, P.M. Butler, J.A. Rossin, and B.E. Hanson, J.Mol.Catal. 1985, 31, 385.
161 H.L. Pelt, G. van der Lee, and J.J.F. Scholten, J.Mol.Catal. 1985, 29, 319.
162 H.L. Pelt, R.P.J. Verburg, and J.J.F. Scholten, J.Mol.Catal. 1985, 32, 77.
163 H.L. Pelt et al., J.Mol.Catal. 1985, 31, 371.
164 H.L. Pelt et al., J.Mol.Catal. 1985, 33, 119.
165 G. Consiglio, F. Morandini, M. Scalone, and P. Pino, J.Organomet.Chem. 1985, 279, 193.
166 P. Haelg, G. Consiglio, and P. Pino, J.Organomet.Chem. 1985, 296, 281.
167 J. Evans, and G. Jingxing, J.Chem.Soc.Chem.Comm. 1985, 39.
168 P.A. Tooley, L.W. Arndt, and M.Y. Darensbourg, J.Am.Chem.Soc. 1985, 107, 2422.
169 M.O. Albers, and E. Singleton, J.Mol.Catal. 1985, 31, 211.
170 F. Bouachir, B. Chaudret, D. Neilbecker, and I. Tkatchenko, Angew.Chem. 1985, 97, 347.
171 J. Lieto, M. Prochazka, D.B. Arnold, and B.C. Gates, J.Mol.Catal. 1985, 31, 89.
172 R.J. McKinney, Organometallics 1985, 4, 1143.
173 K.Tani, T. Yamagata, Y. Tatsuno, Y. Yamagata, K. Tomita, S.Akutagawa, H. Kumobayashi, and S. Otsuka, Angew.Chem. 1985, 97, 232.
174 W. Smadja, J.M. Valery, G. Ville, and J.M. Bernassau, J.Mol.Catal. 1985, 30, 389.
175 K. Inomata, Y. Murata, H. Kato, and Y. Tsukahara, Chem.Lett. 1985, 931.
176 S.I. Murahashi, and Y. Makabe, Tetrahedron Lett. 1985, 26, 5563.
177 T.G. Schenck, and B. Bosnich, J. Am.Chem.Soc. 1985, 107, 2058.
178 S. Matsubara, T. Okazoe, K. Oshima, and K. Takai, Bull.Chem.Soc.Jpn. 1985, 58, 844.
179 J.Mol.Catal. 1985, 28, No. 1-3.
180 R.R. Schrock, J.H. Freudenberger, M.L. Listemann, and L.G. McCullough, J.Mol.Catal. 1985, 28, 1.
181 T.J. Katz, and T.M. Sivavec, J.Am.Chem.Soc. 1985, 107, 737.
182 S.J. Landon, P.M. Shulman, and G.L. Geoffroy, J.Am.Chem.Soc. 1985, 107, 6739.
183 C.C. Han, and T.J. Katz, Organometallics 1985, 4, 2186.
184 T.H. Ho, and T.J. Katz, J.Mol.Catal. 1985, 28, 359.
185 Y.V. Korshak, V.V. Korshak, G. Kanischka, and H. Höcker, Makromol.Chem.,Rapid Comm. 1985, 6, 685.
186 D.A. Straus, and R.H. Grubbs, J.Mol.Catal. 1985, 28, 9.
187 J. Kress, J.A. Osborn, R.M.E. Greene, K.J. Ivin, and J.J. Rooney, J.Chem.Soc.Chem.Comm. 1985, 874.
188 J.G. Hamilton, K.J. Ivin, and J.J. Rooney, J.Mol.Catal. 1985, 28, 255.
189 F. Quignard, M. Leconte, and J.M. Basset, J.Chem.Soc.Chem.Comm. 1985, 1816.
190 L. Bencze, and A. Kraut-Vass, J.Mol.Catal. 1985, 28, 369.

191 L. Bencze, and A. Kraut-Vass, J.Organomet.Chem. 1985, 280, C14.
192 L. Bencze, A. Kraut-Vass, and L. Prókai, J.Chem.Soc.Chem.Comm. 1985, 911.
193 F. Quignard, M. Leconte, and J.M. Basset, J.Mol.Catal. 1985, 28, 27.
194 H.T. Dodd, and K.J. Rutt, J.Mol.Catal. 1985, 28, 33.
195 K. Seyferth, and R. Taube, J.Mol.Catal. 1985, 28, 53.
196 E. Thorn-Csanyi, and H. Timm, J.Mol.Catal. 1985, 28, 37.
197 E. Ceausescu, A. Cornilescu, E. Nicolescu, M. Popescu, S. Coca, M. Dimonie, M. Gherorghui, V. Dragutan, and M. Chipara, J.Mol.Catal. 1985, 28, 337 and 351.
198 W.J. Feast, and K. Harper, J.Mol.Catal. 1985, 28, 293.
199 R.H.A. Bosma, G.C.N. van den Aardweg, and J.C. Mol, J.Organomet.Chem. 1985, 280, 115.
200 J.K. Brunthaler, F. Stelzer, and G. Leising, J.Mol.Catal. 1985, 28, 393.
201 C.T. Thu, T. Bastelberger, and H. Höcker, J.Mol.Catal. 1985, 28, 279.
202 D.S. Banasiak, J.Mol.Catal. 1985, 28, 107.
203 K. Hummel, J.Mol.Catal. 1985, 28, 381.
204 P.S. Kirlin, and B.C. Gates, J.Chem.Soc.Chem.Comm. 1985, 277.
205 L. Clawson, J. Soto, S.I. Buchwald, M.L. Steigerwald, and R.H. Grubbs, J.Am.Chem.Soc. 1985, 107, 3377.
206 J.J. Eisch, A.M. Piotrowski, S.K. Brownstein, E.J. Gabe, and F.L. Lee, J.Am.Chem.Soc. 1985, 107, 7219.
207 G.F. Schmidt, and M. Brookhart, J.Am.Chem.Soc. 1985, 107, 1443.
208 P.D. Smith, J.L. Martin, J.C. Huffman, R.L. Bansemer, and K.G. Caulton, Inorg.Chem. 1985, 24, 2997.
209 C.E. Davies, I.M. Gardiner, J.C. Green, M.L.H. Green, N.J. Hazel, P.D. Grebenik, V.S.B. Mtetwa, and K. Prout, J.Chem.Soc..Dalton Trans. 1985, 669.
210 G. Jeske, H. Lauke, H. Mauermann, P.N. Swepston, H. Schumann, and T.J. Marks, J.Am.Chem.Soc. 1985, 107, 8091.
211 K.A. Ostoja Starzewski, and J. Witte, Angew.Chem. 1985, 97, 610.
212 W. Kaminsky, K.Külper, H.H.Brintzinger, and F.R.W.P.Wild, Angew.Chem. 1985,97, 507.
213 V.M. Möhring, and G. Fink, Angew.Chem. 1985, 97, 982.
214 K. Soga, S.I. Chen, T. Shiono, and Y. Doi, Polymer 1985, 26, 1888.
215 K. Soga, S.I. Chen, T. Shiono, and Y. Doi, Polymer 1985, 26, 1891.
216 W. Skupinski, I. Cieslowska-Glinska, and A. Wasilewski, J.Mol.Catal. 1985, 33, 129.
217 T.J. Katz, S.M. Hacker, R.D. Kendrick, and C.S. Yannoni, J.Am.Chem.Soc. 1985, 107, 2182.
218 T. Matsuda, K. Tamura, and T. Higashimura, J.Chem.Soc.Chem.Comm. 1985, 1615.
219 T. Matsuda, A. Niki, E. Isobe, and T. Higashimura, Macromolecules. 1985, 18, 2109.
220 T. Yamamoto, K. Osakada, T. Wakabayashi, and A. Yamamoto, Makromol.Chem.,Rapid Comm., 1985, 6, 671.
221 S. Sibille, J. Coulombeix, J. Perichon, J.M. Fuchs, A. Mortreux, and F. Petit, J.Mol.Catal. 1985, 32, 239.
222 W.A. Nugent, and R.J. McKinney, J.Mol.Catal. 1985, 29, 65.
223 D.T. Tsou, J.D. Burrington, E.A. Maher, and R.K. Grasselli, J.Mol.Catal. 1985, 30, 219,
224 A. Sen, and J.S. Brumbaugh, J.Organomet.Chem. 1985, 279, C5.
225 K. Jonas, Angew.Chem. 1985, 97, 292.
226 J. Polacek, H. Antropiusová, V. Hanus, L. Petrusová, and K. Mach, J.Mol.Catal. 1985, 29, 165.
227 U. Schuchardt, and F. Santos Dias, J.Mol.Catal. 1985, 29, 145.
228 T. Bartik, P. Heimbach, T. Himmler, and R. Mynott, Angew.Chem. 1985, 97, 345.
229 R. Benn, P.W. Jolly, R. Mynott, and G. Schenker, Organometallics 1985, 4, 1136.
230 A. Goliaszewski, and J. Schwartz, Organometallics. 1985, 4, 415.
231 P. Brun, A. Tenaglia, and B. Waegell, Tetrahedron 1985, 41, 5019.
232 P. Brun, A. Tenaglia, and B. Waegell, Tetrahedron Lett. 1985, 26, 5685.
233 A. Behr, and W. Keim, Angew.Chem. 1985, 97, 326.
234 H. tom Dieck, and J. Dietrich, Angew.Chem. 1985, 97, 795.
235 C.P. Tsonis, J.Mol.Catal. 1985, 33, 61.
236 L. Carlton, J.L. Davidson, P. Ewing, L. Manojlovic-Muir, and K.W. Muir, J.Chem.Soc.Chem.Comm. 1985, 1474.
237 K. Abdulla, B.L. Booth, and C. Stacey, J.Organomet.Chem. 1985, 293, 103.

238 A. Borrini, P.Diversi, G. Ingrosso, A. Lucherini, and G. Serra, J.Mol.Catal. 1985, 30, 181.
239 A. Mantovani, A. Marcomini, and U. Belluco, J.Mol.Catal. 1985, 30, 73.
240 R. Diercks, and H. tom Dieck, Chem.Ber. 1985, 118, 428.
241 M. Kajitani, T. Suetsugo, R. Wakabayashi, A. Igarashi, T. Akiyama, and A. Sugimori, J.Organomet.Chem. 1985, 293, C15.
242 R.L. Halterman, N.H. Nguyen, and K.P.C. Vollhardt, J.Am.Chem.Soc. 1985, 107, 1379.
243 B.C. Berris, G.H. Hovakeemian, Y.H. Lai, H. Mestdagh, and K.P.C. Vollhardt, J.Am.Chem.Soc. 1985, 107, 5670.
244 H. Bönnemann, Angew.Chem. 1985, 97, 264.
245 H. Bönnemann, and W. Brijoux, Bull.Soc.Chim. Belges 1985, 94, 635.
246 Y. Wakatsuki, H. Yamazaki, Bull.Chem.Soc.Jpn. 1985, 58, 2715.
247 C.A. Parnell, and K.P.C. Vollhardt, Tetrahedron 1985, 41, 5791.
248 G. Herrmann, and G. Süss-Fink, Chem.Ber. 1985, 118, 3959.
249 R.J. McKinney, and D.C. Roe, J.Am.Chem.Soc. 1985, 107, 261.
250 E. Puentes, A.F. Noels, R. Warin, A.J. Hubert, P. Teyssié, and D.Y. Waddan, J.Mol.Catal. 1985, 31, 183.
251 G.D. Fallon, N.J. Fitzmaurice, W.R. Jackson, and P. Perlmutter, J.Chem.Soc., Chem.Comm. 1985, 4.
252 N. Chatani, and T. Hanafusa, J.Chem.Soc.,Chem.Comm. 1985, 838.
253 T.N. Mitchell, H. Killing, R. Dicke, and R. Wickenkamp, J.Chem.Soc.,Chem.Comm. 1985, 354.
254 J. Hibino, S. Nagatsukasa, K. Fugami, S. Matsubara, K. Oshima, and H. Nozaki, J.Am.Chem.Soc. 1985, 107, 6416.
255 W.J. Bland, R. Davis, J.L.A. Durrant, J.Organomet.Chem. 1985, 280, 95.
256 J.Tsuji, K. Sato, and H. Nagashima, Tetrahedron 1985, 41, 393.
257 ibid. 1985, 41, 5003.
258 P.W. Jolly, Angew.Chem. 1985, 97, 279.
259 J.E. Bäckvall, J.E. Nyström, and R.E. Nordberg, J.Am.Chem.Soc. 1985, 107, 3676.
260 D.R. Deardorff, D.C. Miles, and K.D. MacFerrin, Tetrahetron Lett. 1985, 26, 5615.
261 T. Antonsson, and C. Moberg, Organometallics 1985, 4, 1083.
262 M. Röper, R. He, and M. Schieren, J.Mol.Catal. 1985, 31, 335.
263 U.M. Dzhemilev, M.M. Sirazova, R.V. Kunakova, A.A. Panasenko, and L.M. Khalilov, Bull.Acad.Sci.USSR, Div.Chem.Sci. 1985, 34, 339.
264 J.P. Bianchini, E.M. Gaydou, B. Waegell, A. Eisenbeis, and W. Keim, J.Mol.Catal. 1985, 30, 197.
265 A. Citterio, P. de Angelis, P. Longo, and A. Musco, J.Chem.Res.(S) 1985, 316.
266 H. Sugimura, and H. Takei, Chem.Lett. 1985, 351.
267 C. Petrier, J.C. de Souza Barbosa, C. Dupuy, and J.L. Luche, J.Org.Chem. 1985, 50, 5761.
268 A.M. Caporusso, F. Da Settimo, and L. Lardicci, Tetrahedron Lett. 1985, 26, 5101.
269 H. Sugimura, and H. Takei, Bull.Chem.Soc.Jpn. 1985, 58, 664.
270 A. Goliaszewski, and J. Schwartz, Organometallics 1985, 4, 417.
271 M.Kosugi, Y.Negishi, M.Kameyama, and T.Migita, Bull.Chem.Soc.Jpn. 1985, 58, 3383.
272 M. Miyaura, K. Yamada, H. Sugiome, and A. Suzuki, J.Am.Chem.Soc. 1985, 107, 972.
273 M.J. Sharp, and V. Snieckus, Tetrahedron Lett. 1985, 26, 5997.
274 K. Karabelas, and A. Halberg, Tetrahedron Lett. 1985, 26, 3131.
275 D. Villemin, and E. Schigeko, J.Organomet.Chem. 1985, 293, C10.
276 H.J. Cristau, B. Chabaud, R. Labaudiniere, and H. Christol, Organometallics 1985, 4, 657.
277 H. Brunner, W. Li, and H. Weber, J.Organomet.Chem. 1985, 288, 359.
278 J.C. Fiaud, and L.Aribi-Zouioueche, J.Organomet.Chem. 1985, 295, 383.
279 P.R. Auburn, P.B. Mackenzie, and B. Bosnich, J.Am.Chem.Soc. 1985, 107, 2033.
280 D.H. Farrar, and N.C. Payne, J.Am.Chem.Soc. 1985, 107, 2054.
281 E. Negishi, D.E. van Horn, and T. Yoshida, J.Am.Chem.Soc. 1985, 107, 6639.
282 U.M.Dzhemilev, O.S.Vostrikova, and R.M.Sultanov, Bull.Acad.Sci.USSR, Div.Chem.Sci., 1985, 34, 1310.
283 K. Wakamatsu, Y. Okuda, K.Oshima, and H. Nozaki, Bull.Chem.Soc.Jpn. 1985, 58, 2425.
284 Y. Tamaru, H. Ochiai, F. Sanda, and Z. Yoshida, Tetrahedron Lett. 1985, 26, 5529.
285 N.A.Bumagin, A.B.Ponomaryov, and I.P.Beletskaya, Tetrahedron Lett. 1985, 26, 4819.

286 T. Kobayashi, T. Sakakura, and M. Tanaka, Tetrahedron Lett. 1985, 26, 3463.
287 C. Cardellichio, V. Fiandanese, G. Marchese, and L. Ronzini, Tetrahedron Lett. 1985, 26, 3595.
288 M. Iyoda, M. Sakaitani, A. Kojima, and M. Oda, Tetrehedron Lett. 1985, 26, 3719.
289 M. Iyoda, K. Sato, and M. Oda, Tetrahedron Lett. 1985, 26, 3829.
290 R. Scheffold, Chimia 1985, 39, 203.
291 Y. Inoue, M. Taguchi, and H. Hashimoto, Bull.Chem.Soc.Jpn. 1985, 58, 2721.
292 K. Takagi, Chem.Lett. 1985, 1307.
293 M. Catellani, and G. Chiusoli, J.Organomet.Chem. 1985, 286, C13.
294 G. Bocelli, M. Catellani, and G. Chiusoli, J.Organomet.Chem. 1985, 279, 225.
295 A. Arcadi, S. Cacchi, and F. Marinelli, Tetrahedron 1985, 41, 5121.
296 H. Hoberg, and E. Hernandez, Angew.Chem. 1985, 97, 987.
297 S.Sato, I. Matsuda, and Y. Izumi, Chem.Lett. 1985, 1875.
298 M.P. Doyle, J.H. Griffin, and J. da Coceicão, J.Chem.Soc..Chem.Comm. 1985, 328.
299 H. Kanai, and H. Matsuda, J.Mol.Catal. 1985, 29, 157.
300 P. Binger, Q.H. Lü, and P. Wedemann, Angew.Chem. 1985, 97, 333.
301 M.G. Vinogradov, A.B. Tuzikov, and G.I. Nikishin, Bull.Acad.Sci.USSR. Div.Chem.Sci. 1985, 34, 325.
302 J.M. Tour, and E. Negishi, J.Am.Chem.Soc. 1985, 107, 289.
303 M. Mori, N. Kanada, I. Oda, and Y. Bau, Tetrahedron 1985, 41, 5465.
304 R. Grigg, P. Stevenson, and T. Worakun, J.Chem.Soc..Chem.Comm. 1985, 971.
305 Y. Inoue, S. Ajioka, M. Toyofuku, A.Mori, T. Fukui, Y. Kawashima, S. Miyano, and H. Hashimoto, J.Mol.Catal. 1985, 32, 91.
306 Y. Wakabayashi, Y. Fukuda, H. Shiragami, K. Utimoto, and H. Nozaki, Tetrehedron 1985, 41, 3655.
307 Y. Tsuji, H. Nishimura, K.T. Huh, and Y. Watanabe, J.Organomet.Chem. 1985, 286, C44.
308 A. Behr, Bull.Soc.Chim. Belges 1985, 94, 671.
309 J.C. Folest, J.M. Duprilot, J.Perichon, Y. Robin, and J.Devynck, Tetrahedron Lett. 1985, 26, 2633.
310 T. Fujinami, T. Suzuki, and M. Kamiya, Chem.Lett. 1985, 199.
311 R.D. Adams, H.S. Kim, and S. Wang, J.Am.Chem.Soc. 1985, 107, 6107.
312 R.B. Wilson, and R.M. Laine, J.Am.Chem.Soc. 1985, 107, 361.
313 R. Sheldon, Bull.Soc.Chim. Belges 1985, 94, 651.
314 B.F.G. Johnson, and M.V. Twigg, Transition Met.Chem. 1985, 10, 439.
315 E.J. Corey, E.P. Barrette, and P.A. Magriotis, Tetrahedron Lett. 1985, 26, 5855.
316 C. Venturello, R. D'Aloisio, J.C.J. Bart, and M. Ricci, J.Mol.Catal.1985, 32, 107.
317 F. Mares, S.E. Diamond, F.J. Regina, and J.P. Solar, J.Am.Chem.Soc. 1985, 107, 3545.
318 G. Cum, R. Gallo, S. Ipsale, and A. Spadaro, J.Chem.Soc..Chem.Comm. 1985, 1571.
319 H. Nagashima, K. Sato, and J. Tsuji, Tetrahedron 1985, 41, 5645.
320 M. Catellani, G.P. Chiusoli, and S. Ricotti, J.Organomet.Chem. 1985, 296, C11.
321 M. Faraj, J. Martin, C. Martin, J.M. Bregeault, and J. Mercier, J.Mol.Catal. 1985, 31, 57.
322 M.Bressan, F.Morandini, A.Morvillo, and P.Rigo, J.Organomet.Chem. 1985, 280, 139.
323 M.J.Nappa, and C.A. Tolman, Inorg.Chem. 1985, 24, 4711.
324 K. Suslick, B. Cook, and M. Fox, J.Chem.Soc..Chem.Comm. 1985, 580.
325 B. de Poorter, M. Ricci, O. Bortolini, and B. Meunier, J.Mol.Catal. 1985, 31, 221.
326 J.P. Colman, J.I. Brauman, B. Meunier, T. Hayashi, T. Kodadek, and S.A. Raybuck, J.Am.Chem.Soc. 1985, 107, 2000.
327 J.A.S.J. Razenberg, A.W. van der Made, J.W.H. Smeets, and R.J.M. Nolte, J.Mol.Catal. 1985, 31, 271.
328 J.P.Renaud, P.Battioni, J.F.Bartoli, and D.Mansuy, J.Chem.Soc..Chem.Comm. 1985, 888.
329 L.C. Yuan, and T.C. Bruce, J.Chem.Soc..Chem.Comm. 1985, 868.
330 J.P. Colman, T. Kodadek, S.A. Raybuck, J.I. Brauman, and L.M. Papazian, J.Am.Chem.Soc. 1985, 107, 4343.
331 J.T. Groves, and R. Quinn, J.Am.Chem.Soc. 1985, 107, 5790.
332 E.G. Samsel, K. Srinivasan, and J.K. Kochi, J.Am.Chem.Soc. 1985, 107, 7606.
333 R.Barret, F.Pautet, M.Daudon, and B.Mathian, Bull.Soc.Chim. Belges 1985, 94, 439.
334 C.Eskénazi, G.Balavoine, F.Meunier, and H.Rivière, J.Chem.Soc..Chem.Comm. 1985, 1111.
335 T.Yokoyama, M. Nishizawa, T.Kimura, and T.M.Suzuki, Bull.Chem.Soc.Jpn. 1985, 58, 3271.

16
Structures of Organometallic Compounds determined by Diffraction Methods

BY D. R. RUSSELL

1 Introduction

This Chapter is arranged in the same format used in Volume 13, and consists of a comprehensive list of organometallic compounds whose structures have been determined by X-ray, neutron or electron diffraction methods and reported during 1985. The definition of metals remains the same, but organic compounds containing $SiMe_3$ groups, and $[AsPh_4]^+$ and $[BPh_4]^-$ salts are excluded this year. The ordering of formulae in the Main Table is based on the modified Hill system as before. Under the Structure heading the line formula is an attempt to describe the structural identity of the compound. A supplementary list of abbreviations used additional to the list at the front of this Volume is given at the end of the Main Table. Mixed metal compounds appear only once in the Main Table. The Metals Cross Reference Table can be used to locate mixed metal compounds in the Main Table which appear alphabetically under another metal.

The growth in numbers of structure determinations remains constant at eleven per cent per year. The Main Table contains 1738 entries, an average of 1.38 structures per citation. The most frequently occurring metal is Fe, found in 225 compounds, with Si(177), Co(144), W(139), Rh(137), Mo(136), and Ru(117) appearing in more than 100 compounds. Compounds 1000 and 1001 contain the largest number of metal atoms(44), compound 136 contains ten different elements, and compound 267 is the largest monomeric species with 301 atoms.

(For references see page 484

Structures Determined by Diffraction Methods   423

2 Main Table

No.	Formula	Structure	Details	Ref.
1	$AgC_{13}H_{20}NO_3$	$AgNO_3(\eta^2,\eta^2,\eta^2-C_{13}H_{20})$		1
2	$AgC_{16}H_{16}ClO_8$	$Ag\{\eta^2-C_6H_4(OCH_2CH_2O)-\underline{o}\}_2(OClO_3)$		2
3	$[AgC_{24}H_{24}]^+$	$[Ag(\eta^2,\eta^2,\eta^2-[2_3](1,4)cyclophane)][tfo]$		3
4	$[AgC_{30}H_{30}]^+$	$[Ag(\eta^1,\eta^1,\eta^1-[2_6](1,2,4,5)cyclophane][tfo]$		3
5	$AgAs_3C_{54}H_{45}Cl$	$AgCl(AsPh_3)_3 \cdot Sv$		4
6	$AgB_{11}C_7H_{18}$	$Ag(CB_{11}H_{12})(\eta^1-PhH)$		5
7	$[AgPtC_{30}H_{15}Cl_2F_{10}P]^-$	$[NBu_4][Ph_3PAgPtCl(C_6F_5)_2(\mu-Cl)] \cdot Sv$		6
8	$AgPtC_{40}H_{23}F_{15}PS$	$Ph_3PAgPt(C_6F_5)_3(SC_4H_8)$		7
9	$[AgPt_2C_{40}H_{10}F_{30}O]^-$	$[NBu_4][Pt_2(C_6F_5)_4(\mu-C_6F_5)_2(\mu-AgOEt_2)] \cdot Et_2O$		7
10	$[AgPt_6C_{60}H_{126}O_6P_6]^+$	$[Ag\{Pt_3(PPr^i_3)_3(\mu-CO)_3\}_2][CF_3SO_3]$		8
11	$AgRh_2C_{36}H_{32}F_2O_3P_3$	$Rh_2(Cp)_2(\mu-AgOPF_2O)(\mu-CO)(\mu-dppm)$		9
12	$Ag_2As_2C_{26}H_{24}N_2O_6$	$(AgNO_3)_2(\mu-dpae)$		4
13	$[Ag_2Pt_2C_{24}Cl_4F_{20}]^{2-}$	$[NBu_4]_2[Ag_2Pt_2(C_6F_5)_4(\mu-Cl)_4]$		6
14	$[Ag_2Re_{14}C_{44}BrO_{42}]^{5-}$	$[NBu_4]_5[\{Re_7C(CO)_{21}Ag\}_2Br]$		10
15	$Ag_2Te_2C_{28}H_{32}F_{10}O_2$	$\{Ag(Cp')_2(\mu-OTeF_5)\}_2$	143	11
16	$[Ag_3Li_2C_{36}H_{30}]^-$	$[Li_6Br_4(OEt_2)_{10}][Li_2Ag_3(\mu-Ph)_6]_2$	171	12
17	$[AlC_2H_5Cl_3]^-$	$[AlCl_2(benzo-15-crown-5)][AlCl_3Et]$		13
	$[AlC_2H_5Cl_3]^-$	$[LAlCl_2][AlCl_3Et]$ (L=12-crown-4,18-crown-6)		14
18	$[AlC_4H_{12}]^-$	$Na[AlMe_4]$		15
19	$[AlC_8H_{20}]^-$	$Na[AlEt_4]$		15
20	$[AlC_{12}H_{28}]^-$	$Na[AlPr_4]$		15
21	$[AlC_{14}H_{16}O_2]^-$	$K[AlMe_2(OPh)_2]$		16
22	$AlLiC_{14}H_{38}N_2P_2$	$(TMED)Li(\mu-CH_2PMe_2)_2AlMe_2$	238	17
23	$AlLiC_{16}H_{41}N_2OP$	$(TMED)(THF)Li(\mu-PMe_2CH_2)AlMe_3$	238	17
24	$AlMgC_{32}H_{49}O_4$	$\{HCAlEt_2O(Et)Mg(THF)_3CH\}(C_6H_4-\underline{o})_2$		18
25	$AlSiC_{10}H_{25}Cl_2N_2$	$Me_2SiNBu^t AlCl_2NHBu^t$		19
26	$AlSi_2C_{26}H_{39}O$	$AlEt(THF)\{C(SiMe_3)_2C_6H_4-\underline{o}\}_2$		20
27	$AlTi_2C_{40}H_{65}$	$\{(Cp^*)_2Ti(\mu-H)_2\}_2AlH$		21
28	$Al_2C_{10}H_{28}P_2$	$\{AlMe_2(\mu-CH_2PMe_2)\}_2$		22

No.	Formula	Structure	Details	Ref.
29	$[Al_2C_{12}H_{23}O]^-$	$[K(db-18-c-6)][(Me_3Al)_2OPh]$		16
30	$Al_2C_{12}H_{32}Cl_2P_4$	$\{AlCl(CH_2PMe_2)(\mu-CH_2PMe_2)\}_2$	233	22
31	$Al_2C_{18}H_{28}N_2$	$[AlMe_2\{\mu-NH(oTol)\}]_2$		23
32	$Al_2C_{26}H_{38}O_2$	$\{AlMe_2(THF)CH\}_2(C_6H_4\text{-}\underline{o})_2$		18
33	$Al_2C_{28}H_{48}N_2O_2$	$\{Al(CH_2Pr^i)_2(\mu-OCH_2C_5H_4N\text{-}2)\}_2$		24
34	$Al_2C_{33}H_{41}ClP_2$	$Et_2Al(\mu-Cl)AlEt_2CH(PPh_2)PPh_2$	233	25
35	$Al_2SiC_{34}H_{44}N_2$	$Me_2Si(NPhAlPh_2)_2$		19
36	$Al_2Si_4YbC_{18}H_{54}N_2$	$Yb\{N(SiMe_3)_2\}_2(AlMe_3)_2$	178	26
37	$Al_2Ti_2C_{40}H_{68}$	$\{(Cp^*)_2Ti(\mu-H)_2AlH(\mu-H)\}_2$		21
38	$Al_2Ti_4C_{40}H_{46}$	$\{Cp_2Ti(\mu-H)_2Al(\mu-\eta^1:\eta^5\text{-}C_5H_4)TiCp(\mu-H)_2\}_2\cdot PhMe$		27
39	$Al_4C_{24}H_{60}O_6$	$[AlMe_3]_4\cdot(18\text{-}c\text{-}6)\cdot\underline{p}\text{-xylene}$		28
40	$Al_6C_{42}H_{48}N_6$	$\{AlMe(\mu_3\text{-}NPh)\}_6$		23
41	$AsC_{10}H_{13}O_4$	$PhAs(OCH_2CH_2O)_2$		29
42	$AsC_{13}H_{27}O_4$	$AsMe(OCMe_2CMe_2O)_2$		30
43	$AsC_{14}H_{17}O_6$	$PhAs\{OC(O)CMe_2O\}_2$		29
44	$AsC_{16}H_{23}O_2$	$PhAs(COBu^t)_2$	203	31
45	$AsC_{18}H_{15}$	$AsPh_3$ complexes		32-34
46	$AsC_{18}H_{33}O_4P_2S_4$	$AsPh\{S_2P(OPr^i)_2\}_2$		35
47	$AsC_{26}H_{17}O_4$	$PhAs(O_2C_{10}H_6)_2$		29
48	$AsC_{29}H_{43}O_4$	$AsMe(O_2C_6H_2Bu^t_2\text{-}3,5)_2$		30
49	$AsBC_3H_9Br_3$	Me_3AsBBr_3		36
50	$AsBC_3H_9Cl_3$	Me_3AsBCl_3		36
51	$AsBC_3H_9I_3$	Me_3AsBI_3		36
52	$AsCoMo_2C_{22}H_{30}O_2S_3$	$(Cp^*)_2Mo_2Co(CO)_2(\mu_3\text{-}S)_2(\mu_3\text{-}\eta^2\text{-}AsS)$		37
53	$AsCr_2C_{13}H_8ClN_2O_{10}S$	$\{Cr(CO)_5\}_2AsClSC(NHMe)_2$	233	38
54	$AsCr_2MnC_{14}O_{14}$	$Cr_2(CO)_8(\mu\text{-}CO)\{\mu\text{-}AsMn(CO)_5\}$	235	39
55	$AsCr_2MnC_{15}O_{15}$	$\{Cr(CO)_5\}_2\{\mu\text{-}AsMn(CO)_5\}$	241	39
56	$AsFeC_{16}H_{13}S_2$	$Fe\{\eta,\eta'\text{-}(C_5H_4S)_2AsPh\}$	163	40
57	$AsFe_2Mo_2C_{21}H_{11}O_{11}$	$\{Mo_2(CO)_4Cp_2\}\{\mu\text{-}AsMo(CHPPh_3)(CO)_2Cp\}$		41
58	$AsMn_2C_{24}H_{31}O_4$	$\{Mn(CO)_2(Cp^*)\}_2AsH$		42

No.	Formula	Structure	Details	Ref.
59	$AsMo_2C_{16}H_{17}O_4$	$\{Mo(CO)_2(Cp)\}_2(\mu-H)(\mu-AsMe_2)$		43
60	$AsMo_3C_{21}H_{15}O_6$	$Mo_3(CO)_6(Cp)_3(\mu_3-As)$		44
61	$AsMo_3C_{39}H_{31}O_6P$	$CpMoFe_2(CO)_8(\mu-H)\{\mu_3-AsMo(CO)_3Cp\}$		41
62	$AsNiC_{44}H_{36}OP$	$Ni(PhOCPhCHAsPh_2)(PPh_3)$		45
63	$AsOs_3C_{26}H_{38}N_2O_{10}PS$	$Os_3H(CO)_{10}(\mu-H)(PBu^t_2NSNAsBu^t_2)$		46
64	$AsRuC_{22}H_{15}O_4$	$\underline{ax}$-$Ru(CO)_4(AsPh_3)$		47
65	$AsSi_2C_8H_{18}Cl_2F_6N$	$AsCl_2(CF_3)_2\{N(SiMe_3)_2\}$		48
66	$AsSi_4C_{20}H_{41}$	$PhAsC(SiMe_3)_2C(SiMe_3)_2$		49
67	$[AsW_7CH_4O_{27}]^{7-}$	$[CN_3H_6]_7[MeAsW_7O_{27}H].3H_2O$		50
68	$[As_2CoC_{54}H_{49}P_3S]^+$	$[Co(SAsCPh_2As)(triphos)][BF_4]$		51
69	$As_2Cr_2Mo_2C_{24}H_{10}O_{14}$	$\{Mo(CO)_2(Cp)\}_2As_2Cr_2(CO)_{10}$	223	52
70	$[As_2FeC_{16}H_{28}O_2P]^+$	$[FeMe(CO)_2(PMe_3)(diars)][BF_4]$		53
71	$As_2Mn_2C_{24}H_{30}O_4$	$\{Mn(CO)_2(Cp^*)\}_2(\mu-\eta^2-As_2)$		42
72	$As_2Mn_4C_{28}H_{20}O_8$	$As_2\{Mn(CO)_2(Cp)\}_4$	237	52
73	$As_2Mn_4C_{30}H_{28}O_6$	$\{Mn_2(CO)_2(Cp')_2\}\{\mu-AsMn(CO)_2(Cp')\}_2$	233	39
74	$As_2Os_3C_{20}H_{28}N_2O_8S$	$Os_3(CO)_8(\mu-H)(\mu-NSNAsBu^t)_2(\mu_3-AsBu^t)$		54
75	$As_2PdC_{43}H_{38}O_5$	$Pd\{CH(CO_2Me)COCH(CO_2Me)\}(AsPh_3)_2.H_2O$		55
76	$[As_2PdRh_2C_{66}H_{58}Cl_3O_2P_4]^+$	$[PdRh_2Cl_3(CO)_2\{\mu_3-AsPh(CH_2PPh_2)_2\}][BPh_4].Sv$	140	56
77	$As_2PtRhC_{51}H_{44}Cl_3OP_2$	$ClPtRhCl_2(CO)(\mu-dapm)_2.2CH_2Cl_2$		57
78	$As_2PtRhC_{51}H_{44}Cl_3OP_2$	$trans$-$RhCl(CO)(\mu-Ph_2AsCH_2PPh_2)_2$-$cis$-$PtCl_2$		58
79	$As_2PtRhC_{51}H_{44}I_3OP_2$	$IPtRhI_2(CO)(\mu-dapm)_2.CH_2Cl_2.2CHCl_3$		57
80	$As_2Rh_2C_{30}H_{63}ClO_2S$	$\{Rh(CO)(AsBu^t_3)\}_2(\mu-Cl)(\mu-SBu^t)$		59
81	$As_2Rh_2C_{44}H_{42}O_8$	$Rh_2(AsPh_3)_2(\mu-OAc)_4$		60
82	$As_2Rh_2C_{66}H_{58}Cl_2O_2P_4$	$Rh_2Cl_2(CO)_2\{\mu-(Ph_2PCH_2)_2AsPh\}.2CH_2Cl_2$	140	56
83	$As_2Si_6C_{20}H_{56}$	$(Me_3Si)_3CAs=AsC(SiMe_3)_3$		61
84	$As_2W_2C_7I_2O_7$	$W_2I(CO)_7(\mu-I)(\mu-\eta^2-As_2)$	243	52
85	$As_3Rh_6C_{31}H_{45}O_{11}$	$Rh_6(CO)_9(\mu-CO)_2(\mu-AsBu^t_2)_2(\mu_4-AsBu^t)$		62
86	$AuC_{15}H_{22}N_7O_4$	$AuMe_2\{(NCHCHNMeC)_3COH\}$		63
87	$[AuC_{16}H_{21}N_5O]^+$	$[AuMe_2\{(C_5H_4N)(mim)_2COH\}][NO_3]$		64
88	$[AuC_{16}H_{22}N_3]^{2+}$	$[Au(C_6H_4CH_2NMe_2-\underline{o})(py)_2][BF_4]_2.Me_2CO$		65

No.	Formula	Structure	Details	Ref.
89	$[AuC_{24}H_{20}]^-$	$[NBu_4][AuPh_4]$		66
90	$AuC_{25}H_{24}OP$	$Ph_3PAuC(CH_2CH_2)COCHCH_2CH_2$	153	67
91	$AuC_{27}H_{25}NOP$	$Au\{C(OMe)=N(pTol)\}(PPh_3)$		68
92	$[AuC_{54}H_{47}OP_3]^+$	$[Rh_2Cp_2(\mu-AuPPh_3)(\mu-CO)(\mu-dppm)][BF_4]$		69
93	$AuBC_{14}H_{18}N_8$	$AuMe_2\{B(pz)_4\}$		64
94	$AuCoC_{39}H_{30}O_3P_2$	$Ph_3PCo(CO)_3AuPPh_3$		70
95	$AuCoRhRuC_{30}H_{15}O_{12}P$	$Ph_3PAuRuCo_xRh_{3-x}(CO)_{12}$ (x=1.6)		71
96	$[AuFeWC_{27}H_{15}O_9]^-$	$[NEt_4][Fe(CO)_4(AuPPh_3)\{W(CO)_5\}]$		72
97	$AuFe_2C_{35}H_{34}O_6PS$	$Fe_2(CO)_6(\mu-AuPPh_3)(\mu-\eta^2-S=CC_6H_7Me_4)$		73
98	$[AuFe_4C_{19}H_{15}O_{13}P]^-$	$[K(18-c-6)][Fe_4(CO)_{12}(AuPEt_3)(\mu_3-\eta^2-CO)].Sv$	163	74
99	$AuFe_4C_{20}H_{18}O_{13}P$	$Fe_4(CO)_{12}(\mu-AuPEt_3)(\mu_4-\eta^2-COMe)$		74,75
100	$AuOs_3C_{36}H_{18}F_5O_{10}P$	$Ph_3PAuOs_3(CO)_{10}(\mu-\eta^2:\eta^1-CHCHC_6F_5)$		76
101	$AuOs_3C_{36}H_{23}O_{10}P$	$Ph_3PAuOs_3(CO)_{10}(\mu-\eta^2:\eta^1-CHCHPh)$		76
102	$AuReC_{31}H_{27}N_2O_2P$	$Ph_3PAuRe(NNC_6H_4OMe-\underline{p})(CO)(Cp)$		77
103	$AuRu_3C_{27}H_{16}O_9PS$	$Ph_3PAuRu_3(CO)_9(\mu-H)(\mu_3-S)$		78
104	$AuRu_3C_{31}H_{24}O_9PS$	$Ph_3PAuRu_3(CO)_9(\mu_3-SBu^t)$		78
105	$AuRu_5C_{20}H_{15}NO_{13}$	$Ru_5C(CO)_{13}(NO)(AuPEt_3)$ (μ_2 & μ_3 isomers)		79
106	$Au_2C_6H_{20}N_2$	$\{AuMe_2(\mu-NHMe)\}_2$	198	80
107	$Au_2C_8H_{22}S_2$	$\{Me_2Au(\mu-SEt)\}_2$		81
108	$Au_2C_{28}H_{28}Br_4P_2$	$[\underline{trans}-AuBr_2\{\mu-(CH_2)_2PPh_2\}]_2.CDCl_3$		82
109	$Au_2C_{28}H_{28}Cl_2P_2$	$ClAuCH_2PPh_2CH_2AuCl\{(CH_2)_2PPh_2\}$		83
110	$Au_2C_{28}H_{28}Cl_4P_2$	$\underline{cis},\underline{trans}-[AuCl_2\{\mu-(CH_2)_2PPh_2\}]_2.2CDCl_3$		82
111	$Au_2C_{28}H_{28}P_2$	$Au_2(\mu-CH_2PPh_2)_2$		84
112	$Au_2C_{29}H_{30}BrClP_2$	$BrAu\{\mu-(CH_2)_2PPh_2\}AuCH_2Cl$		85
113	$Au_2C_{29}H_{31}BrP_2$	$BrAuAuMe(\mu-CH_2PPh_2)_2$		84
114	$Au_2C_{30}H_{28}N_2P_2$	$Au_2(CN)_2\{\mu-(CH_2)_2PPh_2\}_2$		86
115	$Au_2C_{30}H_{30}F_3IP_2$	$Au_2I(CH_2CF_3)\{\mu-(CH_2)_2PPh_2\}_2$		87
116	$Au_2C_{30}H_{33}IP_2$	$Au_2IEt\{\mu-(CH_2)_2PPh_2\}_2$		87
117	$Au_2C_{31}H_{30}N_2P_2$	$[Au_2(CN)_2\{\mu-(CH_2)_2PPh_2\}_2](\mu-CH_2)$		86
118	$[Au_2C_{58}H_{50}O_2P_3]^+$	$[Au_2(PPh_3)_2\{\mu-C(PPh_3)CO_2Et\}][ClO_4]$		88

No.	Formula	Structure	Details	Ref.
119	$Au_2Fe_3C_{45}H_{30}O_9P_2S$	$(Ph_3P)_2Au_2Fe_3(CO)_9(\mu_3-S)$		89
120	$Au_2Os_4C_{25}H_{30}O_{13}P_2$	$(Et_3P)_2Au_2Os_4(CO)_{13}$		90
121	$Au_2Os_4C_{38}H_{26}O_{12}P_2$	$Os_4(CO)_{12}(\mu_3-AuPMePh_2)_2$		90
122	$Au_2Ru_3C_{45}H_{30}O_9P_2S$	$(Ph_3P)_2Au_2Ru_3(CO)_9(\mu_3-S)$		78
123	$Au_2Ru_3C_{51}H_{40}O_9P_2$	$(Ph_3P)_2Au_2Ru_3(CO)_9(\mu_3-\eta^2-C=CHBu^t).Sv$		91
124	$Au_3C_6H_{24}N_3$	$\{AuMe_2(\mu-NH_2)\}_3$	213	92
125	$Au_3C_{42}H_{26}Cl_2F_{10}NP_2$	$ClAu\{Ph_2PCH(Aupy)PPh_2\}AuCl(C_6F_5)_2.xCH_2Cl_2$		93
126	$[Au_3Cu_2C_{48}H_{30}]^-$	$[NBu_4][Cu_2Au_3(\mu-\eta^1:\eta^2-CCPh)_6]$		94
127	$BC_{10}H_{10}N$	$(py)BC_5H_5$		95
128	$BC_{12}H_{22}ClO$	$ClBHOMeCH_2CH_2CHCH(C_5H_6Me_2)$		96
129	$BC_{13}H_{21}O_2$	$B\{CH(CH_2)_3CH\}(acac)$		97
130	$BC_{14}H_7N$	$(py)BC_9H_7$		95
131	$BC_{17}H_{13}ClN$	$BCl\{(\underline{o}-C_6H_4)_2\}(py)$		98
132	$BC_{17}H_{18}NO_3$	$PhBON(CHPh)CMe(CH_2O)_2$		99
133	$BC_{19}H_{11}N_2S_2$	$PhBNMeNMeC(S)S$		100
134	$BC_{19}H_{21}O_2$	$B(pTol)_2(acac)$		97
135	$[BC_{25}H_{17}N]^+$	$[B\{(\underline{o}-C_6H_4)_2\}(9-acridine)][AlCl_4]$		98
136	$[BCoPdC_{22}H_{19}ClFN_6O_3P]^+$	$[PdCl\{P(C_5H_3NCHNO)_3BF\}(\eta^3-CH_2CMeCH_2)][BF_4]$		101
137	$BCrC_{13}H_{10}NO_3$	$Cr(CO)_3\{\eta^6-C_5H_5B(py)\}$		102
138	$BCrC_{17}H_{12}NO_3$	$Cr(CO)_3\{\eta^6-(py)BC_5H_3(3,4-C_6H_4)\}$		102
139	$BCrSnC_{14}H_{24}NO_3$	$Cr(SnMe_3)(CO)_3(\eta^5-MeBNBu^tC_3H_3)$		103
140	$[BFeC_{14}H_{21}N]^+$	$[Fe(\eta^5-C_3H_3BMeNBu^t)(\eta-PhH)]_2[FeX_4]$ (X=Br,Cl)		104
141	$BFeC_{24}H_{28}O_2P$	$Fe(CO)_2(PPh_2BH_3)(Cp^*)$		105
142	$BIr_2C_{20}H_{37}$	$Ir_2H_2(Cp^*)_2(\mu-H)(\mu-BH_4).Sv$	183	106
143	$BMoC_{12}H_{10}BrN_6O_3$	$MoBr(CO)_3(HBpz_3)$		107
144	$BMoC_{13}H_{10}NO_3$	$Mo(CO)_3\{\eta^6-C_5H_5B(py)\}$		102
145	$BMoSnC_{14}H_{24}NO_3$	$Mo(SnMe_3)(CO)_3(\eta^5-MeBNBu^tC_3H_3)$		103
146	$BSi_2C_{28}H_{32}N_3$	$MeN(SiPh_2NMe)_2BMe$		108
147	$BSi_3C_{25}H_{34}FO$	$(Me_2PhSi)_3CBF(OH)$		109
148	$BZrC_{23}H_{39}ClN_2$	$ZrCl(Cp)_2(\mu-H)(\mu-NBu^t)BH(NC_5H_6Me_4)$		110

No.	Formula	Structure	Details	Ref.
149	$[B_2C_{14}H_{19}]^-$	$K[1,9-(Me_2B)_2(H)C_{10}H_6]$		111
150	$B_2C_{14}H_{30}N_2$	$Pr^i_2NBCHCHBNPr^i_2$		112
151	$B_2C_{16}H_{22}N_4$	$PhB(NMeNMe)_2BPh$		100
152	$B_2C_{16}H_{36}N_2$	$BuBNBu^tBBuNBu^t$	118	113
153	$B_2C_{35}H_{28}N_2O_2$	$\{Ph_2B(O)NC_6H_4\}_2C$		114
154	$B_2CoTlC_{10}H_{14}$	$Tl(\mu-\eta^5-MeBCHBMeCHCH)Co(Cp)$		115
155	$B_2Co_2C_{10}H_{12}S_2$	$(CoCp)_2(\mu-\eta^4-SBHBHS)$		116
156	$B_2Mo_2C_{22}H_{20}N_{12}O_4S$	$\{Mo(CO)_2(HBpz_3)\}_2(\mu-S).PhMe$		117
157	$B_2Mo_2C_{34}H_{44}N_{12}O_4S$	$[Mo(CO)_2\{HB(Me_2pz)_3\}]_2(\mu-S)$		117
158	$B_2Mo_2SeC_{22}H_{20}N_{12}O_4$	$\{Mo(CO)_2(HBpz_3)\}_2(\mu-Se).2THF$		117
159	$B_2SiC_{17}H_{25}N_3$	$Ph_2SiNMeBMeNMeBMeNMe$		108
160	$B_2SiC_{23}H_{28}N_2S_2$	$PhBSBPhN(Xy)SNSiMe_3$		118
161	$B_2Si_2TiC_{14}H_{30}Cl_2N_2$	$TiCl_2(\eta^5-Me_3SiNBMeCHCHCH)_2$	253	119
162	$B_2Si_2VC_{14}H_{30}ClN_2$	$VCl(\eta^5-Me_3SiNBMeCHCHCH)_2$		119
163	$B_2SnC_{18}H_{34}N_2$	$Sn(\eta^5-Bu^tNBMeCMeCHCH)_2$	153	120
164	$B_2TiC_{14}H_{32}Cl_4N_2$	$TiCl_4(\eta^4-PrNBu^tBPrNBu^t)$		119
165	$B_3C_{10}H_{25}N_4S_2$	$MeBSBMeNBu^tSNBNMeCH_2CH_2NMe$		118
166	$[B_3C_{24}H_{20}O_3]^-$	$[NMe_4][BPh_2OBPhOBPhO]$		121
167	$B_3Hf_2Si_4C_{20}H_{71}N_2P_4$	$Hf_2(BH_4)_3\{N(SiMe_2CH_2PMe_2)_2\}_2(\mu-H)_3$		122
168	$B_3IrC_{37}H_{38}OP_2$	$IrH(B_3H_7)(CO)(PPh_3)_2$		123
169	$B_3UC_5H_{17}$	$U(BH_4)_3(Cp)$		124
170	$[B_4CH_{10}N]^-$	$[PPN][B_3H_7NCBH_3]$	185	125
171	$B_4CoC_{32}H_{38}ClP_2$	$CoCl(dppe)(\eta^5-Et_2C_2B_4H_4)$		126
172	$B_4CoPtC_{29}H_{51}$	$CpCo(\mu-\eta^5:\eta^5-MeCB_2C_2Et_4)Pt(\eta^5-MeCB_2C_2Et_4)$		127
173	$B_4FeC_{11}H_{16}$	$Fe\{\eta^5,\eta^6-(C_2B_4H_5)CH_2CH_2CH_2Ph\}$		128
174	$B_4FeC_{16}H_{22}$	$Fe(\eta^5-Et_2C_2B_4H_4)(\eta^6-napH)$		129
175	$B_4FeC_{17}H_{32}OS_2$	$Fe(CO)(\eta^5-SBMeCEtCEtBMe)_2$		130
176	$B_4FeC_{18}H_{24}$	$Fe(\eta^5-Et_2C_2B_4H_4)(\eta^6-Ph_2)$		131
177	$B_4FeC_{20}H_{24}$	$Fe(\eta^5-Et_2C_2B_4H_4)(\eta^6-C_{14}H_{10})$		129
178	$B_4FeC_{32}H_{38}ClP_2$	$FeCl(dppe)(\eta^5-Et_2C_2B_4H_4)$		126

No.	Formula	Structure	Details	Ref.
179	$B_4Fe_2PtC_{34}H_{56}$	$Pt\{\mu-\eta^5:\eta^5-MeCB_2C_2Et_4\}FeCp\}_2$		127
180	$B_4NiC_{42}H_{43}BrP_2$	$NiBr(PPh_3)(\eta^5-Et_2C_2B_4H_3PPh_3)$		126
181	$B_4NiPtC_{29}H_{51}$	$CpNi(\mu-\eta^5:\eta^5-MeCB_2C_2Et_4)Pt(\eta^5-MeCB_2C_2Et_4)$		127
182	$B_4Ni_2PtC_{34}H_{56}$	$Pt\{\mu-\eta^5:\eta^5-MeCB_2C_2Et_4\}NiCp\}_2$		127
183	$B_6CoSi_2C_{13}H_{31}$	$1,2,3-(Cp)CoC_2(SiMe_3)_2B_4H_3(5-B_2H_5)$		132
184	$B_7Ni_2C_{11}H_{18}$	$6,7,1-Cp_2Ni_2CB_7H_8$		133
185	$B_7RuC_8H_{11}$	$2,5,6-(\eta-PhH)RuC_2B_7H_{11}$	111	134
186	$B_8Fe_2C_{27}H_{44}$	$\{Fe(\eta^6-PhH)\}_2\{\mu-\eta^5,\eta^{5'}-(Et_2C_2B_4H_3)_2CH_2CHMe\}$		131
187	$B_8Th_2C_{12}H_{58}$	$Th_2(BH_3Me)_6(OEt_2)(\mu-BH_3Me)_2$		135
188	$B_8Th_2C_{16}H_{64}O_2$	$\{Th(BH_3Me)_3(THF)(\mu-BH_3Me)\}_2$		135
189	$[B_9C_9H_{14}INS]^-$	$Cs[\underline{nido}-9-I-7,8-(\underline{p}-SCNC_6H_4)CCB_9H_{10}]$		136
190	$B_9FeC_{11}H_{23}$	$1,2,4-(\eta^6-C_6H_5Me)FeMe_2C_2B_9H_9$	268	137
191	$B_9MoWC_{24}H_{29}O_3$	$(\eta^5-ind)(CO)MoW(CO)_2(\eta^5-Me_2C_2B_9H_9)\{\mu-C(pTol)\}$		138
192	$B_9MoWC_{30}H_{39}O_3$	$MoW(CO)_3(\eta-C_2Et_2)(\eta^5-ind)-$ $\{\mu-\eta^4-CH(pTol)B_9H_8C_2Me_2\}$	190	138
193	$B_9PdC_8H_{27}N_2$	$3,1,2-(TMED)PdC_2B_9H_{11}$		139
194	$B_9PdC_8H_{29}P_2$	$3,1,2-(PMe_3)_2PdC_2B_9H_{11}$		139
195	$B_9ReC_8H_{16}O_3$	$3,1,2-(CO)_3ReC(CMe_2)CB_9H_{10}$		140
196	$B_9RhC_{11}H_{24}$	$1,2,3-(\eta^3,\mu H-nbdH)RhCMeCMeB_9H_9$	115	141
197	$B_9RhC_{11}H_{24}$	$1,2,3-(\eta^3,\eta^2-C_5H_6CHCH_2)RhCMeCMeB_9H_9$		141
198	$B_9RhC_{12}H_{28}$	$3,1,2-(\eta^3-C_8H_{13})RhCMeCMeB_9H_9$		142
199	$[B_9RhC_{38}H_{41}P_2]^-$	$[M][3,1,2-(Ph_3P)_2ReC_2B_9H_{11}] \cdot Sv$ $[M=NMe_4, K(18-c-6) \text{ (2 isomers)}]$		143
200	$B_9RhC_{45}H_{48}P_2$	$2,1,7-Rh(H)(PPh_3)_2CMeCPhB_9H_9$		144
201	$B_9RuC_8H_{17}$	$3,1,2-(\eta^6-PhH)RuC_2B_9H_{11}$		137
202	$B_9W_2C_{24}H_{29}O_3$	$(\eta^5-ind)W_2(CO)_3(\eta^5-Me_2C_2B_9H_9)\{\mu-C(pTol)\}$		138
203	$B_{10}RhC_{18}H_{35}ClOP$	$7,8,11-(Cp^*)Cl(PMe_2Ph)-\underline{nido}-7,12-RhOB_{10}H_9$		145
204	$B_{10}RhC_{18}H_{37}ClP$	$7,8,11-(Cp^*)Cl(PMe_2Ph)-\underline{nido}-7-RhB_{10}H_{11} \cdot CH_2Cl_2$		146
205	$B_{10}TlC_4H_{15}Cl_2$	$TlCl_2(\underline{m}-Me_2C_2B_{10}H_9)$	153	147
206	$B_{10}WC_{18}H_{35}O_2P$	$W(CO)(PMe_3)\{\eta^2-pTol)C_2(B_{10}H_{14})\}(Cp)$	233	148

No.	Formula	Structure	Details	Ref.
207	$B_{12}CoC_{22}H_{49}O$	$Co(\eta^5-Et_2C_2B_4H_8)(\eta^6-Et_4C_4B_8H_7OC_4H_8)$		149
208	$B_{16}CoC_{30}H_{67}$	$CoH(Et_4C_4B_8H_7OCMe_2)_2$		150
209	$B_{18}Rh_2C_{16}H_{50}P_2$	$\{(Et_3P)RhC_2B_9H_{10}\}_2$		151
210	$B_{18}Rh_2C_{24}H_{64}P_2$	$(Et_3P)_2HRh(C_2B_9H_{10})(C_2B_9H_{10})Rh(codH)$		151
211	$B_{18}Rh_2C_{40}H_{52}P_2$	$\{(Ph_3P)RhC_2B_9H_{11}\}_2$	113	151
212	$B_{18}Rh_2C_{44}H_{60}P_2$	$\{(Ph_3P)RhC_2B_9H_{10}Et\}_2$		151
213	BeC_5H_5Cl	$BeCl(\eta^5-Cp)$		152
214	$BiC_2H_6N_3$	Me_2BiN_3	206	153
215	$BiC_{10}H_{13}OS_2$	$PhBiS(CH_2)_2O(CH_2)_2S$		154
216	$BiFeC_{17}H_{25}N_2O_2S_4$	$(Cp)(CO)_2FeBi(detc)_2$		155
217	$Bi_2Fe_3C_9O_9$	$Bi_2Fe_3(CO)_9$		156
218	$Bi_3W_3C_{14}H_3O_{13}$	$W_2(CO)_8(\mu-\eta^2-Bi_2)[\mu-MeBi\{W(CO)_5\}]$		157
219	$[Bi_4Fe_4C_{13}O_{13}]^{2-}$	$[NEt_4]_2[(CO)_4FeBi_4\{\mu_3-Fe(CO)_3\}_3]$		158
220	$CeC_{19}H_{21}N_2$	$Ce(NCMe)_2(Cp)_3$		159
221	CoC_3NO_4	$Co(CO)_3(NO)$	E	160
222	$[CoC_4O_4]^-$	$[Co(CO)_4]^-$ {1324 & Li,Na,K,Rb,SmI$_2$(THF)$_5$}		161-164
223	$CoC_{10}H_{14}N_5O_4$	$CoMe(glyH)_2(py)$		165
224	$CoC_{11}H_{11}O_2$	$Co(\eta^4-OCCMeCMeCO)(Cp)$		166
225	$CoC_{11}H_{16}N_5O_4$	$CoEt(glyH)_2(py)$		165
226	$CoC_{12}H_{18}N_5O_4$	$CoPr^i(glyH)_2(py)$		165
227	$[CoC_{13}H_{23}N_2P]^+$	$[Co\{C(CH_2)NMeCMeNH\}(PMe_3)(Cp)]I$		167
228	$CoC_{13}H_{25}F_3N_4O_7P$	$Co(CH_2CF_3)(dmgH)_2\{P(OMe)_3\}$		168
229	$[CoC_{13}H_{33}P_3]^+$	$[Co(PMe_3)_3(\eta^4-C_4H_6)][BPh_4]$		169
230	$[CoC_{14}H_{26}NOP]^+$	$[Co(CMeNMeCMe_2O)(PMe_3)(Cp)]I \cdot H_2O$		170
231	$[CoC_{14}H_{28}N_4O_3]^+$	$[CoMe(OH_2)(N_4O_2C_{13}H_{23})][PF_6]$		171
232	$CoC_{15}H_{20}ClN$	$CoCl(py)(Cp*)$		172
233	$CoC_{15}H_{20}N_2S$	$Co(SPh)(CNMeCH_2CH_2NMe)(Cp)$		173
234	$CoC_{16}H_{20}F_3N_6O_4$	$Co(CH_2CF_3)(dmgH)_2(NC_5H_4CN-4)$		168
235	$CoC_{16}H_{20}NS$	$Co(NHC_6H_4S-\underline{o})(Cp*)$		174
236	$CoC_{16}H_{21}N_2$	$Co\{(NH)_2C_6H_4-\underline{o}\}(Cp*)$		174

Structures Determined by Diffraction Methods 431

No.	Formula	Structure	Details	Ref.
237	$CoC_{17}H_{25}ClN_5O_6$	$Co\{CHMe(CO_2Me)\}(dmgH)_2(NC_5H_4Cl-4)$	293-353	175
238	$CoC_{17}H_{25}N_6O_4$	$Co(C_2H_4CN)(dmgH)_2(NC_5H_4Me-4)$	293-355	175
239	$CoC_{17}H_{26}N_5O_6$	$Co(CHMe(CO_2Me)(dmgH)_2(py).MeOH$	223-293	175
240	$[CoC_{17}H_{27}N_5O_2]^+$	$[CoMe(py)(N_4O_2C_{11}H_{19})][PF_6].Me_2CO$		176
241	$CoC_{17}H_{28}N_4O_6P$	$CoMe(dmgH)_2\{PPh(OMe)_2\}$		177
242	$CoC_{18}H_{18}IN_2$	$CoI(NC_4H_3CHNCHMePh)(Cp)$		178
243	$[CoC_{18}H_{29}N_6O_6]^{2+}$	$[Co(CNHCH_2CH_2O)_5(C=NCH_2CH_2O)][BPh_4]_2.Me_2CO$		179
244	$CoC_{19}H_{20}IN_2$	$CoI(NC_4H_3CMeNCHMePh)(Cp)$		178
245	$CoC_{19}H_{32}PS$	$Co(PMe_3)(\eta^2-S=CCMe_2(CH_2)_3CMe_2)$		180
246	$[CoC_{20}H_{21}IN_2]^+$	$[CoI(NC_5H_4CMeNCHMePh)]I$		181
247	$CoC_{21}H_{17}N_2O_2$	$Co(saloph)Me.Sv$		182
248	$[CoC_{21}H_{35}N_5O_2]^+$	$[Co(CH_2Bu^t)(py)(N_4O_2C_{11}H_{19})][PF_6]$		176
249	$CoC_{21}H_{37}N_4O_7P$	$Co(adamantyl)(dmgH)_2\{P(OMe)_3\}$		183
250	$CoC_{22}H_{24}N_2OP$	$Co(OCPhNCNPh)(PMe_3)(Cp)$		184
251	$CoC_{22}H_{30}N_4O_5P$	$CoMe(dmgH)_2\{PPh_2(OMe)\}$		177
252	$CoC_{22}H_{34}N_5O_8$	$Co\{CH_2CMe(CO_2Et)_2\}(dmgH)_2(py)$		185
253	$CoC_{23}H_{21}N_2O_2$	$Co(saloph)Pr^i.Sv$		182
254	$CoC_{23}H_{37}P_3$	$Co(PMe_3)_3(\eta^2-anthracene)$	235	186
255	$CoC_{25}H_{27}N_3O_2P$	$Co(CNMeCHMeCH_2NMe)(CO)(NO)(PPh_3)$		187
256	$CoC_{26}H_{27}N_6O_4$	$Co(CH_2CH_2CN)(py)(MePh-glyoxime)_2$		188
257	$CoC_{27}H_{21}F_3N_3O_2$	$Co(saloph)(CH_2CF_3)(py).Sv$		182
258	$CoC_{27}H_{50}N_4O_7P$	$Co(adamantyl)(dmgH)_2\{P(OPr^i)_3\}$		183
259	$CoC_{28}H_{31}F_3N_4O_4P$	$Co(CH_2CF_3)(dmgH)_2(PPh_3)$		168
260	$CoC_{28}H_{34}N_4O_4P$	$CoEt(dmgH)_2(PPh_3)$		185
261	$[CoC_{30}H_{27}IP_2]^+$	$[CoI(dppm)(Cp)][I].CHCl_3$		189
262	$CoC_{30}H_{27}I_2OP_2$	$CoI_2(\eta^1-dppmO)(Cp)$		189
263	$CoC_{30}H_{38}N_4O_4P$	$Co(CHMeEt)(dmgH)_2(PPh_3)$		185
264	$CoC_{33}H_{23}O_7P_2$	$Co(CO)_3(PPh_3)\{P(O_2C_6H_4-\underline{o})_2\}.Et_2O$		190
265	$CoC_{35}H_{44}N_4O_8P$	$Co\{CH_2CMe(CO_2Et)_2\}(dmgH)_2(PPh_3)$		185
266	$CoC_{40}H_{59}O_3P_2$	$Co(CO)_3\{\eta^3-CH(PC_6H_2Bu^t_3-2,4,6)_2\}$		191

No.	Formula	Structure	Details	Ref.
267	$CoC_{62}H_{169}N_{13}O_{55}P$	methylcobalamin.$Me_2CO.4H_2O$		192
268	$CoC_{65}H_{95}N_{13}O_{16}P$	R & S-{$HOCH_2CH(OH)CH_2$}cobalamin.$15H_2O$		193
269	$CoCuC_4O_4$	$[CuCo(CO)_4]_n$		194
270	$CoCuC_{10}H_{16}N_2O_4$	$(TMED)CuCo(CO)_3(\mu\text{-}CO)$		195
271	$CoFeC_{10}H_5Cl_4N_3O_5P_3$	$(Cp)(CO)CoFe(CO)_4(\mu\text{-}P_3N_3Cl_4)$		196
272	$CoFeGeMoC_{17}H_{15}O_8$	$CpMoCoFe(CO)_8(\mu_3\text{-}H)(\mu_3\text{-}GeBu^t)$		197
273	$CoFeNiC_{23}H_{17}O_8$	$(Cp)NiCoFe(CO)_6(\mu_3\text{-}\eta^2\text{-}PhCCCO_2Pr^i)$		198
274	$CoFe_2C_8NO_9S$	$CoFe_2(CO)_8(NO)(\mu_3\text{-}S)$		89
275	$CoFe_2C_{17}H_{13}O_5S$	$(Cp)_2Fe_2Co(CO)_3(\mu\text{-}CO)_2(\mu_3\text{-}CSMe)$		199
276	$CoFe_2C_{34}H_{28}O_4PS$	$(Cp)_2FeCo(CO)_2(PPh_3)(\mu\text{-}CO)_2(\mu_3\text{-}CSMe)$		199
277	$[CoFe_3C_{13}O_{13}]^-$	$[PPN][CoFe_3(CO)_{10}(\mu\text{-}CO)_3]$		200
278	$[CoFe_4C_{15}O_{14}]^-$	$[M][CoFe_4(CO)_{13}(\mu\text{-}CO)]$ (M=PPN,NEt_4 at 153K)		201,202
279	$CoFe_4RhC_{17}O_{16}$	$CoFe_4RhC(CO)_{14}(\mu\text{-}CO)_2$	153	202
280	$CoSiC_{22}H_{33}O$	$Co\{\eta^4\text{-}C_{13}H_{16}(OMe)(SiMe_3)\}(Cp)$		203
281	$CoSnC_{21}H_{26}Cl_3N_2O_3$	$BuSnCl_2(\mu\text{-}OMe)(\mu\text{-}salen)CoCl$		204
282	$CoSnC_{38}H_{43}P_2$	$Co(PMe_3)_2(SnPh_3)(\eta^4\text{-}anthracene)$	233	186
283	$CoSn_3C_{15}H_{27}N_6$	$\{(Me_3Sn)_3Co(CN)_6\}_n$		205
284	$CoVC_{22}H_{42}O_{12}P_3$	$(Cp)Co\{\mu\text{-}P(OEt)_2O\}_3VO(acac)$		206
285	$Co_2C_5H_9F_{12}N_3O_2P_6$	$Co_2(CO)_2(\mu\text{-}PF_2NMePF_2)_3$		207
286	$Co_2C_{12}H_{10}Cl_4N_3O_2P_3$	$Co_2(CO)_2(Cp)_2(\mu\text{-}P_3N_3Cl_4)$		196
287	$Co_2C_{14}H_{12}O_2$	$Co_2(CO)_2(Cp)_2(\mu\text{-}CCH_2)$		208
288	$Co_2C_{14}H_{22}P_2$	$\{Co(Cp)(\mu\text{-}PMe_2)\}_2$		209
289	$[Co_2C_{14}H_{23}P_2]^+$	$[Co_2(Cp)_2(\mu\text{-}H)(\mu\text{-}PMe_2)_2][BPh_4]$		209
290	$Co_2C_{15}H_{24}Br_2P_2$	$\{CoBr(Cp)\}_2(\mu\text{-}PMe_2)(\mu\text{-}CH_2PMe_2)$		210
291	$Co_2C_{16}H_{14}O_8$	$Co_2(CO)_6\{\eta^2\text{-}HCCC(OH)C(COMe)(CH_2)_3CH_2\}$		211
292	$[Co_2C_{16}H_{26}NP_2]^+$	$\{Co(Cp)\}_2(\mu\text{-}PMe_2)_2(\mu\text{-}MeNCH)$		212
293	$Co_2C_{18}H_{30}O_2P_2S$	$Co_2(Cp)_2(\mu\text{-}PEt_2)_2(\mu\text{-}SO_2)$		213
294	$Co_2C_{20}H_{30}S_4$	$\{Co(Cp^*)\}_2(\mu\text{-}\eta^1:\eta^2\text{-}S_2)_2$		214
295	$[Co_2C_{23}H_{40}N_2P_2S_2]^{2+}$	$[\{Cp(Me_3P)Co(S)CMeNMe\}_2C][PF_6]_2$.Sv	253	215
296	$Co_2C_{24}H_{29}O_6P$	$Co_2(CO)_6(\mu\text{-}PC_6H_2Bu^t\text{-}2,4,6)$		216

No.	Formula	Structure	Details	Ref.
297	$[Co_2C_{25}H_{27}P_2]^+$	$[Co_2(Cp)_2(\mu-H)\{\mu-PhP(CH_2)_3PPh\}][BF_4]$	150	213
298	$Co_2C_{25}H_{62}P_6$	$\{Co(PMe_3)_3\}_2(\eta^2,\eta^{2'}-nbd)$		217
299	$[Co_2C_{29}H_{58}O_4P_6]^{2+}$	$[\{Co(CO)_2(PEt_2CH_2CH_2)_2P\}_2CH_2][PF_6]_2$		218
300	$Co_2C_{30}H_{39}O_2P$	$\{Co(CO)(Cp)\}_2PC_6H_2Bu^t_3$		219
301	$Co_2C_{34}H_{30}P_2$	$\{Co(Cp)(\mu-PPh_2)\}_2$		209
302	$Co_2C_{39}H_{40}I_4P_2$	$(CpCoI_2)_2\{\mu-Ph_2P(CH_2)_5PPh_2\}$		189
303	$Co_2C_{52}H_{42}O_4P_4$	$\{Co(CO)_2(PHPh_2)(\mu-PPh_2)\}_2$		220
304	$Co_2CrWC_{16}O_{16}P_2$	$\{Co_2(CO)_6\}\{\mu-\eta^2-(CO)_5CrPPW(CO)_5\}$		221
305	$Co_2Cr_2C_{26}H_{48}O_6P_2S_4$	$(Cp^*)_2Cr_2Co_2\{P(OMe)_3\}_2(\mu_3-S)_4$		222
306	$Co_2FeC_9Cl_4N_3O_9P_3$	$Co_2Fe(CO)_9(\mu_3-P_3N_3Cl_4)$		196
307	$Co_2FeMoC_{25}H_{22}O_7$	$(Cp)_3(Co_2MoFe(CO)_5(\mu_4-CCOOPr^i)$		223
308	$Co_2GeMoWC_{21}H_{10}O_{11}$	$CpMoCo_2(CO)_8\{\mu_3-GeW(CO)_3Cp\}$		197
309	$Co_2IrC_{22}H_{25}O_2$	$(Cp^*)IrCo_2(\mu-CO)_2(Cp)_2$		224
310	$Co_2MnC_{17}H_{13}O_5$	$(CO)_3MnCo_2(\mu-CO)_2(Cp)_2(\mu_3-CMe)$		208
311	$Co_2MoC_{20}H_{18}O_3$	$MoCo_2(CO)_2(\mu-CO)(Cp)_3(\mu_3-CMe)$		208
312	$Co_2Mo_2C_{16}H_{26}N_4O_2S_8$	$(CO)_2Co_2Mo_2(detc)_2(NCMe)_2(\mu_3-S)_4$		225
313	$Co_2NiOs_3C_{36}H_{25}O_{14}P$	$(Cp)NiOs_3(CO)_8PPh_2CCPr^i\{\eta^2-Co_2(CO)_6\}$		226
314	$Co_2ReC_{18}H_7O_{10}$	$Co_2Re(CO)_{10}\{\mu_3-C(pTol)\}$		227
315	$Co_2ReWC_{23}H_7O_{15}$	$\{(CO)_6Co_2WRe(CO)_9\{\mu_3-C(pTol)\}$		227
316	$Co_2RhRuC_{21}H_{10}O_{10}P$	$Co_2RhRu(CO)_{10}(Cp)(\mu_4-PPh)$		228
317	$Co_2Rh_2C_{37}H_{70}O_{21}P_6$	$[CpCo\{\mu-(EtO)_2PO\}_3Rh]_2(\mu-CO)_3$		229
318	$Co_2RuC_{11}O_{11}$	$RuCo_2(CO)_{10}(\mu-CO)$		230
319	$Co_2RuC_{15}H_{10}O_9$	$Co_2Ru(CO)_9(\mu_3-\eta^2-CCHBu^t)$		231
320	$Co_2RuC_{15}H_{10}O_9$	$Co_2Ru(CO)_8(\mu-CO)(\mu_3-\eta^2-HCCBu^t)$		231
321	$Co_2RuC_{17}H_6O_9$	$Co_2Ru(CO)_9(\mu_3-\eta^2-CCHPh)$		231
322	$Co_2RuC_{23}H_{10}O_9$	$Co_2Ru(CO)_8(\mu-CO)(\mu_3-\eta^2-PhCCPh)$		231
323	$Co_2Ru_2C_{13}O_{13}$	$Co_2Ru_2(CO)_9(\mu-CO)_4$		230
324	$Co_2Si_2C_{16}H_{25}O_4P$	$Co_2(CO)_4(Cp)\{\mu-PHCH(SiMe_3)_2\}$		232
325	$Co_2WC_{25}H_{19}O_6P$	$(CO)_4Co_2W(CO)_2(Cp)(\mu-H)(\mu-PPh_2)(\mu_3-CMe)$		233
326	$Co_3C_{13}H_5O_9$	$Co_3(CO)_9(\mu_3-CCHCH_2CH_2)$		234

No.	Formula	Structure	Details	Ref.
327	$Co_3C_{15}H_{15}N_2O_2$	$Co_3(Cp)_3(\mu_3-NO)_2$		235
328	$Co_3C_{18}H_{21}N_2O_2$	$Co_3(Cp')_3(\mu_3-NO)_2$	213	235
329	$Co_3C_{25}H_{21}F_{12}S_2$	$Cp_2Co_2(\mu\text{-SMe})(\mu_3\text{-SMe})Co\{C_4(CF_3)_4\}(Cp)$	153	236
330	$Co_3C_{26}H_{18}O_7PS_3$	$Co_3(CO)_6(PPh_3)(\mu\text{-}S_2COMe)(\mu_3\text{-}S)$		237
331	$Co_3C_{34}H_{25}O_7P_2$	$Co_3(CO)_7(\mu\text{-dppm})(\mu_3\text{-CMe})$	113	238
332	$Co_3C_{42}H_{30}O_6P_3$	$Co_3(CO)_6(\mu\text{-PPh}_2)_3\cdot MeCN$		239
333	$Co_3FeGeC_{16}H_5O_{11}$	$Co_3(CO)_9\{\mu_3\text{-GeFe}(CO)_2(Cp)\}$		240
334	$Co_3Ge_2NiC_{21}H_{23}O_8$	$CpNiCo(CO)_7(\mu\text{-CO})(\mu_4\text{-GeBu}^t)_2$		240
335	$[Co_3Ni_7C_{18}O_{16}]^{2-}$	$[PPh_4]_2[Co_3Ni_7C_2(CO)_6(\mu\text{-CO})_{10}]$		241
336	$Co_3RuC_{46}H_{31}O_{10}P_2$	$RuCo_3H(CO)_7(PPh_3)_2(\mu\text{-CO})_3\cdot CH_2Cl_2$		242
337	$Co_3Si_2C_{16}H_{19}O_9P$	$Co_3(CO)_9\{\mu_3\text{-PCH}(SiMe_3)_2\}$		232
338	$Co_3WC_{14}O_{14}P$	$(CO)_5W(\mu_4\text{-P})Co_3(CO)_9$		243
339	$Co_4C_{12}H_2O_{10}$	$Co_4(CO)_8(\mu\text{-CO})_2(\mu_4\text{-}\eta^2\text{-HCCH})$		244
340	$Co_4C_{19}H_{22}F_{16}N_4P_{10}$	$Co_4(CO)_3\{\mu\text{-}(F_2P)_2NMe\}_4(\mu_4\text{-PPh})_2$		245
341	$Co_4C_{22}H_{20}O_2$	$Co_4(Cp)_4(\mu_3\text{-CO})_2$		246
342	$Co_4C_{29}H_{14}O_9$	$Co_4(CO)_9(\eta^6\text{-triptycene})$		247
343	$Co_4C_{36}H_{25}NO_{14}P_2$	$Co_4(CO)_7(\mu\text{-CO})_3\{\mu\text{-}(PhO)_2PNEtP(OPh)_2\}$		248
344	$Co_4Si_3C_{21}H_{30}O_{10}P_2$	$Co_4(CO)_8(\mu\text{-CO})_2(\mu_4\text{-PCH}_2SiMe_3)(\mu_4\text{-PCH}(SiMe_3)_2)$		232
345	$Co_4Si_3C_{22}H_{18}O_{12}$	$\{Co_2(CO)_6\}_2(\eta^2:\eta^{2'}\text{-Me}_3SiC_4SiMe_3)$	113	249
346	$Co_4Zn_2C_{15}O_{15}$	$Co_2(CO)_7\{\mu\text{-ZnCo}(CO)_4\}_2$		250
347	$[Co_6C_{16}O_{15}]^{2-}$	$[NMe_3Bz]_2[Co_6C(CO)_6(\mu\text{-CO})_9]$		251
348	$Co_6Ge_2C_{19}O_{19}$	$Co_4(CO)_{10}(\mu\text{-CO})\{\mu_4\text{-GeCo}(CO)_4\}_2$		252
349	$Co_6HgRu_2C_{24}O_{24}$	$Hg\{RuCo_3(CO)_9(\mu\text{-CO})_3\}_2$		253
350	$Co_6Hg_9C_{18}O_{18}$	$Hg_9Co_6(CO)_{18}\cdot 2Me_2CO$		254
351	$Co_6Si_2C_{32}H_{18}O_{18}$	$\{Co_2(CO)_6\}_3(\eta^2:\eta^{2'}:\eta^{2''}\text{-Me}_3SiC_8SiMe_3)$	114	249
352	$[Co_{13}C_{26}O_{24}]^{3-}$	$[NMe_3Bz]_3[Co_{13}C_2(CO)_{12}(\mu\text{-CO})_{12}]$		255
353	$[CrC_9H_4O_3S_2]^{2-}$	$[NEt_4]_2[Cr(CO)_3(S_2C_6H_4\text{-}\underline{o})]$		256
354	$CrC_{10}H_{10}N_2O_5S$	$Cr(CO)_5(SCNMeCH_2CH_2NMe)$		257
355	$CrC_{11}H_{10}O_5$	$Cr(CO)_3\{\eta^6\text{-}C_6H_4(OMe)_2\text{-}\underline{o}\}$		258
356	$CrC_{11}H_{10}O_8$	$Cr\{C(OMe)C(OMe)=CHOMe\}(CO)_5$		259

Structures Determined by Diffraction Methods   435

No.	Formula	Structure	Details	Ref.
357	$CrC_{11}H_{12}O_4S_4$	$(CO)_4Cr\{SC(SMe)CMeS(CH_2)_3S\}$		260
358	$CrC_{11}H_{14}ClNO_4$	trans-$CrCl(CO)_4(CNPr^i_2)$	233	261
359	$[CrC_{12}H_{12}]^+$	$[Cr(\eta\text{-PhH})_2]_2[S_2O_6 \cdot 2SO_2]$		262
360	$CrC_{12}H_{19}O_2P$	$Cr(CO)_2(PMe_3)(\eta^4,\eta CH\text{-}C_7H_{10})$		263
361	$CrC_{12}H_{21}O_5P$	$Cr(CO)_2\{P(OMe)_3\}(\eta^4H\text{-}CH_3CMeCHCMe_2)$		264
362	$CrC_{12}H_{27}O_{11}P_3S$	$Cr(CO)_2(CS)\{P(OMe)_3\}_3$		265
363	$CrC_{14}H_{17}N$	$Cr(C_6H_4CH_2NMe_2\text{-}\underline{o})(Cp)$	100	266
364	$CrC_{14}H_{32}O_2P_4$	cis-$Cr(CO)_2(dmpe)_2$		267
365	$CrC_{14}H_{38}P_4$	$CrMe_2(dmpe)_2$		268
366	$CrC_{15}H_{16}O_6$	$Cr(CO)_3\{\eta^6\text{-}C_6H_4C_7H_5(OH)(OMe)_2\}$		269
367	$CrC_{16}H_8N_2O_4$	$Cr(CO)_4(phen)$		270
368	$CrC_{16}H_{10}O_4$	$Cr(CO)_3(\eta^6\text{-}C_6H_5COPh)$		271
369	$CrC_{16}H_{14}$	$Cr(\eta\text{-PhH})(\eta^6\text{-napH})$		272
370	$CrC_{17}H_{15}NO_3$	$Cr(CO)_3\{\eta^5\text{-}[2,2](2,5)\text{pyrroloparacyclophane}\}$		273
371	$CrC_{17}H_{15}NO_3$	$Cr(CO)_3\{\eta^6\text{-}[2,2](2,5)\text{pyrroloparacyclophane}\}$		273
372	$CrC_{17}H_{23}NO_6$	$Cr\{C(OEt)=N\text{-}CBu^t_2\}(CO)_5$		274
373	$[CrC_{18}H_8NO_6]^+$	$[Cr(CO)_5\{CNC(C_6H_4\text{-}\underline{o})_2O\}][BF_4]$	223	275
374	$CrC_{18}H_{14}NO_4PS$	$Cr(CO)_4\{PPh_2C(S)NHMe\} \cdot THF$		276
375	$CrC_{18}H_{21}NO_4S$	$Cr(CO)_2(NO)\{\eta^1,\eta^3\text{-}CHS(O)MeCHCPhCHCHBu^t\}$		277
376	$CrC_{19}H_{20}O_3$	$Cr(CO)_3(\eta^4,\eta^2\text{-}C_{16}H_{20})$		278
377	$CrC_{21}H_{27}N_3O_3$	$Cr(CO)_3\{\eta^6\text{-}C_6H_3(CC_4H_8)_3\text{-}1,3,5\}$	120	279
378	$CrC_{22}H_{15}NO_5$	$Cr(CO)_5(CNMeCHPhC=CHPh)$		280
379	$CrC_{22}H_{24}NO_4P$	$Cr(CO)_4(PPh_2CH_2CH_2NEt_2)$		281
380	$CrC_{22}H_{26}O_6$	$Cr(CO)_3\{\eta^6\text{-}C_6H_3(OMe)(C_{12}H_{20}O_2)\}$ (2 isomers)		282
381	$CrC_{22}H_{29}N_3O_3$	$Cr(CO)_3\{\eta^6\text{-}C_6H_2Me(NC_4H_8)_3\text{-}2,4,6\}$	120	279
382	$CrC_{23}H_{14}O_3$	$Cr(CO)_3(\eta^6\text{-triptycene})$		247
383	$CrC_{23}H_{16}O_3$	$Cr(CO)_3\{\eta^6\text{-}(C_6H_4)_2CHCH_2Ph\}$		283
384	$CrC_{24}H_{21}NO_5S$	$Cr(CO)_5\{S=C(C_6H_4\text{-}\underline{o})CHCMeNPr^i\}$		284
385	$CrC_{25}H_{24}O_7P_2$	trans-$Cr(CO)_4(PPh_3)\{P(OMe)_3\}$		285
386	$CrC_{32}H_{40}O_5P_2$	$Cr(CO)_5\{\underline{Z}\text{-}P(Mes)=P(C_6H_2Bu^t_3\text{-}2,4,6)\}$		286

No.	Formula	Structure	Details	Ref.
387	$CrC_{34}H_{36}P_2$	$Cr(dmpe)(\eta\text{-PhCCPh})_2$	275	267
388	$CrC_{34}H_{42}O_4P_2$	trans-$Cr(CO)_4(PPh_3)(PBu_3)$		285
389	$CrC_{39}H_{58}O_3P_2$	$Cr(CO)_3(\eta^6\text{-}C_6H_2Bu^t_3PPC_6H_2Bu^t_3)$		287
390	$CrC_{40}H_{30}O_7P_2$	trans-$Cr(CO)_4(PPh_3)\{P(OPh)_3\}$		285
391	$[CrC_{40}H_{49}N_4OP_2]^+$	mer-$[Cr(NO)(CNBu^t)_3(dppm)][PF_6]$		288
392	$CrC_{42}H_{35}O_3P_3$	$Cr(CO)_3(CH_2PPh_2CHPPh_2CH_2PPh_2)\cdot PhH$		289
393	$[CrC_{44}H_{30}O_2]^-$	$[K(db\text{-}18\text{-}c\text{-}6)][Cr(CO)_2(\eta\text{-PhCCPh})(\eta^4\text{-}C_4Ph_4)]\cdot THF$		290
394	$CrFeC_{16}H_{28}O_6P_4$	$FeCr(CO)_6(\mu\text{-dmpm})_2\cdot PhMe$	265	291
395	$CrFeC_{17}H_{20}NO_6P$	$(Cp)(CO)FeCr(CO)_4(\mu\text{-CO})(\mu\text{-PHNPr}^i_2)$		292
396	$CrFeC_{28}H_{21}O_5P$	$Fe(\eta\text{-}C_5H_4PMePh_2)\{\eta\text{-}C_5H_4Cr(CO)_5\}$		293
397	$CrSeC_{12}H_{27}O_{11}P_3$	mer-$Cr(CO)_2(CSe)\{P(OMe)_3\}_3$	118	294
398	$CrSiC_{14}H_{18}O_5$	$Cr(CO)_3\{\eta^6\text{-}C_6H_3(SiMe_3)(OMe)_2\text{-}1,2,3\}$		295
399	$CrSiC_{16}H_{12}O_3$	$Cr(CO)_3\{\eta^6\text{-}C_{10}H_6(SiMe_2CH_2)\text{-}1,8\}$		296
400	$CrSiC_{17}H_{14}O_3$	$Cr(CO)_3\{\eta^6\text{-}(C_6H_4)_2SiMe_2\text{-}2,2'\}$		296
401	$CrSiC_{19}H_{20}O_4$	$Cr(CO)_3\{\eta^6\text{-}C_6H_5CHPh(OSiMe_3)\}$		279
402	$CrTeC_{20}H_{36}O_4P_2$	$Cr(CO)_4(PBu^t_2TePBu^t_2)$		297
403	$Cr_2C_{15}H_{12}O_3$	$Cr_2(\eta\text{-PhH})_2(\mu\text{-CO})_3$		298
404	$Cr_2C_{16}H_{20}O_8S_2$	$\{Cr(CO)_4\}_2(\mu\text{-SEt}_2)_2$		299
405	$Cr_2C_{17}H_5O_{10}P$	$\{Cr(CO)_5\}_2(\mu\text{-PO}_2C_7H_5)$	223	300
406	$Cr_2C_{19}H_{11}O_{10}P$	$\{Cr(CO)_5\}_2PMes$	223	301
407	$Cr_2C_{20}H_{30}O_4$	$\{CrO(Cp^*)(\mu\text{-O})\}_2$		302
408	$Cr_2C_{22}H_{10}O_{10}P_2S$	$\{Cr(CO)_5\}_2(\mu\text{-PhPSPPh})$	213	303
409	$Cr_2C_{22}H_{12}O_{11}P_2$	$\{Cr(CO)_5\}_2\{\mu\text{-PhPhP(OH)Ph}\}$	213	303
410	$Cr_2C_{23}H_{12}O_{10}P_2$	$\{Cr(CO)_5\}_2(\mu\text{-PhPCH}_2PPh)$	270	303
411	$Cr_2C_{24}H_{30}O_4S_2$	$\{Cr(CO)_2(Cp^*)\}_2(\mu\text{-}\eta^2\text{-}S_2)$		304
412	$Cr_2C_{26}H_{21}NO_{10}P_2$	$\{Cr(CO)_5\}_2(\mu\text{-PHPhPPhNHBu})$	243	303
413	$Cr_2C_{28}H_{20}O_{10}P_2$	$\{Cr(CO)_5\}_2(\mu\text{-PhPCH}_2CHMeCHMeCH_2PPh)$	228	303
414	$Cr_2C_{32}H_{48}$	$(Cp^*)_2Cr_2\{\mu\text{-}\eta^3,\eta^{3'},\eta^1,\eta^{1'}\text{-}(CMe)_6\}$	100	266
415	$Cr_2C_{34}H_{20}O_{11}P_2$	$\{Cr(CO)_5\}_2(\mu\text{-Ph}_2POPPh_2)$		305
416	$Cr_2C_{34}H_{40}O_8$	$\{Cr(CO)_3\}_2\{\eta^6:\eta^{6'}\text{-}(3,5\text{-Pr}^i_2C_6H_3)_2C_2(OMe)_2\text{-}1,2\}$		306

Structures Determined by Diffraction Methods 437

No.	Formula	Structure	Details	Ref.
417	$Cr_2C_{42}H_{58}O_6P_2$	$\{(\eta^6-C_6H_2Bu^t_3)PCr(CO)_3\}_2-(\underline{P}-\underline{P})$		287
418	$Cr_2Fe_2Sb_2C_{24}H_{10}Br_2O_{14}$	$Sb_2Br_2\{Cr(CO)_5\}_2\{Fe(CO)_2(Cp)\}_2$	228	307
419	$Cr_2SbC_{14}H_8ClO_{11}$	$\{Cr(CO)_5\}_2SbCl(THF)$	208	38
420	$Cr_2Sb_2C_{18}H_{18}Cl_2O_{10}$	$Sb_2Cl_2Bu^t_2\{Cr(CO)_5\}_2$		307
421	$Cr_2SeC_{14}H_{10}O_4$	$\{Cr(CO)_2(Cp)\}_2Se$		304,308
422	$Cr_2Se_2C_{14}H_{10}O_4$	$\{Cr(CO)_2(Cp)\}_2(\mu-Se_2)$		308
423	$Cr_2WC_{17}H_5O_{12}P$	$(CO)_5CrW(CO)_2(Cp)\{\mu-PCr(CO)_5\}$	233	39
424	$Cr_3C_{22}H_{27}O_{10}P_3$	$Cr_3(CO)_9(\mu-CO)(\mu_3-PBu^t)(\mu_3-\eta^2-P_2Bu^t_2)$		309
425	$[Cr_3SbC_{15}O_{15}]^-$	$[Na(THF)][\{Cr(CO)_5\}_3Sb]$	203	39
426	$Cr_3Sb_2C_{23}H_{18}O_{15}$	$\{(CO)_5SbBu^t\}_2\{\mu-Cr(CO)_5\}$		310
427	$[Cr_4C_{12}H_4O_{15}]^{4-}$	$[NEt_4]_4[Cr_4(CO)_{12}(\mu_3-OH)_4]$		311
428	$Cr_4C_{25}H_{24}O_3$	$(Cp)_4Cr_4(\mu_3-O)_3(\mu_3-\eta^2-C_5H_4)$		312
429	$[Cr_4C_{40}H_{66}O_6]^{2+}$	$[(Cp^*)_4Cr_4(\mu-OH)_6][BF_4]_2$		313
430	$[CuC_2H_6]^-$	$Li(12\text{-crown-}4)_2][CuMe_2]$	140	314
431	$[CuC_{12}H_{10}]^-$	$[Li(12\text{-crown-}4)_2][CuPh_2].THF$	140	314
432	$[CuC_{16}H_{19}N_3]^+$	$[Cu(\eta^2-C_6H_{10})\{(NC_5H_4-2)_2NH\}][ClO_4]$	173	315
433	$CuC_{17}H_{16}IN_2O$	$CuI(CO)(BzNCH_2CH_2NBz)$		316
434	$CuC_{24}H_{22}P$	$Cu(PPh_3)(Cp')$		317
435	$CuC_{24}H_{23}IN_3$	$CuI\{CN(pTol)\}(BzNCH_2CH_2NBz)$		316
436	$CuFe_3C_{30}H_{18}O_{10}P$	$(PPh_3)CuFe_3(CO)_9(\mu-CO)(\mu_3-CMe)$		318
437	$[CuFe_4C_{31}H_{15}O_{13}P]^-$	$[PPN][Fe_4(CO)_{12}(\mu-CO)(\mu_3-CuPPh_3)]$		74,75
438	$CuMoC_{14}H_{21}O_3$	$(TMED)CuMo(CO)_3(Cp)$		319
439	$[CuSi_2C_7H_{19}Br]^-$	$[Li(12\text{-crown-}4)_2][CuBrCH(SiMe_3)_2].PhMe$	140	314
440	$CuTeC_2H_5Cl$	$[Et_2TeCuCl]_n$		320
441	$[CuTiC_{38}H_{38}P_2S_2]^+$	$[(Cp)_2Ti(\mu-S)_2Cu\{P(CH_2CH_2)Ph_2\}_2][BF_4]$		321
442	$CuWC_{53}H_{44}ClO_3P_4$	$CuW(CO)_2(\mu-Cl)(\mu-CO)(\mu-dppm)_2.CH_2Cl_2$		322
443	$Cu_2C_{18}H_{10}F_{12}O_4$	$\{Cu(hfacac)\}_2(\mu-\eta^2:\eta^2{'}-cot)$		323
444	$Cu_2C_{32}H_{40}N_4O_4$	$\{Cu_2(\mu-\eta^2-DMAD)\}(C_{26}H_{36}N_4)$		324
445	$Cu_2C_{34}H_{44}N_4O_4$	$\{Cu_2(\mu-\eta^2-EtO_2CC\equiv CCO_2Et)\}(C_{26}H_{36}N_4)$		324
446	$Cu_2FeC_{76}H_{60}O_4P_4$	$Fe(CO)_4\{Cu(PPh_3)_2\}_2$		325

No.	Formula	Structure	Details	Ref.
447	$Cu_2Fe_4C_{52}H_{52}O_{10}$	$(THF)_2Cu_2\{\mu-O_2C(Fc)\}_4 \cdot THF$		326
448	$Cu_2Li_2C_{36}H_{48}N_4$	$Cu_2Li_2(C_6H_4CH_2NMe_2-\underline{o})_4$		327
449	$Cu_2Mn_4C_{28}H_{46}O_{22}P_4$	$Cu_2Mn_4(CO)_{12}(\mu-H)_4(\mu_3-H)_2\{\mu-(EtO)_2POP(OEt)_2\}_2$		328
450	$Cu_3C_{23}H_{21}O_4$	$Cu_3(\mu-O_2CPh)_2(\mu-Mes)$		329
451	$[Cu_3Fe_3C_{12}O_{12}]^{3-}$	$Na_3[Cu_3Fe(CO)_{12}]$		330
452	$[Cu_3Ir_2C_{54}H_{81}N_3P_6]^{3+}$	$[Ir_2Cu_3(\mu-H)_6(NCMe)_3(PMe_2Ph)_6][PF_6]_3$	115	331
453	$Cu_4C_{104}H_{80}P_4$	$\{Ph_3PCu(\mu_3-C\equiv CPh)\}_4$		332
454	$[Cu_4LiC_{36}H_{30}]^-$	$[Li(OEt_2)_4][Cu_4Li(\mu-Ph)_6] \cdot 2Et_2O$		333
455	$Cu_4MgC_{40}H_{40}O$	$(Et_2O)MgCu_4(\mu-Ph)_6 \cdot Et_2O$		333
456	$[Cu_5Fe_4C_{16}O_{16}]^{3-}$	$Na_3[Cu_5Fe_4(CO)_{16}]$		330
457	$[Cu_6Fe_4C_{16}O_{16}]^{2-}$	$[M]_2[Cu_6Fe_4(CO)_{16}]$ {M=Cu(dppe)$_2$, Na}		325,330
458	$ErLiC_{18}H_{32}N_2$	$(TMED)Li(\mu-Me)_2Er(Cp)_2$	138	334
459	$FeC_2N_2O_4$	$Fe(CO)_2(NO)_2$	E	335
460	$FeC_7H_5ClO_2$	$FeCl(CO)_2(Cp)$		336
461	$[FeC_8H_5Cl_2O_2]^+$	$[Fe(CCl_2)(CO)_2(Cp)][BCl_4]$	133	337
462	$FeC_8H_7F_7IO_3PS$	$FeI(C_3F_7)(SPMe_2)(CO)_3$	193	338
463	$FeC_{10}H_9N_3O_2S$	$(Fc)SO_2N_3$		339
464	$FeC_{10}H_{10}F_7O_3PS$	$Fe(C_3F_7)(CO)_3(\eta^2-SPEt_2)$	173	338
465	$FeC_{11}H_{10}O_2$	$(Fc)CO_2H$		340
466	$[FeC_{12}H_{17}N_2O_6S_3]^+$	$[(Fp)(\eta^2-CH_2CHCH_2N(SO_2Me)SNHSO_2Me)][PF_6]$		341
467	$FeC_{12}H_{19}Cl_2N_2$	$FeCl_2\{C_6H_3(CH_2NMe_2)_2-2,6\}$		342
468	$FeC_{13}H_{17}$	$Fe(cod)(Cp)$	100	266
469	$[FeC_{13}H_{33}P_3]^+$	$[Fe(PMe_3)_3\{\eta^4-C(CH_2)_3\}][tfo]$	128	343
470	$FeC_{16}H_{12}O_8$	$Fe(CO)_3\{\eta^4-C_7H_8(C_4HO_3CO_2Me)\}$		344
471	$FeC_{16}H_{18}O_3$	$Fe(CO)_2\{\eta^3,\eta^2-H_2CCHCHCH_2C(pTol)OEt\}$		345
472	$FeC_{16}H_{18}O_3$	$Fe(CO)_2\{\eta^3,\eta^3-H_2CCHCHCH_2C(OEt)(oTol)\}$		345
473	$FeC_{16}H_{19}O_2$	$(Fp)(\eta^2-(Z)-bicyclo[3.3.1]C_9H_{14})$		346
474	$FeC_{16}H_{22}O_4$	$Fe(CO)_4(\eta^4-C_5Me_5Bz)$		347
475	$FeC_{16}H_{24}ClO_2P$	$Fe(PClBu^t)(CO)_2(Cp^*)$		348
476	$[FeC_{16}H_{26}N_2]^{2+}$	$[Fe(\eta-C_5H_4CH_2NHMe_2)_2][B_{12}H_{12}]$		349

No.	Formula	Structure	Details	Ref.
477	$FeC_{17}H_{10}O_4$	$Fe(CO)_3(\eta^4-OCCPh_2)$	N at 15	350
478	$FeC_{17}H_{20}O_2$	$Fe(CO)_2(\eta^1-Cp)(\eta^5-Cp^*)$		351
479	$FeC_{17}H_{20}O_7$	$Fe(CO)_3\{\eta^4-C_6H_6(CH_2CH_2CO_2Me)_2-1,2\}$		352
480	$[FeC_{17}H_{23}]^+$	$[Fe(Cp)(\eta-C_6Me_6)][(TCNQ)_2]$		353
481	$[FeC_{17}H_{26}O_3P]^+$	$[Fe\{C(OMe)=CMePEt_3\}(CO)_2(Cp)][SO_3F]$		354
482	$[FeC_{18}H_{15}O]^+$	$[Fe(Cp)(\eta^6-xanthene)][PF_6]$	163	355
483	$[FeC_{18}H_{15}S_2]^+$	$[Fe(Cp)(\eta^6-C_6H_4S_2C_6H_3Me)][PF_6]$	163	355
484	$FeC_{19}H_{20}Br_2$	$Fe\{\eta,\eta'-(C_5HBr)_2(CH_2CH_2CH_2)_3-2,3,5\}$		356
485	$[FeC_{19}H_{22}]^+$	$[Fe\{\eta:\eta'-(C_5H_2)_2(CH_2CH_2CH_2)_3-1,2,4\}][ClO_4]$		357
486	$FeC_{20}H_{31}N_2O_{10}P$	$\{P(OMe)_3\}(CO)_2Fe\{CRCRCH(CHNPr^i)NPr^iCO\}$ R=CO_2Me		358
487	$[FeC_{22}H_{16}O_7P]^-$	trans-$[NEt_4][Fe(CHO)(CO)_3\{P(OPh)_3\}]$	168	359
488	$FeC_{22}H_{19}NO$	$Fe(Cp)(\eta^5-NC_4H_3CPh_2OH-2)$		360
489	$FeC_{23}H_{14}N_4O_4$	$Fe(CO)_3(\eta^4-C_{20}H_{14}N_4O)$		361
490	$[FeC_{23}H_{18}O_7P]^-$	trans-$[NEt_4][Fe(Ac)(CO)_3\{P(OPh)_3\}]$	203	359
491	$FeC_{23}H_{19}O_3P$	$Fe(CO)_3(PMe_3)(\eta-C_2H_4)$	193	362
492	$[FeC_{23}H_{25}]^+$	$[Fe(Cp)\{\eta^6-MeC_6H_3(CH_2CH_2)_2C_6H_3Me\}][PF_6]$		363
493	$FeC_{24}H_{21}O_2PS_4$	$Fe\{S_2C_2(SMe)_2\}(CO)_2(PPh_3)$		364
494	$FeC_{24}H_{25}O_2P$	$Fe(PPh_2)(CO)_2(Cp^*)$		365
495	$FeC_{25}H_{19}N_3O_6$	$Fe(CO)_3(\eta^4-C_{22}H_{19}N_3O_3)$		366
496	$[FeC_{25}H_{20}F_2OP]^+$	$[Fe(CF_2)(CO)(PPh_3)(Cp)][BF_4]$	233	337
497	$FeC_{25}H_{23}O_3P$	$Fe(CH_2CH_2CH_2CH_2)(CO)_3(PMe_3)$	183	362
498	$FeC_{26}H_{32}F_7O_7PS$	$Fe(C_3F_7)(CO)_3C(CO_2Cy)C(CO_2Cy)PS$	193	338
499	$FeC_{29}H_{29}O_3P$	$Fe\{C(O)CH_2CH_2CH(OH)Me\}(CO)(PPh_3)(Cp)$		367
500	$FeC_{29}H_{29}O_3P$	$Fe\{C(O)CH_2CH(OH)Et\}(CO)(PPh_3)(Cp)$		368
501	$FeC_{32}H_{24}O_3$	$Fe(CO)_3(\eta^1,\eta^3-C_{11}H_9Ph_3)$		369
502	$FeC_{34}H_{42}$	$Fe\{\eta-C_5H_2Me_2(C_6H_2Me_2Et-2,6,4)\}_2$		370
503	$[FeC_{38}H_{30}NO_3P_2]^+$	$[Fe(CO)_2(NO)(PPh_3)_2][BF_4] \cdot CH_2Cl_2$		371
504	$FeC_{38}H_{30}O_{10}P_2S$	$Fe(CO)_2(SO_2)\{P(OPh)_3\}_2$	213	372
505	$FeC_{38}H_{70}NP_3$	$Fe\{\eta-C_5H_3(PBu^t_2)_2-1,3\}\{\eta-C_5H_3(CHMeNMe_2)(PBu^t_2)\}$		373
506	$[FeC_{42}H_{36}P_3]^+$	$[Fe\{(Ph_2P)_3CH\}(Cp)][PF_6]$		374

No.	Formula	Structure	Details	Ref.
507	$[FeC_{42}H_{66}N_6O_6]^{2+}$	$[Fe(CNCMe_2CH_2COMe)_6][BF_4]_2$		375
508	$FeC_{44}H_{48}P_2$	$Fe(\eta^5-PhCCHCBu^tCHCPhPMe_2)_2$		376
509	$FeGeSe_2C_{10}H_8Cl_2$	$Fe\{\mu:\eta'-(C_5H_4Se)_2GeCl_2\}$	163	377
510	$FeMnC_{16}H_{15}N_2O_6$	$(CO)_3FeMn(CO)_3(\mu-H)(\mu-NBu^tCHC_5H_4N-2)$		378
511	$FeMnC_{22}H_{19}N_2O_6$	$FeMn(CO)_6(\mu-H)\{\mu,\mu'-(pTol)NCH_2CH_2N(pTol)\}$		379
512	$FeMn_2C_{24}H_{16}O_7$	$Mn_2(CO)_5(Cp')(\mu-CO)\{\mu-C(CO)(Fc)\}$	233	380
513	$FeMoC_{38}H_{28}O_4P_2$	$Fe\{\eta,\eta'-(C_5H_4PPh_2)_2Mo(CO)_4\}$		381
514	$FeMoNiC_{27}H_{22}O_7$	$(Cp)NiFeMo(CO)_5(Cp)(\mu_3-\eta^2-PhCCCO_2Pr^i)$		198
515	$FeMoTe_2C_{10}H_5BrO_5$	$(CO)_3FeMo(CO)_2(Cp)(\mu-\eta^2-Te_2Br)$		382
516	$FeMoTe_2C_{16}H_{17}NO_5S_2$	$(CO)_3FeMo(CO)_2(Cp')Te_2(\eta^1-detc)$		382
517	$FeMo_2C_{22}H_{30}O_2S_4$	$(Cp^*)_2Mo_2Fe(CO)_2(\mu_3-S)_2(\mu-\eta^2-S_2)$		383
518	$FeMo_2C_{23}H_{30}O_3S_4$	$(Cp^*)_2Mo_2Fe(CO)_2(\mu_3-S)_2(\mu,\mu'-S_2CO)$		383
519	$FeMo_2Te_2C_{17}H_{10}O_7$	$(Cp)_2Mo_2Fe(CO)_7(\mu_3-Te)_2$		384
520	$FeNiC_{34}H_{28}Br_2P_2$	$Fe\{\eta,\eta'-(C_5H_4PPh_2)_2NiBr_2\}$		381
521	$FeNiC_{44}H_{36}$	$Fe(Cp)\{\eta^5-C_4Ph_4Ni(Cp)\}$		385
522	$FeNi_2C_{25}H_{22}O_5$	$(Cp)_2Ni_2Fe(CO)_3\{\mu_3-\eta^2-PhCCCO_2Pr^i\}$		198
523	$FePdC_{18}H_{26}Cl_2S_2$	$Fe\{\eta,\eta'-(C_5H_4SCH_2Pr^i)_2PdCl_2\}$		386
524	$FePdC_{34}H_{28}Cl_2P_2$	$Fe\{\eta,\eta'-(C_5H_4PPh_2)_2PdCl_2\}$		381
525	$FeRhC_{20}H_{28}O_4P$	$(CO)_3FeRh(PPr^i_3)(Cp)(\mu-CCH_2)(\mu-CO)$		387
526	$[FeRhC_{37}H_{44}P_2]^+$	$[(nbd)Rh\{Ph_2P(C_5H_4-\eta)Fe(\eta-C_5H_4)PBu^t_2\}][ClO_4]$		388
527	$[FeRhC_{37}H_{44}P_2]^+$	$[(nbd)Rh\{PhBu^tP(\eta-C_5H_4)\}_2Fe][ClO_4]$		388
528	$[FeRhC_{41}H_{36}P_2]^+$	$[(nbd)Rh\{Ph_2P(\eta-C_5H_4)\}_2Fe][ClO_4]$		388
529	$FeRuC_{14}H_{10}O_4$	$FeRu(CO)_2(\mu-CO)_2(Cp)_2$		389
530	$FeSbC_{16}H_{27}O_4$	$Fe(CO)_4(SbBu^t_3)$		390
531	$FeSb_2Si_4C_{18}H_{38}O_4$	$Fe(CO)_4[\eta^2-Sb_2\{CH(SiMe_3)_2\}_2]$		391
532	$FeSiC_{13}H_{18}O_3P$	$Fe\{PC(OSiMe_3)Bu^t\}(CO)_2(Cp)$		392
533	$FeWC_{17}H_{20}NO_6P$	$(Cp)(CO)FeW(CO)_4(\mu-CO)(\mu-PHNPr^i_2)$		292
534	$FeWC_{28}H_{21}OP$	$Fe\{\eta,\eta'-C_5H_4PPh_2W(CO)_4C(OMe)C_5H_4\}$		293
535	$Fe_2C_8H_4O_6S_2$	$Fe_2(CO)_6(\mu,\mu'-SCH_2CH_2S)$		393
536	$Fe_2C_{10}H_{10}S_4$	$\{Fe(Cp)\}_2(\mu-S_2)(\mu-\eta^2-S_2)$		394

Structures Determined by Diffraction Methods

No.	Formula	Structure	Details	Ref.
537	$Fe_2C_{11}H_{10}O_6S$	$Fe_2(CO)_6(\mu\text{-SEt})(\mu\text{-CH}_2\text{CHCH}_2)$		395
538	$Fe_2C_{13}H_8O_6S_2$	$Fe_2(CO)_6(\mu,\mu'\text{-}S_2C_7H_8)$ (3 isomers)		396
539	$Fe_2C_{13}H_{12}O_6S$	$Fe_2(CO)_6(\mu\text{-SBu}^t)(\mu\text{-}\eta^1:\eta^2\text{-CH=C=CH}_2)$		395
540	$Fe_2C_{14}H_6F_6O_7$	$Fe_2(CO)_6\{\mu\text{-}\eta^3:\eta^1,\eta^2\text{-C(CF}_3)\text{C(CF}_3)\text{CHCHCHOMe}\}$		397
541	$Fe_2C_{14}H_{10}F_{12}S_4$	$[(Cp)Fe\{\mu\text{-SC(CF}_3)\text{C(CF}_3)\text{S}\}]_2$		394
542	$Fe_2C_{14}H_{18}NO_6P$	$Fe_2(CO)_6(\mu\text{-}\eta^2\text{-Bu}^t\text{PNBu}^t)$		398
543	$Fe_2C_{14}H_{30}Cl_2O_4P_4$	$\{FeCl(CO)_2(PMe_3)(\mu\text{-PMe}_2)\}_2$		399
544	$Fe_2C_{15}H_8O_8S$	$(Cp)(CO)_2WFe_2(CO)_6(\mu_3\text{-SCMe})$		400
545	$Fe_2C_{15}H_{28}O_5P_4$	$Fe_2(CO)_4(\mu\text{-CO})(\mu\text{-dmpm})_2$		291
546	$Fe_2C_{16}H_{12}O_6$	$\{Fe(CO)_3\}_2\{\eta^4,\eta^4{}'\text{-C}_6H_2Me_2(=CH_2)_2\text{-}1,4,2,5\}$		401
547	$Fe_2C_{16}H_{18}NO_7P$	$Fe_2(CO)_7(\mu\text{-}\eta^1,\mu P\text{-OCNBu}^t PBu^t)$		398
548	$Fe_2C_{16}H_{19}O_6P$	$Fe_2(CO)_6(\mu\text{-}\eta^3:\eta^1,P\text{-CHCBu}^t PBu^t)$		402
549	$Fe_2C_{18}H_{14}O_7$	$Fe_2(CO)_6(\mu\text{-COEt})(\mu\text{-}\eta^1:\eta^2\text{-CPhCHMe})$		403
550	$Fe_2C_{19}H_{10}O_6$	$Fe_2(CO)_6(\mu\text{-}\eta^3:\eta^3\text{-PhCHC}_6H_4)$	N at 230	350
551	$Fe_2C_{19}H_{28}N_2O_7P_2$	$Fe_2(CO)_6\{\mu,\mu'\text{-}(Pr^iNP)_2CO\}$		404
552	$Fe_2C_{19}H_{30}N_2O_6P_2$	$Fe_2(CO)_6(\mu\text{-}\eta^2\text{-Pr}^i{}_2\text{NPHCHPNPr}^i{}_2)$		404
553	$Fe_2C_{20}H_{12}O_7S$	$(Cp)(CO)WFe_2(CO)_6\{\mu\text{-C(pTol)}\}(\mu_3\text{-S})$		400
554	$[Fe_2C_{20}H_{16}I_2]^+$	$[\{Fe(\eta\text{-C}_5H_4I)\}_2(\eta^5,\eta^5{}'\text{-C}_{10}H_8)][I_3]$		405
555	$[Fe_2C_{20}H_{18}]^+$	<u>trans</u>-$[(Fc)_2][I_3]$		406
556	$Fe_2C_{20}H_{30}S_4$	$\{Fe(Cp^*)\}_2(\mu\text{-S}_2)(\mu\text{-}\eta^1:\eta^2\text{-S}_2)$		214
557	$Fe_2C_{21}H_{16}O_6P_2$	$Fe_2(CO)_6\{\mu,\mu'\text{-PhP(CH}_2)_3\text{PPh}\}$		407
558	$Fe_2C_{22}H_{16}O_6$	$Fe_2(CO)_6(\eta^4:\eta^4{}'\text{-C}_{16}H_{16})$		408
559	$[Fe_2C_{22}H_{16}O_6]^{2+}$	$[Fe_2(CO)_6(\eta^5:\eta^5{}'\text{-C}_{16}H_{16})][PF_6]_2\cdot MeNO_2$		408
560	$Fe_2C_{22}H_{18}O_6$	$\{Fe(CO)_3\}_2\{\eta^4:\eta^4{}'\text{-}(C_8H_9)_2\}$	273	409
561	$Fe_2C_{22}H_{18}O_9$	$(-)\text{-}\{Fe(CO)_3\}_2(\eta^4,\eta^4{}'\text{-C}_{16}H_{18}O_3)$		410
562	$Fe_2C_{23}H_{19}O_3$	$Fe_2(CO)_3(Cp)_2\{\mu\text{-CCH=CH(pTol)}\}$		411
563	$Fe_2C_{23}H_{30}O_3$	$Fe_2(Cp^*)_2(\mu\text{-CO})_3$	223	412
564	$[Fe_2C_{24}H_{24}]^+$	$[Fe_2\{\eta,\eta'\text{-}(C_5H_4)_2CHMe\}_2][I_3]$		413
565	$Fe_2C_{27}H_{31}O_8P$	$Fe(CO)_4\{\eta^2\text{-CH}_2=P(C_6H_2Bu^t{}_3\text{-}2,4,6)Fe(CO)_4\}$		414
566	$Fe_2C_{28}H_{21}F_3O_8$	$Fe_2(CO)_6\{\mu\text{-}\eta^3:\eta^1\text{-PhCCPhC(CF}_3)\text{CHC(OEt)}_2\}$		415

No.	Formula	Structure	Details	Ref.
567	$Fe_2C_{28}H_{21}O_{10}$	$Fe_2(CO)_5(\mu-\eta^2-PhCCHPh)\{\mu-\eta^3-C(OEt)C_2(CO_2Me)_2\}.H_2O$		397
568	$Fe_2C_{28}H_{22}N_2$	(Fc)CHC(CN)CHCHC(CN)CH(Fc)		416
569	$[Fe_2C_{28}H_{34}]^+$	$\{Fe(\eta-C_5H_4Bu)\}_2(\eta^5,\eta^{5'}-fulv)$	150,RT,363	406
570	$Fe_2C_{29}H_{22}O_3$	$\{Fe(CO)(Cp)\}_2(\mu-\eta^2:\eta^1,\eta^1-CH_2=CCPhCPhCO).2MePh$		417
571	$Fe_2C_{30}H_{24}$	1,8-(Fc)$_2$naphthalene		418
572	$Fe_2C_{31}H_{21}NO_4$	(Fp)CCPhC(CN)(CHPh)(Fp)	153	419
573	$Fe_2C_{31}H_{21}NO_4$	(Fp)C(CHPh)CPhC(CN)(Fp)		419
574	$Fe_2C_{35}H_{20}O_7$	$Fe_2(CO)_6(\eta-PhCCPh)_2(\mu-CO)$		420
575	$Fe_2C_{35}H_{33}O_6PS$	$Fe_2(CO)_6(PPh_3)(\mu-\eta^2:S-SCC_6H_6Me_4)$		421
576	$Fe_2C_{35}H_{35}O_6PS$	$Fe_2(CO)_6(PPh_3)(\mu-\eta^2:S-SCHC_6H_7Me_4)$		73
577	$Fe_2C_{36}H_{35}O_6PS$	$Fe_2(CO)_6(CH_2PPh_3)(\mu-\eta^2:S-SCC_6H_6Me_4).Sv$		421
578	$Fe_2C_{41}H_{40}O_5P_2S$	$Fe_2(CO)_5(\mu-dppm)(\mu-\eta^2:S-SCC_6H_6Me_4).CH_2Cl_2$		421
579	$Fe_2C_{41}H_{42}O_5P_2S$	$Fe_2(CO)_5(dppe)(\mu-\eta^2:S-SCCBu^t_2)$		421
580	$Fe_2C_{44}H_{36}O_4P_2S_4$	$\{Fe(CO)_2(PPh_3)\}_2\{\mu,\mu'-S_2C_2(SMe)_2\}$		364
581	$Fe_2MnSeC_{20}H_{15}O_5$	$Fe_2(CO)_3(Cp)_2\{\mu-SeMn(CO)_2(Cp)\}$		422
582	$Fe_2MnSeC_{21}H_{15}O_6$	$Se\{Mn(CO)_2(Cp)\}(Fp)_2$		422
583	$Fe_2Mo_2C_{24}H_{30}O_4S_4$	$(Cp^*)_2Mo_2Fe_2(CO)_4(\mu_3-S)_4$		383
584	$Fe_2Mo_2TeC_{17}H_{10}O_7$	$(Cp)_2Mo_2Fe_2(CO)_6(\mu-CO)(\mu_3-Te)_2$		384
585	$Fe_2Ni_2C_{30}H_{20}O_6$	$(Cp)_2Ni_2Fe_2(CO)_6(\mu_4-\eta^2-PhCCPh)$		423
586	$Fe_2Ni_2C_{35}H_{25}O_5P$	$(Cp)_2Ni_2Fe_2(CO)_5(\mu-PPh_2)(\mu_4-\eta^2-CCPh)$		424
587	$Fe_2PtC_{20}H_{10}N_4O_8$	trans-Pt{Fe(CO)$_3$(NO)}$_2$(NCPh)$_2$		425
588	$Fe_2PtC_{28}H_{26}Cl_4$	$\{Fe(\eta-C_5H_4Cl)(\eta-C_5H_3Cl)\}_2Pt(cod)$		426
589	$[Fe_2Rh_2C_{52}H_{93}P_4]^+$	$[Rh_2H_2(\mu-H)_3(Bu^tPC_5H_4-\eta)_2Fe\}_2][ClO_4]$		427
590	$[Fe_2Rh_2C_{60}H_{77}P_4]^+$	$[Rh_2H_2(\mu-H)_3\{(PhBu^tPC_5H_4-\eta)_2Fe\}_2][ClO_4].MeOH$		427
591	$Fe_2SbSi_2C_{15}H_{19}O_8$	$Fe_2(CO)_8\{\mu-SbCH(SiMe_3)_2\}$		391
592	$Fe_2Sb_2W_2C_{24}H_{10}Br_2O_{14}$	$Sb_2Br_2\{W(CO)_5\}_2(Fp)_2.CH_2Cl_2$		307
593	$Fe_2Se_2C_{19}H_{16}O_7$	$Fe_2(CO)_5(\eta,\eta'-Se_2C_{14}H_{16}O)$		428
594	$Fe_2Si_4C_{20}H_{36}O_6P_2$	$Fe_2(CO)_6\{\mu-PC(SiMe_3)_2\}_2$		429
595	$Fe_2WC_{43}H_{34}O_6P_2$	$Fe_2W(CO)_5(\mu-PPh_2)_2\{\mu_3-OCHCH_2(pTol)\}(Cp)$		430
596	$Fe_2WC_{56}H_{45}O_7P_3$	$(Ph_2HP)Fe_2W(CO)_6(Cp)(\mu-PPh_2)_2\{\mu-OCCH_2(pTol)\}.Sv$		430

No.	Formula	Structure	Details	Ref.
597	$Fe_3C_9O_9S_2$	$Fe_3(CO)_9(\mu-S)_2$		431
598	$Fe_3C_{11}F_2O_9$	$Fe_3(CO)_9(\mu_3-CF)_2$		432
599	$[Fe_3C_{11}HO_{11}]^-$	$[Cp_2MoH(CO)][Fe_3H(CO)_{11}]$		433
600	$[Fe_3C_{13}H_3O_{11}]^-$	$[PPN][Fe_3(CO)_9(\mu_3-\eta^2-CCOCOMe)]$	173	434
601	$Fe_3C_{14}H_9O_{10}P$	$Fe_3(CO)_9(\mu-CO)(\mu_3-PBu^t)$		435
602	$Fe_3C_{16}H_{18}O_9P_2$	$Fe_3(CO)_9(\mu-H)(\mu-PMePr^i)(\mu_3-PPr^i)$		436
603	$Fe_3C_{17}H_7O_{11}PS_2$	$Fe_2(CO)_6\{\mu,\mu'-S_2P(C_6H_4OMe-\underline{p})Fe(CO)_4\}$		437
604	$Fe_3C_{18}H_{14}O_{12}P_2$	$Fe_3(CO)_8\{P(OMe)_3\}(\mu-CO)(\mu_3-PPh)$		435
605	$Fe_3C_{18}H_{19}O_{11}P_3$	$Fe_2(CO)_7\{\mu-Me_2PCH_2PMeCH_2PMe_2Fe(CO)_4\}.CH_2Cl_2$		438
606	$Fe_3C_{19}H_{19}O_{12}P_3$	$(CO)_4FePMe\{CH_2PMe_2Fe(CO)_4\}_2$		438
607	$Fe_3C_{19}H_{24}O_9P_2$	$Fe_3(CO)_9(\mu-H)(\mu_3-Pr^i_2PCH_2PPr^i)$		439
608	$Fe_3C_{21}H_{33}O_7P_3$	$Fe_3(CO)_7(\mu-H)(\mu-PMeBu^t)(\mu,\mu'-Bu^tPCH_2PBu^t)$		436
609	$Fe_3C_{22}H_{25}O_{16}P_3$	$Fe_3(CO)_9\{P(OMe)_3\}_2(\mu_3-PC_6H_4OMe-\underline{p})$		435
610	$Fe_3C_{24}H_{15}O_{10}P_2$	$Fe_3(CO)_9(\mu_3-PPh)(\mu_3-\mu P:C-PC(OEt)Ph)$		440
611	$Fe_3C_{26}H_{23}N_2O_{10}P$	$Fe_2(CO)_6(CNBu^t)_2\{\mu-PPhFe(CO)_4\}$		435
612	$Fe_3C_{27}H_{18}O_9$	$Fe_3(CO)_6(\mu-CO)_2(\mu_3-\eta^1,\eta^1:\eta^4:\eta^4-CPhCPhCMeCOEt)$		441
613	$Fe_3C_{28}H_{20}N_2O_9$	$Fe_3(CO)_8\{C(OEt)Ph\}(\mu_3-NPh)_2$		440
614	$Fe_3C_{32}H_{22}O_8P_2$	$Fe_3(CO)_8(\mu-H)_2(\mu-PPh_2)_2.Sv$		442
615	$Fe_3C_{32}H_{23}O_{10}P$	$Fe_3(CO)_9(PPh_3)(\mu_3-\eta^2-MeCCOEt)$		443
616	$Fe_3C_{33}H_{19}O_9P_3$	$Fe_3(CO)_9(\mu_3-PPhPPhC_6H_4PPh-\underline{o})$	163	444
617	$Fe_3C_{33}H_{19}O_9P_3$	$Fe_3(CO)_9\{\mu_3-(PhP)_2C_6H_4-\underline{o}\}(\mu_3-PPh)$	163	444
618	$Fe_3C_{34}H_{19}O_{10}P_3$	$Fe_2(CO)_6[\mu,\mu'-PhPC_6H_4PPh\{Fe(CO)_4\}PPh]$	163	444
619	$Fe_3C_{37}H_{31}O_8P_3$	$Fe_3(CO)_8(\mu-H)(\mu_3-PBz)(\mu_3-Bz_2PCH_2PBz)$		445
620	$Fe_3C_{38}H_{29}O_9P_3$	$(CO)_6Fe_2(\mu_3-PBz)(\mu_3-BzPCH_2PBzCHPh)Fe(CO)_3$		445
621	$Fe_3Li_6C_{42}H_{56}N_4$	$\{Fe(\eta-C_5H_4Li)_2\}_3(TMED)_2$		446
622	$[Fe_3MoC_6O_6S_6]^{2-}$	$[NEt_4]_2[S_2Mo(\mu-S)_2Fe(\mu_3-S)_2Fe_2(CO)_6]$		447
623	$[Fe_3RhC_{13}O_{12}]^-$	$[PPN][RhFe_3C(CO)_{12}].CH_2Cl_2$		201
624	$Fe_3Si_2C_{21}H_{26}O_9$	$Fe_3(CO)_6(\mu-CO)_2(\mu_3-\eta^4-CMeCOEtCSiMe_3CSiMe_3)$		441
625	$[Fe_4C_{13}O_{13}]^{2-}$	$[PPN]_2[Fe_4(CO)_{12}(\mu_3-CO)]$	115	448
626	$[Fe_4C_{15}H_5O_{11}]^-$	$[PPN][Fe_4(CO)_9(\mu-CO)_2(\mu_4-\eta^3-MeCCHCH)$		449

No.	Formula	Structure	Details	Ref.
627	$Fe_4C_{21}H_{19}O_{14}P_3$	$Fe_3(CO)_{10}\{\mu-Me_2PCH_2PMeCH_2PMe_2Fe(CO)_4\}$		438
628	$Fe_4C_{21}H_{28}O_{10}P_4$	$Fe_4(CO)_{10}(\mu-H)(\mu-PPr^iMe)(\mu,\mu'-Pr^iPCH_2PPr^i)(\mu_4-PH)$		450
629	$Fe_4C_{24}H_{10}O_{12}S_4$	$\{Fe_2(CO)_6(\mu-SPh)\}_2(\mu,\mu'-S_2)$		451
630	$[Fe_4C_{32}H_{29}O_7]^+$	$[\{Fe_2(CO)_3(Cp)_2\}(\mu,\mu'-CHCHCH)][PF_6]$		452
631	$Fe_4C_{35}H_{23}O_{13}P_3$	$Fe_2(CO)_6(\mu_4-P)(\mu_3-BzPCH_2PBz_2)Fe_2(CO)_7 \cdot CH_2Cl_2$		445
632	$[Fe_4HgC_{14}H_3O_{13}]^-$	$[PPN][Fe_4(CO)_{12}(\mu-CO)(\mu_3-HgMe)]$		74
633	$[Fe_5C_{14}NO_{14}]^-$	$[PPN][Fe_5N(CO)_{14}]$		453
634	$[Fe_5C_{15}O_{14}]^{2-}$	$(NBu_4)_2[Fe_5C(CO)_{14}]$		453
635	$Fe_5C_{16}O_{15}$	$Fe_5C(CO)_{15}$		453
636	$Fe_5C_{24}H_{16}N_4O_{13}$	$Fe_5(CO)_{13}(\mu_5-C_{11}H_{16}N_4) \cdot H_2O$	218	454
637	$Fe_6C_{12}N_4O_{15}$	$Fe_6C(CO)_9(\mu-CO)_2(NO)_3(\mu-NO)$		455
638	$[Fe_6C_{16}NO_{16}]^-$	$[PPN][Fe_6C(CO)_{13}(\mu-CO)_2(NO)]$		455
639	$GaC_{15}H_{15}$	$Ga(\eta^1-Cp)_3$		456
640	$GaC_{27}H_{28}ClP_2$	$GaClMe_2(\eta^1-dppm)$	233	457
641	$GaIrC_{16}H_{24}N_4$	$(cod)Ir(\mu-pz)_2GaMe_2$		458
642	$GaReC_{13}H_{12}N_6O_3$	$MeGa(\mu-pz)_3Re(CO)_3$		459
643	$GaReC_{14}H_{23}N_3O_4$	$Re(CO)_3(NMe_2CH_2CH_2OGaMe_2)(\mu-dmpz)$		459
644	$GaReC_{14}H_{23}N_3O_4$	$Re(CO)_3(NH_2CHEtCH_2OGaMe_2)(\mu-dmpz)$		459
645	$GaReC_{18}H_{22}N_2O_4S$	$Re(CO)_3(PhSCH_2CH_2OGaMe_2)(\mu-dmpz)$		459
646	$GaReC_{29}H_{27}N_4O_3P$	$Re(CO)_3(PPh_3)(\mu-pz)_2GaMe_2$		460
647	$GaRhC_8H_{15}N_3O_2$	$Rh(CO)(NH_2CH_2CH_2OGaMe_2)(\mu-pz)$		461
648	$GaRhC_{11}H_{22}IN_3O_2$	$RhI(COMe)(NMe_2CH_2CH_2OGaMe_2)(\mu-pz)$		461
649	$GaRhC_{31}H_{35}N_4OP$	$Rh(CO)(PPh_3)(\mu-pz)_2GaMe_2 \cdot PhMe$		462
650	$Ga_2C_{20}H_{30}Cl_4$	$\{GaCl(\eta^1-Cp^*)(\mu-Cl)\}_2$		463
651	$Ga_2C_{22}H_{38}O_{14}$	$\{GaMe_2(\mu-trimethylcitrate)\}_2$		464
652	$Ga_2C_{24}H_{30}O_2$	$\{Ga(\eta^1-Cp)_2(\mu-OEt)\}_2$		465
653	$Ga_2C_{40}H_{60}Cl_2$	$\{Ga(\eta^1-Cp^*)_2(\mu-Cl)\}_2$		463
654	$Ga_2Si_6C_{30}H_{82}N_2$	$\{Ga(CH_2SiMe_3)_3\}_2(TMED)$		466
655	$Ga_4Tl_4C_{54}H_{72}Br_{16}$	$\{Tl_4(\eta^6-Mes)_6\}(\mu-GaBr_4)_4$	238	467
656	$Gd_2Si_8C_{32}H_{90}N_4S_2$	$[Gd\{N(SiMe_3)_2\}_2(\mu-SBu^t)]_2$		468

Structures Determined by Diffraction Methods

No.	Formula	Structure	Details	Ref.
657	$GeCH_3I_3$	$MeGeI_3$		469
658	$GeC_6H_{10}Cl_3NO$	$Cl_3GeCHMe\{N(CH_2)_3CO\}$		470
659	$GeC_6H_{12}Cl_3NO$	$Cl_3GeCH_2CHMeCONMe_2$		470
660	$GeC_{12}H_{22}N_2O_4$	$Ge\{CHMe(NC_4H_6O)\}(OCH_2CH_2)_3N$		471
661	$GeC_{12}H_{22}S$	$Me_2GeCCCMe_2CH_2SCH_2CMe_2$	153	472
662	$GeC_{16}H_{15}Cl_3O$	$Cl_3GeCMePhCH_2COPh$		473
663	$GeC_{16}H_{22}O_4P_2S_4$	$Ph_2Ge\{SPS(OMe)_2\}_2$		474
664	$GeC_{20}H_{21}O_2PS_2$	$Ph_3Ge\{SPS(OMe)_2\}$		474
665	$GeC_{22}H_{22}ClN$	$PhClGe(1,2-C_6H_3Me-5)_2NEt$		475
666	$GeC_{24}H_{18}O$	$Ph_2Ge(1,2-C_6H_4)_2O$		475
667	$GeC_{26}H_{21}Br_2N$	$Ph_2Ge(1,2-C_6H_3Br-5)_2NEt$		475
668	$GeC_{26}H_{23}N$	$Ph_2Ge(1,2-C_6H_4)_2NEt$		475
669	$GeC_{30}H_{30}$	$Me_2GeCH_2CPh_2CPh_2CH_2$		476
670	$GeC_{80}H_{70}$	$Ge(\eta-C_5Bz_5)_2$	168	477
671	$GeMn_2C_{24}H_{30}O_4$	$Ge\{Mn(CO)_2(Cp^*)\}_2$		478
672	$GeSi_4C_{14}H_{38}$	$Ge\{CH(SiMe_3)_2\}_2$	E	479
673	$GeWC_{15}H_{15}ClO_5$	$(CO)_5WGeCl(\eta^2-Cp^*)$	100	480
674	$GeW_3C_{15}O_{15}$	$\{W(CO)_5\}_3Ge$	230	39
675	$Ge_2C_{10}H_{22}N_2$	$Me_3GeN(CHCH)_2NGeMe_3$	163	481
676	$Ge_2C_{40}H_{52}$	$Ge_2(C_6H_3Et_2-2,6)_4$		482
677	$Ge_2Si_2C_{48}H_{40}O_3$	$Ph_2GeGePh_2OSiPh_2OSiPh_2O$		483
678	$Ge_2Si_4C_{26}H_{50}N_2S_2$	cis-$[Ge(CH_2Ph)\{N(SiMe_3)_2\}(\mu-S)]_2$		484
679	$Ge_3Se_2C_{36}H_{30}$	$Ph_2GeSeGePh_2SeGePh_2$		485
680	$Ge_4C_{48}H_{40}O_2$	$\{(Ph_4Ge_2(\mu-O)\}_2$		485
681	$HfC_{15}H_{20}Cl_2$	$HfCl_2(Cp)(Cp^*)$		486
682	$HfC_{16}H_{20}$	$Hf(\eta^4-CH_2=CMeCMe=CH_2)(Cp)_2$		487
683	$HfC_{18}H_{22}$	$Hf\{\eta^4-(CH_2=)_2C_6H_8-1,2\}(Cp)_2$		487
684	$HfC_{19}H_{33}Cl_2OP$	$HfCl_2(\eta^2-OCPBu^t_2)(Cp^*)$		488
685	$HfC_{24}H_{20}Cl_4O_2$	$Hf(OC_6H_3Cl_2-2,6)_2(Cp')_2$		489
686	$HfC_{31}H_{31}P$	$HfEt(CHPPh_3)(Cp)_2$		490

No.	Formula	Structure	Details	Ref.
687	$HfMoC_{22}H_{30}O_4P_2$	$(Cp)_2Hf(\mu-PPh_2)_2Mo(CO)_4$	173	491
688	$HfSi_2C_{10}H_{28}Cl_3NP_2$	fac-$HfCl_3N(SiMe_2CH_2PMe_2)_2$		492
689	$Hf_2C_{38}H_{74}P_2$	$\{HfMe(Cp^*)(\mu-H)(\mu-PBu^t_2)\}_2$		488
690	$HgC_6H_7N_5$	MeHg(adeninato)		493
691	$HgC_7H_7N_3$	$MeHg(N_3C_6H_4\text{-}\underline{o})$		494
692	$HgC_8H_7NS_2$	$MeHgSCNC_6H_4S$		495
693	$HgC_{11}H_5Cl_5$	$HgPh(\eta^1\text{-}C_5Cl_5)$		496
694	$HgC_{11}H_{16}N_2O_5$	MeHg(thymidinato)$\cdot xH_2O$		497
695	$HgC_{20}H_{10}F_{12}O_{12}S_4$	$Hg\{C(OSO_2CF_3)CPh(OSO_2CF_3)\}_2$		498
696	$HgMn_2SnC_{53}H_{28}N_4O_9$	$(TPP)SnMn(CO)_4HgMn(CO)_5\cdot Sv$		499
697	$HgMoC_9H_7ClO_3$	$Mo(HgCl)(CO)_3(Cp')$		500
698	$HgRuC_{20}H_{28}Cl_2O_4S_2$	$Ru(\eta,\eta'\text{-}C_5H_4S(CH_2CH_2O)_4CH_2CH_2SC_5H_4)HgCl_2$		501
699	$Hg_2C_8H_{11}N_5$	$(MeHg)_2N(C_4N_4H_2Me)$		502
700	$Hg_2C_{18}H_{20}F_6N_2O_4$	$\{MeHg(NC_5H_4Me\text{-}2)\}_2(\mu\text{-}TFAc)_2$		503
701	$[Hg_3C_3H_9S]^+$	$[(MeHg)_3S][ClO_4]$		504
702	$[Hg_4C_{12}H_8Cl_5]^-$	$[PPh_4][\{\underline{o}\text{-}C_6H_4(HgCl)_2\}_2Cl]\cdot CH_2Cl_2$		505
703	$[InC_2H_5I_3]^-$	$[PPh_4][InI_3Et]$		506
704	$InC_8H_{21}Br_2N_2$	$InBr_2Et(TMED)$		506
705	$InC_8H_{21}I_2N_2$	$InI_2Et(TMED)$		506
706	$InC_{13}H_{17}N_4$	$InMe(MeNC_5H_4N\text{-}2)_2$		507
707	$In_2Si_2C_{22}H_{54}N_4$	$\{Me_2Si(\mu\text{-}NBu^t)InMe(\mu_3\text{-}NBu^t)\}_2$		19
708	$In_3C_{13}H_{35}N_4$	$MeIn\{(\mu_3\text{-}TMED)InMe_2\}_2$		507
709	$IrC_{13}H_{16}Cl_2NO$	$IrCl_2(OC_5H_4N\text{-}2)(cod)$		508
710	$IrC_{15}H_{28}P$	$IrH(CH=CH_2)(PMe_3)(Cp^*)$		509
711	$IrC_{19}H_{25}O_3S$	$Ir(C_6H_4COO\text{-}\underline{o})(SOMe_2)(Cp^*)$		510
712	$IrC_{19}H_{42}ClOP_2$	trans-$IrCl(CO)(PPr^i_3)_2$		511
713	$IrC_{26}H_{58}ClP_2$	$IrHCl(CH_2CMe_2CH_2PBu^t_2)(PBu^t_2CH_2Bu^t)$		512
714	$IrC_{27}H_{26}BrOP_2$	$IrH_2Br(CO)(dppe)$		513
715	$[IrC_{27}H_{30}F_4]^+$	$[Ir(\eta^4\text{-}Me_3tfb)(\eta^6\text{-}C_6Me_6)][ClO_4]$		514
716	$[IrC_{34}H_{30}N_2O_6]^+$	$[Ir\{\eta^2,O\text{-}PhCC(CO_2Me)NHCPhO\}_2][BF_4]$		515

Structures Determined by Diffraction Methods

No.	Formula	Structure	Details	Ref.
717	$IrC_{37}H_{31}ClF_2O_3P_3$	$IrHCl(OPOF_2)(CO)(PPh_3)_2$		516
718	$[IrC_{37}H_{33}ClO_2P_2]^+$	$[IrHCl(OH_2)(CO)(PPh_3)_2][BF_4]$		516
719	$IrC_{38}H_{29}Cl_2O_2P_2$	$IrCl_2(COC_6H_4PPh_2\text{-}\underline{o})(CO)(PPh_3)$		517
720	$IrC_{38}H_{29}Cl_2O_2P_2$	$IrCl_2\{C(OH)(C_6H_4PPh_2\text{-}\underline{o})_2\}(CO)$		517
721	$IrC_{38}H_{33}OP_2$	<u>trans</u>-$IrMe(CO)(PPh_3)_2$		518
722	$IrC_{42}H_{39}ClOP_3$	$IrCl(CO)(triphos)$		519
723	$IrC_{43}H_{35}O_2P_2$	<u>trans</u>-$Ir(OPh)(CO)(PPh_3)_2$		520
724	$IrC_{44}H_{44}OP_3$	$IrH(CH_2C_6H_4CO\text{-}\underline{o})\{PhP(CH_2CH_2CH_2PPh_2)_2\}$		521
725	$[IrC_{45}H_{38}P_2]^+$	$[IrH(PPh_3)_2(\eta^5\text{-ind})][BF_4]\cdot CH_2Cl_2$		522
726	$IrSiC_{36}H_{35}O_2P_2$	$Ir(SiMe_2CH_2CH_2PPh_2)(CO)_2(PPh_3)$		523
727	$IrSi_2C_{19}H_{49}INP_2$	$IrHMeI\{NH(SiMe_2CH_2PPr^i_2)_2\}$		524
728	$IrSi_2C_{31}H_{38}NP_2$	$Ir(CH_2)\{N(SiMe_2CH_2PPh_2)_2\}$		525
729	$IrSnC_{28}H_{33}$	$IrH_3(SnPh_3)(Cp^*)$		106
730	$IrWC_{17}H_{15}O_7$	$(CO)_5WIr(CO)_2(Cp^*)$		526
731	$Ir_2C_{10}H_6N_4O_4$	$\{Ir(CO)_2(\mu\text{-}pz)\}_2$		458
732	$Ir_2C_{14}H_{14}N_4O_4$	$\{Ir(CO)_2(\mu\text{-}Me_2pz)\}_2\cdot 0.5XyH$		458
733	$Ir_2C_{16}H_{24}I_4$	$\{IrI(cod)(\mu\text{-}I)\}_2$		527
734	$Ir_2C_{17}H_{38}I_2O_8P_2S_2$	$[IrI(CO)\{P(OMe)_3\}]_2(\mu\text{-}CH_2)(\mu\text{-}SBu^t)_2$		528
735	$[Ir_2C_{22}H_{30}Cl_2N_5O]^+$	$[\{IrCl(cod)\}_2(\mu\text{-}pz)_2(\mu\text{-}NO)][BF_4]\cdot 2CH_2Cl_2$		529
736	$[Ir_2C_{22}H_{30}N_5O]^+$	$[\{Ir(cod)\}\{Ir(cod)(NO)\}(\mu\text{-}pz)_2][BF_4]$		529
737	$Ir_2C_{26}H_{34}N_4O_2$	$\{Ir(cod)(\mu\text{-}pz)\}_2(\mu\text{-}HCCCO_2Me)$		530
738	$[Ir_2C_{26}H_{51}P_2]^+$	$[\{IrH(PMe_3)(Cp^*)\}_2(\mu\text{-}H)][PF_6]$		531
739	$Ir_2C_{28}H_{36}N_2O_2$	$\{Ir(cod)(\mu\text{-}OC_5H_4N\text{-}2)\}_2$		508
740	$Ir_2C_{28}H_{36}N_4O_4$	$\{Ir(cod)(\mu\text{-}pz)\}_2(\mu\text{-}MeCO_2CCCO_2Me)$		530
741	$Ir_2C_{28}H_{42}N_4$	$\{Ir(cod)(\mu\text{-}Me_3pz)\}_2$		532
742	$Ir_2C_{36}H_{42}N_4$	$\{Ir(cod)(\mu\text{-}3,5\text{-}PhMepz)\}_2$		532
743	$Ir_2C_{38}H_{36}N_{18}O_8P_2$	$[Ir(Tcbiim)(CO)(NCMe)\{P(OEt)_3\}]_2\text{-}(\underline{Ir}\text{-}\underline{Ir})\cdot MeCN$		533
744	$Ir_2C_{44}H_{45}ClP_2$	$(cod)Ir(\mu\text{-}H)(\mu\text{-}Cl)IrH_2(PPh_3)_2$		534
745	$Ir_2C_{52}H_{45}ClO_3P_4$	$Ir_2(CO)_2(\mu\text{-}OH.Cl)(\mu\text{-}dppm)_2$		535
746	$Ir_2C_{64}H_{58}Cl_2O_{10}P_4$	$Ir_2Cl_2\{C(CO_2Me)=CHCO_2Me\}_2(\mu\text{-}dppm)_2$		536

No.	Formula	Structure	Details	Ref.
747	$Ir_2W_2C_{25}H_{22}O_{11}$	$(Cp)_2W_2Ir_2(CO)_7(\mu\text{-}CHCO_2Et)(\mu_3\text{-}CHCO_2Et)$		537
748	$Ir_3C_{24}H_{36}IO_2$	$\{Ir(cod)\}_3(\mu\text{-}I)(\mu_3\text{-}O)_2$		527
749	$[Ir_3RuC_{12}O_{12}]^-$	$[PPN][RuIr_3(CO)_9(\mu\text{-}CO)_3]$		538
750	$Ir_4C_{11}O_{13}S$	$Ir_4(CO)_9(\mu\text{-}CO)_2(\mu\text{-}SO_2)$		539
751	$Ir_4C_{59}H_{44}O_7P_4$	$Ir_4H(CO)_7(Ph_2PCHCHPPh_2)(\mu_3\text{-}C_6H_4PPhCHCHPPh_2)\cdot Sv$		540
752	$Ir_4C_{60}H_{44}O_8P_4$	$Ir_4(CO)_8(\mu\text{-}Ph_2PCHCHPPh_2)_2$		540
753	$LaC_{19}H_{21}N_2$	$La(NCMe)_2(Cp)_3$		159
754	$LiC_{11}H_{17}O$	$\{Li(OEt_2)(\mu\text{-}Bz)\}_n$	140	541
755	$LiC_{13}H_{24}N_3$	$(TMED)Li\{NC_5H_3(CH_2)Me\text{-}2,6\}$	119	542
756	$LiC_{15}H_{28}N_3$	$LiPh\{NMe(CH_2CH_2NMe_2)_2\}$		543
757	$LiLuC_{18}H_{32}N_2$	$(TMED)Li(\mu\text{-}CH_2)_2Lu(Cp)_2$	168	544
758	$LiLuC_{36}H_{64}O_2S_2$	$(THF)_2Li(\mu\text{-}SBu^t)_2Lu(Cp^*)_2$	168	545
759	$LiNd_2Si_2C_{48}H_{76}Cl_3O_2$	$(THF)_2Li[(\mu\text{-}Cl)Nd\{\eta,\eta'\text{-}(C_5Me_4)_2SiMe_2\}]_2Cl$	140	546
760	$LiNiC_{14}H_{34}N_3$	$\{MeN(CH_2CH_2NMe_2)_2\}Li(\mu\text{-}Me)Ni(\eta\text{-}C_2H_4)_2$	100	547
761	$LiSi_3C_{20}H_{45}N_2$	$(TMED)Li\{\eta^5\text{-}C_5H_2(SiMe_3)_3\}$		548
762	$LiSi_3C_{21}H_{42}N$	$HC(C_2H_4)_3NLi\{\eta^5\text{-}C_5H_2(SiMe_3)_3\}$	143	548
763	$Li_2C_{16}H_{26}O_4$	<u>trans</u>-$\{(DME)Li\}_2(\mu\text{-}\eta^5:\eta^5{'}\text{-}C_8H_6)$	120	549
764	$Li_2C_{28}H_{46}N_4$	$\{Li(TMED)\}_2(\underline{cis}\text{-}PhCHCHCHCHPh)$	113	550
765	$Li_2C_{32}H_{46}N_4$	$\{(TMED)Li\}_2(\mu\text{-}\eta^2\text{-}Ph_2C_2C_6H_4\text{-}\underline{o})$ (2-isomers)		551
766	$Li_2C_{34}H_{54}O_4$	$\{Li(THF)(\mu\text{-}Mes)\}_2$	140	541
767	$Li_2Si_2C_{26}H_{56}N_4$	$\{(TMED)Li\}_2\{C_6H_4(CHSiMe_3)_2\text{-}\underline{m}\}$		552
768	$Li_2Si_4C_{30}H_{70}O_4S_2$	$Li_2(THF)_4\{SCH(SiMe_3)_2\}_2$		553
769	$Li_2Si_6C_{34}H_{82}O_3S_2$	$Li_2(THF)_{3.5}\{SC(SiMe_3)_3\}_2$		553
770	$Li_2Si_8C_{30}H_{84}O_6$	$\{(DME)LiSi(SiMe_3)_3\}_2(\mu\text{-}DME)$	153	554
771	$Li_4C_{20}H_{48}N_4$	$Li_4(\mu_3\text{-}CH_2CH_2CH_2NMe_2)_4$		555
772	$Li_4C_{40}H_{68}O_4$	$(THF)Li(Bu^tCCCCBu^t)_2Li(THF)\{\eta^4\text{-}Li(THF)\}_2$		556
773	$Li_4C_{44}H_{56}N_4O_4$	$Li_4\{NMe_2CH_2C_6H_4C(=CH_2)O\text{-}\underline{o}\text{-}\mu_3\}_4$		557
774	$Li_4Mn_2C_{36}H_{90}O_2P_8$	$(Et_2O)_2Li_4\{(dmpe\text{-}H)_2MnH(\eta\text{-}C_2H_4)\}_2$		558
775	$Li_6Si_8C_{20}H_{60}N_6$	$(LiNSiMe_3)_6(SiMe_3)_2$		559
776	$Li_6Si_8C_{26}H_{72}N_6$	$(LiNSiMe_3)_6(SiBu^t_3)_2$		559

Structures Determined by Diffraction Methods 449

No.	Formula	Structure	Details	Ref.
777	$Li_6Si_8C_{30}H_{64}N_6$	$\{LiNSiMe_3\}_6(SiPh_3)_2$		559
778	$LuC_{19}H_{23}O$	$Lu(THF)(Cp)_3$		560
779	$LuC_{21}H_{28}NO$	$Lu(C_6H_4NMe_2\text{-}\underline{o})(THF)(\eta^8\text{-}C_8H_8)$		561
780	$MgC_{10}H_{16}$	$Mg(CH_2Bu^t)(Cp)$	E	562
781	$MgC_{26}H_{42}BrNO_4$	$MgBr\{CHN(COBu^t)CH_2CH_2C_6H_4\}(THF)_3$		563
782	$MgC_{28}H_{38}O_3$	$(THF)_3Mg(1,9\eta^2\text{-}1,4\text{-}Me_2\text{anthracene})$		564
783	$MgRe_2C_{24}H_{56}O_6$	$Mg(THF)_4(OReMe_4)_2$		565
784	$MgRe_2Si_8C_{40}H_{104}O_4$	$Mg\{ORe(CH_2SiMe_3)_4\}_2(THF)_2$		565
785	$MgSi_2C_{28}H_{42}O_2$	$(THF)_2Mg\{9,10\text{-}(9,10\text{-}SiMe_3)_2\text{anthracene}\}$		566
786	$Mg_2C_{38}H_{76}N_4O_8$	$\{Mg(CH_2Bu^t)(\mu\text{-}OC_{14}H_{27}N_2O_3)\}_2$		567
787	$Mg_2Si_4C_{32}H_{64}N_4O_2$	$\{Mg(OEt_2)\}_2\{\mu\text{-}N(SiMe_3)C_6H_4NSiMe_3\text{-}\underline{o}\}_2$		568
788	$MnCN_3O_4$	$Mn(CO)(NO)_3$	E	335
789	$MnC_{10}H_{22}$	$Mn(CH_2Bu^t)_2$	E	569
790	$MnC_{12}H_{15}O_3S_2$	$Mn(CO)_2(Cp^*)(\eta^2\text{-}S=SO)$		570
791	$MnC_{14}H_7N_2O_5$	$Mn\{CH(NC_4H_3)_2\}(CO)_5$		571
792	$MnC_{14}H_{10}O_5S$	$\{(Cp)(CO)_2Mn\}_2SO$	193	572
793	$MnC_{14}H_{22}O_2P$	$Mn(CO)_2(PMePr^i_2)(Cp)$		573
794	$MnC_{17}H_{16}O_2P$	$Mn(CCPMe_2Ph)(CO)_2(Cp)$		574
795	$MnC_{17}H_{18}NO$	$Mn(\eta^2\text{-}CH_2CHCH_2NHCPh)(CO)(Cp')$		575
796	$MnC_{17}H_{41}P_4$	$Mn(dmpe)_2(\eta^3\text{-}CH_2CHCHEt)$		576
797	$MnC_{19}H_{20}NO_2S_3$	$Mn(CO)_2(Cp)\{S=CPhSCH_2C(S)NMe_2\}$		284
798	$MnC_{21}H_{18}Cl_2O_4P$	$Mn(CO)_4(CH_2CHPr^iCH_2CH_2PMe_2)$		577
799	$MnC_{21}H_{20}O_4P$	$Mn\{CH_2(CH_2)_4PMe_2\}(CO)_4$		578
800	$[MnC_{21}H_{24}O_3]^+$	$[Mn(CO)_3(\eta^6\text{-}C_{18}H_{24})][BF_4]$		579
801	$[MnC_{26}H_{22}N_4O_2]^+$	<u>cis</u>-<u>cis</u>-$[Mn(CO)_2(CNBu^t)(CNPh)(phen)][ClO_4]$		580
802	$[MnC_{28}H_{23}O_3P]^+$	$[Mn(CO)_3(\eta^5\text{-}C_7H_8PPh_3)][BF_4]$		581
803	$MnC_{34}H_{23}O_{11}P_2$	$Mn\{P(O_2C_6H_4\text{-}\underline{o})_2\}(CO)_4\{P(OPh)_3\}$		582
804	$MnC_{39}H_{35}OP_2$	$Mn(CO)(dppe)(\eta^5\text{-}C_6H_6Ph)$	220	583
805	$[MnC_{39}H_{35}OP_2]^+$	$[Mn(CO)(dppe)(\eta^5\text{-}C_6H_6Ph)][PF_6].Sv$	220	583
806	$MnC_{42}H_{36}O_3P_3$	$Mn\{C(O)CH_2CH_2PPh_2\}(CO)_2(dppm)$		584

No.	Formula	Structure	Details	Ref.
807	$MnMoC_{23}H_{16}O_6P$	$(CO)_4MnMo(\mu-H)(\mu-PPh_2)(CO)_2(Cp)$		585
808	$MnMoC_{25}H_{20}O_4P$	$MnMo(CO)_4Cp(\mu-\eta^2CH,\eta^2:\eta^1,\eta^2PC-CHCHCH_2CHPPh_2)$		585
809	$MnMoC_{26}H_{24}O_4P$	$(CO)_3MnMo(\mu-\eta^2CH:\eta^3-CH_2CMeCHMe)(\mu-PPh_2)(CO)(Cp)$		585
810	$[MnNb_2W_4C_3O_{22}]^{3-}$	$[NBu_4]_3[(CO)_3Mn(Nb_2W_4O_{19})]$		586
811	$MnPtC_{17}H_{31}O_5P_2$	$(CO)_4MnPt(PEt_3)_2(\mu-H)(\mu-CO)$		587
812	$[MnPtC_{53}H_{45}BrO_3P_4]^+$	$[(CO)_3Mn(\mu-dppm)_2PtHBr][BF_4].Et_2O$		588
813	$MnRhC_{23}H_{33}O_2P$	$(Cp)(CO)MnRh(PPr^i_3)(Cp)(\mu-CCH_2)(\eta-CO)$		387
814	$MnSi_6C_{20}H_{54}$	$Mn\{C(SiMe_3)_3\}_2$		589
815	$MnWC_{15}H_9O_8$	$(Cp)(CO)_2MnW(CO)_4(\mu-C=CHCOOMe)$	153	590
816	$[Mn_2C_8O_6S_8]^{2-}$	$[NEt_3Bz]_2[\{Mn(CO)_3(\mu-SSCSS)\}_2]$		591
817	$Mn_2C_{12}H_8O_8S$	$Mn_2(CO)_8(\mu-SC_4H_8)$		592
818	$Mn_2C_{13}H_8O_9S$	$Mn_2(CO)_9(SC_4H_8)$		592
819	$Mn_2C_{14}H_{18}O_{14}P_2$	$Mn_2(CO)_8\{P(OMe)_3\}_2$		593
820	$Mn_2C_{15}H_{14}O_6S$	$Mn_2(CO)_6(Cp)(\mu-SBu^t)$		594
821	$Mn_2C_{18}H_{20}O_4S$	$\{Mn(CO)_2(Cp')\}_2(\mu-SMe_2)$		595
822	$Mn_2C_{24}H_{36}N_4O_8P_4$	$\{Mn(CO)_4\}_2\{\mu-(P_2N_2Bu^t_2)_2\}$		596
823	$Mn_2C_{26}H_{40}N_4O_6$	$Mn_2(CO)_6\{\mu N,\mu N',\eta^1 N-Bu^tNCH(CHNBu^t)\}_2$		597
824	$Mn_2C_{30}H_{60}P_2$	$Mn_2Cy_2(\mu-Cy)_2(\mu-dmpe)$		558
825	$Mn_2C_{35}H_{22}O_6$	$\{Mn(CO)_3\}_2\{\eta,\eta'-(C_5H_4CPh)_2C_5H_4-1,3\}$		598
826	$Mn_2C_{40}H_{26}O_6$	$\{Mn(CO)_3\}_2\{\eta,\eta'-(C_5H_4CPhC_5H_4-1,3)_2\}$	193	598
827	$Mn_2C_{56}H_{44}O_6P_4$	$Mn_2(CO)_6(\mu-dppm)_2$		599
828	$Mn_2MoC_{24}H_{18}O_{10}$	$\{Mn(CO)_2(Cp)\}_2Mo(CO)_2(\mu-C=CHCOOMe)_2$	153	590
829	$Mn_2PbC_{14}H_{10}$	$Pb\{Mn(CO)_2(Cp)\}_2$		600
830	$Mn_2PdC_{48}H_{38}Cl_2N_2O_4P_2$	$PdCl_2[(\mu-N:\eta^5-C_4H_4N)Mn(CO)_2(PPh_3)]_2$		601
831	$Mn_2SeC_{18}H_{20}O_4$	$\{Mn(CO)_2(Cp')\}_2(\mu-SeMe_2)$		595
832	$Mn_2SiC_{24}H_{34}O_4$	$\{Mn(CO)_2(Cp^*)\}_2SiH_2$		602
833	$Mn_3SbC_{24}H_{15}Br_2O_9$	$SbBr_2\{(\eta-C_5H_4)Mn(CO)_3\}_3$		603
834	$[Mn_3Si_9C_{34}H_{89}Cl_4O]^-$	$[Li(THF)_4][Mn_3\{C(SiMe_3)_3\}_3(THF)(\mu-Cl)_4]$		604
835	$MoC_7H_5O_2P_3$	$Mo(CO)_2(\eta^3-P_3)(Cp)$		605
836	$[MoC_8H_5O_3]^-$	$[Cp_2MoH(CO)][CpMo(CO)_3]$		433

Structures Determined by Diffraction Methods 451

No.	Formula	Structure	Details	Ref.
837	$[MoC_8H_5O_5]^-$	$[NEt_4][Mo(O_2CO)(CO)_2(Cp)]$		606
838	$MoC_8H_6O_5S$	$Mo(SO_2H)(CO)_3(Cp)$	243	607
839	$MoC_9H_{10}Cl_2O$	$MoCl_2(CO)(\eta^3-C_3H_5)(Cp)$		608
840	$MoC_9H_{12}O_3S_3$	$Mo(CO)_3(SCH_2CH_2)_3$		609
841	$MoC_{10}H_5O_5P$	$Mo(CO)_5(PC_5H_5)$		610
842	$[MoC_{10}H_{20}FN_2O_2S_4]^-$	$[NEt_4][MoF(detc)_2(CO)_2]$		611
843	$MoC_{11}H_{10}O_2$	$Mo(\eta^2-CO_2)(Cp)_2$		612
844	$[MoC_{11}H_{11}O]^+$	$[Cp_2MoH(CO)]^+$ (see 599 and 836)		433
845	$MoC_{11}H_{12}O$	$Mo(\eta^2-H_2CO)(Cp)_2$		612
846	$MoC_{11}H_{15}$	$Mo(\eta^3-C_3H_5)_2(Cp)$	100	266
847	$MoC_{12}H_{14}INO_2$	$MoI(CO)_2\{C(CH_2)_3NMe\}(Cp)$		613
848	$MoC_{12}H_{14}O$	$MoO(Cp')_2$		614
849	$MoC_{13}H_{19}NO$	$Mo(NO)(\eta^4-Me_2CCHCHCMe_2)(Cp)$		615
850	$MoC_{14}H_{16}$	$Mo(\eta^6-C_5H_4CMe_2)(\eta-PhH)$		616
851	$MoC_{14}H_{30}P_2$	$Mo(PMe_3)_2(\eta^4-C_4H_6)_2$		617
852	$MoC_{15}H_{14}O_2$	$Mo(CO)_2(\eta^3-C_3H_4Me\text{-}\underline{exo})(\eta^5\text{-ind})$		522
853	$MoC_{15}H_{15}O_5P$	$Mo(CO)_4(PMe_2OCHPhCH=CH_2-\eta^2)$		618
854	$MoC_{15}H_{27}F_6O_6P_3$	$Mo(OCOCF_3)_2(CO)_2(PMe_3)_3$		619
855	$MoC_{16}H_8N_2O_4$	$Mo(CO)_4(phen)$	185	620,270
856	$MoC_{16}H_{15}NO_2$	$Mo(CO)_2\{\eta^3-CH_2C(pTol)NH\}$		621
857	$MoC_{16}H_{16}O_2$	$Mo(CO)_2(nbd)_2$		622
858	$[MoC_{17}H_{10}O_6P]^-$	$[NEt_3H][Mo(CO)_5(PPh_2O)]$		623
859	$MoC_{17}H_{18}$	$Mo(\eta^3,\eta^4-C_{12}H_{13})(Cp)$		624
860	$MoC_{17}H_{30}IO_3P$	$MoI\{P(OMe)_3\}(\eta^4-CH_2CHCHCH_2Bu^t)(Cp)$		625
861	$MoC_{18}H_{22}N_2S_2$	$Mo\{\eta^1-C(CN)CH_2\}(\eta^1-detc)(Cp)_2$		626
862	$MoC_{19}H_{19}NS$	$Mo(CHMeCN)(SPh)(Cp)_2$		626
863	$MoC_{19}H_{29}BrN_2O_2$	$MoBr(CO)_2(CyNCHCHNCy)(\eta^3-C_3H_5)$		627
864	$MoC_{20}H_{22}O_5$	$Mo(CO)_2(\eta^3-C_6H_8C(CO_2Me)CH_2CH_2CH_2CO)(Cp)$		628
865	$MoC_{20}H_{23}NO_2S$	$Mo(CO)(NO)(\eta^2-C_8H_{13}SPh)(Cp)$		629
866	$MoC_{20}H_{37}P$	$Mo(PEt_3)(\eta^5-CH_2CMeCHCMeCH_2)_2$		630

No.	Formula	Structure	Details	Ref.
867	$MoC_{21}H_{21}$	$Mo(\eta^4,\eta^2,\eta^2CH-C_{12}H_{14})(\eta^5-ind)$	200	624
868	$MoC_{22}H_{24}NO_4P$	$Mo(CO)_4(PPh_2CH_2CH_2NEt_2)$		281
869	$MoC_{22}H_{40}O_7P_2$	$Mo\{P(OMe)_3\}_2(\eta^3-Bu^tCHCHCMe_2CO)(Cp)$		625
870	$MoC_{23}H_{20}N_2O_6$	$\underline{cis}-Mo(CO)_4(N_2C_{19}H_{20}O_2)\cdot Sv$		631
871	$MoC_{23}H_{38}O_6P_2$	$Mo(\eta^2-Bu^tCCHPh)\{P(OMe)_3\}_2(Cp)$		632
872	$MoC_{23}H_{40}N_4S_{12}$	$Mo_2(detc)_3(\mu-CSCS_2)(\eta-\eta^2-SCSC(S)NEt_2)$		633
873	$MoC_{24}H_{20}$	$Mo(\eta^6-C_5H_4CPh_2)(\eta-PhH)$		616
874	$MoC_{25}H_{33}N_3O_4P_2$	$Mo(CO)_4[\{PPh(NHPr^i)\}_2NPr^i]\cdot Sv$		634
875	$MoC_{25}H_{34}O_6P_2$	$Mo(\eta^2-PhCCHPh)\{P(OMe)_3\}_2(Cp)$		632
876	$MoC_{26}H_{23}O_2$	$Mo(CO)_2(\eta^3-C_3Ph_2Bu^t)(Cp)$		635
877	$MoC_{26}H_{36}O_6P_2$	$Mo(\eta^2-MeCCPh_2)\{P(OMe)_3\}_2(Cp)$		632
878	$[MoC_{27}H_{23}N_2O_2]^+$	$[Mo(NC_5H_4CPh=NCHMePh)(CO)_2(Cp)][PF_6]\cdot 2Me_2CO$		636
879	$MoC_{27}H_{30}F_{12}N_2$	$Mo\{OCH(CF_3)_2\}_2py_2(\eta^3-Bu^tCCCBu^t)$	253	637
880	$MoC_{27}H_{44}O_4$	$Mo(OAc)_2(\eta^5-Bu^tCCBu^tCHCBu^tCHCBu^t)$	223	638
881	$MoC_{28}H_{20}O_2$	$Mo(CO)_2(\eta^3-C_3Ph_3)(Cp)$		635
882	$MoC_{28}H_{26}O_4$	$Mo(\eta^2-PhC=CPhCCOO)(CO)_2(Cp^*)\cdot CH_2Cl_2$		639
883	$MoC_{28}H_{29}NO_3P_2$	$Mo(CO)_3\{NMe(CH_2CH_2CH_2PPh)_2C_6H_4-\underline{o}\}\cdot Sv$	163	640
884	$MoC_{28}H_{35}F_{12}NO_2$	$Mo(\eta^2-Bu^tCCCBu^tCHCBu^t)\{OCH(CF_3)_2\}_2(py)$	223	638
885	$[MoC_{29}H_{31}N_7]^{2+}$	$[Mo(CNEt)_3(bipy)_2][BF_4]_2$		641
886	$MoC_{31}H_{25}O_2P$	$MoPh(CO)_2(PPh_3)(Cp)$		642
887	$MoC_{31}H_{49}O_2P$	$Mo(O)_2(Mes)\{C(Mes)PBu_3\}$		643
888	$MoC_{32}H_{31}Cl_5P_2$	$MoCl_3(dppe)(Cp)\cdot CH_2Cl_2$		644
889	$MoC_{37}H_{26}OS_2$	$Mo(SC_6H_3Ph_2-2,6)(CO)(\eta^6,S-SC_6H_3Ph_2-2,6)$		645
890	$[MoC_{47}H_{69}O_2S_3]^-$	$[PPh_4][Mo(SC_6H_2Pr^i_3-2,4,6)_3(CO)_2]$		646
891	$MoC_{53}H_{36}ClN_4$	$MoClPh(TPP)$		647
892	$MoC_{54}H_{54}N_6$	$Mo(CNC_6H_3Me_2-2,6)_6\cdot PhH$		648
893	$MoNiC_{18}H_{18}O_3$	$(Cp)NiMo(CO)_2(Cp')(\mu-\eta^2:\eta^1,\eta^1-MeCCMeCO)$		649
894	$[MoNiC_{23}H_{33}S_2]^+$	$[(Cp)Ni(\mu-SBu^t)_2Mo(Cp)_2][BF_4]$		650
895	$MoRuC_{56}H_{44}O_6P_4$	$MoRu(CO)_6(\mu-dppm)_2$		651
896	$[MoSiC_3H_9OS_3]^-$	$[PPh_4][MoS_3(OSiMe_3)]$		652

No.	Formula	Structure	Details	Ref.
897	$[MoSiC_{18}H_{15}O_4]^-$	$[NBu_4][MoO_3(OSiPh_3)]$		653
898	$MoSiC_{56}H_{57}BrP_4$	trans-$Mo(CSiMe_3)Br(dppe)_2 \cdot PhMe$	113	654
899	$MoSi_2C_{15}H_{23}O_3P$	$Mo(CO)_3\{P=C(SiMe_3)_2\}(Cp)$		655
900	$MoSi_4C_{19}H_{38}O_5P_2$	$Mo(CO)_5P\{CH(SiMe_3)_2\}=PCH(SiMe_3)_2$		301
901	$MoSn_2C_{14}H_{22}Cl_2$	$Mo(SnClMe_2)(Cp)_2$		656
902	$MoWC_{36}H_{28}O_5$	$Cp(CO)_4WMo\{\eta\text{-}COMeind\}\{\mu\text{-}\eta^2\text{-}C_2(pTol)_2\} \cdot Sv$	200	657
903	$[MoWC_{36}H_{30}NO_8P_2]^-$	$[NEt_4][(CO)_4Mo(\mu\text{-}PPh_2)_2W(CO)_2\{CNEt_2C(O)O\}] \cdot Sv$		658
904	$MoZrC_{19}H_{18}O_3$	$(Cp)(CO)_2MoCOZrMe(Cp)_2$		659
905	$MoZrC_{38}H_{30}O_4P_2$	$(Cp)_2Zr(\mu\text{-}PPh_2)_2Mo(CO)_4$		660
906	$MoZr_2C_{36}H_{42}O_4$	$(Cp)(CO)_2MoCOZr(CMe_2O)(Cp')_2ZrMe(Cp')_2$	233	661
907	$Mo_2C_{12}H_{32}N_4$	$Mo_2(NMe_2)_4\{\mu\text{-}(CH_2)_4\}$	115	662
908	$[Mo_2C_{14}H_{19}S_4]^+$	$[(Cp')_2Mo_2(\mu\text{-}S)(\mu\text{-}SMe)(\mu,\mu'\text{-}S_2CH_2)][HSO_4]$		663
909	$Mo_2C_{18}H_{22}O_2S_2$	$Mo_2(CO)_2(Cp')_2(\mu\text{-}SEt)(\mu\text{-}\eta^2\text{-}SCMe)$		664
910	$Mo_2C_{19}H_{14}O_6$	$Mo_2(CO)_5(CCH_2CH_2O)(\eta^5,\eta^5\text{'-}fulv)$		665
911	$Mo_2C_{19}H_{22}O_4P_2$	$(CO)_3Mo(\eta^5,\eta^5\text{'-}fulv)Mo(CO)(dmpm)$		666
912	$Mo_2C_{19}H_{22}O_5S_2$	$Mo_2(CO)_4(Cp)_2(\mu\text{-}S:\eta^2\text{-}SCMeSEt) \cdot Sv$		664
913	$Mo_2C_{20}H_{22}O_4S$	$Mo(CO)_2(Cp)_2(\mu\text{-}SC_6H_{12})$		667
914	$Mo_2C_{20}H_{30}O_2S_2$	$\{Mo(O)(Cp^*)(\mu\text{-}S)\}_2$ (2 isomers)		668
915	$Mo_2C_{20}H_{30}O_4$	$\{Mo(O)(\mu\text{-}O)(Cp^*)\}_2$		669
916	$Mo_2C_{20}H_{30}P_6$	$Mo_2(Cp^*)_2(\mu\text{-}\eta^6\text{-}P_6)$		670
917	$[Mo_2C_{22}H_{10}O_{10}]^-$	$[PPN][\{Mo(CO)_5\}_2PPh_2]$		671
918	$Mo_2C_{22}H_{14}O_4$	$Mo_2(CO)_4(\eta^5\text{-}ind)_2$		672
919	$Mo_2C_{22}H_{22}O_8$	$[Mo(CO)_3\{\eta\text{-}C_5H_4(CH_2)_3OH\}]_2$		673
920	$Mo_2C_{22}H_{28}O_6$	$[\{Mo(CO)_2\}\{\mu\text{-}\eta^3:\mu O\text{-}CMeCHCMeCH_2CMe(O)CH_2\}]_2$		674
921	$Mo_2C_{22}H_{29}O_4P$	$\{Mo(CO)_2(Cp)\}_2(\mu\text{-}H)(\mu\text{-}PBu^t_2)$		43
922	$Mo_2C_{22}H_{52}O_4P_4$	$\{Mo(PMe_3)_2(\eta\text{-}C_2H_4)(\mu\text{-}\eta^2,O:O\text{-}CH_2CHCO_2H)\}_2$		675
923	$Mo_2C_{25}H_{20}O_{11}$	$Mo_2(CO)_3(\eta\text{-}DMAD)(\mu\text{-}\eta^2\text{-}DMAD)(\mu\text{-}\eta^5:\eta^5\text{'-}fulv)$		676
924	$Mo_2C_{26}H_{30}O_{10}S_2$	$\{Mo(CO)_3(Cp^*)\}_2(\mu\text{-}S_2O_4)$		607
925	$Mo_2C_{28}H_{22}O_{10}P_2$	$\{Mo(CO)_5PMes\}_2\text{-}(\underline{P}\text{-}\underline{P})$		301
926	$Mo_2C_{33}H_{44}NO_5P$	$Mo_2(CO)_3(Cp^*)_2\{PhP(OCH_2CH_2)_2NH\}$		677

No.	Formula	Structure	Details	Ref.
927	$Mo_2C_{42}H_{30}O_6$	$Mo_2(CO)_6(\eta-C_5H_4CHPh_2)_2 \cdot THF$		678
928	$Mo_2Si_2C_{20}H_{56}Br_2P_4$	$Mo_2Br_2(CHSiMe_3)_2(PMe_3)_4$	108	654
929	$Mo_2Te_2C_{20}H_{10}O_8$	$\{Mo(CO)_4(\mu-TePh)\}_2$		679
930	$Mo_2Te_2C_{26}H_{20}O_4$	$\{Mo(CO)_2(Cp)(\mu-TePh)\}_2$		680
931	$Mo_2Ti_2C_{54}H_{70}O_4$	$\{(Cp^*)_2Ti(\mu_3-O:C:C-CO)\}_2\{Mo_2(CO)_4Cp_2\}$	220	681
932	$[Mo_3C_8H_{18}Br_3O_9]^+$	$[\{Mo(OH_2)(\mu-Br)(\mu-OAc)\}_3(\mu_3-CMe)][ClO_4] \cdot 4H_2O$		682
933	$[Mo_3SbC_{15}O_{15}]^-$	$[Na(THF)_2][\{Mo(CO)_5\}_3Sb]$	203	39
934	$[Mo_5TiC_5H_5O_{18}]^{3-}$	$[NBu_4]_3[CpTiMo_5O_{18}]$		683
935	$NaPrC_{28}H_{50}Cl_2O_4$	$(DME)_2Na(\mu-Cl)_2Pr(Cp^*)_2$	138	684
936	$NaRhC_{33}H_{39}O_9P$	$(18-c-6)NaOCRh(CO)_2(PPh_3)$		685
937	$NaRhC_{36}H_{40}O_4P_2$	$(THF)_2NaOCRh(CO)(dppe)$		685
938	$NbC_{10}H_{11}O_3S$	$Nb(CO)_3(SMe_2)(Cp)$		686
939	$NbC_{11}H_{10}ClO$	$NbCl(CO)(Cp)_2$		687
940	$NbC_{11}H_{11}O$	$NbH(CO)(Cp)_2$		687
941	$NbC_{20}H_{37}P$	$Nb(PEt_3)(\eta^5-CH_2CMeCHCMeCH_2)_2$		630
942	$NbC_{22}H_{21}Cl_2$	$NbCl_2\{\eta-C_2(pTol)_2\}(Cp')$		688
943	$NbC_{22}H_{41}O_3P_4$	$Nb(OC_6H_3Me_2-3,5)(CO)_2(dmpe)_2$		689
944	$NbC_{24}H_{22}$	$Nb\{(2-CH_2C_6H_4)_2\}(Cp)_2$		690
945	$[NbC_{24}H_{22}]^-$	$[Na(18-c-6)(THF)_2][Nb\{(2-CH_2C_6H_4)_2\}(Cp)_2]$		690
946	$[NbC_{24}H_{22}]^+$	$[Nb(2-CH_2C_6H_4)_2(Cp)_2][BF_4]$		690
947	$NbC_{30}H_{47}S_2$	$(Cp)_2Nb(SCH=CBu^t_2)(\eta^2CS-SC=CBu^t_2)$		691
948	$NbSi_2C_{18}H_{32}$	$Nb(CH_2SiMe_3)_2(Cp)_2$		692
949	$[Nb_2C_8Cl_3O_8]^-$	$[(THF)_2H][Nb_2(CO)_8(\mu-Cl)_3]$		693
950	$Nb_2C_{16}H_{14}Cl_2O_4$	$\{Nb(CO)_2(Cp')(\mu-Cl)\}_2$		688
951	$Nb_2C_{16}H_{14}Cl_4O_4$	$\{NbCl(CO)_2(Cp')(\mu-Cl)\}_2$		688
952	$Nb_2C_{22}H_{24}O$	$\{Nb(Cp)\}_2(\mu-H)(\mu-OEt)(\eta^5,\eta^{5'}-C_{10}H_8)$		694
953	$Nb_2C_{44}H_{42}Cl_2$	$[Nb\{\eta-C_2(pTol)_2\}(Cp')(\mu-Cl)]_2$		688
954	$[Nb_2UW_{10}C_{15}H_{15}O_{38}]^{5-}$	$[NBu_4]_5[U(Cp)_3(ONbW_5O_{18})_2]$		695
955	$NdSi_2C_{27}H_{49}$	$Nd\{CH(SiMe_3)_2\}(Cp^*)_2$	173	696
956	$NdSi_3C_{27}H_{49}$	$Nd\{CH(SiMe_3)_2\}\{\eta,\eta'-(C_5Me_4)_2SiMe_2\}$	148	546

No.	Formula	Structure	Details	Ref.
957	$[Nd_2C_{30}H_{30}O]^{2-}$	$[(phen)_3Cl][\{Nd(Cp)_3\}_2O]$		697
958	NiC_6H_{10}	$Ni(\eta^3-C_3H_5)_2$	N at 100	698
959	$NiC_{14}H_{12}N_2O_2$	$Ni(CO)_2(4,6-Me_2bipy)$		699
960	$[NiC_{17}H_5F_{10}]^-$	$[NEt_4][Ni(C_6F_5)_2(Cp)]$		700
961	$NiC_{17}H_{40}INOP_2$	trans-$NiI(CONEt_2)(PEt_3)_2$		701
962	$NiC_{19}H_8F_{10}$	$Ni(C_6F_5)_2(nbd)$	193	702
963	$NiC_{20}H_{12}F_{10}$	$Ni(C_6F_5)_2(cod)$	190	702
964	$NiC_{20}H_{26}P_2$	$Ni(dmpe)(\eta^2-PhCCPh)$		703
965	$NiC_{24}H_{12}$	$Ni\{(\eta^2-C\equiv CC_6H_4-\underline{o})_3\}$		704
966	$NiC_{25}H_{25}O_2P$	$Ni(\eta^3-CH_2CMeCMeCH_2COO)(PPh_3)$		705
967	$[NiC_{25}H_{28}Cl_5OP_2]^+$	trans-$[Ni(C_6Cl_5)\{C(OMe)Me\}(PMe_2Ph)_2][BF_4]$		706
968	$NiC_{25}H_{43}O_2P$	$Ni(\eta^3-CH_2CMeCMeCH_2COO)(PCy_3)$		705
969	$NiC_{27}H_{33}Cl_3P_2$	trans-$Ni(CCl=CCl_2)(Mes)(PMe_2Ph)_2$		707
970	$NiC_{32}H_{26}Br_6P_2$	trans-$NiBr(C_6Br_5)(PPh_2Me)_2$		708
971	$[NiC_{32}H_{31}P_2]^+$	$[Ni(dppp)(Cp)]_2[B_{12}H_{12}]$		709
972	$NiC_{32}H_{52}P_2$	$Ni(\eta^2-C_6H_4)(dcpe)$		710
973	$NiC_{34}H_{56}P_2$	$Ni(C_6H_4CH_2CH_2-\underline{o})(dcpe)$		710
974	$NiC_{37}H_{66}O_2P_2$	$Ni(PCy_3)_2(\eta^2-CO_2)$		711
975	$NiC_{38}H_{32}P_2$	$Ni(PPh_3)_2(\eta-HCCH)$		712
976	$NiC_{40}H_{50}P_2$	$Ni(PEt_3)_2(\eta^4-C_4Ph_4)$		713
977	$NiC_{44}H_{36}OP_2$	$NiPh(OCPhCHPPh_2)(PPh_3)$		714
978	$NiC_{54}H_{50}P_2$	$Ni(\eta^3-C_3H_4CH_2CPhC_6H_2Ph_2PEt)PEt(C_6H_2Ph_3-2,4,6)\cdot THF$		715
979	$NiOs_3C_{30}H_{25}O_8P$	$(Cp)NiOs_3(CO)_8(PPh_2CCPr^i)$		226
980	$NiSi_2C_3Cl_6O_3$	$Ni(SiCl_3)_2(CO)_3$		716
981	$NiSi_4C_{15}H_{36}Cl_2OP_2$	$Ni(CO)\{\eta^2-ClP=C(SiMe_3)_2\}_2$		717
982	$NiThC_{46}H_{50}O_2P_2$	$(CO)_2Ni(\mu-PPh_2)_2Th(Cp^*)_2$	208	718
983	$NiWC_{44}H_{35}O_3P_2$	$(Ph_3P)_2NiW(CO)(Cp)(\mu-CO)_2$		719
984	$NiZn_2C_{38}H_{35}$	$(Cp)(Ph_3P)NiZn_2(\eta^1-Cp)_2(\mu-Cp)$		720
985	$NiZrC_{50}H_{51}O_4P_3$	$(CO)Ni(\mu-Ph_2PCH_2O)_3Zr(Cp^*)\cdot MePh$		721
986	$Ni_2C_{36}H_{46}P_2$	$Ni_2(PEt_3)_2\{\mu-\eta^1,\eta^2:\eta^1{'},\eta^2{'}-(C_6H_4)_4\}$		722

No.	Formula	Structure	Details	Ref.
987	$Ni_2C_{36}H_{54}$	$\{(Cp^*)Ni\}_2\{\mu-\eta^3:\eta^3{}'-(C_8H_{12})_2\}$		723
988	$Ni_2C_{44}H_{30}Br_2O_2$	$\{Ni(CO)(\eta^3-C_3Ph_3)(\mu-Br)\}_2$		724
989	$[Ni_2C_{49}H_{47}P_2S]^+$	$[\{(Cp)(Ph_3P)NiCH_2\}_2SMe][PF_6].Me_2CO$		725
990	$[Ni_2C_{58}H_{56}N_4P_4]^{2+}$	$[Ni_2(CNMe)_3(\mu-CNMe)(\mu-dppm)_2][PF_6].CH_2Cl_2$	133	726
991	$Ni_2Si_4C_{40}H_{28}O_2P_4$	$[Ni(CO)\{\mu-P:\eta^2-Ph_2PP=C(SiMe_3)_2\}]_2$		717
992	$Ni_2Zn_4C_{30}H_{30}$	$Ni_2Zn_4(Cp)_6$		727
993	$Ni_3C_{24}H_{51}I_2N_5O_4$	$Ni_3(NHEt_2)(\mu-I)_2(\mu-CONEt_2)_4$		701
994	$Ni_3C_{38}H_{56}O_2P_2$	$Ni_3(CH_2C_6H_4Me\text{-}\underline{o})(PMe_3)_2(\mu_3\text{-}OH)_2$		728
995	$Ni_4Si_6C_{43}H_{99}N_9P_2$	$Ni_4(CNBu^t)_4(\mu_3-\eta^2-CNBu^t)(\mu_3-\eta^2-R_2NPNR_2)$ (R=SiMe$_3$)		729
996	$[Ni_8C_{17}O_{16}]^{2-}$	$[NBu_4]_2[Ni_8C(CO)_8(\mu-CO)_8]$		730
997	$[Ni_9C_{18}O_{17}]^{2-}$	$[NBu_4]_2[Ni_9C(CO)_9(\mu-CO)_4(\mu_3-CO)_4]$		730
998	$[Ni_{10}C_{18}O_{16}]^{2-}$	$[AsPh_4]_2[Ni_{10}C_2(CO)_6(\mu-CO)_{10}]$		731
999	$[Ni_{16}C_{27}O_{23}]^{4-}$	$[NMe_4]_4[Ni_{16}C_2(CO)_{12}(\mu-CO)_{10}(\mu_3-CO)]$		732
1000	$[Ni_{38}Pt_6C_{48}HO_{48}]^{5-}$	$[AsPh_4]_2[NBu_4]_3[Ni_{38}Pt_6H(CO)_{48}]$		733
1001	$[Ni_{38}Pt_6C_{48}H_2O_{48}]^{4-}$	$[AsPh_4]_4[Ni_{38}Pt_6H_2(CO)_{48}]$		733
1002	$[OsC_{19}H_{25}]^+$	$[Os(\eta^2,\eta^3\text{-}CCH\text{-}C_{10}H_{13})(\eta^6\text{-}MesH)][PF_6]$		734
1003	$OsC_{24}H_{44}$	$Os(Cy)_4$		735
1004	$[OsC_{24}H_{46}N_6]^{2+}$	$[Os(CNBu^t)_2(NH_2NCMe_2)_2(cod)][BPh_4]_2.2Me_2CO$		736
1005	$OsC_{28}H_{28}$	$Os(oTol)_4$		735
1006	$[OsC_{28}H_{47}P_4]^+$	$[Os_3(\eta^3\text{-}PhCCC=CHPh)(PMe_3)_4][PF_6]$		737
1007	$[OsC_{38}H_{33}O_3P_2]^+$	$[OsH(OH_2)(CO)_2(PPh_3)_2][BF_4].EtOH$		738
1008	$OsC_{40}H_{30}F_3O_3P_3$	$Os\{P=C(O)CF_3\}(CO)_2(PPh_3)_2$	183	739
1009	$[OsC_{54}H_{49}O_2P_4]^+$	<u>trans</u>-$[Os(CHO)(CO)(dppe)_2][SbF_6].CH_2Cl_2$		740
1010	$OsRhC_{50}H_{40}BrCl_2O_2P_4$	$Cl_2Os(\mu-CO)_2(\mu-dppm)_2RhBr.CH_2Cl_2$		741
1011	$OsSbC_{22}H_{15}O_4$	$\underline{eq}\text{-}Os(CO)_4(SbPh_3)$		47
1012	$[Os_2C_8O_8]^{2-}$	$[PPh_4]_2[Os_2(CO)_8].Sv$		742
1013	$[Os_3C_{11}H_2IO_{10}]^-$	$[PPN][Os_3(CO)_{10}(\mu-CH_2)(\mu-I)]$		743
1014	$Os_3C_{11}H_9NO_8S$	$Os_3(CO)_8(CHNMe_2)(\mu-H)_2(\mu_3-S)$		744
1015	$[Os_3C_{12}H_2NO_{11}]^-$	$[PPN][Os_3(CO)_{10}(\mu-CH_2)(\mu-NCO)]$		743
1016	$Os_3C_{12}H_4O_{12}$	$Os_3(CO)_{10}(\mu-H)(\mu-OAc)$		745

Structures Determined by Diffraction Methods

No.	Formula	Structure	Details	Ref.
1017	$Os_3C_{13}H_8N_2O_{10}$	$Os_3(CO)_{10}(\mu-H)(\mu-NCHNMe_2)$		746
1018	$Os_3C_{14}H_4O_{11}$	$Os_3(CO)_{10}(\mu-H)(\mu-\eta^2:\eta^2-C_4H_3O)$		747
1019	$Os_3C_{15}H_6O_9$	$Os_3(CO)_9(\mu_3-\eta^6-PhH)$		748
1020	$Os_3C_{17}H_6O_{10}$	$Os_3(CO)_{10}(\mu-H)(\mu-CPh)$		749
1021	$Os_3C_{17}H_7NO_{10}$	$Os_3H(CO)_{10}(\mu-\eta^2:\eta^1,N-HC=CHC_5H_4N)$		750
1022	$Os_3C_{19}H_7O_{10}$	$Os_3(CO)_{10}(\mu-H)(\mu-C:N-C_9H_6N)$		751
1023	$Os_3C_{19}H_{12}O_{11}$	$Os_3(CO)_9\{C(OMe)_2\}(\mu-H)(\mu_3-CPh)$		752
1024	$Os_3C_{24}H_{18}NO_9P$	$Os_3H(CO)_9(PMe_2Ph)(\mu-\eta^2:\eta^1,N-HC=CHC_5H_4N)$		750
1025	$Os_3C_{26}H_{14}O_8$	$Os_3(CO)_8(\mu-H)_2(\mu-C:N-C_9H_6N)_2$		751
1026	$Os_3C_{27}H_{17}O_{10}P$	$Os_3(CO)_9(PPh_3)(\mu-H)(\mu-OH)$		753
1027	$Os_3C_{32}H_{22}O_8P_2$	$Os_3(CO)_8(\mu-H)_2(\mu-PPh_2)_2 \cdot Sv$		442
1028	$Os_3C_{33}H_{24}O_8P_2$	$Os_3(\mu-H)_2(CO)_8(\mu-dppm)$		754
1029	$Os_3PtC_{25}H_{22}O_9P_2S_3$	$(Me_2PhP)_2PtOs_3(CO)_9(\mu_3-S)_2$		755
1030	$Os_3PtC_{28}H_{37}O_{10}P$	$Os_3Pt(CO)_{10}(PCy_3)(\mu-H)_4$	210	756
1031	$Os_3PtC_{29}H_{35}O_{10}P$	$(PCy_3)(CO)PtOs_3(CO)_9(\mu-H)_2(\mu_4-C)$	200	757
1032	$Os_3PtC_{29}H_{35}O_{11}P$	$Os_3Pt(CO)_{11}(PCy_3)(\mu-H)_2$		756
1033	$Os_3PtC_{29}H_{37}O_{10}P$	$Os_3Pt(CO)_{10}(PCy_3)(\mu-H)_2(\mu-CH_2)$	220	756
1034	$Os_3Pt_2C_{47}H_{68}O_{10}P_2$	$(PCy_3)_2(CO)Pt_2Os_3(CO)_9(\mu-H)_2(\mu_5-C)$		757
1035	$Os_3Pt_2C_{48}H_{70}O_{11}P_2$	$(PCy_3)_2(CO)Pt_2Os_3(CO)_9(\mu-H)(\mu-OMe)(\mu_5-C)$		757
1036	$Os_3SiC_{16}H_{16}O_{11}$	$Os_3(CO)_{10}(\mu-H)(\mu-OSiEt_3)$		758
1037	$Os_3Si_2C_{16}H_{20}N_2O_{10}S$	$Os_3(CO)_{10}(\mu-H)\{\mu-NHSN(SiMe_3)_2\}$		759
1038	$Os_3Si_4SnC_{24}H_{40}O_{10}$	$Os_3(CO)_9(\mu-H)_2[\mu-Sn\{CH(SiMe_3)_2\}OCCH(SiMe_3)_2]$		760
1039	$Os_3Si_4SnC_{29}H_{46}O_{13}$	$Os_3(CO)_9[\mu-Sn\{CH(SiMe_3)_2\}_2]\{\mu-OC(OMe)CH_2CCO_2Me\}$		760
1040	$Os_3WC_{14}O_{14}S_2$	$Os_3(CO)_9(\mu_3-S)(\mu_4-S)W(CO)_5$		761
1041	$Os_3WC_{23}H_{14}O_{10}$	$(Cp)WOs_3(CO)_9(\mu-O)\{\mu_3-CCH_2(pTol)\}$		762
1042	$Os_3WC_{23}H_{14}O_{10}$	$CpWOs_3(CO)_9(\mu-H)(\mu-O)\{\mu-C=CH(pTol)\}$		763
1043	$Os_3WC_{23}H_{16}O_{10}$	$(Cp)WOs_3(CO)_9(\mu-H)(\mu-O)\{\mu-CHCH_2(pTol)\}$		764
1044	$Os_4C_{11}H_3NO_{12}$	$Os_4(CO)_{11}(NO)(\mu-H)_3$		765
1045	$[Os_4C_{12}H_2IO_{12}]^-$	$[PPN][Os_4I(CO)_{12}(\mu-H)_2]$		766
1046	$Os_4C_{13}H_3BrO_{13}$	$Os_4HBr(CO)_{13}(\mu-H)_2$		767

No.	Formula	Structure	Details	Ref.
1047	$Os_4C_{18}H_9O_{15}P$	$(Me_3P)(CO)_4Os\{Os_3(CO)_{11}\}$		768
1048	$Os_4C_{20}H_6O_{12}S$	$Os_4(CO)_{12}(\mu_3-S)(\mu_4-\mu_3C:\eta^2-CCHPh)$		769
1049	$Os_4C_{20}H_6O_{12}S$	$Os_4(CO)_{12}(\mu_4-\mu S:\eta^2:\eta^1-SCPhCH)$		769
1050	$Os_6C_{21}H_6O_{17}$	$Os_6(CO)_{17}(\mu-H)(\mu_4-\eta^2-CCEt).CH_2Cl_2$		770
1051	$Os_6C_{21}H_6O_{17}$	$Os_6(CO)_{17}(\mu_4-\eta^2,\eta^2-HCCEt)$		770
1052	$Os_6C_{21}H_8O_{16}$	$Os_6(CO)_{16}(\mu_3-\eta^2-MeCCEt)$		770
1053	$Os_6Pt_2C_{32}H_{24}O_{16}$	$(cod)_2Pt_2Os_6(CO)_{16}$		771
1054	$[Os_{10}C_{23}INO_{23}]^{2-}$	$[PPN]_2[Os_{10}Cl(CO)_{22}(NO)]$		772
1055	$[Os_{10}C_{25}IO_{24}]^-$	$[PPN][Os_{10}C(CO)_{24}(\mu-I)]$		772
1056	$Os_{10}C_{25}I_2O_{24}$	$Os_{10}C(CO)_{24}(\mu-I)_2$		772
1057	$Os_{10}C_{27}H_9I_2O_{26}P$	$Os_{10}C(CO)_{23}\{P(OMe)_3\}(\mu-I)_2.CH_2Cl_2$		772
1058	$Os_{10}C_{34}H_{36}O_{33}P_4$	$Os_{10}C(CO)_{21}\{P(OMe)_3\}_4$ (2 isomers)		772
1059	$PbC_3H_9N_3$	Me_3PbN_3	173	153
1060	$PbC_{16}H_{18}OS_2$	$Ph_2Pb(SCH_2CH_2)_2O$	113	773
1061	$PbC_{22}H_{25}O_2PS_2$	$Ph_3PbSPS(OEt)_2$		774
1062	$PbC_{23}H_{23}NS_2$	$PbPh_4(S_2CNC_4H_8)$		775
1063	$PbC_{23}H_{26}INO$	$PbI(pTol)(C_6H_4OMe-\underline{p})(C_6H_4CH_2NMe_2-\underline{o})$		776
1064	$PbC_{24}H_{20}S$	Ph_3PbSPh		774
1065	$PbC_{40}H_{38}O_4P_2S_4$	$Ph_2Pb\{S_2P(OBz)_2\}_2$		774
1066	$Pb_2C_{36}H_{66}$	Pb_2Cy_6	148	777
1067	$Pb_2C_{39}H_{36}$	$Ph_2(pTol)PbPb(pTol)_2Ph$	233	778
1068	$Pb_2Si_{10}C_{32}H_{90}N_2S_2$	$[Pb\{N(SiMe_3)_2\}\{\mu-SC(SiMe_3)_3\}]_2$		484
1069	$PdC_9H_{14}Cl_2$	$PdCl_2(\eta^4-1,4-C_8H_{11}Me-3)$		779
1070	$[PdC_{12}H_{20}N]^+$	$[Pd\{\eta^3,\eta^2-CH_2CHCMe(CH_2)_2CH=CMe_2\}(NCMe)][BF_4]$		780
1071	$[PdC_{12}H_{20}N_4O_4]^{2+}$	$[Pd(CNHCH_2CH_2O)_4]Cl_2$		781
1072	$PdC_{17}H_{16}N_2O_5$	$Pd\{CH(CO_2Me)COCH(CO_2Me)\}(bipy).H_2O$		55
1073	$PdC_{18}H_{19}ClN_2O_4$	$PdCl\{C(CO_2Et)_2CH_2(C_{10}H_7N_2)\}$		782
1074	$PdC_{21}H_{27}ClF_5N_3$	$\underline{trans}-PdCl\{C(NBu^t)C_6F_5\}(CNBu^t)_2$		783
1075	$PdC_{22}H_{22}ClP$	$PdCl(PPh_3)(\eta^3-CH_2CMeCH_2)$		101
1076	$PdC_{26}H_{30}N_2O_8$	$Pd\{C(CO_2Et)_2CH_2C_5H_3N-2,2'\}_2.KNO_3.Sv$		782

No.	Formula	Structure	Details	Ref.
1077	$PdC_{26}H_{31}ClN_2O_8$	$PdCl\{C(CO_2Et)_2CH_2C_{10}H_6N_2CH_2CH(CO_2Et)_2\}$		782
1078	$PdC_{28}H_{34}N_2O_{10}$	$Pd(OAc)\{C(CO_2Et)_2CH_2C_{10}H_6N_2CH_2CH(CO_2Et)_2\}$		782
1079	$PdC_{32}H_{39}ClN_2O_8$	$PdCl\{C(CO_2Pr^i)_2CH_2C_{10}H_6N_2CH_2CH(CO_2Pr^i)_2\}$		782
1080	$PdC_{34}H_{38}O_5P_2$	$Pd\{CH(CO_2Me)COCH(CO_2Me)\}(PPh_3)_2 \cdot H_2O$		55
1081	$PdC_{39}H_{29}F_5O_3P_2$	$Pd(C_6F_5)(\eta^1-C_7H_7O_3)(Ph_2PCHCHPPh_2)$		784
1082	$PdC_{40}H_{35}IO_2P_2$	$PdI(PPh_3)(\eta^2-Ph_3PCHCHCO_2Me)$	153	785
1083	$PdC_{43}H_{43}ClOP_2$	trans-$PdCl\{C(O)(CH_2)_5Me\}(PPh_3)_2$		786
1084	$[PdC_{53}H_{53}P_2]^+$	$[Pd(Me_2dppe)\{\eta^3-PhCHCHC(Xy-3,5)_2\}][BPh_4] \cdot Sv$		787
1085	$PdSiC_{25}H_{30}ClP$	$PdCl(PPh_3)(\eta^3-\underline{anti}-CH_2CHCMeSiMe_3)$		788
1086	$PdSi_{12}Sn_3C_{36}H_{108}N_6$	$Pd[Sn\{N(SiMe_3)_2\}_2]_3$		789
1087	$Pd_2C_{17}H_{35}BrO_2P_2S$	$Pd_2Br(PEt_3)_2(Cp)(\mu-SO_2)$		790
1088	$Pd_2C_{20}H_{28}Cl_2N_2O_2$	$[Pd\{C_6H_3(OMe)CH_2NMe_2\}\{(\mu-Cl)\}]_2$ (3 isomers)		791
1089	$Pd_2C_{20}H_{36}Cl_2N_4$	trans-$Pd_2Cl_2(CNBu^t)_4$		792
1090	$Pd_2C_{26}H_{38}Cl_2N_4$	$\{Pd(CH_2CMe_2CMe=NNMePh)(\mu-Cl)\}_2$		793
1091	$Pd_2C_{26}H_{38}Cl_2N_4$	$\{Pd(CH_2CBu^t=NNMePh)(\mu-Cl)\}_2$		793
1092	$Pd_2C_{31}H_{45}Cl_2OP_3$	$Pd_2Cl_2(PEt_2Ph)_3(\mu-CO)$		794
1093	$Pd_2C_{34}H_{34}N_2O_6$	$[Pd\{C_6H_3(OMe)CH=N(oTol)-4,2\}(\mu-OAc)]_2 \cdot CH_2Cl_2$		795
1094	$Pd_2C_{36}H_{34}Cl_2N_4$	$\{PdCl(NC_5H_4Me-4)(\mu-PhCHC_5H_4N-2)\}_2 \cdot CH_2Cl_2$		796
1095	$Pd_2C_{38}H_{30}Cl_2N_4$	$\{Pd(C_6H_4CPh=NNHPh-\underline{o})(\mu-Cl)\}_2$		797
1096	$Pd_2C_{40}H_{26}Cl_2F_8N_2P_2$	$[PdCl\{\mu N-C(C_6F_4PPh_2-\underline{o})NMe\}]_2$		798
1097	$Pd_2C_{46}H_{38}Cl_2N_2P_2$	trans-$\{PdCl(PPh_3)\}_2(\mu-N:C-C_5H_4N)_2$		799
1098	$[Pd_2C_{51}H_{48}ClP_4]^+$	$MePd(\mu-Cl)(\mu-dppm)_2PdH[BPh_4]$		800
1099	$Pd_2C_{52}H_{50}Cl_2P_4$	$(PdClMe)_2(\mu-dppm)_2$		800
1100	$Pd_2C_{57}H_{49}Cl_2NP_4$	$Pd_2Cl_2(\mu-CNPh)(\mu-dppm)_2$		801
1101	$Pd_2C_{60}H_{55}Cl_2NO_2P_4$	$(PdCl)_2(\mu-CNC_5H_3CHCH_2CO_2Me)(\mu-dppm)_2$		802
1102	$Pd_2Pt_3C_{105}H_{102}P_8$	$Pd_2Pt_3(PPh_3)_5(\mu_4-\eta^2:\mu_3P-PCBu^t)_3$		803
1103	$Pd_2Rh_2C_{14}H_{14}Cl_2N_6O_4$	$(\eta^3-C_3H_5)_2Pd_2(\mu_3-NCHNCHN)Rh_2Cl_2(CO)_4$		804
1104	$[Pd_3C_{76}H_{66}OP_6]^{2+}$	$[Pd_3(\mu-dppm)_3(\mu_3-CO)][Cl][CF_3CO_2] \cdot H_2O$		805
1105	$Pd_3Si_{12}Sn_3C_{39}H_{108}N_6O_3$	$Pd_3(CO)_3[\mu-Sn\{N(SiMe_3)_2\}_2]_3$		806
1106	$Pd_4C_{28}H_{36}N_4O_{16}$	$\{Pd_2(\mu-OAc)_2(\mu-N:\eta^1-NCCHCH_2OAc)_2\}_2$		807

No.	Formula	Structure	Details	Ref.
1107	$Pd_4C_{77}H_{60}O_5P_4$	$Pd_4(PPh_3)_4(\mu-CO)_5$		794
1108	$PrC_{19}H_{21}N_2$	$Pr(NCMe)_2(Cp)_3$		159
1109	$PtC_9H_{19}Cl_2N_3$	trans-$PtCl_2\{N(CHMe_2)CHCHNNMe_2\}(\eta-C_2H_4)$		808
1110	$PtC_{10}H_{22}Br_2N_4S_2$	$PtBr_2\{CH_2NMeC(NMe_2)S\}_2$		809
1111	$PtC_{11}H_{16}Cl_2O_2$	$PtCl_2\{\eta^2,\eta^{2'}-CH_2=CHCH_2)_2C(COMe)_2\}$		810
1112	$PtC_{12}H_{20}N_2O_6$	$Pt\{C(OH)C(O)OCH(CO)CH(O)CH_2OH\}\{(NH_2)_2C_6H_{10}\}\cdot 3H_2O$		811
1113	$PtC_{12}H_{28}Cl_2N_4$	$PtCl_2\{N(NMe_2)CHMeCHMeNNMe_2\}(\eta-C_2H_2Me_2)$		812
1114	$[PtC_{13}H_{19}O_2]^+$	$[Pt(acac)(cod)][BF_4]$		813
1115	$PtC_{14}H_{28}Cl_2N_2$	trans-$PtCl_2(Bu^tNCHCHNBu^t)(\eta-\underline{E}-MeCHCHCMe)$		808
1116	$PtC_{17}H_{20}Cl_3NO_3$	$PtCl_3(\eta^2$-morphinium$)$		814
1117	$PtC_{17}H_{21}IN_2O_2$	$PtIMe_2(O_2Pr^i)(phen)$		815
1118	$[PtC_{20}H_{29}N_2]^+$	$[Pt(oTol)\{C_6H_3(CH_2NMe_2)_2-2,6\}][I]$		816
1119	$PtC_{37}H_{48}OP_2$	$PtH\{\underline{o}-C(O)C_6H_4PPh_2\}(PCy_3)$		817
1120	$PtC_{39}H_{32}O_2P_2$	$Pt(CH=CHCOO)(PPh_3)_2$		818
1121	$PtC_{39}H_{32}O_4P_2$	$Pt\{CH(CO_2H)C(O)O\}(PPh_3)_2$		819
1122	$PtC_{43}H_{38}O_3P_2$	$Pt\{CH(COMe)COCH(COMe)\}(PPh_3)_2$		820
1123	$PtC_{46}H_{39}NO_4P_2$	$Pt\{CH(CONHBz)C(O)OO\}(PPh_3)_2$		821
1124	$PtC_{50}H_{42}P_2$	$Pt\{\eta^1,\eta^{1'}-(C_7H_6)_2\}$		822
1125	$[PtC_{52}H_{43}P_2]^+$	$[Pt(PPh_3)(\eta^3-PhCCPhC=CH_2)][BF_4]$		823
1126	$[PtReC_{53}H_{45}NOP_3]^+$	$[(Ph_3P)_2PtRe(NO)(Cp)(\mu-H)(\mu-PPh_2)][BF_4]$		824
1127	$[PtRhC_{54}H_{47}ClOP_4]^+$	$[ClPtRh(CO)(\mu-\eta^1:\eta^2-CCMe)(\mu-dppm)_2][PF_6]$		825
1128	$[PtRhC_{57}H_{50}OP_4]^+$	$[(MeCC)PtRh(CO)(\mu-\eta^1:\eta^2-CCMe)(\mu-dppm)_2][PF_6]\cdot Sv$		825
1129	$PtSeC_{51}H_{43}P_3S$	$Pt_2(CS)(PPh_3)\{Ph_2PCH_2)_2C_6H_4-\underline{o}\}(\mu-Se)$		826
1130	$PtSe_2C_7H_{17}Cl$	$PtClMe_3(MeSeCH=CHSeMe)$		827
1131	$PtSe_2C_7H_{17}I$	$PtIMe_3(MeSeCH=CHSeMe)$		827
1132	$PtSi_2C_{42}H_{48}N_4P_2S_2$	cis-$[Pt(NSNSiMe_3)_2(PPh_3)_2]$		828
1133	$PtSi_{12}Sn_3C_{36}H_{108}N_6$	$Pt[Sn\{N(SiMe_3)_2\}_2]_3$		829
1134	$PtSnC_{19}H_{35}Cl_3OP_2$	trans-$Pt(SnCl_3)(COPh)(PEt_3)_2$		830
1135	$PtWC_{22}H_{32}O_2P_2$	$(Cp)(CO)_2WPt(PMe_3)_2\{\mu-\eta^2:\eta^1-CH_2C(pTol)\}$		831
1136	$PtWC_{38}H_{30}O_3P_2$	$(Ph_3P)(CO)PtW(CO)_2(Cp)(\mu-PPh_2)$		832

Structures Determined by Diffraction Methods   461

No.	Formula	Structure	Details	Ref.
1137	$PtWC_{42}H_{30}O_5P_2S_2$	$(Ph_3P)_2Pt(\mu-\eta^2:S-CS_2)W(CO)_5$		833
1138	$PtW_4C_{60}H_{112}O_{10}P_2$	trans-$Pt(PMe_2Ph)_2\{\mu_3-\eta^2-C_2W_2(OBu^t)_5\}_2$.Sv	118	834
1139	$Pt_2C_{11}H_{18}Cl_4$	$(\eta-C_2H_4)ClPt(\mu-Cl)_2PtCl(\eta^2,\eta^1,\eta^1-C_9H_{14})$		835
1140	$[Pt_2C_{13}H_{83}IP_4]^+$	$LPt(\mu-I)(\mu-dmpm)_2PtMe_3][I_3]$ (L=0.45I+0.55Me)		836
1141	$[Pt_2C_{14}H_{40}IP_4]^+$	$[Me_3Pt(\mu-I)(\mu-dppm)_2PtMe][I]$		837
1142	$Pt_2C_{16}H_{46}I_2P_4$	$\{Me_3Pt(\mu-dmpm)\}_2(\mu-I)$		838
1143	$Pt_2C_{39}H_{36}I_2P_2S_2$	$Pt_2I_2(PPh_3)_2(\mu-SMe)(\mu-CSMe).Me_2CO$		839
1144	$Pt_2C_{40}H_{42}N_4$	$\{PtPh(NC_5H_4Bu^t-\underline{p})\}_2\{\mu-(bipy-2H)\}$		840
1145	$Pt_2C_{51}H_{46}Cl_2P_4$	$Pt_2Cl_2(\mu-CH_2)(\mu-dppm)_2$		841
1146	$[Pt_2C_{60}H_{57}P_4]^+$	$[Pt_2(dppe)_2(\mu-H)(\mu-CHCH_2Ph)][BF_4]$		842
1147	$Pt_2C_{66}H_{64}P_4$	$Me_2Pt(\mu-dppm)_2Pt(oTol)_2$		843
1148	$[Pt_2C_{70}H_{63}ClP_5]^+$	$[Pt_2Cl(CH_2PPh_3)(\mu-CH_2)(\mu-dppm)_2][PF_6]$		841
1149	$[Pt_2RhC_{80}H_{72}P_4S_2]^+$	$[(Ph_3P)_4Pt_2(\mu_3-S)_2Rh(cod)][PF_6].CH_2Cl_2$		844
1150	$Pt_2Si_8Sn_2C_{36}H_{102}Cl_4N_4P_2$	trans-$[PtSnCl\{N(SiMe_3)_2\}_2(PEt_3)(\mu-Cl)]_2$		829
1151	$Pt_3C_{63}H_{78}N_5P$	$Pt_3(CNR)_2(PCy_3)(\mu-CNR)_3$ (R=2,6-$Me_2C_6H_3$)		845
1152	$Pt_3C_{64}H_{93}N_3OP_2$	$Pt_3(CNR)(PCy_3)_2(\mu-CO)(\mu-CNR)_2$.Sv (R=2,6-$Me_2C_6H_3$)		845
1153	$Pt_3Si_2C_{42}H_{90}N_8P_2$	$Pt_3(CNBu^t)(\mu-CNBu^t)(\mu-PNBu^t(NBu^tSiMe_3)\}_2$		846
1154	$Pt_3Si_3C_{36}H_{81}N_6O_3P_3$	$Pt_3(CO)_3\{\mu-P(NBu^t)(NBu^tSiMe_3)\}_3$		846
1155	$Pt_3Si_{12}Sn_3C_{39}H_{108}N_6O_3$	$Pt_3(CO)_3[\mu-Sn\{N(SiMe_3)_2\}_2]_3$		806
1156	$Pt_4C_{16}H_{48}S_4$	$\{PtMe_3(\mu-SMe)\}_4$		847
1157	$ReC_8H_{12}IO$	$ReOI(\eta-MeCCMe)_2$	173	848
1158	$ReC_{11}H_{13}O_2$	$Re\{CH_2(CH_2)_2CH_2\}(CO)_2(Cp)$		849
1159	$ReC_{11}H_{30}ClN_3O_9P_3$	mer-$ReCl(N_2)(CNMe)\{P(OMe)_3\}_3$		850
1160	$ReC_{13}H_{24}Cl_4P$	$ReCl_4(PMe_3)(Cp^*)$		851
1161	$[ReC_{13}H_{39}NOP_4]^+$	$[ReMe(NO)(PMe_3)_4][Cp]$	183	852
1162	$ReC_{21}H_{15}O_2$	$Re(CO)_2(\eta-PhCCPh)(Cp)$		853
1163	$ReC_{21}H_{20}O_4P$	$Re\{CH_2(CH_2)_4PMe_2\}(CO)_4$		578
1164	$ReC_{22}H_{15}Cl_8O_4$	$Re(O_2C_6Cl_4-\underline{o})_2(Cp^*)$		854
1165	$ReC_{24}H_{25}O_4$	$ReO\{OCPh_2C(O)O\}(Cp^*)$		855
1166	$ReC_{24}H_{35}P_2$	$ReH_2(PMe_2Ph)_2(\eta^5-C_8H_{11})$	113	856

No.	Formula	Structure	Details	Ref.
1167	$ReC_{32}H_{39}P_2$	$ReH_4(dppe)(\eta^3-C_3H_4Pr^i).Sv$		857
1168	$ReC_{33}H_{31}NO_2P$	$Re\{C(O)CHMeCH_2Ph\}(NO)(PPh_3)(Cp).Sv$		858
1169	$ReC_{35}H_{30}NOP_2$	$Re(NO)(PPh_2)(PPh_3)(Cp)$		859
1170	$ReC_{41}H_{39}P_2$	$ReH_3(PPh_3)_2(\eta^4-C_5H_6)$		860
1171	$ReC_{42}H_{35}N_2O_2P_2S$	$Re(NMeCSCHCHN)(CO)_2(PPh_3)_2$		861
1172	$ReC_{44}H_{40}NO_4P_2$	$Re(NHCMeCHC(OEt)O)(CO)_2(PPh_3)_2$		862
1173	$ReC_{48}H_{79}P_2$	$ReH_3(PCy_3)_2(\eta^4-C_{12}H_{10})$	115	863
1174	$ReC_{52}H_{41}O_4P_2$	$Re(O_2CCHPh_2)(CO)_2(PPh_3)_2$		861
1175	$ReC_{57}H_{57}ClNP_4$	<u>trans</u>-$ReCl(CNBu^t)(dppe)_2.THF$		864
1176	$ReSnC_{10}H_{10}Cl_3$	$Re(SnCl_3)(Cp)_2$		865
1177	$ReSnC_{12}H_{16}Cl$	$Re(SnMe_2Cl)(Cp)_2$		865
1178	$ReWC_{21}H_{10}O_9$	$(CO)_5WRe(CO)_4(\mu-PPh_2)$		866
1179	$ReWC_{32}H_{23}O_7P_2$	$(CO)_4WReMe(CO)_3(\mu-PPh_2)_2$		867
1180	$Re_2C_6H_{18}O_3$	$(ReMe_3O)_2O$		565
1181	$[Re_2C_9IO_9]^-$	$[NEt_4][Re_2I(CO)_9]$		868
1182	$Re_2C_{12}O_{12}$	$(CO)_5ReC(O)C(O)Re(CO)_5$	220	869
1183	$Re_2C_{13}H_9N_3O_7$	$Re_2(CO)_6(CNMe)_3$		870
1184	$Re_2C_{14}H_6O_{10}$	<u>trans</u>-$(CO)_5ReCH_2CH=CHCH_2Re(CO)_5$	183	871
1185	$Re_2C_{14}H_9NO_9$	$Re_2(CO)_9(CNBu^t)$		870
1186	$Re_2C_{15}H_8O_8$	$Re_2(CO)_8(\mu-H)(\mu-\eta^1:\eta^2-C_7H_7)$	250	872
1187	$Re_2C_{20}H_{20}O_7P_2$	$Re_2(CCPh)(CO)_7(\mu-H)(\mu-dmpm)$		873
1188	$Re_2C_{23}H_{46}N_2O_6P_4$	$Re_2Me(COMe)(CO)(NO)_2(PMe_3)_4\{\mu-C(C_5H_4)OCO\}$		874
1189	$[Re_2C_{24}H_{15}O_9]^-$	$[NEt_4][Re_2(CO)_6(\mu-OPh)_3]$		875
1190	$Re_2C_{24}H_{22}O_8P_2$	<u>diaxial</u>-$Re_2(CO)_8(PMe_2Ph)_2$		876
1191	$Re_2C_{26}H_{10}N_2O_8$	$Re_2(CO)_8(CNXy-2,6)_2$		870
1192	$Re_2C_{30}H_{36}O_4$	$\{Re(CO)_2(Cp^*)\}_2(\mu-\eta^2:\eta^{2'}-PhH)$		877
1193	$Re_2C_{40}H_{30}O_5P_2$	$Re_2(CO)_5(dppm)(\mu-\eta^1:\eta^2-CHCH_2)(\mu-\eta^1:\eta^2-CCPh)$		873
1194	$Re_2C_{42}H_{36}N_4O_6$	$Re_2(CO)_6(CNXy-2,6)_2$		870
1195	$Re_2C_{44}H_{28}O_8P_2$	$(Ph_3P)(CO)_3Re\{C_6H_3(CO)(PPh_2)\}Re(CO)_4$		878
1196	$Re_2C_{52}H_{44}Cl_4O_2P_4$	$Re_2Cl_3(CO)(\mu-Cl)(\mu-CO)(\mu-dppm)_2$		879

Structures Determined by Diffraction Methods 463

No.	Formula	Structure	Details	Ref.
1197	$[Re_2C_{55}H_{49}Cl_3NO_2P_4]^+$	$[Re_2Cl_2(NCEt)(CO)(\mu-Cl)(\mu-CO)(\mu-dppm)_2][PF_6].Sv$		880
1198	$[Re_2C_{65}H_{72}Cl_3N_3P_4]^+$	$[Re_2Cl_3(CNBu^t)_2(\mu-CNHBu^t)(\mu-dppm)_2][PF_6].Sv$	133	881
1199	$Re_2Si_6C_{24}H_{66}O_3$	${Re(CH_2SiMe_3)_3O}_2O$		565
1200	$[Re_3C_{10}H_4]^-$	$[PPN][Re_3(CO)_{10}(\mu-H)_4]$		882
1201	$[Re_3C_{16}H_3F_5O_{11}]^-$	$[NEt_4][Re_3(CO)_{10}(\mu-H)_3(\mu-OC_6F_5)]$		875
1202	$[Re_3C_{30}H_{45}O_6]^{2+}$	$[Re_3(Cp^*)_3(\mu-O)_6][ReO_4]_2.Sv$		883,884
1203	$[Re_4C_{16}IO_{15}]^-$	$[PPN][Re_4CI(CO)_{15}]$		885
1204	$Re_4C_{18}H_9Cl_2O_{15}P_3$	$Re_4(CO)_{15}(\mu-Cl)_2(\mu_4-MePPMePMe)$		886
1205	$Re_4C_{20}H_{30}O_{11}$	$(Cp^*)OReReO(OReO_3)_2(Cp^*)(\mu-O)_2.CH_2Cl_2$		883
1206	$Re_6C_{21}H_9O_{18}P_3$	$Re_6(CO)_{18}(\mu_4-PMe)_3$		887
1207	$Re_6C_{22}H_9O_{19}P_3$	$Re_5(CO)_{14}(\mu-PMe_2)\{\mu_3-PRe(CO)_5\}(\mu_4-PMe).PhMe$		887
1208	$[RhC_7H_8O_9P_3]^{2-}$	$[NBu_4]_2[Rh(P_3O_9)(nbd)]$		888
1209	$RhC_{13}H_5F_8$	$Rh\{\eta^4-C_6F_6(CFCF)-1,2\}(Cp)$ (2 isomers)	163	889
1210	$RhC_{13}H_8ClN_2O_5$	$Rh(C_6H_4NOO-\underline{o})_2Cl(CO)$		890
1211	$RhC_{13}H_{17}O$	$Rh\{\eta^4-(CH_2=CH)_2CH_2\}(\eta-C_5H_4CHO)$		891
1212	$[RhC_{14}H_{15}INP]^+$	$[RhI(CH_2NC_5H_5)(PMe_3)(Cp)][PF_6]$		892
1213	$RhC_{14}H_{16}F_3O_4$	$Rh(trifluoromenthonato)(CO)_2$		893
1214	$RhC_{15}H_{24}ClN_2$	$RhCl(CNMeCH_2C_4H_7N)(cod)$		187
1215	$RhC_{15}H_{27}BrP$	$RhBr(CH_2CH_2CH_2PMe_2)(Cp^*)$		894
1216	$[RhC_{17}H_{20}O_2]^+$	$Rh(\eta^4-duroquinone)(\eta^6-PhMe)][PF_6]$		895
1217	$RhC_{17}H_{28}ClN_2$	$\underline{cis}-RhCl(CNMeC_6H_{10}NMe)(cod)$		187
1218	$RhC_{19}H_{27}BrP$	$RhBr(C_6H_4CH_2PMe_2-\underline{o})(Cp^*)$		894
1219	$[RhC_{21}H_{18}F_4]^+$	$[Rh(\eta^4-tfb)(\eta^6-MesH-1,2,4)][ClO_4]$		896
1220	$RhC_{21}H_{19}Cl_2N_4$	$RhCl_2(C_6H_4NHC_5H_4N-2)(py)_2$		897
1221	$[RhC_{21}H_{24}N_2O]^+$	$[Rh(cod)\{NC_5H_4C(O)CHNC_5H_4Me-4\}][ClO_4]$		898
1222	$RhC_{21}H_{46}ClP_2$	$RhCl(C=CHMe)(PPr^i_3)_2$		899
1223	$RhC_{22}H_{21}ClN_3O$	$RhClEt\{C(O)C_9H_6N\}(py)_2$		900
1224	$[RhC_{22}H_{26}N_2]^+$	$[Rh\{C_5H_4N(CHNCHMePh)\}(cod)][PF_6]$		901
1225	$RhC_{22}H_{26}N_3O_2S_2$	$Rh(SCN)_2(CNC_5H_3CHCH_2CO_2Me)(Cp^*)$		802
1226	$RhC_{23}H_{35}O$	$Rh(cod)(\eta^6-C_6H_2MeBu^tOH-1,3,5,4).Sv$		902

No.	Formula	Structure	Details	Ref.
1227	$[RhC_{24}H_{24}F_4]^+$	$[Rh(\eta^4-Me_3tfb)(\eta^6-MesH-1,2,4)][ClO_4]$		896
1228	$RhC_{25}H_{19}NO_3P$	$Rh(OCOC_5H_4N-2)(CO)(PPh_3).MeOH$		903
1229	$RhC_{25}H_{25}Cl_2O_4$	$Rh[\{OC(C_6H_4OMe-4)\}_2CH](\eta^4-C_8H_{10}Cl_2-1,6)$		904
1230	$[RhC_{26}H_{26}]^+$	$[Rh(cod)(\eta^5-C_5H_4CPh_2)][ClO_4]$		905
1231	$RhC_{26}H_{38}O_8P$	$Rh(CH_2CMe_2PPhBu^t)(\eta^2-MeO_2CCHCHCO_2Me)_2$		906
1232	$RhC_{27}H_{26}ClOP_2$	$RhCl(CO)(PMePh_2)_2$		907
1233	$RhC_{27}H_{27}N_2$	$Rh\{(PhN)_2CPh\}(cod)$		908
1234	$[RhC_{27}H_{30}F_4]^+$	$[Rh(\eta^4-Me_3tfb)(\eta^6-C_6Me_6)][ClO_4]$		896
1235	$[RhC_{32}H_{28}I_2N_4]^+$	$[RhI_2\{CN(pTol)\}_4][TCNQ]$		909
1236	$RhC_{33}H_{27}NO_2P$	$Rh\{N(oTol)CH_2C_6H_4O-\underline{o}\}(CO)(PPh_3)$		910
1237	$RhC_{34}H_{38}O_2P$	$Rh(CO)(PPh_3)(\eta^6-C_6H_2MeBu^t_2OH-1,3,5,4).PhMe$		902
1238	$RhC_{36}H_{20}Br_2ClF_8OP_2$	$RhBrCl(\underline{o}-C_6F_4PPh_2O)(PPh_2C_6F_4Br-\underline{o})$ (2 isomers)		911
1239	$[RhC_{38}H_{30}O_8P_2]^-$	$[K(18-c-6)][Rh(CO)_2\{P(OPh)_3\}_2]$		685
1240	$[RhC_{43}H_{42}NP_3S_2]^+$	$[Rh(np_3)(\eta^2-CS_2)][BPh_4].Sv$		912
1241	$RhC_{44}H_{39}P_2$	$Rh(PPh_3)_2(\eta^3-CH_2C_6H_4Me-\underline{o})$		913
1242	$[RhC_{44}H_{78}N_2O_{30}]^+$	$[Rh(NH_3)_2(cod).\alpha\text{-cyclodextrin}][PF_6].6H_2O$		914
1243	$RhC_{45}H_{48}ClN_2P_2$	<u>trans</u>-$RhCl\{CNMeCH(CH_2Pr^i)CH_2NMe\}(PPh_3)_2.C_7H_8$		187
1244	$RhC_{50}H_{39}ClF_{12}P_3S_2$	$(triphos)RhCl\{C(CF_3)C(CF_3)CSC(CF_3)C(CF_3)S\}$		915
1245	$RhC_{53}H_{44}O_3P_4$	$Rh_2(CO)_3(\mu\text{-dppm})_2$		916
1246	$[RhC_{54}H_{42}Cl_2O_6P_2]^+$	$[Rh\{P(OPh)_3\}_2\{\eta^5-C_5H_4C(C_6H_4Cl-\underline{p})_2\}][ClO_4]$		905
1247	$RhRuC_{25}H_{23}Cl_2F_4N_2$	$(\eta^6-C_6H_4Me_2-\underline{p})Ru(\mu-Cl)_2(\mu-pz)Rh(\eta^4-tfb)$		917
1248	$RhRuC_{58}H_{57}P_4$	$(cod)RhRuPh(dppm)(\mu-H)(\mu-Ph)(\mu-PhPCH_2PPh_2).Sv$		918
1249	$RhRu_3C_{15}H_7O_{10}$	$(Cp)RhRu_3(CO)_9(\mu-H)_2(\mu-CO)$		919
1250	$RhRu_3C_{19}H_{19}O_9$	$(Cp^*)RhRu_3(CO)_9(\mu-H)_4$		919,920
1251	$RhRu_3C_{20}H_{17}O_{10}$	$(Cp^*)RhRu_3(CO)_9(\mu-H)_2(\mu-CO)$		919
1252	$[RhSeC_8H_{24}N_5]^{2+}$	$[Rh(NH_2CH_2CH_2Se)N(CH_2CH_2NH_2)_3][ClO_4]_2$		921
1253	$RhSeC_{16}H_{30}P$	$Rh(PPr^i_3)(\eta^2-SeCCHMe)(Cp)$		922
1254	$RhSi_4SnC_{27}H_{58}ClN_2$	$Rh[SnCl\{N(SiMe_3)_2\}_2](\eta^2-C_8H_{14})(Cp')$		923
1255	$Rh_2C_7H_3ClN_2O_4$	$\{Rh(CO)_2\}_2(\mu-Cl)(\mu-pz)$		924
1256	$Rh_2C_{12}H_{10}Cl_4N_3O_2P_3$	$Rh_2(CO)_2(Cp)_2(\mu-P_3N_3Cl_4)$		196

Structures Determined by Diffraction Methods 465

No.	Formula	Structure	Details	Ref.
1257	$[Rh_2C_{12}H_{18}O_4]^{2-}$	$[NBu_4]_2[Rh_2(CO)_4(\mu-PBu^t_2)_2]$		925
1258	$Rh_2C_{14}H_{10}O_3$	$Rh_2(CO)_2(\mu-CO)\{\eta,\eta'-(C_5H_4)_2CH_2\}$	125	926
1259	$Rh_2C_{21}H_{15}F_6NO$	$CpRh\{\eta^5-CF_3CC(CF_3)C(O)NPhRhCp\}$		927
1260	$Rh_2C_{21}H_{20}F_6O_4$	$Rh_2Cp_2\{\eta^1,\eta^2:\eta^1,\eta^2-C(CF_3)C(CF_3)OC(OEt)CCO_2Et\}$		928
1261	$[Rh_2C_{22}H_{30}O_2]^{2-}$	$K_2[Rh_2(Cp^*)_2(\mu-CO)_2].2THF$		929
1262	$Rh_2C_{26}H_{30}F_6O_2$	$Rh_2(Cp^*)_2(\mu-CO)\{\mu-\eta^2:\eta^1,\eta^1-F_3CC(CF_3)CO\}$		930
1263	$Rh_2C_{26}H_{36}O_2$	$Rh_2(CO)_2(Cp^*)_2(\mu-C=CMe_2)$		931
1264	$Rh_2C_{27}H_{38}O_2$	$Rh_2(Cp^*)_2(\mu-CO)(\mu-CHCMe_2CHCO)$	230	932
1265	$Rh_2C_{28}H_{26}N_4$	$Rh_2(nbd)_2(\mu-C_7H_5N_2)_2$		933
1266	$[Rh_2C_{28}H_{56}N_4O_4]^{2+}$	$[\{Rh(cod)(NH_3)\}_2\{\mu-N_2O_4C_{12}H_{26}\}][PF_6]_2$		934
1267	$Rh_2C_{32}H_{42}N_8$	$\{Rh(Cp^*)(pz)(\mu-pz)\}_2$		935
1268	$Rh_2C_{33}H_{28}I_2N_4O_2P_2$	$\{RhI(CO)\}_2(\mu-pz)_2(\mu-dppm)$		936
1269	$Rh_2C_{34}H_{34}Cl_2O_4P_2$	$\{Rh(\eta^2-Me_2C=CHCOOPPh_2)(\mu-Cl)\}_2$		937
1270	$Rh_2C_{34}H_{74}P_4$	$[Rh\{Pr^i_2P(CH_2)_3PPr^i_2\}]_2(\eta^3,\eta^3{}'-CH_2CHCHCH_2)$		938
1271	$[Rh_2C_{36}H_{33}OP_2]^+$	$[Rh_2Cp_2(\mu-H)(\mu-CO)(\mu-dppm)][BF_4]$		69
1272	$Rh_2C_{42}H_{44}N_2O_4$	$\{Rh(cod)(\mu-O_2CC_6H_4NHPh-\underline{o})\}_2$		939
1273	$Rh_2C_{43}H_{26}Br_2F_8N_4OP_2$	$Rh_2Br(CO)(\mu-pz)_2(PPh_2C_6F_4Br-\underline{o})(\mu-PPh_2C_6F_4-\underline{o})$		940
1274	$Rh_2C_{44}H_{37}IN_2O_2P_3$	$(CO)Rh(\mu-PPh_2)(\mu-N_2C_3(CH_2PPh_2)_2)RhIMe(CO).Et_2O$		941
1275	$Rh_2C_{44}H_{42}O_8P_2$	$Rh_2(AcOH)_2(\mu-OAc)_2(\mu-C_6H_4PPh_2-\underline{o})_2$		942
1276	$Rh_2C_{48}H_{38}O_2P_2$	$\underline{trans}-Rh_2(CO)_2(PPh_3)_2(\eta^5:\eta^5{}'-fulv).Sv$		943
1277	$[Rh_2C_{48}H_{38}O_2P_2]^{2+}$	$\underline{cis}-[Rh_2(CO)_2(PPh_3)_2(\eta^5:\eta^5{}'-fulv)][PF_6].Sv$		943
1278	$Rh_2C_{50}H_{44}N_2O_4P_2$	$Rh_2(py)_2(\mu-OAc)_2(\mu-C_6H_4PPh_2-\underline{o})_2$		942
1279	$Rh_2C_{56}H_{42}N_4O_2P_4$	$[Rh(CO)\{\mu-PPh(C_6H_4NH-\underline{o})(CH_2)_3PPh(C_6H_4NH_2)\}]_2.Sv$		944
1280	$[Rh_2C_{57}H_{51}I_2N_2O_2P_4]^+$	$[Rh_2I_2(CO)_2(\mu-dmpz)(\mu-dppm)_2][ClO_4]$		945
1281	$[Rh_2C_{57}H_{51}N_2O_2P_4]^+$	$[Rh_2(CO)_2(\mu-dmpz)(\mu-dppm)_2][ClO_4]$		945
1282	$[Rh_2C_{58}H_{53}O_2P_4]^+$	$[Rh_2(CO)_2(\mu-\eta^2,\eta^1-CCBu^t)(\mu-dppm)][ClO_4].Sv$		946
1283	$Rh_2C_{60}H_{50}O_2P_4$	$Rh_2(CO)_2(\mu-C=CHPh)(\mu-dppm)_2$		947
1284	$Rh_2C_{62}H_{42}F_8N_4$	$Rh_2\{\mu-(PhN)_2CPh\}(tfb)_2$		908
1285	$Rh_2C_{63}H_{56}Cl_2OP_4$	$Rh_2Cl_2(\mu-CO)(\mu-dppm)_2.2PhH$		948
1286	$Rh_2RuC_{32}H_{40}O_4$	$RuRh_2(CO)_2(\mu-CO)(\mu_3-CO)(\eta^4-C_8H_{10})(Cp^*)_2$		949

No.	Formula	Structure	Details	Ref.
1287	$Rh_2Sb_2C_{44}H_{42}O_8$	$Rh_2(SbPh_3)_2(\mu\text{-}OAc)_4$		60
1288	$Rh_2Sn_2C_{60}H_{46}Cl_6N_2O_2P_4$	$SnClRh_2Cl_2(CO)2SnCl_3\{\mu_3\text{-}NC_5H_3(PPh_2)_2\}_2.Sv$	140	950
1289	$Rh_2WC_{30}H_{38}O_3$	$(Cp)(CO)_2WRh_2(Cp^*)_2(\mu\text{-}CO)(\mu_3\text{-}CMe)$		951
1290	$Rh_3C_{27}H_{25}F_6O_3$	$Rh_3(CO)(Cp)_2(Cp^*)(\mu\text{-}CO)\{\mu_3\text{-}OCC_2(CF_3)_2\}$		952
1291	$Rh_3C_{32}H_{45}O_2$	$Rh_3(Cp^*)_3(\mu\text{-}CO)_2$		953
1292	$Rh_3C_{32}H_{47}O_2$	$Rh_3(Cp^*)_3(\mu\text{-}H)_2(\mu\text{-}CO)(\mu_3\text{-}CO)$		953
1293	$[Rh_3C_{65}H_{58}I_4OP_6]^+$	$[Rh_3I_4(CO)(\mu_3\text{-}dppmpp)_2][BPh_4].Sv$		954
1294	$[Rh_3C_{66}H_{58}Cl_2O_2P_6]^+$	$[Rh_3Cl_2(CO)_2(\mu_3\text{-}dppmpp)][BPh_4].Sv$	140	955
1295	$[Rh_3C_{66}H_{60}Cl_2O_2P_6]^+$	$[Rh_3H_2(CO)_2(\mu\text{-}Cl)_2(\mu_3\text{-}dppmpp)_2][BPh_4].Sv$	140	956
1296	$Rh_3RuC_{12}HO_{12}$	$RuRh_3H(CO)_9(\mu\text{-}CO)_3$		242
1297	$Rh_3RuC_{35}H_{45}O_5$	$(CO)_3RuRh_3(Cp^*)_3(\mu_3\text{-}CO)_2$		957, 949
1298	$Rh_3RuC_{46}H_{31}O_{10}P_2$	$RuRh_3H(CO)_7(\mu\text{-}CO)_3(PPh_3)_2.hexane$		242
1299	$Rh_4C_{32}H_{26}Cl_2N_4O_4$	$\{Rh_2(CO)(cod)(\mu\text{-}CO)(\mu\text{-}C_7H_5N_2)\}_2(\mu\text{-}Cl)_2$		933
1300	$Rh_4C_{38}H_{22}Cl_4N_2O_6$	$\{Rh_2(CO)_3(\mu\text{-}Cl)_2\{\mu\text{-}\eta^1\text{-}CPh(C_9H_6N)\}\}_2$		958
1301	$[Rh_{12}C_{25}O_{23}]^{4-}$	$[NPr_4][Rh_{12}C_2(CO)_{13}(\mu\text{-}CO)_{10}]$		959
1302	$Rh_{12}C_{25}H_2O_{25}$	$Rh_{12}H_2(CO)_{13}(\mu\text{-}CO)_9(\mu_3\text{-}CO)_3.Sv$		960
1303	$[RuC_3Cl_3O_3]^-$	$[S_5N_5][RuCl_3(CO)_3].Sv$		961
1304	$[RuC_{12}H_{13}O]^+$	$[Ru(CO)(\eta^4\text{-}C_6H_8)(Cp)][BF_4]$		962
1305	$RuC_{12}H_{14}O$	$Ru(CO)(\eta^3\text{-}C_6H_9)(Cp)$		962
1306	$RuC_{13}H_{12}O$	$Ru\{\eta^5:\eta^5{'}\text{-}C_5H_4C(O)CH_2CH_2C_5H_4\}$		963
1307	$RuC_{14}H_{20}O_2$	$Ru(CO)_2\{\eta^3,\eta^3{'}\text{-}(CH_2CMeCMeCH_2)_2\}$		964
1308	$RuC_{16}H_{18}$	$Ru\{\eta^5:\eta^5{'}\text{-}C_5H_4(CH_2)_3C_5H_2(CH_2)_3\}$		963
1309	$RuC_{20}H_{30}$	$Ru(Cp^*)_2$		965
1310	$RuC_{22}H_{33}O_4P$	$Ru[P\{C(O)Bu^t\}_2](CO)_2(Cp^*)$		966
1311	$[RuC_{23}H_{19}]^+$	$[Ru(\eta\text{-}C_5H_3Ph_2\text{-}1,2)(\eta\text{-}PhH)][BF_4]$		967
1312	$RuC_{26}H_{24}N_2O_5S$	$Ru_2(CO)_4(\mu\text{-}CO)\{\mu,\mu'\text{-}NPr^i CH(CHNPr^i)C(S)C_{12}H_8\}$		968
1313	$RuC_{29}H_{23}N_2OPS_2$	$Ru\{S(C_5H_4N)\}_2(CO)(PPh_3)$		969
1314	$RuC_{30}H_{23}N_2O_2PS_2$	$Ru\{S(C_5H_4N)\}_2(CO)_2(PPh_3)$		969
1315	$[RuC_{31}H_{23}N_6O_2]^+$	$[Ru\{C_6H_3(C_5H_4N)(NO_2)\text{-}2,4\}(bipy)_2][BF_4].Sv$		970
1316	$RuC_{32}H_{20}O_4$	$Ru(CO)_3(\eta^4\text{-}C_4Ph_4CO)$		971

Structures Determined by Diffraction Methods 467

No.	Formula	Structure	Details	Ref.
1317	$RuC_{32}H_{35}ClO_2P_2$	$RuCl[C_6H_3(5-Me)\{2-CO(pTol)\}](CO)(PMe_2Ph)_2$		972
1318	$[RuC_{32}H_{44}P_3]^+$	$[Ru(PMe_2Ph)_3(\eta^5-C_8H_{11})][PF_6]$		973
1319	$RuC_{35}H_{33}O_2P$	$Ru\{C(OPr^i)CHPh\}(CO)(PPh_3)(Cp)$		974
1320	$[RuC_{35}H_{36}ClP_2]^+$	$[RuCl\{Ph_2P(CH_2)_4PPh_2\}(\eta^6-PhMe)][PF_6]$		975
1321	$[RuC_{36}H_{41}NOP]^+$	$(\underline{R})-[Ru(CO)(NCMe)(\eta-C_5H_4(neomenthyl))][PF_6].Sv$	273	976
1322	$RuC_{41}H_{32}N_4OP_2S_6$	$Ru\{NNHC(S)SCS\}CO(SC_2HN_2S_2)(PPh_3)_2$		977
1323	$RuC_{41}H_{52}OS_4$	$Ru(SC_6HMe_4-2,3,5,6)_4(CO)$		978
1324	$[RuC_{42}H_{35}OP_2]^+$	$[Ru(CO)(PPh_3)_2(Cp)][X]$ $\{X=Co(CO)_4, BPh_4\}$		162,979
1325	$RuC_{45}H_{35}N_4P_2$	$Ru[C\{C(CN)_2\}CPhC(CN)_2](dppe)(Cp).Sv$		980
1326	$[RuC_{46}H_{37}OP_2]^+$	$[Ru(CO)(PPh_3)_2(\eta^5-ind)][ClO_4].Sv$		981
1327	$[RuC_{46}H_{39}P_2S]^+$	$[Ru\{\eta-C_5H_4CH_2(C_4H_3S)-2\}(PPh_3)_2][BPh_4]$		982
1328	$[RuC_{46}H_{41}P_2]^+$	$[Ru\{C=CPh(C_7H_7)\}(dppe)(Cp)][PF_6].Sv$		983
1329	$[RuC_{48}H_{51}NOP_3]^+$	$(+)-[RuMe(CO)(CNBu^t)(triphos)][PF_6]$		984
1330	$[RuC_{49}H_{39}Br_2P_2]^+$	$[Ru\{C=CBr(C_6H_4Br-\underline{p})\}(PPh_3)_2(Cp)][Br_3].CHCl_3$		985
1331	$RuC_{49}H_{40}P_2$	$Ru(\eta^1-CCPh)(PPh_3)_2(Cp)$	111	979
1332	$[RuC_{57}H_{49}N_2P_2]^+$	$[Ru(C=CPhN=NC_6H_3Me_2-3,4)(PPh_3)_2(Cp)][BF_4].Sv$		983
1333	$RuSbC_7H_9O_4$	$\underline{ax}-Ru(CO)_4(SbMe_3)$		47
1334	$RuSe_4C_{42}H_{55}N$	$Ru(SeC_6HMe_4-2,3,5,6)_4(NCMe)$		978
1335	$RuSiC_{20}H_{33}O_3P$	$Ru\{C(O)PBu^t(SiMe_3)\}(CO)_2(Cp*)$	148	986
1336	$RuThC_{27}H_{35}IO_2$	$(Cp)(CO)_2RuThI(Cp*)_2$	163	987
1337	$RuTiC_{13}H_{23}N_3O_2$	$(Cp)(CO)_2RuTi(NMe_2)_3$		988
1338	$RuWC_{45}H_{38}Cl_2O_2P$	$(PPh_3)_2(CO)RuWCl(CO)(Cp)(\mu-Cl)(\mu-CMe).Sv$	200	989
1339	$RuW_2C_{21}H_{16}O_7$	$Ru_2W(CO)_7(Cp)_2(\mu_3-\eta^2-MeCCMe)$		989
1340	$RuZrC_{18}H_{14}O_3$	$Ru(CO)_2\{\eta-C_5H_4Zr(CO)(Cp)_2\}$		990
1341	$RuZrC_{19}H_{23}O_2P$	$Ru(CO)(PMe_3)\{\eta-C_5H_4Zr(Cp)_2\}(\mu-CO)$		990
1342	$[Ru_2C_8O_8]^{2-}$	$[PPh_4]_2[Ru_2(CO)_8].MeCN$		742
1343	$Ru_2C_{19}H_{10}F_2O_{10}$	$Ru_2(CO)_5(OH_2)(\mu-O_2CC_6H_4F-\underline{p})_2.Sv$		991
1344	$Ru_2C_{25}H_{35}ClN_6$	$Ru_2Cl(pzH)(cod)_2(\mu-H)(\mu-pz)_2.EtOH$		992
1345	$[Ru_2C_{26}H_{35}N_4O]^+$	$[\{Ru(\eta-C_6H_4MePr^i-\underline{p})\}_2(\mu-OH)(\mu-pz)_2][BPh_4]$		993
1346	$[Ru_2C_{28}H_{39}N_8]^+$	$[Ru_2(pzH)_2(cod)_2(\mu-H)(\mu-pz)_2][PF_6]$		992

No.	Formula	Structure	Details	Ref.
1347	$Ru_2C_{30}H_{34}N_4O_5$	$Ru_2(CO)5\{\mu,\mu'-Bu^tN=CHCH(NBu^t)C(=NpTol)N(pTol)\}$		968
1348	$Ru_2C_{41}H_{32}ClO_4P_3$	$Ru_2(CO)_4(\mu-Cl)(\mu-PPh_2)(\mu-dppm)$		994
1349	$Ru_2C_{42}H_{30}O_6P_2$	$Ru_2(CO)_5(PPh_3)(\mu-PPh_2)(\mu-OCPh)$		995
1350	$Ru_2C_{42}H_{31}O_5P_3$	$Ru_2(CO)_4(\mu-H)(\mu-\eta^2PC-Ph_2PCHPPh_2)(\mu-PPhC_6H_4CO-\underline{o})$		994
1351	$Ru_2C_{70}H_{66}N_4O_4P_2$	$Ru_2(pTol)_2(\mu-HNCPhO)_2\{\mu-NCPhOP(pTol)_2\}_2$		996
1352	$Ru_2C_{72}H_{70}N_4O_{12}P_2$	$Ru_2Ph_2(\mu-HNCRO)_2\{\mu-NCROPPh_2\}_2$ R=$C_6H_3(OMe)_2$-3,5		996
1353	$[Ru_2C_{73}H_{64}NP_4]^+$	$[(Cp)(dppe)Ru(\mu-CN)Ru(PPh_3)_2(Cp)][PF_6]$		997
1354	$[Ru_2C_{90}H_{90}N_{10}]^{2+}$	$[Ru_2(CNXy)_{10}][BF_4]_2$		998
1355	$Ru_3C_{10}H_3NO_{10}$	$Ru_3(CO)_{10}(\mu-H)(\mu-NH_2)$		999
1356	$[Ru_3C_{11}O_{10}]^{2-}$	$[PPN]_2[Ru_3(CO)_6(\mu-CO)_3(\mu_3-CCO)]$		1000
1357	$Ru_3C_{12}H_6O_{10}S$	$Ru_3(CO)_{10}(\mu-H)(\mu-SEt)$		1001
1358	$Ru_3C_{13}H_8O_{10}$	$Ru_3(CO)_9(\mu-H)_2(\mu_3-\eta^2-MeCCOMe)$		1002
1359	$Ru_3C_{17}H_{12}N_2O_9$	$Ru_3(CO)_9(py)(\mu-H)(\mu-CNMe_2)$		1003
1360	$Ru_3C_{20}H_8N_2O_{10}$	$Ru_3(CO)_{10}(bipy)$		1004
1361	$Ru_3C_{21}H_{26}N_2O_8$	$Ru_3(CO)_8(\mu-CH_2)\{\mu-\mu N,\mu N',\eta^2-(Bu^tCH_2NCH)_2\}$		1005
1362	$Ru_3C_{23}H_{24}N_2O_9$	$Ru_3(CO)_9(\mu-\eta^4:N,N-CyNCHCHNCy)$		1006
1363	$Ru_3C_{28}H_{16}O_9P_2$	$Ru_3(CO)_9\{\mu_3-\eta^1:P:\mu P-C_6H_4PPhCH_2PPh\}$		1007,1008
1364	$Ru_3C_{28}H_{18}O_9P_2$	$Ru_3(CO)_9(\mu-H)(\mu_3-P:\mu P-Ph_2PCH_2PPh)$	118	1008
1365	$Ru_3C_{29}H_{16}O_{10}P_2$	$Ru_3(CO)_{10}(\mu-PhPCH_2PPhC_6H_4-\underline{o})$	118	1008
1366	$Ru_3C_{30}H_{22}O_8P_2$	$Ru_3(CO)_6(\mu-CO)_2(\mu-\eta^3-C_3H_5)(\mu_3-PhPCH_2PPh_2)$		1009
1367	$Ru_3C_{32}H_{22}O_8P_2$	$Ru_3(CO)_8(\mu-H)_2(\mu-PPh_2)_2$.Sv		442
1368	$[Ru_3C_{35}H_{48}O_3]^+$	$[Ru_3(Cp^*)_3(\mu-CO)_2(\mu_3-CMe)][BF_4]$	200	1010
1369	$Ru_3C_{40}H_{25}O_9P_3$	$Ru_3(CO)_6(\mu-CO)_3\{\mu_3-Ph_2PCH(PPh_2)_2C_6H_4-\underline{o}\}$.$CHCl_3$		994
1370	$Ru_3C_{40}H_{44}O_{16}P_4$	$Ru_3(CO)_6(\mu-CO)_2\{PPh(OMe)_2\}_4$		1011
1371	$Ru_3C_{42}H_{30}NO_9P$	$Ru_3(CO)_9(PPh_3)(\mu-H)(\mu-CNBz_2)$		1003
1372	$[Ru_3Si_2C_{22}H_{31}O_{10}]^-$	$[PPN][Ru_3(SiEt_3)_2(CO)_{10}(\mu-H)]$		1012
1373	$Ru_4C_{12}N_2O_{13}$	$Ru_4(CO)_{12}(\mu-NO)(\mu_3-N)$		1013
1374	$Ru_4C_{13}N_2O_{13}$	$Ru_4(CO)_{12}(\mu-NCO)(\mu_3-N)$		1013
1375	$[Ru_4C_{30}H_{25}O_6]^+$	$[\{Ru_2(CO)_3(Cp)\}_2(\mu_4-\eta^2:\mu\eta^1-MeCCHCH)][BF_4]$.Sv		1014
1376	$Ru_4C_{31}H_{15}O_{11}P$	$Ru_4(CO)_9(\mu-CO)_2(\mu_4-PPh)(\mu_4-\eta^2-PhCCPh)$		1015

Structures Determined by Diffraction Methods 469

No.	Formula	Structure	Details	Ref.
1377	$Ru_4C_{32}H_{15}O_{12}P$	$Ru_4(CO)_{10}(\mu-CO)_2(\mu_4-PhPCPhCPh).Sv$		1015
1378	$Ru_4C_{32}H_{50}N_4O_8$	$Ru_2(CO)_6(\mu-CO)_2\{\eta^5-RNCHCHNRRuH(CO)_2\}_2$ R=CH_2Bu^t		1016
1379	$Ru_4C_{36}H_{20}O_{10}P_2$	$Ru_4(CO)_8(\mu-CO)_2(\mu_4-PPh)(\mu_4-\eta^2-PhCCPPh_2)$		1017
1380	$Ru_4C_{36}H_{24}O_{10}P_2$	$Ru_4(CO)_{10}(\mu-H)_3(\mu-PPh_2)(\mu_4-\eta^2,P-PPh_2CCH)$		1018
1381	$Ru_4C_{62}H_{40}O_{10}P_4$	$\{Ru_2(CO)_5(\mu-PPh_2)(\mu-\eta^2-C\equiv CPPh_2)\}_2$		1017
1382	$[Ru_5C_{14}NO_{14}]^-$	$[NEt_3Bz][Ru_5N(CO)_{13}(\mu-CO)]$		1019
1383	$Ru_5C_{35}H_{28}O_6$	$\{Ru_3(CO)_3(Cp)_3\}(\mu_3-CCH_2CHC-\mu)\{Ru_2(CO)_3(Cp)_2\}.Sv$		1014
1384	$Ru_5C_{37}H_{26}O_{11}P_2$	$Ru_5C(CO)_{11}(PMePh_2)(\mu-H)_3(\mu-PPh_2)$		1020
1385	$Ru_5C_{39}H_{20}O_{13}P_2$	$Ru_5(CO)_{13}(\mu-PPh_2)(\mu_5-\eta^2-CCPPh_2)$		1021,1022
1386	$Ru_5C_{39}H_{22}O_{13}P_2$	$Ru_5(CO)_{13}(\mu-H)(\mu-PPh_2)\mu_5-\mu_3:\eta^2:P-CCHPPh_2)$		1020
1387	$Ru_5C_{41}H_{20}O_{15}P_2$	$Ru_5(CO)_{13}(\mu-PPh_2)(\mu_5-C_2Ph_2)$ (2 isomers)		1023
1388	$Ru_6C_{24}H_{12}O_{11}$	$Ru_6C(CO)_{10}(\mu-CO)(\eta-C_6H_6)(\mu_3-\eta^6-PhH)$		748
1389	$SbC_2H_6N_3$	Me_2SbN_3	206	153
1390	$SbC_{10}H_{13}OS_2$	$PhSb(CH_2CH_2)_2O$		1024
1391	$SbC_{10}H_{13}S_3$	$PhSb(CH_2CH_2)_2S$		1024
1392	$SbC_{11}H_{23}N_2S_4$	$MeSb(detc)_2$		1025
1393	$SbC_{12}H_8Cl$	$ClSb\{(C_6H_4)_2-2,2'\}$		1026
1394	$SbC_{18}H_{13}$	$PhSb\{(C_6H_4)_2-2,2'\}$		1026
1395	$SbC_{18}H_{33}O_4P_2S_4$	$SbPh\{S_2P(OPr^i)_2\}_2$		35
1396	$SbC_{20}H_{20}F_{12}NO_2$	$NSb\{C(CF_3)C(CF_3)CHCOBu^t\}_2$		1027
1397	$SbC_{30}H_{25}O_3S$	$Ph_4SbOSO_2Ph.H_2O$		1028
1398	$SbSi_2C_7H_{19}Cl_2$	$Cl_2SbCH(SiMe_3)_2$		391
1399	$SbSi_4C_{13}H_{39}Cl_2N_2$	$SbCl_2Me\{N(SiMe_3)_2\}_2$	173	1029
1400	$SbWC_{11}H_{11}F_6NO_4P$	$F_5Sb(\mu-F)W(CO)_3(NO)(PMe_2Ph)$	133	1030
1401	$SbW_2C_{29}H_{18}O_{10}P$	$\{W(CO)_5\}_2SbMe(PPh_3)$	238	38
1402	$SbW_3C_{19}H_9O_{15}$	$W_2(CO)_{10}\{\mu-SbBu^tW(CO)_5\}$		310
1403	$Sb_2C_{24}H_{20}Br_2$	$\{Ph_2SbBr(\mu-O)\}_2$		1031
1404	$Sb_2C_{40}H_{40}O_9S_2$	$\{SbPh_3(OSO_2CH_2CH_2OH)\}_2O$		1032
1405	$Sb_3C_5H_9$	$(SbCH_2)_3CMe$		1033
1406	$Sb_6C_{36}H_{30}$	$(SbPh)_6.(1,4-C_4H_8O_2)$		1034

No.	Formula	Structure	Details	Ref.
1407	$Sc_3C_{30}H_{30}F_3$	$\{(Cp)_2Sc(\mu-F)\}_3$		1035
1408	SeC_2F_8	$SeF_2(CF_3)_2$	E	1036
1409	$SeC_9H_{12}N_2O_5$	selenazofurin & α-anomer		1037
1410	$SeC_{16}H_{22}N_4$	$Se=C\{NHNCMe(C_5H_4N)\}(NC_8H_{14})$		1038
1411	$SeC_{19}H_{10}F_{24}N_2O_3$	$Se\{C_4O(CF_3)_4(NEt_2)\}\{C_3NO_2(CF_3)_4\}$		1039
1412	$SeC_{19}H_{17}NO_2S$	$Ph_2Se=NSO_2(pTol).Sv$		1040
1413	$SeTeC_{18}H_{29}BrN_3O_3P$	$TeBrPh\{SeP(NC_4H_8O)_3\}$		1041
1414	$SeTeC_{18}H_{29}ClN_3O_3P$	$TeClPh\{SeP(NC_4H_8O)_3\}$		1041
1415	$SeV_2C_{58}H_{48}O_6P_4$	$\{V(CO)_3(dppe)\}_2Se.3phme$		1042
1416	$Se_2SiC_{16}H_{18}$	$Me_2Si(\underline{o}-C_6H_4CH_2Se_2CH_2CH_2C_6H_4-\underline{o})$		1043
1417	$Se_4C_{10}H_{12}$	$[TMSF]_2[PO_2F_2]$	293 & 125	1044
		$[TMSF]_2[BF_4]$	N at 20	1045
		$[(TMSF)][Ni(S_2C_2S_2CS)_2]$		1046
1418	$Se_4C_{11}H_{12}$	$[3,3-Me_23',3'-(CH_2)_3TSeF]_2[ReO_4]$		1047
1419	$Se_4C_{15}H_{27}P$	$Bu_3PCSeC(Se)SeCSe$	238	1048
1420	$[Se_4W_2C_{10}O_{10}]^{2+}$	$[\{W(CO)_5\}_2(\mu-Se_4)][SbF_6]_2$		1049
1421	$SiC_4H_{11}NO_2$	$NH_2CO_2SiMe_3$		1050
1422	$SiC_4H_{12}O_2$	$Et_2Si(OH)_2$		1051
1423	$SiC_7H_{11}NO_5$	$MeSi\{OC(O)CH_2\}_2NCH_2CH_2O$		1052
1424	$SiC_8H_{19}FO$	$Bu^t_2Si(OH)F$		1053
1425	$SiC_8H_{20}O_2$	$Bu^t_2Si(OH)_2$		1054
1426	$SiC_{13}H_{14}Cl_2N_2$	$SiCl_2Me(1,2,3,4-H_4phen)$		1055
1427	$SiC_{13}H_{19}NO_2$	$PhSiCH_2CH_2CH_2N(CH_2CH_2O)_2$		1056
1428	$SiC_{13}H_{19}NO_3$	$(pTol)Si(OCH_2CH_2)_3N$		1057
1429	$SiC_{14}H_{17}ClN_2$	$SiClMe_2(1,2,3,4-H_4phen)$		1055
1430	$SiC_{15}H_{17}N_2$	$SiMe_3(phen)$		1058
1431	$[SiC_{16}H_{17}O_4]^-$	$[NEt_4][BuSi(O_2C_6H_4-\underline{o})_2]$ (n & t isomers)		1059
1432	$SiC_{16}H_{25}NO_3$	$(pTol)Si(OCHMeCH_2)_3N$		1057
1433	$SiC_{18}H_8F_{12}O_2$	$Si\{C_6H_4C(CF_3)_2O-\underline{o}\}_2$		1060
1434	$[SiC_{18}H_{13}O_4]^-$	$[NHEt_3][PhSi(O_2C_6H_4-\underline{o})]$		1061

No.	Formula	Structure	Details	Ref.
1435	$SiC_{18}H_{14}$	$SiHPh\{(C_6H_4)_2-2,2'\}$		1062
1436	$SiC_{18}H_{29}NO$	rac & R-Si(OH)(Cy)(Ph)(CH$_2$CH$_2$CC$_4$H$_8$)		1063
1437	$[SiC_{22}H_{15}O_4]^-$	[NEt$_4$][(nap)Si(O$_2$C$_6$H$_4$-o)$_2$]		1059
1438	$SiC_{22}H_{20}OS$	Ph$_3$SiC(SEt)CO		1064
1439	$[SiC_{24}H_{13}F_{12}O_2]^-$	[NMe$_4$][Si{C$_6$H$_4$C(CF$_3$)$_2$O-o}$_2$Ph]		1060
1440	$SiC_{24}H_{18}$	$SiPh_2\{(C_6H_4)_2-2,2'\}$		1062
1441	$SiC_{24}H_{20}O_2$	SiPh$_2$(PhO)$_2$		1065
1442	$[SiC_{24}H_{21}O_4]^-$	[NEt$_4$][Bu^{t}Si(O$_2$C$_{10}$H$_6$)$_2$]		1061
1443	$SiC_{25}H_{25}N$	(E)-Ph$_3$SiCH=C(CN)Bu		1066
1444	$[SiC_{34}H_{45}O_4]^-$	[NHEt$_3$][PhSi(O$_2$C$_6$H$_2$Bu$^t_2$-3,5)$_2$]		1061
1445	$SiC_{44}H_{76}P_2$	Bu$^t_2$SiP$_2$(C$_6$H$_2$Bu$^t_3$-2,4,6)$_2$		1067
1446	$SiSnC_{29}H_{41}N_2P$	Me$_2$SiNBu^{t}Sn(CH$_2$PPh$_3$)NBu^t		1068
1447	$SiTaC_{18}H_{23}$	TaH$_2$(SiMe$_2$Ph)(Cp)$_2$		1069
1448	$SiTiC_{12}H_{14}Cl_2$	TiCl$_2\{\eta,\eta'-(C_5H_4)_2SiMe_2\}$		1070
1449	$[SiTiC_{22}H_{27}]^+$	[Ti{E-C(SiMe$_3$)=CMePh}(Cp)$_2$][AlCl$_4$]	115	1071
1450	$SiTiC_{37}H_{57}NO_2$	Tipy{CH$_2$CMe$_2$C$_6$H$_3$Bu^{t}O}(CH$_2$SiMe$_3$)(OC$_6$H$_4$Bu$^t_2$)	108	1072
1451	$SiVC_3H_9Cl_3N$	Me$_3$SiNVCl$_3$		1073
1452	$SiVC_{14}H_{34}Cl_2N_3$	VCl$_2$(NHBu^{t}SiMe$_2$NBu^t)(NBu^t)		1074
1453	$SiWC_9H_{16}O_2$	WO$_2$(CH$_2$SiMe$_3$)(Cp)		1075
1454	$SiZrC_{12}H_{14}Cl_2$	ZrCl$_2\{\eta,\eta'-(C_5H_4)_2SiMe_2\}$		1070
1455	$SiZrC_{14}H_{19}ClO$	ZrCl(η^2-OCSiMe$_3$)(Cp)$_2$		1076
1456	$SiZrC_{17}H_{29}NS_2$	Zr(SiMe$_3$)(detc)(Cp)$_2$		1077
1457	$SiZrC_{29}H_{29}ClO$	(E)-Zr{OC(SiMe$_3$=CH(9-anthryl)}Cl(Cp)$_2$		1078
1458	$Si_2C_3H_{13}N$	NMe(SiH$_2$Me)$_2$	E	1079
1459	$Si_2C_5H_{17}N$	NMe(SiHMe$_2$)$_2$	E	1079
1460	$Si_2C_6H_{16}Cl_2$	Me$_2$ClSiCH$_2$CH$_2$SiMe$_2$Cl	150	1080
1461	$Si_2C_6H_{18}N_2$	Me$_2$SiNMeSiMe$_2$NMe	E	1081
1462	$Si_2C_{10}H_{14}O_3S_2$	{SiMe(OH)(C$_4$H$_3$S)}$_2$O		1082
1463	$Si_2C_{10}H_{22}N_2$	Me$_3$SiN(CHCH)$_2$SiMe$_3$	163	481
1464	$Si_2C_{14}H_{30}N_2$	Me$_3$SiN(CMeCMe)$_2$NSiMe$_3$	163	481

No.	Formula	Structure	Details	Ref.
1465	$Si_2C_{16}H_{24}$	$1,8-(Me_3Si)_2$-naphthalene	183	1083
1466	$Si_2C_{26}H_{42}$	$Bu^t_2HSi_2H(Mes)_2$		1084
1467	$Si_2C_{29}H_{46}O$	$Bu^t_2SiSiBu^t_2CPh_2O$		1085
1468	$Si_2C_{36}H_{30}S$	$(SiPh_3)_2S$		1086
1469	$Si_2C_{36}H_{44}S$	$Si_2(Mes)_4(\mu-S)$	138	1087
1470	$[Si_2SmC_{16}H_{28}]^-$	$[Li(DME)_3][Sm(SiMe_3)_2(Cp)_2]$		1088
1471	$Si_2WC_{13}H_{38}NO$	$W(CH_2SiMe_3)_2(NO)(Cp)$		1089
1472	$Si_2WC_{15}H_{23}O_3P$	$W(CO)_3\{P=C(SiMe_3)_2\}(Cp)$		655
1473	$Si_2W_2C_{20}H_{46}O_4$	$W_2(OPr^i)_4(\mu-CSiMe_3)_2$	204	1090
1474	$Si_2W_2C_{22}H_{48}O_4$	$W_2(OPr^i)_4(\mu-CSiMe_3)(\mu-\eta^3-CHCHCSiMe_3)$	115	1090
1475	$Si_2W_2C_{29}H_{55}NO_4$	$W_2(OPr^i)_4(NXy)(\mu-CSiMe_3)(\mu-\eta^2-CCSiMe_3)$	118	1091
1476	$Si_2W_4C_{35}H_{88}O_{14}$	$\{W_2(OEt)_4(\mu-OEt)_2(\mu-CSiMe_3)\}_2(\mu-OEt)_2$	118	1092
1477	$[Si_2YC_{18}H_{32}]^-$	$[Li_2(DME)_2(THF)][Y(CH_2SiMe_3)_2(Cp)_2]_2$		1093
1478	$Si_2ZrC_{18}H_{44}Cl_3NP_2$	<u>mer</u>-$ZrCl_3N(SiMe_2CH_2PPr^i_2)_2$		492
1479	$Si_3C_{15}H_{36}$	$Me_3Si=C(SiMe_3)SiMeBu^t_2$		1094
1480	$Si_3C_{36}H_{30}O_3$	$Ph_2SiOSiPh_2OSiPh_2O$		1095
1481	$Si_3C_{39}H_{60}$	$Si_3Bu^t_3(Mes)_3$ (2 isomers)		1096
1482	$Si_3WC_{17}H_{38}O$	$WO(CH_2SiMe_3)_3(Cp)$		1075
1483	$Si_4C_{10}H_{30}O_3$	$(Me_3Si)_3CSi(OH)_3$		1097
1484	$Si_4C_{16}H_{34}O_2$	$(OH)_2SiPhC(SiMe_3)_3$		1098
1485	$Si_4C_{16}H_{38}$	<u>trans</u>-$(Me_3Si)_2C=CHCH=C(SiMe_3)_2$		1099
1486	$Si_4C_{17}H_{36}O_2$	$(Me_3Si)_3CSiPh(OMe)(OH)$		1098
1487	$Si_4C_{20}H_{42}O$	$(Me_3Si)_2Si=C(adamantyl)(OSiMe_3)$	223	1100
1488	$Si_4C_{20}H_{58}N_9O_4P_3$	$Me_2Si(OSiMe_2O)_2SiMe(CH_2)_3NHP_3N_3(NMe_2)_5$		1101
1489	$Si_4C_{26}H_{28}O_4$	$Ph_2Si(OSiHMeO)_2SiPh_2$		1102
1490	$Si_4C_{32}H_{72}$	$1,1,2,2,3,3,4,4-(Pr^i)_4(Bu^tCH_2)_4Si_4$		1103
1491	$Si_4C_{38}H_{38}N_2O_2$	$Ph_2SiN(SiPh_2OH)SiMe_2NSiPh_2OH$		1104
1492	$Si_4C_{44}H_{64}O$	$(Mes)_2SiC(SiMe_3)_2CHSi(Mes)_2OH$		1105
1493	$Si_4C_{48}H_{62}O$	$R_2SiSiHBu^tSiR_2SiBu^tOH$ ($R=C_6H_4Et_2-2,6$)		1106
1494	$Si_5C_{28}H_{40}$	$Me_3SiSiMeCPhC(SiMe_3)SiMe_2CPhCSiMe_3$		1107

Structures Determined by Diffraction Methods   473

No.	Formula	Structure	Details	Ref.
1495	$Si_6C_{10}H_{30}S$	$Me_{10}Si_6S$		1108
1496	$Si_6C_{42}H_{84}O_2$	${MeSi(OSiMe_3)C(adamantyl)(SiMe_2Bu^t)}_2$		1109
1497	$Si_6Te_3C_{20}H_{54}$	$(Me_3Si)_3CTeTeTeC(SiMe_3)_3$		1110
1498	$Si_6W_2C_{24}H_{62}Br_2$	${WBr(CH_2SiMe_3)_2}_2(\mu\text{-}CSiMe_3)_2$	116	1111
1499	$Si_6W_2C_{24}H_{62}Cl_2$	${WCl(CH_2SiMe_3)_2}_2(\mu\text{-}CSiMe_3)_2$	113	1111
1500	$Si_6W_2C_{28}H_{68}$	$W_2(CH_2SiMe_3)_4(\mu\text{-}CSiMe_3){\mu\text{-}MeCHC(CH_2)CSiMe_3}$	115	1112
1501	$Si_6W_2C_{33}H_{74}$	$W_2(CH_2SiMe_3)_4(\eta^2\text{-}PhCCMe)(\mu\text{-}CSiMe_3)_2$	116	1090
1502	$Si_6W_2C_{38}H_{72}$	$W_2(CH_2SiMe_3)_4(\mu\text{-}CSiMe_3)(\mu\text{-}\eta^3\text{-}PhCCPhCSiMe_3)$	113	1090
1503	$Si_7C_{14}H_{21}$	$(Me_2Si)_7$	203	1113
1504	$Si_8C_{20}H_{54}F_4O$	${(Me_3Si)_3CSiF_2}_2O$		1114
1505	$Si_8Sn_2C_{28}H_{76}Cl_2$	<u>trans</u>-$[SnCl{CH(SiMe_3)_2}_2]_2$		829
1506	$Si_8WC_{28}H_{64}$	$WH_2{\eta\text{-}C_5H_4Si(SiMe_3)_3}_2$	233	1115
1507	$SmC_{28}H_{46}O_2$	$Sm(THF)_2(Cp^*)_2$		1116
1508	$SmC_{30}H_{43}O$	$SmPh(THF)(Cp^*)_2$		1117
1509	$SmC_{30}H_{43}O$	$Sm(OC_6HMe_4\text{-}2,3,5,6)(Cp^*)_2$		1118
1510	$Sm_2C_{36}H_{62}I_2O_4$	${Sm(THF)_2(Cp^*)(\mu\text{-}I)}_2$		1116
1511	$Sm_2C_{40}H_{60}O$	${Sm(Cp^*)_2}_2O$		1119
1512	$Sm_2C_{78}H_{92}O_4P_2$	${Sm(OPPh_3)(Cp^*)_2}_2(\mu\text{-}OCH=CHO)$ (2 isomers)		1120
1513	$Sm_4C_{94}H_{136}O_8$	$[(Cp^*)_2(THF)Sm{\mu_3\text{-}OC(O)C=C=O}Sm(Cp^*)_2]_2$		1121
1514	$[SnC_2H_5Cl_5]^{2-}$	$[NMe_4]_2[SnEtCl_5]$		1122
1515	$[SnC_2H_6Cl_3]^-$	$[TTF][SnMe_2Cl_3]$		1123
1516	$[SnC_2H_6Cl_4]^{2-}$	$[TTF]_3[SnMe_2Cl_4]$		1123
1517	$SnC_2H_{10}Cl_2O_2$	$SnCl_2Me_2(OH)_2 \cdot 4\text{-purine}$		1124
1518	$[SnC_4H_9Br_5]^{2-}$	$[NMe_4]_2[SnBuBr_5]$		1122
1519	$[SnC_4H_{10}Cl_4]^{2-}$	$[TTF]_3[SnEt_2Cl_4]$		1125
1520	$SnC_6H_{10}N_2O_2S_2$	$Me_2Sn(NCS)_2(OH_2)_2 \cdot (18\text{-crown-}6)$		1126
1521	$[SnC_9H_{13}]^+$	$[Sn(\eta\text{-}C_5H_4Bu^t)][BF_4]$		1127
1522	$SnC_9H_{18}ClN$	$N(CH_2CH_2CH_2)_3SnCl$		1128
1523	$SnC_{10}H_{13}ClS_3$	$SnClPh(SCH_2CH_2)_2S$		1129
1524	$SnC_{10}H_{24}ClN_2P$	$SnClMe_2(NBu^tPNBu^t)$		1130

No.	Formula	Structure	Details	Ref.
1525	$SnC_{11}H_{16}OS_2$	$SnMePh(SCH_2CH_2OCH_2CH_2S)$		1131
1526	$[SnC_{12}H_{10}ClN_2O_6]^-$	$[Ag(AsPh_3)_4][SnPh_2Cl(NO_3)_2]$		34
1527	$[SnC_{12}H_{10}N_3O_9]^-$	$[Ag(AsPh_3)_4][SnPh_2(NO_3)_3]$		34
1528	$SnC_{12}H_{13}I_3N_2$	$SnEtI_3(bipy)$		1132
1529	$SnC_{12}H_{22}Cl_2$	$SnCl_2(Cy)_2$		1133
1530	$SnC_{12}H_{26}N_2S_4$	$SnMe_2(detc)_2$		1134
1531	$SnC_{15}H_{24}ClNS_2$	$SnClBuPh(detc)$		1135
1532	$SnC_{16}H_{19}Cl_2NOS$	$SnCl_2Me_2(O=S=CPhNMePh)$	130	1136
1533	$SnC_{19}H_{25}NS_2$	$SnBuPh_2(dmtc)$		1137
1534	$SnC_{20}H_{26}$	$Bu_2Sn\{(C_6H_4)_2-2,2'\}$		1138
1535	$SnC_{21}H_{20}$	$Ph_3SnCH_2CH=CH_2$		1139
1536	$SnC_{23}H_{23}NS_2$	$SnPh_4(S_2CNC_4H_8)$		775
1537	$SnC_{23}H_{28}$	$SnMePh_2(adamantyl)$		1140
1538	$SnC_{24}H_{18}Cl_2$	$SnCl_2(\underline{o}-C_6H_4Ph)_2$		1141
1539	$[SnC_{25}H_{29}N_7]^{2+}$	$[SnMe_2L][PF_6][Cl]$ L=6,6"-(HNNMe)$_2$-4'-Ph(terpy)		1142
1540	$SnC_{26}H_{21}Br_2N$	$Ph_2Sn(1,2-C_6H_3Br-5)_2NEt$		1143
1541	$SnC_{26}H_{25}I_3O_2S_2$	$SnI_3Et(OSPh_2)_2$		1144
1542	$SnC_{28}H_{22}ClNO_2$	$SnClPh_3(O_2CC_9H_6NH).H_2O$		1145
1543	$SnC_{28}H_{25}Br_2N$	$(oTol)_2Sn(1,2-C_6H_3Br-5)_2NEt$		1143
1544	$SnC_{28}H_{27}N$	$(oTol)_2Sn(1,2-C_6H_4)_2NEt$		1143
1545	$SnC_{30}H_{31}N$	$(oTol)_2Sn(1,2-C_6H_3Me-5)_2NEt$		1143
1546	$SnC_{30}H_{32}Cl_2N_2O_4$	$\underline{trans}$-$SnCl_2Me_2(\underline{o}\text{-}OC_6H_4CNHC_6H_4OMe\text{-}\underline{p})_2$		1146
1547	$SnC_{38}H_{32}N_2O_8P_2$	$SnPh_2(NO_3)_2\{\underline{cis}\text{-}(OPPh_2)_2C_2H_2\}.CHCl_3$		1147
1548	$SnWC_{33}H_{28}O_2$	$W\{=CH(pTol)\}(SnPh_3)(CO)_2(Cp)$	210	1148
1549	$SnW_3C_{15}O_{15}$	$W_2(CO)_{10}\{\mu\text{-}SnW(CO)_5\}$	237	39
1550	$SnW_3C_{19}H_8O_{16}$	$W_2(CO)_{10}\{\mu\text{-}Sn(THF)W(CO)_5\}$		1149
1551	$Sn_2C_{16}H_{38}Br_2O_2$	$\{Bu^t_2SnBr(\mu\text{-}OH)\}_2$	193	1150
1552	$Sn_2C_{16}H_{38}Cl_2O_2$	$\{Bu^t_2SnCl(\mu\text{-}OH)\}_2$		1150
1553	$Sn_2C_{16}H_{38}F_2O_2$	$\{Bu^t_2SnF(\mu\text{-}OH)\}_2$	193	1150
1554	$[Sn_2C_{19}H_{23}O]^+$	$[Sn_2(THF)(Cp)_3][BF_4]$	138	1151

No.	Formula	Structure	Details	Ref.
1555	$Sn_2C_{25}H_{22}Cl_2$	$(Ph_2ClSn)_2CH_2$		1152
1556	$Sn_2C_{50}H_{42}N_4O_{14}P_2$	trans-$C_2H_2\{PPh_2OSnPh_2(NO_3)_2\}_2 \cdot 2H_2O$		1147
1557	$Sn_4C_{32}H_{72}I_2$	$(SnBu^t_2)_4I_2$		1153
1558	$[Sn_5C_{10}H_{30}Cl_5O_3]^-$	$[NHEt_3][(SnMe_2Cl)_5(\mu_3\text{-}O)_3]$		1154
1559	$Sn_6C_{78}H_{96}O_{18}$	$[PhSnO(O_2CCy)]_6$		1155
1560	$[TaC_8H_9O_5P]^-$	$[Ta(CO)_3(PMe_3)_4][Ta(CO)_5(PMe_3)]$	113	1156
1561	$TaC_9H_{11}Cl_2$	$TaCl_2(\eta^4\text{-}C_4H_6)(Cp)$		1157
1562	$TaC_{12}H_{37}ClO_3P_3$	$TaCl(CO)_3(PMe_3)_3$	113	1158
1563	$[TaC_{15}H_{36}O_3P_4]^+$	$[Ta(CO)_3(PMe_3)_4][Ta(CO)_5(PMe_3)]$	113	1156
1564	$TaC_{15}H_{42}P_5$	$Ta(PMe_3)_3(\eta^2\text{-}CH_2PMe_2)(\eta^2\text{-}CHPMe_2)$		1159
1565	$TaC_{17}H_{25}$	$Ta(\eta^4\text{-}CH_2CMeCMeCH_2)_2(Cp)$		1157
1566	$TaC_{22}H_{35}$	$Ta(\eta^4\text{-}CH_2CMeCMeCH_2)_2(Cp^*)$	RT & 213	1157
1567	$TaC_{29}H_{51}Cl_2$	$Cl_2Ta\{\eta^5\text{-}C_5(CH_2CMe_2CH_2)_2(CH_2Bu^t)_2(Bu^t)\}$	220	1160
1568	$TaC_{55}H_{65}O_2$	$Ta(CH_2Bu^t)_3(OC_{20}H_{13})_2$		1161
1569	$[Ta_2C_{20}H_{30}Cl_7]^+$	$[(CpTaCl_2)_2(\mu\text{-}Cl)_3]_2[Cl_5Ta(\mu\text{-}O)TaCl_3\}_2(\mu\text{-}Cl)_2]$		1162
1570	$TcC_{32}H_{35}N_2O_2P_2$	trans-$Tc(PhNCMeNPh)(CO)_2(PMe_2Ph)_2$		1163
1571	$TcC_{32}H_{36}N_3O_2P_2$	trans-$Tc\{(pTol)NNN(pTol)\}(CO)_2(PMe_2Ph)_2$		1163
1572	TeC_2F_8	$\{TeF_3(C_2F_5)\}_n$		1164
1573	$TeC_6H_{12}O_2S_4$	$Me_2Te(S_2COMe)_2$		1165
1574	$TeC_8H_6S_2$	$Te(CCHCHCHS)_2$		1166
1575	$TeC_8H_8I_2$	$TeI_2\{(CH_2)_2C_6H_4\text{-}\underline{o}\}$		1167
1576	$TeC_{11}H_{17}BrN_2S$	$TeBrPh\{SC(NMe_2)_2\}$		1041
1577	$TeC_{11}H_{17}ClN_2S$	$TeClPh\{SC(NMe_2)_2\}$		1041
1578	$TeC_{12}H_9Cl_3N_2$	$TeCl_3(C_6H_4N=NPh\text{-}\underline{o})$		1168
1579	$TeC_{12}H_9Cl_3S$	$TeCl_3(C_6H_4SPh\text{-}\underline{o})$		1169
1580	$TeC_{13}H_9N_3S$	$Te(C_6H_4NNPh\text{-}\underline{o})(SCN)$		1170
1581	$TeC_{14}H_{12}N_2O_2$	$Te(C_6H_4NNPh\text{-}\underline{o})(OAc)$		1170
1582	$TeC_{16}H_{17}N$	$Te\{(C_6H_3Me\text{-}5)\text{-}2\}_2NEt$		1171
1583	$TeV_2C_{58}H_{48}O_6P_4$	$\{V(CO)_3(dppe)\}_2Te \cdot PhMe$		1042
1584	$TeWC_{17}H_{27}O_5P$	$W(CO)_5(TePBu^t_3)$		1172

No.	Formula	Structure	Details	Ref.
1585	$Te_2C_3H_4$	$TeCHCHTeCH_2$		1173
1586	$Te_2C_{24}H_{18}Cl_6O_2$	$\{TeCl_2(C_6H_4OPh\text{-}\underline{p})(\mu\text{-}Cl)\}_2$		1174
1587	$Te_2W_3C_{15}O_{15}$	$\{(CO)_5WTeTeW(CO)_5\}\{(\mu\text{-}W(CO)_5\}$		1149
1588	$[TiC_{12}H_{13}ClN]^+$	$[TiCl(NCMe)(Cp)_2][FeCl_4]$		1175
1589	$TiC_{13}H_{19}ClP$	$TiCl(PMe_3)(Cp)_2$		1176
1590	$TiC_{13}H_{35}ClP_4$	$TiMe_{2-x}Cl_x(dmpe)_2$ (x=0.7)		268
1591	$TiC_{14}H_{19}OP$	$Ti(CO)(PMe_3)(Cp)_2$		1176
1592	$TiC_{15}H_{20}Cl_2$	$TiCl_2(Cp)(Cp^*)$		486
1593	$TiC_{15}H_{28}P_2$	$TiEt(dmpe)(\eta\text{-}C_7H_7)$		1177
1594	$TiC_{16}H_{22}Cl_2$	$TiCl_2(\eta\text{-}C_5H_4Pr^i)_2$		1178
1595	$TiC_{16}H_{28}P_2$	$Ti(PMe_3)_2(Cp)_2$		1179
1596	$TiC_{17}H_{22}$	$Ti(Cp^*)(\eta^7\text{-}C_7H_7)$	183	1180
1597	$TiC_{18}H_{17}NS_2$	$Ti(NCS)(SPh)(Cp)(Cp')$		1181
1598	$TiC_{18}H_{23}$	$Ti(Cp^*)(\eta^8\text{-}C_8H_8)$		1180
1599	$TiC_{20}H_{28}Cl_2$	$TiCl_2\{\eta,\eta'\text{-}(C_5Me_4)_2CH_2CH_2\}$	223	1182
1600	$TiC_{21}H_{28}S$	$Ti\{\eta^2 SC\text{-}SC=CCMe_2(CH_2)_3CMe_2\}(Cp)_2$		1183
1601	$TiC_{27}H_{26}O$	$Ti(CPhCPhCMe_2O)(Cp)_2$	153	1184
1602	$TiC_{28}H_{36}O$	$Ti\{\eta^6\text{-}H_2C=C_5Me_3(CH_2CMePhO)\}(Cp^*)$		1185
1603	$TiC_{32}H_{45}$	$Ti\{CH_2C(CH_2CHPh)CH_2CH_2\}(Cp^*)_2$		1186
1604	$TiC_{36}H_{28}$	$Ti(\eta^6\text{-}C_5H_4CPh_2)_2 \cdot Sv$		616
1605	$TiC_{36}H_{34}P_2$	$Ti(CH_2PPh_2)_2(Cp)_2$		1187
1606	$TiWC_{26}H_{24}O_2$	$(Cp)_2TiW(CO)_2(Cp)\{\mu\text{-}\eta^1:\eta^2\text{-}CH_2C(pTol)\}$	200	831
1607	$Ti_2C_{20}H_{19}Cl$	$Ti_2(Cp)_2(\mu\text{-}H)(\mu\text{-}Cl)(\eta^5,\eta^5{'}\text{-}C_{10}H_8)$		694
1608	$Ti_2C_{22}H_{30}Cl_2O_2$	$\{Ti(THF)(\eta\text{-}C_7H_7)(\mu\text{-}Cl)\}_2$		1177
1609	$Ti_2C_{28}H_{24}O_8$	$\{Cp_2TiOCOCH=CHCOO\}_2$ & $.CHCl_3$		1188
1610	$Ti_2C_{30}H_{44}O_2$	$Ti_2(Cp^*)_2(\mu\text{-}O)_2(\mu\text{-}\eta^1:\eta^5\text{-}CH_2C_5Me_4)$		1189
1611	$[Ti_2U_2W_{10}C_{20}H_{20}O_{38}]^{4-}$	$[NBu_4]_4[\{(Cp_2U)(\mu_3\text{-}O)_2(TiW_5O_{17})\}_2].4MeCN$		1190
1612	$Ti_5C_{25}H_{25}S_6$	$\{Ti(Cp)\}_5(\mu_3\text{-}S)_6$		1191
1613	$UC_{17}H_{23}Br_3O_2$	$UBr_3(THF)_2(\eta^5\text{-}ind)$		1192
1614	$UC_{21}H_{30}P$	$U(PMe_3)(Cp')_3$		1193

Structures Determined by Diffraction Methods 477

No.	Formula	Structure	Details	Ref.
1615	$UC_{22}H_{29}S$	$U(SC_4H_8)(Cp')_3$		1194
1616	$UC_{23}H_{29}N$	$U(\eta^2\text{-MeC=NCy})(Cp)_3$		1195
1617	$UC_{24}H_{20}$	$U(\eta^8\text{-benzo[8]annulene})_2$		1196
1618	$UC_{24}H_{26}N$	$U(NPh)(Cp')_3$	178	1197
1619	$[U_2C_{24}H_{36}Cl_7]^+$	$[\{UCl_2(\eta\text{-}C_6Me_6)\}_2(\mu\text{-}Cl)_3][AlCl_4]$		1198
1620	$U_2C_{43}H_{47}NO$	$\{U(Cp')_3\}_2(\mu\text{-}O:\eta^2 NC\text{-}PhNCO)$		1197
1621	$[VC_6O_6]^-$	$[M][V(CO)_6]$ $\{M=pyH, Cu(PPh_3)_3\}$		1199, 319
1622	$VC_{10}H_{11}O_3S$	$V(CO)_3(SMe_2)(Cp)$		686
1623	$[VC_{10}H_{15}Br_2NO]^-$	$[VBr(NO)(Cp^*)_2][VBr_2(NO)(Cp^*)]$		1200
1624	$VC_{12}H_{24}P_2$	$VMe(dmpe)(Cp)$	115	1201
1625	$VC_{13}H_7O_4$	$V(CO)_4(\eta^5\text{-ind})$	130	1202
1626	$[VC_{20}H_{30}BrNO]^+$	$[VBr(NO)(Cp^*)_2][VBr_2(NO)(Cp^*)]$		1200
1627	$VC_{28}H_{39}N$	$V(NXy)(Cp^*)_2$		1203
1628	$V_2C_{16}H_{16}$	$V_2(Cp)_2(\eta^6,\eta^6\text{'-PhH})$	100	266
1629	$V_2C_{26}H_{30}O_6$	$(Cp^*)_2VOCV(CO)_5$		1204
1630	$V_2C_{38}H_{30}O_8$	$V_2(Cp)_2(\mu\text{-}O_2CPh)_4$		1205
1631	$V_2C_{38}H_{50}N_4O_4$	$\{V(CH_2Ph)(\mu\text{-OCMeCHCMeNCH}_2\text{-})_2\}_2$		1206
1632	$V_2C_{48}H_{50}O_8$	$V_2(Cp^*)_2(\mu\text{-}O_2CPh)_4$		1205
1633	$V_4C_{20}H_{20}I_2N_2O_6$	$\{VI(Cp)\}_2\{V(NO)(Cp)\}_2(\mu\text{-}O)_4$		1207
1634	$V_{13}C_{61}H_{73}N_2O_{18}$	$\{Cp_5V_6(\mu_3\text{-}O)_8\}_2(\mu\text{-}O_2V(NMe_3)_2Cp\}.4PhMe$		1208
1635	$V_{16}C_{70}H_{70}O_{24}$	$\{Cp_5V_6(\mu_3\text{-}O)_8\}_2(\mu_3\text{-}O)_2\{Cp_4V_4(\mu\text{-}O)_6\}.PhMe$		1208
1636	$[WC_5O_5]^{2-}$	$[Na(K\text{-}2.2.1)]_2[W(CO)_5]$		1209
1637	$[WC_6HO_7]^-$	$[Na(K\text{-}2.2.1)][W(OCHO)CO)_5]$		1210
1638	$[WC_8H_{12}O_4P]^-$	$[PPN][\underline{cis}\text{-}WMe(CO)_4(PMe_3)]$	174	1211
1639	$WC_{10}H_{10}N_2O_5S$	$W(CO)_5(SCNMeCH_2CH_2NMe)$		257
1640	$[WC_{11}H_{23}O_4]^-$	$[NEt_4][WO_2(OCMe_2CMe_2O)(CH_2Bu^t)]$		1212
1641	$WC_{12}H_{12}O_5$	$W(CH_2CO_2Et)(CO)_3(Cp)$		862
1642	$WC_{13}H_{14}O_3$	$W(CH_2C(CH_2CH_2)Me)(CO)_3(Cp)$		868
1643	$WC_{14}H_{36}Cl_3P_3$	$WCl_3(CBu^t)(PMe_3)_3$		1213
1644	$WC_{15}H_{13}IO_2$	$W\{CH(pTol)\}I(CO)_2(Cp)$		1214

No.	Formula	Structure	Details	Ref.
1645	$WC_{15}H_{18}$	$W(\eta^6-C_5H_4CMe_2)(\eta-C_6H_5Me)$		616
1646	$WC_{15}H_{27}O_3PS_2$	$W(CO)_3(PPr^i_3)(\eta^3-SCMeSMe)$		1215
1647	$WC_{15}H_{45}P_5$	$WH(PMe_3)_4(\eta^2-CH_2PMe_2)$		1159
1648	$WC_{16}H_{23}Cl$	$WCl(\eta^3-Bu^tC_3Bu^t)(Cp)$		1216
1649	$WC_{16}H_{44}P_4$	trans-$W(PMe_3)_4(\eta-C_2H_4)_2$		1217
1650	$WC_{17}H_{16}$	$W(CHPh)(Cp)_2$	113	1218
1651	$WC_{18}H_{10}BrNO_4$	trans-$WBr(CO)_4(CNCPh_2)$	233	1219
1652	$WC_{18}H_{20}N_2O_3S_4$	$W(dmtc)_2(\eta^2-PhCCH)(\eta^2-CHCHC(O)OCO).Sv$		1220
1653	$[WC_{18}H_{22}O_2P]^+$	$[W(CO)(PMe_3)\{\eta-HOCC(pTol)\}(Cp)][BF_4].Me_2CO$		1214
1654	$WC_{18}H_{24}O_3$	$W(CMe=CMeCOMe)(CO)_2(Cp^*)$	233	1221
1655	$WC_{18}H_{32}O_{13}P_2$	$W(CO)_3\{P(OMe)_3\}\{P(OEt)_3\}\{\eta^2-C_2H_2(CO_2Me)_2\}$	208	1222
1656	$WC_{19}H_{32}O_3$	$W(CPh)(OBu^t)_3$		1223
1657	$[WC_{20}H_{16}O_5P]^-$	$[PPh_4][W\{C(O)CH_2CH_2CH_2PPh_2\}(CO)_4]$		1224
1658	$WC_{20}H_{18}F_{12}S$	$W(\eta^2-C_4F_6)(\eta^4-C(CF_3)C(CF_3)CMeCMeSPr^i)(Cp)$		1225
1659	$WC_{20}H_{22}F_{18}O_3$	$W(\eta^3-Bu^tCCHCBu^t)\{OCH(CF_3)_2\}_3$		1226
1660	$WC_{20}H_{35}Cl_2P$	$WCl_2(PMe_3)(\eta^3-MeCCBu^tCBu^t)(Cp)$		1227
1661	$WC_{21}H_{10}N_4O_2$	$W(CO)_2\{\eta^3-(NC)_2CCPhC(CN)_2\}(Cp).Sv$		1228
1662	$WC_{21}H_{18}O_2$	$W(CO)_2\{\eta^3-CH(pTol)Ph\}(Cp)$		1229
1663	$[WC_{21}H_{33}OP_2]^+$	$[W(PMe_3)_2\{\eta-MeOC\equiv C(pTol)\}(Cp)][BF_4]$	233	1230
1664	$WC_{22}H_{17}ClN_2O_4$	$WCl(CPh)(CO)(py)_2\{\eta^2-CH=CHC(O)OCO\}$		1231
1665	$WC_{22}H_{18}F_{12}O_4S$	$WCp\{\eta^5-C(CF_3)C(CF_3)CRCRC(CF_3)C(CF_3)SPr^i\}$ R=CO_2Me		1225
1666	$WC_{22}H_{18}F_{12}O_4S$	$CpWF\{\eta^4-C(CF_2)C(CF_3)CRCRC(CF_3)C(CF_3)SPr^i\}$ R=CO_2Me		1225
1667	$WC_{24}H_{24}Cl_2$	$W(CPhCBu^tCPh)Cl_2(Cp)$		1232
1668	$WC_{24}H_{30}N_2S_5$	$WS(detc)_2(\eta-PhCCPh)$		1233
1669	$WC_{24}H_{36}N_2O_2S_4$	$W(dmtc)_2(CO)\{\eta^4-(C_8H_{12})_2CO\}$		1234
1670	$WC_{25}H_{22}$	$W(\eta^6-C_5H_4CPh_2)(\eta-tolH)$		616
1671	$WC_{26}H_{15}O_6P$	$W(CO)_5(PPhCPhCPhCO)$	173	1235
1672	$WC_{26}H_{30}F_{12}N_2S$	$W(SPr^i)(CNBu^t)\{\eta^2CN-Bu^tNC_5(CF_3)_4\}(Cp)$		1236
1673	$WC_{31}H_{22}N_2O_5S$	$W(CO)_5\{S=C(NPh)CH_2NPh_2\}$		284
1674	$WC_{33}H_{31}NOS_2$	$WO(detc)(\eta^4-CPhCPhCPhCHPh)$		1237

Structures Determined by Diffraction Methods 479

No.	Formula	Structure	Details	Ref.
1675	$WC_{33}H_{31}O_3P_3$	$W(CO)_3\{PPh(CH_2CH_2CH_2PPh_2)_2C_6H_4-\underline{o}\}$	238	640
1676	$WC_{41}H_{41}NO_2P_2S_2$	$W(\eta^2-OC=CCH_2Ph)(detc)(CO)(dppe)$		1238
1677	$WC_{44}H_{35}O_7P_3$	$(CO)_5W(\mu-PPh_2)OsH(CO)_2(PHPh_2)(PMePh_2).CH_2Cl_2$		1239
1678	$WZrC_{22}H_{14}O_6$	$Cp_2Zr\{C_6H_4C(O)W(CO)_5-\underline{o}\}$		1240
1679	$WZrC_{38}H_{30}O_4P_2$	$(Cp)_2ZrW(CO)_4(\mu-PPh_2)_2.(MeOCH_2CH_2)_2O$		1241
1680	$W_2C_7Br_4O_7$	$W_2Br(CO)_7(\mu-Br)_3$		1242
1681	$W_2C_8Br_4O_8$	$W_2Br_2(CO)_8(\mu-Br)_2$		1242
1682	$[W_2C_{10}HO_{10}]^-$	$[PPh_4][\{W(CO)_5\}_2(\mu-H)]$	X,N at 40	1243
1683	$W_2C_{14}H_{38}Cl_3N_3P_2$	$W_2Cl_3(NMe_2)_2(PMe_3)_2(\mu-NMe_2)(\mu-\eta-C_2H_2))$	111	1244
1684	$W_2C_{16}H_{10}O_6S$	$\{W(CO)_3(Cp)\}_2S$		1245
1685	$W_2C_{18}H_{28}Cl_4N_4$	$W_2Cl_4(py)_2(\mu-NMe_2)_2(\mu-\eta^2-MeCCMe)$	113	1246
1686	$[W_2C_{20}H_{13}O_7]^-$	$[PPN][W_2(CO)_7(Cp)\{\mu-CH(pTol)\}].Sv$	200	1148
1687	$W_2C_{24}H_{30}O_8S_4$	$\{W(CO)_2(Cp^*)(\mu-SSO_2)\}_2$		1245
1688	$W_2C_{24}H_{42}Cl_3N_3P_2$	$W_2Cl_3(NMe_2)_2(PMe_2Ph)_2(\mu-CHCH_2)(\mu-NMeCH_2).Sv$	167	1244
1689	$W_2C_{25}H_{54}O_7$	$W_2(OBu^t)_4(\mu-OBu^t)_2(\mu-CO)$	111	1247
1690	$W_2C_{27}H_{46}O_4$	$W_2H(OPr^i)_4(\mu-CPh)(\mu-\eta^2:\eta^4-C_4Me_4)$	114	1248
1691	$W_2C_{31}H_{59}NO_7$	$W_2(OBu^t)_6(\mu-\eta^2CO:\eta^2CN-PhNCO)$	165	1249
1692	$W_2C_{31}H_{68}N_2O_6$	$\{W(OBu^t)_3\}_2(\eta^2,\eta^2{'}-Pr^iNCNPr^i)$		1250
1693	$W_2C_{34}H_{54}O_4$	$W_2(CH_2Ph)_2(OPr^i)_2(\mu-OPr^i)_2(\eta-C_2Me_2)_2$	114	1248
1694	$W_2C_{34}H_{62}Cl_8P_2$	$\{WCl_4(PMe_3)_2\{\eta,\eta{'}-(C_5Et_4)_2C_2H_4\}.CH_2Cl_2$		1251
1695	$W_2C_{35}H_{42}O_4$	$W_2(CO)_4(Cp)\{\eta-C_5Et_4(pTol)\}(\mu-\eta^2-EtCCEt)$	200	1252
1696	$W_2C_{37}H_{76}N_2O_6$	$\{W(OBu^t)_3\}_2(\eta^2,\eta^2{'}-CyNCNCy)$		1250
1697	$W_2C_{39}H_{28}O_8P_2$	$W_2(CO)_8(\mu-PPh)(\mu-P:C-Ph_2PCC=C=CHCH=CMe_2)$		1253
1698	$W_2C_{39}H_{71}N_2O_6$	$W_2(OBu^t)_6\{\mu-(pTol)NCN(pTol)\}$		1254
1699	$W_2C_{58}H_{42}O_3$	$W_2(CO)_3(\eta-PhCCPh)(Cp)\{\eta-C_5Ph_4(pTol)\}$	200	1252
1700	$W_3C_{12}N_6O_{12}$	$W_3(CO)_9(NO)_3(\mu-C:N-CN)_3$		1255
1701	$[W_3C_{14}H_{27}O_{16}]^{2+}$	$[W_3(OH_2)_3(\mu-OAc)_6(\mu_3-O)(\mu_3-CMe)]Br_2.2H_2O$		1256
1702	$W_3C_{27}H_{10}O_{15}P_2$	$W(CO)_5\{\eta^2-(CO)_5WPPhPPhW(CO)_5\}$		1257
1703	$W_3C_{29}H_{66}O_9$	$W_3(OPr^i)_6(\mu-OPr^i)_3(\mu_3-CMe)$	114	1092
1704	$W_4C_{38}H_{84}O_{14}$	$\{W_2(OPr^i)_5(\mu-OPr^i)(\mu_3-\mu C:O-CO)\}_2$	111	1247

No.	Formula	Structure	Details	Ref.
1705	$W_4C_{38}H_{87}NO_{12}$	$W_4C(OPr^i)_8(\mu\text{-}OPr^i)_4(\mu\text{-}NMe)$		1258
1706	$W_6C_{48}H_{104}O_{16}$	$\{W_3(OBu^t)_5(\mu\text{-}O)_2(\mu\text{-}COPr)\}_2$		1223
1707	$Y_2C_{40}H_{60}Cl_2$	$Y_2Cl(\mu\text{-}Cl)(Cp^*)_4$		1259
1708	$[YbC_{16}H_{16}]^{2-}$	$[K(DME)_2][Yb(\eta^8\text{-}C_8H_8)_2]$		1260
1709	$YbC_{17}H_{29}Br_2O_3$	$YbBr_2(THF)_3(Cp)$		1261
1710	$YbC_{17}H_{29}Cl_2O_3$	$YbCl_2(THF)_3(Cp)$		1262
1711	$ZnC_{10}H_{10}$	$Zn(Cp)_2$		1263
1712	$ZnC_{20}H_{30}$	$Zn(\eta^1\text{-}Cp^*)(\eta^5\text{-}Cp^*)$	E	1264
1713	$ZrC_{10}H_{10}BrNO_3$	$ZrBr(NO_3)(Cp)_2$		1265
1714	$ZrC_{10}H_{10}ClNO_3$	$ZrCl(NO_3)(Cp)_2$		1265
1715	$ZrC_{13}H_{18}ClP$	$ZrCl(CH_2PMe_2)(Cp)_2$	140	1266
1716	$ZrC_{15}H_{20}Cl_2$	$ZrCl_2(Cp)(Cp^*)$		486
1717	$ZrC_{16}H_{20}$	$Zr(\eta^4\text{-}CH_2\text{=}CMeCMe\text{=}CH_2)(Cp)_2$		487
1718	$ZrC_{18}H_{22}$	$Zr\{\eta^4\text{-}(CH_2\text{=})_2C_6H_8\text{-}1,2\}(Cp)_2$		487
1719	$ZrC_{19}H_{21}NO$	$ZrMe(NPhCMeO)(Cp)_2$		1267
1720	$ZrC_{20}H_{24}$	$Zr\{(CH_2)_2C_8H_{10}\}(Cp)_2$		1268
1721	$ZrC_{20}H_{24}Cl_2$	$ZrCl_2\{\eta^5,\eta^5{}'\text{-}(indH_4)CH_2CH_2(indH_4)\}$	228	1269
1722	$ZrC_{20}H_{28}Cl_2$	$ZrCl_2\{\eta,\eta'\text{-}(C_5Me_4)_2CH_2CH_2\}$	216	1182
1723	$ZrC_{20}H_{31}ClO$	$Zr(OH)Cl(Cp^*)_2$		1270
1724	$ZrC_{20}H_{32}O_2$	$Zr(OH)_2(Cp^*)_2$		1270
1725	$ZrC_{20}H_{37}P$	$Zr(PEt_3)(\eta^5\text{-}CH_2CMeCHCMeCH_2)_2$		630
1726	$ZrC_{22}H_{32}$	$Zr(\eta^4\text{-}C_4H_6)(\eta\text{-}C_5H_4Bu^t)_2$		1271
1727	$ZrC_{23}H_{33}ClN_2$	$ZrCl(NCyCHNCy)(Cp)_2$		1272
1728	$ZrC_{26}H_{26}O$	$ZrMe(OCMeCPh_2)(Cp)_2$		1267
1729	$ZrC_{33}H_{42}ClP$	$ZrCl(CH_2PPh_2)(Cp^*)_2$	140	1266
1730	$ZrC_{35}H_{31}P$	$ZrPh(CHPPh_3)(Cp)_2$		1273
1731	$ZrC_{35}H_{39}F_3O$	$Zr(CH_2C_6H_4F\text{-}p)_3(OC_6H_3Bu^t{}_2\text{-}2,6)$	113	1274
1732	$ZrC_{35}H_{42}O$	$ZrBz_3(OC_6H_3Bu^t{}_2\text{-}2,6)$	113	1274
1733	$ZrC_{40}H_{34}$	$Zr(CPhCPhCPhCPh)(Cp')_2 \cdot PhH$		1275
1734	$ZrC_{40}H_{36}O_2$	$Zr(OCMeCPh_2)_2(Cp)_2$		1267

No.	Formula	Structure	Details	Ref.
1735	$ZrC_{44}H_{58}O_2$	$ZrBz_2(OC_6H_3Bu^t_2-2,6)(OC_6H_2Bu^t_2OMe-2,6,4)$	113	1274
1736	$ZrC_{48}H_{66}N_2O_2$	$Zr(\eta^4-XyNCMeCMeNXy)(OC_6H_3Bu^t_2-2,6)_2$	112	1276
1737	$ZrC_{52}H_{65}NO_3$	$Zr(OR)_2(OCBz=CBzNXy)$ $R=C_6H_3Bu^t_2-2,6$	113	1277
1738	$ZrC_{52}H_{74}N_2O_2$	$Zr(OR)_2(\eta^2-Bu^tNCBz)_2$ $R=C_6H_3Bu^t_2-2,6$	116	1277

Additional abbreviations used in Main Table

18-c-6	18-crown-6	Cp'	$\eta^5-C_5H_4Me$
Cp*	$\eta^5-C_5Me_5$	detc	diethydithiocarbamate
dapm	$Ph_2AsCH_2PPh_2$	db-18-c-6	dibenzo-18-crown-6
dcpe	$Cy_2PCH_2CH_2PCy_2$	DMAD	$MeO_2CC\equiv CCO_2Me$
dmpz	3,5-dimethylpyrazol-1-yl	dmtc	dimethyldithiocarbamate
dppmpp	$PhP(CH_2PPh_2)_2$	fulv	fulvalene
$glyH_2$	glyoxime	ind	indenyl
K-2.2.1	4,7,13,16,21-pentaoxa-1,10-diazabicyclo[8.8.5]tricosane		
np_3	$N(CH_2CH_2PPh_2)_3$	oTol	o-tolyl
OTp	9-oxytriptycene	pTol	p-tolyl
Sv	solvated crystal	TFAc	CF_3CO
tfb	5,6,7,8-tetrafluorobarrelene	TMSF	tetramethyltetraselenofulvalene
TSeF	tetraselenofulvalene	Xy	$2,6-Me_2C_6H_4$

3 Metals Cross Reference Table

A list of mixed metal compound numbers which are listed alphabetically in the Main Table under another metal.

Metal	Compound Numbers
As	5, 12
B	6, 52, 53, 54, 96
Co	55, 71, 97, 98, 140, 158, 159, 175, 176, 187, 212, 213
Cr	56, 57, 58, 72, 141, 142, 143, 313, 314
Cu	130, 277, 278
Fe	59, 60, 73, 99, 100, 101, 102, 103, 123, 144, 145, 177, 178, 179, 180, 181, 182, 183, 190, 194, 221, 222, 224, 279, 280, 281, 282, 283, 284, 285, 286, 287, 288, 315, 316, 342, 403, 404, 405, 427, 446, 447, 448, 457, 458, 462, 467, 468, 469
Ge	280, 317, 342, 343, 357, 521
Hf	171
Hg	358, 359, 645
Ir	146, 172, 318, 463, 654
Li	16, 23, 24, 459, 465, 470, 634
Lu	770, 771
Mg	25, 466
Mn	57, 58, 61, 74, 75, 76, 319, 460, 522, 523, 524, 594, 595, 684, 709, 787
Mo	55, 60, 62, 63, 64, 72, 147, 148, 149, 160, 161, 162, 195, 196, 280, 316, 317, 320, 321, 449, 525, 526, 527, 528, 529, 530, 531, 596, 597, 635, 700, 710, 820, 821, 822, 841
Nb	823
Nd	772
Ni	65, 184, 185, 186, 188, 281, 322, 343, 344, 526, 532, 533, 534, 598, 599, 773, 907, 908
Os	66, 77, 104, 105, 124, 125, 322, 994
Pb	842
Pd	78, 79, 140, 197, 198, 535, 536, 843
Pr	949
Pt	7, 8, 9, 10, 13, 80, 81, 82, 176, 183, 185, 186, 600, 601, 824, 825, 1015, 1016, 1045, 1046, 1047, 1048, 1049, 1050, 1051, 1069, 1118
Re	14, 106, 199, 323, 324, 655, 656, 657, 658, 659, 796, 797, 1142
Rh	11, 79, 80, 81, 82, 83, 84, 85, 88, 98, 200, 201, 202, 203, 204, 205, 208, 209, 214, 215, 216, 217, 288, 325, 326, 537, 538, 539, 540, 602, 603, 636, 660, 661, 662, 826, 950, 951, 1025, 1119, 1143, 1144, 1165
Ru	67, 98, 107, 108, 109, 126, 127, 189, 206, 325, 327, 328, 329, 330, 331, 332, 345, 358, 541, 711, 762, 909, 1264, 1265, 1266, 1267, 1268, 1269, 1304, 1314, 1315, 1316, 1317
Sb	427, 428, 429, 435, 436, 542, 543, 604, 605, 846, 947, 1026, 1305, 1355
Se	162, 406, 430, 431, 432, 521, 594, 595, 606, 692, 844, 1145, 1146, 1147, 1270, 1271, 1356
Si	26, 27, 36, 37, 68, 69, 86, 150, 151, 163, 164, 165, 166, 171, 187, 289, 333, 346, 353, 354, 360, 407, 408, 409, 410, 450, 543, 544, 604, 607, 637, 667, 669, 685, 690, 691, 701, 720, 739, 740, 741, 772, 774, 775, 780, 781, 782, 783, 788, 789, 790, 797, 798, 800, 827, 845, 847, 910, 911, 912, 913, 914, 942, 962, 969, 970, 971, 995, 996, 1006, 1010, 1052, 1053, 1054, 1055, 1084, 1101, 1102, 1121, 1148, 1149, 1166, 1169, 1170, 1171, 1215, 1272, 1357, 1395, 1422, 1423, 1440
Sm	1497
Sn	143, 149, 167, 290, 291, 292, 709, 742, 915, 1054, 1055, 1102, 1121, 1149, 1150, 1166, 1171, 1192, 1193, 1272, 1306, 1473, 1532
Ta	1474
Te	15, 411, 451, 527, 528, 531, 597, 943, 944, 1437, 1438, 1524
Th	191, 192, 997, 1358
Ti	28, 38, 39, 165, 168, 452, 945, 948, 1359, 1475, 1476, 1477

Metals Cross Reference Table (continued)

Metal	Compound Numbers
Tl	158, 210, 668
U	173, 968, 1638
V	166, 293, 1439, 1478, 1479, 1610
W	70, 87, 99, 195, 196, 207, 211, 223, 313, 317, 324, 334, 347, 433, 453, 545, 546, 605, 608, 609, 686, 687, 743, 760, 823, 828, 916, 917, 968, 998, 1056, 1057, 1058, 1059, 1151, 1152, 1153, 1154, 1194, 1195, 1307, 1360, 1361, 1424, 1425, 1426, 1446, 1480, 1498, 1499, 1500, 1501, 1502, 1503, 1509, 1525, 1526, 1527, 1528, 1529, 1533, 1575, 1576, 1577, 1611, 1614, 1633, 1638
Y	1504
Yb	37
Zn	355, 999, 1007
Zr	152, 918, 919, 920, 1000, 1362, 1363, 1481, 1482, 1483, 1484, 1505, 1706, 1707

REFERENCES

1. R.Faure, H.Loiseleur, G.Hauffe, H.Trauer, Acta Crystalloqr., 1985, C41, 1593.
2. J.C.Barnes and C.S.Blyth, Inorq. Chim. Acta, 1985, 98, 181.
3. H.C.Kang, A.W.Hanson, B.Eaton, V.Boekelheide, J. Am. Chem. Soc., 1985, 107, 1979.
4. C.Pelizzi, G.Pelizzi, P.Tarasconi, J. Organomet. Chem., 1985, 281, 403.
5. K.Shelly, D.C.Finster, Y.J.Lee, W.R.Scheidt, C.A.Reed, J. Am. Chem. Soc., 1985, 107, 5955.
6. R.Usón, J.Forniés, B.Menjón, F.A.Cotton, L.R.Falvello, M.Tomás, Inorq. Chem., 1985, 24, 4651.
7. R.Usón, J.Forniés, M.Tomás, J.M.Casas, J. Am. Chem. Soc., 1985, 107, 2556.
8. A.Albinati, K.-H.Dahmen, A.Togni, L.M.Venanzi, Angew. Chem., Int. Ed. Engl. 1985, 24, 766.
9. G.Bruno, S.L.Schiavo, P.Piraino, F.Faraone, Organometallics, 1985, 4, 1098.
10. T.Beringhelli, G.D'Alfonso, M.Freni, G.Ciani, A.Sironi, J. Organomet. Chem., 1985, 295, C7.
11. S.H.Strauss, M.D.Noirot, O.P.Anderson, Inorq. Chem., 1985, 24, 4307.
12. M.Y.Chiang, E.Böhlen, R.Bau, J. Am. Chem. Soc., 1985, 107, 1679.
13. S.G.Bott, H.Elgamal, J.L.Atwood, J. Am. Chem. Soc., 1985, 107, 1796.
14. J.L.Atwood, H.Elgamal, G.H.Robinson, S.G.Bott, J.A.Weeks, W.E.Hunter, J. Inclusion Phenom., 1984, 2, 367.
15. J.H.Medley, F.R.Fronczek, N.Ahmad, M.C.Day, R.D.Rogers, C.R.Kerr, J.L.Atwood, J. Crystalloqr. Spectrosc. Res., 1985, 15, 99.
16. M.J.Zaworotko, C.R.Kerr, J.L.Atwood, Organometallics, 1985, 4, 238.
17. H.H.Karsch, A.Appelt, G.Müller, Organometallics, 1985, 4, 1624.
18. H.Lehmkuhl, K.Mehler, A.Shakoor, C.Krüger, Y.-H.Tsay, R.Benn, A.Rufińska, G.Schroth, Chem. Ber., 1985, 118, 4248.
19. M.Veith, H.Lange, O.Recktenwald, W.Frank, J. Organomet. Chem., 1985, 294, 273.
20. H.Lehmkuhl, A.Shakoor, K.Mehler, C.Krüger, Y.-H.Tsay, Z. Naturforsch., 1985, 40B, 1504.
21. V.K.Bel'skii, A.I.Sizov, B.M.Bulychev, G.L.Soloveichik, J. Organomet. Chem., 1985, 280, 67.
22. H.H.Karsch, A.Appelt, F.H.Köhler, G.Müller, Organometallics, 1985, 4, 231.
23. A.-A.I.Al-Wassil, P.B.Hitchcock, S.Sarisaban, J.D.Smith, C.L.Wilson, J. Chem. Soc., Dalton Trans., 1985, 1929.
24. M.R.P.van Vliet, P.Buysingh, G.van Koten, K.Vrieze, B.Kojić-Prodić, A.L.Spek, Organometallics, 1985, 4, 1701.
25. H.Schmidbaur, S.Lauteschläger, G.Müller, J. Organomet. Chem., 1985, 281, 33.
26. J.M.Boncella and R.A.Andersen, Organometallics, 1985, 4, 205.
27. E.B.Lobkovskii, G.L.Soloveichik, A.I.Sizov, B.M.Bulychev, J. Organomet. Chem., 1985, 280, 53.
28. H.Zhang, C.M.Means, N.C.Means, J.L.Atwood, J. Crystalloqr. Spectrosc. Res., 1985, 15, 445.
29. R.R.Holmes, R.O.Day, A.C.Sau, Organometallics, 1985, 4, 714.
30. C.A.Poutasse, R.O.Day, J.M.Holmes, R.R.Holmes, Organometallics, 1985, 4, 708.
31. G.Becker, B.Becker, M.Birkhahn, O.Mundt, R.E.Schmidt, Z. Anorq. Allq. Chem., 1985, 529, 97.
32. M.I.Bruce, P.A.Humphrey, J.M.Patrick, B.W.Skelton, A.H.White, M.L.Williams, Aust. J. Chem., 1985, 38, 1441.
33. M.-U.-Haque, H.A.Tayim, J.Ahmed, W.Horne, J. Crystalloqr. Spectrosc. Res., 1985, 15, 561.
34. M.Nardelli, C.Pelizzi, G.Pelizzi, P.Tarasconi, J. Chem. Soc., Dalton Trans., 1985, 321.
35. R.K.Gupta, A.K.Rai, R.C.Mehrotra, V.K.Jain, B.F.Hoskins, E.R.T.Tiekink, Inorq. Chem., 1985, 24, 3280.
36. R.K.Chadha, J.M.Chehayber, J.E.Drake, J. Crystalloqr. Spectrosc. Res., 1985, 15, 53.
37. H.Brunner, H.Kauermann, U.Klement, J.Wachter, T.Zahn, M.L.Ziegler, Angew. Chem., Int. Ed. Engl., 1985, 24, 132.

38 B.Sigwarth, U.Weber, L.Zsolnai, G.Huttner, Chem. Ber., 1985, 118, 3114.
39 G.Huttner, U.Weber, B.Sigwarth, O.Scheidsteger, H.Lang, L.Zsolnai, J. Organomet. Chem., 1985, 282, 331.
40 A.G.Osborne, R.E.Hollands, R.F.Bryan, S.Lockhart, J. Organomet. Chem., 1985, 288, 207.
41 K.Blechschmitt, T.Zahn, M.L.Ziegler, Angew. Chem., Int. Ed. Engl., 1985, 24, 702.
42 W.A.Herrmann, B.Koumbouris, A.Schäfer, T.Zahn, M.L.Ziegler, Chem. Ber., 1985, 118, 2472.
43 R.A.Jones, S.T.Schwab, A.L.Stuart, B.R.Whittlesey, T.C.Wright, Polyhedron, 1985, 4, 1689.
44 K.Blechschmitt, H.Pfisterer, T.Zahn, M.L.Ziegler, Angew. Chem., Int. Ed. Engl., 1985, 24, 66.
45 Q.Yang, Y.Han, Y.Tang, H.Li, M.Xu, S.Li, W.Xu, Huaxue Xuebao, 1984, 42, 1128.
46 W.Ehrenreich, M.Herberhold, G.Herrmann, G.Süss-Fink, A.Gieren, T.Hübner, J. Organomet. Chem., 1985, 294, 183.
47 L.R.Martin, F.W.B.Einstein, R.K.Pomeroy, Inorg. Chem., 1985, 24, 2777.
48 R.Bohra and H.W.Roesky, J. Fluorine Chem., 1984, 25, 145.
49 R.Appel, T.Gaitzsch, F.Knoch, Angew. Chem., Int. Ed. Engl., 1985, 24, 419.
50 G.B.Jameson, M.T.Pope, S.H.Wasfi, J. Am. Chem. Soc., 1985, 107, 4911.
51 M.Di Vaira, L.Niccolai, M.Peruzzini, P.Stoppioni, Organometallics, 1985, 4, 1888.
52 G.Huttner, B.Sigwarth, O.Scheidsteger, L.Zsolnai, O.Orama, Organometallics, 1985, 4, 326.
53 C.R.Jablonski, Y.-P.Wang, N.J.Taylor, Inorg. Chim. Acta, 1985, 96, L17.
54 G.Süss-Fink, K.Guldner, M.Herberhold, A.Gieren, T.Hübner, J. Organomet. Chem., 1985, 279, 447.
55 R.D.W.Kemmitt, P.McKenna, D.R.Russell, L.J.S.Sherry, J. Chem. Soc., Dalton Trans., 1985, 259.
56 A.L.Balch, L.A.Fossett, M.M.Olmstead, D.E.Oram, P.E.Reedy Jr., J. Am. Chem. Soc., 1985, 107, 5272.
57 A.L.Balch, R.R.Guimerans, J.Linehan, M.M.Olmstead, D.E.Oram, Organometallics, 1985, 4, 1445.
58 A.L.Balch, R.R.Guimerans, J.Linehan, F.E.Wood, Inorg. Chem., 1985, 24, 2021.
59 H.Schumann, S.Jurgis, E.Hahn, J.Pickardt, J.Blum, M.Eisen, Chem. Ber., 1985, 118, 2738.
60 R.J.H.Clark, A.J.Hempleman, H.M.Dawes, M.B.Hursthouse, C.D.Flint, J. Chem. Soc., Dalton Trans., 1985, 1775.
61 A.H.Cowley, N.C.Norman, M.Pakulski, J. Chem. Soc., Dalton Trans., 1985, 383.
62 R.A.Jones and B.R.Whittlesey, J. Am. Chem. Soc., 1985, 107, 1078.
63 P.K.Byers, A.J.Canty, B.W.Skelton, A.H.White, Aust. J. Chem., 1985, 38, 1251.
64 P.K.Byers, A.J.Canty, N.J.Minchin, J.M.Patrick, B.W.Skelton, A.H.White, J. Chem. Soc., Dalton Trans., 1985, 1183.
65 J.Vicente, M.-T.Chicote, M.-D.Bermudez, P.G.Jones, G.M.Sheldrick, J. Chem. Research(S), 1985, 72.
66 A.J.Markwell, J. Organomet. Chem., 1985, 293, 257.
67 E.G.Perevalova, I.G.Bolesov, Yu.T.Struchkov, I.F.Leschova, Ye.S.Kalyuzhnaya, T.I.Voyevodskaya, Yu.L.Slovokhotov, K.I.Grandberg, J. Organomet. Chem., 1985, 286, 129.
68 M.Lanfranchi, M.A.Pellinghelli, A.Tiripicchio, F.Bonati, Acta Crystallogr., 1985, C41, 52.
69 S.L.Schiavo, G.Bruno, F.Nicolò, P.Piraino, F.Faraone, Organometallics, 1985, 4, 2091.
70 J.Bashkin, C.E.Briant, D.M.P.Mingos, R.W.M.Wardle, Transition Met. Chem., 1985, 10, 113.
71 J.Pursiainen, M.Ahlgren, T.A.Pakkanen, J. Organomet. Chem., 1985, 297, 391.
72 L.W.Arndt, M.Y.Darensbourg, J.P.Fackler Jr., R.J.Lusk, D.O.Marler, K.A.Youngdahl, J. Am. Chem. Soc., 1985, 107, 7218.
73 H.Umland and U.Behrens, J. Organomet. Chem., 1985, 287, 109.
74 C.P.Horwitz, E.M.Holt, C.P.Brock, D.F.Shriver, J. Am. Chem. Soc., 1985, 107,

8136.
75 C.P.Horwitz, E.M.Holt, D.F.Shriver, J. Am. Chem. Soc., 1985, 107, 281.
76 M.I.Bruce, E.Horn, J.G.Matisons, M.R.Snow, J. Organomet. Chem., 1985, 286, 271.
77 C.F.Barrientos-Penna, F.W.B.Einstein, T.Jones, D Sutton, Inorg. Chem., 1985, 24, 632.
78 M.I.Bruce, O.bin Shawkataly, B.K.Nicholson, J. Organomet. Chem., 1985, 286, 427.
79 K.Henrick, B.F.G.Johnson, J.Lewis, J.Mace, M.McPartlin, J.Morris, J. Chem. Soc., Chem. Commun., 1985, 1617.
80 U.Grässle, W.Hiller, J.Strähle, Z. Anorg. Allg. Chem., 1985, 529, 29.
81 H.W.Chen, C.Paparizos, J.P.Fackler Jr., Inorg. Chim. Acta, 1985, 96, 137.
82 D.S.Dudis and J.P.Fackler Jr., Inorg. Chem., 1985, 24, 3758.
83 J.P.Fackler Jr. and B.Trzcinska-Bancroft, Organometallics, 1985, 4, 1891.
84 J.D.Basil, H.H.Murray, J.P.Fackler Jr., J.Tocher, A.M.Mazany, B.Trzcinska-Bancroft, H.Knachel, D.Dudis, T.J.Delord, D.O.Marler, J. Am. Chem. Soc., 1985, 107, 6908.
85 H.H.Murray III, J.P.Fackler Jr., D.A.Tocher, J. Chem. Soc., Chem. Commun., 1985, 1278.
86 H.H.Murray, A.M.Mazany, J.P.Fackler Jr., Organometallics, 1985, 4, 154.
87 H.H.Murray, J.P.Fackler Jr., B.Trzcinska-Bancroft, Organometallics, 1985, 4, 1633.
88 J.Vicente, M.T.Chicote, J.A.Cayuelas, J.Fernandez-Baeza, P.G.Jones, G.M.Sheldrick, P.Espinet, J. Chem. Soc., Dalton Trans., 1985, 1163.
89 K.Fischer, W.Deck, M.Schwarz, H.Vahrenkamp, Chem. Ber., 1985, 118, 4946.
90 C.M.Hay, B.F.G.Johnson, J.Lewis, R.C.S.McQueen, P.R.Raithby, R.M.Sorrell, M.J.Taylor, Organometallics, 1985, 4, 202.
91 M.I.Bruce, E.Horn, O.bin Shawkataly, M.R.Snow, J. Organomet. Chem., 1985, 280, 289.
92 U.Grässle and J.Strähle, Z. Anorg. Allg. Chem., 1985, 531, 26.
93 R.Uson, A.Laguna, M.Laguna, B.R.Manzano, P.G.Jones, G.M.Sheldrick, J. Chem. Soc., Dalton Trans., 1985, 2417.
94 O.M.Abu-Salah, A.-R.A.Al-Ohaly, C.B.Knobler, J. Chem. Soc., Chem. Commun., 1985, 1502.
95 R.Boese, N.Finke, J.Henkelmann, G.Maier, P.Paetzold, H.P.Reisenauer, G.Schmid, Chem. Ber., 1985, 118, 1644.
96 C.S.Shiner, C.M.Garner, R.C.Haltiwanger, J. Am. Chem. Soc., 1985, 107, 7167.
97 R.Boese, R.Köster, M.Yalpani, Chem. Ber., 1985, 118, 670.
98 C.K.Narula and H.Nöth, Inorg. Chem., 1985, 24, 2532.
99 W.Kliegel, L.Preu, S.J.Rettig, J.Trotter, Can. J. Chem., 1985, 63, 509.
100 F.Kumpfmüller, D.Nölle, H.Nöth, H.Pommerening, R.Staudigl, Chem. Ber., 1985, 118, 483.
101 J.W.Faller, C.Blankenship, B.Whitmore, S.Sena, Inorg. Chem., 1985, 24, 4483.
102 R.Boese, N.Finke, T.Keil, P.Paetzold, G.Schmid, Z. Naturforsch., 1985, 40B, 1327.
103 G.Schmid, F.Schmidt, R.Boese, Chem. Ber., 1985, 118, 1949.
104 G.Schmid, G.Barbenheim, R.Boese, Z. Naturforsch., 1985, 40B, 787.
105 W.Angerer, W.S.Sheldrick, W.Malisch, Chem. Ber., 1985, 118, 1261.
106 T.M.Gilbert, F.J.Hollander, R.G.Bergman, J. Am. Chem. Soc., 1985, 107, 3508.
107 M.D.Curtis and K.-B.Shiu, Inorg. Chem., 1985, 24, 1213.
108 E.Hanecker and H.Nöth, Z. Naturforsch., 1985, 40B, 717.
109 J.L.Atwood, S.G.Bott, C.Eaborn, M.N.A.El-Kheli, J.D.Smith, J. Organomet. Chem., 1985, 294, 23.
110 D.Männig, H.Nöth, M.Schwartz, S.Weber, U.Wietelmann, Angew. Chem., Int. Ed. Engl., 1985, 24, 998.
111 H.E.Katz, J. Am. Chem. Soc., 1985, 107, 1420.
112 M.Hildenbrand, H.Pritzkow, W.Siebert, Angew. Chem., Int. Ed. Engl., 1985, 24, 759.
113 P.Paetzold, E.Schröder, G.Schmid, R.Boese, Chem. Ber., 1985, 118, 3205.
114 F.Florencio, P.Smith-Verdier, S.Garcia-Blanco, Z. Kristallogr., 1984, 168, 173.
115 K.Stumpf, H.Pritzkow, W.Siebert, Angew. Chem., Int. Ed. Engl., 1985, 24, 71.

116 R.P.Micciche, P.J.Carroll, L.G.Sneddon, Organometallics, 1985, 4, 1619.
117 S.Lincoln, S.-L.Soong, S.A.Koch, M.Sato, J.H.Enemark, Inorg. Chem., 1985, 24, 1355.
118 C.Habben, A.Meller, M.Noltemeyer, G.M.Sheldrick, J. Organomet. Chem., 1985, 288, 1; 295, C45.
119 G.Schmid, D.Kampmann, W.Meyer, R.Boese, P.Paetzold, K.Delpy, Chem. Ber., 1985, 118, 2418.
120 G.Schmid, D.Zaika, R.Boese, Angew. Chem., Int. Ed. Engl., 1985, 24, 602.
121 W.Kliegel, H.-W.Motzkus, S.J.Rettig, J.Trotter, Can. J. Chem., 1985, 63, 3516.
122 M.D.Fryzuk, S.J.Rettig, A.Westerhaus, H.D.Williams, Inorg. Chem., 1985, 24, 4316.
123 J.Bould, N.N.Greenwood, J.D.Kennedy, W.S.McDonald, J. Chem. Soc., Dalton Trans., 1985, 1843.
124 D.Baudry. P.Charpin, M.Ephritikhine, G.Folcher, J.Lambard, M.Lance, M.Nierlich, J.Vigner, J. Chem. Soc., Chem. Commun., 1985, 1553.
125 S.J.Andrews and A.J.Welch, Inorg. Chim. Acta, 1985, 105, 89.
126 H.A.Boyter Jr., R.G.Swisher, E.Sinn, R.N.Grimes, Inorg. Chem., 1985, 24, 3810.
127 H.Wadepohl, H.Pritzkow, W.Siebert, Chem. Ber., 1985, 118, 729.
128 R.G.Swisher, E.Sinn, R.N.Grimes, Organometallics, 1985, 4, 890.
129 R.G.Swisher, E.Sinn, R.N.Grimes, Organometallics, 1985, 4, 896.
130 J.Edwin, W.Siebert, C.Krüger, J. Organomet. Chem., 1985, 282, 297.
131 R.G.Swisher, E.Sinn, R.J.Butcher, R.N.Grimes, Organometallics, 1985, 4, 882.
132 J.J.Briguglio and L.G.Sneddon, Organometallics, 1985, 4, 721.
133 K.A.Solntsev, L.A.Butman, I.Yu.Kuznetsov, N.T.Kuznetsov, B.Štibr, Z.Janoušek, K.Baše, Koord. Khim., 1984, 10, 1132(Engl.Ed. 627).
134 T.P.Hanusa, J.C.Huffman, T.L.Curtis, L.J.Todd, Inorg. Chem., 1985, 24, 787.
135 R.Shinomoto, J.G.Brennan, N.M.Edelstein, A.Zalkin, Inorg. Chem., 1985, 24, 2896.
136 E.A.Mizusawa, M.R.Thompson, M.F.Hawthorne, Inorg. Chem., 1985, 24, 1911.
137 M.P.Garcia, M.Green, F.G.A.Stone, R.G.Somerville, A.J.Welch, C.E.Briant, D.N.Cox, D.M.P.Mingos, J. Chem. Soc., Dalton Trans., 1985, 2343.
138 M.Green, J.A.K.Howard, A.P.James, A.N.de M.Jelfs, C.M.Nunn, F.G.A.Stone, J. Chem. Soc., Chem. Commun., 1985, 1778.
139 H.M.Colquhoun, T.J.Greenhough, M.G.H.Wallbridge, J. Chem. Soc., Dalton Trans., 1985, 761.
140 L.I.Zakharkin, V.V.Kobak, I.V.Pisareva, V.A.Antonovich, V.A.Ol'shevskaya, A.I.Yanovsky, Yu.T.Struchkov, J. Organomet. Chem., 1985, 297, 77.
141 D.M.Speakman, C.B.Knobler, M.F.Hawthorne, Organometallics, 1985, 4, 1692.
142 D.M.Speckman, C.B.Knobler, M.F.Hawthorne, Organometallics, 1985, 4, 426.
143 J.A.Walker, C.B.Knobler, M.F.Hawthorne, Inorg. Chem., 1985, 24, 2688.
144 J.D.Hewes, M.Thompson, M.F.Hawthorne, Organometallics, 1985, 4, 13.
145 X.L.R.Fontaine, H.Fowkes, N.N.Greenwood, J.D.Kennedy, M.Thornton-Pett, J. Chem. Soc., Chem. Commun., 1985, 1722.
146 X.L.R.Fontaine, H.Fowkes, N.N.Greenwood, J.D.Kennedy, M.Thornton-Pett, J. Chem. Soc., Chem. Commun., 1985, 1165.
147 A.Ya.Usyatinskii, V.I.Bregadze, N.N.Godovikov, L.E.Vinogradova, L.A.Leites, A.I.Yanovskii, Yu.T.Struchkov, Izv. Akad. Nauk SSSR, Ser. Khim., 1984, 2009(Engl.Ed. 1832).
148 W.J.Sieber, D.Neugebauer, F.R.Kreissl, Z. Naturforsch., 1985, 40B, 1500.
149 Z.-T.Wang, E.Sinn, R.N.Grimes, Inorg. Chem., 1985, 24, 834.
150 Z.-T.Wang, E.Sinn, R.N.Grimes, Inorg. Chem., 1985, 24, 826.
151 P.E.Behnken, T.B.Marder, R.T.Baker, C.B.Knobler, M.R.Thompson, M.F.Hawthorne, J. Am. Chem. Soc., 1985, 107, 932.
152 R.Goddard, J.Akhtar, K.B.Starowieyski, J. Organomet. Chem., 1985, 282, 149.
153 J.Müller, U.Müller, A.Loss, J.Lorberth, H.Donath, W.Massa, Z. Naturforsch., 1985, 40B, 1320.
154 M.Dräger and B.M.Schmidt, J. Organomet. Chem., 1985, 290, 133.
155 M.Wieber, D.Wirth, C.Burschka, Z. Naturforsch., 1985, 40B, 258.
156 M.R.Churchill, J.C.Fettinger, K.H.Whitmire, J. Organomet. Chem., 1985, 284, 13.

157 A.M.Arif, A.H.Cowley, N.C.Norman, M.Pakulski, J. Am. Chem. Soc., 1985, 107, 1062.
158 K.H.Whitmire, M.R.Churchill, J.C.Fettinger, J. Am. Chem. Soc., 1985, 107, 1056.
159 L.Xing-Fu, S.Eggers, J.Kopf, W.Jahn, R.D.Fischer, C.Apostolidis, B.Kanellakopulos, F.Benetollo, A.Polo, G.Bombieri, Inorg. Chim. Acta, 1985, 100, 183.
160 K.Hedberg, L.Hedberg, K.Hagen, R.R.Ryan, L.H.Jones, Inorg. Chem., 1985, 24, 2771.
161 P.Klüfers, Z. Kristallogr., 1984, 167, 275.
162 G.Doyle and D.van Engen, J. Organomet. Chem., 1985, 280, 253.
163 P.Klüfers, Z. Kristallogr., 1984, 167, 253.
164 W.J.Evans, I.Bloom, J.W.Grate, L.A.Hughes, W.E.Hunter, J.L.Atwood, Inorg. Chem., 1985, 24, 4620.
165 N.Bresciani-Pahor, L.Randaccio, E.Zangrando, P.J.Toscano, Inorg. Chim. Acta, 1985, 96, 193.
166 C.F.Jewell Jr., L.S.Liebeskind, M.Williamson, J. Am. Chem. Soc., 1985, 107, 6715.
167 B.Heiser, A.Kühn, H.Werner, Chem. Ber., 1985, 118, 1531.
168 N.Bresciani-Pahor, M.Calligaris, L.Randaccio, L.G.Marzilli, M.F.Summers, P.J.Toscano, J.Grossman, D.Liotta, Organometallics, 1985, 4, 630.
169 L.C.A.de Carvahlo, Y.Péres, M.Dartiguenave, Y.Dartiguenave, A.L.Beauchamp, Organometallics, 1985, 4, 2021.
170 H.Werner, B.Heiser, U.Schubert, K.Ackermann, Chem. Ber., 1985, 118, 1517.
171 L.G.Marzilli, N.Bresciani-Pahor, L.Randaccio, E.Zangrando, R.G.Finke, S.A.Myers, Inorg. Chim. Acta, 1985, 107, 139.
172 E.Raabe and U.Koelle, J. Organomet. Chem., 1985, 279, C29.
173 D.W.Macomber and R.D.Rogers, Organometallics, 1985, 4, 1485.
174 E.J.Miller, A.L.Rheingold, T.B.Brill, J. Organomet. Chem., 1985, 282, 399.
175 Y.Sasada and Y.Ohashi, J. Mol. Struct., 1985, 126, 477.
176 W.O.Parker Jr., N.Bresciani-Pahor, E.Zangrando, L.Randaccio, L.G.Marzilli, Inorg. Chem., 1985, 24, 3908.
177 N.Bresciani-Pahor, J.D.Orvell, L.Randaccio, Croat. Chem. Acta, 1984, 57, 433.
178 I.Bernal, G.M.Reisner, H.Brunner, G.Riepl, J. Organomet. Chem., 1985, 284, 115.
179 U.Plaia, H.Stolzenberg, W.P.Fehlhammer, J. Am. Chem. Soc., 1985, 107, 2171.
180 H.Werner, O.Kolb, U.Schubert, K.Ackermann, Chem. Ber., 1985, 118, 873.
181 I.Bernal, G.M.Reisner, H.Brunner, G.Riepl, Inorg. Chim. Acta, 1985, 103, 179.
182 L.G.Marzilli, M.F.Summers, N.Bresciani-Pahor, E.Zangrando, J.-P.Charland, L.Randaccio, J. Am. Chem. Soc., 1985, 107, 6880.
183 N.Bresciani-Pahor, L.Randaccio, E.Zangrando, M.F.Summers, J.H.Ramsden Jr., P.Marzilli, L.G.Marzilli, Organometallics, 1985, 4, 2086.
184 H.Werner, B.Heiser, H.Otto, Chem. Ber., 1985, 118, 3932.
185 L.Randaccio, N.Bresciani-Pahor, J.D.Orbell, M.Calligaris, M.F.Summers, B.Snyder, P.J.Toscano, L.G.Marzilli, Organometallics, 1985, 4, 469.
186 H.-F.Klein, K.Ellrich, S.Lamac, G.Lull, L.Zsolnai, G.Huttner, Z. Naturforsch., 1985, 40B, 1377.
187 A.W.Coleman, P.B.Hitchcock, M.F.Lappert, R.K.Maskell, J.H.Müller, J. Organomet. Chem., 1985, 296, 173.
188 A.Uchida, Y.Ohashi, Y.Sasada, Y.Ohgo, Acta Crystallogr., 1985, C41, 25.
189 Q.-B.Bao, S.J.Landon, A.L.Rheingold, T.M.Haller, T.B.Brill, Inorg. Chem., 1985, 24, 900.
190 M.Lattman, S.A.Morse, A.H.Cowley, J.G.Lasch, N.C.Norman, Inorg. Chem., 1985, 24, 1364.
191 R.Appel, W.Schuhn, F.Knoch, Angew. Chem., Int. Ed. Engl., 1985, 24, 420.
192 M.Rossi, J.P.Glusker, L.Randaccio, M.F.Summers, P.J.Toscano, L.G.Marzilli, J. Am. Chem. Soc., 1985, 107, 1729.
193 N.W.Alcock, R.M.Dixon, B.T.Golding, J. Chem. Soc., Chem. Commun., 1985, 603.
194 P.Klüfers, Angew. Chem., Int. Ed. Engl., 1985, 24, 70.
195 G.Doyle, K.A.Eriksen, D.Van Engen, Organometallics, 1985, 4, 877.

196 H.R.Allcock, P.R.Suszko, L.J.Wagner, R.R.Whittle, B.Boso, Organometallics, 1985, 4, 446.
197 P.Gusbeth and H.Vahrenkamp, Chem. Ber., 1985, 118, 1770.
198 M.Mlekuz, P.Bougeard, B.G.Sayer, S.Peng, M.J.McGlinchey, A.Marinetti, Y.-V. Saillard, J.B.Naceur, B.Mentzen, G.Jaouen, Organometallics, 1985, 4, 1123.
199 N.C.Schroeder, J.W.Richardson Jr., S.-L.Wang, R.A.Jacobson, R.J.Angelici, Organometallics, 1985, 4, 1226.
200 C.P.Horwitz, E.M.Holt, D.F.Shriver, Organometallics, 1985, 4, 1117.
201 J.A.Hriljac, P.N.Swepston, D.F.Shriver, Organometallics, 1985, 4, 158.
202 V.E.Lopatin, S.P.Gubin, N.M.Mikova, M.Ts.Tsybenov, Yu.L.Slovokhotov, Yu.T.Struchkov, J. Organomet. Chem., 1985, 292, 275.
203 J.M.Boncella, F.Okino, C.Orvig, R.P.Planalp, G.L.Rosenthal, Acta Crystallogr., 1985, C41, 833.
204 D.Cunningham, T.Higgins, B.Kneafsey, P.McArdle, J.Simmie, J. Chem. Soc., Chem. Commun., 1985, 231.
205 K.Yünlü, N.Höck, R.D.Fischer, Angew. Chem., Int. Ed. Engl., 1985, 24, 879.
206 E.Roman, F.Tapia, M.Barrera, M.-T.Garland, J.-Y.Le Marouille, C.Giannotti, J. Organomet. Chem., 1985, 297, C8.
207 R.B.King, M.Chang, M.G.Newton, J. Organomet. Chem., 1985, 296, 15.
208 E.N.Jacobsen and R.G.Bergman, J. Am. Chem. Soc., 1985, 107, 2023.
209 H.Werner, W.Hofmann, R.Zolk, L.F.Dahl, J.Kocal, A.Kühn, J. Organomet. Chem., 1985, 289, 173.
210 H.Werner and R.Zolk, Organometallics, 1985, 4, 601.
211 G.S.Mikaelyan, V.A.Smit, A.S.Batsanov, Yu.T.Struchkov, Izv. Akad. Nauk SSSR, Ser. Khim., 1984, 2105(Engl.Ed. 1922).
212 R.Zolk and H.Werner, Angew. Chem., Int. Ed. Engl., 1985, 24, 577.
213 L.Chen, D.J.Kountz, D.W.Meek, Organometallics, 1985, 4, 598.
214 H.Brunner, N.Janietz, W.Meier, G.Sergeson, J.Wachter, T.Zahn, M.L.Ziegler, Angew. Chem., Int. Ed. Engl., 1985, 24, 1060.
215 U.Schubert, B.Heiser, L.Hee, H.Werner, Chem. Ber., 1985, 118, 3151.
216 A.M.Arif, A.H.Cowley, M.Pakulski, J. Chem. Soc., Chem. Commun., 1985, 1707.
217 H.-F.Klein, L.Fabry, H.Witty, U.Schubert, H.Lueken, U.Stamm, Inorg. Chem., 1985, 24, 683.
218 F.R.Askham, G.G.Stanley, E.C.Marques, J. Am. Chem. Soc., 1985, 107, 7423.
219 A.M.Arif, A.H.Cowley, N.C.Norman, A.G.Orpen, M.Pakulski, J. Chem. Soc., Chem. Commun., 1985, 1267.
220 G.L.Geoffroy, W.C.Mercer, R.R.Whittle, L.Markó, S.Vastag, Inorg. Chem., 1985, 24, 3771.
221 A.Vizi-Orosz, G.Pályi, L.Markó, R.Boese, G.Schmid, J. Organomet. Chem., 1985, 288, 179.
222 H.Brunner, W.Meier, J.Wachter, H.Pfisterer, M.L.Ziegler, Z. Naturforsch., 1985, 40B, 923.
223 M.Mlekuz, P.Bougeard, B.G.Sayer, R.Faggiani, C.J.L.Lock, M.J.McGlinchey, G.Jaouen, Organometallics, 1985, 4, 2046.
224 W.A.Herrmann, C.E.Barnes, T.Zahn, M.L.Ziegler, Organometallics, 1985, 4, 172.
225 T.R.Halbert, S.A.Cohen, E.I.Stiefel, Organometallics, 1985, 4, 1689.
226 E.Sappa, G.Predieri, A.Tiripicchio, M.T.Camellini, J. Organomet. Chem., 1985, 297, 103.
227 J.C.Jeffery, D.B.Lewis, G.E.Lewis, F.G.A.Stone, J. Chem. Soc., Dalton Trans., 1985, 2001.
228 D.Mani and H.Vahrenkamp, Angew. Chem., Int. Ed. Engl., 1985, 24, 424.
229 W.Kläui, M.Scotti, M.Valderrama, S.Rojas, G.M.Sheldrick, P.G.Jones, T.Schroeder, Angew. Chem., Int. Ed. Engl., 1985, 24, 683.
230 E.Roland and H.Vahrenkamp, Chem. Ber., 1985, 118, 1133.
231 E.Roland, W.Bernhardt, H.Vahrenkamp, Chem. Ber., 1985, 118, 2858.
232 A.M.Arif, A.H.Cowley, M.Pakulski, M.B.Hursthouse, A.Karauloz, Organometallics, 1985, 4, 2227.
233 J.C.Jeffery and J.G.Lawrence-Smith, J. Organomet. Chem., 1985, 280, C34.
234 P.F.Seidler, T.McKenna, M.A.Kulzick, T.M.Gilbert, Acta Crystallogr., 1985, C41, 352.
235 K.A.Kubat-Martin, A.D.Rae, L.F.Dahl, Organometallics, 1985, 4, 2221.

236 F.Y.Pétillon, J.L.Le Quéré, F.le Floch-Pérennou, J.E.Guerchais, P.L'Haridon, J. Organomet. Chem., 1985, 281, 305.
237 L.Markó, G.Gervasio, P.L.Stanghellini, G.Bor, Transition Met. Chem., 1985, 10, 344.
238 G.Balavoine, J.Collin, J.J.Bonnet, G.Lavigne, J. Organomet. Chem., 1985, 280, 429.
239 M.K.Markiewicz, M.E.Fraser, R.K.Ungar, S.Fortier, M.C.Baird, Acta Crystallogr., 1985, C41, 336.
240 P.Gusbeth and H.Vahrenkamp, Chem. Ber., 1985, 118, 1746.
241 A.Arrigoni, A.Ceriotti, R.D.Pergola, G.Longoni, M.Manassero, M.Sansoni, J. Organomet. Chem., 1985, 296, 243.
242 J.Pursiainen, T.A.Pakkanen, J.Jääskeläinen, J. Organomet. Chem., 1985, 290, 85.
243 R.L.De and H.Vahrenkamp, Z. Naturforsch., 1985, 40B, 1250.
244 G.Gervasio, R.Rossetti, P.L.Stanghellini, Organometallics, 1985, 4, 1612.
245 M.G.Richmond, J.D.Korp, J.K.Kochi, J. Chem. Soc., Chem. Commun., 1985, 1102.
246 S.Gambarotta, C.Floriani, A.Chiesi-Villa, C.Guastini, J. Organomet. Chem., 1985, 296, C6.
247 R.A.Gancarz, J.F.Blount, K.Mislow, Organometallics, 1985, 4, 2028.
248 N.J.Bailey, G.de Leeuw, J.S.Field, R.J.Haines, I.C.D.Stuckenberg, R.B.English, S. Afr. J. Chem., 1985, 38, 139.
249 P.Magnus and D.P.Becker, J. Chem. Soc., Chem. Commun., 1985, 640.
250 J.M.Burlitch, S.E.Hayes, J.T.Lemley, Organometallics, 1985, 4, 167.
251 S.Martinengo, D.Strumolo, P.Chini, V.G.Albano, D.Braga, J. Chem. Soc., Dalton Trans., 1985, 35.
252 S.P.Foster, K.M.Mackay, B.K.Nicholson, Inorg. Chem., 1985, 24, 909.
253 P.Braunstein, J.Rosé, A.Tiripicchio, M.T.Camellini, Angew. Chem., Int. Ed. Engl., 1985, 24, 767.
254 J.M.Ragosta and J.M.Burlitch, J. Chem. Soc., Chem. Commun., 1985, 1187.
255 V.G.Albano, D.Braga, A.Fumagalli, S.Martinengo, J. Chem. Soc., Dalton Trans., 1985, 1137.
256 D.Sellmann, W.Ludwig, G.Huttner, L.Zsolnai, J. Organomet. Chem., 1985, 294, 199.
257 T.C.W.Mak, K.S.Jasim, C.Chieh, Inorg. Chim. Acta, 1985, 99, 31.
258 J.-C.Boutonnet, J.Levisalles, F.Rose-Munch, E.Rose, G.Precigoux, F.Leroy, J. Organomet. Chem., 1985, 290, 153.
259 K.H.Dötz, W.Kuhn, U.Thewalt, Chem. Ber., 1985, 118, 1126.
260 S.Lotz, M.Schindehutte, M.M.van Dyk, J.L.M.Dillen, P.H.van Rooyen, J. Organomet. Chem., 1985, 295, 51.
261 H.Fischer, A.Motsch, R,Märkl, K.Ackermann, Organometallics, 1985, 4, 726.
262 C.Elsenbroich, R.Gondrum, W.Massa, Angew. Chem., Int. Ed. Engl., 1985, 24, 967.
263 G.Michael, J.Kaub, C.G.Kreiter, Chem. Ber., 1985, 118, 3944.
264 G.Michael, J.Kaub, C.G.Kreiter, Angew. Chem., Int. Ed. Engl., 1985, 24, 502.
265 P.H.Bird, A.A.Ismail, I.S.Butler, Inorg. Chem., 1985, 24, 2911.
266 K.Angermund, K.H.Claus, R.Goddard, C.Krüger, Angew. Chem., Int. Ed. Engl., 1985, 24, 237.
267 J.E.Salt, G.S.Girolami, G.Wilkinson, M.Motevalli, M.Thornton-Pett, M.B.Hursthouse, J. Chem. Soc., Dalton Trans., 1985, 685.
268 G.S.Girolami, G.Wilkinson, A.M.R.Galas, M.Thornton-Pett, M.B.Hursthouse, J. Chem. Soc., Dalton Trans., 1985, 1339.
269 R.M.Moriarty, S.G.Engerer, O.Prakash, I.Prakash, U.S.Gill, W.A.Freeman, J. Chem. Soc., Chem. Commun., 1985, 1715.
270 J.Huang, Q.Cai, M.Wang, M.He, Jiegou Huaxue, 1985, 4, 66.
271 X.Yang, J.Huang, J.Huang, Jiegou Huaxue, 1985, 4, 52.
272 E.P.Kundig, C.Perret, S.Spichiger, G.Bernardinelli, J. Organomet. Chem., 1985, 286, 183.
273 M.J.Zaworotko, R.J.Stamps, M.T.Ledet, H.Zhang, J.L.Atwood, Organometallics, 1985, 4, 1697.
274 F.Seitz, H.Fischer, J.Riede, J. Organomet. Chem., 1985, 287, 87.
275 H.Fischer, F.Seitz, J.Riede, J.Vogel, Angew. Chem., Int. Ed. Engl., 1985, 24, 121.

276 U.Kunze, H.Jawad, W.Hiller, R.Naumer, Z. Naturforsch., 1985, 40B, 512.
277 L.Weber and R.Boese, Chem. Ber., 1985, 118, 1545.
278 E.Michels, W.S.Sheldrick, C.G.Kreiter, Chem. Ber., 1985, 118, 964.
279 K.Schöllkopf, J.J.Stezowski, F.Effenberger, Organometallics, 1985, 4, 922.
280 A.G.M.Barrett, C.P.Brock, M.A.Sturgess, Organometallics, 1985, 4, 1903.
281 G.R.Dobson, I.Bernal, G.M.Reisner, C.B.Dobson, S.E.Mansour, J. Am. Chem. Soc., 1985, 107, 525.
282 R.C.Cambie, G.R.Clark, A.C.Gourdie, P.S.Rutledge, P.D.Woodgate, J. Organomet. Chem., 1985, 297, 177.
283 N.A.Ustynyuk, L.N.Novikova, V.K.Bel'skii, Yu.F.Oprunenko, S.G.Malyugina, O.I.Trifonova, Yu.A.Ustynyuk, J. Organomet. Chem., 1985, 294, 31.
284 H.G.Raubenheimer, G.J.Kruger, A.van A.Lombard, L.Linford, J.C.Viljoen, Organometallics, 1985, 4, 275.
285 M.J.Wovkulich, J.L.Atwood, L.Canada, J.D.Atwood, Organometallics, 1985, 4, 867.
286 M.Yoshifuji, T.Hashida, N.Inamoto, K.Hirotsu, T.Horiuchi, T.Higuchi, K.Ito, S.Nagase, Angew. Chem., Int. Ed. Engl., 1985, 24, 211.
287 M.Yoshifuji, N.Inamoto, K.Hirotsu, T.Higuchi, J. Chem. Soc., Chem. Commun., 1985, 1109.
288 W.R.Robinson, D.E.Wigley, R.A.Walton, Inorg. Chem., 1985, 24, 918.
289 L.Weber, D.Wewers, R.Boese, Chem. Ber., 1985, 118, 3570.
290 D.J.Wink, J.R.Fox, N.J.Cooper, J. Am. Chem. Soc., 1985, 107, 5012.
291 W.K.Wong, K.W.Chiu, G.Wilkinson, M.Motevalli, M.B.Hursthouse, Polyhedron, 1985, 4, 1231.
292 R.B.King, W.-K.Fu, E.M.Holt, Inorg. Chem., 1985, 24, 3094.
293 I.R.Butler, W.R.Cullen, F.W.B.Einstein, A.C.Willis, Organometallics, 1985, 4, 603.
294 A.A.Ismail, I.S.Butler, J.-J.Bonnet, S.Askenazy, Acta Crystallogr., 1985, C41, 1582.
295 J.-C.Boutonnet, F.Rose-Munch, E.Rose, Y.Jeannin, F.Robert, J. Organomet. Chem., 1985, 297, 185.
296 O.B.Afanasova, Yu.E.Zubarev, V.A.Sharapov, N.I.Kirillova, A.I.Gusev, V.M.Nosova, N.V.Alekseev, E.A.Chernyshev, Yu.T.Struchkov, Dokl. Akad. Nauk SSSR, 1984, 279.
297 R.Hensel, W.-H.du Mont, R.Boese, D.Wewers, L.Weber, Chem. Ber., 1985, 118, 1580.
298 P.Klüfers, L.Knoll, C.Reiners, K.Reiss, Chem. Ber., 1985, 118, 1825.
299 G.Bremer, P.Klüfers, T.Kruck, Chem. Ber., 1985, 118, 4224.
300 H.Lang, L.Zsolnai, G.Huttner, Z. Naturforsch., 1985, 40B, 500.
301 H.Lang, O.Orama, G.Huttner, J. Organomet. Chem., 1985, 291, 293.
302 M.Herberhold, W.Kremnitz, A.Razavi, H.Schöllhorn, U.Thewalt, Angew. Chem., Int. Ed. Engl., 1985, 24, 601.
303 J.Borm, G.Huttner, O.Orama, L.Zsolnai, J. Organomet. Chem., 1985, 282, 53.
304 W.A.Herrmann, J.Rohrmann, H.Nöth, Ch.K.Nanila, I.Bernal, M.Draux, J. Organomet. Chem., 1985, 284, 189.
305 C.Zeiher, J.Mohyla, I.-P.Lorenz, W.Hiller, J. Organomet. Chem., 1985, 286, 159.
306 J.-C.Boutonnet, F.Rose-Munch, E.Rose, G.Precigoux, J. Organomet. Chem., 1985, 284, C25.
307 U.Weber, L.Zsolnai, G.Huttner, Z. Naturforsch., 1985, 40B, 1431.
308 L.Y.Goh, C.Wei, E.Sinn, J. Chem. Soc., Chem. Commun., 1985, 462.
309 J.Borm, G.Huttner, L.Zsolnai, Angew. Chem., Int. Ed. Engl., 1985, 24, 1069.
310 U.Weber, G.Huttner, O.Scheidsteger, L.Zsolnai, J. Organomet. Chem., 1985, 289, 357.
311 T.J.McNeese, T.E.Mueller, D.A.Wierda, D.J.Darensbourg, T.J.Delord, Inorg. Chem., 1985, 24, 3465.
312 F.Bottomley, D.E.Paez, L.Sutin, P.S.White, J. Chem. Soc., Chem. Commun., 1985, 597.
313 D.Wormsbächer, K.M.Nicholas, A.L.Rheingold, J. Chem. Soc., Chem. Commun., 1985, 721.
314 H.Hope, M.M.Olmstead, P.P.Power, J.Sandell, X.Xu, J. Am. Chem. Soc., 1985, 107, 4337.

315 J.S.Thompson, J.C.Calabrese, J.F.Whitney, Acta Crystallogr., 1985, C41, 890.
316 A.Toth, C.Floriani, M.Pasquali, A.Chiesi-Villa, A.Gaetani-Manfredotti, C.Guastini, Inorg. Chem., 1985, 24, 648.
317 T.P.Hanusa, T.A.Ulibarri, W.J.Evans, Acta Crystallogr., 1985, C41, 1036.
318 S.Attali, F.Dahan, R.Mathieu, J. Chem. Soc., Dalton Trans., 1985, 2521.
319 G.Doyle, K.A.Eriksen, D.Van Engen, Organometallics, 1985, 4, 2201.
320 R.K.Chadha and J.E.Drake, J. Organomet. Chem., 1985, 286, 121.
321 G.S.White and D.W.Stephan, Inorg. Chem., 1985, 24, 1499.
322 A.Blagg, A.T.Hutton, B.L.Shaw, M.Thornton-Pett, Inorg. Chim. Acta, 1985, 100, L33.
323 G.Doyle, K.A.Eriksen, D.Van Engen, Organometallics, 1985, 4, 830.
324 G.M.Villacorta, D.Gibson, I.D.Williams, S.J.Lippard, J. Am. Chem. Soc., 1985, 107, 6732.
325 G.Doyle, K.A.Eriksen, D.Van Engen, J. Am. Chem. Soc., 1985, 107, 7914.
326 M.R.Churchill, Y.-J.Li, D.Nalewajek, P.M.Schaber, J.Dorfman, Inorg. Chem., 1985, 24, 2684.
327 G.van Koten, J.T.B.H.Jastrzebski, F.Muller, C.H.Stam, J. Am. Chem. Soc., 1985, 107, 697.
328 V.Riera, M.A.Ruiz, A.Tiripicchio, M.T.Camellini, J. Chem. Soc., Chem. Commun., 1985, 1505.
329 H.L.Aalten, G.van Koten, K.Goubitz, C.H.Stam, J. Chem. Soc., Chem. Commun., 1985, 1252.
330 G.Doyle, B.T.Heaton, E.Occhiello, Organometallics, 1985, 4, 1224.
331 L.F.Rhodes, J.C.Huffman, K.G.Caulton, J. Am. Chem. Soc., 1985, 107, 1759.
332 L.Naldini, F.Demartin, M.Manassero, M.Sansoni, G.Rassu, M.A.Zoroddu, J. Organomet. Chem., 1985, 279, C42.
333 S.I.Khan, P.G.Edwards, H.S.H.Yuan, R.Bau, J. Am. Chem. Soc., 1985, 107, 1682.
334 H.Schumann, H.Lauke, E.Hahn, M.J.Heeg, D.v.d.Helm, Organometallics, 1985, 4, 321.
335 L.Hedberg, K.Hedberg, S.K.Satija, B.I.Swanson, Inorg. Chem., 1985, 24, 2766.
336 K.-J.Jens and E.Weiss, Acta Crystallogr., 1985, C41, 895.
337 A.M.Crespi and D.F.Shriver, Organometallics, 1985, 4, 1830.
338 E.Lindner, C.-P.Krieg, W.Hiller, R.Fawzi, Chem. Ber., 1985, 118, 1398.
339 S.P.McManus, J.A.Knight, E.J.Meehan, R.A.Abramovitch, M.N.Offor, J.L.Atwood, W.E.Hunter, J. Org. Chem., 1985, 50, 2742.
340 F.A.Cotton and A.H.Reid Jr., Acta Crystallogr., 1985, C41, 686.
341 Y.-R.Hu, T.W.Leung, S.-C.H.Su, A.Wojcicki, M.Calligaris, G.Nardin, Organometallics, 1985, 4, 1001.
342 A.de Koster, J.A.Kanters, A.L.Spek, A.A.H.v.d.Zeijden, G.van Koten, K.Vriese, Acta Crystallogr., 1985, C41, 893.
343 J.-M.Grosselin, H.Le Bozec, C.Moinet, L.Toupet, P.H.Dixneuf, J. Am. Chem. Soc., 1985, 107, 2809.
344 Z.Goldschmidt, S.Antebi, H.E.Gottlieb, D.Cohen, U.Shmueli, Z.Stein, J. Organomet. Chem., 1985, 282, 369.
345 C.Jiabi, L.Guixin, X.Weihua, J.Xianglin, S.Meicheng, T.Youqi, J. Organomet. Chem., 1985, 286, 55.
346 R.S.Bly, M.M.Hossain, L.Lebioda, J. Am. Chem. Soc., 1985, 107, 5549.
347 J.P.Blaha and M.S.Wrighton, J. Am. Chem. Soc., 1985, 107, 2694.
348 W.Malisch, W.Angerer, A.H.Cowley, N.C.Norman, J. Chem. Soc., Chem. Commun., 1985, 1811.
349 G.Zhang, Z.Zhang, Y.Zhang, Z.Chen, Wuhan Daxue Xuebao, Ziran Kexueban, 1984, 95(Chem. Abs., 1985, 102:24770b).
350 I.Bkouche-Waksman, J.S.Ricci Jr., T.F.Koetzle, J.Weichmann, W.A.Herrmann, Inorg. Chem., 1985, 24, 1492.
351 M.E.Wright, G.O.Nelson, R.S.Glass, Organometallics, 1985, 4, 245.
352 R.Goddard, F.-W.Grevels, R.Schrader, Angew. Chem., Int. Ed. Engl., 1985, 24, 353.
353 R.-M.Lequan, M.Lequan, G.Jaouen, L.Ouahab, P.Batail, J.Padiou, R.G.Sutherland, J. Chem. Soc., Chem. Commun., 1985, 116.
354 W.Malisch, H.Blau, K.Blank, C.Krüger, L.K.Liu, J. Organomet. Chem., 1985, 296, C32.

355 S.H.Simonsen, V.M.Lynch, R.G.Sutherland, A.Piórko, J. Organomet. Chem., 1985, 290, 387.
356 M.Hillman, J.D.Austin, Å.Kvick, Organometallics, 1985, 4, 316.
357 M.Hillman and A.C.Larson, J. Organomet. Chem., 1985, 280, 389.
358 H.-W.Frühauf, F.Seils, R.J.Goddard, M.J.Romão, Organometallics, 1985, 4, 948.
359 C.P.Casey, M.W.Meszaros, S.M.Neumann, I.G.Cesa, K.J.Haller, Organometallics, 1985, 4, 143.
360 N.I.Pyshnograeva, V.N.Setkina, A.S.Batsanov, Yu.T.Struchkov, J. Organomet. Chem., 1985, 288, 189.
361 F.Pilloud, R.Roulet, G.Chapuis, Acta Crystallogr., 1985, C41, 886.
362 E.Lindner, E.Schauss, W.Hiller, R.Fawzi, Chem. Ber., 1985, 118, 3915.
363 A.R.Koray, T.Zahn, M.L.Ziegler, J. Organomet. Chem., 1985, 291, 53.
364 D.Touchard, J.-L.Fillaut, P.Dixneuf, C.Mealli, M.Sabat, L.Toupet, Organometallics, 1985, 4, 1684.
365 L.Weber, K.Reizig, R.Boese, Chem. Ber., 1985, 118, 1193.
366 E.Tagliaferri, U.Haenisch, R.Roulet, P.Vogel, K.J.Schenk, Helv. Chim. Acta, 1985, 68, 1362.
367 S.L.Brown, S.G.Davies, P.Warner, R.H.Jones, K.Prout, J. Chem. Soc., Chem. Commun., 1985, 1446.
368 S.G.Davies, I.M.Dordor-Hedgecock, P.Warner, R.H.Jones, K.Prout, J. Organomet. Chem., 1985, 285, 213.
369 K.Broadley, N.G.Connelly, P.G.Graham, J.A.K.Howard, W.Risse, M.W.Whiteley, J. Chem. Soc., Dalton Trans., 1985, 777.
370 B.Fuchs, R.Frohlich, H.Musso, Chem. Ber., 1985, 118, 1968.
371 F.R.Ahmed, J.L.A.Rouston, M.Y.Al-Janabi, Inorg. Chem., 1985, 24, 2526.
372 I.-P.Lorenz, W.Hiller, M.Conrad, Z. Naturforsch., 1985, 40B, 1383.
373 T.D.Appleton, W.R.Cullen, S.V.Evans, T.-J.Kim, J.Trotter, J. Organomet. Chem., 1985, 279, 5.
374 J.D.Goodrich and J.P.Selegue, Organometallics, 1985, 4, 798.
375 M.Schaal, W.Weigand, U.Nagel, W.Beck, Chem. Ber., 1985, 118, 2186.
376 G.Baum and W.Massa, Organometallics, 1985, 4, 1572.
377 A.G.Osborne, A.J.Blake, R.E.Hollands, R.F.Bryan, S.Lockhart, J. Organomet. Chem., 1985, 287, 39.
378 J.Keijsper, P.Grimberg, G.van Koten, K.Vrieze, B.Kojić-Prodić, A.L.Spek, Organometallics, 1985, 4, 438.
379 J.Keijsper, P.Grimberg, G.van Koten, K.Vrieze, M.Cristophersen, C.H.Stam, Inorg. Chim. Acta, 1985, 102, 29.
380 E.O.Fischer, J.K.R.Wanner, G.Müller, J.Riede, Chem. Ber., 1985, 118, 3311.
381 I.R.Butler, W.R.Cullen, T.-J.Kim, S.J.Rettig, J.Trotter, Organometallics, 1985, 4, 972.
382 L.E.Bogan Jr., T.B.Rauchfuss, A.L.Rheingold, Inorg. Chem., 1985, 24, 3720.
383 H.Brunner, N.Janietz, J.Wachter, T.Zahn, M.L.Ziegler, Angew. Chem., Int. Ed. Engl., 1985, 24, 133.
384 L.E.Bogan Jr., T.B.Rauchfuss, A.L.Rheingold, J. Am. Chem. Soc., 1985, 107, 3843.
385 S.B.Colbran, B.H.Robinson, J.Simpson, Organometallics, 1985, 4, 1594.
386 B.McCulloch, D.L.Ward, J.D.Woolins, C.H.Brubaker Jr., Organometallics, 1985, 4, 1425.
387 H.Werner, F.J.G.Alonso, H.Otto, K.Peters, H.G.von Schnering, J. Organomet. Chem., 1985, 289, C5.
388 W.R.Cullen, T.-J.Kim, F.W.B.Einstein, T.Jones, Organometallics, 1985, 4, 346.
389 B.P.Gracey, S.A.R.Knox, K.A.Macpherson, A.G.Orpen, S.R.Stobart, J. Chem. Soc., Dalton Trans., 1985, 1935.
390 A.L.Rheingold and M.E.Fountain, Acta Crystallogr., 1985, C41, 1162.
391 R.H.Cowley, N.C.Norman, M.Pakulski, D.L.Bricker, D.H.Russell, J. Am. Chem. Soc., 1985, 107, 8211.
392 L.Weber, K.Reizig, R.Boese, M.Polk, Angew. Chem., Int. Ed. Engl., 1985, 24, 604.
393 J.Messelhäuser, I.-P.Lorenz, K.Haug, W.Hiller, Z. Naturforsch., 1985, 40B, 1064.

394 R.Weberg, R.C.Haltiwanger, M.R.DuBois, Organometallics, 1985, 4, 1315.
395 D.Seyferth, G.B.Womack, J.C.Dewan, Organometallics, 1985, 4, 398.
396 A.I.Nekhaev, S.D.Alekseeva, N.S.Nametkin, V.D.Tyurin, B.I.Kolobov, G.G.Aleksandrov, N.A.Parpiev, M.T.Tashev, H.B.Dustov, J. Organomet. Chem., 1985, 297, C33.
397 J.Ros, G.Commenges, R.Mathieu, X.Solans, M.Font-Altaba, J. Chem. Soc., Dalton Trans., 1985, 1087.
398 A.M.Arif, A.H.Cowley, M.Pakulski, J. Am. Chem. Soc., 1985, 107, 2553.
399 H.Lang, L.Zsolnai, G.Huttner, J. Organomet. Chem., 1985, 282, 23.
400 E.Delgado, A.T.Emo, J.C.Jeffery, N.D.Simmons, F.G.A.Stone, J. Chem. Soc., Dalton Trans., 1985, 1323.
401 A.R.Koray, C.Krieger, H.A.Staab, Angew. Chem., Int. Ed. Engl., 1985, 24, 521.
402 H.Lang, L.Zsolnai, G.Huttner, Chem. Ber., 1985, 118, 4426.
403 X.Solans, M.Font-Altaba, J.Ros, R.Yañez, R.Mathieu, Acta Crystallogr., 1985, C41, 1186.
404 R.B.King, F.-J.Wu, N.D.Sadanani, E.M.Holt, Inorg. Chem., 1985, 24, 4449.
405 T.-Y.Dong, M.J.Cohn, D.N.Hendrickson, C.G.Pierpont, J. Am. Chem. Soc., 1985, 107, 4777.
406 T.-Y.Dong, D.N.Hendrickson, K.Iwai, M.J.Cohn, S.J.Geib, A.L.Rheingold, H.Sano, I.Motoyama, S.Nakashima, J. Am. Chem. Soc., 1985, 107, 7996; J. Chem. Soc., Chem. Commun., 1985, 1095.
407 A.L.Rheingold, Acta Crystallogr., 1985, C41, 1043.
408 N.G.Connelly, M.J.Freeman, A.G.Orpen, J.B.Sheridan, A.N.D.Symonds, M.W.Whiteley, J. Chem. Soc., Dalton Trans., 1985, 1027.
409 N.G.Connelly, A.R.Lucy, R.M.Mills, J.B.Sheridan, P.Woodward, J. Chem. Soc., Dalton Trans., 1985, 699.
410 E.Tagliaferri, P.Campiche, R.Roulet, R.Gabioud, P.Vogel, G.Chapuis, Helv. Chim. Acta, 1985, 68, 126.
411 C.P.Casey, M.S.Konings, R.E.Palermo, R.E.Colborn, J. Am. Chem. Soc., 1985, 107, 5296.
412 J.P.Blaha, B.E.Bursten, J.C.Dewan, R.B.Frankel, C.L.Randolph, B.A.Wilson, M.S.Wrighton, J. Am. Chem. Soc., 1985, 107, 4561.
413 M.F.Moore, S.R.Wilson, M.J.Cohn, T.-Y.Dong, U.T.Mueller-Westerhoff, D.N.Hendrickson, Inorg. Chem., 1985, 24, 4559.
414 R.Appel, C.Casser, F.Knoch, J. Organomet. Chem., 1985, 293, 213.
415 J.Ros, X.Solans, C.Miravitlles, R.Mathieu, J. Chem. Soc., Dalton Trans., 1985, 1981.
416 S.I.Amer, G.Sadler, P.M.Henry, G.Ferguson, B.L.Ruhl, Inorg. Chem., 1985, 24, 1517.
417 C.P.Casey, W.H.Miles, P.J.Fagan, K.J.Haller, Organometallics, 1985, 4, 559.
418 M.-T.Lee, B.M.Foxman, M.Rosenblum, Organometallics, 1985, 4, 539.
419 N.E.Kolobova, T.V.Rozantseva, Yu.T.Struchkov, A.S.Batsanov, V.I.Bakhmutov, J. Organomet. Chem., 1985, 292, 247.
420 J.Huang, S.Li, M.He, M.Wang, Jiegou Huaxue, 1984, 3, 105.
421 H.Umland, D.Wormsbächer, U.Behrens, J. Organomet. Chem., 1985, 284, 353.
422 W.A.Herrmann, J.Rohrmann, M.L.Ziegler, T.Zahn, J. Organomet. Chem., 1985, 295, 175.
423 E.Sappa, A.M.M.Lanfredi, G.Predieri, A.Tiripicchio, A.J.Carty, J. Organomet. Chem., 1985, 288, 365.
424 C.Weatherell, N.J.Taylor, A.J.Carty, E.Sappa, A.Tiripicchio, J. Organomet. Chem., 1985, 291, C9.
425 P.Braunstein, G.Predieri, F.J.Lahoz, A.Tiripicchio, J. Organomet. Chem., 1985, 288, C13.
426 R.E.Hollands, A.G.Osborne, R.H.Whiteley, C.J.Cardin, J. Chem. Soc., Dalton Trans., 1985, 1527.
427 F.W.B.Einstein and T.Jones, Acta Crystallogr., 1985, C41, 365.
428 A.J.Mayr, H.-S.Lien, K.H.Pannell, L.Parkanyi, Organometallics, 1985, 4, 1580.
429 A.M.Arif, A.H.Cowley, S.Quashie, J. Chem. Soc., Chem. Commun., 1985, 428.
430 J.C.Jeffery and J.G.Lawrence-Smith, J. Chem. Soc., Chem. Commun., 1985, 275.
431 A.J.Bard, A.H.Cowley, J.K.Leland, G.J.N.Thomas, N.C.Norman, P.Jutzi,

432 C.P.Morley, E.Schlüter, J. Chem. Soc., Dalton Trans., 1985, 1303.
433 D.Lentz, I.Brüdgam, H.Hartl, Angew. Chem., Int. Ed. Engl., 1985, 24, 119.
   A.S.Antsyshkina, L.M.Dikareva, M.A.Porai-Koshits, V.N.Ostrikova, Yu.V.Skripkin, O.G.Volkov, A.A.Pasynskii, V.T.Kalinnikov, Koord. Khim., 1985, 11, 82(Engl.Ed. 44).
434 J.A.Hriljac and D.F.Shriver, Organometallics, 1985, 4, 2225.
435 K.Knoll, G.Huttner, L.Zsolnai, I.Jibril, M.Wasiucionek, J. Organomet. Chem., 1985, 294, 91.
436 D.J.Brauer, S.Hietkamp, H.Sommer, O.Stelzer, G.Müller, C.Krüger, J. Organomet. Chem., 1985, 288, 35.
437 G.J.Kruger, S.Lotz, L.Linford, M.van Dyk, H.G.Raubenheimer, J. Organomet. Chem., 1985, 280, 241.
438 D.J.Brauer, S.Hietkamp, H.Sommer, O.Stelzer, G.Müller, M.J.Romão, C.Krüger, J. Organomet. Chem., 1985, 296, 411.
439 D.J.Brauer, S.Hietkamp, H.Sommer, O.Stelzer, J. Organomet. Chem., 1985, 281, 187.
440 G.D.Williams, G.L.Geoffroy, R.R.Whittle, J. Am. Chem. Soc., 1985, 107, 729.
441 D.Nuel, F.Dahan, R.Mathieu, J. Am. Chem. Soc., 1985, 107, 1658.
442 V.D.Patel, A.A.Cherkas, D.Nucciarone, N.J.Taylor, A.J.Carty, Organometallics, 1985, 4, 1792.
443 D.Nuel, F.Dahan, R.Mathieu, Organometallics, 1985, 4, 1436.
444 E.P.Kyba, K.L.Hassett, B.Sheikh, J.S.McKennis, R.B.King, R.E.Davis, Organometallics, 1985, 4, 994.
445 D.J.Brauer, S.Hietkamp, H.Sommer, O.Stelzer, Z. Naturforsch., 1985, 40B, 1677.
446 I.R.Butler, W.R.Cullen, J.Ni, S.J.Rettig, Organometallics, 1985, 4, 2196.
447 J.A.Kovacs, J.K.Bashkin, R.H.Holm, J. Am. Chem. Soc., 1985, 107, 1784.
448 G.van Buskirk, C.B.Knobler, H.D.Kaesz, Organometallics, 1985, 4, 149.
449 M.K.Alami, F.Dahan, R.Mathieu, Organometallics, 1985, 4, 2122.
450 D.J.Brauer, G.Hasselkuss, S.Hietkamp, H.Sommer, O.Stelzer, Z. Naturforsch., 1985, 40B, 961.
451 D.Seyferth, A.M.Kiwan, E.Sinn, J. Organomet. Chem., 1985, 281, 111.
452 C.P.Casey, S.R.Mander, A.L.Rheingold, Organometallics, 1985, 4, 762.
453 A.Gourdon and Y.Jeannin, J. Organomet. Chem., 1985, 290, 199.
454 G.Fischer, G.Sedelmeier, H.Prinzbach, K.Knoll, P.Wilharm, G.Huttner, I.Jibril, J. Organomet. Chem., 1985, 297, 307.
455 A.Gourdon and Y.Jeannin, J. Organomet. Chem., 1985, 282, C39.
456 O.T.Beachley Jr., T.D.Getman, R.U.Kirss, R.B.Hallock, W.E.Hunter, J.L.Atwood, Organometallics, 1985, 4, 751.
457 H.Schmidbaur, S.Lauteschläger, G.Müller, J. Organomet. Chem., 1985, 281, 25.
458 S.Nussbaum, S.J.Rettig, A.Storr, J.Trotter, Can. J. Chem., 1985, 63, 692.
459 B.M.Louie, S.J.Rettig, A.Storr, J.Trotter, Can. J. Chem., 1985, 63, 2261.
460 B.M.Louie, S.J.Rettig, A.Storr, J.Trotter, Can. J. Chem., 1985, 63, 703.
461 B.M.Louie, S.J.Rettig, A.Storr, J.Trotter, Can. J. Chem., 1985, 63, 3019.
462 B.M.Louie, S.J.Rettig, A.Storr, J.Trotter, Can. J. Chem., 1985, 63, 503.
463 O.T.Beachley Jr., R.B.Hallock, H.M.Zhang, J.L.Atwood, Organometallics, 1985, 4, 1675.
464 G.A.Banta, S.J.Rettig, A.Storr, J.Trotter, Can. J. Chem., 1985, 63, 2545.
465 A.H.Cowley, S.K.Mehrotra, J.L.Atwood, W.E.Hunter, Organometallics, 1985, 4, 1115.
466 R.B.Hallock, W.E.Hunter, J.L.Atwood, O.T.Beachley Jr., Organometallics, 1985, 4, 547.
467 H.Schmidbaur, W.Bublak, J.Riede, G.Müller, Angew. Chem., Int. Ed. Engl., 1985, 24, 414.
468 H.C.Aspinall, D.C.Bradley, M.B.Hursthouse, K.D.Sales, N.P.C.Walker, J. Chem. Soc., Chem. Commun., 1985, 1585.
469 R.K.Chadha, J.E.Drake, M.K.H.Neo, J. Crystallogr. Spectrosc. Res., 1985, 15, 39.
470 S.N.Gurkova, A.I.Gusev, N.V.Alekseev, T.K.Gar, N.A.Viktorov, Zh. Strukt. Khim., 1984, 25(5), 170(Engl.Ed. 825).
471 S.N.Gurkova, A.I.Gusev, N.V.Alekseev, T.K.Gar, N.A.Viktorov, Zh. Strukt. Khim., 1985, 26(1), 144(Engl.Ed. 124).

472 M.P.Egorov, S.P.Kolesnikov, Yu.T.Struchkov, M.Yu.Antipin, S.V.Sereda, O.M.Nefedov, J. Organomet. Chem., 1985, 290, C27.
473 S.N.Gurkova, A.I.Gusev, N.V.Alekseev, T.K.Gar, N.A.Viktorov, Zh. Strukt. Khim., 1984, 25(5), 174(Engl.Ed. 829).
474 R.K.Chadha, J.E.Drake, A.B.Sarkar, Inorg. Chem., 1985, 24, 3156.
475 V.K.Bel´skii, A.A.Simonenko, I.E.Saratov, V.O.Reikhsfel´d, Kristallografiya, 1985, 30, 304.
476 H.Preut, J.Köcher, W.P.Neumann, Acta Crystallogr., 1985, C41, 912.
477 H.Schumann, C.Janiak, E.Hahn, J.Loebel, J.J.Zuckerman, Angew. Chem., Int. Ed. Engl., 1985, 24, 773.
478 J.D.Korp, I.Bernal, R.Hörlein, R.Serrano, W.A.Herrmann, Chem. Ber., 1985, 118, 340.
479 T.Fjeldberg, A.Haaland, B.E.R.Schilling, H.V.Volden, M.F.Lappert, A.J.Thorne, J. Organomet. Chem., 1985, 280, C43.
480 P.Jutzi, B.Hampel, K.Stroppel, C.Krüger, K.Angermund, P.Hofmann, Chem. Ber., 1985, 118, 2789.
481 H.D.Hausen, O.Mundt, W.Kaim, J. Organomet. Chem., 1985, 296, 321.
482 J.T.Snow, S.Murakami, S.Masamune, D.J.Williams, Tet. Lett., 1984, 25, 4191.
483 H.Puff, T.R.Kök, P.Nauroth, W.Schuh, J. Organomet. Chem., 1985, 281, 141.
484 P.B.Hitchcock, H.A.Jasim, R.E.Kelly, M.F.Lappert, J. Chem. Soc., Chem. Commun., 1985, 1776.
485 M.Dräger and K.Häberle, J. Organomet. Chem., 1985, 280, 183.
486 R.D.Rogers, M.M.Benning, L.K.Kurihara, K.J.Moriarty, M.D.Rausch, J. Organomet. Chem., 1985, 293, 51.
487 C.Krüger, G.Müller, G.Erker, U.Dorf, K.Engel, Organometallics, 1985, 4, 215.
488 D.M.Roddick, B.D.Santarsiero, J.E.Bercaw, J. Am. Chem. Soc., 1985, 107, 4670.
489 S.Dou and S.Chen, Gaodeng Xuexiao Huaxue Xuebao, 1984, 5, 812(Chem. Abs., 1985, 102:158434d).
490 G.Erker, P.Czisch, C.Krüger, J.M.Wallis, Organometallics, 1985, 4, 2059.
491 R.T.Baker, T.H.Tulip, S.S.Wreford, Inorg. Chem., 1985, 24, 1379.
492 M.D.Fryzuk, A.Carter, A.Westerhaus, Inorg. Chem., 1985, 24, 642.
493 J.P.Charland and A.L.Beauchamp, Croat. Chem. Acta, 1984, 57, 679.
494 A.L.Spek, A.R.Siedle, J.Reedijk, Inorg. Chim. Acta, 1985, 100, L15.
495 J.Bravo, J.S.Casas, M.V.Castano, M.Gayoso, Y.P.Mascarenhas, A.Sánchez, C.de O.P.Santos, J.Sordo, Inorg. Chem., 1985, 24, 3435.
496 A.G.Davies, J.P.Goddard, M.B.Hursthouse, N.P.C.Walker, J. Chem. Soc., Dalton Trans., 1985, 471.
497 F.Guay and A.L.Beauchamp, Can. J. Chem., 1985, 63, 3456.
498 G.Maas, R.Brückmann, W.Lorenz, J. Organomet. Chem., 1985, 289, 9.
499 S.Onaka, Y.Kondo, M.Yamashita, Y.Tatematsu, Y.Kato, M.Goto, T.Ito, Inorg. Chem., 1985, 24, 1070.
500 M.Cano, R.Criado, E.Gutierrez-Puebla, A.Monge, M.P.Pardo, J. Organomet. Chem., 1985, 292, 375.
501 S.Akabori, H.Munegumi, Y.Habata, S.Sato, K.Kawazoe, C.Tamura, M.Sato, Bull. Chem. Soc. Jpn., 1985, 58, 2185.
502 J.-P.Charland and A.L.Beauchamp, Acta Crystallogr., 1985, C41, 505.
503 R.D.Bach, H.B.Vardhan, A.F.M.M.Rahman, J.P.Oliver, Organometallics, 1985, 4, 846.
504 B.Kamenar, B.Kaitner, S.Pocev, J. Chem. Soc., Dalton Trans., 1985, 2457.
505 J.D.Wuest and B.Zacharie, Organometallics, 1985, 4, 410.
506 M.A.Khan, C.Peppe, D.G.Tuck, J. Organomet. Chem., 1985, 280, 17.
507 A.M.Arif, D.C.Bradley, D.M.Frigo, M.B.Hursthouse, B.Hussain, J. Chem. Soc., Chem. Commun., 1985, 783.
508 G.S.Redman and K.R.Mann, Inorg. Chem., 1985, 24, 3507.
509 P.O.Stoutland and R.G.Bergman, J. Am. Chem. Soc., 1985, 107, 4581.
510 J.M.Kisenyi, J.A.Cabeza, A.J.Smith, H.Adams, G.J.Sunley, N.J.S.Salt, P.M.Maitlis, J. Chem. Soc., Chem. Commun., 1985, 770.
511 L.Dahlenburg and A.Yardimcioglu, J. Organomet. Chem., 1985, 291, 371.
512 L.Dahlenburg and N.Höck, Inorg. Chim. Acta, 1985, 104, L29.
513 C.E.Johnson and R.Eisenberg, J. Am. Chem. Soc., 1985, 107, 3148.
514 F.H.Cano and C.Foces-Foces, J. Organomet. Chem., 1985, 291, 363.

515 N.W.Alcock, J.M.Brown, A.E.Derome, A.R.Lucy, J. Chem. Soc., Chem. Commun., 1985, 575.
516 H.Bauer, U.Nagel, W.Beck, J. Organomet. Chem., 1985, 290, 219.
517 G.R.Clark, T.R.Greene, W.R.Roper, J. Organomet. Chem., 1985, 293, C25.
518 W.M.Rees, M.R.Churchill, Y.-J.Li, J.D.Atwood, Organometallics, 1985, 4, 1162.
519 P.Janser, L.M.Venanzi, F.Bachechi, J. Organomet. Chem., 1985, 296, 229.
520 W.M.Rees, M.R.Churchill, J.C.Fettinger, J.D.Atwood, Organometallics, 1985, 4, 2179.
521 E.Arpac and L.Dahlenberg, Chem. Ber., 1985, 118, 3188.
522 J.W.Faller, R.H.Crabtree, A.Habib, Organometallics, 1985, 4, 929.
523 M.J.Auburn, S.L.Grundy, S.R.Stobart, M.J.Zaworotko, J. Am. Chem. Soc., 1985, 107, 266.
524 M.D.Fryzuk, P.A.MacNeil, S.J.Rettig, Organometallics, 1985, 4, 1145.
525 M.D.Fryzuk, P.A.MacNeil, S.J.Rettig, J. Am. Chem. Soc., 1985, 107, 6708.
526 F.W.B.Einstein, R.K.Pomeroy, P.Rushman, A.C.Willis, Organometallics, 1985, 4, 250.
527 F.A.Cotton, P.Lahuerta, M.Sanau, W.Schwotzer, J. Am. Chem. Soc., 1985, 107, 8284.
528 M.El Amane, A.Maisonnat, F.Dahan, R.Pince, R.Poilblanc, Organometallics, 1985, 4, 773.
529 D.O.K.Fjeldsted, S.R.Stobart, M.J.Zaworotko, J. Am. Chem. Soc., 1985, 107, 8258.
530 G.W.Bushnell, M.J.Decker, D.T.Eadie, S.R.Stobart, R.Vefghi, J.L.Atwood, M.J.Zaworotko, Organometallics, 1985, 4, 2106.
531 T.M.Gilbert and R.G.Bergman, J. Am. Chem. Soc., 1985, 107, 3502.
532 G.W.Bushnell, D.O.K.Fjeldsted, S.R.Stobart, M.J.Zaworotko, S.A.R.Knox, K.A.Macpherson, Organometallics, 1985, 4, 1107.
533 P.G.Rasmussen, J.E.Anderson, O.H.Bailey, M.Tamres, J. Am. Chem. Soc., 1985, 107, 279.
534 G.G.Hlatky, B.F.G.Johnson, J.Lewis, P.R.Raithby, J. Chem. Soc., Dalton Trans., 1985, 1277.
535 B.R.Sutherland and M.Cowie, Organometallics, 1985, 4, 1637.
536 B.R.Sutherland and M.Cowie, Organometallics, 1985, 4, 1801.
537 M.R.Churchill, L.V.Biondi, J.R.Shapley, C.H.McAteer, J. Organomet. Chem., 1985, 280, C63.
538 A.Fumagalli, F.Demartin, A.Sironi, J. Organomet. Chem., 1985, 279, C33.
539 D.Braga, R.Ros, R.Roulet, J. Organomet. Chem., 1985, 286, C8.
540 V.G.Albano, D.Braga, R.Ros, A.Scrivanti, J. Chem. Soc., Chem. Commun., 1985, 866.
541 M.A.Beno, H.Hope, M.M.Olmstead, P.P.Power, Organometallics, 1985, 4, 2117.
542 P.v.R.Schleyer, R.Hacker, H.Dietrich, W.Mahdi, J. Chem. Soc., Chem. Commun., 1985, 622.
543 U.Schümann, J.Kopf, E.Weiss, Angew. Chem., Int. Ed. Engl., 1985, 24, 215.
544 H.Schumann, F.-W.Reier, E.Hahn, Z. Naturforsch., 1985, 40B, 1289.
545 H.Schumann, I.Albrecht, E.Hahn, Angew. Chem., Int. Ed. Engl., 1985, 24, 985.
546 G.Jeske, L.E.Schock, P.N.Swepston, H.Schumann, T.J.Marks, J. Am. Chem. Soc., 1985, 107, 8103.
547 K.-R.Pörschke, K.Jonas, G.Wilke, R.Benn, R.Mynott, R.Goddard, C.Krüger, Chem. Ber., 1985, 118, 275.
548 P.Jutzi, E.Schlüter, S.Pohl, W.Saak, Chem. Ber., 1985, 118, 1959.
549 J.J.Stezowski, H.Hoier, D.Wilhelm, T.Clark, P.v.R.Schleyer, J. Chem. Soc., Chem. Commun., 1985, 1263.
550 D.Wilhelm, T.Clark, P.v.R.Schleyer, H.Dietrich, W.Mahdi, J. Organomet. Chem., 1985, 280, C6.
551 G.Boche, H.Etzrodt, W.Massa, G.Baum, Angew. Chem., Int. Ed. Engl., 1985, 24, 863.
552 L.M.Engelhardt, W.-P.Leung, C.L.Raston, A.H.White, J. Chem. Soc., Dalton Trans., 1985, 337.
553 M.Aslam, R.A.Bartlett, E.Block, M.M.Olmstead, P.P.Power, G.E.Sigel, J. Chem. Soc., Chem. Commun., 1985, 1674.
554 G.Becker, H.-M.Hartmann, A.Münch, H.Riffel, Z. Anorg. Allg. Chem., 1985,

530, 29.
555 G.W.Klumpp, M.Vos, F.J.J.de Kanter, J. Am. Chem. Soc., 1985, 107, 8292.
556 W.Neugebauer, G.A.P.Geiger, A.J.Kos, J.J.Stezowski, P.v.R.Schleyer, Chem. Ber., 1985, 118, 1504.
557 J.T.B.H.Jastrzebski, G.van Koten, M.J.N.Christopherson, C.H.Stam, J. Organomet. Chem., 1985, 292, 319.
558 G.S.Girolami, C.G.Howard, G.Wilkinson, H.M.Dawes, M.Thornton-Pett, M.Motevalli, M.B.Hursthouse, J. Chem. Soc., Dalton Trans., 1985, 921.
559 D.J.Brauer, H.Bürger, G.R.Liewald, J.Wilke, J. Organomet. Chem., 1985, 287, 305.
560 C.Ni, D.Deng, C.Qian, Inorg. Chim. Acta, 1985, 110, L7.
561 A.L.Wayda and R.D.Rogers, Organometallics, 1985, 4, 1440.
562 R.A.Andersen, R.Blom, A.Haaland, B.E.R.Schilling, H.V.Volden, Acta Chem. Scand., 1985, A39, 563.
563 D.Seebach, J.Hansen, P.Seiler, J.M.Gromek, J. Organomet. Chem., 1985, 285, 1.
564 B.Bogdanović, N.Janke, C.Krüger, R.Mynott, K.Schlichte, U.Westeppe, Angew. Chem., Int. Ed. Engl., 1985, 24, 960.
565 P.Stavropoulos, P.G.Edwards, G.Wilkinson, M.Motevalli, K.M.A.Malik, M.B.Hursthouse, J. Chem. Soc., Dalton Trans., 1985, 2167.
566 H.Lehmkuhl, A.Shakoor, K.Mehler, C.Krüger, K.Angermund, Y.-H.Tsay, Chem. Ber., 1985, 118, 4239.
567 E.P.Squiller, R.R.Whittle, H.G.Richey Jr., Organometallics, 1985, 4, 1154.
568 A.W.Duff, P.B.Hitchcock, M.F.Lappert, R.G.Taylor, J.A.Segal, J. Organomet. Chem., 1985, 293, 271.
569 R.A.Andersen, A.Haaland, K.Rypdal, H.V.Volden, J. Chem. Soc., Chem. Commun., 1985, 1807.
570 M.Herberhold, B.Schmidkonz, M.L.Ziegler, T.Zahn, Angew. Chem., Int. Ed. Engl., 1985, 24, 515.
571 U.Burger, C.Perret, G.Bernardinelli, E.P.Kündig, Helv. Chim. Acta, 1984, 67, 2063.
572 I.-P.Lorenz, J.Messelhäuser, W.Hiller, K.Haug, Angew. Chem., Int. Ed. Engl., 1985, 24, 228.
573 H.Lang, G.Mohr, O.Scheidsteger, G.Huttner, Chem. Ber., 1985, 118, 574.
574 R.E.Cramer, K.T.Higa, J.W.Gilje, Organometallics, 1985, 4, 1140.
575 M.J.McGeary, T.L.Tonker, J.L.Templeton, Organometallics, 1985, 4, 2102.
576 J.R.Bleeke and J.J.Kotyk, Organometallics, 1985, 4, 194.
577 E.Lindner, F.Zinsser, H.A.Mayer, W.Hiller, R.Fawzi, Z. Naturforsch., 1985, 40B, 615.
578 E.Lindner, F.Zinsser, W.Hiller, R.Fawzi, J. Organomet. Chem., 1985, 288, 317.
579 L.H.P.Gommans, L.Main, B.K.Nicholson, J. Organomet. Chem., 1985, 284, 345.
580 M.L.Valin, D.Moreiras, X.Solans, M.Font-Altaba, J.Solans, F.Garćia-Alonso, V.Riera, M.Vivanco, Acta Crystallogr., 1985, C41, 1312.
581 E.D.Honig, M.Quin-jin, W.T.Robinson, P.G.Williard, D.A.Sweigart, Organometallics, 1985, 4, 871.
582 S.K.Chopra, S.S.C.Chu, P.de Meester, D.E.Geyer, M.Lattman, S.A.Morse, J. Organomet. Chem., 1985, 294, 347.
583 N.G.Connelly, M.J.Freeman, A.G.Orpen, A.R.Sheehan, J.B.Sheridan, D.A.Sweigart, J. Chem. Soc., Dalton Trans., 1985, 1019.
584 G.A.Carriedo, J.B.P.Soto, V.Riera, X.Solans, C.Miravitlles, J. Organomet. Chem., 1985, 297, 193.
585 A.D.Horton, M.J.Mays, P.R.Raithby, J. Chem. Soc., Chem. Commun., 1985, 247.
586 C.J.Besecker, V.W.Day, W.G.Klemperer, M.R.Thompson, Inorg. Chem., 1985, 24, 44.
587 P.Braunstein, G.L.Geoffroy, B.Metz, Nouv. J. Chim., 1985, 9, 221.
588 S.W.Carr, B.L.Shaw, M.Thornton-Pett, J. Chem. Soc., Dalton Trans., 1985, 2131.
589 N.H.Buttrus, C.Eaborn, P.B.Hitchcock, J.D.Smith, A.C.Sullivan, J. Chem. Soc., Chem. Commun., 1985, 1380.
590 N.E.Kolobova, L.L.Ivanov, O.S.Zhvanko, A.S.Batsanov, Yu.T.Struchkov, J. Organomet. Chem., 1985, 279, 419.

591 H.Alper, F.Sibtain, F.W.B.Einstein, A.C.Willis, Organometallics, 1985, 4, 604.
592 E.Guggolz, K.Layer, F.Oberdorfer, M.Ziegler, Z. Naturforsch., 1985, 40B, 77.
593 H.Masuda, T.Taga, T.Sowa, T.Kawamura, T.Yonezawa, Inorg. Chim. Acta, 1985, 101, 45.
594 A.Winter, G.Huttner, M.Gottlieb, I.Jibril, J. Organomet. Chem., 1985, 286, 317.
595 A.Belforte, F.Calderazzo, D.Vitali, P.F.Zanazzi, Gazz. Chim. Ital., 1985, 115, 125.
596 D.A.DuBois, E.N.Duesler, R.T.Paine, Inorg. Chem., 1985, 24, 3.
597 J.Keijsper, G.van Koten, K.Vrieze, M.Zoutberg, C.H.Stam, Organometallics, 1985, 4, 1306.
598 A.S.Batsanov, S.P.Dolgova, V.I.Bakhmutov, Yu.T.Struchkov, V.N.Setkina, D.N.Kursanov, J. Organomet. Chem., 1985, 291, 341.
599 F.A.Cotton and K.-B.Shiu, Gazz. Chim. Ital., 1985, 115, 705.
600 W.A.Herrmann, H.-J.Kneuper, E.Herdtwerk, Angew. Chem., Int. Ed. Engl., 1985, 24, 1062.
601 N.I.Pyshnograeva, A.S.Batsanov, Yu.T.Struchkov, A.G.Ginzburg, V.N.Setkina, J. Organomet. Chem., 1985, 297, 69.
602 W.A.Herrmann, E.Voss, E.Guggolz, M.L.Ziegler, J. Organomet. Chem., 1985, 284, 47.
603 Yu.N.Saf´yanov, E.A.Kuz´min, V.V.Sharutin, Kristallografiya, 1984, 29, 928.
604 C.Eaborn, P.B.Hitchcock, J.D.Smith, A.C.Sullivan, J. Chem. Soc., Chem. Commun., 1985, 534.
605 O.J.Scherer, H.Sitzmann, G.Wolmershäuser, Acta Crystallogr., 1985, C41, 1761.
606 M.D.Curtis and K.R.Han, Inorg. Chem., 1985, 24, 378.
607 G.J.Kubas, H.J.Wasserman, R.R.Ryan, Organometallics, 1985, 4, 2012.
608 J.L.Davidson and G.Vasapollo, J. Organomet. Chem., 1985, 291, 43.
609 M.T.Ashby and D.L.Lichtenberger, Inorg. Chem., 1985, 24, 636.
610 A.J.Ashe III, W.Butler, J.C.Colburn, S.Abu-Orabi, J. Organomet. Chem., 1985, 282, 233.
611 S.J.N.Burgmayer and J.L.Templeton, Inorg. Chem., 1985, 24, 2224.
612 S.Gambarotta, C.Floriani, A.Chiesi-Villa, C.Guastini, J. Am. Chem. Soc., 1985, 107, 2985.
613 H.Adams, N.A.Bailey, V.A.Osborn, M.J.Winter, J. Organomet. Chem., 1985, 284, C1.
614 N.D.Silavwe, M.Y.Chiang, D.R.Tyler, Inorg. Chem., 1985, 24, 4219.
615 A.D.Hunter, P.Legzdins, C.R.Nurse, F.W.B.Einstein, A.C.Willis, J. Am. Chem. Soc., 1985, 107, 1791.
616 J.A.Bandy, V.S.B.Mtetwa, K.Prout, J.C.Green, C.E.Davies, M.L.H.Green, N.J.Hazel, A.Izquierdo, J.J.Martin-Polo, J. Chem. Soc., Dalton Trans., 1985, 2037.
617 M.Brookhart, K.Cox, F.G.N.Cloke, J.C.Green, M.L.H.Green, P.M.Hare, J.Bashkin, A.E.Derome, P.D.Grebenik, J. Chem. Soc., Dalton Trans., 1985, 423.
618 R.T.DePue, D.B.Collum, J.W.Ziller, M.R.Churchill, J. Am. Chem. Soc., 1985, 107, 2131.
619 A.L.Beauchamp, F.Belanger-Gariepy, S.Arabi, Inorg. Chem., 1985, 24, 1860.
620 H.J.B.Slot, N.W.Murrall, A.J.Welch, Acta Crystallogr., 1985, C41, 1309.
621 M.Green, R.J.Mercer, C.E.Morton, A.G.Orpen, Angew. Chem., Int. Ed. Engl., 1985, 24, 422.
622 T.J.Chow, M.-Y.Wu, L.-K.Liu, J. Organomet. Chem., 1985, 281, C33.
623 C.Zeiher, W.Hiller, I.-P.Lorenz, Chem. Ber., 1985, 118, 3127.
624 D.G.Bourner, L.Brammer, M.Green, G.Moran, A.G.Orpen, C.Reeve, C.J.Schaverien, J. Chem. Soc., Chem. Commun., 1985, 1409.
625 R.G.Beevor, M.J.Freeman, M.Green, C.E.Morton, A.G.Orpen, J. Chem. Soc., Chem. Commun., 1985, 68.
626 M.M.Kubicki, R.Kergoat, L.C.G.de Lima, M.Cariou, H.Scordia, J.E.Guerchais, Inorg. Chim. Acta, 1985, 104, 191.
627 A.J.Graham, D.Akrigg, B.Sheldrick, Acta Crystallogr., 1985, C41, 995.
628 A.J.Pearson, M.N.I.Khan, J.C.Clardy, H.Cun-heng, J. Am. Chem. Soc., 1985, 107, 2748.

629 W.E.Vanarsdale, R.E.K.Winter, J.K.Kochi, J. Organomet. Chem., 1985, 296, 31.
630 L.Stahl, J.P.Hutchinson, D.R.Wilson, R.D.Ernst, J. Am. Chem. Soc., 1985, 107, 5016.
631 P.Leoni, E.Grilli, M.Pasquali, M.Tomassini, J. Chem. Soc., Dalton Trans., 1985, 2561.
632 S.R.Allen, R.G.Beevor, M.Green, N.C.Norman, A.G.Orpen, I.D.Williams, J. Chem. Soc., Dalton Trans., 1985, 435.
633 S.J.N.Burgmayer and J.L.Templeton, Inorg. Chem., 1985, 24, 3939.
634 T.G.Hill, R.C.Haltiwanger, A.D.Norman, Inorg. Chem., 1985, 24, 3499.
635 R.P.Hughes, J.W.Reich, A.L.Rheingold, Organometallics, 1985, 4, 1754.
636 I.Bernal, W.Ries, H.Brunner, D.K.Rastogi, J. Organomet. Chem., 1985, 290, 353.
637 L.G.McCullough, R.R.Schrock, J.C.Dewan, J.C.Murdzek, J. Am. Chem. Soc., 1985, 107, 5987.
638 H.Strutz, J.C.Dewan, R.R.Schrock, J. Am. Chem. Soc., 1985, 107, 5999.
639 R.P.Hughes, J.W.Reisch, A.L.Rheingold, Organometallics, 1985, 4, 241.
640 E.P.Kyba, R.E.Davis, S.-T.Liu, K.A.Hassett, S.B.Larson, Inorg. Chem., 1985, 24, 4629.
641 M.B.Hursthouse, M.A.Thornton-Pett, J.A.Connor, C.Overton, Acta Crystallogr., 1985, C41, 184.
642 T.Butters, W.Winter, U.Kunze, S.B.Sastrawan, Acta Crystallogr., 1985, C41, 23.
643 H.Arzoumanian, A.Baldy, R.Lai, J.Metzger, M.-L.N.Peh, M.Pierrot, J. Chem. Soc., Chem. Commun., 1985, 1151.
644 K.Starker and M.D.Curtis, Inorg. Chem., 1985, 24, 3006.
645 P.T.Bishop, J.R.Dilworth, J.A.Zubieta, J. Chem. Soc., Chem. Commun., 1985, 257.
646 P.J.Blower, J.R.Dilworth, J.Hutchinson, T.Nicholson, J.A.Zubieta, J. Chem. Soc., Dalton Trans., 1985, 2639.
647 J.Colin and B.Chevrier, Organometallics, 1985, 4, 1090.
648 Y.Yamomoto and H.Yamazaki, J. Organomet. Chem., 1985, 282, 191.
649 M.C.Azar, M.J.Chetcuti, C.Eigenbrot, K.A.Green, J. Am. Chem. Soc., 1985, 107, 7209.
650 H.Werner, B.Ulrich, U.Schubert, P.Hofmann, B.Zimmer-Gasser, J. Organomet. Chem., 1985, 297, 27.
651 B.Chaudret, F.Dahan, S.Sabo, Organometallics, 1985, 4, 1490.
652 Y.Do, E.D.Simhon, R.H.Holm, Inorg. Chem., 1985, 24, 1831.
653 W.G.Klemperer, V.V.Mainz, R.-C.Wang, W.Shum, Inorg. Chem., 1985, 24, 1968.
654 K.J.Ahmed, M.H.Chisholm, J.C.Huffman, Organometallics, 1985, 4, 1168.
655 D.Gudat, E.Niecke, W.Malisch, U.Hofmockel, S.Quashie, A.H.Cowley, A.M.Arif, B.Krebs, M.Dartmann, J. Chem. Soc., Chem. Commun., 1985, 1687.
656 V.K.Bel'skii, A.N.Protskii, B.M.Bulychev, G.L.Soloveichik, J. Organomet. Chem., 1985, 280, 45.
657 J.C.Jeffery, J.C.V.Laurie, F.G.A.Stone, Polyhedron, 1985, 4, 1135.
658 E.O.Fischer, A.C.Filippou, H.G.Alt, U.Thewalt, Angew. Chem., Int. Ed. Engl., 1985, 24, 203.
659 B.Longato, B.D.Martin, J.R.Norton, O.P.Anderson, Inorg. Chem., 1985, 24, 1389.
660 L.Gelmini, L.C.Matassa, D.W.Stephan, Inorg. Chem., 1985, 24, 2585.
661 B.D.Martin, S.A.Matchett, J.R.Norton, O.P.Anderson, J. Am. Chem. Soc., 1985, 107, 7952.
662 M.J.Chetcuti, M.H.Chisholm, H.T.Chiu, J.C.Huffman, Polyhedron, 1985, 4, 1213.
663 C.J.Casewit, R.C.Haltiwanger, J.Noordik, M.R.DuBois, Organometallics, 1985, 4, 119.
664 H.Alper, F.W.B.Einstein, F.W.Hartstock, A.C.Willis, J. Am. Chem. Soc., 1985, 107, 173.
665 J.S.Drage and K.P.C.Vollhardt, Organometallics, 1985, 4, 191.
666 M.Tilset and K.P.C.Vollhardt, Organometallics, 1985, 4, 2230.
667 K.Blechschmitt, E.Guggolz, M.L.Ziegler, Z. Naturforsch., 1985, 40B, 85.
668 X.You, Z.Zhu, J.Huang, R.F.Fenske, L.F.Dahl, New Front. Organomet. Inorg. Chem., Proc. China-Jpn.-U.S.A. Trilateral Semin. 2nd, 1982, 257.(Chem. Abs.

1985, 102:204073k).
669 H.Arzoumanian, A.Baldy, M.Pierrot, J.-F.Petrignani, J. Organomet. Chem., 1985, 294, 327.
670 O.J.Scherer, H.Sitzmann, G.Wolmershäuser, Angew. Chem., Int. Ed. Engl., 1985, 24, 351.
671 D.J.Brauer, G.Hasselkuss, S.Morton, S.Hietkamp, H.Sommer, O.Stelzer, W.S.Sheldrick, Z. Naturforsch., 1985, 40B, 1161.
672 M.A.Greaney, J.S.Merola, T.R.Halbert, Organometallics, 1985, 4, 2057.
673 T.S.Coolbaugh, R.J.Coots, B.D.Santarsiero, R.H.Grubbs, Inorg. Chim. Acta, 1985, 98, 99.
674 M.S.Kralik, J.P.Hutchinson, R.D.Ernst, J. Am. Chem. Soc., 1985, 107, 8296.
675 R.Alvarez, E.Carmona, D.J.Cole-Hamilton, A.Galindo, E.Gutiérrez-Puebla, A.Monge, M.L.Poveda, C.Ruíz, J. Am. Chem. Soc., 1985, 107, 5529.
676 J.Bularzik, K.Kourtakis, J.Nitschke, W.Tötsche, Acta Crystallogr., 1985, C41, 1426.
677 J.G.Reiss, U.Klement, J.Wachter, J. Organomet. Chem., 1985, 280, 215.
678 R.Drews and U.Behrens, Chem. Ber., 1985, 118, 888.
679 T.Vogt and J.Strähle, Z. Naturforsch., 1985, 40B, 1599.
680 P.Jaitner, W.Wohlgenannt, A.Gieren, H.Betz, T.Hübner, J. Organomet. Chem., 1985, 297, 281.
681 E.J.M.de Boer, J.de With, A.G.Orpen, J. Chem. Soc., Chem. Commun., 1985, 1666.
682 A.Birnbaum, F.A.Cotton, Z.Dori, M.Kapon, D.Marler, G.M.Reisner, W.Schwotzer, J. Am. Chem. Soc., 1985, 107, 2405.
683 T.M.Che, V.W.Day, L.C.Francesconi, M.F.Fredrich, W.G.Klemperer, W.Shum, Inorg. Chem., 1985, 24, 4055.
684 I.Albrecht, E.Hahn, J.Pickardt, H.Schumann, Inorg. Chim. Acta, 1985, 110, 145.
685 A.S.C.Chan, H.-S.Shieh, J.R.Hill, J. Organomet. Chem., 1985, 279, 171.
686 A.Belforte, F.Calderazzo, P.F.Zanazzi, Gazz. Chim. Ital., 1985, 115, 71.
687 A.S.Antsyshkina, L.M.Dikareva, M.A.Porai-Koshits, V.N.Ostrikova, Yu.V.Skripkin, A.A.Pasynskii, O.G.Volkov, V.T.Kalinnikov, Koord. Khim., 1984, 10, 1564(Engl.Ed. 872).
688 M.D.Curtis and J.Real, Organometallics, 1985, 4, 940.
689 T.W.Coffindaffer, I.P.Rothwell, K.Folting, J.C.Huffman, W.E.Streib, J. Chem. Soc., Chem. Commun., 1985, 1519.
690 S.I.Bailey, L.M.Engelhardt, W.-P.Leung, C.L.Raston, I.M.Ritchie, A.H.White, J. Chem. Soc., Dalton Trans., 1985, 1747.
691 K.Seitz and U.Behrens, J. Organomet. Chem., 1985, 294, C9.
692 J.Liu and J.L.Peterson, Gaodeng Xuexiao Huaxue Xuebao, 1985, 6, 103.(Chem.Abs., 1985, 103:113685r).
693 F.Calderazzo, M.Castellani, G.Pampaloni, P.F.Zanazzi, J. Chem. Soc., Dalton Trans., 1985, 1989.
694 E.G.Perevalova, I.F.Urazowski, D.A.Lemenovskii, Yu.L.Slovokhotov, Yu.T.Struchkov, J. Organomet. Chem., 1985, 289, 319.
695 V.W.Day, W.G.Klemperer, D.J.Maltbie, Organometallics, 1985, 4, 104.
696 H.Mauermann, P.N.Swepston, T.J.Marks, Organometallics, 1985, 4, 200; with G.Jeske, H.Lauke, H.Schumann, J. Am. Chem. Soc., 1985, 107, 8091.
697 X.Jin and J.Zhou, Jiegou Huaxue, 1984, 3, 235.
698 R.Goddard, C.Krüger, F.Mark, R.Stansfield, X.Zhang, Organometallics, 1985, 4, 285.
699 J.Sieler, N.-N.Than, R.Benedix, E.Dinjus, D.Walther, Z. Anorg. Allg. Chem., 1985, 522, 131.
700 M.M.Brezinski, K.J.Klabunde, S.K.Janikowski, L.J.Radonovich, Inorg. Chem., 1985, 24, 3305.
701 H.Hoberg, F.J.Fañanás, K.Angermund, C.Krüger, M.J.Romão, J. Organomet. Chem., 1985, 281, 379.
702 M.W.Eyring and L.J.Radonovich, Organometallics, 1985, 4, 1841.
703 K.R.Pörschke, R.Mynott, K.Angermund, C.Krüger, Z. Naturforsch., 1985, 40B, 199.
704 J.D.Ferrara, C.Tessier-Youngs, W.J.Youngs, J. Am. Chem. Soc., 1985, 107, 6719.

705 D.Walther, E.Dinjus, H.Gorls, J.Sieler, O.Lindqvist, L.Andersen, J. Organomet. Chem., 1985, 286, 103.
706 D.Xu, K.Miki, Y.Kai, N.Kasai, M.Wada, J. Organomet. Chem., 1985, 287, 265.
707 J.M.Coronas, G.Muller, M.Rocamora, C.Miravitlles, X.Solans, J. Chem. Soc., Dalton Trans., 1985, 2333.
708 Z.Zhang, S.Wang, G.Yang, C.Shen, Y.Fang, Jiegou Huaxue, 1984, 3, 119.
709 L.Chen, K.Pan, X.Chen, L.Zhang, Jiegou Huaxue, 1984, 3, 125.
710 M.A.Bennett, T.W.Hambley, N.K.Roberts, G.B.Robertson, Organometallics, 1985, 4, 1992.
711 A.Döhring, P.W.Jolly, C.Krüger, M.J.Romão, Z. Naturforsch., 1985, 40B, 484.
712 K.R.Pörschke, Y.-H.Tsay, C.Krüger, Angew. Chem., Int. Ed. Engl., 1985, 24, 323.
713 J.J.Eisch, A.M.Piotrowski, A.A.Aradi, C.Krüger, M.J.Romão, Z. Naturforsch., 1985, 40B, 624.
714 H.Qichen, X.Minzhi, Q.Yanlong, X.Weihua, S.Meicheng, T.Youqi, J. Organomet. Chem., 1985, 287, 419.
715 H.Lehmkuhl, J.Elsasser, R.Benn, B.Gabor, A.Rufińska, R.Goddard, C.Krüger, Z. Naturforsch., 1985, 40B, 171.
716 S.K.Janikowski, L.J.Radonovich, T.J.Groshens, K.J.Klabunde, Organometallics, 1985, 4, 396.
717 R.Appel, C.Casser, F.Knoch, J. Organomet. Chem., 1985, 297, 21.
718 J.M.Ritchey, A.J.Zozulin, D.A.Wrobleski, R.R.Ryan, H.J.Wasserman, D.C.Moody, R.T.Paine, J. Am. Chem. Soc., 1985, 107, 501.
719 L.Carlton, W.E.Lindsell, K.J.McCullough, P.N.Preston, Organometallics, 1985, 4, 1138.
720 P.H.M.Budzelaar, J.Boersma, G.J.M.v.d.Kerk, A.L.Spek, A.J.M.Duisenberg, J. Organomet. Chem., 1985, 287, C13.
721 G.S.Ferguson and P.T.Wolczanski, Organometallics, 1985, 4, 1601.
722 J.J.Eisch, A.M.Piotrowski, K.I.Han, C.Krüger, Y.H.Tsay, Organometallics, 1985, 4, 224.
723 B.Fischer, J.Boersma, B.Kojić-Prodić, A.L.Spek, J. Chem. Soc., Chem. Commun., 1985, 1237.
724 C.A.Ghilardi, S.Midollini, A.Orlandini, J. Organomet. Chem., 1985, 295, 377.
725 J.G.Davidson, E.K.Barefield, D.G.V.Derveer, Organometallics, 1985, 4, 1178.
726 D.L.DeLaet, D.R.Powell, C.P.Kubiak, Organometallics, 1985, 4, 954.
727 P.H.M.Budzelaar, J.Boersma, G.J.M.v.d.Kerk, A.L.Spek, A.J.M.Duisenberg, Organometallics, 1985, 4, 680.
728 E.Carmona, J.M.Marín, P.Palma, M.Paneque, M.L.Poveda, Organometallics, 1985, 4, 2053.
729 O.J.Scherer, R.Walter, W.S.Sheldrick, Angew. Chem., Int. Ed. Engl., 1985, 24, 115.
730 A.Ceriotti, G.Longoni, M.Manassero, M.Perego, M.Sansoni, Inorg. Chem., 1985, 24, 117.
731 A.Ceriotti, G.Longoni, M.Manassero, N.Masciocchi, L.Resconi, M.Sansoni, J. Chem. Soc., Chem. Commun., 1985, 181.
732 A.Ceriotti, G.Longoni, M.Manassero, N.Masciocchi, G.Piro, L.Resconi, M.Sansoni, J. Chem. Soc., Chem. Commun., 1985, 1402.
733 A.Ceriotti, F.Demartin, G.Longoni, M.Manssero, M.Marchionna, G.Piva, M.Sansoni, Angew. Chem., Int. Ed. Engl., 1985, 24, 697.
734 M.A.Bennett, I.J.McMahon, S.Pelling, G.B.Robertson, W.A.Wickramasinghe, Organometallics, 1985, 4, 754.
735 R.P.Tooze, P.Stavropoulos, M.Motevalli, M.B.Hursthouse, G.Wilkinson, J. Chem. Soc., Chem. Commun., 1985, 1139.
736 H.E.Oosthuizen, E.Singleton, J.S.Field, G.C.van Niekerk, J. Organomet. Chem., 1985, 279, 433.
737 J.Gotzig, H.Otto, H.Werner, J. Organomet. Chem., 1985, 287, 247.
738 C.Conway, R.D.W.Kemmitt, A.W.G.Platt, D.R.Russell, L.J.S.Sherry, J. Organomet. Chem., 1985, 292, 419.
739 D.S.Bohle, C.E.F.Rickard, W.R.Roper, J. Chem. Soc., Chem. Commun., 1985, 1594.
740 G.Smith, D.J.Cole-Hamilton, M.Thornton-Pett, M.B.Hursthouse, J. Chem. Soc., Dalton Trans., 1985, 387.

741 J.A.Iggo, D.P.Markham, B.L.Shaw, M.Thornton-Pett, J. Chem. Soc., Chem. Commun., 1985, 432.
742 L.-Y.Hsu, N.Bhattacharyya, S.G.Shore, Organometallics, 1985, 4, 1483.
743 E.D.Morrison, G.L.Geoffroy, A.L.Rheingold, W.C.Fultz, Organometallics, 1985, 4, 1413; J. Am. Chem. Soc., 1985, 107, 254.
744 R.D.Adams, H.-S.Kim, S.Wang, J. Am. Chem. Soc., 1985, 107, 6107.
745 P.M.Lausarot, G.A.Vaglio, M.Valle, A.Tiripicchio, M.T.Camellini, P.Gariboldi, J. Organomet. Chem., 1985, 291, 221.
746 J.Banford, M.J.Mays, P.R.Raithby, J. Chem. Soc., Dalton Trans., 1985, 1355.
747 D.Himmelreich and G.Müller, J. Organomet. Chem., 1985, 297, 341.
748 M.P.Gomez-Sal, B.F.G.Johnson, J.Lewis, P.R.Raithby, A.H.Wright, J. Chem. Soc., Chem. Commun., 1985, 1682.
749 W.-Y.Yeh, J.R.Shapley, Y.-J.Li, M.R.Churchill, Organometallics, 1985, 4, 767.
750 K.Burgess, H.D.Holden, B.F.G.Johnson, J.Lewis, M.B.Hursthouse, N.P.C.Walker, A.J.Deeming, P.J.Manning, R.Peters, J. Chem. Soc., Dalton Trans., 1985, 85.
751 A.Eisenstadt, C.M.Giandomenico, M.F.Frederick, R.M.Laine, Organometallics, 1985, 4, 2033.
752 J.R.Shapley, W.-Y.Yeh, M.R.Churchill, Y.-J.Li, Organometallics, 1985, 4, 1898.
753 N.V.Podberezskaya, V.A.Maksakov, L.K.Kedrova, E.D.Korniets, S.P.Gubin, Koord. Khim., 1984, 10, 919(Engl.Ed. 514).
754 J.A.Clucas, M.M.Harding, A.K.Smith, J. Chem. Soc., Chem. Commun., 1985, 1280.
755 R.D.Adams and S.Wang, Inorg. Chem., 1985, 24, 4447.
756 L.J.Farrugia, M.Green, D.R.Hankey, M.Murray, A.G.Orpen, F.G.A.Stone, J. Chem. Soc., Dalton Trans., 1985, 177.
757 L.J.Farrugia, A.D.Miles, F.G.A.Stone, J. Chem. Soc., Dalton Trans., 1985, 2437.
758 L.D´Ornelas, A.Choplin, J.M.Basset, L.Y.Hsu, S.Shore, Nouv. J. Chim., 1985, 9, 155.
759 G.Süss-Fink, W.Bühlmeyer, M.Herberhold, A.Gieren, T.Hübner, J. Organomet. Chem., 1985, 280, 129.
760 C.J.Cardin, D.J.Cardin, J.M.Power, M.B.Hursthouse, J. Am. Chem. Soc., 1985, 107, 505.
761 R.D.Adams, I.T.Horváth, S.Wang, Inorg. Chem., 1985, 24, 1728.
762 M.R.Churchill, J.W.Ziller, L.R.Beanan, J. Organomet. Chem., 1985, 287, 235.
763 M.R.Churchill and Y.-J.Li, J. Organomet. Chem., 1985, 294, 367.
764 M.R.Churchill and Y.-J.Li, J. Organomet. Chem., 1985, 291, 61.
765 J.Puga, R.Sánchez-Delgado, D.Braga, Inorg. Chem., 1985, 24, 3971.
766 J.Puga, R.Sánchez-Delgado, A.Andriollo, J.Ascanio, D.Braga, Organometallics, 1985, 4, 2064.
767 E.J.Ditzel, B.F.G.Johnson, J.Lewis, P.R.Raithby, M.J.Taylor, J. Chem. Soc., Dalton Trans., 1985, 555.
768 F.W.B.Einstein, L.R.Martin, R.K.Pomeroy, P.Rushman, J. Chem. Soc., Chem. Commun., 1985, 345.
769 R.D.Adams and S.Wang, Organometallics, 1985, 4, 1902.
770 M.P.Gomez-Sal, B.F.G.Johnson, R.A.Kamarudin, J.Lewis, P.R.Raithby, J. Chem. Soc., Chem. Commun., 1985, 1622.
771 C.Couture and D.H.Farrar, J. Chem. Soc., Chem. Commun., 1985, 197.
772 R.J.Goudsmit, P.F.Jackson, B.F.G.Johnson, J.Lewis, W.J.H.Nelson, J.Puga, M.D.Vargas, D.Braga, K.Henrick, M.McPartlin, A.Sironi, J. Chem. Soc., Dalton Trans., 1985, 1795.
773 M.Dräger and N.Kleiner, Z. Anorg. Allg. Chem., 1985, 522, 48.
774 M.G.Begley, C.Gaffney, P.G.Harrison, A.Steel, J. Organomet. Chem., 1985, 289, 281.
775 E.M.Holt, F.A.K.Nasser, A.Wilson Jr., J.J.Zuckerman, Organometallics, 1985, 4, 2074.
776 H.O.v.d.Kooi, W.H.den Brinker, A.J.de Kok, Acta Crystallogr., 1985, C41, 869.
777 N.Kleiner and M.Dräger, Z. Naturforsch., 1985, 40B, 477.
778 N.Kleiner and M.Dräger, J. Organomet. Chem., 1985, 293, 323.

779 G.R.Wiger, S.S.Tomita, M.F.Rettig, R.M.Wing, Organometallics, 1985, 4, 1157
780 R.Ciajolo, M.A.Jama, A.Tuzi, A.Vitagliano, J. Organomet. Chem., 1985, 295, 233.
781 W.P.Fehlhammer, K.Bartel, U.Plaia, A.Volkl, A.T.Liu, Chem. Ber., 1985, 118, 2235.
782 G.R.Newkome, W.E.Puckett, G.E.Kiefer, V.K.Gupta, F.R.Fronczek, D.C.Pantaleo, G.L.McClure, J.B.Simpson, W.A.Deutsch, Inorg. Chem., 1985, 24, 811.
783 R.Usón, J.Forniés, P.Espinet, E.Lalinde, P.G.Jones, G.M.Sheldrick, J. Organomet. Chem., 1985, 288, 249.
784 H.Kurosawa, M.Emoto, A.Urabe, K.Miki, N.Kasai, J. Am. Chem. Soc., 1985, 107, 8253.
785 L.V.Rybin, E.A.Petrovskaya, M.I.Rubinskaya, L.G.Kuz'mina, Yu.T.Struchkov, V.V.Kaverin, N.Yu.Koneva, J. Organomet. Chem., 1985, 288, 119.
786 R.Bardi, A.M.Piazzesi, A.D.Pra, G.Cavinato, L.Toniolo, Inorg. Chim. Acta, 1985, 102, 99.
787 D.H.Farrar and N.C.Payne, J. Am. Chem. Soc., 1985, 107, 2054.
788 T.Ohta, T.Hosokawa, S.-I.Murahashi, K.Miki, N.Kasai, Organometallics, 1985, 4, 2080.
789 P.B.Hitchcock, M.F.Lappert, M.C.Misra, J. Chem. Soc., Chem. Commun., 1985, 863.
790 P.Thometzek, K.Zenkert, H.Werner, Angew. Chem., Int. Ed. Engl., 1985, 24, 516.
791 N.Barr, S.F.Dyke, G.Smith, C.H.L.Kennard, V.McKee, J. Organomet. Chem., 1985, 288, 109.
792 Y.Yamamoto and H.Yamazaki, Bull. Chem. Soc. Jpn., 1985, 58, 1843.
793 B.Galli, F.Gasparrini, B.E.Mann, L.Maresca, G.Natile, A.M.Manotti-Lanfredi, A.Tiripicchio, J. Chem. Soc., Dalton Trans., 1985, 1155.
794 R.D.Feltham, G.Elbaze, R.Ortega, C.Eck, J.Dubrawski, Inorg. Chem., 1985, 24, 1503.
795 A.Albinati, P.S.Pregosin, R.Rüedi, Helv. Chim. Acta, 1985, 68, 2046.
796 A.J.Canty, N.J.Minchin, L.M.Engelhardt, A.H.White, Inorg. Chim. Acta, 1985, 102, L29.
797 P.W.Clark, S.F.Dyke, G.Smith, C.H.L.Kennard, A.H.White, Acta Crystallogr., 1985, C41, 1742.
798 R.Usón, J.Forniés, P.Espinet, A.Garcia, M.Tomas, C.Foces-Foces, F.H.Cano, J. Organomet. Chem., 1985, 282, C35.
799 T.A.Anderson, R.J.Barton, B.E.Robertson, K.Venkatasubramanian, Acta Crystallogr., 1985, C41, 1171.
800 B.Kellenberger, S.J.Young, J.K.Stille, J. Am. Chem. Soc., 1985, 107, 6105.
801 M.A.Khan and A.J.McAlees, Inorg. Chim. Acta, 1985, 104, 109.
802 A.W.Hanson, A.J.McAlees, A.Taylor, J. Chem. Soc., Perkin Trans. I, 1985, 441.
803 S.I.Al-Resayes, P.B.Hitchcock, J.F.Nixon, D.M.P.Mingos, J. Chem. Soc., Chem. Commun., 1985, 365.
804 A.Tiripicchio, F.J.Lahoz, L.A.Oro, M.T.Pinillos, C.Tejel, Inorg. Chim. Acta, 1985, 100, L5.
805 Lj.Manojlović-Muir, K.W.Muir, B.R.Lloyd, R.J.Puddephatt, J. Chem. Soc., Chem. Commun., 1985, 536.
806 G.K.Campbell, P.B.Hitchcock, M.F.Lappert, M.C.Misra, J. Organomet. Chem., 1985, 289, C1.
807 M.Lenarda, G.Nardin, G.Pellizer, E.Braye, M.Graziani, J. Chem. Soc., Chem. Commun., 1985, 1536.
808 V.G.Albano, F.Demartin, A.de Renzi, G.Morelli, A.Saporito, Inorg. Chem., 1985, 24, 2032.
809 P.Castan, J.Jaud, N.P.Johnson, R.Soules, J. Am. Chem. Soc., 1985, 107, 5011.
810 R.Grigg, J.F.Malone, T.R.B.Mitchell, A.Ramasubbu, R.M.Scott, J. Chem. Soc., Perkin Trans. I, 1984, 1745.
811 L.S.Hollis, A.R.Amundsen, E.W.Stern, J. Am. Chem. Soc., 1985, 107, 274.
812 V.G.Albano, F.Demartin, B.Di Blasio, G.Morelli, A.Panunzi, Gazz. Chim. Ital., 1985, 115, 361.
813 K.Venkatasubramanian, Indian Acad. Sci., Chem. Sci., 1984, 93, 1237.
814 J.-P.Macquet and A.L.Beauchamp, Acta Crystallogr., 1985, C41, 860.

815 G.Ferguson, P.K.Monaghan, M.Parvez, R.J.Puddephatt, Organometallics, 1985, 4, 1669.
816 J.Terheijden, G.van Koten, I.C.Vinke, A.L.Spek, J. Am. Chem. Soc., 1985, 107, 2891.
817 J.J.Koh, W.H.Lee, P.G.Williard, W.M.Risen Jr., J. Organomet. Chem., 1985, 284, 409.
818 O.J.Scherer, K.Hussong, G.Wolmershäuser, J. Organomet. Chem., 1985, 289, 215.
819 L.Pandolfo, G.Paiaro, G.Valle, P.Ganis, Gazz. Chim. Ital., 1985, 115, 65.
820 A.Imran, R.D.W.Kemmitt, A.J.W.Markwick, P.McKenna, D.R.Russell, L.J.S.Sherry, J. Chem. Soc., Dalton Trans., 1985, 549.
821 L.Pandolfo, G.Paiaro, G.Valle, P.Ganis, Gazz. Chim. Ital., 1985, 115, 59.
822 W.R.Winchester, M.Gawron, W.M.Jones, Organometallics, 1985, 4, 1894.
823 R.P.Hughes, J.M.J.Lambert, A.L.Rheingold, Organometallics, 1985, 4, 2055.
824 J.Powell, J.F.Sawyer, M.V.R.Stainer, J. Chem. Soc., Chem. Commun., 1985, 1314.
825 A.T.Hutton, C.R.Langrick, D.M.McEwan, P.G.Pringle, B.L.Shaw, J. Chem. Soc., Dalton Trans., 1985, 2121.
826 M.Ebner, H.Otto, H.Werner, Angew. Chem., Int. Ed. Engl., 1985, 24, 518.
827 E.W.Abel, S.K.Bhargava, K.G.Orrell, A.W.G.Platt, V.Šik, T.S.Cameron, J. Chem. Soc., Dalton Trans., 1985, 345.
828 N.P.C.Walker, M.B.Hursthouse, C.P.Warrens, J.D.Woollins, J. Chem. Soc., Chem. Commun., 1985, 227.
829 T.A.K.Al-Allaf, C.Eaborn, P.B.Hitchcock, M.F.Lappert, A.Pidcock, J. Chem. Soc., Chem. Commun., 1985, 548.
830 A.Albinati, U.von Gunten, P.S.Pregosin, H.J.Ruegg, J. Organomet. Chem., 1985, 295, 239.
831 M.R.Awang, R.D.Barr, M.Green, J.A.K.Howard, T.B.Marder, F.G.A.Stone, J. Chem. Soc., Dalton Trans., 1985, 2009.
832 J.Powell, J.F.Sawyer, S.J.Smith, J. Chem. Soc., Chem. Commun., 1985, 1312.
833 D.H.Farrar and J.A.Lunniss, Acta Crystallogr., 1985, C41, 1444.
834 R.J.Blau, M.H.Chisholm, K.Folting, R.J.Wang, J. Chem. Soc., Chem. Commun., 1985, 1582.
835 E.J.Parsons, R.D.Larson, P.W.Jennings, J. Am. Chem. Soc., 1985, 107, 1793.
836 L.Manojlović-Muir and K.W.Muir, Croat. Chem. Acta, 1984, 57, 587.
837 S.S.M.Ling, N.C.Payne, R.J.Puddephatt, Organometallics, 1985, 4, 1546.
838 S.S.M.Ling, I.R.Jobe, L.Manojlović-Muir, K.W.Muir, R.J.Puddephatt, Organometallics, 1985, 4, 1198.
839 E.Ma, G.Semelhago, A.Walker, D.H.Farrar, R.R.Gukathasan, J. Chem. Soc., Dalton Trans., 1985, 2595.
840 A.C.Skapski, V.F.Sutcliffe, G.B.Young, J. Chem. Soc., Chem. Commun., 1985, 609.
841 K.A.Azam, A.A.Frew, B.R.Lloyd, L.Manojlovic-Muir, K.W.Muir, R.J.Puddephatt, Organometallics, 1985, 4, 1400.
842 G.Minghetti, A.Albinati, A.L.Bandini, G.Banditelli, Angew. Chem., Int. Ed. Engl., 1985, 24, 120.
843 A.T.Hutton, P.G.Pringle, B.L.Shaw, J. Chem. Soc., Dalton Trans., 1985, 1677.
844 C.E.Briant, D.I.Gilmour, M.A.Luke, D.M.P.Mingos, J. Chem. Soc., Dalton Trans., 1985, 851.
845 C.E.Briant, D.I.Gilmour, D.M.P.Mingos, R.W.M.Wardle, J. Chem. Soc., Dalton Trans., 1985, 1693.
846 O.J.Scherer, R.Konrad, E.Guggolz, M.L.Ziegler, Chem. Ber., 1985, 118, 1.
847 G.Smith, C.H.L.Kennard, T.C.W.Mak, J. Organomet. Chem., 1985, 290, C7.
848 J.M.Mayer, D.L.Thorn, T.H.Tulip, J. Am. Chem. Soc., 1985, 107, 7454.
849 G.K.Yang and R.G.Bergman, Organometallics, 1985, 4, 129.
850 M.F.N.N.Carvalho, A.J.L.Pombeiro, U.Schubert, O.Orama, C.J.Pickett, R.L.Richards, J. Chem. Soc., Dalton Trans., 1985, 2079.
851 W.A.Herrmann, E.Voss, U.Küsthardt, E.Herdtweck, J. Organomet. Chem., 1985, 294, C37.
852 C.P.Casey, J.M.O'Connor, K.J.Haller, J. Am. Chem. Soc., 1985, 107, 1241.
853 F.W.B.Einstein, K.G.Tyers, D.Sutton, Organometallics, 1985, 4, 489.
854 W.A.Herrmann, U.Küsthardt, E.Herdtweck, J. Organomet. Chem., 1985, 294, C33.

855 W.A.Herrmann, U.Küsthardt, M.L.Ziegler, T.Zahn, Angew. Chem., Int. Ed. Engl., 1985, 24, 860.
856 M.C.L.Trimarchi, M.A.Green, J.C.Huffman, K.G.Caulton, Organometallics, 1985, 4, 514.
857 D.Baudry, P.Boydell, M.Ephritikhine, H.Felkin, J.Guilhem, C,Pascard, E.T.H.Dau, J. Chem. Soc., Chem. Commun., 1985, 670.
858 D.E.Smith and J.A.Gladysz, Organometallics, 1985, 4, 1480.
859 W.E.Buhro, S.Georgiou, J.P.Hutchinson, J.A.Gladysz, J. Am. Chem. Soc., 1985, 107, 3346.
860 W.D.Jones and J.A.Maguire, Organometallics, 1985, 4, 951.
861 R.Rossi, A.Duatti, L.Magon, A.Marchi, A.Medici, M.Fogagnolo, U.Casellato, R.Graziani, Transition Met. Chem., 1985, 10, 413.
862 J.J.Doney, R.G.Bergman, C.H.Heathcock, J. Am. Chem. Soc., 1985, 107, 3724.
863 D.G.DeWit, K.Folting, W.E.Streib, K.G.Caulton, Organometallics, 1985, 4, 1149.
864 M.A.A.F.de C.T.Carrondo, A.M.T.S.Domingos, G.A.Jeffrey, J. Organomet. Chem., 1985, 289, 377.
865 V.K.Belsky, A.N.Protsky, I.V.Molodnitskaya, B.M.Bulychev, G.L.Soloveichik, J. Organomet. Chem., 1985, 293, 69.
866 W.C.Mercer, R.R.Whittle, E.W.Burkhardt, G.L.Geoffroy, Organometallics, 1985, 4, 68.
867 W.C.Mercer, G.L.Geoffroy, A.L.Rheingold, Organometallics, 1985, 4, 1418.
868 R.Poli, G.Wilkinson, M.Motevalli, M.B.Hursthouse, J. Chem. Soc., Dalton Trans., 1985, 931.
869 E.J.M.de Boer, J.de With, N.Meijboom, A.G.Orpen, Organometallics, 1985, 4, 259.
870 G.W.Harris, J.C.A.Boeyens, N.J.Coville, Organometallics, 1985, 4, 914.
871 W.Beck, K.Raub, U.Nagel, W.Sacher, Angew. Chem., Int. Ed. Engl., 1985, 24, 505.
872 C.G.Kreiter, K.H.Franzreb, W.Michels, U.Schubert, K.Ackermann, Z. Naturforsch., 1985, 40B, 1188.
873 K.-W.Lee, W.T.Pennington, A.W.Cordes, T.L.Brown, J. Am. Chem. Soc., 1985, 107, 631.
874 C.P.Casey, J.M.O'Connor, K.J.Haller, J. Am. Chem. Soc., 1985, 107, 3172.
875 T.Beringhelli, G.Ciani, G.D'Alfonso, A.Sironi, M.Freni, J. Chem. Soc., Dalton Trans., 1985, 1507.
876 G.W.Harris, J.C.A.Boeyens, N.J.Coville, J. Chem. Soc., Dalton Trans., 1985, 2277.
877 H.v.d.Heijden, A.G.Orpen, P.Pasman, J. Chem. Soc., Chem. Commun., 1985, 1576.
878 H.-J.Haupt, U.Flörke, P.Balsaa, Acta Crystallogr., 1985, C41, 1307.
879 F.A.Cotton, L.M.Daniels, K.R.Dunbar, L.R.Falvello, S.M.Tetrick, R.A.Walton, J. Am. Chem. Soc., 1985, 107, 3524.
880 F.A.Cotton, K.R.Dunbar, L.R.Falvello, R.A.Walton, Inorg. Chem., 1985, 24, 4180.
881 T.J.Barder, D.Powell, R.A.Walton, J. Chem. Soc., Chem. Commun., 1985, 550.
882 T.Beringhelli, G.Ciani, G.D'Alfonso, H.Molinari, A.Sironi, Inorg. Chem., 1985, 24, 2666.
883 W.A.Herrmann, R.Serrano, U.Küsthardt, E.Guggolz, B.Nuber, M.L.Ziegler, J. Organomet. Chem., 1985, 287, 329.
884 W.A.Herrmann, R.Serrano, M.L.Ziegler, H.Pfisterer, B.Nuber, Angew. Chem., Int. Ed. Engl., 1985, 24, 50.
885 T.Beringhelli, G.Ciani, G.D'Alfonso, A.Sironi, M.Freni, J. Chem. Soc., Chem. Commun., 1985, 978.
886 N.J.Taylor, J. Chem. Soc., Chem. Commun., 1985, 476.
887 N.J.Taylor, J. Chem. Soc., Chem. Commun., 1985, 478.
888 C.J.Besecker, V.W.Day, W.G.Klemperer, Organometallics, 1985, 4, 564.
889 N.M.Doherty, B.E.Ewels, R.P.Hughes, D.E.Samkoff, W.D.Saunders, R.E.Davis, B.B.Laird, Organometallics, 1985, 4, 1606.
890 J.Vicente, J.Martin, M.-T.Chicote, X.Solans, C,Miravitlles, J. Chem. Soc., Chem. Commun., 1985, 1004.
891 M.Arthurs, H.Karodia, M.Sedgwick, D.A.Morton-Blake, C.J.Cardin, H.Parge, J.

Organomet. Chem., 1985, 291, 231.
892 H.Werner, W.Paul, R.Feser, R.Zolk, P.Thometzek, Chem. Ber., 1985, 118, 261.
893 V.Schurig, W.Pille, K.Peters, H.G.von Schnering, Mol.Cryst.Liq.Cryst., 1985, 120, 385.
894 W.D.Jones and F.J.Feher, J. Am. Chem. Soc., 1985, 107, 620.
895 M.Valderrama, M.Scotti, R.Ganz, L.A.Oro, F.J.Lahoz, C.Foces-Foces, F.H.Cano, J. Organomet. Chem., 1985, 288, 97.
896 F.H.Cano, C.Foces-Foces, L.A.Oro, J. Organomet. Chem., 1985, 288, 225.
897 A.R.Chakravarty, F.A.Cotton, D.A.Tocher, Organometallics, 1985, 4, 863.
898 G.-E.Matsubayashi and S.Akazawa, Polyhedron, 1985, 4, 419.
899 F.J.G.Alonso, A.Höhn, J.Wolf, H.Otto, H.Werner, Angew. Chem., Int. Ed. Engl., 1985, 24, 406.
900 J.W.Suggs, M.J.Wovkulich, S.D.Cox, Organometallics, 1985, 4, 1101.
901 H.Brunner, P.Beier, G.Riepl, I.Bernal, G.M.Reisner, R.Benn, A.Rufińska, Organometallics, 1985, 4, 1732.
902 L.Dahlenburg and N.Höck, J. Organomet. Chem., 1985, 284, 129.
903 J.G.Leipoldt, G.J.Lamprecht, D.E.Graham, Inorg. Chim. Acta, 1985, 101, 123.
904 J.Ječný and K.Huml, Acta Crystallogr., 1985, C41, 503.
905 Z.Dauter, L.K.Hansen, R.J.Mawby, E.J.Probitts, C.D.Reynolds, Acta Crystallogr., 1985, C41, 850.
906 T.Yoshida, T.Okano, S.Otsuka, I.Miura, T.Kubota, K.Kafuku, K.Nakatsu, Inorg. Chim. Acta, 1985, 100, 7.
907 F.Dahan and R.Choukroun, Acta Crystallogr., 1985, C41, 704.
908 F.J.Lahoz, A.Tiripicchio, M.T.Camellini, L.A.Oro, M.T.Pinillos, J. Chem. Soc., Dalton Trans., 1985, 1487.
909 G.-E.Matsubayashi, T.Iinuma, T.Tanaka, K.Oka, K.Nakatsu, Inorg. Chim. Acta, 1985, 102, 145.
910 J.G.Leipoldt, S.S.Basson, E.C.Grobler, A.Roodt, Inorg. Chim. Acta, 1985, 99, 13.
911 X.Solans, M.Font-Altaba, M.Aguiló, C.Miravitlles, J.C.Besteiro, P.Lahuerta, Acta Crystallogr., 1985, C41, 841.
912 C.Bianchini, D.Masi, C.Mealli, A.Meli, M.Sabat, Organometallics, 1985, 4, 1014.
913 S.D.Chappell, D.J.Cole-Hamilton, A.M.R.Galas, M.B.Hursthouse, N.P.C.Walker, Polyhedron, 1985, 4, 121.
914 D.R.Alston, A.M.Z.Slawin, J.F.Stoddart, D.J.Williams, Angew. Chem., Int. Ed. Engl., 1985, 24, 786.
915 C.Bianchini, C.Mealli, M.Sabat, Organometallics, 1985, 4, 421.
916 C.Woodcock and R.Eisenberg, Inorg. Chem., 1985, 24, 1285.
917 L.A.Oro, D.Carmona, M.P.Garcia, F.J.Lahoz, J.Reyes, C.Foces-Foces, F.H.Cano, J. Organomet. Chem., 1985, 296, C43.
918 B.Delavaux, B.Chaudret, F.Dahan, R.Poilblanc, Organometallics, 1985, 4, 935.
919 W.E.Lindsell, C.B.Knobler, H.D.Kaesz, J. Organomet. Chem., 1985, 296, 209.
920 W.E.Lindsell, C.B.Knobler, H.D.Kaesz, J. Chem. Soc., Chem. Commun., 1985, 1171.
921 K.Nakajima, M.Kojima, J.Fujita, T.Ishii, S.Ohba, M.Ito, Y.Saito, Inorg. Chim. Acta, 1985, 99, 143.
922 H.Werner, L.Hofman, J.Wolf, G.Müller, J. Organomet. Chem., 1985, 280, C55.
923 S.M.Hawkins, P.B.Hitchcock, M.F.Lappert, J. Chem. Soc., Chem. Commun., 1985, 1592.
924 B.M.Louie, S.J.Rettig, A.Storr, J.Trotter, Can. J. Chem., 1985, 63, 688.
925 J.G.Gaudiello, T.C.Wright, R.A.Jones, A.J.Bard, J. Am. Chem. Soc., 1985, 107, 888.
926 H.Werner, H.J.Scholz, R.Zolk, Chem. Ber., 1985, 118, 4531.
927 R.S.Dickson, R.J.Nesbit, H.Pateras, W.Baimbridge, J.M.Patrick, A.H.White, Organometallics, 1985, 4, 2128.
928 R.S.Dickson, G.D.Fallon, R.J.Nesbit, G.N.Pain, Organometallics, 1985, 4, 355.
929 M.J.Krause and R.G.Bergman, J. Am. Chem. Soc., 1985, 107, 2972.
930 R.S.Dickson, G.S.Evans, G.D.Fallon, Aust. J. Chem., 1985, 38, 273.
931 W.A.Herrmann, C.Weber, M.L.Ziegler, O.Serhadli, J. Organomet. Chem., 1985, 297, 245.

932 M.Green, A.G.Orpen, C.J.Schaverien, I.D.Williams, J. Chem. Soc., Dalton Trans., 1985, 2483.
933 L.A.Oro, M.A.Ciriano, B.E.Villarroya, A.Tiripicchio, F.J.Lahoz, J. Chem. Soc., Dalton Trans., 1985, 1891.
934 H.M.Colquhoun, S.M.Doughty, A.M.Z.Slawin, J.F.Stoddart, D.J.Williams, Angew. Chem., Int. Ed. Engl., 1985, 24, 135.
935 L.A.Oro, D.Carmona, M.P.Lamata, C.Foces-Foces, F.H.Cano, Inorg. Chim. Acta, 1985, 97, 19.
936 L.A.Oro, M.T.Pinillos, A.Tiripicchio, M.Tiripicchio-Camellini, Inorg. Chim. Acta, 1985, 99, L13.
937 D.C.Cupertino, M.M.Harding, D.J.Cole-Hamilton, J. Organomet. Chem., 1985, 294, C29.
938 M.D.Fryzuk, W.E.Piers, S.J.Rettig, J. Am. Chem. Soc., 1985, 107, 8259.
939 A.M.Trzeciak, J.J.Ziólkowski, T.Lis, A.Borowski, Polyhedron, 1985, 4, 1677.
940 F.Barceló, P.Lahuerta, M.A.Ubeda, C.Foces-Foces, F.H.Cano, M.Martinez-Ripoll, J. Chem. Soc., Chem. Commun., 1985, 43.
941 T.G.Schenck, C.R.C.Milne, J.F.Sawyer, B.Bosnich, Inorg. Chem., 1985, 24, 2338.
942 A.R.Chakravarty, F.A.Cotton, D.A.Tocher, J.H.Tocher, Organometallics, 1985, 4, 8.
943 M.J.Freeman, A.G.Orpen, N.G.Connelly, I.Manners, S.J.Raven, J. Chem. Soc., Dalton Trans., 1985, 2283.
944 C.W.G.Ansell, M.K.Cooper, K.P.Dancey, P.A.Duckworth, K.Henrick, M.McPartlin, G.Organ, P.A.Tasker, J. Chem. Soc., Chem. Commun., 1985, 437.
945 L.A.Oro, D.Carmona, P.L.Pérez, M.Esteban, A.Tiripicchio, M.Tiripicchio-Camelloni, J. Chem. Soc., Dalton Trans., 1985, 973.
946 M.Cowie and S.J.Loeb, Organometallics, 1985, 4, 852.
947 D.H.Berry and R.Eisenberg, J. Am. Chem. Soc., 1985, 107, 7181.
948 L.Gelmini, D.W.Stephan, S.J.Loeb, Inorg. Chim. Acta, 1985, 98, L3.
949 L.J.Farrugia, J.C.Jeffery, C.Marsden, F.G.A.Stone, J. Chem. Soc., Dalton Trans., 1985, 645.
950 A.L.Balch, H.Hope, F.E.Wood, J. Am. Chem. Soc., 1985, 107, 6936.
951 J.C.Jeffery, C.Marsden, F.G.A.Stone, J. Chem. Soc., Dalton Trans., 1985, 1315.
952 R.S.Dickson, G.S.Evans, G.D.Fallon, G.N.Pain, J. Organomet. Chem., 1985, 295, 109.
953 A.C.Bray, M.Green, D.R.Hankey, J.A.K.Howard, O.Johnson, F.G.A.Stone, J. Organomet. Chem., 1985, 281, C12.
954 A.L.Balch and M.M.Olmstead, Isr. J. Chem., 1985, 25, 189.
955 A.L.Balch, L.A.Fossett, R.R.Guimerans, M.M.Olmstead, Organometallics, 1985, 4, 781.
956 A.L.Balch, J.C.Linehan, M.M.Olmstead, Inorg. Chem., 1985, 24, 3975.
957 R.P.Hughes, A.L.Rheingold, W.A.Herrmann, J.L.Hubbard, J. Organomet. Chem., 1985, 286, 361.
958 J.W.Suggs, M.J.Wovkulich, K.S.Lee, J. Am. Chem. Soc., 1985, 107, 5546.
959 V.G.Albano, D.Braga, D.Strumolo, C.Seregni, S.Martinengo, J. Chem. Soc., Dalton Trans., 1985, 1309.
960 G.Ciani, A.Sironi, S.Martinengo, J. Chem. Soc., Chem. Commun., 1985, 1757.
961 A.Berg, K.Dehnicke, D.Fenske, Z. Anorg. Allg. Chem., 1985, 527, 111.
962 M.Crocker, M.Green, C.E.Morton, K.R.Nagle, A.G.Orpen, J. Chem. Soc., Dalton Trans., 1985, 2145.
963 S.Ohba, Y.Saito, T.Ishii, S.Kamiyama, A.Kasahara, Acta Crystallogr., 1985, C41, 709.
964 D.N.Cox, R.Roulet, G.Chapuis, Organometallics, 1985, 4, 2001.
965 D.C.Liles, A.Shaver, E.Singleton, M.B.Wiege, J. Organomet. Chem., 1985, 288, C33.
966 L.Weber, K.Reizig, R.Boese, Organometallics, 1985, 4, 1890.
967 Z.L.Lutsenko, G.G.Aleksandrov, P.V.Petrovskii, E.S.Shubina, V.G.Andrianov, Yu.T.Struchkov, A.Z.Rubezhov, J. Organomet. Chem., 1985, 281, 349.
968 J.Keijsper, L.Polm, G.van Koten, K.Vrieze, C.H.Stam, J.-D.Schagen, Inorg. Chim. Acta, 1985, 103, 137.
969 P.Mura, B.G.Olby, S.D.Robinson, J. Chem. Soc., Dalton Trans., 1985,

2101;Inorg. Chim. Acta, 1985, 98, L21.
970 P.Reveco, R.H.Schmehl, W.R.Cherry, F.R.Fronczek, J.Selbin, Inorg. Chem., 1985, 24, 4078.
971 Y.Blum, Y.Shvo, D.F.Chodosh, Inorg. Chim. Acta, 1985, 97, L25.
972 Z.Dauter, R.J.Mawby, C.D.Reynolds, D.R.Saunders, J. Chem. Soc., Dalton Trans., 1985, 1235.
973 T.V.Ashworth, A.A.Chalmers, D.C.Liles, E.Meintjies, H.E.Oosthuizen, E.Singleton, J. Organomet. Chem., 1985, 284, C19.
974 M.I.Bruce, D.N.Duffy, M.G.Humphrey, A.G.Swincer, J. Organomet. Chem., 1985, 282, 383.
975 I.S.Thorburn, S.J.Rettig, B.R.James, J. Organomet. Chem., 1985, 296, 103.
976 E.Cesarotti, M.Angoletta, N.P.C.Walker, M.B.Hursthouse, R.Vefghi, P.A.Schofield, C.White, J. Organomet. Chem., 1985, 286, 343.
977 P.Mura, B.G.Olby, S.D.Robinson, Inorg. Chim. Acta, 1985, 97, 45.
978 M.M.Millar, T.O'Sullivan, N.de Vries, J. Am. Chem. Soc., 1985, 107, 3714.
979 J.M.Wisner, T.J.Bartczak, J.A.Ibers, Inorg. Chim. Acta, 1985, 100, 115.
980 M.I.Bruce, T.W.Hambley, M.R.Snow, A.G.Swincer, Organometallics, 1985, 4, 501.
981 L.A.Oro, M.A.Ciriano, M.Campo, C.Foces-Foces, F.H.Cano, J. Organomet. Chem., 1985, 289, 117.
982 M.Draganjac, C.J.Ruffing, T.B.Rauchfuss, Organometallics, 1985, 4, 1909.
983 M.I.Bruce, C.Dean, D.N.Duffy, M.G.Humphrey, G.A.Koutsantonis, J. Organomet. Chem., 1985, 295, C40.
984 S.I.Hommeltoft, A.D.Cameron, T.A.Shackleton, M.E.Fraser, S.Fortier, M.C.Baird, J. Organomet. Chem., 1985, 282, C17.
985 M.I.Bruce, M.G.Humphrey, G.A.Koutsantonis, B.K.Nicholson, J. Organomet. Chem., 1985, 296, C47.
986 L.Weber, K.Reizig, R.Boese, Organometallics, 1985, 4, 2097.
987 R.S.Sternal, C.P.Brock, T.J.Marks, J. Am. Chem. Soc., 1985, 107, 8270.
988 W.J.Sartain and J.P.Selegue, J. Am. Chem. Soc., 1985, 107, 5818.
989 J.A.K.Howard, J.C.V.Laurie, O.Johnson, F.G.A.Stone, J. Chem. Soc., Dalton Trans., 1985, 2017.
990 C.P.Casey, R.E.Palermo, R.F.Jordan, A.L.Rheingold, J. Am. Chem. Soc., 1985, 107, 4597.
991 M.Rotem, I.Goldberg, Y.Shvo, Inorg. Chim. Acta, 1985, 97, L27.
992 T.V.Ashworth, D.C.Liles, E.Singleton, Inorg. Chim. Acta, 1985, 98, L65.
993 L.A.Oro, M.P.Garcia, D.Carmona, C.Foces-Foces, F.H.Cano, Inorg. Chim. Acta, 1985, 96, L21.
994 J.A.Clucas, M.M.Harding, B.S.Nicholls, A.K.Smith, J. Chem. Soc., Dalton Trans., 1985, 1835.
995 A.Basu, S.Bhaduri, H.Khwaja, P.G.Jones, T.Schroeder, G.M.Sheldrick, J. Organomet. Chem., 1985, 290, C19.
996 A.R.Chakravarty and F.A.Cotton, Inorg. Chim., 1985, 24, 3584.
997 G.J.Baird, S.G.Davies, S.D.Moon, S.J.Simpson, R.H.Jones, J. Chem. Soc., Dalton Trans., 1985, 1479.
998 A.A.Chalmers, D.C.Liles, E.Meintjies, H.E.Oosthuizen, J.A.Pretorius, E.Singleton, J. Chem. Soc., Chem. Commun., 1985, 1340.
999 J.E.Smieja, R.E.Stevens, D.E.Fjare, W.L.Gladfelter, Inorg. Chem., 1985, 24, 3206.
1000 M.J.Sailor and D.F.Shriver, Organometallics, 1985, 4, 1476.
1001 M.R.Churchill, J.W.Ziller, J.B.Keister, J. Organomet. Chem., 1985, 297, 93.
1002 M.R.Churchill, J.C.Fettinger, J.B.Keister, R.F.See, J.W.Ziller, Organometallics, 1985, 4, 2112.
1003 M.R.Churchill, J.C.Fettinger, J.B.Keister, Organometallics, 1985, 4, 1867.
1004 T.Venalainen, J.Pursiainen, T.A.Pakkanen, J. Chem. Soc., Chem. Commun., 1985, 1348.
1005 J.Keijsper, L.H.Polm, G.van Koten, K.Vrieze, K.Goubitz, C.H.Stam, Organometallics, 1985, 4, 1876.
1006 J.Keijsper, L.H.Polm, G.van Koten, K.Vrieze, P.F.A.B.Seignette, C.H.Stam, Inorg. Chem., 1985, 24, 518.
1007 M.I.Bruce, P.A.Humphrey, B.W.Skelton, A.H.White, M.L.Williams, Aust. J. Chem., 1985, 38, 1301.

1008 N.Lugan, J.-J.Bonnet, J.A.Ibers, J. Am. Chem. Soc., 1985, 107, 4484.
1009 M.I.Bruce and M.L.Williams, J. Organomet. Chem., 1985, 288, C55.
1010 N.G.Connelly, N.J.Forrow, S.A.R.Knox, K.A.Macpherson, A.G.Orpen, J. Chem. Soc., Chem. Commun., 1985, 16.
1011 M.I.Bruce, J.G.Matisons, J.M.Patrick, A.H.White, A.C.Willis, J. Chem. Soc., Dalton Trans., 1985, 1223.
1012 H.-P.Klein, U.Thewalt, G.Herrmann, G.Süss-Fink, C.Moinet, J. Organomet. Chem., 1985, 286, 225.
1013 J.P.Attard, B.F.G.Johnson, J.Lewis, J.M.Mace, P.R.Raithby, J. Chem. Soc., Chem. Commun., 1985, 1526.
1014 D.L.Davies, J.A.K.Howard, S.A.R.Knox, K.Marsden, K.A.Mead, M.J.Morris, M.C.Rendle, J. Organomet. Chem., 1985, 279, C37.
1015 J.Lunniss, S.A.MacLaughlin, N.J.Taylor, A.J.Carty, E.Sappa, Organometallics, 1985, 4, 2066.
1016 J.Keijsper, L.H.Polm, G.van Koten, K.Vrieze, E.Nielsen, C.H.Stam, Organometallics, 1985, 4, 2006.
1017 J.-C.Daran, Y.Jeannin, O.Kristiansson, Organometallics, 1985, 4, 1882.
1018 M.I.Bruce, M.L.Williams, B.W.Skelton, A.H.White, J. Organomet. Chem., 1985, 282, C53.
1019 M.L.Blohm and W.L.Gladfelter, Organometallics, 1985, 4, 45.
1020 M.I.Bruce, B.W.Skelton, A.H.White, M.L.Williams, J. Chem. Soc., Chem. Commun., 1985, 744.
1021 J.C.Daran, O.Kristiansson, Y.Jeannin, C.R.Acad.Sci., Ser.2, 1985, 300, 943.
1022 M.I.Bruce, M.L.Williams, J.M.Patrick, A.H.White, J. Chem. Soc., Dalton Trans., 1985, 1229.
1023 M.I.Bruce and M.L.Williams, J. Organomet. Chem., 1985, 282, C11.
1024 H.M.Hoffmann and M.Dräger, J. Organomet. Chem., 1985, 295, 33.
1025 M.Wieber, D.Wirth, J.Metter, Ch.Burschka, Z. Anorg. Allg. Chem., 1985, 520, 65.
1026 V.K.Bel'skii, Zh. Strukt. Khim., 1984, 25(6), 138(Engl.Ed. 965).
1027 C.A.Stewart, R.L.Harlow, A.J.Arduengo III, J. Am. Chem. Soc., 1985, 107, 5543.
1028 R.Rüther, F.Huber, H.Preut, J. Organomet. Chem., 1985, 295, 21.
1029 W.Kolondra, W.Schwarz, J.Weidlein, Z. Naturforsch., 1985, 40B, 872.
1030 W.H.Hersh, J. Am. Chem. Soc., 1985, 107, 4599.
1031 D.M.Wesolek, D.B.Sowerby, M.J.Begley, J. Organomet. Chem., 1985, 293, C5.
1032 H.Preut, R.Rüther, F.Huber, Acta Crystallogr., 1985, C41, 358.
1033 J.Ellermann, E.Köck, H.Burzlaff, Acta Crystallogr., 1985, C41, 1437.
1034 H.J.Breunig, K.Häberle, M.Dräger, T.Severengiz, Angew. Chem., Int. Ed. Engl., 1985, 24, 72.
1035 F.Bottomley, D.E.Paez, P.S.White, J. Organomet. Chem., 1985, 291, 35.
1036 P.L.Baxter, A.J.Downs, A.M.Forster, M.J.Goode, D.W.H.Rankin, H.E.Robertson, J. Chem. Soc., Dalton Trans., 1985, 941.
1037 B.M.Goldstein, F.Takusagawa, H.M.Berman, P.C.Srivastava, R.K.Robins, J. Am. Chem. Soc., 1985, 107, 1394.
1038 N.N.Dhaneshwar, S.S.Tavale, T.N.G.Row, Y.K.Bhoon, Acta Crystallogr., 1985, C41, 1188.
1039 H.W.Roesky, J.Lucas, K.-L.Weber, H.Djarrah, E.Egert, M.Noltemeyer, G.M.Sheldrick, Chem. Ber., 1985, 118, 2396.
1040 H.W.Roesky, K.-L.Weber, U.Seseke, W.Pinkert, M.Noltemeyer, W.Clegg, G.M.Sheldrick, J. Chem. Soc., Dalton Trans., 1985, 565.
1041 S.Hauge and O.Vikane, Acta Chem. Scand., 1985, A39, 553.
1042 N.Albrecht, P.Hübener, U.Behrens, E.Weiss, Chem. Ber., 1985, 118, 4059.
1043 Yu.E.Ovchinnikov, V.E.Shklover, Yu.T.Struchkov, V.A.Palyulin, V.I.Rokitskaya, M.Yu.Aismont, V.F.Traven', J. Organomet. Chem., 1985, 290, 25.
1044 K.Eriks, H.H.Wang, P.E.Reed, M.A.Beno, E.H.Appelman, J.M.Williams, Acta Crystallogr., 1985, C41, 257.
1045 T.J.Emge, J.M.Williams, P.C.Leung, A.J.Schultz, M.A.Beno, H.H.Wang, Mol. Cryst. Liq. Cryst., 1985, 119, 237.
1046 H.Kobayashi, R.Kato, A.Kobayashi, Y.Sasaki, Chem. Lett., 1985, 535.
1047 K.Kikuchi, K.Yakushi, H.Kuroda, K.Kobayashi, M.Honda, C.Katayama, J.Tanaka, Chem. Lett., 1984, 1885.

1048 H.P.Fritz, G.Müller, G.Reber, M.Weis, Angew. Chem., Int. Ed. Engl., 1985, 24, 1058.
1049 C.Belin, T.Makani, J.Rozière, J. Chem. Soc., Chem. Commun., 1985, 118.
1050 V.D.Sheludyakov, A.B.Dmitrieva, A.I.Gusev, G.M.Apal'kova, A.D.Kirilin, Zh. Obshch. Khim., 1984, 54, 2298(Engl.Ed. 2056).
1051 P.E.Tomlins, J.E.Lydon, D.Akrigg, B.Sheldrick, Acta Crystallogr., 1985, C41, 941.
1052 A.Kemme, J.Bleidelis, A.Lapsina, M.Fleisher, G.Zelcans, E.Lukevics, Latv. PSR Zinat. Akad. Vestis, Kim. Ser., 1985, 242(Chem. Abs. 1985, 103:87961f).
1053 N.H.Buttrus, C.Eaborn, P.B.Hitchcock, A.K.Saxena, J. Organomet. Chem., 1985, 287, 157.
1054 N.H.Buttrus, C.Eaborn, P.B.Hitchcock, A.K.Saxena, J. Organomet. Chem., 1985, 284, 291.
1055 G.Klebe, J.W.Bats, K.Hensen, J. Chem. Soc., Dalton Trans., 1985, 1.
1056 P.Hencsei, I.Kovács, L.Párkányi, J. Organomet. Chem., 1985, 293, 185.
1057 L.Párkányi, P.Hencsei, L.Bihátsi, I.Kovács, Polyhedron, 1985, 4, 243.
1058 G.Klebe, J. Organomet. Chem., 1985, 293, 147.
1059 R.R.Holmes, R.O.Day, V.Chandrasekhar, J.J.Harland, J.M.Holmes, Inorg. Chem., 1985, 24, 2016.
1060 W.H.Stevenson III, S.Wilson, J.C.Martin, W.B.Farnham, J. Am. Chem. Soc., 1985, 107, 6340.
1061 R.R.Holmes, R.O.Day, V.Chandrasekhar, J.M.Holmes, Inorg. Chem., 1985, 24, 2009.
1062 V.K.Bel'skii and A.V.Dzyabchenko, Zh. Strukt. Khim., 1985, 26(1), 94(Engl.Ed. 78).
1063 W.S.Sheldrick, H.Linoh, R.Tacke, G.Lambrecht, U.Moser, E.Mutschler, J. Chem. Soc., Dalton Trans., 1985, 1743.
1064 H.Hörnig, E.Walther, U.Schubert, Organometallics, 1985, 4, 1905.
1065 S.N.Gurkova, Yu.M.Varezhkin, A.I.Gusev, N.V.Alekseev, A.N.Mikhailova, Zh. Strukt. Khim., 1985, 26(1), 150(Engl.Ed. 130).
1066 G.D.Fallon, N.J.Fitzmaurice, W.R.Jackson, P.Perlmutter, J. Chem. Soc., Chem. Commun., 1985, 4.
1067 M.Weidenbruch, M.Herrndorf, A.Schäfer, K.Peters, H.G.von Schnering, J. Organomet. Chem., 1985, 295, 7.
1068 M.Veith and V.Huch, J. Organomet. Chem., 1985, 293, 161.
1069 M.D.Curtis, L.G.Bell, W.M.Butler, Organometallics, 1985, 4, 701.
1070 C.S.Bajgur, W.R.Tikkanen, J.L.Petersen, Inorg. Chem., 1985, 24, 2539.
1071 J.J.Eisch, A.M.Piotrowski, S.K.Brownstein, E.J.Gabe, F.L.Lee, J. Am. Chem. Soc., 1985, 107, 7219.
1072 S.L.Latesky, A.K.McMullen, I.P.Rothwell, J.C.Huffman, J. Am. Chem. Soc., 1985, 107, 5981.
1073 E.Schweda, K.D.Scherfise, K.Dehnicke, Z. Anorg. Allg. Chem., 1985, 528, 117.
1074 F.Preuss, E.Fuchslocher, W.S.Sheldrick, Z. Naturforsch., 1985, 40B, 1040.
1075 P.Legzdins, S.J.Rettig, L.Sánchez, Organometallics, 1985, 4, 1470.
1076 T.D.Tilley, J. Am. Chem. Soc., 1985, 107, 4084.
1077 T.D.Tilley, Organometallics, 1985, 4, 1452.
1078 M.F.Lappert, C.L.Raston, L.M.Engelhardt, A.H.White, J. Chem. Soc., Chem. Commun., 1985, 521.
1079 G.Gundersen, D.W.H.Rankin, H.E.Robertson, J. Chem. Soc., Dalton Trans., 1985, 191.
1080 Yu.E.Ovchinnikov, V.E.Shklover, Yu.T.Struchkov, Yu.P.Polyakov, L.E.Guselnikov, Acta Crystallogr., 1985, C41, 1055.
1081 É.Gergö, G.Schultz, I.Hargittai, J. Organomet. Chem., 1985, 292, 343.
1082 L.M.Khananashvili, Ts.N.Vardosanidze, G.V.Gridunova, V.E.Shklover, Yu.T.Struchkov, E.G.Markarashvili, M.Sh.Tsutsunava, Izv. Akad. Nauk Gruz. SSR, Ser. Khim., 1985, 10, 262(Chem. Abs., 1985, 103:87950b).
1083 R.Sooriyakumaran, P.Boudjouk, R.G.Garvey, Acta Crystallogr., 1985, C41, 1348.
1084 M.Weidenbruch, K.Kramer, K.Peters, H.G.von Schnering, Z. Naturforsch., 1985, 40B, 601.
1085 A.Schäfer, M.Weidenbruch, S.Pohl, J. Organomet. Chem., 1985, 282, 305.
1086 W.Wojnowski, K.Peters, E.-M.Peters, H.G.von Schnering, Z. Anorg. Allg.

1087 R.West, D.J.De Young, K.J.Haller, J. Am. Chem. Soc., 1985, 107, 4942.
1088 H.Schumann, S.Nickel, E.Hahn, M.J.Heeg, Organometallics, 1985, 4, 800.
1089 P.Legzdins, S.J.Rettig, L.Sánchez, B.E.Bursten, M.G.Gatter, J. Am. Chem. Soc., 1985, 107, 1411.
1090 M.H.Chisholm, J.C.Huffman, J.A.Heppert, J. Am. Chem. Soc., 1985, 107, 5116.
1091 M.H.Chisholm, J.A.Heppert, J.C.Huffman, W.E.Streib, J. Chem. Soc., Chem. Commun., 1985, 1771.
1092 M.H.Chisholm, K.Folting, J.A.Heppert, D.M.Hoffman, J.C.Huffman, J. Am. Chem. Soc., 1985, 107, 1234.
1093 W.J.Evans, R.Dominguez, K.R.Levan, R.J.Doedens, Organometallics, 1985, 4, 1836.
1094 N.Wiberg, G.Wagner, G.Müller, Angew. Chem., Int. Ed. Engl., 1985, 24, 229.
1095 P.E.Tomlins, J.E.Lydon, D.Akrigg, B.Sheldrick, Acta Crystallogr., 1985, C41, 292.
1096 J.C.Dewan, S.Murakami, J.T.Snow, S.Collins, S.Masamune, J. Chem. Soc., Chem. Commun., 1985, 892.
1097 N.H.Buttrus, R.I.Damja, C.Eaborn, P.B.Hitchcock, P.D.Lickiss, J. Chem. Soc., Chem. Commun., 1985, 1385.
1098 Z.H.Aiube, N.H.Buttrus, C.Eaborn, P.B.Hitchcock, J.A.Zora, J. Organomet. Chem., 1985, 292, 177.
1099 N.S.Hosmane, M.N.Mollenhauer, A.H.Cowley, N.C.Norman, Organometallics, 1985, 4, 1194.
1100 S.C.Nyburg, A.G.Brook, F.Abdesaken, G.Gutekunst, W.Wong-Ng, Acta Crystallogr., 1985, C41, 1632.
1101 Yu.E.Ovchinnikov, V.E.Shklover, Yu.T.Struchkov, A.A.Remizova, V.M.Kopylov, V.A.Kovyazin, V.V.Kireyev, Z. Anorg. Allg. Chem., 1985, 523, 14.
1102 Yu.E.Ovchinnikov, V.E.Shklover, Yu.T.Struchkov, A.A.Remizova, B.D.Lavrukhin, T.V.Astapova, A.A.Zhdanov, Z. Anorg. Allg. Chem., 1985, 524, 75.
1103 H.Matsumoto, K.Takatsuna, M.Minemura, Y.Nagai, M.Goto, J. Chem. Soc., Chem. Commun., 1985, 1366.
1104 B.Chen, Z.Xie, J.Wang, Jieqou Huaxue, 1984, 3, 113.
1105 M.Ishikawa, S.Matsuzawa, H.Sugisawa, F.Yano, S.Kamitori, T.Higuchi, J. Am. Chem. Soc., 1985, 107, 7706.
1106 S.Masamune, Y.Kabe, S.Collins, D.J.Williams, R.Jones, J. Am. Chem. Soc., 1985, 107, 5552.
1107 M.Ishikawa, S.Matsuzawa, T.Higuchi, S.Kamitori, K.Hirotsu, Organometallics, 1985, 4, 2040.
1108 W.Wojnowski, B.Dreczewski, A.Herman, K.Peters, E.-M.Peters, H.G.von Schnering, Angew. Chem., Int. Ed. Engl., 1985, 24, 992.
1109 A.G.Brook, K.D.Safa, P.D.Lickiss, K.M.Baines, J. Am. Chem. Soc., 1985, 107, 4338.
1110 F.Sladky, B.Bildstein, C.Rieker, A.Gieren, H.Betz, T.Hübner, J. Chem. Soc., Chem. Commun., 1985, 1800.
1111 M.H.Chisholm, J.A.Heppert, J.C.Huffman, P.Thornton, J. Chem. Soc., Chem. Commun., 1985, 1466.
1112 M.H.Chisholm, K.Folting, J.A.Heppert, W.E.Streib, J. Chem. Soc., Chem. Commun., 1985, 1755.
1113 F.Shafiee, J.R.Damewood Jr., K.J.Haller, R.West, J. Am. Chem. Soc., 1985, 107, 6950.
1114 U.Klingebiel, S.Pohlmann, L.Skoda, C.Lensch, G.M.Sheldrick, Z. Naturforsch., 1985, 40B, 1023.
1115 U.Schubert, A.Schenkel, J.Müller, J. Organomet. Chem., 1985, 292, C11.
1116 W.J.Evans, J.W.Grate, H.W.Choi, I.Bloom, W.E.Hunter, J.L.Atwood, J. Am. Chem. Soc., 1985, 107, 941.
1117 W.J.Evans, I.Bloom, W.E.Hunter, J.L.Atwood, Organometallics, 1985, 4, 112.
1118 W.J.Evans, T.P.Hanusa, K.R.Levan, Inorg. Chim. Acta, 1985, 110, 191.
1119 W.J.Evans, J.W.Grate, I.Bloom, W.E.Hunter, J.L.Atwood, J. Am. Chem. Soc., 1985, 107, 405.
1120 W.J.Evans, J.W.Grate, R.J.Doedens, J. Am. Chem. Soc., 1985, 107, 1671.
1121 W.J.Evans, J.W.Grate, L.A.Hughes, H.Zhang, J.L.Atwood, J. Am. Chem. Soc., 1985, 107, 3728.

1122 K.A.Paseshnitchenko, L.A.Aslanov, A.V.Jatsenko, S.V.Medvedev, J. Organomet. Chem., 1985, 287, 187.
1123 G.Matsubayashi, K.Ueyama, T.Tanaka, J. Chem. Soc., Dalton Trans., 1985, 465.
1124 G.Valle, G.Plazzogna, R.Ettorre, J. Chem. Soc., Dalton Trans., 1985, 1271.
1125 K.Ueyama, G.-E.Matsubayashi, R.Shimizu, T.Tanaka, Polyhedron, 1985, 4, 1783.
1126 G.Valle, G.Ruisi, U.Russo, Inorg. Chim. Acta, 1985, 99, L21.
1127 R.Hani and R.A.Geanangel, J. Organomet. Chem., 1985, 293, 197.
1128 K.Jurkschat, A.Tzschach, J.Meunier-Piret, M.van Meerssche, J. Organomet. Chem., 1985, 290, 285.
1129 M.Dräger, Z. Anorg. Allg. Chem., 1985, 527, 169.
1130 M.Bürklin, E.Hanecker, H.Nöth, W.Storch, Angew. Chem., Int. Ed. Engl., 1985, 24, 999.
1131 M.Dräger, Z. Naturforsch., 1985, 40B, 1511.
1132 K.A.Paseshnichenko, L.A.Aslanov, A.V.Yatsenko, S.V.Medvedev, Koord. Khim,, 1984, 10, 1279(Engl.Ed. 706).
1133 K.C.Molloy, K.Quill, I.W.Nowell, J. Organomet. Chem., 1985, 289, 271.
1134 T.P.Lockhart, W.F.Manders, E.O.Schlemper, J. Am. Chem. Soc., 1985, 107, 7451.
1135 C.Wei, V.G.K.Das, E.Sinn, Inorg. Chim. Acta, 1985, 100, 245.
1136 R.J.F.Jans, G.van Koten, K.Vrieze, B.Kojić-Prodić, A.L.Spek, J.L.de Boer, Inorg. Chim. Acta, 1985, 105, 193.
1137 V.G.K.Das, C.Wei, E.Sinn, J. Organomet. Chem., 1985, 290, 291.
1138 V.K.Bel'skii, Zh. Strukt. Khim., 1984, 25(6), 136(Engl.Ed. 963).
1139 P.Ganis, D.Furlani, D.Marton, G.Tagliavini, G.Valle, J. Organomet. Chem., 1985, 293, 207.
1140 C.S.Frampton, R.M.G.Roberts, J.Silver, J. Organomet. Chem., 1985, 297, 273.
1141 J.L.Baxter, E.M.Holt, J.J.Zuckerman, Organometallics, 1985, 4, 255.
1142 E.C.Constable, F.K.Khan, J.Lewis, M.C.Liptrot, P.R.Raithby, J. Chem. Soc., Dalton Trans., 1985, 333.
1143 V.K.Bel'skii, A.A.Simonenko, I.E.Saratov, V.O.Reikhsfel'd, Kristallografiya, 1985, 30, 297.
1144 A.V.Jatsenko, S.V.Medvedev, K.A.Paseshnitchenko, L.A.Aslanov, J. Organomet. Chem., 1985, 284, 181.
1145 E.J.Gabe, F.J.Lee, L.E.Khoo, F.E.Smith, Inorg. Chim. Acta, 1985, 105, 103.
1146 M.E.Kamwaya and L.E.Khoo, Acta Crystallogr., 1985, C41, 1027.
1147 S.Dondi, M.Nardelli, C.Pelizzi, G.Pelizzi, G.Predieri, J. Chem. Soc., Dalton Trans., 1985, 487.
1148 D.Hodgson, J.A.K.Howard, F.G.A.Stone, M.J.Went, J. Chem. Soc., Dalton Trans., 1985, 1331.
1149 O.Scheidsteger, G.Huttner, K.Dehnicke, J.Pebler, Angew. Chem., Int. Ed. Engl., 1985, 24, 428.
1150 H.Puff, H.Hevendehl, K.Höfer, H.Reuter, W.Schuh, J. Organomet. Chem., 1985, 287, 163.
1151 T.S.Dory, J.J.Zuckerman, C.L.Barnes, J. Organomet. Chem., 1985, 281, C1.
1152 J.Meunier-Piret, M.van Meerssche, K.Jurkschat, M.Gielen, J. Organomet. Chem., 1985, 288, 139.
1153 S.Adams and M.Dräger, J. Organomet. Chem., 1985, 288, 295.
1154 N.W.Alcock, M.Pennington, G.R.Willey, J. Chem. Soc., Dalton Trans., 1985, 2683.
1155 V.Chandrasekhar, R.O.Day, R.R.Holmes, Inorg. Chem., 1985, 24, 1970.
1156 M.L.Luetkens Jr., J.C.Huffman, A.P.Sattelberger, J. Am. Chem. Soc., 1985, 107, 3361.
1157 H.Yasuda, K.Tatsumi, T.Okamoto, K.Mashima, K.Lee, A.Nakamura, Y.Kai, N.Kanehisa, N.Kasai, J. Am. Chem. Soc., 1985, 107, 2410.
1158 M.L.Luetkens Jr., D.J.Santure, J.C.Huffman, A.P.Sattelberger, J. Chem. Soc., Chem. Commun., 1985, 552.
1159 V.C.Gibson, C.E.Graimann, P.M.Hare, M.L.H.Green, J.A.Bandy, P.D.Grebenik, K.Prout, J. Chem. Soc., Dalton Trans., 1985, 2025.
1160 H.v.d.Heijden, A.W.Gal, P.Pasman, A.G.Orpen, Organometallics, 1985, 4, 1847.
1161 R.E.LaPointe, P.T.Wolczanski, G.D.Van Duyne, Organometallics, 1985, 4, 1810.
1162 J.M.Marín, V.S.B.Mtetwa, K.Prout, Acta Crystallogr., 1985, C41, 55.
1163 A.Marchi, R.Rossi, A.Duatti, L.Magon, V.Bertolasi, V.Ferretti, G.Gilli,

Inorg. Chem., 1985, 24, 4744.
1164 C.Lau, J.Passmore, E.K.Richardson, T.K.Whidden, P.S.White, Can. J. Chem., 1985, 63, 2273.
1165 M.Wieber, E.Schmidt, Ch.Burschka, Z. Anorg. Allg. Chem., 1985, 525, 127.
1166 G.Bandoli, J.Bergman, K.J.Irgolic, A.Grassi, G.C.Pappalardo, Z. Naturforsch., 1985, 40B, 1157.
1167 J.D.McCullough, C.Knobler, R.F.Ziolo, Inorg. Chem., 1985, 24, 1814.
1168 M.A.K.Ahmed, W.R.McWhinnie, T.A.Hamor, J. Organomet. Chem., 1985, 281, 205.
1169 R.Chakravorty, K.J.Irgolic, E.A.Meyers, Acta Crystallogr., 1985, C41, 1545.
1170 M.A.K.Ahmed, W.R.McWhinnie, T.A.Hamor, J. Organomet. Chem., 1985, 293, 219.
1171 N.G.Furmanova, N.I.Sorokina, G.M.Abakarov, I.D.Sadekov, Zh. Strukt. Khim., 1985, 26(1), 120(Engl.Ed. 100).
1172 N.Kuhn, H.Schumann, G.Wolmershäuser, J. Chem. Soc., Chem. Commun., 1985, 1595.
1173 M.R.Detty, N.F.Haley, R.S.Eachus, J.W.Hassett, H.R.Luss, M.G.Mason, J.M.McKelvey, A.A.Wernberg, J. Am. Chem. Soc., 1985, 107, 6298.
1174 R.K.Chadha and J.E.Drake, J. Organomet. Chem., 1985, 293, 37.
1175 U.Thewalt, K.Berhalter, E.W.Neuse, Transition Met. Chem., 1985, 10, 393.
1176 L.B.Kool, M.D.Rausch, H.G.Alt, M.Herberhold, B.Wolf, U.Thewalt, J. Organomet. Chem., 1985, 297, 159.
1177 C.E.Davies, I.M.Gardiner, J.C.Green, M.L.H.Green, N.J.Hazel, P.D.Grebenik, V.S.B.Mtetwa, K.Prout, J. Chem. Soc., Dalton Trans., 1985, 669.
1178 R.A.Howie, G.P.McQuillan, D.W.Thompson, Acta Crystallogr., 1985, C41, 1045.
1179 L.B.Kool, M.D.Rausch, H.G.Alt, M.Herberhold, U.Thewalt, B.Wolf, Angew. Chem., Int. Ed. Engl., 1985, 24, 394.
1180 L.B.Kool, M.D.Rausch, R.D.Rogers, J. Organomet. Chem., 1985, 297, 289.
1181 J.Besancon, D.Camboli, B.Trimaille, Y.Dusausoy, C.R.Acad.Sci., Ser.2, 1985, 301, 83.
1182 F.Wochner, L.Zsolnai, G.Huttner, H.H.Brintzinger, J. Organomet. Chem., 1985, 288, 69.
1183 K.Seitz and U.Behrens, J. Organomet. Chem., 1985, 288, C47.
1184 V.B.Shur, V.V.Burlakov, A.I.Yanovsky, P.V.Petrovsky, Yu.T.Struchkov, M.E.Vol'pin, J. Organomet. Chem., 1985, 297, 51.
1185 J.W.Pattiasina, C.E.Hissink, J.L.De Boer, A.Meetsma, J.H.Teuben, J. Am. Chem. Soc., 1985, 107, 7758.
1186 K.Mashima and H.Takaya, Organometallics, 1985, 4, 1464.
1187 M.Etienne, R.Choukroun, M.Basso-Bert, F.Dahan, D.Gervais, Nouv. J. Chim., 1984, 8, 531.
1188 H.-P.Klein, K.Döppert, U.Thewalt, J. Organomet. Chem., 1985, 280, 203.
1189 F.Bottomley, G.O.Egharevba, I.J.B.Lin, P.S.White, Organometallics, 1985, 4, 550.
1190 V.W.Day, C.W.Earley, W.G.Klemperer, D.J.Maltbie, J. Am. Chem. Soc., 1985, 107, 8261.
1191 F.Bottomley, G.O.Egharevba, P.S.White, J. Am. Chem. Soc., 1985, 107, 4353.
1192 J.Rebizant, M.R.Spirlet, J.Goffart, Acta Crystallogr., 1985, C41, 334.
1193 J.G.Brennan and A.Zalkin, Acta Crystallogr., 1985, C41, 1038.
1194 A.Zalkin and J.G.Brennan, Acta Crystallogr., 1985, C41, 1295.
1195 P.Zanella, G.Paolucci, G.Rossetto, F.Benetollo, A.Polo, R.D.Fischer, G.Bombieri, J. Chem. Soc., Chem. Commun., 1985, 96.
1196 A.Zalkin, D.H.Templeton, R.Kluttz, A.Streitwieser Jr., Acta Crystallogr., 1985, C41, 327.
1197 J.G.Brennan and R.A.Andersen, J. Am. Chem. Soc., 1985, 107, 514.
1198 F.A.Cotton and W.Schrotzer, Organometallics, 1985, 4, 942.
1199 F.Calderazzo, G.Pampaloni, M.Lanfranchi, G.Pelizzi, J. Organomet. Chem., 1985, 296, 1.
1200 F.Bottomley, J.Darkwa, P.S.White, Organometallics, 1985, 4, 961.
1201 T.H.Lemmen, J.C.Huffman, K.G.Caulton, Organometallics, 1985, 4, 946.
1202 R.M.Kowaleski, D.O.Kipp, K.J.Stauffer, P.N.Swepston, F.Basolo, Inorg. Chem., 1985, 24, 3750.
1203 J.H.Osborne, A.L.Rheingold, W.C.Trogler, J. Am. Chem. Soc., 1985, 107, 7945.
1204 J.H.Osborne, A.L.Rheingold, W.C.Trogler, J. Am. Chem. Soc., 1985, 107, 6292.
1205 F.A.Cotton, S.A.Duraj, W.J.Roth, Organometallics, 1985, 4, 1174.

1206 S.Gambarotta, M.Mazzanti, C.Floriani, A.Chiesi-Villa, C.Guastini, J. Chem. Soc., Chem. Commun., 1985, 829.
1207 F.Bottomley, J.Darkwa, P.S.White, J. Chem. Soc., Dalton Trans., 1985, 1435.
1208 F.Bottomley, D.E.Paez, P.S.White, J. Am. Chem. Soc., 1985, 107, 7226.
1209 J.M.Maher, R.P.Beatty, N.J.Cooper, Organometallics, 1985, 4, 1354.
1210 D.J.Darensbourg and M.Pala, J. Am. Chem. Soc., 1985, 107, 5687.
1211 D.J.Darensbourg, R.K.Hanckel, C.G.Bauch, M.Pala, D.Simmons, J.N.White, J. Am. Chem. Soc., 1985, 107, 7463.
1212 I.Feinstein-Jaffe, J.C.Dewan, R.R.Schrock, Organometallics, 1985, 4, 1189.
1213 M.R.Churchill and Y.-J.Li, J. Organomet. Chem., 1985, 282, 239.
1214 J.A.K.Howard, J.C.Jeffery, J.C.V.Laurie, I.Moore, F.G.A.Stone, A.Stringer, Inorg. Chim. Acta, 1985, 100, 23.
1215 W.A.Schenk, D.Rüb, C.Burschka, Angew. Chem., Int. Ed. Engl., 1985, 24, 971.
1216 M.R.Churchill and J.W.Ziller, J. Organomet. Chem., 1985, 281, 237.
1217 E.Carmona, A.Galindo, M.L.Poveda, R.D.Rogers, Inorg. Chem., 1985, 24, 4033.
1218 J.A.Marsella, J.C.Huffman, K.Folting, K.G.Caulton, Inorg. Chim. Acta, 1985, 96, 161.
1219 H.Fischer, F.Seitz, J.Riede, J. Chem. Soc., Chem. Commun., 1985, 537.
1220 J.R.Morrow, T.L.Tonker, J.L.Templeton, J. Am. Chem. Soc., 1985, 107, 6956.
1221 H.G.Alt, H.E.Engelhardt, U.Thewalt, J.Riede, J. Organomet. Chem., 1985, 288, 165.
1222 H.Berke, G.Huttner, C.Sontag, L.Zsolnai, Z. Naturforsch., 1985, 40B, 799.
1223 F.A.Cotton, W.Schwotzer, E.S.Shamshoum, J. Organomet. Chem., 1985, 296, 55.
1224 D.J.Darensbourg, R.Kudaroski, T.Delord, Organometallics, 1985, 4, 1094.
1225 L.Carlton, J.L.Davidson, P.Ewing, Lj.Manojlović-Muir, K.W.Muir, J. Chem. Soc., Chem. Commun., 1985, 1474.
1226 M.R.Churchill and J.W.Ziller, J. Organomet. Chem., 1985, 286, 27.
1227 M.R.Churchill and J.C.Fettinger, J. Organomet. Chem., 1985, 290, 375.
1228 M.I.Bruce, T.W.Hambley, M.R.Snow, A.G.Swincer, Organometallics, 1985, 4, 494.
1229 J.C.Jeffery, A.L.Ratermann, F.G.A.Stone, J. Organomet. Chem., 1985, 289, 367.
1230 F.R.Kreissl, W.J.Sieber, P.Hofmann, J.Reide, M.Wolfgruber, Organometallics, 1985, 4, 788.
1231 A.Mayr, A.M.Dorries, G.A.McDermott, S.J.Geib, A.L.Rheingold, J. Am. Chem. Soc., 1985, 107, 7775.
1232 M.R.Churchill and J.W.Ziller, J. Organomet. Chem., 1985, 279, 403.
1233 J.R.Morrow, T.L.Tonker, J.L.Templeton, Organometallics, 1985, 4, 745.
1234 M.A.Bennett, I.W.Boyd, G.B.Robertson, W.A.Wickramasinghe, J. Organomet. Chem., 1985, 290, 181.
1235 A.Marinetti, J.Fischer, F.Mathey, J. Am. Chem. Soc., 1985, 107, 5001.
1236 J.L.Davidson, W.F.Wilson, K.W.Muir, J. Chem. Soc., Chem. Commun., 1985, 460.
1237 J.R.Morrow, T.L Tonker, J.L.Templeton, J. Am. Chem. Soc., 1985, 107, 5004.
1238 K.R.Birdwhistell, T.L.Tonker, J.L.Templeton, J. Am. Chem. Soc., 1985, 107, 4474.
1239 S.Rosenberg, G.L.Geoffroy, A.L.Rheingold, Organometallics, 1985, 4, 1184.
1240 G.Erker, U.Dorf, R.Mynott, Y.-H.Tsay, C.Krüger, Angew. Chem., Int. Ed. Engl., 1985, 24, 584.
1241 T.S.Targos, R.P.Rosen, R.R.Whittle, G.L.Geoffroy, Inorg. Chem., 1985, 24, 1375.
1242 F.A.Cotton, L.R.Falvello, J.H.Meadows, Inorg. Chem., 1985, 24, 514.
1243 D.W.Hart, R.Bau, T.F.Koetzle, Organometallics, 1985, 4, 1590.
1244 K.J.Ahmed, M.H.Chisholm, K.Folting, J.C.Huffman, J. Chem. Soc., Chem. Commun., 1985, 152.
1245 G.J.Kubas, H.J.Wasserman, R.R.Ryan, Organometallics, 1985, 4, 419.
1246 K.J.Ahmed, M.H.Chisholm, J.C.Huffman, Organometallics, 1985, 4, 1312.
1247 M.H.Chisholm, D.M.Hoffman, J.C.Huffman, Organometallics, 1985, 4, 986.
1248 M.H.Chisholm, B.W.Eichhorn, J.C.Huffman, J. Chem. Soc., Chem. Commun., 1985, 861.
1249 F.A.Cotton and E.S.Shamshoum, J. Am. Chem. Soc., 1985, 107, 4662.
1250 F.A.Cotton and E.S.Shamshoum, Polyhedron, 1985, 4, 1727.
1251 S.A.MacLaughlin, R.C.Murray, J.C.Dewan, R.R.Schrock, Organometallics, 1985,

1252 G.A.Carriedo, J.A.K.Howard, D.B.Lewis, G.E.Lewis, F.G.A.Stone, J. Chem. Soc., Dalton Trans., 1985, 905.
1253 J.Levisalles, F.Rose-Munch, H.Rudler, J.C.Daran, Y.Jeannin, J. Organomet. Chem., 1985, 279, 413.
1254 F.A.Cotton, W.Schwotzer, E.S.Shamshoum, Organometallics, 1985, 4, 461.
1255 H.M.Dawes, M.B.Hursthouse, A.A.del Paggio, E.L.Muetterties, A.W.Parkins, Polyhedron, 1985, 4, 379.
1256 F.A.Cotton, Z.Dori, M.Kapon, D.O.Marler, G.M.Reisner, W.Schwotzer, M.Shaia, Inorg. Chem., 1985, 24, 4381.
1257 A.Marinetti, C.Charrier, F.Mathey, J.Fischer, Organometallics, 1985, 4, 2134.
1258 M.H.Chisholm, K.Folting, J.C.Huffman, J.Leonelli, N.S.Marchant, C.A.Smith, L.C.E.Taylor, J. Am. Chem. Soc., 1985, 107, 3722.
1259 W.J.Evans, T.T.Peterson, M.D.Rausch, W.E.Hunter, H.Zhang, J.L.Atwood, Organometallics, 1985, 4, 554.
1260 S.A.Kinsley, A.Streitwieser Jr., A.Zalkin, Organometallics, 1985, 4, 52.
1261 G.B.Deacon, G.D.Fallon, D.L.Wilkinson, J. Organomet. Chem., 1985, 293, 45.
1262 M.Adam, X.-F.Li, W.Oroschin, R.D.Fischer, J. Organomet. Chem., 1985, 296, C19.
1263 P.H.M.Budzelaar, J.Boersma, G.J.M.v.d.Kerk, A.L.Spek, A.J.M.Duisenberg, J. Organomet. Chem., 1985, 281, 123.
1264 R.Blom, A.Haaland, J.Weidlein, J. Chem. Soc., Chem. Commun., 1985, 266.
1265 L.G.Kuz´mina, A.I.Yanovskii, Yu.T.Struchkov, M.Kh.Minacheva, E.M.Brainina, Koord. Khim., 1985, 11, 116.
1266 S.J.Young, M.M.Olmstead, M.J.Knudsen, N.E.Schore, Organometallics, 1985, 4, 1432.
1267 S.Gambarotta, S.Strologo, C.Floriani, A.Chiesi-Villa, C.Guastini, Inorg. Chem., 1985, 24, 654.
1268 G.Erker, K.Engel, C.Krüger, Y.-H.Tsay, E.Samuel, P.Vogel, Z. Naturforsch., 1985, 40B, 150.
1269 F.R.W.P.Wild, M.Wasiucionek, G.Huttner, H.H.Brintzinger, J. Organomet. Chem., 1985, 288, 63.
1270 R.Bortolin, V.Patel, I.Munday, N.J.Taylor, A.J.Carty, J. Chem. Soc., Chem. Commun., 1985, 456.
1271 G.Erker, T.Mühlenbernd, R.Benn, A.Rufińska, Y.-H.Tsay, C.Krüger, Angew. Chem., Int. Ed. Engl., 1985, 24, 321.
1272 S.Gambarotta, S.Strologo, C.Floriani, A.Chiesi-Villa, C.Guastini, J. Am. Chem. Soc., 1985, 107, 6278.
1273 G.Erker, P.Czisch, R.Mynott, Y.-H.Tsay, C.Krüger, Organometallics, 1985, 4, 1310.
1274 S.L.Latesky, A.K.McMullen, G.P.Niccolai, I.P.Rothwell, J.C.Huffman, Organometallics, 1985, 4, 902.
1275 S.B.Jones and J.L.Petersen, Organometallics, 1985, 4, 966.
1276 S.L.Latesky, A.K.McMullen, G.P.Niccolai, I.P.Rothwell, J.C.Huffman, Organometallics, 1985, 4, 1896.
1277 A.K.McMullen, I.P.Rothwell, J.C.Huffman, J. Am. Chem. Soc., 1985, 107, 1072.